Karl J. Smith

For the past 22 years, Karl Smith has been teaching at Santa Rosa Junior College (for 7 of those years, he was department chairman) and has gained a reputation as a master teacher. From 1988 to 1989 he was President of the American Mathematical Association of Two-Year Colleges and on the list of lecturers for the American Mathematical Association. He has served on numerous boards, including the Conference Board of Mathematical Sciences (which serves as a liaison between the National Academy of Sciences, the federal government, and the mathematical community in the United States), the Council of Scientific Society Presidents, and the Board of the California Mathematics Council for Community Colleges. He received his B.A. and M.A. from the University of California at Los Angeles and his Ph.D. from Southeastern University. In his spare time he enjoys running and swimming.

One of the most important goals of this book is to make mathematics real and alive for the students. For that reason, personal notes of many mathematicians are included in the book.

Other Brooks/Cole Titles by Karl J. Smith

Algebra and Trigonometry
Trigonometry for College Students, 4th Edition
Essentials of Trigonometry, 2nd Edition
The Nature of Mathematics, 5th Edition
Mathematics: Its Power and Utility, 3rd Edition

The Smith Business Series

College Mathematics and Calculus
Calculus with Applications
Finite Mathematics, 2nd Edition

The Precalculus Series by Karl J. Smith and Patrick J. Boyle

Beginning Algebra for College Students, 4th Edition
Intermediate Algebra for College Students, 4th Edition
Primer for College Algebra
College Algebra, 4th Edition
Study Guide for Algebra

Conic Sections

$y = mx + b$ is the **line** with slope m and y intercept b.

$y - k = m(x - h)$ is the **line** with slope m passing through (h, k).

$y - k = a(x - h)^2$ is the **parabola** with vertex (h, k) opening upward if $a > 0$ and downward if $a < 0$.

$x - h = a(y - k)^2$ is the **parabola** with vertex (h, k) opening to the right if $a > 0$ and to the left if $a < 0$.

$(x - h)^2 + (y - k)^2 = r^2$ is the **circle** with center (h, k) and radius r. This is the special case of an ellipse, where $a = b = r$.

Translation substitution:

$x' = x - h$
$y' = y - k$

Rotation substitution:

$x = x' \cos\theta - y' \sin\theta$
$y = x' \sin\theta + y' \cos\theta$
where $\cot 2\theta = (A - C)/B$

The following standard conic sections can be translated:

$\dfrac{x^2}{a^2} + \dfrac{y^2}{b^2} = 1$ is the **ellipse** with center at $(0, 0)$, x intercepts $\pm a$, y intercepts $\pm b$, and foci $(\pm c, 0)$ with constant sum $2a$; $(a^2 - b^2 = c^2)$.

$\dfrac{y^2}{a^2} + \dfrac{x^2}{b^2} = 1$ is the **ellipse** with center at $(0, 0)$, x intercepts $\pm b$, y intercepts $\pm a$, and foci $(0, \pm c)$ with constant sum $2a$; $(a^2 - b^2 = c^2)$.

$\dfrac{x^2}{a^2} - \dfrac{y^2}{b^2} = 1$ is the **hyperbola** with center at $(0, 0)$, x intercepts $\pm a$, foci $(\pm c, 0)$, and constant difference $2a$; $(a^2 + b^2 = c^2)$.

$\dfrac{y^2}{a^2} - \dfrac{x^2}{b^2} = 1$ is the **hyperbola** with center at $(0, 0)$, y intercepts $\pm a$, foci $(0, \pm c)$, and constant difference $2a$; $(a^2 + b^2 = c^2)$.

$y = \pm \dfrac{b}{a} x$ are the **asymptotes** for the hyperbola.

Graphing Procedure

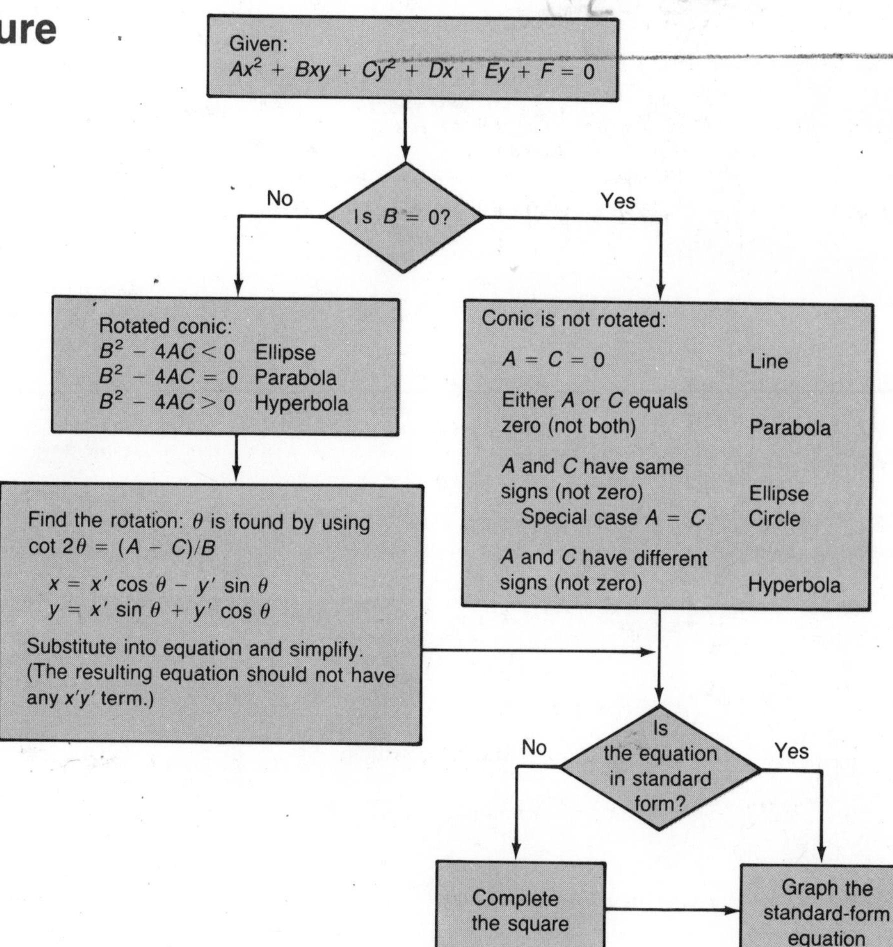

Formulas

Distance between (x_1, y_1) and (x_2, y_2): $d = \sqrt{(\Delta x)^2 + (\Delta y)^2}$, where $\Delta x = x_2 - x_1$ and $\Delta y = y_2 - y_1$.

Slope of the line passing through (x_1, y_1) and (x_2, y_2): $m = \Delta y / \Delta x$.

Pythagorean Theorem: The sum of the squares of the lengths of the legs of a right triangle is equal to the square of the length of the hypotenuse.

Quadratic Formula: If $ax^2 + bx + c = 0$, $a \neq 0$, then $x = \dfrac{-b \pm \sqrt{b^2 - 4ac}}{2a}$.

Discriminant: $b^2 - 4ac$
$b^2 - 4ac < 0 \rightarrow$ *no real* solutions
$b^2 - 4ac = 0 \rightarrow$ *one real* solution
$b^2 - 4ac > 0 \rightarrow$ *two real* solutions

Arithmetic Sequence: $a_n = a_1 + (n-1)d$

Geometric Sequence: $g_n = g_1 r^{n-1}$

Arithmetic Series: $A_n = \dfrac{n}{2}[2a_1 + (n-1)d]$ or $n\left(\dfrac{a_1 + a_n}{2}\right)$

Geometric Series: $G_n = \dfrac{g_1(1 - r^n)}{1 - r}$; $\quad G = \dfrac{g_1}{1 - r}$ if $|r| < 1$

Binomial Expansion: $(a + b)^n = \sum_{k=0}^{n} \binom{n}{k} a^{n-k} b^k$, where $\binom{n}{k} = \dfrac{n!}{k!(n-k)!}$

Forms of a Complex Number z:

Rectangular form: $z = a + bi$
Polar form: $z = r \operatorname{cis} \theta$
$= r(\cos \theta + i \sin \theta)$
Conversions: $a = r \cos \theta \qquad r = \sqrt{a^2 + b^2}$
$b = r \sin \theta \qquad \theta' = \tan^{-1}\left|\dfrac{b}{a}\right|$, where θ' is the reference angle for θ

De Moivre's Theorem: If n is a natural number, $(r \operatorname{cis} \theta)^n = r^n \operatorname{cis} n\theta$.

nth Root Theorem: If n is a positive integer, $(r \operatorname{cis} \theta)^{1/n} = \sqrt[n]{r} \operatorname{cis} \dfrac{1}{n}(\theta + 360°k)$, where $k = 0, 1, 2, \ldots, n - 1$.

Cramer's Rule: If

$$\begin{cases} a_{11}x + a_{12}y + a_{13}z = b_1 \\ a_{21}x + a_{22}y + a_{23}z = b_2 \\ a_{31}x + a_{32}y + a_{33}z = b_3 \end{cases} \quad \text{then} \quad x = \frac{D_x}{D}, \quad y = \frac{D_y}{D}, \quad z = \frac{D_z}{D} \quad (D \neq 0)$$

D is the determinant of the coefficients and the constants b_1, b_2, b_3 replace the coefficients x, y, z for D_x, D_y, and D_z, respectively.

Matrices

$A + B = [a_{ij}]_{m \times n} + [b_{ij}]_{m \times n} = [a_{ij} + b_{ij}]_{m \times n}$

$cA = c[a_{ij}]_{m \times n} = [ca_{ij}]_{m \times n}$

$I = [a_{ij}]_{n \times n}$, where $a_{ij} = \begin{cases} 1 & \text{if } i = j \\ 0 & \text{if } i \neq j \end{cases}$

$AB = [a_{ij}]_{m \times n}[b_{ij}]_{n \times p} = [a_{i1}b_{1j} + a_{i2}b_{2j} + \cdots + a_{in}b_{nj}]_{m \times p}$

$A^{-1}A = AA^{-1} = I$

$IA = AI = A$

Principle of Mathematical Induction: If a given proposition $P(n)$ is true for $P(1)$ and if the truth of $P(k)$ implies its truth for $P(k+1)$, then $P(n)$ is true for all positive integers n.

FOURTH EDITION

Precalculus Mathematics

A Functional Approach

FOURTH EDITION

Precalculus Mathematics

A Functional Approach

Karl J. Smith, Ph.D.

Brooks/Cole Publishing Company
Pacific Grove, California

Brooks/Cole Publishing Company
A Division of Wadsworth, Inc.

Printed in the United States of America

10 9 8 7 6 5 4 3 2 1

Library of Congress Cataloging-in-Publication Data

Smith, Karl J.
Precalculus mathematics: a functional approach / Karl J. Smith.—
—4th ed.
p. cm.
ISBN 0-534-11922-0
1. Functions. I. Title.
QA331.S617 1990
512'.1—dc20 89-22059
ISBN 0-534-11922-0 CIP

Sponsoring Editor: *Paula-Christy Heighton*
Editorial Assistant: *Sarah Wilson*
Production Editor: *Phyllis Niklas*
Manuscript Editor: *Phyllis Niklas*
Permissions Editor: *Carline Haga*
Interior Design: *Vernon T. Boes/John Edeen*
Cover Design: *Lisa Thompson*
Cover Neon Art: *Brian Coleman*
Interior Illustration: *Carl Brown*
Typesetting: *Polyglot Compositors*
Cover Printing: *Lehigh Press*
Printing and Binding: *Arcata Graphics, Hawkins*

**This book is dedicated to my daughter,
Melissa Ann Smith**

Missy is a student at the University of California at Los Angeles and after graduation wishes to enter a program to teach in elementary school. She loves people, especially children, and has just completed a course in signing for the deaf. I am proud of her, and dedicate this book to her with love.

Preface

As I began writing this fourth edition of a very successful precalculus textbook I sorted through the ideas I wished to emphasize:

- The book must be designed to prepare students for the study of calculus.
- Precalculus is a difficult subject, and there is no magic key to success; so this book should create no problems or awkward moments for the instructor, and at the same time must make it clear to the student what is required and how to proceed. Therefore, this material must be clearly presented without intimidation. The discussion and examples must be straightforward, and the problems and examples should amplify and illustrate the objectives.
- The book must present the best set of problems (both drill and applied) that are available in the marketplace. The problems should offer the instructor choices in creating the assignment and should be graded from easy to difficult.
- I want the student to realize that mathematics is alive and created by real people, so the book should include biographical sketches and historical questions.
- The book must be free of errors.

How did I propose to accomplish these goals? Let me consider them one at a time.

Preparation for Calculus

The selection of topics in this book is, of course, designed to prepare the student for calculus. In addition, at the beginning of each section I have included not only a chapter overview but also a chapter perspective. This perspective points out how the techniques learned in that chapter will be used in calculus. The definitions and formulas given on the inside front and back covers summarize the ideas presented in this book that will be needed in calculus.

The main concept unifying the material of this book is the one that is most needed for the study of calculus—namely, the notion of a function. This concept is defined and discussed in Chapter 2 and is then used as the unifying idea for the rest of the material. Although it is assumed that students have had high school algebra, the central ideas from algebra (factoring, solving equations, simplifying algebraic expressions, the laws of exponents) are integrated into the text where appropriate, since these topics are often the very ones that cause difficulty for the beginning calculus student. It is not necessary for students to have had trigonometry, since an entire trigonometry course is presented in Chapters 6, 7, and 8. I do not believe that the right triangle approach is the best way to introduce trigonometry to students who are about to take calculus. Therefore, I use reference angles and the ratio definition for the trigonometric functions, and quickly introduce radian measure of angles.

In addition to the usual precalculus material, I have included problems that are not typical in precalculus courses but will be particularly useful in the study of calculus. A list of these problems is included in the Applications Index on page 588.

Techniques for Success

Success in this course is a joint effort by the student, the instructor, and the author. The student must be willing to attend class and devote time to it on a daily basis. There is no substitute for working problems in mathematics. The problem sets in this book are divided into A, B, and C problems according to the level of difficulty. A word of warning is necessary here. The C problems are generally extraordinary problems and should be considered difficult and beyond the scope of the usual precalculus course. A *Student Solutions Manual*, which shows the complete solutions to all the odd-numbered problems is available.

I have tried to do my part in writing this material so that the student can master the material necessary for the study of calculus:

- Important terms are presented in **boldface type** where they are defined.
- Important ideas are set off in boxes.
- Common pitfalls, helpful hints, and explanations are shown in *italics*.
- WARNINGS are given to call attention to common mistakes.
- Color is used in a functional way to help the student see what to do next.

In addition to these pedagogical aids, I have included entire sections to provide built-in redundancy and help students review for exams:

- The goals of each section are listed as objectives at the end of each chapter.
- Review problems are provided for each chapter objective.
- Answers to the odd-numbered problems are provided in the back of the book to let the students know if they are on the right track. Many solutions and hints are also provided, and more than 350 graphs are presented in the answer section alone to provide additional examples.
- Problems are graded and presented in sets so that students can work some of the simpler problems before progressing to the more challenging ones.
- Cumulative reviews are given as practice tests.

Fully Developed Problem Sets

I do not believe that students completely learn the material by working a particular type of problem only once. However, as a college instructor, I am well aware of time constraints and the amount of material that must be covered. I have therefore developed the idea of uniform problem sets for this book. This means that the instructor can make "standard assignments" consisting of the same problem being assigned from section to section. This makes it easy for the instructor and student to work problems from more than one section at a time. For example, a typical spiral assignment would be given as follows:

First day: 5–60 multiples of 5 from Problem Set x
Second day: 5–60 multiples of 5 from Problem Set $x + 1$
7–28 multiples of 7 from Problem Set x
Third day: 5–60 multiples of 5 from Problem Set $x + 2$
7–28 multiples of 7 from Problem Set $x + 1$
6–24 multiples of 6 from Problem Set x

This means that the standard assignment would be 5–60 multiples of 5, multiples of 7 from the previous set, and multiples of 6 from the assignment two class meetings past. On the other hand, if the instructor does not wish to give spiral assignments, the same assignment may be given from section to section, as follows:

Standard 1–2 hour assignment: Problems 5–50, multiples of 5
Standard 2–3 hour assignment: Problems 3–60, multiples of 3

If any of these standard assignments or spiral assignments are given, all the important ideas in this book will be covered thoroughly. (These standard assignments do not apply to the review problems at the end of each chapter.)

In addition to a large number of drill problems, extensive application problems are provided. The Applications Index at the back of the book lists applications from agriculture, archaeology, architecture, astronomy, aviation, ballistics, business, chemistry, earth science, economics, engineering, medicine, navigation, physics, police science, psychology, social science, space science, sports, surveying, as well as consumer and general interest applications.

Finally, at the end of each part of the book, I have included an extended application introduced by a news clipping that examines a topic in more than the usual depth. These include population growth, solar power, and planetary orbits.

Accuracy of the Material

It is extremely important that the material be accurately presented. Not only has this book been reviewed by the usual number of persons, but all the problems have been checked and rechecked. Karen Sharp of Mott Community College and Donna Szott of Allegheny Community College have worked all the problems. In addition, there were three independent accuracy checks of all the problems by Gary Gislason, Steve Simonds, and Mary Ellen White. If you find *any* errors in this book, please call me at home to let me know where they are. My phone number is (707) 829-0606.

Organization of the Material

A great deal of flexibility is possible in the instructor's selection of topics presented in this book. I have written the book so that each section represents approximately one day of class material. I have provided more material than can be used in a single semester or quarter so that you can select material appropriate for your school or class.

Some of the material that I have included may not be considered typical. Included is a chapter on graphing techniques to allow a consideration of graphing, not only of polynomial and rational functions, but also radical and absolute functions. Also included are discussions of the ambiguous case in solving triangles; De Moivre's Theorem; the Fundamental Theorem of Algebra; linear programming; partial fractions; translations and rotations of the conic sections; three-dimensional coordinate systems; vectors (in two and three dimensions); parametric equations; polar coordinates; and intersection of polar-form curves.

Calculator Usage

We are now seeing a new generation of calculators that are able to do symbolic manipulation and graphing. These calculators are characterized by the Hewlett-Packard 28S, Sharp EL-5200, and Casio fx-7000G. Even though these calculators and computers may revolutionize the way precalculus and calculus are taught in the future, their use is still rather limited. Most students and instructors do not yet use calculators that do symbolic manipulation and graphing, so my discussion of these calculators is limited to a general overview in Appendix A and a few examples scattered throughout the book (and clearly set off in boxes).

In the general text of this book, I assume only that the student has an inexpensive scientific calculator. Procedures for using inexpensive calculators are presented as they occur in a natural way without any special designation. This is in recognition of the assumption that most students will have such calculators. For example, the computation aspects of the logarithmic function in Chapter 5 have been minimized, and the introduction to the evaluation of the trigonometric functions in Chapter 6 assumes that each student has a calculator. In writing this book, I have been using a Sharp EL-531A (I paid $5.95) to illustrate algebraic logic and an HP-35 to illustrate RPN logic. You should always consult the owner's manual for the calculator you own. Before beginning, you should know the type of logic used by your calculator. There are three types of logic: arithmetic, algebraic, and RPN. Arithmetic logic works from left to right, algebraic logic recognizes the order of operations agreement, and RPN logic is characterized by an [ENTER] or [SAVE] key. In order to determine the type of logic used by a calculator, you should consider the following test problem: $2 + 3 \cdot 4$. If there is a [SAVE] or [ENTER] key, the calculator uses RPN logic; otherwise, press [2] [+] [3] [×] [4] [=]. If the output is 20, then your calculator uses arithmetic logic; if the output is 14, then your calculator uses algebraic logic. It is suggested that you use a calculator with algebraic or RPN logic. The correct sequence of steps for this test problem is shown below for each type of calculator:

ALGEBRAIC LOGIC:	[2] [+] [3] [×] [4] [=]
RPN LOGIC:	[2] [ENTER] [3] [+] [4] [×]
ARITHMETIC LOGIC:	[3] [×] [4] [+] [2] [=]

The correct output is 14.

Acknowledgments

I would like to thank the reviewers of the previous editions of this book:

James Bailey, University of Toledo
Roy Bergstrom, University of the Pacific
Virginia Buchanan, Southwest Texas State University
David Bush, Shasta College
Lee R. Clancy, Golden West College
John Cross, University of Northern Iowa
Joel Cunningham, Susquehanna University

Daniel Drucker, Wayne State University
Vivian Fielder, Fisk University
Marjorie S. Freeman, University of Houston, Downtown College
Sheldon Gordon, Suffolk Community College
Thomas Green, Contra Costa College
Steve Hinthorne, Central Washington University
James Householder, California State University, Humboldt
John Kuisti, Michigan Technological University at South Range
Laurence P. Maher, Jr., North Texas State University
Joan Mahmud, Bergen Community College
Raymond McGivney, University of Hartford
James McMurdo, Shasta College
Gordon D. Mock, Western Illinois University
Bill Orr, San Bernardino Valley College
Bill Pelletier, Wayne State University
Hubert J. Pollick, Killgore College
Richard Redner, University of Tulsa
Thomas Strommer, Oxford College of Emery University
David Tabor, University of Texas
Donald Taranto, California State University, Sacramento
Robert Webber, Longwood College
Terry Wilson, San Jacinto College

I would also like to thank the reviewers of the fourth edition:

Gary A. Gislason, University of Alaska
Herbert E. Kasube, Bradley University
Robert Keicher, Delta College
Anna Penk, Western Oregon State University
Kenneth M. Shiskowski, Eastern Michigan University
John R. Unbehaun, University of Wisconsin, La Crosse
Howard L. Wilson, Oregon State University

Special thanks go to Dr. Charles Ray and his students for all their helpful suggestions on how to improve the third edition as well as to the problem checkers of this edition: Gary A. Gislason, Steve Simonds, and Mary Ellen White. George E. Morris of Scientific Illustrators was most helpful in preparing the material on graphing calculators.

Finally, I would like to thank those who worked on the production of this book. It is always a pleasure to work with Phyllis Niklas, whose attention to detail and dedication to the project is unsurpassed; special thanks to Joan Marsh, whose consistent support is greatly appreciated; thanks to Paula-Christy Heighton, whose leadership is superb; and thanks to Carl Brown, John Edeen, Lisa Thompson, and Carline Haga, who all contributed to the success of this book.

Karl J. Smith
Sebastopol, California

Contents

INTRODUCTION

J. J. Sylvester

1 Fundamental Concepts 2

FUNCTIONS AND GRAPHING

Leonhard Euler

2 Functions 44

* Optional section

Karl Gauss

†3 Graphing Techniques 88

Amalie (Emmy) Noether

4 Polynomial and Rational Functions 118

John Napier

5 Exponential and Logarithmic Functions 164

* Optional section
† Chapter 3 is not required for the development of the other chapters in this book.

TRIGONOMETRY

Hipparchus

Nicholas Copernicus

Benjamin Bannecker

* Optional section

ADVANCED ALGEBRA TOPICS AND ANALYTIC GEOMETRY

Gottfried Wilhelm von Leibniz

9 Systems and Matrices 328

Sir Isaac Newton

10 Additional Topics in Algebra 386

René Descartes

11 Analytic Geometry— Conic Sections 422

* Optional section

Charlotte Angas Scott

12 Additional Topics in Analytic Geometry 460

APPENDIX

* Optional section

FOURTH EDITION

Precalculus Mathematics

A Functional Approach

J. J. Sylvester (1814–1897)

There are three ruling ideas, three so to say, spheres of thought, which pervade the whole body of mathematical science, to some one or other of which, or to two or all three of them combined, every mathematical truth admits of being referred; these are the three cardinal notions, of Number, Space, and Order. Arithmetic has for its object the properties of number in the abstract. In algebra, viewed as a science of operations, order is the predominating idea. The business of geometry is with the evolution of the properties of space, or of bodies viewed as existing in space.

J. J. Sylvester
Philosophical Magazine, vol. 24 (1844), p. 285

HISTORICAL NOTE

It is appropriate to begin our study of precalculus with a quotation from J. J. Sylvester, one of the most colorful men in the history of mathematics. This quotation of Sylvester ties together the ideas of number, space, and order. As we look at the topics of precalculus we see that those very notions form the basis not only for this chapter, but also for the entire course.

Sylvester's writings are flowery and eloquent. He was able to make the dullest subject bright, fresh, and interesting. His enthusiasm and zest for life were evident in every line of his writings. He was, however, a perfect fit for the stereotype of an absent-minded mathematics professor.

Born and educated in England, he accepted an appointment as Professor of Mathematics at the University of Virginia in 1841. After only 3 months, he resigned because of an altercation with a student. He returned to England penniless, but in 1876 he came back to America to accept a position at Johns Hopkins University. The 7 years he spent at Johns Hopkins were the happiest and most productive in his life. During his stay there, he founded the *American Journal of Mathematics* (in 1878).

It is in the spirit of Sylvester that I write this book for you, the student. I have tried to deal with your frustration and concerns by writing a book that is bright, fresh, and interesting. I have tried to make the material easy to read and understand. If I have overlooked anything, I hope that you will not dismiss this as the work of a typical absent-minded professor; please take time to write me a letter or a postcard to communicate your thoughts.

As you begin your study of precalculus, I want to remind you that learning mathematics requires systematic and regular study. Do not expect it to come to you in a flash, and do not expect to cram just before an examination. Take it in small steps, 1 day at a time, and you will find success at your doorstep.

1

Fundamental Concepts

CONTENTS

* Optional section

PREVIEW

This chapter reviews many of the preliminary ideas from previous courses and sets the foundation for the remainder of this course. Numbers, linear and quadratic equations and inequalities, as well as a two-dimensional system are introduced in this chapter. This chapter has 25 objectives, which are listed at the end of the chapter on pages 41–43.

PERSPECTIVE

Calculus is divided into two parts: differential and integral calculus. Both the definition of a derivative and that of an integral are dependent on a notion defined in calculus and called the *limit*. The formal definition of a limit involves both an absolute value and an inequality,

$$0 < |t - c| < \delta$$

as you can see in the portion of a page from a calculus book shown below. In the middle of the page you can also see that intervals are used. We introduce interval notation and discuss absolute value inequalities in Section 1.2.

We need to require that for *every* interval about L, no matter how small, we can find an interval of numbers about c all of whose F-values lie within that interval about L. In other words, given *any* positive radius ε about L, there exists some positive radius δ about c such that for all t within δ units of c (except $t = c$ itself) the values of $F(t)$ lie within ε units of L:

Thus, the closer t stays to c without equaling c, the closer $F(t)$ must stay to L.

DEFINITION

Limit

The *limit* of $F(t)$ as t approaches c is the number L if:

Given any radius $\varepsilon > 0$ about L there exists a radius $\delta > 0$ about c such that for all t

$$0 < |t - c| < \delta \quad \text{implies} \quad |F(t) - L| < \varepsilon. \qquad (5)$$

From George Thomas and Ross Finney, *Calculus and Analytic Geometry*, 7th ed. (Reading, Mass.: Addison-Wesley), pp. 57–58.

1.1 REAL NUMBERS

You are, no doubt, familiar with various sets of numbers, as well as with certain properties of those numbers. Table 1.1 gives a brief summary of some of the more common subsets of the set of **real numbers**.

TABLE 1.1
Sets of numbers

Name	Symbol	Set	Examples
Counting numbers or natural numbers	$\mathbb{N}$	$\{1, 2, 3, 4, \ldots\}$	$86; 1{,}986{,}412; \sqrt{16}; \sqrt{1}; \sqrt{100}; \ldots$ $\sqrt{a}$ is the nonnegative real number b so that $b^2 = a$. It is called the *principal square root* of a. Some square roots are natural numbers.
Whole numbers	$\mathbb{W}$	$\{0, 1, 2, 3, 4, \ldots\}$	$0; 86; 49; \frac{8}{4}; \frac{16{,}425}{25}; \ldots$ $\frac{a}{b}$ means $a \div b$. It is called the *quotient* of a and b. Some quotients are whole numbers.
Integers	$\mathbb{Z}$	$\{\ldots, -2, -1, 0, 1, 2, 3, \ldots\}$	$-8; 0; \frac{-16}{4}; 43{,}812; -96; -\sqrt{25}; \ldots$
Rational numbers	$\mathbb{Q}$	Numbers that can be written in the form p/q, where p and q are integers with $q \neq 0$; they are characterized as numbers whose decimal representations either terminate or repeat.	$\frac{2}{3}; \frac{4}{17}; 0.863214; 0.866\ldots; 5; 0; \frac{-19}{10}; -16; \sqrt{\frac{1}{4}};$ $3.1416; 8.\overline{6}; 0.1\overline{56}; \ldots$ An overbar indicates repeating decimals. Some square roots are also rational.
Irrational numbers	$\mathbb{Q}'$	Numbers whose decimal representations do not terminate and do not repeat	$4.1234567891011\ldots; 6.31331333133331\ldots;$ $\sqrt{2}; \sqrt{3}; \sqrt{5}; \pi; \frac{\pi}{2}; \ldots$ π is the ratio of the circumference of a circle to its diameter. It is sometimes approximated by 3.1416, but this is a rational approximation of an irrational number. We write $\pi \approx 3.1416$ to mean the numbers are *approximately* equal. Square roots of most numbers are irrational.
Real numbers	$\mathbb{R}$	Numbers that are either rational or irrational	All examples listed above are real numbers. Not all numbers are real numbers, however. Some of these will be considered in Section 1.3; they are called *complex numbers*.

The real numbers can most easily be visualized by using a **one-dimensional coordinate system** called a **real number line** (Figure 1.1). A **one-to-one correspondence**

Figure 1.1 A real number line

is established between all real numbers and all points on a real number line:

1. Every point on the line corresponds to precisely one real number.
2. Every real number corresponds to precisely one point.

A point associated with a particular number is called the **graph** of that number. Numbers associated with points to the right of the **origin** are called **positive real numbers** and those to the left are called **negative real numbers**. Numbers are called **opposites** if they are plotted an equal distance from the origin. The **opposite of a real number a** is denoted by $-a$. Notice that if a is positive, then $-a$ is negative; and if a is negative, then $-a$ is positive. It is a common error to think of $-a$ as a negative number; it might be negative, but it might also be positive. If, say, $a = -5$, then $-a = -(-5) = 5$, which is positive.

EXAMPLE 1 Graph the following numbers on a real number line:

$$5;\ -2;\ 2.5;\ 1.31331133311\ldots;\ \tfrac{2}{3};\ \pi;\ -\sqrt{2}$$

Solution When graphing, the exact positions of the points are usually approximated.

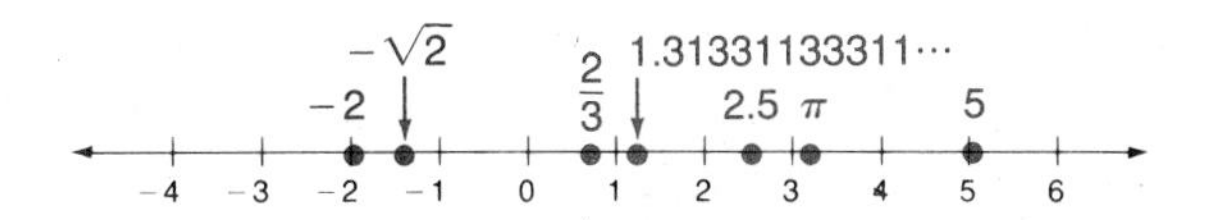

■

There are certain relationships between real numbers with which you should be familiar:

Less Than $a < b$, read "*a is less than b*," means $b - a$ is positive. On a number line, it means the graph of a is to the left of the graph of b.

Greater Than $a > b$, read "*a is greater than b*," means $b < a$ or $a - b$ is positive. On a number line, it means the graph of a is to the right of the graph of b.

Equal to $a = b$, read "*a is equal to b*," means that a and b represent the same real number. On a number line, the graphs of both a and b are the same point.

If a and b represent any real numbers, then they are related by a property called **Trichotomy** or **Property of Comparison**.

Property of Comparison

Given any two real numbers a and b, exactly one of the following holds:

1. $a = b$ 2. $a > b$ 3. $a < b$

EXAMPLE 2 Illustrate the Property of Comparison by replacing the □ by =, >, or <.

a. $-3 \square -7$ **b.** $\frac{2}{5} \square 0.4$ **c.** $\pi \square 3.1416$ **d.** $\sqrt{2} \square 1.4142$

Solution **a.** $-3 > -7$ Think of a number line. The graph of -3 is to the right of the graph of -7; we see that -3 is greater than -7 (so "$>$" is the answer).

b. $\frac{2}{5} = 0.4$ $.4 = \frac{4}{10} = \frac{2}{5}$, so the numbers are equal.

c. $\pi < 3.1416$ $\pi \approx 3.1415926$, so use "$<$," less than. (A good approximation of π is found by pressing [π] on a calculator.)

d. $\sqrt{2} > 1.4142$ $(1.4142)^2 = 1.99996164$, which is less than 2. Alternatively, you can press [2] [√] on a calculator for a display of 1.4142136, so $1.4142136 > 1.4142$. ■

The Property of Comparison establishes the order on a real number line. If you let $b = 0$, for example, then $a = 0$, $a < 0$, or $a > 0$. Using the definition of less than and greater than, it follows that

$a > 0$ if and only if a is positive.

$a < 0$ if and only if a is negative.

In addition to the Property of Comparison, there are four properties of equality that are used in mathematics:

Properties of Equality

Let a, b, and c be real numbers.

REFLEXIVE PROPERTY:	1. $a = a$.
SYMMETRIC PROPERTY:	2. If $a = b$, then $b = a$.
TRANSITIVE PROPERTY:	3. If $a = b$ and $b = c$, then $a = c$.
SUBSTITUTION PROPERTY:	4. If $a = b$, then a may be replaced by b (or b by a) throughout any statement without changing the truth or falsity of the statement.

EXAMPLE 3 Identify the property of equality illustrated by each:

a. $(a + b)(c + d) = (a + b)(c + d)$.

b. If $(a + b)(c + d) = (a + b)c + (a + b)d$ and $(a + b)c + (a + b)d = ac + bc + ad + bd$, then $(a + b)(c + d) = ac + bc + ad + bd$.

c. If $a(b + c) = ab + ac$, then $ab + ac = a(b + c)$.

d. If $a = 3$ and $(a + 2)(a + 5) = 40$, then $(3 + 2)(3 + 5) = 40$.

Solution **a.** Reflexive **b.** Transitive **c.** Symmetric **d.** Substitution ■

Finally, consider some properties of the real numbers. When we are adding real numbers the result is called the **sum** and the numbers added are called **terms**. When multiplying real numbers the result is called the **product** and the numbers multiplied are called the **factors**. The result from subtraction is called the **difference**; the result from division is the **quotient**. The real numbers, together with the relation of equality and the operations of addition and multiplication, satisfy what are called **field properties**.

Field Properties for the Set $\mathbb{R}$ of Real Numbers

Let a, b, and c be real numbers.

	Addition Properties	*Multiplication Properties*
CLOSURE:	$a + b$ is a unique real number.	ab is a unique real number.
COMMUTATIVE:	$a + b = b + a$	$ab = ba$
ASSOCIATIVE:	$(a + b) + c = a + (b + c)$	$(ab)c = a(bc)$
IDENTITY:	There exists a unique real number zero, denoted by 0, such that $a + 0 = 0 + a = a$.	There exists a unique real number one, denoted by 1, such that $a \cdot 1 = 1 \cdot a = a$.
INVERSE:	For each real number a, there is a unique real number $-a$ such that $a + (-a) = (-a) + a = 0$.	For each *nonzero* real number a, there is a unique real number $1/a$ such that $a\left(\frac{1}{a}\right) = \left(\frac{1}{a}\right)a = 1$
DISTRIBUTIVE:	$a(b + c) = ab + ac$	

A set is said to be **closed** if it satisfies the closure property, as shown in Example 4.

EXAMPLE 4 The set $\{-1, 0, 1\}$ is closed for multiplication since

$$\begin{array}{ll} (-1)(-1) = 1 & (0)(1) = 0 \\ (-1)(0) \quad = 0 & (0)(0) = 0 \\ (-1)(1) \quad = -1 & (1)(1) = 1 \end{array}$$

But it is not closed for addition since

$$1 + 1 = 2$$

which is not in the set. ■

EXAMPLE 5 Identify the field property illustrated.

a. $4 \cdot 3$ is a real number.

b. $\frac{1}{4}(4 \cdot 3) = (\frac{1}{4} \cdot 4)3$

c. $(\frac{1}{4} \cdot 4)3 = 1 \cdot 3$
d. $1 \cdot 3 = 3$
e. $4 + (3 + 5) = (3 + 5) + 4$
f. $4 + (3 + 5) = (4 + 3) + 5$

Solution a. Closure property
b. Associative property for multiplication
c. Inverse for multiplication
d. Identity for multiplication
e. Commutative property for addition
f. Associative property for addition ■

PROBLEM SET 1.1

A

Classify each example in Problems 1–6 as a natural number, whole number, integer, rational number, irrational number, real number, or none of these. Notice from Table 1.1 that each number listed may be in more than one of these sets.

1. a. -5 **b.** $\frac{13}{2}$ **c.** $\sqrt{25}$
d. $\sqrt{20}$ **e.** 2π

2. a. 9 **b.** $\sqrt{16}$ **c.** $\frac{0}{5}$
d. $\frac{5}{0}$ **e.** $0.\overline{3}$

3. a. $0.\overline{5}$ **b.** $0.\overline{463}$ **c.** $\sqrt{1000}$
d. 0.281 **e.** $\frac{\pi}{6}$

4. a. $\sqrt{\frac{1}{8}}$ **b.** $\sqrt{\frac{1}{9}}$ **c.** 0.5656
d. $0.5656\ldots$ **e.** $\frac{\pi}{3}$

5. a. $\sqrt{169}$ **b.** $\sqrt{200}$ **c.** π
d. 3.14192 **e.** -12.5

6. a. $-\sqrt{4}$ **b.** $\frac{3}{5}$ **c.** $\frac{4}{9}$
d. $\sqrt{\frac{4}{9}}$ **e.** $\frac{\pi}{4}$

Graph each number given in Problems 7–12 on a real number line.

7. $-3; -\sqrt{3}; \frac{5}{4}; 2; 0.1234567891011\ldots$

8. $-1; \frac{4}{5}; -\sqrt{5}; 2\frac{1}{8}; 1.03469217$

9. $-4; -\frac{-5}{3}; 0; \sqrt{2}; -1.343443444\ldots$

10. $0; \frac{\pi}{6}; \frac{\pi}{-3}; \frac{\pi}{2}; \pi$

11. $-\pi; -\frac{\pi}{2}; \frac{\pi}{3}; 0; 0.5$

12. $-\pi; 0.25; 0; \frac{\pi}{4}; 1$

Illustrate the Property of Comparison in Problems 13–18 by replacing the □ *by* $=$, $>$, *or* $<$.

13. a. $-10 \square -3$ **b.** $6 - 9 \square 9 - 6$

14. a. $-4 \square -8$ **b.** $\frac{5}{4} \square 1.2$

15. a. $-\frac{8}{3} \square -2.66$ **b.** $\frac{35}{21} \square \frac{-30}{-18}$

16. a. $\sqrt{3} \square 1.7$ **b.** $\frac{1}{3} \square 0.33333333$

17. a. $\frac{3}{4} + \frac{4}{5} \square \frac{31}{20}$ **b.** $\pi \square \frac{22}{7}$

18. a. $\frac{2}{3} + \frac{5}{2} \square \frac{19}{6}$ **b.** $\frac{2}{3} \square 0.66666667$

Identify the property of equality or field property illustrated by Problems 19–36.

19. $14x + 8x = 14x + 8x$

20. $10y + 4y = 10y + 4y$

21. $14x + 8x = (14 + 8)x$

22. $10y + 4y = (10 + 4)y$

23. $(14 + 8)x = 22x$

24. $(10 + 4)y = 14y$

25. If $14x + 8x = (14 + 8)x$ and $(14 + 8)x = 22x$, then $14x + 8x = 22x$.

26. If $10y + 4y = (10 + 4)y$ and $(10 + 4)y = 14y$, then $10y + 4y = 14y$.

27. If $14x + 8y = 22$ and $y = b$, then $14x + 8b = 22$.

28. If $22 = 14x + 8y$, then $14x + 8y = 22$.

29. 4π is a real number.

30. $\frac{1}{6}$ is a real number.

31. $\pi \cdot \frac{1}{\pi} = 1$

32. $\frac{1}{\frac{1}{6}}$ is a real number.

33. $5x(a + b) = 5xa + 5xb$

34. $(a + b)5x = 5x(a + b)$

35. $\frac{\sqrt{3} + 2}{\sqrt{5} + 1} = \frac{\sqrt{3} + 2}{\sqrt{5} + 1} \cdot \frac{\sqrt{5} - 1}{\sqrt{5} - 1}$

36. $\frac{4}{5} = \frac{4}{5} \cdot \frac{3}{3}$

B

37. Is the set $\{0, 1\}$ closed for addition?

38. Is the set $\{0, 1\}$ closed for multiplication?

39. Is the set $\{-1, 0, 1\}$ closed for subtraction?

40. Is the set $\{-1, 0, 1\}$ closed for nonzero division?

41. Is the set $\{0, 3, 6, 9, 12, \ldots\}$ closed for addition?

42. Is the set $\{0, 3, 6, 9, 12, \ldots\}$ closed for multiplication?

43. Is the set $\{0, 3, 6, 9, 12, \ldots\}$ closed for nonzero division?

44. Find an example showing that the operation of subtraction on $\mathbb{R}$ is not commutative.

45. Find an example showing that the operation of division on the set of nonzero real numbers is not commutative.

46. Is the operation of subtraction on $\mathbb{R}$ associative?

47. Is the operation of division on the set of nonzero real numbers associative?

C

48. What is the additive inverse of the real number $2 + \sqrt{3}$?

49. What is the additive inverse of the real number $\frac{\pi}{3} + 1$?

50. What is the multiplicative inverse of the real number $2 + \sqrt{3}$?

51. What is the multiplicative inverse of the real number $\frac{\pi}{3} + 1$?

52. What is the multiplicative inverse of the real number $\sqrt{5} + 1$?

53. Which of the field properties are satisfied by the set of natural numbers?

54. Which of the field properties are satisfied by the set of whole numbers?

55. Which of the field properties are satisfied by the set of integers?

56. Which of the field properties are satisfied by the set of rational numbers?

57. Which of the field properties are satisfied by the set of irrational numbers?

58. Which of the field properties are satisfied by the set of real numbers?

59. Which of the field properties are satisfied by the set $H = \{1, -1, i, -i\}$, where i is a number such that $i^2 = -1$?

60. Which of the field properties are satisfied by the set of nonnegative multiples of 5, namely $F = \{0, 5, 10, 15, 20, 25, \ldots\}$?

1.2 INTERVALS, INEQUALITIES, AND ABSOLUTE VALUES

A **linear equation in one variable** is an equation that can be written in the form

$$ax + b = 0 \qquad (a \neq 0)$$

where x is a **variable** and a and b are any real numbers. An **open** or **conditional equation** is an equation containing a variable that may be either true or false, depending on the replacement for the variable. A **root** or a **solution** is a replacement for the variable that makes the equation true. We also say that the root **satisfies** the equation. The **solution set** of an open equation is the set of all solutions of the equation. To **solve an equation** means to find its solution set. If there are no values for the variable that satisfy an equation, then the solution set is said to be **empty** and is denoted by ∅. If every replacement of the variable makes the equation true, then the equation is called an **identity**.

You will need to know how to solve linear inequalities involving variables. There are other inequality relationships besides less than and greater than. For example:

Less Than or Equal to $a \leq b$, read "*a is less than or equal to b*," means that either $a < b$ or $a = b$ (but not both).

Greater Than or Equal to $a \geq b$, read "*a is greater than or equal to b*," means that either $a > b$ or $a = b$ (but not both).

Between $b < a < c$, read "*a is between b and c*," means *both* $b < a$ and $a < c$. Additional between relationships are also used:

$b \leq a \leq c$ means $b \leq a$ and $a \leq c$
$b \leq a < c$ means $b \leq a$ and $a < c$
$b < a \leq c$ means $b < a$ and $a \leq c$

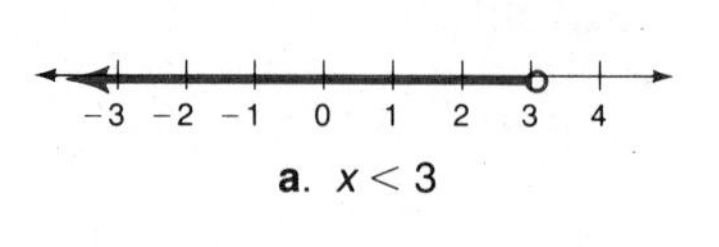

a. $x < 3$

-3 -2 -1 0 1 2 3 4
b. $x \leq 3$

-3 -2 -1 0 1 2 3 4
c. $x > 3$

-3 -2 -1 0 1 2 3 4
d. $x \geq 3$

Figure 1.2 Graph of inequality statements

In this book, when we use the word *between* we mean strictly between, namely $b < a < c$.

Graphs of linear inequality statements with a single variable are drawn on a one-dimensional coordinate system. For example, $x < 3$ denotes the interval shown in Figure 1.2a. Notice that $x \neq 3$, and this fact is shown by an open circle as the endpoint of the ray. Compare this with $x \leq 3$ (Figure 1.2b) in which the endpoint $x = 3$ is included. Notice that to sketch the graph you darken (or color) the appropriate portion.

The between relationships define **intervals** on a number line. An interval is said to be **closed** if it includes both endpoints; it is **open** if it does not include either endpoint. Study the terminology shown in Figure 1.3. In all cases, a and b are called the **endpoints** of the interval.

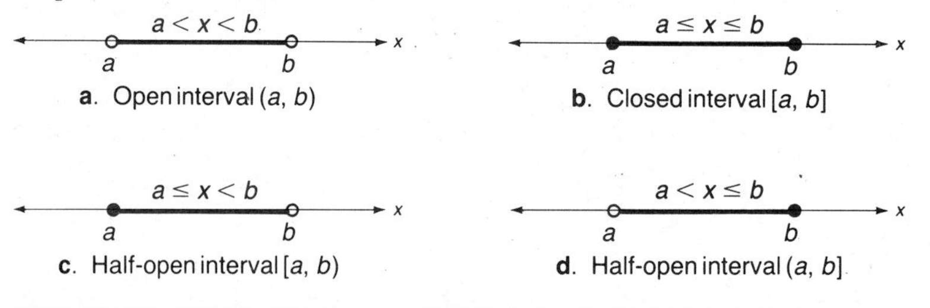

a. Open interval (a, b)
b. Closed interval $[a, b]$
c. Half-open interval $[a, b)$
d. Half-open interval $(a, b]$

Figure 1.3 Interval notation

Note the correct use of **interval notation**: (a, b), $[a, b]$, $[a, b)$, and $(a, b]$

EXAMPLE 1 Write interval notation for each given inequality statement. Also graph each of the given inequalities.

a. $-6 \leq x < 5$; $[-6, 5)$

-6 5

b. $7 < x < 8$; $(7, 8)$

7 8

c. $-9 \leq x \leq -2$; $[-9, -2]$

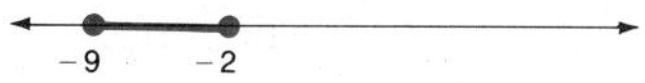

d. $-5 < x \leq 0$; $(-5, 0]$

EXAMPLE 2 Graph each of the given intervals.

a. $[1, 3]$

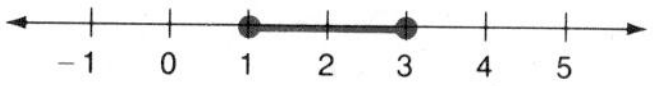

b. $(2, 5]$

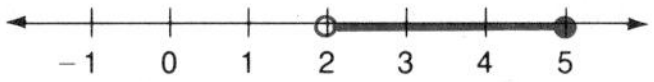

c. $[0, 4)$

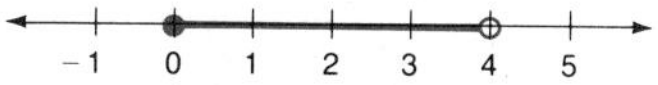

d. $(-1, 4)$

Sometimes intervals of unlimited extent in one or both directions are also denoted by using interval notation and the symbol ∞ as shown in Example 3. Do not interpret ∞ (infinity) as a number; as illustrated by Example 3, it is merely a notational device.

EXAMPLE 3 Write interval notation for each given inequality statement. Also graph each of the given inequalities.

a. $x > 2$; $(2, \infty)$

b. $x \geq 2$; $[2, \infty)$

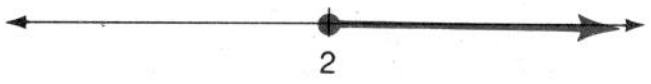

c. $x < 2$; $(-\infty, 2)$

d. $x \leq 2$; $(-\infty, 2]$

e. x is any real number; $(-\infty, \infty)$

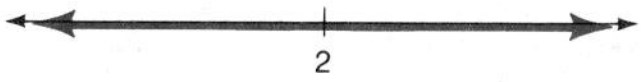

A **linear inequality in one variable** is an inequality that can be written in one of the following forms:

$$ax + b < 0 \qquad ax + b \leq 0 \qquad ax + b > 0 \qquad ax + b \geq 0$$

where x is a variable and a is a nonzero real number and b is a real number. Linear inequalities are solved by using the following principles stated for less than, but they are also true for $\leq$, $>$, and $\geq$. These symbols are referred to as the **order** of the inequality.

Properties of Inequality

TRANSITIVITY:
If $a < b$ and $b < c$, then $a < c$ for any real numbers a, b, and c.

ADDITION PRINCIPLE:
If $a < b$, then $a + c < b + c$ for any real numbers a, b, and c.

POSITIVE MULTIPLICATION PRINCIPLE:
If $a < b$, and c is a positive real number then $ac < bc$.

NEGATIVE MULTIPLICATION PRINCIPLE:
If $a < b$, and c is a negative real number then $ac > bc$.

Procedure for Solving Linear Inequalities

The procedure and terminology for solving linear inequalities are identical to the procedure and terminology for solving linear equations except for one fact: If you multiply or divide by a negative number, the order of the inequality is reversed. And remember, from the definition of greater than, that if you reverse the left and right sides of an inequality, the order is also reversed.

EXAMPLE 4 Solve the inequality $5x - 3 \geq 7$ and state the solution set using interval notation.

Solution

$$5x - 3 \geq 7$$
$$5x - 3 + 3 \geq 7 + 3$$
$$5x \geq 10$$
$$x \geq 2$$

Solution: $[2, \infty)$ ■

EXAMPLE 5 Solve the given inequality and write the solution set using interval notation.

Solution

$$2(4 - 3t) < 5t - 14$$ Eliminate parentheses.

$$8 - 6t < 5t - 14$$ Isolate the variable on one side.

$$-6t < 5t - 22$$
$$-11t < -22$$
$$t > 2$$ Reverse the order of the inequality when you divide by a negative number.

Solution: $(2, \infty)$ ■

Sometimes two inequality statements are combined, as with

$$-3 \leq 2x + 1 \leq 5$$

This means that $-3 \leq 2x + 1$ *and* $2x + 1 \leq 5$. It also means that $2x + 1$ is between -3 and 5. When solving an inequality of this type, you can frequently work both inequalities simultaneously as shown in Examples 6 and 7.

EXAMPLE 6

$$-3 \leq 2x + 1 \leq 5$$
$$-3 - 1 \leq 2x + 1 - 1 \leq 5 - 1$$
$$-4 \leq 2x \leq 4$$
$$\frac{-4}{2} \leq \frac{2x}{2} \leq \frac{4}{2}$$
$$-2 \leq x \leq 2$$

Your goal is now to isolate the variable in the middle of the between statement. Each inequality statement is equivalent to the preceding one. Whatever you do to one part, you do to all three parts.

Solution: $[-2, 2]$ ■

EXAMPLE 7

$$-5 \leq 1 - 3x < 10$$
$$-6 \leq -3x < 9$$
$$\frac{-6}{-3} \geq \frac{-3x}{-3} > \frac{9}{-3}$$ Inequality reverses when you multiply or divide by a negative.

$$2 \geq x > -3$$

However, $2 \geq x > -3$ is not convenient for converting to interval notation, so it is rewritten as

$$-3 < x \leq 2$$

Solution: $\mathbf{(-3, 2]}$ ■

A very important idea in mathematics involves the notion of **absolute value**. If a is a real number, then its graph on a number line is some point, call it A. The distance between A and the origin is the geometric interpretation of the **absolute value of *a***.

Absolute Value

The absolute value of a real number a is denoted by $|a|$ and is defined by

$$|a| = a \qquad \text{if } a \geq 0$$
$$|a| = -a \qquad \text{if } a < 0$$

Thus $|5| = 5$, $|-5| = 5$, $|0| = 0$, and $|-\pi| = \pi$. Notice that $|a|$ is nonnegative for all values of a.

EXAMPLE 8 $|\pi - 3| = \boldsymbol{\pi - 3}$ since $\pi - 3 > 0$; use the first part of the definition of absolute value ■

EXAMPLE 9 $|\pi - 4| = -(\pi - 4)$
$= \boldsymbol{4 - \pi}$ since $\pi - 4 < 0$; use the second part of the definition of absolute value ■

EXAMPLE 10 $|w^2 + 1| = \boldsymbol{w^2 + 1}$ since $w^2 + 1$ is positive for all real values of w ■

EXAMPLE 11 $|-4 - t^2| = -(-4 - t^2)$
$= \boldsymbol{t^2 + 4}$ since $-4 - t^2$ is negative for all real values of t ■

Since $|a|$ can be interpreted as the distance between the point A whose coordinate is a and the origin, it is a straightforward derivation to show that the distance between any two points on a number line can be expressed as an absolute value.

Distance on a Number Line

P_1 P_2
x_1 x_2
d

Let x_1 and x_2 be the coordinates of two points P_1 and P_2, respectively, on a number line. The distance d between P_1 and P_2 is

$$d = |x_2 - x_1|$$

Notice that if a is the coordinate of P_1 and 0 is the coordinate of P_2, then $d = |a - 0| = |a|$; and if 0 is the coordinate of P_1 and a is the coordinate of P_2, then $d = |0 - a| = |-a|$. Since d is the same distance for both these calculations,

$$|a| = |-a|$$

It also follows that $d = |x_2 - x_1| = |x_1 - x_2|$ and that $|x^2| = |x|^2$.

EXAMPLE 12 Find the distance between the points whose coordinates are given.

a. (-8) and (2); $d = |2 - (-8)| = |10| = \mathbf{10}$
b. (108) and (-34); $d = |-34 - 108| = |-142| = \mathbf{142}$
c. $(-\pi)$ and $(-\sqrt{2})$; $d = |-\sqrt{2} + \pi| = \boldsymbol{\pi - \sqrt{2}}$
d. $(-\pi)$ and $(-\sqrt{10})$; $d = |-\sqrt{10} + \pi| = -(-\sqrt{10} + \pi) = \boldsymbol{\sqrt{10} - \pi}$ ■

Since $|x - a|$ can be interpreted as the distance between x and a on the number line, an equation of the form

$$|x - a| = b$$

has two values of x that are a given distance from a when represented on a number line. For example, $|x - 5| = 3$ states that x is 3 units from 5 on a number line. Thus x is either 2 or 8.

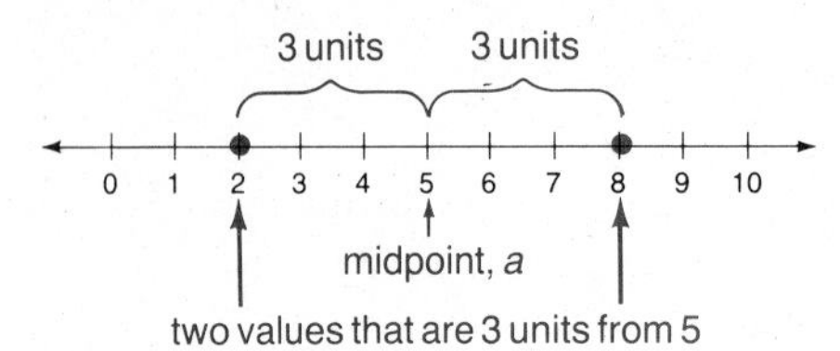

You can verify this conclusion by proving the following property.

Absolute Value Equations

$|a| = b$ where $b \geq 0$ if and only if $a = b$ or $a = -b$.

Proof There are two parts to a proof using the words "if and only if." **(1)** If $|a| = b$, then $a = b$ or $a = -b$. **(2)** If $a = b$ or $a = -b$, then $|a| = b$. We will prove the first part here and leave the second part as a problem. That is, suppose $|a| = b$; we now wish to show that $a = b$ or $a = -b$. Begin with the Property of Comparison to compare the real number a with the real number zero:

$$a > 0 \qquad a = 0 \qquad \text{or} \qquad a < 0$$

If $a > 0$ or if $a = 0$, then $|a| = a$ and

$\lvert a\rvert = b$	Given.
$a = b$	Substitution of a for $\lvert a\rvert$.

If $a < 0$, then $|a| = -a$ and

$\lvert a\rvert = b$	Given.
$-a = b$	Substitution of $-a$ for $\lvert a\rvert$.
$a = -b$	Multiplication of both sides by -1. □

EXAMPLE 13 Solve $|x + 5| = 2$.

Solution

$$x + 5 = 2 \quad \text{or} \quad x + 5 = -2$$
$$x = -3 \quad \text{or} \quad x = -7$$

Solution: $\{\mathbf{-3, -7}\}$ ■

EXAMPLE 14 Solve $|x + 5| = -2$.

Solution The absolute value of every real number is nonnegative, so the solution set is empty. ■

If you are solving an absolute value equation with absolute values on both sides of the equation, the following property may be useful.

Procedure for Solving Absolute Value Equations

$|a| = |b|$ if and only if $a = b$ or $a = -b$.

You are asked to prove this in Problem 70 of Problem Set 1.2.

EXAMPLE 15 Solve $|x + 5| = |3x - 4|$.

Solution Use the given property of absolute value.

$$x + 5 = 3x - 4 \qquad \text{or} \qquad x + 5 = -(3x - 4)$$
$$-2x = -9 \qquad\qquad x + 5 = -3x + 4$$
$$x = \frac{9}{2} \qquad\qquad 4x = -1$$
$$x = -\frac{1}{4}$$

Solution: $\{\frac{9}{2}, -\frac{1}{4}\}$ ■

Since $|x - 5| = 3$ states that x is 3 units from 5, the inequality $|x - 5| < 3$ states that x is any number less than 3 units from 5.

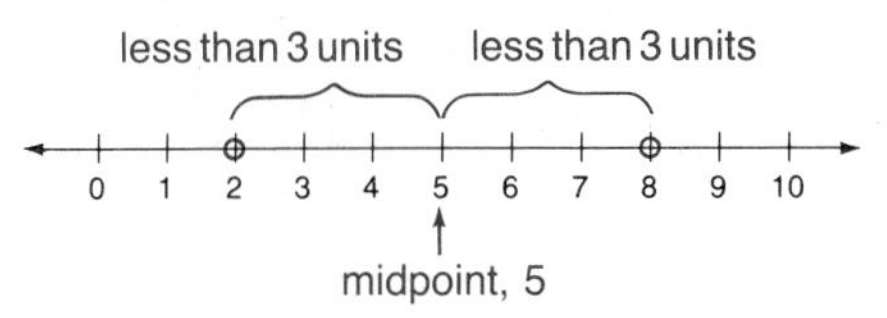

On the other hand, $|x - 5| > 3$ states that x is any number greater than 3 units from 5.

These properties are summarized by four absolute value inequality properties where a and b are real numbers and $b > 0$.

Absolute Value Inequalities

$\lvert a\rvert < b$	if and only if	$-b < a < b$
$\lvert a\rvert \le b$	if and only if	$-b \le a \le b$
$\lvert a\rvert > b$	if and only if	$a > b$ or $a < -b$
$\lvert a\rvert \ge b$	if and only if	$a \ge b$ or $a \le -b$

Proof *Prove:* If $|a| < b$, then $-b < a < b$. By definition:

$$|a| = a \quad \text{if } a \geq 0$$
$$|a| = -a \quad \text{if } a < 0$$

Case i: If $a \geq 0$

$\lvert a\rvert < b$	Given.
$a < b$	Substitute a for $\lvert a\rvert$.

Case ii: If $a < 0$

$\lvert a\rvert < b$	Given.
$-a < b$	Substitute $-a$ for $\lvert a\rvert$.
$0 < a + b$	Add a to both sides.
$-b < a$	Subtract b from both sides.

Therefore $-b < a$ and $a < b$ as seen from cases *i* and *ii*. This is the same as saying $-b < a < b$. The proofs of the other parts are similar and are left for the exercises. □

EXAMPLE 16 Solve $|2x - 3| \leq 4$.

Solution $-4 \leq 2x - 3 \leq 4$. Now solve the *between* relationship:

$$-4 + 3 \leq 2x - 3 + 3 \leq 4 + 3$$
$$-1 \leq 2x \leq 7$$
$$\frac{-1}{2} \leq \frac{2x}{2} \leq \frac{7}{2}$$
$$-\frac{1}{2} \leq x \leq \frac{7}{2}$$

Solution: $[-\frac{1}{2}, \frac{7}{2}]$ ■

EXAMPLE 17 Solve $|4 - x| < -3$.

Solution Absolute values must be nonnegative, so the solution set is empty. ■

EXAMPLE 18 Solve $|x + 3| > 4$.

Solution

$$x + 3 > 4 \quad \text{or} \quad x + 3 < -4$$
$$x > 1 \quad \text{or} \quad x < -7$$

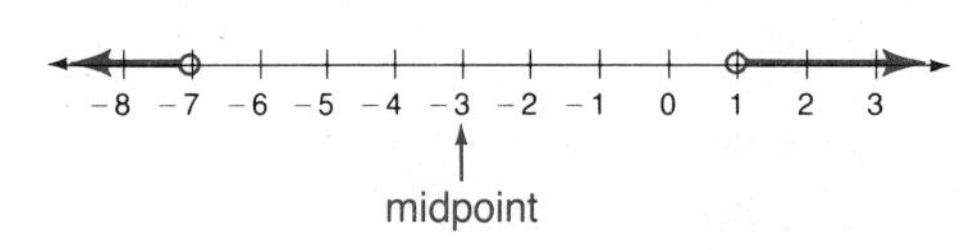

Do not attempt to write this as $1 < x < -7$ since transitivity implies $1 < -7$ (which is not true). Remember: $a < x < c$ means that x satisfies *both* the conditions $a < x$ and $x < c$, and transitivity implies that $a < c$.

Interval notation may be used to express the solution set when more than one part of a number line is included if you use the union symbol $\cup$. The solution is

$$(-\infty, -7) \cup (1, \infty)$$ ■

PROBLEM SET 1.2

A

Write interval notation and graph each of the inequalities in Problems 1–4.

1. a. $3 < x < 7$ **b.** $-4 < x < -1$
c. $-2 \leq x \leq 6$ **d.** $-3 < x \leq 0$

2. a. $-6 \leq x \leq 0$ **b.** $-3 < x \leq 3$
c. $-2 \leq x \leq 2$ **d.** $-5 \leq x < 1$

3. a. $x \leq -3$ **b.** $x \geq -2$
c. $x < 0$ **d.** $2 < x$

4. a. $3 < x \leq 6$ **b.** $5 < x$
c. $10 > x$ **d.** $-0.01 \leq x \leq 0.01$

Graph each of the intervals in Problems 5–8.

5. a. $[-3, 2]$ **b.** $(-2, 2)$
c. $(-\infty, 3]$ **d.** $[-2, \infty)$

6. a. $(-3, 4]$ **b.** $[-3, 4)$
c. $[-4, 3)$ **d.** $(2, \infty)$

7. a. $(-2, 0) \cup (3, 5)$ **b.** $[-8, 3) \cup [5, \infty)$
c. $(-\infty, -3] \cup (0, 3]$ **d.** $[-5, -1) \cup (0, 5]$

8. a. $(-\infty, 0) \cup (0, \infty)$ **b.** $(-\infty, 2) \cup (2, \infty)$
c. $(-\infty, 1) \cup (1, \infty)$ **d.** $(-\infty, 6) \cup (6, \infty)$

Write each of the intervals in Problems 9–12 as an inequality. Use x as the variable.

9. a. $[-4, 2]$ **b.** $[-1, 2]$
c. $(0, 8)$ **d.** $(-5, 3]$

10. a. $(-3, -1)$ **b.** $[-5, 2)$
c. $[5, 9]$ **d.** $(-6, -2)$

11. a. $(-\infty, 2)$ **b.** $(6, \infty)$
c. $(-1, \infty)$ **d.** $(-\infty, 3]$

12. a. $(-\infty, -6]$ **b.** $(-\infty, 0)$
c. $[5, \infty)$ **d.** $(-1, \infty)$

Write each expression in Problems 13–15 without using the symbol for absolute value.

13. a. $|\pi - 2|$ **b.** $|\pi - 5|$
c. $|2\pi - 6|$ **d.** $|2\pi - 7|$

14. a. $|x^2|$ **b.** $|x^2 + 4|$
c. $|-3 - x^2|$ **d.** $|-2 + \sqrt{5}|$

15. a. $|\sqrt{2} - 1|$ **b.** $|\sqrt{2} - 2|$
c. $\left|\frac{\pi}{6} - 1\right|$ **d.** $\left|\frac{2\pi}{3} - 1\right|$

Find the distance between the points whose coordinates are given in Problems 16–18.

16. a. (8) and (15) **b.** (-103) and (6)
c. (-143) and (-120)

17. a. $(-\pi)$ and (-3) **b.** (23) and (-96)
c. $(-\sqrt{5})$ and (-3)

18. a. (π) and (-4) **b.** $(\sqrt{3})$ and $(-\pi)$
c. $(-\pi)$ and $(-\sqrt{2})$

Solve the equations in Problems 19–30.

19. $|x| = 5$ **20.** $|x| = 10$
21. $|x| = -1$ **22.** $|x| = -4$
23. $|x - 3| = 4$ **24.** $|x + 9| = 3$
25. $|x - 9| = 15$ **26.** $|2x + 4| = -8$
27. $|2x + 4| = -12$ **28.** $|3 - 2x| = 7$
29. $|5x + 4| = 6$ **30.** $|5 - 3x| = 14$

B

Solve the inequalities in Problems 31–54 and write each solution in interval notation.

31. $3x - 9 \geq 12$ **32.** $-x > -36$
33. $-3x \geq 123$ **34.** $3(2 - 4x) \leq 0$
35. $5(3 - x) > 3x - 1$ **36.** $-5 \leq 5x \leq 25$
37. $-8 < 5x < 0$ **38.** $3 \leq -x < 8$
39. $-3 \leq -x < -1$ **40.** $-5 < 3x + 2 \leq 5$
41. $-7 \leq 2x + 1 < 5$ **42.** $-4 < 1 - 5x \leq 11$
43. $-5 \leq 3 - 2x < 18$ **44.** $|x - 5| \leq 1$
45. $|x - 3| \leq 5$ **46.** $|x - 8| \leq 0.01$
47. $|x - 3| < 0.001$ **48.** $|3x + 2| < -3$
49. $|2x + 1| < -1$ **50.** $|3 - x| < 5$
51. $|4 - 3x| < 3$ **52.** $|x + 1| > 3$
53. $|2x + 7| > 5$ **54.** $|3 - 2x| > 5$

55. ***Business*** If Amex Automobile Rental charges \$35 per day and 45¢ per mile, how many miles can you drive and keep the cost under \$125 per day?

56. ***Business*** A saleswoman is paid a salary of \$300 plus a 40% commission on sales. How much does she need to sell in order to have an income of at least \$2000?

57. ***Chemistry*** A certain experiment requires that the temperature be between 20° and 30° C. If Fahrenheit and Celsius degrees are related by the formula $C = \frac{5}{9}(F - 32)$, what are the permissible temperatures in Fahrenheit?

58. ***Economics*** An economist estimates that the consumer price index will grow by 14%, give or take 1%. Let c represent the growth rate of the consumer price index.
a. Write the condition of this problem as an inequality.
b. Write the condition of this problem as an absolute value statement.

59. ***Business*** The management of a certain company needs to monitor the activities of salespersons whose sales are not the usual \$15,000 per week. That is, if the sales of a certain employee are less than \$10,000 or greater than \$20,000, the company needs to monitor the amount of time spent on the job. Let s represent the amount of weekly sales.

   a. Write the condition of this problem as an inequality.
   b. Write the condition of this problem as an absolute value statement.

C

Solve the inequalities in Problems 60–65 and write the solution in interval notation.

60. $\left|\frac{5-x}{2}\right| > 1$ **61.** $\left|\frac{3-x}{2}\right| > 3$

62. $\frac{5}{|x-2|} \le 1$ **63.** $\frac{3}{|x+1|} \ge 3$

64. $|2x-1| + |x| - 3 = 0$ **65.** $|3x+1| + |x| - 5 = 0$

66. Prove that if b is a nonnegative real number, and if $a = b$ or $a = -b$, then $|a| = b$.

67. If $|a| \le b$, show that $-b \le a \le b$.

68. If $|a| > b$, show that either $a > b$ or $a < -b$.

69. If $|a| \ge b$, show that either $a \ge b$ or $a \le -b$.

70. Prove that $|a| = |b|$ if and only if $a = b$ or $a = -b$.

1.3 COMPLEX NUMBERS*

To find the roots of certain equations, you must sometimes consider the square roots of negative numbers. Since the set of real numbers does not allow for such a possibility, a number that is *not a real number* is defined. This number is denoted by the symbol i.

The Imaginary Unit

The number i, called the **imaginary unit**, is defined as a number with the following properties:

$$i^2 = -1 \quad \text{and} \quad \sqrt{-a} = i\sqrt{a} \quad (a > 0)$$

With this number you can write the square roots of any negative numbers as the product of a real number and the number i. Thus $\sqrt{-9} = i\sqrt{9} = 3i$. For any positive real number b, $\sqrt{-b} = i\sqrt{b}$. We now consider a new set of numbers.

Complex Numbers

The set of all numbers of the form

$$a + bi$$

with real numbers a and b and i the imaginary unit is called the set of **complex numbers**.

If $b = 0$, then $a + bi = a + 0i = a$, which is a **real number**; thus the real numbers form a subset of the complex numbers. If $a \ne 0$ and $b \ne 0$, then $a + bi$ is called an **imaginary number**; if $a = 0$ and $b \ne 0$, then $a + bi = 0 + bi = bi$ is called a **pure imaginary number**. If $a = 0$ with $b = 1$, then $a + bi = 0 + 1 \cdot i = i$, which is the imaginary unit.

* Complex numbers can be omitted from this text. All parts of the book requiring the use of complex numbers will be marked as optional.

A complex number is **simplified** if it is written in the form $a + bi$, where a and b are simplified real numbers and i is the imaginary unit. In order to work with complex numbers, you will need definitions of equality along with the usual arithmetic operations. Let $a + bi$ and $c + di$ be complex numbers. Values that could cause division by zero are excluded.

Operations with Complex Numbers

EQUALITY: $a + bi = c + di$ if and only if $a = c$ and $b = d$

ADDITION: $(a + bi) + (c + di) = (a + c) + (b + d)i$

SUBTRACTION: $(a + bi) - (c + di) = (a - c) + (b - d)i$

MULTIPLICATION: $(a + bi)(c + di) = (ac - bd) + (ad + bc)i$

DIVISION: $\dfrac{a + bi}{c + di} = \dfrac{(ac + bd) + (bc - ad)i}{c^2 + d^2} = \dfrac{ac + bd}{c^2 + d^2} + \dfrac{bc - ad}{c^2 + d^2}i$

It is not necessary to memorize these definitions, since you can deal with two complex numbers as you would any binomials as long as you remember that $i^2 = -1$. Also notice $i^3 = i \cdot i^2 = -i$ and $i^4 = i^2 \cdot i^2 = 1$.

EXAMPLE 1 Simplify each expression.

a. $(4 + 5i) + (3 + 4i) = \mathbf{7 + 9i}$

b. $(2 - i) - (3 - 5i) = \mathbf{-1 + 4i}$

c. $(5 - 2i) + (3 + 2i) = \mathbf{8}$ or $8 + 0i$

d. $(4 + 3i) - (4 + 2i) = \mathbf{i}$ or $0 + i$

e. $(2 + 3i)(4 + 2i) = 8 + 16i + 6i^2$
$= 8 + 16i - 6$
$= \mathbf{2 + 16i}$

f. $i^{94} = i^{4 \cdot 23 + 2}$
$= (i^4)^{23}(i^2)$
$= 1 \cdot i^2$
$= \mathbf{-1}$

g. $i^{125} = i^{4 \cdot 31 + 1}$
$= (i^4)^{31}(i^1)$
$= \mathbf{i}$

h. $i^{1883} = i^{4 \cdot 470 + 3}$
$= (i^4)^{470}(i^3)$
$= i^3$
$= \mathbf{-i}$ ■

EXAMPLE 2 Verify that multiplying $(a + bi)(c + di)$ in the usual algebraic way gives the same result as that shown in the definition of multiplication of complex numbers.

Solution

$$\begin{aligned}(a + bi)(c + di) &= ac + adi + bci + bdi^2 \\ &= ac + (ad + bc)i - bd \\ &= (ac - bd) + (ad + bc)i\end{aligned}$$ ■

EXAMPLE 3 Simplify $(4 - 3i)(4 + 3i)$.

Solution

$$\begin{aligned}(4 - 3i)(4 + 3i) &= 16 - 9i^2 \\ &= 16 + 9 \\ &= \mathbf{25}\end{aligned}$$ ■

Example 3 gives a clue for dividing complex numbers. The definition would be difficult to remember, so instead we use the idea of *conjugates*. The complex numbers $a + bi$ and $a - bi$ are called **complex conjugates**, and each is the conjugate of the other:

$$(a + bi)(a - bi) = a^2 - b^2i^2$$
$$= a^2 + b^2$$

which is a real number. Thus you can simplify a quotient by using the conjugate of the denominator, as illustrated by Examples 4 to 6.

EXAMPLE 4 Simplify $\dfrac{15 - 5i}{2 - i}$.

Solution

$$\frac{15 - 5i}{2 - i} = \frac{15 - 5i}{2 - i} \cdot \overbrace{\frac{\mathbf{2 + i}}{\mathbf{2 + i}}}^{\text{multiply by 1}}$$

(conjugates: $2 - i$ and $2 + i$)

$$= \frac{30 + 5i - 5i^2}{4 - i^2}$$

$$= \frac{35 + 5i}{5}$$

$$= \mathbf{7 + i}$$

You can check this by multiplying $(2 - i)(7 + i)$:

$$(2 - i)(7 + i) = 14 - 5i - i^2$$
$$= 15 - 5i$$

■

EXAMPLE 5

$$\frac{6 + 5i}{2 + 3i} = \frac{6 + 5i}{2 + 3i} \cdot \frac{\mathbf{2 - 3i}}{\mathbf{2 - 3i}}$$

$$= \frac{12 - 8i - 15i^2}{4 - 9i^2}$$

$$= \mathbf{\frac{27}{13} - \frac{8}{13}i}$$

■

EXAMPLE 6 Verify that the conjugate method gives the same result as the definition of division of complex numbers.

Solution

$$\frac{a + bi}{c + di} = \frac{a + bi}{c + di} \cdot \frac{\mathbf{c - di}}{\mathbf{c - di}}$$

$$= \frac{ac - adi + bci - bdi^2}{c^2 - d^2i^2}$$

$$= \frac{(ac + bd) + (bc - ad)i}{c^2 + d^2}$$

$$= \frac{ac + bd}{c^2 + d^2} + \frac{bc - ad}{c^2 + d^2}i$$

■

PROBLEM SET 1.3

A

Simplify the expressions in Problems 1–36.

1. $\sqrt{-36}$
2. $\sqrt{-100}$
3. $\sqrt{-49}$
4. $\sqrt{-8}$
5. $\sqrt{-20}$
6. $\sqrt{-24}$
7. $(3 + 3i) + (5 + 4i)$
8. $(6 - 2i) + (5 + 3i)$
9. $(5 - 3i) - (5 + 2i)$
10. $(3 + 4i) - (7 + 4i)$
11. $(4 - 2i) - (3 + 4i)$
12. $5i - (5 + 5i)$
13. $5 - (2 - 3i)$
14. $-2(-4 + 5i)$
15. $6(3 + 2i) + 4(-2 - 3i)$
16. $4(2 - i) - 3(-1 - i)$
17. $i(5 - 2i)$
18. $i(2 + 3i)$
19. $(3 - i)(2 + i)$
20. $(4 - i)(2 + i)$
21. $(5 - 2i)(5 + 2i)$
22. $(8 - 5i)(8 + 5i)$
23. $(3 - 5i)(3 + 5i)$
24. $(7 - 9i)(7 + 9i)$
25. $-i^2$
26. $-i^3$
27. i^3
28. i^4
29. $-i^4$
30. $-i^5$
31. $-i^6$
32. i^6
33. i^{11}
34. i^{236}
35. $-i^{1980}$
36. i^{1982}

B

Simplify the expressions in Problems 37–57.

37. $(6 - 2i)^2$
38. $(3 + 3i)^2$
39. $(4 + 5i)^2$
40. $(1 + i)^3$
41. $(3 - 5i)^3$
42. $(2 - 3i)^3$
43. $\dfrac{-3}{1 + i}$
44. $\dfrac{5}{4 - i}$
45. $\dfrac{2}{1 - i}$
46. $\dfrac{5}{i}$
47. $\dfrac{2}{i}$
48. $\dfrac{3}{-i}$
49. $\dfrac{-2i}{3 + i}$
50. $\dfrac{3i}{5 - 2i}$
51. $\dfrac{-i}{2 - i}$
52. $\dfrac{1 - 6i}{1 + 6i}$
53. $\dfrac{4 - 2i}{3 + i}$
54. $\dfrac{5 + 3i}{4 - i}$
55. $\dfrac{1 + 3i}{1 - 2i}$
56. $\dfrac{3 - 2i}{5 + i}$
57. $\dfrac{2 + 7i}{2 - 7i}$

C

58. Simplify $(1.9319 + 0.5176i)(2.5981 + 1.5i)$.
59. Simplify $\dfrac{-3.2253 + 8.4022i}{3.4985 + 1.9392i}$.

Let $z_1 = (1 + \sqrt{3}) + (2 + \sqrt{3})i$ *and* $z_2 = (2 + \sqrt{12}) + \sqrt{3}i$. *Perform the indicated operations in Problems 60–65.*

60. $z_1 + z_2$
61. $z_1 - z_2$
62. $z_1 z_2$
63. z_1/z_2
64. $(z_1)^2$
65. $(z_2)^2$

For each of Problems 66–68 let $z_1 = a + bi$, $z_2 = c + di$, *and* $z_3 = e + fi$.

66. Prove the commutative laws for complex numbers. That is, prove that:

$$z_1 + z_2 = z_2 + z_1$$
$$z_1 z_2 = z_2 z_1$$

67. Prove the associative laws for complex numbers. That is, prove that:

$$z_1 + (z_2 + z_3) = (z_1 + z_2) + z_3$$
$$z_1(z_2 z_3) = (z_1 z_2) z_3$$

68. Prove the distributive law for complex numbers. That is, prove that:

$$z_1(z_2 + z_3) = z_1 z_2 + z_1 z_3$$

1.4 QUADRATIC EQUATIONS

A **quadratic equation in one variable** is an equation that can be written in the form

$$ax^2 + bx + c = 0 \qquad (a \neq 0)$$

where x is a variable and a, b, and c are real numbers. Quadratic equations can be solved by several methods. The simplest method can be used if the quadratic expression $ax^2 + bx + c$ is factorable over the integers.* The solution depends on the following property of zero.

Property of Zero

$AB = 0$ if and only if $A = 0$ or $B = 0$ (or both).

* In this section we will consider only simple factoring types from beginning algebra. For a comprehensive review of factoring, see Section 4.1.

Thus, if the product of two factors is zero, then at least one of the factors is zero. If a quadratic is factorable, this property provides a method of solution.

EXAMPLE 1 Solve $x^2 = 2x + 15$.

Solution

$$x^2 - 2x - 15 = 0$$
$$(x + 3)(x - 5) = 0$$
$$x + 3 = 0 \quad \text{or} \quad x - 5 = 0$$

Since the product is zero, one of the factors must be zero.

$$x = -3 \quad \text{or} \quad x = 5$$

Solution: $\{-3, 5\}$ ■

Solution of Quadratic Equations by Factoring

To solve a quadratic equation that can be expressed as a product of linear factors:

1. rewrite all nonzero terms on one side of the equation;
2. factor the expression;
3. set each of the factors equal to zero;
4. solve each of the linear equations;
5. write the solution set which is the union of the solution sets of the linear equations.

Completing the Square

When the quadratic is not factorable, other methods must be employed. One such method depends on the square-root property.

Square-Root Property

If $P^2 = Q$, then $P = \pm\sqrt{Q}$.

The equation $x^2 = 4$ could be rewritten as $x^2 - 4 = 0$, factored, and solved. However, the square-root property can be used, as illustrated below.

Using the square-root property:

$$x^2 = 4$$
$$x = \pm\sqrt{4}$$
$$x = \pm 2$$

Using factoring:

$$x^2 - 4 = 0$$
$$(x + 2)(x - 2) = 0$$
$$x = -2 \quad \text{or} \quad x = 2$$

The square-root property can be derived by using the following property of square roots and an absolute value equation.

For all real numbers x,

$$\sqrt{x^2} = |x|$$

This means, if $x \geq 0$, then $\sqrt{x^2} = x$ and if $x < 0$, then $\sqrt{x^2} = -x$. For example, $\sqrt{2^2} = |2| = 2$ and $\sqrt{(-2)^2} = |-2| = 2$. The importance of this property, however, is that it can be applied to any quadratic! This is because every quadratic may be expressed in the form $P^2 = Q$ by isolating the variable terms and **completing the square**, as illustrated in the following examples.

EXAMPLE 2 Solve $x^2 = 2x + 15$.

Solution

$$x^2 - 2x = 15$$
Isolate the variable.

$$x^2 - 2x + 1 = 15 + 1$$
Complete the square by adding the square of half the coefficient of x to both sides.

$$(x - 1)^2 = 16$$
Factor.

$$x - 1 = \pm 4$$
Use the square-root property.

$$x = 1 \pm 4$$
Solve for x.

$$x = 1 + 4 \quad \text{or} \quad x = 1 - 4$$
$$x = 5 \quad \text{or} \quad x = -3$$

Solution: $\{\mathbf{5, -3}\}$ ■

EXAMPLE 3 Solve $4x^2 - 4x - 7 = 0$.

Solution

$$4x^2 - 4x = 7$$
Isolate the variable.

$$x^2 - x = \frac{7}{4}$$
Divide by 4 so that the coefficient of the squared term is 1.

$$x^2 - x + \left(\frac{1}{2}\right)^2 = \left(\frac{1}{2}\right)^2 + \frac{7}{4}$$
Complete the square by adding the square of half the coefficient of x to both sides.

$$\left(x - \frac{1}{2}\right)^2 = 2$$
Factor.

$$x - \frac{1}{2} = \pm\sqrt{2}$$
Remember $\pm$.

$$x = \frac{1 \pm 2\sqrt{2}}{2}$$
Solve for x. Note $\pm$ means two different solutions.

Solution: $\left\{\frac{\mathbf{1 + 2\sqrt{2}}}{\mathbf{2}}, \frac{\mathbf{1 - 2\sqrt{2}}}{\mathbf{2}}\right\}$ ■

Quadratic Formula

The process of completing the square is often cumbersome. However, if *any* quadratic $ax^2 + bx + c = 0$, $a \neq 0$, is considered, completing the square can be used to derive a formula for x in terms of the coefficients a, b, and c. The formula can then be used to solve all quadratics, even those that are nonfactorable.

$$ax^2 + bx + c = 0$$
$$ax^2 + bx = -c$$
Isolate the variable.

$$x^2 + \frac{b}{a}x = -\frac{c}{a}$$ Divide by a.

$$x^2 + \frac{b}{a}x + \left(\frac{b}{2a}\right)^2 = \left(\frac{b^2}{4a^2}\right) - \frac{c}{a}$$ Complete the square; $\frac{1}{2}$ of $\frac{b}{a}$ is $\frac{b}{2a}$.

$$\left(x + \frac{b}{2a}\right)^2 = \frac{b^2 - 4ac}{4a^2}$$ Factor.

$$x + \frac{b}{2a} = \pm\sqrt{\frac{b^2 - 4ac}{4a^2}}$$ Use the square-root property.

$$x + \frac{b}{2a} = \pm\frac{\sqrt{b^2 - 4ac}}{2a}$$

$$x = -\frac{b}{2a} \pm \frac{\sqrt{b^2 - 4ac}}{2a}$$ Solve for x.

$$x = \frac{-b \pm \sqrt{b^2 - 4ac}}{2a}$$

Quadratic Formula

> If $ax^2 + bx + c = 0$, $a \neq 0$, then
>
> $$x = \frac{-b \pm \sqrt{b^2 - 4ac}}{2a}$$

EXAMPLE 4 Solve $5x^2 + 2x - 2 = 0$.

Solution

$$x = \frac{-2 \pm \sqrt{2^2 - 4(5)(-2)}}{2(5)}$$

$$= \frac{-2 \pm 2\sqrt{1 + 10}}{2(5)}$$

$$= \frac{-1 \pm \sqrt{11}}{5}$$

Solution: $\left\{\frac{-1 \pm \sqrt{11}}{5}\right\}$

To approximate these roots using a calculator, press:

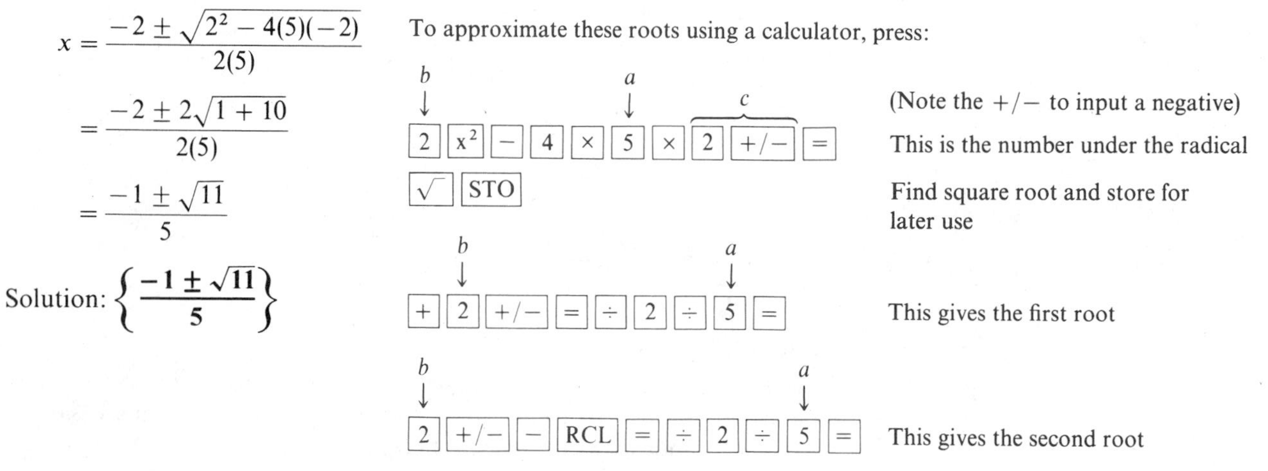

Some of the steps shown here could be combined because these are simple numbers. These steps will give the numerical approximation for a quadratic equation with real roots. For this quadratic equation these roots are (to four decimal places) 0.4633 and −0.8633. ■

HP-28S ACTIVITY*

Quadratic equations can be solved by any four-function calculator, as shown in Example 4. However, these calculators are best suited to finding decimal approximations of the real roots. In order to work with irrational or imaginary roots you will need to break down the quadratic formula into its component parts. However, graphic calculators can do a better job in solving quadratic equations. Let us reconsider Example 4 using the HP-28S calculator.

First enter the equation, and then the variable:

Next, press [SOLV] to gain access to the QUADratic routine, as shown at the right.

When we press the [QUAD] key, we see a symbolic expression representing the solution to the quadratic equation. The display shows a symbol s1, which represents the $\pm$ symbol used in algebra. You can use either $+1$ or -1 for s1 in order to see the two solutions.

Next, press the [STEQ] key, and the symbolic equation will be stored. Then press [SOLVR], which activates the program to solve this stored equation. The display after pressing these keys is shown at the right.

Now, substitute 1 and -1 for the s1 key and evaluate. This is done by pressing the following keys (the display is shown at the right):

[1] [s1] [EXPR =] This substitutes 1 for s1

[1] [CHS] [s1] [EXPR =] This substitutes -1 for s1

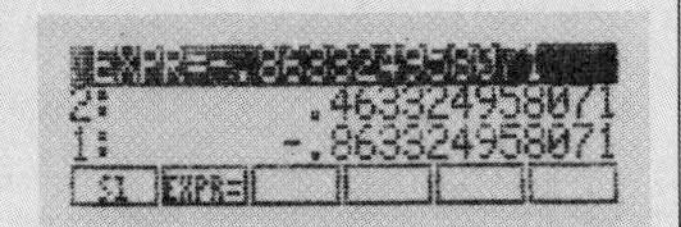

The procedure in solving a quadratic equation with discriminant less than zero is identical to that used in solving any other quadratic equation. Let us solve Example 6 using the HP-28S.

['] [5] [×] [X] [^] [2] [+] [2] [×] [X] [+] [2] [=] [0] [ENTER]

['] [X] [ENTER]

[SOLV] [QUAD] [STEQ] [SOLVR]

After executing these keystrokes, the menus shown at the right will result. As in Example 4 above, we enter -1 and 1 on successive passes through the solver routine to arrive at the roots

$$-0.2 - 0.6i \qquad \text{and} \qquad -0.2 + 0.6i$$

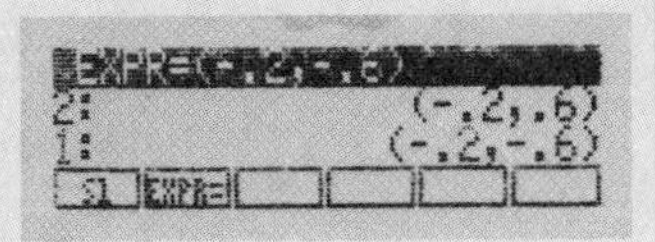

Notice that on the HP-28S these complex roots are shown using ordered pair notation where $(a, b) = a + bi$.

* For a discussion of the HP-28S, along with other graphic calculators, see Appendix A.

EXAMPLE 5 Solve for x:

$$5x^2 + 2x - (w + 4) = 0$$

Solution

$$x = \frac{-2 \pm \sqrt{4 - 4(5)(-w - 4)}}{2(5)}$$

$$= \frac{-2 \pm 2\sqrt{5w + 21}}{2(5)}$$

There are some steps not shown here. Can you fill in the details?

$$= \frac{\mathbf{-1 \pm \sqrt{5w + 21}}}{\mathbf{5}}$$

When solving equations, it is not necessary to recopy the answer using solution set notation. If you leave your answer as shown in this example, it will be understood that the solution set is

$$\left\{\frac{-1 \pm \sqrt{5w + 21}}{5}\right\}$$

■

EXAMPLE 6 Solve $5x^2 + 2x + 2 = 0$.

Solution

$$x = \frac{-2 \pm \sqrt{4 - 4(5)(2)}}{2(5)} = \frac{-2 \pm 2\sqrt{-9}}{10}$$

Since the square root of a negative number is not a real number, the solution set is **empty over the reals**. On the other hand, if you are using the domain consisting of the complex numbers (Section 1.3), then you can solve this equation.

$$x = \frac{-2 \pm 2(3i)}{10} = \frac{2(-1 \pm 3i)}{10} = \frac{-1 \pm 3i}{5} = \mathbf{-\frac{1}{5} \pm \frac{3}{5}i}$$

■

Since the quadratic formula contains a radical, the sign of the radicand will determine whether the roots will be real or nonreal. This radicand is called the **discriminant** of the quadratic, and its properties are summarized in the box.

The Discriminant of the Quadratic

If $ax^2 + bx + c = 0$, $a \neq 0$, then $b^2 - 4ac$ is called the ***discriminant***.
If $b^2 - 4ac < 0$, there are *no real* solutions.
If $b^2 - 4ac = 0$, there is *one real* solution.
If $b^2 - 4ac > 0$, there are *two real* solutions.

Suppose you wish to check your answers for the above examples, say Example 4. One method is to substitute the roots into the original equation. This may involve considerable effort:

$$x = \frac{-1 \pm \sqrt{11}}{5}$$

Substituting,

$$5\left(\frac{-1+\sqrt{11}}{5}\right)^2 + 2\left(\frac{-1+\sqrt{11}}{5}\right) - 2 \stackrel{?}{=} 0 \qquad \text{and}$$

$$5\left(\frac{-1-\sqrt{11}}{5}\right)^2 + 2\left(\frac{-1-\sqrt{11}}{5}\right) - 2 \stackrel{?}{=} 0$$

Instead, suppose we represent the roots by r_1 and r_2. Then:

$$x = r_1 \quad \text{or} \quad x = r_2$$

$$x - r_1 = 0 \quad \text{or} \quad x - r_2 = 0$$

$$(x - r_1)(x - r_2) = 0$$

$$x^2 - r_1x - r_2x + r_1r_2 = 0$$

$$x^2 - (r_1 + r_2)x + r_1r_2 = 0$$

Sum and Product of the Roots of a Quadratic Equation

$$x^2 - \begin{pmatrix}\text{sum of}\\ \text{the roots}\end{pmatrix}x + \begin{pmatrix}\text{product of}\\ \text{the roots}\end{pmatrix} = 0$$

Comparing this result to $ax^2 + bx + c = 0$ or

$$x^2 + \frac{b}{a}x + \frac{c}{a} = 0,$$

we can see that

$$\textbf{sum of roots} = -\frac{b}{a} \qquad \text{and} \qquad \textbf{product of roots} = \frac{c}{a}$$

For Example 4:

$$5x^2 + 2x - 2 = 0$$

$$x^2 + \frac{2}{5}x - \frac{2}{5} = 0$$

sum of roots:

$$\frac{-1+\sqrt{11}}{5} + \frac{-1-\sqrt{11}}{5} = -\frac{2}{5}$$

product of roots:

$$\left(\frac{-1+\sqrt{11}}{5}\right)\left(\frac{-1-\sqrt{11}}{5}\right) = \frac{1-11}{25} = \frac{-10}{25} = -\frac{2}{5}$$

The answer checks since

$$x^2 + \frac{2}{5}x - \frac{2}{5} = 0$$

opposite of the sum of the roots (points to $\frac{2}{5}$ in $\frac{2}{5}x$)

same as the product of the roots (points to $-\frac{2}{5}$)

EXAMPLE 7 Check the answer for Example 5 using the sum and product method.

Solution

$$5x^2 + 2x - (w + 4) = 0$$

$$x^2 + \frac{2}{5}x + \frac{-w-4}{5} = 0$$

$\frac{2}{5}$: opposite of sum of roots; $\frac{-w-4}{5}$: product of roots

$$r_1 = \frac{-1 + \sqrt{5w + 21}}{5} \qquad r_2 = \frac{-1 - \sqrt{5w + 21}}{5}$$

$$r_1 + r_2 = \frac{-1 + \sqrt{5w + 21} - 1 - \sqrt{5w + 21}}{5} = -\frac{2}{5}$$ Checks

$$r_1 r_2 = \left(\frac{-1 + \sqrt{5w + 21}}{5}\right)\left(\frac{-1 - \sqrt{5w + 21}}{5}\right)$$

$$= \frac{1 - (5w + 21)}{25}$$

$$= \frac{-5w - 20}{25}$$

$$= \frac{5(-w - 4)}{25}$$

$$= \frac{-w - 4}{5}$$ Checks ■

PROBLEM SET 1.4

A

Solve each equation in Problems 1–12 by factoring.

1. $x^2 + 2x - 15 = 0$
2. $x^2 - 8x + 12 = 0$
3. $x^2 + 7x - 18 = 0$
4. $2x^2 + 5x - 12 = 0$
5. $10x^2 - 3x - 4 = 0$
6. $6x^2 + 7x - 10 = 0$
7. $9x^2 - 34x - 8 = 0$
8. $4x^2 + 12x + 9 = 0$
9. $9x^2 - 24x + 16 = 0$
10. $6x^2 = 5x$
11. $6x^2 = 12x$
12. $12x^2 = 48x$

Solve each equation in Problems 13–24 by completing the square.

13. $x^2 + 4x - 5 = 0$
14. $x^2 - x - 6 = 0$
15. $x^2 + 2x - 8 = 0$
16. $x^2 + x - 6 = 0$
17. $x^2 + 7x + 12 = 0$
18. $x^2 + 5x + 6 = 0$
19. $x^2 - 10x - 2 = 0$
20. $x^2 + 2x - 15 = 0$
21. $x^2 - 3x = 1$
22. $x^2 - 4x = 2$
23. $6x^2 = x + 2$
24. $x^2 = 4x + 2$

*Solve each equation in Problems 25–45 over the set of real numbers.**

25. $x^2 + 5x - 6 = 0$
26. $x^2 + 5x + 6 = 0$
27. $x^2 - 10x + 25 = 0$
28. $x^2 + 6x + 9 = 0$
29. $12x^2 + 5x - 2 = 0$
30. $2x^2 - 6x + 5 = 0$
31. $5x^2 - 4x + 1 = 0$
32. $2x^2 + x - 15 = 0$
33. $4x^2 - 5 = 0$
34. $3x^2 - 1 = 0$
35. $3x^2 = 7x$
36. $7x^2 = 3$
37. $3x^2 = 5x + 2$
38. $3x^2 - 2 = 5x$
39. $5x = 3 - 4x^2$
40. $4x^2 = 12x - 9$

* If you have covered Section 1.3, you can solve these over the set of complex numbers.

41. $3x = 1 - 2x^2$

42. $5x^2 = 3x - 4$

43. $\sqrt{5} - 4x^2 = 3x$

44. $x = \sqrt{2} - 2x^2$

45. $3x^2 - 4x = \sqrt{5}$

B

Solve the equations in Problems 46–55 for x in terms of the other variable.

46. $2x^2 + x - w = 0$

47. $2x^2 + wx + 5 = 0$

48. $3x^2 + 2x + (y + 2) = 0$

49. $3x^2 + 5x + (4 - y) = 0$

50. $4x^2 - 4x + (1 - t^2) = 0$

51. $4x^2 - (3t + 10)x + (6t + 4) = 0, t > 2$

52. $2x^2 + 3x + 4 - y = 0$

53. $y = 2x^2 + x + 6$

54. $(x - 3)^2 + (y - 2)^2 = 4$

55. $x^2 - 6x + y^2 - 4y + 9 = 0$

56. ***Business*** A small manufacturer of citizens' band radios determines that the price of each item is related to the number of items produced per day. The manufacturer knows that (a) the maximum number that can be produced is 10 items; (b) the price should be $400 - 25x$ dollars; (c) the overhead (the cost of producing x items) is $5x^2 + 40x + 600$ dollars; and (d) the daily profit is then found by subtracting the overhead from the revenue:

$$\begin{aligned} \text{profit} &= \text{revenue} - \text{cost} \\ &= (\text{number of items})(\text{price per item}) - \text{cost} \\ &= x(400 - 25x) - (5x^2 + 40x + 600) \\ &= 400x - 25x^2 - 5x^2 - 40x - 600 \\ &= -30x^2 + 360x - 600 \end{aligned}$$

What is the number of radios produced if the profit is zero?

57. ***Physics*** Suppose you throw a rock at 48 ft/sec from the top of the Sears Tower in Chicago and the height in feet, h, from the ground after t sec is given by

$$h = -16t^2 + 48t + 1454$$

a. What is the height of the Sears Tower?

b. How long will it take (to the nearest tenth of a second) for the rock to hit the ground?

58. ***Physics*** If an object is shot up from the ground with an initial velocity of 256 ft/sec, its distance in feet above the ground at the end of t sec is given by $d = 256t - 16t^2$ (neglecting air resistance). Find the length of time for which $d \geq 240$.

59. ***Physics*** Find the length of time the projectile described in Problem 58 will be in the air.

C

60. ***Engineering*** Many materials, such as brick, steel, aluminum, and concrete, expand due to increases in temperature. This is why fillers are placed between the cement slabs in sidewalks. Suppose you have a 100-ft roof truss securely fastened at both ends, and assume that the buckle is linear. (It is not, but this assumption will serve as a worthwhile approximation.) Let the height of the buckle be x ft. If the percentage of swelling is y, then, for each half of the truss,

$$\begin{aligned} \text{new length} &= \text{old length} + \text{change in length} \\ &= 50 + (\text{percentage})(\text{length}) \\ &= 50 + \left(\frac{y}{100}\right)50 \\ &= 50 + \frac{y}{2} \end{aligned}$$

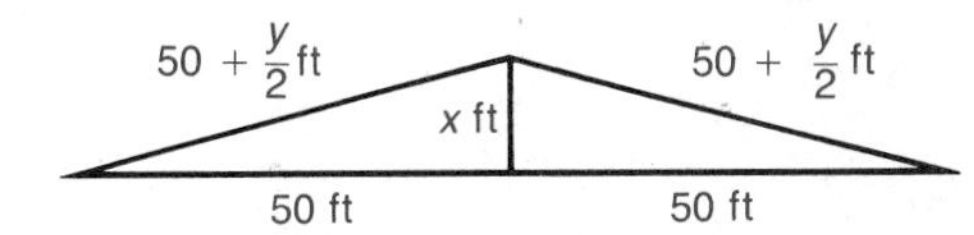

These relationships are shown in the figure. Then, by the Pythagorean Theorem (see page 34),

$$x^2 + 50^2 = \left(50 + \frac{y}{2}\right)^2$$

$$x^2 + 50^2 = \frac{(100 + y)^2}{4}$$

$$4x^2 + 4 \cdot 50^2 = 100^2 + 200y + y^2$$

$$4x^2 - y^2 - 200y = 0$$

Solve this equation for x and then calculate the amount of buckling (to the nearest inch) for the following materials:

a. Brick; $y = 0.03$

b. Steel; $y = 0.06$

c. Aluminum; $y = 0.12$

d. Concrete; $y = 0.05$

61. ***Space Science*** Suppose a model rocket weighs $\frac{1}{4}$ lb. Its engine propels it vertically to a height of 52 ft and a speed of 120 ft/sec at burnout. If its parachute fails to open, determine the approximate time to fall to earth according to the following equation for free fall in a vacuum:

$$h = h_0 + v_0 t - \tfrac{1}{2}gt^2$$

where h is the height (in feet) at time t, h_0 and v_0 are the height (in feet) and velocity (in feet per second) at the time selected at $t = 0$, and g is approximately 32 ft/sec². (For this problem, $h_0 = 52$ ft and $v_0 = 120$ ft/sec.)

62. If $ax^2 + bx + c = 0$, show that the following are roots:

$$r_1 = \frac{2c}{-b + \sqrt{b^2 - 4ac}} \quad \text{and} \quad r_2 = \frac{2c}{-b - \sqrt{b^2 - 4ac}}$$

1.5 QUADRATIC INEQUALITIES

A **quadratic inequality in one variable** is an inequality that can be written in the form

$$ax^2 + bx + c < 0 \qquad (a \neq 0)$$

where x is a variable and a, b, and c are any real numbers. The symbol $<$ can be replaced by $\leq$, $>$, or $\geq$ and it is still called a quadratic inequality in one variable.

The procedure for solving a quadratic inequality is similar to that for solving a quadratic equality. First, use the properties of inequality to obtain a zero on one side of the inequality. The next step is to factor, if possible, the quadratic expression on the left. For example:

$$x^2 - 4 \geq 0$$

$$(x-2)(x+2) \geq 0$$

Figure 1.4 Number line with critical values

Find the values of x that make the inequality valid. A value for which a factor is zero is called a **critical value** of x. The critical values for this example are 2 and -2. For *every other value of* x the inequality is either positive or negative. Next, plot the critical values on a number line as in Figure 1.4. These critical values divide the number line into three intervals. Choose a sample value from each interval. Evaluate each factor to determine the sign only—it is not necessary to complete the arithmetic to find its sign.

Sample value	Factor	Sign of factor	Sign of product
This is *your* choice ↓	This is done mentally ↓		
$x = -100$	$x - 2 = -100 - 2$ ········>	$-$	positive product
	$x + 2 = -100 + 2$ ········>	$-$	
$x = 0$	$x - 2 = 0 - 2$ ········>	$-$	negative product
	$x + 2 = 0 + 2$ ········>	$+$	
$x = 100$	$x - 2 = 100 - 2$ ········>	$+$	positive product
	$x + 2 = 100 + 2$ ········>	$+$	

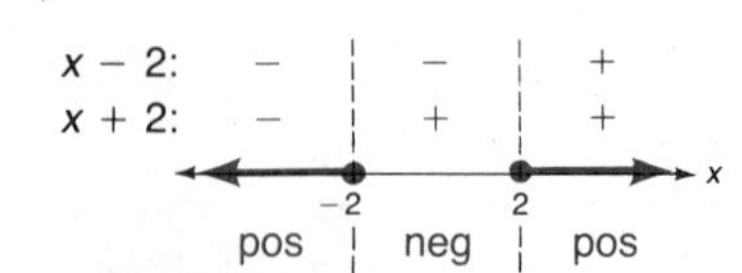

Figure 1.5 Procedure for solving $x^2 - 4 \geq 0$

This procedure can be summarized as in Figure 1.5. You want

$$(x-2)(x+2) \geq 0$$

Since this is positive or zero, you pick out the parts of Figure 1.5 labeled positive and also include those that are zero (namely the critical values). If the given inequality is of the form $\leq$ or $\geq$, the endpoints are included; if it has the form $<$ or $>$, the endpoints are excluded. The solution set is $x \leq -2$ or $x \geq 2$ or, using interval notation, $(-\infty, -2] \cup [2, \infty)$.

EXAMPLE 1 Solve the inequality $2x^2 < 5 - 9x$.

Solution

$$2x^2 + 9x - 5 < 0$$
$$(2x - 1)(x + 5) < 0$$

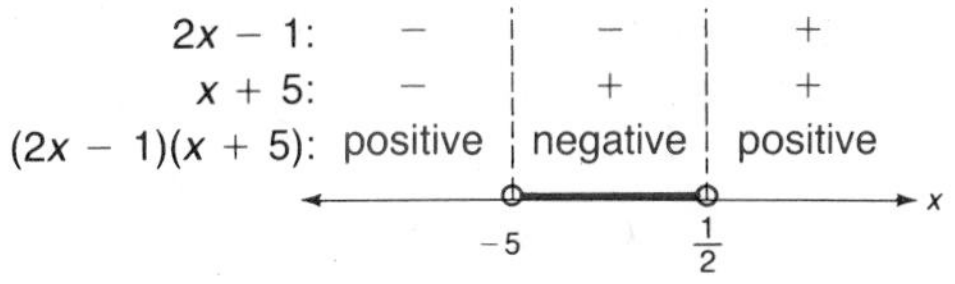

Solution: $-5 < x < \frac{1}{2}$ or $\mathbf{(-5, \frac{1}{2})}$ ■

EXAMPLE 2 Solve the inequality $x^2 + 2x - 4 < 0$.

Solution The term on the left is in simplified form and cannot be easily factored. Therefore proceed by considering $x^2 + 2x - 4$ as a single factor. To find the critical values, find the values for which the factor $x^2 + 2x - 4$ is zero.

$$x^2 + 2x - 4 = 0$$

$$x = \frac{-2 \pm \sqrt{4 - 4(1)(-4)}}{2} = \frac{-2 \pm 2\sqrt{5}}{2} = -1 \pm \sqrt{5}$$

Use the quadratic formula (as shown here) whenever you cannot factor the quadratic expression.

Plot the critical values and check the sign of the factor in each interval:

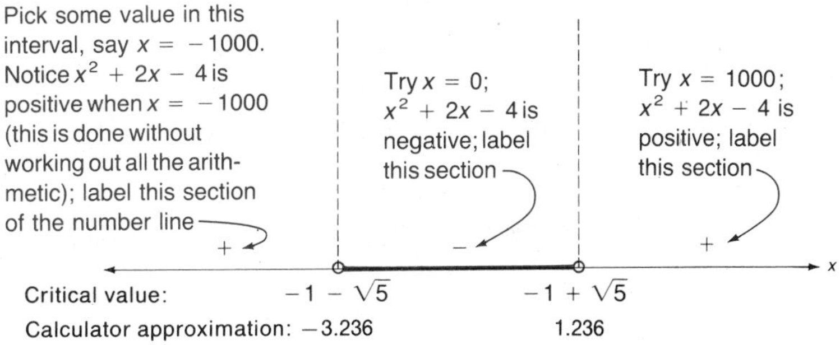

The solution is $-1 - \sqrt{5} < x < -1 + \sqrt{5}$ or, using interval notation,

$$\mathbf{(-1 - \sqrt{5}, -1 + \sqrt{5})}$$ ■

The factoring procedure used in this section for solving quadratic inequalities can be used for other inequalities that are in factored form, even though they may not be quadratic.

EXAMPLE 3 Solve the inequality $(x - 5)(2 - x)(2 - 3x) > 0$.

Solution

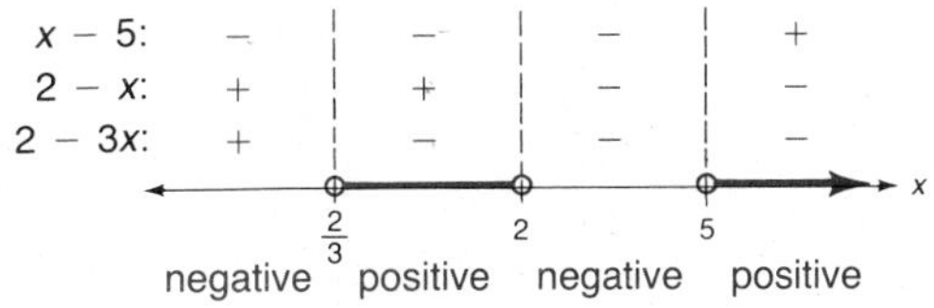

Solution: $\mathbf{(\frac{2}{3}, 2) \cup (5, \infty)}$ ■

EXAMPLE 4 Solve

$$\frac{3x(x+1)(x-2)}{(x-1)(x+3)} \geq 0$$

Solution Plot the critical values of 0, -1, 2, 1, and -3. These are included (because it is $\geq$), but *values that cause division by zero* ($x = 1, x = -3$) need to be **excluded**.

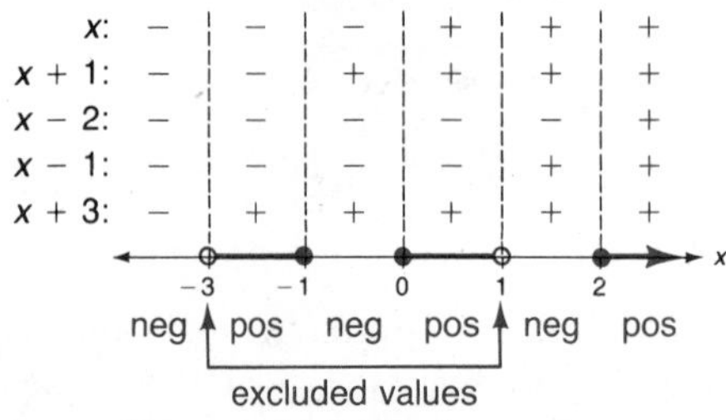

Solution: $\mathbf{(-3, -1] \cup [0, 1) \cup [2, \infty)}$ ■

Be careful about following the procedure described above. It is tempting to attempt these solutions without testing intervals on a number line. You must also take care not to divide both sides by a variable. For example, if

$$x^2 > x$$

and you divide both sides by x to obtain $x > 1$, you will have made a very common mistake, which can be shown by a counterexample. Suppose $x = -2$. Then

$$x^2 > x$$
$$(-2)^2 > -2$$
$$4 > -2$$

which is true. But if you divide both sides by $x = -2$ you obtain

$$\frac{4}{-2} > \frac{-2}{-2}$$

$$-2 > 1 \qquad \text{False!}$$

Using numbers you can easily see the mistake; the order of inequality was not reversed as it should have been.

PROBLEM SET 1.5

A

Solve the inequalities in Problems 1–54 and write each answer using interval notation. Graph each interval.

1. $x(x+3) < 0$

2. $x(x-3) \geq 0$

3. $(x-6)(x-2) \geq 0$

4. $(y+2)(y-8) \leq 0$

5. $(x-8)(x+7) < 0$

6. $(x+1)(x+6) > 0$

7. $(x+2)(2x-1) \leq 0$

8. $(x-2)(2x+1) \leq 0$

9. $(3x+2)(x-3) > 0$

10. $(3x-2)(x+2) > 0$

11. $(x+2)(8-x) \leq 0$

12. $(2-x)(x+8) \geq 0$

13. $(1-3x)(x-4) < 0$

14. $(2x+1)(3-x) > 0$

15. $x(x-3)(x+4) \leq 0$

16. $x(x+3)(x-4) \geq 0$

17. $(x-2)(x+3)(x-4) \geq 0$

18. $(x+1)(x-2)(x+3) \leq 0$

19. $(x+1)(2x+5)(7-3x) > 0$

20. $(x-2)(3x+2)(3-2x) < 0$

21. $\dfrac{x+2}{x} < 0$ **22.** $\dfrac{x}{x+3} < 0$ **23.** $\dfrac{x}{x-8} > 0$

24. $\dfrac{x-8}{x} > 0$ **25.** $\dfrac{x-2}{x+5} \leq 0$ **26.** $\dfrac{x+2}{4-x} \geq 0$

B

27. $\dfrac{x(2x-1)}{5-x} > 0$ **28.** $\dfrac{x}{(2x+3)(x-2)} < 0$

29. $\dfrac{1}{x(x-3)(x+2)} \leq 0$ **30.** $\dfrac{1}{x(x+1)(5-x)} \geq 0$

31. $x^2 \geq 9$ **32.** $x^2 \geq 4$

33. $x^2 + 9 \geq 0$ **34.** $x^2 + 2x - 3 < 0$

35. $x^2 - x - 6 > 0$ **36.** $x^2 - 7x + 12 > 0$

37. $5x - 6 \geq x^2$ **38.** $4 \geq x^2 + 3x$

39. $5 - 4x \geq x^2$ **40.** $x^2 + 2x - 1 < 0$

41. $x^2 - 2x - 2 < 0$ **42.** $x^2 - 8x + 13 > 0$

43. $2x^2 + 4x + 5 \geq 0$ **44.** $x^2 - 2x - 6 \leq 0$

45. $x^2 + 3x - 7 \geq 0$ **46.** $\dfrac{(x-3)(x+1)}{(x-2)(x-1)} \leq 0$

47. $\dfrac{x(x+5)(x-3)}{(x+3)(x-4)} \geq 0$ **48.** $\dfrac{x(x-6)(2-x)}{(x-4)(3-x)} \leq 0$

C

49. $\dfrac{2}{x-2} \leq \dfrac{3}{x+3}$ **50.** $\dfrac{1}{x+1} \geq \dfrac{2}{x-1}$

51. $\dfrac{x-3}{3x-1} \geq \dfrac{x+3}{2x+1}$ **52.** $\dfrac{x-7}{x^2-4x-21} < 0$

53. $\dfrac{x-4}{x^2-5x+6} \leq 0$ **54.** $\dfrac{x^2+4x+3}{x+2} \geq 0$

55. The product of two numbers is at least 340. One of the numbers is three less than the other. What are the possible values of the larger number?

56. The product of two numbers is no larger than 300. One number is five larger than the other. What are the possible values of the smaller number?

57. The quotient of two numbers is positive. The divisor is three larger than the dividend. What are the possibilities for the smaller number?

58. Two numbers have a negative quotient. What are the possibilities for the dividend if it is five larger than the divisor?

59. A rectangular area is to be fenced. If the space is twice as long as it is wide, for what dimensions is the area numerically greater than the perimeter?

60. A rectangular area three times as long as it is wide is to be fenced. For what dimensions is the perimeter numerically greater than the area?

1.6 TWO-DIMENSIONAL COORDINATE SYSTEM AND GRAPHS

There is a one-to-one correspondence between the real numbers and points on a real number line, called a one-dimensional coordinate system. A **two-dimensional coordinate system** can be introduced by considering two perpendicular coordinate lines in a plane. Usually one of the coordinate lines is horizontal with the positive direction to the right; the other is vertical with the positive direction upward. These coordinate lines are called **coordinate axes** and the point of intersection is called the **origin**. Notice from Figure 1.6 that the axes divide the plane into four parts called the **first, second, third,** and **fourth quadrants**. This two-dimensional coordinate system is also called a **Cartesian coordinate system** in honor of René Descartes, who was the first to describe such a coordinate system in mathematical detail.

Points of a plane are denoted by ordered pairs. The term **ordered pair** refers to two real numbers represented by (a, b), where a is called the **first component** and b is the **second component**. The order in which the components are listed is important, since $(a, b) \neq (b, a)$ if $a \neq b$.

y-axis
Quadrant II (−, +)
Quadrant I (+, +)
Coordinates of a point (a, b)
b
Origin
a
x-axis
Quadrant III (−, −)
Quadrant IV (+, −)

Figure 1.6 Cartesian coordinate system

Equality of Ordered Pairs

$(a, b) = (c, d)$ if and only if $a = c$ and $b = d$

The notation (a, b) was used to denote an interval on a number line in the last section. You should know from the discussion whether a one-dimensional or a two-dimensional coordinate system is involved.

The horizontal number line is called the $\boldsymbol{x}$ **axis** (sometimes called the *axis of abscissas*), and x represents the first component of the ordered pair. The vertical number line is called the $\boldsymbol{y}$ **axis** (sometimes called the *axis of ordinates*), and y represents the second component of the ordered pair. The plane determined by the x and y axes is called a **coordinate plane, Cartesian plane,** or $\boldsymbol{xy}$ **plane.** When we refer to a point (x, y), we are referring to a point in the coordinate plane whose abscissa is x and whose ordinate is y. To **plot a point** (x, y) means to locate the point with coordinates (x, y) in the plane and represent its location by a dot.

Finding the distance between points in two dimensions requires the Pythagorean Theorem.

Pythagorean Theorem

A triangle with sides a, b, and c is a right triangle if and only if

$$a^2 + b^2 = c^2$$

Let $P_1(x_1, y_1)$ and $P_2(x_2, y_2)$ be any two distinct points in a plane. If $x_1 = x_2$, then P_1P_2 is a *vertical line segment*; if $y_1 = y_2$, then P_1P_2 is a *horizontal line segment* as shown in Figure 1.7.

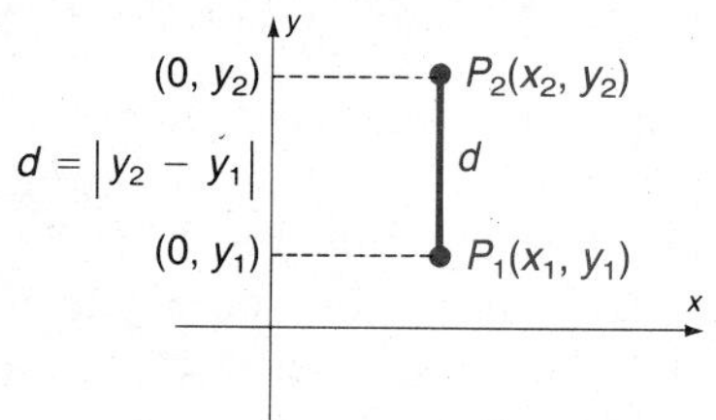

a. Vertical segment, $x_1 = x_2$

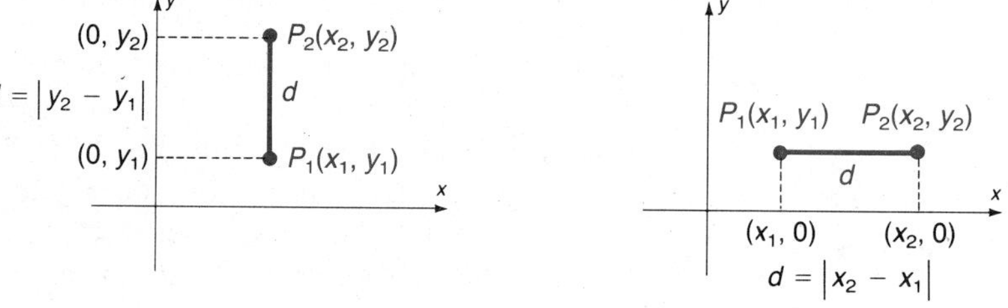

b. Horizontal segment, $y_1 = y_2$

Figure 1.7 Vertical and horizontal line segments

With these special cases, it is easy to find the distance d between P_1 and P_2 because these distances correspond directly to distances on a one-dimensional coordinate system as discussed in Section 1.2. Study Figure 1.7 and see Problems 68 and 69 of Problem Set 1.6 for the details.

Since these special cases are considered in the problem set, we will focus our attention on the general case in which P_1 and P_2 do not lie on the same horizontal or

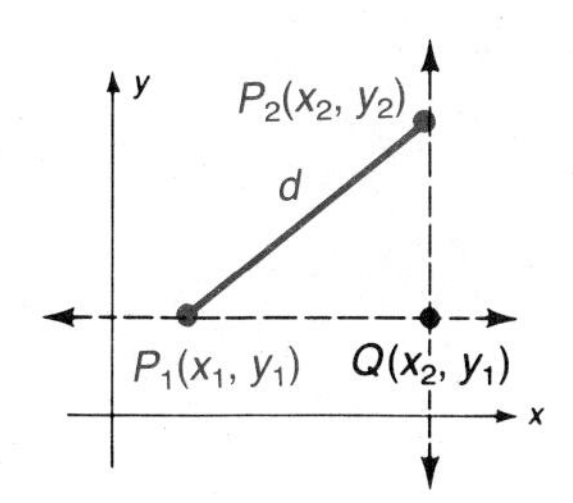

Figure 1.8 Distance between points

vertical line. Draw a line through P_1 parallel to the x axis and another through P_2 parallel to the y axis. These lines intersect at a point Q with coordinates (x_2, y_1) as shown in Figure 1.8. The distance P_1Q is $|x_2 - x_1|$; the distance QP_2 is $|y_2 - y_1|$. By the Pythagorean Theorem,

$$d^2 = |x_2 - x_1|^2 + |y_2 - y_1|^2$$

Thus, since d is nonnegative and $|a|^2 = a^2$ for every real number a, the distance from P_1 to P_2 is given by the following formula.

Distance Formula

If $P_1(x_1, y_1)$ and $P_2(x_2, y_2)$ are any two points, then the distance d from P_1 to P_2 is

$$d = \sqrt{(x_2 - x_1)^2 + (y_2 - y_1)^2}$$

EXAMPLE 1 Find the distance between $(-3, 2)$ and $(-1, -6)$.

Solution

$$\begin{aligned} d &= \sqrt{[(-1) - (-3)]^2 + (-6 - 2)^2} \\ &= \sqrt{4 + 64} \\ &= \mathbf{2\sqrt{17}} \end{aligned}$$

■

To find the **midpoint** M of a segment $P_1(x_1, y_1)$ and $P_2(x_2, y_2)$, you simply average the coordinates of the two endpoints.

Midpoint Formula

The midpoint M between point $P_1(x_1, y_1)$ and $P_2(x_2, y_2)$ is

$$M = \left(\frac{x_1 + x_2}{2}, \frac{y_1 + y_2}{2}\right)$$

EXAMPLE 2 Find the midpoint of the segment connecting $(-3, 2)$ and $(-1, -6)$.

Solution

$$\begin{aligned} M &= \left(\frac{(-3) + (-1)}{2}, \frac{(2) + (-6)}{2}\right) \\ &= \mathbf{(-2, -2)} \end{aligned}$$

■

One of the main topics of this course is the graphing of certain relations. Consider the formula for the volume V of a right circular cone whose radius is one-half its height h:

$$V = \frac{\pi h^3}{12}$$

A table of values showing the volumes for different heights is given here:

Height	1	2	3	4	5	6
Volume	$\frac{\pi}{12}$	$\frac{2\pi}{3}$	$\frac{9\pi}{4}$	$\frac{16\pi}{3}$	$\frac{125\pi}{12}$	18π

Do you see how these values were obtained? If $h = 1$, then

$$V = \frac{\pi(1)^3}{12} = \frac{\pi}{12}$$

If $h = 2$, then

$$V = \frac{\pi(2)^3}{12} = \frac{8\pi}{12} = \frac{2\pi}{3}$$

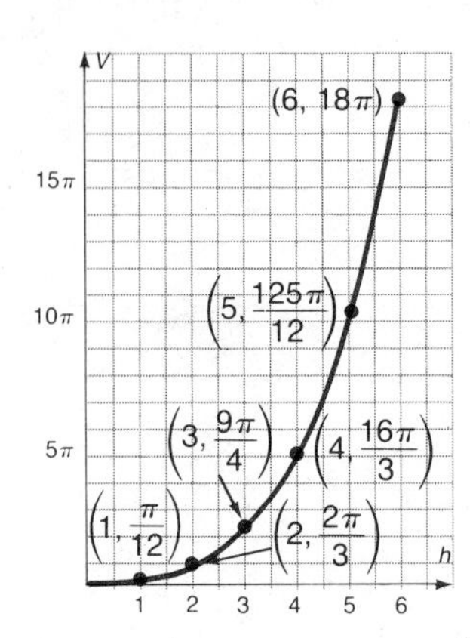

Figure 1.9 Graph of $V = \dfrac{\pi h^3}{12}$ for $0 \leq h \leq 6$

And so on. This table can be written as a set of ordered pairs: $(1, \pi/12)$, $(2, 2\pi/3)$, $(3, 9\pi/4), \ldots$. A set of ordered pairs is called a **relation.** In this example, the ordered pairs are (h, V) so that $V = \pi h^3/12$. If an ordered pair has components that, when substituted for their corresponding variables, yield a true equation, we say that the ordered pair **satisfies** the equation. For example, $(h, V) = (1, \pi/12)$ satisfies the equation $V = \pi h^3/12$. By plotting the points whose coordinates are shown in the table and then connecting them with a smooth curve, you arrive at the **graph** of this relation, a portion of which is shown in Figure 1.9.

In general, when we speak of a **graph of a relation** we mean there is a one-to-one correspondence between the set of all ordered pairs (x, y) in the relation and the set of all points with coordinates (x, y) that lie on the curve.

EXAMPLE 3 Graph $2x + 3y + 6 = 0$ by plotting points.

Solution One of the most efficient methods for graphing a curve by plotting points is to solve the equation for y and then substitute values for x, arranging the ordered pairs in table form.

$$2x + 3y + 6 = 0$$
$$3y = -2x - 6$$
$$y = -\tfrac{2}{3}x - 2$$

x	y	
0	-2	You choose the x values.
3	-4	Do you see why we chose 3 and not 1 or 2?
-3	0	
-6	2	This table entry represents the point $(-6, 2)$.

Plot the points $(0, -2)$, $(3, -4), \ldots$ and draw a smooth curve connecting them:

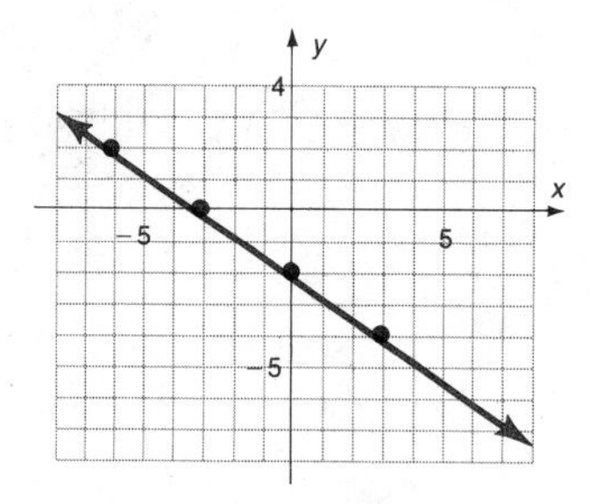

■

EXAMPLE 4 Graph $y = x^2$ by plotting points.

Solution

x	y
0	0
1	1
2	4
-1	1
-2	4

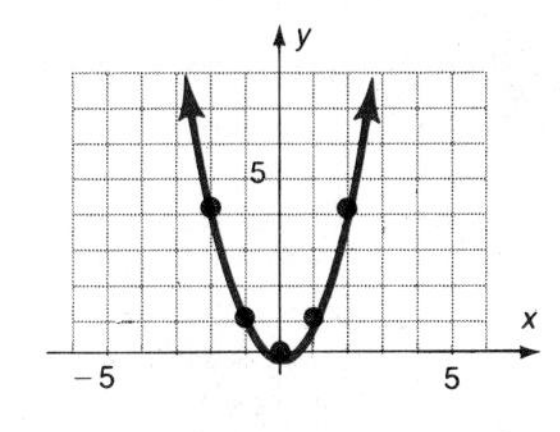

■

EXAMPLE 5 Graph $y = |x|$ by plotting points.

Solution

x	y
0	0
1	1
-1	1
2	2
-2	2

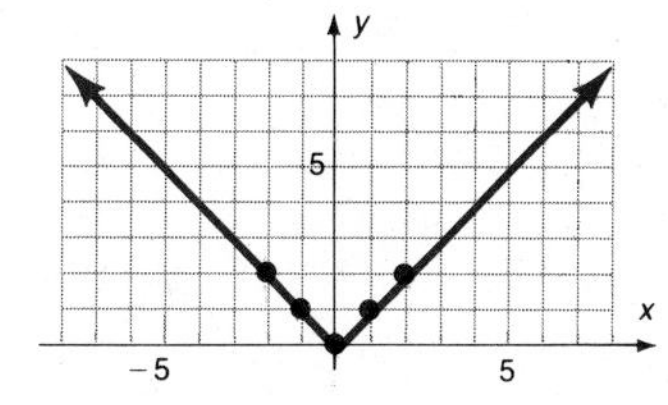

■

The method of drawing the graph of a relation as shown here—that of plotting points—is very primitive, and one of the primary purposes of this book is to develop more efficient methods of graphing curves than simply plotting points.

The first property of a graph we will consider is called **symmetry**. The *idea* of symmetry is the *idea* of mirror images. A graph or curve is **symmetric with respect to a line**, for example, if the graph is the same on both sides of that line as shown in Figures 1.10 and 1.11.

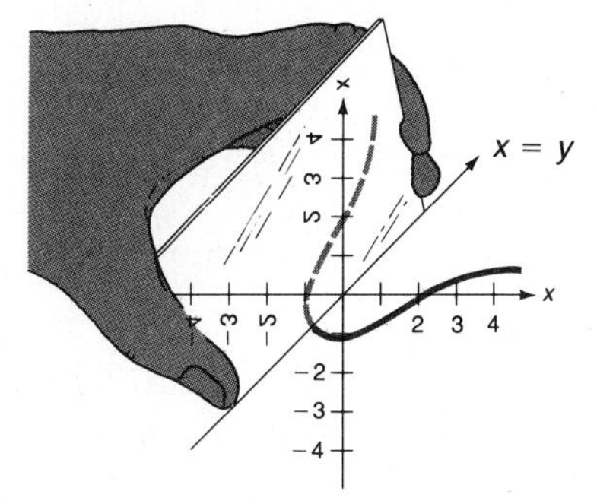

Curve is symmetric with respect to given line.

Figure 1.10 Symmetry with respect to a given line

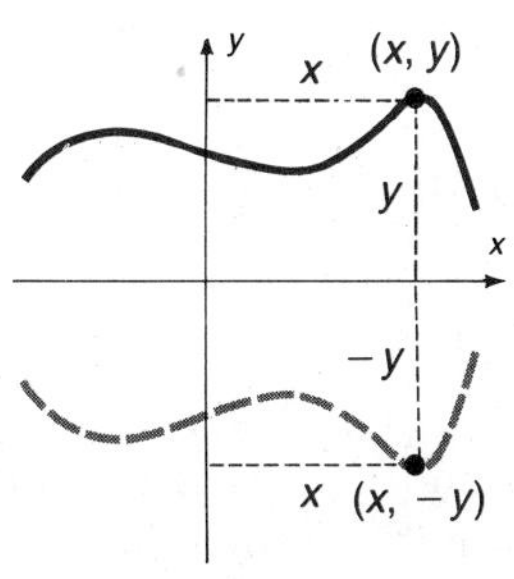

a. Symmetry with respect to the x axis

b. Symmetry with respect to the y axis

Figure 1.11 Symmetry with respect to the x axis and the y axis

EXAMPLE 6 Draw the reflection of the given curve as it would appear in the mirror.

Solution The answer is shown as a dashed curve. Your paper would look like the graph on the right.

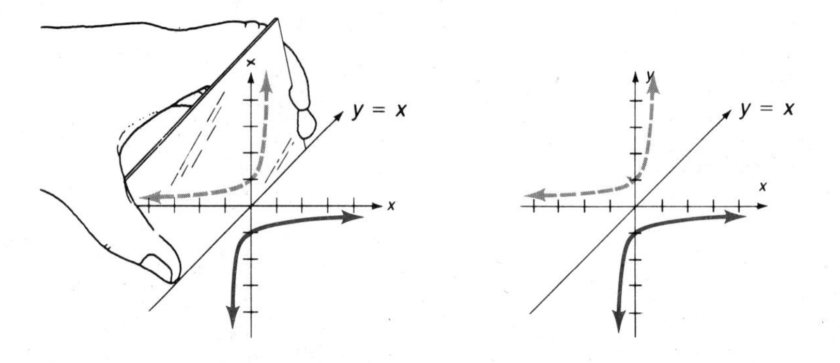

■

Symmetry with Respect to a Coordinate Axis

> The graph of a relation is **symmetric with respect to the x axis** if substitution of $-y$ for y does not change the set of coordinates satisfying the relation. The graph of a relation is **symmetric with respect to the y axis** if substitution of $-x$ for x does not change the set of coordinates satisfying the relation.

EXAMPLE 7 Draw the given curve so that it is: (**a**) symmetric with respect to the x axis, and (**b**) symmetric with respect to the y axis.

Solution

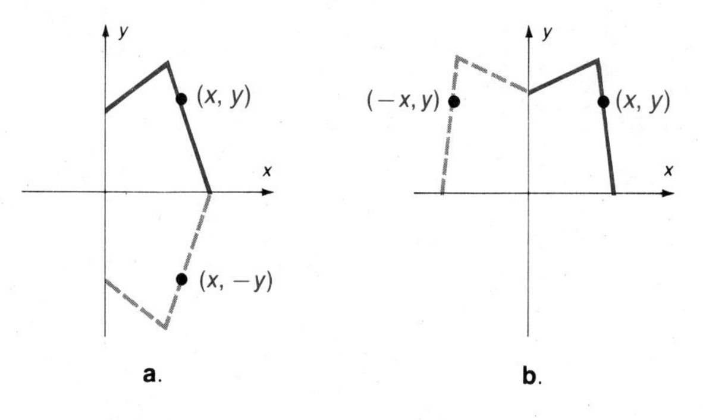

■

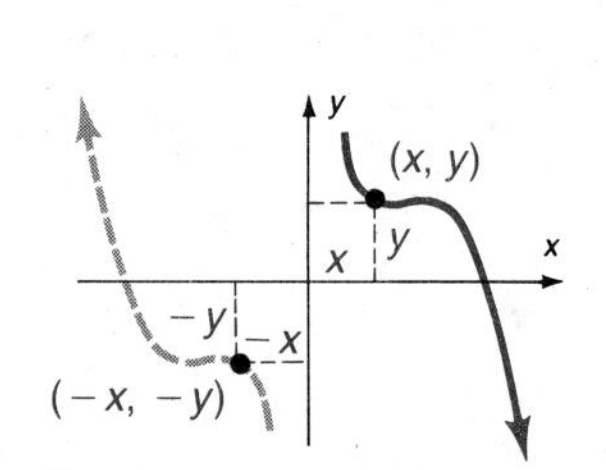

Figure 1.12 Symmetry with respect to the origin

Some graphs possess another type of symmetry called symmetry with respect to the origin (Figure 1.12).

Symmetry with Respect to the Origin

> The graph of a relation is **symmetric with respect to the origin** if the simultaneous substitution of $-x$ for x and $-y$ for y does not change the set of coordinates satisfying the relation.

EXAMPLE 8 Draw the given curve so that it is symmetric with respect to the origin.

Solution

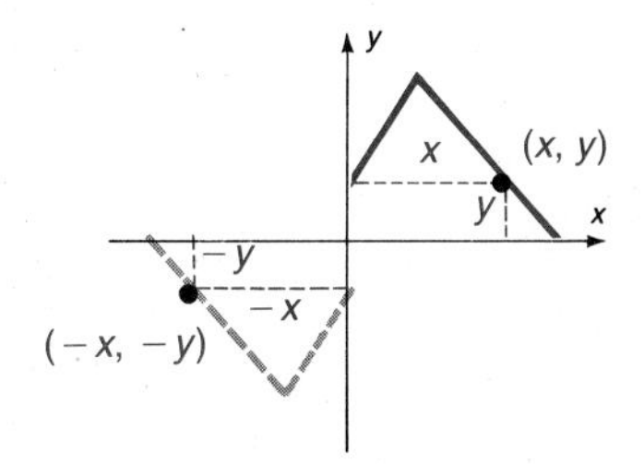

■

PROBLEM SET 1.6

A

Plot the points whose coordinates are given in Problems 1–6.

1. $A(3,4)$; $B(6,2)$; $C(-3,-5)$; $D(-4,6)$; $E(-3,0)$
2. $F(4,-3)$; $G(7,2)$; $H(0,5)$; $I(-3,0)$; $J(-5,-2)$
3. $A(1,175)$; $B(-3,-125)$; $C(-2,-150)$; $D(2,25)$; $E(0,100)$
4. $F(250,-3)$; $G(-350,-2)$; $H(-150,3)$; $I(300,4)$; $J(-50,-1)$
5. $A(\frac{\pi}{6},1)$; $B(\frac{-2\pi}{3},-2)$; $C(-\pi,-1)$; $D(\frac{\pi}{4},3)$; $E(-\frac{\pi}{4},-2)$
6. $F(-\frac{\pi}{3},\frac{1}{2})$; $G(\frac{\pi}{4},\frac{3}{4})$; $H(\frac{3\pi}{2},-\frac{1}{2})$; $I(\pi,1)$; $J(-\pi,-1)$
7. Plot five points (x,y) whose second component is the opposite of the first component.
8. Plot five points (x,y) whose first and second components are equal.
9. Plot five points (x,y) whose second component is twice the first component.
10. Plot five points (x,y) whose first component is 3.
11. Plot five points (x,y) whose second component is -2.
12. Plot five points (x,y) whose second component is the opposite of twice the first component.

Find the distance between the points whose coordinates are given in Problems 13–21.

13. $(5,1)$ and $(8,5)$
14. $(1,4)$ and $(13,9)$
15. $(-2,4)$ and $(0,0)$
16. $(0,0)$ and $(5,-2)$
17. $(4,5)$ and $(3,-1)$
18. $(-2,1)$ and $(-1,-5)$
19. $(7x,5x)$ and $(3x,2x)$, $x < 0$
20. $(x,5x)$ and $(-3x,2x)$, $x < 0$
21. $(x,5x)$ and $(-3x,2x)$, $x > 0$

Find the midpoint of the segment connecting the points whose coordinates are given in Problems 22–30.

22. $(5,1)$ and $(8,5)$
23. $(1,4)$ and $(13,9)$
24. $(-2,4)$ and $(0,0)$
25. $(0,0)$ and $(5,-2)$
26. $(4,5)$ and $(3,-1)$
27. $(-2,1)$ and $(-1,-5)$
28. $(x,5x)$ and $(3x,2x)$, $x < 0$
29. $(x,5x)$ and $(-3x,2x)$, $x < 0$
30. $(x,5x)$ and $(-3x,2x)$, $x > 0$

Draw a reflection of each curve given in Problems 31–34. To do this, draw coordinate axes on your paper. Next draw the line $x = y$ and the curve as shown. Finally imagine a mirror as shown here and draw the curve as it would look in this mirror.

31.

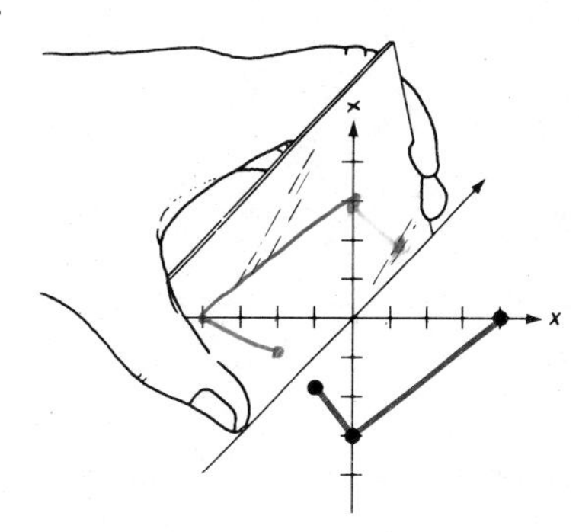

32.

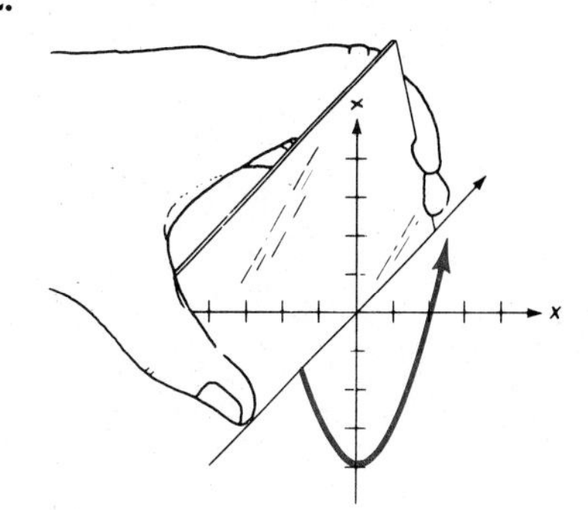

33.

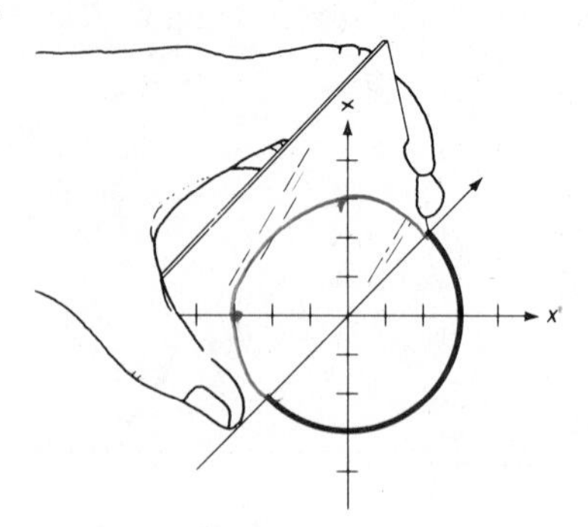

34.

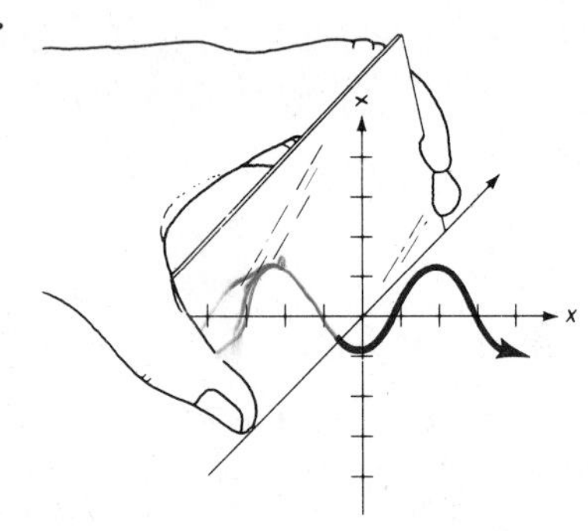

Graph each relation in Problems 35–46 by plotting points.

35. $x + y + 3 = 0$
36. $3x - y + 1 = 0$
37. $3x + 2y + 6 = 0$
38. $4x - 5y + 2 = 0$
39. $y = -x^2$
40. $2x^2 + y = 0$
41. $x^2 + 3y = 0$
42. $x^2 - 2y = 0$
43. $y = -|x|$
44. $y = -|x + 2|$
45. $y = 2|x|$
46. $y = |x| + 2$

B

In Problems 47–52, draw a curve so that it is symmetric to the given curve with respect to: (a) the x axis; (b) the y axis; and (c) the origin.

47.

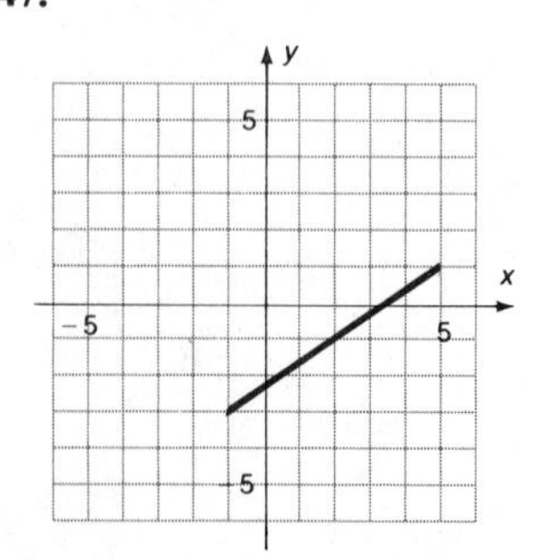

48.

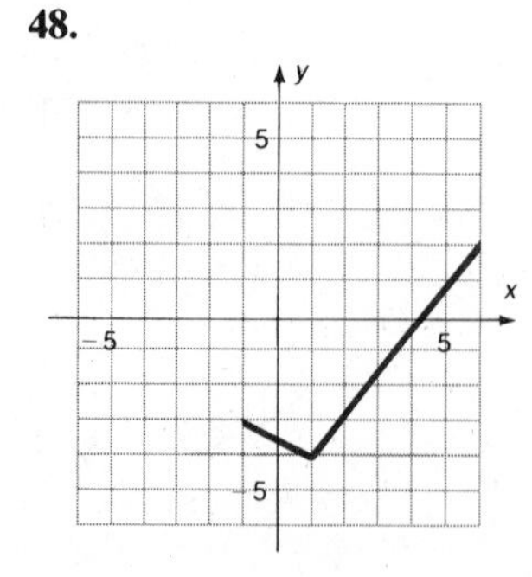

49.

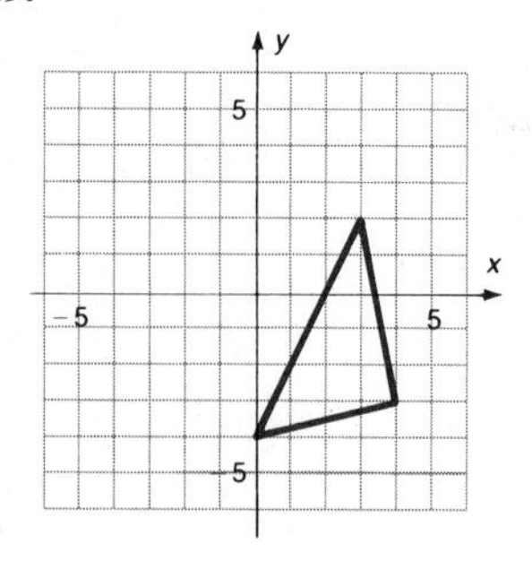

50.

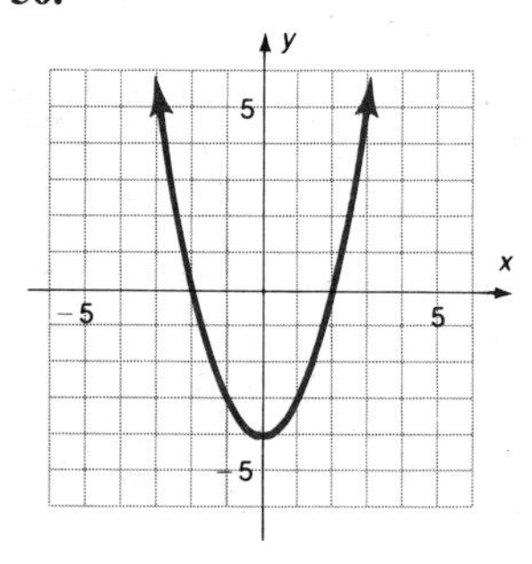

51.

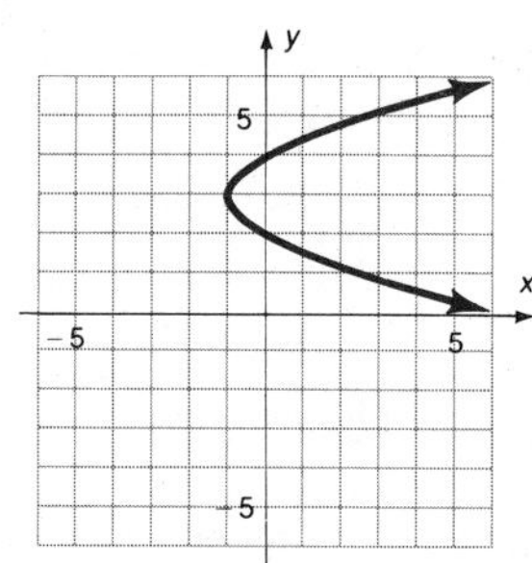

52.

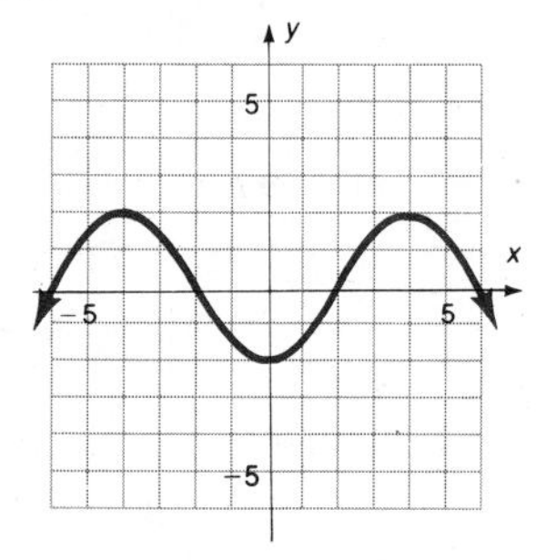

The sides of a triangle satisfy $a^2 + b^2 = c^2$ if and only if it is a right triangle. Which of the triangles whose vertices are given in Problems 53–58 are right triangles?

53. (1, 3), (7, 1), (7, 10)
54. (0, 0), (6, 4), (2, 10)
55. (0, 0), (4, 3), (−3, 8)
56. (−6, 0), (−5, −6), (0, −5)
57. (3, 2), (7, −4), (4, −6)
58. (8, −1), (14, 2), (6, 3)

59. Find a formula for the set of points (x, y) for which the distance from (x, y) to (2, 3) is 7.
60. Find a formula for the set of points (x, y) for which the distance from (x, y) to (−3, −5) is 5.
61. Find a formula for the set of points (x, y) for which the distance from (x, y) to (−4, 1) is 3.
62. Find all points on the x axis that are 8 units from the point (2, 4).
63. Find all points on the y axis that are 8 units from the point (2, 4).
64. Find all points on the y axis that are 5 units from the point (−3, −1).
65. Draw the graph of the relation $A = \pi r^2$.
66. Draw the graph of the relation $V = \frac{2}{3}\pi R^2$.
67. Draw the graph of the relation $h = \frac{a}{2}\sqrt{3}$.
68. Let $P_1(x_1, y_1)$ and $P_2(x_2, y_2)$ be two points such that $y_1 = y_2$. Show that the distance from P_1 to P_2 is $|x_2 - x_1|$.

69. Let $P_1(x_1, y_1)$ and $P_2(x_2, y_2)$ be two points such that $x_1 = x_2$. Show that the distance from P_1 to P_2 is $|y_2 - y_1|$.

70. Find a formula for the set of points (x, y) for which the distance from (x, y) to $(3, 0)$ plus the distance from (x, y) to $(-3, 0)$ equals 10.

71. Find a formula for the set of points (x, y) for which the distance from (x, y) to $(4, 0)$ plus the distance from (x, y) to $(-4, 0)$ equals 10.

72. Find a formula for the set of points (x, y) for which the distance from (x, y) to $(5, 0)$ minus the distance from (x, y) to $(-5, 0)$, in absolute value, is 6.

73. Find a formula for the set of points (x, y) for which the distance from (x, y) to $(0, 5)$ minus the distance from (x, y) to $(0, -5)$, in absolute value, is 8.

1.7 CHAPTER 1 SUMMARY

The material of this chapter is reviewed in the following list of objectives. After each objective there are some practice questions. For a sample test, select the first question of each set and check your answers with the answer section. For a sample test without answers, use the second question of each set. Additional practice is given by the other questions in each set. If you are having trouble with a particular type of problem, look back to that section for extra help.

1.1 REAL NUMBERS

Objective 1 Be familiar with the counting numbers, whole numbers, integers, rational numbers, irrational numbers and real numbers. Classify each of the following numbers using the above listed sets.

1. $\frac{14}{7}, \sqrt{144}, 6.\bar{2}, \pi$

2. $\sqrt{2.25}, \sqrt{30}, 6.545454\ldots, \frac{\pi}{6}$

3. $3.\bar{1}, \frac{5\pi}{6}, \frac{22}{7}, \sqrt{10}$

4. $3.1416, 4.513, \sqrt{1.69}, \sqrt{12}$

Objective 2 Graph numbers on a number line. Graph each of the following sets of numbers on the same real number line.

5. $\frac{14}{7}, \sqrt{144}, 6.\bar{2}, \pi$

6. $\sqrt{2.25}, \sqrt{30}, 6.545454\ldots, \frac{\pi}{6}$

7. $3.\bar{1}, \frac{5\pi}{6}, \frac{22}{7}, \sqrt{10}$

8. $3.1416, 4.513, \sqrt{1.69}, \sqrt{12}$

Objective 3 Use $<$, $>$, and $=$ relationships. Illustrate the Property of Comparison by replacing □ by $=$, $<$, or $>$.

9. $\frac{5}{8}$ □ 0.625

10. $\frac{5}{9}$ □ 0.555

11. $\frac{5}{7}$ □ $\frac{8}{11}$

12. $\sqrt{2}$ □ 1.414

Objective 4 Know the reflexive, symmetric, transitive, and substitution properties. Complete the given statement so that the requested property of equality is demonstrated.

13. $a(b + c) =$ ____________ (reflexive property)
14. If $a(b + c) = ab + ac$, then ____________ (symmetric property)
15. If $a(b + c) = ab + ac$ and $ab + ac = 5$, then ____________ (transitive property)
16. If $a(b + c) = 5$ and $a = 3$, then ____________ (substitution property)

Objective 5 Be familiar with the real number properties. Complete the given statement so that the requested field property is demonstrated.

17. $a(b + c) =$ ____________ (commutative property for multiplication)
18. $a(b + c) =$ ____________ (commutative property for addition)
19. $a(b + c) =$ ____________ (distributive property)
20. $a(b + c) =$ ____________ (identity property for multiplication)

1.2 INTERVALS, INEQUALITIES, AND ABSOLUTE VALUES

Objective 6 *Write interval notation and graph inequalities.*

21. $-4 < x < 2$ **22.** $-12 \le x < -8$ **23.** $x > -3$ **24.** $4 \ge x$

Objective 7 *Graph inequalities given interval notation.*

25. $[-5, -2]$ **26.** $[-3, 0)$ **27.** $[-1, \infty)$ **28.** $(-\infty, 4) \cup (4, \infty)$

Objective 8 *Write interval notation using inequality notation.* Write each of the intervals as an inequality. Use x as the variable.

29. $[-8, -5)$ **30.** $[-2, 0)$ **31.** $(-\infty, 3)$ **32.** $[3, \infty)$

Objective 9 *Solve linear inequalities.* Leave your answer in interval notation.

33. $3x + 2 \le 14$ **34.** $-9 \le -x$
35. $5 \le 1 - x < 9$ **36.** $-0.001 \le x + 2 \le 0.001$

Objective 10 *Know the definition of absolute value.* Write expressions without using absolute value notation.

37. $|-\sqrt{11}|$ **38.** $|4 - \sqrt{11}|$ **39.** $|3 - \sqrt{11}|$ **40.** $|2\pi - 9|$

Objective 11 *Find the distance between points on a number line.*

41. (3) and (−5) **42.** (−5) and (−1) **43.** $(-\pi)$ and (2) **44.** (4) and $(-\sqrt{5})$

Objective 12 *Solve absolute value equations.*

45. $|x| = 8$ **46.** $|x| = -5$ **47.** $|2x + 3| = 8$ **48.** $|2 - 3x| = 11$

Objective 13 *Solve absolute value inequalities.*

49. $|x - 4| < 5$ **50.** $|3x + 1| \le 5$ **51.** $|5 - 2x| \le 25$ **52.** $|x - 5| \ge 3$

*1.3 COMPLEX NUMBERS

Objective 14 *Define a complex number. Simplify expressions involving complex numbers.*

53. $-i^7$ **54.** $(2 - 3i) - (5 + 6i)$ **55.** $(2 + 5i)(2 - 5i)$ **56.** $\dfrac{1 - 8i}{3 + 2i}$

1.4 QUADRATIC EQUATIONS

Objective 15 *Solve quadratic equations by factoring.*

57. $x^2 - x - 12 = 0$ **58.** $x^2 - 10x + 24 = 0$ **59.** $(3 - x)(5 + 2x) = 0$
60. $x^2 - 100 = 0$

Objective 16 *Solve quadratic equations by completing the square.*

61. $x^2 - 2x - 15 = 0$ **62.** $x^2 + 6x + 8 = 0$ **63.** $x^2 + 9x + 20 = 0$
64. $x^2 - 3x + 1 = 0$

Objective 17 *Know the quadratic formula. Solve quadratic equations over the set of real numbers.***

65. $x^2 - 5x + 3 = 0$ **66.** $2x^2 - 5x - 3 = 0$ **67.** $x^2 + 2x - 5 = 0$
68. $3x^2 + 2x + 1 = 0$

1.5 QUADRATIC INEQUALITIES

Objective 18 *Solve quadratic inequalities.* Write your answer using interval notation.

69. $3x^2 - 2x - 1 < 0$ **70.** $3 + 5x \ge 2x^2$ **71.** $x^2 + 2x + 1 \ge 0$ **72.** $x^2 - x - 1 \le 0$

Objective 19 *Solve inequalities in factored form.* Write your answer using interval notation.

73. $x(3 - x)(x + 1) < 0$ **74.** $\dfrac{x + 5}{x - 9} \le 0$

---

* Optional section.
** Find the solution over the set of complex numbers if you covered Section 1.3.

75. $x(x-2)^2(x+1) > 0$

76. $\dfrac{(2x-1)(x-2)}{(x+1)(x+2)} \geq 0$

1.6 TWO-DIMENSIONAL COORDINATE SYSTEM AND GRAPHS

Objective 20 Plot points on a Cartesian coordinate system.

77. $\left(\dfrac{\pi}{2}, 0\right)$

78. $\left(\dfrac{2\pi}{3}, -\dfrac{1}{2}\right)$

79. $\left(\dfrac{\pi}{4}, \dfrac{\sqrt{2}}{2}\right)$

80. $\left(\dfrac{5\pi}{6}, \dfrac{-\sqrt{3}}{2}\right)$

Objective 21 Know the Pythagorean Theorem and the distance formula. Find the distance between points in a plane. Find the distance between each pair of points.

81. (α, β) and (γ, δ)

82. (x, x) and $(5x, 4x)$, where $x > 0$

83. (x, x) and $(5x, 4x)$, where $x < 0$

84. $(-3, -2)$ and $(1, -4)$

Objective 22 Find the midpoint of a segment. Find the midpoint of the segment connecting the given points.

85. (α, β) and (γ, δ)

86. (x, x) and $(5x, 4x)$, where $x > 0$

87. (x, x) and $(5x, 4x)$, where $x < 0$

88. $(-3, -2)$ and $(1, -4)$

Objective 23 Know the definition of a relation and the terminology of graphing in two dimensions. Fill in the blanks.

89. A relation is ______________________________.

90. By a graph of a relation we mean ______________________________.

91. If an ordered pair has components that, when substituted for their corresponding variables, yield a true equation, we say that the ordered pair ______________ the equation.

92. Draw a Cartesian coordinate system, label the axes, origin, and quadrants by number.

Objective 24 Graph relations specified by an equation by plotting points.

93. $2x - y + 3 = 0$

94. $y = -\frac{2}{3}x$

95. $y = -\frac{2}{3}x^2$

96. $y = -\frac{2}{3}|x|$

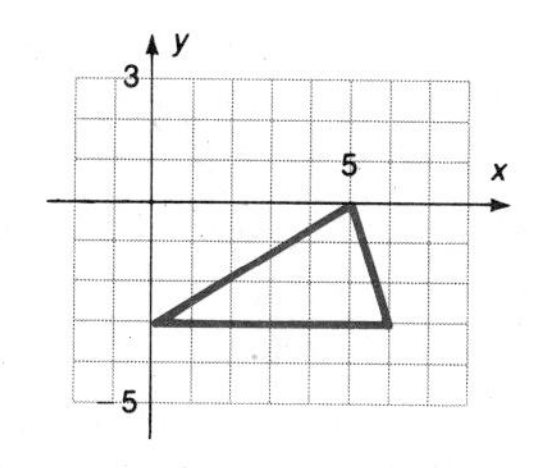

Objective 25 Draw curves showing symmetry with respect to a line; the x axis; the y axis; and the origin. Given the curve at the left, draw a curve so that it has the indicated symmetry.

97. x axis

98. y axis

99. origin

100. the line $x = y$

Leonhard Euler (1707–1783)

Nature herself exhibits to us measurable and observable quantities in definite mathematical dependence; the conception of a function is suggested by all the processes of nature where we observe natural phenomena varying according to distance or to time. Nearly all the "known" functions have presented themselves in an attempt to solve geometrical, mechanical, or physical problems.

J. T. Mertz
History of European Thought in the Nineteenth Century

Euler calculated without any apparent effort, just as men breathe and as eagles sustain themselves in the air.

F. Arago in
Howard Eves, *In Mathematical Circles*, p. 47

HISTORICAL NOTE

The word *function* was used as early as 1694 by the universal genius of the seventeenth century, Gottfried Wilhelm von Leibniz (1646–1716), to denote any quantity connected with a curve. The notion was generalized and modified by Johann Bernoulli (1667–1748) and by Leonhard Euler, the most prolific mathematical writer in history. Throughout his life, Euler was both a frequent contributor to research journals and a superb textbook writer who was widely known for his clarity, detail, and completeness. Euler's work with functions was later expanded by the mathematician P. G. Lejeune-Dirichlet (1805–1859). Around 1815, functions were being considered that were not "nice" and it was Dirichlet who, in 1837, suggested a very broad definition of a function, the one, in fact, that leads to the definition used in this chapter.

An interesting story about Euler is found in a book by Howard Eves:*

"In 1735, the year after his wedding and when he was in Russia, Euler received a problem in celestial mechanics from the French Academy. Though other mathematicians had required several months to solve this problem, Euler, using improved methods of his own and by devoting intense concentration to it, solved it in three days and the better part of the two intervening nights. The strain of the effort induced a fever from which Euler finally recovered, but with the loss of the sight of his right eye. Stoically accepting the misfortune, he commented, 'Now I will have less distraction.' Thirty-one years later, in 1766, when he was again in Russia, Euler developed a cataract in his remaining eye and went completely blind. Now blindness would seem to be an insurmountable barrier to a mathematician, but, like Beethoven's loss of hearing, Euler's loss of sight in no way impaired his amazing productivity. He continued his creative work by dictating to a secretary and by writing formulas in chalk on a large slate for his secretary to copy down. In 1771, after five years in darkness, Euler underwent an operation to remove the cataract from his left eye, and for a brief period he was able to see again. But within a few weeks a very painful infection set in, and when it was over Euler was once again totally blind—so to continue for the remaining twelve years of his life."

* From Howard Eves, *In Mathematical Circles* (Boston: Prindle, Weber, & Schmidt), p. 48.

2

Functions

CONTENTS

PREVIEW

The central idea for this course is the notion of a function, and you are introduced to this concept here in Chapter 2. Sections 2.3 and 2.5 give you specific examples of functions; the other sections present essential properties of functions in general that will be used throughout the course. This chapter has 24 objectives, which are listed on pages 85–87.

PERSPECTIVE

The derivative in calculus involves, among other things, the evaluation of a function. In this section you will learn the definition of a function, notation for a function, and the evaluation of a function. The definition of derivative involves finding the limit of this expression:

$$\frac{f(x+h)-f(x)}{h}$$

Functional notation is discussed in Section 2.2, and you might want to compare Example 3 and Problems 20–28 in that section with what you see on the page below, which is reproduced from a leading calculus book. You will not only need to thoroughly understand the functional concept but also need to be at ease with functional notation in order to succeed in calculus.

the graph of f, and the secant line through P and Q will approach the tangent line at P. Thus, the slope m_{sec} of the secant line approaches the slope m_{tan} of the tangent line as x_1 approaches x_0. Therefore, from (1)

$$m_{\tan} = \lim_{x_1 \to x_0} \frac{f(x_1) - f(x_0)}{x_1 - x_0} \tag{2}$$

For many purposes it is desirable to rewrite this expression in an alternative form by letting

$$h = x_1 - x_0$$

Thus (see Figure 3.1.2*b*), $x_1 = x_0 + h$ and $h \to 0$ as $x_1 \to x_0$, so (2) can be rewritten as

$$m_{\tan} = \lim_{h \to 0} \frac{f(x_0 + h) - f(x_0)}{h}$$

Motivated by the foregoing discussion, we make the following definition.

3.1.1 DEFINITION. If $P(x_0, y_0)$ is a point on the graph of a function f, then the ***tangent line*** to the graph of f at P is defined to be the line through P with slope

$$m_{\tan} = \lim_{h \to 0} \frac{f(x_0 + h) - f(x_0)}{h} \tag{3}$$

From Howard Anton, *Calculus*, 3rd ed. (New York: Wiley), p. 140.

2.1 FUNCTIONS

The material presented in this book is designed to prepare you for the study of calculus. As the title suggests, the main thread that will lead you through this book is the concept of a *function*. You have, no doubt, been introduced to this idea before, probably in algebra. The notion of a correspondence between sets is a common idea. The price of a stock, for example, can be determined by looking at the daily quote in the newspaper; the height of a bridge can be determined by dropping a rock and measuring the time it takes to hit the bottom; and the surface area of a balloon can be determined if you know its radius. All of these are everyday examples of functions.

In the last chapter a relation was defined as a set of ordered pairs (x, y). For a function, consider a correspondence, or a mapping, between two sets X and Y so that x is a member of X and y is a member of Y.

Function as a Mapping

A **function** f is a mapping that assigns to each element x of X a unique element y of Y. The element y is called the **image** of x under f and is denoted by $f(x)$. The set X is called the **domain** of the function. The set of all images of elements of X is called the **range** of the function.

If a function f maps a set X *into* a set Y, it can be illustrated as shown in Figure 2.1.

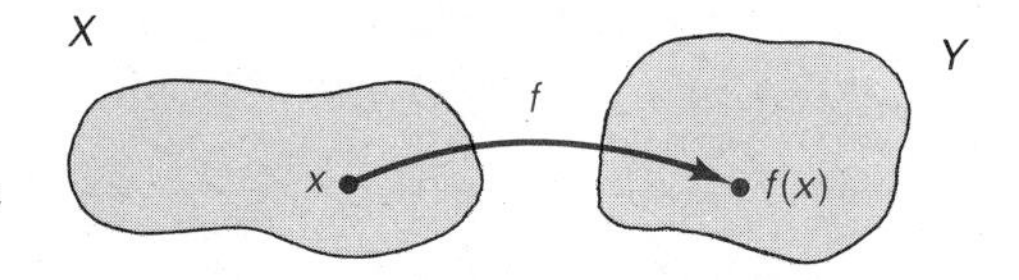

Figure 2.1 A function as a mapping

The sets X and Y are represented by points within regions in a plane. We have shown them as different sets, but they may have elements in common or may even be equal. Notice that the range of f is a subset of Y and does not have to include all of Y. If the range and Y are equal, however, then X maps *onto* Y. The symbol $f(x)$ is read "f of x" and does not mean multiplication; it represents that unique element of Y which corresponds to the element x in X.

EXAMPLE 1

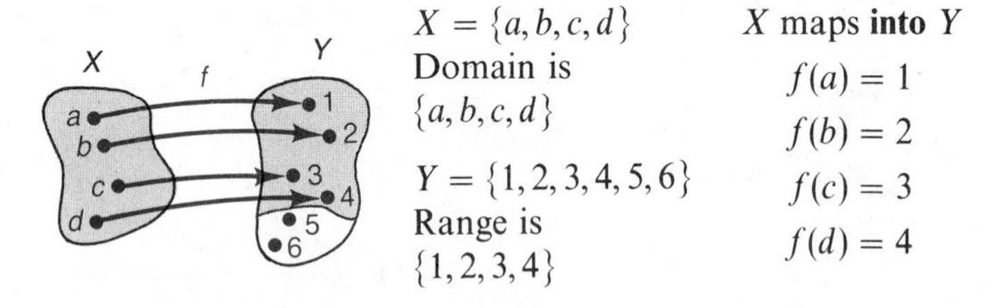

$X = \{a, b, c, d\}$
Domain is $\{a, b, c, d\}$

$Y = \{1, 2, 3, 4, 5, 6\}$
Range is $\{1, 2, 3, 4\}$

X maps **into** Y

$f(a) = 1$

$f(b) = 2$

$f(c) = 3$

$f(d) = 4$

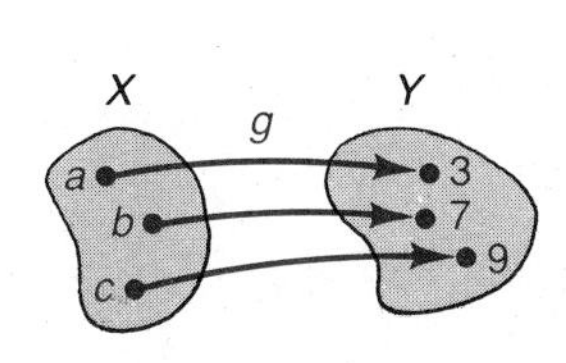

$X = \{a, b, c\}$ $\quad Y = \{3, 7, 9\}$
Domain is $\{a, b, c\}$ $\quad$ Range is $\{3, 7, 9\}$

X maps **onto** Y

$g(a) = 3$
$g(b) = 7$
$g(c) = 9$ ■

EXAMPLE 2 It is possible that different elements in the domain have the same image as shown by this example.

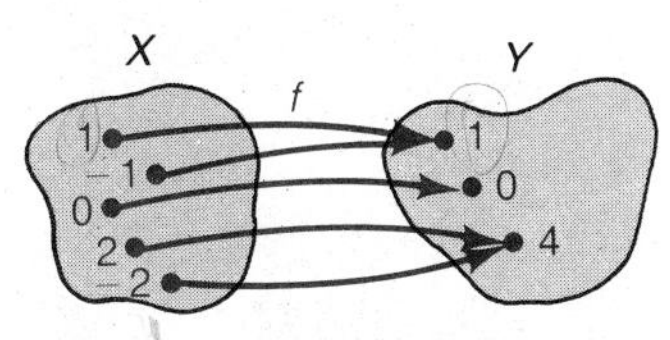

$D = \{1, -1, 0, 2, -2\}$ $\quad R = \{0, 1, 4\}$

$f(0) = 0$
$f(1) = 1$
$f(-1) = 1$
$f(2) = 4$
$f(-2) = 4$ ■

If the images are always different (as they were in Example 1), then the function is called *one-to-one*.

One-to-One Function

> If f maps X into Y so that for any distinct elements x_1 and x_2 of X, $f(x_1) \neq f(x_2)$, then f is a **one-to-one function** of X into Y.

Example 3 illustrates a mapping that is not a function.

EXAMPLE 3

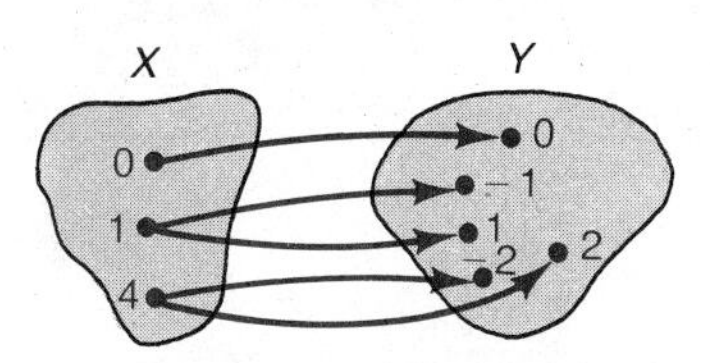

$D = \{0, 1, 4\}$ $\quad R = \{-2, -1, 0, 1, 2\}$

This is not a function because 1 is associated with more than one image (so is 4). ■

It is perhaps more common to describe a function as a set of ordered pairs than it is to describe it as a mapping. It must, however, possess a certain property as described in the following alternative definition of a function.

Function as a Set of Ordered Pairs

> A **function** is a set of ordered pairs for which each member x of the domain is associated with exactly one member $f(x)$ of the range.

EXAMPLE 4 $f = \{(0,2), (1,5), (2,8), (3,11)\}$
$\boldsymbol{D} = \{0,1,2,3\}$ $\quad \boldsymbol{R} = \{2,5,8,11\}$

$\boldsymbol{D}$	$\boldsymbol{R}$	
$0 \rightarrow$	2	$f(0) = 2$
$1 \rightarrow$	5	$f(1) = 5$
$2 \rightarrow$	8	$f(2) = 8$
$3 \rightarrow$	11	$f(3) = 11$

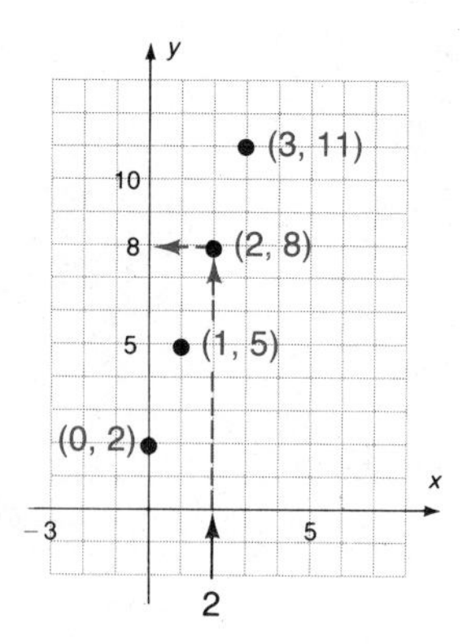

Notice that each value in the domain is associated with a single value in the range. ■

EXAMPLE 5 $g = \{(0,0), (1,1), (-1,1), (2,4), (-2,4)\}$
$\boldsymbol{D} = \{0,1,-1,2,-2\}$ $\quad \boldsymbol{R} = \{0,1,4\}$

$\boldsymbol{D}$	$\boldsymbol{R}$	
$0 \rightarrow$	0	$g(0) = 0$
$1 \searrow$	1	$g(1) = 1$
$-1 \nearrow$		$g(-1) = 1$
$2 \searrow$	4	$g(2) = 4$
$-2 \nearrow$		$g(-2) = 4$

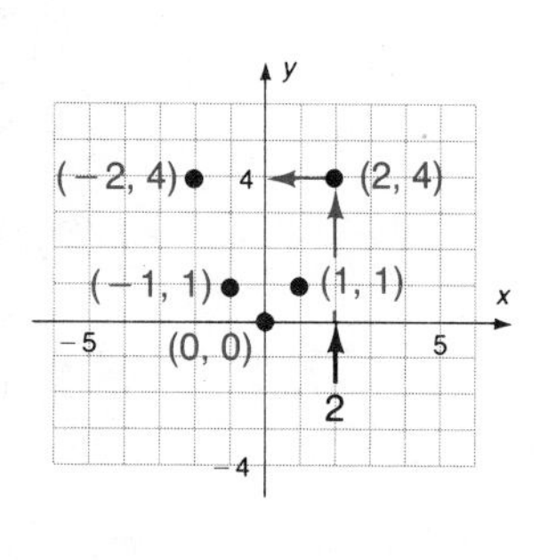

Notice that in Example 4 f is one-to-one while in this example g is not. ■

Not all sets of ordered pairs are functions, as illustrated in Example 6. Do not use $f(x)$ notation unless the set of ordered pairs is a function.

EXAMPLE 6 $\{(0,0), (1,1), (1,-1), (4,2), (4,-2)\}$
$\boldsymbol{D} = \{0,1,4\}$ $\quad \boldsymbol{R} = \{0,1,-1,2,-2\}$

$\boldsymbol{D}$	$\boldsymbol{R}$
$0 \rightarrow$	0
$1 \nearrow$	1
$1 \searrow$	-1
$4 \nearrow$	2
$4 \searrow$	-2

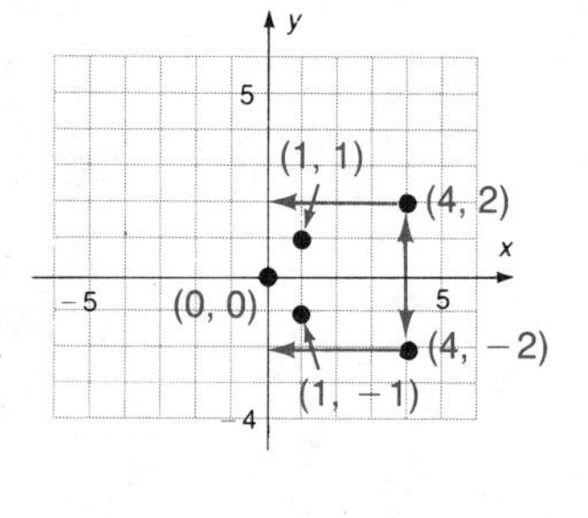

This is not a function because both 1 and 4 are associated with more than one second component. ■

The **graph of a function** f is the set of all points $(x, f(x))$ in a coordinate plane, where x is in the domain of f. That is, the graph of f can be described as the set of all points $P(x, y)$ such that $y = f(x)$. Since a function is a special type of relation, the graphing of functions is a special case of the graphing of relations (discussed in Chapter 1). The graph of a typical function f is shown in Figure 2.2. Notice that the graph of a function is such that for each a in the domain there is only *one point* $(a, f(a))$ on the graph. This means that every vertical line passes through the graph of a function in at most one point. This is the so-called **vertical line test** for the graphs of functions as illustrated in Example 7.

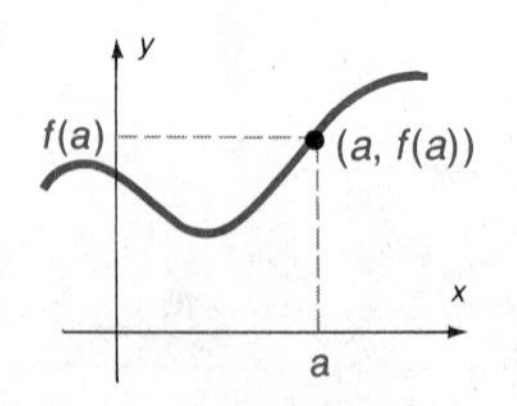

Figure 2.2 Graph of a function

EXAMPLE 7 Which of the following are graphs of functions?

Solution

Function	Function	Not a function	Not a function	Function

Imagine a vertical line sweeping from left to right across the plane—if it passes through more than one point of the graph at one time, then the graph does not represent a function. ■

There are several points of the graph of a function that will be of particular importance to us. The first are the intercepts. These points are called intercepts because they are the places where a curve intercepts the coordinate axes.

Intercepts

If the number zero is in the domain of f, then $f(0)$ is called the **y intercept** of the graph of f and is the point $(0, f(0))$. If a is a real number in the domain such that $f(a) = 0$, then a is called an **x intercept** and is the point $(a, 0)$. Any number x such that $f(x) = 0$ is called a **zero of the function.**

EXAMPLE 8 Find the domain, range, and intercepts for g defined by the graph.

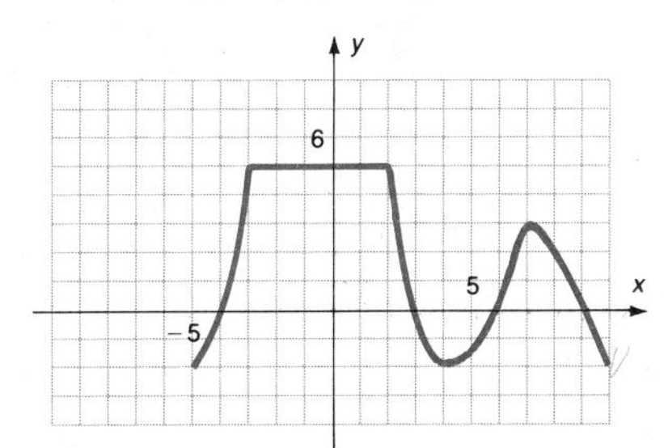

Domain: $-5 \le x \le 10$
Range: $-2 \le y \le 5$
y intercept: $(0, 5)$; we usually say simply that the y intercept is 5.
x intercepts: $(-4, 0)$, $(3, 0)$, $(6, 0)$, and $(9, 0)$; the zeros of the function g are -4, 3, 6, and 9.

■

Notice that a function might have several x intercepts, or several zeros, but a function can have at most only one y intercept. As you progress through this book, you will encounter a variety of special types of functions. The first types we will mention are functions that are increasing, decreasing, or constant.

Increasing, Decreasing, and Constant Functions

Let S be a subset of the domain of a function f. Then:
f is **increasing** on S if $f(x_1) < f(x_2)$ whenever $x_1 < x_2$ in S.
f is **decreasing** on S if $f(x_1) > f(x_2)$ whenever $x_1 < x_2$ in S.
f is **constant** on S if $f(x_1) = f(x_2)$ for every x_1 and x_2 in S.

EXAMPLE 9 Consider the function g graphed in Example 8. Note that g is increasing on $(-5, -3)$; g is constant on $(-3, 2)$; g is decreasing on $(2, 4)$; g is increasing on $(4, 7)$; and g is decreasing on $(7, 10)$. Notice that these terms apply only to open intervals (points in S) and not at the endpoints. ■

PROBLEM SET 2.1

A

Let $X = \{1, 2, 3\}$ and $Y = \{1, 2, 3\}$. Classify the sets in Problems 1–6 as onto, one-to-one, function, or not a function. More than one, or none, of these terms may apply.

1. $\{(1,1), (2,2)\ (3,3)\}$
2. $\{(1,1), (2,1), (3,1)\}$
3. $\{(1,1), (1,2), (1,3)\}$
4. $\{(2,3), (2,1), (2,2)\}$
5. $\{(2,3), (1,2), (3,3)\}$
6. $\{(1,3), (2,1), (3,2)\}$

Let $X = \{a, b, c, d, e\}$ and $Y = \{1, 2, 3, 4, 5\}$. Classify the sets of Problems 7–12 as onto, one-to-one, function, or not a function. More than one, or none, of these terms may apply.

7. $f = \{(a,3), (b,5), (c,2), (d,1), (e,3)\}$
8. $F = \{(a,1), (b,2), (c,3), (d,4), (e,5)\}$
9. $g = \{(a,4), (b,4), (d,4), (e,4)\}$
10. $h = \{(a,3), (b,2), (c,1), (d,4), (a,4)\}$
11. $G = \{(a,5), (a,4), (a,3), (a,2), (a,1)\}$
12. $H = \{(a,5), (b,4), (c,3), (d,4), (e,5)\}$

State whether each set in Problems 13–29 is or is not a function.

13. $\{(8,2), (7,1), (6,3), (5,1)\}$
14. $\{(5,2), (7,3), (1,6), (7,4)\}$
15. $\{1, 2, 3, 4\}$
16. $\{6, 9, 12, 15\}$
17. $\{(x,y) \mid y = 4x + 3\}$*
18. $\{(x,y) \mid y \le 4x + 3\}$

19.

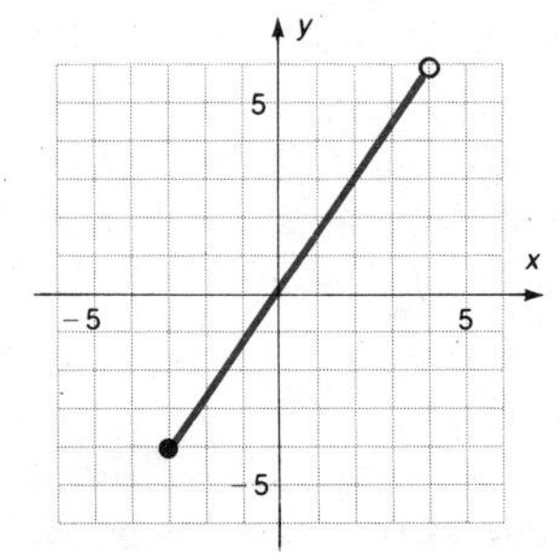

20.

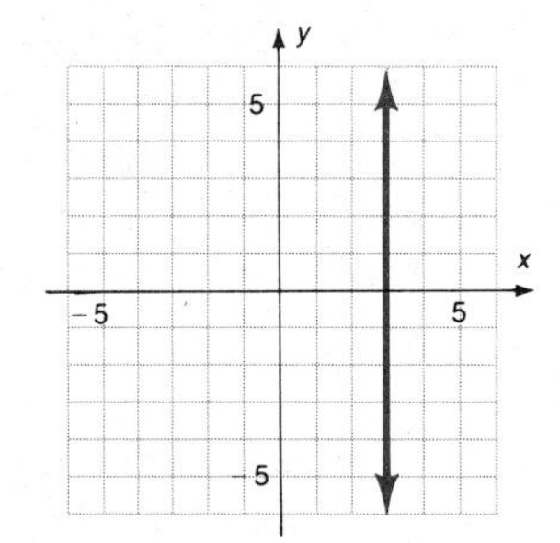

21.

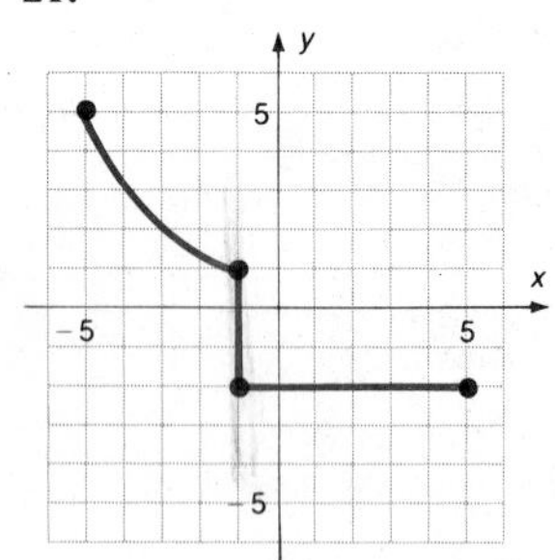

22.

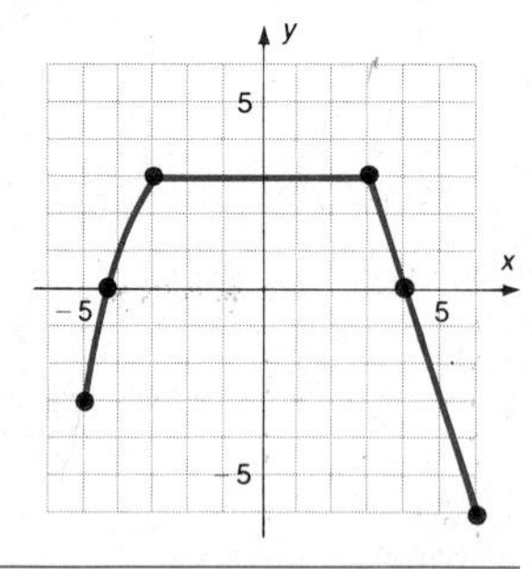

* This is called set builder notation and is read as "the set of all ordered pairs (x, y) such that $y = 4x + 3$."

23.

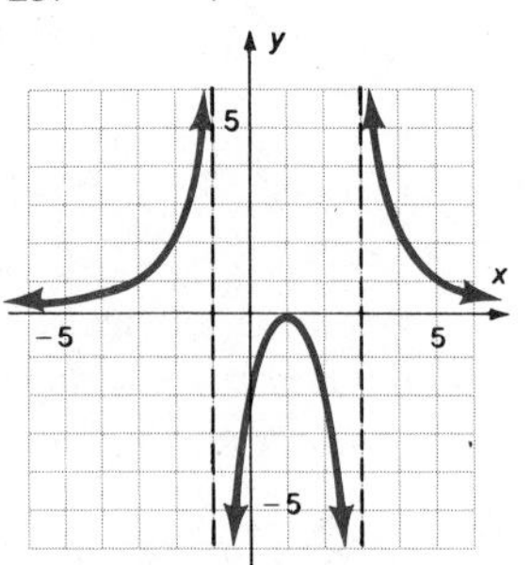

24.

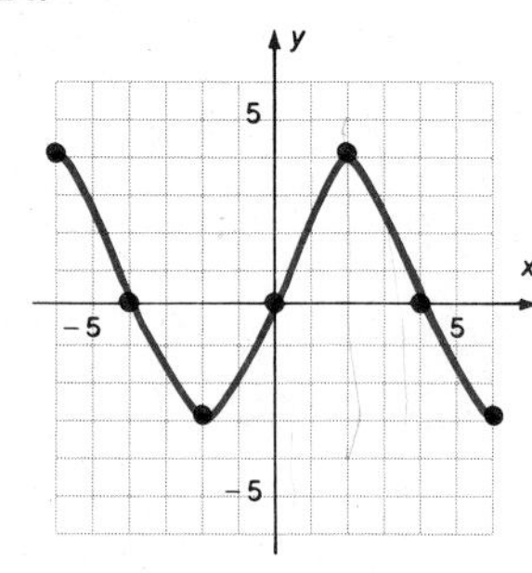

25. Let $y = -1$ if x is a rational number, and let $y = 1$ if x is an irrational number.
26. Let $y = 1$ if x is positive and $y = -1$ if x is negative.
27. Suppose that x is the closing price of Xerox stock on July 1 of year y.
28. Suppose that y is the closing price of IBM stock on January 2 of year x.
29. Let A be the area of a cross-sectional slice of an orange whose circumference is C.
30. See Figure 2.3a. If point A has coordinates $(2, f(2))$, what are the coordinates of P and Q?

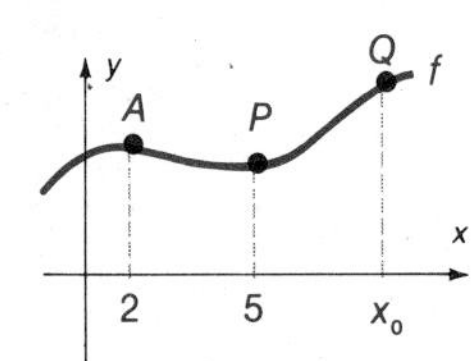

(a) Graph of f

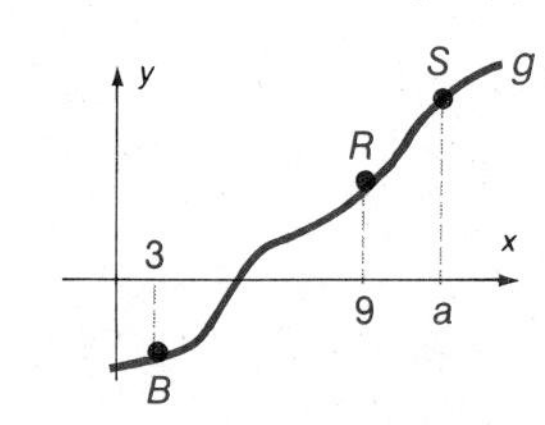

(b) Graph of g

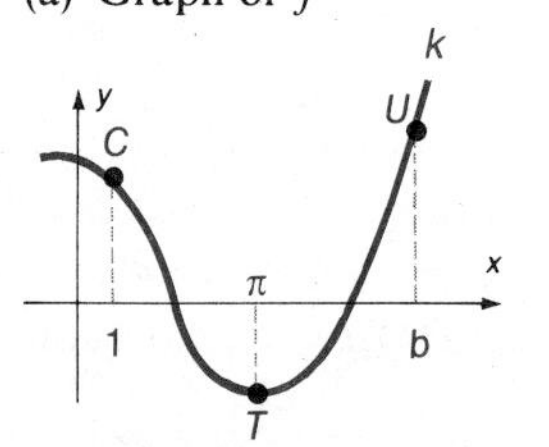

(c) Graph of k

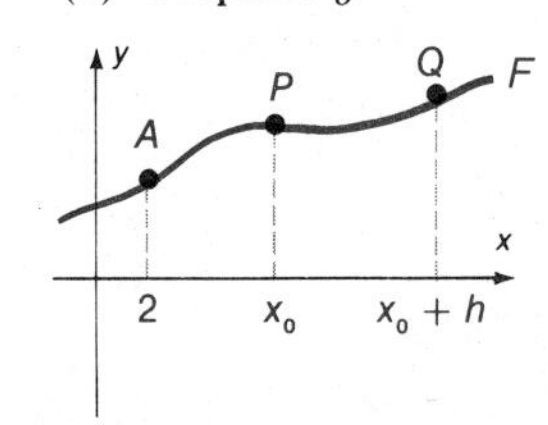

(d) Graph of F

Figure 2.3

31. See Figure 2.3b. If point B has coordinates $(3, g(3))$, what are the coordinates of R and S?
32. See Figure 2.3c. If point C has coordinates $(1, k(1))$, what are the coordinates of T and U?
33. See Figure 2.3d. If point A has coordinates $(2, F(2))$, what are the coordinates of P and Q?

34. See Figure 2.3e. If point B has coordinates $(3, G(3))$, what are the coordinates of R and S?

35. See Figure 2.3f. If point C has coordinates $(1, K(1))$, what are the coordinates of T and U?

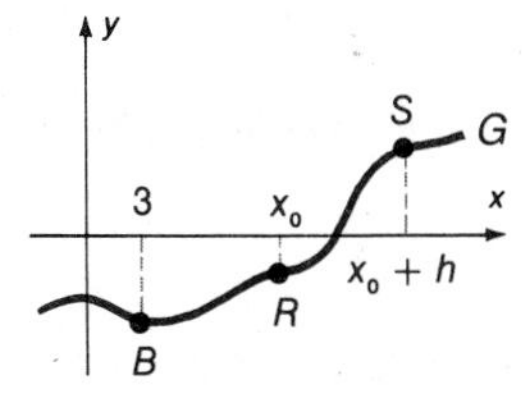

(e) Graph of G

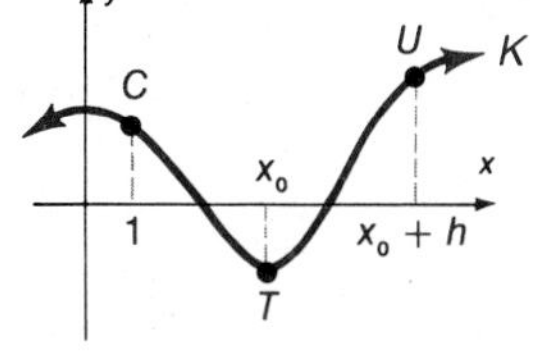

(f) Graph of K

Figure 2.3

B

Find the domain, range, and intercepts of the functions defined by the graphs indicated in Problems 36–44. Also tell where the function is increasing, decreasing, and constant.

36. Graph of Problem 19

37. Graph of Problem 22

38. Graph of Problem 24

39.

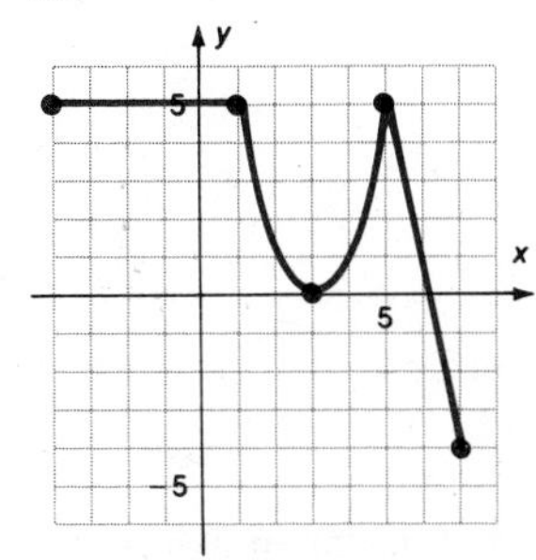

40.

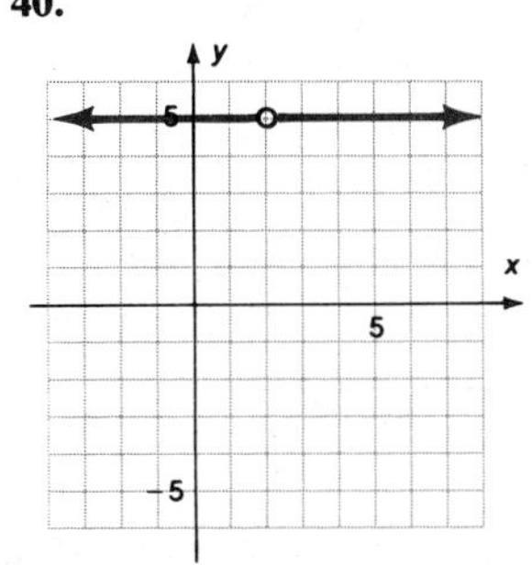

41.

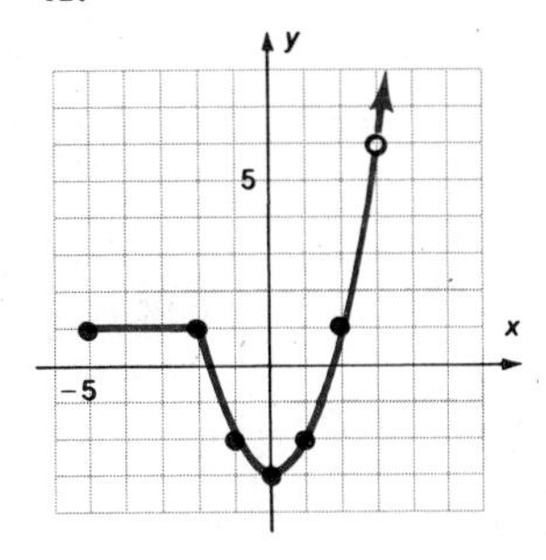

42.

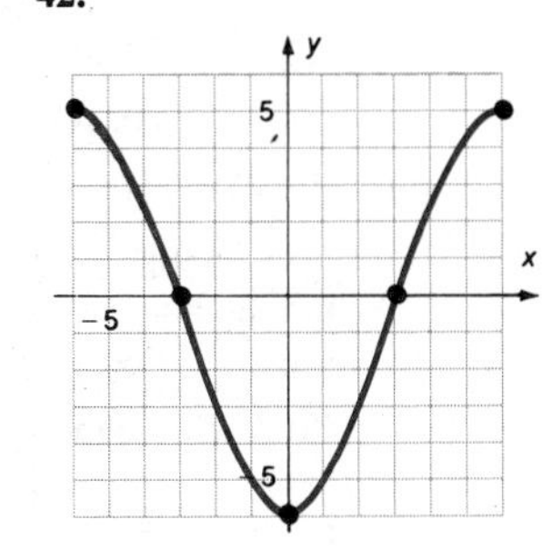

43.

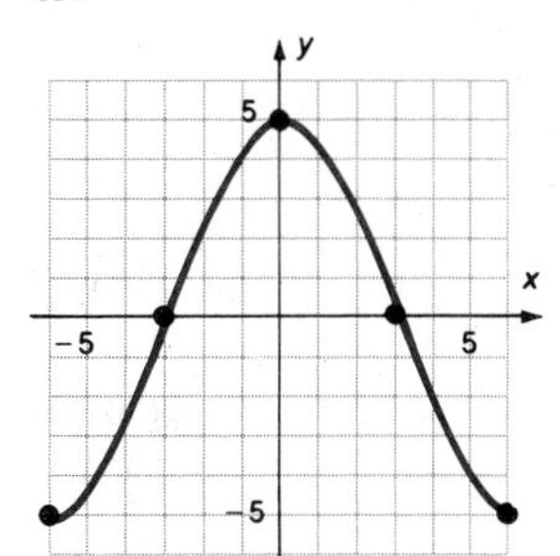

44.

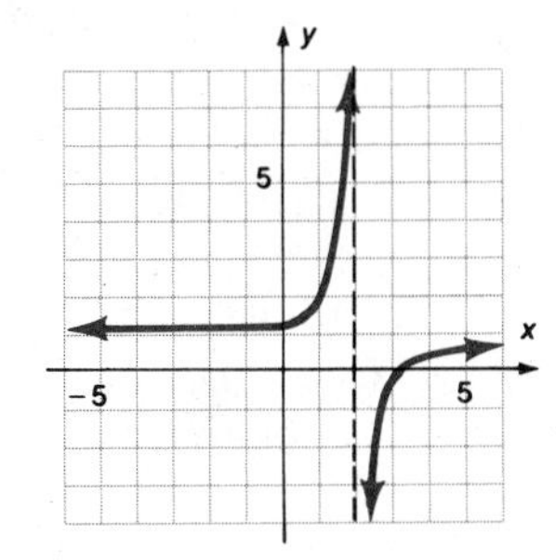

For Problems 45–54, use the following table, which reflects the purchasing power of the dollar from October 1944 to October 1984 (Source: Bureau of Labor Statistics, Consumer Division). Let x represent the year, let the domain be the set $\{1944, 1954, 1964, 1974, 1984\}$, and let

$r(x) =$ *price of 1 lb of round steak*

$s(x) =$ *price of a 5-lb bag of sugar*

$b(x) =$ *price of a loaf of bread*

$c(x) =$ *price of 1 lb of coffee*

$e(x) =$ *price of a dozen eggs*

$m(x) =$ *price of $\frac{1}{2}$ gal of milk*

$g(x) =$ *price of 1 gal of gasoline*

Year	Round steak (1 lb)	Sugar (5-lb bag)	Bread (loaf)	Coffee (1 lb)	Eggs (1 dozen)	Milk ($\frac{1}{2}$ gal)	Gasoline (1 gal)
1944	$0.45	$0.34	$0.09	$0.30	$0.64	$0.29	$0.21
1954	0.92	0.52	0.17	1.10	0.60	0.45	0.29
1964	1.07	0.59	0.21	0.82	0.57	0.48	0.30
1974	1.78	2.08	0.36	1.31	0.84	0.78	0.53
1984	2.15	1.49	1.29	2.69	1.15	1.08	1.10

45. Find: a. $r(1954)$ b. $m(1954)$

46. Find: a. $g(1944)$ b. $c(1984)$

47. Find $s(1984) - s(1944)$.

48. Find $b(1984) - b(1944)$.

49. a. Find the change in the price of eggs from 1944 to 1984.
    b. Write the change in the price of eggs using functional notation.

50. a. Find the change in the price of round steak from 1944 to 1984.
    b. Write the change in the price of round steak using functional notation.

51. a. Find $[g(1944 + 40) - g(1944)]/40$.
    b. In words, attach some meaning to the expression given in part **a**.

52. a. Find $[m(1944 + 40) - m(1944)]/40$.
    b. In words, attach some meaning to the expression given in part **a**.

53. a. What is the average increase in the price of sugar per year from 1944 to 1954? Write this in functional notation.
    b. What is the average increase in the price of sugar per year from 1944 to 1964? Write this in functional notation.
    c. What is the average increase in the price of sugar per year from 1944 to 1974? Write this in functional notation.
    d. What is the average increase in the price of sugar per year from 1944 to 1984? Write this in functional notation.
    e. What is the average increase in the price of sugar per year from 1944 to $1944 + h$, where h is an unspecified number of years? Write this in functional notation.

54. Repeat Problem 53 for coffee instead of sugar.

55. According to the U.S. Public Health Service, the number of marriages in the United States was about 2,176,000 in 1977 and about 2,495,000 in 1982. Let $M(x)$ represent the number of marriages in year x.
    a. Find $[M(1982) - M(1977)]/5$.
    b. Give a verbal description for the following functional expression:

$$[M(1977 + h) - M(1977)]/h$$

56. According to the U.S. Public Health Service, the number of divorces in the United States was about 1,097,000 in 1977 and about 1,180,000 in 1982. Let $D(x)$ represent the number of divorces in year x.
    a. Find $[D(1982) - D(1977)]/5$.
    b. Give a verbal description for the following functional expression:

$$[D(1977 + h) - D(1977)]/h$$

C

57. If F is a one-to-one function mapping X onto Y, and the domain of F contains exactly five elements, what can you conclude about the set Y?

58. If a function f is increasing throughout its domain, prove that f is one-to-one.

59. If a function f is decreasing throughout its domain, prove that f is one-to-one.

60. If g maps X into Y, and the range of g is a set S (where S is a subset of Y), explain why g maps X onto S.

2.2 FUNCTIONAL NOTATION

The notation for functions introduced in the last section is fundamental for this course and for calculus. This section provides additional examples and practice in using functional notation. Remember:

The function is denoted by f

x is a member of the domain

$$f(x)$$

$f(x)$ is a member of the range

The *function* is denoted by f; $f(x)$ is the *number* associated with x. Sometimes functions are defined by expressions such as

$$f(x) = 3x + 2 \qquad \text{or} \qquad g(x) = x^2 + 4x + 3$$

To emphasize the difference between f and $f(x)$, some books use $f: x \to 3x + 2$ to define functions, but this book simply uses the notation $f(x) = 3x + 2$ to mean the set of all ordered pairs (x, y) such that $y = 3x + 2$.

EXAMPLE 1 Given f and g defined by $f(x) = 3x + 2$ and $g(x) = x^2 + 4x + 3$, find the indicated values: **(a)** $f(1)$; **(b)** $g(2)$; **(c)** $g(-3)$; **(d)** $f(-3)$.

Solution **a.** The symbol $f(1)$ represents the second component of the ordered pair of the function f with first component 1. Replace x by 1 in the expression

$$\begin{aligned} f(x) &= 3x + 2 \\ &\updownarrow \quad \updownarrow \\ f(\mathbf{1}) &= 3(\mathbf{1}) + 2 \\ &= 5 \end{aligned}$$

b. $g(2)$:
$$\begin{aligned} g(x) &= x^2 + 4x + 3 \\ &\updownarrow \quad \updownarrow \quad \updownarrow \\ g(\mathbf{2}) &= (\mathbf{2})^2 + 4(\mathbf{2}) + 3 \\ &= 4 + 8 + 3 \\ &= 15 \end{aligned}$$

c.
$$\begin{aligned} g(-3) &= (-3)^2 + 4(-3) + 3 \\ &= 9 - 12 + 3 \\ &= 0 \end{aligned}$$

d.
$$\begin{aligned} f(-3) &= 3(-3) + 2 \\ &= -9 + 2 \\ &= -7 \end{aligned}$$

■

The members of the domain of a function may also be represented by variables, as shown in Example 2.

EXAMPLE 2 Let F and G be defined by $F(x) = x^2 + 1$ and $G(x) = (x + 1)^2$.

a. $F(w) = w^2 + 1$

b.
$$\begin{aligned} G(t) &= (t + 1)^2 \\ &= t^2 + 2t + 1 \end{aligned}$$

c.
$$\begin{aligned} F(w + 3) &= (w + 3)^2 + 1 \\ &= w^2 + 6w + 9 + 1 \\ &= w^2 + 6w + 10 \end{aligned}$$

d.
$$\begin{aligned} G(x - 2) &= [(x - 2) + 1]^2 \\ &= (x - 1)^2 \\ &= x^2 - 2x + 1 \end{aligned}$$

e.
$$\begin{aligned} F(w + h) &= (w + h)^2 + 1 \\ &= w^2 + 2wh + h^2 + 1 \end{aligned}$$

f. $G(x + h) = [(x + h) + 1]^2$
$$= x^2 + xh + x + xh + h^2 + h + x + h + 1$$
$$= x^2 + 2xh + h^2 + 2x + 2h + 1$$
■

In calculus, functional notation is used to carry out manipulations such as those shown in Example 3.

EXAMPLE 3 Find $\dfrac{f(x + h) - f(x)}{h}$ for each function.

a. $f(x) = x^2$, where $x = 5$

$$\frac{f(5 + h) - f(5)}{h} = \frac{(5 + h)^2 - 5^2}{h}$$
$$= \frac{25 + 10h + h^2 - 25}{h}$$
$$= \frac{(10 + h)h}{h}$$
$$= 10 + h$$

b. $f(x) = 2x^2 + 1$, where $x = 1$

$$\frac{f(1 + h) - f(1)}{h} = \frac{[2(1 + h)^2 + 1] - [2(1)^2 + 1]}{h}$$
$$= \frac{[2(1 + 2h + h^2) + 1] - (2 + 1)}{h}$$
$$= \frac{2h^2 + 4h + 3 - 3}{h}$$
$$= 2h + 4$$

c. $f(x) = x^2 + 3x - 2$

$$\frac{f(x + h) - f(x)}{h} = \frac{[(x + h)^2 + 3(x + h) - 2] - [x^2 + 3x - 2]}{h}$$
$$= \frac{x^2 + 2xh + h^2 + 3x + 3h - 2 - x^2 - 3x + 2}{h}$$
$$= \frac{2xh + h^2 + 3h}{h}$$
$$= 2x + 3 + h$$
■

Functional notation can be used to work a wide variety of applied problems, as shown by Example 4 and again in the problem set.

EXAMPLE 4 If an object is dropped from a certain height, it is known that it will fall a distance of s ft in t sec according to the formula

$$s = 16t^2$$

This formula can be represented by $f(t) = 16t^2$.

a. How far will the object fall in the first second?

$$\begin{aligned} f(1) &= 16 \cdot 1^2 \\ &= 16 \quad \text{or} \quad \mathbf{16\ ft} \end{aligned}$$

b. How far will it fall in the *next* 2 sec?

$$\begin{aligned} f(1 + 2) &= 16 \cdot 3^2 \\ &= 144 \quad \text{or} \quad 144 \text{ ft in 3 sec} \end{aligned}$$

So the answer to the question is

$$\begin{aligned} f(3) - f(1) &= 144 - 16 \\ &= 128 \quad \text{or} \quad \mathbf{128\ ft} \end{aligned}$$

c. How far will it fall during the time $t = 1$ sec to $t = 1 + h$ sec?

$$\begin{aligned} f(1 + h) - f(1) &= 16(1 + h)^2 - 16 \\ &= 16 + 32h + 16h^2 - 16 \\ &= \mathbf{(32h + 16h^2)\ ft} \end{aligned}$$

d. What is the average rate of change of distance (in feet per second, fps) during the time $t = 1$ sec to $t = 3$ sec?

$$\frac{f(3) - f(1)}{3 - 1} = \frac{128}{2} = \mathbf{64\ fps}$$

e. What is the average rate of change of distance during the time $t = 1$ sec to $t = 1 + h$ sec?

$$\begin{aligned} \frac{f(1 + h) - f(1)}{h} &= \frac{32h + 16h^2}{h} \\ &= \mathbf{(32 + 16h)\ fps} \end{aligned}$$

f. What is the average rate of change of distance during the time $t = x$ sec to $t = x + h$ sec?

$$\begin{aligned} \frac{f(x + h) - f(x)}{(x + h) - x} &= \frac{f(x + h) - f(x)}{h} \qquad \text{Does this look familiar?} \\ &= \frac{16(x + h)^2 - 16x^2}{h} \\ &= \frac{16x^2 + 32xh + 16h^2 - 16x^2}{h} \\ &= \mathbf{(32x + 16h)\ fps} \end{aligned}$$

■

The variable t in Example 4 represents an arbitrary number from the domain of f and is often called the **independent variable**. The variable s, which represents a number from the range of f, is called a **dependent variable** since its value depends on the value assigned to t.

It is also important to make note of the domain of a function. In this book, unless otherwise specified, the domain is the set of real numbers for which the given function is meaningful. For Example 4, the domain is the set of all nonnegative real numbers, since the formula is meaningless for negative time. If a function f is **undefined at** x, it means that x is not in the domain of f.

EXAMPLE 5 Find the domain for the given functions:

a. $f(x) = 2x - 1$ **b.** $g(x) = \dfrac{(2x-1)(x+3)}{x+3}$

c. $h(x) = \sqrt{x+1}$ **d.** $F(x) = \sqrt{2 - 3x - 2x^2}$

e. $G(x) = 2x - 1,\ x \neq -3$ **f.** $r(x) = 2 - \dfrac{x}{x}$

Solution **a.** All real numbers; $(-\infty, \infty)$

b. All real numbers except $x = -3$ because if $x = -3$, then the expression is meaningless. That is, set the denominator $(x + 3)$ equal to zero $(x + 3 = 0)$, and solve $(x = -3)$. The domain is $(-\infty, -3) \cup (-3, \infty)$. This domain is usually denoted simply by $x \neq -3$.

c. Here h has meaning if $x + 1$ is nonnegative. That is,

$$x + 1 \geq 0$$
$$x \geq -1$$

This domain can be described by writing $[-1, \infty)$.

d. F has meaning if $2 - 3x - 2x^2$ is nonnegative.

$$2 - 3x - 2x^2 \geq 0$$
$$(1 - 2x)(2 + x) \geq 0$$

Domain: $[-2, \frac{1}{2}]$

+ − neg | + + pos | − + neg

-2 $\frac{1}{2}$

e. The domain has $x = -3$ explicitly eliminated. This means that the domain is $(-\infty, -3) \cup (-3, \infty)$ or simply $x \neq -3$.

f. $r(0)$ is meaningless, but $r(x) = 1$ for $x \neq 0$, so the domain is all real numbers except 0. ■

Equality of Functions

Two functions f and g are **equal** if and only if

1. f and g have the same domain.
2. $f(x) = g(x)$ for all x in the domain.

Compare Examples 5a and 5b where $f(x) = 2x - 1$ and $g(x) = \dfrac{(2x-1)(x+3)}{x+3}$.

In algebra you wrote

$$\frac{(2x-1)(x+3)}{x+3} = 2x - 1$$

but it is *not* true that $f = g$, since their domains are not the same. However, $g = G$ (from Example 5e, $G(x) = 2x - 1$, $x \neq -3$) because both conditions of the definition are met.

EXAMPLE 6 $f(x) = \dfrac{(x-3)(x+5)}{x+5}$; $\quad g(x) = x - 3$

$f \neq g$ since the domain of f is all reals except -5 and the domain of g is all real numbers. ■

EXAMPLE 7 $f(x) = \dfrac{(2x-5)(x+1)}{x+1}$; $\quad g(x) = 2x - 5, x \neq -1$

$f = g$ since the domain of both f and g is all reals except $x = -1$ and $f(x) = g(x)$ for all x in the domain. ■

Functions are sometimes classified as *even* or *odd.*

Even and Odd Functions

A function f is called

even if $f(-x) = f(x)$ and

odd if $f(-x) = -f(x)$.

Just as not every real number is even or odd (2 is even, 3 is odd, but 2.5 is neither), not every function is even or odd.

EXAMPLE 8 Classify the given functions as even, odd, or neither.

a. $f(x) = x^2$ is **even** since
$$\begin{aligned} f(-x) &= (-x)^2 \\ &= x^2 \\ &= f(x) \end{aligned}$$

b. $g(x) = x^3$ is **odd** since
$$\begin{aligned} g(-x) &= (-x)^3 \\ &= -x^3 \\ &= -[x^3] \\ &= -g(x) \end{aligned}$$

c. $h(x) = x^2 + x$ is **neither** since
$$\begin{aligned} h(-x) &= (-x)^2 + (-x) \\ &= \underbrace{x^2 - x} \end{aligned}$$

This is neither $h(x)$ nor $-h(x)$. ■

PROBLEM SET 2.2

A

In Problems 1–12, let $f(x) = 2x + 1$ and $g(x) = 2x^2 - 1$. Find the requested values.

1. a. $f(0)$ b. $f(2)$ c. $f(-3)$ d. $f(\sqrt{5})$ e. $f(\pi)$
2. a. $f(1)$ b. $g(1)$ c. $f(\sqrt{3})$ d. $g(\sqrt{3})$ e. $g(\pi)$
3. a. $f(w)$ b. $g(w)$ c. $g(t)$ d. $g(v)$ e. $f(m)$
4. a. $f(t)$ b. $f(p)$ c. $f(t + 1)$ d. $g(t + 1)$ e. $f(t^2)$
5. a. $f(1 + \sqrt{2})$ b. $g(1 + \sqrt{2})$ c. $g(t + 3)$ d. $f(t^2 + 2t + 1)$ e. $g(m - 1)$
6. a. $f(x + 2)$ b. $g(x + 2)$ c. $f(t + h)$ d. $g(t + h)$ e. $f(x + h)$
7. $\dfrac{f(t + 3) - f(t)}{3}$
8. $\dfrac{f(t + h) - f(t)}{h}$
9. $\dfrac{f(x + h) - f(x)}{h}$
10. $\dfrac{g(t + 2) - g(t)}{2}$
11. $\dfrac{g(t + h) - g(t)}{h}$
12. $\dfrac{g(x + h) - g(x)}{h}$

In Problems 13–19, compute the given value where $f(x) = x^2 - 1$ and $g(x) = 2x + 5$.

13. a. $f(w)$ b. $f(h)$ c. $f(w + h)$ d. $f(w)+f(h)$
14. a. $g(s)$ b. $g(t)$ c. $g(s + t)$ d. $g(s) + g(t)$
15. a. $f(x^2)$ b. $f(\sqrt{x})$ c. $f(x + h)$ d. $f(-x)$
16. a. $g(x^2)$ b. $g(\pi)$ c. $g(x + \pi)$ d. $g(-x)$
17. $\dfrac{g(x + h) - g(x)}{h}$
18. $\dfrac{f(t + h) - f(t)}{h}$
19. $\dfrac{f(x + h) - f(x)}{h}$

B

In Problems 20–28, find $\dfrac{f(x + h) - f(x)}{h}$ *for the given function f.*

20. $f(x) = 9x + 3$
21. $f(x) = 5 - 2x$
22. $f(x) = |x|$
23. $f(x) = |2x + 1|$
24. $f(x) = 5x^2$
25. $f(x) = 3x^2 + 2x$
26. $f(x) = 2x^2 + 3x - 4$
27. $f(x) = \dfrac{1}{x}$
28. $f(x) = \dfrac{x + 1}{x - 1}$

Find the domain for the functions defined by the equations in Problems 29–38 and leave your answer in interval notation.

29. $f(x) = 3x + 1$
30. $g(x) = 3x + 1, x \neq 2$
31. $h(x) = \dfrac{(3x + 1)(x + 2)}{x + 2}$
32. $F(x) = \dfrac{(2x + 1)(x - 1)}{x^2 + 1}$
33. $G(x) = \sqrt{2x + 1}$
34. $H(x) = \sqrt{1 - 3x}$
35. $f(x) = \sqrt{2 - x - x^2}$
36. $g(x) = \sqrt{2 + x - x^2}$
37. $h(x) = \dfrac{3x^2 - 4x - 4}{x^2 - 4}$
38. $h(x) = \dfrac{x^2 - 3x + 2}{x^2 + 2x - 3}$

State whether the functions f and g are equal in Problems 39–44.

39. $f(x) = \dfrac{2x^2 + x}{x}$; $g(x) = 2x + 1$
40. $f(x) = \dfrac{2x^2 + x}{x}$; $g(x) = 2x + 1, x \neq 0$
41. $f(x) = \dfrac{2x^2 + x - 6}{x - 2}$; $g(x) = 2x + 3, x \neq 2$
42. $f(x) = \dfrac{3x^2 - 7x - 6}{x - 3}$; $g(x) = 3x + 2, x \neq 3$
43. $f(x) = \dfrac{3x^2 - 5x - 2}{x - 2}$; $g(x) = 3x + 1$
44. $f(x) = \dfrac{(3x + 1)(x - 2)}{x - 2}, x \neq 6$;
    $g(x) = \dfrac{(3x + 1)(x - 6)}{x - 6}, x \neq 2$

Classify the functions defined in Problems 45–53 as even, odd, or neither.

45. $f_1(x) = x^2 + 1$
46. $f_2(x) = \sqrt{x^2}$
47. $f_3(x) = \dfrac{1}{3x^3 - 4}$
48. $f_4(x) = x^3 + x$
49. $f_5(x) = \dfrac{1}{(x^3 + 3)^2}$
50. $f_6(x) = \dfrac{1}{(x^3 + x)^2}$
51. $f_7(x) = |x|$
52. $f_8(x) = |x| + 3$
53. $f_9(x) = 5$
54. ***Business*** A firm determines that the total cost C (in dollars) of producing x units of a certain product is given by
$$C(x) = -0.02x^2 + 4x + 500 \qquad (0 \leq x \leq 150)$$
Find $C(50)$ and $C(100)$.
55. ***Business*** What is the average cost per unit in Problem 54 if 50 units are produced? Repeat for 100 units. What is the

per-unit increase in cost for the increase from 50 to 100 units?

56. ***Business*** What is the per-unit increase in cost in Problem 54 for an increase from 50 units to 51 units? Compare this answer with the answer to Problem 55.

57. ***Business*** What is the per-unit increase in cost in Problem 54 for an h-unit increase in production above a level of x units?

58. ***Physics*** Let d be a function that represents the distance an object falls (neglecting air resistance) in t sec. It can be shown that $d(t) = 16t^2$. Find the average rate that the object falls for the intervals of time given:

a. From $t = 2$ to $t = 6$ **b.** From $t = 2$ to $t = 4$
c. From $t = 2$ to $t = 3$ **d.** From $t = 2$ to $t = 2 + h$
e. From $t = x$ to $t = x + h$

59. ***Physics*** In Problem 58, give a physical interpretation for

$$\frac{d(x + h) - d(x)}{h}$$

C

60. If $f(x) = x^2$, then

$$f\left(\frac{1}{x}\right) = \left(\frac{1}{x}\right)^2 = \frac{1}{x^2} = \frac{1}{f(x)}$$

Give an example of a function for which

$$f\left(\frac{1}{x}\right) \neq \frac{1}{f(x)}$$

61. If $f(x) = x$, then $f(x^2) = [f(x)]^2$. Give an example of a function for which $f(x^2) \neq [f(x)]^2$.

62. If f is an even function, show that the graph of f is symmetric with respect to the y axis.

63. If f is an odd function, show that the graph of f is symmetric with respect to the origin.

2.3 LINEAR FUNCTIONS

The first type of function we will consider in this book is one with which you have had some experience in beginning algebra.

Linear Function

> A function f is a **linear function** if
>
> $$f(x) = mx + b$$
>
> where m and b are real numbers.

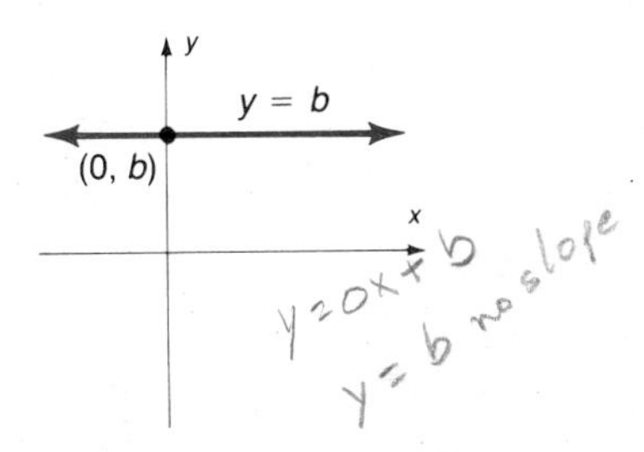

Figure 2.4 A horizontal line

Notice that if $m = 0$, then $f(x) = b$, which we called a constant function in Section 2.1. If the domain of a constant function is the set of real numbers, then the graph of $f(x) = b$ is a **horizontal line** as shown in Figure 2.4.

Let $P_1(x_1, y_1)$ and $P_2(x_2, y_2)$ be any points on a line, and suppose that $x_1 = x_2$. Then the line is parallel to the y axis and is called a **vertical line** (Figure 2.5). Notice that vertical lines are not functions.

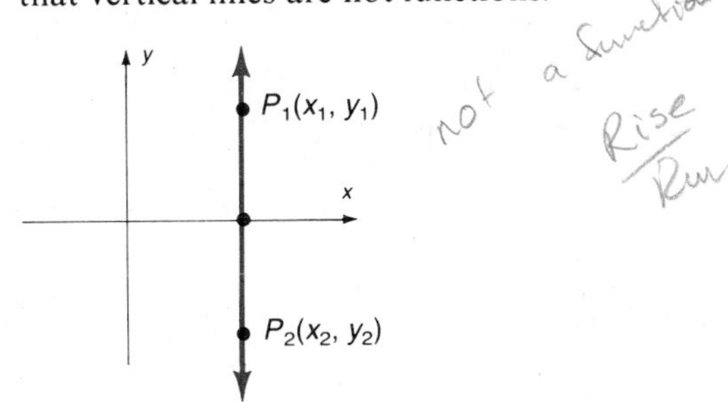

Figure 2.5 A vertical line is not a function

The steepness of a line is specified by using the idea of slope. If $x_1 \neq x_2$, then the slope is defined as follows.

Slope of a Line

> Let $P_1(x_1, y_1)$ and $P_2(x_2, y_2)$ be distinct points on a line such that $x_1 \neq x_2$. Then
>
> $$\textbf{Slope} = \frac{\text{vertical change}}{\text{horizontal change}} = \frac{y_2 - y_1}{x_2 - x_1}$$

The numerator $y_2 - y_1$ is often called the **rise** and the denominator $x_2 - x_1$ the **run** from P_1 to P_2. If you use functional notation for the points $P_1(x_1, f(x_1))$ and $P_2(x_2, f(x_2))$, then the slope is found by

$$\text{Slope} = \frac{f(x_2) - f(x_1)}{x_2 - x_1}$$

EXAMPLE 1 Sketch the line passing through the points whose coordinates are given. Then find the slope of each line.

a. $(2, -3)$ and $(-1, 2)$ **b.** $(-4, -1)$ and $(1, 3)$
c. $(-3, 4)$ and $(5, 4)$ **d.** $(-3, 2)$ and $(-3, 4)$
e. $(3, f(3))$ and $(3 + h, f(3 + h))$ **f.** $(a, f(a))$ and $(a + h, f(a + h))$

Solution **a.** $m = \dfrac{2 - (-3)}{-1 - 2} = \dfrac{5}{-3} = -\dfrac{5}{3}$

negative slope

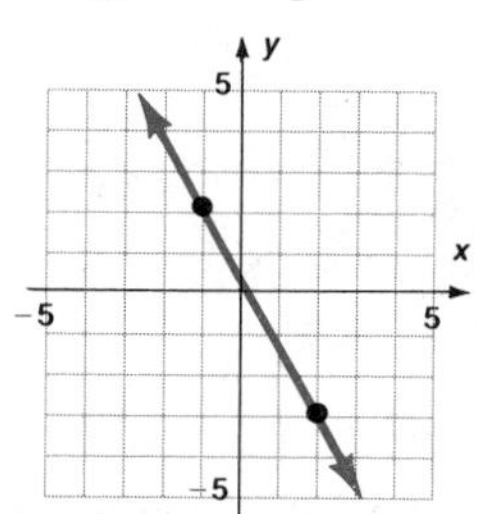

b. $m = \dfrac{3 - (-1)}{1 - (-4)} = \dfrac{4}{5}$

positive slope

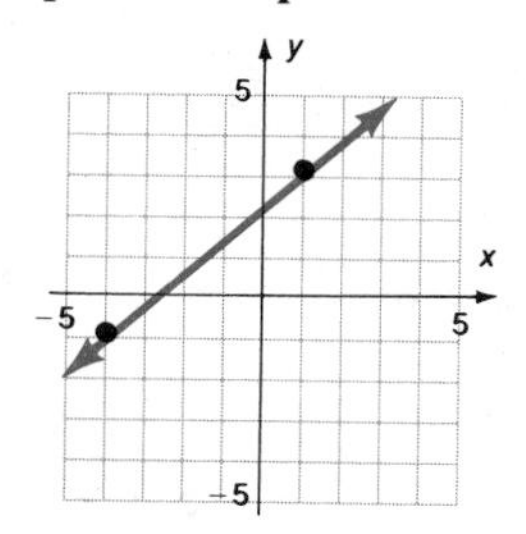

c. $m = \dfrac{4 - 4}{5 + 3} = 0$

0 slope; horizontal line

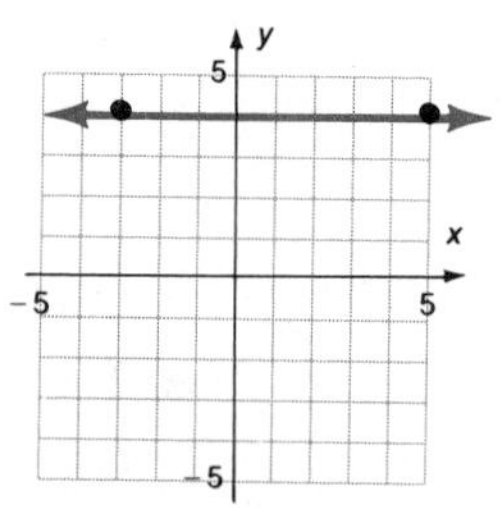

d. $m = \dfrac{4 - 2}{-3 + 3}$ is undefined

undefined slope; vertical line

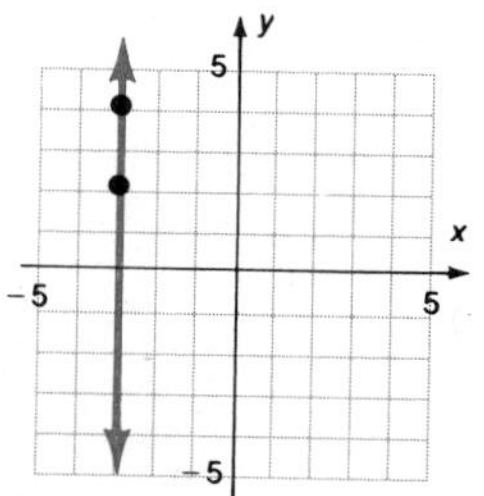

e. $m = \dfrac{f(3+h) - f(3)}{3+h-3} = \dfrac{f(3+h) - f(3)}{h}$

arbitrary slope

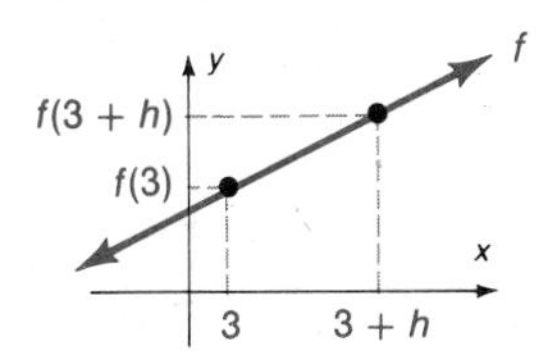

f. $m = \dfrac{f(a+h) - f(a)}{h}$

arbitrary slope

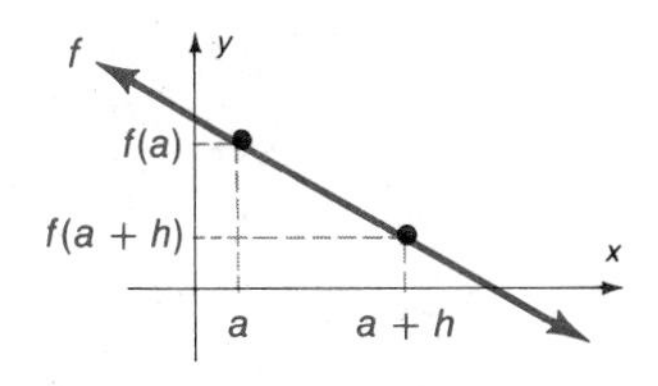

■

Two properties relating lines and slopes can be proved geometrically.

Parallel and Perpendicular Lines

Let L_1 and L_2 be two nonvertical lines with slopes m_1 and m_2, respectively. Then:

1. L_1 and L_2 are parallel if and only if $m_1 = m_2$.
2. L_1 and L_2 are perpendicular if and only if $m_1 m_2 = -1$.

EXAMPLE 2 Show that the points $Q(-3, 7)$, $U(8, 2)$, $A(4, -3)$, and $D(-7, 2)$ are the corners of a parallelogram $QUAD$.

Solution

$$m_{QU} = \frac{2-7}{8+3} = \frac{-5}{11}$$

$$m_{DA} = \frac{-3-2}{4+7} = \frac{-5}{11}$$

Thus $\overline{QU}$ is parallel to $\overline{DA}$.

$$m_{UA} = \frac{-3-2}{4-8} = \frac{-5}{-4} = \frac{5}{4}$$

$$m_{QD} = \frac{2-7}{-7+3} = \frac{-5}{-4} = \frac{5}{4}$$

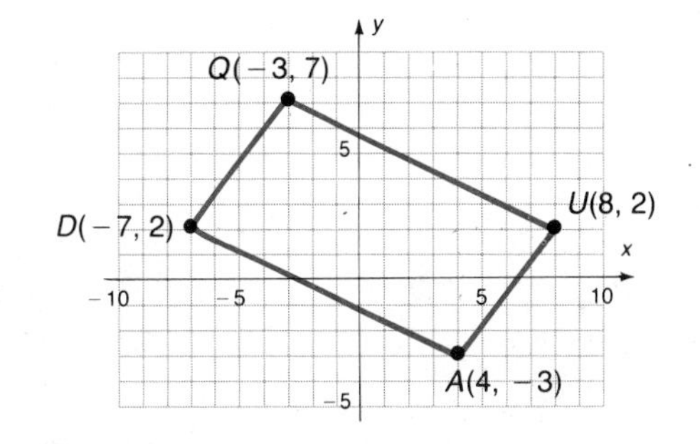

Thus $\overline{UA}$ is parallel to $\overline{QD}$. Since opposite sides of quadrilateral $QUAD$ are parallel, it is a parallelogram. Notice that even though it is helpful to draw the graph, you cannot use the figure to prove your argument. ■

EXAMPLE 3 Are the diagonals of $QUAD$ in Example 2 perpendicular?

Solution

$$m_{QA} = \frac{-3-7}{4+3} = \frac{-10}{7}$$

$$m_{DU} = \frac{2-2}{8+7} = 0$$

Since $m_{QA} m_{DU} \neq -1$, the segments $\overline{QA}$ and $\overline{DU}$ are not perpendicular. ■

Consider the linear function $f(x) = mx + b$. Then

$$\text{Slope} = \frac{\text{rise}}{\text{run}} = \frac{f(x_2) - f(x_1)}{x_2 - x_1}$$

$$= \frac{(mx_2 + b) - (mx_1 + b)}{x_2 - x_1}$$

$$= \frac{m(x_2 - x_1)}{x_2 - x_1} = m$$

Thus the slope of the graph of a linear function is m.

The y intercept for the linear function is found when $x = 0$:

$$f(0) = m(0) + b = b$$

Thus the y intercept of a linear function is b.

Slope-Intercept Form of the Equation of a Line

> The graph of the equation $y = mx + b$ is a line having slope m and having y intercept b.

This form of the equation of a line can be used for graphing certain lines for which it is not convenient to plot points. This procedure is summarized in Figure 2.6.

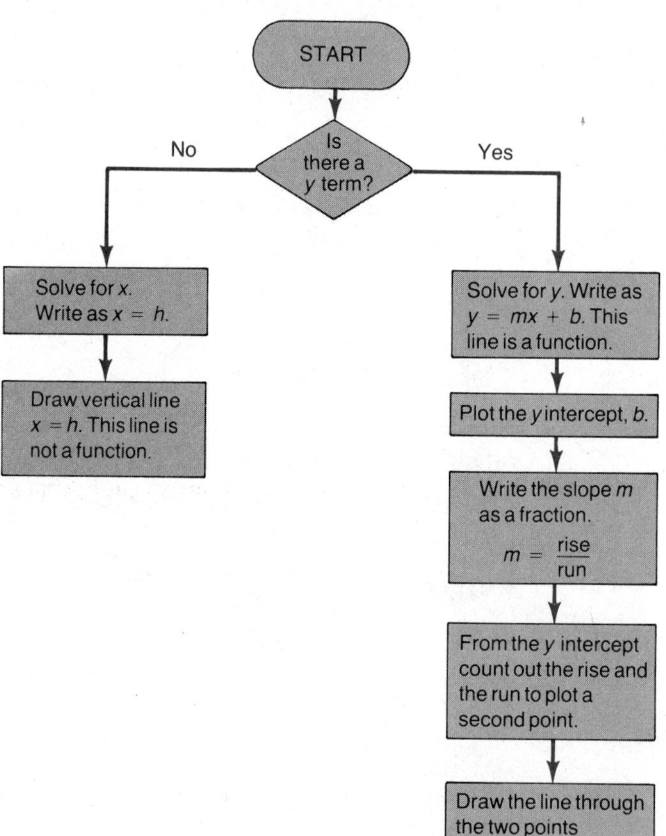

Figure 2.6 Procedure for graphing a line by using the slope-intercept form.

EXAMPLE 4 Graph $y = \frac{1}{2}x + 3$.

Solution By inspection, the y intercept is 3 and the slope is $\frac{1}{2}$; the line is graphed by first plotting the y intercept $(0, 3)$, then finding a second point by counting out the slope: up 1 and over 2.

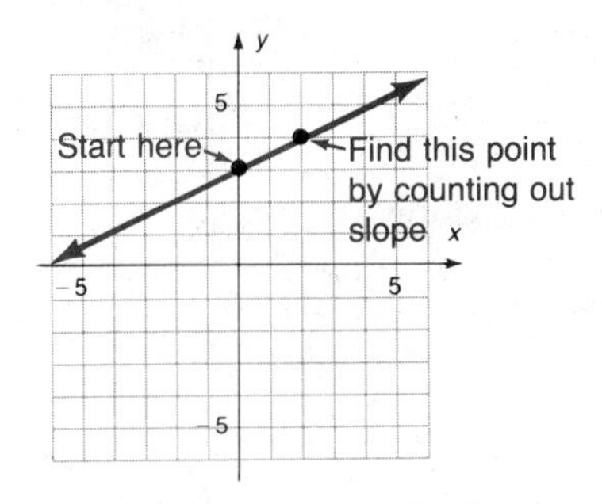

■

EXAMPLE 5 Graph $2x + 3y - 6 = 0$.

Solution Solve for y:

$$3y = -2x + 6$$

$$y = -\frac{2}{3}x + 2$$

The y intercept is 2 and the slope is $-\frac{2}{3}$; the line is graphed as shown. ■

EXAMPLE 6 Graph $4x + 2y - 5 = 0$ for $-1 \le x \le 3$.

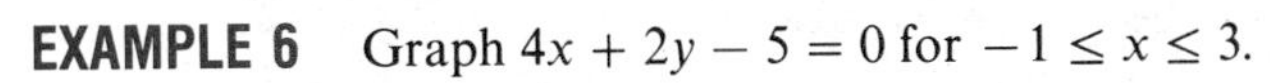

Solution Solve for y:

$$2y = -4x + 5$$

$$y = -2x + \frac{5}{2}$$

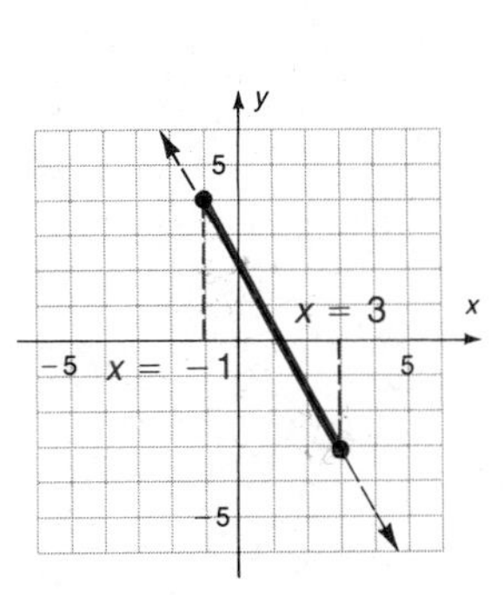

The y intercept is $\frac{5}{2}$ and the slope is -2; this line is shown as a black dashed line. Because of the restriction on the domain you draw that part of the line with x values between -1 and 3 (inclusive) as shown by the colored line segment. ■

A variation of graphing linear functions is seen when we graph absolute value functions.

EXAMPLE 7 Graph $y = |x|$

Solution First apply the definition of absolute value:

$$y = x \quad \text{if} \quad x \ge 0$$
$$y = -x \quad \text{if} \quad x < 0$$

Now graph these two equations with restrictions as you did in Example 6. The graph of $y = |x|$ is the part shown in color.

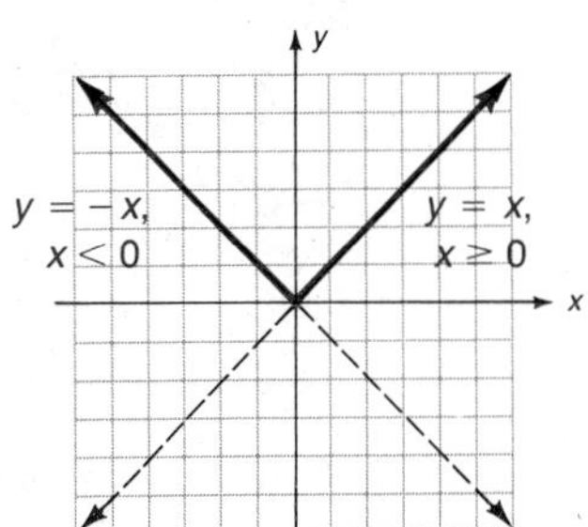

■

In algebra you studied several forms of the equation of a line. The derivation of some of these is reviewed in the problems, and the forms are stated below for review.

Forms of a Linear Equation

STANDARD FORM: $\mathbf{Ax + By + C = 0}$, where (x, y) is any point on the line and A, B, and C are constants, A and B not both zero.

SLOPE-INTERCEPT FORM: $\mathbf{y = mx + b}$, where m is the slope and b is the y intercept.

POINT-SLOPE FORM: $\mathbf{y - k = m(x - h)}$, where m is the slope and (h, k) is a point on the line.

TWO-POINT FORM: $\mathbf{y - y_1 = \left(\dfrac{y_2 - y_1}{x_2 - x_1}\right)(x - x_1)}$, where (x_1, y_1) and (x_2, y_2) are points on the line.

INTERCEPT FORM: $\mathbf{\dfrac{x}{a} + \dfrac{y}{b} = 1}$, where $(a, 0)$ and $(0, b)$ are the x and y intercepts, respectively.

HORIZONTAL LINE passing through the point (h, k): $\mathbf{y = k}$.

VERTICAL LINE passing through the point (h, k): $\mathbf{x = h}$.

EXAMPLE 8 Find the equation of the line using the given information. Leave your answer in standard form.

a. y intercept 4; slope 6
b. Slope 4; passing through $(-3, 2)$
c. Passing through $(2, 3)$ and $(5, 7)$
d. No slope; passing through $(7, -3)$

Solution **a.** Since you are given the slope and the intercept, use the slope-intercept form: $y = mx + b$, where $b = 4$ and $m = 6$:

$$y = 6x + 4$$

In standard form, $\mathbf{6x - y + 4 = 0}$.

b. Use the point-slope form, where $h = -3$, $k = 2$, and $m = 4$:

$$y - k = m(x - h)$$
$$y - 2 = 4(x + 3)$$
$$y - 2 = 4x + 12$$

In standard form, $\mathbf{4x - y + 14 = 0}$.

c. Use the two-point form:

$$y - 3 = \left(\frac{7 - 3}{5 - 2}\right)(x - 2)$$

This is the point (2, 3), but you could also use (5,7) to obtain the same result.

$$y - 3 = \frac{4}{3}(x - 2)$$

$$3y - 9 = 4x - 8$$

In standard form, $\mathbf{4x - 3y + 1 = 0}$.

d. Do not confuse no slope (vertical line) with zero slope (horizontal line). This is a vertical line, so the equation has the form $x = h$ when it passes through (h, k). Thus

$$x = 7$$

In standard form, $\mathbf{x - 7 = 0}$. ■

PROBLEM SET 2.3

A

Sketch the line passing through the points whose coordinates are given in Problems 1–9. Also find the slope of each line.

1. $(2, 3)$ and $(5, 6)$
2. $(0, 7)$ and $(3, 0)$
3. $(-1, -2)$ and $(4, 11)$
4. $(4, -2)$ and $(7, -3)$
5. $(-6, -4)$ and $(-9, 3)$
6. $(6, 0)$ and $(-3, 0)$
7. $(0, 0)$ and $(0, 3)$
8. $(4, -3)$ and $(4, 1)$
9. $(-1, 2)$ and $(3, 2)$

Graph the lines whose equations are given in Problems 10–21 by finding the slope and y intercept.

10. $y = 3x + 3$
11. $y = -4x - 1$
12. $y = \frac{2}{3}x + \frac{4}{3}$
13. $y = \frac{1}{5}x - \frac{6}{5}$
14. $y = 40x$
15. $y = 300x$
16. $x - 4 = 0$
17. $y + 2 = 0$
18. $5x - 4y - 8 = 0$
19. $x - 3y + 2 = 0$
20. $100x - 250y + 500 = 0$
21. $2x - 5y - 1200 = 0$

Graph the line segments or the absolute value functions in Problems 22–27.

22. $3x + y - 2 = 0, -7 \le x \le 1$
23. $2x - 2y + 6 = 0, 5 \le x \le 9$
24. $5x - 3y - 9 = 0, -3 \le x \le 1$
25. $y = 2|x|$
26. $y = -3|x|$
27. $y = |4x|$

Use slopes to decide whether the coordinates given in Problems 28–30 are vertices of a right triangle.

28. $T(-6, -4), R(6, 12), I(-4, 7)$
29. $A(1, -1), N(4, 1), G(0, 7)$
30. $L(-4, 6), E(-10, 2), S(-3, -1)$

Use slopes to decide whether the coordinates given in Problems 31–34 are vertices of a parallelogram.

31. $R(3, 0), E(6, 7), C(2, 9), T(-3, 3)$
32. $A(-1, 5), N(2, 3), G(-3, -4), E(-6, -2)$
33. $P(1, 10), A(-4, 5), R(-3, -2), L(2, 3)$
34. $E(-3, 6), L(9, 11), G(14, 1), R(2, -4)$
35. Are the diagonals of the quadrilateral in Problem 33 perpendicular?
36. Are the diagonals of the quadrilateral in Problem 34 perpendicular?

B

Find the equation of the line satisfying the given conditions in Problems 37–52. Give your answer in standard form.

37. y intercept 6; slope 5
38. y intercept -3; slope -2
39. y intercept 0; slope 0
40. y intercept 5; slope 0
41. Slope 3; passing through $(2, 3)$
42. Slope -1; passing through $(-4, 5)$
43. Slope $\frac{1}{2}$; passing through $(3, 3)$
44. Slope $\frac{2}{5}$; passing through $(5, -2)$
45. Passing through $(-4, -1)$ and $(4, 3)$
46. Passing through $(4, -2)$ and $(4, 5)$
47. Passing through $(5, 6)$ and $(1, -2)$
48. Passing through $(5, 6)$ and $(7, 6)$
49. Passing through $(2, 4)$ parallel to $2x + 3y - 6 = 0$
50. Passing through $(-1, -2)$ parallel to $x - 2y + 4 = 0$
51. Passing through $(-1, -2)$ perpendicular to $x - 2y + 4 = 0$
52. Passing through $(2, 4)$ perpendicular to $2x + 3y - 6 = 0$
53. Consider Figure 2.7a.
  a. What are the coordinates of A and B?
  b. What is the slope of the line passing through A and B?

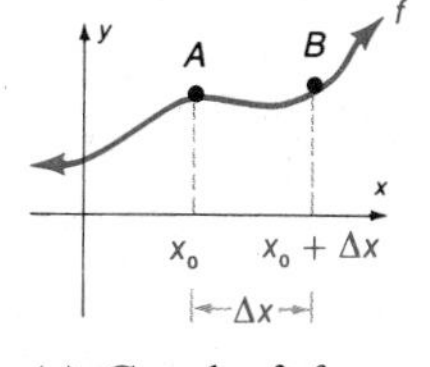

(a) Graph of f

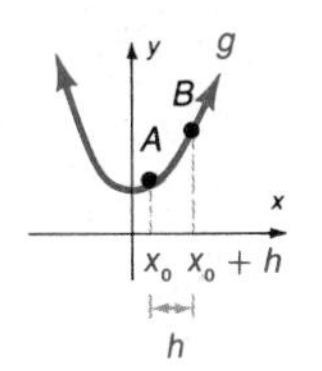

(b) Graph of g

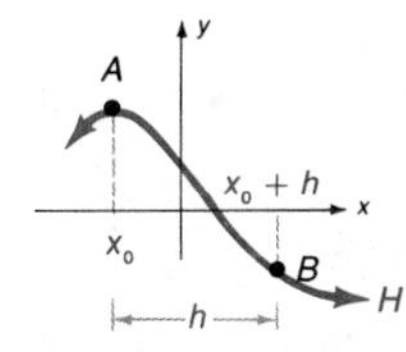

(c) Graph of H

Figure 2.7

54. Consider Figure 2.7b.
  a. What are the coordinates of A and B?
  b. What is the slope of the line passing through A and B?

55. Consider Figure 2.7c.
  a. What are the coordinates of A and B?
  b. What is the slope of the line passing through A and B?

Problems 56–62 provide some real-world examples of line graphs. One way of finding the equation of the line is to write two data points from the given information, and then use those two points to write the equation. Use the given information to write a standard-form equation of the line described by the problem.

56. The demand for a certain product is related to the price of the item. Suppose a new line of stationery is tested at two stores. It is found that 25 boxes are sold within a month if they are priced at $1 and 15 boxes priced at $2 are sold in the same time. Let x be the price and y be the number of boxes sold.

57. An important factor that is related to the demand for a product is the supply. The amount of the stationery in Problem 56 that can be supplied is also related to the price. At $1 each 10 boxes can be supplied and at $2 each 20 boxes can be supplied. Let x be the price and y be the number of boxes sold.

58. The population of Florida was roughly 6.8 million in 1970, and 9.7 million in 1980. Let x be the year (let 1950 be the base year; that is, $x = 0$ represents 1950 and $x = 10$ represents 1960) and y be the population. Use this equation to predict the population in 1990.

59. The population of Texas was roughly 11.2 million in 1970, and 14.2 million in 1980. Let x be the year (let 1950 be the base year; that is, $x = 0$ represents 1950 and $x = 10$ represents 1960) and y be the population. Use this equation to predict the population in 1990.

60. It costs $90 to rent a car if you drive 100 miles and $140 if you drive 200 miles. Let x be the number of miles driven and y the total cost of the rental. Use this equation to find how much it would cost if you drove 394 miles.

61. It costs $60 to rent a car if you drive 50 miles and $60 if you drive 260 miles. Let x be the number of miles and y be the total cost of the rental. Use this equation to find how much it would cost if you drove 394 miles.

62. Suppose it costs $100 for maintenance and repairs to drive a three-year-old car 1000 miles and $650 for maintenance and repairs to drive it 6500 miles. Let x be the number of miles and y be the cost for repairs and maintenance.

63. Begin with the slope-intercept form and derive the point-slope form:

$$y - k = m(x - h)$$

64. Begin with the point-slope form and derive the two-point form:

$$y - y_1 = \left(\frac{y_2 - y_1}{x_2 - x_1}\right)(x - x_1)$$

65. Begin with the two-point form and derive the intercept form:

$$\frac{x}{a} + \frac{y}{b} = 1$$

66. Prove that if $m > 0$, then the linear function is an increasing function throughout its domain.

67. Prove that if $m < 0$, then the linear function is a decreasing function throughout its domain.

2.4 TRANSLATION OF FUNCTIONS

In this section we look at a technique for graphing that will pay dividends in the amount of work you will need to do when graphing certain functions. For example, to graph the functions $y = x^2$ and $y - 3 = (x - 2)^2$ by plotting points, first set up tables of values and then draw the graphs, as shown on the next page.*

* You will not really appreciate what is being done here unless *you* carry through the arithmetic shown in the tables. We are introducing a method that makes *the arithmetic* easier, but you will not see why it is better unless you actually do the arithmetic.

Function $y = x^2$

x	y
0	0
1	1
−1	1
2	4
−2	4
3	9
−3	9

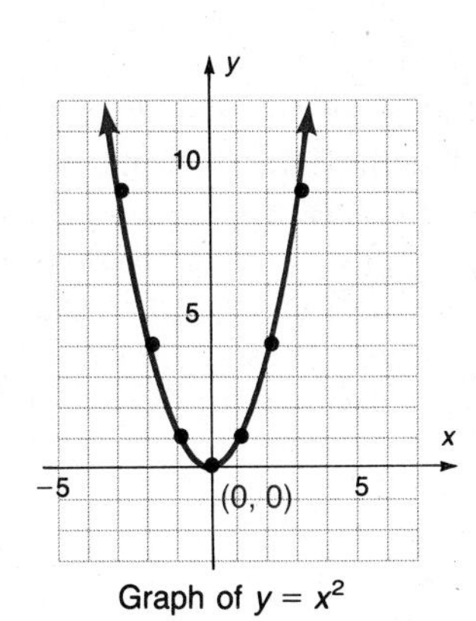

Graph of $y = x^2$

Function $y - 3 = (x - 2)^2$

x	y
0	7
1	4
−1	12
2	3
3	4
4	7
5	12

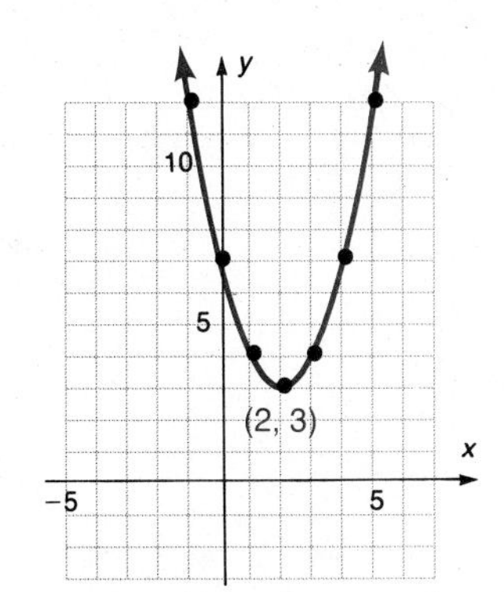

Graph of $y - 3 = (x - 2)^2$

Notice that the graphs are the same, but they are in different locations. You also should have noticed (if you did the arithmetic) that the first table of values was much easier to calculate than the second. When two curves are congruent (have the same size and shape) and have the same orientation, we say that one can be found from the other by a **shift** or **translation.**

Consider the function f defined by the graph in Figure 2.8. It is possible to shift the entire curve up, down, right, or left, as shown in Figure 2.9.

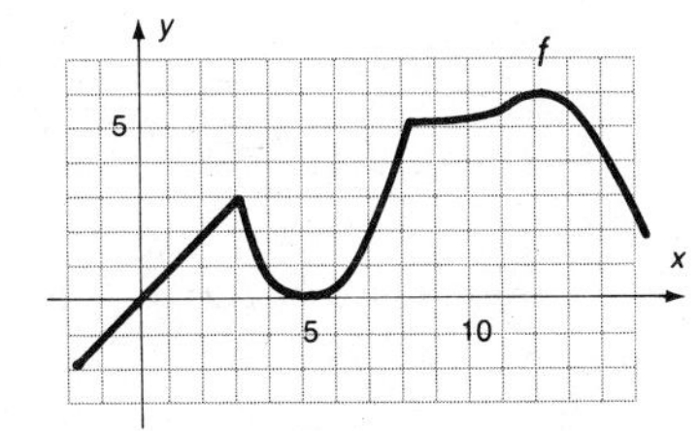

Figure 2.8 Graph of f

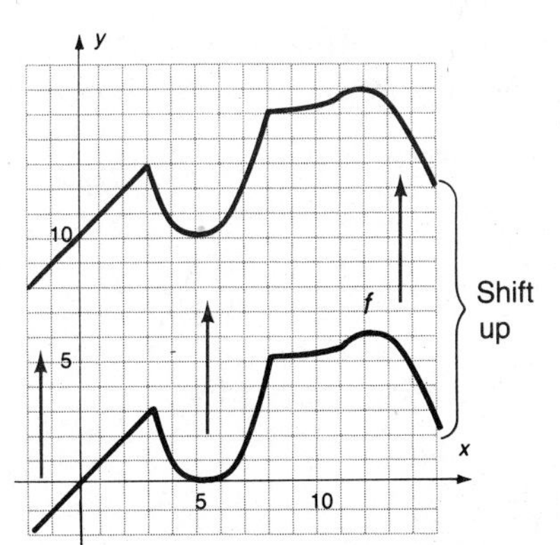

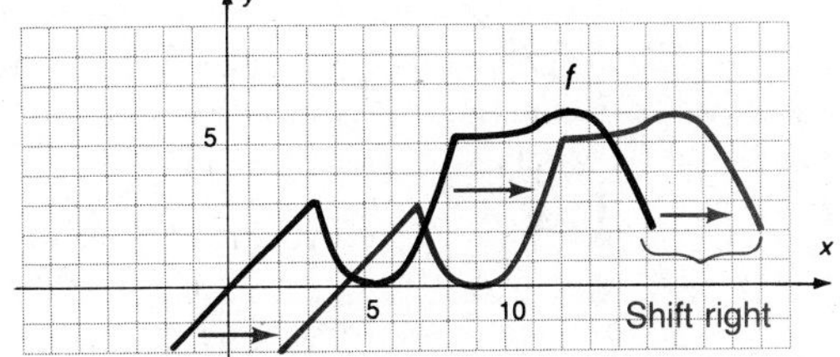

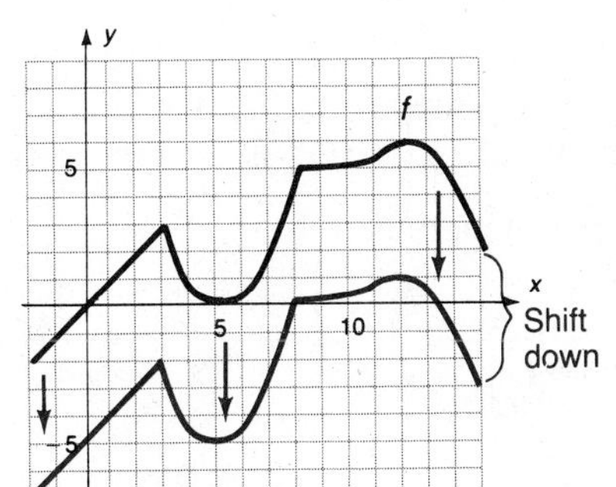

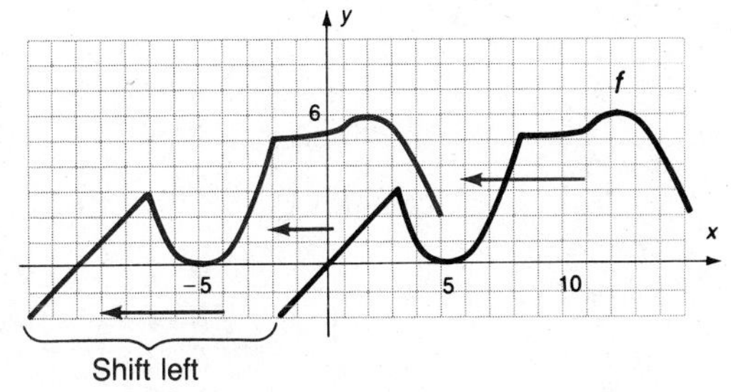

Figure 2.9 Shifting the graph of f

Instead of considering the curve shifting relative to fixed axes, consider the effect of shifting the axes. If the coordinate axes are shifted up k units, the origin of this new coordinate system would correspond to the point $(0, k)$ in the old coordinate system. If the axes are shifted to the right h units, the origin would correspond to the point $(h, 0)$ in the old system. A horizontal shift of h units followed by a vertical shift of k units would shift the new coordinate axes so that the origin corresponds to a point (h, k) on the old axes. Suppose a *new* coordinate system with origin at (h, k) is drawn and the new axes are labeled x' and y', as shown in Figure 2.10.

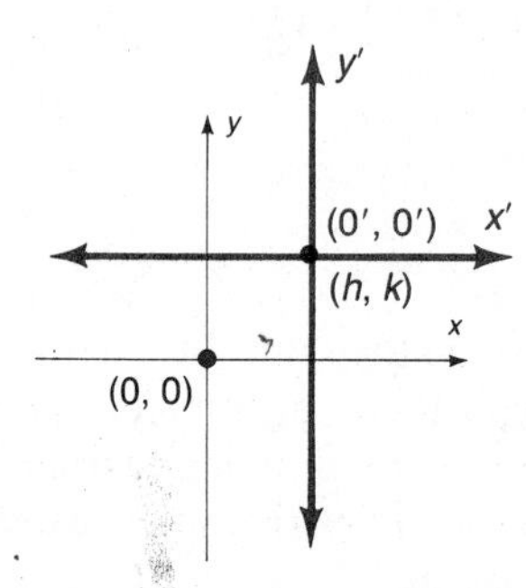

Figure 2.10 Shifting the axes to (h, k)

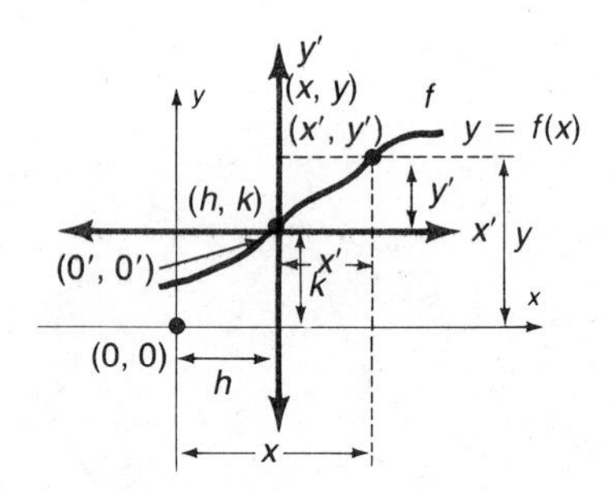

Figure 2.11 Comparison of coordinate axes

Every point on a given curve can now be denoted in two ways (see Figure 2.11):

1. As (x, y) measuring from the old origin
2. As (x', y') measuring from the new origin (color)

To find the relationship between (x, y) and (x', y'), consider the graph shown in Figure 2.11:

$$x = x' + h \qquad\qquad x' = x - h$$
$$\text{or}$$
$$y = y' + k \qquad\qquad y' = y - k$$

This says that if you are given any function

$$y - k = f(x - h)$$

the graph of this function is the same as

$$y' = f(x')$$

where (x', y') are measured from the new origin located at (h, k). This can greatly simplify our work since $y' = f(x')$ is usually easier to graph than $y - k = f(x - h)$.

EXAMPLE 1 Find (h, k) for each equation.

a. $y - 5 = f(x - 7)$
b. $y + 6 = f(x - 1)$
c. $y + 1 = f(x + 3)$
d. $y = f(x)$
e. $y - 6 = f(x) + 15$

Solution
a. $(h, k) = \mathbf{(7, 5)}$
b. Notice that $y + 6$ can be written as $y - (-6)$; $(h, k) = \mathbf{(1, -6)}$.
c. $(h, k) = \mathbf{(-3, -1)}$
d. This indicates no shift; $(h, k) = \mathbf{(0, 0)}$.
e. Write the equation as $y - 21 = f(x)$; thus $(h, k) = \mathbf{(0, 21)}$. ■

EXAMPLE 2 If $f(x) = x^2$, write $y - k = f(x - h)$ for $(h, k) = (3, -2)$.

Solution
$$\mathbf{y + 2 = (x - 3)^2}$$ ■

EXAMPLE 3 If $f(x) = 3x^2 + 5x$, write $y - k = f(x - h)$ for $(h, k) = (-\sqrt{2}, \pi)$.

Solution
$$\mathbf{y - \pi = 3(x + \sqrt{2})^2 + 5(x + \sqrt{2})}$$ ■

Translation of Axes

> The graph of the equation $y - k = f(x - h)$ is the same as the graph of $y = f(x)$ on a system of coordinate axes that has been **translated** h units horizontally and k units vertically.

EXAMPLE 4 Given f defined by $y = f(x)$ as shown by the graph in Figure 2.12, graph the following functions.

a. $y = f(x - 3)$ **b.** $y + 2 = f(x)$ **c.** $y - 4 = f(x + 5)$

Solution
a. Since $(h, k) = (3, 0)$, the shift is 3 units to the right.
b. Since $(h, k) = (0, -2)$, the shift is 2 units down.
c. Since $(h, k) = (-5, 4)$, the shift is 5 units to the left and 4 units up.

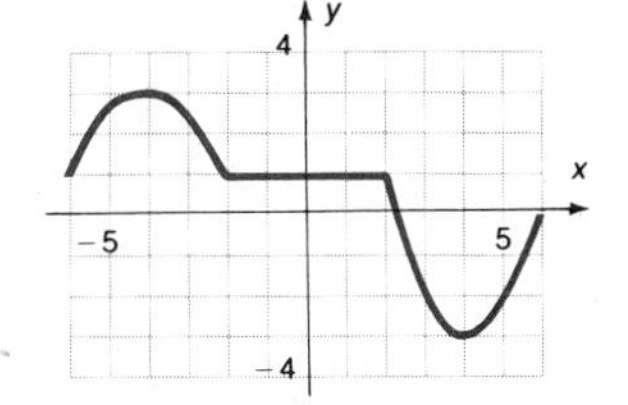

Figure 2.12 Graph of f

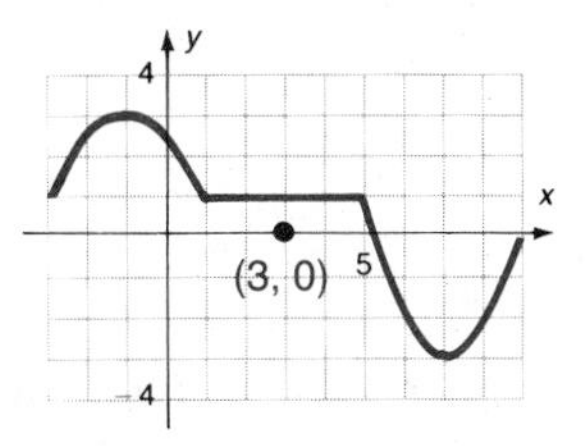

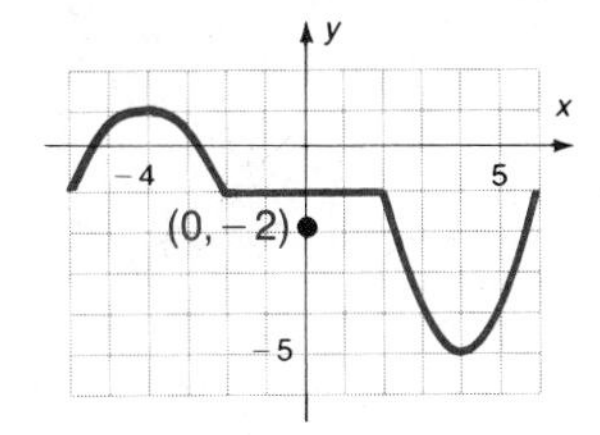

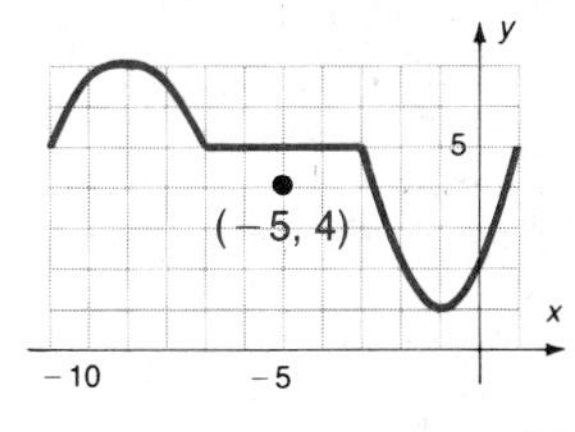

■

EXAMPLE 5 Substitute $x' = x + 3$ and $y' = y - 6$ into the given equations.

a. $y - 6 = (x + 3)^2$ **b.** $y = 5(x + 3)^2 + 6$
c. $y = 6 - 2(x + 3)^2$ **d.** $y - 6 = 7(x + 3)^2 + 2(x + 3)$

Solution
a. $\mathbf{y' = x'^2}$
b. Rewrite as $y - 6 = 5(x + 3)^2$; thus $\mathbf{y' = 5x'^2}$.
c. Rewrite as $y - 6 = -2(x + 3)^2$; thus $\mathbf{y' = -2x'^2}$.
d. $\mathbf{y' = 7x'^2 + 2x'}$ ■

Translations are useful because they sometimes allow us to take a computationally difficult equation and rewrite it in terms of a simpler equation. For example, recall the two previously considered curves shown in Figure 2.13.

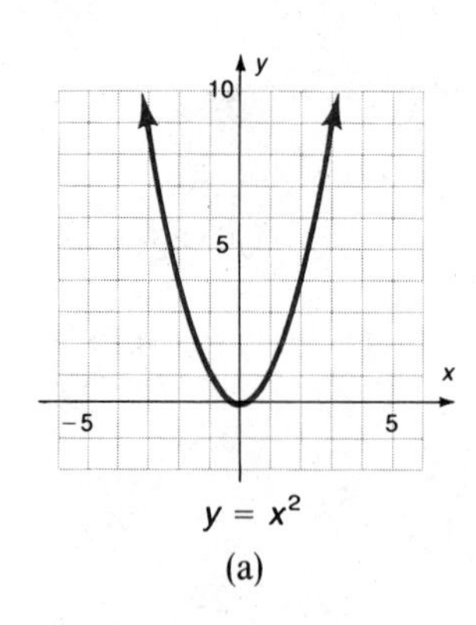

(a)

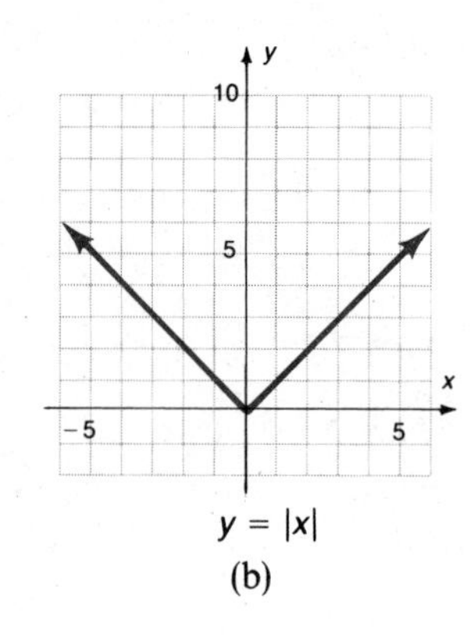

(b)

Figure 2.13 Standard parabola and absolute value curves

We can now draw the graphs of some rather complicated equations by using these curves and the idea of translation.

EXAMPLE 6 Graph $y - \frac{1}{2} = (x - \frac{3}{2})^2$.

Solution First, plot $(\frac{3}{2}, \frac{1}{2})$. Next, let $y' = y - \frac{1}{2}$ and $x' = x - \frac{3}{2}$. *Imagine* the new origin at $(h, k) = (\frac{3}{2}, \frac{1}{2})$ and plot values from the new origin using the equation $y' = x'^2$.

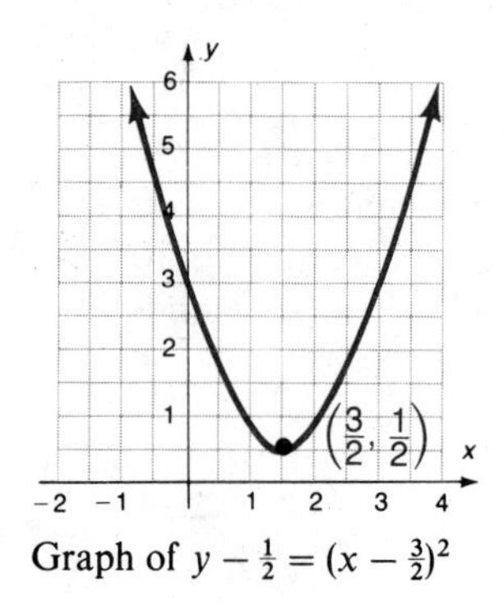

Graph of $y - \frac{1}{2} = (x - \frac{3}{2})^2$ ■

As you can see, the graph of $y - k = f(x - h)$ is done in two steps:

1. Plot (h, k).
2. Graph the simpler curve $y' = f(x')$ by using (h, k) as the new origin.

EXAMPLE 7 Graph $f(x) = |x - 3| + 2$.

Solution This can be rewritten as $y - 2 = |x - 3|$, which is the graph of $y = |x|$ (see Figure 2.13b) on the system of coordinate axes that is translated to $(h, k) = (3, 2)$.

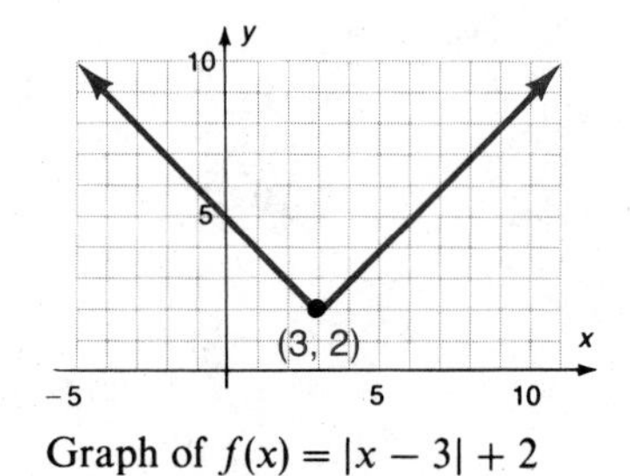

Graph of $f(x) = |x - 3| + 2$ ■

PROBLEM SET 2.4

A

Find (h, k) for each of the functions given in Problems 1–12.

1. $y - 3 = f(x - 6)$
2. $y - 5 = f(x + 3)$
3. $y + 1 = f(x - 6)$
4. $y = f(x - 4)$
5. $y - \sqrt{2} = f(x)$
6. $y = f(x) + 6$
7. $y = f(x - 3) - 4$
8. $y = f(x + 1) + 2$
9. $y = 4f(x)$
10. $y = 2f(x + \sqrt{2})$
11. $6y = f(x + 3)$
12. $y = 2f\left(x + \frac{\pi}{2}\right) - \frac{\pi}{3}$

If $f(x) = x^2$, write $y - k = f(x - h)$ for the (h, k) given in Problems 13–17.

13. $(2, 3)$
14. $(-3, 0)$
15. $(0, -1)$
16. $\left(-\pi, \frac{\pi}{4}\right)$
17. $(-\sqrt{3}, -2)$

If $f(x) = |x|$, write $y - k = f(x - h)$ for the (h, k) given in Problems 18–22.

18. $(6, 1)$
19. $(0, -6)$
20. $(-\pi, 0)$
21. $(\sqrt{2}, -\sqrt{3})$
22. $\left(-\frac{\pi}{2}, -\frac{\sqrt{2}}{2}\right)$

Substitute $x' = x - 5$ and $y' = y + 1$ into the equations given in Problems 23–31.

23. $y + 1 = (x - 5)^2$
24. $y + 1 = \frac{2}{3}(x - 5)^2$
25. $y + 1 = -5(x - 5)^2$
26. $y = (x - 5)^2 - 1$
27. $y = -2(x - 5)^2 - 1$
28. $y = -1 + 3(x - 5)^2$
29. $y + 1 = |x - 5|$
30. $y + 1 = \frac{2}{3}|x - 5|$
31. $y = -2|x - 5| - 1$

B

Let f, g, and h be the functions whose graphs are shown in Figure 2.14. Graph the functions indicated by the equations in Problems 32–49.

32. $y + 3 = f(x)$
33. $y + 2 = g(x)$
34. $y + \frac{1}{2} = h(x)$
35. $y - 5 = g(x)$
36. $y - 1 = h(x)$
37. $y + \pi = f(x)$
38. $y = h\left(x - \frac{\pi}{2}\right)$
39. $y = g(x + 3)$
40. $y = f(x - \sqrt{5})$
41. $y + 4 = f(x - 3)$
42. $y + 1 = g(x - 4)$
43. $y - 2 = h(x + \pi)$
44. $y - 3 = g(x + 5)$
45. $y - 2 = h(x + 3)$
46. $y + 2 = f(x - 4)$
47. $y - \frac{21}{5} = g\left(x - \frac{15}{2}\right)$
48. $y - \sqrt{3} = f(x + \sqrt{2})$
49. $y + \pi = h\left(x - \frac{3\pi}{2}\right)$

Graph the functions given in Problems 50–66.

50. $y - 3 = (x - 2)^2$
51. $y = (x + 3)^2$
52. $y = x^2 - 1$
53. $y - 1 = |x - 7|$
54. $y = |x + \pi|$
55. $y = |x| + 6$
56. $y + \sqrt{3} = |x - \sqrt{2}|$
57. $y - \frac{\pi}{4} = (x + \pi)^2$
58. $y + 2 = (x + \sqrt{3})^2$
59. $y - \sqrt{2} = (x + \sqrt{5})^2$
60. $y - \sqrt{2} = (x - \sqrt{3})^2$
61. $y - \pi = \left|x + \frac{\pi}{6}\right|$

C

62. $y + 3 = (x + 8)^2$, such that $-14 \le x \le -8$
63. $y - 2 = (x + 3)^2$, such that $-7 \le x \le -2$
64. $y + 12 = \left(x + \frac{25}{2}\right)^2$, such that $y > -10$
65. $y + 3 = (x + 3)^2$, such that $y < 6$
66. $y + 12 = (x - 8)^2$, such that $y < 4$

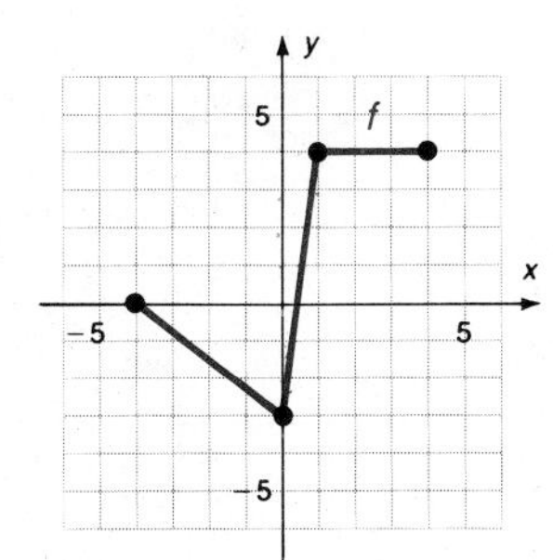

(a) Graph of f

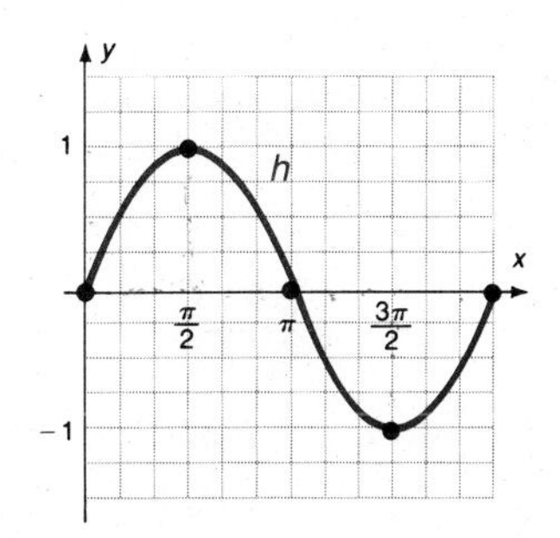

(b) Graph of g

(c) Graph of h

Figure 2.14

2.5 QUADRATIC FUNCTIONS

The second type of function to be considered in this chapter is the quadratic function.

Quadratic Function

> A function f is a **quadratic function** if
>
> $f(x) = ax^2 + bx + c$
>
> where a, b, and c are real numbers and $a \neq 0$.

If $b = c = 0$, however, the quadratic function has the form

$$y = ax^2$$

and has a graph called a **standard-position parabola**. You graphed functions of this type by plotting points in Section 1.6 and by translations in the last section. We begin by sketching two additional standard-position parabolas in order to make some generalizations.

EXAMPLE 1 Sketch the graph of $y = 3x^2$.

Solution Find some ordered pairs satisfying the equation:

x	0	1	2	3	...
y	0	3	12	27	...

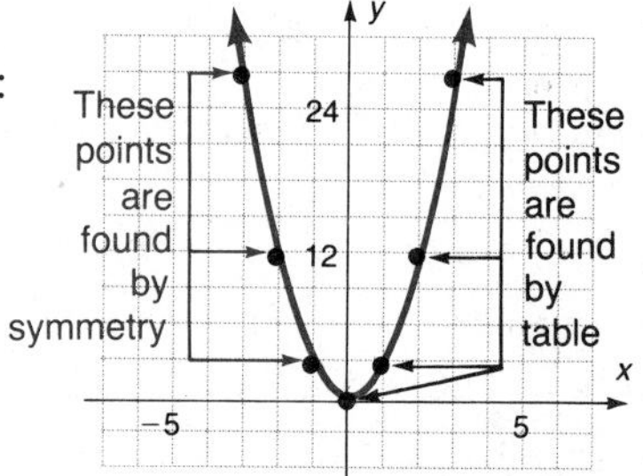

Figure 2.15 Graph of $y = 3x^2$

Also, if $y = f(x)$ then

$$f(-x) = 3(-x)^2 = 3x^2$$

so the graph is symmetric with respect to the y axis. Plot the points represented by the ordered pairs, use symmetry, and draw a smooth graph as shown in Figure 2.15. ■

EXAMPLE 2 Sketch the graph of $y = -\frac{1}{2}x^2$.

x	0	1	2	3	4
y	0	$-\frac{1}{2}$	-2	$-\frac{9}{2}$	-8

Plot these points and use symmetry as shown in Figure 2.16. ■

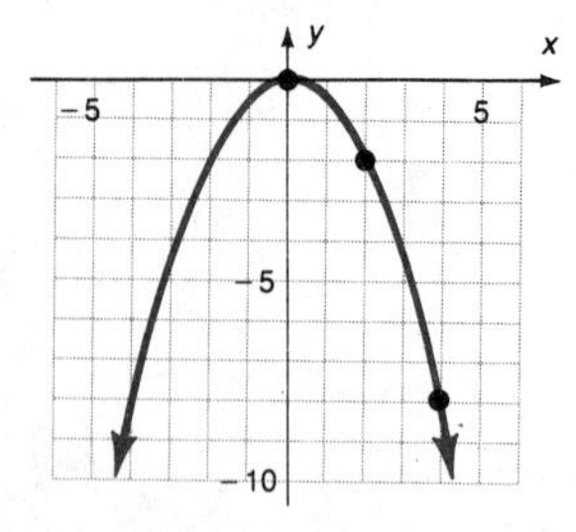

Figure 2.16 Graph of $y = -\frac{1}{2}x^2$

We will now make some general observations based on the special case $y = ax^2$:

1. The graph has a shape of a curve which is called a **parabola**.
2. The point $(0, 0)$ is the lowest point if the parabola opens up $(a > 0)$; $(0, 0)$ is the highest point if the parabola opens down $(a < 0)$. This highest or lowest point is called the **vertex**.
3. The parabola is **symmetric** with respect to the vertical line passing through the vertex.
4. Relative to a fixed scale, the magnitude of a determines the "wideness" of the parabola: Small values of $|a|$ yield "wide" parabolas; large values of $|a|$ yield "narrow" parabolas.

For graphs of parabolas of the form

$$y - k = a(x - h)^2$$

you simply translate the axes to the point (h, k) and then graph the parabola $y' = ax'^2$, as shown in Example 3. The point (h, k) is the vertex. Using functional notation, this can be written

$$f(x) = a(x - h)^2 + k$$

EXAMPLE 3 Sketch the graph of $y + 5 = 3(x + 2)^2$.

Solution By inspection, $(h, k) = (-2, -5)$ and the standard-position parabola is $y' = 3x'^2$. The table of values (or points to plot) is the same as shown for Example 1, and is repeated below; the only difference here is that you count out these points from $(-2, -5)$ instead of from the origin.

Table from Example 1

x	y
0	0
1	3
2	12
3	27

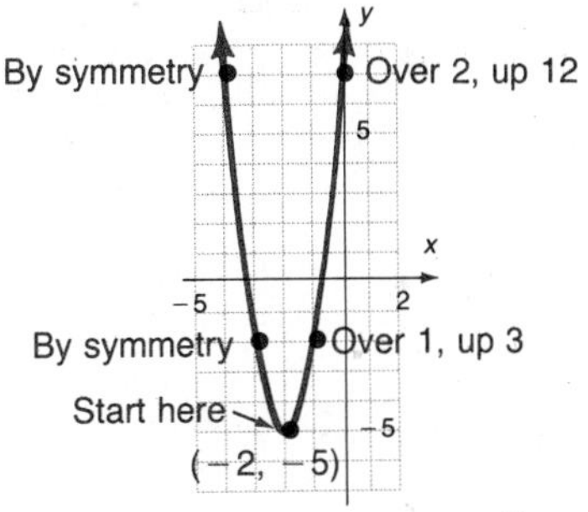

Graph of $y + 5 = 3(x + 2)^2$ ■

Now you are ready to consider the general quadratic function

$$y = ax^2 + bx + c \qquad (a \neq 0)$$

Consider Example 3 above: $y + 5 = 3(x + 2)^2$. This was graphed by translating the axes to $(-2, -5)$ and then considering $y' = 3x'^2$. Suppose you rewrite the given equation as

$$y + 5 = 3(x + 2)^2$$
$$y + 5 = 3(x^2 + 4x + 4)$$
$$y + 5 = 3x^2 + 12x + 12$$
$$y = 3x^2 + 12x + 7$$

The last form is the general quadratic form, where $a = 3, b = 12$, and $c = 7$. Suppose you are given this form and told to graph the curve. Then, to reverse the process, you must **complete the square** (review Section 1.4).

To complete the square, follow these steps:

Step 1 Subtract c (the constant term) from both sides:

General Form	*Example*
$y = ax^2 + bx + c$	$y = 3x^2 + 12x + 7$
$y - c = ax^2 + bx$	$y - 7 = 3x^2 + 12x$

Step 2 Factor the a term from the expression on the right (remember $a \neq 0$):

$$y - c = a\left(x^2 + \frac{b}{a}x\right) \qquad y - 7 = 3(x^2 + 4x)$$

Step 3 To complete the square on the number inside parentheses:

$$x^2 + \frac{b}{a}x + ? = (x + ?)^2 \qquad x^2 + 4x + ? = (x + ?)^2$$

you square one-half the coefficient of the x term:

$$\left(\frac{b}{2a}\right)^2 = \frac{b^2}{4a^2} \qquad \left(\frac{4}{2}\right)^2 = 4$$

Then add a times this number to both sides of the original equation:

$$y - c + \frac{b^2}{4a} = a\left(x^2 + \frac{b}{a}x + \frac{b^2}{4a^2}\right) \qquad y - 7 + 12 = 3(x^2 + 4x + 4)$$

$\frac{1}{2} \cdot \frac{b}{a}$ squared

distributive property

$$a \cdot \frac{b^2}{4a^2} = \frac{b^2}{4a}$$

Add $\frac{b^2}{4a}$ to both sides

$\frac{1}{2} \cdot 4$ squared

distributive property

$3 \cdot 4 = 12$

Add 12 to both sides

Step 4 Factor the right-hand side as a perfect square and simplify the left-hand side:

$$y + \frac{-4ac}{4a} + \frac{b^2}{4a} = a\left(x + \frac{b}{2a}\right)^2 \qquad y - 7 + 12 = 3(x + 2)^2$$

common denominator if fractions are involved

$$\left(y + \frac{b^2 - 4ac}{4a}\right) = a\left(x + \frac{b}{2a}\right)^2 \qquad y + 5 = 3(x + 2)^2$$

This is of the form

$$y - k = a(x - h)^2$$

and is called the **general form of a parabola**. This equation can be graphed by doing a translation as shown in Example 4.

EXAMPLE 4 Sketch the graph of $y = x^2 + 6x + 10$.

Solution Complete the square:

$$y - 10 = x^2 + 6x$$

$$y - 10 + \mathbf{9} = x^2 + 6x + \mathbf{9}$$

Since $\frac{1}{2} \cdot 6 = 3$ and $3^2 = 9$, you add 9 to both sides.

$$y - 1 = (x + 3)^2$$

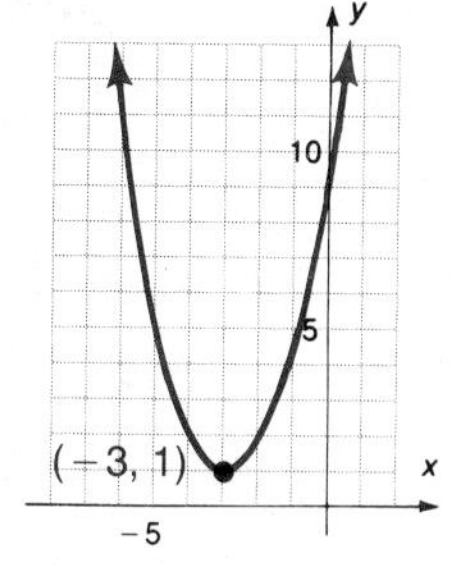

Figure 2.17 Graph of $y = x^2 + 6x + 10$

The vertex is at $(-3, 1)$, the parabola opens up, and the graph is shown in Figure 2.17. ■

Notice from Example 4 above that $c = 10$ (compare the given equation with the form $y = ax^2 + bx + c$) and that the y intercept is $(0, 10)$. In general, if $x = 0$ then

$$y = a(0)^2 + b(0) + c$$
$$= c$$

which shows that the **y intercept for a quadratic function is $(0, c)$**. This often serves as a checkpoint for graphs such as the one shown in Example 4.

EXAMPLE 5 Sketch the graph of $y = 1 - 5x - 2x^2$

Solution

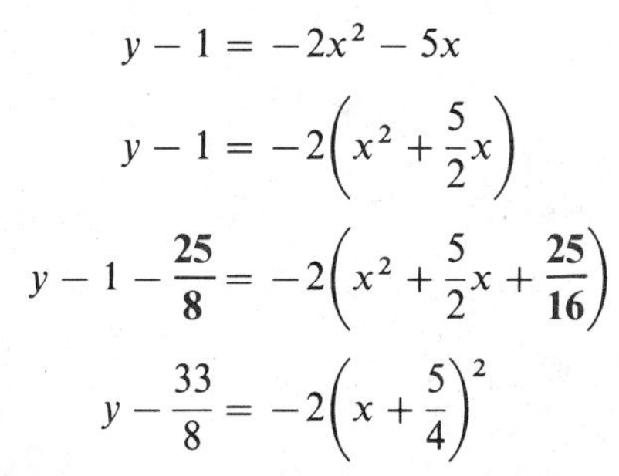

$$y - 1 = -2x^2 - 5x$$

$$y - 1 = -2\left(x^2 + \frac{5}{2}x\right)$$

$$y - 1 - \frac{\mathbf{25}}{\mathbf{8}} = -2\left(x^2 + \frac{5}{2}x + \frac{\mathbf{25}}{\mathbf{16}}\right)$$

$$y - \frac{33}{8} = -2\left(x + \frac{5}{4}\right)^2$$

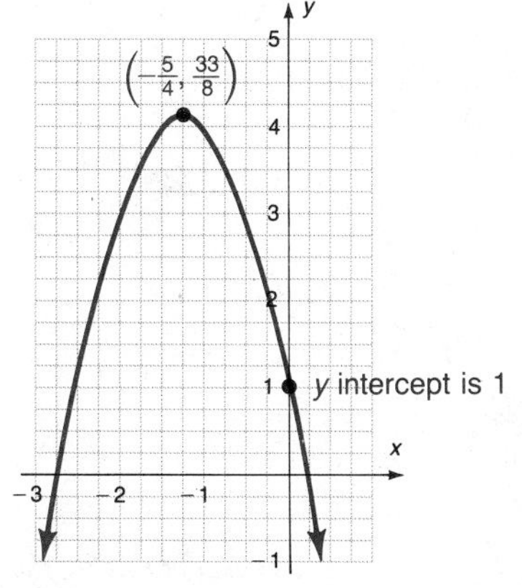

Figure 2.18 Graph of $y = 1 - 5x - 2x^2$

The vertex is $(-\frac{5}{4}, \frac{33}{8})$, the parabola opens down, and the graph is shown in Figure 2.18. For fractions, you can sometimes choose a scale that is more convenient than one square per unit. Notice in Figure 2.18 that there are four squares per unit. Also, remember that once you have found the vertex, you simply have to graph $y' = -2x'^2$ translated to the point $(-\frac{5}{4}, \frac{33}{8})$. ■

Many applications are concerned with finding the maximum or minimum value of a function f. If f is a quadratic function defined by

$$y - k = a(x - h)^2$$

then the **maximum or minimum value is at the vertex**. That is, if $x = h$ then the maximum value of $f(x) = a(x - h)^2 + k$ is $y = k$. You can see this is true because $(x - h)^2$ is necessarily nonnegative for all x and zero if and only if $x = h$. For Example 5, the maximum value of $y = \frac{33}{8}$, which occurs for $x = -\frac{5}{4}$.

EXAMPLE 6 A small manufacturer of CB radios determines that the price of each item is related to the number of items produced. Suppose that x items are produced per day where the maximum number that can be produced is 10 items, and that the profit, in dollars, is determined to be

$$P(x) = -30(x - 6)^2 + 480$$

How many radios should be produced in order to maximize the profit?

Solution The profit function has a graph which opens down. The maximum profit is found at the vertex of this parabola as shown in Figure 2.19.

The vertex of the function

$$P(x) - 480 = -30(x - 6)^2$$

is (6, 480) so the **maximum profit of $480 is obtained when 6 radios are manufactured**. Notice also from the graph that if there were a strike and no radios could be produced, the daily profit would be −600 (this is a $600 per day *loss*).

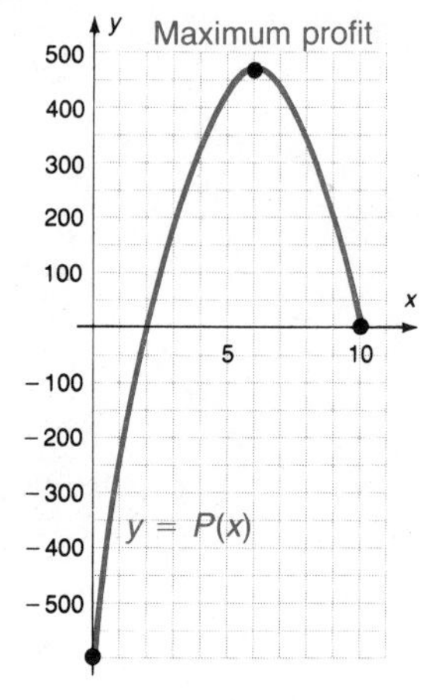

Figure 2.19 Graph of $P(x) = -30(x - 6)^2 + 480$

■

PROBLEM SET 2.5

A

Sketch the graph of each equation given in Problems 1–24.

1. $y = x^2$
2. $y = -x^2$
3. $y = -2x^2$
4. $y = 2x^2$
5. $y = -5x^2$
6. $y = 5x^2$
7. $y = \frac{1}{3}x^2$
8. $y = -\frac{1}{3}x^2$
9. $y = \frac{1}{10}x^2$
10. $y = -\frac{1}{10}x^2$
11. $y = \pi x^2$
12. $y = -2\pi x^2$
13. $y = (x - 3)^2$
14. $y = (x - 1)^2$
15. $y = (x + 2)^2$
16. $y = -2(x - 1)^2$
17. $y = \frac{1}{4}(x - 1)^2$
18. $y = \frac{1}{2}(x + 1)^2$
19. $y = \frac{1}{3}(x + 2)^2$
20. $y - 2 = (x - 1)^2$
21. $y - 2 = 3(x + 2)^2$
22. $y - 2 = -\frac{3}{5}(x - 1)^2$
23. $y + 3 = \frac{2}{3}(x + 2)^2$
24. $y - 1 = \frac{1}{3}(x - 4)^2$

B

Sketch the graph of each equation given in Problems 25–40.

25. $y = x^2 + 4x + 4$
26. $y = x^2 + 6x + 9$
27. $y = x^2 + 2x - 3$
28. $y = 2x^2 - 4x + 5$
29. $y = 2x^2 - 4x + 4$
30. $y = 2x^2 + 8x + 5$
31. $y = 3x^2 - 12x + 10$
32. $y = 3x^2 - 30x + 76$
33. $y = \frac{1}{2}x^2 + 2x - 1$
34. $y = \frac{1}{2}x^2 + 4x + 10$
35. $y = \frac{1}{2}x^2 - x + \frac{5}{2}$
36. $y = \frac{1}{2}x^2 - 3x + \frac{3}{2}$
37. $x^2 - 6x - 2y - 1 = 0$
38. $x^2 + 2x + 2y - 3 = 0$
39. $x^2 - 6x - 3y - 3 = 0$
40. $2x^2 - 4x - 3y + 11 = 0$

Find the maximum value of y for the functions defined by the equations in Problems 41–46.

41. $y = -4x^2 - 8x - 1$
42. $y = -3x^2 + 6x - 5$
43. $10x^2 - 160x + y + 655 = 0$

44. $6x^2 + 84x + y + 302 = 0$ **45.** $9x^2 + 6x + 81y - 53 = 0$

46. $100x^2 - 120x + 25y + 41 = 0$

47. A manufacturer produces high-quality boats at a profit, in dollars, that is determined to be

$$P(x) = -10(x - 375)^2 + 1{,}156{,}250$$

a. How many boats should be produced in order to maximize the profit?
b. What is the profit (or loss) if no boats are produced?
c. What is the maximum profit?

48. A profit function, P, is

$$P(x) = -10(x - 75)^2 + 3750$$

Find the maximum profit.

49. The profit function for a ratchet flange is

$$P(x) = -2(x - 25)^2 + 650$$

What is the maximum profit?

50. An arch has an equation

$$y - 18 = -\frac{2}{81}x^2$$

a. What is the maximum height?
b. What is the width of the arch?

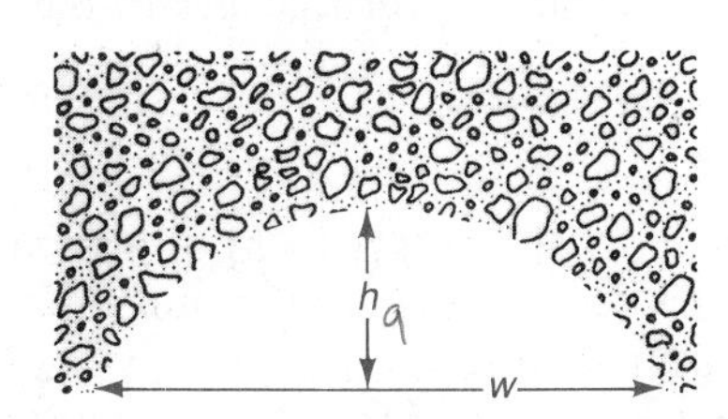

51. What is the height of the arch in Problem 50 at the following places?
a. Nine feet from the center
b. Eighteen feet from the center

52. One side of a storage yard is against a building. The other three sides of the rectangular yard are to be fenced with 36 ft of fencing. How long should the sides be to produce the greatest area with the given length of fence? What is the area obtained?

53. The sum of the length and the width of a rectangular area is 50 ft. Find the greatest area possible, and find the dimensions of the figure.

54. ***Physics*** The highest bridge in the world is the bridge over the Royal Gorge of the Arkansas River in Colorado. It is 1053 ft above the water. If a rock is projected vertically upward from this bridge with an initial velocity of 64 fps, the height h of the object above the river at time t is described by the function

$$h(t) = -16t^2 + 64t + 1053$$

What is the maximum height possible for a rock projected vertically upward from the bridge with an initial velocity of 64 fps? After how many seconds does it reach that height?

55. ***Physics*** In 1974, Evel Knievel attempted a skycycle ride across the Snake River. Suppose the path of the skycycle is given by the equation

$$d(x) = -0.0005x^2 + 2.39x + 600$$

where $d(x)$ is the height in feet above the canyon floor for a horizontal distance of x units from the launching ramp. What was Knievel's maximum height?

C

Sketch the graph of each equation given in Problems 56–61.

56. $y = x^2 - 5x + 2$

57. $2x^2 - x - y + 3 = 0$

58. $2x^2 - 8x + 3y + 20 = 0$

59. $4x^2 - 20x - 16y + 33 = 0$

60. $4x^2 + 24x - 27y - 17 = 0$

61. $25x^2 - 30x - 5y + 2 = 0$

62. ***Business*** A small manufacturer of CB radios determines that the price of each item is related to the number of items produced. If x items are produced per day, and the maximum number that can be produced is 10 items, then the price should be

$$400 - 25x \text{ dollars}$$

It is also determined that the overhead (the cost of producing x items) is

$$5x^2 + 40x + 600 \text{ dollars}$$

The daily profit is then found by subtracting the overhead from the revenue:

$$\begin{aligned} \text{profit} &= \text{revenue} - \text{cost} \\ &= (\text{number of items})(\text{price per item}) - \text{cost} \\ &= x(400 - 25x) - (5x^2 + 40x + 600) \\ &= 400x - 25x^2 - 5x^2 - 40x - 600 \\ &= -30x^2 + 360x - 600 \end{aligned}$$

A negative profit is called a loss.

a. What is the domain for the profit function?
b. For what values of the domain does the manufacturer show a positive profit?
c. Does the manufacturer ever show a loss? For what values?

d. If there were a strike and production were brought to a halt, what would be the value of x? What would be the profit (or loss) for this situation?

e. How many items should the manufacturer produce per day, and what should be the expected daily profit in order to maximize the profit?

63. ***Physics (General Interest)***

a. In most states, drivers are required to know approximately how long it takes to stop their cars at various speeds. Suppose you estimate three car lengths for 30 mph and six car lengths for 60 mph. One car length per 10 mph assumes a linear relationship between speed and distance covered. Write a linear equation where x is the speed of the car and y is the distance traveled by the car in feet. (Assume that one car length = 20 ft.)

b. The scheme for the stopping distance of a car given in part **a** is convenient but not accurate. The stopping distance is more accurately approximated by the quadratic equation

$$y = 0.071x^2$$

This says that your car requires four times as many feet to stop at 60 mph as at 30 mph. Doubling your speed quadruples the braking distance. Graph this quadratic equation and the linear equation you found in part **a** on the same coordinate axes.

c. Comment on the results from part **b**. At about what speed are the two measures the same?

2.6 COMPOSITE AND INVERSE FUNCTIONS

Suppose a farmer sells eggs to Safeway. If the farmer's price for a dozen eggs is x dollars, then there is a function f that can be used to describe $f(x)$, the total cost of those eggs to Safeway. Note that x and $f(x)$ are not the same because Safeway must pay for ordering, shipping, and distributing the eggs. Moreover, in order to determine the price to the consumer, Safeway must add an appropriate markup. Suppose this markup function is called g. Then the price of the eggs to the consumer is

$$g[f(x)] \quad \text{and not} \quad g(x)$$

since the markup must be on Safeway's *total* cost of the eggs and not just on the price the farmer charges for the eggs. This process of evaluating a function of a function illustrates the idea of composition of functions.

Consider two functions f and g such that f is a function from X to Y and g is a function from Y to Z as illustrated by Figure 2.20.

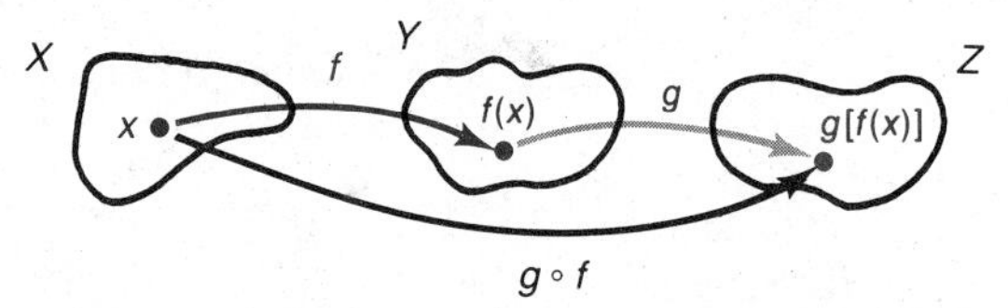

Figure 2.20 Composition of f and g

The image of x in the set Z is the number $g[f(x)]$ and defines a function from X to Z called the **composition of functions f and g**.

Composite Function

Let X, Y, and Z be sets of real numbers. Let f be a function from X to Y and g be a function from Y to Z. Then the **composite function $g \circ f$** is the function from X to Z defined by

$$(g \circ f)(x) = g[f(x)]$$

EXAMPLE 1 If $f(x) = x^2$ and $g(x) = x + 4$, find:

a. $(g \circ f)(x)$ **b.** $(f \circ g)(x)$ **c.** $(g \circ f)(-1)$ **d.** $(f \circ g)(5)$

Solution

a. $$(g \circ f)(x) = g[f(x)] = g[x^2] = x^2 + 4$$

b. $$(f \circ g)(x) = f[g(x)] = f[x + 4] = (x + 4)^2 = x^2 + 8x + 16$$

c. $$(g \circ f)(-1) = (-1)^2 + 4 = 5$$

d. $$(f \circ g)(5) = 5^2 + 8(5) + 16 = 81$$

■

Notice from Examples 1a and 1b that $g \circ f \neq f \circ g$.

Some attention also must be paid to the domain of $g \circ f$. Let X be the domain of a function f, and let Y be the range of f. The situation can be viewed as shown in Figure 2.21.

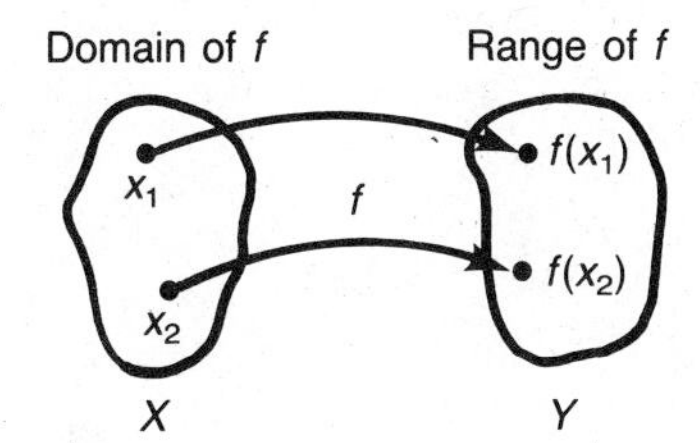

Figure 2.21 Function f

Now the function g may have as a domain the set Y, but not necessarily. Suppose the domain of g is a subset of Y, as shaded in Figure 2.22.

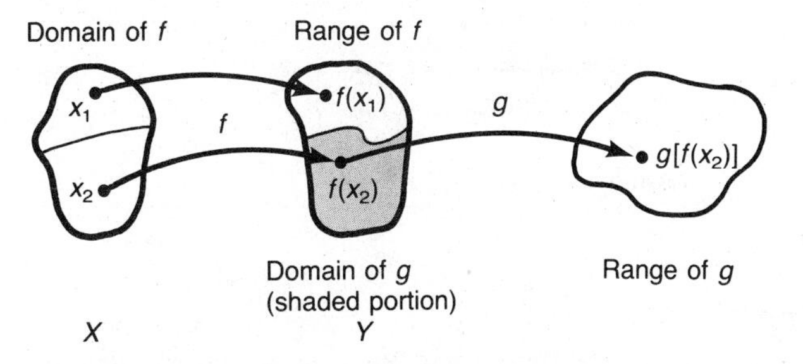

Figure 2.22 Function g

If part of f maps into the shaded portion of Y and part of f maps into the portion that is not shaded, as indicated in Figure 2.22, then the domain of $g \circ f$ is just that part of X that maps into the shaded portion of Y, as shown in Figure 2.23.

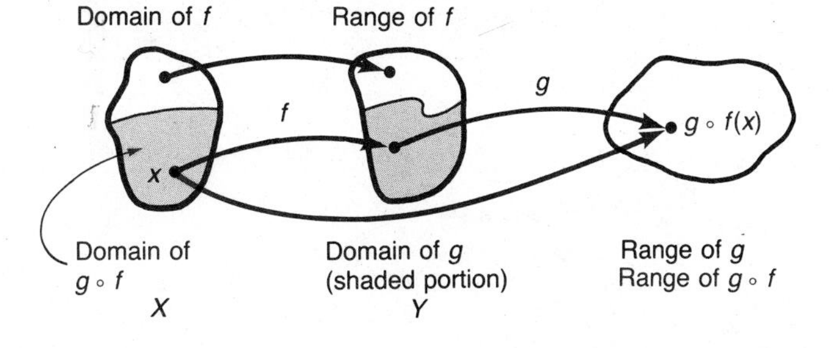

Figure 2.23 Composition of two functions, $g \circ f$; notice that the domain of $g \circ f$ is the subset of X for which $g \circ f$ is defined (shaded portion of Y)

EXAMPLE 2

$$f = \{(0,0), (-1,1), (-2,4), (-3,9), (5,25)\}$$
$$g = \{(0,-5), (-1,0), (2,-3), (4,-2), (5,-1)\}$$

a. Find $g \circ f$. **b.** Find $f \circ g$.

Solution **a.**

f	g
$0 \to 0 \cdots \to 0 \to -5$	
$-1 \to 1 \cdots \to$ Not defined, so go back to domain of f and exclude -1	
$-2 \to 4 \cdots \to 4 \to -2$	
$-3 \to 9 \cdots \to$ Not defined; exclude -3 from the domain of $g \circ f$	
$5 \to 25 \cdots \to$ Not defined; exclude 5 from the domain of $g \circ f$	

$$g \circ f = \{(0,-5), (-2,-2)\}$$

Notice that -1, -3, and 5 are excluded from the domain of $g \circ f$ even though they are in the domain of f.

b.

g	f
$0 \to -5 \cdots \to$ Not defined	
$-1 \to 0 \cdots \to 0 \to 0$	
$2 \to -3 \cdots \to -3 \to 9$	
$4 \to -2 \cdots \to -2 \to 4$	
$5 \to -1 \cdots \to -1 \to 1$	

$$f \circ g = \{(-1,0), (2,9), (4,4), (5,1)\}$$

■

Inverse Functions

If f is a one-to-one function from X onto Y, then a function g from Y to X is called the **inverse function of f** whenever

$$(g \circ f)(x) = x \qquad \text{for every } x \text{ in } X$$

and

$$(f \circ g)(x) = x \qquad \text{for every } x \text{ in } Y$$

EXAMPLE 3 Show that f and g defined by

$$f(x) = 5x + 4 \qquad \text{and} \qquad g(x) = \frac{x-4}{5}$$

are inverse functions.

Solution You must show that f and g are inverse functions in two parts:

$$(g \circ f)(x) = g(5x + 4) = \frac{(5x + 4) - 4}{5} = \frac{5x}{5} = x$$

and

$$(f \circ g)(x) = f\left(\frac{x-4}{5}\right) = 5\left(\frac{x-4}{5}\right) + 4 = (x - 4) + 4 = x$$

Thus $(g \circ f)(x) = (f \circ g)(x) = x$, so f and g are inverse functions. ■

Once you are certain that a function g is the inverse of a function f, you can denote it by f^{-1} so that:

Notation for Inverse Functions

$(f^{-1} \circ f)(x) = x$ for x in the domain of f
$(f \circ f^{-1})(x) = x$ for x in the domain of f^{-1}

Be careful about this notation. The symbol f^{-1} means the *inverse* of f and does not mean 1 *divided* by f.

Now consider a function defined by a set of ordered pairs $y = f(x)$. The image of x is y, as shown in Figure 2.24:

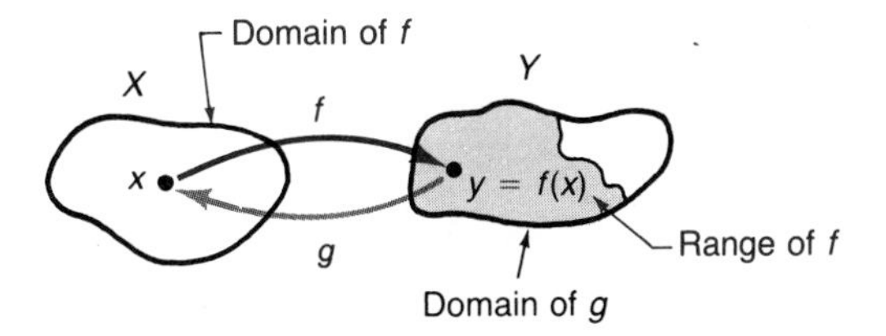

Figure 2.24 Inverse functions

If y is a member of the domain of the function g, then $g(y) = x$, as illustrated in Figure 2.24. This means that if you think of a function as a set of ordered pairs (x, y), the inverse of f is the set of ordered pairs with the components (y, x).

EXAMPLE 4 Let $f = \{(0, 3), (1, 5), (3, 9), (5, 13)\}$. Find the inverse of f.

Solution The inverse simply reverses the ordered pairs: $f^{-1} = \{(3, 0), (5, 1), (9, 3), (13, 5)\}$ ■

Example 5, on the next page, tells us how to find the inverse if the function is defined by a graph.

EXAMPLE 5 Consider the function f defined by the graph in Figure 2.25.

a. Find $f(5)$. **b.** Find $f^{-1}(6)$.

Solution **a.** Use the graph as shown in Figure 2.25:

This is a member of the domain of f—locate on x axis.

$f(5) = 3$

This is found by following dotted lines in Figure 2.25.

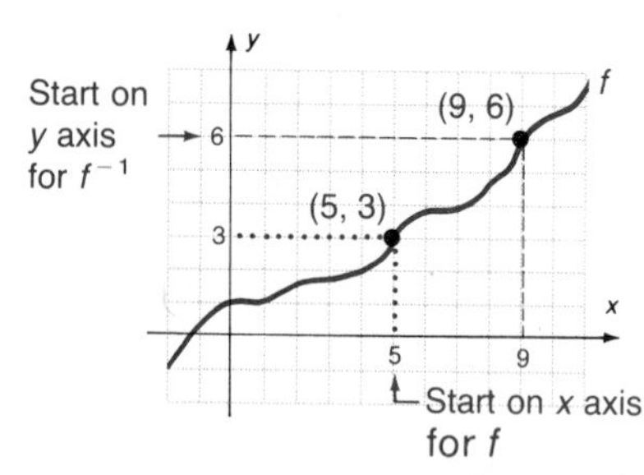

Figure 2.25 Graph of f

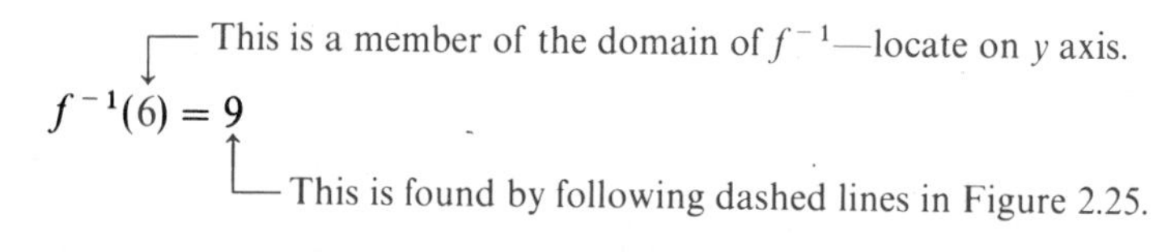

Although Example 5 illustrates a convenient method for finding ordered pairs in f or f^{-1}, it is not the usual way for representing f^{-1} since the first component of a function is usually plotted on the axis of abscissas. That is, to draw the graph of f^{-1}, *each* ordered pair of f—(5, 3) for example—must be rewritten as (3, 5) and plotted in the usual fashion (see Figure 2.26).

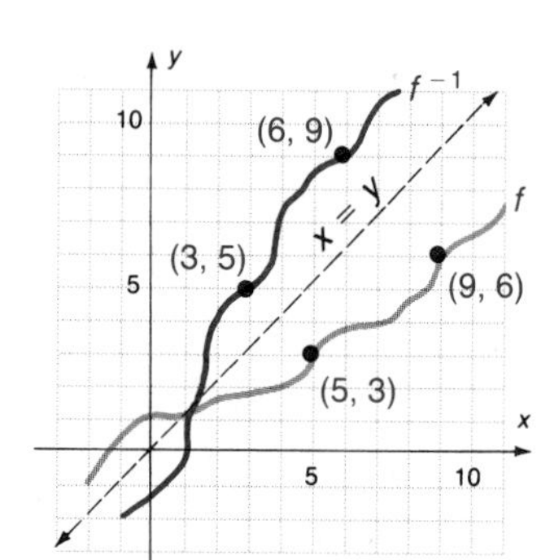

Figure 2.26 Graphs of f and f^{-1}

The graph of f and its inverse f^{-1} are symmetric with respect to the line $x = y$. If a function f is defined by an equation such as

$$f(x) = 4x - 3$$

you can find the inverse by first writing this equation using (x, y) notation:

$$y = 4x - 3$$

Next find the inverse by interchanging the x and y components:

$$x = 4y - 3 \qquad \text{Interchange } x \text{ and } y.$$

Finally, solve for y:

$$x + 3 = 4y \qquad \text{Solve for } y.$$

$$y = \frac{x + 3}{4}$$

The result is an equation for the inverse of f:

$$f^{-1}(x) = \frac{x + 3}{4}$$

EXAMPLE 6 Let f be defined by $f(x) = x^2 - 1$ on the interval $[0, \infty)$. Find f^{-1}.

Solution Do you see why some restriction on the domain is necessary? The interval given in this example forces f to be increasing and therefore one-to-one. Without this restriction you have

$$f(3) = 9 - 1 \qquad \text{and} \qquad f(-3) = 9 - 1$$

which means that $(3, 8)$ and $(-3, 8)$ both belong to f—which shows that even though f is a function, it is not one-to-one. If x is a member of the domain $[0, \infty)$, however, then f is one-to-one and the inverse function is defined. Now let $y = f(x)$ so that

	Domain	*Range*
$y = x^2 - 1$	$x \geq 0$	$y \geq -1$

Interchange x and y for the inverse:

$x = y^2 - 1$	$y \geq 0$	$x \geq -1$
	This one is now called the range because it shows the restriction on y.	This one is now called the domain because it shows the restriction on x.

Solve for y:

$y^2 = x + 1$

$y = \pm\sqrt{x + 1}$ Reject $y = -\sqrt{x + 1}$ since y is nonnegative.

$y = \sqrt{x + 1}$ $x \geq -1$

Thus

$f^{-1}(x) = \sqrt{x + 1}$ on $[-1, \infty)$ ■

PROBLEM SET 2.6

A

In Problems 1–12, find $(f \circ g)(x)$ and $(g \circ f)(x)$.

1. $f(x) = 2x - 3; \quad g(x) = x + 6$

2. $f(x) = 2x - 3; \quad g(x) = x^2 + 1$

3. $f(x) = 5x - 1; \quad g(x) = \dfrac{x + 1}{5}$

4. $f(x) = x^2; \quad g(x) = x^2 - x$

5. $f(x) = \dfrac{x - 2}{x + 1}$ and $g(x) = x^2 - x$

6. $f(x) = 4x + 1$ and $g(x) = x^3 + 3$

7. $f(x) = \dfrac{1}{x - 1}$ and $g(x) = x^2 - 1$

8. $f(x) = |x|$ and $g(x) = x^2$

9. $f(x) = 3$ and $g(x) = 5x^2$

10. $f = \{(0, 1), (1, 4), (2, 7), (3, 10)\}$ and $g = \{(0, -3), (1, -1), (2, 1), (3, 3)\}$

11. $f = \{(5, 3), (6, 2), (7, 9), (8, 12)\}$ and $g = \{(5, 8), (6, 5), (7, 4), (8, 3)\}$

12. $f = \{(5, 9), (10, 29), (15, 39), (20, 49)\}$ and $g = \{(5, 4), (10, 5), (15, 6), (20, 9)\}$

Determine which pairs of functions defined by the equations in Problems 13–18 are inverses.

13. $f(x) = 5x + 3; g(x) = \dfrac{x - 3}{5}$

14. $f(x) = \frac{2}{3}x + 2; g(x) = \frac{3}{2}x + 3$

15. $f(x) = \frac{4}{5}x + 4; g(x) = \frac{5}{4}x + 3$

16. $f(x) = \dfrac{1}{x}, x \neq 0; g(x) = \dfrac{1}{x}, x \neq 0$

17. $f(x) = x^2, x < 0; g(x) = \sqrt{x}, x > 0$

18. $f(x) = x^2, x \geq 0; g(x) = \sqrt{x}, x \geq 0$

Find the inverse function, if it exists, of each function given in Problems 19–34.

19. $f = \{(4, 5), (6, 3), (7, 1), (2, 4)\}$

20. $f = \{(1, 4), (6, 1), (4, 5), (3, 4)\}$

21. $f(x) = x + 3$

22. $f(x) = 2x + 3$

23. $g(x) = 5x$

24. $g(x) = \frac{1}{5}x$

25. $h(x) = x^2 - 5$

26. $h(x) = \sqrt{x} + 5$

27. $f(x) = x$

28. $f(x) = \dfrac{1}{x}$

29. $f(x) = 6$

30. $g(x) = -2$

31. $f(x) = \dfrac{1}{x - 2}$

32. $f(x) = \dfrac{2x + 1}{x}$

33. $f(x) = \dfrac{2x - 6}{3x + 3}$

34. $f(x) = \dfrac{3x + 1}{2x - 3}$

B

In Problems 35–42 find **a.** $f \circ g$ **b.** $g \circ h$ **c.** $(f \circ g) \circ h$ **d.** $f \circ (g \circ h)$

35. $f(x) = x^2, g(x) = 2x - 1, h(x) = 3x + 2$

36. $f(x) = x^2, g(x) = 3x - 2, h(x) = x^2 + 1$
37. $f(x) = 2x + 4, g(x) = \frac{1}{2}x - 2, h(x) = x^2 + 1$
38. $f(x) = \sqrt{x}, g(x) = x^2, h(x) = x + 2; x > 0$ for f, g, and h
39. $f(x) = 3x + 2, g(x) = 2x - 5, h(x) = x + 1$
40. $f(x) = g(x) = h(x) = x$
41. $f(x) = g(x) = h(x) = 2x$
42. $f(x) = x, g(x) = x^2, h(x) = x^3$

If f is defined by the graph in Figure 2.27, find the values requested in Problems 43–48.

43. a. $f(3)$ b. $f(4)$ c. $f(7)$ d. $f(11)$ e. $f(14)$
44. a. $f(0)$ b. $f^{-1}(4)$ c. $f^{-1}(5)$ d. $f(15)$ e. $f^{-1}(9)$
45. a. $f^{-1}(0)$ b. $f^{-1}(1)$ c. $f^{-1}(6)$ d. $f^{-1}(-2)$ e. $f^{-1}(-6)$
46. a. $f(1)$ b. $f^{-1}(-1)$ c. $f^{-1}(-4)$ d. $f(2)$ e. $f^{-1}(2)$
47. a. $f^{-1}(-5)$ b. $f(5)$ c. $f^{-1}(3)$ d. $f^{-1}(8)$ e. $f(13)$

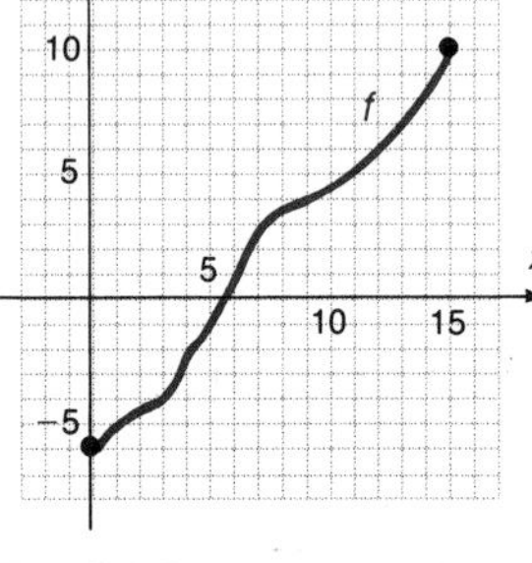

Figure 2.27 Graph of f

48. a. $f^{-1}(4)$ b. $f(8)$ c. $f^{-1}(-3)$ d. $f(9)$ e. $f^{-1}(7)$
49. a. What is the domain and range of the function f defined by the graph in Figure 2.27?
    b. What is the domain and range of the function f^{-1}?
50. Graph f^{-1} for the function f defined by the graph in Figure 2.27.

Determine which pairs of functions defined by the equations in Problems 51–54 are inverses.

51. $f(x) = 2x^2 + 1, x \geq 0;$ $g(x) = \frac{1}{2}\sqrt{2x - 2}, x \geq 1$
52. $f(x) = 2x^2 + 1, x \leq 0;$ $g(x) = -\frac{1}{2}\sqrt{2x - 2}, x \geq 1$
53. $f(x) = (x + 1)^2, x \geq 1;$ $g(x) = -1 - \sqrt{x}, x \geq 0$
54. $f(x) = (x + 1)^2, x \geq -1;$ $g(x) = -1 + \sqrt{x}, x \geq 0$
55. ***Business*** Suppose that a store sells calculators by marking up the price 20%. That is, the price of an item costing c dollars is

$$p(c) = c + 0.20c$$

Also suppose that the cost c of manufacturing n calculators is $50n + 200$ (dollars). This means that the cost of each calculator depends on the number manufactured according to the function

$$c(n) = \frac{50n + 200}{n} \qquad (n \geq 1)$$

a. Find the price for one calculator if only one calculator is manufactured.
b. Find the price for one calculator if 10 calculators are manufactured.
c. Express the price as a function of the number of calculators produced by finding $p \circ c$.

56. ***Physics*** Suppose that the volume of a certain cone is expressed as a function of its height so that it is defined by

$$V(h) = \frac{\pi h^3}{12}$$

Suppose the height is expressed as a function of time by $h(t) = 2t$.
a. Find the volume for $t = 2$.
b. Express the volume as a function of time by finding $V \circ h$.
c. If the domain of V is $\{h \mid 0 < h \leq 6\}$, find the domain of h. That is, what are the permissible values for t?

57. ***Physics*** The surface area of a spherical balloon is given by

$$S(r) = 4\pi r^2$$

Suppose the radius is expressed as a function of time by $r(t) = 3t$.
a. Find the surface area for $t = 2$.
b. Express the surface area as a function of time by finding $S \circ r$.
c. If the domain of S is $\{r \mid 0 < r \leq 8\}$, find the domain of r. That is, what are the permissible values for t?

C

Find the inverse of each function given in Problems 58–67.

58. $f(x) = x^2$ on $[0, \infty)$
59. $f(x) = x^2$ on $(-\infty, 0]$
60. $f(x) = x^2 + 1$ on $(-\infty, 0]$
61. $f(x) = x^2 + 1$ on $[0, \infty)$
62. $f(x) = 2x^2$ on $[2, 10]$
63. $f(x) = 2x^2$ on $[-10, -1]$
64. $f(x) = \dfrac{1}{3x + 1}$ on $(-\frac{1}{3}, \infty)$
65. $f(x) = \dfrac{1}{2x - 1}$ on $(\frac{1}{2}, \infty)$
66. $f(x) = x$ on $[0, \infty)$
67. $f(x) = x + 1$ on $[-1, \infty)$
68. If $f(x) = 1 + \dfrac{1}{x}$, find:
    a. $(f \circ f)(x)$ b. $(f \circ f \circ f)(x)$ c. $(f \circ f \circ f \circ f)(x)$

69. a. Let $f(x) = \sqrt{x}$. Choose *any* positive x. Find a numerical value for $(f \circ f)(x)$, $(f \circ f \circ f)(x)$, and $(f \circ f \circ f \circ f)(x)$. Suppose this procedure is repeated a large number of times:

$$(f \circ f \circ f \circ \ldots \circ f)(x)$$

Can you predict the outcome for *any* x?

b. Let $f(x) = 2\sqrt{x}$. Answer the questions of part **a**.

c. Let $f(x) = 3\sqrt{x}$. Answer the questions of part **a**.

d. Let $f(x) = k\sqrt{x}$, where k is a positive integer. Answer the questions of part **a**.

2.7 CHAPTER 2 SUMMARY

The material of this chapter is reviewed in the following list of objectives. After each objective there are some practice questions. For a sample test, select the first question of each set and check your answers with the answer section. For a sample test without answers, use the second question of each set. Additional practice is given by the other questions in each set. If you are having trouble with a particular type of problem, look back to that section for extra help.

2.1 FUNCTIONS

Objective 1 Be able to classify examples as onto, one-to-one, functions, or not functions. Let X and Y be sets of real numbers such that $x \in X$ and $y \in Y$. (The symbol $\in$ means "is a member of.")

1. $y = 3x + 3$ **2.** $y = 2x^2 + 3$ **3.** $y \leq 2x + 3$

4. The set of ordered pairs defined by the graph at the left.

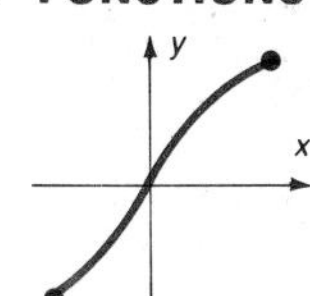

Objective 2 Name subsets of the domain of a function in which the function is increasing, decreasing, or constant. Find the intercepts. Consider the graph shown in the following figure.

5. What is the domain and range of f?

6. Name the subsets of the domain for which f is increasing, f is decreasing, and f is constant.

7. What are the intercepts?

8. What are the coordinates of A and B?

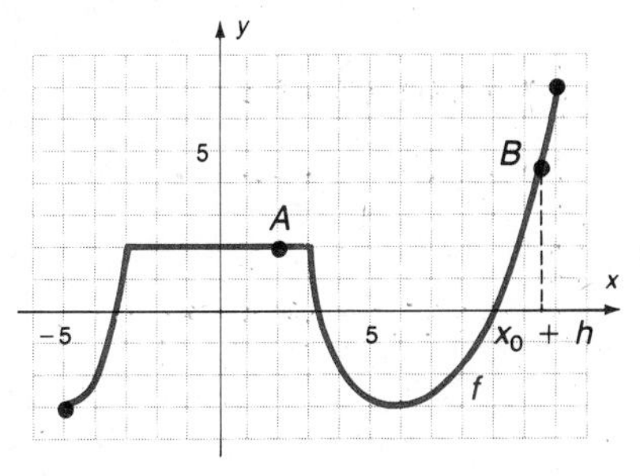

2.2 FUNCTIONAL NOTATION

Objective 3 Distinguish between the f and f(x) notation. Use functional notation. Let $f(x) = 3x - 1$ and $g(x) = 5 - x^2$. Find the requested values.

9. a. $f(4)$ **b.** $g(4)$

10. a. $f(-6)$ **b.** $g(-6)$

11. a. $f(w)$ **b.** $g(w + h)$

12. a. $3f(t)$ **b.** $g(\sqrt{t})$

Objective 4 Find $\dfrac{f(x+h) - f(x)}{h}$ *for a given function f.*

13. $f(x) = 5 - x^2$ **14.** $f(x) = 5x + 3$ **15.** $f(x) = 5$ **16.** $f(x) = 3x^2 - 1$

Objective 5 Find the domain of a given function over the set of real numbers.

17. $f(x) = \sqrt{x^2 - 7x + 6}$

18. $f(x) = \dfrac{(3x+1)(2x+5)}{2x+5}$

19. $f(x) = 3x^2 - 5x + 1$

20. $f(x) = \dfrac{6x^2 + 17x + 15}{3x + 1}, x \neq 0$

Objective 6 Determine when functions are equal.

21. $f(x) = \dfrac{x^2 + x}{x + 1}$; $g(x) = x$

22. $f(x) = \dfrac{(x + 3)(x - 2)}{x - 2}$, $x \neq 2$; $g(x) = \dfrac{(x + 3)(x - 2)}{x + 3}$, $x \neq -3$

23. $f(x) = \sqrt{x^2}$; $g(x) = x$

24. $f(x) = \dfrac{(2x + 1)(3x - 2)}{3x - 2}$; $g(x) = 2x + 1$, $x \neq 0$

Objective 7 Classify functions as even, odd, or neither.

25. $f(x) = 3x^2 + 2x - 5$

26. $f(x) = 2x^3 + 5x$

27. $f(x) = \dfrac{1}{(x + 4)^2}$

28. $f(x) = 7$

2.3 LINEAR FUNCTIONS

Objective 8 Define a linear function. Graph linear functions.

29. $y = 5x - 3$

30. $2x + 3y + 6 = 0$

31. $5x - 2y + 8 = 0$

32. $y + 3 = \frac{3}{5}(x - 4)$

Objective 9 Find the slope of a line.

33. The line passing through $(2, 2)$ and $(6, -3)$
34. The line passing through $(a, f(a))$ and $(b, f(b))$
35. The line $7x - 5y + 3 = 0$
36. A horizontal line

Objective 10 Graph line segments.

37. $4x - 3y + 9 = 0$ for $-3 < x \leq 4$

38. $y = \frac{3}{5}x - 3$ for $-5 \leq x \leq 5$

39. $2x = y - 5$ for $-4 \leq x < 2$

40. $y = 5$ for $-4 < x < 4$

Objective 11 Graph absolute value relations.

41. $y = |-x|$

42. $y = -2|x|$

43. $y = \frac{2}{3}|x|$

44. $y = |4x|$

Objective 12 Determine when given lines or line segments are parallel or perpendicular.

45. Use slopes to decide whether the triangle with vertices $A(12, -7)$, $B(4, 3)$, $C(-3, -2)$ is a right triangle.
46. What is the slope of a line parallel to $3x - 4y + 7 = 0$?
47. What is the slope of a line perpendicular to $6x - 4y + 3 = 0$?
48. Are the diagonals of the quadrilateral $ABCD$ with vertices $A(3, 4)$, $B(-2, 5)$, $C(-4, -2)$, $D(3, -5)$ perpendicular?

Objective 13 Know and use the various forms of a linear equation.

49. State the standard-form equation and define the variables.
50. State the point-slope-form equation and define the variables.
51. State the slope-intercept-form equation and define the variables.
52. State the equation of a vertical line and define the variables.

Objective 14 Given the graph, or information about the graph, of a line, write the standard-form equation of the line.

53. y intercept -9; slope $\frac{2}{3}$

54. Slope -3; passing through $(3, -2)$

55. Passing through $(-3, -1)$ and $(5, -6)$

56. Passing through $(4, 5)$ and perpendicular to the line $2x - 3y + 8 = 0$

2.4 TRANSLATION OF FUNCTIONS

Objective 15 *Find the shift (h, k) when given an equation $y - k = f(x - h)$.*

57. $y - 6 = f(x + \pi)$

58. $y - 1 = 3(x + 3)^2$

59. $y = 5(x - 4)^2$

60. $y = \frac{2}{3}|x - \sqrt{2}| + 5$

Objective 16 *Given $y = f(x)$ and (h, k), write $y - k = f(x - h)$.*

61. $f(x) = 9x^2; (h, k) = (-\sqrt{2}, 3)$

62. $f(x) = 2|x|; (h, k) = (4, -3)$

63. $f(x) = -2x^2; (h, k) = (\pi, 0)$

64. $f(x) = 3x^2 + 2x + 5; (h, k) = (1, 2)$

Objective 17 *Substitute $x' = x - h$ and $y' = y - k$ into an equation $y - k = f(x - h)$.* Substitute $x' = x + 3$ and $y' = y - 1$ into the following equations.

65. $y - 1 = 3(x + 3)^2$

66. $y - 1 = 2|x + 3|$

67. $y = 5(x + 3)^2 + 1$

68. $y - 1 = (x + 3)^2 + (x + 3) - 5$

Objective 18 *Given a function defined by $y = f(x)$ and shown by a graph, draw the graph of $y - k = f(x - h)$.*

Figure 2.28 Graph for Problems 69–72

69. $y - 2 = f(x + 1)$

70. $y + 1 = f(x - 3)$

71. $y = f(x - \pi)$

72. $y + \sqrt{2} = f(x - \sqrt{3})$

2.5 QUADRATIC FUNCTIONS

Objective 19 *Define and graph quadratic functions of the form $y - k = a(x - h)^2$.*

73. $y - 1 = \frac{3}{4}(x + 3)^2$

74. $y + 1 = -2(x + 2)^2$

75. $y = 10x^2$

76. $y - 1 = 3(x + 3)^2$

Objective 20 *Graph quadratic functions by completing the square.*

77. $x^2 + 2x - y - 1 = 0$

78. $x^2 + 4x - 2y - 2 = 0$

79. $y = x^2 - 6x + 10$

80. $x^2 - 4x + 5y + 9 = 0$

Objective 21 *Find the maximum or minimum value for a quadratic function.* Be sure to state whether your answer is a maximum or a minimum value.

81. $2x^2 + 24x + y = 178$

82. $2x^2 + 20x + y + 190 = 0$

83. $x^2 - 10x - y - 825 = 0$

84. $x^2 - 8x - 3y + 3616 = 0$

2.6 COMPOSITE AND INVERSE FUNCTIONS

Objective 22 *Find the composition of functions.* Let $f(x) = 3x - 1$, $g(x) = 5 - x^2$, and $h(x) = 6$. Find the requested values.

85. $(f \circ g)(x)$

86. $(g \circ f)(x)$

87. $(f \circ h)(x)$

88. $(h \circ f)(x)$

Objective 23 *Given two functions, decide whether or not they are inverses.* Decide if f and g are inverses.

89. $f(x) = 3x + 6; g(x) = x - 2$

90. $f(x) = 2x + 1; g(x) = x - 2$

91. $f(x) = 3x - 2; g(x) = \frac{x - 2}{3}$

92. $f(x) = 2x - 3; g(x) = \frac{x + 2}{3}$

Objective 24 *Given a one-to-one function, find its inverse.*

93. $f(x) = 3x - 1$

94. $g(x) = 5 - x^2$ on $[0, \infty)$

95. $f(x) = \frac{1}{2}x + 5$

96. $g(x) = \sqrt{2x}$ on $[0, \infty)$

Karl Gauss (1777–1885)

Mathematics is the queen of the sciences and arithmetic the queen of mathematics. She often condescends to render service to astronomy and other natural sciences, but in all relations she is entitled to the first rank.

Karl Gauss

Gauss is sometimes described as the last mathematician to know everything in his subject. Such a generalization is bound to be inexact, but it does emphasize the breadth of interests Gauss displayed.

Carl B. Boyer in
A History of Mathematics,
pp. 561–562

HISTORICAL NOTE

Karl Gauss is considered to be one of the three greatest mathematicians of all time, along with Archimedes and Newton. Gauss had a great admiration for Archimedes, but could not understand how Archimedes failed to invent the positional numeration system. Howard Eves quotes Gauss as saying, "To what heights would science now be raised if Archimedes had made that discovery!" Because of the greatness of Gauss, many stories and anecdotes about him have survived. In all the history of mathematics there is nothing approaching the precocity of Gauss as a child. He showed his caliber before he was 3 years old! In later life, Gauss joked that he knew how to reckon before he could talk. It seems that when he was only 3 he corrected his father's computations on a payroll report.

Gauss graduated from college at the age of 15 and proved what was to become the Fundamental Theorem of Algebra for his doctoral thesis at the age of 22. He published only a small portion of the ideas that seemed to storm his mind because he felt that each published result had to be complete, concise, polished, and convincing. His motto was "Few, but ripe."

In this chapter you begin your study of one of the most important tools in mathematics—namely, the ability to draw a graph corresponding to a given equation. There is a quotation of Gauss in which he described the joy of learning. This quotation is from a letter dated September 2, 1808 to his friend Wolfgang Bolyai:

"It is not knowledge, but the act of learning, not possession, but the act of getting there, which grants the greatest enjoyment. When I have clarified and exhausted a subject, then I turn away from it, in order to go into darkness again; the never-satisfied man is so strange—if he has completed a structure, then it is not in order to dwell in it peacefully, but in order to begin another. I imagine the world conqueror must feel thus, who, after one kingdom is scarcely conquered, stretches out his arms for another."

3

Graphing Techniques

CONTENTS

PREVIEW

The ability to sketch a curve quickly is a very important skill in mathematics. In the last chapter, we looked at the graphs of lines and certain parabolas. In this chapter, we will tie some of the techniques you have already learned into a general plan for graphing, which you will be able to use not only when you study calculus, but whenever you need to sketch the graph of a given equation quickly and efficiently. There are 7 objectives in this chapter, which are listed on pages 116–117.

PERSPECTIVE

Graphing of functions is an essential concept for the study of calculus. You can open a calculus book to almost any page at random and you will find some function whose graph is shown. For example, when finding the area bounded by two functions in calculus, it is desirable to first graph the functions in order to see which one is above and which is below and where the boundaries begin and end. The example shown below finds the area of the region bounded by the two curves

$$y = x^4 \qquad \text{and} \qquad y = 2x - x^2$$

We will discuss techniques for curve sketching in this chapter.

EXAMPLE 5 Find the area of the region between the curves $y = x^4$ and $y = 2x - x^2$.

Solution We start by finding where the two curves intersect and sketching the required region. This means that we need to solve $2x - x^2 = x^4$, a fourth-degree equation, which would usually be difficult to solve. However, in this case, $x = 0$ and $x = 1$ are rather obvious solutions. Our sketch of the region, together with the appropriate approximation and the corresponding integral, is shown in Figure 8.

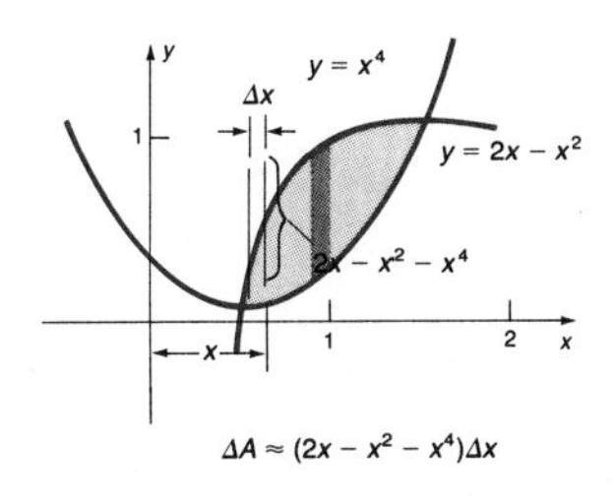

FIGURE 8

3.1 PLOTTING POINTS, INTERCEPTS, AND SYMMETRY

This chapter considers a very important topic in mathematics, namely the relationship between an algebraic equation and a geometric curve. In the last chapter, we introduced you to the Cartesian coordinate system, the graphs of lines and certain parabolas, and the ideas of translation and symmetry. In this chapter, we will elaborate on those ideas and add many more, so that you will begin to feel confident of your ability to graph an equation.*

It is important that you clearly understand the relationship between an equation and its graph. This relationship is very explicit: There is a one-to-one correspondence between ordered pairs satisfying an equation and coordinates of points on the graph.

Relationship between an Equation and Its Graph

The **graph of an equation**, or the **equation of a graph**, means:

1. Every point on the graph has coordinates that satisfy the equation.
2. Every ordered pair satisfying the equation has coordinates that lie on the graph.

This fundamental relationship between an equation and its graph tells you that to graph an equation (or check to see if a graph is correct) you can find ordered pairs satisfying an equation, and then plot those ordered pairs. After plotting many points you can see the graph "take shape." You used this idea to graph some of the lines and parabolas of the last chapter. Computer programs for graphing curves are readily available and, for the most part, rely on plotting points to determine the shape of a graph. However, this procedure of plotting a great number of points can be very tedious without a computer. Graphic calculators also can be used for graphing many curves (see the HP28S Activity on page 136, for example). In this section, however, we will investigate other properties of graphs that give us a great deal of information about a wide variety of curves and, at the same time, minimize the amount of actual work we need to do.

Rather than randomly plotting points, we begin by looking for the intercepts—those places where the curve passes through the coordinate axes.

Intercepts

The **x intercepts** are those points where a curve passes through the x axis. They are found by setting $y = 0$ and solving the resulting equation for x.

The **y intercepts** are those points where a curve passes through the y axis. They are found by setting $x = 0$ and solving the resulting equation for y.

---

* Remember our agreement from Chapter 2 about variables in this book. Unless otherwise stated, the domain for all variables is the set of real numbers for which the given function is meaningful.

Remember, if a graph is a function, then, unless stated otherwise, for each x value there can be associated only one y value. This means that if a curve represents a function, there can be at most one y intercept, but if the graph is not a function, there might be more than one y intercept. In either case, there may be several x intercepts. We usually begin by noting whether or not we are dealing with a function. Next, we find the y intercept(s) and finally the x intercepts (if any).

EXAMPLE 1 Find the intercepts and determine whether each is a function.

a. $y = \dfrac{1}{x^2 + 1}$ **b.** $y^2 = \dfrac{x^2 - 1}{x - 4}$ **c.** $|x| + |y| = 2$ **d.** $x^{2/3} + y^{2/3} = 9$

Solution **a.** This is a **function.**

***y* intercept:** Let $x = 0$. $y = \dfrac{1}{0^2 + 1} = 1$ Point: **(0, 1)**

***x* intercepts:** Let $y = 0$. $0 = \dfrac{1}{x^2 + 1}$

$0 = 1$ Multiply both sides by $x^2 + 1$.

A false equation means that there is no point; in this example, there are **no *x* intercepts.**

b. This is **not a function.**

***y* intercepts:** Let $x = 0$. $y^2 = \dfrac{0^2 - 1}{0 - 4} = \dfrac{1}{4}$

$y = \pm\dfrac{1}{2}$ Points: $(\mathbf{0}, \tfrac{\mathbf{1}}{\mathbf{2}}), (\mathbf{0}, -\tfrac{\mathbf{1}}{\mathbf{2}})$

x intercepts: Let $y = 0$. $0 = \dfrac{x^2 - 1}{x - 4}$ Multiply both sides by $x - 4$.

$0 = x^2 - 1$

$= (x - 1)(x + 1)$

$x = \pm 1$ Points: **(1, 0), (−1, 0)**

c. This is **not a function.**

***y* intercepts:** Let $x = 0$. $|0| + |y| = 2$

$y = \pm 2$ Points: **(0, 2), (0, −2)**

***x* intercepts:** Let $y = 0$. $|x| + |0| = 2$

$x = \pm 2$ Points: **(2, 0), (−2, 0)**

d. This is **not a function.**

***y* intercepts:** Let $x = 0$. $0^{2/3} + y^{2/3} = 9$

$(y^{2/3})^{3/2} = 9^{3/2}$

$y = 27$ Point: **(0, 27)**

x intercepts: Let $y = 0$. $x^{2/3} + 0^{2/3} = 9$

$x = 27$ Point: **(27, 0)** ■

Sometimes when you are graphing a curve, you want to find a point in a certain region or with certain properties. For example, if you want to know one point on the line $2x + 3y - 4 = 0$ where $x > 5$, then you can choose an x value satisfying $x > 5$ and find the corresponding y value. Consider the following example.

EXAMPLE 2 Find a point on each of the given curves that satisfies the given conditions.

a. $y = \dfrac{2x^2 - 3x + 5}{x^2 - x - 2}$ Find a point where $x > 5$.

You choose some value of x, say $x = 10$:

$$y = \frac{2(10)^2 - 3(10) + 5}{10^2 - 10 - 2} = \frac{175}{88} \approx 2 \qquad \text{Point: } (10, \tfrac{175}{88}) \approx \mathbf{(10, 2)}$$

b. Find a point (if it exists) where the curve in part a passes through the line $y = 2$. This means that you should substitute the value 2 for y in the given equation:

$$2 = \frac{2x^2 - 3x + 5}{x^2 - x - 2}$$

$$2(x^2 - x - 2) = 2x^2 - 3x + 5 \qquad \text{Multiply both sides by } x^2 - x - 2.$$

$$2x^2 - 2x - 4 = 2x^2 - 3x + 5$$

$$x = 9$$

The point is **(9, 2)**.

c. Given

$$y = \frac{2x^2 - 5x + 1}{x - 3}$$

find a point (if any exists) where this curve passes through the line $y = 2x + 1$. This means substitute $2x + 1$ for y in the given equation and solve for x:

$$2x + 1 = \frac{2x^2 - 5x + 1}{x - 3}$$

$$(2x + 1)(x - 3) = 2x^2 - 5x + 1$$

$$2x^2 - 5x - 3 = 2x^2 - 5x + 1$$

$$-3 = 1$$

This is a false equation, so there is **no point of intersection.** ■

In Section 1.6, we introduced the idea of symmetry, but there we focused on what symmetry means. We now revisit this idea, this time from an algebraic standpoint. Later in this chapter we will use these algebraic tests to determine the types of symmetry present so that we can use symmetry, along with plotting points, as an aid to curve sketching. We now repeat the symmetry tests from Section 1.6.

Symmetry

With respect to the x axis:	The equation is unchanged when y is replaced by $-y$.
With respect to the y axis:	The equation is unchanged when x is replaced by $-x$.
With respect to the origin:	The equation is unchanged when x and y are simultaneously replaced by $-x$ and $-y$, respectively.

If a curve has any two of the three types of symmetry, it will also have the third. (You are asked to show this in Problem 64.) If you find that a curve has one type of symmetry but not a second, then it cannot have the third type either.

EXAMPLE 3 Test for symmetry with respect to the x axis, y axis, and origin.

a. $y = \dfrac{3x^2 + 1}{x^4}$ **b.** $xy = 2$ **c.** $9x^2 - 16y^2 = 144$

Solution **a. x axis:** $-y = \dfrac{3x^2 + 1}{x^4}$ Substitute $-y$ for y. This is a different equation.

y axis: $y = \dfrac{3(-x)^2 + 1}{(-x)^4}$ Substitute $-x$ for x.

$= \dfrac{3x^2 + 1}{x^4}$ This is the same as the original equation.

Because one symmetry holds and the other does not, the graph will not be symmetric with respect to the third type of symmetry. Therefore, this curve is **symmetric with respect to the y axis.**

b. x axis: $x(-y) = 2$ This is a different equation.
y axis: $(-x)y = 2$ This is a different equation.
origin: $(-x)(-y) = 2$ Substitute $-x$ for x and $-y$ for y.
$xy = 2$ This is the same equation.

This curve is **symmetric with respect to the origin.**

c. x axis: $9x^2 - 16(-y)^2 = 144$ This is the same equation.
y axis: $9(-x)^2 - 16y^2 = 144$ This is the same equation.
origin: Since two symmetries hold, the third symmetry also must hold.

This curve is **symmetric with respect to the x axis, the y axis, and the origin.** ■

PROBLEM SET 3.1

A

In Problems 1–23, find all intercepts.

1. $y = x + 5$
2. $y = x - 2$
3. $y = x^2 - 4$
4. $y = x^2 + 3$
5. $y = x^2$
6. $y = x^3$
7. $y = \dfrac{3}{x^2 + 1}$
8. $y = \dfrac{x^2 - 4}{x + 1}$

9. $y = \frac{x^2 - 9}{x - 2}$

10. $y = \frac{x + 1}{x^2 - 16}$

11. $y = \frac{x^3 - 8}{x^2 + 2x + 4}$

12. $y = \frac{x^3 - 1}{x^2 + x + 1}$

13. $y = \frac{x^3 - 1}{x - 1}$

14. $y = \frac{x^3 + 8}{x + 2}$

15. $y = -\frac{\sqrt{x + 3}}{x - 1}$

16. $y = -\frac{\sqrt{9 - x^2}}{x + 3}$

17. $2|x| - |y| = 5$

18. $|y| = 5 - 3|x|$

19. $xy = 1$

20. $xy + 6 = 0$

21. $x^2 + 2xy + y^2 = 4$

22. $x^{1/2} + y^{1/2} = 4$

23. $x^2y + 2x^2 - 3x + 2y + 6 = 0$

In Problems 24–46, test for symmetry with respect to both axes and the origin.

24. $y = x + 5$

25. $y = x - 2$

26. $y = x^2 - 4$

27. $y = x^2 + 3$

28. $y = x^2$

29. $y = x^3$

30. $y = \frac{3}{x^2 + 1}$

31. $y = \frac{x^2 - 4}{x + 1}$

32. $y = \frac{x^2 - 9}{x - 2}$

33. $y = \frac{x + 1}{x^2 - 16}$

34. $y = \frac{x^3 - 8}{x^2 + 2x + 4}$

35. $y = \frac{x^3 - 1}{x^2 + x + 1}$

36. $y = \frac{x^3 - 1}{x - 1}$

37. $y = \frac{x^3 + 8}{x + 2}$

38. $y = -\frac{\sqrt{x + 3}}{x - 1}$

39. $y = -\frac{\sqrt{9 - x^2}}{x + 3}$

40. $2|x| - |y| = 5$

41. $|y| = 5 - 3|x|$

42. $xy = 1$

43. $xy + 6 = 0$

44. $x^2 + 2xy + y^2 = 4$

45. $x^{1/2} + y^{1/2} = 4$

46. $x^2y + 2x^2 - 3x + 2y + 6 = 0$

B

Find a point on each of the given curves satisfying the given conditions.

47. Find the points (if any) where the curve

$$y = \frac{5x^2 - 8x}{2x + 1}$$

passes through the line $y = 3$.

48. Find the points (if any) where the curve

$$y = \frac{2x^3 + 2x}{x^2 + 1}$$

passes through the line $y = 1$.

49. Find the points (if any) where the curve

$$y = \frac{5x^2 - 8x}{2x + 1}$$

has a value $x = -4$.

50. Find the points (if any) where the curve

$$y = \frac{2x^3 + 2x}{x^2 - 2}$$

has a value $x = -1$.

51. Find the points (if any) where the curve

$$y = \frac{x^3 + 2x^2 - 2x}{x^2 - 2}$$

passes through the line $y = x + 1$.

52. Find the points (if any) where the curve

$$y = \frac{3x^3 + 4x^2 + 3}{3x^2 + 1}$$

passes through the line $y = x + 2$.

In Problems 53–62, find the intercepts and test for symmetry.

53. $x^2 - xy + x + y + 1 = 0$

54. $2x^2 - xy + x - 2y - 2 = 0$

55. $9x^2 + 4y^2 = 36$

56. $4x^2 - 9y^2 = 36$

57. $y = \sqrt{\frac{x^3 + 1}{x^2 - x + 1}}$

58. $y = -\sqrt{\frac{x^3 - 1}{x^2 + x + 1}}$

59. $y^2 = \frac{x^2 - 4}{x^2 + 1}$

60. $y^2 = \frac{x^2 - 9}{x^2 - 4}$

61. $xy^2 - x + 4y^2 + 16 = 0$

62. $x^2y^2 - y + 3x^2 + 2x - 1 = 0$

C

63. Show that if a curve is symmetric with respect to both the x axis and the y axis, then it must also be symmetric with respect to the origin.

64. Show that if a curve has any two of the three types of symmetry discussed in this section, then it will also have the third.

65. Show that if a curve is symmetric with respect to the x axis but not with respect to the y axis, then it cannot be symmetric with respect to the origin.

66. Show that if a curve has one type of symmetry discussed in this section but not a second, then it cannot have the third type either.

3.2 DOMAIN AND RANGE

Earlier, we mentioned that a curve can be sketched by plotting points satisfying the equation for that curve. We also noted that it is usually much too tedious to find those points by making random choices. A valuable tool in curve sketching is to find **excluded values** and **excluded regions**. These are points or regions in the plane in which a given curve cannot lie. These points and regions are found by noting where either x or y does not exist in the set of real numbers. Exclusion will occur in the following situations:

1. Division by zero
2. Negative under a radical with an even index
3. Even-powered variables equal to negative numbers

EXAMPLE 1 Graph $y = \dfrac{2x^2 + 5x - 3}{x + 3}$

Solution Notice that there is a value of x that would cause division by zero ($x = -3$). This value is excluded from the domain, and is therefore called a **deleted point**.

$$y = \frac{2x^2 + 5x - 3}{x + 3}$$
$$= \frac{(2x - 1)(x + 3)}{x + 3}$$
$$= 2x - 1 \qquad (x \neq -3)$$

The graph is the same as for the linear function $y = 2x - 1$, with the point at $x = -3$ deleted from the domain, as shown in Figure 3.1.

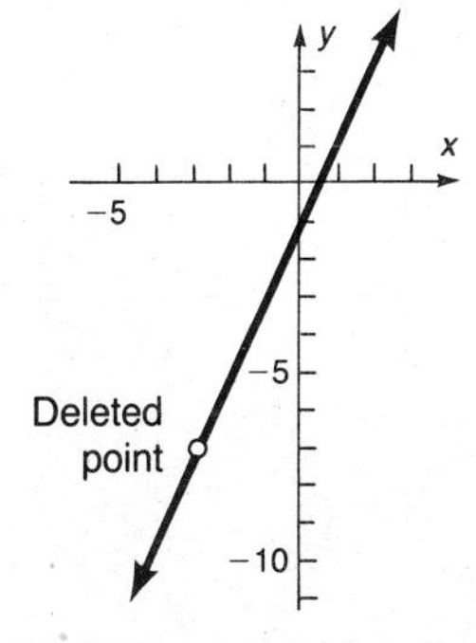

Figure 3.1 Graph of $f(x) = \dfrac{2x^2 + 5x - 3}{x + 3}$ ■

EXAMPLE 2 Graph $y = \dfrac{x^3 + 4x^2 + 7x + 6}{x + 2}$

Solution Simplify, if possible:

$$y = \frac{x^3 + 4x^2 + 7x + 6}{x + 2}$$
$$= x^2 + 2x + 3 \qquad (x \neq -2)$$

This simplification can be done by long division:

$$\begin{array}{r} x^2 + 2x + 3 \\ x + 2 \overline{) x^3 + 4x^2 + 7x + 6} \\ \underline{x^3 + 2x^2} \qquad\qquad \\ 2x^2 \qquad\qquad \\ \underline{2x^2 + 4x} \qquad \\ 3x \qquad \\ \underline{3x + 6} \\ 0 \end{array}$$

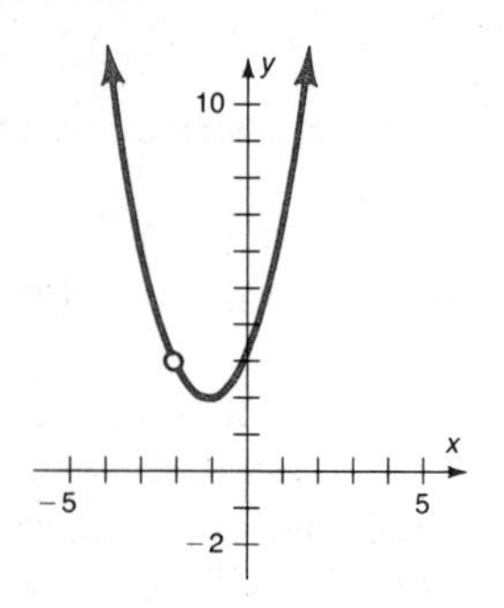

Figure 3.2 Graph of $f(x) = \dfrac{x^3 + 4x^2 + 7x + 6}{x + 2}$

The graph of this curve is the same as for the quadratic function $y = x^2 + 2x + 3$, with the point at $x = -2$ deleted from the domain. To sketch this parabola, complete the square:

$$\begin{aligned} y &= x^2 + 2x + 3 \\ y - 3 &= x^2 + 2x \\ y - 3 + 1 &= x^2 + 2x + 1 \\ y - 2 &= (x + 1)^2 \end{aligned}$$

The vertex is at $(-1, 2)$, and the parabola opens upward. It is drawn with the point at $x = -2$ deleted, as in Figure 3.2. ■

The second situation to give excluded values is that of negative values under radical symbols with an even index. Remember that $\sqrt[n]{b}$ is defined for all b if n is odd, but only for nonnegative b if n is even. Therefore, when sketching an equation with a radical symbol, first look to see if it is a square root, fourth root, or other even-indexed root; if it is, you will find the excluded regions by finding and then stating the domain and range as illustrated by Examples 3–7.

EXAMPLE 3 Find the domain and range for the curve $y = \sqrt{3 - x}$.

Solution There is a square root, so the domain requires that

$$\begin{aligned} 3 - x &\geq 0 \\ 3 &\geq x \end{aligned}$$

Therefore, $\boldsymbol{D = (-\infty, 3]}$.

To find the range, solve for x:

$$x = 3 - y^2 \quad (\text{where } y \geq 0)$$ Since y is equal to a square root, it is nonnegative.

Thus, x is defined for all positive y, so $\boldsymbol{R = [0, \infty)}$. ■

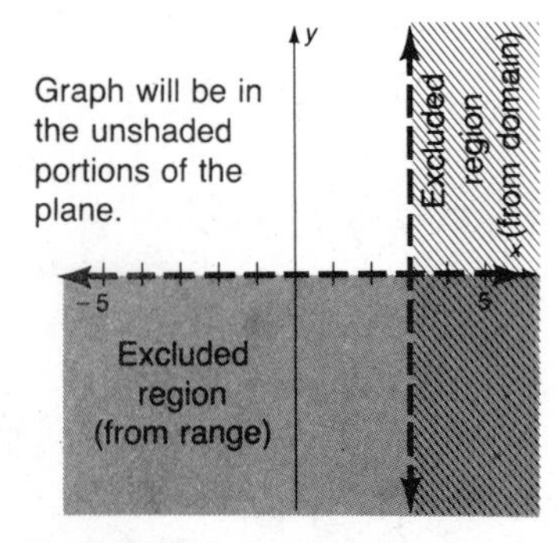

Figure 3.3 Excluded regions for Example 3

As an aid to graphing in Section 3.4, we will use the domain and range to shade in **excluded regions**. We will then focus our attention on the unshaded portions of the plane. For Example 3, we would shade the plane as shown in Figure 3.3. For convenience, you can draw the boundaries of all excluded regions as dashed lines.

EXAMPLE 4 Find the domain and range for the curve $y^2 + x - 3 = 0$, and shade in excluded regions.

Solution To find the domain, solve for y:

$$\begin{aligned} y^2 &= 3 - x \\ y &= \pm\sqrt{3 - x} \end{aligned}$$ Use the square-root property.

At first glance, this example looks just like Example 3, but it is not. The domain is the same, but notice that the range is no longer restricted to positive values. Solve for x:

$$x = 3 - y^2$$

Thus, x is defined for all y, so $\boldsymbol{D = (-\infty, 3]}$ and $\boldsymbol{R = (-\infty, \infty)}$; see Figure 3.4.

Figure 3.4 Excluded region for Example 4. Notice that there are no exclusions for the range.

■

You must be careful when finding both the domain and the range. For example, consider the following variations of curves similar to those in Examples 3 and 4:

Equation	*Domain*	*Range*
$y = \sqrt{3 - x^2}$	Solve $3 - x^2 \geq 0$ to find $D = (-\sqrt{3}, \sqrt{3})$.	Solve $y^2 = 3 - x^2$, where $y \geq 0$, to find $x = \pm\sqrt{3 - y^2}$, so that $R = (0, \sqrt{3})$.
$x^2 + y^2 = 3$	Solve $y = \pm\sqrt{3 - x^2}$ to find $D = (-\sqrt{3}, \sqrt{3})$.	Solve $y^2 = 3 - x^2$ to find $x = \pm\sqrt{3 - y^2}$, so that $R = (-\sqrt{3}, \sqrt{3})$.

EXAMPLE 5 Find the domain and range for the curve $y = \sqrt{x^2 - 2x - 3}$, and shade the excluded regions.

Solution The domain is found by making sure the radicand is nonnegative:

$$x^2 - 2x - 3 \geq 0$$
$$(x + 1)(x - 3) \geq 0$$

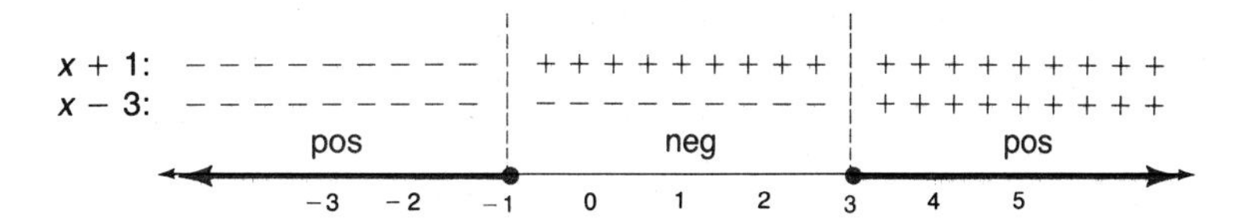

$$D = (-\infty, -1] \cup [3, \infty)$$

For the range, we see that $y \geq 0$ and $y^2 = x^2 - 2x - 3$. Solve for x to find

$$y^2 + 3 = x^2 - 2x$$
$$y^2 + 3 + 1 = x^2 - 2x + 1$$
$$y^2 + 4 = (x - 1)^2$$
$$\pm\sqrt{y^2 + 4} = x - 1$$
$$x = 1 \pm \sqrt{y^2 + 4}$$

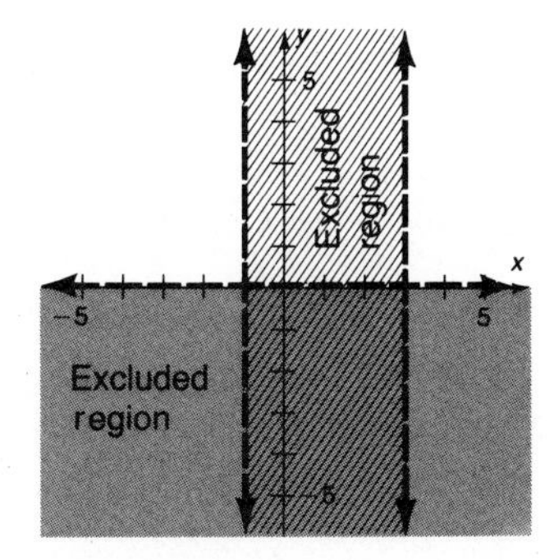

Figure 3.5 Excluded regions for Example 5

We see that x is defined for all values of y, so (since $y \geq 0$) the range is $\boldsymbol{R = [0, \infty)}$. See Figure 3.5. ■

EXAMPLE 6 Find the domain and range for the curve $xy^2 - y^2 - 1 = 0$.

Solution For the domain, solve for y:

$$y^2(x-1) - 1 = 0$$
$$y^2(x-1) = 1$$
$$y^2 = \frac{1}{x-1}$$
$$y = \frac{\pm 1}{\sqrt{x-1}}$$

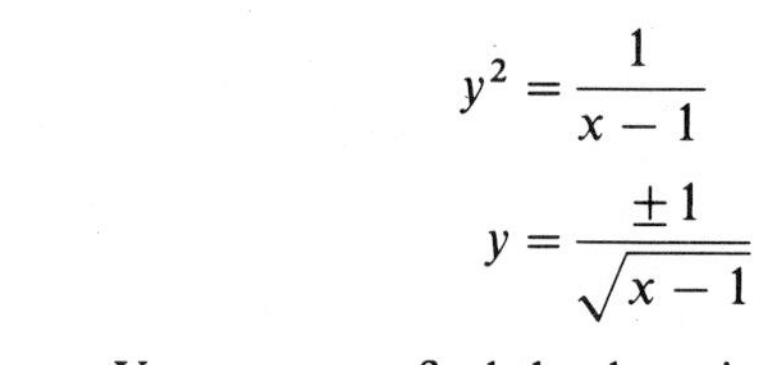

Figure 3.6 Excluded regions for Example 6

You can now find the domain by inspection: $\boldsymbol{D = (1, \infty)}$. See Figure 3.6.

For the range, solve for x:

$$xy^2 = y^2 + 1$$
$$x = \frac{y^2 + 1}{y^2}$$

From this equation, we see that x is real for all y except $y = 0$ (see the dashed line in Figure 3.6). Therefore, $\boldsymbol{R = (-\infty, 0) \cup (0, \infty)}$. ■

EXAMPLE 7 Find the domain and range for the curve $y^2x^2 - x^2 - 4y^2 + 9 = 0$.

Solution For the domain, solve for y (if possible):

$$y^2(x^2 - 4) - x^2 + 9 = 0$$
$$y^2(x^2 - 4) = x^2 - 9$$
$$y^2 = \frac{x^2 - 9}{x^2 - 4}$$

From this equation we see that

$$\frac{x^2 - 9}{x^2 - 4} \geq 0$$
$$\frac{(x-3)(x+3)}{(x-2)(x+2)} \geq 0$$

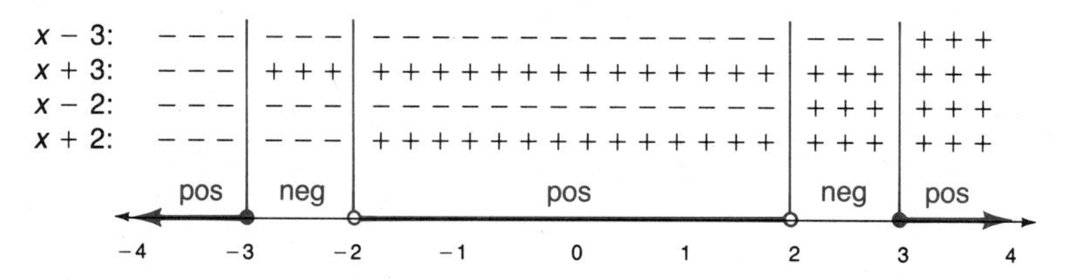

$$\boldsymbol{D = (-\infty, -3] \cup (-2, 2) \cup [3, \infty)}$$

For the range, solve for x:

$$y^2x^2 - x^2 - 4y^2 + 9 = 0$$
$$x^2(y^2 - 1) = 4y^2 - 9$$
$$x^2 = \frac{(2y-3)(2y+3)}{(y-1)(y+1)}$$

Thus,

$$\frac{(2y-3)(2y+3)}{(y-1)(y+1)} \geq 0$$

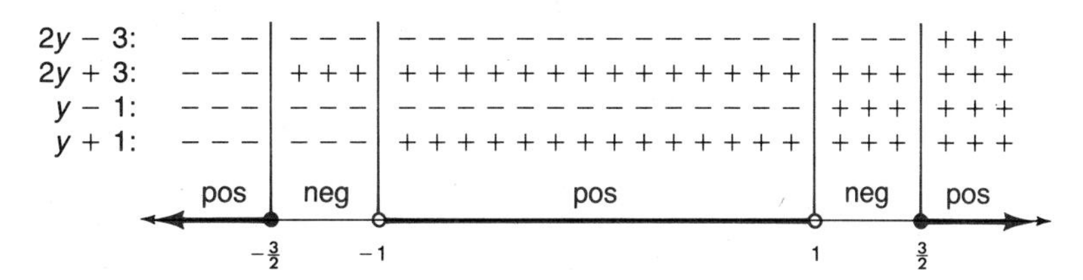

$$R = (-\infty, -\tfrac{3}{2}] \cup (-1, 1) \cup [\tfrac{3}{2}, \infty)$$

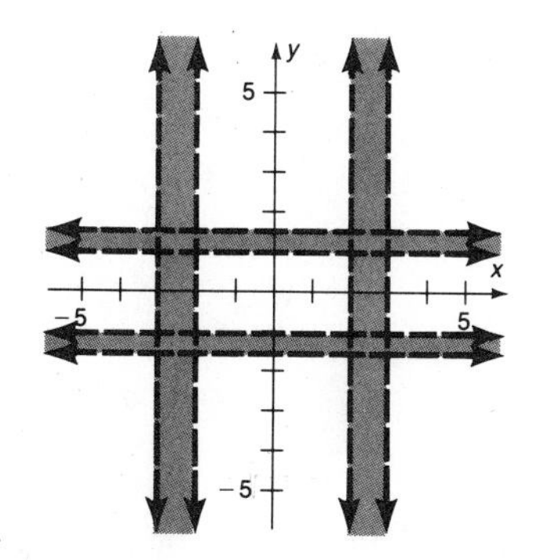

Figure 3.7 Excluded regions for Example 7

The excluded regions are shown in Figure 3.7. Notice in this example that some boundaries for the domain and range are included and others are not. However, since we are now showing only the **excluded regions**, we will agree to show these using only dashed lines. Then, in the next section, we will pay attention to included and excluded boundaries when actually sketching the graph. Remember, we want to use excluded regions as an **aid** to sketching the graph. ■

PROBLEM SET 3.2

A

Graph the curves defined by the equations given in Problems 1–22.

1. $y = \dfrac{(x+1)(x+2)}{x+2}$

2. $y = \dfrac{(x+3)(x-2)}{x-2}$

3. $y = \dfrac{(2x+1)(x+1)}{x+1}$

4. $y = \dfrac{(3x-1)(x-1)}{x-1}$

5. $y = \dfrac{(x+1)(x-2)(x+3)}{(x+1)(x-2)}$

6. $y = \dfrac{(x-3)(x+1)(x+5)}{(x+1)(x+5)}$

7. $y = \dfrac{(x+2)(x-1)(3x+2)}{x^2+x-2}$

8. $y = \dfrac{(x-3)(2x+3)(x+2)}{x^2-x-6}$

9. $y = \dfrac{(x+1)(x+2)(x-1)}{x+2}$

10. $y = \dfrac{(x-1)(x-2)(x+2)}{x-1}$

11. $y = \dfrac{(x-1)(x+2)(x-3)(x+4)}{(x+2)(x-1)}$

12. $y = \dfrac{(x+3)(x-3)(x+1)(2x-3)}{(x-3)(2x-3)}$

13. $y = \dfrac{x^2-x-12}{x+3}$

14. $y = \dfrac{x^2+x-2}{x-1}$

15. $y = \dfrac{x^2-x-6}{x+2}$

16. $y = \dfrac{2x^2-13x+15}{x-5}$

17. $y = \dfrac{6x^2-5x-4}{2x+1}$

18. $y = \dfrac{15x^2+13x-6}{3x-1}$

19. $y = \dfrac{x^3+6x^2+10x+4}{x+2}$

20. $y = \dfrac{x^3+9x^2+15x-9}{x+3}$

21. $y = \dfrac{x^3+12x^2+40x+40}{x+2}$

22. $y = \dfrac{x^3+5x^2+6x}{x+3}$

B

Find the domain and range for the curves defined by the equations in Problems 23–62, and shade the excluded regions.

23. $y = \sqrt{x}$

24. $y = \sqrt{2x}$

25. $y = \sqrt{x-2}$

26. $y = \sqrt{x+2}$

27. $y = \sqrt{2x-1}$

28. $y = \sqrt{2x+10}$

29. $y = x\sqrt{x}$

30. $y = 2x\sqrt{x}$

31. $y = \sqrt{x^3}$

32. $y = x\sqrt{x^3}$

33. $y^2 = x^3$

34. $y^2 = 8x^3$

35. $y^2 = x^2 - 9$

36. $y^2 = x^2 - 4$

37. $y = \sqrt{x^2-4}$

38. $y = \sqrt{x^2-x-6}$

39. $y = \sqrt{x^2+x-12}$

40. $y = \sqrt{x^3-9x}$

41. $y = \sqrt{\dfrac{x^2 - 1}{x - 2}}$
42. $y = \sqrt{\dfrac{x + 3}{x^2 - 4}}$
43. $y = \dfrac{x}{\sqrt{1 - x^2}}$
44. $y = \dfrac{x}{\sqrt{16 - x^2}}$
45. $\dfrac{x^2}{9} + \dfrac{y^2}{16} = 1$
46. $\dfrac{x^2}{9} - \dfrac{y^2}{16} = 1$
47. $(x - 1)^2 + (y + 3)^2 = 9$
48. $(x - 2)^2 + (y + 1)^2 = 4$
49. $y^2 = \dfrac{3}{x^2 - 1}$
50. $y^2 = \dfrac{1}{1 - x^2}$
51. $x^2 + y^2 - 4x + 2y + 1 = 0$
52. $x^2 - y^2 - 4x - 2y - 1 = 0$
53. $y^2 = \dfrac{x^2 + 1}{x}$
54. $y^2 = \dfrac{x^2 + 4}{x}$
55. $x^2 = \dfrac{y}{3 - y}$
56. $x^2 = \dfrac{y - 1}{y + 3}$
57. $x^2y^2 - 4y^2 - x^2 = 0$
58. $x^2y^2 - 4x^2 - y^2 = 0$
59. $x^2y - x - y = 0$
60. $xy^2 - x - y = 0$

C

61. $y^2x^2 - x^2 - 4y^2 + x = 0$
62. $x^2y^2 - 9x^2 - y^2 + y = 0$

3.3 ASYMPTOTES

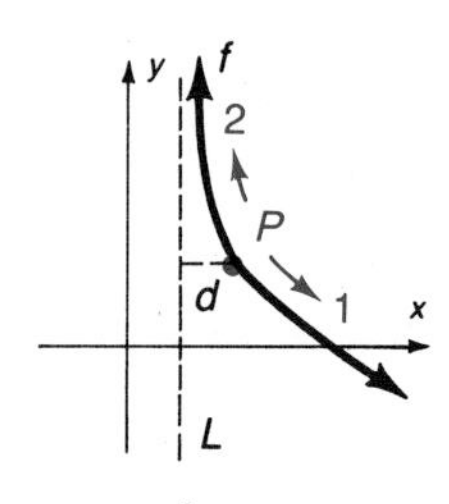

a.

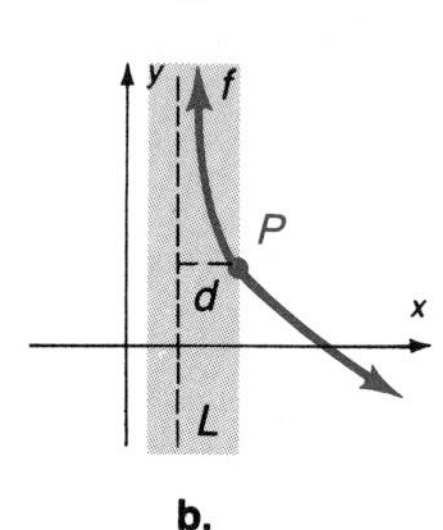

b.

Figure 3.8 An asymptote to a curve

As an additional aid to sketching certain functions, consider the notion of an asymptote. An **asymptote** is a line having the property that the distance from a point P on the curve to the line approaches zero as the distance from P to the origin increases without bound, and P is *on a suitable part of the curve.* This last phrase (in italics) is best illustrated by considering Figure 3.8, where L is an asymptote for the function f. Consider P and d, the distance from P to the line L, as shown in Figure 3.8a. Now the distance from P to the origin can increase in two ways, depending on whether P moves along the curve in direction 1 or direction 2. In direction 1 the distance d increases without bound, but in direction 2 the distance d approaches zero. Thus, if you consider the portion of the curve in the shaded region of Figure 3.8b, you see that the conditions of the definition of an asymptote apply. There are three types of asymptotes that occur frequently enough when sketching curves to merit consideration. These are **vertical**, **horizontal**, and **slant asymptotes**. An example of each is shown in Figure 3.9. Notice that a curve may pass through a horizontal or slant asymptote.

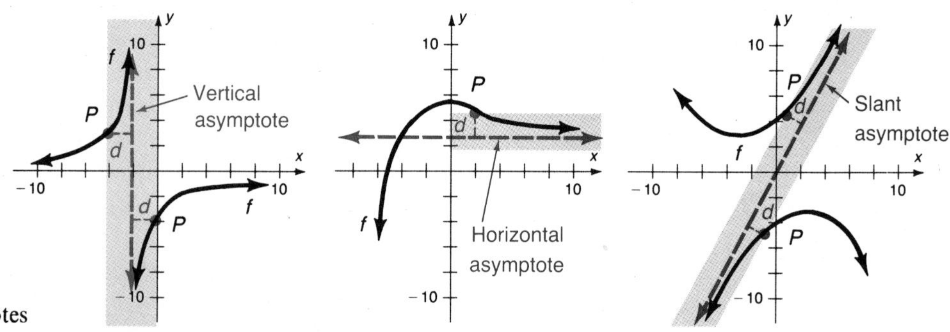

Figure 3.9 Asymptotes

The easiest asymptotes to find are the vertical asymptotes of a function. If $f(x) = P(x)/D(x)$, where $P(x)$ and $D(x)$ have no common factors, then $|f(x)|$ must get large as x gets close to any value for which $D(x) = 0$ and $P(x) \neq 0$. This means that **if r is a root of $D(x) = 0$, then $x = r$ is the equation of a vertical asymptote.**

EXAMPLE 1 Let $y = \dfrac{1}{x-3}$. Find any vertical asymptotes.

Solution Notice that $x - 3 = 0$ if $\boldsymbol{x = 3}$, which is the equation of a vertical line. This line is a vertical asymptote. ■

EXAMPLE 2 Let $y = \dfrac{2x^2 - 3x + 5}{x^2 - x - 2}$. Find any vertical asymptotes.

Solution There are no common factors in the numerator and denominator of f, so solve

$$\begin{aligned} x^2 - x - 2 &= 0 \\ (x-2)(x+1) &= 0 \\ x &= 2 \text{ or } -1 \end{aligned}$$

The vertical asymptotes are given by $\boldsymbol{x = 2}$ and $\boldsymbol{x = -1}$ since neither of these values also causes the numerator to be zero. ■

EXAMPLE 3 Let $y = \dfrac{x^2 + 2x - 8}{x^2 - 4}$. Find any vertical asymptotes.

Solution Be careful with functions defined by equations like this one. If you do not notice the common factor, you might be led to the incorrect conclusion that $x = 2$ and $x = -2$ are the equations of the vertical asymptotes. Instead, factor *both* numerator and denominator as shown:

$$\begin{aligned} f(x) &= \frac{x^2 + 2x - 8}{x^2 - 4} \\ &= \frac{(x-2)(x+4)}{(x-2)(x+2)} \\ &= \frac{x+4}{x+2} \qquad (x \neq 2) \end{aligned}$$

Do you see why we wrote $x \neq 2$? You might think it should be $x \neq -2$.
WARNING: $x \neq -2$ is implied by the statement

$$\frac{x+4}{x+2}$$

because it would cause division by zero. The condition $x \neq 2$ is not implied, but it is necessary because of the *previous* step (before reducing).

From the simplified form

$$\frac{x+4}{x+2}$$

you can see that $\boldsymbol{x = -2}$ is the equation of the vertical asymptote. ■

If the equation can be solved easily for x so that

$$x = \frac{p(y)}{d(y)}$$

where $p(y)$ and $d(y)$ have no common factors, and if r is a root of $d(y) = 0$, then the line with the equation $y = r$ is a horizontal asymptote. Unfortunately, it is not always easy to solve for x, so we need to set some preliminary groundwork to find horizontal asymptotes.

To discuss horizontal asymptotes, we need to investigate the behavior of the curve as we permit x to become large (either positively or negatively). We symbolize this by writing $|x| \to \infty$. If we want to look at x becoming large positively, we write $x \to \infty$, and for x becoming large negatively we write $x \to -\infty$. There are three possibilities for what happens to y:

1. y approaches zero (written $y \to 0$), in which case there is a horizontal asymptote $y = 0$.
2. y approaches a nonzero number b (written $y \to b$), in which case there is a horizontal asymptote $y = b$.
3. y becomes numerically large (written $y \to \infty$ or $y \to -\infty$), in which case there is no horizontal asymptote.

EXAMPLE 4 Let $y = 1/x$. Find the horizontal asymptote.

Solution We need to find the value of $1/x$ as x increases without bound. This is written as

$$\lim_{|x| \to \infty} \frac{1}{x}$$

Consider the following table of values for $1/x$ as x increases without bound:

x	$1/x$	x	$1/x$
1	1	-1	-1
2	0.5	-2	-0.5
10	0.1	-10	-0.1
100	0.01	-100	-0.01
1,000	0.001	$-1,000$	-0.001
10,000	0.0001	$-10,000$	-0.0001
$x \to \infty$	$1/x \to 0$	$x \to -\infty$	$1/x \to 0$

We say that $1/x \to 0$ as $|x|$ increases without bound and symbolize this by

$$\frac{1}{x} \to 0 \qquad \text{as } |x| \to \infty$$

Thus, the horizontal asymptote is $\mathbf{y = 0}$. This example illustrates the first possibility for horizontal asymptotes. ■

The table and results of Example 4 lead to a useful theorem about limits that can help us find horizontal asymptotes.

Limit Theorem

> Let x be any real number, k any constant, and n a natural number. Then as $|x| \to \infty$:
>
> 1. $\frac{1}{x} \to 0$ 2. $\frac{1}{x^n} \to 0$ 3. $\frac{k}{x^n} \to 0$

EXAMPLE 5 Find any horizontal asymptotes for the curve $y = \frac{x}{2x+1}$.

Solution We need to consider

$$\lim_{|x| \to \infty} \frac{x}{2x+1}$$

We could set up a table of values, but instead, multiply the rational expression by 1, written as $\frac{1/x}{1/x}$, and use the limit theorem:

$$\frac{x}{2x+1} \cdot \frac{1/x}{1/x} = \frac{1}{2 + \frac{1}{x}}$$

Since $1/x \to 0$ as $x \to \infty$, you can see that

$$\frac{1}{2 + \frac{1}{x}} \to \frac{1}{2+0} \quad \text{as } |x| \to \infty$$

Thus,

$$\frac{x}{2x+1} \to \frac{1}{2} \quad \text{as } |x| \to \infty$$

The horizontal asymptote is $\boldsymbol{y = \frac{1}{2}}$. ■

EXAMPLE 6 Find any horizontal asymptotes for the curve $y = \frac{3x^2 - 7x + 2}{7x^2 + 2x + 5}$.

Solution Notice that the largest power of x in the expression is x^2, so multiply the numerator and denominator by $1/x^2$:

$$\frac{3x^2 - 7x + 2}{7x^2 + 2x + 5} = \frac{3x^2 - 7x + 2}{7x^2 + 2x + 5} \cdot \frac{1/x^2}{1/x^2} = \frac{3 - \frac{7}{x} + \frac{2}{x^2}}{7 + \frac{2}{x} + \frac{5}{x^2}}$$

Since k/x and k/x^2 both approach zero as x increases without bound,

$$\frac{3 - \frac{7}{x} + \frac{2}{x^2}}{7 + \frac{2}{x} + \frac{5}{x^2}} \to \frac{3 - 0 + 0}{7 + 0 + 0} = \frac{3}{7} \qquad \text{as } |x| \to \infty$$

The horizontal asymptote is $y = \frac{3}{7}$. ■

Examples 5 and 6 illustrate the second possibility listed on page 102 for horizontal asymptotes.

EXAMPLE 7 Find any horizontal asymptotes for the curve

$$y = \frac{6x^3 + 7x^2 - 5}{3x^2 + 2}$$

Solution The highest power of x is x^3, so multiply the numerator and denominator by $1/x^3$:

$$\frac{6x^3 + 7x^2 - 5}{3x^2 + 2} \cdot \frac{1/x^3}{1/x^3} = \frac{6 + \frac{7}{x} - \frac{5}{x^3}}{\frac{3}{x} + \frac{2}{x^3}}$$

As $|x| \to \infty$ the denominator approaches zero, so y gets numerically large. Thus, y does not approach a constant, and there is no horizontal asymptote. This example illustrates the third possibility for horizontal asymptotes. ■

The last type of asymptote we will consider is a slant asymptote. Suppose the degree of $P(x)$ is one more than the degree of $D(x)$. Then

$$f(x) = \frac{P(x)}{D(x)} = mx + b + \frac{R(x)}{D(x)}$$

where the degree of $R(x)$ is less than the degree of $D(x)$. Then

$$\lim_{|x| \to \infty} \frac{R(x)}{D(x)} = 0$$

which means that for large values of $|x|$, $f(x)$ is near the line given by $y = mx + b$. This says that the line with the equation $y = mx + b$ is a slant asymptote for the curve given by $y = f(x)$.

We can now summarize the finding of asymptotes, as given in the box at the top of the next page.

Procedure for Finding Asymptotes

Let

$$f(x) = \frac{P(x)}{D(x)}$$

where $P(x)$ and $D(x)$ are polynomial functions and $P(x)$ and $D(x)$ have no common factors. Let $P(x)$ have degree m with leading coefficient p, and let $D(x)$ have degree n with leading coefficient d. Then:

Vertical asymptote: $x = r$ where $D(r) = 0$

Horizontal asymptote: $y = 0$ if $m < n$

$y = \frac{p}{d}$ if $m = n$

None if $m > n$

Slant asymptote: $y = mx + b$ if $m = n + 1$ and

$f(x) = mx + b + \frac{R(x)}{D(x)}$ (by long division)

EXAMPLE 8 Find the slant asymptote for $y = \dfrac{3x^3 - 2x^2 + x - 5}{x^2 + 3}$.

Solution Divide:

$$\begin{array}{r|l} & 3x - 2 \\ \hline x^2 + 3 & 3x^3 - 2x^2 + x - 5 \\ & 3x^3 \qquad\quad + 9x \\ \hline & -2x^2 - 8x - 5 \\ & -2x^2 \qquad\quad - 6 \\ \hline & -8x + 1 \end{array}$$

Thus,

$$y = 3x - 2 + \frac{-8x + 1}{x^2 + 3}$$

and the slant asymptote is given by $\mathbf{y = 3x - 2}$. ■

The last two examples of this section illustrate how you can find asymptotes for a curve whose equation cannot be written as a quotient of polynomials.

EXAMPLE 9 Find all asymptotes for $y = \dfrac{x}{\sqrt{x^2 - 4}}$.

Solution 1. **Vertical:** Determine whether the denominator can be zero:

$$x^2 - 4 = 0$$
$$(x - 2)(x + 2) = 0$$
$$x = 2, -2$$

The vertical asymptotes are $\mathbf{x = 2}$ and $\mathbf{x = -2}$.

WARNING: If the denominator had been $\sqrt{x^2 + 4}$, there would have been no vertical asymptotes because $x^2 + 4 \neq 0$ for all real values of x.

2. **Horizontal:** Find $\lim_{x \to \infty} \dfrac{x}{\sqrt{x^2 - 4}}$:

$$\frac{x}{\sqrt{x^2 - 4}} = \frac{x}{\sqrt{x^2\left(1 - \frac{4}{x^2}\right)}} = \frac{x}{|x|\sqrt{1 - \frac{4}{x^2}}}$$

As $x \to \infty$, $|x| = x$, and

$$\frac{x}{|x|\sqrt{1 - \frac{4}{x^2}}} = \frac{1}{\sqrt{1 - \frac{4}{x^2}}} \to 1$$

Thus, $y \to 1$ as $x \to \infty$, and the line $\mathbf{y = 1}$ is a horizontal asymptote.
As $x \to -\infty$, $|x| = -x$, and

$$\frac{x}{|x|\sqrt{1 - \frac{4}{x^2}}} = \frac{-1}{\sqrt{1 - \frac{4}{x^2}}} \to -1$$

Thus, $y \to -1$ as $x \to -\infty$, and the line $\mathbf{y = -1}$ is also a horizontal asymptote.
WARNING: If the denominator had been $\sqrt{4 - x^2}$, there would have been no horizontal asymptote because the domain is

$$4 - x^2 \geq 0$$
$$(2 - x)(2 + x) \geq 0$$

$2 - x$:	+ + + + + + +	+ + + + + + + + + + + + + +	− − − − − − −
$2 + x$:	− − − − − − −	+ + + + + + + + + + + + + +	+ + + + + + +
	neg	pos	neg

−4 −3 −2 −1 0 1 2 3 4

$$D = [-2, 2]$$

It is impossible for $|x| \to \infty$ and still remain in the domain $[-2, 2]$.

3. **Slant:** None ■

EXAMPLE 10 Find all asymptotes for $4x^2 - 9y^2 = 36$.

Solution Solve for y:

$$9y^2 = 4x^2 - 36$$
$$y^2 = \frac{4}{9}(x^2 - 9)$$
$$|y| = \sqrt{\frac{4}{9}(x^2 - 9)}$$
$$y = \pm\frac{2}{3}\sqrt{x^2 - 9}$$

1. **Vertical:** There are no values of x that cause division by zero. There are no vertical asymptotes.
2. **Horizontal:** As $|x| \to \infty$, y increases without bound, so there are no horizontal asymptotes.
3. **Slant:** As $|x| \to \infty$, $\sqrt{x^2 - 9} \approx \sqrt{x^2} = |x|$. Thus, $y \to \pm\frac{2}{3}x$ as $x \to \infty$. The slant asymptotes are $\boldsymbol{y = \frac{2}{3}x}$ and $\boldsymbol{y = -\frac{2}{3}x}$, or (in standard form) $\boldsymbol{2x - 3y = 0}$ and $\boldsymbol{2x + 3y = 0}$. ■

PROBLEM SET 3.3

A

Find the limits in Problems 1–30.

1. $\lim\limits_{x\to\infty} x^3$
2. $\lim\limits_{x\to\infty} (x^2 - 4)$
3. $\lim\limits_{x\to\infty} \dfrac{1}{x - 3}$
4. $\lim\limits_{x\to\infty} \dfrac{1}{x - 3}$
5. $\lim\limits_{x\to\infty} \dfrac{x}{2x + 1}$
6. $\lim\limits_{x\to\infty} \dfrac{1}{x^2 + 1}$
7. $\lim\limits_{x\to\infty} 2x$
8. $\lim\limits_{x\to\infty} (3x - 4)$
9. $\lim\limits_{x\to\infty} \dfrac{12x - 3}{x - 2}$
10. $\lim\limits_{x\to\infty} \dfrac{x^2 + 3x - 10}{x^2 - 2}$
11. $\lim\limits_{x\to\infty} \dfrac{x^2 - 8x + 15}{x - 3}$
12. $\lim\limits_{x\to\infty} \dfrac{x^2 + 3x - 10}{2x^2 + 5}$
13. $\lim\limits_{x\to\infty} \dfrac{x^2 - 1}{x - 2}$
14. $\lim\limits_{x\to\infty} \dfrac{2x^2 - 5x - 12}{x - 4}$
15. $\lim\limits_{x\to\infty} \dfrac{x^2 - 8}{x^3 + 2x + 4}$
16. $\lim\limits_{x\to\infty} \dfrac{x^2 + 2x + 4}{x^2 - 8}$
17. $\lim\limits_{x\to\infty} \dfrac{x^3 + 2}{4x^3 + 5}$
18. $\lim\limits_{x\to\infty} \dfrac{6 - x}{2x - 15}$
19. $\lim\limits_{|x|\to\infty} \dfrac{2x^2 - 5x - 3}{x^2 - 9}$
20. $\lim\limits_{|x|\to\infty} \dfrac{3x - 1}{2x + 3}$
21. $\lim\limits_{|x|\to\infty} \dfrac{x^2 + 6x + 9}{x + 3}$
22. $\lim\limits_{|x|\to\infty} \dfrac{6x^2 - 5x + 2}{2x^2 + 5x + 1}$
23. $\lim\limits_{x\to\infty} \dfrac{5x + 10{,}000}{x - 1}$
24. $\lim\limits_{x\to-\infty} \dfrac{4x + 10^5}{x + 1}$
25. $\lim\limits_{x\to\infty} \left(x + 2 + \dfrac{3}{x - 1}\right)$
26. $\lim\limits_{x\to-\infty} \left(2x - 3 + \dfrac{4}{x + 2}\right)$
27. $\lim\limits_{x\to-\infty} \dfrac{4x^4 - 3x^3 + 2x + 1}{3x^4 - 9}$
28. $\lim\limits_{x\to\infty} \dfrac{x^4 + 1}{x^2 - 1}$
29. $\lim\limits_{x\to\infty} \dfrac{3x^3 - 2x^2 + 1}{5x^3 + 3x - 100}$
30. $\lim\limits_{x\to\infty} \dfrac{x^2 + x + 1}{x^3 - 1}$

B

Find the horizontal, vertical, and slant asymptotes, if any exist, for the functions given in Problems 31–60.

31. $y = \dfrac{1}{x}$
32. $y = \dfrac{1}{x} + 2$
33. $y = -\dfrac{1}{x} + 1$
34. $y = \dfrac{4}{x^2}$
35. $y = \dfrac{2x^2 + 2}{x^2}$
36. $y = \dfrac{1}{x - 4}$
37. $y = \dfrac{-1}{x + 3}$
38. $y = \dfrac{4x}{x^2 - 2}$
39. $y = \dfrac{x^2}{x - 4}$
40. $y = \dfrac{x^3}{(x - 1)^2}$
41. $y = \dfrac{-x^2}{x - 1}$
42. $y = \dfrac{x^2 + x - 2}{x - 1}$
43. $y = \dfrac{x}{x^2 + x - 6}$
44. $y = \dfrac{-x}{x^2 + x - 6}$
45. $y = \dfrac{x^2}{x^3 - x^2 - 20x}$
46. $y = \dfrac{x^2}{20x - x^2 - x^3}$
47. $y = \dfrac{x^2 + x - 6}{x + 3}$
48. $y = \dfrac{x^2 + 3x - 2}{x^2 + 2x - 8}$
49. $y = \dfrac{x^3 - 2x^2 + x - 2}{(x - 2)(x^2 + 1)}$
50. $y = \dfrac{(x - 3)(x^2 + 1)}{x^3 + 3x^2 + x + 3}$
51. $y = \dfrac{(15x^2 + 13x - 6)(x - 1)}{3x^2 - 4x + 1}$

C

52. $y = \dfrac{2x^3 - 3x^2 - 32x - 15}{x^2 - 2x - 15}$
53. $y = \dfrac{3x^3 + 5x^2 - 26x + 8}{x^2 + 2x - 8}$
54. $y = \dfrac{x^3 + 6x^2 + 10x + 4}{x + 2}$
55. $y = \dfrac{x^3 + 9x^2 + 15x - 9}{x + 3}$
56. $y = \dfrac{x^3 + 12x^2 + 40x + 40}{x + 2}$
57. $y = \dfrac{x^3 + 5x^2 + 6x}{x + 3}$
58. $y = \dfrac{x^2}{x^3 - x^2 - 20x}$
59. $y = \dfrac{x^2 + 3x - 2}{x^2 + 2x - 8}$
60. $y = \dfrac{x^2}{20x - x^2 - x^3}$

3.4 CURVE SKETCHING

In this chapter we have discussed plotting points, intercepts, symmetry, domain, range, excluded points and regions, and asymptotes. You are now ready to use these concepts to sketch a graph. This section consists of several examples of graphing curves that are functions as well as some that are not functions. But first, consider the following ideas, which you should keep in mind as you work through the examples:

1. In most advanced work, we are interested in sketching a graph in the quickest, most efficient way. Usually, we need to know only a curve's general shape and location.
2. Sometimes the effort in finding some information about a curve is greater than the effort necessary to plot some points. You should exercise common sense in knowing when to abandon the effort of finding out a particular bit of information about a curve.
3. Being able to classify a curve by inspection of the equation is extremely important. You can now do this for lines and parabolas, but later in this book we will consider general characteristics of the graphs of polynomial, rational, trigonometric, exponential, and logarithmic functions, as well as various types of polar forms. The more you know about the general characteristics of a curve, the less work you will need to do in sketching that curve.
4. In later courses, you will add even more types of curves to your general knowledge. The most useful information about curve sketching will be discussed in a calculus course.
5. When in doubt about what a curve looks like in a certain region, you can substitute values and plot some points. In addition to knowing the intercepts, additional points that are often useful are the endpoints of the domain or range and places where a curve passes through an asymptote. Curves may intersect horizontal and slant asymptotes, but may never intersect vertical asymptotes.

In the examples that follow, we will discuss the steps that you would do as well as show you the graph, but limitations of space will not permit all the algebraic details to be shown. You should read these examples with a paper and pencil at hand so you can fill in these details.

EXAMPLE 1 Discuss and sketch $y = \dfrac{x^2 - 1}{x^2 - 4}$.

Solution **Domain:** Values that cause division by zero are excluded, so we need to determine when $x^2 - 4 = 0$. We find that $x^2 - 4 = 0$ when $x = 2, -2$. Thus, $D = (-\infty, -2) \cup (-2, 2) \cup (2, \infty)$.

Range: Solve for x:

$$x^2 = \frac{4y - 1}{y - 1}$$

$$x = \pm\sqrt{\frac{4y - 1}{y - 1}}$$

Graphing is done in stages, which is difficult to show in a textbook. I have attempted to show you these steps in the margins alongside the examples of this section. Note also that we show the boundaries of all excluded regions as dashed lines for convenience.

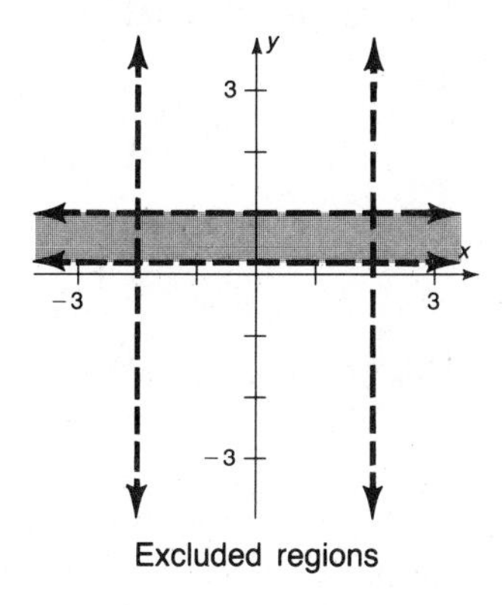

Excluded regions

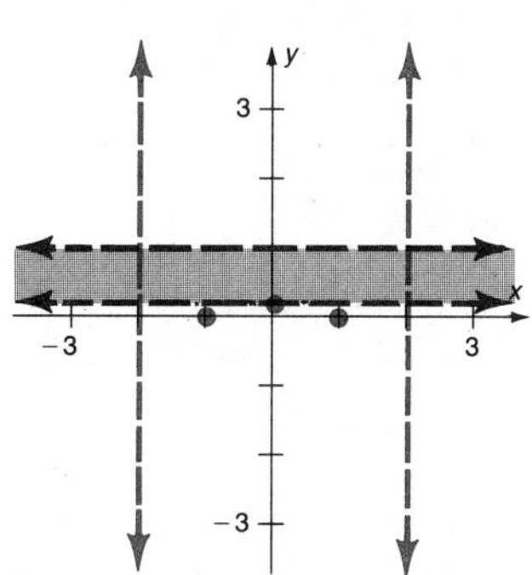

Asymptotes and intercepts

We need to find out when

$$\frac{4y-1}{y-1} \geq 0$$

We find critical values $y = \frac{1}{4}, 1$.

$4y-1$:	− − − − − − − − −	+ + + + + + + + +	+ + + + + + + + +
$y-1$:	− − − − − − − − −	− − − − − − − − −	+ + + + + + + + +
	pos	neg	pos

(critical points on the number line: $\frac{1}{4}$, 1)

Thus, $R = (-\infty, \frac{1}{4}] \cup (1, \infty)$.

Symmetry: It is symmetric with respect to the y axis only.

Asymptotes: $x = 2$, $x = -2$, and $y = 1$

Plotting points:

x intercepts: Solve $\frac{x^2-1}{x^2-4} = 0$ to obtain $x = 1, -1$.
Plot the points $(1, 0)$ and $(-1, 0)$.

y intercepts: Substitute $x = 0$ to obtain $y = \frac{1}{4}$.
Plot the point $(0, \frac{1}{4})$.

Asymptote intercepts: $x = 2$, $x = -2$ are excluded values.
Substitute $y = 1$:

$$1 = \frac{x^2-1}{x^2-4}$$

Solving this equation, you get $-4 = -1$, which tells you the graph does not cross this asymptote.

Additional points: Now look at the information you have and plot some additional points using symmetry. The graph is shown in Figure 3.10. Plot as many points as you feel you need to be confident of your graph, but you do not need to plot many. Notice that this curve is a function, so each x value is associated with exactly one y value. (It will pass the vertical line test.)

To complete the graph shown in Figure 3.10, only one additional point was calculated: If $x = 3$, then calculate $y = \frac{8}{5}$.

By symmetry — Plotted point

Figure 3.10 Graph of $y = \dfrac{x^2-1}{x^2-4}$

■

EXAMPLE 2 Discuss and sketch $x^2 + x^2y^2 - 4y^2 - 1 = 0$.

Solution This curve is not a function.

Domain: Solve for y:

$$y^2 = \frac{1 - x^2}{x^2 - 4} \qquad (x \neq 2, -2)$$

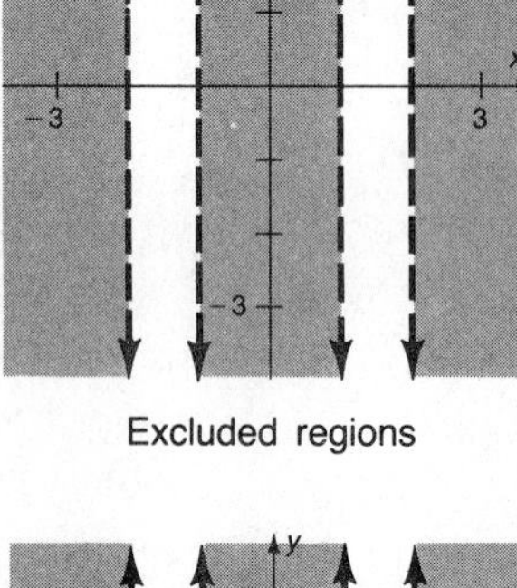
Excluded regions

Also, $\dfrac{1 - x^2}{x^2 - 4} \geq 0$ since y^2 is nonnegative. Solving this inequality, you will find the domain to be $D = (-2, -1] \cup [1, 2)$.

Range: Solve for x:

$$x^2 = \frac{4y^2 + 1}{1 + y^2}$$

Also, $\dfrac{4y^2 + 1}{1 + y^2} \geq 0$ since x^2 is nonnegative. This inequality is true for all values of y, so $R = (-\infty, \infty)$.

Asymptotes and intercepts

Symmetry: The graph is symmetric with respect to the x axis, y axis, and origin.

Asymptotes: $x = 2$, $x = -2$; notice that as $x \to \infty$, we have $y^2 \to -1$, which is impossible (a square number approaching a negative number), so there are no horizontal asymptotes; no slant asymptotes.

Plotting points:

x intercepts: Solve $0 = \dfrac{1 - x^2}{x^2 - 4}$ to obtain $x = 1, -1$.
Plot the points $(1, 0)$ and $(-1, 0)$.

y intercepts: Solve $0 = \dfrac{4y^2 + 1}{1 + y^2}$ (no solution).
The curve does not cross the y axis.

Asymptote intercepts: $x = 2$, $x = -2$ are excluded values, so the curve does not cross its asymptotes.

Additional points: If $x = 1.5$, $y \approx 0.71$; plot $(1.5, 0.7)$ as well as three other symmetric points.
If $x = 1.3$, $y \approx 0.3$; plot $(1.3, 0.3)$ as well as three other symmetric points.

The completed graph is shown in Figure 3.11.

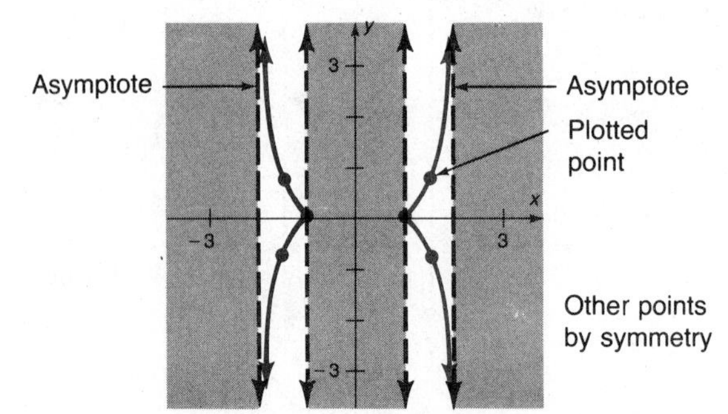

Figure 3.11 Graph of $x^2 + x^2y^2 - 4y^2 - 1 = 0$

■

EXAMPLE 3 Discuss and sketch $2|x| - |3y| = 9$.

Solution This curve is not a function.

Domain: Solve for y:

$$|3y| = 2|x| - 9$$

This requires that $2|x| - 9 \geq 0$. Solving,

$$|x| \geq \tfrac{9}{2}$$
$$x \geq \tfrac{9}{2} \text{ or } x \leq -\tfrac{9}{2}$$

$$D = (-\infty, -\tfrac{9}{2}] \cup [\tfrac{9}{2}, \infty)$$

Range: Solve for x:

$$|x| = \frac{9 + |3y|}{2}$$

This requires that $\dfrac{9 + |3y|}{2} \geq 0$. Solving,

$$|3y| \geq -9$$

This is true for all values of y, so $R = (-\infty, \infty)$.

Symmetry: The graph is symmetric with respect to the x axis, y axis, and origin.

Asymptotes: None

Plotting points:

x intercepts: If $y = 0$, then $2|x| = 9$, so $x = \pm\frac{9}{2}$.

y intercepts: If $x = 0$, then $|3y| = -9$ (no values).

Additional points: Use your knowledge of lines:

If $x \geq 0$ and $y \geq 0$, sketch $2x - 3y = 9$.
If $x \geq 0$ and $y < 0$, sketch $2x + 3y = 9$.
If $x < 0$ and $y \geq 0$, sketch $-2x - 3y = 9$.
If $x < 0$ and $y < 0$, sketch $-2x + 3y = 9$.

The graph is shown in Figure 3.12.

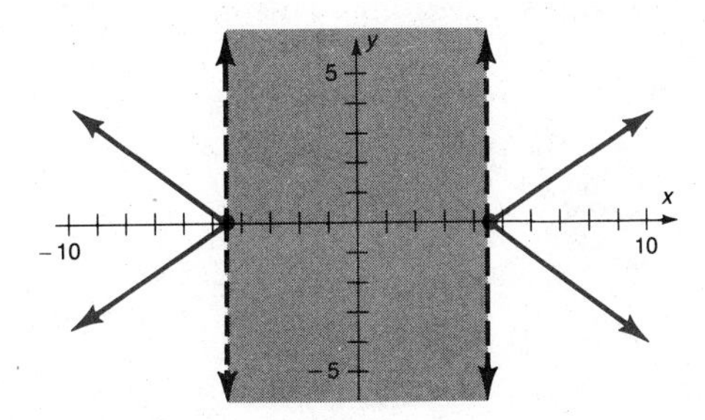

Figure 3.12 Graph of $2|x| - |3y| = 9$

■

EXAMPLE 4 Discuss and sketch $\sqrt{x} + \sqrt{y} = 2$.

Solution This curve is a function.

Domain: Solve for y:

$$\sqrt{y} = 2 - \sqrt{x}$$

$x \geq 0$ and $2 - \sqrt{x} \geq 0$ or $x \leq 4$, so $D = [0, 4]$.

Range: Same analysis as for the domain; $R = [0, 4]$.

Symmetry: The curve is not symmetric with respect to the x axis, the y axis, or the origin.

Asymptotes: None

Plotting points:

x intercepts: If $y = 0$, then $x = 4$; plot $(4, 0)$.

y intercepts: If $x = 0$, then $y = 4$; plot $(0, 4)$.

Additional points: If $x = 1$, then $y = 1$.
If $x = 2$, then $y \approx 0.34$.
If $x = 3$, then $y \approx 0.07$.

The graph is shown in Figure 3.13.

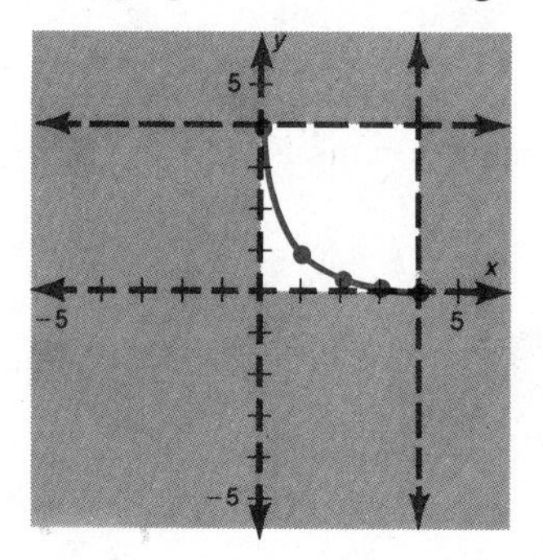

Figure 3.13 Graph of $\sqrt{x} + \sqrt{y} = 2$

■

EXAMPLE 5 Discuss and graph $y = \dfrac{2x^2 - 3x + 5}{x^2 - x - 2}$.

Solution This curve is a function, so it will pass the vertical line test.

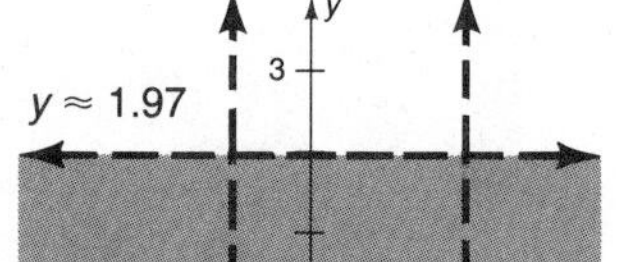

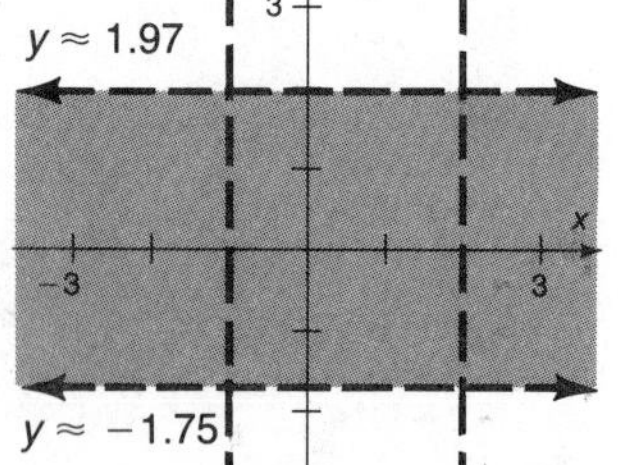

Excluded regions

Domain: $x^2 - x - 2 \neq 0$. Solving the equation $x^2 - x - 2 = 0$, you obtain $x = 2$, $x = -1$. Thus, $D = (-\infty, -1) \cup (-1, 2) \cup (2, \infty)$.

Range: Solve for x (you may wish to skip this step because of the amount of work involved). Multiply both sides by $x^2 - x - 2$ and simplify to obtain

$$(2 - y)x^2 + (y - 3)x + (2y + 5) = 0$$

Use the quadratic equation to solve for x (if $y \neq 2$)

$$x = \frac{-y + 3 \pm \sqrt{(y - 3)^2 - 4(2 - y)(2y + 5)}}{2(2 - y)}$$

From this we see that $(y - 3)^2 - 4(2 - y)(2y + 5) \geq 0$. Solving *this* inequality, you first find the critical values:

$$(y-3)^2 - 4(2-y)(2y+5) = 0$$

$$9y^2 - 2y - 31 = 0 \qquad \text{Expand and simplify.}$$

$$y = \frac{1 \pm \sqrt{280}}{9} \qquad \text{By the quadratic formula}$$

$$\approx -1.75, 1.97$$

Finally, conclude that the range is

$$R = \left(-\infty, \frac{1-\sqrt{280}}{9}\right] \cup \left[\frac{1+\sqrt{280}}{9}, \infty\right)$$

Symmetry: Not symmetric with respect to the x axis, the y axis, or the origin.

Asymptotes: $x = 2$, $x = -1$, and $y = 2$

Plotting points:

x intercepts: If $y = 0$, then $0 = \dfrac{2x^2 - 3x + 5}{x^2 - x - 2}$, which has no real roots; there are no x intercepts.

y intercepts: If $x = 0$, then $y = -\frac{5}{2}$. Plot the y intercept, $(0, -\frac{5}{2})$.

Asymptote intercepts: $x = 2$ and $x = -1$ are excluded values. If $y = 2$ (notice that we considered the possibility $y \neq 2$ above), then

$$2 = \frac{2x^2 - 3x + 5}{x^2 - x - 2}$$

$$2x^2 - 2x - 4 = 2x^2 - 3x + 5 \qquad (x \neq 2, -1)$$

$$-2x - 4 = -3x + 5$$

$$x = 9$$

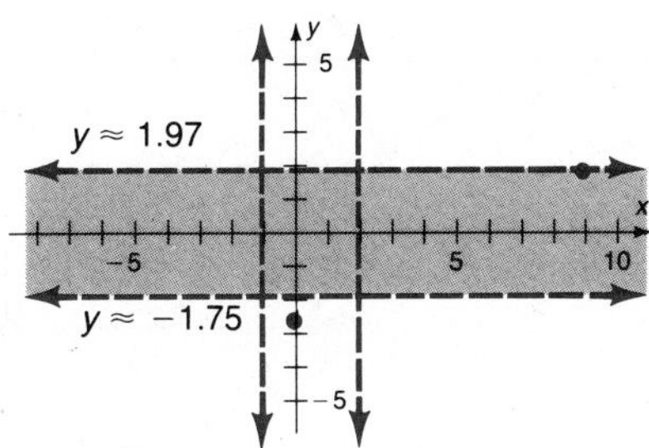

Asymptotes and intercepts

Note: The asymptote is the line $y = 2$, but the excluded region is bounded by the line $y = (1 + \sqrt{280})/9 \approx 1.97$. You cannot see both of these lines (nor should you try to draw them both in your work).

The curve passes through $(9, 2)$. That is, $(9, 2)$ is on the curve and is also on the horizontal asymptote.

Additional points: If $x = -4$, then $y \approx 2.72$.
If $x = -2$, then $y \approx 4.75$.
If $x = 0.5$, then $y \approx -1.8$.
If $x = 1$, then $y = -2$.
If $x = 1.5$, then $y = -4$.
If $x = 3$, then $y \approx 3.5$.
If $x = 4$, then $y \approx 2.5$.
If $x = 12$, then $y \approx 1.98$.

The graph is shown in Figure 3.14.

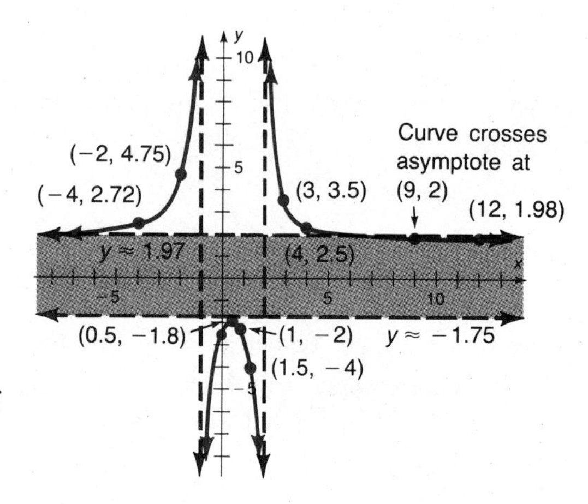

Figure 3.14 Graph of $y = \dfrac{2x^2 - 3x + 5}{x^2 - x - 2}$

■

EXAMPLE 6 Discuss and sketch $y = \sqrt{x(x^2 - 4)}$.

Solution This curve is a function, so it will pass the vertical line test.

Domain:
$$x(x^2 - 4) \geq 0$$
$$x(x - 2)(x + 2) \geq 0$$

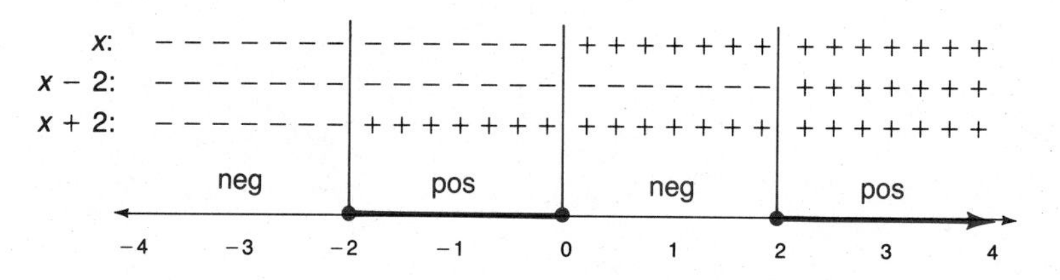

$$D = [-2, 0] \cup [2, \infty)$$

Range: $y \geq 0$ (by inspection; it is too difficult to solve for x).

Symmetry: None

Asymptotes: None

Plotting points:

x intercepts: $\sqrt{x(x^2 - 4)} = 0$ if $x = 0, 2, -2$. Plot the points $(0, 0)$, $(2, 0)$, and $(-2, 0)$.

y intercepts: If $x = 0$, then $y = 0$.

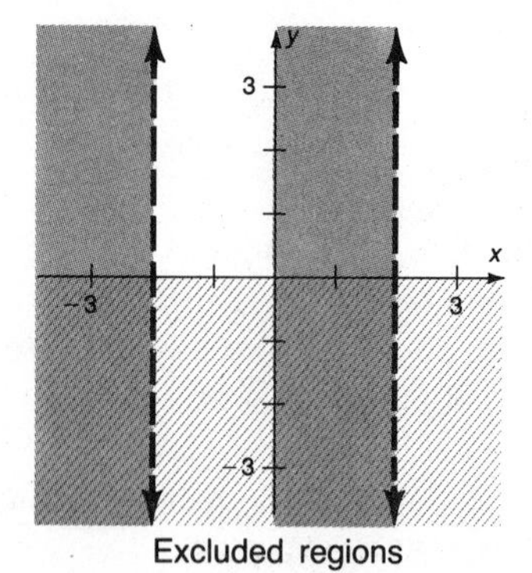

Excluded regions

Additional points: If $x = -1$, then $y = \sqrt{3}$.
If $x = 3$, then $y = \sqrt{15}$.

The graph is shown in Figure 3.15.

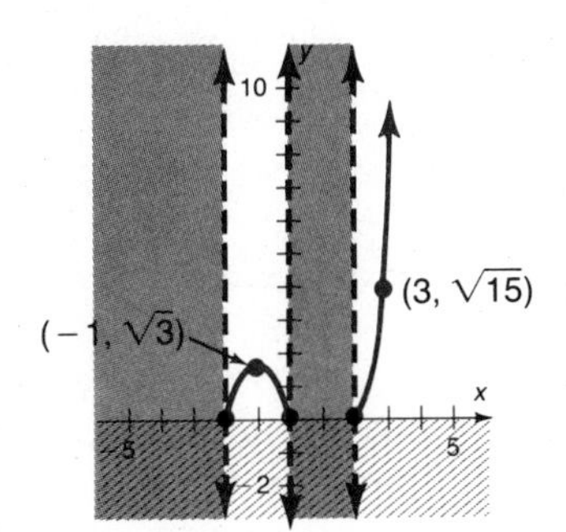

Figure 3.15 Graph of $y = \sqrt{x(x^2 - 4)}$

PROBLEM SET 3.4

Determine whether each is a function; then find the domain, range, intercepts, symmetry, and asymptotes for the graphs whose equations are given in Problems 1–30.

A

1. $xy = 2$

2. $xy = 6$

3. $y = \dfrac{x+1}{x}$

4. $y = \dfrac{x+1}{x+2}$

5. $y = \sqrt{4-x}$

6. $y = \sqrt{x-4}$

7. $y = -\sqrt{x-6}$

8. $y = -\sqrt{6-x}$

9. $y = \dfrac{1}{x^2-4}$

10. $y = \dfrac{1}{x^2+1}$

11. $y = \dfrac{2x^2+x-10}{x+2}$

12. $y = \dfrac{3x^2+5x-2}{x+2}$

13. $y = \dfrac{2x^3-3x^2-2x}{2x+1}$

14. $y = \dfrac{x^3+6x^2+15x+14}{x+2}$

B

15. $y = \sqrt{x} - x$

16. $y = x - \sqrt{x}$

17. $y = \sqrt{x^2+2x-3}$

18. $y = \sqrt{x^2+x-6}$

19. $y = \dfrac{-x}{\sqrt{4-x^2}}$

20. $y = \dfrac{9}{\sqrt{x^2-9}}$

21. $|x| + |y| = 5$

22. $|2x| - 3|y| = 9$

23. $9x^2 + 16y^2 = 144$

24. $9x^2 - 16y^2 = 144$

25. $y = \dfrac{x^2+1}{x}$

26. $y = \dfrac{5x^2+2}{x}$

27. $y^2 = \dfrac{x^2-1}{x-4}$

28. $y^2 = \dfrac{x^2-4}{x-2}$

29. $y^2 = \dfrac{x^2-9}{x^2-2x-15}$

30. $y^2 = \dfrac{x^2-4}{x^2-3x+2}$

Sketch the graphs of the curves whose equations are given in Problems 31–60. Notice that these are the same equations as those in Problems 1–30.

31. $xy = 2$

32. $xy = 6$

33. $y = \dfrac{x+1}{x}$

34. $y = \dfrac{x+1}{x+2}$

35. $y = \sqrt{4-x}$

36. $y = \sqrt{x-4}$

37. $y = -\sqrt{x-6}$

38. $y = -\sqrt{6-x}$

39. $y = \dfrac{1}{x^2-4}$

40. $y = \dfrac{1}{x^2+1}$

41. $y = \dfrac{2x^2+x-10}{x+2}$

42. $y = \dfrac{3x^2+5x-2}{x+2}$

43. $y = \dfrac{2x^3-3x^2-2x}{2x+1}$

44. $y = \dfrac{x^3+6x^2+15x+14}{x+2}$

45. $y = \sqrt{x} - x$

46. $y = x - \sqrt{x}$

47. $y = \sqrt{x^2+2x-3}$

48. $y = \sqrt{x^2+x-6}$

49. $y = \dfrac{-x}{\sqrt{4-x^2}}$

50. $y = \dfrac{9}{\sqrt{x^2-9}}$

51. $|x| + |y| = 5$

52. $|2x| - 3|y| = 9$

53. $9x^2 + 16y^2 = 144$

54. $9x^2 - 16y^2 = 144$

55. $y = \dfrac{x^2 + 1}{x}$

56. $y = \dfrac{5x^2 + 2}{x}$

57. $y^2 = \dfrac{x^2 - 1}{x - 4}$

58. $y^2 = \dfrac{x^2 - 4}{x - 2}$

59. $y^2 = \dfrac{x^2 - 9}{x^2 - 2x - 15}$

60. $y^2 = \dfrac{x^2 - 4}{x^2 - 3x + 2}$

C

Discuss and sketch the curves whose equations are given in Problems 61–66.

61. $x^{2/3} + y^{2/3} = 4$

62. $x^{2/3} + y^{2/3} = 1$

63. $x^2y^2 - x^2 - 9y^2 + 4 = 0$

64. $y^2x^2 - x^2 - 3y^2 + x = 0$

65. $x^4 + y^2 - 4x^2 = 0$

66. $x^4 + 4x^2 + y^2 = 0$

3.5 CHAPTER 3 SUMMARY

The material of this chapter is reviewed in the following list of objectives. After each objective there are some practice questions. For a sample test, select the first question of each set and check your answers with the answer section. For a sample test without answers, use the second question of each set. Additional practice is given by the other questions in each set. If you are having trouble with a particular type of problem, look back to that section for extra help.

3.1 PLOTTING POINTS, INTERCEPTS, AND SYMMETRY

Objective 1 *Be able to find the x and y intercepts for a curve whose equation is given.*

1. $y = (x - 2)(3x + 5)(x + 2)$

2. $y = \dfrac{5x^2 - 16x + 3}{x - 3}$

3. $2x^2 + xy - 4y^2 = 1$

4. $\dfrac{(x - 1)^2}{4} + \dfrac{(y + 1)^2}{9} = 1$

Objective 2 *Be able to check the symmetry of a given curve with respect to the x axis, y axis, and origin.*

5. $y^2 = x^2 - x$

6. $y = \sqrt{x^2 - x}$

7. $y = \dfrac{x^3 - x}{9x^4}$

8. $y^2 - 2y - 3 = |x|$

3.2 DOMAIN AND RANGE

Objective 3 *Graph curves with deleted points.*

9. $y = \dfrac{x^3 - 8x^2 + 20x - 16}{x - 4}$

10. $y = \dfrac{(2x - 3)^2(x + 1)}{(x + 1)(3 - 2x)}$

11. $y = \dfrac{x^3 - 7x^2 + 16x - 12}{x^2 - 5x + 6}$

12. $2x^2 + 9x - xy - 5y - 5 = 0$

Objective 4 *Find the domain and range for a curve whose equation is given.*

13. $x^2 - y + 5 = 0$

14. $y = \sqrt{10 - 2x}$

15. $y = \sqrt{\dfrac{x + 5}{1 - x}}$

16. $16x^2 - x^2y^2 + y^2 - 1 = 0$

3.3 ASYMPTOTES

Objective 5 *Find limits as $|x| \to \infty$.*

17. $\displaystyle\lim_{|x| \to \infty} \frac{3x^5 - 2x^3 + 1}{4x^5 - 1}$

18. $\displaystyle\lim_{|x| \to \infty} \frac{3x^5 - 2x^3 + 1}{5x^4 - 1}$

19. $\displaystyle\lim_{|x| \to \infty} \frac{5x^3 - 2x^2 + 1}{4x^5 - 1}$

20. $\displaystyle\lim_{|x| \to \infty} \frac{(x - 1)(3x - 2)(5x - 3)}{x^3}$

Objective 6 Find vertical, horizontal, and slant asymptotes.

21. $y = \dfrac{6x^2 - 11x}{2x - 1}$

22. $y = \dfrac{2x^2 - 5x - 3}{3x^2 - 7x - 6}$

23. $y = \dfrac{3x}{\sqrt{9 - x^2}}$

24. $x^2 - x^2y^2 + 9y^2 - 4 = 0$

3.4 CURVE SKETCHING

Objective 7 Discuss and sketch curves, given an equation.

25. $x^2 - x^2y^2 - y^2 + 25 = 0$

26. $|y| - |x| = 5$

27. $x^{2/3} + y^{2/3} = 1$

28. $y = \sqrt{x(x^2 - 9)}$

NOTE: For additional practice with curve sketching, you can discuss and sketch the curves whose equations are given in Problems 1–8, 13–16, and 21–24.

Amalie (Emmy) Noether (1882–1935)

Women who have left behind a name in mathematics are very few in number—three or four, perhaps five. Does this say, as a common prejudice would tend to persuade us, that mathematics, so very abstract, is not congenial to the feminine disposition?... The growth of female education, the overthrow of prejudices, the profound changes in the kind of life and in the role assigned to woman during the last few years will doubtless bring about a revision of her position in science. Then we shall see in what measure she can, as the equal of man, emerge from the role of the excellent pupil or the perfect collaborator, and join with those scientists whose work has opened new paths and bears the mark of genius.

Marie-Louise Dubreil-Jacotin
Great Currents of Mathematical Thought

HISTORICAL NOTE

The greatest woman mathematician of all time was Emmy Noether, who is famous for her work in physics and algebra. She obtained her degrees at a time when it was unusual for a woman to attend college. In fact, a leading historian of the day wrote that talk of "surrendering our universities to the invasion of women ... is a shameful display of moral weakness." Nevertheless, Noether did receive her degrees and significantly influenced the development of abstract algebra. In 1935, Albert Einstein wrote the following tribute:*

"In the judgement of the most competent living mathematicians, Fraulein Noether was the most significant creative mathematical genius thus far produced since the higher education of women began. In the realm of algebra in which the most gifted of mathematicians have been busy for centuries, she discovered methods which have proved of enormous importance in the development of the present day younger generation of mathematicians."

In the nineteenth century, women were not allowed to enroll in many universities. At one school, which will remain nameless, a talented woman had persuaded the administration to let her audit a mathematics course. However, when a professor walked in and found four men and one woman in the classroom, he let her know in no uncertain terms that despite the administration's weakness and willingness to allow her in the class, he was not changing his attitude. Forthwith he began to lecture and lecture he did—with a vengeance. Week by week the lectures took their toll and by midterm only the woman and one struggling male student remained. The professor did not reduce the pace, and within another week the woman student found that she was the only person in the classroom when the professor walked in to begin a lecture. Glancing around, the professor announced, "Since there are no students here, I will not give a lecture," and walked out of the room. This story is somewhat depressing in its implication that unknown genius has been wasted or suppressed by societies that have not recognized mathematics as an appropriate activity for women.**

* As quoted in Howard Eves, *In Mathematical Circles Revisited* (Boston: Prindle, Weber, & Schmidt), p. 125.

** From Marie-Louise Dubreil-Jacotin, "Women Mathematicians," in Douglas M. Campbell and John C. Higgins, eds., *Mathematics: People, Problems, Results*, vol. I (Belmont, Ca.: Wadsworth), p. 166.

4

Polynomial and Rational Functions

CONTENTS

* Optional sections

PREVIEW

Equation solving is one of the most important topics of elementary algebra. In high school you learned about first-degree equations, and in Chapter 1 we reviewed second-degree equations. In this chapter we consider the solution of polynomial equations of degree higher than two. There are 17 objectives in this chapter, which are listed on pages 161–163.

PERSPECTIVE

In Chapter 3 we considered techniques of graphing, and in this chapter we consider graphing polynomial and rational functions. However, you will have to wait until you take a course in calculus to be able to find the turning points and maximum and minimum values for a graph. In order to find these values in calculus you will need to analyze equations such as

$$y = x^3 - 3x^2 + 1$$

as shown on the page below from a leading calculus book. Notice that the author assumes that you know how to graph this curve (no discussion). You will learn how to do this in Section 4.3.

Solution Since f is continuous on $[-\frac{1}{2}, 4]$ we can use the procedure outlined in (4.8):

$$f(x) = x^3 - 3x^2 + 1$$
$$f'(x) = 3x^2 - 6x = 3x(x - 2)$$

Since $f'(x)$ exists for all x, the only critical numbers of f occur when $f'(x) = 0$, that is, $x = 0$ or $x = 2$. Notice that each of these critical numbers lies in the interval $[-\frac{1}{2}, 4]$. The values of f at these critical numbers are

$$f(0) = 1 \qquad f(2) = -3$$

The values of f at the endpoints of the interval are

$$f(-\tfrac{1}{2}) = \tfrac{1}{8} \qquad f(4) = 17$$

Comparing these four numbers, we see that the absolute maximum value is $f(4) = 17$ and the absolute minimum value is $f(2) = -3$.

Note that in this example the absolute maximum occurs at an endpoint, whereas the absolute minimum occurs at a critical number. The graph of f is sketched in Figure 4.12. ●

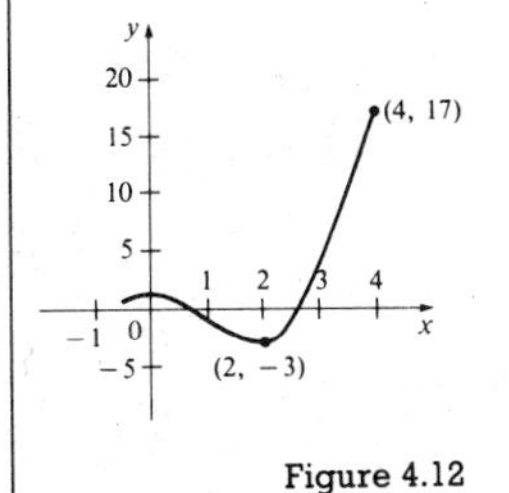

Figure 4.12

From James Stewart, *Calculus* (Pacific Grove, Ca.: Brooks/Cole), p. 180.

4.1 POLYNOMIAL FUNCTIONS

The key to working with polynomials is the proper handling of exponents. You have used the laws of exponents extensively in algebra, and they must also be fully understood for the study of calculus.

Definition of Exponents

If b is any real number and n is any natural number, then

$$b^n = \underbrace{b \cdot b \cdot b \cdots b}_{n \text{ factors}}$$

And if $b \neq 0$, then

$$b^0 = 1 \qquad b^{-n} = \frac{1}{b^n}$$

The number b is called the **base**, n is called the **exponent**, and b^n is called the **nth power** of b. The five laws of exponents can now be summarized.

Laws of Exponents

Let a and b be real numbers and let m and n be any integers. Then the five rules listed below govern the use of exponents except that the form 0^0 and division by zero are excluded.

First law: $b^m \cdot b^n = b^{m+n}$ Fourth law: $(ab)^m = a^m b^m$

Second law: $\dfrac{b^m}{b^n} = b^{m-n}$ Fifth law: $\left(\dfrac{a}{b}\right)^m = \dfrac{a^m}{b^m}$

Third law: $(b^n)^m = b^{mn}$

The linear and quadratic functions defined in Chapter 2 are special cases of a more general function called a polynomial function.

Polynomial Function

A function P is a **polynomial function** in x if

$$P(x) = a_n x^n + a_{n-1}x^{n-1} + a_{n-2}x^{n-2} + \cdots + a_1 x + a_0$$

where n is an integer greater than or equal to zero and the coefficients $a_0, a_1, a_2, \ldots, a_{n-1}, a_n$ are real numbers.

In an expression of the form $a_n x^n$, a_n is called the **coefficient** of x^n. And if a_n is the coefficient of the highest power of x, it is called the **leading coefficient** of the polynomial. If a_n and x are nonzero, the polynomial function is said to have **degree** n. Notice that if $n = 2$ (degree 2), then $P(x)$ is a *quadratic function*; if $n = 1$ (degree 1), then $P(x)$ is a *linear function*; and if $n = 0$ (degree 0), then $P(x)$ is a *constant function.* If all the coefficients of a polynomial are 0, it is called the **zero polynomial** and is denoted by 0. The zero polynomial is not assigned a degree.

Each $a_k x^k$ of a polynomial function is called a **term** of the polynomial, and it is customary to arrange the terms of a polynomial in order of decreasing powers of the variable. **Similar terms** are terms with the same variable and the same degree. Two polynomials are **equal** if and only if coefficients of similar terms are the same. If some of the numerical coefficients are negative, they are usually written as subtractions of terms. Thus

$$6 + (-3)x + x^3 + (-2)x^2$$

would customarily be written as

$$x^3 - 2x^2 - 3x + 6$$

A polynomial is said to be **simplified** if it is written with all similar terms combined and is expressed in the form of a polynomial function, as given in the preceding box.

Examples 1–4 provide some practice with polynomial notation and operations. Let $P(x) = 4x - 5$, $Q(x) = 5x^2 + 2x + 1$, and $R(x) = 3x^3 - 4x^2 + 3x - 2$.

EXAMPLE 1 Find $P(2)$, $Q(2)$, and $R(2)$.

Solution

$$\begin{aligned} P(2) &= 4(2) - 5 \\ &= 8 - 5 \\ &= \mathbf{3} \end{aligned} \qquad \begin{aligned} Q(2) &= 5(2)^2 + 2(2) + 1 \\ &= 20 + 4 + 1 \\ &= \mathbf{25} \end{aligned}$$

$$\begin{aligned} R(2) &= 3(2)^3 - 4(2)^2 + 3(2) - 2 \\ &= 24 - 16 + 6 - 2 \\ &= \mathbf{12} \end{aligned}$$

■

EXAMPLE 2 Find $Q(x) + R(x)$.

Solution Use the distributive law to combine similar terms:

$$\begin{aligned} Q(x) + R(x) &= (5x^2 + 2x + 1) + (3x^3 - 4x^2 + 3x - 2) \\ &= 3x^3 + \underbrace{5x^2 + (-4)x^2}_{\text{similar terms}} + \underbrace{2x + 3x}_{\text{similar terms}} + \underbrace{1 + (-2)}_{\text{similar terms}} \\ &= \mathbf{3x^3 + x^2 + 5x - 1} \end{aligned}$$

This step is usually done mentally.

■

EXAMPLE 3 Find $Q(x) - R(x)$.

Solution Be careful when subtracting polynomials; to subtract $R(x)$ you must subtract *each* term of $R(x)$. That is, remember that $Q(x) - R(x) = Q(x) + (-1)R(x)$. Thus

$$\begin{aligned} Q(x) - R(x) &= (5x^2 + 2x + 1) - (3x^3 - 4x^2 + 3x - 2) \\ &= 5x^2 + 2x + 1 + (-3x^3) + 4x^2 + (-3x) + 2 \\ &= \mathbf{-3x^3 + 9x^2 - x + 3} \end{aligned}$$

■

EXAMPLE 4 Find $P(x)R(x)$.

Solution Use the distributive property:

$$\begin{aligned} P(x)R(x) &= (4x - 5)(3x^3 - 4x^2 + 3x - 2) \\ &= (4x - 5)(3x^3) + (4x - 5)(-4x^2) + (4x - 5)(3x) + (4x - 5)(-2) \\ &= 12x^4 - 15x^3 - 16x^3 + 20x^2 + 12x^2 - 15x - 8x + 10 \\ &= \mathbf{12x^4 - 31x^3 + 32x^2 - 23x + 10} \end{aligned}$$

■

If a polynomial is written as the product of other polynomials, each polynomial in the product is called a **factor** of the original polynomial. **Factoring** a polynomial

TABLE 4.1 Factoring types

Type	Form	Comments
1. Common factors	$ax + ay + az = a(x + y + z)$	This simply uses the distributive property. It can be applied with any number of terms.
2. Difference of squares	$x^2 - y^2 = (x - y)(x + y)$	The *sum* of two squares cannot be factored in the set of real numbers.
3. Difference of cubes	$x^3 - y^3 = (x - y)(x^2 + xy + y^2)$	This is similar to the difference of squares and can be proved by multiplying the factors.
4. Sum of cubes	$x^3 + y^3 = (x + y)(x^2 - xy + y^2)$	Unlike the sum of squares, the sum of cubes can be factored.
5. Perfect square	$x^2 + 2xy + y^2 = (x + y)^2$ $x^2 - 2xy + y^2 = (x - y)^2$	The middle term is twice the product of xy.
6. Perfect cube	$x^3 + 3x^2y + 3xy^2 + y^3 = (x + y)^3$ $x^3 - 3x^2y + 3xy^2 - y^3 = (x - y)^3$	The numerical coefficients of the terms are: 1 3 3 1 or 1 −3 3 −1
7. Binomial product	$acx^2 + (ad + bc)xy + bdy^2$ $= (ax + by)(cx + dy)$ See Examples 12–14 in the text.	This procedure is used to factor a trinomial into binomial factors. It should be used after types 1–6 have been checked.
8. Grouping	See Examples 16 and 17 in the text.	After types 1–7 have been checked, you can factor some expressions by grouping.
9. Irreducible (cannot be factored over the set of integers)	Examples arise in every factoring situation. Expressions such as $x^2 + 4$, $x^2 + xy + y^2$, and $x^2 + y^2$ cannot be factored over the set of integers.	When factoring, you are not through until all the factors are irreducible.

means to break up the polynomial into the product of individual factors. The types of factoring problems usually encountered are summarized in Table 4.1. For example, the expression $6x^2 + 3x - 18$ is properly factored by *first* finding the common factor (type 1) and then noting that types 2–6 do not apply. Finally, use type 7 (binomial factors) to write

$$\begin{aligned} 6x^2 + 3x - 18 &= 3(2x^2 + x - 6) \\ &= 3(2x - 3)(x + 2) \end{aligned}$$

We usually factor **over the set of integers**, which means that all the numerical coefficients are integers. If the original polynomial has fractional coefficients, you should factor out the fractional part first, as shown in Example 5.

EXAMPLE 5

$$\begin{aligned} \frac{1}{36}x^2 - 9y^4 &= \frac{1}{36}(x^2 - 6^2 \cdot 3^2 y^4) \\ &= \frac{1}{36}[x^2 - (18y^2)^2] \\ &= \mathbf{\frac{1}{36}(x - 18y^2)(x + 18y^2)} \end{aligned}$$

■

Factoring over the set of integers rules out factoring

$$x^2 - 3 = (x - \sqrt{3})(x + \sqrt{3})$$

WARNING: Note the agreement about factoring in this book.

since the factors do not have integer coefficients. In this book, a polynomial is **completely factored** if all fractions are eliminated by common factoring (as shown in Example 5) and if no further factoring is possible *over the set of integers*. If, after fractional common factors have been removed, a polynomial cannot be factored further, then it is said to be **irreducible**. Examples 6–17 review various factoring techniques.

EXAMPLE 6 $5x + 7xy = \mathbf{x(5 + 7y)}$ Common factor x ■

EXAMPLE 7 $7 - 35x = \mathbf{7(1 - 5x)}$ Common factor 7 (note the 1) ■

EXAMPLE 8 $5x(3a - 2b) + y(3a - 2b) = \mathbf{(5x + y)(3a - 2b)}$ Common factor $(3a - 2b)$ ■

EXAMPLE 9

$$\begin{aligned} 3x^2 - 75 &= 3(x^2 - 25) && \text{Common factor first} \\ &= \mathbf{3(x - 5)(x + 5)} && \text{Difference of squares} \end{aligned}$$

■

EXAMPLE 10

$$\begin{aligned} \frac{9a^2}{b^2} - (a + 3b)^2 &= \frac{1}{b^2}[9a^2 - b^2(a + 3b)^2] && \text{Common factor to eliminate fractions} \\ &= \frac{1}{b^2}\{(3a)^2 - [b(a + 3b)]^2\} && \text{This step can be done mentally.} \\ &= \frac{1}{b^2}[3a - b(a + 3b)][3a + b(a + 3b)] && \text{Difference of squares} \\ &= \mathbf{\frac{1}{b^2}(3a - ab - 3b^2)(3a + ab + 3b^2)} && \text{Simplify} \end{aligned}$$

■

EXAMPLE 11 $(x+3y)^3+8 = [(x+3y)+2][(x+3y)^2-(x+3y)(2)+(2)^2]$ Sum of cubes

$$= \mathbf{(x+3y+2)(x^2+6xy+9y^2-2x-6y+4)}$$ ■

EXAMPLE 12 $x^2-8x+15 = \mathbf{(x-5)(x-3)}$ ■

EXAMPLE 13 $6x^2+x-12 = \mathbf{(2x+3)(3x-4)}$ ■

Binomial product is sometimes called FOIL; you may have to try several possibilities before you find the correct binomial factors.

EXAMPLE 14 $6x^2-9x-15 = 3(2x^2-3x-5)$

$$= \mathbf{3(2x-5)(x+1)}$$ ■

Remember to factor completely; several types may be combined in one problem.

EXAMPLE 15 $4x^4-13x^2y^2+9y^4 = (x^2-y^2)(4x^2-9y^2)$

$$= \mathbf{(x-y)(x+y)(2x-3y)(2x+3y)}$$ ■

EXAMPLE 16 $9x^3+18x^2-x-2 = (9x^3+18x^2)-(x+2)$

$$= 9x^2(x+2)-(x+2)$$
$$= (9x^2-1)(x+2)$$
$$= \mathbf{(3x-1)(3x+1)(x+2)}$$

None of the types 1–7 from Table 4.1 seem to apply. Thus try grouping the terms. Some groupings may lead to a factorable form; others may not. ■

EXAMPLE 17 $a^2-b^2-8b-16 = a^2-(b^2+8b+16)$

$$= a^2-(b+4)^2$$

Try grouping only after you have tried types 1–7 on Table 4.1.

$$= [a-(b+4)][a+(b+4)]$$ Difference of squares

$$= \mathbf{(a-b-4)(a+b+4)}$$ ■

An important application of factoring in algebra, and in calculus, is the **Zero Factor Theorem**. It is simply the principle of zero products of Section 1.4 stated for polynomials.

Zero Factor Theorem

> If $P(x)$ and $Q(x)$ are polynomials such that
>
> $$P(x)Q(x) = 0$$
>
> then either
>
> $$P(x) = 0 \quad \text{or} \quad Q(x) = 0 \quad \text{(perhaps both)}$$

You can apply this theorem to graph equations expressible in factored form. Suppose you wish to graph the curve whose equation is

$$x^2-y^2 = 0$$

Since this equation can be factored as

$$(x-y)(x+y) = 0$$

you can use the Zero Factor Theorem to write

$$x-y=0 \quad \text{or} \quad x+y=0$$

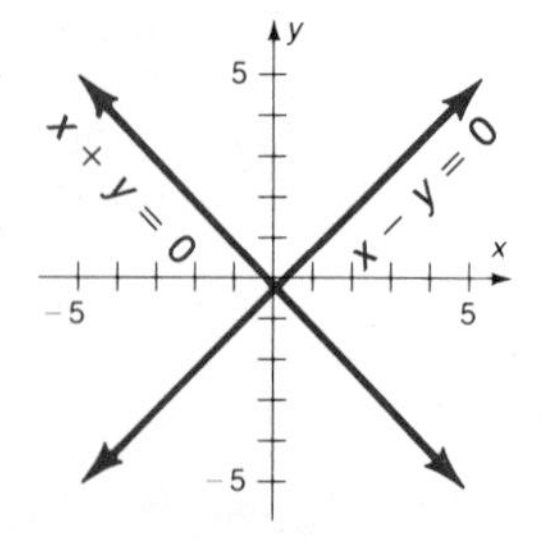

Figure 4.1 Graph of $x^2-y^2=0$

Thus the graph of $x^2-y^2=0$ is found by graphing both $x-y=0$ and $x+y=0$, as shown in Figure 4.1. You might look at Figure 4.1 and say it is two graphs, but the graphs of the two lines give the *single* curve representing the graph of $x^2 - y^2 = 0$. That is, not all curves must be continuous.

EXAMPLE 18 Graph $3x^3 - 3xy - 2x^2y + 2y^2 = 0$.

Solution First factor by grouping:

$$3x(x^2 - y) - 2y(x^2 - y) = 0$$
$$(x^2 - y)(3x - 2y) = 0$$

Next use the Zero Factor Theorem:

$$x^2 - y = 0$$

or

$$3x - 2y = 0$$

The graph is shown in Figure 4.2. Note that both parts (the parabola and the line) form what is called *the* graph of the equation

$$3x^3 - 3xy - 2x^2y + 2y^2 = 0$$

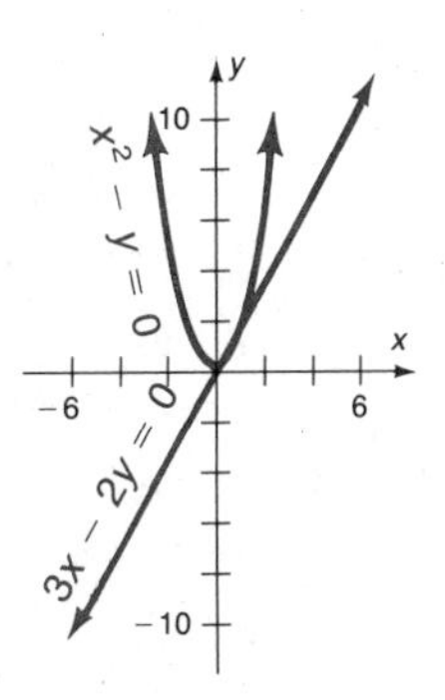

Figure 4.2 Graph of $3x^3 - 3xy - 2x^2y + 2y^2 = 0$ ■

PROBLEM SET 4.1

A

Let $P(x) = 5x + 1$, $Q(x) = 3x^2 - 5x + 2$, *and* $R(x) = x^3 - 4x^2 + x - 4$ *in Problems 1–6. Find the required result.*

1. a. $P(2)$ **b.** $Q(2)$ **c.** $R(2)$

2. a. $P(3)$ **b.** $Q(0)$ **c.** $R(1)$

3. a. $P(5)$ **b.** $Q(1)$ **c.** $R(0)$

4. a. $P(x) + R(x)$ **b.** $Q(x) - P(x)$

5. a. $R(x) - Q(x)$ **b.** $P(x)Q(x)$

6. a. $P(x)Q(x) - R(x)$ **b.** $3Q(x) - 4R(x)$

Write the expressions in Problems 7–12 in simplified polynomial form.

7. a. $(x + 2)(x + 1)$ **b.** $(y - 2)(y + 3)$ **c.** $(x + 1)(x - 2)$ **d.** $(y - 3)(y + 2)$

8. a. $(a - 5)(a - 3)$ **b.** $(b + 3)(b - 4)$ **c.** $(c + 1)(c - 7)$ **d.** $(z - 3)(z + 5)$

9. a. $(2x + 1)(x - 1)$ **b.** $(2x - 3)(x - 1)$ **c.** $(x + 1)(3x + 1)$ **d.** $(x + 1)(3x + 2)$

10. a. $(x + y)(x - y)$ **b.** $(a + b)(a - b)$ **c.** $(5x - 4)(5x + 4)$ **d.** $(3y - 2)(3y + 2)$

11. a. $(a + 2)^2$ **b.** $(b - 2)^2$ **c.** $(x + 4)^2$ **d.** $(y - 3)^2$

12. a. $(u + v)^3$ **b.** $(s - 2t)^3$ **c.** $(1 - 3n)^3$ **d.** $(2x - y)^3$

Factor completely, if possible, the expressions in Problems 13–18.

13. a. $me + mi + my$ **b.** $a^2 - b^2$ **c.** $a^2 + b^2$ **d.** $a^3 - b^3$

14. a. $a^3 + b^3$ **b.** $s^2 + 2st + t^2$ **c.** $m^2 - 2mn + n^2$ **d.** $u^2 + 2uv + v^2$

15. a. $a^3 + 3a^2b + 3ab^2 + b^3$ **b.** $p^3 - 3p^2q + 3pq^2 - q^3$ **c.** $-c^3 + 3c^2d - 3cd^2 + d^3$ **d.** $x^2y + xy^2$

16. a. $(a + b)x + (a + b)y$ **b.** $(4x - 1)x + (4x - 1)3$ **c.** $x^2 - 2x - 35$ **d.** $2x^2 + 7x - 15$

17. a. $3x^2 - 5x - 2$ **b.** $6y^2 - 7y + 2$ **c.** $8a^2b + 10ab - 3b$ **d.** $2s^2 - 10s - 48$

18. a. $4y^3 + y^2 - 21y$ **b.** $12m^2 - 7m - 12$ **c.** $12p^4 + 11p^3 - 15p^2$ **d.** $9x^2y + 15xy - 14y$

B

Write the expressions in Problems 19–24 in simplified polynomial form.

19. a. $(3x - 1)(x^2 + 3x - 2)$ **b.** $(2x + 1)(x^2 + 2x - 5)$

20. a. $(5x + 1)(x^3 - 2x^2 + 3x - 5)$ **b.** $(4x - 1)(x^3 + 3x^2 - 2x - 4)$

21. a. $(x + 1)(x - 3)(2x + 1)$ **b.** $(2x - 1)(x + 3)(3x + 1)$

22. a. $(x - 2)(x + 3)(x - 4)$ **b.** $(x + 1)(2x - 3)(2x + 3)$

23. a. $(x - 2)^2(x + 1)$ **b.** $(x - 2)(x + 1)^2$

24. a. $(x - 5)^2(x + 2)$ **b.** $(x - 3)^2(3x + 2)$

Factor completely, if possible, the expressions in Problems 25–51.

25. $(x - y)^2 - 1$

26. $(2x + 3)^2 - 1$

27. $(5a - 2)^2 - 9$

28. $(3p - 2)^2 - 16$

29. $\frac{4}{25}x^2 - (x + 2)^2$

30. $\frac{4x^2}{9} - (x + y)^2$

31. $\frac{x^6}{y^8} - 169$

32. $\frac{9x^{10}}{4} - 144$

33. $(a + b)^2 - (x + y)^2$

34. $(m - 2)^2 - (m + 1)^2$

35. $2x^2 + x - 6$

36. $3x^2 - 11x - 4$

37. $6x^2 + 47x - 8$

38. $6x^2 - 47x - 8$

39. $6x^2 + 49x + 8$

40. $6x^2 - 49x + 8$

41. $4x^2 + 13x - 12$

42. $9x^2 - 43x - 10$

43. $9x^2 - 56x + 12$

44. $12x^2 + 12x - 25$

45. $4x^4 - 17x^2 + 4$

46. $4x^4 - 45x^2 + 81$

47. $x^6 + 9x^3 + 8$

48. $x^6 - 6x^3 - 16$

49. $(x^2 - \frac{1}{4})(x^2 - \frac{1}{9})$

50. $(x^2 - \frac{1}{9})(x^2 - \frac{1}{16})$

51. $(x^3 + \frac{1}{8})(x^2 - \frac{1}{4})$

Sketch the curves whose equations are given in Problems 52–61.

52. $\frac{x^2}{9} - \frac{y^2}{4} = 0$

53. $\frac{x^2}{16} - \frac{y^2}{25} = 0$

54. $\frac{y^2}{49} = \frac{x^2}{36}$

55. $\frac{y^2}{25} - \frac{x^2}{36} = 0$

56. $2x^3 - 2xy + x^2y - y^2 = 0$

57. $4x^3 - 8xy + 3x^2y - 6y^2 = 0$

58. $(x - y - 1)(3x + 2y - 4) = 0$

59. $(5x - 2y - 6)(x + y - 2) = 0$

60. $(x - y)(x + 2y)(x^2 - y) = 0$

61. $(x^2 - y^2)(2x + y - 3) = 0$

C

Write the expressions in Problems 62–67 in simplified polynomial form.

62. $(2x - 1)^2(3x^4 - 2x^3 + 3x^2 - 5x + 12)$

63. $(x - 3)^3(2x^3 - 5x^2 + 4x - 7)$

64. $(2x + 1)^2(5x^4 - 6x^3 - 3x^2 + 4x - 5)$

65. $(3x^2 + 4x - 3)(2x^2 - 3x + 4)$

66. $(2x^3 + 3x^2 - 2x + 4)^3$

67. $(x^3 - 2x^2 + x - 5)^3$

Factor completely, if possible, the expressions in Problems 68–83.

68. $x^{2n} - y^{2n}$

69. $x^{3n} - y^{3n}$

70. $x^{3n} + y^{3n}$

71. $x^{2n} - 2x^ny^n + y^{2n}$

72. $(x - 2)^2 - (x - 2) - 6$

73. $(x + 3)^2 - (x + 3) - 6$

74. $z^5 - 8z^2 - 4z^3 + 32$

75. $x^5 + 8x^2 - x^3 - 8$

76. $x^2 - 2xy + y^2 - a^2 - 2ab - b^2$

77. $x^2 + 2xy + y^2 - a^2 - 2ab - b^2$

78. $x^2 + y^2 - a^2 - b^2 - 2xy + 2ab$

79. $x^3 + 3x^2y + 3xy^2 + y^3 + a^3 + 3a^2b + 3ab^2 + b^3$

80. $(x + y + 2z)^2 - (x - y + 2z)^2$

81. $(x^2 - 3x - 6)^2 - 4$

82. $2(x + y)^2 - 5(x + y)(a + b) - 3(a + b)^2$

83. $2(s + t)^2 + 3(s + t)(s + 2t) - 2(s + 2t)^2$

The following problems require common factoring and are the type of problems you will encounter in calculus.

84. $3(x + 1)^2(x - 2)^4 + 4(x + 1)^3(x - 2)^3$

85. $4(x - 5)^3(x + 3)^2 + 2(x - 5)^4(x + 3)$

86. $3(2x - 1)^2(2)(3x + 2)^2 + 2(2x - 1)^3(3x + 2)(3)$

87. $(2x - 3)^3(3)(1 - x)^2(-1) + 3(2x - 3)^2(1 - x)^3(2)$

88. $4(x + 5)^3(x^2 - 2)^3 + (x + 5)^4(3)(x^2 - 2)^2(2x)$

89. $5(x - 2)^4(x^2 + 1)^3 + (x - 2)^5(3)(x^2 + 1)^2(2x)$

4.2 SYNTHETIC DIVISION

In the previous section we considered sums, differences, and products of polynomials. In this section we will consider quotients of polynomials and see how this division can be quickly and easily accomplished using a process called *synthetic division.* This will, in turn, lead us in the next sections to some methods for finding the roots of certain polynomials as well as to some techniques for graphing polynomial functions.

Consider positive integers P and D. If the result of P divided by D is an integer Q, then D is a factor of P. For example, if $P = 15$ and $D = 3$, then $P/D = 5$, and 5 is called the *quotient*. If D is not a factor of P, we will obtain a quotient Q and a remainder R that is less than D so that

$$\frac{P}{D} = Q + \frac{R}{D}$$

For example:

$$\text{If } \frac{15}{3} = 5, \qquad \text{then} \qquad 15 = 5 \cdot 3$$

$$\text{If } \frac{17}{3} = 5 + \frac{2}{3}, \qquad \text{then} \qquad 17 = 5 \cdot 3 + 2$$

Notice how division may be checked by multiplying:

$$\text{If } \frac{P}{D} = Q, \qquad \text{then} \qquad P = QD$$

$$\text{If } \frac{P}{D} = Q + \frac{R}{D}, \qquad \text{then} \qquad P = QD + R \qquad (R < D)$$

Division of polynomials is similar and leads to a result called the **Division Algorithm**.

Division Algorithm

If $P(x)$ and $D(x)$ are polynomials $[D(x) \neq 0]$, then there exist unique polynomials $Q(x)$ and $R(x)$ such that

$$\frac{P(x)}{D(x)} = Q(x) + \frac{R(x)}{D(x)}$$

where $Q(x)$ is a unique polynomial and $R(x)$ is a polynomial such that the degree of $R(x)$ is less than the degree of $D(x)$. The polynomial $Q(x)$ is called the **quotient** and $R(x)$ the **remainder**.

1. If $R(x) = 0$, then $D(x)$ is a factor of $P(x)$.
2. If the degree of $D(x)$ is greater than the degree of $P(x)$, then $Q(x) = 0$.
3. If both sides are multiplied by $D(x)$, the Division Algorithm is then stated in product form:

$$P(x) = Q(x)D(x) + R(x)$$

The question to be considered is how to *find* $Q(x)$ and $R(x)$. The first method is by long division. Let $P(x) = 3x^4 + 7x^3 + 2x^2 + 3x + 5$ and $D(x) = x + 1$. Then:

Multiply $3x^3(x + 1) = 3x^4 + 3x^3$ and write the answer so that similar terms are aligned.

$$\begin{array}{r} \mathbf{3x^3} \\ x + 1 \overline{)3x^4 + 7x^3 + 2x^2 + 3x + 5} \\ \mathbf{3x^4 + 3x^3} \\ \hline \end{array}$$

$3x^3$ was picked so that the terms shown by the arrow are identical.

$$
\begin{array}{r}
3x^3 + \mathbf{4x^2} \\
x + 1\overline{)3x^4 + 7x^3 + 2x^2 + 3x + 5} \\
\underline{3x^4 + 3x^3} \\
4x^3 + 2x^2 + 3x + 5 \\
\underline{\mathbf{4x^3 + 4x^2}}
\end{array}
$$

Multiply $4x^2(x + 1)$ and align similar terms.

Next subtract (or add the opposite).

This must be zero since the terms were identical.

$4x^2$ was picked so that these terms are identical.

Now subtract and repeat the procedure:

$$
\begin{array}{r}
3x^3 + 4x^2 - 2x + 5 \\
x + 1\overline{)3x^4 + 7x^3 + 2x^2 + 3x + 5} \\
\underline{3x^4 + 3x^3} \\
4x^3 + 2x^2 + 3x + 5 \\
\underline{4x^3 + 4x^2} \\
-2x^2 + 3x + 5 \\
\underline{-2x^2 - 2x} \\
5x + 5 \\
\underline{5x + 5} \\
0
\end{array}
$$

Do not forget to subtract: $2x^2 - 4x^2 = -2x^2$

Subtract $3x - (-2x) = 5x$

$0 \leftarrow 5 - 5 = 0$

The remainder is zero, so $x + 1$ is a factor of $3x^4 + 7x^3 + 2x^2 + 3x + 5$. You can check this by verifying that $P(x) = Q(x)D(x)$:

$$
\begin{aligned}
Q(x)D(x) &= (3x^3 + 4x^2 - 2x + 5)(x + 1) \\
&= 3x^4 + 4x^3 - 2x^2 + 5x + 3x^3 + 4x^2 - 2x + 5 \\
&= 3x^4 + 7x^3 + 2x^2 + 3x + 5
\end{aligned}
$$

Since this is $P(x)$, the result checks.

EXAMPLE 1 Let $P(x) = 4x^4 - 6x^2 - 10x + 3$ and $D(x) = x^2 + x - 2$. Find $P(x)/D(x)$.

Solution

$$
\begin{array}{r}
4x^2 - 4x + 6 \\
x^2 + x - 2\overline{)4x^4 + 0x^3 - 6x^2 - 10x + 3} \\
\underline{4x^4 + 4x^3 - 8x^2} \\
-4x^3 + 2x^2 - 10x + 3 \\
\underline{-4x^3 - 4x^2 + 8x} \\
6x^2 - 18x + 3 \\
\underline{6x^2 + 6x - 12} \\
-24x + 15
\end{array}
$$

Notice that there is no x^3 term. Leave a space for this "missing" term, or write $0x^3$.

The remainder is $-24x + 15$. Thus

$$\frac{P(x)}{D(x)} = \mathbf{4x^2 - 4x + 6 + \frac{-24x + 15}{x^2 + x - 2}}$$

■

The process of long division is indeed *long* because of the duplication of symbols when carrying out this process. Consider the following example for dividing by a polynomial of the form $x - r$:

$$\begin{array}{r} 2x^3 - 4x^2 - x + 3 \\ x - 1\overline{)2x^4 - 6x^3 + 3x^2 + 4x - 3} \\ \mathbf{2x^4} - 2x^3 \qquad\qquad\qquad \\ \hline - 4x^3 + \mathbf{3x^2} + \mathbf{4x} - \mathbf{3} \\ \mathbf{- 4x^3} + 4x^2 \qquad\qquad \\ \hline - x^2 + \mathbf{4x} - \mathbf{3} \\ \mathbf{- x^2} + x \qquad \\ \hline 3x - 3 \\ 3x - 3 \\ \hline 0 \end{array}$$

This is first degree, so the quotient is 1 degree less than the given polynomial.

Notice that each term in color is a repetition of the term directly above, so we could eliminate writing it down.

Also, the position of the term indicates the degree, so it is not even necessary to write down the variable:

The degree of the first term is 1 less than the degree of the given polynomial.

$$\begin{array}{r} 2 - 4 - 1 + 3 \\ -1\overline{)2 - 6 + 3 + 4 - 3} \\ -2 \qquad\qquad\quad \\ \hline -4 \qquad\qquad\quad \\ +4 \qquad\qquad \\ \hline -1 \qquad\qquad \\ +1 \qquad \\ \hline 3 \qquad \\ -3 \\ \hline 0 \end{array}$$

Interpret this answer by recognizing that it is of degree 1 less than the degree of the dividend. This means the answer is $2x^3 - 4x^2 - x + 3$. That is, these numbers are the coefficients of the quotient polynomial.

$0 \longleftarrow$ This is the remainder.

There is no reason to spread out the array; it can be compressed into a more efficient form:

$$\begin{array}{rrrrrr} & \textcircled{2} & -4 & -1 & +3 & \\ \hline -1) & 2 & -6 & +3 & +4 & -3 \\ & & -2 & +4 & +1 & -3 \\ \hline & \uparrow & -4 & -1 & +3 & 0 \end{array}$$

$\longleftarrow$ Delete these terms.

$\longleftarrow$ This line is no longer needed.

This is the remainder.

The top line is the same as the bottom line if you bring down the leading coefficient

$$\begin{array}{r|rrrrr} -1 & 2 & -6 & +3 & +4 & -3 \\ & & -2 & +4 & +1 & -3 \\ \hline & 2 & -4 & -1 & +3 & 0 \end{array}$$

These are the coefficients of the quotient (which begins with a degree 1 less than that of the given polynomial).

This is the remainder.

The process is now fairly compact, but the most common error in this procedure is in the subtraction. Remember: It is usually easier to add the opposite than to subtract, so change the sign of the divisor (-1 to 1 in this example) and add:

$$\begin{array}{r|rrrrr} 1 & 2 & -6 & 3 & 4 & -3 \\ & & 2 & -4 & -1 & 3 \\ \hline & 2 & -4 & -1 & 3 & 0 \end{array}$$

Change the sign of this number, and **add**.

The condensed form of the division of a polynomial by $x - r$ is called **synthetic division**. **Notice synthetic division is used only when the divisor is of the form $x - r$** (*r positive or negative*). **If the divisor is not linear, then long division must be used.**

EXAMPLE 2 Divide $x^4 + 3x^3 - 12x^2 + 5x - 2$ by $x - 2$.

Solution

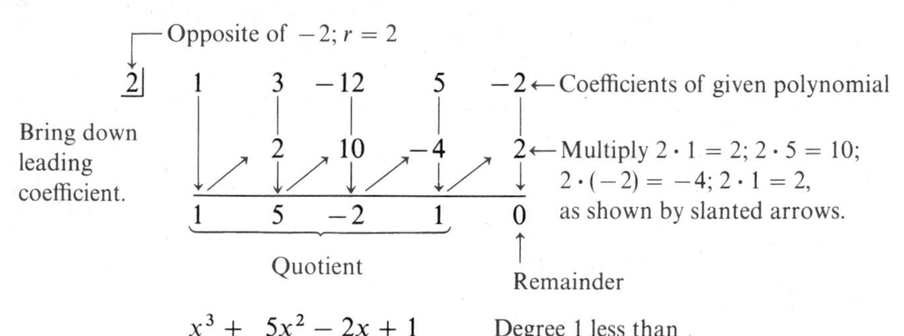

$$\frac{x^4 + 3x^3 - 12x^2 + 5x - 2}{x - 2} = \mathbf{x^3 + 5x^2 - 2x + 1}$$ ■

EXAMPLE 3 Divide $x^4 - 2x^3 - 10x^2 + 26x + 3$ by $x + 3$.

Solution

Opposite of $+3$

$$\begin{array}{r|rrrrr} -3 & 1 & -2 & -10 & 26 & 3 \\ & & -3 & 15 & -15 & -33 \\ \hline & 1 & -5 & 5 & 11 & -30 \end{array}$$

$3 \leftarrow$ Coefficients; -30: Remainder

The answer is $\mathbf{x^3 - 5x^2 + 5x + 11 + \dfrac{-30}{x + 3}}$ ■

EXAMPLE 4 Divide $x^5 - 3x^2 + 1$ by $x - 2$.

Solution

$$\begin{array}{r|rrrrrr} 2 & 1 & 0 & 0 & -3 & 0 & 1 \\ & & 2 & 4 & 8 & 10 & 20 \\ \hline & 1 & 2 & 4 & 5 & 10 & 21 \end{array} \leftarrow (R)$$

Be sure to include zero coefficients for missing terms.

The answer is $\mathbf{x^4 + 2x^3 + 4x^2 + 5x + 10 + \dfrac{21}{x - 2}}$ ■

PROBLEM SET 4.2

A

Fill in the blanks for the synthetic divisions in Problems 1–10.

1. $\dfrac{3x^3 - 9x^2 + 11x - 10}{x - 2}$

2	3	**b**	11	−10
		6	**c**	10
	a	−3	5	0

The answer is **d**.

2. $\dfrac{4x^3 - 6x^2 - 8x - 5}{x - 3}$

a	4	−6	**c**	**e**
		b	**d**	**f**
	4	6	10	25

The answer is **g**.

3. $\dfrac{2x^3 - 4x^2 - 10x + 12}{x + 2}$

a	2	−4	**c**	**e**
		b	**d**	**f**
	2	−8	6	0

The answer is **g**.

4. $\dfrac{5x^3 + 12x^2 + 2x - 4}{x + 3}$

a	5	12	**d**	**f**
		−15	9	−33
	b	**c**	**e**	**g**

The answer is **h**.

5. $\dfrac{x^3 + 7x^2 + 8x - 20}{x + 4}$

a	1	7	**d**	**f**
		−4	−12	16
	b	**c**	**e**	**g**

The answer is **h**.

6. $\dfrac{x^4 - 3x^2 + 2x - 7}{x - 1}$

a	**b**	**c**	**d**	**e**	**g**
		1	1	−2	0
	1	1	−2	**f**	**h**

The answer is **i**.

7. $\dfrac{x^4 + 2x^3 - 5x + 2}{x - 1}$

a	**b**	**c**	**d**	**e**	**g**
		1	3	3	−2
	1	3	3	**f**	**h**

The answer is **i**.

8. $\dfrac{x^5 - 1}{x - 1}$

a	**b**	**c**	**d**	**e**	**f**	**h**
		1	1	1	1	1
	1	1	1	1	**g**	**i**

The answer is **j**.

9. $\dfrac{x^5 - 32}{x + 2}$

a	**b**	**c**	**d**	**e**	**g**	**j**
		−2	4	−8	**h**	**k**
	1	−2	4	**f**	**i**	**l**

The answer is **m**.

10. $\dfrac{x^4 + 3x^3 - 2x^2}{x + 3}$

a	b	c	e	h	k
		−3	f	i	l
	1	d	g	j	m

The answer is ___n.

Use synthetic division to find the quotients in Problems 11–30.

11. $\dfrac{3x^3 - 2x^2 + 4x - 75}{x - 3}$

12. $\dfrac{3x^3 + 2x^2 - 4x + 8}{x + 2}$

13. $\dfrac{x^4 - 6x^3 + x^2 - 8}{x + 1}$

14. $\dfrac{x^4 - 3x^3 + x + 6}{x - 2}$

15. $\dfrac{2x^4 - 15x^2 + 8x - 3}{x + 3}$

16. $\dfrac{2x^4 - 15x - 2}{x - 2}$

17. $\dfrac{4x^5 - 3x^4 - 5x^3 + 4}{x - 1}$

18. $\dfrac{4x^3 - 3x^2 - 5x + 2}{x + 1}$

19. $\dfrac{3x^3 - 2x^2 + 4x - 24}{x - 2}$

20. $\dfrac{3x^3 + 2x^2 + 4x + 24}{x + 2}$

21. $\dfrac{x^3 + 5x^2 - 2x - 24}{x + 4}$

22. $\dfrac{x^3 + 3x^2 - 6x - 8}{x - 2}$

23. $\dfrac{x^3 - 4x^2 - 17x + 60}{x - 5}$

24. $\dfrac{x^5 - 3x^4 + x - 3}{x - 3}$

25. $\dfrac{4x^4 + 4x^3 - 15x^2 + 7}{x - 1}$

26. $\dfrac{5x^3 - 21x^2 - 13x - 35}{x - 5}$

27. $\dfrac{x^4 - x^3 + x^2 - x - 4}{x + 1}$

28. $\dfrac{2x^4 + 6x^3 - 4x^2 - 11x + 3}{x + 3}$

29. $\dfrac{3x^4 + 10x^3 - 8x^2 - 5x - 20}{x + 4}$

30. $\dfrac{x^4 - 9x^3 + 20x^2 - 15x + 18}{x - 6}$

B

In Problems 31–36, use long division to find $Q(x)$ and $R(x)$ if $P(x)$ is divided by $D(x)$.

31. $P(x) = 4x^4 - 14x^3 + 10x^2 - 6x + 2$; $D(x) = 2x - 1$
32. $P(x) = x^5 - x^3 + x^2 + x - 1$; $D(x) = x^2 + x$
33. $P(x) = x^5 - x^3 + x^2 + 1$; $D(x) = x^2 + x$
34. $P(x) = x^2 + 2x + 1$; $D(x) = x^3$
35. $P(x) = 6x^4 - 11x^3 + 6x^2 - 2x + 5$; $D(x) = 2x^2 + x + 1$
36. $P(x) = 2x^4 + 5x^3 - 16x^2 - 45x - 18$; $D(x) = x^2 + x - 2$

In Problems 37–56, use synthetic division to find $Q(x)$.

37. $\dfrac{3x^3 - 7x^2 - 5x + 2}{x - 3}$

38. $\dfrac{2x^3 - 3x^2 + 4x - 10}{x - 2}$

39. $\dfrac{x^4 - 3x^3 - 4x^2 + 2x - 5}{x - 4}$

40. $\dfrac{x^4 - 5x^3 + 2x^2 - x + 3}{x - 2}$

41. $\dfrac{5x^4 + 10x^3 - 20x^2 - 12x - 2}{x + 3}$

42. $\dfrac{2x^4 + 5x^3 + 2x^2 + 5x + 2}{x + 2}$

43. $\dfrac{x^5 - 3x^4 + 2x^2 - 5}{x + 2}$

44. $\dfrac{x^4 - 20x^2 - 10x - 50}{x - 5}$

45. $\dfrac{4x^3 - x^2 + 2x - 1}{x + 1}$

46. $\dfrac{6x^4 - x^2 + 1}{x - 3}$

47. $\dfrac{5x^5 - 2x + 1}{x + 2}$

48. $\dfrac{5x^4 + 7x^3 - 27x^2 + 14x - 12}{x - 1}$

49. $\dfrac{x^5 + 3x^3 - 3x^4 - 16x^2 + 21x - 6}{x - 3}$

50. $\dfrac{5x^4 - x + x^5 - 1}{x + 5}$

51. $\dfrac{x^4 - 12x^2 + 4x + 15}{(x + 1)(x - 3)}$

52. $\dfrac{x^4 - x^3 - 12x^2 + 28x - 16}{(x - 2)(x + 4)}$

53. $\dfrac{2x^3 - 3x^2 - 11x + 6}{(x - 3)(x + 2)}$

54. $\dfrac{2x^3 - 3x^2 - 2x + 3}{(2x - 3)(x + 1)}$

55. $\dfrac{2x^4 + 5x^3 - 25x^2 - 40x + 48}{(x - 3)(x + 4)}$

56. $\dfrac{2x^4 + 5x^3 - 16x^2 - 49x - 30}{(x + 1)(x + 3)}$

C

57. Devise a procedure for using synthetic division on

$$\frac{x^4 + 7x^3 + 5x^2 - 23x + 10}{x^2 + 4x - 5}$$

58. Devise a procedure for using synthetic division on

$$\frac{2x^4 - 7x^3 - 4x^2 + 27x - 18}{x^2 - x - 6}$$

59. Find K so that $x^4 + Kx^3 + 7x^2 - 2x + 8$ has no remainder when divided by $x + 2$.

60. Find h and k so that $x^4 + hx^3 - kx + 15$ has no remainder when divided by $x - 1$ and $x + 3$.

61. Let $P(x) = 2x^3 - x^2 + 5x + 3$.

a. Divide $P(x)$ by $2x + 1$ using long division.

b. Divide $P(x)$ by $x + \frac{1}{2}$ using synthetic division.

c. Compare your answers to parts **a** and **b**.

d. Notice that, in part **b**, the divisor of part **a** was divided by 2. If a divisor in any division problem is divided by 2, how will the quotient be affected?

e. Write out a general procedure for dividing $P(x)$ by $ax + b$ synthetically.

62. Check the procedure you outlined in Problem 61**e** by applying it to the following problems (divide synthetically):

a. $2x^3 + 7x^2 + x - 1$ divided by $2x + 1$

b. $6x^3 - 13x^2 + 14x - 12$ divided by $2x - 3$

c. $6x^3 + x^2 + 3x + 1$ divided by $2x + 1$

d. $3x^3 - 10x^2 - 19x + 5$ divided by $3x + 5$

4.3 GRAPHING POLYNOMIAL FUNCTIONS

Graphs of linear and quadratic functions were discussed in Chapter 2, so we begin this section by considering a process for graphing a cubic function. Suppose you wish to graph the cubic function

$$f(x) = 2x^3 - 3x^2 - 12x + 17$$

Since there is a one-to-one correspondence between points on the graph and ordered pairs satisfying this equation, you can begin by plotting points:

$$\begin{aligned} f(0) &= 2 \cdot 0^3 - 3 \cdot 0^2 - 12 \cdot 0 + 17 \\ &= 17 && (0, 17) \text{ is on the graph.} \\ f(1) &= 2 \cdot 1^3 - 3 \cdot 1^2 - 12 \cdot 1 + 17 \\ &= 2 - 3 - 12 + 17 \\ &= 4 && (1, 4) \text{ is on the graph.} \\ f(-1) &= 2(-1)^3 - 3(-1)^2 - 12(-1) + 17 \\ &= -2 - 3 + 12 + 17 \\ &= 24 && (-1, 24) \text{ is on the graph.} \end{aligned}$$

But you can see that this procedure could be very tedious by the time you find enough points to determine the shape of the curve. Instead, consider the **Remainder Theorem**.

Remainder Theorem

When a polynomial $f(x)$ is divided by $x - r$, the remainder is equal to $f(r)$.

To verify this theorem, recall the Division Algorithm:

$$\frac{P(x)}{D(x)} = Q(x) + \frac{R(x)}{D(x)} \quad \text{or} \quad P(x) = Q(x)D(x) + R(x)$$

In this context, $P(x) = f(x)$, $D(x) = x - r$, and $R(x)$ is a constant since the degree of $R(x)$ must be less than the degree of $D(x)$, which is 1. The Division Algorithm can be restated as

$$f(x) = Q(x)(x - r) + R$$

where R represents the remainder. Now

$$\begin{aligned} f(r) &= Q(r)(r - r) + R \\ &= R \end{aligned}$$

since $Q(r)(r - r) = Q(r) \cdot 0 = 0$. Points on the curve can therefore be found by synthetic division.

EXAMPLE 1 Find several points of the curve defined by the function $f(x) = 2x^3 - 3x^2 - 12x + 17$ by using synthetic division.

Solution Since the same polynomial is to be evaluated repeatedly, it is not necessary to recopy it each time. The work can be arranged as shown below:

Try to do the work mentally: → 1
$1 \cdot 2 + (-3) = -1$;
$1 \cdot (-1) + (-12) = -13$;
and so on.

	2	**−3**	**−12**	**17**	point
1	2	−1	−13	4	(1, 4)
−1	2	−5	−7	24	(−1, 24)
2	2	1	−10	−3	(2, −3)
−2	2	−7	2	13	(−2, 13)
0				17	(0, 17) ← Why is $f(0)$ always equal to the constant term?
3	2	3	−3	8	(3, 8)
−3	2	−9	15	−28	(−3, −28)
4	2	5	8	49	(4, 49)
−4	2	−11	32	−111	(−4, −111)

■

After you have found enough points, which is a much less tedious procedure when you use synthetic division, you can connect these points to draw a smooth curve. This property, called **continuity**, is studied extensively in calculus, but for our purposes we can state the following useful theorem even though its proof requires calculus:

Intermediate-Value Theorem for Polynomial Functions

If f is a polynomial function on $[a, b]$ such that $f(a) \neq f(b)$, then $f(x)$ takes on every value between $f(a)$ and $f(b)$ over the interval $[a, b]$.

This theorem allows you to connect the points found by synthetic division in order to draw a smooth curve.

EXAMPLE 2 Use the information in Example 1 as well as the Intermediate-Value Theorem to sketch the graph of the polynomial function

$$f(x) = 2x^3 - 3x^2 - 12x + 17$$

Solution Plot the points from Example 1, paying attention to the scales on the axes so that they accommodate most of the values obtained. Connect the points to draw a smooth curve as shown in Figure 4.3.

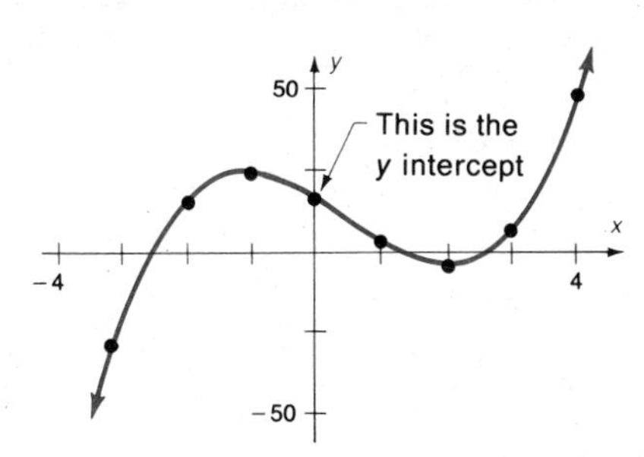

Figure 4.3 Graph of $f(x) = 2x^3 - 3x^2 - 12x + 17$

■

EXAMPLE 3 Sketch the graph of $f(x) = 3x^4 - 8x^3 - 30x^2 + 72x + 47$.

Solution *You* select the integers that are convenient.

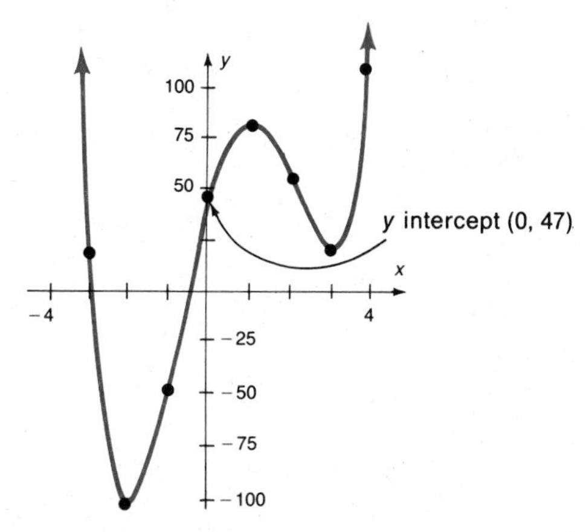

Figure 4.4 Graph of $f(x) = 3x^4 - 8x^3 - 30x^2 + 72x + 47$

	3	**−8**	**−30**	**72**	**47**	
−4	3	−20	50	−128	559	(−4, 559)
−3	3	−17	21	9	20	(−3, 20)
−2	3	−14	−2	76	−105	(−2, −105)
−1	3	−11	−19	91	−44	(−1, −44)
0					47	(0, 47)
1	3	−5	−35	37	84	(1, 84)
2	3	−2	−34	4	55	(2, 55)
3	3	1	−27	−9	20	(3, 20)
4	3	4	−14	16	111	(4, 111)

Plot the points and draw a smooth curve as in Figure 4.4. ■

The graphs of polynomial functions have *turning points* where the functions change from increasing to decreasing or from decreasing to increasing. A polynomial of degree n has at most $n - 1$ turning points. In calculus you will find how to locate the exact position of these turning points. For now, however, you will need to rely on synthetic division and plotting points to sketch graphs. If a polynomial function can be factored into linear factors, the number of plotted points can be reduced as illustrated by Example 4.

HP-28S ACTIVITY

A graphic calculator can be invaluable if we wish to have a graphic representation of the algebraic equation we are solving or investigating. Many techniques which up to now required calculus (maximum or minimum applications, for example) can now be done by using a graphic calculator and looking at the graph. Here, we will demonstrate the power of the HP-28S by reworking Example 3.

The first step is to key in the function:

['] [3] [×] [X] [^] [4] [−] [8] [×] [X] [^] [3] [−] [30]

[×] [X] [^] [2] [+] [72] [×] [X] [+] [47] [ENTER]

The default extremes of the graphing area for the HP-28S are $P_{min}(-6.8, -1.5)$ and $P_{max}(6.8, 1.6)$. A quick check shows the y intercept of 47 when $x = 0$. Since default Y_{max} is only 1.6, we see that we must alter the value for Y_{max} or we will see none of the curve. As a starting point, expand the y axis to a maximum of 125, or roughly 75 times. Entering the plot menu, use the [*H] menu key to expand the y axis the amount you wish (75 times for this example):

[PLOT] [STEQ] [NEXT] [75] [*H] [PREV] [DRAW]

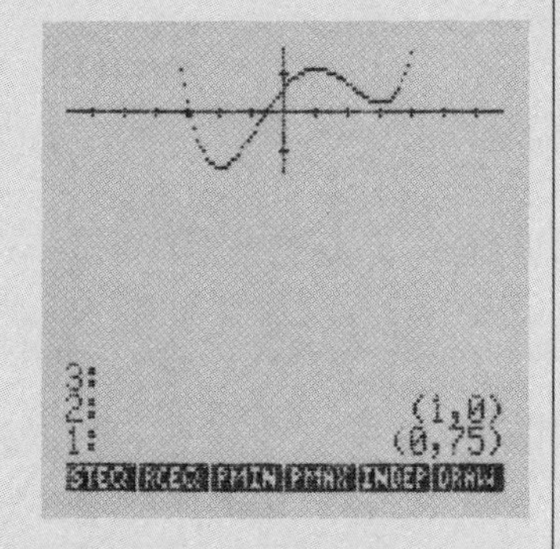

Notice that the scale is not shown. Using the [▶] key, move the cursor so that it aligns with the first tick mark on the positive x axis and press the [INS] key. In the same fashion, you can find the coordinates (this is called *digitizing*) of the first tick mark on the positive y axis. Next, press [ON] to return to the main menu. The two digitized points in stack registers 1 and 2 are $(0, 75)$ and $(1, 0)$, respectively, as shown at the right.

In the next section we will solve polynomial equations. You can use the HP-28S to find the points at which a curve crosses the x axis (the *zeros* of the function). First press the menu key [DRAW] once more, to redraw the curve. Since the equation was stored earlier, it is still in memory. Move the cursor (use the [▲] [▼] [◀] [▶] keys) and then the [INS] key to digitize the approximate points of crossing. For this example, these are $(-2.9, 0)$ and $(-0.6, 0)$. Press the [ON] key to return to the menu. Use the SOLVR function:

[SOLV] [SOLVR]

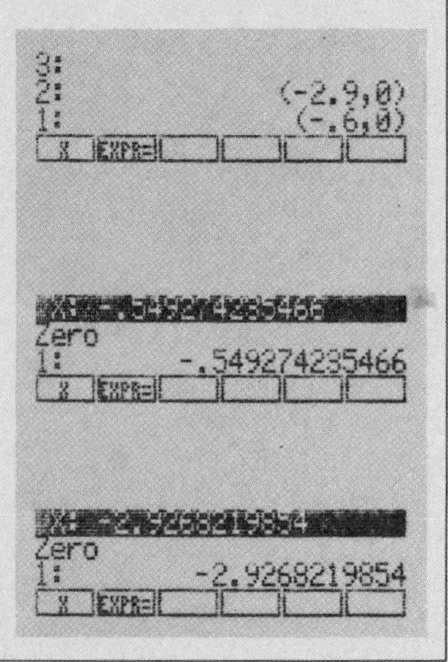

Pressing the [SOLVR] key brings us to the equation solving menu. In addition to the menu, we have the two digitized points in stack registers 1 and 2. Pressing the [X] key in the menu, load the contents of register 1 into the solver routine as a first guess. By pressing the shift key (the red button) and then the [X] key, the equation is solved. The result is given in the highlighted top row in the display:

$$x = -0.549274235466$$

Using the same technique again, we find that the second zero point is $x = -2.9268219854$.

EXAMPLE 4 Sketch the graph of $y = (x + 1)(x - 2)(3x - 2)$.

Solution Begin by locating the *critical values* for the polynomial function. Recall that these are the x values for which $y = 0$ (the x intercepts):

$$(x + 1)(x - 2)(3x - 2) = 0 \quad \leftarrow y = 0$$

$$x = -1, 2, \frac{2}{3}$$

These critical values divide the x axis into four regions:

$x < -1 \qquad -1 < x < \frac{2}{3} \qquad \frac{2}{3} < x < 2 \qquad x > 2$

Since $y = 0$ for $x = -1$, $x = \frac{2}{3}$, and $x = 2$, it follows that y must be either positive or negative in each of the four regions. That is, you need to determine whether y is positive or negative in *each* of the four regions. To do this, select an x value in *each* region and substitute that value into the function, as shown below:

	$x<-1$	$-1<x<\frac{2}{3}$	$\frac{2}{3}<x<2$	$x>2$
$x + 1$:	−	+	+	+
$x - 2$:	−	−	−	+
$3x - 2$:	−	−	+	+
	neg	pos	neg	pos

Wherever you have labeled the number line as *neg*, the graph is below the axis; and wherever you have labeled it *pos*, the graph is above the axis.

Plot some points (to the nearest tenth) using synthetic division to draw the graph, as shown in Figure 4.5. Synthetic division requires the coefficients in polynomial (not factored) form. Thus,

$$\begin{aligned} y &= (x + 1)(x - 2)(3x - 2) \\ &= 3x^3 - 5x^2 - 4x + 4 \end{aligned}$$

	3	**−5**	**−4**	**4**		
−2	3	−11	18	−32	(−2, −32)	Need a point to the left of the critical value −1.
−0.3	3	−5.9	−2.2	4.7	(−0.3, 4.7)	Need a point between −1 and 0.
0				4	(0, 4)	
1	3	−2	−6	−2	(1, −2)	
1.5	3	−0.5	−4.75	−3.125	(1.5, −3.1)	Need a point between 1 and 2.
3	3	4	8	28	(3, 28)	Need a point to the right of 2.

■

Figure 4.5 Graph of $y = (x + 1)(x - 2)(3x - 2)$ or $y = 3x^3 - 5x^2 - 4x + 4$

PROBLEM SET 4.3

A

Use synthetic division to find the values specified for the functions given in Problems 1–12.

1. $f(x) = 5x^3 - 7x^2 + 3x - 4$
   a. $f(1)$ **b.** $f(-1)$ **c.** $f(0)$
   d. $f(6)$ **e.** $f(-4)$
2. $g(x) = 4x^3 + 10x^2 - 120x - 350$
   a. $g(1)$ **b.** $g(-1)$ **c.** $g(0)$
   d. $g(5)$ **e.** $g(-5)$
3. $P(x) = x^4 - 10x^3 + 20x^2 - 23x - 812$
   a. $P(1)$ **b.** $P(-1)$ **c.** $P(0)$
   d. $P(3)$ **e.** $P(7)$
4. $f(t) = 3t^4 + 5t^3 - 8t^2 - 3t$
   a. $f(0)$ **b.** $f(1)$ **c.** $f(2)$
   d. $f(-2)$ **e.** $f(-4)$
5. $g(t) = 4t^4 - 3t^3 + 5t - 10$
   a. $g(0)$ **b.** $g(-1)$ **c.** $g(-2)$
   d. $g(5)$ **e.** $g(-3)$
6. $P(t) = t^4 - t^3 - 39t - 70$
   a. $P(0)$ **b.** $P(1)$ **c.** $P(-1)$
   d. $P(-5)$ **e.** $P(7)$
7. $f(h) = 8h^4 - 6h^3 + 5h^2 + 4h - 3$
   a. $f(0)$ **b.** $f(1)$ **c.** $f(\frac{1}{2})$
   d. $f(-\frac{1}{2})$ **e.** $f(-3)$
8. $g(h) = 16h^4 + 64h^3 + 19h^2 - 81h + 18$
   a. $g(0)$ **b.** $g(-2)$ **c.** $g(-3)$
   d. $g(\frac{1}{4})$ **e.** $g(\frac{3}{4})$
9. $P(x) = 4x^4 - 8x^3 - 43x^2 + 29x + 60$
   a. $P(-1)$ **b.** $P(1)$ **c.** $P(4)$
   d. $P(\frac{3}{2})$ **e.** $P(-\frac{5}{2})$
10. $f(x) = (x - 2)(x + 3)(2x - 5)$
    a. $f(2)$ **b.** $f(-3)$ **c.** $f(\frac{5}{2})$
    d. $f(1)$ **e.** $f(-2)$
11. $g(x) = (x - 1)(x - 4)(2x + 1)$
    a. $g(1)$ **b.** $g(4)$ **c.** $g(-\frac{1}{2})$
    d. $g(-2)$ **e.** $g(-1)$
12. $h(x) = (x + 1)(x + 2)(3x - 1)$
    a. $h(-1)$ **b.** $h(1)$ **c.** $h(-2)$
    d. $h(2)$ **e.** $h(\frac{1}{3})$

Sketch the graph of each polynomial function in Problems 13–18.

13. $f(x) = x^3 - 3x^2 + 10$
14. $f(x) = x^3 + 3x^2 + 11$
15. $f(x) = 2x^3 - 3x^2 - 12x + 3$
16. $f(x) = 2x^3 + 3x^2 - 12x + 48$
17. $f(x) = x^3 - 2x^2 + x - 5$
18. $f(x) = x^3 + 4x^2 - 3x + 2$

B

Sketch the graph of each polynomial function in Problems 19–46.

19. $f(x) = x^3 - 6x^2 + 9x - 9$
20. $f(x) = 2x^3 - 3x^2 - 36x + 78$
21. $f(x) = 4x^4 - 8x^3 - 43x^2 + 29x + 60$
22. $f(x) = 16x^4 + 64x^3 + 19x^2 - 81x + 18$
23. $f(x) = x^4 - 7x^2 - 2x + 2$
24. $f(x) = x^4 - 14x^3 + 58x^2 - 46x - 9$
25. $f(x) = x^6 - 4x^4 - 4x^2 + 4$
26. $f(x) = x^5 + 2x^4 - 5x^3 - 10x^2 + 4x + 8$
27. $y = 3x^4 - x^3 - 14x^2 + 4x + 8$
28. $y = 5x^4 + 3x^3 - 22x^2 - 12x + 8$
29. $y = x^4 - x^3 - 3x^2 + 2x + 4$
30. $y = x^4 - 2x^2 - 4x + 3$
31. $y = x^4 - 7x^2 - 2x + 2$
32. $y = x^3 + 3x^2 - x - 3$
33. $y = (x - 1)(x + 1)(x + 3)$
34. $y = (x - 1)(x - 4)(x + 3)$
35. $y = (x + 1)(x + 3)(2x - 5)$
36. $y = (x - 1)(x - 4)(2x + 1)$
37. $y = (x + 1)(x + 2)(3x - 1)$
38. $y = x(x - 3)(x + 3)$
39. $y = 3x^2(x - 3)(x + 1)$
40. $y = 5x^2(x - 4)(x + 2)$
41. $y = x^2(x^2 - 1)$
42. $y = x^2(x^2 - 4)$
43. $f(x) = 3x^4 - 7x^3 + 5x^2 + x - 10$
44. $f(x) = 8x^4 + 12x^3 - 3x^2 + 4x + 20$
45. $f(x) = x^5 - 3x^4 + 2x^3 - 7x + 15$
46. $f(x) = x^5 - 5x^4 + 3x^3 + x^2$
47. Graph $y = x^3$ and $y = -x^3$ on the same axes.
48. Graph $y = x^3$, $y = \frac{1}{2}x^3$, and $y = \frac{1}{100}x^3$ on the same axes.
49. Graph $y = x^3$ and $y - 2 = (x - 1)^3$ on the same axes.
50. Graph $y = x^4$ and $y = -x^4$ on the same axes.
51. Graph $y = x^4$, $y = \frac{1}{10}x^4$, and $y = \frac{1}{100}x^4$ on the same axes.
52. Graph $y = x^4$ and $y + 4 = (x - 2)^4$ on the same axes.

C

53. Using Problems 47–49, discuss the graph of $y - k = a(x - h)^3$ by comparing it with the graph of $y = x^3$.

54. Using Problems 50–52, discuss the graph of $y - k = a(x - h)^4$ by comparing it with the graph of $y = x^4$.

55. On the same axes, graph $y = x^n$ for $-1 \le x \le 1$, where n is equal to:
a. 0 **b.** 2 **c.** 4 **d.** 6

56. Repeat Problem 55 for $-4 \le x \le 4$ and $-500 \le y \le 500$.

57. On the same axes, graph $y = x^n$ for $-1 \le x \le 1$, where n is equal to:
a. 1 **b.** 3 **c.** 5 **d.** 7

58. Repeat Problem 57 for $-4 \le x \le 4$ and $-2000 \le y \le 2000$.

59. Using the results from Problems 55–58, make some conjectures concerning the graph of $y = x^n$.

60. Using the results from Problems 55–59, make a conjecture about the graph of $y - k = (x - h)^n$.

4.4 REAL ROOTS OF POLYNOMIAL EQUATIONS

If

$$P(x) = a_n x^n + a_{n-1}x^{n-1} + \cdots + a_2x^2 + a_1x + a_0 \qquad (a_n \neq 0)$$

then the *roots* or *solutions* of $P(x) = 0$ are values of x that satisfy this equation. Such an equation is called a **polynomial equation**. Recall from Chapter 1 that a is called a *zero* of a function P if $P(a) = 0$. Thus we speak of the **roots** of a polynomial equation and the **zeros** of a polynomial function. If a zero is a real number, then it is an ***x* intercept** of the graph of the polynomial function.

Consider the polynomial equation

$$4x^4 - 8x^3 + 43x^2 + 29x + 60 = 0$$

To solve this equation, find the values of x that make it true. As a first step, write the equation in factored form if possible. From the Division Algorithm, we know that if $R = 0$ and

$$P(x) = Q(x)(x - r) + R$$

then $P(x) = Q(x)(x - r)$. This says that $x - r$ is a factor of $P(x)$. But since you are looking for values of x such that $P(x) = 0$,

$$0 = Q(x)(x - r)$$

Notice that this equation is satisfied by $x = r$ and leads to a result called the **Factor Theorem**. The polynomial equation $Q(x) = 0$ is called a **depressed equation** of the polynomial equation $P(x) = 0$.

Factor Theorem

If r is a root of the polynomial equation $P(x) = 0$, then $x - r$ is a factor of $P(x)$. Moreover, if $x - r$ is a factor of $P(x)$, then r is a root of the polynomial equation $P(x) = 0$.

The Factor Theorem is a generalization of the method of solving quadratic equations by factoring. Remember: If $P \cdot Q = 0$, then $P = 0$ or $Q = 0$ (perhaps both). Suppose you wish to solve

$$x^2 - x - 2 = 0$$

You can do so by factoring:

$$(x - 2)(x + 1) = 0$$

Then each factor is set equal to zero to find

$$x = 2, -1$$

Suppose further that the same factor appears more than once in the factoring process, as in

$$(x - 3)(x - 3) = 0$$

Then there is a single root (which occurs twice) to give

$$x = 3$$

In this chapter it will be useful to attach some terminology to a repeated root. If a factor $x - r$ occurs exactly k times in the factorization of $P(x)$, then r is called a **zero of multiplicity k**. If a factor $x - r$ occurs exactly k times in the factorization of $P(x)$ in the equation $P(x) = 0$, then r is called a **root of multiplicity k**. In the preceding illustration, 3 is a root of multiplicity 2. If

$$f(x) = (x - 1)(x - 3)(x - 4)(x - 1)(x - 3)(x - 1)$$

the **zeros** are 1 (multiplicity 3), 3 (multiplicity 2), and 4. Notice that this is a function, not an equation, so we use the word *zero*. On the other hand,

$$(x - 1)(x - 3)(x - 4)(x - 1)(x - 3)(x - 1) = 0$$

has **roots** 1 (multiplicity 3), 3 (multiplicity 2), and 4. This is an equation, so we use the word *root* (or *solution*).

The relationship between roots of the polynomial equation $f(x) = 0$ and factors of $f(x)$, as described by the Factor Theorem, leads to a statement concerning the number of roots to expect for a polynomial equation of degree n. Suppose $P(x) = 0$ is a polynomial equation in which $P(x)$ is a third-degree polynomial. It is impossible for $P(x) = 0$ to have more than three roots. If it had more, say four roots (r_1, r_2, r_3, r_4), then the Factor Theorem provides

$$P(x) = a_4(x - r_1)(x - r_2)(x - r_3)(x - r_4)$$

so that $P(x)$ is a *fourth-degree* polynomial. Thus a third-degree polynomial equation cannot have more than three roots.

The related question—Does it necessarily have three zeros?—is answered by the Fundamental Theorem of Algebra, which is considered in the next section.

Root Limitation Theorem

A polynomial function f of degree n has, at most, n distinct zeros.

As you saw in the previous section, there is a close relationship between the zeros of a polynomial function and its graph. Suppose, by synthetic division, you find for two values a and b that $P(a)$ and $P(b)$ are opposite in sign. Then the Intermediate-Value Theorem for polynomial functions tells you that there is a value r such that $a < r < b$ and $P(r) = 0$.

Location Theorem

If f is a polynomial function such that $f(a)$ and $f(b)$ are opposite in sign, then there is at least one real zero on the interval between a and b.

Next you need some reasonable method for finding the zeros, since you cannot simply find them by trial and error, even with the Location Theorem and synthetic division. The best that mathematicians can offer is a theorem that provides a list of *possible* rational roots of the polynomial equation $f(x) = 0$. Not every number on the list will be a root, but every rational root of the polynomial equation will appear someplace on the list. Given this *finite* list (which admittedly might be large), you can check values from this list using synthetic division and the Location Theorem.

Rational Root Theorem

> If $P(x) = a_n x^n + a_{n-1}x^{n-1} + \cdots + a_1 x + a_0$ has integer coefficients and p/q (where p/q is reduced) is a rational zero, then p is a factor of a_0 and q is a positive factor of a_n.

To use this theorem, make a list of *all* factors of a_0 and divide these integers by the factors of a_n. Notice also, if $a_n = 1$, then all rational zeros are integers that divide a_0. The procedure for finding all possible rational roots is not very difficult if you work systematically.

EXAMPLE 1 List all possible rational roots of $x^3 - x^2 - 4x + 4 = 0$.

Solution

p: $a_0 = 4$, with factors $1, -1, 2, -2, 4, -4$

q: $a_n = 1$, with factor 1

Shorten these lists by using the $\pm$ sign: p: $\pm 1, \pm 2, \pm 4$

Form all possible fractions:

$$\frac{p}{q}: \frac{1}{1}, \frac{-1}{1}, \frac{2}{1}, \frac{-2}{1}, \frac{4}{1}, \frac{-4}{1}$$

Simplifying and not bothering to rewrite those that are repeated, we find that the possible rational roots are $\pm \mathbf{1}, \pm \mathbf{2}, \pm \mathbf{4}$. ■

EXAMPLE 2 List all possible rational zeros of $P(x) = 4x^4 - 8x^3 - 43x^2 + 29x + 60$.

Solution

$p(a_0 = 60)$: $\pm 1, \pm 2, \pm 3, \pm 4, \pm 5, \pm 6, \pm 10, \pm 12, \pm 15, \pm 20, \pm 30, \pm 60$

$q(a_n = 4)$: 1, 2, 4

$$\frac{p}{q}: \pm 1, \pm\frac{1}{2}, \pm\frac{1}{4}, \pm 2, \pm 3, \pm\frac{3}{2}, \pm\frac{3}{4}, \pm 4, \pm 5, \pm\frac{5}{2}, \pm\frac{5}{4},$$

$$\pm 6, \pm 10, \pm 12, \pm 15, \pm\frac{15}{2}, \pm\frac{15}{4}, \pm 20, \pm 30, \pm 60$$

Note: If a factor is already listed, it is not repeated. For example, $\pm\frac{2}{4}$ is not listed separately from $\pm\frac{1}{2}$. ■

You can see from the examples that the list of possible rational zeros may be quite large, but it is a *finite* list, so with enough time and effort the entire list could be checked. Usually this is not necessary if you first pick the values that are easiest to check. (That is, do not start with the fractions or large numbers.) There is also another theorem, called the **Upper and Lower Bound Theorem**, that helps to rule out many of the values listed with the Rational Root Theorem. If f is a polynomial function, then a real number b is called an **upper bound** for the polynomial equation

$f(x) = 0$ if there is no root, or solution, larger than b. A real number a is a **lower bound** if there is no solution less than a.

Upper and Lower Bound Theorem

If $a > 0$ and, in the synthetic division of $P(x)$ by $x - a$, all the numbers in the last row have the *same sign*, then a is an *upper bound* for the roots of $P(x) = 0$. If $b < 0$ and, in the synthetic division of $P(x)$ by $x - b$, the numbers in the last row *alternate in sign*, then b is a *lower bound* for the roots of $P(x) = 0$.

EXAMPLE 3 Solve $2x^4 - 5x^3 - 8x^2 + 25x - 10 = 0$.

Solution

$p(a_0 = -10)$: $\pm 1, \pm 2, \pm 5, \pm 10$

$q(a_n = 2)$: 1, 2

$\frac{p}{q}$: $\pm 1, \pm\frac{1}{2}, \pm 2, \pm 5, \pm\frac{5}{2}, \pm 10$

	2	**−5**	**−8**	**25**	**−10**
1	2	−3	−11	14	4
−1	2	−7	−1	26	−36
2	2	−1	−10	5	0

Begin with the values that are easiest to check. In Examples 3–5 we will use this shaded portion to tell you the value checked is not a root.

$x - 2$ is a factor.

Since $x - 2$ is a factor, you can write the polynomial equation as

$$(x - 2)(2x^3 - x^2 - 10x + 5) = 0$$

Now focus your attention on the depressed equation

$$2x^3 - x^2 - 10x + 5 = 0$$

by using these coefficients in the synthetic division process. It is not necessary to recopy these coefficients in your work.

	2	−1	−10	5	
−2	2	−5	0	5	
5	2	9	35	180	← All sums are positive, so 5 is an upper bound. No larger values need to be checked.
−5	2	−11	45	−220	← Sums have alternating signs, so −5 is a lower bound. No smaller values need be checked.
$\frac{1}{2}$	2	0	−10	0	← $x - \frac{1}{2}$ is a factor. The resulting depressed equation is now quadratic, so you can stop the synthetic division.

$$(x - 2)(x - \tfrac{1}{2})(2x^2 - 10) = 0$$

$$2(x - 2)(x - \tfrac{1}{2})(x^2 - 5) = 0$$

Note: The resulting quadratic may not have rational roots. In fact, you may need the quadratic formula for its solution.

The roots are **$2, \frac{1}{2}, \sqrt{5}, -\sqrt{5}$**. ■

EXAMPLE 4 Solve $4x^4 - 8x^3 - 43x^2 + 29x + 60 = 0$.

Solution The list of possible rational roots is shown in Example 2.

	4	**−8**	**−43**	**29**	**60**	
1	4	−4	−47	−18	42	
−1	4	−12	−31	60	0	⟵ $x + 1$ is a factor, so -1 is a root. Do not recopy, but use these *coefficients* as you continue synthetic division.
2	4	−4	−39	−18		
−2	4	−20	9	42		
−3	4	−24	41	−63		⟵ -3 is a lower bound; since the Location Theorem says there is a root between -2 and -3, look on the list of possible rational roots for the next number to try.
$-\frac{5}{2}$	4	−22	24	0		

$x + \frac{5}{2}$ is a factor. The resulting depressed equation is quadratic, so now you can write the given polynomial equation in factored form and complete the factoring process directly.

$$4x^4 - 8x^3 - 43x^2 + 29x + 60 = 0$$

$$(x + 1)\left(x + \frac{5}{2}\right)(4x^2 - 22x + 24) = 0$$

Do you see where these factors come from in the synthetic division process?

$$(x + 1)\left(x + \frac{5}{2}\right)(2)(2x^2 - 11x + 12) = 0$$

$$(x + 1)\left(x + \frac{5}{2}\right)(2)(2x - 3)(x - 4) = 0$$

Now use the Factor Theorem.

The roots are $\mathbf{-1, -\frac{5}{2}, \frac{3}{2}}$**, and 4.** ■

EXAMPLE 5 Solve $8x^5 - 44x^4 + 86x^3 - 73x^2 + 28x - 4 = 0$.

Solution

$p(a_0 = -4)$: $\pm 1, \pm 2, \pm 4$

$q(a_n = 8)$: $1, 2, 4, 8$

$\frac{p}{q}$: $\pm 1, \pm\frac{1}{2}, \pm\frac{1}{4}, \pm\frac{1}{8}, \pm 2, \pm 4$

Note: Do not forget possible multiple roots. Just because $x - 2$ was a factor once does not mean it cannot be again.

	8	**−44**	**86**	**−73**	**28**	**−4**	
1	8	−36	50	−23	5	1	
−1	8	−52	138	−211	239	−243	⟵ Lower bound
2	8	−28	30	−13	2	0	⟵ $x - 2$ is a factor.
4	8	4	46	171	686		⟵ Upper bound
2	8	−12	6	−1	0		⟵ $x - 2$ is a factor.
2	8	4	14	27			⟵ Upper bound; this is a new upper bound because we are working with the depressed equation.
$\frac{1}{2}$	8	−8	2	0			⟵ $x - \frac{1}{2}$ is a factor.

$$(x-2)(x-2)\left(x-\frac{1}{2}\right)(8x^2-8x+2)=0$$

$$2(x-2)(x-2)\left(x-\frac{1}{2}\right)(4x^2-4x+1)=0 \qquad \text{Common factor}$$

$$2(x-2)(x-2)\left(x-\frac{1}{2}\right)(2x-1)(2x-1)=0$$

The roots are 2 and $\frac{1}{2}$; 2 is a root of multiplicity 2 and $\frac{1}{2}$ is a root of multiplicity 3. ■

In Examples 3 and 4 the degree was 4 and in both cases you found four roots. The Root Limitation Theorem tells you that there cannot be more roots. Even when you find multiple roots as illustrated by Example 5, you know you have found all the real roots. Suppose, however, you attempt to solve a polynomial equation that has fewer *real* roots than the degree. In such a situation you will not know whether you cannot find the real roots because although they *are not* rational they are still real (that is, are x intercepts) or because they are simply not real numbers. The following theorem will help you to answer this dilemma by telling you when you have found all the positive or negative real roots. When applying this theorem, remember that a root of multiplicity m is counted as m roots.

WARNING: Note the correct way to count multiple roots.

Descartes' Rule of Signs

Let $P(x)$ be a polynomial with real coefficients written in descending powers of x. Count the number of sign changes in the signs of the coefficients.

1. The number of positive real zeros is equal to the number of sign changes or is equal to that number decreased by an even integer.
2. The number of negative real zeros is equal to the number of sign changes in $P(-x)$ or is equal to that number decreased by an even integer.

EXAMPLE 6

$$f(x)=\underbrace{2x^4 - 5x^3}_{1} \underbrace{- 8x^2 + 25x}_{2} \underbrace{+ 25x - 10}_{3} \qquad \text{From Example 3}$$

There are three sign changes, so there are three or one positive real zeros. Next calculate $f(-x)$:

$$\begin{aligned} f(-x) &= 2(-x)^4 - 5(-x)^3 - 8(-x)^2 + 25(-x) - 10 \\ &= 2x^4 \underbrace{+ 5x^3 - 8x^2}_{1} - 25x - 10 \end{aligned}$$

There is one sign change on the coefficients of $f(-x)$, so there is one negative real zero. Compare this result with the answer you found in Example 3:

$\underbrace{2, \frac{1}{2}, \sqrt{5}}_{\text{3 positive zeros}}, \; \underbrace{-\sqrt{5}}_{\text{1 negative zero}}$ ■

EXAMPLE 7

$$f(x) = 8x^5 - 44x^4 + 86x^3 - 73x^2 + 28x - 4 = 0 \qquad \text{From Example 5}$$

(sign changes numbered 1, 2, 3, 4, 5)

There are five sign changes here, so there are five, three, or one positive real roots.

$$f(-x) = -8x^5 - 44x^4 - 86x^3 - 73x^2 - 28x - 4 = 0$$

Here there are no sign changes, so there are no negative real roots. Compare this result with the answer you found in Example 5: $2, \frac{1}{2}$. There seems to be a discrepancy, but remember that 2 has multiplicity 2 and $\frac{1}{2}$ has multiplicity 3, *so think*:

$$\underbrace{2, 2, \frac{1}{2}, \frac{1}{2}, \frac{1}{2}}$$

5 positive real roots
and no negative real roots ■

The following example ties together some of the ideas of this section and the preceding section.

EXAMPLE 8 Solve $x^4 - 3x^2 - 6x - 2 = 0$.

Solution

$p(a_0 = -2)$: $\pm 1, \pm 2$ $\quad \frac{p}{q}$: $\pm 1, \pm 2$

$q(a_n = 1)$: 1

	1	**0**	**−3**	**−6**	**−2**	
1	1	1	−2	−8	−10	
2	1	2	1	−4	−10	
−1	1	−1	−2	−4	2	
−2	1	−2	1	−8	14	← Lower bound

There are no rational roots (we have tried all the numbers on our list). Next verify the type of roots by using Descartes' Rule of Signs:

$$f(x) = x^4 - 3x^2 - 6x - 2$$

(sign change numbered 1)

One positive root

$$f(-x) = x^4 - 3x^2 + 6x - 2$$

(sign changes numbered 1, 2, 3)

Three or one negative root(s)

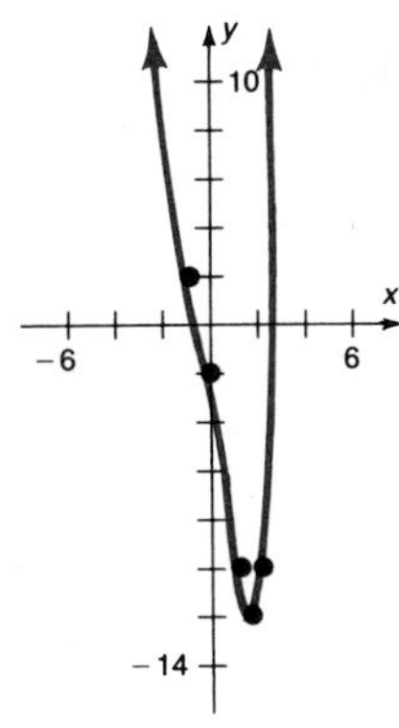

Figure 4.6 Graph of $y = x^4 - 3x^2 - 6x - 2$

Using synthetic division to find some additional points: $(0, -2)$ for the y intercept; $(1.5, -12.6875)$; and $(3, 34)$ for the upper bound. Plot the known points as shown in Figure 4.6. You would expect roots between -1 and 0 as well as between 2 and 3. Moreover, because of the upper and lower bounds, as well as Descartes' Rule of Signs, you would expect these to be the only real roots. Since these roots are not rational you will not be able to find them exactly. You can, however, approximate them to any desired degree of accuracy by using synthetic division. For the root between -1 and 0:

A calculator would be very helpful

	1	0	-3	-6	-2	
-0.5	1	-0.5	-2.75	-4.625	0.3125	Root between -0.5 and -0.4 since one is negative and the other is positive
-0.4	1	-0.4	-2.84	-4.864	-0.0544	
-0.41	1	-0.41	-2.8319	-4.8389	-0.0161	Root between -0.41 and -0.42
-0.42	1	-0.42	-2.8236	-4.8141	0.0219	

Continue in this fashion to approximate the root to any degree of accuracy desired. (This would be a good problem for a computer.) Repeat the procedure for the root between 2 and 3 (it is about 2.41). Thus the real roots (to the nearest tenth) are **-0.4, 2.4.** ■

HP-28S ACTIVITY

As we saw in the previous HP-28S Activity section (page 136), the graphic calculator can be an extremely powerful tool in visualizing a function. In this example, we will use the HP-28S to solve the polynomial equation given in Example 8.

If you examine this polynomial, you will find that you do not obtain an acceptable graph using the default scale values. So, we will expand the scale in the y direction 10 times as a first try to obtain a "good graph." First key in the expression:

['] [X] [^] [4] [−] [3] [×] [X] [^] [2] [−] [6] [×] [X] [−] [2] [ENTER]

Then enter the PLOT menu and store the equation (STEQ command). Finally, expand the height by a factor of 10:

[PLOT] [STEQ] [NEXT] [10] [*H] [PREV] [DRAW]

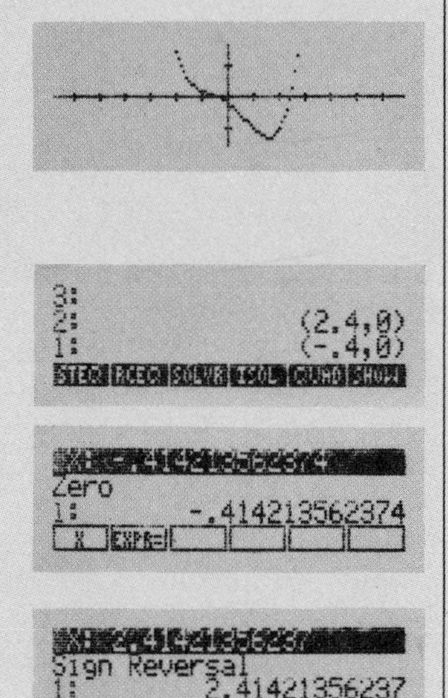

This is an acceptable graph. Using the cursor keys, move along the positive x axis to approximate the point at which the curve crosses the axis. Digitize this point using the [INS] key. Repeat the process for the point at which the curve crosses the negative x axis. Finally, use the [SOLV] key to enter the SOLVR menu:

[SOLV]

Note that the two digitized points are in registers 1 and 2. We will use them next. Now press the [SOLVR] menu key and you see a new menu. By pressing the [X] menu key, you can load the contents of register 1 into the SOLVR routine and proceed to find a closer approximation to the answer by pressing the shift key and then the [X] menu key. In the same fashion, we find a closer approximation to the answer for the second digitized point. Compare this with the synthetic division in Example 8 and notice the degree of accuracy we obtain from the HP-28S.

The roots are $\{-0.414213562374, 2.41421356237\}$.

PROBLEM SET 4.4

A

In Problems 1–12, find the zeros of the polynomial and state the multiplicity of each zero.

1. $f(x) = (x - 2)(x + 3)^2$
2. $f(x) = (x + 1)^2(x - 3)^2$
3. $f(x) = x^3(2x - 3)^2$
4. $f(x) = x^2(3x + 1)^3$
5. $f(x) = (x^2 - 1)^2(x + 2)$
6. $f(x) = (x - 5)(x^2 - 4)^3$
7. $f(x) = (x^2 + 2x - 15)^2$
8. $f(x) = (6x^2 + 7x - 3)^2$
9. $f(x) = x^4 - 8x^3 + 16x^2$
10. $f(x) = x^4 + 6x^3 + 9x^2$
11. $f(x) = (x^3 - 9x)^2$
12. $f(x) = (x^3 - 25x)^2$

In Problems 13–24, use Descartes' Rule of Signs to state the number of possible positive and negative real roots.

13. $x^4 - 3x^3 + 7x^2 - 19x + 15 = 0$
14. $3x^3 - 7x^2 + 5x + 7 = 0$
15. $2x^5 + 6x^4 - 3x + 12 = 0$
16. $x^3 + 2x^2 - 5x - 6 = 0$
17. $x^3 + 3x^2 - 4x - 12 = 0$
18. $2x^3 + x^2 - 13x + 6 = 0$
19. $2x^3 - 3x^2 - 32x - 15 = 0$
20. $x^4 - 12x^3 + 54x^2 - 108x + 81 = 0$
21. $x^4 + 3x^3 - 20x^2 - 3x + 18 = 0$
22. $x^4 - 13x^2 + 36 = 0$
23. $2x^2 + 6x - 3x^3 - 4 = 0$
24. $5x^3 - 2x^4 + x^2 - 7 = 0$

Find the possible rational roots for the polynomial equations in Problems 25–36.

25. $x^4 - 3x^3 + 7x^2 - 19x + 15 = 0$
26. $3x^3 - 7x^2 + 5x + 7 = 0$
27. $2x^5 + 6x^4 - 3x + 12 = 0$
28. $x^3 + 2x^2 - 5x - 6 = 0$
29. $x^3 + 3x^2 - 4x - 12 = 0$
30. $2x^3 + x^2 - 13x + 6 = 0$
31. $2x^3 - 3x^2 - 32x - 15 = 0$
32. $x^4 - 12x^3 + 54x^2 - 108x + 81 = 0$
33. $x^4 + 3x^3 - 20x^2 - 3x + 18 = 0$
34. $x^4 - 13x^2 + 36 = 0$
35. $5x^2 + 6x^3 - 2x - 1 = 0$
36. $10x^2 - 8x^3 + 17x - 10 = 0$

B

Solve the polynomial equations in Problems 37–56.

37. $x^3 - x^2 - 4x + 4 = 0$
38. $2x^3 - x^2 - 18x + 9 = 0$
39. $x^3 - 2x^2 - 9x + 18 = 0$
40. $x^3 + 2x^2 - 5x - 6 = 0$
41. $x^3 + 3x^2 - 4x - 12 = 0$
42. $2x^3 + x^2 - 13x + 6 = 0$
43. $2x^3 - 3x^2 - 32x - 15 = 0$
44. $x^4 - 12x^3 + 54x^2 - 108x + 81 = 0$
45. $x^4 + 3x^3 - 19x^2 - 3x + 18 = 0$
46. $x^4 - 13x^2 + 36 = 0$
47. $x^3 + 15x^2 + 71x + 105 = 0$
48. $x^3 - 15x^2 + 74x - 120 = 0$
49. $8x^3 - 12x^2 - 66x + 35 = 0$
50. $12x^3 + 16x^2 - 7x - 6 = 0$
51. $x^5 + 8x^4 + 10x^3 - 60x^2 - 171x - 108 = 0$
52. $x^5 + 6x^4 + x^3 - 48x^2 - 92x - 48 = 0$
53. $x^7 + 2x^6 - 4x^5 - 2x^4 + 3x^3 = 0$
54. $x^6 - 3x^4 + 3x^2 - 1 = 0$
55. $x^6 - 12x^4 + 48x^2 - 64 = 0$
56. $x^7 + 3x^6 - 4x^5 - 16x^4 - 13x^3 - 3x^2 = 0$
57. Does there exist a real number that exceeds its cube by 1?
58. ***Consumer*** The dimensions of a rectangular box are consecutive integers, and its volume is 720 cm^3. What are the dimensions of the box?
59. ***Engineering*** A 2-cm-thick slice is cut from a cube, leaving a volume of 384 cm^3. What is the length of a side of the original cube?
60. ***Engineering*** A rectangular sheet of tin with dimensions 3×5 m has equal squares cut from its four corners. The resulting sheet is folded up on the sides to form a topless box. Find all possible dimensions of the cutout square to the nearest 0.1 m such that the box has a volume of 1 m^3.

C

Italian mathematicians discovered the algebraic solution of cubic and quartic equations in the sixteenth century. At that time they would challenge one another to solve certain equations. Problems 61–64 were such challenge problems. Find the real roots in each problem to the nearest tenth.

61. In 1515, Scipione del Ferro solved the cubic equation $x^3 + mx + n = 0$ and revealed his secret to his pupil Antonio Fior. At about the same time, Tartaglia solved the equation $x^3 + px^2 = n$. Fior thought Tartaglia was bluffing and challenged him to a public contest of solving cubic equations. According to the historian Howard Eves, Tartaglia triumphed completely in this contest. Solve the cubic $x^3 + px^2 = n$, where $p = 5$ and $n = 21$.
62. Girolamo Cardano stole the solution of the cubic equation from Tartaglia and published it in his *Ars Magna.*

Tartaglia protested, but Cardano's pupil, Ludovico Ferrari (who solved the biquadratic equation), claimed that both Cardano and Tartaglia stole it from del Ferro. According to the historian Howard Eves, there was a dispute from which Tartaglia was lucky to have escaped alive. One of the problems in *Ars Magna* was $x^3 - 63x = 162$. Solve this cubic.

63. Cardano solved the quartic $13x^2 = x^4 + 2x^3 + 2x + 1$. Find the real roots for this equation.

64. In 1540, Cardano was given the problem "Divide 10 into three parts such that they shall be in continued proportion and that the product of the first two shall be 6." Let x, y, and z be the three parts. Then $x + y + z = 10$. Also,

$$\frac{x}{y} = \frac{y}{z} \qquad \text{and} \qquad xy = 6$$

Eliminating x and z, you obtain $y^4 + 6y^2 + 36 = 60y$. Find the real roots for this equation.

4.5 THE FUNDAMENTAL THEOREM OF ALGEBRA*

In 1799, a 22-year-old graduate student named Karl Gauss proved in his doctoral thesis that every polynomial equation has at least one solution in the complex numbers. This, of course, is an assumption that you have made throughout your study of algebra—from the time you solved first-degree equations in beginning algebra until now. It is an idea so basic to algebra that it is called the Fundamental Theorem of Algebra. This theorem strengthens the Root Limitation Theorem of the last section. However, in order to prove this theorem it is necessary to allow the domain to be the set of complex numbers. Remember that the set of complex numbers has the set of real numbers as a subset so that when we speak of complex coefficients of a polynomial equation, we are including all those polynomial equations previously considered in this chapter.

Fundamental Theorem of Algebra

If $P(x)$ is a polynomial of degree $n \geq 1$ with complex coefficients, then $P(x) = 0$ has at least one complex root.

If an equation has one solution, a depressed equation of 1 degree less may be obtained. That new equation, according to the Fundamental Theorem, has a root. This root may now be used to obtain an equation of lower degree. The result of this process suggests the following theorem.

Number of Roots Theorem

If $P(x)$ is a polynomial of degree $n \geq 1$ with complex coefficients, then $P(x) = 0$ has exactly n roots (if roots are counted according to their multiplicity).

Of course, the roots need not be distinct or real. Consider the following example.

EXAMPLE 1 Show that $x^6 - 2x^3 + x^2 - 2x + 2 = 0$ has at least two nonreal complex roots.

---

* This section is optional and requires complex numbers.

Solution Check Descartes' Rule of Signs:

$$P(x) = x^6 - 2x^3 + x^2 - 2x + 2$$

1 2 3 4

4, 2, or 0 positive real roots

$$P(-x) = x^6 + 2x^3 + x^2 + 2x + 2$$

0 negative real roots

The polynomial equation has six roots, and at most four of these are real (positive). Thus, at least two roots are complex and nonreal. ■

If synthetic division is used on the equation in Example 1, two positive real roots are quickly found:

	1	0	0	−2	1	−2	2	
1	1	1	1	−1	0	−2	0	
1	1	2	3	2	2	0		All positive, upper bound

Notice that $x = 1$ is a root of *multiplicity 2.* When counting the number of roots using Descartes' Rule, multiple roots are *not* counted as a single root. A root of multiplicity 2 counts as two roots and roots of multiplicity n count as n roots.

Note further that the depressed equation produced by the synthetic division has all positive coefficients; thus, all the positive roots have been found. Hence, the polynomial equation $x^6 - 2x^3 + x^2 - 2x + 2 = 0$ has one real root (with multiplicity 2) and four nonreal complex roots.

We can now distinguish between roots and x intercepts of polynomial equations. There are n roots of an nth-degree polynomial equation $P(x) = 0$. The **real roots** correspond to the x intercepts of the graph of $y = P(x)$ whereas the **imaginary roots** do not.

EXAMPLE 2 Solve $x^4 - 2x^3 + x^2 - 8x - 12 = 0$, and draw the graph of $y = x^4 - 2x^3 + x^2 - 8x - 12$.

Solution Check Descartes' Rule of Signs:

$$P(x) = x^4 - 2x^3 + x^2 - 8x - 12$$

1 2 3

3 or 1 positive real roots

$$P(-x) = x^4 + 2x^3 + x^2 + 8x - 12$$

1

1 negative real root

p: $\pm 1, \pm 2, \pm 3, \pm 4, \pm 6, \pm 12$ $\quad \frac{p}{q}$: $\pm 1, \pm 2, \pm 3, \pm 4, \pm 6, \pm 12$

q: 1

	1	−2	1	−8	−12	
1	1	−1	0	−8	−20	
−1	1	−3	4	−12	0	Use this depressed equation.
−1	1	−4	8	−20		Since −1 could be a multiple root, try it again.
3	1	0	4	0		

The depressed equation is

$$x^2 + 4 = 0$$

$$x = \pm 2i$$

The roots are $-1, 3, 2i, -2i$; this is consistent with the results from Descartes' Rule of Signs (one positive and one negative real root). It is also consistent with the Number of Roots Theorem (degree 4, four roots). Now you would also expect the graph to have two x intercepts corresponding to the real roots. The graph is shown in Figure 4.7.

	1	−2	1	−8	−12
0					−12
1	1	−1	0	−8	−20
−1	1	−3	4	−12	0
2	1	0	1	−6	−24
−2	1	−4	9	−26	40
3	1	1	4	4	0
4	1	2	9	28	100

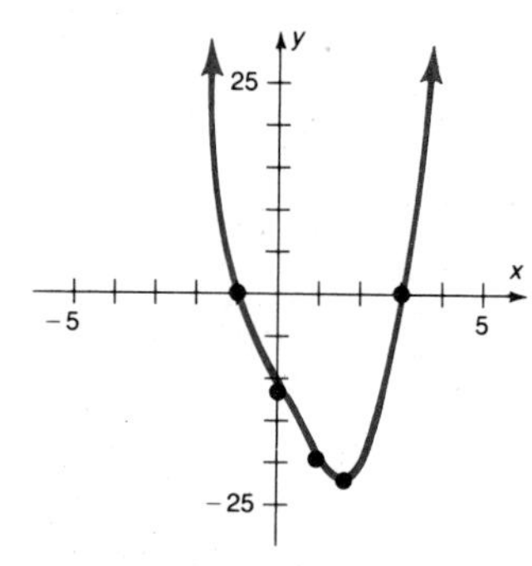

Figure 4.7 Graph of $y = x^4 - 2x^3 + x^2 - 8x - 12$ ■

Another result that is sometimes helpful when finding roots is the **Conjugate Pair Theorem.**

Conjugate Pair Theorem

1. If $P(x) = 0$ is a polynomial equation with real coefficients, then when $a + bi$ is a root, $a - bi$ is also a root (a and b are real numbers).
2. If $P(x) = 0$ is a polynomial equation with rational coefficients, then when $m + \sqrt{n}$ is a root, $m - \sqrt{n}$ is also a root (m and n are rational numbers and $\sqrt{n}$ is irrational).

EXAMPLE 3 Solve $x^4 - 4x - 1 = 4x^3$ given that $2 + \sqrt{5}$ is a root.

Solution

$$x^4 - 4x^3 - 4x - 1 = 0$$

$2+\sqrt{5}$	1	-4	0	-4	-1
		$2+\sqrt{5}$	1	$2+\sqrt{5}$	1
	1	$-2+\sqrt{5}$	1	$-2+\sqrt{5}$	0

And $2 - \sqrt{5}$ must be a root also:

$$\begin{array}{r|rrrr} 2-\sqrt{5} & 1 & -2+\sqrt{5} & 1 & -2+\sqrt{5} \\ & & 2-\sqrt{5} & 0 & 2-\sqrt{5} \\ \hline & 1 & 0 & 1 & 0 \end{array}$$

The depressed equation is

$$x^2 + 1 = 0$$
$$x = \pm i$$

The roots are $\mathbf{2 \pm \sqrt{5}, \pm i}$. ■

Alternate Solution The synthetic division solution shown above is rather cumbersome, so we offer an alternate procedure. Since $2 + \sqrt{5}$ is a root, then $2 - \sqrt{5}$ must also be a root, which (by the Factor Theorem) says that the original polynomial has factors $x - (2 + \sqrt{5})$ and $x - (2 - \sqrt{5})$. Multiply these factors:

$$\begin{aligned}[x - (2 + \sqrt{5})][x - (2 - \sqrt{5})] &= x^2 - (2 - \sqrt{5})x - (2 + \sqrt{5})x + (2 - \sqrt{5})(2 + \sqrt{5}) \\ &= x^2 - 4x - 1\end{aligned}$$

You could also use the formula for the sum and product of the roots of a quadratic equation (page 27)

$$(2 + \sqrt{5}) + (2 - \sqrt{5}) = 4$$
$$(2 + \sqrt{5})(2 - \sqrt{5}) = 4 - 5 = -1$$

Thus, the quadratic is

$$x^2 - 4x - 1$$

↑ opposite of the sum of the roots (coefficient -4); ↑ product of the roots (constant -1)

Now divide (using long division) this factor into the original polynomial:

$$\begin{array}{r} x^2 + 1 \\ x^2 - 4x - 1 \overline{)\, x^4 - 4x^3 + 0x^2 - 4x - 1} \\ \underline{x^4 - 4x^3 - x^2} \\ x^2 \\ \underline{x^2 - 4x - 1} \\ 0 \end{array}$$

Thus, $x^4 - 4x^3 - 4x - 1 = \underbrace{(x^2 - 4x - 1)}\underbrace{(x^2 + 1)}$

$$x^2 - 4x - 1 = 0 \qquad\qquad x^2 + 1 = 0$$
$$x = 2 \pm \sqrt{5} \qquad\qquad x = \pm i$$

Now set each factor equal to zero and solve.

The roots are $\mathbf{2 \pm \sqrt{5}, \pm i}$. ■

EXAMPLE 4 Solve $x^4 - 3x^2 - 6x - 2 = 0$ given that $-1 + i$ is a root.

Solution You can divide by $-1 + i$ synthetically, but instead we will multiply together the known factors:

$$\begin{aligned}[x-(-1-i)][x-(-1+i)] &= x^2-(-1+i)x-(-1-i)x \\ &\quad +(-1-i)(-1+i) \\ &= x^2+x-ix+x+ix+(1-i^2) \\ &= x^2+2x+2\end{aligned}$$

opposite of sum ↑ (the $2x$) ↑ product of roots (the 2)

Find the other factor(s) by long division:

$$\begin{array}{r} x^2-2x-1 \\ x^2+2x+2\overline{)x^4+0x^3-3x^2-6x-2} \\ \underline{x^4+2x^3+2x^2} \\ -2x^3-5x^2 \\ \underline{-2x^3-4x^2-4x} \\ -x^2-2x \\ \underline{-x^2-2x-2} \\ 0 \end{array}$$

Thus, $x^4-3x^2-6x-2=(x^2+2x+2)(x^2-2x-1)$.

$$x^2+2x+2=0 \qquad\qquad x^2-2x-1=0$$

$$x=\frac{-2\pm\sqrt{4-8}}{2} \qquad\qquad x=\frac{2\pm\sqrt{4+4}}{2}$$

$$x=-1\pm i \qquad\qquad x=1\pm\sqrt{2}$$

The roots are $\mathbf{-1\pm i, 1\pm\sqrt{2}}$.

You might wish to try using synthetic division to check this answer. ■

PROBLEM SET 4.5

A

In Problems 1–6 let $f(x)=x^4-6x^3+15x^2-2x-10$ *to find the requested value.*

1. $f(i)$ **2.** $f(-i)$ **3.** $f(\sqrt{2})$

4. $f(-\sqrt{3})$ **5.** $f(2+i)$ **6.** $f(1-\sqrt{3})$

In Problems 7–12 let $f(x)=x^4-8x^3+21x^2-14x-10$ *to find the requested value.*

7. $f(-i)$ **8.** $f(i)$ **9.** $f(\sqrt{3})$

10. $f(-\sqrt{2})$ **11.** $f(3-i)$ **12.** $f(1+\sqrt{2})$

In Problems 13–18 let $f(x)=x^4-10x^3+36x^2-58x+35$ *to find the requested value.*

13. $f(i)$ **14.** $f(-i)$ **15.** $f(2-i)$

16. $f(2+i)$ **17.** $f(3+\sqrt{2})$ **18.** $f(3-\sqrt{2})$

In Problems 19–24 let $f(x)=x^4-6x^3+18x^2-30x+25$ *to find the requested value.*

19. $f(2)$ **20.** $f(3)$ **21.** $f(2+i)$

22. $f(2-i)$ **23.** $f(1+2i)$ **24.** $f(1-2i)$

In Problems 25–30 let $f(x)=2x^4-x^3-13x^2+5x+15$ *to find the requested value.*

25. $f(2)$ **26.** $f(-2)$ **27.** $f(\sqrt{5})$

28. $f(-\sqrt{5})$ **29.** $f(i)$ **30.** $f(-i)$

31. Is $1+\sqrt{2}$ a root of $x^3-2x^2+1=0$? If it is, name another root.

32. Is $1+i$ a root of $x^3-4x^2+6x-4=0$? If it is, name another root.

33. Is $1-2i$ a root of $x^3-x^2+3x+5=0$? If it is, name another root.

34. Is $1 - 2i$ a root of $x^4 - 2x^3 + 4x^2 + 2x - 5 = 0$? If it is, name another root.

35. Is $1 + 2i$ a root of $x^4 - 7x^3 + 14x^2 + 2x - 20 = 0$? If it is, name another root.

36. Is $2 + \sqrt{5}$ a root of $x^4 - 4x^3 - 5x^2 + 16x + 4 = 0$? If it is, name another root.

B

Solve the polynomial equations $P(x) = 0$ and graph the curve $y = P(x)$ in Problems 37–55.

37. $P(x) = x^3 - 8$

38. $P(x) = x^3 - 64$

39. $P(x) = x^3 - 125$

40. $P(x) = x^4 - 64$

41. $P(x) = x^4 - 81$

42. $P(x) = x^4 - 625$

43. $P(x) = x^4 + 9x^2 + 20$

44. $P(x) = x^4 + 10x^2 + 9$

45. $P(x) = x^4 + 13x^2 + 36 = 0$

46. $P(x) = (x^2 - 4x - 1)(x^2 - 6x + 7)$

47. $P(x) = (x^2 - 6x + 10)(x^2 - 8x + 17)$

48. $P(x) = (x^2 + 2x + 5)(x^2 - 3x + 5)$

49. $P(x) = (x^2 - 4x - 1)(x^2 - 3x + 5)$

50. $P(x) = 2x^4 - x^3 + 2x - 1$

51. $P(x) = 2x^3 - 3x^2 + 4x + 3 = 0$

52. $P(x) = x^4 - 7x^3 + 14x^2 + 2x - 20$

53. $P(x) = x^4 - 2x^3 + 4x^2 + 2x - 5$

54. $P(x) = 4x^4 - 10x^3 + 10x^2 - 5x + 1$

55. $P(x) = 27x^4 - 180x^3 + 213x^2 + 62x - 10$

C

Assuming the given value is a root, solve the equations in Problems 56–65.

56. $x^3 - 2x^2 + 4x - 8 = 0;\ 2i$

57. $x^4 + 13x^2 + 36 = 0;\ -3i$

58. $x^4 - 6x^2 + 25 = 0;\ 2 + i$

59. $x^4 - 4x^3 + 3x^2 + 8x - 10 = 0;\ 2 + i$

60. $2x^4 - 5x^3 + 9x^2 - 15x + 9 = 0;\ i\sqrt{3}$

61. $2x^4 - x^3 - 13x^2 + 5x + 15 = 0;\ -\sqrt{5}$

62. $3x^5 + 10x^4 - 8x^3 + 12x^2 - 11x + 2 = 0;\ -2 + \sqrt{5}$

63. $2x^5 + 9x^4 - 3x^2 - 8x - 42 = 0;\ i\sqrt{2}$

64. $x^5 - 11x^4 + 24x^3 + 16x^2 - 17x + 3 = 0;\ 2 + \sqrt{3}$

65. $x^6 + x^5 - 3x^4 - 4x^3 + 4x + 4 = 0;\ -\sqrt{2}$ is a multiple root.

4.6 RATIONAL FUNCTIONS

To evaluate and graph polynomial functions, we used synthetic division and the Division Algorithm. That is, we considered $P(x)/D(x)$ for polynomials $P(x)$ and $D(x)$, where $D(x) \neq 0$. Now consider this quotient from another viewpoint. An expression of the form $P(x)/D(x)$ is called a *rational function.*

Rational Function

A **rational function** f is the quotient of polynomial functions $P(x)$ and $D(x)$; that is,

$$f(x) = \frac{P(x)}{D(x)} \qquad \text{where } D(x) \neq 0$$

Figure 4.8 $y = 1/x$

If you studied Chapter 3, you have already graphed quite a few rational functions by looking at the domain, range, asymptotes, symmetry, and intercepts. However, in this section (which does not require Chapter 3), we will focus on some of the general properties of rational functions. We begin with the simplest rational function: $y = 1/x$, which is graphed by plotting points as shown in Figure 4.8.

Each of the graphs in Figure 4.9 shows a variation of the graph of $1/x$.

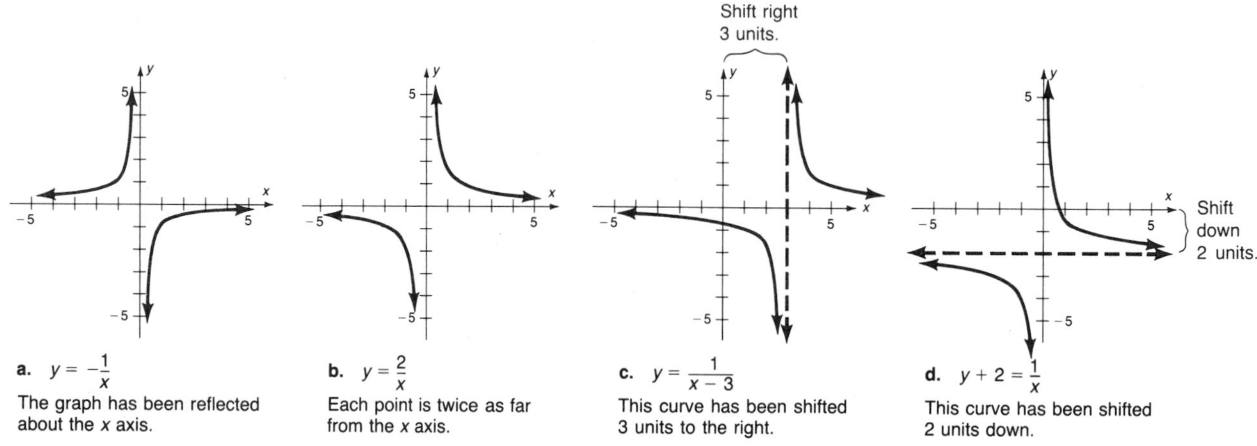

Figure 4.9 Variations on the graph of the rational function $1/x$.

EXAMPLE 1 Graph $y + 1 = \dfrac{-3}{x-2}$.

Solution We can sketch this graph by comparing it to $y = 1/x$. There are four changes: It is reflected about the x axis (negative); each point is three times as far from the x axis; and it has been shifted 2 units to the *right* and 1 unit *down*. The graph is shown in Figure 4.10.

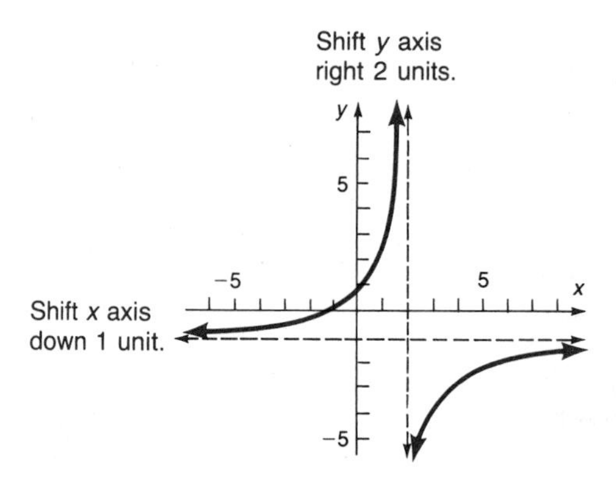

Figure 4.10 Graph of $y + 1 = \dfrac{-3}{x-2}$

■

You may have noticed that, unlike polynomial functions, rational functions cannot be drawn as a single curve. In Section 4.3 we saw that the graph of a polynomial function was continuous. However, we cannot say the same for rational functions. If $f(x) = P(x)/D(x)$ where there are no common factors for $P(x)$ and $D(x)$, we say that there is a **discontinuity** at a value of $x = b$ for which $D(b) = 0$. If you graph the line $x = b$, you will see that the graph of f will get closer and closer to this line without ever touching it. We call such a line an **asymptote**. If you studied Chapter 3, you did a lot of work with asymptotes, but in this section we will limit our work to finding vertical asymptotes.

EXAMPLE 2 Sketch the graph of $f(x) = \dfrac{x}{(x-2)(x+3)}$.

Solution The rational expression is reduced, so we see that there are two vertical asymptotes ($x = 2$ and $x = -3$). Remember that $x = b$ is an asymptote if substitution of b for x results in an undefined expression. Begin by sketching the asymptotes. (If you studied Chapter 3, you can use some of the analysis discussed there.) Then plot some points to obtain the graph shown in Figure 4.11.

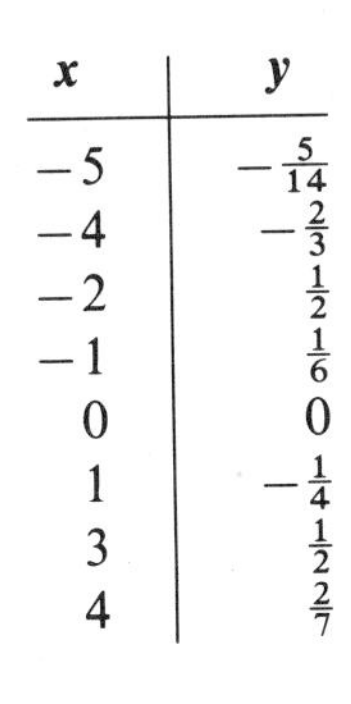

x	y
-5	$-\frac{5}{14}$
-4	$-\frac{2}{3}$
-2	$\frac{1}{2}$
-1	$\frac{1}{6}$
0	0
1	$-\frac{1}{4}$
3	$\frac{1}{2}$
4	$\frac{2}{7}$

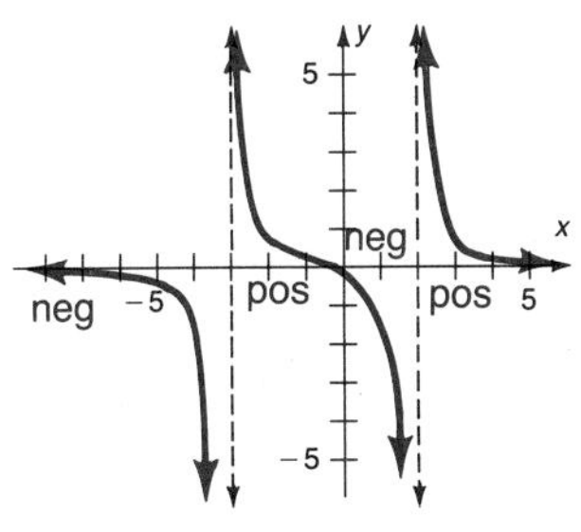

Figure 4.11 Graph of $f(x) = \dfrac{x}{(x-2)(x+3)}$ ■

If you compare Example 2 with Example 3, you should see the effect of changing the degree of the numerator.

EXAMPLE 3 Graph $f(x) = \dfrac{x^2}{(x-2)(x+3)}$.

Solution The graph is shown in Figure 4.12.*

x	y
-5	$\frac{25}{14} \approx 1.78$
-4	$\frac{8}{3} \approx 2.67$
-3	undefined; vertical asymptote at $x = -3$
-2	-1
-1	$-\frac{1}{6} \approx 0.17$
0	0
1	$-\frac{1}{4} = -0.25$
2	undefined; vertical asymptote at $x = 2$
3	$\frac{3}{2}$
4	$\frac{8}{7} \approx 1.14$
5	$\frac{25}{24} \approx 1.04$

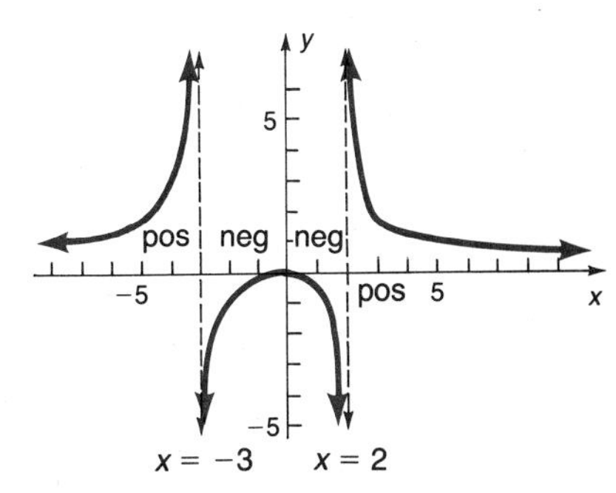

Figure 4.12 Graph of $f(x) = \dfrac{x^2}{(x-2)(x+3)}$ ■

* If you studied Chapter 3, you may find it helpful to find that a horizontal asymptote is $y = 1$ and that the curve crosses this asymptote at $(6, 1)$.

PROBLEM SET 4.6

A

Graph the function defined by each equation given in Problems 1–60.

1. $y = \dfrac{x^2 - x - 12}{x - 4}$
2. $y = \dfrac{2x^2 + x - 6}{x + 2}$
3. $y = \dfrac{3x^2 - 16x + 5}{x - 5}$
4. $y = \dfrac{2x^2 - 11x - 6}{x - 6}$
5. $y = \dfrac{3x^2 - 11x - 4}{3x + 1}$
6. $y = \dfrac{10x^2 + x - 2}{5x - 2}$
7. $y = \dfrac{3}{x}$
8. $y = -\dfrac{3}{x}$
9. $y = \dfrac{-2}{x}$
10. $y = \dfrac{1}{x} + 2$
11. $y = \dfrac{1}{x} + 1$
12. $y = \dfrac{1}{x} - 3$
13. $y = -\dfrac{1}{x} + 2$
14. $y = -\dfrac{1}{x} + 1$
15. $y = -\dfrac{1}{x} - 3$
16. $y = \dfrac{1}{x + 2}$
17. $y = \dfrac{1}{x - 3}$
18. $y = \dfrac{1}{x - 4}$
19. $y = \dfrac{-1}{x + 3}$
20. $y = \dfrac{-1}{x - 2}$
21. $y = \dfrac{-2}{x - 1}$
22. $y - 3 = \dfrac{-1}{x + 1}$
23. $y + 1 = \dfrac{-1}{x + 2}$
24. $y = \dfrac{4}{x + 2} - 3$
25. $y = \dfrac{4}{x^2}$
26. $y = \dfrac{-2}{x^2}$
27. $y = \dfrac{-3}{x^2}$
28. $y = \dfrac{2}{(x - 1)^2}$
29. $y = \dfrac{-1}{(x - 1)^2} + 3$
30. $y = \dfrac{1}{(x + 1)^2} - 2$

B

31. $y = \dfrac{2x^3 - 3x^2 - 32x - 15}{x^2 - 2x - 15}$
32. $y = \dfrac{3x^3 + 5x^2 - 26x + 8}{x^2 + 2x - 8}$
33. $y = \dfrac{x^3 + 6x^2 + 10x + 4}{x + 2}$
34. $y = \dfrac{x^3 + 9x^2 + 15x - 9}{x + 3}$
35. $y = \dfrac{x^3 + 12x^2 + 40x + 40}{x + 2}$
36. $y = \dfrac{x^3 + 5x^2 + 6x}{x + 3}$
37. $y = \dfrac{2x^2 + 2}{x^2}$
38. $y = \dfrac{2x^2 - 1}{x^2}$
39. $y = \dfrac{x^2}{x - 4}$
40. $y = \dfrac{4x}{x^2 - 2}$
41. $y = \dfrac{-x^2}{x - 1}$
42. $y = \dfrac{x^2}{x - 1}$
43. $y = \dfrac{x^2}{x - 2}$
44. $y = \dfrac{x^2}{2 - x}$
45. $y = \dfrac{2x + 1}{3x - 2}$
46. $y = \dfrac{4x + 3}{3x - 1}$
47. $y = \dfrac{3x - 1}{2x + 3}$
48. $y = \dfrac{2x^2 + 3x - 1}{x - 1}$
49. $y = \dfrac{3x^2 - 2x + 1}{x - 2}$
50. $y = \dfrac{x^2 + 3x - 2}{x + 1}$
51. $y = \dfrac{40{,}000x^2}{110x - x^2}$
52. $y = \dfrac{18{,}000x^2}{100x - x^2}$
53. $y = \dfrac{20{,}000x^3}{120x^2 - x^3}$

C

54. $y = \dfrac{x^2}{x^3 - x^2 - 20x}$
55. $y = \dfrac{x^2}{20x - x^2 - x^3}$
56. $y = \dfrac{x^2 + 3x - 2}{x^2 + 2x - 8}$
57. $y = \dfrac{x}{x^2 - 1}$
58. $y = \dfrac{x^2}{x^2 - 1}$
59. $y = \dfrac{x^3}{x^2 - 1}$
60. $y = \dfrac{x^4}{x^2 - 1}$
61. Can you see and describe a pattern in Problems 57–60?

4.7 PARTIAL FRACTIONS*

In algebra, rational expressions are added by finding common denominators; for example:

$$\frac{5}{x - 2} + \frac{3}{x + 1} = \frac{5(x + 1) + 3(x - 2)}{(x - 2)(x + 1)} = \frac{8x - 1}{(x - 2)(x + 1)}$$

* Optional section

In calculus, however, it is sometimes necessary to break apart the expression

$$\frac{8x-1}{(x-2)(x+1)}$$

into two fractions with denominators that are linear. The technique for doing this is called the **method of partial fractions.**

The rational expression

$$f(x)=\frac{P(x)}{D(x)}$$

can be **decomposed** into partial fractions if there are no common factors and if the degree of $P(x)$ is less than the degree of $D(x)$. If the degree of $P(x)$ is greater than or equal to the degree of $D(x)$, then use either long division or synthetic division to obtain a polynomial plus a proper fraction. For example,

$$\frac{x^4+2x^3-4x^2+x-3}{x^2-x-2}=x^2+3x+1+\underbrace{\frac{8x-1}{x^2-x-2}}$$

This was found by long division.

Proper fraction: This is the part that is decomposed into partial fractions.

Now look at the proper fraction. There is a theorem that says this proper fraction can be written as a sum,

$$F_1+F_2+\cdots+F_j$$

where *each* F_i is of the form

$$\frac{A}{(x-r)^n} \quad \text{or} \quad \frac{Ax+B}{(x^2+sx+t)^n}$$

We begin by focusing on the first form.

Partial Fraction Decomposition—Linear Factors

Let $f(x)=P(x)/D(x)$, where $P(x)$ and $D(x)$ have no common factors and the degree of $P(x)$ is less than the degree of $D(x)$. Also suppose that $D(x)=(x-r)^n$. Then $f(x)$ can be decomposed into partial fractions:

$$\frac{A_1}{x-r}+\frac{A_2}{(x-r)^2}+\cdots+\frac{A_n}{(x-r)^n}$$

EXAMPLE 1 Decompose $\dfrac{8x-1}{x^2-x-2}$ into partial fractions.

Solution

$$\frac{8x-1}{x^2-x-2}=\frac{8x-1}{(x-2)(x+1)}$$

First, factor the denominator, if possible, and make sure there are no common factors.

$$=F_1+F_2$$

Break up the fraction into parts, each with a linear factor.

$$F_1 + F_2 = \frac{A}{x-2} + \frac{B}{x+1}$$ The task is to find A and B.

$$= \frac{A(x+1) + B(x-2)}{(x-2)(x+1)}$$ Obtain a common denominator on the right.

Now, multiply both sides of this equation by the least common denominator, which is $(x-2)(x+1)$ for this example:

$$8x - 1 = A(x+1) + B(x-2)$$

Substitute, one at a time, the values that cause each of the factors in the least common denominator to be zero.

Let $x = -1$:

$$8x - 1 = A(x+1) + B(x-2)$$
$$8(-1) - 1 = A(-1+1) + B(-1-2)$$
$$-9 = 0 + B(-3)$$
$$-9 = -3B$$
$$\mathbf{3 = B}$$

Let $x = 2$:

$$8x - 1 = A(x+1) + B(x-2)$$
$$8(2) - 1 = A(2+1) + B(2-2)$$
$$15 = 3A$$
$$\mathbf{5 = A}$$

If $A = 5$ and $B = 3$, then

$$\frac{8x-1}{(x-2)(x+1)} = \frac{5}{x-2} + \frac{3}{x+1}$$ ■

Example 2 illustrates the process if there is a repeated linear factor.

EXAMPLE 2 Decompose $\frac{x^2 - 6x + 3}{(x-2)^3}$ by using the method of partial fractions.

Solution

$$\frac{x^2 - 6x + 3}{(x-2)^3} = \frac{A}{x-2} + \frac{B}{(x-2)^2} + \frac{C}{(x-2)^3}$$

Multiply both sides by $(x-2)^3$:

$$x^2 - 6x + 3 = A(x-2)^2 + B(x-2) + C$$

Let $x = 2$:

$$(2)^2 - 6(2) + 3 = A(2-2)^2 + B(2-2) + C$$
$$4 - 12 + 3 = 0 + 0 + C$$
$$\mathbf{-5 = C}$$

Notice that with repeated factors we cannot find all the numerators as we did in

Example 1. Now substitute $C = -5$ into the original equation and simplify by combining terms on the right side:

$$\begin{aligned} x^2 - 6x + 3 &= A(x-2)^2 + B(x-2) - 5 \\ &= A(x^2 - 4x + 4) + B(x-2) - 5 \\ &= Ax^2 - 4Ax + Bx + 4A - 2B - 5 \\ &= Ax^2 + (-4A + B)x + (4A - 2B - 5) \end{aligned}$$

If the polynomials on the left and right sides of the equality are equal, then the coefficients of the like terms must be equal. That is,

$$x^2 - 6x + 3 = Ax^2 + (-4A + B)x + (4A - 2B - 5)$$

$$A = 1$$

$$-4A + B = -6$$

$$4A - 2B - 5 = 3$$

If $A = 1$, then

$$\begin{aligned} -4A + B &= -6 \\ -4(1) + B &= -6 \\ \mathbf{B} &= \mathbf{-2} \end{aligned}$$

Check: If $A = 1$ and $B = -2$, then

$$\begin{aligned} 4A - 2B - 5 &= 4(1) - 2(-2) - 5 \\ &= 4 + 4 - 5 \\ &= 3 \end{aligned}$$

Thus,

$$\frac{\mathbf{x^2 - 6x + 3}}{\mathbf{(x-2)^3}} = \frac{\mathbf{1}}{\mathbf{x-2}} + \frac{\mathbf{-2}}{\mathbf{(x-2)^2}} + \frac{\mathbf{-5}}{\mathbf{(x-2)^3}}$$

■

We will now consider quadratic factors.

Partial Fraction Decomposition—Quadratic Factors

Let $f(x) = P(x)/D(x)$, where $P(x)$ and $D(x)$ have no common factors and the degree of $P(x)$ is less than the degree of $D(x)$. If $D(x) = (x^2 + sx + t)^m$, then $f(x)$ can be decomposed into partial fractions:

$$\frac{A_1x + B_1}{x^2 + sx + t} + \frac{A_2x + B_2}{(x^2 + sx + t)^2} + \cdots + \frac{A_mx + B_m}{(x^2 + sx + t)^m}$$

EXAMPLE 3 Decompose $f(x) = \dfrac{2x^3 + 3x^2 + 3x + 2}{(x^2 + 1)^2}$

Compare the denominator with the expression in the box above: $s = 0$, $t = 1$, and $m = 2$

Solution

$$\frac{2x^3 + 3x^2 + 3x + 2}{(x^2 + 1)^2} = \frac{Ax + B}{x^2 + 1} + \frac{Cx + D}{(x^2 + 1)^2}$$

Multiply by $(x^2+1)^2$:

$$2x^3+3x^2+3x+2=(Ax+B)(x^2+1)+Cx+D$$

This time, $x^2+1\neq 0$ in the set of real numbers, so multiply out the right side:

$$\begin{aligned}2x^3+3x^2+3x+2&=Ax^3+Bx^2+Ax+B+Cx+D\\&=Ax^3+Bx^2+\underbrace{(A+C)}x+\underbrace{(B+D)}\end{aligned}$$

Equate the coefficients of the similar terms on the left and right:

$$\mathbf{A=2}$$
$$\mathbf{B=3}$$
$A+C=3$ If $A=2$, then $2+C=3$ and $\mathbf{C=1}$.

$B+D=2$ If $B=3$, then $3+D=2$ and $\mathbf{D=-1}$.

Thus,

$$\frac{2x^3+3x^2+3x+2}{(x^2+1)^2}=\frac{2x+3}{x^2+1}+\frac{x-1}{(x^2+1)^2}$$

■

PROBLEM SET 4.7

A

List the factors of each denominator in the decomposition of the rational expressions in Problems 1–24. For Example 1 these factors are $(x-2)$ *and* $(x+1)$; *for Example 2 they are* $(x-2)$, $(x-2)^2$, *and* $(x-2)^3$.

1. $\dfrac{2x+10}{x^2+7x+12}$ 2. $\dfrac{2x-14}{x^2+x-6}$ 3. $\dfrac{7x-7}{2x^2-5x-3}$

4. $\dfrac{4(x-1)}{x^2-4}$ 5. $\dfrac{34-5x}{48-14x+x^2}$ 6. $\dfrac{x-7}{20-9x+x^2}$

7. $\dfrac{5x^2-5x-4}{x^3-x}$ 8. $\dfrac{4x^2-7x-3}{x^3-x}$ 9. $\dfrac{2x^2-18x-12}{x^3-4x}$

10. $\dfrac{2x-1}{(x-2)^2}$ 11. $\dfrac{4x-22}{(x-5)^2}$ 12. $\dfrac{x^2+5x+1}{x(x+1)^2}$

13. $\dfrac{5x^2-2x+2}{x(x-1)^2}$ 14. $\dfrac{2x^2+7x+2}{(x+1)^3}$ 15. $\dfrac{7x-3x^2}{(x-2)^3}$

16. $\dfrac{x}{x^2+4x-5}$ 17. $\dfrac{x}{x^2-2x-3}$ 18. $\dfrac{7x-1}{x^2-x-2}$

19. $\dfrac{10x^2-11x-6}{x^3-x^2-2x}$ 20. $\dfrac{-17x-6}{x^3+x^2-6x}$

21. $\dfrac{12+9x-6x^2}{x^3-5x^2+4x}$ 22. $\dfrac{x^3}{(x+1)^2(x-2)}$

23. $\dfrac{2x^3-3x^2+6x-1}{1-x^4}$ 24. $\dfrac{2x^3-7x^2+8x-7}{x^2-4x+4}$

B

Decompose each fraction in Problems 25–60 by using the method of partial fractions.

25. $\dfrac{x^2+2x+5}{x^3}$ 26. $\dfrac{3x^2-2x+1}{x^3}$ 27. $\dfrac{2x^2-5x+4}{x^3}$

28. $\dfrac{1}{(x+2)(x+3)}$ 29. $\dfrac{1}{(x+4)(x+5)}$ 30. $\dfrac{1}{(x+3)(x+2)}$

31. $\dfrac{7x-10}{(x-2)(x-1)}$ 32. $\dfrac{11x-1}{(x-1)(x+1)}$ 33. $\dfrac{7x+2}{(x+2)(x-4)}$

34. $\dfrac{2x+10}{x^2+7x+12}$ 35. $\dfrac{2x-14}{x^2+x-6}$ 36. $\dfrac{7x-7}{2x^2-5x-3}$

37. $\dfrac{4(x-1)}{x^2-4}$ 38. $\dfrac{34-5x}{48-14x+x^2}$

39. $\dfrac{x-7}{20-9x+x^2}$ 40. $\dfrac{5x^2-5x-4}{x^3-x}$

41. $\dfrac{4x^2-7x-3}{x^3-x}$ 42. $\dfrac{2x^2-18x-12}{x^3-4x}$

43. $\dfrac{2x-1}{(x-2)^2}$ 44. $\dfrac{4x-22}{(x-5)^2}$

45. $\dfrac{x^2+5x+1}{x(x+1)^2}$ 46. $\dfrac{5x^2-2x+2}{x(x-1)^2}$

47. $\dfrac{2x^2+8x+3}{(x+1)^3}$ 48. $\dfrac{7x-3x^2}{(x-2)^3}$

49. $\dfrac{x}{x^2 + 4x - 5}$

50. $\dfrac{x}{x^2 - 2x - 3}$

51. $\dfrac{7x - 1}{x^2 - x - 2}$

52. $\dfrac{10x^2 - 11x - 6}{x^3 - x^2 - 2x}$

53. $\dfrac{-17x - 6}{x^3 + x^2 - 6x}$

54. $\dfrac{12 + 9x - 6x^2}{x^3 - 5x^2 + 4x}$

C

55. $\dfrac{5x^2 - 6x + 7}{(x - 1)(x^2 + 1)}$

56. $\dfrac{x^2}{(x + 1)(x^2 + 1)}$

57. $\dfrac{x^3}{(x - 1)^2}$

58. $\dfrac{x^3}{(x + 1)^2}$

59. $\dfrac{2x^3 - 3x^2 + 6x - 1}{1 - x^4}$

60. $\dfrac{2x^3 - 7x^2 + 8x - 7}{x^2 - 4x + 4}$

4.8
CHAPTER 4 SUMMARY

The material of this chapter is reviewed in the following list of objectives. After each objective there are some practice questions. For a sample test, select the first question of each set and check your answers with the answer section. For a sample test without answers, use the second question of each set. Additional practice is given by the other questions in each set. If you are having trouble with a particular type of problem, look back to that section for extra help.

4.1 POLYNOMIAL FUNCTIONS

Objective 1 Be familiar with the terminology of polynomials, including the definition and laws of exponents. Fill in the blanks.

1. A function P is a polynomial function in x if ______
2. If b is any real number and n is ______, then $b^n =$ ______
3. $b^m \cdot b^n =$ ______
4. $(ab)^m =$ ______

Objective 2 Add, subtract, and multiply polynomials. Let $P(x) = 3x^3 + 4x^2 - 35x - 12$ and $D(x) = 3x + 1$. Find the requested expressions.

5. $P(x)D(x)$
6. $P(t) + D(t)$
7. $D(w) - P(w)$
8. $[D(s)]^3$

Objective 3 Factor polynomials.

9. $\dfrac{4x^2}{y^2} - (2x + y)^2$
10. $x^4 - 26x^2 + 25$
11. $(x^3 - \frac{1}{8})(8x^3 + 8)$
12. $4x^3 + 8x^2 - x - 2$

Objective 4 Graph polynomial forms that can be factored.

13. $\dfrac{x^2}{4} - \dfrac{y^2}{9} = 0$
14. $(x^2 - y)(x^2 - y^2) = 0$
15. $(2y + x^2)(y - x) = 0$
16. $x^3 + xy + x^2y + y^2 = 0$

4.2 SYNTHETIC DIVISION

Objective 5 Be familiar with the Division Algorithm and do long division of polynomials with and without remainders.

17. $\dfrac{2x^3 + 9x^2 + 5x - 6}{x^2 + 3x - 2}$
18. $\dfrac{12x^4 + 22x^3 + 7x + 4}{3x + 1}$
19. $\dfrac{6x^4 - x^3 + 4x^2 + 3x - 2}{2x + 1}$
20. $\dfrac{3x^3 + 2x^2 - 12x - 8}{x^2 - 4}$

Objective 6 *Use synthetic division.*

21. $\dfrac{x^4 + 2x^2 - x - 26}{x + 2}$

22. $\dfrac{6x^4 - x^3 + 4x^2 + 3x - 2}{x + \frac{1}{2}}$

23. $\dfrac{3x^3 + 2x^2 - 12x - 8}{x - 2}$

24. $\dfrac{4x^5 - 3x^3 + 2x - 5}{x + 1}$

4.3 GRAPHING POLYNOMIAL FUNCTIONS

Objective 7 *Use synthetic division and the Remainder Theorem to find points on the graph of a polynomial function.* Let $P(x) = 3x^3 + 4x^2 - 35x - 12$. Find the specified values.

25. $P(2)$
26. $P(-2)$
27. $P(3)$
28. $P(-3)$

Objective 8 *Use the Intermediate-Value Theorem for polynomial functions to sketch the graph of a polynomial function.*

29. $P(x) = 3x^3 + 4x^2 - 35x - 12$
30. $f(x) = 3x^4 - 8x^3 - 48x^2 + 492$
31. $Q(x) = 3x^3 + 2x^2 - 12x - 8$
32. $g(x) = 6x^4 - x^3 + 3x^2 + 3x - 2$

4.4 REAL ROOTS OF POLYNOMIAL EQUATIONS

Objective 9 *Find the zeros of polynomial functions by using the Factor Theorem, and state the multiplicity of each zero.*

33. $y = (x - 1)^3(x + 2)^2$
34. $y = (x^2 - 1)^2(x + 1)^3$
35. $y = (x^3 - 16x)^2$
36. $y = (x^2 - 4)^3$

Objective 10 *List the possible rational roots of a polynomial equation.*

37. $3x^3 + 4x^2 - 35x - 12 = 0$
38. $6x^4 - 13x^3 + 3x^2 + 3x - 2 = 0$
39. $3x^3 + 2x^2 - 12x - 8 = 0$
40. $x^4 + 2x^2 - x - 26 = 0$

Objective 11 *Use Descartes' Rule of Signs to determine the number of positive and negative real roots.*

41. $3x^3 + 4x^2 - 35x - 12 = 0$
42. $6x^4 - 13x^3 + 3x^2 + 9x - 5 = 0$
43. $3x^3 + 2x^2 - 12x - 8 = 0$
44. $x^4 + 11x^3 - 25x^2 + 11x - 26 = 0$

Objective 12 *Solve polynomial equations over the set of real numbers.*

45. $3x^3 + 4x^2 - 35x - 12 = 0$
46. $6x^4 - 13x^3 + 3x^2 + 9x - 5 = 0$
47. $3x^3 + 2x^2 - 12x - 8 = 0$
48. $x^4 + 11x^3 - 25x^2 + 11x - 26 = 0$

*4.5 THE FUNDAMENTAL THEOREM OF ALGEBRA

Objective 13 *Evaluate polynomial functions over the set of complex numbers.* Let $f(x) = x^5 + x^4 - 2x^3 - 2x^2 - 3x - 3$. Find the requested values.

49. $f(i)$
50. $f(-i)$
51. $f(\sqrt{3})$
52. $f(-\sqrt{3})$

Objective 14 *Solve polynomial equations over the set of complex numbers and graph the corresponding polynomial functions.*

53. Solve $x^4 - 3x^3 - 9x^2 + 25x - 6 = 0$
54. Graph $f(x) = x^4 - 3x^3 - 9x^2 + 25x - 6$
55. Solve $x^4 - 14x^2 + x^3 - 14x = 0$
56. Graph $g(x) = x^4 - 14x^2 + x^3 - 14x$

---

* Optional section

Objective 15 Use the Conjugate Pair Theorem to solve polynomial equations. Solve the equations, assuming the given value is a root.

57. $x^4 - 2x^3 + 5x^2 - 6x + 6 = 0;\ 1 + i$
58. $x^4 + 2x^2 - 63 = 0;\ -3i$
59. $x^4 - 2x^3 - x^2 + 10x - 20 = 0;\ 1 + \sqrt{3}i$
60. $x^4 - 6x^3 - 8x^2 + 62x + 15 = 0;\ 2 + \sqrt{5}$

4.6 RATIONAL FUNCTIONS

Objective 16 Graph rational functions.

61. $f(x) = \dfrac{1}{x - 2} + 2$

62. $f(x) = \dfrac{3x^2 + 2x - 5}{x - 1}$

63. $f(x) = \dfrac{2x^2 - 3x - 1}{x^2 - x - 2}$

64. $f(x) = \dfrac{x^3 - x - 6}{x - 2}$

*4.7 PARTIAL FRACTIONS

Objective 17 Decompose rational expressions by using the method of partial fractions.

65. $\dfrac{5x^2 - 19x + 17}{(x - 1)(x - 2)^2}$

66. $\dfrac{7x - 7}{2x^2 - 9x + 4}$

67. $\dfrac{2x^2 + 13x - 9}{x^2 + 2x - 15}$

68. $\dfrac{2x^3 + 2x + 3}{(x^2 + 1)^2}$

* Optional section

John Napier (1550–1617)

The invention of logarithms and the calculation of the earlier tables form a very striking episode in the history of exact science, and with the exception of the "Principia" *of Newton, there is no mathematical work published in the country which has produced such important consequences, or to which so much interest attaches as to Napier's* "Description."

J. W. L. Glaisher
Encyclopedia Britannica

The advancement and perfection of mathematics are intimately connected with the prosperity of the State.

Napoleon I

HISTORICAL NOTE

John Napier was the Isaac Asimov of his day, having envisioned the tank, the machine gun, and the submarine. He also predicted that the end of the world would occur between 1688 and 1700. He is best known today as the inventor of logarithms. Napier, however, believed that his reputation would rest ultimately on his predictions about the end of the world. He considered logarithms merely an interesting recreational diversion.

The word *logarithm* means "ratio number" and was adopted by Napier after he had first chosen the term *artificial number*. As you will see in this chapter, today we define a logarithm as an exponent, but historically logarithms were discovered before exponents were in use. Napier's original tables were very cumbersome. They did not give the logarithm of x but $10^7 \ln(10^7 x^{-1})$. Common logarithms, or logs to the base 10, were introduced by a professor at Oxford named Henry Briggs (1516–1631). Briggs saw Napier's book, immediately recognized the importance of logarithms, and began working on them himself. Within months he joined Napier in Scotland and convinced him that tables for $\log x$ would be more convenient.

*"Today, when logarithms appear so easy, the student may have trouble understanding how they could ever have been difficult. But in Napier's time the notion of a function was unknown, and even the simple exponential notation $a^n = a \cdot a \cdot a \cdots a$, introduced later by René Descartes, did not exist. Napier invented logarithms to ease computation, and by so doing became known as the founder of applied mathematics. Although the computational efficiency of calcutors is now greater than that of logarithms, Napier's place in mathematics remains secure, for from his work came the logarithm functions and their inverses, the exponentials. These functions occupy a central role in the mathematical sciences."**

Until the advent of the low-cost calculator, logarithms were used extensively in complicated mathematical calculation. The mathematician F. Cajori said the powers of modern calculations are due to three inventions: "arabic notation, decimal fractions, and logarithms." Today we would have to add another to those three inventions: the hand-held calculator. This, however, does not minimize the importance of the logarithm. It simply changes the emphasis from calculation to application. This chapter reflects the recent change in emphasis brought about by the calculator.

* From R. A. Bonic, G. Hajian, E. DuCasse, and M. M. Lipschutz, *Freshman Calculus*, 2nd ed. (Lexington, Mass.: D.C. Heath).

5

Exponential and Logarithmic Functions

CONTENTS

PREVIEW

The important ideas of this chapter are solving exponential and logarithmic equations. In order to do this, two new and important functions are defined and discussed. The 16 objectives of this chapter are listed on pages 197–198. Following these objectives is a Cumulative Review for Chapters 2–5. It is a natural breaking point for the functions and graphing in the first part of the book. Following the Cumulative Review you will find the first Extended Application in the book. This is introduced with a newspaper article, and a mathematical discussion follows, in which the techniques of this part of the book are used to answer questions related to the article. This application concerns population growth.

PERSPECTIVE

Transcendental functions form a major category of functions discussed and developed in calculus. The logarithmic and exponential functions introduced in this chapter fall into this category, and are used in calculus in many applications including growth, decay, population growth, evaluation of integrals, and solving equations called *differential equations*. The following excerpt is from the introduction to transcendental functions in a calculus book.

6.0 Introduction

A function f is *algebraic* if its formula involves only the operations of addition, subtraction, multiplication, division, absolute value, and exponentiation to rational powers. Otherwise the function is called *transcendental*. The six trigonometric functions, which we have been working with since Chapter 2, are familiar transcendental functions. This chapter develops the calculus of additional transcendental functions.

The first five sections are devoted to logarithmic and exponential functions. Following a review of the basic properties of logarithms, Section 1 defines the natural logarithm function *ln* as an antiderivative. Properties of integration are then used to show that this function has all the usual logarithmic properties, and to derive its calculus. The next section defines and studies the natural exponential function *exp* as the inverse function of *ln*. Section 3 uses *ln* and *exp* to analyze several important types of growth and decay. The following section discusses further exponential and logarithmic functions, which arise in several areas outside mathematics.

Section 5 introduces the inverse trigonometric functions and their calculus. These functions are at the heart of one of the techniques of integration presented in the next chapter. Section 6 is devoted to hyperbolic functions, a class of exponential functions with properties similar to those of the trigonometric functions. Section 7 discusses an important application of the hyperbolic functions to physics and engineering, as well as some applications of trigonometric functions to motion problems.

From James Hurley, *Calculus* (Belmont, Ca.: Wadsworth), p. 335.

5.1 EXPONENTIAL FUNCTIONS

The linear, quadratic, polynomial, and rational functions considered in the first part of this book are all called **algebraic functions**. An algebraic function is a function that can be expressed in terms of algebraic operations alone. If a function is not algebraic, it is called a **transcendental function**. In this chapter, two examples of transcendental functions, *exponential* and *logarithmic* functions, are considered. In the next chapter, other types of transcendental functions will be considered.

Exponential Function

> The function f is an **exponential function** if
>
> $$f(x) = b^x$$
>
> where b is a positive constant other than 1 and x is any real number. The number x is called the **exponent** and b is called the **base**.

Recall that if n is a natural number, then

$$b^n = \underbrace{b \cdot b \cdot b \cdots b}_{n \text{ factors}}, \quad b^0 = 1, \quad b^{-n} = \frac{1}{b^n}$$

This definition of b^n is used in conjunction with the five laws of exponents stated in Section 4.1 on page 120.

The next step is to extend the definition of an exponent to include rational exponents. To do this, you need to recall the definition of roots. If n is a natural number, then an ***n*th root** of a number b is a only if $a^n = b$. If $n = 2$, the root is called a **square root**; if $n = 3$, it is called a **cube root**. The number $\sqrt[n]{a}$ is called the **principal *n*th root** of a. If $n = 2$, it is customary to write $\sqrt{a}$ instead of $\sqrt[2]{a}$. The symbol $\sqrt[n]{a}$ is also called a **radical**; the number a is the **radicand**, and the **index** is the number n. To relate this discussion to exponents, consider $\sqrt{2}$. Suppose we wish to find an x so that

$$\sqrt{2} = 2^x$$

From the definition of square root,

$$2 = (2^x)^2$$

If the properties of exponents are to hold, then

$$2^1 = 2^{2x}$$

We would now like to conclude that the exponents are equal. In this case, it is indeed true. But under what conditions does $x = y$ when

$$b^x = b^y$$

If $b = 1$, you cannot conclude that $x = y$ since

$$1^5 = 1^4$$

but $5 \neq 4$. If $b \neq 1$, however, it can be proved true for all positive real numbers b and is called the **exponential property of equality**.

Exponential Property of Equality

> For positive real b ($b \neq 1$):
>
> If $b^x = b^y$, then $x = y$.

We now use this property to conclude that if

$$2^1 = 2^{2x}$$

then $1 = 2x$ and $x = \frac{1}{2}$. This shows that $2^{1/2} = \sqrt{2}$. Another way of showing this same fact is to notice that

$$2^{1/2} \cdot 2^{1/2} = 2^1 = 2$$

and

$$\sqrt{2}\sqrt{2} = 2$$

so $2^{1/2} = \sqrt{2}$.

The use of rational numbers as exponents preserves the laws of exponents and, more important, gives an alternative choice of notation for roots as shown by the following definition.

Rational Exponents

> For $b > 0$ and m, n positive integers where m/n is reduced,
>
> $$b^{1/n} = \sqrt[n]{b} \qquad \text{and} \qquad b^{m/n} = (b^{1/n})^m = \sqrt[n]{b^m} = (\sqrt[n]{b})^m$$

EXAMPLE 1 Simplify the given expressions.

a. $16^{1/2} = (4^2)^{1/2}$
$= 4^1$
$= \mathbf{4}$

b. $-16^{1/2} = -(4^2)^{1/2}$
$= \mathbf{-4}$

c. $(-16)^{1/2} = \sqrt{-16}$ is not defined (b must be greater than zero by definition).

d. $(343)^{2/3} = (7^3)^{2/3}$
$= 7^2$
$= \mathbf{49}$

e. $25^{-3/2} = \dfrac{1}{25^{3/2}}$ First use the definition $b^{-p} = \dfrac{1}{b^p}$.

$= \dfrac{1}{(5^2)^{3/2}}$

$= \dfrac{1}{5^3}$

$= \mathbf{\dfrac{1}{125}}$ ■

EXAMPLE 2 Use the ordinary rules of algebra to simplify the given expressions.

a. $x(x^{2/3} + x^{1/2}) = x^1x^{2/3} + x^1x^{1/2}$

$= x^{1+(2/3)} + x^{1+(1/2)}$ Recall that $x^px^q = x^{p+q}$, so $x^1x^{2/3} = x^{1+(2/3)}$.

$= \mathbf{x^{5/3} + x^{3/2}}$

b. $(x^{1/2} + y^{1/2})(x^{1/2} - y^{1/2}) = x^{1/2}x^{1/2} - x^{1/2}y^{1/2} + x^{1/2}y^{1/2} - y^{1/2}y^{1/2}$

$= \mathbf{x - y}$ ■

The next step in enlarging the domain for x in $f(x) = b^x$ requires the following property:

Squeeze Theorem for Exponents

Suppose b is a real number greater than 1. Then for any real number x there is a unique real number b^x. Moreover, if h and k are any two rational numbers such that $h < x < k$, then

$$b^h < b^x < b^k$$

The Squeeze Theorem will give meaning to expressions such as $2^{\sqrt{3}}$. Consider the graph of the function $f(x) = 2^x$ by plotting the points shown in the table as in Figure 5.1.

x	$y = f(x) = 2^x$
-3	$2^{-3} = \frac{1}{8}$
-2	$2^{-2} = \frac{1}{4}$
-1	$2^{-1} = \frac{1}{2}$
0	$2^0 = 1$
1	$2^1 = 2$
2	$2^2 = 4$
3	$2^3 = 8$

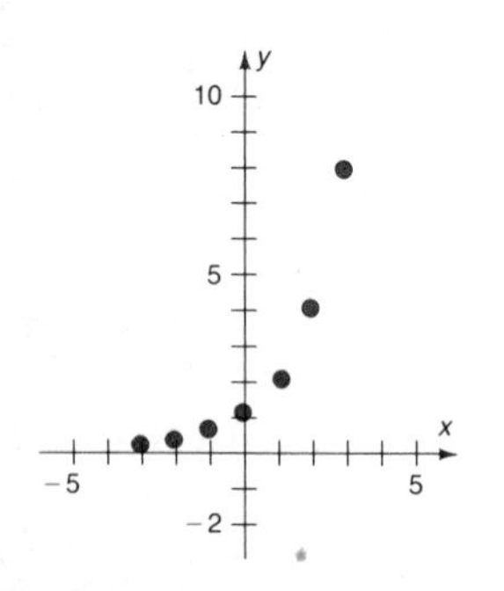

Figure 5.1 Selected points that satisfy $f(x) = 2^x$

If these points are connected with a smooth curve, as shown in Figure 5.2, you can see that $2^{\sqrt{3}}$ is defined and is between 2^1 and 2^2.

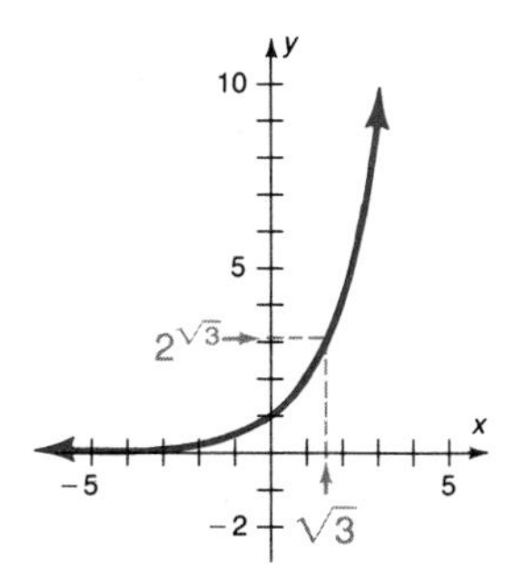

Figure 5.2 Graph of $f(x) = 2^x$

The number $2^{\sqrt{3}}$ can be approximated to any desired degree of accuracy:

$$\begin{array}{llll} & 1 < \sqrt{3} < 2 & & 2^1 < 2^{\sqrt{3}} < 2^2 \\ & 1.7 < \sqrt{3} < 1.8 & & 2^{1.7} < 2^{\sqrt{3}} < 2^{1.8} \\ \text{Since} & 1.73 < \sqrt{3} < 1.74 & \text{we have} & 2^{1.73} < 2^{\sqrt{3}} < 2^{1.74} \\ & 1.732 < \sqrt{3} < 1.733 & & 2^{1.732} < 2^{\sqrt{3}} < 2^{1.733} \\ & \vdots & & \vdots \end{array}$$

Base with irrational exponent is squeezed between same base with rational exponents.

Even using a calculator,

[2] [y^x] [3] [$\sqrt{\ }$] [=]

gives a *rational* approximation of $\sqrt{3} \approx 1.732050808$ and does not find $2^{\sqrt{3}}$ but rather $2^{1.732050808}$. This process can be visualized by looking at the portions of Figure 5.2 shown in Figure 5.3.

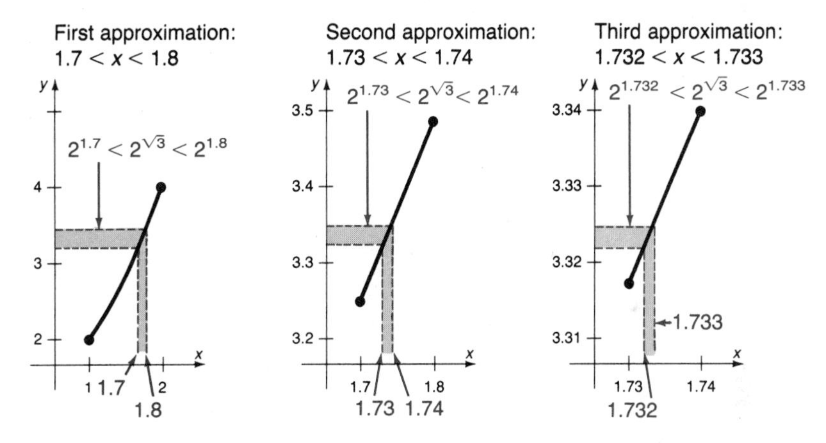

Figure 5.3 Successive approximations for locating $2^{\sqrt{3}}$

Because it is beyond the scope of this book to prove that the usual laws of exponents hold for all real exponents in exponential functions, we will accept them as axioms. Because of the restrictions on b, however, the hypotheses for these laws of exponents with any real exponents must be changed so that they apply only when the bases are positive numbers not equal to 1. Suppose we sketch several exponential functions and observe their behavior.

EXAMPLE 3 Sketch $f(x) = (\frac{1}{2})^x$.

Solution Notice that

$$y = \left(\frac{1}{2}\right)^x$$
$$= (2^{-1})^x$$

The points are plotted in Figure 5.4 and connected by a smooth curve.

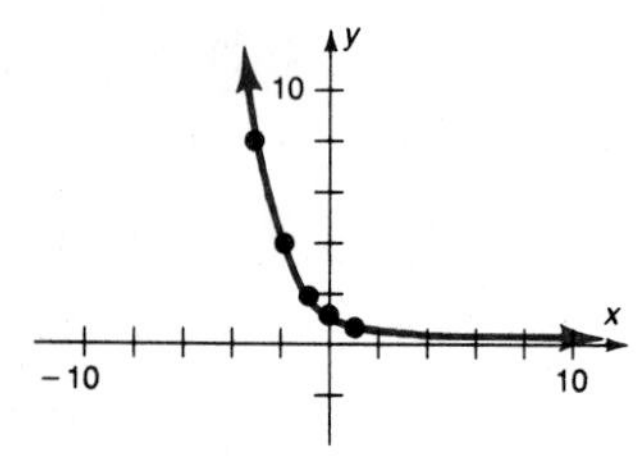

Figure 5.4 Graph of $f(x) = (\frac{1}{2})^x$

x	$f(x) = (\frac{1}{2})^x$
-3	$(2^{-1})^{-3} = 8$
-2	$(2^{-1})^{-2} = 4$
-1	$(2^{-1})^{-1} = 2$
0	$(2^{-1})^{0} = 1$
1	$(\frac{1}{2})^1 = \frac{1}{2}$
2	$(\frac{1}{2})^2 = \frac{1}{4}$
3	$(\frac{1}{2})^3 = \frac{1}{8}$

f is a decreasing function*

■

EXAMPLE 4 Sketch $f(x) = 10^x$.

Solution

x	$f(x) = 10^x$
-3	$10^{-3} = \frac{1}{1000}$
-2	$10^{-2} = \frac{1}{100}$
-1	$10^{-1} = \frac{1}{10}$
0	$10^0 = 1$
1	$10^1 = 10$
2	$10^2 = 100$
3	$10^3 = 1000$

f is an increasing function*

Figure 5.5 Graph of $f(x) = 10^x$

Notice that it is often necessary to alter the scale for exponential functions. ■

By looking at Figures 5.2, 5.4, and 5.5, we can make some observations regarding the graph of $f(x) = b^x$:

1. It passes through the point $(0, 1)$.
2. $f(x) > 0$ for all x.
3. If $b > 1$, f is an increasing function
4. If $b < 1$, f is a decreasing function

EXAMPLE 5 Sketch $f(x) = 10^{x-3} + 50$.

Solution Write this as

$$y = 10^{x-3} + 50$$

$$y - 50 = 10^{x-3}$$

or

$$y' = 10^{x'} \qquad (x' = x - 3;\ y' = y - 50)$$

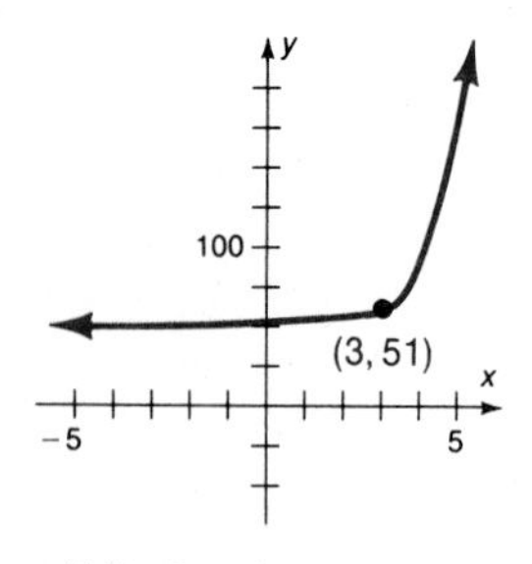

Figure 5.6 Graph of $f(x) = 10^{x-3} + 50$

Thus, this is the graph shown in Figure 5.5 translated to $(h, k) = (3, 50)$, as shown in Figure 5.6. Notice that the curve passes through the point $(3, 51)$. ■

---

* If you studied Chapter 3, you can see that there is a horizontal asymptote at $y = 0$.

EXAMPLE 6 Sketch $f(x) = 2^{-x^2}$.

Solution

x	$f(x) = 2^{-x^2}$
-3	$2^{-9} = \frac{1}{512}$
-2	$2^{-4} = \frac{1}{16}$
-1	$2^{-1} = \frac{1}{2}$
0	$2^0 = 1$
1	$2^{-1} = \frac{1}{2}$
2	$2^{-4} = \frac{1}{16}$
3	$2^{-9} = \frac{1}{512}$

A calculator could be used to estimate additional points.*

x	$f(x) = 2^{-x^2}$
-1.5	0.21
-0.5	0.84
0.5	0.84
1.5	0.21

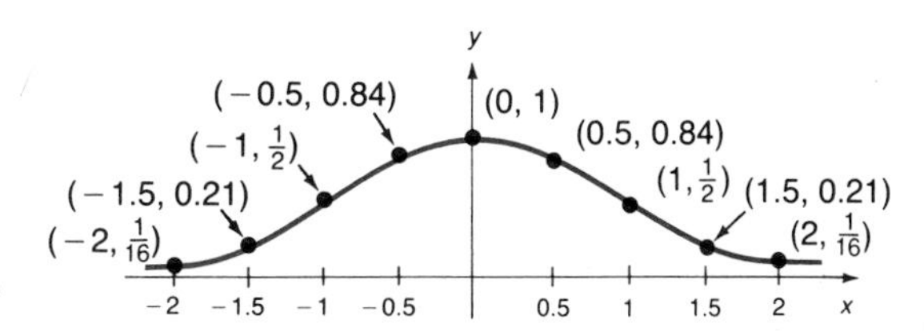

Figure 5.7 Graph of $f(x) = 2^{-x^2}$ ■

PROBLEM SET 5.1

A

Simplify the expressions in Problems 1–20. Eliminate negative exponents from your answers.

1. $25^{1/2}$ **2.** $-25^{1/2}$ **3.** $-27^{1/3}$ **4.** $-216^{1/3}$
5. $216^{1/3}$ **6.** $8^{2/3}$ **7.** $64^{2/3}$ **8.** $64^{3/2}$
9. $-64^{3/2}$ **10.** $-64^{2/3}$ **11.** $(-64)^{3/2}$ **12.** $7^{1/3} \cdot 7^{2/3}$
13. $8^{4/3} \cdot 8^{-1/3}$ **14.** $10^{4/3}/10^{1/3}$
15. $(2^{1/3} \cdot 3^{1/2})^6$ **16.** $(2^6 \cdot 3^{12})^{1/6}$
17. $1000^{-2/3}$ **18.** $0.001^{-2/3}$
19. $100^{-3/2}$ **20.** $0.01^{-3/2}$

B

Simplify the expressions in Problems 21–30.

21. $x^{1/2}(x^{1/2} + x^{-1/2})$ **22.** $x(x^{1/2} + x^{-1/2})$
23. $x^{2/3}(x^{-2/3} + x^{1/3})$ **24.** $x^{1/4}(x^{3/4} + x^{-1/4})$
25. $(x^{2/3}y^{-1/3})^3$ **26.** $(x^{2^2+1}x^5)^{1/10}$
27. $(x^{1/2} + y^{1/2})^2$ **28.** $(x^{1/2} - y^{1/2})^2$
29. $(x^{1/3} + y^{1/3})(x^{2/3} - x^{1/3}y^{1/3} + y^{2/3})$
30. $(x^{1/3} - y^{1/3})(x^{2/3} + x^{1/3}y^{1/3} + y^{2/3})$

Sketch the graph of each function given in Problems 31–54.

31. $y = 3^x$ **32.** $y = 4^x$ **33.** $y = 5^x$ **34.** $y = (\frac{1}{3})^x$
35. $y = 4^{-x}$ **36.** $y = 5^{-x}$ **37.** $y = 2^{x-2}$ **38.** $y = 2^{x-1}$
39. $y = 2^{x+3}$ **40.** $y - 2 = 2^x$
41. $y - 3 = 2^x$ **42.** $y + 5 = 2^x$
43. $y - 5 = 2^{x+4}$ **44.** $y - 10 = 2^{x+3}$
45. $y = 2^{x-3} - 10$ **46.** $y = 2^{|x|}$
47. $y = 3^{|x|}$ **48.** $y = 2^{-|x|}$ **49.** $y = 3^{x^2}$ **50.** $y = 2^{x^2}$
51. $y = 10^{x^2}$ **52.** $y = 3^{-x^2}$ **53.** $y = 5^{-x^2}$ **54.** $y = 10^{-x^2}$

55. Use graphical methods to estimate the value of $2^{\sqrt{2}}$.

56. Use graphical methods to estimate the value of $3^{\sqrt{2}}$.

57. Use graphical methods to estimate the value of $10^{\sqrt{2}}$.

58. Graph $y = 10^x$, $-1 \le x \le 1$ and approximate x if:
a. $10^x = 5$ **b.** $10^x = 0.5$ **c.** $10^x = 2$
d. $10^x = 8.4$ **e.** $10^x = -1$

59. In the definition of the exponential function $f(x) = b^x$, we require that b is a positive constant.
a. What happens if $b = 1$? Draw the graph of $f(x) = b^x$, where $b = 1$. Is this an algebraic or a transcendental function?
b. What happens if $b = 0$? Draw the graph of $f(x) = b^x$, where $b = 0$. Is this an algebraic or a transcendental function?

60. In the definition of the exponential function $f(x) = b^x$, we require that b is a positive constant. What happens if $b < 0$, say $b = -2$? For what values of x is f defined? Describe the graph of $f(x)$ in this case.

* If you studied Chapter 3, notice that $2^{-x^2} \to 0$ as $|x| \to \infty$, so $y = 0$ is the equation of a horizontal asymptote.

C

61. ***Physics*** Radioactive argon-39 has a half-life of 4 min. This means that the time required for half of the argon-39 to decompose is 4 min. If we start with 100 milligrams (mg) of argon-39, the amount (A) left after t min is given by

$$A = 100\left(\frac{1}{2}\right)^{t/4}$$

Graph this function.

62. ***Earth Science*** Carbon-14, used for archaeological dating, has a half-life of 5700 years. This means that the time required for half of the carbon-14 to decompose is 5700 years. If we start with 100 mg of carbon-14, the amount (A) left after t years is given by

$$A = 100\left(\frac{1}{2}\right)^{t/5700}$$

Graph this function for $t \geq 0$.

63. ***Social Science*** In 1982 the world population was about 4.56 billion. If we assume a growth rate of 2%, the formula expressing the world population for t years after 1982 is given by

$$\begin{aligned} P &= 4.56(1 + 0.02)^t \\ &= 4.56(1.02)^t \end{aligned}$$

where P is the population in billions. Graph this function for 1982–1992

64. Use graphical methods to determine which is larger:
   a. $(\sqrt{3})^\pi$ or $\pi^{\sqrt{3}}$ **b.** $(\sqrt{5})^\pi$ or $\pi^{\sqrt{5}}$ **c.** $(\sqrt{6})^\pi$ or $\pi^{\sqrt{6}}$
   d. Consider $(\sqrt{N})^\pi = \pi^{\sqrt{N}}$, where N is a positive real number. From parts **a**–**c**, notice that $(\sqrt{N})^\pi$ is larger for some values of N and $\pi^{\sqrt{N}}$ is larger for others. For $N = \pi^2$,

$$(\sqrt{N})^\pi = \pi^{\sqrt{N}}$$

is obviously true. Using a graphic method, find another value (approximately) for which the given statement is true.

5.2 INTRODUCTION TO LOGARITHMS

Many natural phenomena follow patterns of exponential growth or decay. Human population growth exhibits exponential growth, for example, and is considered at length in a special Extended Application following this chapter. Compound interest provides another application of exponential growth that is very important in business. If a sum of money, called the **principal**, is denoted by P and invested at an annual interest rate of r for t years, then the amount of money present is denoted by A and is found by

$$A = P + I$$

where I denotes the interest. **Interest** is an amount of money paid for the use of another's money. **Simple interest** is found by multiplication:

$$I = Prt$$

For example, \$1000 invested for 3 years at 15% simple interest would generate an interest of \$1000(0.15)(3) = \$450, so the amount present after 3 years is \$1450.

Most businesses, however, pay interest on the interest, as well as the principal, after a certain period of time. When this is done, it is called **compound interest**. For example, \$1000 invested at 15% annual interest compounded annually for 3 years can be found as follows:

$$\begin{aligned} \text{First year: } A &= P + I \\ &= P + Pr \qquad\qquad I = Prt \text{ and } t = 1 \\ &= P(1 + r) \\ &= \$1000(1 + 0.15) \\ &= \$1150 \end{aligned}$$

Second year: The amount from the first year becomes the principal for the second year:

$$A = \$1150(1 + 0.15)$$
$$= \$1322.50$$

Third year: $A = \$1322.50(1 + 0.15)$

$= \$1520.88$ Rounded to the nearest cent

Notice that the amount with simple interest is \$1450, whereas with compound interest it is \$1520.88. The calculation for compound interest, shown above, can become very tedious (especially for large t), so it is desirable to derive a general formula:

First year: $A = P + I$

$= P + Pr$ $\quad I = Prt$ and $t = 1$

$= P(1 + r)$

Second year: $A = P(1 + r) + I$

$= P(1 + r) + P(1 + r)r$

$= P(1 + r)(1 + r)$ Common factor $P(1 + r)$

$= P(1 + r)^2$

Third year: $A = P(1 + r)^2 + P(1 + r)^2 r$

$= P(1 + r)^2(1 + r)$ Common factor $P(1 + r)^2$

$= P(1 + r)^3$

$\vdots$

This pattern leads to the compound interest formula. The proof requires mathematical induction, which is discussed in Chapter 10.

Compound Interest

If a principal (or present value) of P dollars is invested at an interest rate of r per period for a total of t periods, then the amount present (or future value) after t periods, A, is given by the formula

$$\mathbf{A = P(1 + r)^t}$$

EXAMPLE 1 If \$12,000 is invested for 5 years at 18% compounded annually, what is the amount present at the end of 5 years?

Solution $P = \$12{,}000$; $r = 0.18$; and $t = 5$. Then

$A = \$12{,}000(1 + 0.18)^5$

$\approx \$12{,}000(2.2877578)$ Use a calculator.

$\approx \mathbf{\$27{,}453.09}$

■

EXAMPLE 2 If the interest in Example 1 is compounded monthly, find the amount present.

Solution

$P = \$12{,}000$

$r = \dfrac{0.18}{12}$ — r is the rate per period; in this case the period is monthly, so divide by 12.

$= 0.015$

$t = 5(12)$ — t is the number of periods.

$= 60$

$A = \$12{,}000(1 + 0.015)^{60}$

$\approx \$12{,}000(2.4432198)$ — Use a calculator.

$\approx$ **\$29,318.64** ■

Instead of finding the amount present in Example 2, suppose we want to know how long it will take for \$12,000 to grow to some specified amount. This question gives rise to an equation for which the variable or unknown value is an exponent. Such equations are called **exponential equations.** Consider

$$A = b^x$$

where $b > 1$. How can you solve this equation for x? Notice that

x is the exponent of b that yields A

This can be rewritten

x = exponent of b to get A

It appears that the equation is now solved for x, but this is simply a notational change. The expression "exponent of b to get A" is called, for historical reasons, "the log of A to the base b." That is,

x = log A to the base b

But this phrase is shortened to the notation

$$x = \log_b A$$

The term **log** is an abbreviation for **logarithm,** but even the introduction of this notation does not give us the numerical answer we are looking for. It does solve for x algebraically, however, which is a first step in solving exponential equations. It is important to recognize this as a notational change only:

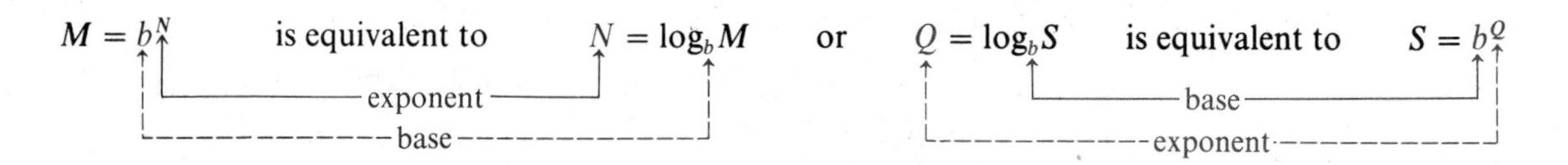

EXAMPLE 3 Change from exponential form to logarithmic form.

a. Remember: The log form solves for the exponent.

$5^2 = 25 \Leftrightarrow \log_5 25 = 2$ Use the symbol $\Leftrightarrow$ to mean "is equivalent to."

Remember: This is the base.

b. $3^2 = 9 \Leftrightarrow \log_3 9 = 2$

base exponent

c. $\frac{1}{8} = 2^{-3} \Leftrightarrow \log_2 \frac{1}{8} = -3$

d. $\sqrt{16} = 4 \Leftrightarrow \log_{16} 4 = \frac{1}{2}$ Remember: $\sqrt{16} = 16^{1/2}$. ∎

EXAMPLE 4 Change from logarithmic form to exponential form.

a. $\log_{10} 100 = 2 \Leftrightarrow 10^2 = 100$

base exponent

b. $\log_{10} \frac{1}{1000} = -3 \Leftrightarrow 10^{-3} = \frac{1}{1000}$

c. $\log_3 1 = 0 \Leftrightarrow 3^0 = 1$ ∎

To **evaluate** a logarithm means to find a numerical value for the given logarithm. The first ones you are asked to find use the definition of logarithm and the exponential property of equality.

EXAMPLE 5 Evaluate the given logarithms.

a. $\log_2 64$. Since it is usually necessary to supply a variable to convert to exponential form, we will use N in these examples.

$$\log_2 64 = N \quad \text{or} \quad 2^N = 64$$

$$2^N = 2^6$$

$$N = 6$$

Thus $\mathbf{\log_2 64 = 6}$.

b. $\log_3 \frac{1}{9}$

$$3^N = \frac{1}{9}$$

$$3^N = 3^{-2}$$

$$N = -2$$

Thus $\mathbf{\log_3 \frac{1}{9} = -2}$.

c. $\log_9 27$

$$9^N = 27$$

$$3^{2N} = 3^3$$

$$2N = 3$$

$$N = \frac{3}{2}$$

Thus $\mathbf{\log_9 27 = \frac{3}{2}}$.

d. $\log_{10} 1 = \mathbf{0}$ Can you do this mentally?
e. $\log_{10} 10 = \mathbf{1}$
f. $\log_{10} 100 = \mathbf{2}$
g. $\log_{10} 0.1 = \mathbf{-1}$ ■

Consider Example 5d–g; suppose you want to find $\log_{10} 5.03$.

$$\log_{10} 5.03 = x \Leftrightarrow 10^x = 5.03$$

Since 5.03 is between 1 and 10 and

$10^0 = 1$

$10^x = 5.03$ You want to find this x.

$10^1 = 10$

the number x should be between 0 and 1 by the Squeeze Theorem for Exponents. Although you *still* do not have the value of x, all is not lost because tables showing approximations for these exponents have been prepared. Base 10 is fairly common, and if a logarithm is to the base 10 it is called a **common logarithm** and written without the subscript 10. Table C.II in Appendix C at the back of the book is a table of common logarithms. To find log 5.03, locate 5.0 (the first two digits) in the column headed N and then read over to the column headed 3. The result is

$\log 5.03 \approx 0.7016$ Table values are approximate.

This means that $10^{0.7016} \approx 5.03$.

Calculators have, to a large extent, eliminated the need for extensive log tables. To find x on a calculator that has logarithmic keys, you must push the keys in the indicated order:

[5.03] [log]
↑ ↑
Number first, then log key

Notice that the key for $\log_{10}$ on a calculator is simply labeled *log*. You can always assume that a logarithm is to the base 10 unless it is otherwise specified.

EXAMPLE 6 Evaluate log 7.68 correct to four significant digits using Table C.II or a calculator.*

Solution **a.** From Table C.II, $\log 7.68 \approx \mathbf{0.8854}$.
b. By calculator,

[7.68] [log] DISPLAY: .88536122

To four significant digits, $x = \mathbf{0.8854}$. ■

Notice that Table C.II is limited to logarithms of numbers between 1.00 and 9.99. Other logarithms can be found by using essential properties of logarithms. Let A and B be positive real numbers and let b be a positive real number other than 1.

---

* Significant digits are discussed in Appendix B.

Then:

First Law of Logarithms	$\log_b AB = \log_b A + \log_b B$	The log of the product of two numbers is the sum of the logs of those numbers.
Second Law of Logarithms	$\log_b \frac{A}{B} = \log_b A - \log_b B$	The log of the quotient of two numbers is the log of the numerator minus the log of the denominator.
Third Law of Logarithms	$\log_b A^p = p \log_b A$	The log of the pth power of a number is p times the log of that number.

The proofs of these laws of logarithms are easy if you remember that logarithmic equations are equivalent to exponential equations and that the properties of exponents can be applied. The first law of logarithms is a restatement of the first law of exponents:

$$b^x b^y = b^{x+y}$$

Let $A = b^x$ and $B = b^y$. Then $\log_b A = x$ and $\log_b B = y$. The first law concerns the product of A and B, so

$$\begin{aligned} AB &= b^x b^y \\ &= b^{x+y} \end{aligned}$$

Therefore, by changing to logarithmic form,

$$\begin{aligned} \log_b AB &= x + y \\ &= \log_b A + \log_b B \qquad \text{By substitution} \end{aligned}$$

The second law concerns the quotient A/B $(B \neq 0)$, so

$$\frac{A}{B} = \frac{b^x}{b^y}$$

$$\frac{A}{B} = b^{x-y} \qquad \text{Second law of exponents}$$

Thus

$$\begin{aligned} \log_b \frac{A}{B} &= x - y \\ &= \log_b A - \log_b B \qquad \text{By substitution} \end{aligned}$$

The proof of the third law of logarithms follows from the third law of exponents and is left as an exercise.

EXAMPLE 7 Evaluate $\log 852$ correct to three significant digits by using Table C.II or a calculator.

Solution **a.** From Table C.II: 852 is not found in the table, so rewrite it as an integral power of 10 times a number between 1 and 10. This form of a number is called **scientific**

notation of the number:

$$852 = \underbrace{8.52}_{\uparrow} \times \overset{\uparrow}{10^2}$$

This number is found in Table C.II. — This part is handled by using the definition of log and the laws of exponents.

$$\begin{aligned} \log 852 &= \log(8.52 \times 10^2) \\ &= \log 8.52 + \log 10^2 \\ &= \log 8.52 + 2\log 10 \\ &\approx 0.9304 + 2(1) \\ &= 2.9304 \\ &\approx \mathbf{2.93} \end{aligned}$$

These steps are usually done mentally, but this first example will show you why it works.

log 8.52 from Table C.II; log 10 from the definition (found by inspection)

Rounded to three significant digits

b. By calculator:

[852] [log]   DISPLAY: 2.930439595

To three significant digits, $\log 852 = \mathbf{2.93}$. ■

If a number n is written in scientific notation as

$$n = M \cdot 10^C \qquad (1 \le M < 10)$$

then

$$\begin{aligned} \log n &= \log(M \cdot 10^C) \\ &= \log M + C \log 10 \\ &= C + \log M \end{aligned}$$

Since $1 \le M < 10$, $\log M$ can be found in Table C.II. Log M is called the **mantissa** and C is called the **characteristic**; C can be found by inspection.

EXAMPLE 8 Find x, where $x = \log 2420$.

Solution **a.** By Table C.II: The mantissa is $\log 2.420 \approx 0.3838$ and the characteristic is 3, so

$$x = \log 2420 \approx \mathbf{3.3838}$$

b. By calculator:

$x =$ [2420] [log]   DISPLAY: 3.383815366

To five significant digits, $\boldsymbol{x} = \mathbf{3.3838}$. ■

EXAMPLE 9 Find x, where $\log 2426 = x$.

Solution **a.** By table: The mantissa is $\log 2.426$ and the characteristic is 3. But 2.426 is not listed in the table so proceed as follows:

$$6\left[\begin{matrix} 2.420 \\ 2.426 \end{matrix}\right. \qquad \left.\begin{matrix} 0.3838 \\ ? \\ 0.3856 \end{matrix}\right] \text{Difference is } 0.0018$$

$$\ 2.430$$

Take a proportion of the difference between the two logs we know and add it to the smaller one. In this case, we add $\frac{6}{10}$ of 0.0018 or 0.00108. Thus $\log 2426 \approx 3.3849$. This procedure is called **linear interpolation** and is explained more fully in the Extended Application following Chapter 8. In practice, linear interpolation for logarithms is no longer necessary. In this book we will normally require only the accuracy found in a four-place table, such as Table C.II; since 2.426 is closer to 2.430 than to 2.420, the answer $\boldsymbol{x = \log 2426 \approx 3.3856}$ is acceptable. Five- or six-place tables could be used for greater accuracy, but calculators have made them effectively obsolete.

b. By calculator:

$x =$ [2426] [log] DISPLAY: 3.3848907965

To four significant digits, $\boldsymbol{x = 3.385}$. ■

EXAMPLE 10 Find x, where $\log 0.00728 = x$.

Solution **a.** By table: The mantissa is $\log 7.28 \approx 0.8621$ and the characteristic is -3, so

$$\begin{aligned} x = \log 0.00728 &\approx 0.8621 - 3 \\ &= \mathbf{-2.1379} \end{aligned}$$

Since mantissas are always positive but characteristics are negative for any number less than 1, the answer is sometimes left in the form

$$0.8621 - 3$$

or

$$\begin{aligned} 0.8621 - 3 &= (0.8621 + 10) - 10 - 3 \\ &= \mathbf{10.8621 - 13} \end{aligned}$$

b. By calculator:

[0.00728] [log] DISPLAY: −2.137868621

To three significant digits, $\boldsymbol{x = -2.14}$. ■

Let us now return to the compound interest formula used in Examples 1 and 2. If \$12,000 is deposited at 18% and is compounded annually for 5 years the amount present is \$27,453.09 (Example 1), and if it is compounded monthly the amount present is \$29,318.64 (Example 2). A reasonable extension is to ask what happens if the interest is compounded even more frequently than monthly. Can we compound daily, hourly, every minute, or every split second? We certainly can; in fact money can be compounded **continuously**, which means that every instant the newly accumulated interest is used as part of the principal for the next instant. In order to understand these concepts consider the following contrived example. Suppose \$1 is invested at 100% interest for 1 year compounded at different intervals. The compound interest formula for this example is

$$A = \left(1 + \frac{1}{n}\right)^n$$

where n is the number of times of compounding in one year. The calculations of this formula for different values of n are shown in Table 5.1.

TABLE 5.1
Effect of compounding on $1 investment

Number of periods	Formula	Amount
Annually, $n = 1$	$\left(1 + \frac{1}{1}\right)^1$	$2.00
Semiannually, $n = 2$	$\left(1 + \frac{1}{2}\right)^2$	2.25
Quarterly, $n = 4$	$\left(1 + \frac{1}{4}\right)^4$	2.44
Monthly, $n = 12$	$\left(1 + \frac{1}{12}\right)^{12}$	2.61
Daily, $n = 360$	$\left(1 + \frac{1}{360}\right)^{360}$	2.715
Hourly, $n = 8640$	$\left(1 + \frac{1}{8640}\right)^{8640}$	2.7181

If you continue these calculations for even larger n, you will obtain the following results:

$n = 10{,}000$	the formula yields	2.718145926
$n = 100{,}000$		2.718268237
$n = 1{,}000{,}000$		2.718280469
$n = 10{,}000{,}000$		2.718281828
$n = 100{,}000{,}000$		2.718281828

The calculator can no longer distinguish the values of $(1 + 1/n)^n$ for larger n. These values are approaching a particular number. This number, it turns out, is an irrational number so it does not have a convenient decimal representation. (That is, its decimal representation does not terminate and does not repeat.) Mathematicians, therefore, have agreed to denote this number by the symbol e, which is defined as a limit.

The Number e

$$\left(1 + \frac{1}{n}\right)^n \to e \qquad \text{as } n \to \infty$$

For interest compounded continuously, the following formula is used:

Continuous Interest

$$A = Pe^{rt}$$

You can find e, as well as powers of e, by using Table C.I in Appendix C or by using a calculator.

EXAMPLE 11 Find e, e^2, and e^{-3}.

Solution **a.** From Table C.I,

$$e \approx 2.718 \qquad e = e^1$$
$$e^2 \approx 7.389$$
$$e^{-3} \approx 0.050$$

b. On a calculator, locate a key labeled e^x. First enter the value for x, then press [e^x]:

$e \approx 2.718$ [1] [e^x] DISPLAY: 2.718281828

$e^2 \approx 7.389$ [2] [e^x] DISPLAY: 7.389056099

$e^{-3} \approx 0.050$ [3] [+/−] [e^x] DISPLAY: .0497870684 ■

EXAMPLE 12 If the interest in Example 1 is compounded continuously, find the amount present.

Solution Since $P = \$12{,}000$, $r = 0.18$, and $t = 5$,

$$A = \$12{,}000e^{0.18(5)}$$
$$\approx \$12{,}000(2.4596031) \qquad \text{PRESS: [.18] [×] [5] [=] [}e^x\text{]}$$
$$\approx \mathbf{\$29{,}515.24}$$ ■

You should memorize at least the first six digits of e.

Approximation of e

$$e \approx 2.71828$$

Logarithms to the base e are called **natural logarithms** and are denoted by

WARNING: Notice the notation for natural logarithm.

$$\log_e x = \ln x$$

The expression $\ln x$ is often pronounced "lon x." Tables for natural logarithms are usually given as powers of e and are used as shown in Example 13.

EXAMPLE 13 Find $\ln 3.49$.

Solution **a.** By table: *Think* $\ln 3.49 = \log_e 3.49 = x$ and write this as an exponential equation:

$$e^x = 3.49$$

Now look at Table C.I (found in Appendix C at the back of the book). Look down the e^x columns until you find an entry equal to (or close to) 3.49; you will see $x = 1.25$ in the column headed x. This means

$$e^{1.25} \approx 3.49$$

Thus $\ln 3.49 \approx \mathbf{1.25}$.

b. By calculator: If your calculator has a [log] key, chances are it also has an [ln] key:

[3.49] [ln] DISPLAY: 1.249901736

Calculator answers are more accurate than table answers. However, it is important to realize that any answer (whether from the table or a calculator) is only as accurate as the input number (3.49 in this problem). So the answer is $x = \mathbf{1.25}$ to three significant digits. ■

EXAMPLE 14 Find ln 0.403.

Solution **a.** By table: Find 0.403 in Table C.I; it is found for $x = 0.91$ in the column headed e^{-x}. Thus

$$e^{-0.91} \approx 0.403$$

and $\ln 0.403 \approx \mathbf{-0.91}$.

b. By calculator:

[.403] [ln]   DISPLAY: -.908818717

Thus, $x = \mathbf{-0.909}$ to three significant digits. ■

EXAMPLE 15 Find ln 15.

Solution **a.** By table: Since there is no entry in Table C.I, you can use linear interpolation or find the entry closest to 15 in the column headed e^x. It is found for $x = 2.71$. Thus $\ln 15 \approx \mathbf{2.71}$.

b. By calculator:

[15] [ln]   DISPLAY: 2.708050201

Thus, $x = \mathbf{2.71}$ to three significant digits. ■

Tables for natural logs (or for e^x) are more restrictive than tables for common logs because it is not convenient to use the idea of characteristic and mantissa when working with natural logs. Notice that the limitations on Table C.I are for

$$0.00 \le x \le 3.00 \qquad \text{or} \qquad 0.00 \le e^x \le 20.086$$

Other values of $\ln x$ are considered in the next section.

PROBLEM SET 5.2

A

Write the equations in Problems 1–12 in logarithmic form.

1. $64 = 2^6$ **2.** $100 = 10^2$ **3.** $1000 = 10^3$ **4.** $64 = 8^2$
5. $125 = 5^3$ **6.** $a = b^c$ **7.** $m = n^p$ **8.** $1 = e^0$
9. $9 = (\frac{1}{3})^{-2}$ **10.** $8 = (\frac{1}{2})^{-3}$ **11.** $\frac{1}{3} = 9^{-1/2}$ **12.** $\frac{1}{2} = 4^{-1/2}$

Write the equations in Problems 13–28 in exponential form.

13. $\log_{10} 10{,}000 = 4$ **14.** $\log 0.01 = -2$
15. $\log 1 = 0$ **16.** $\log x = 2$ **17.** $\log_e e^2 = 2$
18. $\ln e^3 = 3$ **19.** $\ln x = 5$ **20.** $\ln x = 0.03$
21. $\log_2 \frac{1}{8} = -3$ **22.** $\log_2 32 = 5$ **23.** $\log_4 2 = \frac{1}{2}$
24. $\log_{1/2} 16 = -4$ **25.** $\log_m n = p$ **26.** $\log_a b = c$
27. $\log_x 8 = 3$ **28.** $\log_x 52 = 2$

Use the definition of logarithm, Tables C.I and C.II, or a calculator to evaluate the expressions in Problems 29–52.

29. $\log_b b^2$ **30.** $\log_t t^3$ **31.** $\log_e e^4$ **32.** $\log_\pi \sqrt{\pi}$
33. $\log_\pi (1/\pi)$ **34.** $\log_2 8$ **35.** $\log_3 9$ **36.** $\log_{19} 1$
37. $\log 4.27$ **38.** $\log 1.08$ **39.** $\log 8.43$ **40.** $\log 9760$
41. $\log 71{,}600$ **42.** $\log 0.042$ **43.** $\log 0.321$ **44.** $\log 0.0532$

45. $\ln 2.27$ **46.** $\ln 16.77$ **47.** $\ln 2$ **48.** $\ln \frac{1}{8}$

49. $\ln 13$ **50.** $\ln 0.15$ **51.** $\ln 7.3$ **52.** $\ln 10.57$

B

53. Prove the third law of logarithms, $\log_b A^p = p \log_b A$.

54. If \$1000 is invested at 7% compounded annually, how much money will there be in 25 years?

55. If \$1000 is invested at 12% compounded semiannually, how much money will there be in 10 years?

56. If \$1000 is invested at 16% interest compounded continuously, how much money will there be in 25 years?

57. If \$1000 is invested at 14% interest compounded continuously, how much money will there be in 10 years?

58. If \$8500 is invested at 18% interest compounded monthly, how much money will there be in 3 years?

59. If \$3600 is invested at 15% interest compounded daily, how much money will there be in 7 years? (Use a 365-day year; this is called *exact* interest.)

60. If \$10,000 is invested at 14% interest compounded daily, how much money will there be in 6 months? (Use a 360-day year; this is called *ordinary* interest.)

61. ***Business*** An advertising agency conducted a survey and found that the number of units sold, N, is related to the amount a spent on advertising (in dollars) by the following formula:

$$N = 1500 + 300 \ln a \qquad (a \geq 1)$$

a. How many units are sold after spending \$1000?
b. How many units are sold after spending \$50,000?
c. If the company wants to sell 5000 units, how much money will it have to spend?

62. ***Chemistry*** The pH of a substance measures its acidity or alkalinity. It is found by the formula

$$\text{pH} = -\log[\text{H}^+]$$

where $[\text{H}^+]$ is the concentration of hydrogen ions in an aqueous solution given in moles per liter.

a. What is the pH (to the nearest tenth) of a lemon for which $[\text{H}^+] = 2.86 \times 10^{-4}$?
b. What is the pH (to the nearest tenth) of rainwater for which $[\text{H}^+] = 6.31 \times 10^{-7}$?
c. If a shampoo advertises that it has a pH of 7, what is $[\text{H}^+]$?

63. ***Earth Science*** The Richter scale for measuring earthquakes was developed by Gutenberg and Richter. It relates the energy E (in ergs) to the magnitude of the earthquake, M, by the formula

$$M = \frac{\log E - 11.8}{1.5}$$

a. A small earthquake is one that releases 15^{15} ergs of energy. What is the magnitude of such an earthquake on the Richter scale?
b. A large earthquake is one that releases 10^{25} ergs of energy. What is the magnitude of such an earthquake on the Richter scale?

C

64. ***Physics*** The intensity of sound is measured in decibels D and is given by

$$D = \log\left(\frac{I}{I_0}\right)^{10}$$

where I is the power of the sound in watts per cubic centimeter (W/cm^3) and $I_0 = 10^{-16}\ \text{W/cm}^3$ (the power of sound just below the threshold of hearing). Find the intensity in decibels of the given sound:

a. A whisper, $10^{-13}\ \text{W/cm}^3$
b. Normal conversation, $3.16 \cdot 10^{-10}\ \text{W/cm}^3$
c. The world's loudest shout by Skipper Kenny Leader, $10^{-5}\ \text{W/cm}^3$
d. A rock concert, $5.23 \cdot 10^{-6}\ \text{W/cm}^3$

5.3 LOGARITHMIC FUNCTIONS AND EQUATIONS

The notion of a logarithm, which was introduced in the previous section, can also be considered as a special type of function.

Logarithmic Function

> The function f defined by
>
> $$f(x) = \log_b x$$
>
> where $b > 0$, $b \neq 1$, is called the **logarithmic function with base *b*.**

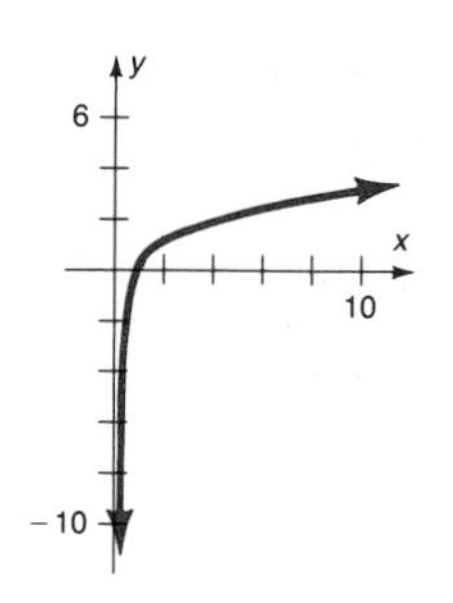

Figure 5.8 Graph of $y = \log_2 x$

By relating logarithmic and exponential functions, you will notice that you already know a great deal about the logarithmic function. For example, the graph of $y = \log_2 x$ is the same as the graph of $2^y = x$, as shown in Figure 5.8.

y	x
-3	$2^{-3} = \frac{1}{8}$
-2	$2^{-2} = \frac{1}{4}$
-1	$2^{-1} = \frac{1}{2}$
0	$2^0 = 1$
1	$2^1 = 2$
2	$2^2 = 4$
3	$2^3 = 8$

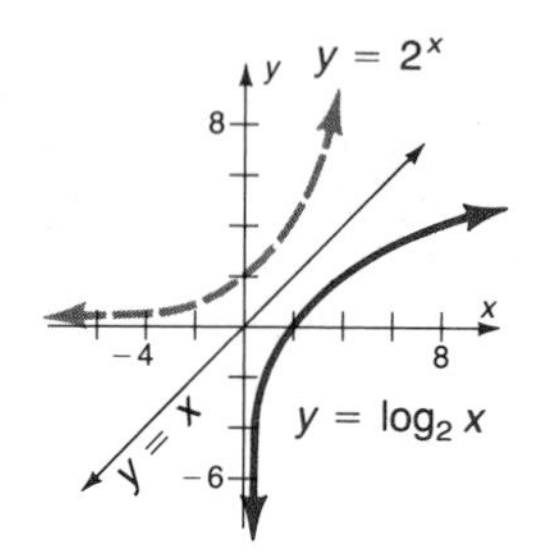

Figure 5.9 Graphs of $y = 2^x$, $y = \log_2 x$, and $y = x$

Compare the graph in Figure 5.8 with the graph of $y = 2^x$, as shown along with $y = x$ in Figure 5.9.

The functions $f(x) = 2^x$ and $g(x) = \log_2 x$ appear to be inverse functions. To prove this, we will need to consider two properties of logarithms that follow from the definition of logarithm:

1. $\log_b b^x = x$ 2. $b^{\log_b x} = x$

Both of these properties follow directly from the definition of logarithms:

$$b^M = N \Leftrightarrow \log_b N = M$$

exponent

Definition: $b^M = N \Leftrightarrow \log_b N = M$
Property 1: $b^x = b^x \Leftrightarrow \log_b b^x = x$

exponent

Property 1 is an exact statement of the definition, only with $M = x$ and $N = b^x$. Since $b^x = b^x$ is true, then $\log_b b^x = x$ is true by the definition of logarithm.

exponent

Definition: $b^M = N \Leftrightarrow \log_b N = M$
Property 2: $b^{\log_b x} = x \Leftrightarrow \log_b x = \log_b x$

exponent

Property 2 is an exact statement of the definition. Let $M = \log_b x$ and $N = x$. Since $\log_b x = \log_b x$, then $b^{\log_b x} = x$ is true by the definition of logarithm.

To show that $f(x) = \log_b x$ and $g(x) = b^x$ are inverse functions, show $(f \circ g)(x) = (g \circ f)(x) = x$:

$$\begin{aligned}(f \circ g)(x) &= f[g(x)] \\ &= f(b^x) \\ &= \log_b b^x \\ &= x \quad \text{By Property 1}\end{aligned}$$

and

$$\begin{aligned}(g \circ f)(x) &= g[f(x)] \\ &= g(\log_b x) \\ &= b^{\log_b x} \\ &= x \quad \text{By Property 2}\end{aligned}$$

This proves the following theorem.

Exponential and Logarithmic Functions Are Inverses

> The exponential and logarithmic functions with base b are inverse functions of one another.

This relationship of inverse functions is needed to find e on several brands of calculators. If a calculator has

[ln x] and [INV]

keys, but no e^x key, how can you find e or e^x? Since

$$y = \ln x \qquad \text{and} \qquad y = e^x$$

are inverse functions, to find e (or e^1) you can press

[1] [INV] [ln x] DISPLAY: 2.718281828

This gives the inverse of the ln x function; that is, it is e^x where the x value is input just prior to pressing these buttons.

If you need $e^{5.2}$ on such a calculator, press

[5.2] [INV] [ln x] DISPLAY: 181.2722419

Properties 1 and 2 also lead to another theorem that allows us to solve logarithmic equations.

Log of Both Sides Theorem

> If A, B, and b are positive real numbers with $b \neq 1$, then
>
> $$\log_b A = \log_b B \Leftrightarrow A = B$$

Proof If $A = B$, then

$$\log_b A = \log_b A$$
$$= \log_b B \qquad \text{By substitution}$$

If $\log_b A = \log_b B$, then

$$b^{\log_b B} = A \qquad \text{By definition of logarithm}$$
$$b^{\log_b B} = B \qquad \text{By Property 2}$$
$$B = A \qquad \text{By substitution}$$

□

We now use this theorem and the definition of logarithm to solve exponential equations.

EXAMPLE 1 Solve $\log_2 \sqrt{2} = x$ for x.

Solution For this logarithmic equation, the exponent is the unknown. Apply the definition of logarithm:

$$2^x = \sqrt{2}$$
$$2^x = 2^{1/2}$$
$$x = \tfrac{1}{2} \qquad \text{Exponential property of equality}$$

■

EXAMPLE 2 Solve $\log_x 25 = 2$.

Solution For this logarithmic equation, the base is the unknown. Apply the definition of logarithm:

$$x^2 = 25$$
$$x = \pm 5$$

Be sure the values you obtain are permissible values for the definition of a logarithm. In this case, $x = -5$ is not a permissible value since a logarithm with a negative base is not defined. Therefore, the solution is $\mathbf{x = 5}$. ■

EXAMPLE 3 Solve $\ln x = 5$.

Solution The power itself is the unknown in this logarithmic equation. Apply the definition of logarithm (to the base e):

$$e^5 = x$$

$\mathbf{x \approx 148.41}$ PRESS: [5] [e^x] DISPLAY: 148.4131591 ■

EXAMPLE 4 Solve $\log_5 x = \log_5 72$.

Solution Use the Log of Both Sides Theorem: $\mathbf{x = 72}$. ■

Basically, all logarithmic equations fall into one of the four types illustrated by Examples 1–4. For problems like those in Examples 1–3, apply the definition of logarithm and for those like Example 4, apply the Log of Both Sides Theorem. More difficult examples require algebraic simplification using the laws of logarithms (see page 177) in order to write a single logarithmic function on either one or both sides, as illustrated by Examples 5–7.

EXAMPLE 5

$\log_8 3 + \frac{1}{2}\log_8 25 = \log_8 x$

$\log_8 3 + \log_8 25^{1/2} = \log_8 x$ Third law of logarithms

$\log_8 3 + \log_8 5 = \log_8 x$ $25^{1/2} = \sqrt{25} = 5$

$\log_8(3 \cdot 5) = \log_8 x$ First law of logarithms

$15 = x$ Log of Both Sides Theorem

The solution is **15**. ■

EXAMPLE 6

$$5\log_x 2 - \tfrac{1}{2}\log_x 8 = 2 - \tfrac{1}{2}\log_x 2$$

$$\log_x 2^5 - \log_x \sqrt{8} = 2 - \log_x \sqrt{2}$$
$$\log_x 32 - \log_x 2\sqrt{2} + \log_x \sqrt{2} = 2$$
$$\log_x\left(\frac{32}{2\sqrt{2}} \cdot \sqrt{2}\right) = 2$$
$$\log_x 16 = 2$$

$x^2 = 16$ Definition of logarithm

$$x = \pm 4$$

By the definition of a logarithm, x must be positive, so the solution is **4**. ■

EXAMPLE 7

$$\ln x - \tfrac{1}{2}\ln 2 = \tfrac{1}{2}\ln(x + 4)$$
$$\ln x - \ln\sqrt{2} = \ln\sqrt{x + 4}$$
$$\ln\left(\frac{x}{\sqrt{2}}\right) = \ln\sqrt{x + 4}$$
$$\frac{x}{\sqrt{2}} = \sqrt{x + 4}$$
$$\frac{x^2}{2} = x + 4$$
$$x^2 - 2x - 8 = 0$$
$$(x - 4)(x + 2) = 0$$
$$x = 4, -2$$

Notice that $\ln(-2)$ is not defined, so $x = -2$ is an extraneous root. Therefore the solution is **4**. ■

For many years logarithms were taught in elementary mathematics primarily as an aid to computation. As you have seen in this chapter, logarithms are useful as functions in mathematics, but with the widespread use of calculators, they are no longer necessary for complicated calculations. In the remaining part of this section we will consider some problems and give their logarithmic and calculator solutions so you can compare the different methods. If you do not have a calculator available, you can work the problems using the four-place log tables in the back of the book.

EXAMPLE 8 Find $25{,}000(1.15)^{30}$ to two significant digits.

Solution By calculator with algebraic logic:

[1.15] [y^x] [30] [×] [25000] [=] DISPLAY: 1655294.299

By calculator with RPN logic:

[30] [ENTER] [1.15] [x^y] [25000] [×] DISPLAY: 1655294.299

By logarithms:

$$A = 25{,}000(1.15)^{30}$$
$$\log A = \log[25{,}000(1.15)^{30}]$$
$$= \log 25{,}000 + 30\log 1.15$$
$$\approx 4.3979 + 30(0.0607)$$
$$= 6.2189$$

The problem now is to use Table C.II in reverse by finding the mantissa 0.2189 in the table; it is closest to 1.66. (You can use linear interpolation for a more accurate answer.) Thus if $\log A = 6.2189$, then

$$A = 1.66 \times 10^6$$
$$\approx 1{,}660{,}000$$

To two significant digits the answer is **1,700,000**. ■

EXAMPLE 9 Find $13{,}250e^{1.35}$ to three significant digits.

Solution By calculator with algebraic logic:

[1.35] [e^x] [×] [13250] [=] DISPLAY: 51110.88828

By calculator with RPN logic:

[1.35] [e^x] [ENTER] [13250] [×] DISPLAY: 51110.88828

By logarithms:

$$\begin{aligned} A &= 13{,}250e^{1.35} \\ &= (13{,}250)(3.857) \qquad \text{By Table C.I} \\ \log A &= \log 13{,}250 + \log 3.857 \\ &\approx 4.1222 + 0.5862 \\ &= 4.7085 \\ A &\approx 5.11 \times 10^4 \end{aligned}$$

To three significant digits, the answer is **51,000**. ■

PROBLEM SET 5.3

A

Solve the equations in Problems 1–15.

1. a. $\log_5 25 = x$ b. $\log_2 128 = x$
2. a. $\log_3 81 = x$ b. $\log_4 64 = x$
3. a. $\log \frac{1}{10} = x$ b. $\log 10{,}000 = x$
4. a. $\log 1000 = x$ b. $\log \frac{1}{1000} = x$
5. a. $\log_x 28 = 2$ b. $\log_x 81 = 4$
6. a. $\log_x 50 = 2$ b. $\log_x 84 = 2$
7. a. $\log_x e = 2$ b. $\log_x e = 1$
8. a. $\ln x = 2$ b. $\ln x = 3$
9. a. $\ln x = 4$ b. $\ln x = \ln 14$
10. a. $\ln 9.3 = \ln x$ b. $\ln 109 = \ln x$
11. a. $\log_3 x^2 = \log_3 125$ b. $\ln x^2 = \ln 12$
12. a. $\log x^2 = \log 70$ b. $\log_2 8\sqrt{2} = x$
13. a. $\log_3 27\sqrt{3} = x$ b. $\log_x 1 = 0$
14. a. $\log_x 10 = 0$ b. $\log_e x = 3$
15. a. $\log_2 x = 5$ b. $\log_{10} x = 5$

Use a calculator or logarithms to evaluate the expressions in Problems 16–22. Be sure to round off your answers to the appropriate number of significant digits. (See Appendix B for a discussion of significant digits.)

16. a. (14)(351) b. (218)(263)
17. a. $(2.00)^4(1245)(277)$ b. $(3.00)^3(182)$
18. a. $\dfrac{(1979)(1356)}{452}$ b. $\dfrac{(515)(20{,}600)}{200}$
19. a. $(990)(1117)(342) - 89$ b. $[0.14 + (197)(25.08)](19)$
20. a. $\dfrac{4(25)}{10(0.05)}$ b. $\dfrac{20}{(5.0)(0.20)}$
21. a. $\dfrac{1.00}{0.005 + 0.020}$ b. $\dfrac{1.00}{8.21 + 2.45}$
22. a. $\dfrac{17^2 + 34^2 - 12^2}{2.0(17)(34)}$ b. $\dfrac{241^2 + 568^2 - 351^2}{2.00(241)(568)}$

B

Graph the functions given in Problems 23–38.

23. $f(x) = e^{-x}$
24. $f(x) = e^x$
25. $y = e^{x+1}$
26. $y = e^x + 1$
27. $y - 2 = e^{x+2}$
28. $y + 1 = e^{x+3}$
29. $y = \log_3 x$
30. $y = \log_{1/3} x$
31. $y = \log_\pi x$
32. $y = \log|x|$
33. $y = \log(|x| + 1)$
34. $y - 1 = \log x^2$
35. $y = \log(x - 4)$
36. $y - 3 = \log x$
37. $y - 1 = \log(x + 2)$
38. $y + 2 = \log(x - 1)$

Solve the logarithmic equations in Problems 39–55.

39. $\log 2 = \frac{1}{4}\log 16 - x$
40. $\frac{1}{2}\log x - \log 100 = 2$

41. $\ln 2 + \frac{1}{2}\ln 3 + 4\ln 5 = \ln x$
42. $\ln x = \ln 5 - 3\ln 2 + \frac{1}{4}\ln 3$
43. $1 = 2\log x - \log 1000$
44. $\log_8 5 + \frac{1}{2}\log_8 9 = \log_8 x$
45. $\log_7 x - \frac{1}{2}\log_7 4 = \frac{1}{2}\log_7(2x - 3)$
46. $\ln 10 - \frac{1}{2}\ln 25 = \ln x$
47. $\frac{1}{2}\ln x = 3\ln 5 - \ln x$
48. $\ln x - \frac{1}{2}\ln 3 = \frac{1}{2}\ln(x + 6)$
49. $2\ln x - \frac{1}{2}\ln 9 = \ln 3(x - 2)$
50. $\log 2 - \frac{1}{2}\log 9 + 3\log 3 = x$
51. $3\log 3 - \frac{1}{2}\log 3 = \log\sqrt{x}$

52. $3\ln\dfrac{e}{\sqrt[3]{5}} = 3 - \ln x$

53. $5\ln\dfrac{e}{\sqrt[5]{5}} = 3 - \ln x$

54. $2\ln\dfrac{e}{\sqrt{7}} = 2 - \ln x$

55. $\log_x(x + 6) = 2$

Business *If P dollars are borrowed for n months at a monthly interest rate of r, then the monthly payment is found by the formula*

$$M = \frac{Pr}{1 - (1 + r)^{-n}}$$

To use this formula to find M with a calculator with algebraic logic, press:

$\boxed{r}\ \boxed{+}\ \boxed{1}\ \boxed{=}\ \boxed{y^x}\ \boxed{n}\ \boxed{+/-}\ \boxed{=}$

$\boxed{+/-}\ \boxed{+}\ \boxed{1}\ \boxed{=}\ \boxed{1/x}\ \boxed{\times}\ \boxed{P}\ \boxed{\times}\ \boxed{r}\ \boxed{=}$

Use this information in Problems 56–59.

56. What is the monthly car payment for a new car costing \$12,487 with a down payment of \$2487? The car is financed for 4 years at 12%. (*Hint:* $P = \$10{,}000$ and $r = 0.01$.)

57. A home loan is made for \$110,000 at 12% interest for 30 years. What is the monthly payment? (*Hint:* $P = \$110{,}000$ and $r = 0.01$.)

58. A purchase of \$2430 is financed at 23% for 3 years. What is the monthly payment?

59. A home is financed at $14\frac{1}{2}\%$ for 30 years. If the amount financed is \$125,000, what is the monthly payment?

60. ***Chemistry*** The pH (hydrogen potential) of a solution is given by

$$\text{pH} = \log\frac{1}{[\text{H}^+]}$$

where $[\text{H}^+]$ is the concentration of hydrogen ions in a water solution in moles per liter. Find the pH (to the nearest tenth) for the substances with the $[\text{H}^+]$ given.

a. Lemon juice, $5.01 \cdot 10^{-3}$
b. Milk, $3.98 \cdot 10^{-7}$
c. Vinegar, $7.94 \cdot 10^{-4}$
d. Rainwater, $5.01 \cdot 10^{-7}$
e. Seawater, $4.35 \cdot 10^{-9}$

61. ***Psychology*** A learning curve describes the rate at which a person learns certain tasks. If a person sets a goal of typing N words per minute (wpm), the length of time t, in days, to achieve this goal is given by

$$t = -62.5\ln\left(1 - \frac{N}{80}\right)$$

a. How long would it take to learn to type 30 wpm?
b. If we accept this formula, is it possible to learn to type 80 wpm?
c. Solve for N.

62. ***Psychology*** In Problem 61 an equation for learning was given. Psychologists are also concerned with forgetting. In a certain experiment, students were asked to remember a set of nonsense syllables, such as "htm." The students then had to recall the syllables after t seconds. The equation that was found to describe forgetting was

$$R = 80 - 27\ln t \qquad (t \geq 1)$$

where R is the percentage of students who remember the syllables after t sec.

a. What percentage of the students remembered the syllables after 3 sec?
b. In how many seconds would only 10% of the students remember?
c. Solve for t.

63. The equation for the Richter scale relating energy E (in ergs) to the magnitude of the earthquake, M, is given by the formula

$$M = \frac{\log E - 11.8}{1.5}$$

a. Solve for E.
b. What was the energy of the 1988 Armenian earthquake, which measured 7.5 on the Richter scale?

C

64. Consider the following argument:

$4 > 3$	Obviously true
$4\log_{10}\frac{1}{3} > 3\log_{10}\frac{1}{3}$	Multiply both sides by $\log_{10}\frac{1}{3}$.
$\log_{10}(\frac{1}{3})^4 > \log_{10}(\frac{1}{3})^3$	Property of exponents
$(\frac{1}{3})^4 > (\frac{1}{3})^3$	Theorem of logarithms
$\frac{1}{81} > \frac{1}{27}$	Obviously false

Can you find the error?

65. Graph $y = \left(1 + \frac{1}{x}\right)^x$ for $x > 0$.

66. Graph: **a.** $c(x) = \frac{e^x + e^{-x}}{2}$ **b.** $s(x) = \frac{e^x - e^{-x}}{2}$

67. ***Engineering*** The functions c and s of Problem 66 are called the *hyperbolic cosine* and *hyperbolic sine* functions. These are defined by

$$\cosh x = \frac{e^x + e^{-x}}{2} \quad \text{and} \quad \sinh x = \frac{e^x - e^{-x}}{2}$$

a. Show that $\cosh^2 x - \sinh^2 x = 1$.
b. Show that the hyperbolic cosine is an even function and that the hyperbolic sine is an odd function.
c. Show that $\sinh 2x = 2 \sinh x \cosh x$.
d. Graph $y = \cosh x + \sinh x$.
e. Define the *hyperbolic tangent* function as

$$\tanh x = \frac{\sinh x}{\cosh x}$$

Graph $\tanh x$.

5.4 EXPONENTIAL EQUATIONS

In this section, we return to the question that prompted our discussion of logarithms in the first place—How do you solve an exponential equation? The most straightforward method for solving exponential equations applies the exponential property of equality stated in Section 5.1. Recall that if equal bases are raised to some power and the results are equal, then the exponents must also be equal.

EXAMPLE 1 Solve $7^x = 343$ for x.

Solution Write both sides with the same base if possible:

$$7^x = 7^3$$
$$\mathbf{x = 3} \quad \text{Exponential property of equality}$$

■

EXAMPLE 2 Solve $25^x = 125$.

Solution

$$(5^2)^x = 5^3$$
$$5^{2x} = 5^3$$
$$2x = 3 \quad \text{Exponential property of equality}$$
$$\mathbf{x = \tfrac{3}{2}}$$

■

EXAMPLE 3 Solve $36^x = \frac{1}{6}$.

Solution

$$(6^2)^x = 6^{-1}$$
$$6^{2x} = 6^{-1}$$
$$2x = -1$$
$$\mathbf{x = -\tfrac{1}{2}}$$

■

Not all exponential equations are as easy to solve as those in Examples 1–3. If the bases cannot be forced to be the same, then the exponential equations will fall into one of three types:

Base:	10 (common log)	e (natural log)	b (arbitrary base)
Example:	$10^x = 5$	$e^t = 3.456$	$7^x = 3$

We will work two examples of each of these types. First, consider base 10.

EXAMPLE 4 Solve $10^x = 5$.

Solution Use the definition of logarithm.

$x = \log 5$ Exact answer

$\mathbf{x \approx 0.699}$ Approximate answer, found by using tables or a calculator ∎

EXAMPLE 5 Solve $10^{5x+3} = 195$.

Solution Rewrite in logarithmic form: $\log 195 = 5x + 3$. Solve for x:

$$\log 195 - 3 = 5x$$

$$x = \frac{1}{5}(\log 195 - 3) \qquad \text{Exact answer}$$

By table: $\log 195 \approx 2.2900$, so

$$x \approx \frac{1}{5}(2.2900 - 3)$$

$$= \frac{1}{5}(-0.71)$$

$$= \mathbf{-0.142}$$

By calculator: You can multiply by $\frac{1}{5}$, but it is more common to think of this as division by 5. *Think:*

$$x = \frac{\log 195 - 3}{5}$$

[195] [log] [−] [3] [=] [÷] [5] [=]

↑

If you do not press this equals key, your calculator may assume you want $3 \div 5$ and not the whole quantity divided by 5.

DISPLAY: −.1419930777

RPN logic: [195] [ln] [ENTER] [3] [−] [5] [÷] ∎

The next two examples involve base e.

EXAMPLE 6 Solve $e^{0.06t} = 3.456$ for t.

Solution **a.** By table: Find $e^x = 3.456$; $x = 1.24$. Thus

$$e^{0.06t} \approx e^{1.24}$$

By the exponential property of equality,

$$0.06t = 1.24$$

$$\mathbf{t \approx 20.67}$$

b. By calculator: Write in logarithmic form:

$$e^{0.06t} = 3.456$$
$$0.06t = \ln 3.456$$
$$t = \frac{\ln 3.456}{0.06}$$

Algebraic logic:

[3.456] [ln] [÷] [.06] [=]   DISPLAY: 20.66853085

RPN logic:

[3.456] [ln] [ENTER] [.06] [÷]   DISPLAY: 20.66853085 ■

EXAMPLE 7 Solve $\frac{1}{2} = e^{-0.000425t}$.

Solution **a.** By table: Find $e^x = \frac{1}{2} = 0.5$ in Table C.I. Notice that this is found in the column headed e^{-x} for $x \approx 0.69$. (You can interpolate if better accuracy is needed.) Thus, since $e^{-0.69} \approx \frac{1}{2}$,

$$e^{-0.69} \approx e^{-0.000425t}$$
$$-0.69 \approx -0.000425t$$
$$t \approx \frac{0.69}{0.000425}$$
$$\approx \mathbf{1624}$$

b. By calculator: $\frac{1}{2} = e^{-0.000425t}$ in logarithmic form is

$$\ln 0.5 = -0.000425t$$
$$t = \frac{\ln 0.5}{-0.000425}$$

Algebraic logic:

[.5] [ln] [÷] [.000425] [+/−] [=]   DISPLAY: 1630.934542

RPN logic:

[.5] [ln] [ENTER] [.000425] [+/−] [÷]   DISPLAY: 1630.934542 ■

Notice that there is a considerable discrepancy between the answers in Example 7 depending on whether you use Table C.I or a calculator. Remember, however, that only two significant digits were used in finding the result from Table C.I, so to two significant digits both of these answers are 1600.

Since you do not have tables or calculator keys for bases other than 10 and e, you must proceed differently for an arbitrary base b. You should use the Log of Both Sides Theorem and the procedure for solving logarithmic equations. You generally can elect to work with base 10 or base e.

EXAMPLE 8 Solve $7^x = 3$.

Solution Using base e:

$$\ln 7^x = \ln 3$$
$$x \ln 7 = \ln 3$$
$$x = \frac{\ln 3}{\ln 7}$$
$$\approx \frac{1.099}{1.946} \quad \text{By Table C.I or by calculator}$$
$$\approx \mathbf{0.5646}$$

Using base 10:

$$\log 7^x = \log 3$$
$$x \log 7 = \log 3$$
$$x = \frac{\log 3}{\log 7}$$
$$\approx \frac{0.4771}{0.8451} \quad \text{By Table C.II or by calculator}$$
$$\approx \mathbf{0.5646}$$

Same answer (you use only one of these) ■

There is another method for solving the exponential equation of Example 8. This one directly applies the definition of logarithm:

$$7^x = 3 \Leftrightarrow \log_7 3 = x$$

If you had a $\log_7$ key on your calculator or a $\log_7$ table, you would have the answer. Since you do not, you need one final logarithm theorem that changes logarithms from one base to another.

Change of Base Theorem

$$\log_a x = \frac{\log_b x}{\log_b a}$$

Notice that to change from base a to another (possibly more familiar) base b, you simply change the base on the given logarithm from a to b and then divide by the logarithm to the base b of the old base a.

Proof Let $y = \log_a x$

$$a^y = x \quad \text{Definition of logarithm}$$
$$\log_b a^y = \log_b x \quad \text{Log of Both Sides Theorem}$$
$$y \log_b a = \log_b x \quad \text{Third law of logarithms}$$
$$y = \frac{\log_b x}{\log_b a} \quad \text{Divide both sides by } \log_b a \ (\log_b a \neq 0).$$

□

EXAMPLE 9 Change $\log_7 3$ to logarithms with base 10 and evaluate.

Solution

$$\log_7 3 = \frac{\log 3}{\log 7}$$
$$\approx \frac{0.4771}{0.8451}$$
$$\approx \mathbf{0.5646} \quad \text{Compare this result with the one from Example 8.}$$

■

EXAMPLE 10 Change $\log_3 3.84$ to logarithms with base e and evaluate.

Solution

$$\log_3 3.84 = \frac{\ln 3.84}{\ln 3} \approx \frac{1.345}{1.099} \approx \mathbf{1.225}$$

■

We are now ready to consider the second of our examples of exponential equations with an arbitrary base.

EXAMPLE 11 Solve $6^{3x+2} = 200$.

Solution Method 1:

$$\log_6 200 = 3x + 2$$
$$\log_6 200 - 2 = 3x$$
$$x = \frac{\log_6 200 - 2}{3}$$
$$= \frac{\frac{\ln 200}{\ln 6} - 2}{3}$$

Algebraic calculator:

[200] [ln] [÷] [6] [ln] [=] [−] [2] [=] [÷] [3] [=] DISPLAY: .3190157417

RPN calculator:

[200] [ln] [ENTER] [6] [ln] [÷] [2] [−] [3] [÷] DISPLAY: .3190157417

Method 2:

$$\log 6^{3x+2} = \log 200 \quad \text{Log of Both Sides Theorem}$$
$$(3x + 2)\log 6 = \log 200$$
$$3x \log 6 + 2 \log 6 = \log 200$$
$$(3 \log 6)x = \log 200 - 2 \log 6$$
$$x = \frac{\log 200 - 2 \log 6}{3 \log 6}$$

Algebraic calculator:

[200] [log] [−] [2] [×] [6] [log] [=] [÷] [(] [3] [×] [6] [log] [)] [=]

DISPLAY: .3190157417

RPN calculator:

[200] [log] [ENTER] [2] [ENTER] [6] [log] [×] [−] [3] [ENTER] [6] [log] [×] [÷]

DISPLAY: .3190157417

Answer: $x \approx 0.32$ ■

In Section 5.2 we considered continuous compounding of interest. This is an example of continuous *growth*. A similiar application involves continuous *decay*.

Decay Formula

$$A = A_0 e^{-kt}$$

where an initial quantity A_0 decays to an amount A after a time t. The positive constant k depends on the substance.

EXAMPLE 12 If 100.0 mg of neptunium-239 (^{239}Np) decays to 73.36 mg after 24 hr, find the value of k (to four significant digits) in the decay formula for t expressed as days.

Solution Since $A = 73.36$, $A_0 = 100$, and $t = 1$, we have

$$73.36 = 100e^{-k(1)}$$
$$0.7336 = e^{-k}$$
$$k = -\ln 0.7336$$
$$\approx 0.30979136$$
$$\mathbf{k = 0.3098} \quad \text{Rounded to four significant digits}$$ ■

Radioactive decay is usually specified in terms of its **half-life**, which means the amount of time necessary for one-half of its substance to disintegrate into another substance. The decay formula for half-life is

Half-Life Decay

$$\tfrac{1}{2} = e^{-kt}$$

EXAMPLE 13 What is the half-life of ^{239}Np? Use the value of k found in Example 12 ($k = 0.3098$).

Solution

$$\tfrac{1}{2} = e^{-0.3098t}$$
$$\ln 0.5 = -0.3098t$$
$$t = \frac{\ln 0.5}{-0.3098}$$
$$\approx 2.2374021$$

The half-life is about **2.24 days**. ■

A very important application of both growth and decay is the prediction of the size of various populations. Population growth is considered at length in a special Extended Application following the Cumulative Review at the end of this chapter.

PROBLEM SET 5.4

A

Solve the exponential equations in Problems 1–16.

1. $2^x = 128$ 2. $3^x = 243$ 3. $8^x = 32$ 4. $9^x = 27$
5. $125^x = 25$ 6. $4^x = \frac{1}{16}$ 7. $27^x = \frac{1}{81}$ 8. $(\frac{1}{2})^x = 8$
9. $(\frac{1}{2})^x = \frac{1}{8}$ 10. $(\frac{2}{3})^x = \frac{9}{4}$ 11. $(\frac{3}{4})^x = \frac{16}{9}$ 12. $2^{3x+1} = \frac{1}{2}$
13. $3^{4x-3} = \frac{1}{9}$ 14. $27^{2x+1} = 3$
15. $8^{5x+2} = 16$ 16. $125^{2x+1} = 25$

Evaluate the logarithms in Problems 17–22 by using the Change of Base Theorem.

17. $\log_5 30$ 18. $\log_2 15$ 19. $\log_6 0.1$ 20. $\log_4 0.05$
21. $\log_\pi 10$ 22. $\log_\pi 25$

B

Solve the exponential equations in Problems 23–46.

23. $10^x = 42$ 24. $10^x = 126$
25. $10^x = 0.00325$ 26. $10^x = 0.0234$
27. $10^{x+3} = 214$ 28. $10^{x-5} = 0.036$
29. $10^{x-1} = 0.613$ 30. $10^{4x+1} = 719$
31. $10^{5-3x} = 0.041$ 32. $10^{2x-1} = 515$
33. $e^{2x} = 10$ 34. $e^{5x} = \frac{1}{4}$
35. $e^{4x} = \frac{1}{10}$ 36. $e^{2x+1} = 5.474$
37. $e^{1-2x} = 3$ 38. $e^{1-5x} = 15$
39. $8^x = 300$ 40. $2^x = 1000$
41. $5^x = 10$ 42. $9^x = 0.045$
43. $2^{-x} = 5$ 44. $4^x = 0.82$
45. $5^{-x} = 8$ 46. $7^{-x} = 125$

47. If \$1000 is invested at 12% compounded semiannually, how long will it take (to the nearest year) for the money to double?
48. If \$1000 is invested at 16% interest compounded annually, how long will it take (to the nearest year) for the money to quadruple?
49. If \$1000 is invested at 12% interest compounded quarterly, how long will it take (to the nearest quarter) for the money to reach \$2500?
50. If \$1000 is invested at 15% interest compounded continuously, how long will it take (to the nearest year) for the money to triple?
51. If the half-life of cesium-137 is 30 years, find the constant k for which $A = A_0e^{-kt}$ where t is expressed in years.
52. Find the half-life of strontium-90 if $k = 0.0246$, where $A = A_0e^{-kt}$ and t is expressed in years.
53. Find the half-life of krypton if $k = 0.0641$, where $A = A_0e^{-kt}$ and t is expressed in years.
54. The formula used for carbon-14 dating in archaeology is

$$A = A_0\left(\frac{1}{2}\right)^{t/5700} \quad \text{or} \quad P = \left(\frac{1}{2}\right)^{t/5700}$$

where P is the percentage of carbon-14 present after t years. Solve for t.
55. Some bone artifacts were found at the Lindenmeier site in northeastern Colorado and tested for their carbon-14 content. If 25% of the original carbon-14 was still present, what is the probable age of the artifacts? Use the formula given in Problem 54.
56. An artifact was discovered at the Debert site in Nova Scotia. Tests showed that 28% of the original carbon-14 was still present. What is the probable age of the artifact? Use the formula given in Problem 54.
57. An artifact was found and tested for its carbon-14 content. If 12% of the original carbon-14 was still present, what is its probable age? Use the formula given in Problem 54.
58. An artifact was found and tested for its carbon-14 content. If 85% of the original carbon-14 was still present, what is its probable age? Use the formula given in Problem 54.
59. ***Physics*** The atmospheric pressure P in pounds per square inch (psi) is given by

$$P = 14.7e^{-0.21a}$$

where a is the altitude above sea level (in miles). If a city has an atmospheric pressure of 13.23 psi, what is its altitude?
60. ***Earth Science*** In 1985 ($t = 0$), the world use of petroleum, P_0, was 19,473 million barrels of oil. If the world reserves are 584,600 million barrels and the growth rate for the use of oil is k, then the total amount A used during a time interval $t > 0$ is given by

$$A = \frac{P_0}{k}(e^{kt} - 1)$$

How long will it be before the world reserves are depleted if:

a. $k = 0.08$ (8%)? **b.** $k = 0.052$ (5.2%)?
61. ***Space Science*** A satellite has an initial radioisotope power supply of 50 watts (W). The power output in watts is given by the equation

$$P = 50e^{-t/250}$$

where t is the time in days. Solve for t.

C

62. ***Physics*** Newton's law of cooling states that an object at temperature B surrounded by air temperature A will cool to a temperature T after t min according to the equation

$$T = A + (B - A)e^{-kt}$$

where k is a constant depending on the item being cooled.
   a. If you draw a tub of 120°F water for a bath and let it stand in a 75°F room, what is the temperature of the water after 30 min if $k = 0.01$?
   b. What is k for an apple pie taken from a 375°F oven and cooled to 75°F after it is left in a 72°F room for 1 hr?
   c. Solve the given equation for t.

63. In calculus it is shown that

$$e^x = 1 + x + \frac{x^2}{2} + \frac{x^3}{2 \cdot 3} + \frac{x^4}{2 \cdot 3 \cdot 4} + \cdots$$

   a. What are the next two terms?
   b. What is the rth term?
   c. Calculate e correct to the nearest thousandth using this equation.
   d. Calculate $\sqrt{e} = e^{0.5}$ correct to the nearest thousandth using this equation.

64. Solve $10^{5x+1} = e^{2-3x}$.

65. Solve $5^{2+x} = 6^{3x+2}$.

66. Solve $e^x + e^{-x} = 10$. (*Hint:* Multiply both sides by e^x and use the quadratic formula.)

67. Solve $e^x - e^{-x} = 100$.

5.5 CHAPTER 5 SUMMARY

The material of this chapter is reviewed in the following list of objectives. After each objective there are some practice questions. For a sample test, select the first question of each set and check your answers with the answer section. For a sample test without answers, use the second question of each set. Additional practice is given by the other questions in each set. If you are having trouble with a particular type of problem, look back to that section for extra help.

5.1 EXPONENTIAL FUNCTIONS

Objective 1 Define an exponential function and a rational exponent. Apply the five laws of exponents to rational exponents. Simplify the given expressions (assume that the variables are positive).

1. $125^{2/3}$
2. $(2^{1/2} \cdot 3^{1/3})^6$
3. $\frac{27^{2/3}}{27^{1/2}}$
4. $(x^{1/2} - y^{1/2})(x^{1/2} + y^{1/2})$

Objective 2 Sketch exponential functions.

5. $y = (\frac{1}{2})^x$
6. $y = -2^x$
7. $y = 2^{-x}$
8. $y = e^{-x/2}$

5.2 INTRODUCTION TO LOGARITHMS

Objective 3 Write an exponential equation in logarithmic form.

9. $10^{0.5} = \sqrt{10}$
10. $e^0 = 1$
11. $9^3 = 729$
12. $(\sqrt{2})^3 = 2\sqrt{2}$

Objective 4 Write a logarithmic equation in exponential form.

13. $\log 1 = 0$
14. $\ln \frac{1}{e} = -1$
15. $\log_2 64 = 6$
16. $\log_\pi \pi = 1$

Objective 5 Evaluate common logarithms.

17. $\log 3$
18. $\log 451$
19. $\log 0.0021$
20. $\log 3^4$

Objective 6 Evaluate natural logarithms.

21. $\ln 3$
22. $\ln 451$
23. $\ln 0.013$
24. $\ln 3^4$

Objective 7 Evaluate logarithms using the definition of logarithm.

25. $\log_3 27$
26. $\log_2 0.125$
27. $\ln e^5$
28. $\log 0.01$

Objective 8 *State and prove the laws of exponents. Fill in the blanks.*

29. $\log_b AB =$ ______________.
30. $\log_b A - \log_b B =$ ______________.
31. $\log_b A^p =$ ______________.
32. If $A = b^x$, then $x =$ ______________.

5.3 LOGARITHMIC FUNCTIONS AND EQUATIONS

Objective 9 *Graph logarithmic functions.*

33. $y = \log x$
34. $y = \ln x$
35. $y = \log_3 x$
36. $y + 3 = \log(x - 2)$

Objective 10 *Solve logarithmic equations. Solve for x.*

37. $\log_5 25 = x$
38. $\log_x (x + 6) = 2$
39. $3 \log 3 - \frac{1}{2} \log 3 = \log \sqrt{x}$
40. $2 \ln \dfrac{e}{\sqrt{7}} = 2 - \ln x$

Objective 11 *Use logarithms or a calculator to carry out complicated calculations. Simplify the given expressions.*

41. $(3450)(241)$
42. $\dfrac{689}{14}$
43. 19^6
44. $\left(1 + \dfrac{0.12}{360}\right)^{720}$

5.4 EXPONENTIAL EQUATIONS

Objective 12 *Solve exponential equations with base* 10.

45. $10^{x+2} = 125$
46. $10^{2x-3} = 0.5$
47. $10^{-x^2} = 0.45$
48. $10^{4-3x} = 15$

Objective 13 *Solve exponential equations with base e.*

49. $e^{3x} = 50$
50. $e^{x-5} = 0.49$
51. $e^{1-2x} = 690$
52. $e^{x^2} = 9$

Objective 14 *Solve exponential equations with arbitrary bases.*

53. $5^x = 125$
54. $5^{2x+1} = 0.2$
55. $3^x = 7$
56. $7^{1-3x} = 0.048$

Objective 15 *Change logarithms from one base to another. Evaluate the given expressions.*

57. $\log_2 818$
58. $\log_5 4.51$
59. $\log_3 100$
60. $\log_4 \sqrt[3]{4}$

Objective 16 *Solve applied problems involving exponential equations.*

61. The half-life of arsenic-76 is 26.5 hr. Find the constant k for which $A = A_0 e^{-kt}$ where t is expressed in hours.

62. If \$1500 is placed in a $2\frac{1}{2}$-year time certificate paying 13.5% compounded monthly, what is the amount in the account when the certificate matures in $2\frac{1}{2}$ years?

63. If a person's present salary is \$20,000 per year, use the formula $A = P(1 + r)^n$ to determine the salary necessary to equal this salary in 15 years if you assume the 1980 U.S. inflation rate of 13.4%.

64. Solve the formula given in Problem 63 for n.

65. Repeat Problem 63 for the 1984 rate of 4%.

CUMULATIVE REVIEW I

Suggestions for study of Chapters 2–5:

Make a list of important ideas from Chapters 2–5. Study this list. Use the objectives at the end of each chapter to help you make up this list.

Work some practice problems. A good source of problems is the set of chapter objectives at the end of each chapter. You should try to work at least one problem from each objective:

Chapter 2, pp. 85–87; work 24 problems
*Chapter 3, pp. 116–117; work 7 problems
Chapter 4, pp. 161–163; work 17 problems
Chapter 5, pp. 197–198; work 16 problems

Check the answers for the practice problems you worked. The first and third problems of each objective have their answers listed in the back of the book.

Additional problems. Work additional odd-numbered problems (answers in the back of the book) from the problem sets as needed. Focus on the problems you missed in the chapter summaries.

Work the problems in the following cumulative review. These problems should be done after you have studied the material. They should take you about 1 hour and will serve as a sample test for Chapters 2–5. Assume that all variables are restricted so that each expression is defined. All the answers for these questions are provided in the back of the book for self-checking.

PRACTICE TEST FOR CHAPTERS 2–5

1. If $f(x) = 3x^2 + 2$ and $g(x) = 2x - 1$, find:
a. $f(-2)$ **b.** $g(t)$ **c.** $f(t + h)$
d. $f[g(x)]$ **e.** $(g \circ f)(x)$ **f.** $\dfrac{f(x + h) - f(x)}{h}$

2. Find the inverse of the given functions, if possible.
a. $f(x) = 3x^2 + 2$ **b.** $g(x) = 2x - 1$

3. Graph the given functions.
a. $y + 2 = -\frac{1}{2}(x + 3)$ **b.** $y + 2 = -\frac{1}{2}(x + 3)^2$

4. Let $F(x) = 4x^3 - 20x^2 + 3x + 27$. Find:
a. $F(-1)$ **b.** $F(0)$ **c.** $F(3)$ **d.** $F(\frac{3}{2})$
e. the zeros for F

5. Graph $F(x) = 4x^3 - 20x^2 + 3x + 27$. You might want to look at your work for Problem 4.

6. Solve $4x^4 - 20x^3 - 19x^2 + 26x - 6 = 0$, given that $3 + \sqrt{7}$ is a root.

***7.** Let $G(x) = \dfrac{2x^2 - 3x + 1}{x^2 - x - 2}$.
a. What is the domain?
b. Is the curve defined by $G(x)$ symmetric with respect to the x axis, y axis, or origin?
c. Are there any horizontal, vertical, or slant asymptotes?
d. What are the x intercepts and y intercepts?
e. Graph G.

8. Graph the following rational functions.
a. $y = \dfrac{-10}{x}$ **b.** $y = \dfrac{-3x}{x + 4}$ **c.** $y = \dfrac{x^2}{x(x - 3)}$

9. Solve the given equations.
a. $8^x = 32$ **b.** $\log_3 x = 4.12$
c. $\log_8 x = \log_8(4 - 3x)$ **d.** $2 \log 3 = \frac{1}{2} \log x$

10. If the original quantity of a radioactive substance is A_0, then the amount present, A, after t years is given by

$$A = A_0 2^{-kt}$$

a. Carbon-14 has a half-life of 5700 years. Find k.
b. If an ancient scroll is found and it is estimated that 72.4% of the original carbon-14 is still present, how old is the scroll? (Use the k you found in part **a.**)
c. Solve the equation for t.

* Optional

Extended Application: POPULATION GROWTH

Population Growth

World Population 4 Billion Tonight

BY EDWARD K. DeLONG

WASHINGTON (UPI)—By midnight tonight, the Earth's population will reach the 4 billion mark, twice the number of people living on the planet just 46 years ago, the Population Reference Bureau said Saturday.

The bureau expressed no joy at the new milestone.

It said global birth rates are too high, placing serious pressures on all aspects of future life and causing "major concern" in the world scientific community, and more than one-third of the present population has yet to reach child-bearing age.

The PRB found cause for optimism, however, in that some governments are stressing birth control to blunt the impact of "explosive growth" and the population growth rate dropped slightly in the past year.

"In 1976, each new dawn brings a formidable increase of approximately 195,000 newborn infants to share the resources of our finite world," it said.

One expert warned that a lack of jobs, rather than too little food, may be the "ultimate threat" facing society as the planet becomes more and more crowded.

It took between two and three million years for the human race to hit the one billion mark in 1850, the PRB said. By 1930, 80 years later, the population stood at 2 billion. A mere 31 years after that, in 1961, it was 3 billion. The growth from 3 to the present 4 billion took just 15 years.

The world could find it has 5 billion people by 1989—just 13 years from now—if population growth continues at the present rate of 1.8 per cent a year, said Dr. Leon F. Bouvier, vice president of the private, nonprofit PRB.

Bouvier said the newly calculated growth rate is a little lower than the 1.9 per cent estimated last year. Thanks to that slowdown, the passing of the 4 billion milestone came a year later than some demographers had predicted.

"I really think the rate of growth is going to start declining ever so slightly because of declining fertility," Bouvier said. "I think there is some evidence of progress—ever so slow, much too slow."

The new PRB figures show there were 3,982,815,000 people on Earth on Jan. 1. By March 1 the number had grown to 3,994,812,000, the organization said, and by April 1 the total will be 4,000,824,000.

The bureau said its calculations are based on estimates of 328,000 live births per day minus 133,000 deaths.

A growing number of governments are taking steps to slow growth rates, the PRB said.

Singapore appears likely to meet the goal of the two-child family "well before the target date of 1980," it said, and several states in India, which yearly adds the equivalent of the population of Australia, are considering financial incentives to birth control and mandatory sterilization after the birth of two children.

Dr. Paul Ehrlich of Stanford University, one of several population experts contacted by PRB, said he was sad to realize at the age of 44 he had lived through a doubling of Earth's population. He expressed fear the next 44 years could see population growth halted "by a horrifying increase in death rates."

"At this point, hunger does not seem the greatest issue presented by the ever growing number of people," said Dr. Louis M. Hellman, chief of population staff at the Health, Education and Welfare Department.

"Rather, the threat appears to lie in the increasing numbers who can find no work. As these masses of unemployed migrate toward the cities, they create a growing impetus toward political unrest and instability."

SOURCE: Courtesy of *The Press Democrat*, Santa Rosa, California, March 28, 1976. Reprinted by permission.

Human populations grow according to the exponential equation

$$P = P_0 e^{rt}$$

where P_0 is the size of the initial population, r is the growth rate, t is the length of time, and P is the size of the population after time t.

EXAMPLE 1 On March 28, 1976, the world population reached 4 billion. If the annual growth rate is 2%, what was the expected population on March 28, 1990?

Solution The initial population P_0 is 4 (billion), $r = 0.02$, and $t = 14$. Then

$$\begin{aligned} P &= 4e^{0.02(14)} \\ &= 4e^{0.28} \\ &\approx 4(1.323) \quad \text{From Table C.I or by calculator} \\ &\approx 5.29 \end{aligned}$$

The expected population was **5.3 billion.** ■

EXAMPLE 2 When is the world population expected to reach 6 billion, given a growth rate of 2%?

Solution $r = 0.02$, $P_0 = 4$, $P = 6$

$$\begin{aligned} 6 &= 4e^{0.02t} \\ e^{0.02t} &= 1.5 \\ 0.02t &= \ln 1.5 \\ t &= 50 \ln 1.5 \quad \text{Divide both sides by 0.02 } (\tfrac{1}{0.02} = 50). \\ &\approx 20.2733 \quad \text{By Table C.I or a calculator} \end{aligned}$$

This is 20 years, 100 days after March 28, 1976, or **July 6, 1996.** ■

If you can find the present world population (consult a recent almanac or call your library), you can compare the *predicted* population from Example 1 (using 1976 figures) with the actual figures. Suppose they differ; what conclusion can you make?

If you know the actual population of a city, state, country, or even the world for two dates, you can use the population formula to find the **growth rate** over that period. If factors influencing population growth do not change significantly, this growth rate can be used to predict future population growth. The question asked at the end of the preceding paragraph would be answered by saying that the *actual* growth rate was not the 2% assumed in Example 1.

EXAMPLE 3 The population actually reached 5 billion on July 7, 1986. Find the growth rate since it was reported in the news article.

Solution $P_0 = 4$, $P = 5$, and $t \approx 10.2767$

To find t, it is 10 years from March 28, 1976 to March 28, 1986; from March 28 to July 7 is 101 days; $101/365 \approx 0.2767123$. Now solve $P = P_0 e^{rt}$ for r:

$$\begin{aligned} \frac{P}{P_0} &= e^{rt} \\ rt &= \ln \frac{P}{P_0} \\ r &= \frac{1}{t} \ln \frac{P}{P_0} \\ &= \frac{1}{10.2767123} \ln \tfrac{5}{4} \quad \text{PRESS: } \boxed{5}\boxed{\div}\boxed{4}\boxed{=}\boxed{\ln}\boxed{\div}\boxed{10.2767123}\boxed{=} \\ &\approx 0.0217135 \end{aligned}$$

The growth rate is about **2.2%**. ■

EXAMPLE 4 The 1970 population of San Antonio, Texas, was 654,153 and the 1980 population was 783,296. What was the growth rate of San Antonio for this period?

Solution Since $P_0 = 654{,}153$, $P = 783{,}296$, and $t = 10$, we have

$$r = \frac{1}{t}\ln\frac{P}{P_0} \qquad \text{From Example 3}$$
$$= \frac{1}{10}\ln\frac{783{,}296}{654{,}153}$$

[783296] [÷] [654153] [=] [ln] [÷] [10] [=] DISPLAY: .0180169389

RPN logic:

[783296] [ENTER] [654153] [÷] [ln] [10] [÷]

The growth rate is about **1.8%**. ■

EXAMPLE 5 Using the growth rate found in Example 4, predict the population of San Antonio in 1990.

Solution Find P when $P_0 = 783{,}296$ and $r = 0.0180169389$

$$P = P_0 e^{rt}$$
$$= 783{,}296e^{0.0180169389}$$
$$\approx 937{,}934$$

The predicted population of San Antonio in 1990 is **937,934**. ■

Extended Application Problems—Population Growth

1. The world population was 1 billion in 1850 and it took 80 years to reach the 2 billion mark. What was the annual growth rate for this period from 1850 to 1930?
2. The world population was 2 billion in 1930 and it took 31 years to reach 3 billion. What was the annual growth rate for this period from 1930 to 1961?
3. The world population was 3 billion in 1961 and it took 15 years to reach 4 billion. What was the annual growth rate for this period from 1961 to 1976?
4. The world population was 4 billion in 1976 and it took 10 years to reach 5 billion. What is the growth rate for this period from 1976 to 1986?
5. On July 7, 1986, the world population reached 5 billion. If the annual growth rate is 2.2%, when would you expect the population to reach 6 billion?
6. On July 7, 1986, the world population reached 5 billion. If the annual growth rate is 2.2%, when would you expect the population to reach 7 billion?
7. On July 7, 1986, the world population reached 5 billion. If the annual growth rate is 2.2%, when would you expect the population to reach 8 billion?
8. **a.** If San Jose, California, grew from 459,913 in 1970 to 625,763 in 1980, what was the growth rate for this period?

   b. If you use the growth rate found in part **a**, what was the expected population in 1990?

9. **a.** If the population of Boston declined from 641,071 in 1970 to 562,118 in 1980, what was the growth rate for this period? (*Note:* A negative growth rate means the population declined.)
   b. If you use the growth rate found in part **a**, what was the expected population in 1990?
10. **a.** If the population of Denver declined from 514,678 in 1970 to 488,765 in 1980, what was the growth rate for this period? (*Note:* A negative growth rate means the population declined.)
    b. If you use the growth rate found in part **a**, what was the expected population in 1990?
11. ***Research Problem*** Use the information from the news article and Problems 1–7 to graph the world population from 1800 to 2020.
12. ***Research Problem*** From your local Chamber of Commerce, obtain the population figures for your city for 1960, 1970, and 1980.
    a. Find the rate of growth for each period.
    b. Forecast the population of your city for the year 2000. Which of the rates you obtained in part **a** is the most accurate for this purpose?
13. ***Research Problem*** Use the table of petroleum consumption given in the margin for this problem.
    a. Establish that the world consumption of petroleum for 1915 to 1975 grew exponentially by drawing a graph.
    b. Using the graph from part **a**, estimate the worldwide use of petroleum for 1980, 1985, 1990, and 1995.
    c. The actual consumption of petroleum for 1980 was 5.896 billion barrels and for 1985, 5.331 billion barrels. Plot these, points on the graph you used to estimate production in part **b**. Reevaluate your answer for part **b** using this new information.
    d. How do the figures of this problem compare with the estimated world petroleum reserves? Graph use and remaining reserves on one grid. Try to reach some conclusions.
14. List some factors, such as new zoning laws, that could change the growth rate of your city.
15. List some factors, such as a change in the tax laws of your state, that could change the growth rate of your state.
16. List some factors, such as war, that could change the growth rate of a country or the world.

World consumption of petroleum products

Year	Billions of barrels
1915	0.20
1920	0.48
1925	0.77
1930	1.04
1935	1.19
1940	1.54
1945	1.82
1950	2.38
1955	3.10
1960	3.61
1965	4.20
1970	5.36
1975	5.64

Hipparchus (c. 180–125 B.C.)

Trigonometry contains the science of continually undulating magnitude: meaning magnitude which becomes alternately greater and less, without any termination to succession of increase or decrease .... All trigonometric functions are not undulating; but it may be stated that in common algebra nothing but infinite series undulate; in trigonometry nothing but infinite series do not undulate.

Augustus De Morgan
Trigonometry and Double Algebra

$\sin^2\phi$ is odious to me, even though Laplace made use of it; should it be feared that $\sin\phi^2$ might become ambiguous, which would perhaps never occur, or at most very rarely when speaking of $\sin(\phi^2)$, well then, let us write $(\sin\phi)^2$, but not $\sin^2\phi$, which by analogy should signify $\sin(\sin\phi)$.

Karl Gauss
Gauss–Schumacher Briefweches, vol. 3, p. 292

We might add that the notation we use today does not conform to Gauss' suggestion, because $\sin^2\phi$ is used to mean $(\sin\phi)^2$.

HISTORICAL NOTE

The origins of trigonometry are obscure, but we do know that it began more than 2000 years ago with the Mesopotamian, Babylonian, and Egyptian civilizations. Much of the knowledge from these civilizations was passed on to the Greeks, who formally developed many of the ideas of this chapter. The word *trigonometry* comes from the words *trigonon* (triangle) and *metron* (measurement), and the ancients used trigonometry in a very practical way to measure triangles, as we will in Chapter 8.

During the second half of the second century B.C., the astronomer Hipparchus of Nicaea compiled the first trigonometric tables in 12 books. It was he who used the 360° of the Babylonians and thus introduced trigonometry with the degree angle measure that we still use today. Hipparchus' work formed the foundation for Ptolemy's *Mathematical Syntaxis*, the most significant early work in trigonometry.

The Greek discoveries were later lost in Europe. Fortunately, they had been translated into Arabic, and the Arabs took them as far as India, which accounts for the interesting origin of the word *sine*. The Hindu mathematician Aryabhata (c. A.D. 476–550) called it *jyā-adhā* (chord half) and abbreviated it *jyā*, which the Arabs wrote as *jiba* but shortened to *jb*. Later writers erroneously interpreted *jb* to stand for *jaib*, which means cove or bay. When Europeans rediscovered trigonometry through the Arab tradition, they translated the texts into Latin. The Latin equivalent for *jaib* is *sinus*, from which our present word *sine* is derived. Several of our mathematical terms, including the word *algebra*, are based on misunderstandings of Arabic words—but in this case the Arabs themselves were confused. The words *tangent* and *secant* come from the relationship of the ratios to the lengths of the tangent and secant lines drawn on a circle. Finally, Edmund Gunter (1581–1626) first used the prefix *co* to invent the words *cosine*, *cotangent*, and *cosecant*. In 1620, Gunter published a seven-place table of the common logarithms of the sines and tangents of angles for intervals of one minute of an arc. Gunter originally entered the ministry, but later decided on astronomy as a career. He was such a poor preacher that historian Howard Eves stated that Gunter left the ministry in 1619 to the "benefit of both occupations."

6

Trigonometric Functions

CONTENTS

PREVIEW

This chapter introduces you to six very important functions in mathematics, called the trigonometric functions. After considering the definition we consider their evaluation as well as their graphs. Finally, we look at the inverse trigonometric functions. You will need to be very familiar with these functions in order to continue with your studies in mathematics. The 25 objectives of this chapter are listed on pages 243–245.

PERSPECTIVE

Trigonometric functions are tools in calculus just as are linear or quadratic functions. For example, trigonometric substitutions form a common method of integration, which is an operation considered in integral calculus. In the page below, taken from a calculus book, notice that the example uses the definitions of both the secant and the tangent functions as we introduce them in Section 6.2. You may also notice, toward the bottom of this extract, that the range of the inverse function is used, so you would need the material we introduce in Section 6.5. (Notice also the use of ln, which we introduced in Chapter 5.)

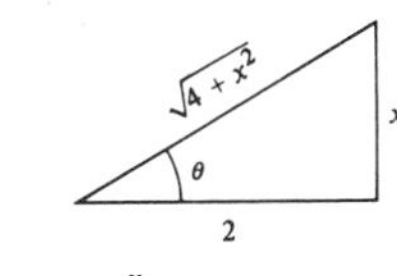

Figure 9.3 $\frac{x}{2} = \tan\theta$

Since $\tan\theta = x/2$ we see from the triangle in Figure 9.3 that

$$\sec\theta = \frac{\sqrt{4+x^2}}{2}$$

and hence

$$\int \frac{1}{\sqrt{4+x^2}}\,dx = \ln\left|\frac{\sqrt{4+x^2}}{2} + \frac{x}{2}\right| + C.$$

The expression on the right may be written

$$\ln\left|\frac{\sqrt{4+x^2}+x}{2}\right| + C = \ln\left|\sqrt{4+x^2}+x\right| - \ln 2 + C.$$

Since $\sqrt{4+x^2} + x > 0$ for all x, the absolute value sign is unnecessary. If we also let $D = -\ln 2 + C$, then

$$\int \frac{1}{\sqrt{4+x^2}}\,dx = \ln\left(\sqrt{4+x^2}+x\right) + D. \quad ■$$

For integrands containing $\sqrt{x^2 - a^2}$ we substitute $x = a\sec\theta$, where θ is chosen in the range of the inverse secant function; that is, either $0 \le \theta < \pi/2$ or $\pi \le \theta < 3\pi/2$.

From Earl Swokowski, *Calculus with Analytic Geometry*, 3rd ed. (Boston: Prindle, Weber, & Schmidt), p. 416.

6.1 ANGLES AND THE UNIT CIRCLE

The circle is of primary importance in the study of trigonometry. Although you are no doubt familiar with a circle, it is worthwhile to review circles briefly at this time. A **circle** is the set of points a given distance, called the **radius**, from a given point called the **center**. Since a circle is not the graph of a function, we will delay discussion of the general equation and properties of a circle until Chapter 11. We do, however, have need for the equation of a circle with radius r and center at the origin. The equation of this circle is derived by using the distance formula (from Section 1.6). Let (x, y) be any point on a circle of radius r. Then

$$(x_2 - x_1)^2 + (y_2 - y_1)^2 = d^2$$ This is the distance formula with both sides squared.

$$(x - 0)^2 + (y - 0)^2 = r^2$$ The center is $(0, 0)$, so let $(x_1, y_1) = (0, 0)$; let $(x_2, y_2) = (x, y)$ and $d = r$.

$$x^2 + y^2 = r^2$$

If $r = 1$, this is called the *unit circle.*

Unit Circle

The **unit circle** is the circle with radius 1 and center at the origin. The equation of the unit circle is

$$x^2 + y^2 = 1$$

In mathematics it is often useful to consider functions of angles; however, the definition of an angle depends on the context in which it is being used. In geometry, an angle is usually defined as the union of two rays with a common endpoint. In advanced mathematics courses, a more general definition is used.

Angle

An **angle** is formed by rotating a ray about its endpoint (called the **vertex**) from some initial position (called the **initial side**) to some terminal position (called the **terminal side**). The measure of an angle is the amount of rotation. An angle is also formed if a line segment is rotated about one of its endpoints.

If the rotation of the ray is in a counterclockwise direction, the measure of the angle is called **positive**. If the rotation is in a clockwise direction, the measure is called **negative**. The notation $\angle ABC$ means the measure of an angle with vertex B and points A and C (different from B) on the sides; $\angle B$ denotes the measure of an angle with vertex at B, and a curved arrow is used to denote the direction and amount of rotation, as shown in Figure 6.1. If no arrow is shown, the measure of the angle is considered to be the smallest positive rotation. Lowercase Greek letters are also used to denote the angles as well as the measure of angles. For example, θ may

Commonly used Greek letters

Symbol	Name
α	alpha
β	beta
γ	gamma
δ	delta
θ	theta
λ	lambda
ϕ or φ	phi
ω	omega

π (pi) is a lowercase Greek letter that will not be used to represent an angle. It denotes an irrational number approximately equal to 3.141592654.

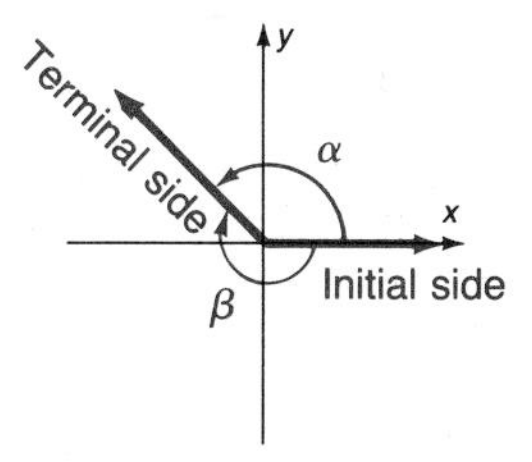

Figure 6.2 Standard-position angles α and β; α is a positive angle and β is a negative angle; α and β are coterminal angles.

represent the angle or the measure of the angle called θ; you will know which is meant by the context in which it is used. Some examples are shown in Figure 6.1.

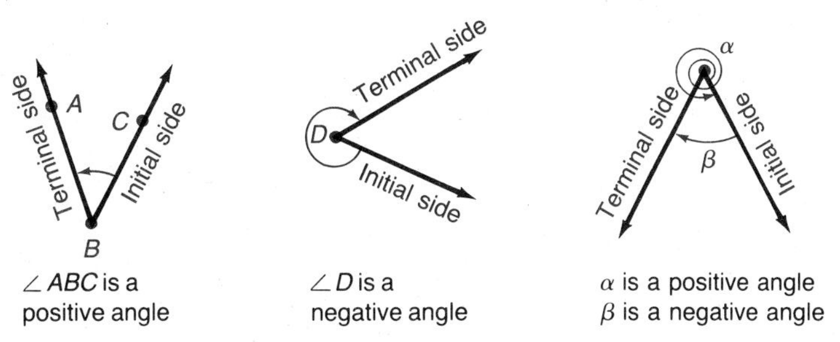

Figure 6.1 Examples of angles

A Cartesian coordinate system may be superimposed on an angle so that the vertex is at the origin and the initial side is along the positive x axis. In this case the angle is in **standard position**. Angles in standard position having the same terminal sides are **coterminal angles**. Given any angle α, there is an unlimited number of coterminal angles (some positive and some negative). In Figure 6.2, β is coterminal with α. Can you find other angles coterminal with α?

Several units of measurement are used for measuring angles. Let α be an angle in standard position with a point P not the vertex but on the terminal side. As this side is rotated through one revolution, the trace of the point P forms a circle. The measure of the angle is one revolution, but since much of our work will be with amounts less than one revolution, we need to define measures of smaller angles. Historically the most common scheme divides one revolution into 360 equal parts with each part called a **degree**. Sometimes even finer divisions are necessary, so a degree is divided into 60 equal parts, each called a **minute** ($1° = 60'$). Furthermore, a minute is divided into 60 equal parts, each called a **second** ($1' = 60''$). For most applications, we will write decimal parts of degrees instead of minutes and seconds. That is, $32.5°$ is preferred over $32° \, 30'$.

In calculus and scientific work, another measure for angles is defined. This method uses real numbers to measure angles. Draw a circle with any nonzero radius r. Next measure out an arc with length r. Figure 6.3a shows the case in which $r = 1$ and Figure 6.3b shows $r = 2$. Regardless of your choice for r, the angle determined by this arc of length r is the same. (It is labeled θ in Figure 6.3.)

This angle is used as a basic unit of measurement and is called a **radian**. Notice that the circumference C generates an angle of one revolution. Since $C = 2\pi r$, and

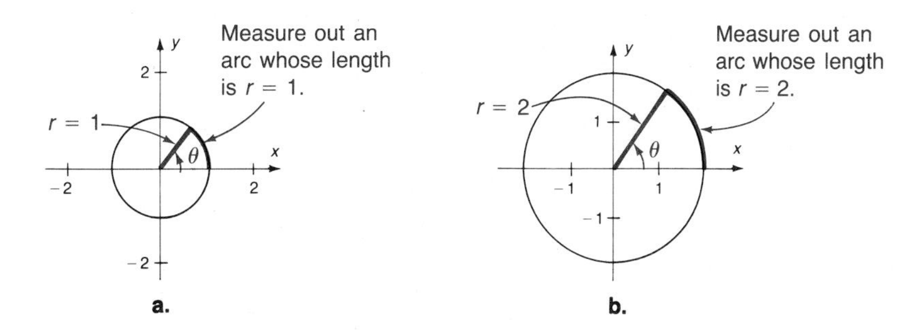

Figure 6.3 Radian measure

since the basic unit of measurement on the circle is r,

$$\text{One revolution} = \frac{C}{r}$$

$$= \frac{2\pi r}{r}$$

$$= 2\pi$$

Thus $\frac{1}{2}$ revolution is $\frac{1}{2}(2\pi) = \pi$ radians; $\frac{1}{4}$ revolution is $\frac{1}{4}(2\pi) = \frac{\pi}{2}$ radians.

WARNING: Note agreement in this book.

Notice that when measuring angles in radians, you are using *real numbers.* Because radian measure is used so frequently, we agree that **radian measure is understood** when **no units of measure** for an angle are indicated. Figure 6.4 shows a protractor for measuring angles using radian measure.

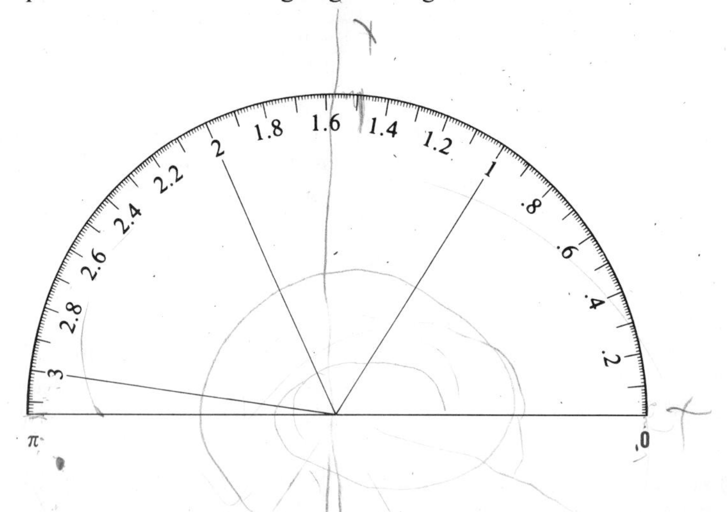

Figure 6.4 Protractor for radian measure

Example 1 asks you to draw several angles using radian measure. Do not try to change these angles to degree measure. Work them directly as shown in the examples. You will have to work at thinking in terms of radian measure. You should memorize the approximate size of an angle of measure 1 radian in much the same way you have memorized the approximate size of an angle of measure 45°.

EXAMPLE 1 Let $r = 1$ and draw the angles with the given measures.

a. $\theta = 2$
(2 radians is understood)

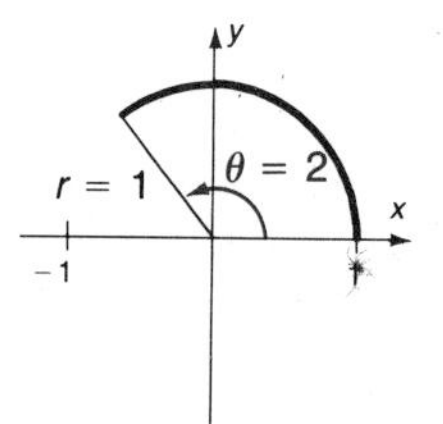

b. $\theta = 3$
(3 radians is understood)

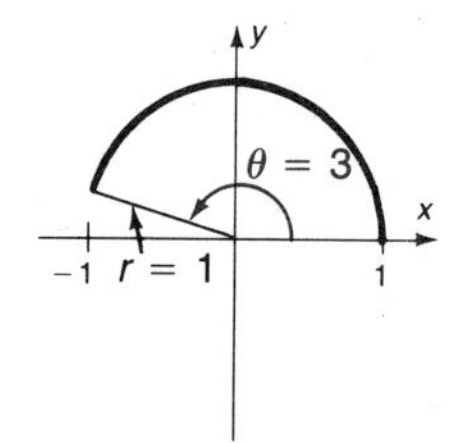

c. $\theta = \sqrt{19}$
$\sqrt{19} \approx 4.36$; since a straight angle is $\pi \approx 3.14$, use the radian protractor for $\sqrt{19} - \pi \approx 1.22$.

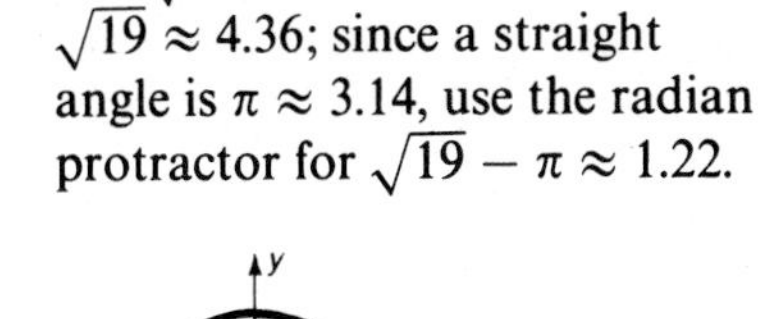

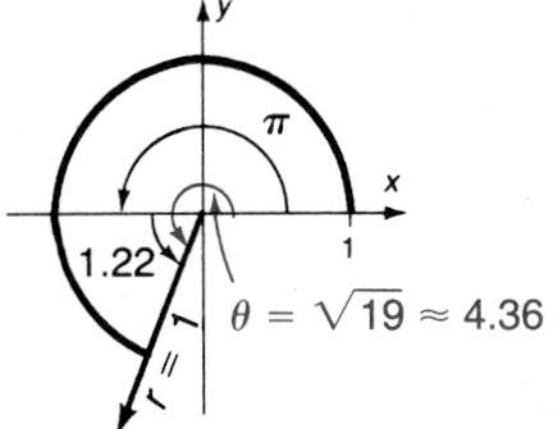

d. $\theta = \sqrt{50}$
$\sqrt{50} \approx 7.07$; since one revolution is $2\pi \approx 6.28$, use the radian protractor for $\sqrt{50} - 2\pi \approx 0.79$.

e. $\theta = 0.5$

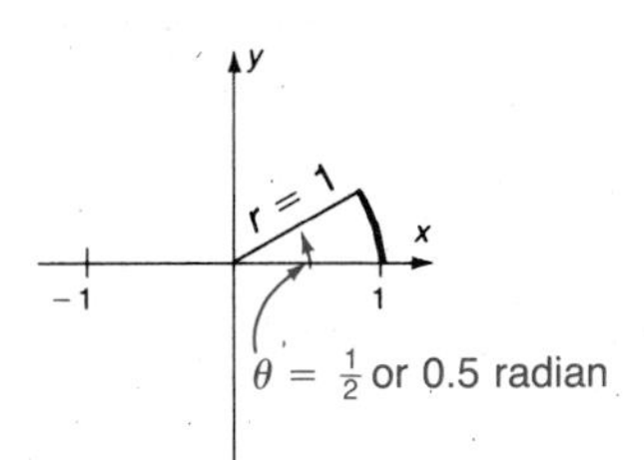

f. $\theta = -\dfrac{\pi}{4}$

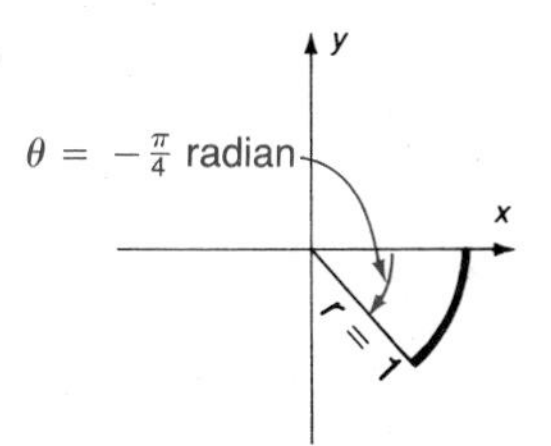

Next consider the relationship between degree and radian measure:

One revolution is measured by 360° or by 2π radians

Then

$$\text{Number of revolutions} = \frac{\text{angle in degrees}}{360}$$

$$\text{Number of revolutions} = \frac{\text{angle in radians}}{2\pi}$$

Therefore:

Relationship Between Degree and Radian Measure

$$\frac{\text{angle in degrees}}{360} = \frac{\text{angle in radians}}{2\pi}$$

EXAMPLE 2 **a.** Change 45° to radians.

$$\frac{45}{360} = \frac{\theta}{2\pi}$$

$$\frac{90\pi}{360} = \theta$$

$$\frac{\pi}{4} = \theta$$

An alternative method is to remember that π radians is 180°. That is, since 45° is $\frac{1}{4}$ of 180°, you know that the radian measure is $\frac{\pi}{4}$. As a decimal, θ can be approximated by calculator:

$$\boldsymbol{\theta \approx 0.78539816}$$

b. Change 2.30 to degrees.

$$\theta = \frac{180}{\pi}(2.30)$$

$$\approx (57.296)(2.30)$$
$$= 131.7808$$

To the nearest hundredth, the angle is 131.78°. If you have a calculator, you can obtain a much more accurate answer by using a better approximation for $\frac{180}{\pi}$. For example, if 2.30 is exact, then by calculator:

Algebraic logic: $\boxed{180}\ \boxed{\div}\ \boxed{\pi}\ \boxed{\times}\ \boxed{2.3}\ \boxed{=}$

RPN logic: $\boxed{180}\ \boxed{\text{ENTER}}\ \boxed{\pi}\ \boxed{\div}\ \boxed{2.3}\ \boxed{\times}$

The result is approximately **131.7802929°.**

c. Change 1° to radians.

$$\frac{1}{360} = \frac{\theta}{2\pi}$$

$$\frac{\pi}{180} = \theta$$

Even though this solution is in the desired form, you might be interested in performing the division on a calculator:

1° ≈ 0.0174532925 radian

d. Change 1 to degrees.

$$\frac{\theta}{360} = \frac{1}{2\pi}$$

$$\theta = \frac{180}{\pi}$$

By calculator:

$$\boldsymbol{\theta \approx 57.29577951° \text{ or } 57°17'45''}$$

Decimal degrees is the preferred form. ■

For the more common measures of angles, it is a good idea to memorize the equivalent degree and radian measures. If you keep in mind that 180° in radian measure is π, the rest of the values will be easy to remember.

Commonly Used Degree and Radian Measures

Degrees	Radians
0°	0
30°	$\frac{\pi}{6}$
45°	$\frac{\pi}{4}$
60°	$\frac{\pi}{3}$
90°	$\frac{\pi}{2}$
180°	π
270°	$\frac{3\pi}{2}$
360°	2π

90° or $\frac{\pi}{2}$
60° or $\frac{\pi}{3}$
45° or $\frac{\pi}{4}$
30° or $\frac{\pi}{6}$
180° or π
0° or 0
270° or $\frac{3\pi}{2}$

We now relate the radian measure of an angle to a circle to find the arc length. An **arc** is part of a circle; thus **arc length** is the distance around part of a circle. The arc length corresponding to one revolution is the **circumference** of the circle. Let s be the length of an arc and let θ be the angle measured in radians. Then

$$\text{Angle in revolutions} = \frac{s}{2\pi r}$$

since one revolution has an arc length (circumference of the circle) of $2\pi r$. Also,

$$\text{Angle in radians} = (\text{angle in revolutions})(2\pi)$$

Substituting,

$$\theta = \frac{s}{2\pi r}(2\pi)$$

$$= \frac{s}{r}$$

From this result, we derive the following formula.

Arc Length Formula

The **arc length** cut by a central angle θ (measured in radians) from a circle of radius r is denoted by s and is found by

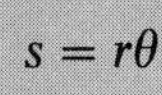

$$s = r\theta$$

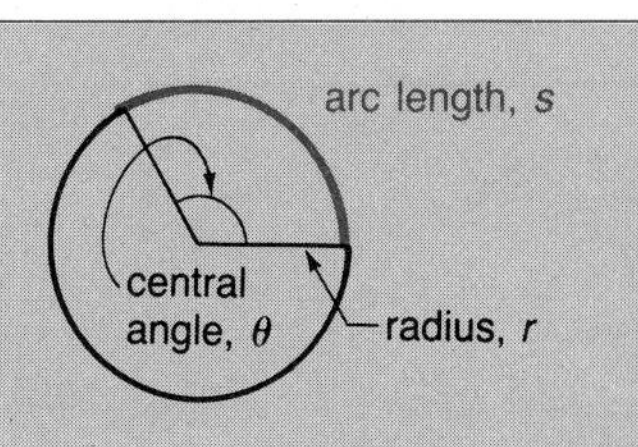

EXAMPLE 3 The length of the arc subtended (cut off) by a central angle of 36° in a circle with a radius of 20 centimeters (cm) is found as follows. First change 36° to radians so that you can use the formula given above.

$$\frac{36}{360} = \frac{\theta}{2\pi}$$

36°
s is the arc length

Solving for θ,

$$\frac{\pi}{5} = \theta$$

Thus

$$\begin{aligned} s &= 20\left(\frac{\pi}{5}\right) \\ &= 4\pi \end{aligned}$$

The length of the arc is **4π cm**. This is about 12.6 cm. ■

If the terminal side of an angle coincides with a coordinate axis, the angle is called a **quadrantal angle**. If the angle θ is not a quadrantal angle then we refer to its *reference angle*, which is denoted by θ' throughout this book.

Reference Angle

> Given an angle θ in standard position the **reference angle** θ' is defined as the acute angle the terminal side makes with the x axis.

The procedure for finding the reference angle depends on the quadrant of θ. One example for each quadrant is shown in Figure 6.5.

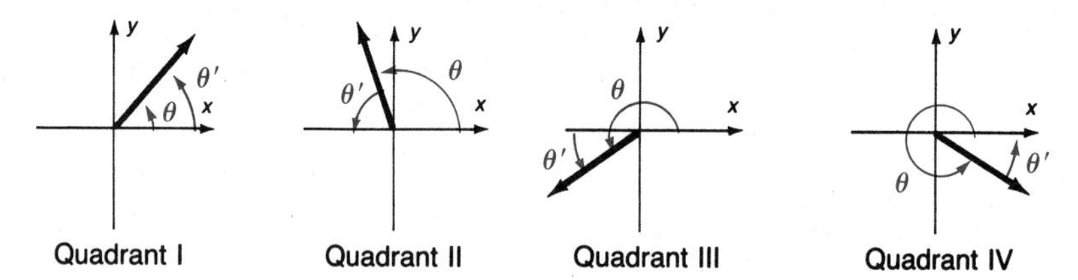

Figure 6.5 Reference angles

EXAMPLE 4 Find the reference angle and draw both the given angle and the reference angle.

a. 210°

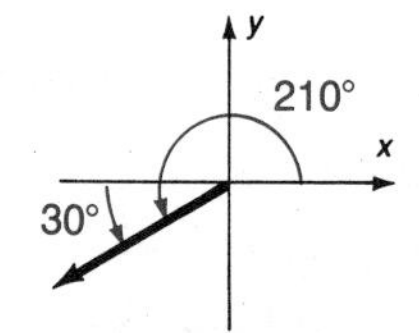

$210° - 180° = 30°$
Reference angle is **30°**.

b. 150°

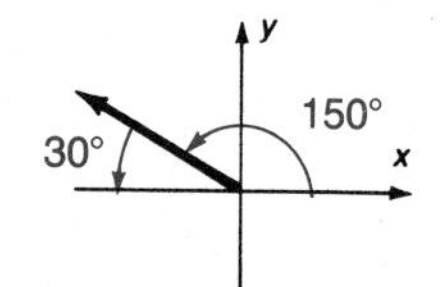

$180° - 150° = 30°$
Reference angle is **30°**.

c. $-\dfrac{5\pi}{3}$

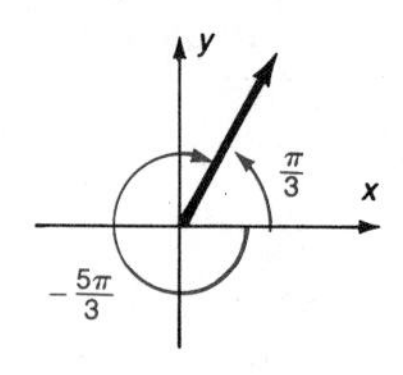

$$2\pi - \frac{5\pi}{3} = \frac{\pi}{3}$$

Reference angle is $\frac{\pi}{3}$.

d. 2.5

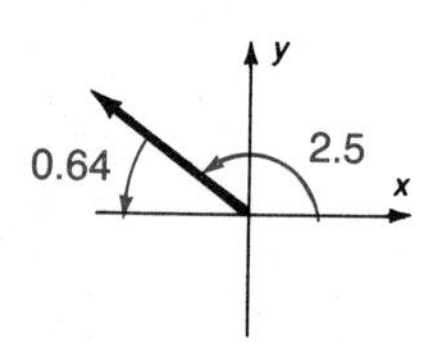

$\pi - 2.5 \approx 0.64$
Reference angle is approximately **0.64**.

e. 812°. If the angle is more than one revolution, first find a nonnegative coterminal angle that is less than one revolution. 812° is coterminal with 92°.

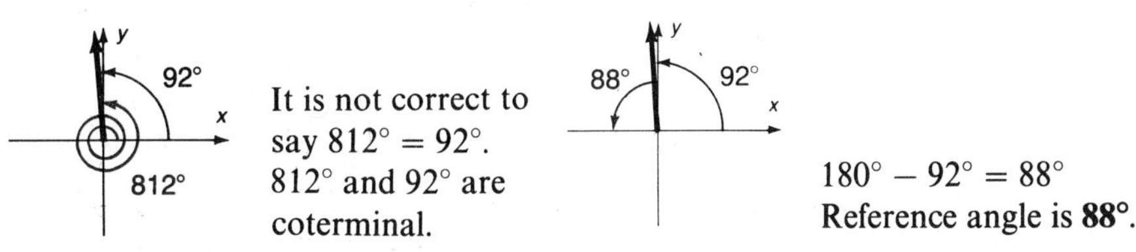

It is not correct to say 812° = 92°. 812° and 92° are coterminal.

$180° - 92° = 88°$
Reference angle is **88°**.

f. The angle whose measure is 30; 30 is coterminal with 4.867 ($30 - 8\pi \approx 4.867$).

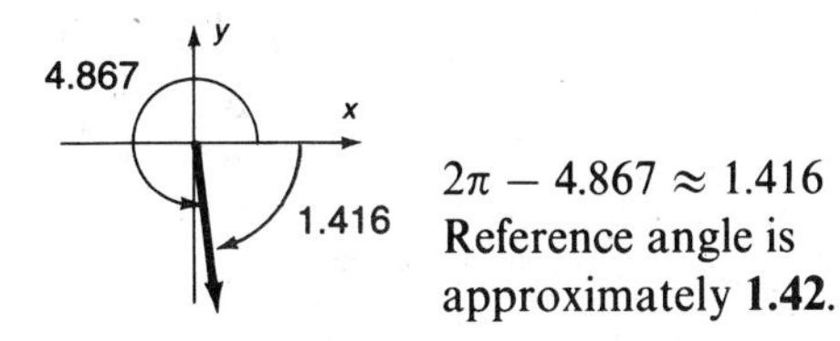

$2\pi - 4.867 \approx 1.416$
Reference angle is approximately **1.42**.

The purpose of this example is to remind you that $30 \neq 30°$. ■

Keep in mind the differences between coterminal angles and reference angles. Note also that reference angles are always between 0 and 90° or 0 and $\frac{\pi}{2}$ (about 1.57).

PROBLEM SET 6.1

A

From memory, give the radian measure for each of the angles whose degree measure is stated, and the degree measure for each of the angles whose radian measure is stated, in Problems 1–4.

1. a. 30° **b.** 90° **c.** 270° **d.** 45°

2. a. 360° **b.** 60° **c.** 180° **d.** 0°

3. a. π **b.** $\frac{\pi}{4}$ **c.** $\frac{\pi}{3}$ **d.** 2π

4. a. $\frac{\pi}{2}$ **b.** 0 **c.** $\frac{3\pi}{2}$ **d.** $\frac{\pi}{6}$

Use the radian protractor in Figure 6.4 to help you sketch each of the angles in Problems 5–10.

5. a. $\frac{\pi}{2}$ **b.** $\frac{\pi}{6}$ **c.** -2.5 **d.** $\sqrt{17}$

6. a. $\frac{2\pi}{3}$ **b.** $\frac{3\pi}{4}$ **c.** -1 **d.** $\sqrt{95}$

7. a. $-\frac{\pi}{4}$ **b.** $\frac{7\pi}{6}$ **c.** -2.76 **d.** $\sqrt{115}$

8. a. $-\frac{3\pi}{2}$ **b.** $\frac{13\pi}{3}$ **c.** -1.2365 **d.** $\sqrt{23}$

9. a. $\frac{\pi}{15}$ **b.** $-\frac{9\pi}{7}$ **c.** -4 **d.** $\sqrt{10}$

10. a. $-\frac{5\pi}{6}$ **b.** $\frac{5\pi}{4}$ **c.** -3 **d.** $\sqrt{19}$

Find the exact value of a positive angle less than one revolution that is coterminal with each of the angles in Problems 11–16.

11. a. 400° **b.** 540° **c.** 750° **d.** 1050°

12. a. −30° **b.** −200° **c.** −55° **d.** −320°

13. a. −120° **b.** 500° **c.** −180° **d.** 1000°

14. a. 3π **b.** $\frac{13\pi}{6}$ **c.** $-\pi$ **d.** 7

15. a. $-\frac{\pi}{4}$ **b.** $\frac{17\pi}{4}$ **c.** $\frac{11\pi}{3}$ **d.** -2

16. a. $-\frac{\pi}{6}$ **b.** $-\frac{5\pi}{4}$ **c.** $\frac{15\pi}{6}$ **d.** 8

Find a positive angle less than one revolution correct to four decimal places so that it is coterminal with each of the angles in Problems 17–19.

17. a. 9 **b.** -5 **c.** $\sqrt{50}$ **d.** -6

18. a. 6.2832 **b.** -3.1416 **c.** 30 **d.** $3\sqrt{5}$

19. a. 6.8068 **b.** -0.7854 **c.** 150 **d.** 9.4247

Find the reference angle for the angles given in Problems 20–27. Use the unit of measurement (degrees or radians) given in the problem.

20. a. 150° **b.** 210° **c.** 240° **d.** 330°

21. a. 60° **b.** 120° **c.** 300° **d.** 135°

22. a. $\frac{5\pi}{3}$ **b.** $\frac{7\pi}{6}$ **c.** $\frac{4\pi}{3}$ **d.** $\frac{5\pi}{4}$

23. a. $\frac{11\pi}{12}$ **b.** $\frac{2\pi}{3}$ **c.** $\frac{11\pi}{6}$ **d.** $\frac{\pi}{4}$

24. a. −30° **b.** −200° **c.** −55° **d.** −320°

25. a. $-\frac{\pi}{4}$ **b.** $-\pi$ **c.** $-\frac{13\pi}{6}$ **d.** $-\frac{5\pi}{3}$

26. a. 7 **b.** 9 **c.** -5 **d.** -6

27. a. $\sqrt{50}$ **b.** $3\sqrt{5}$ **c.** 6.8068 **d.** -0.7854

B

Change the angles in Problems 28–39 to decimal degrees correct to the nearest hundredth.

28. $\frac{2\pi}{9}$ **29.** $\frac{\pi}{10}$ **30.** $\frac{\pi}{30}$ **31.** $\frac{5\pi}{3}$

32. $-\frac{11\pi}{12}$ **33.** $\frac{3\pi}{18}$ **34.** 2 **35.** -3

36. -0.25 **37.** -2.5 **38.** 0.4 **39.** 0.51

Change the angles in Problems 40–45 to radians using exact values.

40. 40° **41.** 20° **42.** $-64°$ **43.** $-220°$

44. 254° **45.** 85°

Change the angles in Problems 46–51 to radians correct to the nearest hundredth.

46. 112° **47.** 314° **48.** $-62.8°$ **49.** 350°

50. $-480°$ **51.** 985°

In Problems 52–59, find the intercepted arc to the nearest hundredth if the central angle and radius are as given.

52. Angle 1, radius 1 m **53.** Angle 2.34, radius 6 cm

54. Angle 3.14, radius 10 m **55.** Angle $\frac{\pi}{3}$, radius 4 m

56. Angle $\frac{3\pi}{2}$, radius 15 cm **57.** Angle 40°, radius 7 ft

58. Angle 72°, radius 10 ft **59.** Angle 112°, radius 7.2 cm

60. How far does the tip of an hour hand on a clock move in 3 hr if the hour hand is 2.00 cm long?

61. A 50-cm pendulum on a clock swings through an angle of 100°. How far does the tip travel in one arc?

C

62. ***Surveying*** In about 230 B.C., a mathematician named Eratosthenes estimated the radius of the earth using the following information: Syene and Alexandria in Egypt are on the same line of longitude. They are also 800 km apart. At noon on the longest day of the year, when the sun was directly overhead in Syene, Eratosthenes measured the sun to be 7.2° from the vertical in Alexandria. Because of the distance of the earth from the sun, he assumed that the rays were parallel. Thus he concluded that the arc from Syene to Alexandria is subtended by a central angle of 7.2° measured at the center of the earth. Using this information, find the approximate radius of the earth.

63. ***Geography*** Omaha, Nebraska, is located at approximately 97° west longitude, 41° north latitude; Wichita, Kansas, is located at approximately 97° west longitude, 37° north latitude. Notice that these two cities have about the same longitude. If we know that the radius of the earth is about 6370 kilometers (km), what is the distance between these cities to the nearest 10 km?

64. ***Geography*** Entebbe, Uganda, is located at approximately 33° east longitude, and Stanley Falls in Zaire is located at 25° east longitude. Both these cities lie approximately on the equator. If we know that the radius of the earth is about 6370 km, what is the distance between the cities to the nearest 10 km?

65. ***Astronomy*** Suppose it is known that the moon subtends an angle of 45.75′ at the center of the earth (see Figure 6.6). It is also known that the center of the moon is 384,417 km from the surface of the earth. What is the diameter of the moon to the nearest 10 km?

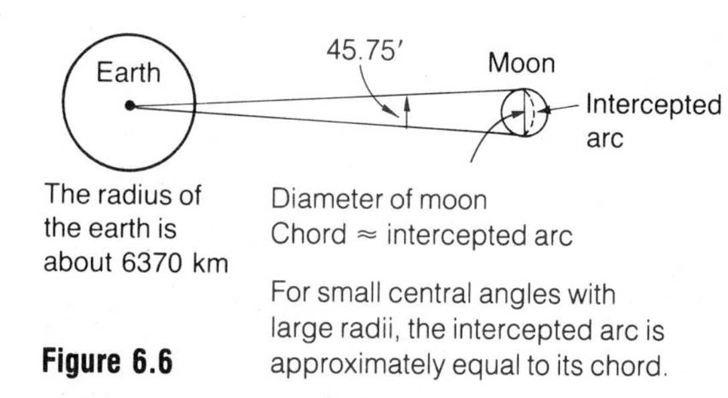

Figure 6.6

66. One side of a triangle is 20 cm longer than another, and the angle between them is 60°. If two circles are drawn with these sides as diameters, one of the points of intersection of the two circles is the common vertex. How far from the third side is the other point of intersection?

6.2 TRIGONOMETRIC FUNCTIONS

To introduce you to the trigonometric functions, we will consider a relationship between angles and circles. Draw a unit circle with an angle θ in standard position, as in Figure 6.7.

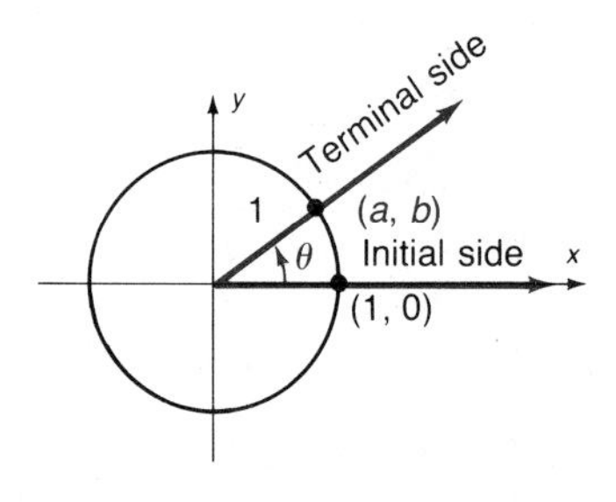

Figure 6.7 The unit circle $x^2 + y^2 = 1$ with an angle θ in standard position

The initial side of θ intersects the unit circle at $(1, 0)$, and the terminal side intersects the unit circle at (a, b). Define functions of θ as follows:

$$c(\theta) = a \quad \text{and} \quad s(\theta) = b$$

EXAMPLE 1 Find: **a.** $s(90°)$ **b.** $c(90°)$ **c.** $s(-270°)$ **d.** $s(-\frac{\pi}{2})$ **e.** $c(-3.1416)$

Solution **a.** $s(90°)$. This is the second component of the ordered pair (a, b), where (a, b) is the point of intersection of the terminal side of a 90° standard-position angle and the unit circle. By inspection, it is 1. Thus, $s(90°) = \mathbf{1}$.

b. $c(90°) = \mathbf{0}$.

c. $s(-270°) = \mathbf{1}$. This is the same as part a because the angles 90° and $-270°$ are coterminal.

d. $s(-\frac{\pi}{2}) = \mathbf{-1}$.

e. $c(-3.1416) \approx c(-\pi) = \mathbf{-1}$. ■

The function $c(\theta)$ is called the **cosine function**, and the function $s(\theta)$ is called the **sine function**. These functions, along with four others, make up the **trigonometric functions** or **trigonometric ratios**, which are further examples of *transcendental functions*. One way to define the trigonometric functions is as follows.

Unit Circle Definition of the Trigonometric Functions

Let θ be an angle in standard position with the point (a, b) the intersection of the terminal side of θ and the unit circle. Then the six trigonometric functions, with their standard abbreviations, are defined as follows:

cosine:	$\cos\theta = a$	**secant:**	$\sec\theta = \frac{1}{a} \quad (a \neq 0)$
sine:	$\sin\theta = b$	**cosecant:**	$\csc\theta = \frac{1}{b} \quad (b \neq 0)$
tangent:	$\tan\theta = \frac{b}{a} \quad (a \neq 0)$	**cotangent:**	$\cot\theta = \frac{a}{b} \quad (b \neq 0)$

Notice the condition on the tangent, secant, cosecant, and cotangent functions. These exclude division by 0; for example, $a \neq 0$ means that θ cannot be 90°, 270°, or any angle coterminal to these angles. We summarize this condition by saying that the *tangent and secant are not defined for 90° or 270°*. If $b \neq 0$, then $\theta \neq 0°, 180°$, or any angle coterminal to these angles; thus, *cosecant and cotangent are not defined for 0° or 180°*.

The angle θ in the above definition is called the **argument** of the function. The argument, of course, does not need to be the same as the variable. For example, in $\cos(2\theta + 1)$ the argument is $2\theta + 1$ and the variable is θ.

In many applications, you will know the angle measure and will want to find one or more of its trigonometric functions. To do this you will carry out a process called **evaluation of the trigonometric functions**. In order to help you see the relationship between the angle and the function, consider Figure 6.8 and Example 2.

EXAMPLE 2 Evaluate the trigonometric functions of $\theta = 110^\circ$ by drawing the unit circle and approximating the point (a, b).

Solution Draw a unit circle as shown in Figure 6.8.

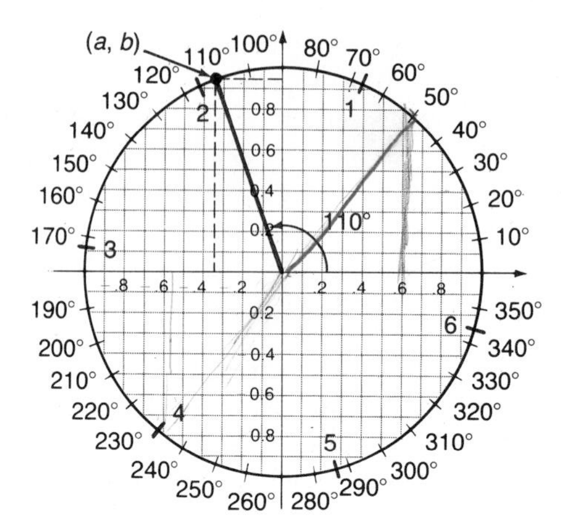

Figure 6.8 Approximate values of circular functions

Next draw the terminal side of the angle 110°. Estimate $a \approx -0.35$ and $b \approx 0.95$. Thus

$$\cos 110^\circ \approx \mathbf{-0.35} \qquad \sec 110^\circ \approx \frac{1}{-0.35} \approx \mathbf{-2.9}$$

$$\sin 110^\circ \approx \mathbf{0.95} \qquad \csc 110^\circ \approx \frac{1}{0.95} \approx \mathbf{1.1}$$

$$\tan 110^\circ \approx \frac{0.95}{-0.35} \approx \mathbf{-2.7} \qquad \cot 110^\circ \approx \frac{-0.35}{0.95} \approx \mathbf{-0.37}$$

■

Notice from Example 2 that some of the trigonometric functions of 110° are positive and others are negative. We can make certain predictions about the signs of these functions, as summarized in Table 6.1.

TABLE 6.1 Signs of trigonometric functions

	Quadrant I a pos b pos	*Quadrant II* a neg b pos	*Quadrant III* a neg b neg	*Quadrant IV* a pos b neg
$\cos\theta = a$	pos	neg	neg	pos
$\sin\theta = b$	pos	pos	neg	neg
$\tan\theta = b/a$	pos	neg	pos	neg
Summary	**All pos**	**Sine pos**	**Tangent pos**	**Cosine pos**

Table 6.1 can be summarized by remembering the following:

Sine positive	All positive
Tangent positive	Cosine positive

Easy-to-remember form:

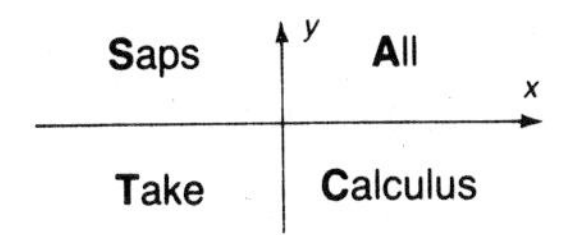

Proof In Quadrant I, a and b are both positive, so all six trigonometric functions must be positive. In Quadrant II, a is negative and b is positive, so from the definition of trigonometric functions, all are negative except the sine and cosecant. In Quadrant III, a and b are both negative, so all the functions are negative except tangent and cotangent because those are ratios of two negatives, which are positive. In Quadrant IV, a is positive and b is negative, so all are negative except cosine and secant. □

You may have noticed certain relationships among these functions. The first is that since the tangent is the ratio of b to a, it can be found by dividing the sine by the cosine. That is,

$$\tan\theta = \frac{\sin\theta}{\cos\theta}$$

whenever $\cos\theta \neq 0$. Note also that the secant, cosecant, and cotangent functions are reciprocals of the cosine, sine, and tangent functions. This means that if values of θ which cause division by zero are excluded, then

$$\sec\theta = \frac{1}{\cos\theta} \qquad \csc\theta = \frac{1}{\sin\theta} \qquad \cot\theta = \frac{1}{\tan\theta}$$

These are called the **reciprocal** relationships, or **identities**. The term *identity* is defined and discussed in the next chapter, so for now we will just remember which pairs of functions are called reciprocals. This can be confusing because the functions are also paired according to their names: sine and cosine, secant and cosecant, and tangent and cotangent. These pairs are called **cofunctions**. Study Table 6.2 until this terminology is clear to you.

TABLE 6.2
Reciprocal and cofunction relationships

Reciprocals	Cofunctions
cosine and secant	cosine and sine
sine and cosecant	cosecant and secant
tangent and cotangent	cotangent and tangent

Since the method shown in Example 2 for evaluating the trigonometric functions is not very practical, many additional procedures for finding them have been discovered. The most common method today, however, is to use a calculator. The procedure is straightforward, but you must note several details:

1. Note the unit of measure used: degree or radian. Calculators have a variety of ways of changing from radian to degree format, so consult your owner's manual to find out how your particular calculator does it. Most, however, simply have a switch (similar to an on/off switch) that sets the calculator in either degree or radian mode. From now on, we will assume that you are working in the appropriate radian/degree mode. Remember: If later in the course you suddenly

start obtaining strange answers and have no idea what you are doing wrong, double-check to make sure you are using the proper mode.

2. With most calculators you enter the angle first and then press the button corresponding to the trigonometric function.
3. You must remember which functions are reciprocals since a normal scientific calculator does not have sec θ, csc θ, or cot θ keys. It does, however, have a reciprocal key, which is labeled

   [1/x]

EXAMPLE 3 Find the trigonometric functions of 110° by calculator.

Solution

cos 110° ≈ **−0.34202014** PRESS: [110] [cos]

sin 110° ≈ **0.93969262** PRESS: [110] [sin]

tan 110° ≈ **−2.74747742** PRESS: [110] [tan]

sec 110° ≈ **−2.9238044** PRESS: [110] [cos] [1/x]

csc 110° ≈ **1.06417777** PRESS: [110] [sin] [1/x]

cot 110° ≈ **−0.36397023** PRESS: [110] [tan] [1/x] ■

EXAMPLE 4 Find csc(π/12) by calculator.

Solution

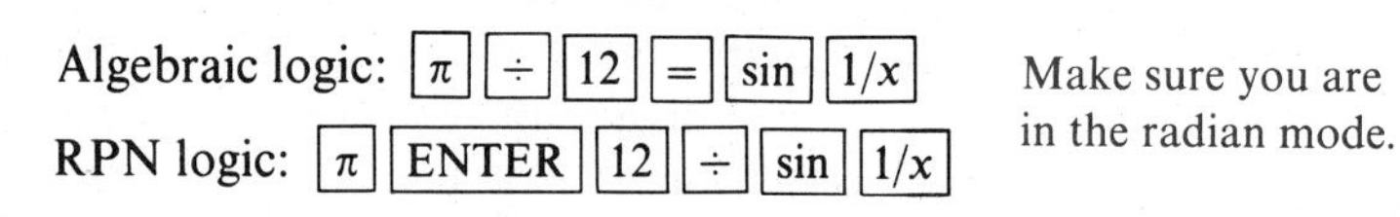

The answer now displayed is **3.8637033**. ■

EXAMPLE 5 Find tan 70°23′40″.

Solution Switch key to degrees; this measure must first be changed to a decimal degree measure:

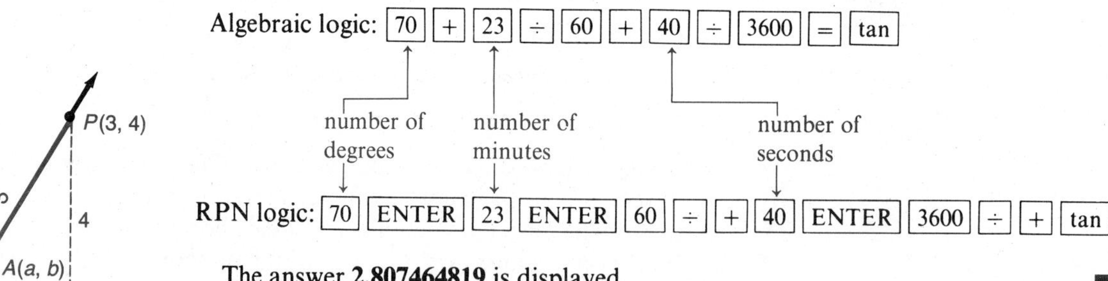

The answer **2.807464819** is displayed. ■

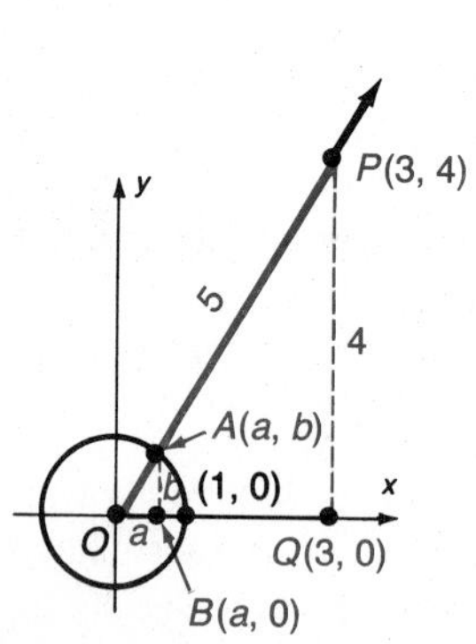

Figure 6.9 Finding (a, b) when a point on the terminal side is known

Suppose you want to find the trigonometric functions of an angle whose terminal side passes through some known point, say (3, 4). To apply the definition, you need to find the point (a, b), as shown in Figure 6.9.

Let (3, 4) be denoted by P and (a, b) by A. Let B be the point $(a, 0)$ and Q be the point (3, 0) ($OA = 1$, $OP = \sqrt{3^2 + 4^2} = 5$). Now consider $\triangle AOB$ and $\triangle POQ$.

Recall from geometry that two triangles are **similar** if two angles of one are congruent to two angles of the other. For these triangles, $\angle OBA$ is congruent to $\angle OQP$ since they are both right angles; also $\angle O$ is congruent to $\angle O$ since equal angles are congruent. Thus these triangles are similar, which is denoted by

$$\triangle AOB \sim \triangle POQ$$

The important property of similar triangles is that corresponding parts of similar triangles are proportional. Thus

$$b = \frac{b}{1} = \frac{4}{5} \qquad \text{and} \qquad a = \frac{a}{1} = \frac{3}{5}$$

Thus $\cos\theta = \frac{3}{5}$, $\sin\theta = \frac{4}{5}$, and $\tan\theta = \frac{4/5}{3/5} = \frac{4}{3}$. The reciprocals are $\sec\theta = \frac{5}{3}$, $\csc\theta = \frac{5}{4}$, and $\cot\theta = \frac{3}{4}$.

If you carry out these steps for a point $P(x, y)$ instead of $(3, 4)$, as shown in Figure 6.10, it is still true that

$$\triangle AOB \sim \triangle POQ$$

Let r be the distance from O to P. That is, let

$$r = \sqrt{x^2 + y^2}$$

Then

$$a = \frac{a}{1} = \frac{x}{r} \qquad \frac{1}{a} = \frac{r}{x}$$

$$b = \frac{b}{1} = \frac{y}{r} \qquad \frac{1}{b} = \frac{r}{y}$$

$$\frac{b}{a} = \frac{y/r}{x/r} = \frac{y}{x} \qquad \frac{a}{b} = \frac{x}{y}$$

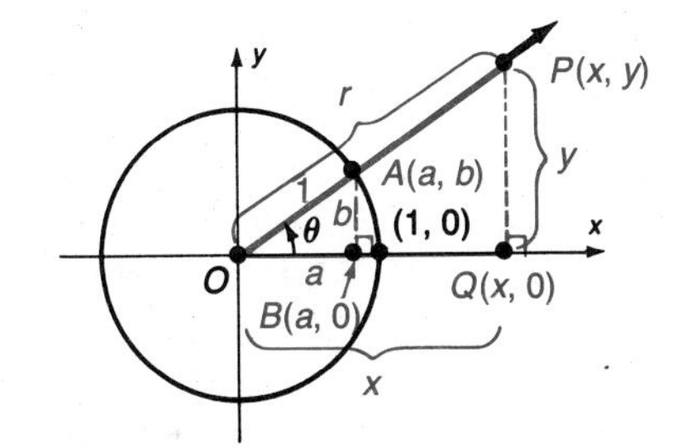

Figure 6.10

These ratios lead to an alternative definition of the trigonometric functions that allows you to choose *any* point (x, y). In practice, it is this definition that is most frequently used.

Ratio Definition of the Trigonometric Functions

Let θ be an angle in standard position with any point $P(x, y)$ on the terminal side a distance r from the origin $(r \neq 0)$. Then the six trigonometric functions are defined as follows:

$$\cos\theta = \frac{x}{r} \qquad \sec\theta = \frac{r}{x} \quad (x \neq 0)$$

$$\sin\theta = \frac{y}{r} \qquad \csc\theta = \frac{r}{y} \quad (y \neq 0)$$

$$\tan\theta = \frac{y}{x} \quad (x \neq 0) \qquad \cot\theta = \frac{x}{y} \quad (y \neq 0)$$

EXAMPLE 6 Find the values of the six trigonometric functions for an angle θ in standard position with the terminal side passing through $(-5, -2)$.

Solution $x = -5$, $y = -2$, and $r = \sqrt{25 + 4} = \sqrt{29}$. Thus

$$\cos\theta = \frac{-5}{\sqrt{29}} = \frac{-5}{29}\sqrt{29} \qquad \sec\theta = \frac{-\sqrt{29}}{5}$$

$$\sin\theta = \frac{-2}{\sqrt{29}} = \frac{-2}{29}\sqrt{29} \qquad \csc\theta = \frac{-\sqrt{29}}{2}$$

$$\tan\theta = \frac{-2/\sqrt{29}}{-5/\sqrt{29}} = \frac{2}{5} \qquad \cot\theta = \frac{5}{2}$$

■

Both the unit circle and ratio definitions of the trigonometric functions were given using angle domains. We can extend the definition to include real number domains since radian measure is in terms of real numbers. That is, for

$$\cos\frac{\pi}{2} \qquad \text{or} \qquad \cos 2$$

it does not matter whether $\frac{\pi}{2}$ and 2 are considered as radian measures of angles or simply as real numbers—the functional values are the same.

Trigonometric Functions of Real Numbers

For any real number x, let $x = \theta$, where θ is a standard-position angle measured in radians. Then

$$\cos x = \cos\theta \qquad \sin x = \sin\theta \qquad \tan x = \tan\theta$$

$$\sec x = \sec\theta \qquad \csc x = \csc\theta \qquad \cot x = \cot\theta$$

PROBLEM SET 6.2

A

From the unit circle definition of the trigonometric functions, estimate to one decimal place the numbers in Problems 1–8.

1. $\cos 50°$ **2.** $\sin 20°$ **3.** $\sin 320°$ **4.** $\tan 80°$

5. $\tan(-20°)$ **6.** $\cos(-340°)$ **7.** $\sec 70°$ **8.** $\csc 190°$

Use a calculator to evaluate the functions given in Problems 9–27.

9. a. $\cos 50°$ **b.** $\sin 20°$

10. a. $\sec 70°$ **b.** $\csc 150°$

11. a. $\cot 250°$ **b.** $\sec 135°$

12. a. $\cos(-34°)$ **b.** $\sin(-95°)$

13. a. $-\cot(-18°)$ **b.** $-\sec(-213°)$

14. a. $\tan 56.2°$ **b.** $\cot 78.4°$

15. a. $\sin 1$ **b.** $\sin 1°$

16. a. $\tan 15$ **b.** $\tan 15°$

17. a. $\cot 2.5$ **b.** $-\sec 1.5$

18. a. $\cos(-0.48)$ **b.** $\sec(-21.3°)$

19. a. $\tan 129°9'12''$ **b.** $\cos 240°8''$

20. a. $\dfrac{3}{5\sin 2}$ **b.** $\frac{3}{5}\sin 2$

21. a. $\frac{3}{5}\sin\frac{1}{2}$ **b.** $\dfrac{3\sin 2}{5}$

22. a. $\dfrac{1}{2\sec 3}$ **b.** $\frac{1}{2}\sec 3$

23. **a.** $2 \sec \frac{1}{3}$ **b.** $\dfrac{2}{\sec \frac{1}{3}}$

24. **a.** $(-3)\cos 2$ **b.** $3\cos(-2)$

25. **a.** $3\cos(-\frac{1}{2})$ **b.** $\dfrac{-3}{\cos \frac{1}{2}}$

26. **a.** $(-3)\cot 2$ **b.** $3\cot(-2)$

27. **a.** $3\cot(-\frac{1}{2})$ **b.** $\dfrac{-3}{\cot \frac{1}{2}}$

Tell whether each of the functions in Problems 28–39 is positive or negative. You should be able to do this without tables or a calculator.

28. sine, Quadrant I
29. cosine, Quadrant I
30. tangent, Quadrant II
31. secant, Quadrant II
32. cosecant, Quadrant III
33. cotangent, Quadrant IV
34. $\sin 1$
35. $\cos 2$
36. $\tan 3$
37. $\sec 4$
38. $\sin(-1)$
39. $\cos(-2)$

Tell in which quadrant(s) a standard-position angle θ could lie if the conditions in Problems 40–45 are true.

40. $\sin\theta > 0$
41. $\cos\theta < 0$
42. $\tan\theta < 0$
43. $\sin\theta < 0$ and $\tan\theta > 0$
44. $\sin\theta > 0$ and $\tan\theta < 0$
45. $\cos\theta < 0$ and $\sin\theta < 0$

B

Find the values of the six trigonometric functions for an angle θ in standard position with terminal side passing through the points given in Problems 46–54. Draw a picture showing θ and the reference angle θ'.

46. $(3, 4)$
47. $(-3, 4)$
48. $(-3, -4)$
49. $(5, 12)$
50. $(-5, -12)$
51. $(5, -12)$
52. $(2, -5)$
53. $(-6, 1)$
54. $(-4, -5)$

In the next section you will need to simplify some radical expressions. Recall that $\sqrt{x^2} = |x|$. This means that $\sqrt{x^2} = x$ if x is positive and $-x$ if x is negative. Simplify the expressions in Problems 55–63.

55. $\sqrt{2x^2}$
56. $\sqrt{9x^2}$
57. $\sqrt{2x^2}$ if x is negative
58. $\sqrt{2x^2}$ if x is positive
59. $\sqrt{9x^2}$ if x is positive
60. $\sqrt{9x^2}$ if x is negative
61. $\dfrac{x}{\sqrt{4x^2}}$ if x is positive
62. $\dfrac{x}{\sqrt{4x^2}}$ if x is negative
63. $\dfrac{\sqrt{16x^2}}{x}$ if x is negative

C

64. **a.** Let $P(x, y)$ be any point in the plane. Show that $P(r\cos\theta, r\sin\theta)$ is a representation for P, where θ is the standard-position angle formed by drawing ray $\overrightarrow{OP}$.
b. Let $A(\cos\alpha, \sin\alpha)$ and $B(\cos\beta, \sin\beta)$ be any two points on a unit circle. Use the distance formula to show that

$$|AB| = \sqrt{2 - 2(\cos\alpha\cos\beta + \sin\alpha\sin\beta)}$$

65. You will learn in calculus that

$$\sin x = x - \frac{x^3}{3!} + \frac{x^5}{5!} - \frac{x^7}{7!} + \cdots$$

where $n! = n(n-1)(n-2)\cdots 3 \cdot 2 \cdot 1$. Find $\sin 1$ correct to four decimal places by using this equation. (Remember that the 1 in $\sin 1$ refers to radian measure.)

66. You will learn in calculus that

$$\cos x = 1 - \frac{x^2}{2!} + \frac{x^4}{4!} - \frac{x^6}{6!} + \cdots$$

Use this equation to find $\cos 1$ correct to four decimal places.

6.3 VALUES OF THE TRIGONOMETRIC FUNCTIONS

If an angle has a terminal side that coincides with one of the coordinate axes, it is easy to evaluate the trigonometric functions by using the unit circle definition.

EXAMPLE 1 Evaluate the trigonometric functions for $-\frac{5\pi}{2}$.

Solution Since $-\frac{5\pi}{2}$ has a terminal side coinciding with the negative y axis, the intersection of this terminal side and the unit circle is $(0, -1)$. This means that $a = 0$ and $b = -1$.

Hence

$$\cos\frac{-5\pi}{2} = a = \mathbf{0} \qquad \sin\frac{-5\pi}{2} = b = \mathbf{-1}$$

$$\csc\frac{-5\pi}{2} = \frac{1}{b} = \mathbf{-1} \qquad \cot\frac{-5\pi}{2} = \frac{a}{b} = \mathbf{0}$$

$\tan\frac{-5\pi}{2} = \frac{b}{a}$ and $\sec\frac{-5\pi}{2} = \frac{1}{a}$ are **undefined** since $a = 0$. ■

There are times when you will not be able to rely on calculator approximations for the trigonometric functions but instead will need to find **exact values.**

EXAMPLE 2 Evaluate the trigonometric functions for $\frac{\pi}{4}$.

Solution

$$\cos\frac{\pi}{4} = \frac{x}{r} \qquad \text{From the ratio definition (this is true for any angle)}$$

$$= \frac{x}{\sqrt{x^2 + y^2}} \qquad r = \sqrt{x^2 + y^2} \text{ for any angle}$$

If $\theta = \frac{\pi}{4}$, then $x = y$ since $\frac{\pi}{4}$ bisects Quadrant I. By substitution,

$$\cos\frac{\pi}{4} = \frac{x}{\sqrt{x^2 + x^2}}$$

$$= \frac{x}{\sqrt{2x^2}}$$

$$= \frac{x}{x\sqrt{2}} \qquad \sqrt{x^2} = |x| = x \text{ since } x \text{ is positive in Quadrant I}$$

$$= \frac{1}{\sqrt{2}}$$

$$= \mathbf{\frac{1}{2}\sqrt{2}} \qquad \frac{1}{\sqrt{2}} = \frac{1}{\sqrt{2}} \cdot \frac{\sqrt{2}}{\sqrt{2}} = \frac{\sqrt{2}}{2} = \frac{1}{2}\sqrt{2}$$

Similarly, $\sin\frac{\pi}{4} = \mathbf{\frac{\sqrt{2}}{2}}$, $\tan\frac{\pi}{4} = \mathbf{1}$, $\sec\frac{\pi}{4} = \mathbf{\sqrt{2}}$, $\csc\frac{\pi}{4} = \mathbf{\sqrt{2}}$, and $\cot\frac{\pi}{4} = \mathbf{1}$. ■

EXAMPLE 3 Find the exact values for the trigonometric functions of 30°.

Solution Consider not only the standard-position angle 30° but also the standard-position angle −30°. Choose $P_1(x, y)$ and $P_2(x, -y)$, respectively, on the terminal sides. (See Figure 6.11.)

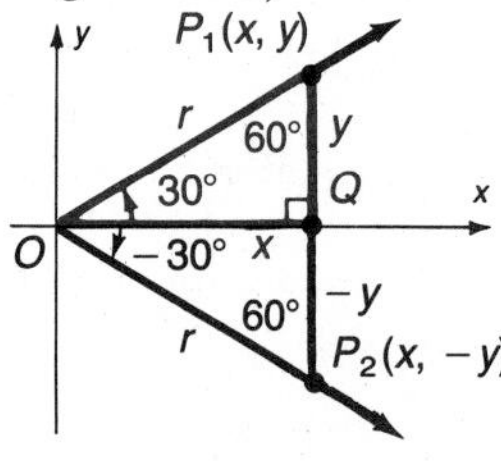

Figure 6.11

Angles OQP_1 and OQP_2 are right angles, so $\angle OP_1Q = 60°$ and $\angle OP_2Q = 60°$. Thus $\triangle OP_1P_2$ is an equiangular triangle. (All angles measure 60°.) From geometry we know that an equiangular triangle has sides the same length. Thus $2y = r$. Notice the following relationship between x and y:

$$r^2 = x^2 + y^2$$
$$(2y)^2 = x^2 + y^2 \qquad 2y = r$$
$$3y^2 = x^2$$
$$\sqrt{3}|y| = |x|$$

For 30°, x and y are both positive, so $x = \sqrt{3}y$.

$$\cos 30° = \frac{x}{r} = \frac{\sqrt{3}y}{2y} = \frac{\sqrt{3}}{2} \qquad \sec 30° = \frac{2}{\sqrt{3}} = \frac{2}{3}\sqrt{3}$$

$$\sin 30° = \frac{y}{r} = \frac{y}{2y} = \frac{1}{2} \qquad \csc 30° = 2$$

$$\tan 30° = \frac{y}{x} = \frac{y}{\sqrt{3}y} = \frac{1}{\sqrt{3}} = \frac{\sqrt{3}}{3} \qquad \cot 30° = \sqrt{3}$$ ■

The derivation in Example 3 leads to the following result from plane geometry.

30°–60°–90° Triangle Theorem

> In a 30°–60°–90° triangle, the leg opposite the 30° angle equals one-half the hypotenuse and the leg opposite the 60° angle equals one-half the hypotenuse times the square root of 3.

EXAMPLE 4 Evaluate the trigonometric functions for 60°.

Solution Using the 30°–60°–90° triangle theorem, the hypotenuse r is twice the length of the shorter leg, as shown in Figure 6.12.

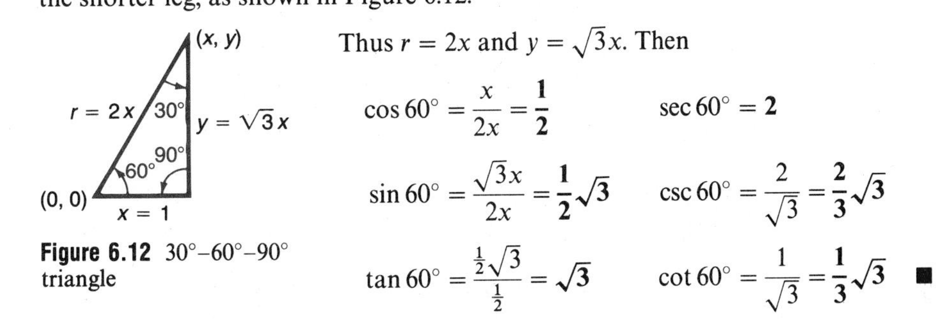

Figure 6.12 30°–60°–90° triangle

Thus $r = 2x$ and $y = \sqrt{3}x$. Then

$$\cos 60° = \frac{x}{2x} = \frac{1}{2} \qquad \sec 60° = 2$$

$$\sin 60° = \frac{\sqrt{3}x}{2x} = \frac{1}{2}\sqrt{3} \qquad \csc 60° = \frac{2}{\sqrt{3}} = \frac{2}{3}\sqrt{3}$$

$$\tan 60° = \frac{\frac{1}{2}\sqrt{3}}{\frac{1}{2}} = \sqrt{3} \qquad \cot 60° = \frac{1}{\sqrt{3}} = \frac{1}{3}\sqrt{3}$$ ■

In a manner similar to Examples 1 to 4, a **table of exact values** is constructed (Table 6.3, page 224). Since this table of values is used extensively, you should memorize at least the values for $\cos\theta$, $\sin\theta$, and $\tan\theta$ as you did multiplication tables in elementary school.

The values for $\sec\theta$, $\csc\theta$, and $\cot\theta$ do not need to be memorized as separate entries because they are simply the reciprocals of $\cos\theta$, $\sin\theta$, and $\tan\theta$.

TABLE 6.3 Exact values

Function \ Angle θ	$0 = 0°$	$\frac{\pi}{6} = 30°$	$\frac{\pi}{4} = 45°$	$\frac{\pi}{3} = 60°$	$\frac{\pi}{2} = 90°$	$\pi = 180°$	$\frac{3\pi}{2} = 270°$
$\cos\theta$	1	$\frac{\sqrt{3}}{2}$	$\frac{\sqrt{2}}{2}$	$\frac{1}{2}$	0	-1	0
$\sin\theta$	0	$\frac{1}{2}$	$\frac{\sqrt{2}}{2}$	$\frac{\sqrt{3}}{2}$	1	0	-1
$\tan\theta$	0	$\frac{\sqrt{3}}{3}$	1	$\sqrt{3}$	undef.	0	undef.
$\sec\theta$	1	$\frac{2}{\sqrt{3}} = \frac{2}{3}\sqrt{3}$	$\frac{2}{\sqrt{2}} = \sqrt{2}$	$\frac{2}{1} = 2$	undef.	$\frac{1}{-1} = -1$	undef.
$\csc\theta$	undef.	$\frac{2}{1} = 2$	$\frac{2}{\sqrt{2}} = \sqrt{2}$	$\frac{2}{\sqrt{3}} = \frac{2}{3}\sqrt{3}$	1	undef.	-1
$\cot\theta$	undef.	$\frac{3}{\sqrt{3}} = \sqrt{3}$	1	$\frac{1}{\sqrt{3}} = \frac{\sqrt{3}}{3}$	0	undef.	0

These are the reciprocals (which is why the exact values are given in reciprocal form as well as in simplified form). In a problem you would use the rationalized form. The reciprocal form makes it easy to remember them.

You can find exact values of the trigonometric functions that are multiples of those in Table 6.3 by using the idea of a reference angle and the ***reduction principle***:

Reduction Principle

> If t represents any of the six trigonometric functions, then
>
> $$t(\theta) = \pm t(\theta')$$
>
> where θ' is the reference angle of θ and the sign plus or minus depends on the quadrant of the terminal side of the angle θ.

EXAMPLE 5 $\tan 210° = + \tan 30° = \frac{\sqrt{3}}{3}$ ■

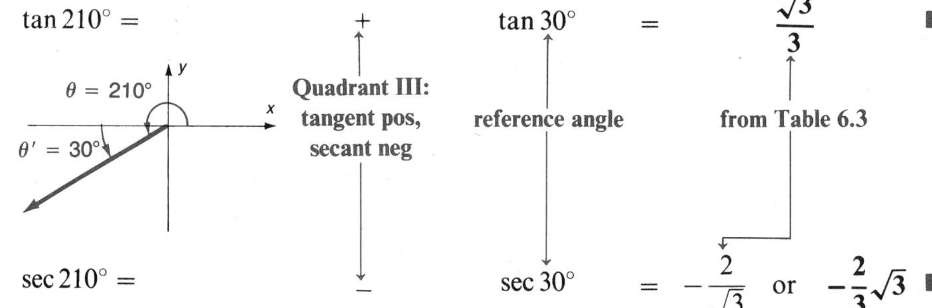

EXAMPLE 6 $\sec 210° = - \sec 30° = -\frac{2}{\sqrt{3}}$ or $-\frac{2}{3}\sqrt{3}$ ■

EXAMPLE 7 $\csc\frac{3\pi}{2} = -1$

If it is a quadrantal angle, then it comes directly from the memorized table. ■

EXAMPLE 8 $\cos 405° = \cos 45° = \frac{1}{2}\sqrt{2}$

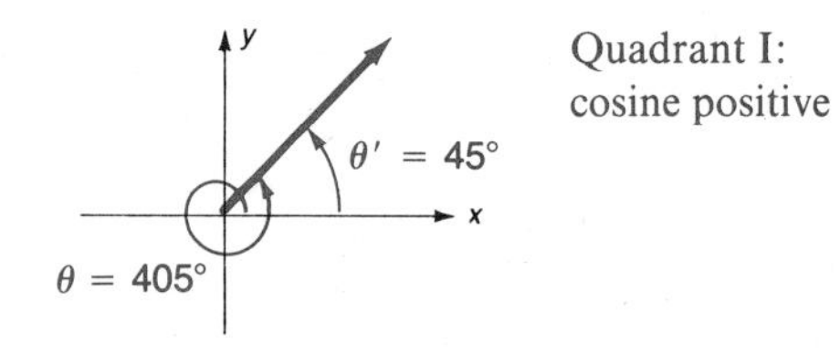

Sketch the angle if necessary to find the quadrant and the reference angle. ■

EXAMPLE 9 $\cot\left(-\frac{7\pi}{6}\right) = -\cot\frac{\pi}{6} = -\sqrt{3}$

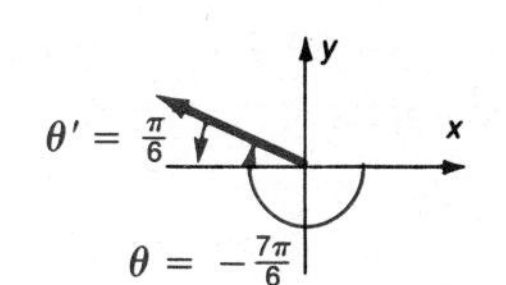

Quadrant II: cotangent negative ■

EXAMPLE 10 $\sec\frac{5\pi}{3} = +\sec\frac{\pi}{3} = 2$

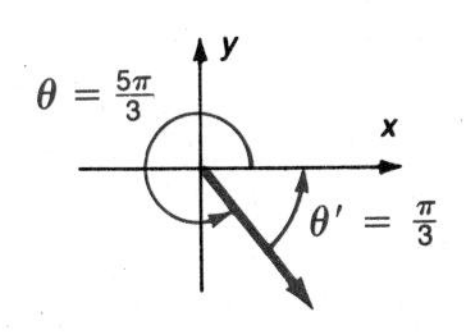

Quadrant IV: secant positive ■

The reduction principle is also needed when using Table III in Appendix C. If you look at Table C.III, you will see that only first quadrant angles are given, so the reduction principle is necessary.

EXAMPLE 11 Find cos 186° using Table C.III or a calculator.

Solution To find the table value, first use the reduction principle:

$$\cos 186° = -\cos 6° = 0.9945$$

The reduction principle is not needed when evaluating with a calculator:

$\cos 186° = -0.9945219$ PRESS: [186] [cos] ■

PROBLEM SET 6.3

A

In Problems 1–10, give the exact values in simplified form.

1. **a.** $\tan\frac{\pi}{4}$ **b.** $\cos 0$ **c.** $\sin 60°$ **d.** $\cos 30°$
2. **a.** $\cos 270°$ **b.** $\tan\frac{\pi}{6}$ **c.** $\tan 180°$ **d.** $\sin 45°$
3. **a.** $\sin\pi$ **b.** $\sin\frac{\pi}{2}$ **c.** $\tan 0$ **d.** $\cos\frac{\pi}{4}$
4. **a.** $\sec\frac{\pi}{6}$ **b.** $\csc 0$ **c.** $\sec\frac{\pi}{4}$ **d.** $\sec 0°$
5. **a.** $\csc\frac{\pi}{4}$ **b.** $\cot\pi$ **c.** $\sec\frac{\pi}{3}$ **d.** $\cot\frac{\pi}{6}$
6. **a.** $\tan 90°$ **b.** $\tan 60°$ **c.** $\cos\frac{\pi}{3}$ **d.** $\sec\frac{\pi}{3}$
7. **a.** $\cot 45°$ **b.** $\cos\pi$ **c.** $\sin\frac{3\pi}{2}$ **d.** $\sin 0°$
8. **a.** $\sec\pi$ **b.** $\tan 270°$ **c.** $\sin\frac{\pi}{6}$ **d.** $\csc\frac{3\pi}{2}$
9. **a.** $\cos(-300°)$ **b.** $\sin 390°$ **c.** $\sin\frac{17\pi}{4}$ **d.** $\cos(-6\pi)$
10. **a.** $\cos\frac{9\pi}{2}$ **b.** $\sin(-765°)$ **c.** $\tan(-765°)$ **d.** $\cos 495°$

Use a calculator or Table C.III to evaluate the functions in Problems 11–26. Round your answers to four decimal places.

11. **a.** $\sin 34.4°$ **b.** $\cos 54.2°$
12. **a.** $\tan 70.2°$ **b.** $\cot 46.7°$
13. **a.** $\cos 50°$ **b.** $\cot 80°$
14. **a.** $\sin 70°$ **b.** $\tan 20°$
15. **a.** $\tan(-20°)$ **b.** $\sin 190°$
16. **a.** $\cos(-340°)$ **b.** $\cot(-213°)$
17. **a.** $\sin 132.8°$ **b.** $\tan(-25.6°)$
18. **a.** $\cot(-125.6°)$ **b.** $\cos 163.4°$
19. **a.** $\sin 1.20$ **b.** $\cos 0.65$
20. **a.** $\tan 0.51$ **b.** $\cot 1.85$
21. **a.** $\tan 1$ **b.** $\cot 1.5$
22. **a.** $\sin 0.8$ **b.** $\cos 0.5$
23. **a.** $\tan 2.5$ **b.** $\sin 3$
24. **a.** $\cos 4.5$ **b.** $\cot 6$
25. **a.** $\cos(-0.45)$ **b.** $\tan(-2.8)$
26. **a.** $\sin(-3.9)$ **b.** $\cot 10$

B

27. Verify the entries in Table 6.3 for the angle $\frac{\pi}{3}$.
28. Find $\cos\frac{3\pi}{4}$ by using the procedure illustrated in Example 2.
29. Find $\cos\frac{5\pi}{4}$ by using the procedure illustrated in Example 2.

30. Find $\cos 135°$ by choosing an arbitrary point (x, y) on the terminal side of $135°$ and applying the ratio definition of the trigonometric functions.

31. Find $\sin(-\frac{\pi}{4})$ by choosing an arbitrary point (x, y) on the terminal side of $-\frac{\pi}{4}$ and applying the ratio definition of the trigonometric functions.

32. Find $\sin 210°$ by choosing an arbitrary point (x, y) on the terminal side of $210°$ and applying the ratio definition of the trigonometric functions.

33. Find $\cos 210°$ by choosing an arbitrary point (x, y) on the terminal side of $210°$ and applying the ratio definition of the trigonometric functions.

Substitute the exact values for the trigonometric functions in the expressions in Problems 34–59 and simplify. When a trigonometric function is raised to a power, such as $(\sin x)^2$, *it is written as* $\sin^2 x$.

34. $\sin 30° + \cos 0°$ **35.** $\sin\frac{\pi}{2} + 3\cos\frac{\pi}{2}$

36. $2\cos\frac{\pi}{2}$ **37.** $\cos\frac{2\pi}{2}$

38. $\sin\frac{2\pi}{4}$ **39.** $2\sin\frac{\pi}{4}$

40. $\sin^2 60°$ **41.** $\cos^2\frac{\pi}{4}$

42. $\sin^2\frac{\pi}{6} + \cos^2\frac{\pi}{2}$ **43.** $\sin^2\frac{\pi}{2} + \cos^2\frac{\pi}{2}$

44. $\sin^2\frac{\pi}{3} + \cos^2\frac{\pi}{3}$ **45.** $\sin^2\frac{\pi}{6} + \cos^2\frac{\pi}{3}$

46. $\sin\frac{\pi}{6}\csc\frac{\pi}{6}$ **47.** $\csc\frac{\pi}{2}\sin\frac{\pi}{2}$

48. $\cos(\frac{\pi}{4} - \frac{\pi}{2})$ **49.** $\cos\frac{\pi}{4} - \cos\frac{\pi}{2}$

50. $\tan(2 \cdot 30°)$ **51.** $2\tan 30°$

52. $\csc(\frac{1}{2} \cdot 60°)$ **53.** $\dfrac{\csc 60°}{2}$

54. $\cos(\frac{1}{2} \cdot 60°)$ **55.** $\sqrt{\dfrac{1 + \cos 60°}{2}}$

56. $\tan(2 \cdot 60°)$ **57.** $\dfrac{2\tan 60°}{1 - \tan^2 60°}$

58. $\cos(\frac{\pi}{2} - \frac{\pi}{6})$ **59.** $\cos\frac{\pi}{2}\cos\frac{\pi}{6} + \sin\frac{\pi}{2}\sin\frac{\pi}{6}$

C

60. What is the smaller angle between the hands of a clock at 12:25 P.M.?

61. **a.** If θ is in Quadrant I, then $\theta + \pi$ is in Quadrant III with a reference angle θ. Use this fact and the reduction principle to show that $\sin(\theta + \pi) = -\sin\theta$ if θ is in Quadrant I.
b. Show that $\sin(\theta + \pi) = -\sin\theta$ if θ is in Quadrant II.
c. Show that $\sin(\theta + \pi) = -\sin\theta$ if θ is in Quadrant III.
d. Show that $\sin(\theta + \pi) = -\sin\theta$ if θ is in Quadrant IV.
e. By considering parts **a**–**d**, show that $\sin(\theta + \pi) = -\sin\theta$ for any angle θ.

62. Show that $\cos(\theta + \pi) = -\cos\theta$ for any angle θ. (*Hint:* See Problem 61.)

63. ***Computer*** If you have access to a computer, write a program that will output a table of trigonometric values for the sine, cosine, and tangent for every degree from 0° to 45°.

6.4 GRAPHS OF THE TRIGONOMETRIC FUNCTIONS

As with the polynomial and rational functions, we are interested in the graphs of the trigonometric functions. We will first determine the general shape of the trigonometric functions by plotting points and then generalize so we can graph the functions without too many calculations concerning points.

To graph $y = \sin x$, begin by plotting familiar values for the sine:

x = real number	0	$\frac{\pi}{6}$	$\frac{\pi}{4}$	$\frac{\pi}{3}$	$\frac{\pi}{2}$	π	$\frac{3\pi}{2}$
$y = \sin x$	0	$\frac{1}{2}$	$\frac{\sqrt{2}}{2}$	$\frac{\sqrt{3}}{2}$	1	0	-1
y (approximate)	0	0.5	0.71	0.87	1	0	-1

We are using exact values here, but you could also use Table C.III or a calculator to generate these values. (See Problems 3 and 4 in the problem set.) The hardest part of graphing the sine function is deciding on the scales to use for the x and y axes. You may find it convenient to choose 12 intervals on the x axis for π units and 10 intervals on the y axis for 1 unit. You can then plot additional values by using the reduction principle. Continue the table to include x in Quadrants II, III and IV and plot the points (x, y) as shown in Figure 6.13. The smooth curve that connects these points is called the **sine curve**.

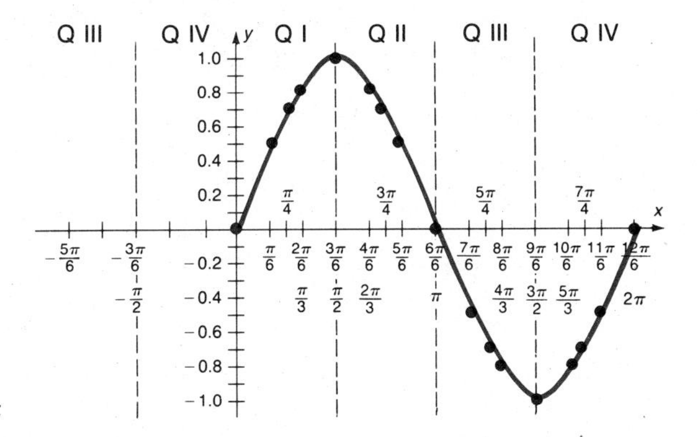

Figure 6.13 Graph of $y = \sin x$ for $0 \le x \le 2\pi$

Notice that when x is in Quadrant I, then $0 < x < \frac{\pi}{2}$, which does *not* correspond to the first quadrant of the graph $y = \sin x$. Figure 6.13 shows the intervals corresponding to the quadrants of the angle x.

If you plot values for $y = \sin x$ outside the interval $[0, 2\pi]$, you will see that this curve repeats the curve already plotted (Figure 6.14). Since $\sin(\theta + 2\pi) = \sin\theta$, we say the **period of the sine is 2π.**

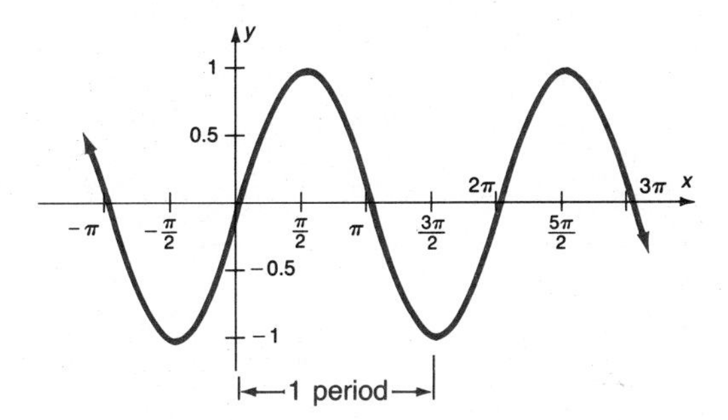

Figure 6.14 Graph of $y = \sin x$

Notice that for the base period shown the sine curve starts at $(0, 0)$, goes *up* to $(\frac{\pi}{2}, 1)$, then *down* to $(\frac{3\pi}{2}, -1)$ passing through $(\pi, 0)$, and then back *up* to $(2\pi, 0)$, which completes one period. The procedure for sketching the sine curve is shown in Figure 6.15.

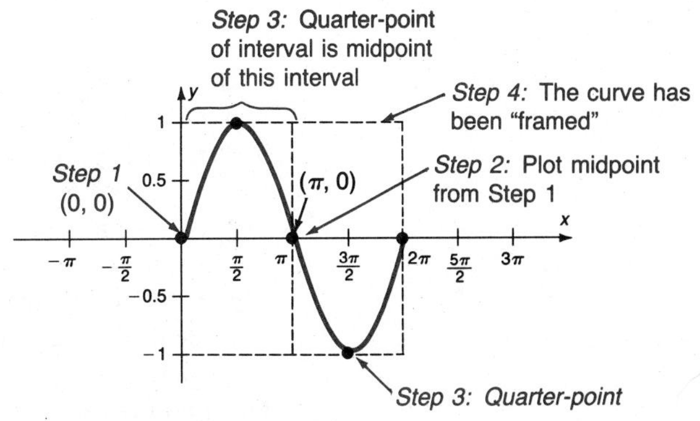

Figure 6.15 Procedure for framing the sine curve

Procedure for framing a sine curve:

1. Plot the endpoints of the base period—namely $(0, 0)$ and $(2\pi, 0)$.
2. Plot the midpoint $(\pi, 0)$.
3. Halfway between these points, plot the highest (up) point $(\frac{\pi}{2}, 1)$, and the lowest (down) point $(\frac{3\pi}{2}, -1)$; these are called the quarter-points. This is easy to remember—the sine curve "goes up and down."
4. Now the sine curve is framed; draw the curve through the plotted points, remembering the shape of the sine curve.

The cosine curve, like the sine curve, can be graphed by plotting points. We will leave the details of plotting these points as an exercise and summarize the results by "framing" the cosine curve (see Figure 6.16).

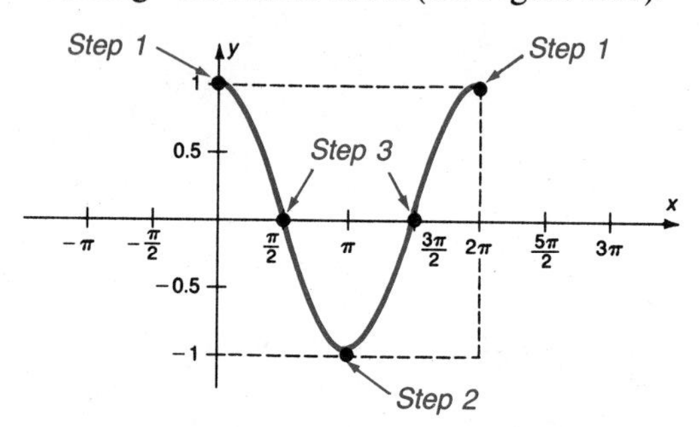

Figure 6.16 Procedure for framing the cosine curve

1. Plot the endpoints of the base period—namely $(0, 1)$ and $(2\pi, 1)$.
2. Plot the midpoint $(\pi, -1)$.
3. Halfway between these points, plot the points $(\frac{\pi}{2}, 0)$ and $(\frac{3\pi}{2}, 0)$.
4. Now the cosine curve is framed; draw the curve through the plotted points.

Since values for x greater than 2π or less than zero are coterminal with those already considered, the **period of the cosine is 2π**. The cosine curve is shown in Figure 6.17.

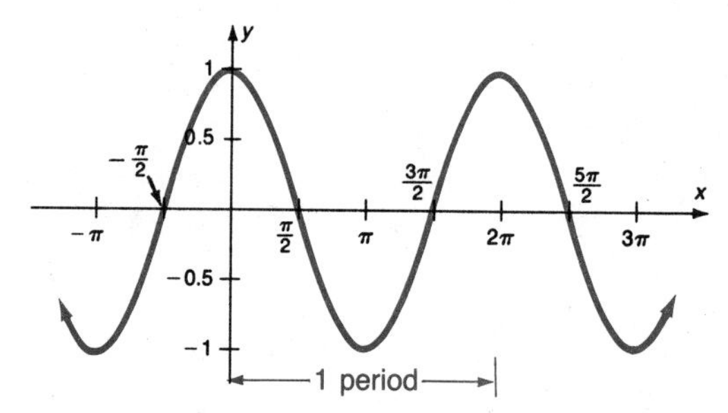

Figure 6.17 Graph of $y = \cos x$

By setting up a table of values and plotting points (the details are left as an exercise), notice that $y = \tan x$ does not exist at $\frac{\pi}{2}, \frac{3\pi}{2}$, or $\frac{\pi}{2} \pm n\pi$ for any integer n. Thus, the lines $x = \frac{\pi}{2}, x = \frac{3\pi}{2}, \ldots, x = \frac{\pi}{2} \pm n\pi$ for which the tangent is not defined are vertical asymptotes. We now frame the tangent curve as in Figure 6.18.

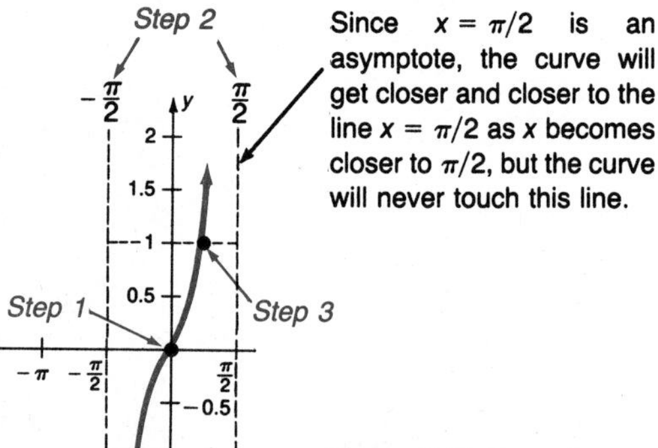

Figure 6.18 Procedure for framing a tangent curve

Procedure for framing a tangent curve:

1. Plot the midpoint $(0, 0)$.
2. Draw a pair of adjacent asymptotes—say, $-\frac{\pi}{2}$ and $\frac{\pi}{2}$.
3. Halfway between the midpoint and the asymptotes, plot the points $(\frac{\pi}{4}, 1)$ and $(-\frac{\pi}{4}, -1)$.
4. The tangent curve is now framed. Draw the curve through the plotted points; it does not cross any of the asymptotes.

The tangent curve is indicated in Figure 6.19. Notice that the curve repeats for values of x greater than 2π or less than zero, but it also repeats after it has passed through an interval with length π. This result can be shown algebraically if we use the answers to Problems 61 and 62 of Section 6.3. Since $\sin(\theta + \pi) = -\sin\theta$ and $\cos(\theta + \pi) = -\cos\theta$,

$$\tan(\theta + \pi) = \frac{\sin(\theta + \pi)}{\cos(\theta + \pi)} = \frac{-\sin\theta}{-\cos\theta} = \frac{\sin\theta}{\cos\theta} = \tan\theta$$

Since $\tan(\theta + \pi) = \tan\theta$, then $\tan(\theta + n\pi) = \tan\theta$ for any integer n, and we see that the **tangent has a period of π.**

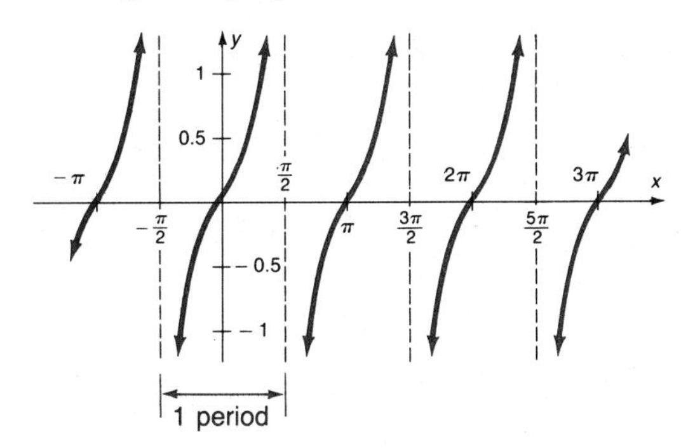

Figure 6.19 Graph of $y = \tan x$

The graphs of the other three trigonometric functions can be done in the same fashion. Instead, however, we will make use of the reciprocal relationships and graph them as shown in Example 1.

EXAMPLE 1 Sketch $y = \sec x$ by first sketching the reciprocal $y = \cos x$.

Solution Begin by sketching the reciprocal, $y = \cos x$ (black curve in Figure 6.20). Wherever $\cos x = 0$, $\sec x$ is undefined; draw asymptotes at these places. Now plot points by finding the reciprocals of the ordinates of points previously plotted. When

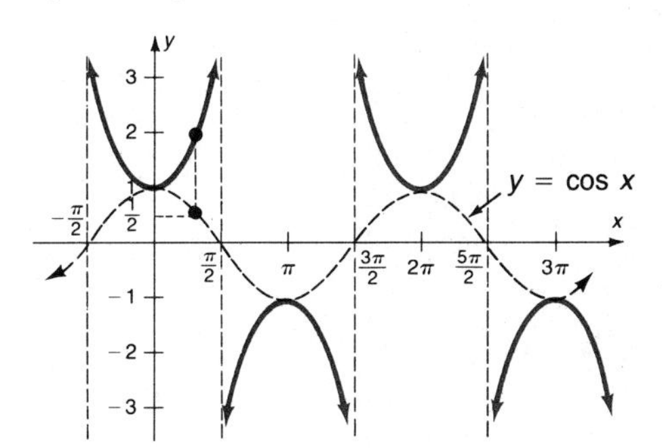

Figure 6.20 Graph of $y = \sec x$

$y = \cos x = \frac{1}{2}$, for example, the reciprocal is

$$y = \sec x = \frac{1}{\cos x} = \frac{1}{\frac{1}{2}} = 2$$

The completed graph is shown in Figure 6.20. ■

The graphs of the other reciprocal trigonometric functions are shown in Figure 6.21.

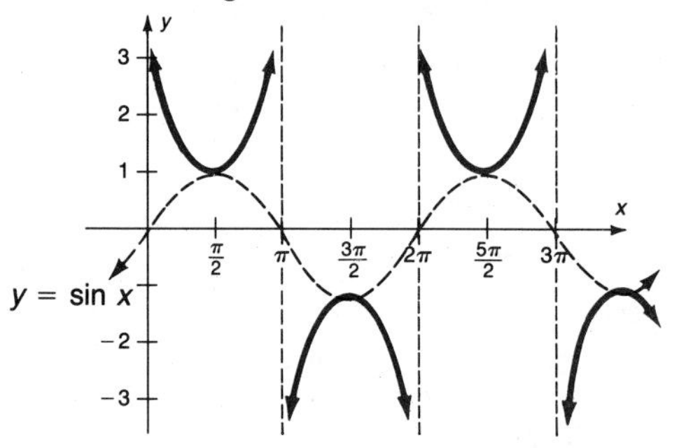

Figure 6.21 a. Graph of $y = \csc x$ b. Graph of $y = \cot x$

In Section 2.4 we saw that $y - k = f(x - h)$ can be sketched by translating the coordinate axes to a point (h, k) and then graphing the related function $y = f(x)$ on this new coordinate system. Thus if $f(x) = \sin x$, then $f(x - h) = \sin(x - h)$ is a sine curve shifted h units to the right. This shifting is called a **phase shift** but, in this book, we will treat it as a **translation**.

EXAMPLE 2 Graph one period of $y = \sin(x + \frac{\pi}{2})$.

Solution *Step 1* Frame the curve as in Figure 6.22.

a. Plot $(h, k) = (-\frac{\pi}{2}, 0)$.

b. The period of the sine curve is 2π, and it has a high point up 1 unit and a low point down 1 unit.

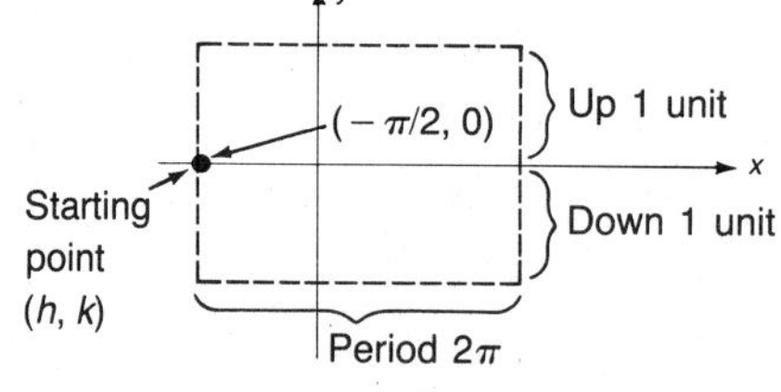

Figure 6.22 Framing the curve: this step is the same regardless of whether you are graphing a sine or a cosine

Step 2 Plot the five critical points (two endpoints, the midpoint, and two quarter-points). For the sine curve, plot the endpoint (h, k) and use the frame to plot the other endpoint and the midpoint. For the quarter-points, remember that the sine curve is "up–down"; use the frame to plot the quarter-points as shown in Figure 6.23 on the next page.

Step 3 Remembering the shape of the sine curve, sketch one period of $y = \sin(x + \frac{\pi}{2})$ using the frame and the five critical points. If you want to show more than one period, just repeat the same pattern.

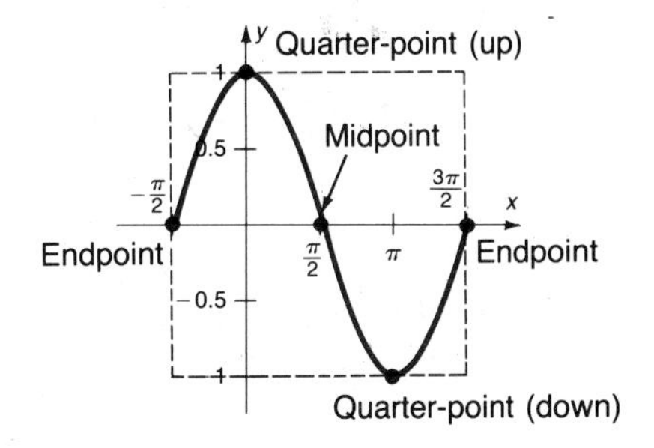

Figure 6.23 Graph of one period of $y = \sin(x + \frac{\pi}{2})$

■

Notice from Figure 6.23 that the graph of $y = \sin(x + \frac{\pi}{2})$ is the same as the graph of $y = \cos x$. Thus

$$\sin\left(x + \frac{\pi}{2}\right) = \cos x$$

EXAMPLE 3 Graph one period of $y - 2 = \cos(x - \frac{\pi}{6})$.

Solution *Step 1* Frame the curve as shown in Figure 6.24. Notice that $(h, k) = (\frac{\pi}{6}, 2)$.

Step 2 Plot the five critical points. For the cosine curve the left and right endpoints are at the upper corners of the frame; the midpoint is at the bottom of the frame; the quarter-points are on a line through the middle of the frame.

Step 3 Draw one period of the curve, as shown in Figure 6.24.

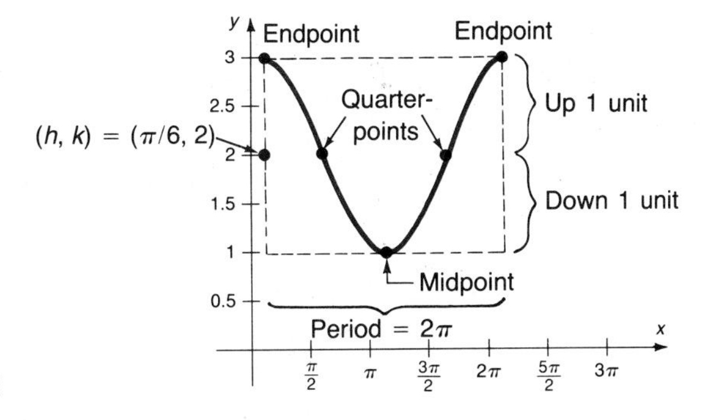

Figure 6.24 Graph of one period of $y - 2 = \cos(x - \frac{\pi}{6})$

■

EXAMPLE 4 Sketch one period of $y + 3 = \tan(x + \frac{\pi}{3})$.

Solution *Step 1* Frame the curve as shown in Figure 6.25, page 232. Notice that $(h, k) = (-\frac{\pi}{3}, -3)$, and remember that the period of the tangent is π.

Step 2 For the tangent curve, (h, k) is the midpoint of the frame. The endpoints, which are each a distance of one-half the period from the midpoint, determine the location of the asymptotes. The top and bottom of the frame are one unit from (h, k). Locate the quarter-points at the top and bottom of the frame as shown in Figure 6.25.

Step 3 Sketch one period of the curve as shown in Figure 6.25. Remember that the tangent curve is not contained entirely within the frame.

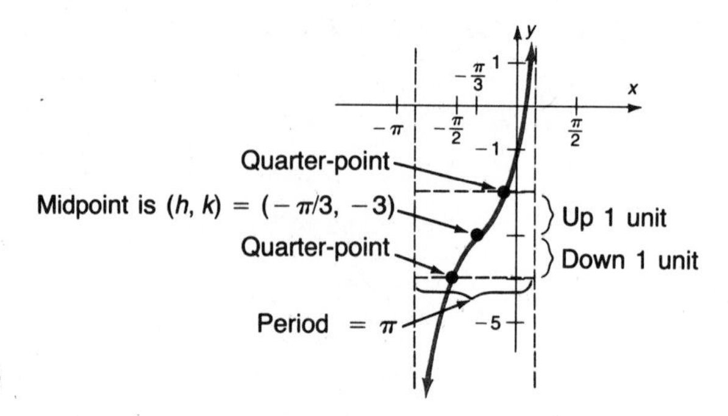

Figure 6.25 Graph of one period of $y + 3 = \tan(x + \frac{\pi}{3})$

■

We will now discuss two additional changes for the function defined by $y = f(x)$. The first, $y = af(x)$, changes the scale on the y axis; the second, $y = f(bx)$, changes the scale on the x axis.

For a function $y = af(x)$, it is clear that the y value is a times the corresponding value of $f(x)$, which means that $f(x)$ is stretched or shrunk in the y direction by the factor of a. For example, if $y = f(x) = \cos x$, then $y = 3f(x) = 3\cos x$ is the graph of $\cos x$ that has been stretched so that the high point is at 3 units and the low point is at negative 3 units. In general, given

$$y = af(x)$$

where f represents a trigonometric function, $2|a|$ gives the height of the frame for f. To graph $y = 3\cos x$, frame the cosine curve using 3 units rather than 1 (see Figure 6.26). For the sine and cosine curves, $|a|$ is the **amplitude** of the function. When $a = 1$, the amplitude is 1, so $y = \cos x$ and $y = \sin x$ are said to have amplitude 1.

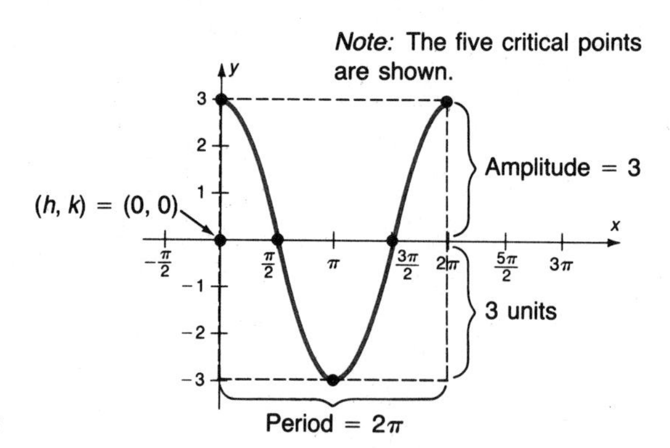

Figure 6.26 Graph of one period of $y = 3\cos x$

For a function $y = f(bx)$, $b > 0$, b affects the scale on the x axis. Recall that $y = \sin x$ has a period of 2π ($f(x) = \sin x$, so $b = 1$). A function $y = \sin 2x$ ($f(x) = \sin x$ and $f(2x) = \sin 2x$) must complete one period as $2x$ varies from zero to 2π. This means that one period is completed as x varies from zero to π. (Remember that for each value of x the result is doubled *before* we find the sine of that number.) In general, the period of $y = \sin bx$ is $\frac{2\pi}{b}$ and the period of $y = \cos bx$ is $\frac{2\pi}{b}$. Since the period of $y = \tan x$ is π, however, $y = \tan bx$ has a period of $\frac{\pi}{b}$. Therefore, when framing the curve, use $\frac{2\pi}{b}$ for the sine and cosine and $\frac{\pi}{b}$ for the tangent.

EXAMPLE 5 Graph one period of $y = \sin 2x$.

Solution The period is $\frac{2\pi}{2} = \pi$; thus the endpoints of the frame are $(0,0)$ and $(\pi, 0)$, as shown in Figure 6.27.

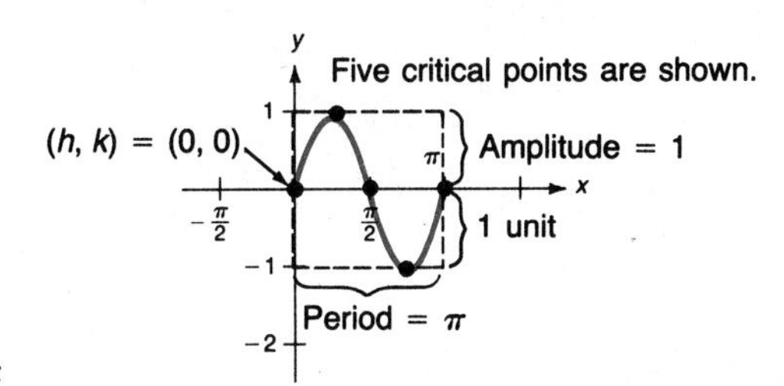

Figure 6.27 Graph of one period of $y = \sin 2x$

■

Summarizing all these results, we have the *general* cosine, sine, and tangent curves:

General Cosine, Sine, and Tangent Curves

$$y - k = a \cos b(x - h)$$
$$y - k = a \sin b(x - h)$$
$$y - k = a \tan b(x - h)$$

1. The origin has been translated or shifted to the point (h, k).
2. The height (width) of the frame is $2|a|$.
3. The length of the frame is $\frac{2\pi}{b}$ for the cosine and sine curves and $\frac{\pi}{b}$ for the tangent curve.
4. The curves are sketched by translating the origin to the point (h, k) and then framing the curve to complete the graph.

EXAMPLE 6 Graph $y + 1 = 2 \sin \frac{2}{3}(x - \frac{\pi}{2})$.

Solution Notice that $(h, k) = (\frac{\pi}{2}, -1)$ and that the amplitude is 2; the period is $2\pi/(\frac{2}{3}) = 3\pi$. Now plot (h, k) and frame the curve. Then plot the five critical points (two endpoints, the midpoint, and two quarter-points). Finally, after sketching one period, draw the other periods as in Figure 6.28.

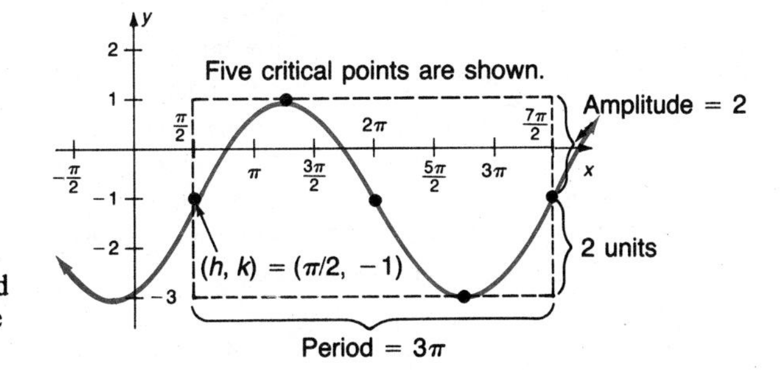

Figure 6.28 Graph of $y + 1 = 2 \sin \frac{2}{3}(x - \frac{\pi}{2})$; one period inside the frame is drawn first, and then the curve is extended outside the frame

■

EXAMPLE 7 Graph $y = 3\cos(2x + \frac{\pi}{2}) - 2$.

Solution Rewrite in standard form to obtain $y + 2 = 3\cos 2(x + \frac{\pi}{4})$
Notice that $(h, k) = (-\frac{\pi}{4}, -2)$; the amplitude is 3 and the period is $\frac{2\pi}{2} = \pi$. Plot (h, k) and frame the curve as shown in Figure 6.29.

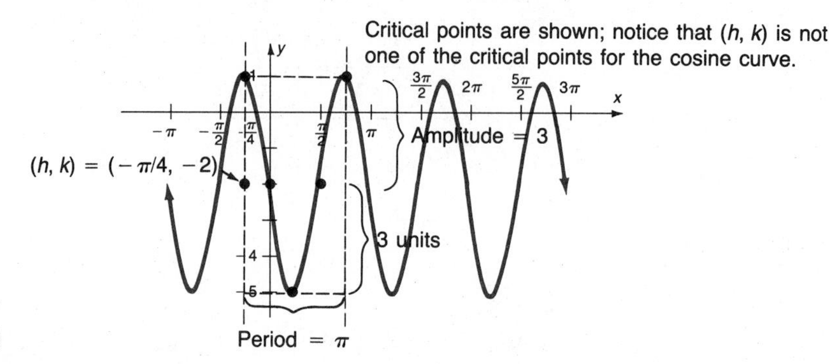

Figure 6.29 Graph of $y + 2 = 3\cos 2(x + \frac{\pi}{4})$ ■

EXAMPLE 8 Graph $y - 2 = 3\tan\frac{1}{2}(x - \frac{\pi}{3})$.

Solution Notice that $(h, k) = (\frac{\pi}{3}, 2)$, $a = 3$, and the period is $\pi/(\frac{1}{2}) = 2\pi$. Plot (h, k) and frame the curve as shown in Figure 6.30.

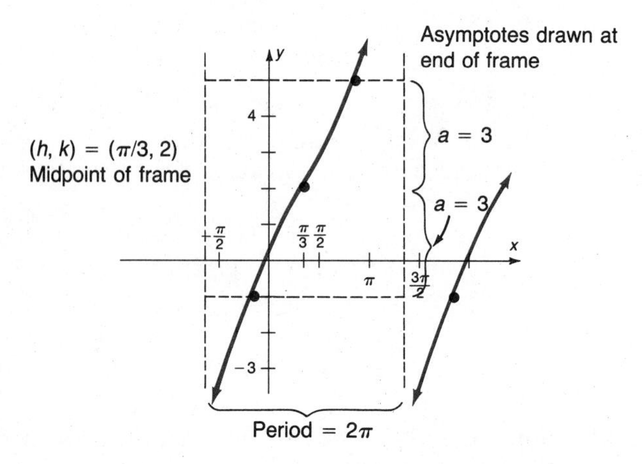

Figure 6.30 Graph of $y - 2 = 3\tan\frac{1}{2}(x - \frac{\pi}{3})$ ■

PROBLEM SET 6.4

A

1. Complete the following table of values for $y = \cos x$:

x = angle	$\frac{2\pi}{3}$	$\frac{3\pi}{4}$	$\frac{5\pi}{6}$	$\frac{7\pi}{6}$	$\frac{5\pi}{4}$	$\frac{4\pi}{3}$	$\frac{7\pi}{4}$	$\frac{11\pi}{6}$
Quadrant; sign of $\cos x$								
$y = \cos x$								
y (approximate)								

Use this table, along with other values if necessary, to plot $y = \cos x$.

2. Complete a table of values like the one in Problem 1 for $y = \tan x$. Use this table, along with other values if necessary, to plot $y = \tan x$.

3. We have emphasized the fact that the sine function can be considered as a function of a real number, x. Instead of using units of π, graph the sine function by finding additional values for the table shown at the top of the next page.

x = real number	0	1	2	3	4	5	6
$y = \sin x$	0	0.84	0.91	0.14	-0.76	-0.96	-0.28

Plot the ordered pairs, $(0, 0)$, $(1, 0.84)$, $(2, 0.91), \ldots$, and complete the graph of $y = \sin x$.

4. We have emphasized the fact that the cosine function can be considered as a function of a real number, x. Instead of using units of π, graph the cosine function by finding additional values for the table shown here:

x = real number	0	1	2	3	4	5	6
$y = \cos x$	1	0.54	-0.42	-0.99	-0.65	0.28	0.96

Plot the ordered pairs, $(0, 1)$, $(1, 0.54)$, $(2, -0.42), \ldots$, and complete the graph of $y = \cos x$.

Graph one period of each function given in Problems 5–19.

5. $y = \sin(x + \pi)$
6. $y = \cos(x + \frac{\pi}{2})$
7. $y = \tan(x + \frac{\pi}{3})$
8. $y = 3 \sin x$
9. $y = 2 \cos x$
10. $y = \frac{1}{2} \sin x$
11. $y = \sin 3x$
12. $y = \cos 2x$
13. $y = \cos \frac{1}{2}x$
14. $y = \tan(x - \frac{3\pi}{2})$
15. $y = \tan(x + \frac{\pi}{6})$
16. $y = \frac{1}{2} \sin x$
17. $y = \frac{1}{3} \tan x$
18. $y = 4 \tan x$
19. $y = 5 \sin x$

B

Graph one period of each function given in Problems 20–31.

20. $y - 2 = \sin(x - \frac{\pi}{2})$
21. $y + 1 = \cos(x + \frac{\pi}{3})$
22. $y - 3 = \tan(x + \frac{\pi}{6})$
23. $y - \frac{1}{2} = \frac{1}{2} \cos x$
24. $y - 1 = 2 \sin x$
25. $y + 2 = 3 \cos x$
26. $y - 1 = 2 \cos(x - \frac{\pi}{4})$
27. $y - 1 = \cos 2(x - \frac{\pi}{4})$
28. $y + 2 = 3 \sin(x + \frac{\pi}{6})$
29. $y + 2 = \sin 3(x + \frac{\pi}{6})$
30. $y = 1 + \tan 2(x - \frac{\pi}{4})$
31. $y + 2 = \tan(x - \frac{\pi}{4})$

Graph the curves given in Problems 32–46.

32. $y = \sin(4x + \pi)$
33. $y = \sin(3x + \pi)$
34. $y = \tan(2x - \frac{\pi}{2})$
35. $y = \tan(\frac{x}{2} + \frac{\pi}{3})$
36. $y = \frac{1}{2} \cos(x + \frac{\pi}{6})$
37. $y = \cos(\frac{1}{2}x + \frac{\pi}{12})$
38. $y = 3 \cos(3x + 2\pi) - 2$
39. $y = 4 \sin(\frac{1}{2}x + 2)$
40. $y = \sqrt{2} \cos(x - \sqrt{2}) - 1$
41. $y = \sqrt{3} \sin(\frac{1}{3}x - \sqrt{\frac{1}{3}})$
42. $y = 2 \sin(2\pi x)$
43. $y = 3 \cos(3\pi x)$
44. $y = 4 \tan(\frac{\pi x}{5})$
45. $y + 2 = \frac{1}{2} \cos(\pi x + 2\pi)$
46. $y - 3 = 3 \cos(2\pi x + 4)$

Use the technique of plotting points to graph the functions in Problems 47–55.

47. $y = \sec x$
48. $y = 2 \sec x$
49. $y = \csc x$
50. $y = \cot x$
51. $y = \csc 2x$
52. $y = 2 \cot x - 1$
53. $y = \sin x + \cos x$
54. $y = \sin 2x + \cos x$
55. $y = 2 \cos x + \sin 2x$

So far we have limited ourselves to $a > 0$. If $a < 0$, the curve is reflected through the x axis. Graph the curves in Problems 56–61.

56. $y = -\sin x$
57. $y = -\cos x$
58. $y = -\tan x$
59. $y = -3 \sin x$
60. $y = -2 \cos x$
61. $y = -\sin 3x$

C

62. ***Electrical Engineering*** The current I (in amperes) in a certain circuit is given by

$$I = 60 \cos(120\pi t - \pi)$$

where t is time in seconds. Graph this equation for $0 \le t \le \frac{1}{30}$.

63. ***Engineering*** Suppose a point P on a waterwheel with a 30-ft radius is d units from the water as shown in Figure 6.31. If it turns at 6 revolutions per minute, then

$$d = 29 + 30 \cos(\tfrac{\pi}{5}t - \pi)$$

Graph this equation for $0 \le t \le 20$.

Figure 6.31

64. ***Space Science*** The distance (in kilometers) that a certain satellite is north or south of the equator is given by

$$y = 3000 \cos(\tfrac{\pi}{60}t + \tfrac{\pi}{5})$$

where t is the number of minutes that have elapsed since liftoff.

a. Graph the equation for $0 \le t \le 120$.
b. What is the farthest distance that the satellite ever reaches north of the equator?
c. How long does it take to complete one orbit?

65. Plot points to graph $y = (\sin x)/x$.

6.5 INVERSE TRIGONOMETRIC FUNCTIONS

In Section 2.6 the notion of inverse functions was introduced. In this section, that idea is applied to the trigonometric functions. Recall that a function f must be one-to-one in order to have an inverse function. This means that the trigonometric functions do not have inverse functions. We can, however, restrict the domains of the trigonometric functions so that they become one-to-one. We will illustrate with the sine function. Figure 6.32 shows the graph of $y = \sin x$.

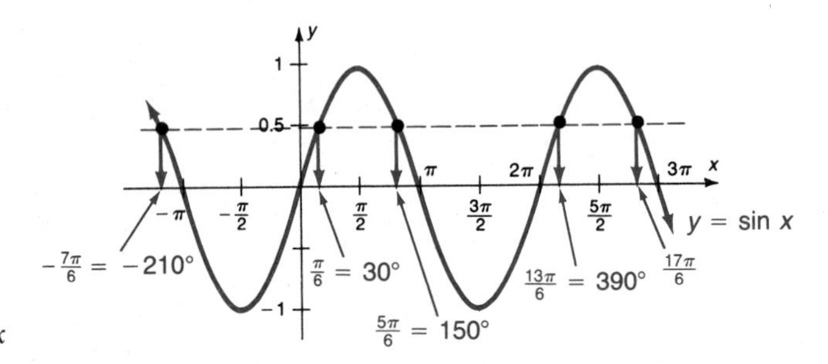

Figure 6.32 Graph of $y = \sin x$

Notice that if $y = \sin x$, then each x, say $\frac{5\pi}{6}$, is associated with exactly one y value, $\frac{1}{2}$ in this case. But it is *not* one-to-one because for a given y value, say $\frac{1}{2}$, there are infinitely many x values as shown in Figure 6.32.

Define a new function related to $y = \sin x$ but having the property that it is one-to-one. We do this simply by restricting its domain to the first and fourth quadrants. That is, define

$$y = \operatorname{Sin} x \text{ so that } x \text{ is on the interval } \left[-\frac{\pi}{2}, \frac{\pi}{2}\right]$$

WARNING: The capital S in $y = \operatorname{Sin} x$ is important; $\operatorname{Sin} x \neq \sin x$ because of their different domains. This function is shown in color in Figure 6.33a; notice that this function *is* one-to-one.

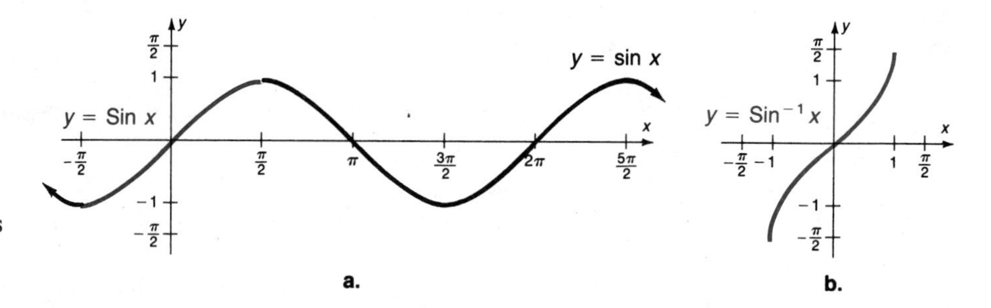

Figure 6.33 Comparison graphs of $y = \sin x$, $y = \operatorname{Sin} x$, and $y = \operatorname{Sin}^{-1} x$

Now we can define the inverse by using the notation introduced in Section 2.6. Remember: The inverse of $y = \operatorname{Sin} x$ is found by interchanging the x and y components to $x = \operatorname{Sin} y$.

Inverse Sine

The **inverse sine function**, denoted by $\text{Sin}^{-1} x$ or by $\text{Arcsin}\, x$ and called "arcsine of x," is defined by

$$y = \text{Sin}^{-1} x \quad \text{if and only if} \quad x = \sin y$$

where $-1 \le x \le 1$ and $-\frac{\pi}{2} \le y \le \frac{\pi}{2}$. See Figure 6.33b.

EXAMPLE 1 Find $\text{Sin}^{-1}(\frac{1}{2}\sqrt{3})$.

Solution Let $\theta = \text{Sin}^{-1}(\frac{1}{2}\sqrt{3})$. (Remember: An inverse sine is an angle, so we denote it by θ.) Find the angle or real number θ with sine equal to $\frac{1}{2}\sqrt{3}$ so that $-\frac{\pi}{2} \le \theta \le \frac{\pi}{2}$. From the memorized table of exact values you know that $\sin(\frac{\pi}{3}) = \frac{1}{2}\sqrt{3}$. And since $\frac{\pi}{3}$ is between $-\frac{\pi}{2}$ and $\frac{\pi}{2}$, you have $\text{Sin}^{-1}(\frac{1}{2}\sqrt{3}) = \frac{\pi}{3}$. ■

EXAMPLE 2 Find $\text{Sin}^{-1}(-\frac{1}{2}\sqrt{3})$.

Solution You will find it easier to work with reference angles when finding inverse trigonometric functions. That is, because the table of exact values was memorized for the first quadrant, work with the reference angle. Let

$$\theta' = \text{Sin}^{-1}(\tfrac{1}{2}\sqrt{3})$$

reference angle ↑ (points to θ'); ↑ absolute value of the given number (points to $\frac{1}{2}\sqrt{3}$)

$$\theta' = \frac{\pi}{3} \qquad \text{From Example 1}$$

Now place θ in the appropriate quadrant. The sine is negative in both the third and fourth quadrants, but you choose the fourth quadrant because of the restrictions on $y = \text{Sin}\, x$. Thus θ is the fourth-quadrant angle with its reference angle $\theta' = \frac{\pi}{3}$. Therefore $\text{Sin}^{-1}(-\frac{1}{2}\sqrt{3}) = -\frac{\pi}{3}$. ■

The other trigonometric functions are handled similarly:

Given function	Inverse	Other notations for inverse	
$y = \text{Cos}\, x$	$x = \text{Cos}\, y$	$y = \text{Cos}^{-1} x$	$y = \text{Arccos}\, x$
$y = \text{Sin}\, x$	$x = \text{Sin}\, y$	$y = \text{Sin}^{-1} x$	$y = \text{Arcsin}\, x$
$y = \text{Tan}\, x$	$x = \text{Tan}\, y$	$y = \text{Tan}^{-1} x$	$y = \text{Arctan}\, x$
$y = \text{Sec}\, x$	$x = \text{Sec}\, y$	$y = \text{Sec}^{-1} x$	$y = \text{Arcsec}\, x$
$y = \text{Csc}\, x$	$x = \text{Csc}\, y$	$y = \text{Csc}^{-1} x$	$y = \text{Arccsc}\, x$
$y = \text{Cot}\, x$	$x = \text{Cot}\, y$	$y = \text{Cot}^{-1} x$	$y = \text{Arccot}\, x$

These are the same. (Inverse, and both other notations for inverse)

Consider the graphs in Figure 6.34. We have to restrict each trigonometric function so it is one-to-one, but we also want to include all possible values in the range of the original function. For the sine curve, x was restricted so that $-\frac{\pi}{2} \le x \le \frac{\pi}{2}$. Then the inverse is the function

$$y = \text{Sin}^{-1} x \qquad \text{where } -\tfrac{\pi}{2} \le y \le \tfrac{\pi}{2}$$

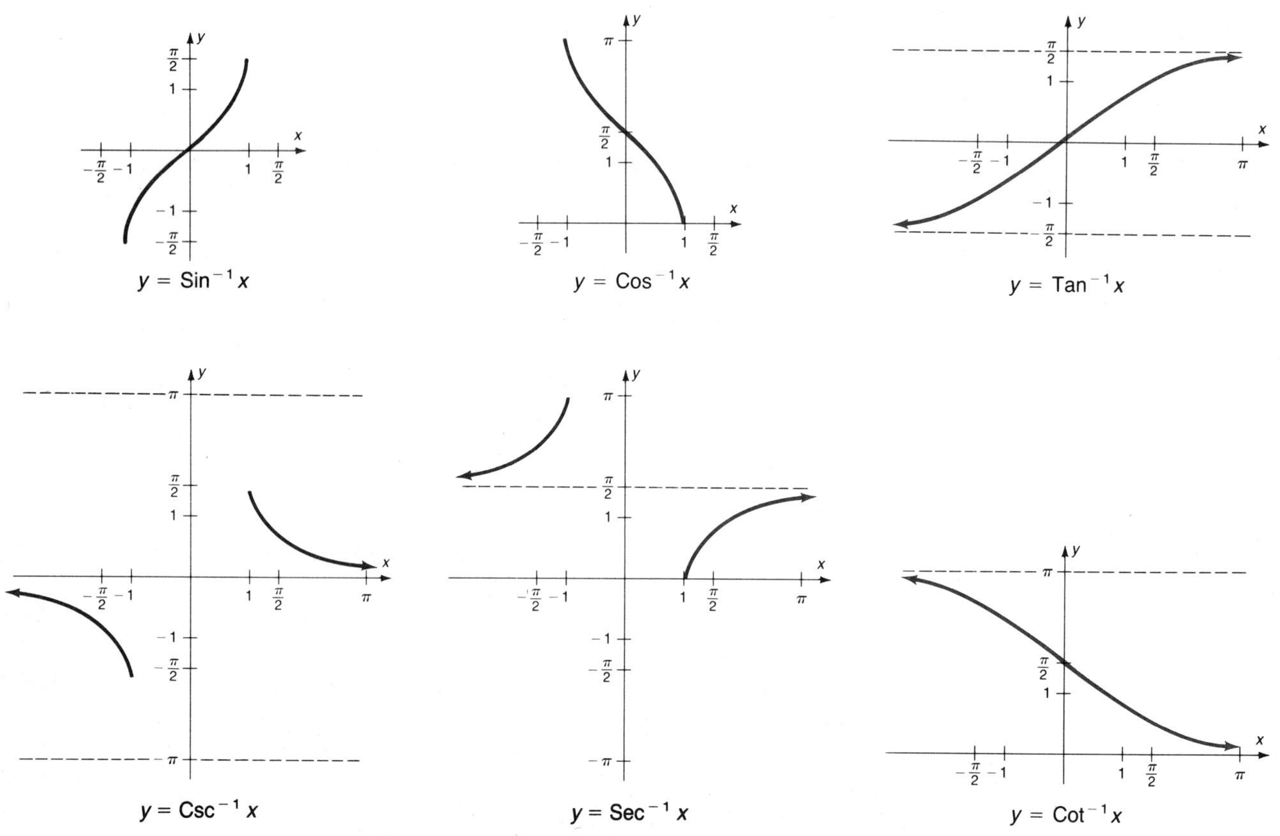

Figure 6.34 Graphs of inverse trigonometric functions

The same restrictions (leaving out the values $-\frac{\pi}{2}$ and $\frac{\pi}{2}$) apply for the tangent and arctangent curves. For the cosine function, however, notice that by restricting x to the same interval you obtain only positive values for $y = \cos x$. Thus, to include the entire range of the cosine curve, x can be restricted so that $0 \le x \le \pi$. Then the inverse is the function

$$y = \text{Cos}^{-1} x \qquad \text{where } 0 \le y \le \pi$$

The cotangent function is restricted in almost the same way, and the results are summarized in the following definitions.

The following box lists the domain, range, and quadrants for the range of the six inverse trigonometric functions.

Inverse Trigonometric Functions

Inverse function	Domain	Range	Quadrants of the range
$y = \text{Arcsin}\, x$ or $y = \text{Sin}^{-1} x$	$-1 \le x \le 1$	$-\frac{\pi}{2} \le y \le \frac{\pi}{2}$	I and IV
$y = \text{Arccos}\, x$ or $y = \text{Cos}^{-1} x$	$-1 \le x \le 1$	$0 \le y \le \pi$	I and II
$y = \text{Arctan}\, x$ or $y = \text{Tan}^{-1} x$	All reals	$-\frac{\pi}{2} < y < \frac{\pi}{2}$	I and IV
$y = \text{Arccot}\, x$ or $y = \text{Cot}^{-1} x$	All reals	$0 < y < \pi$	I and II
$y = \text{Arcsec}\, x$ or $y = \text{Sec}^{-1} x$	$x \ge 1$ or $x \le -1$	$0 \le y \le \pi$, $y \ne \frac{\pi}{2}$	I and II
$y = \text{Arccsc}\, x$ or $y = \text{Csc}^{-1} x$	$x \ge 1$ or $x \le -1$	$-\frac{\pi}{2} \le y \le \frac{\pi}{2}$, $y \ne 0$	I and IV

EXAMPLE 3 Find Arctan 1.

Solution Let Arctan 1 $= \theta$. You are looking for an angle θ with tangent equal to 1. Since this is an exact value, you know that $\theta = \frac{\pi}{4}$ **or 45°**. ■

EXAMPLE 4 Find $\text{Arccot}(-\sqrt{3})$.

Solution Let $\text{Arccot}(-\sqrt{3}) = \theta$. Find θ so that $\cot\theta = -\sqrt{3}$; the reference angle is 30°, and the cotangent is negative in Quadrants II and IV. Since Arccot x is defined in Quadrant II but not in Quadrant IV, $\text{Arccot}(-\sqrt{3}) = \frac{5\pi}{6}$ **or 150°**. ■

EXAMPLE 5 Find Arcsin(−0.4695).

Solution Let Arcsin(−0.4695) $= \theta$. Find θ so that $\sin\theta = -0.4695$; from Table C.III we find that the reference angle is 28°. Since Arcsin x is defined for values between −90° and 90°, Arcsin(−0.4695) = **−28°**. If you want to solve this problem on your calculator, check the algorithm used on your own calculator. You may have an [arc] button instead of an [inv] button, but in the text we will indicate inverse by showing an [inv] button. For this example:

PRESS: [.4695] [+/−] [inv] [sin]

The display: **−28.00184535** is the decimal representation of the angle in degrees (if the calculator is set to degrees) or the angle in radians (if it is set to radians). ■

EXAMPLE 6 Find Arccot(−2.747).

Solution Let Arccot(−2.747) $= \theta$. You can use Table C.III to find the reference angle 20°. Since $0 < \cot^{-1} x < \pi$, you must place this angle in Quadrant II, so $\theta =$ **160°**. Because calculators have no cotangent function, note that if $\cot y = x$ then $\tan y = \frac{1}{x}$. Thus

$$y = \cot^{-1} x \text{ and } y = \tan^{-1}\left(\frac{1}{x}\right) \qquad \text{so} \qquad \cot^{-1} x = \tan^{-1}\left(\frac{1}{x}\right)$$

This tells you that to find the inverse cotangent on a calculator you must first take the reciprocal of the given value and then complete the problem.

Another way of looking at this is to consider the keys pressed on a calculator to find a cotangent. For example, find cot 30°:

PRESS: [30] [tan] [1/x] DISPLAY: 1.73205081

Thus to find the arccot 1.73205081, just reverse the steps:

PRESS: [1/x] [inv] [tan] DISPLAY: 30

For this example:

PRESS: [2.747] [1/x] [inv] [tan] DISPLAY: 20.00320032

This result is correct for the inverse tangent since it is in Quadrant IV. However, the inverse cotangent must be in Quadrant I or II, so it is necessary to add 180° (or π if working in radians) for the proper placement of the angle.

PRESS: [2.747] [+/−] [1/x] [inv] [tan] [+] [180] [=]

DISPLAY: 159.9967997

Thus, the answer is approximately **160°**. ■

We can now summarize the calculator steps illustrated in Examples 5 and 6.

Inverse function	**Calculator** Enter the value of x, then press:
$y = \text{Arccos } x$	[inv] [cos]
$y = \text{Arcsin } x$	[inv] [sin]
$y = \text{Arctan } x$	[inv] [tan]
$y = \text{Arcsec } x$	[1/x] [inv] [cos]
$y = \text{Arccsc } x$	[1/x] [inv] [sin]
$y = \text{Arccot } x$	If $x > 0$: [1/x] [inv] [tan] If $x < 0$: [1/x] [inv] [tan] [+] [π] [=]*

* Use 180 instead of π if working in degrees.

EXAMPLE 7 Find θ (in radians) using a calculator.

a. $\text{Arcsec}(-3) = \theta$ **b.** $\text{Arccsc } 7.5 = \theta$

c. $\text{Arccot } 2.4747 = \theta$ **d.** $\text{Arccot}(-4.852) = \theta$

Solution Once again, we remind you to make sure your calculator is in the proper mode. These are all in radian mode.

a. Arcsec$(-3) = \theta$.

PRESS: [3] [+/−] [1/x] [inv] [cos] DISPLAY: 1.910633236

b. Arccsc $7.5 = \theta$.

PRESS: [7.5] [1/x] [inv] [sin] DISPLAY: .1337315894

c. Arccot $2.4747 = \theta$. Since 2.4747 is positive,

PRESS: [2.4747] [1/x] [inv] [tan] DISPLAY: .3840267299

d. Arccot(-4.852). Since -4.852 is negative,

PRESS: [4.852] [+/−] [1/x] [inv] [tan] [+] [π] [=]

DISPLAY: 2.938338095 ■

A final word of caution is in order regarding the inverse trigonometric functions, especially when you are using a calculator. Recall from Section 2.6 that

$$(f^{-1} \circ f)(x) = (f \circ f^{-1})(x) = x$$

In the context of this section, this means

$$\begin{aligned} \text{Cos}^{-1}(\text{Cos}\, x) &= x \quad \text{and} \quad \text{Cos}(\text{Cos}^{-1} x) = x \\ \text{Sin}^{-1}(\text{Sin}\, x) &= x \quad \text{and} \quad \text{Sin}(\text{Sin}^{-1} x) = x \\ \text{Tan}^{-1}(\text{Tan}\, x) &= x \quad \text{and} \quad \text{Tan}(\text{Tan}^{-1} x) = x \\ \text{Cot}^{-1}(\text{Cot}\, x) &= x \quad \text{and} \quad \text{Cot}(\text{Cot}^{-1} x) = x \end{aligned}$$

However, you should not forget the appropriate restrictions that are also part of the definition. Consider Examples 8 to 13.

EXAMPLE 8 $\text{Cos}^{-1}(\cos 2.2) = \mathbf{2.2}$

Calculator check: [2.2] [cos] [inv] [cos] DISPLAY: 2.2

(Some calculators may show a display such as 2.1999997; this should be considered a proper check.) ■

EXAMPLE 9 $\text{Sin}^{-1}(\sin 2.2) \approx \mathbf{0.9}$ (Not 2.2)

The reason for this answer is that Sin x is defined in Quadrants I and IV, whereas 2.2 is not an angle in these quadrants.

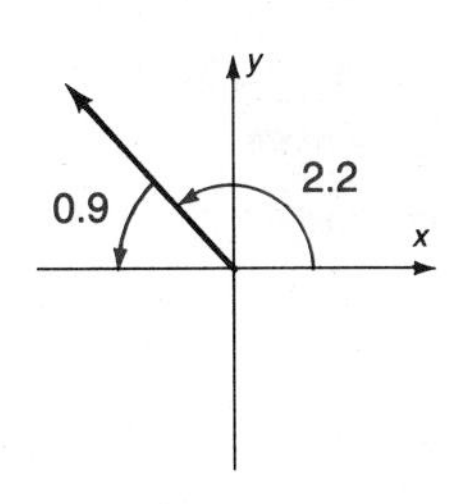

Since $\sin 2.2 \approx \sin 0.9$ (by the reduction principle) and since 0.9 is an angle in Quadrants I or IV,

$$\begin{aligned} \text{Sin}^{-1}(\sin 2.2) &\approx \text{Sin}^{-1}(\sin 0.9) \\ &= 0.9 \end{aligned}$$

Calculator check: [2.2] [sin] [inv] [sin] DISPLAY: .94159265 ■

EXAMPLE 10 $\text{Arccos}(\cos 4) \approx \mathbf{2.3}$

Since the angle 4 radians is in Quadrant III, cos 4 is negative. The Arccosine of a negative angle will be a Quadrant II angle. This is the Quadrant II angle having the

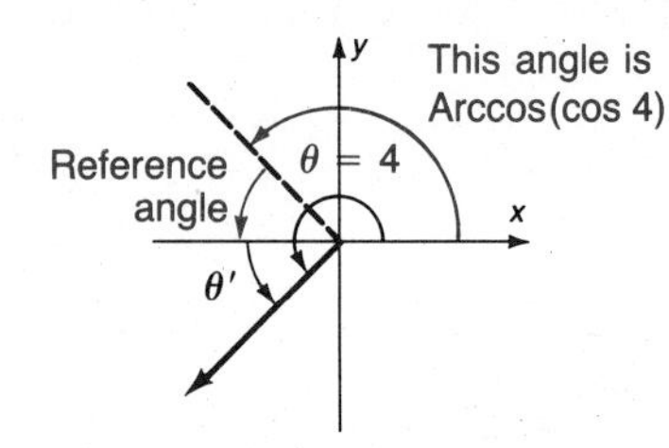

same reference angle as 4. The reference angle is $4 - \pi \approx 0.86$. In Quadrant II, π − reference angle ≈ 2.28. Therefore

$$\text{Arccos}(\cos 4) \approx 2.28$$

By calculator: [4] [cos] [inv] [cos] DISPLAY: 2.283185307 ■

EXAMPLE 11 $\sin(\text{Sin}^{-1} 0.463) = \mathbf{0.463}$ ■

EXAMPLE 12 $\sin(\text{Sin}^{-1} 2.463)$ is **not defined** since 2.463 is not between -1 and $+1$. ■

EXAMPLE 13 $\tan(\text{Tan}^{-1} 2.463) = \mathbf{2.463}$ ■

PROBLEM SET 6.5

A

In Problems 1–7, obtain the given angle from memory.

1. a. Arcsin 0 b. $\text{Tan}^{-1}(\frac{\sqrt{3}}{3})$ c. Arccot $\sqrt{3}$ d. Arccos 1
2. a. $\text{Cos}^{-1}(\frac{\sqrt{3}}{2})$ b. Arcsin $\frac{1}{2}$ c. $\text{Tan}^{-1} 1$ d. $\text{Sin}^{-1} 1$
3. a. Arctan $\sqrt{3}$ b. $\text{Cos}^{-1}(\frac{\sqrt{2}}{2})$ c. $\text{Arcsin}(\frac{1}{2}\sqrt{2})$ d. Arccot 1
4. a. Arcsin(-1) b. $\text{Cot}^{-1}(-1)$ c. $\text{Arcsin}(-\frac{\sqrt{3}}{2})$ d. $\text{Cos}^{-1}(-1)$
5. a. $\text{Cot}^{-1}(-\sqrt{3})$ b. Arctan(-1) c. $\text{Sin}^{-1}(-\frac{1}{2}\sqrt{2})$ d. $\text{Cos}^{-1}(-\frac{1}{2})$
6. a. $\text{Arccos}(-\frac{\sqrt{2}}{2})$ b. $\text{Cot}^{-1}(-\frac{\sqrt{3}}{3})$ c. $\text{Sin}^{-1}(-\frac{1}{2})$ d. $\text{Arctan}(-\frac{\sqrt{3}}{3})$
7. a. $\text{Tan}^{-1} 0$ b. $\text{Arccot}(\frac{\sqrt{3}}{3})$ c. Arccos $\frac{1}{2}$ d. $\text{Sin}^{-1}(\frac{\sqrt{3}}{2})$

Use Table C.III or a calculator to find the values (in radians correct to the nearest hundredth) for the functions in Problems 8–25.

8. Arcsin 0.20846
9. $\text{Cos}^{-1} 0.83646$
10. Arctan 1.1156
11. $\text{Cot}^{-1}(-0.08097)$
12. $\text{Tan}^{-1}(-3.7712)$
13. Arccos(-0.94604)
14. $\text{Sin}^{-1} 0.75$
15. Arccos 0.25
16. Arctan 2
17. $\text{Tan}^{-1} 1.489$
18. $\text{Cot}^{-1} 3.451$
19. $\text{Sec}^{-1} 4.315$
20. $\text{Csc}^{-1} 5.791$
21. Arccsc 2.985
22. Arccot(-3)
23. Arccot(-4)
24. Arctan(-2)
25. Arcsec(-5)

Use Table C.III or a calculator to find the values (in degrees) given in Problems 26–43.

26. $\text{Sin}^{-1} 0.3584$
27. $\text{Cos}^{-1} 0.3584$
28. Arccos 0.9455
29. $\text{Sin}^{-1}(-0.4695)$
30. $\text{Tan}^{-1} 2.050$
31. Arctan 1.036
32. $\text{Tan}^{-1}(-3.732)$
33. $\text{Cot}^{-1} 0.0875$
34. Arcsin(-0.9135)
35. Arccot 0.7265
36. $\text{Cot}^{-1}(-0.3249)$
37. Arccot(-1.235)
38. $\text{Csc}^{-1} 2.816$
39. Arccot(-1)
40. Arccsc 3.945
41. Arccot(-2)
42. Arcsec(-6)
43. Arctan(-3)

B

Simplify the expressions in Problems 44–58.

44. cot(Arccot 1)
45. $\text{Arccos}[\cos(\frac{\pi}{6})]$
46. $\sin(\text{Arcsin } \frac{1}{3})$
47. $\text{Tan}^{-1}[\tan(\frac{\pi}{15})]$
48. $\cos(\text{Arccos } \frac{2}{3})$
49. $\text{Arcsin}[\sin(\frac{2\pi}{15})]$
50. Arccot(cot 35°)
51. tan(Arctan 0.4163)
52. Arcsin(sin 4)
53. Arccos(cos 5)
54. sin(Arcsin 0.7568)
55. cos(Arccos 0.2836)
56. tan(Arctan 0.2910)
57. Arctan(tan 2.5)
58. Arctan(Tan 2.5)

In Problems 59–61 graph the given pair of curves on the same axes.

59. $y = \text{Sin } x$; $y = \text{Sin}^{-1} x$
60. $y = \text{Cos } x$; $y = \text{Cos}^{-1} x$
61. $y = \text{Tan } x$; $y = \text{Tan}^{-1} x$

C

Graph the curves given in Problems 62–67.

62. $y + 2 = \text{Arctan } x$

63. $y - 1 = \text{Arcsin } x$

64. $y = 2\,\text{Cos}^{-1} x$

65. $y = 3\,\text{Sin}^{-1} x$

66. $y = \text{Arcsin}(x - 2)$

67. $y = \text{Arcsin}(x + 1)$

6.6 CHAPTER 6 SUMMARY

The material of this chapter is reviewed in the following list of objectives. After each objective there are some practice questions. For a sample test, select the first question of each set and check your answers with the answer section. For a sample test without answers, use the second question of each set. Additional practice is given by the other questions in each set. If you are having trouble with a particular type of problem, look back to that section for extra help.

6.1 ANGLES AND THE UNIT CIRCLE

Objective 1 Know the definition and equation of a unit circle. Know the definition and notation of an angle, including positive and negative angles. Know the Greek letters. Know what it means for an angle to be in standard position.

1. Name the Greek letter used for each angle: **a.** λ **b.** θ **c.** ϕ **d.** α **e.** β

Fill in the blanks.

2. A unit circle is ______________________________.

3. The equation of a unit circle is ______________________________.

4. An angle is in standard position if ______________________________.

Objective 2 Find angles coterminal with a given angle. Find the positive angle coterminal with the given angle and less than one revolution.

5. $-215°$ **6.** $\frac{11\pi}{3}$ **7.** $-\frac{5\pi}{6}$ **8.** $1000°$

Objective 3 Be familiar with the degree measure of an angle and be able to approximate the angle associated with a given degree measure without using any measuring devices. Draw the indicated angles.

9. $180°$ **10.** $120°$ **11.** $-30°$ **12.** $135°$

Objective 4 Be familiar with the radian measure of an angle and be able to approximate the angle associated with a given radian measure without using any measuring devices. Draw the indicated angles.

13. $\frac{\pi}{3}$ **14.** $\frac{\pi}{4}$ **15.** $\frac{5\pi}{6}$ **16.** 2

Objective 5 Change from radian to degree measure; know the commonly used degree and radian measure equivalences.

17. $\frac{3\pi}{2}$ **18.** 2 **19.** $-\frac{7\pi}{4}$ **20.** $\frac{5\pi}{6}$

Objective 6 Change from degree measure to radian measure; know the commonly used radian and degree measure equivalences. Use exact values when possible.

21. $300°$ **22.** $-45°$ **23.** $54°$ **24.** $-210°$

Objective 7 Know the arc length formula and be able to apply it.

25. The arc length formula is _______________ where s is the arc length, r is the _______________ and θ is _______________.

26. If the radius is 1 and the angle is 1, then what is the arc length?

27. If the minute hand on a clock is 15 cm long, how far does the tip move in 10 minutes?
28. A curve on a highway is laid out as the arc of a circle of radius 500 m. If the curve subtends a central angle of 18°, what is the distance around this section of road? Give the exact answer and an answer rounded off to the nearest meter.

Objective 8 Find the reference angle θ for a given angle θ.

29. 300° **30.** $\frac{11\pi}{3}$ **31.** -4 **32.** $-215°$

6.2 TRIGONOMETRIC FUNCTIONS

Objective 9 Know the unit circle definition of the trigonometric functions. Fill in the blanks.

33. Let θ be ____________. Then the trigonometric functions are defined as follows:
34. ____________ ____________
35. ____________ ____________
36. ____________ ____________

Objective 10 Know the signs of the six trigonometric functions in each of the four quadrants. Name the function(s) which is (are) positive in the given quadrant.

37. I **38.** II **39.** III **40.** IV

Objective 11 Know that $\tan\theta = \dfrac{\sin\theta}{\cos\theta}$*; know the reciprocal relationships.*

41. If $\sin\theta = \frac{4}{5}$ and $\cos\theta = \frac{3}{5}$, then what is $\tan\theta$?
42. What is the reciprocal function of tangent?
43. What is the reciprocal function of cosine?
44. What is the reciprocal function of sine?

Objective 12 Be able to evaluate the trigonometric functions using tables or a calculator.

45. sec 23.4° **46.** cot 2.5 **47.** csc 43.28° **48.** sin 7

Objective 13 Know the ratio definition of the trigonometric functions.

49. If ____________________________________, then
50. ____________ ____________
51. ____________ ____________
52. ____________ ____________

Objective 14 Use the definition of the trigonometric functions to approximate their values for a given angle or for an angle passing through a given point. Assume that the terminal side passes through the given point.

53. $(5, -12)$ **54.** $(3, -4)$ **55.** $(-5, 2)$ **56.** $(4, 5)$

6.3 VALUES OF THE TRIGONOMETRIC FUNCTIONS

Objective 15 Know and be able to derive the table of exact values. Complete the table.

	Function	0	$\frac{\pi}{6}$	$\frac{\pi}{4}$	$\frac{\pi}{3}$	$\frac{\pi}{2}$	π	$\frac{3\pi}{2}$
57.	$\cos\theta$	a.	b.	c.	d.	e.	f.	g.
58.	$\csc\theta$	a.	b.	c.	d.	e.	f.	g.
59.	$\tan\theta$	a.	b.	c.	d.	e.	f.	g.
60.	$\sin\theta$	a.	b.	c.	d.	e.	f.	g.

Objective 16 Use the reduction principle, along with the table of exact values, to evaluate certain trigonometric functions.

61. $\cos(-\frac{5\pi}{3})$ **62.** $\sin(\frac{11\pi}{6})$ **63.** $\tan 135°$ **64.** $\cos(-210°)$

Objective 17 Use the reduction principle, along with tables, to approximate values of the trigonometric functions.

65. $\csc 43.28°$ **66.** $\sin 9$ **67.** $\sec 23.4°$ **68.** $\cot 2.5$

6.4 GRAPHS OF THE TRIGONOMETRIC FUNCTIONS

Objective 18 Graph the trigonometric functions, or variations, by plotting points.

69. $y = 2\cot\theta$ **70.** $y = 2\sec\theta$
71. $y = \frac{1}{2}\csc\theta$ **72.** $y = 2\cos\theta + \sin 2\theta$

Objective 19 Sketch $y = \cos x$, $y = \sin x$, and $y = \tan x$ from memory; know their periods and amplitudes.

73. $y = \sin x$ **74.** $y = \cos x$ **75.** $y = \tan x$

76. Fill in the blanks:

Function	Period	Amplitude
$\sin x$	a. ______	b. ______
$\cos x$	c. ______	d. ______
$\tan x$	e. ______	f. ______

Objective 20 Graph the general cosine, sine, and tangent curves.

77. $y = 2\cos\frac{2}{3}x$ **78.** $y = \cos(x + \frac{\pi}{4})$
79. $y - 2 = \sin(x - \frac{\pi}{6})$ **80.** $y = \tan(x - \frac{\pi}{3}) - 2$

6.5 INVERSE TRIGONOMETRIC FUNCTIONS

Objective 21 Know the definition of the inverse cosine, sine, tangent, and cotangent functions, especially the range of each. Fill in the blanks.

	Inverse Function	*Domain*	*Range*
81.	$y = \text{Arctan}\, x$	All reals	______
82.	$y = \text{Cos}^{-1} x$	$-1 \le x \le 1$	______
83.	$y = \text{Cot}^{-1} x$	All reals	______
84.	$y = \text{Arcsin}\, x$	$-1 \le x \le 1$	______

Objective 22 Evaluate inverse cosine, sine, tangent, and cotangent functions using exact values.

85. $\text{Arcsin}\,\frac{1}{2}$ **86.** $\text{Cot}^{-1}(\frac{1}{3}\sqrt{3})$ **87.** $\text{Arccos}(\frac{\sqrt{3}}{2})$ **88.** $\text{Tan}^{-1}(-\sqrt{3})$

Objective 23 Evaluate inverse cosine, sine, tangent, and cotangent functions using tables or a calculator. Answer in radians.

89. Arcsin 0.3140 **90.** Arccos(−0.6494) **91.** Arctan 3.271 **92.** Arccot 2

Objective 24 Graph the inverse cosine, sine, and tangent functions.

93. $y = \text{Sin}^{-1} x$ **94.** $y = \text{Arccos}\, x$ **95.** $y = \text{Arctan}\, x$ **96.** $y - 1 = \text{Sin}^{-1} x$

Objective 25 Simplify a function and its inverse function.

97. Arccos(cos 1) **98.** Arcsin(sin 1) **99.** Arcsin(cos 2) **100.** Arcsin(sin 2)

Nicholas Copernicus (1473–1543)

There is perhaps nothing which so occupies, as it were, the middle position of mathematics, as trigonometry.

J. F. Herbart
Idee eines ABC der Anshauung

HISTORICAL NOTE

Trigonometry was invented by Ptolemy, known as Claudius Ptolemy, who worked with Hipparchus and Menelaus. Their goal was to build a quantitative astronomy that could be used to predict the paths and positions of the heavenly bodies and to aid in telling time, calendar reckoning, navigating, and studying geography. In his book *Syntaxis Mathematica*, Ptolemy derived many of the trigonometric identities of this chapter. Although his purposes were more related to solving triangles (Chapter 8), today a major focus of trigonometry is on the relationships of the functions themselves, and it is important to be able to change the form of a trigonometric expression in many different ways.

Nicholas Copernicus is probably best known as the astronomer who revolutionized the world with his heliocentric theory of the universe, but in his book *De revolutionibus orbium coelestium* he also developed a substantial amount of trigonometry. This book was published in the year of his death; as a matter of fact, the first copy off the press was rushed to him as he lay on his deathbed. It was on Copernicus' work that his student Rheticus based his ideas, which soon brought trigonometry into full use. In a two-volume work, *Opus palatinum de triangulis*, Rheticus used and calculated elaborate tables for all six trigonometric functions.

The transition of trigonometry from the Renaissance to the modern world is due to a Frenchman, François Viète (1540–1603). He was a lawyer, not a mathematician, and served as a member of the king's council under Henry III and Henry IV. Viète spent his leisure time working on mathematics. He was the first to link trigonometry to the solution of algebraic problems. In 1593, a Belgian ambassador to the court of Henry IV boasted that France had no mathematician capable of solving the equation

$$x^{45} - 45x^{43} + 945x^{41} - \cdots - 3795x^3 + 45x = K$$

Viète used trigonometry to find that this equation results when expressing $K = \sin 45\theta$ in terms of $x = 2\sin\theta$. He was then able to find the positive roots of this equation. It was during this period that the word *trigonometry* was first used.

7

Trigonometric Equations and Identities

CONTENTS

* Optional section

PREVIEW

Equation solving is an important skill to be learned in mathematics. In this chapter we consider solving equations in which the unknown is related to the angle of a trigonometric function. A **solution** of a trigonometric equation has the same meaning as the solution of any equation, namely value(s) that make a given equation true. Remember, **to solve an equation** means to find all replacements for the variable that make the equation true.

This chapter also introduces you to *eight fundamental identities* that can be used to change the form of a trigonometric expression or equation. These identities will then be used to derive a variety of other useful identities in the next chapter. The 17 objectives of this chapter are listed on pages 287–289.

PERSPECTIVE

Trigonometric identities are frequently used in a variety of calculus applications. In the example shown on this page the application is finding the length of part of a curve. Notice that first the identity $\cos^2 t + \sin^2 t = 1$ is used (without a remark or warning), and then some half-angle identities are used. We introduce the fundamental identities in Section 7.2 and the half-angle identities in Section 7.5.

Solution First we notice that

$$\frac{dx}{dt} = r - r\cos t \quad \text{and} \quad \frac{dy}{dt} = r\sin t$$

Therefore by (4) we have

$$\begin{aligned}\mathscr{L} &= \int_0^{2\pi} \sqrt{(r - r\cos t)^2 + (r\sin t)^2}\,dt \\ &= \int_0^{2\pi} \sqrt{r^2 - 2r^2\cos t + r^2\cos^2 t + r^2\sin^2 t}\,dt \\ &= \int_0^{2\pi} \sqrt{2r^2 - 2r^2\cos t}\,dt \\ &= r\int_0^{2\pi} \sqrt{2(1 - \cos t)}\,dt\end{aligned}$$

By the half-angle formula for $\sin t/2$,

$$\frac{1 - \cos t}{2} = \sin^2\frac{t}{2}, \quad \text{so that} \quad \sqrt{2(1 - \cos t)} = \sqrt{4\sin^2\frac{t}{2}}$$

Since $\sin t/2 \geq 0$ for $0 \leq t \leq 2\pi$, we conclude that

From Fig. 8-29, "Solution" from *Calculus and Analytic Geometry*, Third Edition, by Robert Ellis and Denny Gulick, copyright © 1986 by Harcourt Brace Jovanovich, Inc., reprinted by permission of the publisher.

7.1 TRIGONOMETRIC EQUATIONS

Section 6.5 introduced notation for inverse functions. In this section we will use a similar notation to solve equations.

If $\cos x = \frac{1}{2}$, then write $x = \cos^{-1}(\frac{1}{2})$ to mean that x is *any* angle or real number whose cosine is $\frac{1}{2}$. Note the use of the small letter on $\cos^{-1}(\frac{1}{2})$ rather than the capital C we used for the inverse cosine function. The procedure for finding $\cos^{-1}(\frac{1}{2})$ relies on knowing $\text{Cos}^{-1}(\frac{1}{2})$. These steps are summarized:

1. Find $\text{Cos}^{-1}|y|$; this will give you the reference angle. It can be found by using the table of exact values, Appendix Table C.III, or a calculator.

2. Find the principal values; use the sign of y to determine the proper quadrant placement. Use reference angles for finding the values of x less than one revolution that satisfy the equation.

   For cosine:
   y positive: Quadrants I and IV
   y negative: Quadrants II and III

   For sine:
   y positive: Quadrants I and II
   y negative: Quadrants III and IV

   For tangent:
   y positive: Quadrants I and III
   y negative: Quadrants II and IV

   Use:

Sine pos	All pos
Tangent pos	Cosine pos

3. For the entire solution, use the period of the function:

   For cosine and sine: add multiples of 2π or 360°
   For tangent: add multiples of π or 180°

EXAMPLE 1 Solve $\cos\theta = \frac{1}{2}$.

Solution *Step 1* $\theta' = \text{Cos}^{-1}|\frac{1}{2}| = 60°$ or $\frac{\pi}{3}$.

Step 2 $\frac{1}{2}$ is positive; cosine is positive in Quadrants I and IV; find the angles less than one revolution whose reference angle is 60° or $\frac{\pi}{3}$.

	Degrees	*Radians*
Quadrant I:	60°	$\frac{\pi}{3}$
Quadrant IV:	300°	$\frac{5\pi}{3}$

Step 3 Add multiples (let k be any integer):

$$\theta = \begin{cases} 60^\circ + 360^\circ k \\ 300^\circ + 360^\circ k \end{cases} \quad \text{or} \quad \theta = \begin{cases} \dfrac{\pi}{3} + 2k\pi \\ \dfrac{5\pi}{3} + 2k\pi \end{cases}$$

The solution is infinite. To check, select *any* integral value of k, say $k = 5$. From the solution, $300^\circ + 360^\circ(5) = 2100^\circ$. If this is a solution, it must satisfy the given equation: $\cos 2100^\circ = \frac{1}{2}$, which checks. ■

EXAMPLE 2 Solve $\cos\theta = -\frac{1}{2}$.

Solution $\theta' = \text{Cos}^{-1}|-\frac{1}{2}| = 60^\circ$ or $\frac{\pi}{3}$; this time the reference angle is in Quadrants II and III (since cosine is negative):

$$\theta = \begin{cases} 120^\circ + 360^\circ k \\ 240^\circ + 360^\circ k \end{cases} \quad \text{or} \quad \theta = \begin{cases} \dfrac{2\pi}{3} + 2k\pi \\ \dfrac{4\pi}{3} + 2k\pi \end{cases}$$

■

In trigonometry it is customary to delete the step in which multiples of 2π (for cosine and sine) or π (for tangent) are added to the principal values. In the examples that follow, you need only find the positive solutions less than one revolution. Give your solutions in radians unless specifically asked for degrees. The reason for this is that radians occur in calculus more often than degrees.

EXAMPLE 3 Solve $\sin\theta = -0.4446$ for $0^\circ \le \theta < 360^\circ$ (in degrees).

Solution **a.** From the table, $\theta' \approx 26.4^\circ$. Sine is negative in Quadrants III and IV:

206.4°, 333.6°

b. By calculator: [0.4446] [inv] [sin]

gives the reference angle 26.397749720; when placed in the proper quadrants, the solution is

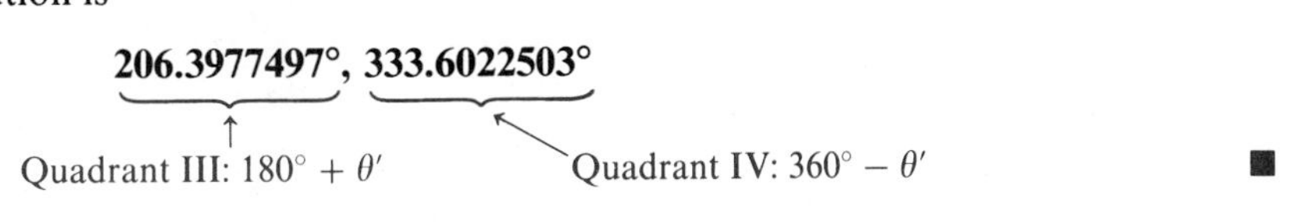

■

EXAMPLE 4 Solve $\tan\theta = -0.66956$ for $0 \le \theta < 2\pi$.

Solution **a.** From the table, $\theta' \approx 0.59$. (Work in radians if degrees are not specified.) Tangent is negative in Quadrants II and IV:

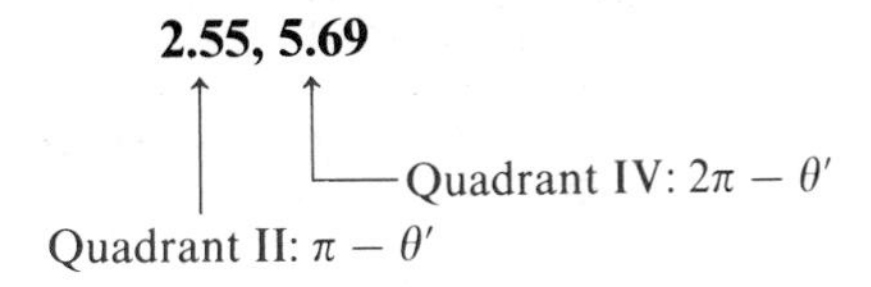

b. By calculator: .66956 inv tan

gives the reference angle 0.59000301; when placed in the proper quadrants, the solution is

$$\underset{\pi - \theta'}{\mathbf{2.5515896}}, \quad \underset{2\pi - \theta}{\mathbf{5.6931823}}$$

■

If the angle is a multiple of θ, the following inequalities hold true:

$$\begin{array}{ll} 0^\circ \le \theta < 360^\circ & 0 \le \theta < 2\pi \\ 0^\circ \le 2\theta < 720^\circ & 0 \le 2\theta < 4\pi \\ 0^\circ \le 3\theta < 1080^\circ & 0 \le 3\theta < 6\pi \\ 0^\circ \le 4\theta < 1440^\circ & 0 \le 4\theta < 8\pi \\ \vdots & \vdots \end{array}$$

EXAMPLE 5 Solve $\sin 2\theta = \frac{1}{2}$ for $0 \le \theta < 2\pi$.

Solution Since $0 \le \theta < 2\pi$, solve for 2θ such that $0 \le 2\theta < 4\pi$. Thus $\theta' = \frac{\pi}{6}$.

$$2\theta = \frac{\pi}{6}, \frac{5\pi}{6}, \frac{13\pi}{6}, \frac{17\pi}{6}$$

- $\frac{\pi}{6}$: Quadrant I
- $\frac{5\pi}{6}$: Quadrant II (sine is positive in Quadrants I and II)
- $\frac{13\pi}{6}$: $\frac{\pi}{6} + 2\pi$
- $\frac{17\pi}{6}$: $\frac{5\pi}{6} + 2\pi$

Add 2π to each of the principal values.

Mentally solve each equation for θ:

$$2\theta = \frac{\pi}{6} \qquad 2\theta = \frac{5\pi}{6} \qquad 2\theta = \frac{13\pi}{6} \qquad 2\theta = \frac{17\pi}{6}$$

As you solve these equations, notice that in each case θ is between zero and 2π. The solution is

$$\boldsymbol{\frac{\pi}{12}, \frac{5\pi}{12}, \frac{13\pi}{12}, \frac{17\pi}{12}}$$

■

EXAMPLE 6 Solve $\cos 3\theta = 1.2862$.

Solution The solution is **empty** since it is not true that $-1 \le \cos 3\theta \le 1$, which is why $\cos 3\theta \ne 1.2862$. ■

EXAMPLE 7 Solve $\cos 3\theta = -0.68222$ for $0 \le \theta < 2\pi$.

Solution By table, $\theta' \approx 0.82$; 3θ is in Quadrants II and III and $0 \le 3\theta < 6\pi$.

$$3\theta \approx 2.32, 3.96, 8.60, 10.24, 14.89, 16.53$$

Divide by 3 to find the solution:

0.77, 1.32, 2.87, 3.41, 4.96, 5.51

By calculator: [.68222] [inv] [cos]

The reference angle is 0.82000165. To two decimal places the solution is the same as shown above. ■

Remember,

$$\underbrace{\cos}_{\text{The function is cosine.}}\underbrace{(2x+1)}_{\text{The angle, or argument, is } 2x+1.} = 0$$

The unknown is x.

The steps in solving a trigonometric equation are now given.

Procedure for Solving Trigonometric Equations

1. Solve for a single trigonometric function. You may use identities, factoring, or the quadratic formula.
2. Solve for the argument (angle). You will use the definition of the inverse trigonometric functions for this step.
3. Solve for the unknown.

EXAMPLE 8 Solve $2\cos\theta\sin\theta = \sin\theta$ for $0 \le \theta < 2\pi$.

Solution This problem is solved by factoring:

$$2\cos\theta\sin\theta - \sin\theta = 0$$
$$\sin\theta(2\cos\theta - 1) = 0$$
$$\sin\theta = 0 \qquad 2\cos\theta - 1 = 0$$
$$\theta = 0, \pi \qquad \cos\theta = \frac{1}{2}$$
$$\theta = \frac{\pi}{3}, \frac{5\pi}{3}$$

Solution: $\mathbf{0, \pi, \frac{\pi}{3}, \frac{5\pi}{3}}$ ■

EXAMPLE 9 Solve $2\sin^2\theta = 1 + 2\sin\theta$ for $0 \le \theta < 2\pi$.

Solution This problem is solved by the quadratic formula: If $ax^2 + bx + c = 0$, $a \ne 0$, then

$$x = \frac{-b \pm \sqrt{b^2 - 4ac}}{2a}$$

Since $2\sin^2\theta - 2\sin\theta - 1 = 0$, let $x = \sin\theta$ in the quadratic formula:

$$\sin\theta = \frac{2 \pm \sqrt{4 - 4(2)(-1)}}{2(2)}$$
$$= \frac{1 \pm \sqrt{3}}{2}$$
$$\approx 1.366, \ -0.366025$$

(1.366: Reject since $-1 \le \sin\theta \le 1$.)

Solve $\sin\theta \approx -0.366025$ by using Table C.III or a calculator to find a reference angle of 0.3747. Since the sine is negative in Quadrants III and IV, the solutions are $\pi + 0.3747 \approx 3.5163$ and $2\pi - 0.3747 \approx 5.9085$. To four decimal places: **3.5163, 5.9085.** ■

PROBLEM SET 7.1

A

Solve each of the equations in Problems 1–12. Use exact values.

1. $\sin x = \frac{1}{2}$ $(0 \le x < \frac{\pi}{2})$
2. $\sin x = \frac{1}{2}$ $(0° \le x < 90°)$
3. $\sin x = -\frac{1}{2}$ $(-\frac{\pi}{2} \le x \le \frac{\pi}{2})$
4. $\sin x = -\frac{1}{2}$ $(0 \le x \le \pi)$
5. $\sin x = \frac{1}{2}$
6. $\sin x = -\frac{1}{2}$
7. $\cos x = \frac{1}{2}$ $(0 \le x < \frac{\pi}{2})$
8. $\cos x = \frac{1}{2}$ $(0° \le x < 90°)$
9. $\cos x = -\frac{1}{2}$ $(0° \le x \le 180°)$
10. $\cos x = -\frac{1}{2}$ $(0 \le x \le \pi)$
11. $\cos x = \frac{1}{2}$
12. $\cos x = -\frac{1}{2}$

Solve each of the equations in Problems 13–27 for exact values such that $0 \le x < 2\pi$.

13. $\cos 2x = \frac{1}{2}$
14. $\cos 3x = \frac{1}{2}$
15. $\cos 2x = -\frac{1}{2}$
16. $\sin 2x = \frac{\sqrt{2}}{2}$
17. $\sin 2x = -\frac{\sqrt{3}}{2}$
18. $\sin 3x = \frac{\sqrt{2}}{2}$
19. $\tan 3x = 1$
20. $\tan 3x = -1$
21. $\sec 2x = -\frac{2\sqrt{3}}{3}$
22. $(\sin x)(\cos x) = 0$
23. $(\sec x)(\tan x) = 0$
24. $(\sin x)(\cot x) = 0$
25. $(\cot x)(\cos x) = 0$
26. $(\csc x - 2)(2\cos x - 1) = 0$
27. $(\sec x - 2)(2\sin x - 1) = 0$

B

Solve each of the equations in Problems 28–41 for $0 \le x < 2\pi$. Use exact values where possible, but state approximate answers to four decimal places.

28. $\tan^2 x = \sqrt{3}\tan x$
29. $\tan^2 x = \tan x$
30. $\sin^2 x = \frac{1}{2}$
31. $\cos^2 x = \frac{1}{2}$
32. $3\sin x\cos x = \sin x$
33. $2\cos x\sin x = \sin x$
34. $\sin^2 x - \sin x - 2 = 0$
35. $\cos^2 x - 1 - \cos x = 0$
36. $4\cot^2 x - 8\cot x + 3 = 0$
37. $\tan^2 x - 3\tan x + 1 = 0$
38. $\sec^2 x - \sec x - 1 = 0$
39. $\csc^2 x - \csc x - 1 = 0$
40. $\cos x + 1 = \sqrt{3}$
41. $\sin x + 1 = \sqrt{3}$

Solve each of the equations in Problems 42–55 for $0 \le x < 2\pi$ correct to two decimal places.

42. $2\cos 2x\sin 2x = \sin 2x$
43. $\sin 2x + 2\cos x\sin 2x = 0$
44. $\cos 3x + 2\sin 2x\cos 3x = 0$
45. $\sin 2x + 1 = \sqrt{3}$
46. $\cos 3x - 1 = \sqrt{2}$
47. $\tan 2x + 1 = \sqrt{3}$
48. $1 - 2\sin^2 x = \sin x$
49. $2\cos^2 x - 1 = \cos x$
50. $2\cos x\sin x + \cos x = 0$
51. $1 - \sin x = 1 - 2\sin^2 x$
52. $\cos x = 2\sin x\cos x$
53. $\sin^2 3x + \sin 3x + 1 = 1 - \sin^2 3x$
54. $3\sin 4x - \sin 3x + \sqrt{3} = 2\sin 2x - \sin 3x + 3\sin 4x$
55. $3\cos x - \cos 3x + 2\cos 5x = 1 + 3\cos x - \cos 3x$

Find all solutions in radian measure of the equations given in Problems 56–61.

56. $\tan x = -\sqrt{3}$
57. $\cos x = -\frac{\sqrt{3}}{2}$
58. $\sin x = -\frac{\sqrt{2}}{2}$
59. $\sin x = 0.3907$
60. $\cos x = 0.2924$
61. $\tan x = 1.376$

C

62. ***Sound Waves*** A tuning fork vibrating at 264 Hz (frequency $f = 264$) with an amplitude of 0.0050 cm produces C on the musical scale and can be described by an equation

of the form

$$y = 0.0050 \sin 528\pi x$$

Find the smallest positive value of x (correct to four decimal places) for which $y = 0.0020$.

63. ***Electrical*** In a certain electric circuit, the electromotive force V in volts and the time t (in seconds) are related by an equation of the form

$$V = \cos 2\pi t$$

Find the smallest positive value for t (correct to three decimal places) for which $V = 0.400$.

64. ***Space Science*** The orbit of a certain satellite alternates above and below the equator according to the equation

$$y = 4000 \sin\left(\frac{\pi}{45}t + \frac{5\pi}{18}\right)$$

where t is the time in minutes and y is the distance (in kilometers) from the equator. Find the times at which the satellite crosses the equator during the first hour and a half (that is, for $0 \le t \le 90$).

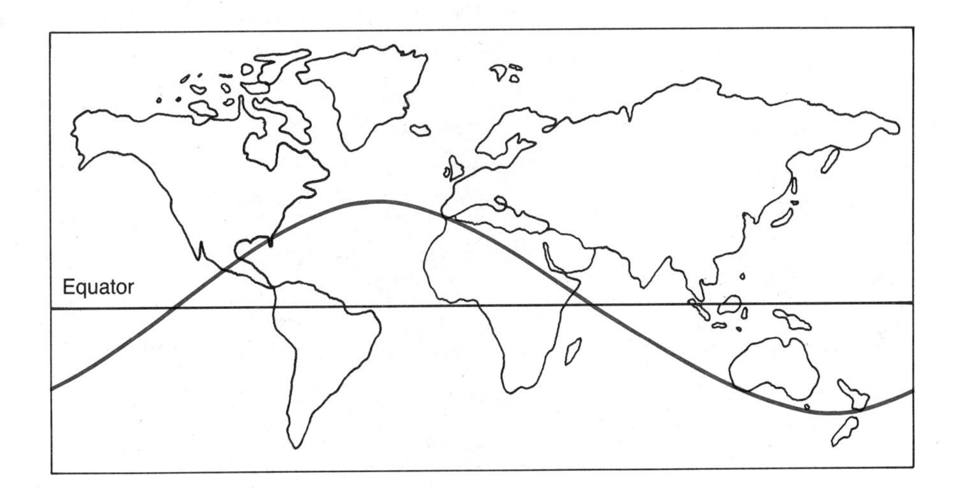

7.2 FUNDAMENTAL IDENTITIES

In the last section we solved some trigonometric equations. In this section we focus on trigonometric identities. The procedure for proving identities is quite different from the procedure for solving equations. Compare the following two examples from algebra:

SOLVE:	$2x + 3x = 5$	
SOLUTION:	$2x + 3x = 5$	Given
	$5x = 5$	Combining similar terms
	$x = 1$	Multiply both sides by $\frac{1}{5}$

PROVE:	$2x + 3x = 5x$	
SOLUTION:	$2x + 3x = (2 + 3)x$	Distributive property
	$= 5x$	Closure
	$2x + 3x = 5x$	Transitive property

Notice that when *solving* the equation, $2x + 3x = 5$ was given and *used as a starting point*. On the other hand, when *proving* the identity, $2x + 3x = 5x$ *could not be used as a starting point*. Indeed $2x + 3x = 5x$ was the *last step*, not the first step.

The reason for this difference in procedure is apparent if you look at the addition and multiplication principles:

Addition principle: If $a = b$, then $a + c = b + c$.

Multiplication principle: If $a = b$, then $ac = bc$.

In both cases you must *know* that $a = b$ before you can use the addition or multiplication principles. If you are asked to *prove* that $a = b$, you cannot assume $a = b$ to work the problem. You must begin with what is known to be true and *end* with the given identity.

All our work with trigonometric identities is ultimately based on eight basic identities called the **fundamental identities.** Notice that these identities are classified into three categories and numbered for later reference. Values of θ that cause division by zero are excluded.

Fundamental Identities

RECIPROCAL IDENTITIES:

1. $\sec\theta = \dfrac{1}{\cos\theta}$ 2. $\csc\theta = \dfrac{1}{\sin\theta}$ 3. $\cot\theta = \dfrac{1}{\tan\theta}$

RATIO IDENTITIES:

4. $\tan\theta = \dfrac{\sin\theta}{\cos\theta}$ 5. $\cot\theta = \dfrac{\cos\theta}{\sin\theta}$

PYTHAGOREAN IDENTITIES:

6. $\sin^2\theta + \cos^2\theta = 1$ 7. $1 + \tan^2\theta = \sec^2\theta$ 8. $\cot^2\theta + 1 = \csc^2\theta$

The proofs of these identities follow directly from the definitions of the trigonometric functions. Some proofs were given in the last chapter, but they are repeated here and in Problem Set 7.2.

Let θ be an angle in standard position with point $P(x, y)$ on the terminal side a distance of r from the origin, with $r \neq 0$.

By definition, $\cos\theta = \frac{x}{r}$; thus

$$\frac{1}{\cos\theta} = \frac{1}{x/r}$$

$$= 1 \cdot \frac{r}{x} \qquad \text{Division of fractions}$$

$$= \frac{r}{x} \qquad \text{Multiplication of fractions}$$

$$= \sec\theta \qquad \text{By definition of } \sec\theta$$

Therefore $\dfrac{1}{\cos\theta} = \sec\theta.$ □

Identities 2 and 3 are proved in precisely the same way and are left as problems.

Proof of Identity 4

$$\frac{\sin\theta}{\cos\theta} = \frac{y/r}{x/r}$$ By definition of $\sin\theta$ and $\cos\theta$

$$= \frac{y}{r}\cdot\frac{r}{x}$$ Division of fractions

$$= \frac{y}{x}$$ Multiplication and simplification of fractions

$$= \tan\theta$$ By definition of $\tan\theta$

Therefore $\dfrac{\sin\theta}{\cos\theta} = \tan\theta.$ □

The proof of Identity 5 is similar to the proof of Identity 4 and is left as a problem. For Identities 6, 7, and 8, begin with the Pythagorean Theorem (which is why these are called the *Pythagorean identities*):

$$x^2 + y^2 = r^2$$ By the Pythagorean Theorem

To prove Identity 6, divide both sides by r^2; for Identity 7 divide by x^2; and for Identity 8, divide by y^2. We will show the details for Identity 6 and leave Identities 7 and 8 as problems.

Proof of Identity 6

$$x^2 + y^2 = r^2$$ Pythagorean Theorem

$$\frac{x^2}{r^2} + \frac{y^2}{r^2} = \frac{r^2}{r^2}$$ Dividing both sides by r^2 $(r \neq 0)$

$$\left(\frac{x}{r}\right)^2 + \left(\frac{y}{r}\right)^2 = 1$$ Properties of exponents

$$(\cos\theta)^2 + (\sin\theta)^2 = 1$$ Definition of $\cos\theta$ and $\sin\theta$

$$\sin^2\theta + \cos^2\theta = 1$$ Commutative property □

EXAMPLE 1 Write all six trigonometric functions in terms of $\sin\theta$.

Solution

a. $\sin\theta = \sin\theta$

b. $\cos\theta = \pm\sqrt{1 - \sin^2\theta}$ From Identity 6

c. $\tan\theta = \dfrac{\sin\theta}{\cos\theta}$ Identity 4

$= \dfrac{\sin\theta}{\pm\sqrt{1 - \sin^2\theta}}$ From part b

d. $\cot\theta = \dfrac{1}{\tan\theta}$ Identity 3

$= \dfrac{\pm\sqrt{1 - \sin^2\theta}}{\sin\theta}$ From part c

e. $\csc\theta = \dfrac{1}{\sin\theta}$ From Identity 2

f. $\sec\theta = \dfrac{1}{\cos\theta}$ From Identity 1

$= \dfrac{1}{\pm\sqrt{1-\sin^2\theta}}$ From part b ■

The $\pm$ sign we have been using, as in

$$\cos\theta = \pm\sqrt{1-\sin^2\theta}$$

means that $\cos\theta$ is positive for some values of θ and negative for other values of θ. The plus or the minus sign is chosen by determining the proper quadrant, as shown in Example 2.

EXAMPLE 2 Given $\sin\theta = \frac{3}{5}$ and $\tan\theta < 0$, find the other functions of θ.

Solution Since the tangent is negative and the sine is positive, the quadrant is II. Thus

$$\begin{aligned}\cos\theta &= -\sqrt{1-\sin^2\theta} && \text{Since the cosine is negative in Quadrant II}\\ &= -\sqrt{1-(\tfrac{3}{5})^2}\\ &= -\sqrt{1-\tfrac{9}{25}}\\ &= -\sqrt{\tfrac{16}{25}}\\ &= -\tfrac{4}{5}\end{aligned}$$

Also,

$$\tan\theta = \frac{\sin\theta}{\cos\theta} = \frac{3/5}{-4/5} = -\frac{3}{4}$$

Using the reciprocal identities, $\cot\theta = -\frac{4}{3}$, $\sec\theta = -\frac{5}{4}$, and $\csc\theta = \frac{5}{3}$. ■

PROBLEM SET 7.2

A

1. State from memory the eight fundamental identities.

In Problems 2–9, state the quadrant or quadrants in which θ may lie to make the expression true.

2. $\sin\theta = \sqrt{1-\cos^2\theta}$

3. $\sin\theta = -\sqrt{1-\cos^2\theta}$

4. $\sec\theta = -\sqrt{1+\tan^2\theta}$

5. $\sec\theta = \sqrt{1+\tan^2\theta}$

6. $\csc\theta = \sqrt{1+\cot^2\theta}$; $\tan\theta < 0$

7. $\cos\theta = -\sqrt{1-\sin^2\theta}$; $\sin\theta > 0$

8. $\tan\theta = \sqrt{\sec^2\theta - 1}$; $\cos\theta < 0$

9. $\csc\theta = \sqrt{1+\cot^2\theta}$; $\cos\theta > 0$

Write each of the expressions in Problems 10–18 as a single trigonometric function of some angle by using one of the eight fundamental identities.

10. $\dfrac{\sin 50^\circ}{\cos 50^\circ}$

11. $\dfrac{\cos(A+B)}{\sin(A+B)}$

12. $\dfrac{1}{\sec 75^\circ}$

13. $\dfrac{1}{\cot(\frac{\pi}{15})}$

14. $\tan 42^\circ \cos 42^\circ$

15. $\cot\frac{\pi}{8}\sin\frac{\pi}{8}$

16. $1-\cos^2 18^\circ$

17. $-\sqrt{1-\sin^2 127^\circ}$

18. $\sec^2(\frac{\pi}{6}) - 1$

Evaluate the expressions in Problems 19–24 by using one of the eight fundamental identities.

19. $\cos 128° \sec 128°$

20. $\sin^2 \frac{\pi}{3} + \cos^2 \frac{\pi}{3}$

21. $\sec^2 \frac{\pi}{6} - \tan^2 \frac{\pi}{6}$

22. $\cot^2 45° - \csc^2 45°$

23. $\tan^2 135° - \sec^2 135°$

24. $\csc 85° \sin 85°$

25. Prove that $\csc \theta = 1/\sin \theta$.

26. Prove that $\cot \theta = 1/\tan \theta$.

27. Prove that $\cot \theta = \dfrac{\cos \theta}{\sin \theta}$.

28. Prove that $\cot^2 \theta + 1 = \csc^2 \theta$.

29. Prove that $1 + \tan^2 \theta = \sec^2 \theta$.

B

In Problems 30–33, write all the trigonometric functions in terms of the given function.

30. $\tan \theta$ **31.** $\cot \theta$ **32.** $\sec \theta$ **33.** $\csc \theta$

In Problems 34–45, find the other functions of θ using the given information.

34. $\cos \theta = \frac{3}{5}$; $\tan \theta > 0$

35. $\cos \theta = \frac{3}{5}$; $\csc \theta < 0$

36. $\cos \theta = \frac{5}{13}$; $\tan \theta < 0$

37. $\cos \theta = \frac{5}{13}$; $\tan \theta > 0$

38. $\tan \theta = \frac{5}{12}$; $\sin \theta > 0$

39. $\tan \theta = \frac{5}{12}$; $\sin \theta < 0$

40. $\sin \theta = \frac{2}{3}$; $\sec \theta > 0$

41. $\sin \theta = \frac{2}{3}$; $\sec \theta < 0$

42. $\sec \theta = \frac{\sqrt{34}}{5}$; $\tan \theta < 0$

43. $\sec \theta = \frac{\sqrt{34}}{5}$; $\tan \theta > 0$

44. $\csc \theta = -\frac{\sqrt{10}}{3}$; $\cos \theta > 0$

45. $\csc \theta = -\frac{\sqrt{10}}{3}$; $\cos \theta < 0$

Simplify the expressions in Problems 46–54 using only sines, cosines, and the fundamental identities.

46. $\dfrac{1 - \sin^2 \theta}{\cos \theta}$

47. $\dfrac{1 - \cos^2 \theta}{\sin \theta}$

48. $\dfrac{\sin \theta}{\cos \theta} + \dfrac{\cos \theta}{\sin \theta}$

49. $\dfrac{1}{1 + \cos \theta} + \dfrac{1}{1 - \cos \theta}$

50. $\dfrac{\dfrac{\sin \theta}{\cos \theta} + \dfrac{\cos \theta}{\sin \theta}}{\dfrac{1}{\sin \theta \cos \theta}}$

51. $\sin \theta + \dfrac{\cos^2 \theta}{\sin \theta}$

52. $\dfrac{\cos \theta + \dfrac{\sin^2 \theta}{\cos \theta}}{\sin \theta}$

53. $\dfrac{\sin \theta - \dfrac{\cos^2 \theta}{\sin \theta}}{\cos \theta}$

54. $\dfrac{\dfrac{\cos^4 \theta}{\sin^2 \theta} + \cos^2 \theta}{\dfrac{\cos^2 \theta}{\sin^2 \theta}}$

Reduce the expressions in Problems 55–62 so that they involve only sines and cosines, and then simplify.

55. $\sin \theta + \cot \theta$

56. $\sec \theta + \tan \theta$

57. $\dfrac{\tan \theta + \cot \theta}{\sec \theta \csc \theta}$

58. $\dfrac{\sec \theta + \csc \theta}{\tan \theta \cot \theta}$

59. $\sec^2 \theta + \tan^2 \theta$

60. $\csc^2 \theta + \cot^2 \theta$

61. $(\cot \theta - \sec \theta)(\sin \theta \cos \theta)$

62. $(\tan \theta - \csc \theta)(\cos \theta \sin \theta)$

7.3 PROVING IDENTITIES

In the last section we considered eight fundamental identities, which are used to simplify and change the form of a variety of trigonometric expressions. Suppose you are given a trigonometric equation such as

$$\tan \theta + \cot \theta = \sec \theta \csc \theta$$

and are asked to show that it is an identity. You must be careful not to treat this problem as though it were an algebraic equation. When asked to prove an identity, do *not* start with the given expression, since you cannot assume it is true. You should *begin* with what you know is true and *end* with the given identity. There are three ways to proceed:

1. Reduce the left-hand side to the right-hand side by using algebra and the fundamental identities.
2. Reduce the right-hand side to the left-hand side.
3. Reduce both sides independently to the same expression.

EXAMPLE 1 Prove that $\tan\theta + \cot\theta = \sec\theta\csc\theta$.

Solution Begin with either the left- or right-hand side:

$$\tan\theta + \cot\theta = \frac{\sin\theta}{\cos\theta} + \frac{\cos\theta}{\sin\theta}$$

$$= \frac{\sin^2\theta + \cos^2\theta}{\cos\theta\sin\theta}$$

$$= \frac{1}{\cos\theta\sin\theta}$$

This is algebraically simplified, so return to the other side and begin anew:

$$\sec\theta\csc\theta = \frac{1}{\cos\theta}\cdot\frac{1}{\sin\theta}$$

$$= \frac{1}{\cos\theta\sin\theta}$$

This too is simplified, but notice that the simplified forms for both the left and right sides are the same. Therefore

$$\tan\theta + \cot\theta = \sec\theta\csc\theta$$

■

Usually it is easier to begin with the more complicated side and try to reduce it to the simpler side. If both sides seem equally complex, you might change all the functions to sines and cosines and then simplify.

EXAMPLE 2 Prove that $2\csc^2\theta = \dfrac{1}{1+\cos\theta} + \dfrac{1}{1-\cos\theta}$.

Solution Begin with the more complicated side:

$$\frac{1}{1+\cos\theta} + \frac{1}{1-\cos\theta} = \frac{(1-\cos\theta) + (1+\cos\theta)}{(1+\cos\theta)(1-\cos\theta)}$$

$$= \frac{2}{1-\cos^2\theta}$$

$$= \frac{2}{\sin^2\theta}$$

$$= 2\csc^2\theta$$

■

EXAMPLE 3 Prove that $\dfrac{\sec 2\lambda + \cot 2\lambda}{\sec 2\lambda} = 1 + \csc 2\lambda - \sin 2\lambda$.

Solution Begin with the left-hand side. When working with a fraction consisting of a single function as a denominator, it is often helpful to separate the fraction into the sum of several fractions:

$$\frac{\sec 2\lambda + \cot 2\lambda}{\sec 2\lambda} = \frac{\sec 2\lambda}{\sec 2\lambda} + \frac{\cot 2\lambda}{\sec 2\lambda}$$

$$= 1 + \cot 2\lambda \cdot \frac{1}{\sec 2\lambda}$$

$$= 1 + \frac{\cos 2\lambda}{\sin 2\lambda} \cdot \cos 2\lambda$$

$$= 1 + \frac{\cos^2 2\lambda}{\sin 2\lambda}$$

$$= 1 + \frac{1 - \sin^2 2\lambda}{\sin 2\lambda}$$

$$= 1 + \frac{1}{\sin 2\lambda} - \frac{\sin^2 2\lambda}{\sin 2\lambda}$$

$$= 1 + \csc 2\lambda - \sin 2\lambda$$

■

EXAMPLE 4 Prove that $\dfrac{\cos\theta}{1 - \sin\theta} = \dfrac{1 + \sin\theta}{\cos\theta}$.

Solution Sometimes, when there is a binomial in the numerator or denominator, the identity can be proved by multiplying one side by 1, where 1 is written in the form of the conjugate of the binomial. When changing one side, keep a sharp eye on the other side, since it often gives a clue about what to do. Thus in this example we can multiply the numerator and denominator of the left-hand side by $1 + \sin\theta$:

$$\frac{\cos\theta}{1 - \sin\theta} = \frac{\cos\theta}{1 - \sin\theta} \cdot \frac{1 + \sin\theta}{1 + \sin\theta}$$

$$= \frac{\cos\theta(1 + \sin\theta)}{1 - \sin^2\theta}$$

$$= \frac{\cos\theta(1 + \sin\theta)}{\cos^2\theta}$$

$$= \frac{1 + \sin\theta}{\cos\theta}$$

■

EXAMPLE 5 Prove that $\dfrac{\sec^2 2\theta - \tan^2 2\theta}{\tan 2\theta + \sec 2\theta} = \dfrac{\cos 2\theta}{1 + \sin 2\theta}$.

Solution Sometimes the identity can be proved by factoring:

$$\frac{\sec^2 2\theta - \tan^2 2\theta}{\tan 2\theta + \sec 2\theta} = \frac{(\sec 2\theta + \tan 2\theta)(\sec 2\theta - \tan 2\theta)}{\tan 2\theta + \sec 2\theta}$$

$$= \sec 2\theta - \tan 2\theta$$

$$= \frac{1}{\cos 2\theta} - \frac{\sin 2\theta}{\cos 2\theta}$$

$$\frac{\sec^2 2\theta - \tan^2 2\theta}{\tan 2\theta + \sec 2\theta} = \frac{1 - \sin 2\theta}{\cos 2\theta}$$

$$= \frac{1 - \sin 2\theta}{\cos 2\theta} \cdot \frac{1 + \sin 2\theta}{1 + \sin 2\theta}$$

$$= \frac{1 - \sin^2 2\theta}{\cos 2\theta(1 + \sin 2\theta)}$$

$$= \frac{\cos^2 2\theta}{\cos 2\theta(1 + \sin 2\theta)}$$

$$= \frac{\cos 2\theta}{1 + \sin 2\theta}$$

■

EXAMPLE 6 Prove that $\dfrac{-2\sin\theta\cos\theta}{1 - \sin\theta - \cos\theta} = 1 + \sin\theta + \cos\theta$.

Solution Sometimes, when there is a fraction on one side, the identity can be proved by multiplying the other side by 1 written so that the desired denominator is obtained. Thus, for this example,

$$1 + \sin\theta + \cos\theta = (1 + \sin\theta + \cos\theta) \cdot \frac{1 - \sin\theta - \cos\theta}{1 - \sin\theta - \cos\theta}$$

$$= \frac{(1 + \sin\theta + \cos\theta)(1 - \sin\theta - \cos\theta)}{1 - \sin\theta - \cos\theta}$$

$$= \frac{1 - \sin\theta - \cos\theta + \sin\theta - \sin^2\theta - \sin\theta\cos\theta + \cos\theta - \cos\theta\sin\theta - \cos^2\theta}{1 - \sin\theta - \cos\theta}$$

$$= \frac{1 - (\sin^2\theta + \cos^2\theta) - 2\sin\theta\cos\theta}{1 - \sin\theta - \cos\theta}$$

$$= \frac{-2\sin\theta\cos\theta}{1 - \sin\theta - \cos\theta}$$

■

In summary, there is no single method that is best for proving identities. However, the following hints should help:

Procedures for Proving Identities

1. If one side contains one function only, write all the trigonometric functions on the other side in terms of that function.
2. If the denominator of a fraction consists of only one function, break up the fraction.
3. Simplify by combining fractions.
4. Factoring is sometimes helpful.
5. Change all trigonometric functions to sines and cosines and simplify.
6. Multiply by the conjugate of either the numerator or the denominator.
7. If there are squares of functions, look for alternate forms of the Pythagorean identities.
8. Avoid the introduction of radicals.
9. Keep your destination in sight. Watch where you are going, and know when you are finished.

PROBLEM SET 7.3

A

Prove that the equations in Problems 1–38 are identities.

1. $\sin\theta = \sin^3\theta + \cos^2\theta\sin\theta$
2. $\sec\theta = \sec\theta\sin^2\theta + \cos\theta$
3. $\tan\theta = \cot\theta\tan^2\theta$
4. $\dfrac{\sin\theta\cos\theta + \sin^2\theta}{\sin\theta} = \cos\theta + \sin\theta$
5. $\tan^2\theta - \sin^2\theta = \tan^2\theta\sin^2\theta$
6. $\cot^2\theta\cos^2\theta = \cot^2\theta - \cos^2\theta$
7. $\tan A + \cot A = \sec A\csc A$
8. $\cot A = \csc A\sec A - \tan A$
9. $\sin x + \cos x = \dfrac{\sec x + \csc x}{\csc x\sec x}$
10. $\dfrac{\cos\gamma + \tan\gamma\sin\gamma}{\sec\gamma} = 1$
11. $\dfrac{1 - \sec^2 t}{\sec^2 t} = -\sin^2 t$
12. $\dfrac{1 + \cot^2 t}{\cot^2 t} = \sec^2 t$
13. $(\sec\theta - \cos\theta)^2 = \tan^2\theta - \sin^2\theta$
14. $\dfrac{\sin\theta}{\csc\theta} + \dfrac{\cos\theta}{\sec\theta} = 1$
15. $1 - \sin 2\theta = \dfrac{1 - \sin^2 2\theta}{1 + \sin 2\theta}$
16. $\dfrac{1 - \tan^2 3\theta}{1 - \tan 3\theta} = 1 + \tan 3\theta$
17. $\sin\lambda = \dfrac{\sin^2\lambda + \sin\lambda\cos\lambda + \sin\lambda}{\sin\lambda + \cos\lambda + 1}$
18. $\dfrac{1 + \cot 2\lambda\sec 2\lambda}{\tan 2\lambda + \sec 2\lambda} = \cot 2\lambda$
19. $\sin 2\alpha\cos 2\alpha(\tan 2\alpha + \cot 2\alpha) = 1$
20. $(\sin\beta - \cos\beta)^2 + (\sin\beta + \cos\beta)^2 = 2$
21. $\csc 3\beta - \cos 3\beta\cot 3\beta = \sin 3\beta$
22. $\dfrac{1 + \cot^2 A}{1 + \tan^2 A} = \cot^2 A$
23. $\dfrac{\sin^2 B - \cos^2 B}{\sin B + \cos B} = \sin B - \cos B$
24. $\dfrac{\tan^2\gamma - \cot^2\gamma}{\tan\gamma + \cot\gamma} = \tan\gamma - \cot\gamma$
25. $\tan^2 2\gamma + \sin^2 2\gamma + \cos^2 2\gamma = \sec^2 2\gamma$
26. $\cot^2 C + \cos^2 C + \sin^2 C = \csc^2 C$
27. $\dfrac{\tan\theta + \cot\theta}{\sec\theta\csc\theta} = 1$
28. $\dfrac{\tan\theta - \cot\theta}{\sec\theta\csc\theta} = \sin^2\theta - \cos^2\theta$
29. $1 + \sin^2\lambda = 2 - \cos^2\lambda$
30. $2 - \sin^2 3\lambda = 1 + \cos^2 3\lambda$
31. $\dfrac{\sin\alpha}{\tan\alpha} + \dfrac{\cos\alpha}{\cot\alpha} = \cos\alpha + \sin\alpha$
32. $\dfrac{1}{1 + \cos 2\alpha} + \dfrac{1}{1 - \cos 2\alpha} = 2\csc^2 2\alpha$
33. $\sec\beta + \cos\beta = \dfrac{2 - \sin^2\beta}{\cos\beta}$
34. $2\sin^2 3\beta - 1 = 1 - 2\cos^2 3\beta$
35. $\dfrac{\tan 2\theta + \cot 2\theta}{\sec 2\theta} = \csc 2\theta$
36. $\dfrac{\tan 3\theta + \cot 3\theta}{\csc 3\theta} = \sec 3\theta$
37. $\dfrac{\sec\lambda + \tan^2\lambda}{\sec\lambda} = 1 + \sec\lambda - \cos\lambda$
38. $\dfrac{\sin 2\lambda}{\tan 2\lambda} + \dfrac{\cos 2\lambda}{\cot 2\lambda} = \cos 2\lambda + \sin 2\lambda$

B

Prove that the equations in Problems 39–60 are identities.

39. $\dfrac{1 + \tan C}{1 - \tan C} = \dfrac{\sec^2 C + 2\tan C}{2 - \sec^2 C}$
40. $(\cot x + \csc x)^2 = \dfrac{\sec x + 1}{\sec x - 1}$
41. $\dfrac{\sin^3 x - \cos^3 x}{\sin x - \cos x} = 1 + \sin x\cos x$
42. $\dfrac{\tan^3 t - \cot^3 t}{\tan t - \cot t} = \sec^2 t + \cot^2 t$
43. $\dfrac{1 - \cos\theta}{1 + \cos\theta} = \left(\dfrac{1 - \cos\theta}{\sin\theta}\right)^2$
44. $\dfrac{(\sec^2\gamma + \tan^2\gamma)^2}{\sec^4\gamma - \tan^4\gamma} = 1 + 2\tan^2\gamma$
45. $\dfrac{(\cos^2\gamma - \sin^2\gamma)^2}{\cos^4\gamma - \sin^4\gamma} = 2\cos^2\gamma - 1$
46. $(\sec 2\theta + \csc 2\theta)^2 = \dfrac{1 + 2\sin 2\theta\cos 2\theta}{\cos^2 2\theta\sin^2 2\theta}$
47. $\dfrac{1}{\sec\theta + \tan\theta} = \sec\theta - \tan\theta$

48. $\csc\theta + \cot\theta = \dfrac{1}{\csc\theta - \cot\theta}$

49. $\sec^2 2\lambda + \csc^2 2\lambda = \csc^2 2\lambda \sec^2 2\lambda$

50. $\dfrac{1 + \tan^3\theta}{1 + \tan\theta} = \sec^2\theta - \tan\theta$

51. $\dfrac{1 - \sec^3\theta}{1 - \sec\theta} = \tan^2\theta + \sec\theta + 2$

52. $\dfrac{\cos^2\theta - \cos\theta\csc\theta}{\cos^2\theta\csc\theta - \cos\theta\csc^2\theta} = \sin\theta$

53. $\dfrac{\tan^2\theta - 2\tan\theta}{2\tan\theta - 4} = \dfrac{1}{2}\tan\theta$

54. $\dfrac{\tan\theta}{\cot\theta} - \dfrac{\cot\theta}{\tan\theta} = \sec^2\theta - \csc^2\theta$

55. $\sqrt{(3\cos\theta - 4\sin\theta)^2 + (3\sin\theta + 4\cos\theta)^2} = 5$

56. $\dfrac{\cos\theta + \cos^2\theta}{\cos\theta + 1} = \dfrac{\cos\theta\sin\theta + \cos^2\theta}{\sin\theta + \cos\theta}$

57. $\sec^2\lambda - \csc^2\lambda = (2\sin^2\lambda - 1)(\sec^2\lambda + \csc^2\lambda)$

58. $2\csc A = 2\csc A - \cot A\cos A + \cos^2 A\csc A$

59. $\dfrac{\csc\theta + 1}{\cot^2\theta + \csc\theta + 1} = \dfrac{\sin^2\theta + \sin\theta\cos\theta}{\sin\theta + \cos\theta}$

60. $\dfrac{\cos^4\theta - \sin^4\theta}{(\cos^2\theta - \sin^2\theta)^2} = \dfrac{\cos\theta}{\cos\theta + \sin\theta} + \dfrac{\sin\theta}{\cos\theta - \sin\theta}$

C

Prove that the equations in Problems 61–72 are identities.

61. $(\cos\alpha - \cos\beta)^2 + (\sin\alpha - \sin\beta)^2$
$= 2 - 2(\cos\alpha\cos\beta + \sin\alpha\sin\beta)$

62. $(\sec\alpha + \sec\beta)^2 - (\tan\alpha - \tan\beta)^2$
$= 2 + 2(\sec\alpha\sec\beta + \tan\alpha\tan\beta)$

63. $\tan A + \cot B = (\sin A\sin B + \cos A\cos B)\sec A\csc B$

64. $\sec A + \csc B = (\cos A\sin^2 B + \sin B\cos^2 A)\sec^2 A\csc^2 B$

65. $(\sin A\cos A\cos\beta + \sin B\cos B\cos A)\sec A\sec B$
$= \sin A + \sin B$

66. $(\cos A\cos B\tan A + \sin A\sin B\cot B)\csc A\sec B = 2$

67. $\sin\theta + \cos\theta + 1 = \dfrac{2\sin\theta\cos\theta}{\sin\theta + \cos\theta - 1}$

68. $\dfrac{2\tan^2\theta + 2\tan\theta\sec\theta}{\tan\theta + \sec\theta - 1} = \tan\theta + \sec\theta + 1$

69. $\dfrac{\csc\theta + 1}{\csc\theta - 1} - \dfrac{\sec\theta - \tan\theta}{\sec\theta + \tan\theta} = 4\tan\theta\sec\theta$

70. $\dfrac{\cos\theta + \sin\theta}{\cos\theta - \sin\theta} + \dfrac{\cot\theta - 1}{\cot\theta + 1} = \dfrac{-2}{\sin^2\theta - \cos^2\theta}$

71. $\dfrac{\cos\theta + 1}{\cos\theta - 1} + \dfrac{1 - \sec\theta}{1 + \sec\theta} = -2\cot^2\theta - 2\csc^2\theta$

72. $\dfrac{\sin\theta}{1 - \cos\theta} + \dfrac{\cos\theta}{1 - \sin\theta} = (1 + \sin\theta + \cos\theta)(\sec\theta\csc\theta)$

7.4 ADDITION LAWS

When proving identities, it is sometimes necessary to simplify the functional value of the sum or difference of two angles. If α and β represent any two angles,

$$\cos(\alpha - \beta) \neq \cos\alpha - \cos\beta$$

For example, if $\alpha = 60°$ and $\beta = 30°$ then

$$\cos(60° - 30°) = \cos 30° = \frac{\sqrt{3}}{2} \qquad \text{and} \qquad \cos 60° - \cos 30° = \frac{1}{2} - \frac{\sqrt{3}}{2} = \frac{1 - \sqrt{3}}{2}$$

Thus $\cos(60° - 30°) \neq \cos 60° - \cos 30°$.

In this section, we discuss twelve more identities. First we consider $\cos(\alpha - \beta)$ expanded in terms of trigonometric functions of single angles. We shall later list this as Identity 16 but it is convenient to discuss it first because it provides the cornerstone for building a great many additional identities.

Difference of Angles Identity

$$\cos(\alpha - \beta) = \cos\alpha\cos\beta + \sin\alpha\sin\beta$$

Proof Find the length of any chord in a unit circle with a corresponding arc intercepted by the central angle θ, where θ is in standard position. Let A be the point $(1, 0)$ and P be the point on the intersection of the terminal side of angle θ and the unit circle. This means that the coordinates of P are $(\cos\theta, \sin\theta)$. Now find the length of the chord AP (see Figure 7.1) by using the distance formula:

$$\begin{aligned} |AP| &= \sqrt{(1 - \cos\theta)^2 + (0 - \sin\theta)^2} \\ &= \sqrt{1 - 2\cos\theta + \cos^2\theta + \sin^2\theta} \\ &= \sqrt{1 - 2\cos\theta + 1} \\ &= \sqrt{2 - 2\cos\theta} \end{aligned}$$

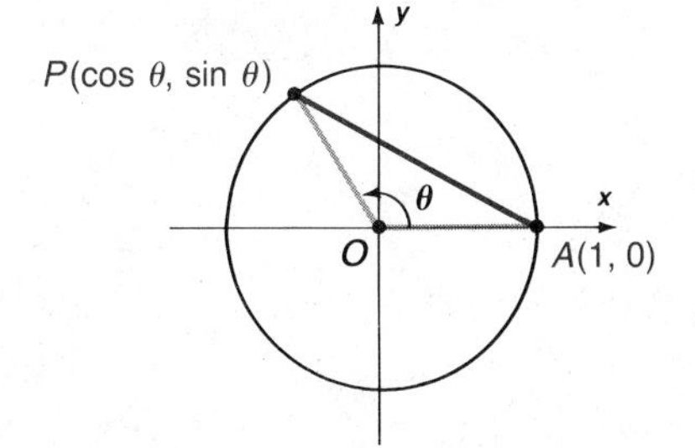

Figure 7.1 Length of a chord determined by an angle θ

Next apply this result to a chord determined by any two angles α and β, as shown in Figure 7.2. Let P_α and P_β be the points on the unit circle determined by the angles α and β, respectively. By the previous result,

$$|P_\alpha P_\beta| = \sqrt{2 - 2\cos(\alpha - \beta)}$$

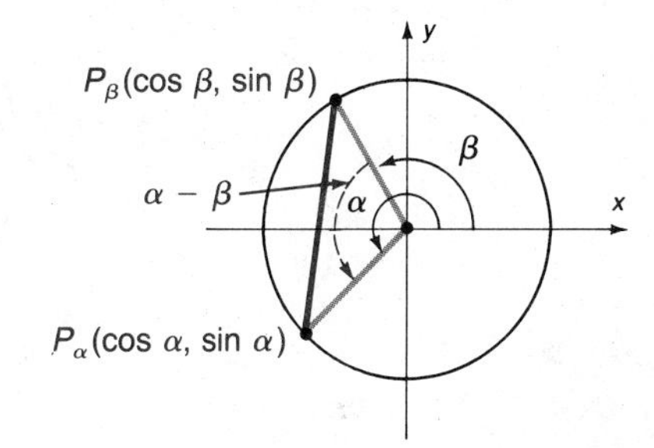

Figure 7.2 Distance between P_α and P_β

But you could also have found this distance directly via the distance formula:

$$\begin{aligned} |P_\alpha P_\beta| &= \sqrt{(\cos\beta - \cos\alpha)^2 + (\sin\beta - \sin\alpha)^2} \\ &= \sqrt{\cos^2\beta - 2\cos\alpha\cos\beta + \cos^2\alpha + \sin^2\beta - 2\sin\alpha\sin\beta + \sin^2\alpha} \\ &= \sqrt{(\cos^2\beta + \sin^2\beta) + (\cos^2\alpha + \sin^2\alpha) - 2(\cos\alpha\cos\beta + \sin\alpha\sin\beta)} \\ &= \sqrt{2 - 2(\cos\alpha\cos\beta + \sin\alpha\sin\beta)} \end{aligned}$$

Finally, equate these quantities since they both represent the distance between P_α and P_β:

$$\begin{aligned} \sqrt{2 - 2\cos(\alpha - \beta)} &= \sqrt{2 - 2(\cos\alpha\cos\beta + \sin\alpha\sin\beta)} \\ 2 - 2\cos(\alpha - \beta) &= 2 - 2(\cos\alpha\cos\beta + \sin\alpha\sin\beta) \\ -2\cos(\alpha - \beta) &= -2(\cos\alpha\cos\beta + \sin\alpha\sin\beta) \\ \boldsymbol{\cos(\alpha - \beta)} &\boldsymbol{= \cos\alpha\cos\beta + \sin\alpha\sin\beta} \end{aligned}$$

□

You can use this identity to find the exact values of functions of angles that are multiples of 15° as shown by Example 1.

EXAMPLE 1

$$\begin{aligned}\cos 345^\circ &= \cos 15^\circ && \text{Reduction principle}\\ &= \cos(45^\circ - 30^\circ) && \text{Since } 45^\circ - 30^\circ = 15^\circ\\ &= \cos 45^\circ \cos 30^\circ + \sin 45^\circ \sin 30^\circ\\ &= \frac{1}{2}\sqrt{2}\cdot\frac{1}{2}\sqrt{3} + \frac{1}{2}\sqrt{2}\cdot\frac{1}{2} && \text{Using exact values}\\ &= \frac{\sqrt{6}}{4} + \frac{\sqrt{2}}{4}\\ &= \frac{\sqrt{6}+\sqrt{2}}{4}\end{aligned}$$

■

Even though this identity is helpful for making evaluations (as in the preceding example), its real value lies in the fact that it is true for *any* choice of α and β. By making some particular choices for α and β, we find several useful special cases of this identity.

EXAMPLE 2 Prove $\cos(\frac{\pi}{2} - \theta) = \sin\theta$.

Solution This proof is based on the identity

$$\cos(\alpha - \beta) = \cos\alpha\cos\beta + \sin\alpha\sin\beta$$

Let $\alpha = \frac{\pi}{2}$ and $\beta = \theta$:

$$\begin{aligned}\cos\left(\frac{\pi}{2} - \theta\right) &= \cos\frac{\pi}{2}\cos\theta + \sin\frac{\pi}{2}\sin\theta\\ &= 0\cdot\cos\theta + 1\cdot\sin\theta\\ &= \sin\theta\end{aligned}$$

■

Example 2 is one of three identities known as the **cofunction identities**. (Remember that Identities 1–8 are the fundamental identities discussed in Section 7.2.)

Cofunction Identities

For any real number (or angle) θ,

9. $\cos\left(\frac{\pi}{2} - \theta\right) = \sin\theta$ 10. $\sin\left(\frac{\pi}{2} - \theta\right) = \cos\theta$ 11. $\tan\left(\frac{\pi}{2} - \theta\right) = \cot\theta$

The proof of Identity 9 was done in Example 2, and the proof of Identity 10 depends on Identity 9, and is shown below.

Proof of Identity 10

$$\begin{aligned}\cos\theta &= \cos\left[\frac{\pi}{2} - \left(\frac{\pi}{2} - \theta\right)\right]\\ &= \sin\left(\frac{\pi}{2} - \theta\right) && \text{This is Identity 9.}\end{aligned}$$

Therefore $\sin\left(\frac{\pi}{2} - \theta\right) = \cos\theta$. □

Identities involving the tangent are usually proved after proving similar identities for cosine and sine. The fundamental identity $\tan\theta = \sin\theta/\cos\theta$ is applied first, allowing you then to use the appropriate identities for cosine and sine. This process is illustrated with the following proof.

Proof of Identity 11

$$\tan\left(\frac{\pi}{2} - \theta\right) = \frac{\sin(\frac{\pi}{2} - \theta)}{\cos(\frac{\pi}{2} - \theta)} = \frac{\cos\theta}{\sin\theta} = \cot\theta$$

□

The cofunction identities allow us to change a trigonometric function to the cofunction of its complement.

EXAMPLE 3 Write each function in terms of its cofunction.

a. $\sin 28^\circ$ **b.** $\cos 43^\circ$ **c.** $\cot 9^\circ$ **d.** $\sin\frac{\pi}{6}$

Solution

a. $\sin 28^\circ = \cos(90^\circ - 28^\circ) = \cos 62^\circ$

b. $\cos 43^\circ = \sin(90^\circ - 43^\circ) = \sin 47^\circ$

c. $\cot 9^\circ = \tan(90^\circ - 9^\circ) = \tan 81^\circ$

d. $\sin\frac{\pi}{6} = \cos(\frac{\pi}{2} - \frac{\pi}{6}) = \cos\frac{2\pi}{6} = \cos\frac{\pi}{3}$

■

Suppose the given angle in Example 3 is larger than 90°; for example,

$$\cos 125^\circ = \sin(90^\circ - 125^\circ) = \sin(-35^\circ)$$

This result can be further simplified using the following **opposite-angle identities**.

Opposite-Angle Identities

For any real number (or angle) θ,

12. $\cos(-\theta) = \cos\theta$ 13. $\sin(-\theta) = -\sin\theta$ 14. $\tan(-\theta) = -\tan\theta$

Proof of Identity 12 Let $\alpha = 0$ and $\beta = \theta$ in the identity $\cos(\alpha - \beta) = \cos\alpha\cos\beta + \sin\alpha\sin\beta$.

$$\cos(0 - \theta) = \cos 0\cos\theta + \sin 0\sin\theta = 1 \cdot \cos\theta + 0 \cdot \sin\theta = \cos\theta$$

But, if you simplify directly,

$$\cos(0 - \theta) = \cos(-\theta)$$

Therefore $\cos(-\theta) = \cos\theta$ □

The proofs of Identities 13 and 14 are left for the reader.

EXAMPLE 4 Write each as a function of a positive angle or number.

a. $\cos(-19°)$ **b.** $\sin(-19°)$ **c.** $\tan(-2)$

Solution **a.** $\cos(-19°) = \cos 19°$ **b.** $\sin(-19°) = -\sin 19°$ **c.** $\tan(-2) = -\tan 2$

■

EXAMPLE 5 Write the given functions in terms of their cofunctions.

a. $\cos 125°$ **b.** $\sin 102°$ **c.** $\cot 2.5$

Solution **a.**
$$\begin{aligned}\cos 125° &= \sin(90° - 125°)\\ &= \sin(-35°)\\ &= -\sin 35°\end{aligned}$$

b.
$$\begin{aligned}\sin 102° &= \cos(90° - 102°)\\ &= \cos(-12°)\\ &= \cos 12°\end{aligned}$$

c.
$$\begin{aligned}\cot 2.5 &= \tan(\tfrac{\pi}{2} - 2.5)\\ &\approx \tan(-0.9292)\\ &= -\tan 0.9292\end{aligned}$$

■

You may also need to use opposite-angle identities together with other identities. For example, you know from algebra that

$$a - b \qquad \text{and} \qquad b - a$$

are opposites. This means that $a - b = -(b - a)$. In trigonometry you often see angles like $\frac{\pi}{2} - \theta$ and want to write $\theta - \frac{\pi}{2}$. This means that $\frac{\pi}{2} - \theta = -(\theta - \frac{\pi}{2})$. In particular,

$$\begin{aligned}\cos\left(\frac{\pi}{2} - \theta\right) &= \cos\left[-\left(\theta - \frac{\pi}{2}\right)\right]\\ &= \cos\left(\theta - \frac{\pi}{2}\right) \qquad \text{By Identity 12}\end{aligned}$$

EXAMPLE 6 Write the given functions using the opposite-angle identities.

a. $\sin(\frac{\pi}{2} - \theta)$ **b.** $\tan(\frac{\pi}{2} - \theta)$ **c.** $\cos(\pi - \theta)$

Solution **a.**
$$\begin{aligned}\sin(\tfrac{\pi}{2} - \theta) &= \sin[-(\theta - \tfrac{\pi}{2})]\\ &= -\sin(\theta - \tfrac{\pi}{2})\end{aligned}$$

b.
$$\begin{aligned}\tan(\tfrac{\pi}{2} - \theta) &= \tan[-(\theta - \tfrac{\pi}{2})]\\ &= -\tan(\theta - \tfrac{\pi}{2})\end{aligned}$$

c.
$$\begin{aligned}\cos(\pi - \theta) &= \cos[-(\theta - \pi)]\\ &= \cos(\theta - \pi)\end{aligned}$$

■

The procedure for graphing $y = -\cos\theta$ is identical to the procedure for graphing $y = \cos\theta$, except that, after the frame is drawn, the endpoints are at the bottom of the frame instead of at the top of the frame, and the midpoint is at the top, as shown in Figure 7.3.

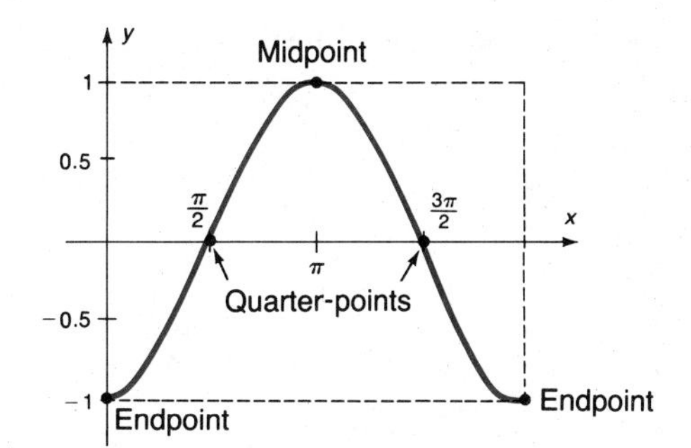

Figure 7.3 One period of $y = -\cos\theta$

EXAMPLE 7 Graph $y = \sin(-\theta)$.

Solution First use an opposite-angle identity, if necessary: $\sin(-\theta) = -\sin\theta$. To graph $y = -\sin\theta$, build the frame as before; the endpoints and midpoints are the same, but the quarter-points are reversed as shown in Figure 7.4.

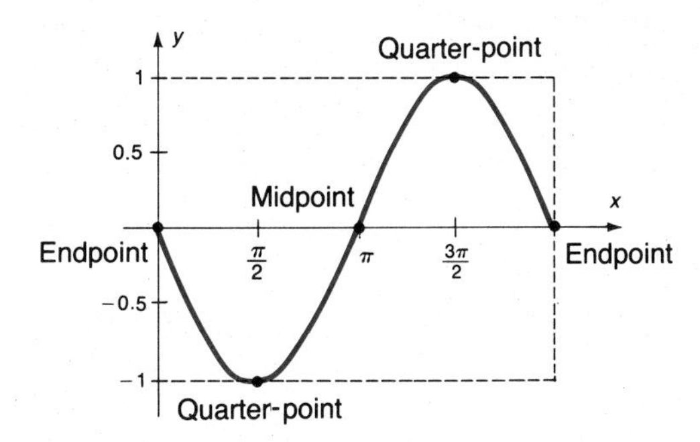

Figure 7.4 One period of $y = -\sin\theta$

■

The difference-of-angles identity proved at the beginning of this section is one of six identities known as the **addition laws**. Since subtraction can easily be written as a sum, the designation *addition laws* refers to both addition and subtraction.

Addition Laws

15. $\cos(\alpha + \beta) = \cos\alpha\cos\beta - \sin\alpha\sin\beta$
16. $\cos(\alpha - \beta) = \cos\alpha\cos\beta + \sin\alpha\sin\beta$
17. $\sin(\alpha + \beta) = \sin\alpha\cos\beta + \cos\alpha\sin\beta$
18. $\sin(\alpha - \beta) = \sin\alpha\cos\beta - \cos\alpha\sin\beta$
19. $\tan(\alpha + \beta) = \dfrac{\tan\alpha + \tan\beta}{1 - \tan\alpha\tan\beta}$
20. $\tan(\alpha - \beta) = \dfrac{\tan\alpha - \tan\beta}{1 + \tan\alpha\tan\beta}$

EXAMPLE 8 Write $\cos(\frac{2\pi}{3} + \theta)$ as a function of θ only.

Solution Use Identity 15:

$$\cos\left(\frac{2\pi}{3} + \theta\right) = \cos\frac{2\pi}{3}\cos\theta - \sin\frac{2\pi}{3}\sin\theta$$

$$= \left(-\frac{1}{2}\right)\cos\theta - \left(\frac{\sqrt{3}}{2}\right)\sin\theta$$ Substitute exact values where possible.

$$= -\frac{1}{2}(\cos\theta + \sqrt{3}\sin\theta)$$ Simplify. ■

EXAMPLE 9 Evaluate

$$\frac{\tan 18^\circ - \tan 40^\circ}{1 + \tan 18^\circ \tan 40^\circ}$$

using Table C.III or a calculator.

Solution You can do a lot of arithmetic, or use Identity 20:

$$\frac{\tan 18^\circ - \tan 40^\circ}{1 + \tan 18^\circ \tan 40^\circ} = \tan(18^\circ - 40^\circ)$$

$$= \tan(-22^\circ)$$

$$= -\tan 22^\circ$$ Identity 14

$$\approx -0.4040$$ By Table C.III or a calculator ■

We will conclude this section by proving some of the addition laws; others are left as problems.

Proof of Identity 15

$$\cos(\alpha + \beta) = \cos[\alpha - (-\beta)]$$

$$= \cos\alpha\cos(-\beta) + \sin\alpha\sin(-\beta)$$ By Identity 16, proved earlier

$$= \cos\alpha\cos\beta - \sin\alpha\sin\beta$$ By Identities 12 and 13

□

Identity 16 is the main cosine identity proved at the beginning of this section. It is numbered and included here for the sake of completeness.

Proof of Identity 17

$$\sin(\alpha + \beta) = \cos\left[\frac{\pi}{2} - (\alpha + \beta)\right]$$

$$= \cos\left[\left(\frac{\pi}{2} - \alpha\right) - \beta\right]$$

$$= \cos\left(\frac{\pi}{2} - \alpha\right)\cos\beta + \sin\left(\frac{\pi}{2} - \alpha\right)\sin\beta$$

$$= \sin\alpha\cos\beta + \cos\alpha\sin\beta$$ □

To prove Identity 18, replace β by $-\beta$ in Identity 17; the details are left as an exercise.

$$\sin 2\theta = \sin(\theta + \theta) = \sin\theta\cos\theta + \cos\theta\sin\theta$$
$$= \mathbf{2\sin\theta\cos\theta}$$

$$\tan 2\theta = \tan(\theta + \theta) = \frac{\tan\theta + \tan\theta}{1 - \tan\theta\tan\theta}$$
$$= \frac{\mathbf{2\tan\theta}}{\mathbf{1 - \tan^2\theta}}$$

□

EXAMPLE 1 $\cos 100x = \cos(2 \cdot 50x) = \cos^2 50x - \sin^2 50x$ ■

EXAMPLE 2 $\sin 120° = \sin 2(60°) = 2\sin 60° \cos 60°$ ■

EXAMPLE 3
$$\cos 3\theta = \cos(2\theta + \theta) = \cos 2\theta\cos\theta - \sin 2\theta\sin\theta$$
$$= (\cos^2\theta - \sin^2\theta)\cos\theta - (2\sin\theta\cos\theta)\sin\theta$$
$$= \cos^3\theta - \sin^2\theta\cos\theta - 2\sin^2\theta\cos\theta$$
$$= \cos^3\theta - 3\sin^2\theta\cos\theta$$
$$= \cos^3\theta - 3(1 - \cos^2\theta)\cos\theta$$
$$= \cos^3\theta - 3\cos\theta + 3\cos^3\theta$$
$$= 4\cos^3\theta - 3\cos\theta$$
■

EXAMPLE 4 Evaluate $\dfrac{2\tan\frac{\pi}{16}}{1 - \tan^2\frac{\pi}{16}}$.

Solution Notice this is the right-hand side of Identity 23 so it is the same as $\tan(2 \cdot \frac{\pi}{16}) = \tan\frac{\pi}{8}$. Now use tables or a calculator to find $\tan\frac{\pi}{8} \approx \mathbf{0.4142}$. ■

EXAMPLE 5 If $\cos\theta = \frac{3}{5}$ and θ is in Quadrant IV, find $\cos 2\theta$, $\sin 2\theta$, and $\tan 2\theta$.

Solution Since $\cos 2\theta = 2\cos^2\theta - 1$,

$$\cos 2\theta = 2\left(\frac{3}{5}\right)^2 - 1$$
$$= 2\left(\frac{9}{25}\right) - 1$$
$$= \mathbf{-\frac{7}{25}}$$

For the other functions of 2θ, you need to know $\sin\theta$. Begin with the fundamental identity relating cosine and sine:

$$\sin^2\theta = 1 - \cos^2\theta$$
$$\sin\theta = -\sqrt{1 - \cos^2\theta} \qquad \text{Negative since the sine is negative in Quadrant IV}$$
$$= -\sqrt{1 - \left(\frac{3}{5}\right)^2}$$

$$\begin{aligned}\sin\theta &= -\sqrt{1-\frac{9}{25}}\\ &= -\sqrt{\frac{16}{25}}\\ &= -\frac{4}{5}\end{aligned}$$

Now,

$$\begin{aligned}\sin 2\theta &= 2\sin\theta\cos\theta\\ &= 2\left(-\frac{4}{5}\right)\left(\frac{3}{5}\right)\\ &= \mathbf{-\frac{24}{25}}\end{aligned}$$

Finally,

$$\begin{aligned}\tan 2\theta &= \frac{\sin 2\theta}{\cos 2\theta}\\ &= \frac{-\frac{24}{25}}{-\frac{7}{25}}\\ &= \mathbf{\frac{24}{7}}\end{aligned}$$

■

Identity 21 leads us to the second important special case of the addition laws, called the **half-angle identities**. We wish to solve $\cos 2\alpha = 2\cos^2\alpha - 1$ for $\cos^2\alpha$.

$$\begin{aligned}2\cos^2\alpha - 1 &= \cos 2\alpha\\ 2\cos^2\alpha &= 1 + \cos 2\alpha\\ \cos^2\alpha &= \frac{1+\cos 2\alpha}{2}\end{aligned}$$

Now, if $\alpha = \frac{1}{2}\theta$, then $2\alpha = \theta$ and

$$\cos^2\frac{1}{2}\theta = \frac{1+\cos\theta}{2}$$

If $\frac{1}{2}\theta$ is in Quadrant I or IV, then

$$\cos\frac{1}{2}\theta = \sqrt{\frac{1+\cos\theta}{2}}$$

If $\frac{1}{2}\theta$ is in Quadrant II or III, then

$$\cos\frac{1}{2}\theta = -\sqrt{\frac{1+\cos\theta}{2}}$$

These results are summarized by writing

$$\cos\frac{1}{2}\theta = \pm\sqrt{\frac{1+\cos\theta}{2}}$$

WARNING: Notice $\pm$ usage and meaning.

However, *you must be careful.* The sign $+$ or $-$ is chosen according to which quadrant $\frac{1}{2}\theta$ is in. The formula requires either $+$ or $-$, but not both. This use of $\pm$ is different from the use of $\pm$ in algebra. For example, when using $\pm$ in the quadratic formula, we are indicating *two* possible correct roots. In this trigonometric identity we will obtain *one* correct value depending on the quadrant of $\frac{1}{2}\theta$.

For the sine, solve $\cos 2\alpha = 1 - 2\sin^2\alpha$ for $\sin^2\alpha$.

$$\cos 2\alpha = 1 - 2\sin^2\alpha$$
$$2\sin^2\alpha = 1 - \cos 2\alpha$$
$$\sin^2\alpha = \frac{1-\cos 2\alpha}{2}$$

Replace $\alpha = \frac{1}{2}\theta$, and

$$\sin^2\frac{1}{2}\theta = \frac{1-\cos\theta}{2}$$

or

$$\sin\frac{1}{2}\theta = \pm\sqrt{\frac{1-\cos\theta}{2}}$$

where the sign depends on the quadrant of $\frac{1}{2}\theta$. If $\frac{1}{2}\theta$ is in Quadrant I or II, you use $+$; if it is in Quadrant III or IV, you use $-$.

Finally, to find the half-angle identity for the tangent, write

$$\tan\frac{1}{2}\theta = \frac{\sin\frac{1}{2}\theta}{\cos\frac{1}{2}\theta}$$

$$= \frac{\pm\sqrt{\frac{1-\cos\theta}{2}}}{\pm\sqrt{\frac{1+\cos\theta}{2}}}$$

$$= \pm\sqrt{\frac{1-\cos\theta}{1+\cos\theta}}$$

$$= \pm\sqrt{\frac{1-\cos\theta}{1+\cos\theta}\cdot\frac{1-\cos\theta}{1-\cos\theta}}$$

$$= \pm\sqrt{\frac{(1-\cos\theta)^2}{\sin^2\theta}}$$

Remember that $1 - \cos^2\theta = \sin^2\theta$.

$$= \frac{1-\cos\theta}{\sin\theta}$$

Notice that $1 - \cos\theta$ is positive. Also, since $\tan\frac{1}{2}\theta$ and $\sin\theta$ have the same sign regardless of the quadrant of θ, the desired result follows.

You can also show that $\tan\frac{1}{2}\theta = \dfrac{\sin\theta}{1+\cos\theta}$.

Half-Angle Identities

24. $\cos\dfrac{1}{2}\theta = \pm\sqrt{\dfrac{1+\cos\theta}{2}}$

25. $\sin\dfrac{1}{2}\theta = \pm\sqrt{\dfrac{1-\cos\theta}{2}}$

26. $\tan\dfrac{1}{2}\theta = \dfrac{1-\cos\theta}{\sin\theta}$

$\qquad = \dfrac{\sin\theta}{1+\cos\theta}$

To help you remember the correct sign between the first two half-angle identities, remember "*sinus-minus*"—the sine is minus.

EXAMPLE 6 Find the exact value of $\cos\frac{9\pi}{8}$.

Solution

$$\cos\frac{9\pi}{8} = \cos\left(\frac{1}{2}\cdot\frac{9\pi}{4}\right) = -\sqrt{\frac{1+\cos\frac{9\pi}{4}}{2}}$$

Choose a negative sign, since $\frac{9\pi}{8}$ is in Quadrant III and the cosine is negative in this quadrant.

$$= -\sqrt{\frac{1+\cos\frac{\pi}{4}}{2}}$$

$$= -\sqrt{\frac{1+\frac{\sqrt{2}}{2}}{2}}$$

$$= -\sqrt{\frac{2+\sqrt{2}}{4}}$$

$$= -\frac{1}{2}\sqrt{2+\sqrt{2}}$$

■

EXAMPLE 7 If $\cot 2\theta = \frac{3}{4}$, find $\cos\theta$, $\sin\theta$, and $\tan\theta$, where 2θ is in Quadrant I.

Solution You need to find $\cos 2\theta$ so that you can use it in the half-angle identities. To do this, first find $\tan 2\theta$:

$$\tan 2\theta = \frac{1}{\cot 2\theta} = \frac{4}{3}$$

Next, find $\sec 2\theta$:

$$\sec 2\theta = \pm\sqrt{1+\tan^2 2\theta}$$ From Identity 7

$$= \sqrt{1+\frac{16}{9}}$$ It is positive because 2θ is in Quadrant I.

$$= \frac{5}{3}$$

Finally, $\cos 2\theta$ is the reciprocal of $\sec 2\theta$: $\quad \cos 2\theta = \frac{3}{5}$

Next use the half-angle identities:

$$\cos\theta = \pm\sqrt{\frac{1+\cos 2\theta}{2}} \qquad \text{and} \qquad \sin\theta = \pm\sqrt{\frac{1-\cos 2\theta}{2}}$$

Do you see that θ is one-half of 2θ in these formulas?

$$\cos\theta = +\sqrt{\frac{1+\frac{3}{5}}{2}} \qquad \sin\theta = +\sqrt{\frac{1-\frac{3}{5}}{2}}$$

Positive value chosen because θ is in Quadrant I.

$$\cos\theta = \frac{2}{\sqrt{5}} \qquad \sin\theta = \frac{1}{\sqrt{5}}$$

$$\begin{aligned}\tan\theta &= \frac{\sin\theta}{\cos\theta}\\ &= \frac{1/\sqrt{5}}{2/\sqrt{5}}\\ &= \frac{1}{2}\end{aligned}$$

■

EXAMPLE 8 Prove that $\sin\theta = \dfrac{2\tan\frac{1}{2}\theta}{1+\tan^2\frac{1}{2}\theta}$.

Solution When proving identities involving functions of different angles, you should write all the trigonometric functions in the problem as functions of a single angle.

$$\begin{aligned}\frac{2\tan\frac{1}{2}\theta}{1+\tan^2\frac{1}{2}\theta} &= \frac{2\dfrac{\sin\frac{1}{2}\theta}{\cos\frac{1}{2}\theta}}{\sec^2\frac{1}{2}\theta}\\ &= 2\frac{\sin\frac{1}{2}\theta}{\cos\frac{1}{2}\theta}\cdot\cos^2\tfrac{1}{2}\theta\\ &= 2\sin\tfrac{1}{2}\theta\cos\tfrac{1}{2}\theta\\ &= \sin\theta \qquad \text{From Identity 22 (double-angle identity)}\end{aligned}$$

■

It is sometimes convenient, or even necessary, to write a trigonometric sum as a product or a product as a sum. We conclude this section with eight additional identities.

Product Identities

27. $2\cos\alpha\cos\beta = \cos(\alpha-\beta) + \cos(\alpha+\beta)$
28. $2\sin\alpha\sin\beta = \cos(\alpha-\beta) - \cos(\alpha+\beta)$
29. $2\sin\alpha\cos\beta = \sin(\alpha+\beta) + \sin(\alpha-\beta)$
30. $2\cos\alpha\sin\beta = \sin(\alpha+\beta) - \sin(\alpha-\beta)$

The proofs of the product identities involve systems of equations, and will therefore be delayed until Chapter 9, Problem Set 9.1.

EXAMPLE 9 Write $2\sin 3\sin 1$ as the sum of two functions.*

Solution Use Identity 28, where $\alpha = 3$ and $\beta = 1$:

$$\begin{aligned}2\sin 3\sin 1 &= \cos(3-1) - \cos(3+1)\\ &= \cos 2 - \cos 4\end{aligned}$$

■

EXAMPLE 10 Write $\sin 40^\circ \cos 12^\circ$ as the sum of two functions.

Solution Use Identity 29 where $\alpha = 40^\circ$ and $\beta = 12^\circ$:

$$\begin{aligned}2\sin 40^\circ \cos 12^\circ &= \sin(40^\circ + 12^\circ) + \sin(40^\circ - 12^\circ)\\ &= \sin 52^\circ + \sin 28^\circ\end{aligned}$$

But what about the coefficient 2? Since you know that the preceding is an *equation* that is true, you can divide both sides by 2 to obtain:

$$\sin 40^\circ \cos 12^\circ = \frac{1}{2}(\sin 52^\circ + \sin 28^\circ)$$

■

By making appropriate substitutions and again using systems (as shown in Chapter 9), Identities 27–30 can be rewritten in a form known as the **sum identities**.

Sum Identities

31. $\cos x + \cos y = 2\cos\left(\frac{x+y}{2}\right)\cos\left(\frac{x-y}{2}\right)$

32. $\cos x - \cos y = -2\sin\left(\frac{x+y}{2}\right)\sin\left(\frac{x-y}{2}\right)$

33. $\sin x + \sin y = 2\sin\left(\frac{x+y}{2}\right)\cos\left(\frac{x-y}{2}\right)$

34. $\sin x - \sin y = 2\sin\left(\frac{x-y}{2}\right)\cos\left(\frac{x+y}{2}\right)$

EXAMPLE 11 Write $\sin 35^\circ + \sin 27^\circ$ as a product.

Solution $x = 35^\circ$, $y = 27^\circ$, and

$$\frac{x+y}{2} = \frac{35^\circ + 27^\circ}{2} = 31^\circ; \qquad \frac{x-y}{2} = 4^\circ$$

Therefore, $\sin 35^\circ + \sin 27^\circ = 2\sin 31^\circ \cos 4^\circ$. ■

You sometimes will use these product and sum identities to prove other identities.

---

* Remember that sum also includes difference, because $a - b = a + (-b)$.

EXAMPLE 12 Prove $\dfrac{\sin 7\gamma + \sin 5\gamma}{\cos 7\gamma - \cos 5\gamma} = -\cot\gamma$.

Solution

$$\frac{\sin 7\gamma + \sin 5\gamma}{\cos 7\gamma - \cos 5\gamma} = \frac{2\sin\left(\frac{7\gamma + 5\gamma}{2}\right)\cos\left(\frac{7\gamma - 5\gamma}{2}\right)}{-2\sin\left(\frac{7\gamma + 5\gamma}{2}\right)\sin\left(\frac{7\gamma - 5\gamma}{2}\right)}$$

$$= \frac{2\sin 6\gamma\cos\gamma}{-2\sin 6\gamma\sin\gamma}$$

$$= -\frac{\cos\gamma}{\sin\gamma}$$

$$= -\cot\gamma$$

■

In calculus you frequently need to solve trigonometric equations that are sums or differences. The procedure is to change a sum to a product and then use the Zero Factor Theorem as illustrated by Example 13.

EXAMPLE 13 Solve $\cos 5x - \cos 3x = 0$ $(0 \leq x < 2\pi)$.

Solution

$$\cos 5x - \cos 3x = 0$$

$$-2\sin\left(\frac{5x + 3x}{2}\right)\sin\left(\frac{5x - 3x}{2}\right) = 0 \qquad \text{Use Identity 32.}$$

$$\sin 4x \sin x = 0 \qquad \text{Divide both sides by } -2 \text{ and simplify the argument.}$$

$\sin 4x = 0$ $\qquad$ $\sin x = 0$ $\qquad$ Use the Zero Factor Theorem.

$4x = 0, \pi, 2\pi, 3\pi, 4\pi, 5\pi, 6\pi, 7\pi$ $\qquad$ $x = 0, \pi$

$x = 0, \frac{\pi}{4}, \frac{\pi}{2}, \frac{3\pi}{4}, \pi, \frac{5\pi}{4}, \frac{3\pi}{2}, \frac{7\pi}{4}$

The solution is: $\mathbf{0, \frac{\pi}{4}, \frac{\pi}{2}, \frac{3\pi}{4}, \pi, \frac{5\pi}{4}, \frac{3\pi}{2}, \frac{7\pi}{4}}$ ■

PROBLEM SET 7.5

A

Use the double-angle or half-angle identities to evaluate each of Problems 1–9 using exact values.

1. $2\cos^2 22.5° - 1$

2. $\dfrac{2\tan\frac{\pi}{8}}{1 - \tan^2\frac{\pi}{8}}$

3. $\sqrt{\dfrac{1 - \cos 60°}{2}}$

4. $\cos^2 15° - \sin^2 15°$

5. $1 - 2\sin^2 90°$

6. $-\sqrt{\dfrac{1 - \cos 420°}{2}}$

7. $\sin 22.5°$

8. $\cos\frac{\pi}{8}$

9. $\tan 22.5°$

In each of Problems 10–15, find the exact values of cosine, sine, and tangent of 2θ.

10. $\sin\theta = \frac{3}{5}$; θ in Quadrant I

11. $\sin\theta = \frac{5}{13}$; θ in Quadrant II

12. $\tan\theta = -\frac{5}{12}$; θ in Quadrant IV

13. $\tan\theta = -\frac{3}{4}$; θ in Quadrant II

14. $\cos\theta = \frac{5}{9}$; θ in Quadrant I
15. $\cos\theta = -\frac{5}{13}$; θ in Quadrant III

In each of Problems 16–21, find the exact values of cosine, sine, and tangent of $\frac{1}{2}\theta$.

16. $\sin\theta = \frac{3}{5}$; θ in Quadrant I
17. $\sin\theta = \frac{5}{13}$; θ in Quadrant II
18. $\tan\theta = -\frac{5}{12}$; θ in Quadrant IV
19. $\tan\theta = -\frac{3}{4}$; θ in Quadrant II
20. $\cos\theta = \frac{5}{9}$; θ in Quadrant I
21. $\cos\theta = -\frac{5}{13}$; θ in Quadrant III

Write each of the expressions in Problems 22–29 as the sum of two functions.

22. $2\sin 35^\circ \sin 24^\circ$
23. $2\cos 46^\circ \cos 18^\circ$
24. $\sin 41^\circ \cos 19^\circ$
25. $\cos 115^\circ \sin 200^\circ$
26. $\sin 225^\circ \sin 300^\circ$
27. $\sin 2\theta \sin 5\theta$
28. $\cos\theta \cos 3\theta$
29. $\cos 3\theta \sin 2\theta$

Write each of the expressions in Problems 30–37 as a product of functions.

30. $\sin 43^\circ + \sin 63^\circ$
31. $\sin 22^\circ - \sin 6^\circ$
32. $\cos 81^\circ - \cos 79^\circ$
33. $\cos 78^\circ + \cos 25^\circ$
34. $\sin 215^\circ + \sin 300^\circ$
35. $\cos 25^\circ - \cos 100^\circ$
36. $\sin 3x - \sin 2x$
37. $\cos 6x + \cos 2x$

Solve each of the equations in Problems 39–45 for $0 \le x < 2\pi$.

38. $\sin x - \sin 3x = 0$
39. $\cos 5x - \cos 3x = 0$
40. $\sin x + \sin 3x = 0$
41. $\cos 3\theta + \cos\theta = 0$
42. $\cos 5y + \cos 3y = 0$
43. $\sin 3z - \sin 5z = 0$
44. $\cos 3y + \cos 3y = 0$
45. $\sin 4\theta + \sin 4\theta = 0$

B

In each of Problems 46–51, find $\cos\theta$, $\sin\theta$, *and* $\tan\theta$ *when* θ *is in Quadrant I and* $\cot 2\theta$ *is given.*

46. $\cot 2\theta = -\frac{3}{4}$
47. $\cot 2\theta = 0$
48. $\cot 2\theta = \frac{1}{\sqrt{3}}$
49. $\cot 2\theta = -\frac{1}{\sqrt{3}}$
50. $\cot 2\theta = -\frac{4}{3}$
51. $\cot 2\theta = \frac{4}{3}$

52. ***Aviation*** An airplane flying faster than the speed of sound is said to have a speed greater than Mach 1. The Mach number is the ratio of the speed of the plane to the speed of sound and is denoted by M. When a plane flies faster than the speed of sound, a sonic boom is heard, created by sound waves that form a cone with a vertex angle θ, as shown in Figure 7.5. It can be shown that, if $M > 1$, then

Figure 7.5 Pattern of sound waves creating a sonic boom

$$\sin\frac{\theta}{2} = \frac{1}{M}$$

a. If $\theta = \frac{\pi}{6}$, find the Mach number to the nearest tenth.
b. Find the exact Mach number for part **a**.

Prove each of the identities in Problems 53–61.

53. $\sin\alpha = 2\sin\frac{\alpha}{2}\cos\frac{\alpha}{2}$
54. $\cos 4\theta = \cos^2 2\theta - \sin^2 2\theta$
55. $\sin 2\theta = \dfrac{2\tan\theta}{1+\tan^2\theta}$
56. $\tan\frac{3}{2}\beta = \dfrac{2\tan\frac{3\beta}{4}}{1-\tan^2\frac{3\beta}{4}}$
57. $\tan\frac{1}{2}\theta = \dfrac{1-\cos\theta}{\sin\theta}$
58. $\tan\frac{1}{2}\theta = \dfrac{\sin\theta}{1+\cos\theta}$
59. $\dfrac{\sin 5\theta + \sin 3\theta}{\cos 5\theta + \cos 3\theta} = \tan 4\theta$
60. $\dfrac{\cos 5w + \cos w}{\cos w - \cos 5w} = \dfrac{\cot 2w}{\tan 3w}$
61. $\dfrac{\cos 3\theta - \cos\theta}{\sin\theta - \sin 3\theta} = \tan 2\theta$

C

Prove each of the identities in Problems 62–66.

62. $\sin 3\theta = 3\sin\theta - 4\sin^3\theta$
63. $\tan\frac{B}{2} = \csc B - \cot B$
64. $\sin 4\theta = 4\sin\theta\cos\theta - 8\sin^3\theta\cos\theta$
65. $\frac{1}{2}\cot x - \frac{1}{2}\tan x = \cot 2x$
66. $\cos^4\theta = \frac{1}{8}(3 + 4\cos 2\theta + \cos 4\theta)$

7.6 DE MOIVRE'S THEOREM*

Consider a graphical representation of a complex number. To give a graphic representation of complex numbers, such as

$$2 + 3i, \qquad -i, \qquad -3 - 4i, \qquad 3i, \qquad -2 + \sqrt{2}i, \qquad \frac{3}{2} - \frac{5}{2}i,$$

a two-dimensional coordinate system is used. The horizontal axis represents the **real axis** and the vertical axis is the **imaginary axis**, so that $a + bi$ is represented by the ordered pair (a, b). Remember that a and b represent real numbers, so we plot (a, b) in the usual manner, as shown in Figure 7.6. The coordinate system in Figure 7.6 is called the **complex plane** or the **Gaussian plane**, in honor of Karl Friedrich Gauss.

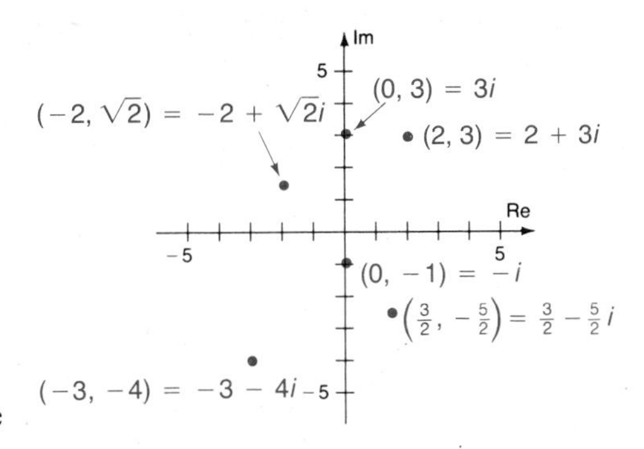

Figure 7.6 Complex plane

The **absolute value** of a complex number z is, graphically, the distance between z and the origin (just as it is for real numbers). The absolute value of a complex number is also called the **modulus**. The distance formula leads to the following definition.

Absolute Value, or Modulus, of a Complex Number

> If $z = a + bi$, then the *absolute value*, or *modulus*, of z is denoted by $|z|$ and defined by
>
> $$|z| = \sqrt{a^2 + b^2}$$

EXAMPLE 1 Find the absolute value.

a. $3 + 4i$

$$\text{Absolute value: } |3 + 4i| = \sqrt{3^2 + 4^2} = \sqrt{25} = \mathbf{5}$$

---

* This is an optional section which requires complex numbers from Section 1.3.

b. $-2 + \sqrt{2}i$

$$\text{Absolute value: } |-2 + \sqrt{2}i| = \sqrt{4 + 2} = \sqrt{6}$$

c. -3

$$\text{Absolute value: } |-3 + 0i| = \sqrt{9 + 0} = 3$$

This example shows that the definition of absolute value for complex numbers is consistent with the definition of absolute value given for real numbers. ■

The form $a + bi$ is called the **rectangular form**, but another useful representation uses trigonometry. Consider the graphical representation of a complex number $a + bi$, as shown in Figure 7.7. Let r be the distance from the origin to (a, b) and let θ be the angle the segment makes with the real axis. Then

$$r = \sqrt{a^2 + b^2}$$

Figure 7.7 Trigonometric form and rectangular form of a complex number

and θ, called the **argument**, is chosen so that it is the smallest nonnegative angle the terminal side makes with the positive real axis. From the definition of the trigonometric functions,

$$\cos\theta = \frac{a}{r} \qquad \sin\theta = \frac{b}{r} \qquad \tan\theta = \frac{b}{a}$$

$$a = r\cos\theta \qquad b = r\sin\theta$$

Therefore

$$a + bi = r\cos\theta + ir\sin\theta = r(\cos\theta + i\sin\theta)$$

Sometimes $r(\cos\theta + i\sin\theta)$ is abbreviated by

$$r(\cos\theta + i\sin\theta) \rightarrow r(\text{c}\ \ i\text{s}\ \theta) \quad \text{or } r\,\text{cis}\,\theta$$

Trigonometric Form of a Complex Number

> The **trigonometric form** of a complex number $z = a + bi$ is
>
> $$r(\cos\theta + i\sin\theta) = r\,\text{cis}\,\theta$$
>
> where $r = \sqrt{a^2 + b^2}$; $\tan\theta = b/a$ if $a \neq 0$; $a = r\cos\theta$; $b = r\sin\theta$. This representation is unique for $0^\circ \leq \theta < 360^\circ$ for all z except $0 + 0i$.

The placement of θ in the proper quadrant is an important consideration because there are two values of $0^\circ \leq \theta < 360^\circ$ that will satisfy the relationship

$$\tan\theta = \frac{b}{a}$$

For example, compare the following:

1. $-1 + i \qquad a = -1, b = 1 \qquad \tan\theta = \frac{1}{-1} \quad \text{or} \quad \tan\theta = -1$

2. $1 - i \qquad a = 1, b = -1 \qquad \tan\theta = \frac{-1}{1} \quad \text{or} \quad \tan\theta = -1$

Notice the same trigonometric equation for both complex numbers, even though $-1 + i$ is in Quadrant II and $1 - i$ is in Quadrant IV. This consideration of quadrants is even more important when you are doing the problem on a calculator, since the proper sequence of steps for this example is

[1] [+/−] [inv] [tan]

This gives the result $-45°$, which is not true for either example since $0° \le \theta < 360°$. The entire process can be dealt with quite simply if you let θ' be the reference angle for θ. Then find the reference angle

$$\theta' = \tan^{-1}\left|\frac{b}{a}\right|$$

After you know the reference angle and the quadrant, it is easy to find θ. For these examples,

$$\theta' = \tan^{-1}|-1| \qquad \text{On a calculator: [1] [inv] [tan]}$$
$$= 45°$$

For Quadrant II, $\theta = 135°$; for Quadrant IV, $\theta = 315°$.
$-1 + i$ is in Quadrant II (plot it), so $\theta = 135°$; $1 - i$ is in Quadrant IV, so $\theta = 315°$.

EXAMPLE 2 Change the complex numbers to trigonometric form.

a. $1 - \sqrt{3}i$

$a = 1$ and $b = -\sqrt{3}$; the number is in Quadrant IV.

$$r = \sqrt{1^2 + (-\sqrt{3})^2} \qquad \theta' = \tan^{-1}\left|\frac{-\sqrt{3}}{1}\right|$$
$$= \sqrt{4} \qquad = 60°$$
$$= 2$$

On a calculator: [3] [$\sqrt{x}$] [inv] [tan]

The reference angle is 60°; in Quadrant IV: $\theta = 300°$.

Thus $\mathbf{1 - \sqrt{3}i = 2 \operatorname{cis} 300°}$.

b. $6i$

$a = 0$ and $b = 6$; notice that $\tan\theta$ is not defined for $\theta = 90°$. By inspection, $6i = \mathbf{6 \operatorname{cis} 90°}$.

c. $4.310 + 5.516i$

$a = 4.310$ and $b = 5.516$; the number is in Quadrant I.

$$r = \sqrt{(4.310)^2 + (5.516)^2} \qquad \theta = \tan^{-1}\left|\frac{5.516}{4.310}\right|$$
$$\approx \sqrt{49} \qquad \approx \tan^{-1}(1.2798)$$
$$= 7 \qquad \approx 52°$$

Thus $\mathbf{4.310 + 5.516i \approx 7 \operatorname{cis} 52°}$. ■

HP-28S ACTIVITY

In general, changing a complex number in rectangular form to trigonometric form is not a major undertaking. However, the procedure is particularly easy on an HP-28S. Consider the conversion shown in Example 2c.

Obtain the MODE menu, fix the number of decimal points to 4 (4 FIX), and set the calculator to degrees (DEG).

For Example 2c, press

4.31 ENTER 5.516 ENTER

Now obtain the TRIG menu and press NEXT and then R → C (which means rectangular to complex form). This converts the two real numbers entered above to complex number notation in the calculator.

Next, press the R → P key, which converts to the trigonometric (polar) form. Notice that the output is in ordered pair notation, so (7.0002, 51.9972) means 7.0002 cis 51.9972°.

EXAMPLE 3 Change the complex numbers to rectangular form.

a. 4 cis 330°

$r = 4$ and $\theta = 330°$

$$a = 4\cos 330° = 4\left(\frac{\sqrt{3}}{2}\right) = 2\sqrt{3} \qquad b = 4\sin 330° = 4\left(-\frac{1}{2}\right) = -2$$

Thus **4 cis 330° = $2\sqrt{3} - 2i$.**

b. $5(\cos 38° + i\sin 38°)$

$r = 5$ and $\theta = 38°$

$$a = 5\cos 38° \approx 3.94 \qquad b = 5\sin 38° \approx 3.08$$

Thus **$5(\cos 38° + i\sin 38°) \approx 3.94 + 3.08i$.** ■

The great advantage of trigonometric form over rectangular form is the ease with which you can multiply and divide complex numbers.

Products and Quotients of Complex Numbers in Trigonometric Form

> Let $z_1 = r_1 \operatorname{cis} \theta_1$ and $z_2 = r_2 \operatorname{cis} \theta_2$ be nonzero complex numbers. Then
>
> $$z_1 z_2 = r_1 r_2 \operatorname{cis}(\theta_1 + \theta_2) \qquad \frac{z_1}{z_2} = \frac{r_1}{r_2} \operatorname{cis}(\theta_1 - \theta_2)$$

Proof

$$\begin{aligned} z_1 z_2 &= (r_1 \operatorname{cis} \theta_1)(r_2 \operatorname{cis} \theta_2) \\ &= [r_1(\cos\theta_1 + i\sin\theta_1)][r_2(\cos\theta_2 + i\sin\theta_2)] \\ &= r_1 r_2(\cos\theta_1 + i\sin\theta_1)(\cos\theta_2 + i\sin\theta_2) \\ &= r_1 r_2(\cos\theta_1\cos\theta_2 + i\cos\theta_1\sin\theta_2 + i\sin\theta_1\cos\theta_2 - \sin\theta_1\sin\theta_2) \\ &= r_1 r_2[(\cos\theta_1\cos\theta_2 - \sin\theta_1\sin\theta_2) + i(\cos\theta_1\sin\theta_2 + \sin\theta_1\cos\theta_2)] \\ &= r_1 r_2[\cos(\theta_1 + \theta_2) + i\sin(\theta_1 + \theta_2)] \\ &= r_1 r_2 \operatorname{cis}(\theta_1 + \theta_2) \end{aligned}$$

The proof of the quotient form is similar and is left as a problem. □

EXAMPLE 4 Simplify:

a. $5 \operatorname{cis} 38° \cdot 4 \operatorname{cis} 75° = 5 \cdot 4 \operatorname{cis}(38° + 75°)$
$= 20 \operatorname{cis} 113°$

b. $\sqrt{2} \operatorname{cis} 188° \cdot 2\sqrt{2} \operatorname{cis} 310° = 4 \operatorname{cis} 498°$
$= 4 \operatorname{cis} 138°$

c. $(2 \operatorname{cis} 48°)^3 = (2 \operatorname{cis} 48°)(2 \operatorname{cis} 48°)^2$
$= (2 \operatorname{cis} 48°)(4 \operatorname{cis} 96°)$
$= 8 \operatorname{cis} 144°$

Notice that this result is the same as $(2 \operatorname{cis} 48°)^3 = 2^3 \operatorname{cis}(3 \cdot 48°) = 8 \operatorname{cis} 144°$ ■

EXAMPLE 5 Find $\dfrac{15(\cos 48° + i\sin 48°)}{5(\cos 125° + i\sin 125°)}$.

Solution

$$\begin{aligned} \frac{15 \operatorname{cis} 48°}{5 \operatorname{cis} 125°} &= 3 \operatorname{cis}(48° - 125°) \\ &= 3 \operatorname{cis}(-77°) \\ &= 3 \operatorname{cis} 283° \end{aligned}$$

Remember that arguments should be between 0° and 360°. ■

EXAMPLE 6 Simplify $(1 - \sqrt{3}i)^5$.

Solution First change to trigonometric form:

$a = 1; b = -\sqrt{3}$; Quadrant IV

$r = \sqrt{1 + 3} \qquad \theta' = \tan^{-1}\left|-\frac{\sqrt{3}}{1}\right|$

$= 2 \qquad\qquad = 60°$

$\theta = 300° \qquad$ (Quadrant IV)

$$\begin{aligned}(1-\sqrt{3}i)^5 &= (2\operatorname{cis}300^\circ)^5\\ &= 2^5\operatorname{cis}(5\cdot 300^\circ)\\ &= 32\operatorname{cis}1500^\circ\\ &= 32\operatorname{cis}60^\circ\end{aligned}$$

If you want the answer in rectangular form, you can now change back:

$$\begin{aligned} a &= 32\cos 60^\circ & b &= 32\sin 60^\circ\\ &= 32\left(\frac{1}{2}\right) & &= 32\left(\frac{1}{2}\sqrt{3}\right)\\ &= 16 & &= 16\sqrt{3}\end{aligned}$$

Thus $(1-\sqrt{3}i)^5 = 16 + 16\sqrt{3}i$. ■

As you can see from Example 6, multiplication in trigonometric form extends quite nicely to any positive integral power in a result called *De Moivre's Theorem.* This theorem is proved by mathematical induction (Chapter 10).

De Moivre's Theorem

If n is a natural number, then

$$(r\operatorname{cis}\theta)^n = r^n\operatorname{cis}n\theta$$

for a complex number $r\operatorname{cis}\theta = r(\cos\theta + i\sin\theta)$.

Although De Moivre's Theorem is useful for powers as illustrated by Example 6, its real usefulness is in finding the complex roots of numbers. Recall from algebra that $\sqrt[n]{r} = r^{1/n}$ is used to denote the principal nth root of r. However, $r^{1/n}$ is only *one* of the nth roots of r. How do you find *all* nth roots of r? To find the principal root, you can use a calculator or logarithms along with the following theorem, which follows directly from De Moivre's Theorem.

nth Root Theorem

If n is any positive integer, then the nth roots of $r\operatorname{cis}\theta$ are given by

$$\sqrt[n]{r}\operatorname{cis}\left(\frac{\theta+360^\circ k}{n}\right) \quad\text{or}\quad \sqrt[n]{r}\operatorname{cis}\left(\frac{\theta+2\pi k}{n}\right)$$

for $k = 0, 1, 2, 3, \ldots, n-1$.

The proof of this theorem is left as a problem.

EXAMPLE 7 Find the square roots of $-\frac{9}{2}+\frac{9}{2}\sqrt{3}i$.

Solution First change to trigonometric form:

$$r = \sqrt{\left(-\frac{9}{2}\right)^2 + \left(\frac{9}{2}\sqrt{3}\right)^2} \qquad \theta' = \tan^{-1}\left|\frac{\frac{9}{2}\sqrt{3}}{-\frac{9}{2}}\right|$$

$$= \sqrt{\frac{81}{4} + \frac{81 \cdot 3}{4}} \qquad = \tan^{-1}(\sqrt{3})$$

$$= \sqrt{81\left(\frac{1}{4} + \frac{3}{4}\right)} \qquad = 60^\circ$$

$$\theta = 120^\circ \qquad \text{(Quadrant II)}$$

$$= 9$$

By the nth root theorem, the square roots of $9 \text{ cis } 120^\circ$ are

$$9^{1/2} \text{ cis}\left(\frac{120^\circ + 360^\circ k}{2}\right) = 3 \text{ cis}(60^\circ + 180^\circ k)$$

$$k = 0: \quad 3 \text{ cis } 60^\circ = \frac{3}{2} + \frac{3}{2}\sqrt{3}i$$

$$k = 1: \quad 3 \text{ cis } 240^\circ = -\frac{3}{2} - \frac{3}{2}\sqrt{3}i$$

All other integral values of k repeat one of the previously found roots. For example,

$$k = 2: \quad 3 \text{ cis } 420^\circ = \frac{3}{2} + \frac{3}{2}\sqrt{3}i$$

Check:

$$\left(\frac{3}{2} + \frac{3}{2}\sqrt{3}i\right)^2 = \frac{9}{4} + \frac{9}{2}\sqrt{3}i + \frac{9}{4} \cdot 3i^2 \qquad \left(-\frac{3}{2} - \frac{3}{2}\sqrt{3}i\right)^2 = \frac{9}{4} + \frac{9}{2}\sqrt{3}i + \frac{9}{4} \cdot 3i^2$$

$$= -\frac{9}{2} + \frac{9}{2}\sqrt{3}i \qquad = -\frac{9}{2} + \frac{9}{2}\sqrt{3}i$$

■

EXAMPLE 8 Find the fifth roots of 32.

Solution Begin by writing 32 in trigonometric form: $32 = 32 \text{ cis } 0^\circ$. The fifth roots are found by

$$32^{1/5} \text{ cis}\left(\frac{0^\circ + 360^\circ k}{5}\right) = 2 \text{ cis } 72^\circ k$$

$k = 0$: $2 \text{ cis } 0^\circ = 2$ ← The first root, which is located so that its argument is θ/n, is called the **principal *n*th root**.

$k = 1$: $2 \text{ cis } 72^\circ = 0.6180 + 1.9021i$

$k = 2$: $2 \text{ cis } 144^\circ = -1.6180 + 1.1756i$

$k = 3$: $2 \text{ cis } 216^\circ = -1.6180 - 1.1756i$

$k = 4$: $2 \text{ cis } 288^\circ = 0.6180 - 1.9021i$

All other integral values for k repeat those listed here. ■

If all the fifth roots of 32 are represented graphically, as shown in Figure 7.8, notice that they all lie on a circle of radius 2 and are equally spaced.

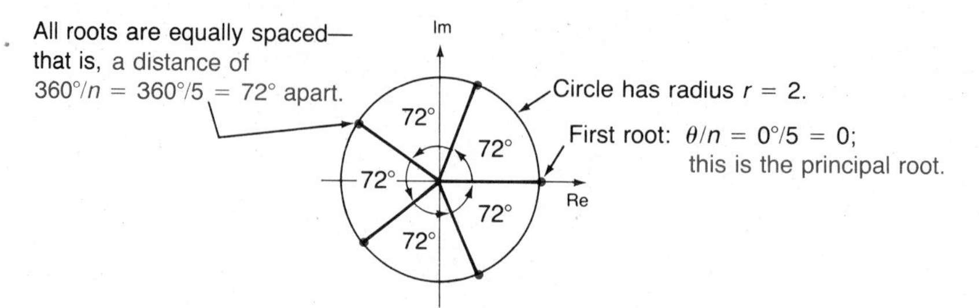

Figure 7.8 Graphical representation of the fifth roots of 32

If n is a positive integer, then the nth roots of a complex number $a + bi = r \operatorname{cis} \theta$ are equally spaced on the circle of radius r centered at the origin.

PROBLEM SET 7.6

A

Plot the complex numbers given in Problems 1–3. Find the modulus of each.

1. a. $3 + i$ **b.** $7 - i$ **c.** $3 + 2i$ **d.** $-3 - 2i$

2. a. $-1 + 3i$ **b.** $2 + 4i$ **c.** $5 + 6i$ **d.** $2 - 5i$

3. a. $-2 + 5i$ **b.** $-5 + 4i$ **c.** $4 - i$ **d.** $-1 + i$

Plot and then change to trigonometric form each of the numbers in Problems 4–15.

4. $1 + i$ **5.** $1 - i$ **6.** $\sqrt{3} - i$ **7.** $\sqrt{3} + i$

8. $1 - \sqrt{3}i$ **9.** $-1 - \sqrt{3}i$ **10.** 1 **11.** 5

12. $-4i$ **13.** $5.7956 - 1.5529i$

14. $-0.6946 + 3.9392i$ **15.** $1.5321 - 1.2856i$

Plot and then change to rectangular form each of the numbers in Problems 16–27. Use exact values whenever possible.

16. $2(\cos 45° + i \sin 45°)$ **17.** $3(\cos 60° + i \sin 60°)$

18. $4(\cos 315° + i \sin 315°)$ **19.** $5 \operatorname{cis}(\frac{4\pi}{3})$

20. $\operatorname{cis}(\frac{5\pi}{6})$ **21.** $5 \operatorname{cis}(\frac{3\pi}{2})$

22. $4 \operatorname{cis} 30°$ **23.** $2 \operatorname{cis} \pi$

24. $10 \operatorname{cis} 65°$ **25.** $8 \operatorname{cis} 24°$

26. $6 \operatorname{cis} 247°$ **27.** $9 \operatorname{cis} 190°$

B

Perform the indicated operations in Problems 28–39.

28. $2 \operatorname{cis} 60° \cdot 3 \operatorname{cis} 150°$ **29.** $3 \operatorname{cis} 48° \cdot 5 \operatorname{cis} 92°$

30. $4(\cos 65° + i \sin 65°) \cdot 12(\cos 87° + i \sin 87°)$

31. $\dfrac{5(\cos 315° + i \sin 315°)}{2(\cos 48° + i \sin 48°)}$

32. $\dfrac{8 \operatorname{cis} 30°}{4 \operatorname{cis} 15°}$ **33.** $\dfrac{12 \operatorname{cis} 250°}{4 \operatorname{cis} 120°}$

34. $(2 \operatorname{cis} 50°)^3$ **35.** $(3 \operatorname{cis} 60°)^4$

36. $(\cos 210° + i \sin 210°)^5$ **37.** $(2 - 2i)^4$

38. $(1 + i)^6$ **39.** $(\sqrt{3} - i)^8$

Find the indicated roots of the numbers in Problems 40–57. Leave your answers in trigonometric form.

40. Square roots of $16 \operatorname{cis} 100°$ **41.** Cube roots of $8 \operatorname{cis} 240°$

42. Fourth roots of $81 \operatorname{cis} 88°$ **43.** Fifth roots of $32 \operatorname{cis} 200°$

44. Cube roots of $64 \operatorname{cis} 216°$ **45.** Fifth roots of $32 \operatorname{cis} 160°$

46. Cube roots of -1 **47.** Cube roots of 27

48. Cube roots of 8 **49.** Fourth roots of i

50. Fourth roots of $1 + i$ **51.** Fourth roots of $-1 - i$

52. Sixth roots of -64 **53.** Sixth roots of $64i$

54. Ninth roots of 1 **55.** Ninth roots of $-1 + i$

56. Tenth roots of i **57.** Tenth roots of 1

Find the indicated roots of the numbers in Problems 58–63. Leave your answers in rectangular form. Show the roots graphically.

58. Cube roots of 1 **59.** Fourth roots of 1

60. Cube roots of -8 **61.** Cube roots of $4\sqrt{3} - 4i$

62. Square roots of $\dfrac{25}{2} - \dfrac{25\sqrt{3}}{2}i$

63. Fourth roots of $12.2567 + 10.2846i$

C

64. Find the cube roots of $(4\sqrt{2} + 4\sqrt{2}i)^2$.

65. Find the fifth roots of $(-16 + 16\sqrt{3}i)^3$.

66. Solve $x^5 - 1 = 0$.

67. Solve $x^4 + x^3 + x^2 + x + 1 = 0$.

68. Prove that

$$\frac{r_1 \operatorname{cis} \theta_1}{r_2 \operatorname{cis} \theta_2} = \frac{r_1}{r_2} \operatorname{cis}(\theta_1 - \theta_2)$$

69. Prove that

$$\sqrt[n]{r} \operatorname{cis}\left(\frac{\theta + 360°k}{n}\right) = (r \operatorname{cis} \theta)^{1/n}$$

70. If $z_1 = a + bi$ and $z_2 = c + di$, show that
$|z_1 + z_2| \le |z_1| + |z_2|$.
This relationship is called the *triangle inequality*.

71. If

$$\cos \theta = 1 - \frac{\theta^2}{2!} + \frac{\theta^4}{4!} - \frac{\theta^6}{6!} + \cdots + \frac{(-1)^n \theta^{2n}}{(2n)!} + \cdots$$

$$\sin \theta = \theta - \frac{\theta^3}{3!} + \frac{\theta^5}{5!} - \frac{\theta^7}{7!} + \cdots + \frac{(-1)^n \theta^{2n+1}}{(2n+1)!} + \cdots$$

and

$$e^{i\theta} = 1 + (i\theta) + \frac{(i\theta)^2}{2!} + \frac{(i\theta)^3}{3!} + \frac{(i\theta)^4}{4!} + \cdots + \frac{(i\theta)^n}{n!} + \cdots$$

show that $e^{i\theta} = \cos \theta + i \sin \theta$. This equation is called *Euler's formula*.

72. Using Problem 71, show that $e^{i\pi} = -1$.

7.7 CHAPTER 7 SUMMARY

The material of this chapter is reviewed in the following list of objectives. After each objective there are some practice questions. For a sample test, select the first question of each set and check your answers with the answer section. For a sample test without answers, use the second question of each set. Additional practice is given by the other questions in each set. If you are having trouble with a particular type of problem, look back to that section for extra help.

7.1 TRIGONOMETRIC EQUATIONS

Objective 1 *Solve first-degree trigonometric equations.* Solve for $0 \le \theta < 2\pi$.

1. $3 \tan 2\theta - \sqrt{3} = 0$

2. $\sin^2 \theta + 2 \cos \theta = 1 + 3 \cos \theta + \sin^2 \theta$

3. $2 \sin 3\theta = \sqrt{2}$

4. $3 \sec 2\theta - 2 = 0$

Objective 2 *Solve second-degree trigonometric equations by factoring or by using the quadratic formula.* Solve for $0 \le \theta < 2\pi$.

5. $4 \cos^2 \theta = 1$

6. $3 \cos^2 \theta = 1 + \cos \theta$

7. $\frac{1}{2} \cos^2 \theta = 1$

8. $\tan^2 2\theta = 3 \tan 2\theta$

7.2 FUNDAMENTAL IDENTITIES

Objective 3 *State and prove the eight fundamental identities.*

9. State the reciprocal identities.

10. State the ratio identities.

11. State the Pythagorean identities.

12. Prove one of the Pythagorean identities.

Objective 4 *Use the fundamental identities to find the other trigonometric functions if you are given the value of one function and want to find the others.*

13. Find the other trigonometric functions so that $\sin \delta = \frac{3}{5}$ when $\tan \delta < 0$.

14. Find the other trigonometric functions so that $\cos \beta = \frac{3}{5}$ and $\tan \beta < 0$.

15. Find the other trigonometric functions so that $\sin \omega = -\frac{3}{5}$ when $\tan \omega > 0$.

16. Find the other trigonometric functions so that $\cos \omega = -\frac{3}{5}$ when $\tan \omega < 0$.

Objective 5 Algebraically simplify expressions involving the trigonometric functions. Leave your answer in terms of sines and cosines.

17. $\dfrac{\sin\theta}{\cos\theta}+\dfrac{1}{\sin\theta}$

18. $\dfrac{1}{\sin\theta+\cos\theta}+\dfrac{1}{\sin\theta-\cos\theta}$

19. $\tan^2\theta+\sec^2\theta$

20. $\dfrac{\tan\theta+\cot\theta}{\sec\theta}$

7.3 PROVING IDENTITIES

Objective 6 Prove identities using algebraic simplification and the eight fundamental identities.

21. $\dfrac{\csc^2\alpha}{1+\cot^2\alpha}=1$

22. $\dfrac{1+\tan^2\theta}{\csc\theta}=\sec\theta\tan\theta$

23. $\dfrac{\cos\theta}{\sec\theta}-\dfrac{\sin\theta}{\cot\theta}=\dfrac{\cos\theta\cot\theta-\tan\theta}{\csc\theta}$

24. $\tan\beta+\sec\beta=\dfrac{1+\csc\beta}{\cos\beta\csc\beta}$

Objective 7 Prove a given identity by using various "tricks of the trade."

25. $\dfrac{\sin^2\theta-\cos^2\theta}{\sin\theta+\cos\theta}=\sin\theta-\cos\theta$

26. $(\tan\theta+\cot\theta)^2=\sec^2\theta+\csc^2\theta$

27. $\cos^2 x\tan x\csc x\sec x=1$

28. $\dfrac{1}{\sin\theta+\cos\theta}+\dfrac{1}{\sin\theta-\cos\theta}=\dfrac{2\sin\theta}{\sin^4\theta-\cos^4\theta}$

7.4 ADDITION LAWS

Objective 8 Use the cofunction identities.

29. Write $\sin 38^\circ$ in terms of its cofunction.

30. Write $\tan\frac{\pi}{8}$ in terms of its cofunction.

31. Write $\cos 0.456$ in terms of its cofunction.

32. Prove $\tan(\frac{\pi}{2}-\theta)=\cot\theta$.

Objective 9 Use the opposite-angle identities.

33. Write $\cos(\frac{\pi}{6}-\theta)$ as a function of $(\theta-\frac{\pi}{6})$.

34. Write $\tan(1-\theta)$ as a function of $(\theta-1)$.

35. Graph $y-1=2\sin(\frac{\pi}{6}-\theta)$.

36. Graph $y+1=2\cos(2-\theta)$.

Objective 10 Use the addition laws.

37. Write $\cos(\theta-30^\circ)$ as a function of $\cos\theta$ and $\sin\theta$.

38. Find the exact value of $\sin 105^\circ$.

39. Evaluate $\dfrac{\tan 23^\circ-\tan 85^\circ}{1+\tan 23^\circ\tan 85^\circ}$.

40. Prove $\cos(\alpha-\beta)=\cos\alpha\cos\beta+\sin\alpha\sin\beta$.

7.5 DOUBLE-ANGLE AND HALF-ANGLE IDENTITIES

Objective 11 Use the double-angle identities. Let θ be a positive angle.

41. Evaluate $\dfrac{2\tan\frac{\pi}{6}}{1-\tan^2\frac{\pi}{6}}$ using exact values.

42. If $\cos\theta=-\frac{4}{5}$ and 2θ is in Quadrant IV, find the exact value of $\cos 2\theta$.

43. If $\cos\theta=\frac{4}{5}$ and 2θ is in Quadrant IV, find the exact value of $\sin 2\theta$.

44. If $\cos\theta=-\frac{4}{5}$ and 2θ is in Quadrant IV, find the exact value of $\tan 2\theta$.

Objective 12 Use the half-angle identities.

45. Evaluate $-\sqrt{\dfrac{1+\cos 240^\circ}{2}}$ using exact values.

46. If $\cot 2\theta=-\frac{4}{3}$, find the exact value of $\cos\theta$ (θ in Quadrant I).

47. If $\cot 2\theta = \frac{4}{3}$, find the exact value of $\sin\theta$ (θ in Quadrant I).
48. If $\cot 2\theta = -\frac{4}{3}$, find the exact value of $\tan\theta$ (θ in Quadrant I).

Objective 13 Use the product and sum identities.

49. Write $\sin 3\theta \cos\theta$ as a sum.
50. Write $\sin(x + h) - \sin x$ as a product.
51. Solve $\sin x - \sin 3x = 0$ (for $0 \le x \le 2\pi$).
52. Prove $\dfrac{\sin 5\theta + \sin 3\theta}{\cos 5\theta - \cos 3\theta} = -\cot\theta$.

*7.6 DE MOIVRE'S THEOREM

Objective 14 Change rectangular-form complex numbers to trigonometric form.

53. $7 - 7i$
54. $-3i$
55. $\frac{7}{2}\sqrt{3} - \frac{7}{2}i$
56. $2 + 3i$

Objective 15 Change trigonometric-form complex numbers to rectangular form.

57. $4(\cos\frac{7\pi}{4} + i\sin\frac{7\pi}{4})$
58. $2 \operatorname{cis} 150°$
59. $5 \operatorname{cis} 270°$
60. $5 \operatorname{cis} 25°$

Objective 16 Multiply and divide complex numbers. Perform the operations and leave your answer in the form indicated.

61. $(\sqrt{12} - 2i)^4$; rectangular
62. $\dfrac{(3 + 3i)(\sqrt{3} - i)}{1 + i}$; rectangular
63. $\dfrac{2 \operatorname{cis} 158° \cdot 4 \operatorname{cis} 212°}{(2 \operatorname{cis} 312°)^3}$; trigonometric
64. $2i(-1 + i)(-2 + 2i)$; trigonometric

Objective 17 State De Moivre's Theorem and the nth root theorem, and find all roots of a complex number.

65. State the nth root theorem.
66. State De Moivre's Theorem.
67. Find and plot the square roots of $\frac{7}{2}\sqrt{3} - \frac{7}{2}i$. Leave your answer in rectangular form.
68. Find the cube roots of i. Leave your answer in trigonometric form.

---

* Optional section

Benjamin Bannecker (1731–1806)

I apprehend you will embrace every opportunity to eradicate that train of absurd and false ideas and opinions which so generally prevail with respect to us [black people] and that your sentiments are concurrent with mine, which are: that one universal Father hath given being to us all; and He not only made us all of one flesh, but that He hath also without partiality afforded us all with these same faculties and that, however diversified in situation or color, we are all the same family and stand in the same relation to Him.

Benjamin Banneker in
Black Mathematicians and Their Works

HISTORICAL NOTE

Analytic trigonometry has its roots in geometry and in the practical application of surveying. Surveying is the technique of locating points on or near the surface of the earth. The Egyptians are credited with being the first to do surveying. The annual flooding of the Nile and its constant destruction of property markers led the Egyptians to the principles of surveying, as noted by the Greek historian Herodotus:

"Sesostris . . . made a division of the soil of Egypt among the inhabitants . . . . If the river carried away any portion of a man's lot . . . the king sent persons to examine, and determine by measurement the exact extent of the loss . . . . From this practice, I think, geometry first came to be known in Egypt, whence it passed into Greece."

The first recorded measurements to determine the size of the earth were also made in Egypt when Eratosthenes measured a meridian arc in 230 B.C. (See Problem 62 of Section 6.1.) A modern-day example of a surveyor's dream is the city of Washington D.C. The center of the city is the Capitol, and the city is divided into four sections: Northwest, Northeast, Southwest, and Southeast. These sections are separated by North, South, East, and Capitol streets, which all converge at the Capitol. The initial surveying of this city was done by a group of mathematicians and surveyors who included the first distinguished black mathematician, Benjamin Banneker.

*"Benjamin Banneker, born in 1731 in Maryland, was the first black to be recognized as a mathematician and astronomer. He exhibited unusual talent in mathematics, and, with little formal education, produced an almanac and invented a clock. Bannecker was born free in a troubled time of black slavery. The right to vote was his by birth. It is important to note that this right to vote was denied him as well as other free black Americans by 1803 in his native state of Maryland. Yet, history must acknowledge him as the first American black man of science."**

* From Virginia K. Newell, Joella H. Gipson, L. Waldo Rich, and Beauregard Stubblefield, *Black Mathematicians and Their Works* (Admore: Dorrance), p. xiv.

8

Analytic Trigonometry

CONTENTS

PREVIEW

In this chapter, we are concerned with another use of trigonometry—solving triangles to find unknown parts. We begin by solving right triangles using the Pythagorean Theorem. We then turn to the solution of *oblique* triangles (triangles that are not right triangles). We will derive two trigonometric laws, the *Law of Cosines* and the *Law of Sines*, which allow us to solve certain triangles using a calculator or trigonometric tables and logarithms. There is a third trigonometric law that lends itself to logarithmic calculation, the *Law of Tangents*, which is useful for times when no calculator is at hand. This chapter has 6 objectives, which are listed at the end of the chapter on pages 318–320. Following these objectives is a Cumulative Review for Chapters 6–8 and an Extended Application about solar power.

PERSPECTIVE

The first step in carrying out the operation of integration in calculus is frequently solving a triangle. This technique, called *trigonometric substitution*, involves expressions of the form $a^2 + u^2$, $\sqrt{a^2 - u^2}$, $\sqrt{a^2 + u^2}$, and $\sqrt{u^2 - a^2}$. In the page below, taken from a calculus book, two reference triangles are shown in Figures 7.1 and 7.2.

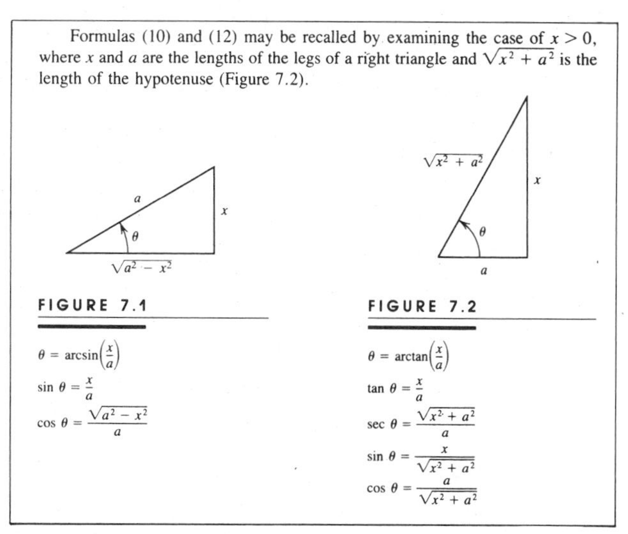

Formulas (10) and (12) may be recalled by examining the case of $x > 0$, where x and a are the lengths of the legs of a right triangle and $\sqrt{x^2 + a^2}$ is the length of the hypotenuse (Figure 7.2).

FIGURE 7.1

$\theta = \arcsin\left(\frac{x}{a}\right)$

$\sin\theta = \frac{x}{a}$

$\cos\theta = \frac{\sqrt{a^2 - x^2}}{a}$

FIGURE 7.2

$\theta = \arctan\left(\frac{x}{a}\right)$

$\tan\theta = \frac{x}{a}$

$\sec\theta = \frac{\sqrt{x^2 + a^2}}{a}$

$\sin\theta = \frac{x}{\sqrt{x^2 + a^2}}$

$\cos\theta = \frac{a}{\sqrt{x^2 + a^2}}$

From *Calculus and Analytic Geometry*, 4th Ed., by A. Shenk, p. 451.

8.1 RIGHT TRIANGLES

One of the most important uses of trigonometry is in solving triangles. Recall from geometry that every **triangle** has three sides and three angles, which are called the six *parts* of the triangle. We say that a **triangle is solved** if all six parts are known. Typically three parts will be given, or known, and you will want to find the other three parts. Label a triangle as shown in Figure 8.1. The vertices are labeled A, B, and C, with the sides opposite those vertices a, b, and c, respectively. The angles are labeled α, β, and γ, respectively. In this section the examples are limited to right triangles in which γ denotes the right angle and c the **hypotenuse**.

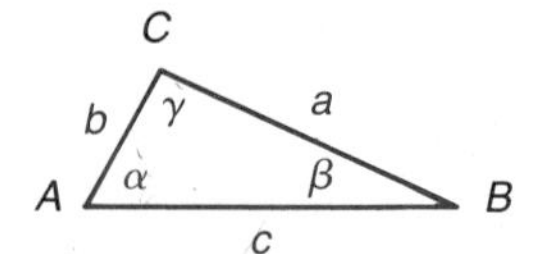

Figure 8.1 Correctly labeled triangle

According to the definition of the trigonometric functions, the angle under consideration must be in standard position. This requirement is sometimes inconvenient, so we use that definition to create a special case that applies to any acute angle θ of a right triangle. Notice that in Figure 8.1, θ might be α or β, but it would not be γ since γ is not an acute angle. Also notice that the hypotenuse is one of the sides of both acute angles. The other side making up the angle is called the **adjacent side**. Thus side a is adjacent to β and side b is adjacent to α. The third side of the triangle (the one not making up the angle) is called the **opposite side**. Thus side a is opposite α and side b is opposite β.

Right-Triangle Definition of the Trigonometric Functions

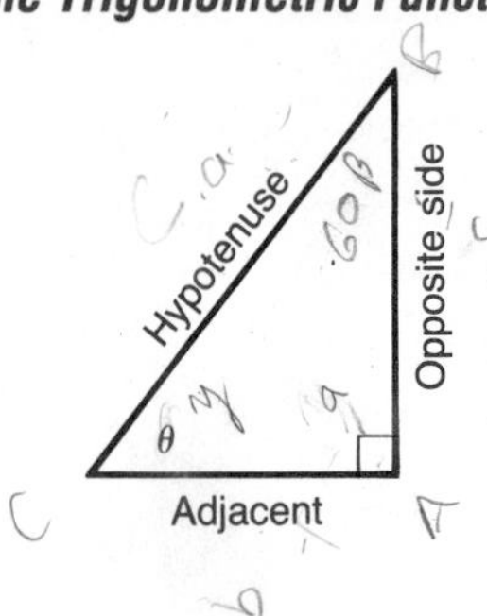

If θ is an acute angle in a right triangle, then

$$\cos\theta = \frac{\text{adjacent side}}{\text{hypotenuse}}$$

$$\sin\theta = \frac{\text{opposite side}}{\text{hypotenuse}}$$

$$\tan\theta = \frac{\text{opposite side}}{\text{adjacent side}}$$

The other trigonometric functions are the reciprocals of these relationships.

We can now use this definition to solve some given triangles.

EXAMPLE 1 Solve the triangle with $a = 50$, $\alpha = 35°$, and $\gamma = 90°$. (*Note:* $\alpha = 35°$ means the measure of angle α is 35°.)

Solution

$\alpha = 35°$	Given
$\beta = 55°$	Since $\alpha + \beta = 90°$ for any right triangle with right angle at C
$\gamma = 90°$	Given
$a = 50$	Given

b: $\tan 35^\circ = \dfrac{50}{b}$

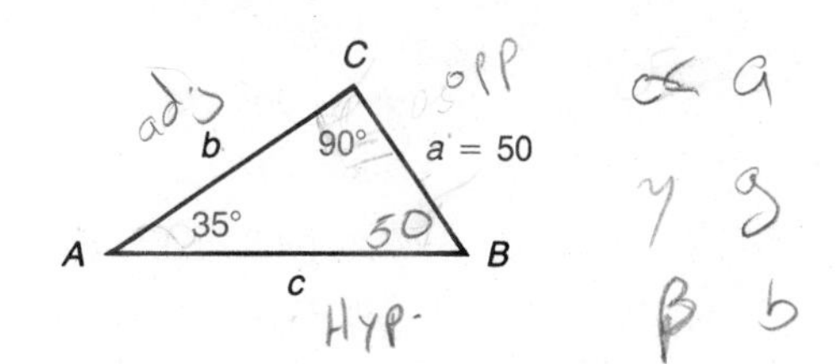

$$b = \frac{50}{\tan 35^\circ}$$

By table:

$$b = \frac{50}{\tan 35^\circ} \approx \frac{50}{0.7002} \qquad \text{From Table C.III, } \tan 35^\circ \approx 0.7002$$

$$\approx 71.4082$$

or

$$b = 50 \cot 35^\circ \qquad \text{This avoids division.}$$
$$\approx 50(1.4281) \qquad \text{From Table C.III}$$
$$= 71.405$$

By calculator with algebraic logic:

[50] [÷] [35] [tan] [=] DISPLAY: 71.40740034

By calculator with RPN logic:

[50] [ENTER] [35] [tan] [÷] DISPLAY: 71.40740034

Notice that some of the answers above differ. However, to two significant digits, $\boldsymbol{b = 71}$.

c: $\sin 35^\circ = \dfrac{50}{c}$

$$c = \frac{50}{\sin 35^\circ}$$

$$\approx 87.1723 \qquad \text{Use a calculator or Table C.III.}$$

To two significant digits, $\boldsymbol{c = 87}$. ■

EXAMPLE 2 Solve the triangle with $a = 32$, $b = 58$, and $\gamma = 90^\circ$.

Solution α: $\tan \alpha = \dfrac{32}{58}$ or $\alpha = \text{Tan}^{-1}\left(\dfrac{32}{58}\right)$

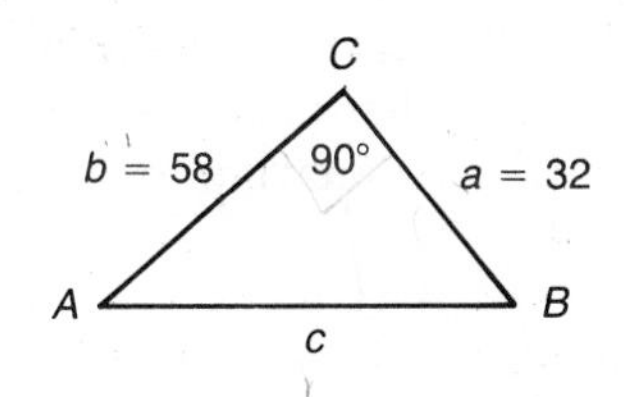

1. By table: $\frac{32}{58} \approx 0.5517$, so from Table C.III, $\alpha \approx 28.9^\circ$.
2. By calculator with algebraic logic:

3. By calculator with RPN logic:

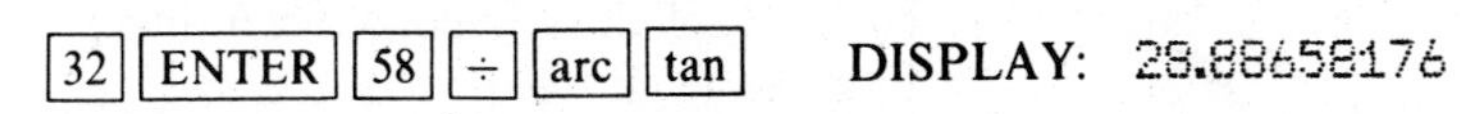

To the nearest degree,

$\alpha = 29°$

$\beta = 61°$ $\quad \beta = 90° - \alpha$

$\gamma = 90°$ $\quad$ Given

$a = 32$ $\quad$ Given

$b = 58$ $\quad$ Given

c: To find c by table:

$$\sin \alpha = \frac{32}{c}$$

$$c = \frac{32}{\sin \alpha}$$

$$\approx \frac{32}{0.4833} \qquad \text{By table, } \sin 28.9° = .4833$$

$$\approx 66.2139$$

To find c by calculator, use the Pythagorean Theorem:

$$c = \sqrt{a^2 + b^2}$$

Algebraic logic: [32] [x^2] [+] [58] [x^2] [=] [$\sqrt{x}$] DISPLAY: 66.24198065

RPN logic: [32] [ENTER] [×] [58] [ENTER] [×] [+] [$\sqrt{x}$]

DISPLAY: 66.24198067

To two significant digits, $\mathbf{c = 66.}$ ■

As you can see, there are many ways to solve a triangle; the method you choose will depend on the accuracy you want and the type of table or calculator you have. In the rest of this chapter, it is assumed that you have access to a calculator.

When solving triangles, you are dealing with measurements that are never exact. In real life, measurements are made with a certain number of digits of precision, and results are not claimed to have more digits of precision than the least accurate number in the input data. There are rules for working with significant digits (see Appendix B), but it is important for you to focus first on the trigonometry. Just remember that your answers should not be more accurate than any of the given measurements. However, be careful not to round your answers when doing your calculations; you should round only when stating your answers.

WARNING: Note rounding agreement.

The solution of right triangles is necessary in a variety of situations. The first one we will consider concerns an observer looking at an object. The **angle of depression** is the acute angle measured down from the horizontal line to the line of sight whereas the **angle of elevation** is the acute angle measured up from a horizontal line to the line of sight. These ideas are illustrated in Examples 3 and 4.

EXAMPLE 3 The angle of elevation of a tree from a point on the ground 42 m from its base is 33°. Find the height of the tree.

Solution Let θ = angle of elevation and h = height of tree. Then

$$\tan\theta = \frac{h}{42}$$

$$h = 42\tan 33^\circ$$
$$\approx 42(0.6494)$$
$$\approx 27.28$$

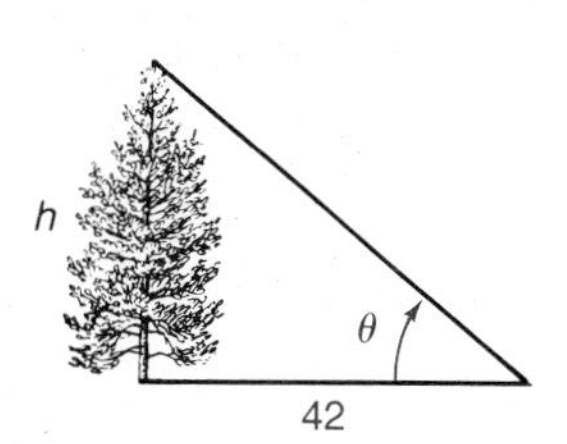

The tree is 27 m tall. ■

EXAMPLE 4 The distance across a canyon can be determined from an airplane. Suppose the angles of depression to the two sides of the canyon are 43° and 55° as shown in Figure 8.2. If the altitude of the plane is 20,000 ft, how far is it across the canyon?

Solution Label parts x, y, θ, and ϕ as shown in Figure 8.2.

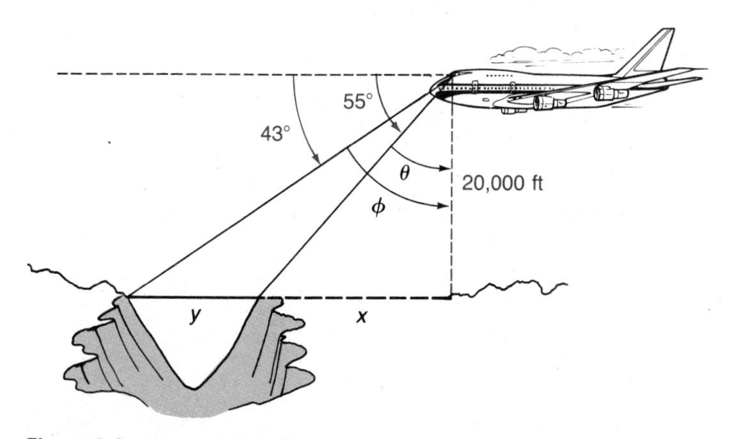

Figure 8.2 Determining the distance across a canyon from the air

Remember, when you are working these problems on your calculator, do not work with the rounded results here, but with the entire accuracy possible with your calculator. You should round only once and that is with your final answer.

$$\theta = 90^\circ - 55^\circ = 35^\circ \quad \text{and} \quad \phi = 90^\circ - 43^\circ = 47^\circ$$

First find x:

$$\tan 35^\circ = \frac{x}{20{,}000}$$

$$x = 20{,}000\tan 35^\circ$$
$$\approx 14{,}004$$

Next find $x + y$:

$$\tan 47^\circ = \frac{x+y}{20{,}000}$$

$$x + y = 20{,}000\tan 47^\circ$$
$$\approx 21{,}447$$

Thus

$$y \approx 21{,}447 - 14{,}004$$
$$= 7443$$

To two significant digits, **the distance across the canyon is 7400 ft.** ■

A second application of the solution of right triangles involves the **bearing** of a line, which is defined as an acute angle made with a north–south line. When giving the bearing of a line, first write N or S to determine whether to measure the angle from the north or the south side of a point on the line. Then give the measure of the angle followed by E or W, denoting on which side of the north–south line the angle is to be measured. Some examples are shown in Figure 8.3, page 296.

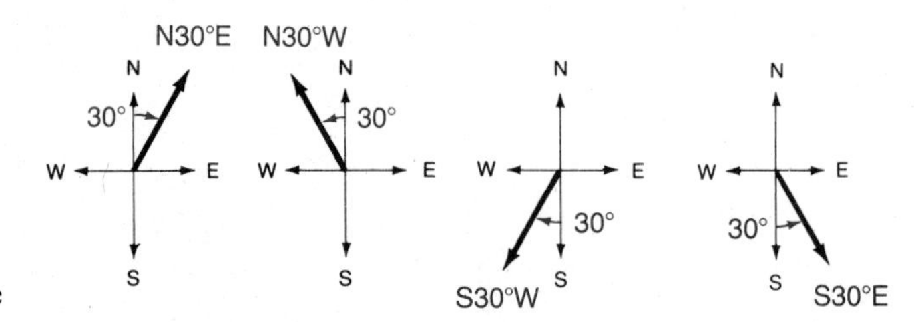

Figure 8.3 Bearing of an angle

EXAMPLE 5 To find the width AB of a canyon, a surveyor measures 100 m from A in the direction of N42.6°W to locate point C. The surveyor then determines that the bearing of CB is N73.5°E. Find the width of the canyon if point B is situated so that $\angle BAC = 90.0°$.

Solution Let $\theta = \angle BCA$ in Figure 8.4.

$\angle BCE' = 16.5°$ — Complementary angles

$\angle ACS' = 42.6°$ — Alternate interior angles

$\angle E'CA = 47.4°$ — Complementary angles

$$\theta = \angle BCA = \angle BCE' + \angle E'CA = 16.5° + 47.4° = 63.9°$$

Since $\tan\theta = \dfrac{AB}{AC}$,

$$AB = AC\tan\theta = 100\tan\theta = 100\tan 63.9° \approx 204.125$$

The canyon is 204 m across.

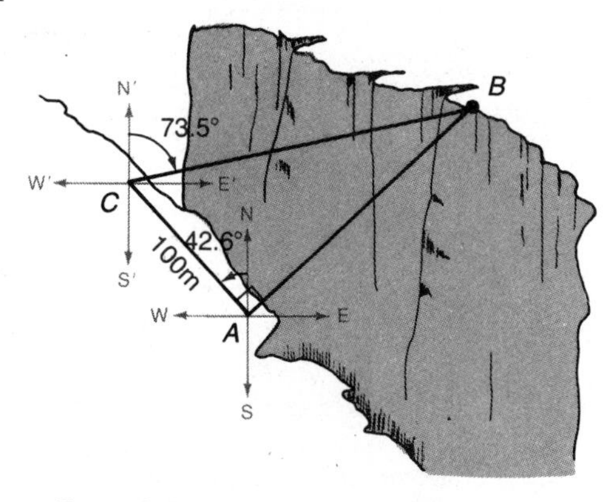

Figure 8.4 Surveying problem ■

PROBLEM SET 8.1

A

Solve the right triangles ($\gamma = 90°$) in Problems 1–30.

1. $a = 80;\ \beta = 60°$
2. $a = 18;\ \beta = 30°$
3. $a = 9.0;\ \beta = 45°$
4. $b = 37;\ \alpha = 65°$
5. $b = 15;\ \alpha = 37°$
6. $b = 50;\ \alpha = 53°$
7. $a = 69;\ c = 73$
8. $b = 23;\ c = 45$
9. $b = 13;\ c = 22$
10. $a = 68;\ b = 83$
11. $a = 24;\ b = 29$
12. $a = 12;\ b = 8.0$
13. $a = 29;\ \alpha = 76°$
14. $a = 93;\ \alpha = 26°$
15. $a = 49;\ \beta = 45°$
16. $b = 13;\ \beta = 65°$
17. $b = 90;\ \beta = 13°$
18. $b = 47;\ \beta = 108°$
19. $b = 82;\ \alpha = 50°$
20. $c = 28.3;\ \alpha = 69.2°$
21. $c = 36;\ \alpha = 6°$
22. $\beta = 57.4°;\ a = 70.0$
23. $\alpha = 56.00°;\ b = 2350$
24. $\beta = 23°;\ a = 9000$
25. $b = 3100;\ c = 3500$
26. $\beta = 16.4°;\ b = 2580$
27. $\alpha = 42°;\ b = 350$
28. $b = 3200;\ c = 7700$
29. $b = 4100;\ c = 4300$
30. $c = 75.4;\ \alpha = 62.5°$

B

31. ***Surveying*** The angle of elevation of a building from a point on the ground 30 m from its base is 38°. Find the height of the building.

32. ***Surveying*** Repeat Problem 31 for 150 ft instead of 30 m.

33. ***Surveying*** From a cliff 150 m above the shoreline, the angle of depression of a ship is 37°. Find the distance of the ship from a point directly below the observer.

34. ***Surveying*** Repeat Problem 33 for 450 ft instead of 150 m.

35. From a police helicopter flying at 1000 ft, a stolen car is sighted at an angle of depression of 71°. Find the distance of the car from a point directly below the helicopter.

36. Repeat Problem 35 for an angle of depression of 64° instead of 71°.

37. ***Surveying*** To find the east–west boundary of a piece of land, a surveyor must divert his path from point C on the boundary by proceeding due south for 300 ft to point A. Point B, which is due east of point C, is now found to be in the direction of N49°E from point A. What is the distance CB?

38. ***Surveying*** To find the distance across a river that runs east–west, a surveyor locates points P and Q on a north–south line on opposite sides of the river. She then paces out 150 ft from Q due east to a point R. Next she determines that the bearing of RP is N58°W. How far is it across the river?

39. A 16-ft ladder on level ground is leaning against a house. If the angle of elevation of the ladder is 52°, how far above the ground is the top of the ladder?

40. How far is the base of the ladder in Problem 39 from the house?

41. If the ladder in Problem 39 is moved so that the bottom is 9 ft from the house, what will be the angle of elevation?

42. Find the height of the Barrington Space Needle if the angle of elevation at 1000 ft from a point on the ground directly below the top is 58.15°.

43. The world's tallest chimney is the stack of the International Nickel Company. Find its height if the angle of elevation at 1000 ft from a point on the ground directly below the top of the stack is 51.36°.

44. In the movie *Close Encounters of the Third Kind*, there was a scene in which the star, Richard Dreyfuss, was approaching Devil's Tower in Wyoming. He could have determined his distance from Devil's Tower by first stopping at a point P and estimating the angle P, as shown in Figure 8.5. After moving 100 m toward Devil's Tower, he could have estimated the angle N, as shown in Figure 8.5. How far away from Devil's Tower is point N?

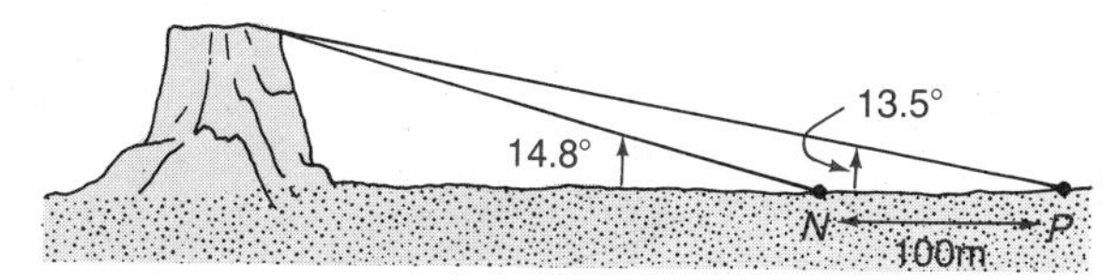

Figure 8.5 Procedure for determining the height of Devil's Tower

45. Determine the height of Devil's Tower in Problem 44.

46. ***Surveying*** To find the boundary of a piece of land, a surveyor must divert his path from a point A on the boundary for 500 ft in the direction S50°E. He then determines that the bearing of a point B located directly south of A is S40°W. Find the distance AB.

47. ***Surveying*** To find the distance across a river, a surveyor locates points P and Q on either side of the river. Next she measures 100 m from point Q in the direction S35°E to point R. Then she determines that point P is now in the direction of N25.0°E from point R and that $\angle PQR$ is a right angle. Find the distance across the river.

48. ***Surveying*** If the Empire State Building and the Sears Tower were situated 1000 ft apart, the angle of depression from the top of the Sears Tower to the top of the Empire State Building would be 11.53°, and the angle of depression to the foot of the Empire State Building would be 55.48°. Find the heights of the buildings.

49. ***Surveying*** On the top of the Empire State Building is a television tower. From a point 1000 ft from a point on the ground directly below the top of the tower, the angle of elevation to the bottom of the tower is 51.34° and to the top of the tower is 55.81°. What is the length of the television tower?

50. ***Physics*** A wheel 5.00 ft in diameter rolls up a 15.0° incline. What is the height of the center of the wheel above the base of the incline after the wheel has completed one revolution?

51. ***Physics*** What is the height of the center of the wheel in Problem 50 after three revolutions?

52. ***Astronomy*** If the distance from the earth to the sun is 92.9 million mi and the angle formed between Venus, the

earth, and the sun (as shown in Figure 8.6) is 47.0°, find the distance from the sun to Venus.

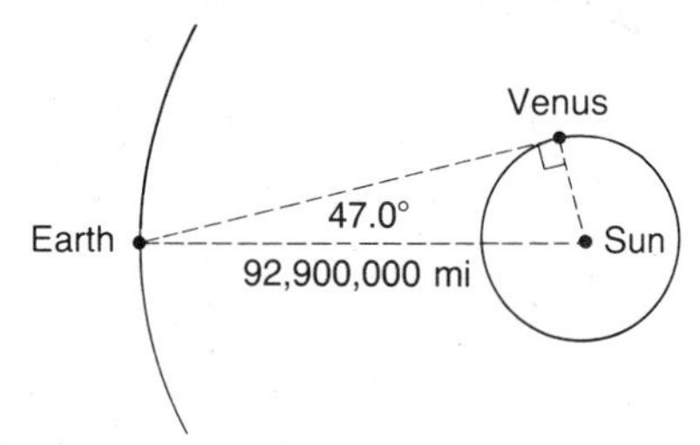

Figure 8.6

53. ***Astronomy*** Use the information in Problem 52 to find the distance from the earth to Venus.

C

54. The largest ground area covered by any office building is that of the Pentagon in Arlington, Virginia. If the radius of the circumscribed circle is 783.5 ft, find the length of one side of the Pentagon.
55. Use the information in Problem 54 to find the radius of the circle inscribed in the Pentagon.
56. ***Surveying*** To determine the height of the building shown in Figure 8.7, we select a point P and find that the angle of elevation is 59.64°. We then move out a distance of 325.4 ft (on a level plane) to point Q and find that the angle of elevation is now 41.32°. Find the height h of the building.

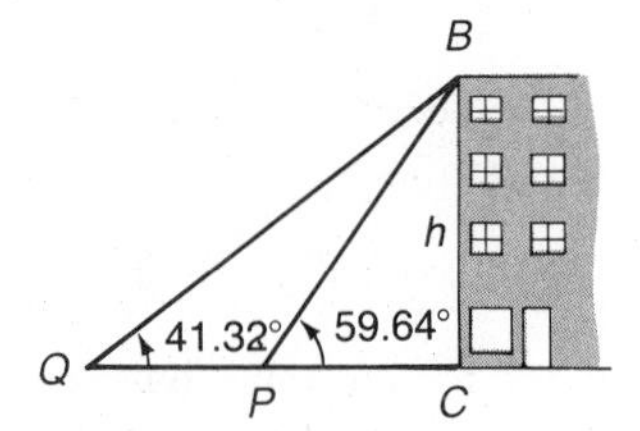

Figure 8.7

57. Using Figure 8.7, let the angle of elevation at P be α and at Q be β, and let the distance from P to Q be d. If h is the height of the building, show that
$$h = \frac{d \sin \alpha \sin \beta}{\sin(\alpha - \beta)}$$
58. A 6.0-ft person is casting a shadow of 4.2 ft. What time of the morning is it if the sun rose at 6:15 A.M. and is directly overhead at 12:15 P.M.?
59. How long will the shadow of the person in Problem 58 be at 8:00 A.M.?
60. ***Surveying*** From the top of a tower 100 ft high, the angles of depression to two landmarks on the plane upon which the tower stands are 18.5° and 28.4°.
    a. Find the distance between the landmarks when they are on the same side of the tower.
    b. Find the distance between the landmarks when they are on opposite sides of the tower.
61. Show that in every right triangle the value of c lies between $(a + b)/\sqrt{2}$ and $a + b$.

8.2 LAW OF COSINES

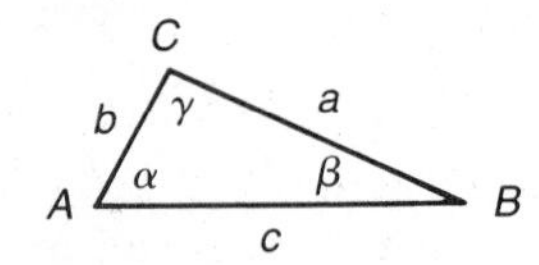

Figure 8.8 Correctly labeled triangle

In the last section we solved right triangles. We now extend that study to triangles with no right angles. Such triangles are called **oblique triangles.** Consider a triangle labeled as in Figure 8.8, and notice that γ is not now restricted to 90°.

In general, you will be given three parts of a triangle and be asked to find the remaining three parts. But can you do so given *any* three parts? Consider the possibilities:

1. SSS: By SSS we mean that you are given three sides and want to find the three angles.
2. SAS: You are given two sides and an included angle.
3. AAA: You are given three angles.
4. ASA or AAS: You are given two angles and a side.
5. SSA: You are given two sides and the angle opposite one of them.

We will consider these possibilities one at a time.

SSS

To solve a triangle given SSS, it is necessary for the sum of the lengths of the two smaller sides to be greater than the length of the largest side. In this case, we use a generalization of the Pythagorean Theorem called the **Law of Cosines**:

Law of Cosines

In triangle ABC,

$$c^2 = a^2 + b^2 - 2ab\cos\gamma$$

Proof Let γ be an angle in standard position with A on the positive x axis, as shown in Figure 8.9. The coordinates of the vertices are as follows:

$C(0,0)$	Since C is in standard position.
$A(b,0)$	Since A is on the x axis a distance of b units from the origin.
$B(a\cos\gamma, a\sin\gamma)$	Let $B = (x, y)$; then by definition of the trigonometric functions, $\cos\gamma = x/a$ and $\sin\gamma = y/a$. Thus, $x = a\cos\gamma$ and $y = a\sin\gamma$.

Figure 8.9

Use the distance formula for the distance between the points $A(b,0)$ and the points $B(a\cos\gamma, a\sin\gamma)$:

$$\begin{aligned} c^2 &= (a\cos\gamma - b)^2 + (a\sin\gamma - 0)^2 \\ &= a^2\cos^2\gamma - 2ab\cos\gamma + b^2 + a^2\sin^2\gamma \\ &= a^2(\cos^2\gamma + \sin^2\gamma) + b^2 - 2ab\cos\gamma \\ &= a^2 + b^2 - 2ab\cos\gamma \end{aligned}$$

□

Notice that for a right triangle, $\gamma = 90^\circ$. This means that

$$c^2 = a^2 + b^2 - 2ab\cos 90^\circ \qquad \text{or} \qquad c^2 = a^2 + b^2$$

since $\cos 90^\circ = 0$. But this last equation is simply the Pythagorean Theorem.

By letting A and B, respectively, be in standard position, it can also be shown that

$$a^2 = b^2 + c^2 - 2bc\cos\alpha \qquad \text{and} \qquad b^2 = a^2 + c^2 - 2ac\cos\beta$$

To find the angles when you are given three sides, solve for α, β, or γ.

Law of Cosines, Alternate Forms

$$a^2 = b^2 + c^2 - 2bc\cos\alpha \qquad \cos\alpha = \frac{b^2 + c^2 - a^2}{2bc}$$

$$b^2 = a^2 + c^2 - 2ac\cos\beta \qquad \cos\beta = \frac{a^2 + c^2 - b^2}{2ac}$$

$$c^2 = a^2 + b^2 - 2ab\cos\gamma \qquad \cos\gamma = \frac{a^2 + b^2 - c^2}{2ab}$$

EXAMPLE 1 What is the smallest angle of a triangular patio whose sides measure 25, 18, and 21 ft?

Solution If γ represents the smallest angle, then c (the side opposite γ) must be the smallest side, so $c = 18$. Then:

$$\cos\gamma = \frac{a^2 + b^2 - c^2}{2ab}$$

$$= \frac{25^2 + 21^2 - 18^2}{2(25)(21)}$$ Use this number and trigonometric tables if you have only a four-function calculator.

$$\gamma = \cos^{-1}\left[\frac{25^2 + 21^2 - 18^2}{2(25)(21)}\right]$$

1. By calculator with algebraic logic:

   [25] [x^2] [+] [21] [x^2] [−] [18] [x^2] [=] [÷] [2] [÷] [25] [÷] [21] [=] [inv] [cos]

   DISPLAY: 45.03565072

2. By calculator with RPN logic:

   [25] [ENTER] [×] [21] [ENTER] [×] [+] [18] [ENTER] [×] [−] [2] [÷] [25] [÷] [21] [÷] [arc] [cos]

   DISPLAY: 45.03565071

To two significant digits, the answer is **45°**. ■

SAS

The second possibility listed for solving oblique triangles is that of being given two sides and an included angle. It is necessary that the given angle be less than 180°. Again use the Law of Cosines for this possibility, as shown by Example 2.

EXAMPLE 2 Find c where $a = 52.0$, $b = 28.3$, and $\gamma = 28.5°$.

Solution By the Law of Cosines:

$$c^2 = a^2 + b^2 - 2ab\cos\gamma$$
$$= (52.0)^2 + (28.3)^2 - 2(52.0)(28.3)\cos 28.5°$$

By calculator:

$$c^2 \approx 918.355474$$
$$c \approx 30.30438044$$

To three significant digits, $\boldsymbol{c = 30.3}$. ■

AAA

The third case supposes that three angles are given. However, from what you know of similar triangles (see Figure 8.10)—that they have the same shape but not necessarily the same size, and thus their corresponding angles have equal measure—you can conclude that the triangle cannot be solved without knowing the length of at least one side.

We will discuss ASA and AAS in the next section. The last possibility, SSA, is discussed in Section 8.4.

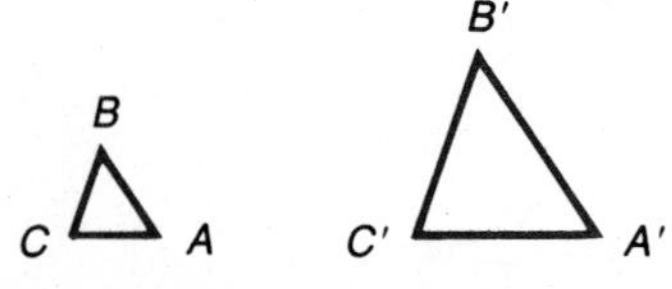

Figure 8.10 Similar triangles

PROBLEM SET 8.2

A

Solve $\triangle ABC$ for the parts requested in Problems 1–24. If the triangle cannot be solved, tell why.

1. $a = 7.0$; $b = 8.0$; $c = 2.0$. Find α.

2. $a = 7.0$; $b = 5.0$; $c = 4.0$. Find β.

3. $a = 11$; $b = 9.0$; $c = 8.0$. Find α.

4. $a = 4.0$; $b = 5.0$; $c = 6.0$. Find α.

5. $a = 15$; $b = 8.0$; $c = 20$. Find β.

6. $a = 25$; $b = 18$; $c = 40$. Find γ.

7. $a = 18$; $b = 25$; $\gamma = 30°$. Find c.

8. $a = 18$; $c = 11$; $\beta = 63°$. Find b.

9. $a = 15$; $b = 8.0$; $\gamma = 38°$. Find c.

10. $b = 21$; $c = 35$; $\alpha = 125°$. Find a.

11. $b = 14$; $c = 12$; $\alpha = 82°$. Find a.

12. $a = 31$; $b = 24$; $\gamma = 120°$. Find c.

13. $b = 18$; $c = 15$; $\alpha = 50°$. Find a.

14. $a = 20$; $c = 45$; $\beta = 85°$. Find b.

15. $a = 241$; $b = 187$; $c = 100$. Find β.

16. $a = 38.2$; $b = 14.8$; $\gamma = 48.2°$. Find c.

17. $b = 123$; $c = 485$; $\alpha = 163.0°$. Find a.

18. $a = 48.3$; $c = 35.1$; $\beta = 215.0°$. Find b.

19. $a = 11$; $b = 9.0$; $c = 8.0$. Find the largest angle.

20. $a = 12$; $b = 6.0$; $c = 15$. Find the smallest angle.

21. $a = 123$; $b = 310$; $c = 250$. Find the smallest angle.

22. $a = 123$; $b = 310$; $c = 250$. Find the largest angle.

23. $a = 38$; $b = 41$; $c = 25$. Find the largest angle.

24. $a = 45$; $b = 92$; $c = 41$. Find the smallest angle.

Solve $\triangle ABC$ in Problems 25–36. If the triangle cannot be solved, tell why.

25. $b = 5.2$; $c = 3.4$; $\alpha = 54.4°$

26. $a = 81$; $b = 53$; $\gamma = 85.2°$

27. $a = 214$; $b = 320$; $\gamma = 14.8°$

28. $a = 18$; $b = 12$; $c = 23.4$

29. $a = 140$; $b = 85.0$; $c = 105$

30. $\alpha = 83°$; $\beta = 52°$; $\gamma = 45°$

31. $\alpha = 35°$; $\beta = 123°$; $\gamma = 22°$

32. $a = 4.82$; $b = 3.85$; $\gamma = 34.2°$

33. $b = 520$; $c = 235$; $\alpha = 110.5°$

34. $a = 31.2$; $c = 51.5$; $\beta = 109.5°$

35. $a = 341$; $b = 340$; $\gamma = 23.4°$

36. $b = 1234$; $c = 3420$; $\alpha = 24.58°$

Problems 37–42 refer to the baseball field shown in Figure 8.11. Find the requested distances to the nearest foot.

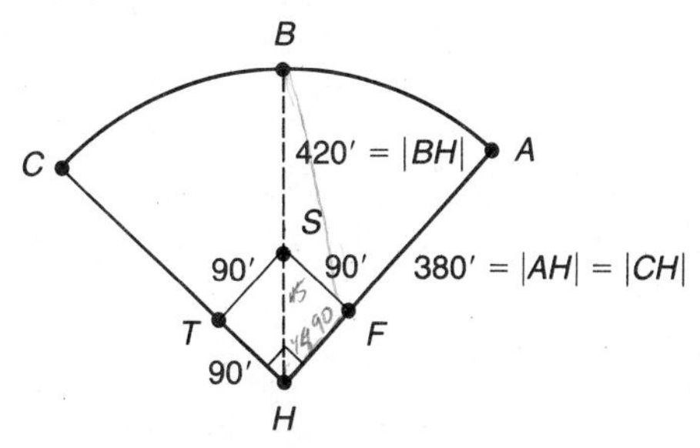

Figure 8.11 Baseball field

37. Right field (point A) to left field (point C)

38. Right field (point A) to center field (point B)

39. First base (point F) to center field (point B)

40. Third base (point T) to right field (point A)

41. Second base (point S) to center field (point B)

42. First base (point F) to third base (point T)

B

43. Prove that $a^2 = b^2 + c^2 - 2bc\cos\alpha$.

44. Prove that $b^2 = a^2 + c^2 - 2ac\cos\beta$.

45. Prove that $c^2 = a^2 + b^2 - 2ab\cos\gamma$.

46. Prove that $\cos\alpha = \dfrac{b^2 + c^2 - a^2}{2bc}$.

47. Prove that $\cos\beta = \dfrac{a^2 + c^2 - b^2}{2ac}$.

48. Prove that $\cos\gamma = \dfrac{a^2 + b^2 - c^2}{2ab}$.

49. ***Aviation*** New York City is approximately 600 mi N9°E of Washington, D.C., and Buffalo, New York, is N49°W of Washington, D.C. How far is Buffalo from New York City if the distance from Buffalo to Washington, D.C., is approximately 800 mi?

50. ***Aviation*** New Orleans, Louisiana, is approximately 3000 mi S56°E of Denver, Colorado, and Chicago, Illinois, is N76°E of Denver. How far is Chicago from New Orleans if it is approximately 2500 mi from Denver to Chicago?

51. ***Aviation*** An airplane flies N51°W for a distance of 345 mi and then alters course and flies 150 mi in the direction N25°W. How far is the airplane from the starting point?

52. ***Aviation*** An airplane flies due south for a distance of 135 mi and then alters course and flies S46°W for 82 mi before landing. How far from the starting point is the plane upon landing?

53. ***Geometry*** A dime, a penny, and a quarter are placed on a table so that they just touch each other, as shown in Figure 8.12. Let D, P, and Q be the respective centers of the coins.) Solve $\triangle DPQ$. (*Hint:* If you measure the coins, the diameters are 1.75 cm, 2.00 cm, and 2.50 cm.)

Figure 8.12

54. ***Geometry*** An equilateral triangle is inscribed in a circle of radius 5.0 in., as shown in Figure 8.13. Find the perimeter of the triangle.

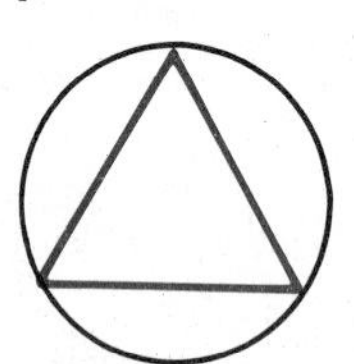

Figure 8.13

55. ***Geometry*** An equilateral triangle is inscribed in a circle of radius 3.0 in., as shown in Figure 8.13. Find the perimeter of the triangle.

56. ***Geometry*** A square is inscribed in a circle of radius 5.0 in. Find the area of the square.

57. ***Geometry*** A square is inscribed in a circle of radius 15.0 in. Find the area of the square.

58. In $\triangle ABC$, use the Law of Cosines to prove
$$a = b\cos\gamma + c\cos\beta$$

59. In $\triangle ABC$, use the Law of Cosines to prove
$$b = a\cos\gamma + c\cos\alpha$$

60. In $\triangle ABC$, use the Law of Cosines to prove
$$c = a\cos\beta + b\cos\alpha$$

C

61. Using the Law of Cosines, show that
$$\cos\alpha + 1 = \frac{(b + c - a)(a + b + c)}{2bc}$$

62. Using the Law of Cosines, show that
$$\cos\beta + 1 = \frac{(a + c - b)(a + b + c)}{2ac}$$

63. Using the Law of Cosines, show that
$$\cos\gamma + 1 = \frac{(a + b - c)(a + b + c)}{2ab}$$

8.3 LAW OF SINES

In the last section, we listed five possibilities for given information in solving oblique triangles: SSS, SAS, AAA, ASA or AAS, and SSA. We discussed three of those cases: SSS, SAS, and AAA. In this section we will consider the fourth case, ASA or AAS. Notice that ASA and AAS both give us the same information about a triangle, because knowing any two angles is the same as knowing all three angles (the sum of the angles of any triangle is always 180°).

ASA or AAS

Case 4 supposes that two angles and a side are given. For a triangle to be formed, the sum of the two given angles must be less than 180°, and the given side must be greater than 0. The Law of Cosines is not sufficient in this case because at least two sides are needed for the application of that law. So we state and prove a result called the **Law of Sines**.

Law of Sines

> In any $\triangle ABC$,
>
> $$\frac{\sin\alpha}{a} = \frac{\sin\beta}{b} = \frac{\sin\gamma}{c}$$

The equation in the Law of Sines means that you can use any of the following equations to solve a triangle:

$$\frac{\sin\alpha}{a} = \frac{\sin\beta}{b} \qquad \frac{\sin\alpha}{a} = \frac{\sin\gamma}{c} \qquad \frac{\sin\beta}{b} = \frac{\sin\gamma}{c}$$

Proof Consider any oblique triangle, as shown in Figure 8.14.

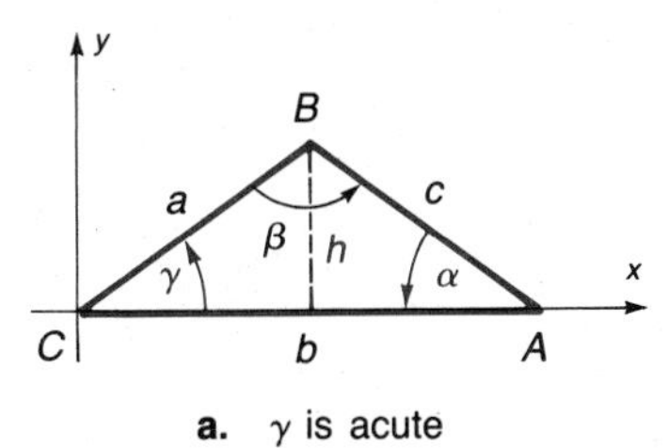

a. γ is acute

b. γ is obtuse

Figure 8.14 Oblique triangles for Law of Sines

Let h = height of the triangle with base CA. Then

$$\sin\alpha = \frac{h}{c} \qquad \text{and} \qquad \sin\gamma = \frac{h}{a}$$

Solving for h,

$$h = c\sin\alpha \qquad \text{and} \qquad h = a\sin\gamma$$

Thus,

$$c\sin\alpha = a\sin\gamma$$

Dividing by ac,

$$\frac{\sin\alpha}{a} = \frac{\sin\gamma}{c}$$

Repeat these steps for the height of the same triangle with base AB to get

$$\frac{\sin\alpha}{a} = \frac{\sin\beta}{b}$$

□

EXAMPLE 1 Solve the triangle in which $a = 20$, $\alpha = 38°$, and $\beta = 121°$.

Solution

$\boldsymbol{\alpha = 38°}$	Given
$\boldsymbol{\beta = 121°}$	Given
$\boldsymbol{\gamma = 21°}$	Since $\alpha + \beta + \gamma = 180°$, then $\gamma = 180° - 38° - 121° = 21°$
$\boldsymbol{a = 20}$	Given

b: Use the Law of Sines:

$$\frac{\sin 38^\circ}{20} = \frac{\sin 121^\circ}{b}$$

$$b = \frac{20 \sin 121^\circ}{\sin 38^\circ}$$ Use tables or a calculator.

$$\approx \frac{20(0.8572)}{0.6157}$$ Use logarithms or a calculator.

$$\approx 27.8454097$$ Give answer to two significant digits.

$$\boldsymbol{b \approx 28}$$

c: Use the Law of Sines:

$$\frac{\sin 38^\circ}{20} = \frac{\sin 21^\circ}{c}$$

$$c \approx \frac{20 \sin 21^\circ}{\sin 38^\circ}$$

$$\approx 11.64172078$$ Give answer to two significant digits.

$$c \approx 12$$ ■

EXAMPLE 2 A boat traveling at a constant rate due west passes a buoy that is 1.0 kilometer from a lighthouse. The lighthouse is N30°W of the buoy. After the boat has traveled for $\frac{1}{2}$ hour, its bearing to the lighthouse is N74°E. How fast is the boat traveling?

Solution The angle at the lighthouse (see Figure 8.15) is $180^\circ - 60^\circ - 16^\circ = 104^\circ$.

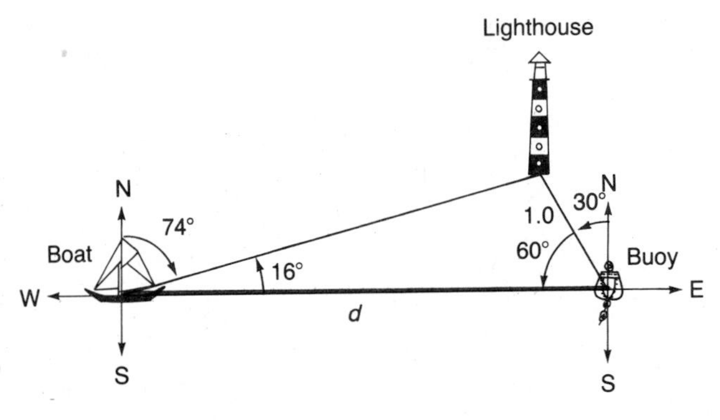

Figure 8.15 Finding the rate of travel

Therefore, by the Law of Sines,

$$\frac{\sin 104^\circ}{d} = \frac{\sin 16^\circ}{1.0}$$

$$d = \frac{\sin 104^\circ}{\sin 16^\circ}$$

$$\approx 3.520189502$$ Do not round answers in your work; round only when stating the answer

After 1 hour the distance is $2d$, so the rate of the boat is about **7.0 kph**. ■

PROBLEM SET 8.3

A

Solve $\triangle ABC$ with the parts given in Problems 1–40. If the triangle cannot be solved, tell why.

1. $a = 10$; $\alpha = 48°$; $\beta = 62°$

2. $a = 18$; $\alpha = 65°$; $\gamma = 15°$

3. $a = 30$; $\beta = 50°$; $\gamma = 100°$

4. $b = 23$; $\alpha = 25°$; $\beta = 110°$

5. $b = 40$; $\alpha = 50°$; $\gamma = 60°$

6. $b = 90$; $\alpha = 85°$; $\gamma = 25°$

7. $c = 43$; $\alpha = 120°$; $\gamma = 7°$

8. $c = 115$; $\beta = 81.0°$; $\gamma = 64.0°$

9. $a = 107$; $\alpha = 18.3°$; $\beta = 54.0°$

10. $b = 223$; $\beta = 85°$; $\gamma = 24°$

11. $a = 85$; $\alpha = 48.5°$; $\gamma = 72.4°$

12. $a = 183$; $\alpha = 65.0°$; $\gamma = 105.2°$

13. $a = 105$; $\alpha = 48.5°$; $\beta = 62.7°$

14. $b = 128$; $\alpha = 165°$; $\beta = 15°$

15. $b = 10.3$; $\alpha = 78°$; $\beta = 102°$

16. $c = 110$; $\beta = 105.0°$; $\gamma = 15.5°$

17. $c = 105$; $\beta = 148.0°$; $\gamma = 22.5°$

18. $c = 1823$; $\beta = 65.25°$; $\gamma = 15.55°$

19. $a = 41.0$; $\alpha = 45.2°$; $\beta = 21.5°$

20. $b = 55.0$; $c = 92.0°$; $\alpha = 98.0°$

21. $b = 58.3$; $\alpha = 120°$; $\gamma = 68.0°$

22. $c = 123$; $\alpha = 85.2°$; $\beta = 38.7°$

23. $a = 26$; $b = 71$; $c = 88$

24. $a = 38$; $b = 82$; $c = 115$

25. $\alpha = 48°$; $\beta = 105°$; $\gamma = 27°$

26. $\alpha = 38°$; $\beta = 100°$; $\gamma = 42°$

27. $a = 80.6$; $b = 23.2$; $\gamma = 89.2°$

28. $b = 1234$; $\alpha = 85.26°$; $\beta = 24.45°$

29. $c = 28.36$; $\beta = 42.10°$; $\gamma = 102.30°$

30. $a = 481$; $\beta = 28.6°$; $\gamma = 103.0°$

31. $a = 10$; $b = 48$; $c = 52$

32. $a = 58$; $b = 165$; $c = 180$

33. $a = 48.1$; $\alpha = 4.8°$; $\beta = 82.0°$

34. $c = 101$; $\beta = 35.0°$; $\gamma = 25.85°$

35. $c = 83.1$; $\beta = 81.5°$; $\gamma = 162°$

36. $a = 381$; $\alpha = 25°$; $\gamma = 90.00°$

37. $a = 8.1$; $\alpha = 28.5°$; $\gamma = 90.00°$

38. $b = 23.1$; $c = 16.5$; $\alpha = 15.25°$

39. $b = 10.9$; $c = 4.45$; $\alpha = 16.2°$

40. $a = 18.55$; $b = 16.54$; $c = 51.45$

B

41. ***Surveying*** In San Francisco, a certain hill makes an angle of 20° with the horizontal and has a tall building at the top. At a point 100 ft down the hill from the base of the building, the angle of elevation to the top of the building is 72°. What is the height of the building?

42. ***Surveying*** A certain hill makes an angle of 15° with the horizontal and has a tall building at the top. At a point 100 ft down the hill from the base of the building, the angle of elevation to the top of the building is 68°. What is the height of the building?

43. ***Surveying*** A hill makes an angle of 8° with the horizontal and has a tower at the top. At a point 500 ft down the hill from the base of the building, the angle of elevation to the top of the tower is 35°. What is the height of the tower?

44. ***Navigation*** In the movie *Star Wars*, the hero, Luke, must hit a small target on the Death Star by flying a horizontal distance to reach the target. When the target is sighted, the onboard computer calculates the angle of depression to be 28.0°. If, after 150 km, the target has an angle of depression of 42.0°, how far is the target from Luke's spacecraft at that instant?

45. ***Navigation*** The Galactic Empire's computers on the Death Star are monitoring the positions of the invading forces (see Problem 44). At a particular instant, two observation points 2500 m apart make a fix on Luke's spacecraft, which is between the observation points and in the same vertical plane. If the angle of elevation from the first observation point is 3.00° and the angle of elevation from the second is 1.90°, find the distance from Luke's spacecraft to each of the observation points.

46. ***Navigation*** Solve Problem 45 if both observation points are on the same side of the spacecraft and all the other information is unchanged.

47. Mr. T, whose eye level is 6 ft, is standing on a hill with an inclination of 5° and needs to throw a rope to the top of a nearby building. If the angle of elevation to the top of the building is 20° and the angle of depression to the base of the building is 11°, how tall is the building and how far is it from Mr. T to the top of the building? See Figure 8.16, page 306.

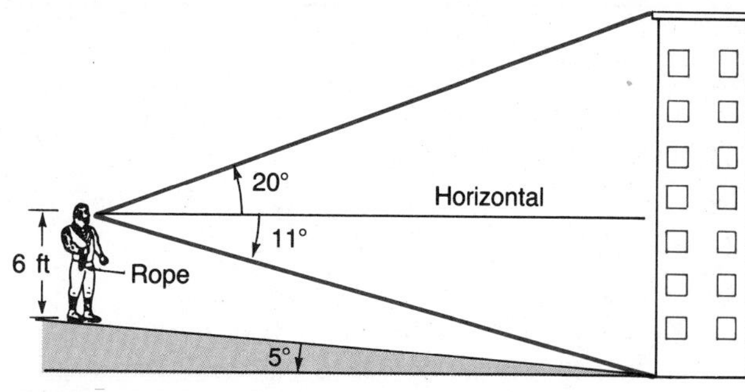

Figure 8.16

48. Repeat Problem 47 with all the information unchanged, except assume that the hill has an inclination of 9°.

49. Johnny Carson is standing on the deck of his Malibu home and is looking down a stairway with an inclination of 35° to a boat on the beach. If the angles of depression to the bow (point B) and stern (point S) are $\beta = 32°$ and $\theta = 28°$, how long is the boat if it is known that the distance JF is 50 ft, and F, B, and S are on level ground? See Figure 8.17.

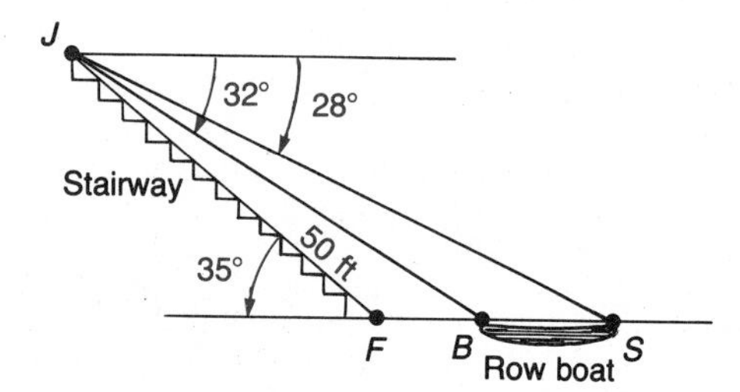

Figure 8.17

50. Repeat Problem 49, except assume that JF is 100 ft and all of the other information is unchanged.

51. Given $\triangle ABC$, show that

$$\frac{\sin\alpha}{a} = \frac{\sin\beta}{b}$$

52. Given $\triangle ABC$, show that

$$\frac{\sin\beta}{b} = \frac{\sin\gamma}{c}$$

53. Given $\triangle ABC$, show that

$$\frac{\sin\alpha}{\sin\beta} = \frac{a}{b}$$

C

54. Using Problem 53, show that

$$\frac{\sin\alpha - \sin\beta}{\sin\alpha + \sin\beta} = \frac{a-b}{a+b}$$

55. Using Problem 54 and the formulas for the sum and difference of sines, show that

$$\frac{2\cos\frac{1}{2}(\alpha+\beta)\sin\frac{1}{2}(\alpha-\beta)}{2\sin\frac{1}{2}(\alpha+\beta)\cos\frac{1}{2}(\alpha-\beta)} = \frac{a-b}{a+b}$$

56. Using Problem 55, show that

$$\frac{\tan\frac{1}{2}(\alpha-\beta)}{\tan\frac{1}{2}(\alpha+\beta)} = \frac{a-b}{a+b}$$

This result is known as the **Law of Tangents.**

57. Another form of the Law of Tangents (see Problem 56) is

$$\frac{\tan\frac{1}{2}(\beta-\gamma)}{\tan\frac{1}{2}(\beta+\gamma)} = \frac{b-c}{b+c}$$

Derive this formula.

58. A third form of the Law of Tangents (see Problems 56 and 57) is

$$\frac{\tan\frac{1}{2}(\alpha-\gamma)}{\tan\frac{1}{2}(\alpha+\gamma)} = \frac{a-c}{a+c}$$

Derive this formula.

59. Newton's formula involves all six parts of a triangle. It is not useful in solving a triangle, but it is helpful in checking your results after you have done so. Show that

$$\frac{a+b}{c} = \frac{\cos\frac{1}{2}(\alpha-\beta)}{\sin\frac{1}{2}\gamma}$$

60. Mollweide's formula involves all six parts of a triangle. It is not useful in solving a triangle, but it is helpful in checking your results after you have done so. Show that

$$\frac{a-b}{c} = \frac{\sin\frac{1}{2}(\alpha-\beta)}{\cos\frac{1}{2}\gamma}$$

61. The letter to the editor shown on the facing page appeared in the February 1977 issue of *Popular Science.* See if you can solve this puzzle. As a hint, you might guess from its placement in this book that it uses the Law of Cosines or the Law of Sines.

Thirty-Seven-Year-Old Puzzle*

For more than 37 years I have tried, on and off, to solve a problem that appeared in *Popular Science*.

All right, I give up. What's the answer? In the September, 1939 issue (page 14) the following letter appeared:

"Have enjoyed monkeying with the problems you print, for the last couple of years. Here's one that may make the boys scratch their heads a little:

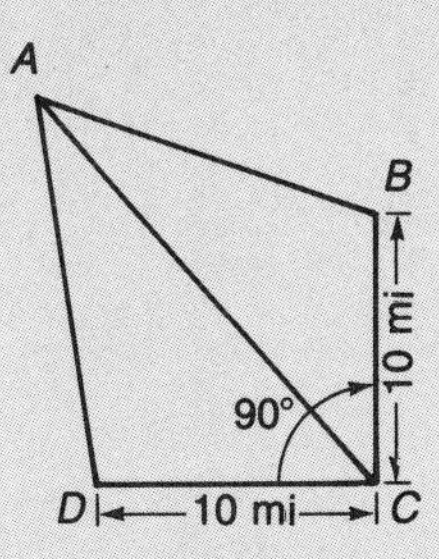

A man always drives at the same speed (his car probably has a governor on it). He makes it from A direct to C in 30 minutes; from A through B to C in 35 minutes; and from A through D to C in 40 minutes. How fast does he drive?

D.R.C., Sacramento, Calif."

It certainly looked easy, and I started to work on it. By 3 a.m. I had filled a lot of sheets of paper, both sides, with notations. In subsequent issues of *Popular Science*, all I could find on the subject was the following, in the November, 1939 issue:

"In regard to the problem submitted by D.R.C. of Sacramento, Calif. in the September issue. According to my calculations, the speed of the car must be 38.843 miles an hour. W.L.B., Chicago, Ill."

But the reader gave no hint of how he had arrived at that figure. It may or may not be correct. Over the years, I probably have shown the problem to more than 500 people. The usual reaction was to nod the head knowingly and start trying to find out what the hypotenuse of the right angle is. But those few who remembered the formula then found there is little you can do with it when you know the hypotenuse distance.

Now I'd like to challenge *Popular Science*. If your readers of today cannot solve this little problem (and prove the answer) how about your finding the answer, if there is one, editors?

R. F. Davis, Sun City, Ariz.

Darrell Huff, master of the pocket calculator and frequent PS contributor [Feb. '76, June '75, Dec. '74] has come up with a couple of solutions. We'd like to see what you readers can do before we publish his answers.

8.4 AMBIGUOUS CASE: SSA

The remaining case of solving oblique triangles as given in Section 8.2 is case 5, in which we are given two sides and an angle that is not an included angle. Call the given angle θ (which may be α, β, or γ, depending on the problem). Since we are given SSA, one of the given sides must not be one of the sides of θ; that is, it is opposite θ. Call this side OPP. The given side that is one of the sides of θ is called ADJ. (You might find it helpful to refer to Table 8.1 on page 311 as you read this section.)

a. Suppose that $\theta > 90°$. There are two possibilities:

i. OPP $\leq$ ADJ

No triangle is formed (see Figure 8.18).

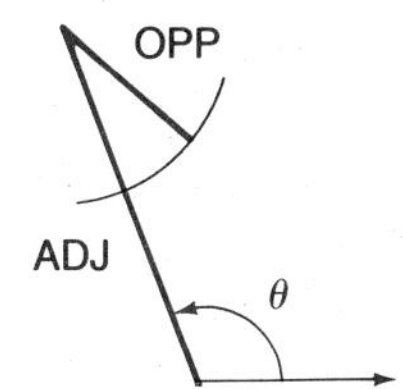

Figure 8.18 $\theta > 90°$, OPP $\leq$ ADJ

* Reprinted with permission from *Popular Science*. © 1977, Times Mirror Magazines, Inc.

ii. OPP > ADJ
One triangle is formed (see Figure 8.19). Use the Law of Sines as shown in Example 1.

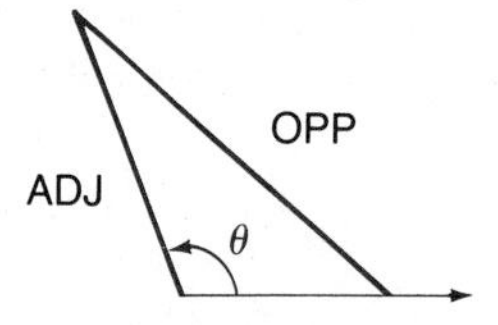

Figure 8.19 $\theta > 90°$, OPP > ADJ

EXAMPLE 1 Let $a = 3.0$, $b = 2.0$, and $\alpha = 110°$. Solve the triangle.

Solution

$\boldsymbol{\alpha = 110°}$ Given

$\boldsymbol{\beta = 39°}$ Use the Law of Sines:

$$\frac{\sin \alpha}{a} = \frac{\sin \beta}{b}$$

$$\frac{\sin 110°}{3.0} = \frac{\sin \beta}{2.0}$$

$$\sin \beta = \frac{2.0}{3.0} \sin 110°$$

$$\beta \approx 38.78955642°$$

$\boldsymbol{\gamma = 31°}$

$$\gamma = 180° - 110° - \beta$$
$$\approx 31.2104436°$$

$\boldsymbol{a = 3.0}$ Given

$\boldsymbol{b = 2.0}$ Given

$\boldsymbol{c = 1.7}$ Use the Law of Sines:

$$\frac{\sin 110°}{3.0} = \frac{\sin \gamma}{c}$$

$$c = \frac{3.0 \sin \gamma}{\sin 110°}$$
$$\approx 1.654$$

■

Remember: Do not work with rounded results.

Notice, when finding c in Example 1, if you work with

$$\gamma \approx 31.2104436°$$

you obtain $c \approx 1.654$, or 1.7, to two significant digits. However, if you work with $\gamma \approx 31°$, you obtain $c \approx 1.644$, or 1.6, to two significant digits. The proper procedure is to round only when stating answers.

b. Suppose that $\theta < 90°$. There are four possibilities. Let h be the altitude of the triangle drawn from the vertex connecting the OPP and ADJ sides. Now, to find h use the right-triangle definition of sine:

$$h = (\text{ADJ}) \sin \theta$$

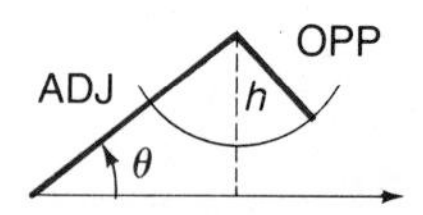

Figure 8.20 $\theta < 90°$, OPP $< h <$ ADJ

i. OPP $< h <$ ADJ
No triangle is formed (see Figure 8.20).

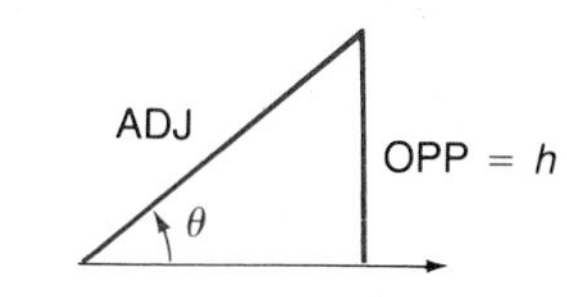

Figure 8.21 $\theta < 90°$, OPP $= h <$ ADJ

ii. OPP $= h <$ ADJ
A right triangle is formed (see Figure 8.21). Use the methods of Section 8.1 to solve the triangle.

iii. $h <$ OPP $<$ ADJ
This situation is called the **ambiguous case** and is really the only special case you must watch for. All the other cases can be determined from the calculations without any special consideration. Notice from Figure 8.22 that two *different* triangles are formed with the given information. The process for finding both solution is shown in Example 2.

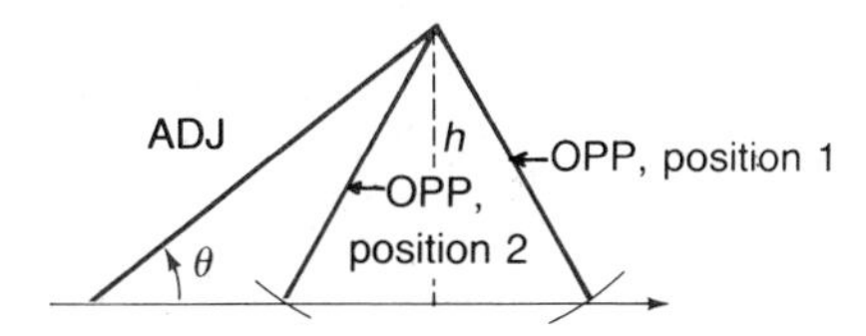

Figure 8.22 The ambiguous case

EXAMPLE 2 Solve the triangle with $a = 1.50$, $b = 2.00$, and $\alpha = 40.0°$.

Solution

$$\frac{\sin \alpha}{a} = \frac{\sin \beta}{b}$$

$$\frac{\sin 40.0°}{1.50} = \frac{\sin \beta}{2.00}$$

$$\sin \beta = \frac{2.00 \sin 40.0°}{1.50}$$

$$\approx \frac{4}{3}(0.6428)$$

$$\approx 0.8571$$

$$\beta \approx 59.0°$$

Figure 8.23

But from Figure 8.23, you can see that this is only the acute angle solution. For the obtuse angle solution—call it β'—find

$$\beta' = 180° - \beta$$

$$\approx 121.0°$$

Finish the problem by working two calculations:

Solution 1

$\alpha = 40.0°$	Given	$c = 2.30$	$\frac{\sin \alpha}{a} = \frac{\sin \gamma}{c}$
$\beta = 59.0°$	See above		
$\gamma = 81.0°$	$\gamma = 180° - \alpha - \beta$		$c = \frac{1.50 \sin \gamma}{\sin 40.0°}$
$a = 1.50$	Given		≈ 2.3049
$b = 2.00$	Given		

Solution 2:

$\alpha = 40.0°$	Given	$c' = 0.759$	$\dfrac{\sin\alpha}{a} = \dfrac{\sin\gamma'}{c'}$
$\beta' = 121.0°$	See calculations, page 309		
$\gamma' = 19.0°$	$\gamma' = 180° - \alpha - \beta'$		$c' = \dfrac{1.50\sin\gamma'}{\sin 40.0°}$
$a = 1.50$	Given		
$b = 2.00$	Given		≈ 0.7592

■

iv. OPP ≥ ADJ
One triangle is formed, as shown in Figure 8.24.

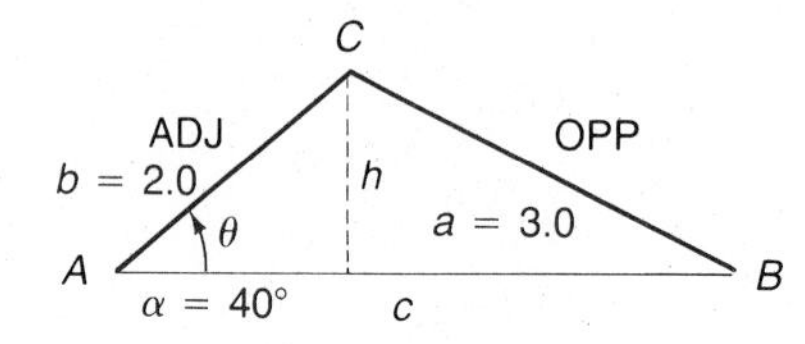

Figure 8.24 $\theta < 90°$, OPP ≥ ADJ

EXAMPLE 3 Solve the triangle given by $a = 3.0$, $b = 2.0$, and $\alpha = 40°$, as shown in Figure 8.24.

Solution

$$\frac{\sin\alpha}{a} = \frac{\sin\beta}{b}$$

$$\frac{\sin 40°}{3.0} = \frac{\sin\beta}{2.0}$$

$$\sin\beta = \frac{2}{3}\sin 40°$$

$$\approx \frac{2}{3}(0.6428)$$

$$\approx 0.4285$$

$$\beta \approx 25.374$$

$\alpha = 40°$	Given	
$\beta = 25°$	See work shown above.	
$\gamma = 115°$	$\gamma = 180° - \alpha - \beta$	
$a = 3.0$	Given	
$b = 2.0$	Given	
$c = 4.2$	$\dfrac{\sin 40°}{3.0} = \dfrac{\sin\gamma}{c}$	
	$c = \dfrac{3.0\sin\gamma}{\sin 40°}$	
	≈ 4.2427	

■

The most important skill to be learned from this section is the ability to select the proper trigonometric law when given a particular problem. In the rest of this section, you will encounter applications of right triangles, the Law of Cosines, and the Law of Sines. A review of various types of problems may be helpful. Table 8.1 summarizes the solution of oblique triangles.

TABLE 8.1 Summary for solving triangles

Given	Conditions on given information	Law to use for solution
1. **SSS**	*a.* The sum of the lengths of the two smaller sides is less than or equal to the length of the larger side.	No solution
	b. The sum of the lengths of the two smaller sides is greater than the length of the larger side.	**Law of Cosines**
2. **SAS**	*a.* The angle is greater than or equal to 180°.	No solution
	b. The angle is less than 180°.	**Law of Cosines**
3. **AAA**		**No solution**
4. **ASA** or **AAS**	*a.* The sum of the angles is greater than or equal to 180°.	No solution
	b. The sum of the angles is less than 180°.	**Law of Sines**
5. **SSA**	Let θ be the given angle with adjacent (ADJ) and opposite (OPP) sides given; the height h is found by $h = (\text{ADJ})\sin\theta$	
	a. $\theta > 90°$	
	i. OPP $\leq$ ADJ	No solution
	ii. OPP $>$ ADJ	**Law of Sines**
	b. $\theta < 90°$	
	i. OPP $< h <$ ADJ	No solution
	ii. OPP $= h <$ ADJ	Right-triangle solution
	iii. $h <$ OPP $<$ ADJ	***Ambiguous case:*** Use Law of Sines to find two solutions
	iv. OPP $\geq$ ADJ	**Law of Sines**

Remember: When given two sides and an angle that is not an included angle, check to see whether one side is between the height of the triangle and the length of the other side. If it is, there will be two solutions.

EXAMPLE 4 An airplane is 100 km N40°E of a Loran station, traveling due west at 240 kph. How long will it be (to the nearest minute) before the plane is 90 km from the Loran station? (See Figure 8.25, page 312.)

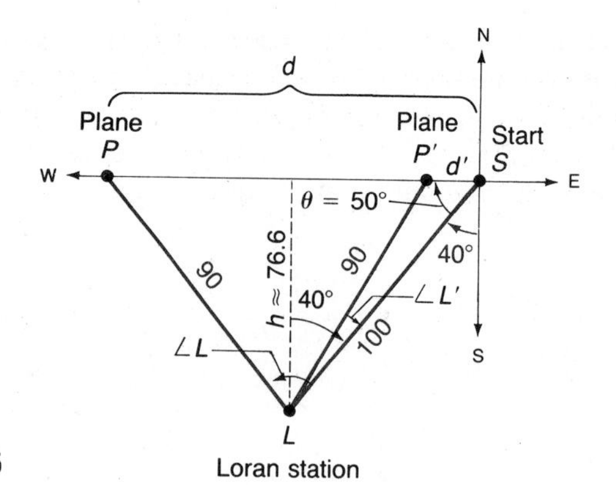

Figure 8.25

Solution You are given SSA; check the other conditions for the ambiguous case.

Since the heading is 40°, $\angle PSL = 50°$; call this angle θ. Notice that 100 km is the side adjacent to θ; 90 km is the side opposite to θ; thus, we have

$$\begin{aligned} h &= (\text{ADJ})\sin\theta \\ &= 100\sin 50° \\ &\approx 76.6 \end{aligned}$$

$$\begin{aligned} h &< \text{OPP} < \text{ADJ} \\ 76.6 &< \ 90 \ < 100 \end{aligned}$$

which is the ambiguous case.

Solution 1

$\angle S$: 50°

$\angle P$: $\dfrac{\sin 50°}{90} = \dfrac{\sin P}{100}$

$\sin P \approx 0.8512$

$P \approx 58.34°$

$\angle L$: $L = 180° - S - P$

$\approx 71.66°$

d: $\dfrac{\sin L}{d} = \dfrac{\sin 50°}{90}$

$d = \dfrac{90\sin L}{\sin 50°}$

≈ 111.52

Solution 2

$\angle S$: 50° — This is θ.

$\angle P'$: $P' = 180° - P$

$\approx 121.66°$

$\angle L'$: $L' = 180 - S - P'$

$\approx 8.34°$

d: $\dfrac{\sin L'}{d'} = \dfrac{\sin 50°}{90}$

$d' = \dfrac{90\sin L'}{\sin 50°}$

≈ 17.04

At 240 kph, the times are

$$\frac{d}{240} \approx 0.4647 \quad \text{and} \quad \frac{d'}{240} \approx 0.0710$$

To convert these to minutes, multiply by 60 and round to the nearest minute. The times are **28 min** and **4 min**. ■

In elementary school you learned that the area, K, of a triangle is $K = \frac{1}{2}bh$.* Now, you sometimes need to find the area of a triangle given the measurements for the sides and angles but not the height. Then you need trigonometry to find the area.

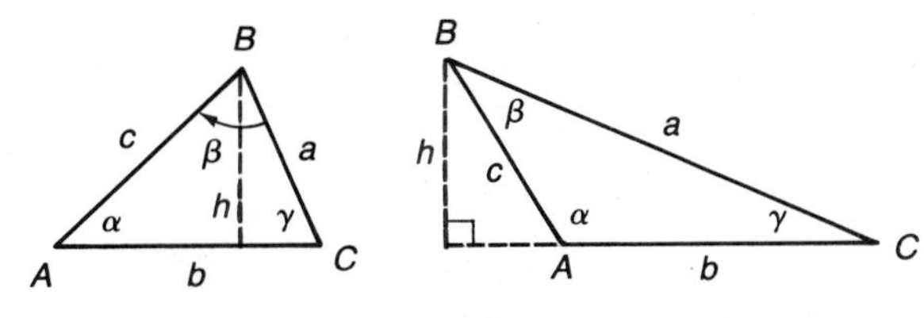

Figure 8.26 Oblique triangles

Using the orientation shown in Figure 8.26,

$$\sin\alpha = \frac{h}{c} \qquad \text{or} \qquad h = c\sin\alpha$$

Thus,

$$\begin{aligned} K &= \frac{1}{2}bh \\ &= \frac{1}{2}bc\sin\alpha \end{aligned}$$

We could use any other pair of sides to derive the following area formulas:

Area of a Triangle (Two Sides and an Included Angle Are Known)

$$K = \frac{1}{2}bc\sin\alpha \qquad K = \frac{1}{2}ac\sin\beta \qquad K = \frac{1}{2}ab\sin\gamma$$

EXAMPLE 5 Find the area of $\triangle ABC$ where $\alpha = 18.4°$, $b = 154$ ft, and $c = 211$ ft.

Solution

$$\begin{aligned} K &= \tfrac{1}{2}(154)(211)\sin 18.4° \\ &\approx 5128.349903 \end{aligned}$$

To three significant figures, the area is **5130 ft²**. ■

The area formulas given above are useful for finding the area of a triangle when two sides and an included angle are known. Suppose, however, that you know the angles but only one side. Say you know side a. Then you can use either formula involving a and replace the other variable by using the Law of Sines:

$$\frac{\sin\alpha}{a} = \frac{\sin\beta}{b}$$

so

$$b = \frac{a\sin\beta}{\sin\alpha}$$

* In elementary school you no doubt used A for area. In trigonometry, we use K for area because we have already used A to represent a vertex of a triangle.

Therefore,

$$K = \frac{1}{2}ab\sin\gamma = \frac{1}{2}a\frac{a\sin\beta}{\sin\alpha}\sin\gamma = \frac{a^2\sin\beta\sin\gamma}{2\sin\alpha}$$

The other area formulas shown in the following box can be similarly derived:

Area of a Triangle (Two Angles and an Included Side Are Known)

$$K = \frac{a^2\sin\beta\sin\gamma}{2\sin\alpha} \qquad K = \frac{b^2\sin\alpha\sin\gamma}{2\sin\beta} \qquad K = \frac{c^2\sin\alpha\sin\beta}{2\sin\gamma}$$

EXAMPLE 6 Find the area of a triangle with angles 20°, 50°, and 110° if the side opposite the 50° angle is 24 in. long.

Solution

$$K = \frac{24^2\sin 20°\sin 110°}{2\sin 50°}$$
$$\approx 120.8303469$$

By calculator:

[24] [x^2] [×] [20] [sin] [×] [110] [sin] [=] [÷] [2] [÷] [50] [sin] [=]

To two significant figures, the area is **120 m²**. ■

If three sides (but none of the angles) are known, you will need another formula to find the area of a triangle. The formula is derived from the Law of Cosines; this derivation is left as an exercise but is summarized in the following box. The result is known as **Heron's** (or **Hero's**) **Formula**.

Area of a Triangle (Three Sides Known)

$$K = \sqrt{s(s-a)(s-b)(s-c)} \qquad \text{where } s = \tfrac{1}{2}(a+b+c)$$

EXAMPLE 7 Find the area of a triangle having sides of 43 ft, 89 ft, and 120 ft.

Solution Let $a = 43$, $b = 89$, and $c = 120$. Then $s = \frac{1}{2}(43 + 89 + 120) = 126$. Thus,

$$K = \sqrt{126(126-43)(126-89)(126-120)}$$
$$\approx 1523.704696$$

To two significant digits, the area is **1500 ft²**. ■

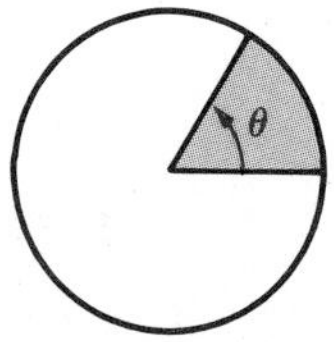

Figure 8.27 Sector of a circle (shaded portion)

Another type of area problem that requires trigonometry is the procedure for finding the area of a sector of a circle. A **sector of a circle** is the portion of the interior of a circle cut by a central angle, θ (Figure 8.27). Since the area of a circle is $A = \pi r^2$, where r is the radius, the area of a sector is the fraction $\theta/2\pi$ of the entire circle. In general,

$$\text{Area of sector} = (\text{fractional part of circle})(\text{area of circle})$$
$$= \frac{\theta}{2\pi} \cdot \pi r^2$$

Area of a Sector

> Area of sector $= \frac{1}{2}\theta r^2$ where θ is measured in radians

EXAMPLE 8 What is the area of the sector of a circle of radius 12 in. whose central angle is 2?

Solution

$$\text{Area of sector} = \frac{1}{2} \cdot 2 \cdot (12 \text{ in.})^2$$
$$= \mathbf{144 \text{ in.}^2}$$

■

EXAMPLE 9 What is the area of the sector of a circle of radius 420 m whose central angle is 2°?

Solution First change 2° to radians:

$$2° = 2\left(\frac{\pi}{180}\right) = \frac{\pi}{90} \text{ radian}$$

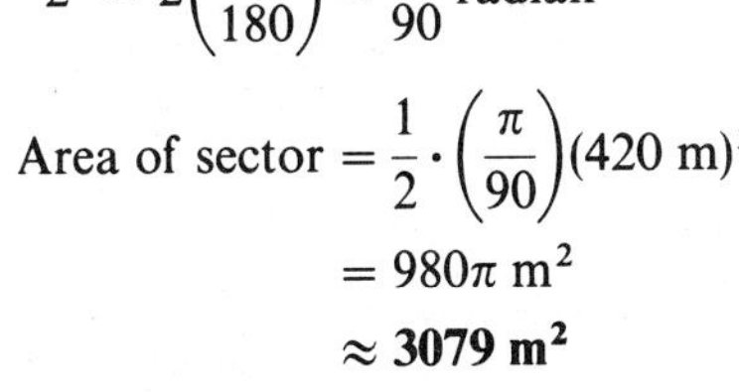

$$\text{Area of sector} = \frac{1}{2} \cdot \left(\frac{\pi}{90}\right)(420 \text{ m})^2$$
$$= 980\pi \text{ m}^2$$
$$\approx \mathbf{3079 \text{ m}^2}$$

■

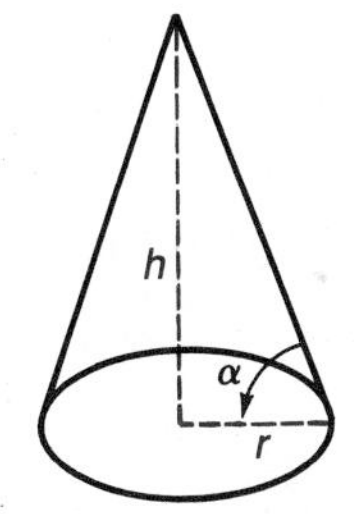

Figure 8.28

A third application in this section is finding the volume of a cone when you do not know its height. From geometry, the volume is

$$V = \frac{\pi r^2 h}{3}$$

where r is the radius of the base and h is the height of the cone. Now suppose you know the radius of the base but not the height. If you know the angle of elevation α (see Figure 8.28), then

$$\tan\alpha = \frac{h}{r} \qquad \text{or} \qquad h = r\tan\alpha$$

Thus, the volume is

$$V = \frac{\pi r^3 \tan\alpha}{3} = \frac{1}{3}\pi r^3 \tan\alpha$$

EXAMPLE 10 If sand is dropped from the end of a conveyor belt, the sand will fall in a conical heap such that the angle of elevation α is about 33°. Find the volume of sand when the radius r is exactly 10 ft. (See Figure 8.29, page 316.)

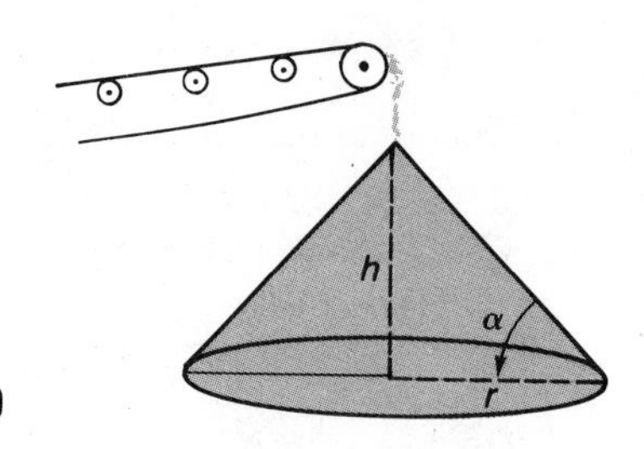

Figure 8.29

Solution $V = \frac{1}{3}\pi(10)^3 \tan 33°$ PRESS: [π] [×] [10] [y^x] [3] [×] [33] [tan] [÷] [3] [=]

≈ 680 DISPLAY: 680.0580413

The volume (to two significant figures) is **680 ft³**. ■

PROBLEM SET 8.4

A

Solve △ABC in Problems 1–12. If two triangles are possible, give both solutions. If the triangle does not have a solution, tell why.

1. $a = 3.0$; $b = 4.0$; $\alpha = 125°$
2. $a = 3.0$; $b = 4.0$; $\alpha = 80°$
3. $a = 5.0$; $b = 7.0$; $\alpha = 75°$
4. $a = 5.0$; $b = 7.0$; $\alpha = 135°$
5. $a = 5.0$; $b = 4.0$; $\alpha = 125°$
6. $a = 5.0$; $b = 4.0$; $\alpha = 80°$
7. $a = 9.0$; $b = 7.0$; $\alpha = 75°$
8. $a = 9.0$; $b = 7.0$; $\alpha = 135°$
9. $a = 7.0$; $b = 9.0$; $\alpha = 52°$
10. $a = 12.0$; $b = 9.00$; $\alpha = 52.0°$
11. $a = 8.629973679$; $b = 11.8$; $\alpha = 47.0°$
12. $a = 10.2$; $b = 11.8$; $\alpha = 47.0°$

Solve △ABC with the parts given in Problems 13–24. If the triangle cannot be solved, tell why.

13. $a = 14.2$; $b = 16.3$; $\beta = 115.0°$
14. $a = 14.2$; $c = 28.2$; $\gamma = 135.0°$
15. $\beta = 15.0°$; $\gamma = 18.0°$; $b = 23.5$
16. $b = 45.7$; $\alpha = 82.3°$; $\beta = 61.5°$
17. $b = 82.5$; $c = 52.2$; $\gamma = 32.1°$
18. $a = 151$; $b = 234$; $c = 416$
19. $a = 68.2$; $\alpha = 145°$; $\beta = 52.4°$
20. $\alpha = 82.5°$; $\beta = 16.9°$; $\gamma = 80.6°$
21. $a = 123$; $b = 225$; $c = 351$
22. $a = 27.2$; $c = 35.7$; $\alpha = 43.7°$
23. $c = 196$; $\alpha = 54.5°$; $\gamma = 63.0°$
24. $b = 428$; $c = 395$; $\gamma = 28.4°$

Find the area of each triangle in Problems 25–36.

25. $a = 15$; $b = 8.0$; $\gamma = 38°$
26. $a = 18$; $c = 11$; $\beta = 63°$
27. $b = 14$; $c = 12$; $\alpha = 82°$
28. $b = 21$; $c = 35$; $\alpha = 125°$
29. $a = 30$; $\beta = 50°$; $\gamma = 100°$
30. $b = 23$; $\alpha = 25°$; $\beta = 110°$
31. $b = 40$; $\alpha = 50°$; $\gamma = 60°$
32. $b = 90$; $\beta = 85°$; $\gamma = 25°$
33. $a = 7.0$; $b = 8.0$; $c = 2.0$
34. $a = 10$; $b = 4.0$; $c = 8.0$
35. $a = 11$; $b = 9.0$; $c = 8.0$
36. $a = 12$; $b = 6.0$; $c = 15$

B

Find the area of each triangle in Problems 37–48. If two triangles are formed, give the area of each, and if no triangle is formed, so state.

37. $a = 12.0$; $b = 9.00$; $\alpha = 52.0°$
38. $a = 7.0$; $b = 9.0$; $\alpha = 52°$
39. $a = 10.2$; $b = 11.8$; $\alpha = 47.0°$
40. $a = 8.629973679$; $b = 11.8$; $\alpha = 47.0°$
41. $b = 82.5$; $c = 52.2$; $\gamma = 32.1°$
42. $a = 352$; $b = 230$; $c = 418$

43. $\beta = 15.0°$; $\gamma = 18.0°$; $b = 23.5$

44. $b = 45.7$; $\alpha = 82.3°$; $\beta = 61.5°$

45. $a = 68.2$; $\alpha = 145°$; $\beta = 52.4°$

46. $a = 151$; $b = 234$; $c = 416$

47. $a = 124$; $b = 325$; $c = 351$

48. $a = 27.2$; $c = 35.7$; $\alpha = 43.7°$

49. ***Ballistics*** An artillery-gun observer must determine the distance to a target at point T. He knows that the target is 5.20 mi from point I on a nearby island. He also knows that he (at point H) is 4.30 mi from point I. If $\angle HIT$ is 68.4°, how far is he from the target? (See Figure 8.30.)

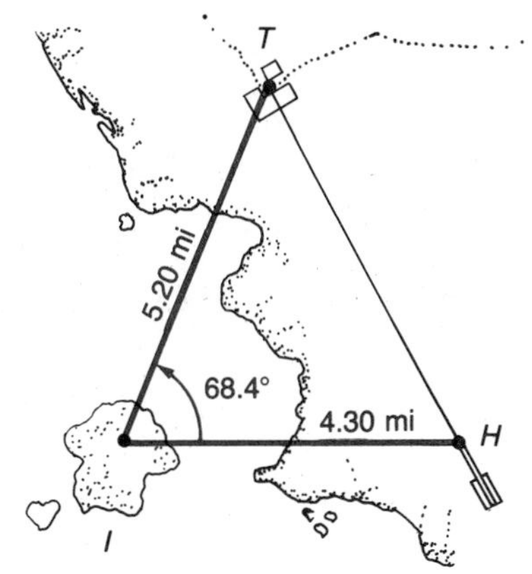

Figure 8.30 Determining the distance to a target

50. ***Consumer*** A buyer is interested in purchasing a triangular lot with vertices LOT, but unfortunately, the marker at point L has been lost. The deed indicates that TO is 453 ft and LO is 112 ft and that the angle at L is 82.6°. What is the distance from L to T?

51. ***Navigation*** A UFO is sighted by people in two cities 2.300 mi apart. The UFO is between and in the same vertical plane as the two cities. The angle of elevation of the UFO from the first city is 10.48° and from the second is 40.79°. At what altitude is the UFO flying? What is the actual distance of the UFO from each city?

52. ***Engineering*** At 500 ft in the direction that the Tower of Pisa is leaning, the angle of elevation is 20.24°. If the tower leans at an angle of 5.45° from the vertical, what is the length of the tower?

53. ***Engineering*** What is the angle of elevation of the leaning Tower of Pisa (described in Problem 52) if you measure from a point 500 ft in the direction exactly opposite from the way it is leaning?

54. ***Surveying*** The world's longest deepwater jetty is at Le Havre, France. Since access to the jetty is restricted, it was necessary for me to calculate its length by noting that it forms an angle of 85.0° with the shoreline. After pacing out 1000 ft along the line making an 85.0° angle with the jetty, I calculated the angle to the end of the jetty to be 83.6°. What is the length of the jetty?

55. If the central angle subtended by the arc of a segment of a circle is 1.78 and the area is 54.4 cm^2, what is the radius of the circle?

56. If the area of a sector of a circle is 162.5 cm^2 and the angle of the sector is 0.52, what is the radius of the circle?

57. A field is in the shape of a sector of a circle with a central angle of 20° and a radius of 320 m. What is the area of the field?

58. ***Consumer*** A level lot has the dimensions shown in Figure 8.31. What is the total cost of treating the area for poison oak if the fee is \$45 per acre (1 acre = 43,560 ft^2)?

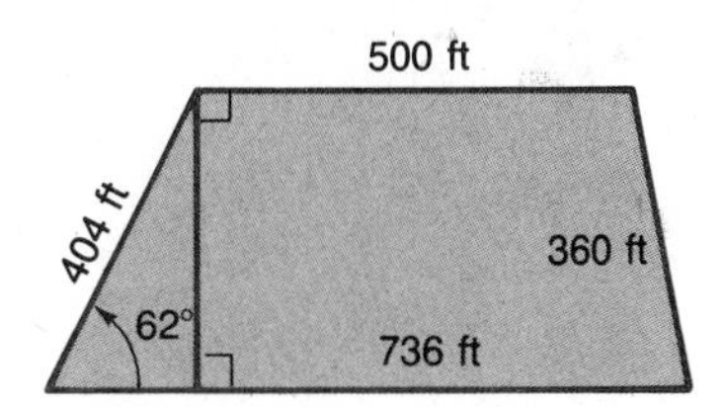

Figure 8.31 Area of a lot

59. If vulcanite is dropped from a conveyor belt, it will fall in a conical heap such that the angle of elevation is about 36°. Find the volume of vulcanite when the radius is 30 ft.

60. The volume of a slice cut from a cylinder is found from the formula

$$V = hr^2\left(\frac{\theta}{2} - \sin\frac{\theta}{2}\cos\frac{\theta}{2}\right)$$

for θ measured in radians such that $0 \le \theta \le \pi$. See Figure 8.32. Find the volume of a slice cut from a log with a 6.0-in. radius that is 3.0 ft long when $\theta = \pi/3$.

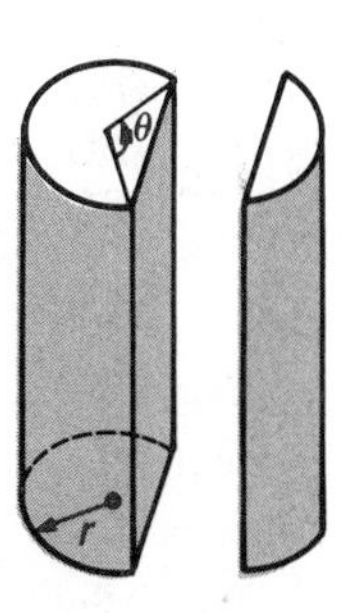

Figure 8.32

C

61. ***Navigation*** From a blimp, the angle of depression to the top of the Eiffel Tower is 23.2° and to the bottom is 64.6°. After flying over the tower at the same height and at a distance of 1000 ft from the first location, you determine that the angle of depression to the top of the tower is now 31.4°. What is the height of the Eiffel Tower given that these measurements are in the same vertical plane? (See Figure 8.33.)

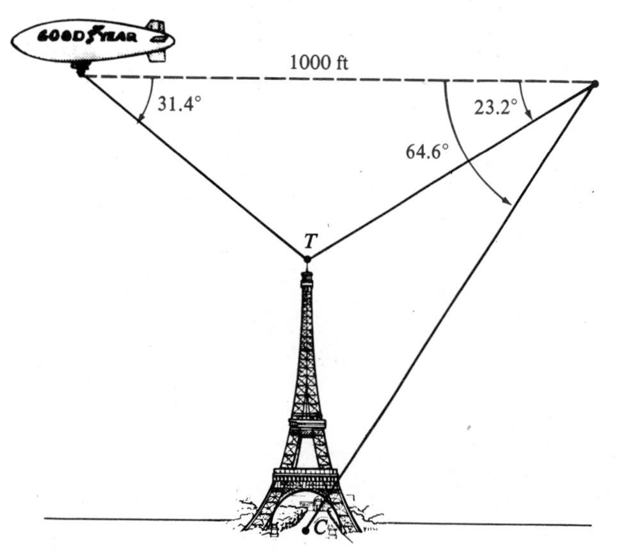

Figure 8.33 Determining the height of the Eiffel Tower

62. ***Historical Question*** In January 1978, Martin Cohen, Terry Goodman, and John Benard published an article, "SSA and the Law of Cosines," in *The Mathematics Teacher*. In this article, they used the quadratic formula to solve the Law of Cosines for c:

$$c = b\cos\alpha \pm \sqrt{a^2 - b^2\sin^2\alpha}$$

Show this derivation.

63. Suppose $\alpha < 90°$. If $d = a^2 - b^2\sin^2\alpha$ in Problem 62, show that if $d < 0$, there is no solution and this equation corresponds to no triangle; if $d = 0$, there is one solution, which corresponds to one triangle; if $d > 0$, there are two solutions and two triangles are formed.

64. ***Construction*** A 50-ft culvert carries water under a road. If there is 2 ft of water in a culvert with a 3-ft radius, use Problem 60 to find the volume of water in the culvert.

65. Derive the formula for finding the area when you know three sides of a triangle.

8.5 CHAPTER 8 SUMMARY

The material of this chapter is reviewed in the following list of objectives. After each objective there are some practice questions. For a sample test, select the first question of each set and check your answers with the answer section. For a sample test without answers, use the second question of each set. Additional practice is given by the other questions in each set. If you are having trouble with a particular type of problem, look back to that section for extra help.

8.1 RIGHT TRIANGLES

Objective 1 Know the right-triangle definition of the trigonometric functions and solve right triangles.

1. State the right-triangle definition of the trigonometric functions.
2. Solve $\triangle ABC$, where $a = 7.3$, $c = 15$, and $\gamma = 90°$.
3. Solve $\triangle ABC$, where $b = 678$, $\beta = 55.0°$, and $\gamma = 90.0°$.
4. Solve $\triangle ABC$, where $a = 3.0$, $b = 4.0$, and $c = 5.0$.

8.2 LAW OF COSINES
8.3 LAW OF SINES

Objective 2 Know the laws of cosines and sines, as well as the proof for each.

5. State the Law of Cosines.
6. State the Law of Sines.
7. Prove the Law of Cosines.
8. Prove the Law of Sines.

8.4 AMBIGUOUS CASE: SSA

Objective 3 Know when to apply the laws of cosines and sines, as well as recognizing the ambiguous case. Complete the table for $\triangle ABC$ labeled in the usual fashion.

	To find	Known	Procedure	Solution
	β	a, b, α	$\frac{\sin\alpha}{a} = \frac{\sin\beta}{b}$	$\beta = \sin^{-1}\left(\frac{b\sin\alpha}{a}\right)$
9.	α	a, b, c		
10.	α	a, b, β		
11.	α	a, β, γ		
12.	b	a, α, β		

Objective 4 Solve oblique triangles.

13. $a = 24; c = 61; \beta = 58^\circ$
14. $a = 6.8; b = 12.2; c = 21.5$
15. $b = 34; c = 21; \gamma = 16^\circ$
16. $b = 4.6; \alpha = 108^\circ; \gamma = 38^\circ$

Objective 5 Find the area of a given triangle, including the ambiguous case.

17. $\alpha = 48.0^\circ; b = 25.5$ ft; $c = 48.5$ ft
18. $a = 275$ ft; $b = 315$ ft; $\alpha = 50.0^\circ$
19. $a = 14.5$ in.; $b = 17.2$ in.; $\alpha = 35.5^\circ$
20. $a = 6.50$ m; $b = 8.30$ m; $c = 12.6$ m

Objective 6 Solve applied problems using either right triangles or oblique triangles.

21. A mine shaft is dug into the side of a sloping hill. The shaft is dug horizontally for 485 ft. Next, a turn is made so that the angle of elevation of the second shaft is 58.0°, thus forming a 58.0° angle between the shafts. The shaft is then continued for 382 ft before exiting, as shown in Figure 8.34. How far is it along a straight line from the entrance to the exit, assuming that all tunnels are in a single plane? If the slope of the hill follows the line from the entrance to the exit, what is the angle of elevation from the entrance to the exit?

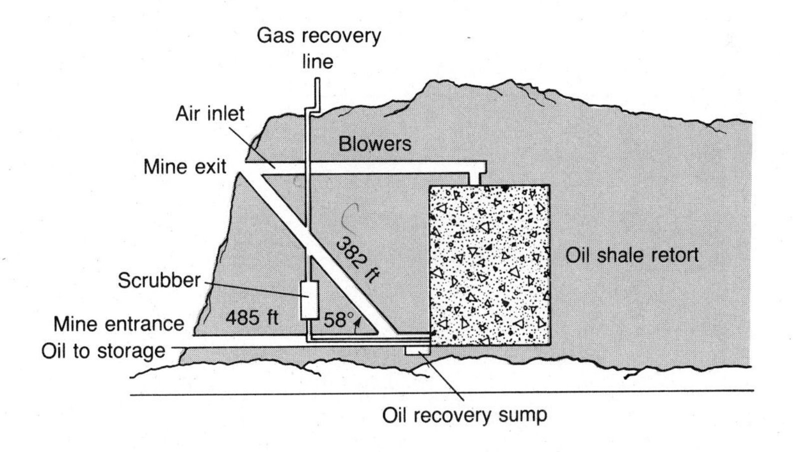

Figure 8.34 Determining the exit of a mine shaft.

22. Ferndale is 7 mi N50°W of Fortuna. If I leave Fortuna at noon and travel due west at 2 mph, when will I be exactly 6 mi from Ferndale?

23. To measure the span of the Rainbow Bridge in Utah, a surveyor selected two points, P and Q, on either end of the bridge. From point Q, the surveyor measured 500 ft in the direction N38.4°E to point R. Point P was then determined to be in the direction

S67.5°W. What is the span of the Rainbow Bridge if all the preceding measurements are in the same plane and $\angle PQR$ is a right angle?

24. When viewing Angel Falls (the world's highest waterfall) from Observation Platform *A*, located on the same level as the bottom of the falls, we calculate the angle of elevation to the top of the falls to be 69.30°. From Observation Platform *B*, which is located on the same level exactly 1000 ft from the first observation point, we calculate the angle of elevation to the top of the falls to be 52.90°. How high are the falls?

CUMULATIVE REVIEW II

Suggestions for study of Chapters 6–8:

Make a list of important ideas from Chapters 6–8. Study this list. Use the objectives at the end of each chapter to help you make up this list.

Work some practice problems. A good source of problems is the set of chapter objectives at the end of each chapter. You should try to work at least one problem from each objective:

Chapter 6, pp. 243–245; work 25 problems
Chapter 7, pp. 287–289; work 17 problems
Chapter 8, pp. 318–320; work 6 problems

Check the answers for the practice problems you worked. The first and third problems of each objective have their answers listed in the back of the book.

Additional problems. Work additional odd-numbered problems (answers in the back of the book) from the problem sets as needed. Focus on the problems you missed in the chapter summaries.

Work the problems in the following cumulative review. These problems should be done after you have studied the material. They should take you about 1 hour and will serve as a sample test for Chapters 6–8. Assume that all variables are restricted so that each expression is defined. All the answers for these questions are provided in the back of the book for self-checking.

PRACTICE TEST FOR CHAPTERS 6–8

1. Define the six trigonometric functions of any angle θ.

2. Evaluate the given functions (use exact values wherever possible).

a. $\sin 300^\circ$ **b.** $\cos\frac{5\pi}{6}$
c. $\tan 3\pi$ **d.** $\sin^{-1}\left(-\frac{\sqrt{3}}{2}\right)$
e. $\operatorname{arccos}\frac{1}{2}$ **f.** $\operatorname{Arccos}(-4.521)$
g. $\operatorname{Tan}^{-1} 2.310$ **h.** $\tan(\operatorname{Sec}^{-1} 3)$

3. Graph each curve.
a. $y = \sqrt{2}\cos\frac{1}{2}x$
b. $y + 3 = 2\sin(3x + \pi)$
c. $y - 1 = \tan\left(x - \frac{\pi}{3}\right)$

4. a. State and prove one of the reciprocal identities.
b. State and prove one of the ratio identities
c. State and prove one of the Pythagorean identities

5. Solve the given equations for $0 \le \theta < 2\pi$.
a. $3\tan 3\theta = \sqrt{3}$ (exact values)
b. $4\sin^2\theta + 8\sin\theta - 1 = 0$ (four decimal places)

6. Prove any *two* of the following identities:
a. $\frac{\sin 3\theta}{\sin\theta} - \frac{\cos 3\theta}{\cos\theta} = 2$
b. $\frac{\sin\theta}{\csc\theta - \cot\theta} = 1 + \cos\theta$
c. $\cos 2\theta = 1 - 2\sin^2\theta$
(Do not assume the identity for $\cos 2\theta$.)

7. For the given triangle, state (one form for each is sufficient):

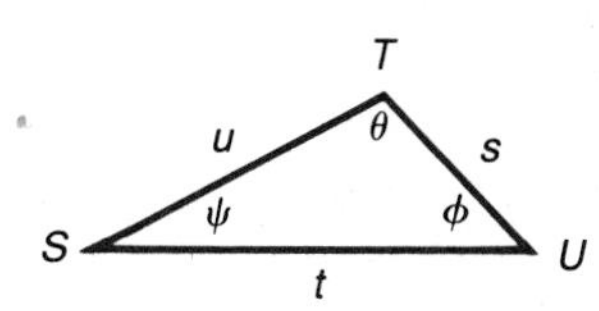

a. Law of Sines
b. Law of Cosines

8. Solve each of the given triangles.

a. $a = 14$; $b = 27$; $c = 19$

b. $b = 7.2$; $c = 15$; $\alpha = 113°$

c. $a = 35$; $b = 45$; $\beta = 35°$

d. $c = 85.7$; $\alpha = 50.8°$; $\beta = 83.5°$

9. Find the area of each of the given triangles.

a. $b = 16$ ft, $c = 43$ ft, $\alpha = 113°$

b. $\alpha = 40.0°$, $\beta = 51.8°$, $c = 14.3$ in.

c. $a = 121$ cm, $b = 46$ cm, $c = 92$ cm

10. a. The world's largest pyramid is located 63 mi S45°E of Mexico City. If I leave Mexico City in a jeep and travel due east, how far from Mexico City will I be when I am 50 mi from the pyramid?

b. The longest bridge in the world is over the Humber Estuary in England. To measure its length I paced out 100.0 ft in a direction that formed a right angle with one end. I then determined that the angle formed when sighting the other end was 88.76°. How long is the bridge?

Extended Application: SOLAR POWER

Solar Power

Energy from the sun provides enough power to heat water or even a home

BY NEALE LESLIE A revolutionary solar-powered water heater that aligns itself facing directly into the sun now is available on the market.

Unlike flat plate collectors which have to be installed in a southward direction, usually on roofs of the buildings being served, this device can be installed on any part of a north or south-facing roof—or any portion of the yard—as long as the sun's rays reach it.

This device, called the SAV, manufactured by SAV Solar Systems Inc. of Los Angeles, follows the sun. It is both a parabolic reflector and storage tank combined.

The tank—riding piggyback above the parabolic mirror—is a collector itself. It is coated with black chrome, one of the most superior materials known for absorbing heat. Actually, it absorbs 95 percent of all incoming radiation, and only emits 10 percent of the heat collected. The result: A trapping and holding of 85 percent of the solar radiation.

The tank is further encased in cylindrical double glazing made of clear plastic, with one inch of air space between the tank and inner glazing and another inch of air space between the inner and outer glazing.

This provides a "glass house" heating and insulating effect. At the same time the tank is tilted to the most direct angle to the sun, an angle the homeowner can adjust to conform to his latitude and the change of the season.

The tank, designed by Jon Makeever, SAV Solar Systems president, permits the sun's heat to set up "thermosyphonic" circulation patterns, greatly increasing the efficiency of the unit.

Makeever also incorporated a system of parabolic surfaces concentrating the energy of the sun directly onto the solar collector and storage tank. And working in conjunction with this is his tracking system which causes the entire mechanism, including the tank, to follow the sun across the sky.

Makeever chose Freon 12, available anywhere in refrigeration service shops, as the agent to provide the power for the tracking mechanism. The Freon 12 is energized by the sun. It is maintenance free, requires no wires, no electricity, timers or other apparatus. It is non-explosive, providing enough pressure to operate a couple of hydraulic pistons, but not enough to burst any of the tracker's components. The Freon 12 has to be replaced once every four years.

The unit, approximately 4 by 4 feet, provides up to 98 percent more heat per collector area than flat plate systems used in a test, Makeever said.

One unit would supply the hot water needs of a typical family, but three would be needed for space heating a house. Empty, the unit weighs 140 pounds; filled another 100, but way under the minimal weight standard for roofs.

The unit is said to be adaptable to any part of a sloping or flat roof, and capable of withstanding wind velocities up to 120 m.p.h.

Retail price of the unit is about $2,500 installed, available through plumbing, heating and air conditioning firms and solar equipment dealers.

The efficient use of a solar collector requires knowledge of the length of daylight and the angle of the sun throughout the year at the location of the collector. Table 1 on page 324 gives the times of sunrise and sunset for various latitudes. Figure 1 shows a graph of the sunrise and sunset times for latitude 35°N. These are definitely not graphs of sine or cosine functions. If you plot the length of daylight (see Problem 7), however, you will obtain a curve that is nearly a sine curve.

TABLE 1
Times of sunrise and sunset for various latitudes

Date	Time of Sunrise						Time of Sunset					
	20°N. Latitude (Hawaii)	30°N. Latitude (New Orleans)	35°N. Latitude (Albuquerque)	40°N. Latitude (Philadelphia)	45°N. Latitude (Minneapolis)	60°N. Latitude (Alaska)	20°N. Latitude (Hawaii)	30°N. Latitude (New Orleans)	35°N. Latitude (Albuquerque)	40°N. Latitude (Philadelphia)	45°N. Latitude (Minneapolis)	60°N. Latitude (Alaska)
	h m	h m	h m	h m	h m	h m	h m	h m	h m	h m	h m	h m
Jan. 1	6 35	6 56	7 08	7 22	7 38	9 03	17 31	17 10	16 58	16 44	16 28	15 03
Jan. 15	6 38	6 57	7 08	7 20	7 35	8 48	17 41	17 22	17 12	16 59	16 44	15 31
Jan. 30	6 36	6 52	7 01	7 11	7 23	8 19	17 51	17 35	17 27	17 17	17 05	16 09
Feb. 14	6 30	6 41	6 48	6 55	7 03	7 42	17 59	17 48	17 42	17 34	17 26	16 48
Mar. 1	6 20	6 26	6 30	6 34	6 39	6 59	18 05	17 59	17 56	17 52	17 47	17 27
Mar. 16	6 08	6 09	6 10	6 11	6 11	6 15	18 10	18 09	18 08	18 08	18 07	18 04
Mar. 31	5 55	5 51	5 49	5 46	5 43	5 29	18 14	18 18	18 20	18 23	18 26	18 41
Apr. 15	5 42	5 34	5 28	5 23	5 16	4 44	18 18	18 27	18 32	18 38	18 45	19 18
Apr. 30	5 32	5 18	5 11	5 02	4 51	4 01	18 23	18 37	18 44	18 53	19 04	19 55
May 15	5 24	5 07	4 57	4 45	4 31	3 23	18 29	18 46	18 56	19 08	19 22	20 31
May 30	5 20	5 00	4 48	4 34	4 18	2 53	18 35	18 55	19 07	19 21	19 38	21 04
June 14	5 20	4 58	4 45	4 30	4 13	2 37	18 40	19 02	19 15	19 30	19 48	21 24
June 29	5 23	5 02	4 49	4 34	4 16	2 40	18 43	19 05	19 18	19 33	19 51	21 26
July 14	5 29	5 08	4 56	4 43	4 26	3 01	18 43	19 03	19 15	19 29	19 45	21 09
July 29	5 34	5 17	5 07	4 55	4 41	3 33	18 39	18 56	19 06	19 17	19 31	20 38
Aug. 13	5 39	5 26	5 18	5 09	4 59	4 09	18 30	18 43	18 51	19 00	19 10	19 59
Aug. 28	5 43	5 35	5 29	5 24	5 17	4 45	18 19	18 27	18 33	18 38	18 45	19 16
Sept. 12	5 47	5 43	5 40	5 38	5 35	5 20	18 06	18 10	18 12	18 14	18 17	18 31
Sept. 27	5 50	5 51	5 51	5 52	5 53	5 55	17 52	17 51	17 50	17 49	17 49	17 45
Oct. 12	5 54	6 00	6 03	6 07	6 11	6 31	17 39	17 33	17 29	17 26	17 21	17 01
Oct. 22	5 57	6 06	6 12	6 18	6 25	6 56	17 32	17 22	17 17	17 11	17 04	16 32
Nov. 6	6 04	6 18	6 26	6 35	6 45	7 35	17 23	17 10	17 02	16 52	16 42	15 52
Nov. 21	6 12	6 30	6 40	6 52	7 05	8 12	17 19	17 02	16 51	16 40	16 26	15 19
Dec. 6	6 22	6 42	6 54	7 07	7 23	8 44	17 20	17 00	16 48	16 35	16 19	14 58
Dec. 21	6 30	6 52	7 04	7 18	7 35	9 02	17 26	17 05	16 52	16 38	16 21	14 54

Table courtesy of U.S. Naval Observatory. This table of sunrise and sunset may be used in any year of the 20th century with an error not exceeding two minutes and generally less than one minute. It may also be used anywhere in the vicinity of the stated latitude with an additional error of less than one minute for each 9 miles.

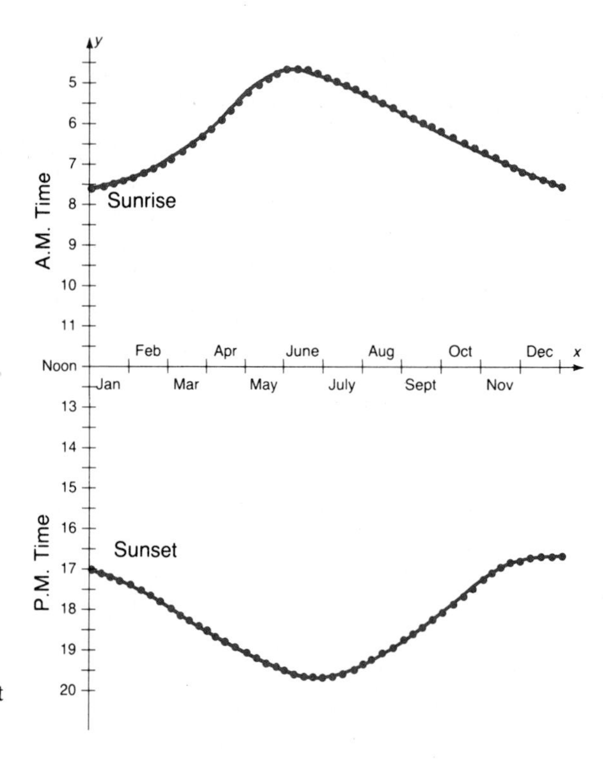

Figure 1 Sunrise and sunset times for latitude 35°N

If you wish to use Table 1 for your own town's latitude and it is not listed, you can use a procedure called **linear interpolation**. Your local Chamber of Commerce will probably be able to tell your town's latitude. The latitude for Santa Rosa, California, is about 38°N, which falls between 35° and 40° on the table (see Figure 2).

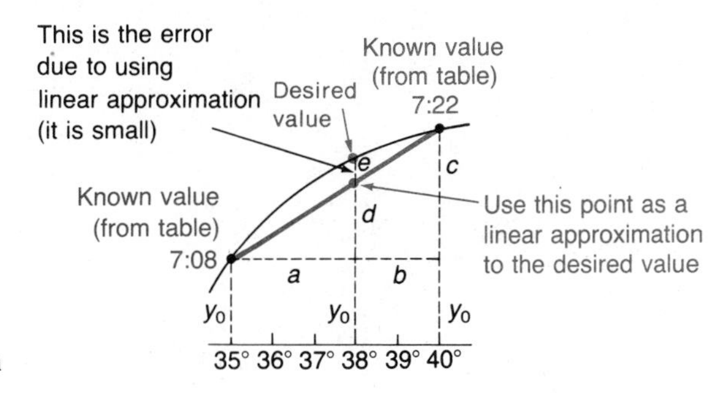

Figure 2 Linear interpolation

Angle	*Time*	
35°	7:08	Known, from Table 1
38°	x	Unknown; this is the latitude of the town for which the times of sunrise and sunset are not available.
40°	7:22	Known, from Table 1

Use the known information and a proportion to find x:

$$5\left[\begin{matrix} 3\left[\begin{matrix}35^\circ \\ 38^\circ\end{matrix}\right. \\ 40^\circ \end{matrix}\right. \quad \left.\begin{matrix} 7{:}08 \\ x \\ 7{:}22 \end{matrix}\right] 14$$

Since 38° is $\frac{3}{5}$ of the way between 35° and 40°, the sunrise time will be $\frac{3}{5}$ of the way between 7:08 and 7:22.

$$\frac{3}{5} = \frac{x}{14}$$

$$x = \frac{3}{5}(14)$$

$$= 8.4$$ To the nearest minute this is :08.

The sunrise time is about 7:08 + :08 = 7:16.

This solar tank riding piggyback above a parabolic mirror is called the SAV. It is pictured with its designer, Jon Makeever, president of SAV Solar Systems. (Courtesy of *The Press Democrat*.)

One type of solar collector requires the construction of a central cylinder. The construction of a template for cutting a 45° angle in a cardboard tube to make a rotating mirror is described in the May 1978 issue of *Byte* magazine. The template is shown in Figure 3, and the details of construction are given in Problem 12.

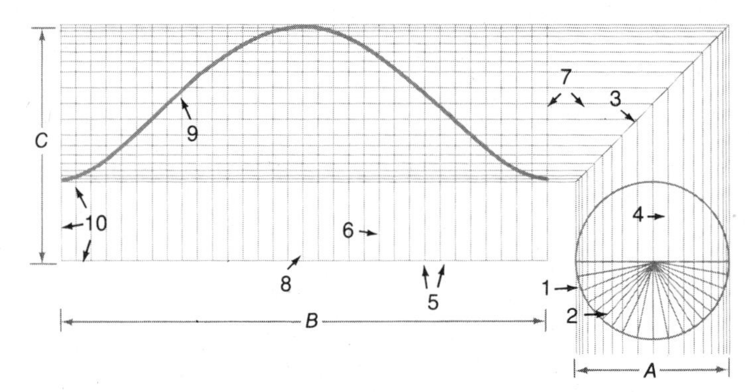

Figure 3 Construction of a template for cutting a 45° angle in a cardboard tube to make a rotating mirror

Extended Application Problems—Solar Power

Use linear interpolation and Table 1 to approximate the times requested in each of Problems 1–6.

1. Sunset for Santa Rosa (lat. 38°N) on January 1.
2. Sunrise for Tampa, Florida (lat. 28°N), on June 29.

3. Sunset for Tampa, Florida (lat. 28°N), on June 29.
4. Sunset for Winnipeg, Canada (lat. 50°N), on October 12.
5. Sunrise for Winnipeg, Canada (lat. 50°N), on October 12.
6. Sunrise for Juneau, Alaska (lat. 58°N), on March 1.
7. Plot the length of daylight in Chattanooga, Tennessee (lat. 35°N).
8. Plot the time of sunrise and sunset for Seward, Alaska (lat. 60°).
9. Plot the length of daylight for Seward, Alaska (lat. 60°N).
10. Plot the sunrise and sunset for your town's latitude by drawing a graph similar to that in Figure 1.
11. Plot the length of daylight for your town from the data in Problem 10.
12. Carry out the following steps from *Byte* magazine (May 1978, p. 129) to cut a 45° angle in a tube of diameter A, circumference B, and finished height C:*

1. Draw a circle the same size as the outside diameter of the tube.
2. Divide the circle into 32 equal parts of $11\frac{1}{4}$ in. each.
3. Draw a 45° angle above the circle.
4. Carry the points at the outside of the circle straight up to the 45° angle.
5. Find the circumference of the circle, draw a straight line, and divide it into 32 equal parts.
6. Carry lines from those divisions straight upward.
7. Bring lines straight across from the intersections at the 45° angle line and intersect with the vertical lines.
8. Starting at the centerline, mark points of intersection.
9. Fill in between points of intersection.
10. Cut out pattern and wrap around tube, lining up straight circumference side of pattern with (cut) end of tube square.
11. Carefully trace curved line end of pattern onto tube.
12. Remove pattern from tube and save for making black paper cover for tube.
13. Cut along traced line on tube with an X-acto (or similar) knife.
14. Tape large sheet of sandpaper to a table top or other flat surface.
15. Sand end of tube until flat.
16. Tube is now ready for application of Mylar mirror surface.

---

* From "How to Multiply in a Wet Climate," by J. Bryant and M. Swasdee, May, 1978, p. 129.

Gottfried Wilhelm von Leibniz (1646–1716)

I have so many ideas that may perhaps be of some use in time if others more penetrating than I go deeply into them some day and join the beauty of their minds to the labor of mine.

Leibniz

For what is the theory of determinants? It is an algebra upon algebra; a calculus which enables us to combine and foretell the results of algebraical operations, in the same way as special operations of arithmetic. All analysis must ultimately clothe itself under this form.

J. J. Sylvester
Philosophical Magazine, vol. 1 (1851)

HISTORICAL NOTE

Gottfried Wilhelm von Leibniz has been described as a true example of a "universal genius." Morris Kline described Leibniz as a philosopher, lawyer, historian, philologist, and pioneer geologist who did important work in logic, mechanics, optics, mathematics, hydrostatics, pneumatics, nautical science, and calculating machines. He was a professional lawyer who was employed as a diplomat and historian. By the time he was 20, he had mastered most of the important works on mathematics and had set the stage for modern logic. At 30, he invented calculus, and at 36, he invented a reckoning machine, a forerunner of the computer. He tried, without success, to reconcile the Catholic and Protestant faiths. Throughout his life he was searching for a *universal mathematics* in which he aimed to create "a general method in which all truths of the reason would be reduced to a kind of calculation." Because of his vast talent, his search for a universal mathematics led him to many applied problems.

In this chapter, we investigate an important tool in solving a wide variety of applied problems. That tool—*matrices*—will provide procedures for solving *systems of equations and inequalities.* Systems of equations were solved by the Babylonians as early as 1600 B.C. By 1683, the Japanese mathematician Seki Kōwa had updated an old Chinese method of solving simultaneous linear equations that involved rearrangements of rods similar to the way we simplify *determinants* in this chapter. It was Leibniz, however, who originated the notation of determinants.

Leibniz' later years were dimmed by a bitter controversy with Isaac Newton concerning whether he had discovered calculus independently of Newton. When he died, in 1716, his funeral was attended only by his secretary. Today we ascribe the independent discovery of calculus to both Leibniz and Newton.

9

Systems and Matrices

CONTENTS

* Optional section

PREVIEW

This chapter introduces a very important tool in mathematics, namely the matrix. We are motivated by our desire to solve systems of equations, but we soon learn that matrices have many other applications and uses. This chapter concludes with an introduction to a very significant new branch of mathematics, linear programming. There are 14 specific objectives in this chapter, which are listed on pages 382–385.

PERSPECTIVE

We live in a three-dimensional world, and surfaces in three dimensions are studied in calculus. In order to analyze a surface you need to look at what are called *tangent planes*. In the discussion taken from a leading calculus book shown below, the following determinant is used to find the tangent plane $\mathbf{T}_x \times \mathbf{T}_y$:

$$\begin{vmatrix} \mathbf{i} & \mathbf{j} & \mathbf{k} \\ 1 & 0 & \dfrac{\partial f}{\partial x} \\ 0 & 1 & \dfrac{\partial f}{\partial y} \end{vmatrix}$$

We introduce determinants in Section 9.2.

since $\sin\phi \geqslant 0$ for $0 \leqslant \phi \leqslant \pi$. Therefore the area of the sphere is

$$A = \iint_D |\mathbf{T}_\phi \times \mathbf{T}_\theta|\,dA = \int_0^{2\pi}\int_0^{\pi} a^2 \sin\phi\,d\phi\,d\theta$$

$$= a^2 \int_0^{2\pi} d\theta \int_0^{\pi} \sin\phi\,d\phi = a^2(2\pi)2 = 4\pi a^2 \quad \bullet$$

For the special case of a surface S with equation $z = f(x, y)$, where (x, y) lies in D and f has continuous partial derivatives, we take x and y as parameters. The parametric equations are

$$x = x \qquad y = y \qquad z = f(x, y)$$

so
$$\mathbf{T}_x = \mathbf{i} + \left(\frac{\partial f}{\partial x}\right)\mathbf{k} \qquad \mathbf{T}_y = \mathbf{j} + \left(\frac{\partial f}{\partial y}\right)\mathbf{k}$$

and

(14.47)
$$\mathbf{T}_x \times \mathbf{T}_y = \begin{vmatrix} \mathbf{i} & \mathbf{j} & \mathbf{k} \\ 1 & 0 & \dfrac{\partial f}{\partial x} \\ 0 & 1 & \dfrac{\partial f}{\partial y} \end{vmatrix} = -\frac{\partial f}{\partial x}\mathbf{i} - \frac{\partial f}{\partial y}\mathbf{j} + \mathbf{k}$$

From James Stewart, *Calculus* (Pacific Grove, Ca.: Brooks/Cole), p. 895.

9.1 SYSTEMS OF EQUATIONS

Solving systems of equations is a procedure that arises throughout mathematics. We will begin our study of this topic by considering an arbitrary system of two equations with two variables:

$$\begin{cases} a_{11}x_1 + a_{12}x_2 = b_1 \\ a_{21}x_1 + a_{22}x_2 = b_2 \end{cases}$$

The notation may seem strange, but it will prove to be useful. The variables are x_1 and x_2; the constants are $a_{11}, a_{12}, a_{21}, a_{22}, b_1$, and b_2. By a **system** of two equations with two variables, we mean any two equations in those variables. The **simultaneous solution** of a system is the intersection of the solution sets of the individual equations. The brace is used to show that this intersection is desired. If all the equations in a system are linear, it is called a **linear system.**

You solved linear systems with two variables in elementary algebra, so we will begin by reviewing the methods used there. Then we will generalize first to nonlinear systems and finally to more complicated linear systems.

Since the graph of each equation in a system of linear equations in two variables is a line, the solution set is the intersection of two lines. In two dimensions, two lines must be related to each other in one of three possible ways:

1. They intersect at a single point.
2. The graphs are parallel lines. In this case, the solution set is empty and the system is called *inconsistent.* In general, any system that has an empty solution set is referred to as an **inconsistent system.**
3. The graphs are the same line. In this case, there are infinitely many points in the solution set and any solution of one equation is also a solution of the other. Such a system is called **dependent**. This word is also used to describe nonlinear systems.

EXAMPLE 1 Relate the system

$$\begin{cases} 2x - 3y = -8 \\ x + y = 6 \end{cases}$$

to the general system

$$\begin{cases} a_{11}x_1 + a_{12}x_2 = b_1 \\ a_{21}x_1 + a_{22}x_2 = b_2 \end{cases}$$

and solve by graphing.

Solution The variables are $x_1 = x$ and $x_2 = y$. The subscripts in a_{11}, a_{12}, a_{21}, and a_{22} are called double subscripts and indicate *position.* That is, a_{12} should not be read "*a* sub

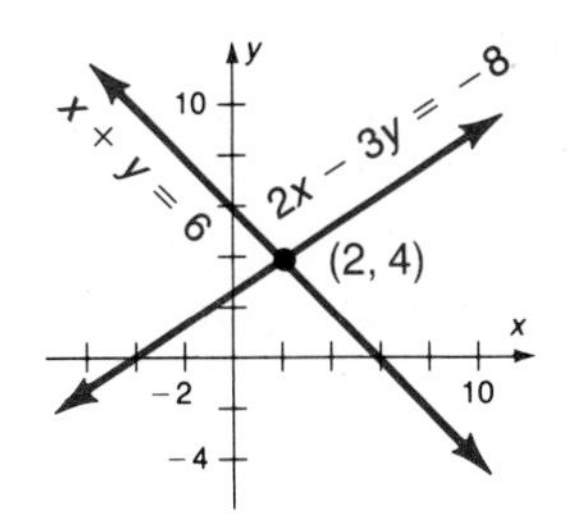

Figure 9.1 Graph of the system

$$\begin{cases} 2x - 3y = -8 \\ x + y = 6 \end{cases}$$

twelve" but rather "a sub one two." It represents the second constant in the first equation:

$$a_{12}$$

equation number ↗ ↖ constant number in equation

This effort in using notation may seem unnecessarily complicated when dealing with only two equations and two variables, but we want to develop a notation that can be used with many variables and many equations. For this example,

$$a_{11} = 2 \qquad a_{12} = -3 \qquad b_1 = -8$$
$$a_{21} = 1 \qquad a_{22} = 1 \qquad b_2 = 6$$

The solution is $x = 2$, $y = 4$, or **(2, 4)** as shown in Figure 9.1. ■

EXAMPLE 2 Relate the system

$$\begin{cases} 2x - 3y = -8 \\ 4x - 6y = 0 \end{cases}$$

to the general system and solve by graphing.

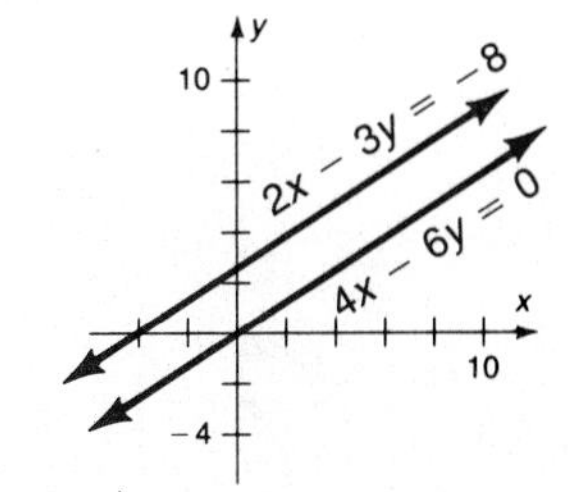

Figure 9.2 Graph of the system

$$\begin{cases} 2x - 3y = -8 \\ 4x - 6y = 0 \end{cases}$$

Solution The variables are $x_1 = x$, $x_2 = y$, and the constants are

$$a_{11} = 2 \qquad a_{12} = -3 \qquad b_1 = -8$$
$$a_{21} = 4 \qquad a_{22} = -6 \qquad b_2 = 0$$

The graphs of the lines are parallel since there is no point of intersection (Figure 9.2). This is an **inconsistent system**. ■

EXAMPLE 3 Relate the system

$$\begin{cases} 2x - 3y = -8 \\ y = \frac{2}{3}x + \frac{8}{3} \end{cases}$$

to the general system and solve by graphing.

Solution To relate to the general system, both equations must be algebraically in the usual form; the second equation needs to be rewritten:

$$y = \frac{2}{3}x + \frac{8}{3}$$
$$3y = 2x + 8$$
$$-2x + 3y = 8$$

This means that the variables are $x_1 = x$, $x_2 = y$, and the constants are

$$a_{11} = 2 \qquad a_{12} = -3 \qquad b_1 = -8$$
$$a_{21} = -2 \qquad a_{22} = 3 \qquad b_2 = 8$$

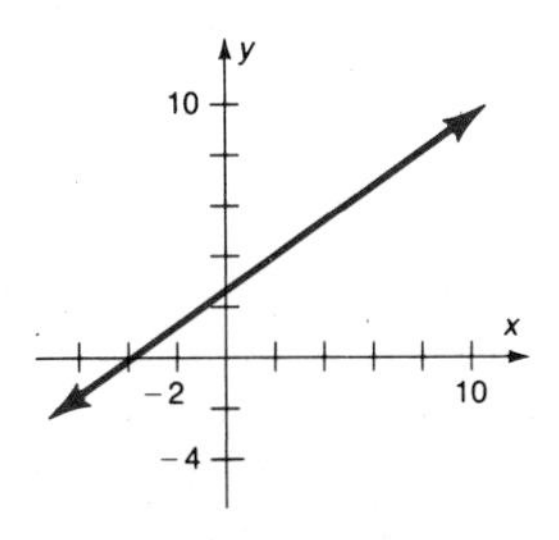

Figure 9.3 Graph of the system

$$\begin{cases} 2x - 3y = -8 \\ y = \frac{2}{3}x + \frac{8}{3} \end{cases}$$

The equations represent the same line as shown in Figure 9.3. This is a **dependent system**. ■

EXAMPLE 4 Solve the system $\begin{cases} x - y = 3 \\ 6x - y = x^2 + 7 \end{cases}$

Solution This is a nonlinear system. The graphs of $y = x - 3$ and $y = -x^2 + 6x - 7$ are shown in Figure 9.4. Remember that to graph this second-degree equation you need to complete the square:

$$y = -x^2 + 6x - 7$$

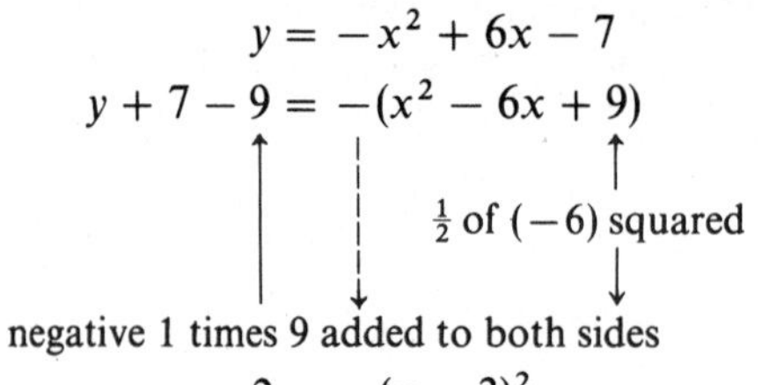

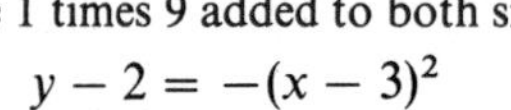

$$y - 2 = -(x - 3)^2$$

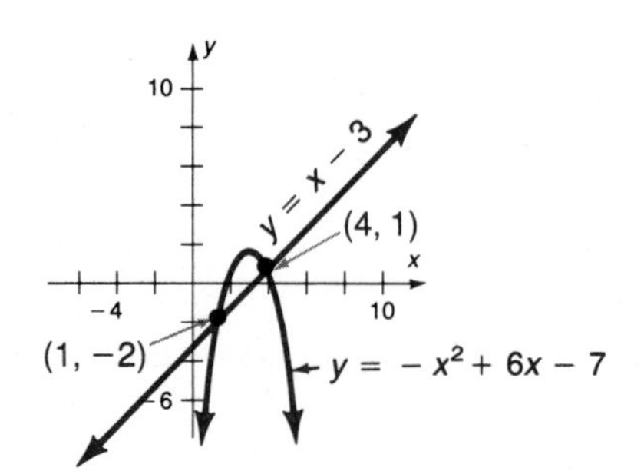

Figure 9.4 Graph of $\begin{cases} x - y = 3 \\ 6x - y = x^2 + 7 \end{cases}$

This is a parabola that opens downward with vertex at $(3, 2)$. By inspection, the solution is **(4, 1) and (1, −2)**.

To check an answer, make sure that every member of the solution set satisfies all the equations of the system:

Check $(4, 1)$: $x - y = 4 - 1 = 3$ (checks)

$6x - y = 6(4) - 1 = 23$ and $x^2 + 7 = (4)^2 + 7 = 23$ — checks

Check $(1, -2)$: $x - y = 1 - (-2) = 3$ (checks)

$6x - y = 6(1) - (-2) = 8$ and $x^2 + 7 = (1)^2 + 7 = 8$ — checks

Both $(4, 1)$ and $(1, -2)$ check. ■

The graphing method can give solutions only as accurate as the graphs you can draw, and consequently it is inadequate for most applications. There is therefore a need for more efficient methods.

In general, given a system, the procedure is to write a simpler equivalent system. Two systems are said to be **equivalent** if they have the same solution set. In this chapter we will limit ourselves to finding only real roots. There are several ways to go about writing equivalent systems. The first nongraphical method we will consider comes from the substitution property of real numbers and leads to a **substitution method** for solving systems.

Substitution Method for Solving Systems of Equations

1. *Solve* one of the equations for one of the variables.
2. *Substitute* the expression that you obtain into the other equation.
3. *Solve* the resulting equation in a single variable for the value of that variable.
4. *Substitute* that value into either of the original equations to determine the value of the other variable.
5. *State* the solution.

EXAMPLE 5 Solve $\begin{cases} 2p + 3q = 5 \\ q = -2p + 7 \end{cases}$ by substitution.

Solution Since $q = -2p + 7$, substitute $-2p + 7$ for q in the other equation:

$$\begin{aligned} 2p + \quad 3q \quad &= 5 \\ 2p + 3(\mathbf{-2p + 7}) &= 5 \\ 2p - 6p + 21 &= 5 \\ -4p &= -16 \\ p &= 4 \end{aligned}$$

Substitute 4 for p in either of the given equations:

$$\begin{aligned} q &= -2p \quad + 7 \\ &= -2(\mathbf{4}) + 7 \\ &= -1 \end{aligned}$$

The solution is $\mathbf{(p, q) = (4, -1)}$. ■

EXAMPLE 6 Solve $\begin{cases} x - y + 7 = 0 \\ y = x^2 + 4x + 3 \end{cases}$

Solution Solve one of the equations for one of the variables. The second equation is solved for y, so substitute $x^2 + 4x + 3$ for y in the first equation:

$$\begin{aligned} x - \quad y \quad + 7 &= 0 \\ x - (\mathbf{x^2 + 4x + 3}) + 7 &= 0 \\ -x^2 - 3x + 4 &= 0 \\ x^2 + 3x - 4 &= 0 \\ (x + 4)(x - 1) &= 0 \\ x &= -4, 1 \end{aligned}$$

If $x = -4$, then

$$\begin{aligned} y &= (-4)^2 + 4(-4) + 3 \\ &= 3 \end{aligned}$$

If $x = 1$, then

$$\begin{aligned} y &= 1^2 + 4(1) + 3 \\ &= 8 \end{aligned}$$

Solution: $\mathbf{(-4, 3)}$ **and** $\mathbf{(1, 8)}$. ■

EXAMPLE 7 Solve $\begin{cases} x^2 + 2xy = 0 \\ x^2 - 5xy = 14 \end{cases}$

Solution You want to solve the first equation for y, but that requires $x \neq 0$. If $x = 0$, then the first equation is satisfied but the second is not. Thus you can say that $x \neq 0$. Then

$$y = \frac{-x^2}{2x} = \frac{-x}{2}$$

Substitute into the second equation:

$$\begin{aligned} x^2 - 5xy &= 14 \\ x^2 - 5x\left(\frac{-x}{2}\right) &= 14 \\ 2x^2 + 5x^2 &= 28 \\ 7x^2 &= 28 \\ x^2 &= 4 \\ x^2 - 4 &= 0 \\ (x-2)(x+2) &= 0 \\ x &= 2, -2 \end{aligned}$$

If $x = 2$, then $y = \frac{-2}{2} = -1$; if $x = -2$, then $y = \frac{-(-2)}{2} = 1$. Thus the solution is **(2, −1) and (−2, 1)**. ■

EXAMPLE 8 Solve $\begin{cases} y = x^2 - x - 1 \\ 4x = 7 + 2y - y^2 \end{cases}$

Solution Substitute the first equation into the second one:

$$\begin{aligned} 4x &= 7 + 2y - y^2 \\ 4x &= 7 + 2(x^2 - x - 1) - (x^2 - x - 1)^2 \\ 4x &= 7 + 2x^2 - 2x - 2 - x^4 + 2x^3 + x^2 - 2x - 1 \\ 4x &= 4 + 3x^2 - 4x - x^4 + 2x^3 \\ x^4 - 2x^3 &- 3x^2 + 8x - 4 = 0 \end{aligned}$$

The possible rational roots are: $\pm 1, \pm 2, \pm 4$.

	1	−2	−3	8	−4
1	1	−1	−4	4	0
1	1	0	−4	0	

The depressed equation is

$$\begin{aligned} x^2 - 4 &= 0 \\ (x-2)(x+2) &= 0 \\ x &= 2, -2 \end{aligned}$$

The roots of this fourth-degree equation are $x = 1, 2$, and -2.

If $x = 1$, then $y = 1^2 - 1 - 1 = -1$.

If $x = 2$, then $y = 2^2 - 2 - 1 = 1$.

If $x = -2$, then $y = (-2)^2 - (-2) - 1 = 5$.

The solution is **(1, −1), (2, 1), and (−2, 5)**. ■

A third method for solving systems is called the **linear combination method**. It involves substitution and the idea that if equal quantities are added to equal quantities, the resulting equation is equivalent to the original system. In general, such addition will not simplify matters unless the numerical coefficients of one or more terms are opposites. However, you can often force them to be opposites by multiplying one or both of the given equations by nonzero constants.

Linear Combination Method for Solving Systems of Equations

1. *Multiply* one or both of the equations by a constant or constants, so that the coefficients of one of the variables become opposites.
2. *Add* corresponding members of the equations to obtain a new equation in a single variable.
3. *Solve* the derived equation for that variable.
4. *Substitute* the value of the found variable into either of the original equations and solve for the second variable.
5. *State* the solution.

EXAMPLE 9 Solve $\begin{cases} 3x + 5y = -2 \\ 2x + 3y = 0 \end{cases}$

Solution Multiply both sides of the first equation by 2 and both sides of the second equation by -3. This procedure, denoted as shown below, forces the coefficients of x to be opposites:

$$\begin{matrix} 2 \\ -3 \end{matrix}\begin{cases} 3x + 5y = -2 \\ 2x + 3y = 0 \end{cases}$$

This means you should add the equations of the system.

$$+\begin{cases} 6x + 10y = -4 \\ -6x - 9y = 0 \end{cases}$$

$$[6x + (-6x)] + [10y + (-9y)] = -4 + 0 \leftarrow$$ This step should be done mentally.

$$y = -4$$

If $y = -4$, then $2x + 3y = 0$ means $2x + 3(-4) = 0$, or $x = 6$. The solution is **(6, −4)**. ■

EXAMPLE 10 Solve $\begin{cases} y^2 - 5xy = 3 \\ 5xy + 4 = y^2 \end{cases}$

Solution

$$-1\begin{cases} y^2 - 5xy = 3 \\ y^2 - 5xy = 4 \end{cases}$$

$$+\begin{cases} -y^2 + 5xy = -3 \\ y^2 - 5xy = 4 \end{cases}$$

$$0 = 1$$

This is an **inconsistent system**. ■

PROBLEM SET 9.1

A

Solve the systems in Problems 1–9 by graphing.

1. $\begin{cases} y = 3x - 7 \\ y = -2x + 8 \end{cases}$

2. $\begin{cases} x - y = -1 \\ 3x - y = 5 \end{cases}$

3. $\begin{cases} y = \frac{2}{3}x - 7 \\ 2x + 3y = 3 \end{cases}$

4. $\begin{cases} y = \frac{3}{5}x + 2 \\ 3x - 5y = -10 \end{cases}$

5. $\begin{cases} 2x - 3y = 9 \\ y = \frac{2}{3}x - 3 \end{cases}$

6. $\begin{cases} x + y = 0 \\ y = -x^2 - 6x - 4 \end{cases}$

7. $\begin{cases} x + y = 4 \\ y = x^2 + 6x + 14 \end{cases}$

8. $\begin{cases} y = x^2 + 8x + 11 \\ x + y = -7 \end{cases}$

9. $\begin{cases} y = x^2 - 4x \\ y = x^2 - 4x + 8 \end{cases}$

Solve the systems in Problems 10–18 by substitution. Relate each system to the general system

$$\begin{cases} a_{11}x_1 + a_{12}x_2 = b_1 \\ a_{21}x_1 + a_{22}x_2 = b_2 \end{cases}$$

10. $\begin{cases} a = 3b - 7 \\ a = -2b + 8 \end{cases}$

11. $\begin{cases} s + t = 1 \\ 3s + t = -5 \end{cases}$

12. $\begin{cases} m = \frac{2}{3}n - 7 \\ 2n + 3m = 3 \end{cases}$

13. $\begin{cases} v = \frac{3}{5}u + 2 \\ 3u - 5v = 10 \end{cases}$

14. $\begin{cases} 2p - 3q = 9 \\ q = \frac{2}{3}p - 3 \end{cases}$

15. $\begin{cases} 3t_1 + 5t_2 = 1541 \\ t_2 = 2t_1 + 160 \end{cases}$

16. $\begin{cases} \alpha = -7\beta - 3 \\ 2\alpha + 5\beta = 3 \end{cases}$

17. $\begin{cases} \gamma = 3\delta - 4 \\ 5\gamma - 4\delta = -9 \end{cases}$

18. $\begin{cases} \theta + 3\phi = 0 \\ \theta = 5\phi + 16 \end{cases}$

Solve the systems in Problems 19–27 by linear combinations. Relate each system to the general system

$$\begin{cases} a_{11}x_1 + a_{12}x_2 = b_1 \\ a_{21}x_1 + a_{22}x_2 = b_2 \end{cases}$$

19. $\begin{cases} c + d = 2 \\ 2c - d = 1 \end{cases}$

20. $\begin{cases} 2s_1 + s_2 = 10 \\ 5s_1 - 2s_2 = 16 \end{cases}$

21. $\begin{cases} 3q_1 - 4q_2 = 3 \\ 5q_1 + 3q_2 = 5 \end{cases}$

22. $\begin{cases} 9x + 3y = 5 \\ 3x + 2y = 2 \end{cases}$

23. $\begin{cases} 7x + y = 5 \\ 14x - 2y = -2 \end{cases}$

24. $\begin{cases} 2x + 3y = 1 \\ 3x - 2y = 0 \end{cases}$

25. $\begin{cases} \alpha + \beta = 12 \\ \alpha - 2\beta = -4 \end{cases}$

26. $\begin{cases} 2\gamma - 3\delta = 16 \\ 5\gamma + 2\delta = 21 \end{cases}$

27. $\begin{cases} 2\theta + 5\phi = 7 \\ 3\theta + 4\phi = 0 \end{cases}$

B

Solve the systems in Problems 28–45 for x and y by any method. Limit your answers to the set of real numbers.

28. $\begin{cases} 5x + 4y = 5 \\ 15x - 2y = 8 \end{cases}$

29. $\begin{cases} 3x + 2y = 1 \\ 6x + 4y = 2 \end{cases}$

30. $\begin{cases} 4x - 2y = -28 \\ y = \frac{1}{2}x + 5 \end{cases}$

31. $\begin{cases} 12x - 5y = -39 \\ y = 2x + 9 \end{cases}$

32. $\begin{cases} y = 2x - 1 \\ y = -3x - 9 \end{cases}$

33. $\begin{cases} y = \frac{2}{3}x - 5 \\ y = -\frac{4}{3}x + 7 \end{cases}$

34. $\begin{cases} x + y = \alpha \\ x - y = \beta \end{cases}$

35. $\begin{cases} x - y = \gamma \\ x - 2y = \delta \end{cases}$

36. $\begin{cases} x + y = 2\alpha \\ x - y = 2\beta \end{cases}$

37. $\begin{cases} x + y = a \\ x - y = b \end{cases}$

38. $\begin{cases} x + 2y = a \\ x - 3y = b \end{cases}$

39. $\begin{cases} 2x - y = c \\ 3x + y = d \end{cases}$

40. $\begin{cases} x^2 - 10x = -5 - 2y \\ y = x + 1 \end{cases}$

41. $\begin{cases} y = 2x + 6 \\ 2x^2 + 8x = -2 - 3y \end{cases}$

42. $\begin{cases} 3x^2 + 4y^2 = 12 \\ x^2 + y^2 = -8 \end{cases}$

43. $\begin{cases} 3x^2 + 4y^2 = 19 \\ x^2 + y^2 = 5 \end{cases}$

44. $\begin{cases} x^2 + y^2 = 25 \\ y^2 = 5 - x \end{cases}$

45. $\begin{cases} x^2 - 12y - 1 = 0 \\ x^2 - 4y^2 - 9 = 0 \end{cases}$

46. Use the system

$$\begin{cases} \cos(x + y) = \cos x \cos y - \sin x \sin y \\ \cos(x - y) = \cos x \cos y + \sin x \sin y \end{cases}$$

to prove the identity

$$2 \cos x \cos y = \cos(x + y) + \cos(x - y)$$

47. Use the system

$$\begin{cases} \sin(x + y) = \sin x \cos y + \cos x \sin y \\ \sin(x - y) = \sin x \cos y - \cos x \sin y \end{cases}$$

to prove the identity

$$2 \sin x \cos y = \sin(x + y) + \sin(x - y)$$

48. Use the system in Problem 46 to prove the identity

$$2 \sin x \sin y = \cos(x - y) - \cos(x + y)$$

49. Use the system in Problem 47 to prove the identity

$$2 \cos x \sin y = \sin(x + y) - \sin(x - y)$$

50. Find two numbers whose sum is 9 and whose product is 18.

51. Find two numbers whose difference is 4 and whose product is -3.

52. Find the lengths of the legs of a right triangle whose area is 60 ft^2 and whose hypotenuse is 17 ft.

53. Find the length and width of a rectangle whose area is 60 ft^2 with a diagonal of length 13 ft.

C

Solve the systems in Problems 54–63 for x and y. For Problems 54–56 give only those solutions between 0 and 6.

54. $\begin{cases} \sin x + \cos y = 1 \\ \sin x - \cos y = 0 \end{cases}$

55. $\begin{cases} \cos x - \sin y = 1 \\ \cos x + \sin y = 0 \end{cases}$

56. $\begin{cases} x + 3y = \cos^2 60° \\ x + \ \ y = -\sin^2 60° \end{cases}$

57. $\begin{cases} x^2 - xy + y^2 = 21 \\ xy - y^2 = 15 \end{cases}$

58. $\begin{cases} x^2 - xy + y^2 = 3 \\ x^2 + y^2 = 6 \end{cases}$

59. $\begin{cases} x^2 + 2xy = 8 \\ x^2 - 4y^2 = 8 \end{cases}$

60. $\begin{cases} \dfrac{3}{4} + \dfrac{4}{y} = 3 \\ \dfrac{9}{x} - \dfrac{2}{y} = 2 \end{cases}$

61. $\begin{cases} \dfrac{2}{x-1} - \dfrac{5}{y+2} = 12 \\ \dfrac{4}{x-1} - \dfrac{2}{y+2} = -12 \end{cases}$

62. $\begin{cases} y^2 - 5xy = 1 \\ y^2 - 4xy = 2y \end{cases}$

63. $\begin{cases} x^2 - 3xy = 2 \\ x^2 - 4xy = 3x \end{cases}$

64. If the parabola $(x - h)^2 = y - k$ passes through the points $(-2, 6)$ and $(-4, 2)$, find (h, k).

65. If the parabola $-3(x - h)^2 = y - k$ passes through the points $(2, 5)$ and $(-1, -4)$, find (h, k).

66. ***Historical Question*** The Babylonians knew how to solve systems of equations as early as 1600 B.C. One tablet, called the Yale Tablet, shows a system equivalent to

$$\begin{cases} xy = 600 \\ (x + y)^2 - 150(x - y) = 100 \end{cases}$$

Find a solution for this system correct to the nearest tenth.

67. ***Historical Question*** The Louvre Tablet from the Babylonian civilization is dated at about 1500 B.C. It shows a system equivalent to

$$\begin{cases} xy = 1 \\ x + y = a \end{cases}$$

Solve this system for x and y in terms of a.

9.2 CRAMER'S RULE

As you encounter systems more difficult than those considered in the last section, you will need more efficient methods for solving them. The first new method we will consider is a solution by *formula* in a procedure known as *Cramer's Rule.* Consider

$$\begin{cases} a_{11}x_1 + a_{12}x_2 = b_1 \\ a_{21}x_1 + a_{22}x_2 = b_2 \end{cases}$$

Solve this system by using linear combinations.

$$\begin{matrix} a_{22} \\ -a_{12} \end{matrix} \begin{cases} a_{11}x_1 + a_{12}x_2 = b_1 \\ a_{21}x_1 + a_{22}x_2 = b_2 \end{cases}$$

$$+\begin{cases} a_{11}a_{22}x_1 + a_{12}a_{22}x_2 = \quad b_1a_{22} \\ -a_{12}a_{21}x_1 - a_{12}a_{22}x_2 = -b_2a_{12} \end{cases}$$

$$(a_{11}a_{22} - a_{12}a_{21})x_1 = b_1a_{22} - b_2a_{12}$$

If $a_{11}a_{22} - a_{12}a_{21} \neq 0$, then

$$x_1 = \frac{b_1a_{22} - b_2a_{12}}{a_{11}a_{22} - a_{12}a_{21}}$$

Similarly, solving for x_2,

$$x_2 = \frac{b_2a_{11} - b_1a_{21}}{a_{11}a_{22} - a_{12}a_{21}}$$

The difficulty with using this formula is that it is next to impossible to remember. To help with this matter, the following definition is given.

Determinant of Order 2

The **determinant** of order 2 is denoted by

$$\begin{vmatrix} a & b \\ c & d \end{vmatrix}$$

and is defined to be the real number $ad - bc$.

EXAMPLE 1 Evaluate the given determinants.

a. $\begin{vmatrix} 1 & 2 \\ 3 & 4 \end{vmatrix} = (1)(4) - (2)(3)$
$= -2$

b. $\begin{vmatrix} 3 & -2 \\ 4 & 1 \end{vmatrix} = 3 + 8$
$= 11$

c. $\begin{vmatrix} \sin\theta & \cos\theta \\ -\cos\theta & \sin\theta \end{vmatrix} = \sin^2\theta + \cos^2\theta$
$= 1$

d. $\begin{vmatrix} a_{11} & a_{12} \\ a_{21} & a_{22} \end{vmatrix} = a_{11}a_{22} - a_{12}a_{21}$

e. $\begin{vmatrix} b_1 & a_{12} \\ b_2 & a_{22} \end{vmatrix} = b_1a_{22} - b_2a_{12}$

f. $\begin{vmatrix} a_{11} & b_1 \\ a_{21} & b_2 \end{vmatrix} = b_2a_{11} - b_1a_{21}$ ■

The solution to the general system

$$\begin{cases} a_{11}x_1 + a_{12}x_2 = b_1 \\ a_{21}x_1 + a_{22}x_2 = b_2 \end{cases}$$

can now be stated using determinant notation:

$$x_1 = \frac{\begin{vmatrix} b_1 & a_{12} \\ b_2 & a_{22} \end{vmatrix}}{\begin{vmatrix} a_{11} & a_{12} \\ a_{21} & a_{22} \end{vmatrix}} \qquad x_2 = \frac{\begin{vmatrix} a_{11} & b_1 \\ a_{21} & b_2 \end{vmatrix}}{\begin{vmatrix} a_{11} & a_{12} \\ a_{21} & a_{22} \end{vmatrix}}$$

Notice that the denominator of both variables is the same. This determinant is called the **determinant of the coefficients** for the system and is denoted by $|D|$. The numerators are also found by looking at $|D|$. For x_1, replace the first column of $|D|$ by the constant numbers b_1 and b_2, respectively. Denote this by $|D_1|$. Similarly, let $|D_2|$ be the determinant formed by replacing the coefficients of x_2 in $|D|$ by the constant numbers b_1 and b_2, respectively. The solution to the system can now be stated with a result called **Cramer's Rule**.

Cramer's Rule (Two Unknowns)

Let $|D|$ be the determinant of the coefficients ($|D| \neq 0$) of a system of linear equations, and let $|D_1|$ and $|D_2|$ be the determinants where the coefficients of x_1 and x_2 are replaced respectively by b_1 and b_2. Then

$$x_1 = \frac{|D_1|}{|D|} \quad \text{and} \quad x_2 = \frac{|D_2|}{|D|}$$

EXAMPLE 2 Solve $\begin{cases} 2x - 3y = -8 \\ x + y = 6 \end{cases}$ using Cramer's Rule.

$$x = \frac{\begin{vmatrix} -8 & -3 \\ 6 & 1 \end{vmatrix}}{\begin{vmatrix} 2 & -3 \\ 1 & 1 \end{vmatrix}} = \frac{-8 + 18}{2 + 3} = \frac{10}{5} = 2 \qquad y = \frac{\begin{vmatrix} 2 & -8 \\ 1 & 6 \end{vmatrix}}{5} = \frac{12 + 8}{5} = \frac{20}{5} = 4$$

The solution is **(2, 4)**. ■

One advantage of Cramer's Rule is the ease with which its form can be used when working with more complicated solutions.

EXAMPLE 3 Solve $\begin{cases} 14x - 3y = 1 \\ 5x + 7y = -2 \end{cases}$ using Cramer's Rule.

Solution

$$x = \frac{\begin{vmatrix} 1 & -3 \\ -2 & 7 \end{vmatrix}}{\begin{vmatrix} 14 & -3 \\ 5 & 7 \end{vmatrix}} = \frac{7 - 6}{98 + 15} = \frac{1}{113}$$

$$y = \frac{\begin{vmatrix} 14 & 1 \\ 5 & -2 \end{vmatrix}}{113} = \frac{-28 - 5}{113} = \frac{-33}{113}$$

The solution is $\left(\frac{1}{113}, \frac{-33}{113}\right)$. ■

EXAMPLE 4 Solve $\begin{cases} 2x + 3y = 9 \\ y = -\frac{2}{3}x + 1 \end{cases}$ using Cramer's Rule.

Solution You must first put both equations into the usual form:

$$\begin{cases} 2x + 3y = 9 \\ 2x + 3y = 3 \end{cases}$$

$$x = \frac{\begin{vmatrix} 9 & 3 \\ 3 & 3 \end{vmatrix}}{\begin{vmatrix} 2 & 3 \\ 2 & 3 \end{vmatrix}} = \frac{27 - 9}{6 - 6} = \frac{18}{0}$$

Since division by zero is not defined, **Cramer's Rule fails.** ■

Notice, however, from Example 4 that if $a_{11}a_{22} - a_{12}a_{21} = 0$ and the numerators are not zero, the system is inconsistent. To show this, note that

$$a_{11}a_{22} = a_{12}a_{21}$$

since $a_{11}a_{22} - a_{12}a_{21} = 0$. Divide both sides by $a_{12}a_{22}$. (If $a_{12}a_{22} = 0$, then $a_{12}a_{21} = 0$ and the problem would be trivial.) Assume, therefore, that $a_{12}a_{22} \neq 0$:

$$\frac{a_{11}}{a_{12}} = \frac{a_{21}}{a_{22}}$$

Now multiply both sides by -1:

$$-\frac{a_{11}}{a_{12}} = -\frac{a_{21}}{a_{22}}$$

But $-a_{11}/a_{12}$ is the slope of the line represented by the first equation of the system, and $-a_{21}/a_{22}$ is the slope of the line represented by the second equation. This means that if $a_{11}a_{22} - a_{12}a_{21} = 0$, then the slopes of the lines are the same. Thus the lines are either parallel or coincident. Next we show that they are not coincident by considering the numerators:

$$b_1a_{22} - b_2a_{12} \neq 0 \qquad\qquad a_{11}b_2 - a_{21}b_1 \neq 0$$

$$b_1a_{22} \neq b_2a_{12} \qquad\qquad a_{11}b_2 \neq a_{21}b_1$$

$$\frac{b_1}{b_2} \neq \frac{a_{12}}{a_{22}} \qquad\qquad \frac{b_1}{b_2} \neq \frac{a_{11}}{a_{21}}$$

This shows that the lines differ in at least one point, but since they are parallel or coincident they must be parallel. Therefore, if Cramer's Rule gives a solution of the form where the denominator is zero and the numerators are not zero, the system is *inconsistent*. (That is, the lines are parallel.) If Cramer's Rule yields a form $\frac{0}{0}$, the system is *dependent*.

PROBLEM SET 9.2

A

Evaluate the determinants given in Problems 1–15.

1. $\begin{vmatrix} 3 & 1 \\ 2 & 4 \end{vmatrix}$ **2.** $\begin{vmatrix} 6 & 3 \\ 2 & 2 \end{vmatrix}$ **3.** $\begin{vmatrix} -2 & 4 \\ 1 & 3 \end{vmatrix}$

4. $\begin{vmatrix} -5 & -3 \\ 4 & 2 \end{vmatrix}$ **5.** $\begin{vmatrix} 6 & -3 \\ 2 & 5 \end{vmatrix}$ **6.** $\begin{vmatrix} -3 & 2 \\ -4 & 5 \end{vmatrix}$

7. $\begin{vmatrix} 6 & 8 \\ -2 & -1 \end{vmatrix}$ **8.** $\begin{vmatrix} 6 & -3 \\ 4 & -2 \end{vmatrix}$ **9.** $\begin{vmatrix} 8 & -3 \\ 4 & -5 \end{vmatrix}$

10. $\begin{vmatrix} \sqrt{3} & \sqrt{5} \\ \sqrt{5} & \sqrt{3} \end{vmatrix}$ **11.** $\begin{vmatrix} \sqrt{2} & 3 \\ 1 & \sqrt{2} \end{vmatrix}$ **12.** $\begin{vmatrix} \pi & 3 \\ 0 & 1 \end{vmatrix}$

13. $\begin{vmatrix} \sin\theta & -\cos\theta \\ \cos\theta & \sin\theta \end{vmatrix}$ **14.** $\begin{vmatrix} \tan\theta & 1 \\ -1 & \tan\theta \end{vmatrix}$

15. $\begin{vmatrix} \csc\theta & 1 \\ 1 & \csc\theta \end{vmatrix}$

Solve the systems in Problems 16–30 by using Cramer's Rule.

16. $\begin{cases} x + y = 1 \\ 3x + y = -5 \end{cases}$ **17.** $\begin{cases} 2s + t = 7 \\ 3s + t = 12 \end{cases}$

18. $\begin{cases} c + d = 2 \\ 2c - d = 1 \end{cases}$ **19.** $\begin{cases} 2x + 3y = 1 \\ 3x - 2y = 0 \end{cases}$

20. $\begin{cases} 2s_1 + 3s_2 = 3 \\ 5s_1 - 2s_2 = 1 \end{cases}$ **21.** $\begin{cases} 3q_1 - 4q_2 = 3 \\ 5q_1 + 3q_2 = 4 \end{cases}$

22. $\begin{cases} 4x + 3y = 5 \\ 3x + 2y = 2 \end{cases}$ **23.** $\begin{cases} 2x - 3y = 6 \\ 5x + 2y = 3 \end{cases}$

24. $\begin{cases} y = \frac{2}{3}x - 7 \\ 2x + 3y = 3 \end{cases}$ **25.** $\begin{cases} y = \frac{3}{5}x + 2 \\ 3x - 5y = 10 \end{cases}$

26. $\begin{cases} 2x - 3y = 9 \\ y = \frac{2}{3}x - 3 \end{cases}$ **27.** $\begin{cases} 2p - 3q = 9 \\ q = \frac{2}{3}p - 3 \end{cases}$

28. $\begin{cases} y = \frac{2}{3}x + 5 \\ x + 7y = 1 \end{cases}$ **29.** $\begin{cases} 3x + 5y + 6 = 0 \\ y = -\frac{5}{3}x + 2 \end{cases}$

30. $\begin{cases} 2x - 3y - 4 = 0 \\ y = \frac{3}{2}x - 8 \end{cases}$

B

Solve the systems in Problems 31–51 for (x, y) *by using Cramer's Rule.*

31. $\begin{cases} 2x - 3y - 5 = 0 \\ 3x - 5y + 2 = 0 \end{cases}$ **32.** $\begin{cases} y = \frac{2}{3}x + 3 \\ y = -\frac{3}{4}x - 5 \end{cases}$

33. $\begin{cases} y = \frac{1}{2}x - 4 \\ y = \frac{2}{3}x + 5 \end{cases}$ **34.** $\begin{cases} x + y = a \\ x - y = b \end{cases}$

35. $\begin{cases} 2x + 3y = a \\ x - 5y = b \end{cases}$ **36.** $\begin{cases} 2x + y = \alpha \\ x - 3y = \beta \end{cases}$

37. $\begin{cases} ax + by = 1 \\ bx + ay = 0 \end{cases}$ **38.** $\begin{cases} ax + by = 0 \\ bx + ay = 1 \end{cases}$

39. $\begin{cases} cx + dy = s \\ ex + fy = t \end{cases}$

40. $\begin{cases} y = m_1x + b_1 \\ y = m_2x + b_2 \end{cases}$

41. $\begin{cases} ax + by = \alpha \\ cx + dy = \beta \end{cases}$

42. $\begin{cases} ax - by = \gamma \\ cx + dy = \delta \end{cases}$

43. $\begin{cases} 0.12x + 0.06y = 108 \\ x + y = 1000 \end{cases}$

44. $\begin{cases} 0.12x + 0.06y = 210 \\ x + y = 2000 \end{cases}$

45. $\begin{cases} 0.12x + 0.06y = 228 \\ x + y = 2000 \end{cases}$

46. $\begin{cases} 0.12x + 0.06y = 1140 \\ x + y = 10{,}000 \end{cases}$

47. $\begin{cases} x + y = 147 \\ 0.25x + 0.1y = 24.15 \end{cases}$

48. $\begin{cases} x + y = 10 \\ 0.4x + 0.9y = 0.5 \end{cases}$

49. $\begin{cases} 2x + 3y = \cos^2 45° \\ x - y = -\sin^2 45° \end{cases}$

50. $\begin{cases} 3x - y = \sec^2 30° \\ x - y = \tan^2 30° \end{cases}$

51. $\begin{cases} x + 3y = \csc^2 60° \\ x - 2y = \cot^2 60° \end{cases}$

For Problems 52–56 let $|A| = \begin{vmatrix} a & b \\ c & d \end{vmatrix}$

52. Show that $\begin{vmatrix} c & d \\ a & b \end{vmatrix} = -|A|$

53. Show that $\begin{vmatrix} a & b \\ a + c & b + d \end{vmatrix} = |A|$

54. Show that $\begin{vmatrix} a & a + b \\ c & c + d \end{vmatrix} = |A|$

55. Show that for any constant k, $\begin{vmatrix} ak & bk \\ c & d \end{vmatrix} = k|A|$

56. Show that $\begin{vmatrix} a & b \\ c + ka & d + kb \end{vmatrix} = \begin{vmatrix} a + kc & b + kd \\ c & d \end{vmatrix} = |A|$

C

Solve the systems in Problems 57–60 by using Cramer's Rule.

57. $\begin{cases} 3x^2 + 4y^2 = 16 \\ x^2 + y^2 = 5 \end{cases}$

58. $\begin{cases} 13x^2 + 12y^2 = 169 \\ x^2 + y^2 = 12 \end{cases}$

59. $\begin{cases} \dfrac{3}{x} + \dfrac{4}{y} = 3 \\ \dfrac{9}{x} - \dfrac{2}{y} = 2 \end{cases}$

60. $\begin{cases} \dfrac{2}{x - 1} - \dfrac{5}{y + 2} = 12 \\ \dfrac{4}{x - 1} - \dfrac{2}{y + 2} = -12 \end{cases}$

9.3 PROPERTIES OF DETERMINANTS

The strength of Cramer's Rule is that it can be applied to linear systems of n equations with n unknowns. However, to apply Cramer's Rule to these systems, some additional notation and definitions are required. Since the number of rows and columns in a determinant is the same, we call the size of either the **order** or **dimension** of the determinant. The determinants of the last section were of order 2. Consider a determinant of order 3:

$$\begin{vmatrix} a_{11} & a_{12} & a_{13} \\ a_{21} & a_{22} & a_{23} \\ a_{31} & a_{32} & a_{33} \end{vmatrix}$$

3 rows

3 columns

The entry a_{ij} refers to the entry in the ith row and jth column. That is, a_{23} refers to the entry in the second row and third column. The **minor** of an element in a determinant is the (smaller) determinant that remains after deleting all entries in its row and column.

EXAMPLE 1 Consider

$$\begin{vmatrix} 2 & -4 & 0 \\ 1 & -3 & -1 \\ 6 & 5 & 3 \end{vmatrix}$$

a. The a_{13} entry is 0; the a_{31} entry is 6; and the a_{23} entry is -1.
b. The minor of -3 is

$$\begin{vmatrix} 2 & -4 & 0 \\ 1 & \circled{-3} & -1 \\ 6 & 5 & 3 \end{vmatrix} \quad \text{or} \quad \begin{vmatrix} 2 & 0 \\ 6 & 3 \end{vmatrix} = 6$$

Delete shaded portions.

c. The minor of 0 is

$$\begin{vmatrix} 1 & -3 \\ 6 & 5 \end{vmatrix} = 5 + 18 = 23$$

d. The minor of 2 is

$$\begin{vmatrix} -3 & -1 \\ 5 & 3 \end{vmatrix} = -9 + 5 = -4$$

■

The **cofactor** of an entry a_{ij} is $(-1)^{i+j}$ times the minor of the a_{ij} entry. This says that if the sum of the row and column number is even, the cofactor is the same as the minor. If the sum of the row and column of an entry is odd, the cofactor of that entry is the opposite of its minor.

EXAMPLE 2 Consider

$$\begin{vmatrix} 2 & -4 & 0 \\ 1 & -3 & -1 \\ 6 & 5 & 3 \end{vmatrix}$$

a. The cofactor of 1 is

$$-\underbrace{\begin{vmatrix} -4 & 0 \\ 5 & 3 \end{vmatrix}}_{\text{minor}} = -(-12 - 0) = 12$$

Entry is in row 2, column 1, and 2 + 1 is odd so use the negative sign.
Notice that this sign is independent of whether the entry is positive or negative or whether the minor is positive or negative.

b. The cofactor of -3 is

$$\underset{\text{sign}}{+}\underbrace{\begin{vmatrix} 2 & 0 \\ 6 & 3 \end{vmatrix}}_{\text{minor}} = 6$$

c. The cofactor of -1 is

$$-\begin{vmatrix} 2 & -4 \\ 6 & 5 \end{vmatrix} = -(10 + 24) = -34$$

■

EXAMPLE 3 Consider

$$\begin{vmatrix} 1 & 32 & -3 & 4 & 5 \\ -6 & 7 & 8 & 9 & 10 \\ 11 & -10 & -9 & -8 & -7 \\ -6 & -5 & -4 & 3 & \textcircled{2} \\ 1 & 0 & 3 & 5 & 12 \end{vmatrix}$$

The a_{45} entry is 2 and its cofactor is

$$-\underbrace{\begin{vmatrix} 1 & 32 & -3 & 4 \\ -6 & 7 & 8 & 9 \\ 11 & -10 & -9 & -8 \\ 1 & 0 & 3 & 5 \end{vmatrix}}_{\text{minor}}$$

sign: Row number is 4; column number is 5; the sum is 9, which is an odd number, so use the negative sign. ■

Determinant of Order n

> A determinant of order n is a real number whose value is the sum of the products obtained by multiplying each element of a row (or column) by its cofactor.

EXAMPLE 4 Expand $\begin{vmatrix} 1 & -2 & -5 \\ 2 & -1 & 0 \\ -4 & 5 & 6 \end{vmatrix}$ about the second row.

Solution

$$-(2)\begin{vmatrix} -2 & -5 \\ 5 & 6 \end{vmatrix} + (-1)\begin{vmatrix} 1 & -5 \\ -4 & 6 \end{vmatrix} - (0)\begin{vmatrix} 1 & -2 \\ -4 & 5 \end{vmatrix}$$

signs of cofactors

$$= -2(-12 + 25) - (6 - 20) - 0$$
$$= -26 - (-14)$$
$$= \mathbf{-12}$$
■

EXAMPLE 5 Expand the determinant of Example 4 about the first column.

Solution

$$+(1)\begin{vmatrix} -1 & 0 \\ 5 & 6 \end{vmatrix} - (2)\begin{vmatrix} -2 & -5 \\ 5 & 6 \end{vmatrix} + (-4)\begin{vmatrix} -2 & -5 \\ -1 & 0 \end{vmatrix}$$
$$= (-6 - 0) - 2(-12 + 25) - 4(0 - 5)$$
$$= -6 - 2(13) + 20$$
$$= \mathbf{-12}$$
■

Notice from Examples 4 and 5 that the same value is obtained regardless of the row or column chosen. You are asked to prove this for order 3 in the problem set.

The method of evaluating determinants by rows or columns as shown in Examples 4 and 5 is not very efficient for higher-order determinants. The following theorem considerably simplifies the work in evaluating determinants.

Determinant Reduction Theorem

If $|A'|$ is a determinant obtained from a determinant $|A|$ by multiplying any row by a constant k and adding the result to any other row (entry by entry), then $|A'| = |A|$. The same result holds for columns.

Proof We will prove this theorem for the determinant of order 2; in the problems you are asked to prove it for order 3. For order 2, let

$$|A| = \begin{vmatrix} a & b \\ c & d \end{vmatrix} = ad - bc$$

Multiply row 1 by k and add it to row 2:

$$\begin{aligned} |A'| = \begin{vmatrix} a & b \\ c + ak & d + bk \end{vmatrix} &= a(d + bk) - b(c + ak) \\ &= ad + abk - bc - abk \\ &= ad - bc \\ &= |A| \end{aligned}$$

Thus $|A'| = |A|$. Next multiply row 2 by k and add it to row 1 to obtain the same result. □

EXAMPLE 6 Expand

$$\begin{vmatrix} 1 & -2 & -5 \\ 2 & -1 & 0 \\ -4 & 5 & 6 \end{vmatrix}$$

by using the determinant reduction theorem.

Solution *You want to obtain some row or column with two zeros to simplify the arithmetic.* Add twice the second column to the first column:

$$\begin{vmatrix} 1 & -2 & -5 \\ 2 & -1 & 0 \\ -4 & 5 & 6 \end{vmatrix} = \begin{vmatrix} -3 & -2 & -5 \\ 0 & -1 & 0 \\ 6 & 5 & 6 \end{vmatrix}$$

× 2 These columns are unchanged.

Now expand about the second row (do not even bother to write down the products that are zero):

positive understood

$$\begin{aligned} &= (-1)\begin{vmatrix} -3 & -5 \\ 6 & 6 \end{vmatrix} \\ &= (-1)(-18 + 30) \\ &= \mathbf{-12} \end{aligned}$$

■

EXAMPLE 7 Expand

$$\begin{vmatrix} 2 & -3 & 2 & 5 & 0 \\ 4 & 2 & -1 & 4 & 0 \\ 5 & 1 & 0 & -2 & 0 \\ 6 & 2 & 3 & 6 & 0 \\ 3 & 4 & 6 & 1 & -2 \end{vmatrix}$$

by using the determinant reduction theorem.

Solution First notice that all the entries in the fifth column except one are zeros, so begin by expanding about the fifth column:

$$+(-2)\begin{vmatrix} 2 & -3 & 2 & 5 \\ 4 & 2 & -1 & 4 \\ 5 & 1 & 0 & -2 \\ 6 & 2 & 3 & 6 \end{vmatrix}$$

Row 5, column 5; 5 + 5 is even, so this sign is positive.

Next obtain a row or column with all entries zero except one; row 3 or column 3 seem to be likely candidates. We will work toward obtaining zeros in column 3:

$$(-2)\begin{vmatrix} 2 & -3 & 2 & 5 \\ 4 & 2 & -1 & 4 \\ 5 & 1 & 0 & -2 \\ 6 & 2 & 3 & 6 \end{vmatrix} \times 2 = (-2)\begin{vmatrix} 10 & 1 & 0 & 13 \\ 4 & 2 & -1 & 4 \\ 5 & 1 & 0 & -2 \\ 6 & 2 & 3 & 6 \end{vmatrix} \times 3$$

$$= (-2)\begin{vmatrix} 10 & 1 & 0 & 13 \\ 4 & 2 & -1 & 4 \\ 5 & 1 & 0 & -2 \\ 18 & 8 & 0 & 18 \end{vmatrix} = (-2)\left[-(-1)\begin{vmatrix} 10 & 1 & 13 \\ 5 & 1 & -2 \\ 18 & 8 & 18 \end{vmatrix}\right]$$

$\times (-5)$

Finished when all entries but one are zero; now reduce the order of the determinant.

Row 2, column 3; 2 + 3 is odd.

$$= (-2)\begin{vmatrix} 5 & 1 & 13 \\ 0 & 1 & -2 \\ -22 & 8 & 18 \end{vmatrix} = (-2)\begin{vmatrix} 5 & 1 & 15 \\ 0 & 1 & 0 \\ -22 & 8 & 34 \end{vmatrix}$$

$\times 2$

$$= (-2)\left[+(1)\begin{vmatrix} 5 & 15 \\ -22 & 34 \end{vmatrix}\right]$$

Row 2, column 2; 2 + 2 is even.

$= -2(170 + 330) = -2(500) = \mathbf{-1000}$ ■

We can now state Cramer's Rule for n linear equations with n unknowns. Consider the general n by n system

$$\begin{cases} a_{11}x_1 + a_{12}x_2 + a_{13}x_3 + \cdots + a_{1n}x_n = b_1 \\ a_{21}x_1 + a_{22}x_2 + a_{23}x_3 + \cdots + a_{2n}x_n = b_2 \\ a_{31}x_1 + a_{32}x_2 + a_{33}x_3 + \cdots + a_{3n}x_n = b_3 \\ \vdots \\ a_{n1}x_1 + a_{n2}x_2 + a_{n3}x_3 + \cdots + a_{nn}x_n = b_n \end{cases}$$

The unknowns are $x_1, x_2, x_3, \ldots, x_n$; a_{ij} are the coefficients and $b_1, b_2, b_3, \ldots, b_n$ are the constants. When written in this form, the system is said to be in **standard form**.

Cramer's Rule

Let $|D|$ be the determinant of the coefficients ($|D| \neq 0$) of a system of n linear equations with n unknowns. Let $|D_i|$ be the determinant where the coefficients of x_i have been replaced by the constants $b_1, b_2, b_3, \ldots, b_n$, respectively. Then

$$x_i = \frac{|D_i|}{|D|}$$

As before, if $|D_i|$ and $|D|$ are zero, the system is dependent; if at least one of the $|D_i|$ is not zero when $|D|$ is zero, the system is inconsistent.

EXAMPLE 8 Solve

$$\begin{cases} x - 2y - 5z = -12 \\ 2x - y = 7 \\ 5y + 6z = 4x + 1 \end{cases}$$

Solution Rewrite the system in standard form:

$$\begin{cases} x - 2y - 5z = -12 \\ 2x - y \qquad = 7 \\ -4x + 5y + 6z = 1 \end{cases}$$

Be sure to consider "missing terms" as terms with zero coefficient.

$$|D| = \begin{vmatrix} 1 & -2 & -5 \\ 2 & -1 & 0 \\ -4 & 5 & 6 \end{vmatrix} = -12 \qquad \text{From Example 6}$$

$$|D_1| = \begin{vmatrix} -12 & -2 & -5 \\ 7 & -1 & 0 \\ 1 & 5 & 6 \end{vmatrix} = \begin{vmatrix} -26 & -2 & -5 \\ 0 & -1 & 0 \\ 36 & 5 & 6 \end{vmatrix} = (-1)\begin{vmatrix} -26 & -5 \\ 36 & 6 \end{vmatrix}$$

$\times 7$

$$= -(-156 + 180)$$
$$= -24$$

$$|D_2| = \begin{vmatrix} 1 & -12 & -5 \\ 2 & 7 & 0 \\ -4 & 1 & 6 \end{vmatrix} \times 1 = \begin{vmatrix} 1 & -12 & -5 \\ 2 & 7 & 0 \\ -3 & -11 & 1 \end{vmatrix} \times 5$$

$$= \begin{vmatrix} -14 & -67 & 0 \\ 2 & 7 & 0 \\ -3 & -11 & 1 \end{vmatrix} = \begin{vmatrix} -14 & -67 \\ 2 & 7 \end{vmatrix} = -98 + 134 = 36$$

$$|D_3| = \begin{vmatrix} 1 & -2 & -12 \\ 2 & -1 & 7 \\ -4 & 5 & 1 \end{vmatrix} = \begin{vmatrix} -3 & -2 & -26 \\ 0 & -1 & 0 \\ 6 & 5 & 36 \end{vmatrix} = (-1)\begin{vmatrix} -3 & -26 \\ 6 & 36 \end{vmatrix}$$

$\times 2 \quad \times 7$

$$= -(-108 + 156) = -48$$

Thus

$$x = \frac{|D_1|}{|D|} = \frac{-24}{-12} = 2 \qquad y = \frac{|D_2|}{|D|} = \frac{36}{-12} = -3 \qquad z = \frac{|D_3|}{|D|} = \frac{-48}{-12} = 4$$

The solution is $\mathbf{(x, y, z) = (2, -3, 4)}$. ■

PROBLEM SET 9.3

A

Evaluate the determinants in Problems 1–15. If you want to check, evaluate the determinant by using a different row or column.

1. $\begin{vmatrix} 3 & 0 & 0 \\ 2 & 1 & 4 \\ 3 & 6 & -1 \end{vmatrix}$

2. $\begin{vmatrix} 1 & -2 & 3 \\ 0 & 4 & 0 \\ 3 & -1 & -3 \end{vmatrix}$

3. $\begin{vmatrix} 4 & -2 & 6 \\ 3 & 1 & 4 \\ 0 & 0 & -2 \end{vmatrix}$

4. $\begin{vmatrix} 0 & 1 & 1 \\ -3 & 2 & 4 \\ 0 & -2 & -3 \end{vmatrix}$

5. $\begin{vmatrix} 1 & -2 & 3 \\ -2 & 0 & 4 \\ 3 & 0 & 5 \end{vmatrix}$

6. $\begin{vmatrix} 3 & 1 & 0 \\ -2 & 4 & 0 \\ -3 & 5 & -4 \end{vmatrix}$

7. $\begin{vmatrix} 1 & 1 & 1 \\ 1 & 3 & 2 \\ 1 & -2 & 1 \end{vmatrix}$

8. $\begin{vmatrix} 4 & 1 & 1 \\ 4 & 3 & 2 \\ 7 & -2 & 1 \end{vmatrix}$

9. $\begin{vmatrix} 1 & 4 & 1 \\ 1 & 4 & 2 \\ 1 & 7 & 1 \end{vmatrix}$

10. $\begin{vmatrix} -3 & -1 & 1 \\ 14 & 2 & -3 \\ 12 & 3 & -1 \end{vmatrix}$

11. $\begin{vmatrix} 2 & -3 & 1 \\ 1 & 14 & -3 \\ 3 & 12 & -1 \end{vmatrix}$

12. $\begin{vmatrix} 2 & -1 & -3 \\ 1 & 3 & 14 \\ 3 & 3 & 12 \end{vmatrix}$

13. $\begin{vmatrix} 2 & 4 & 3 \\ -2 & 3 & -2 \\ 4 & 3 & 5 \end{vmatrix}$

14. $\begin{vmatrix} 6 & 3 & -3 \\ 2 & 0 & 5 \\ 3 & 5 & -2 \end{vmatrix}$

15. $\begin{vmatrix} 4 & 8 & 5 \\ 3 & 2 & 3 \\ 5 & 5 & 4 \end{vmatrix}$

B

Evaluate the determinants in Problems 16–30.

16. $\begin{vmatrix} 5 & 2 & 6 & -11 \\ -3 & 0 & 3 & 1 \\ 4 & 0 & 0 & 6 \\ 5 & 0 & 0 & -1 \end{vmatrix}$

17. $\begin{vmatrix} 4 & 3 & 2 & 1 \\ -5 & 0 & 0 & 0 \\ 11 & -4 & 0 & 0 \\ 9 & 6 & 3 & -5 \end{vmatrix}$

18. $\begin{vmatrix} 7 & 2 & -5 & 3 \\ 1 & 0 & -3 & 2 \\ 4 & 0 & 1 & -5 \\ 0 & 0 & 3 & 0 \end{vmatrix}$

19. $\begin{vmatrix} 2 & 1 & -1 & 3 \\ 4 & 0 & 0 & 0 \\ 2 & 1 & -2 & 3 \\ 1 & 4 & 3 & 5 \end{vmatrix}$

20. $\begin{vmatrix} 6 & 3 & 0 & -2 \\ 4 & 3 & 4 & -1 \\ 1 & 2 & 0 & 5 \\ 3 & -2 & 0 & 5 \end{vmatrix}$

21. $\begin{vmatrix} 1 & 6 & 2 & -1 \\ 0 & -5 & 0 & 0 \\ 3 & 4 & 5 & -3 \\ 2 & 1 & 4 & 0 \end{vmatrix}$

22. $\begin{vmatrix} 2 & 1 & 2 & 4 \\ 3 & -1 & 2 & 5 \\ -3 & 2 & 3 & -4 \\ -3 & 2 & 8 & -4 \end{vmatrix}$

23. $\begin{vmatrix} 3 & 1 & -1 & 2 \\ 4 & 0 & 3 & 0 \\ 2 & 4 & 3 & -3 \\ 6 & 1 & 4 & 0 \end{vmatrix}$

24. $\begin{vmatrix} 2 & 1 & 3 & 1 \\ 6 & 3 & -3 & 2 \\ 2 & 0 & 5 & 1 \\ 3 & 5 & -2 & -1 \end{vmatrix}$

25. $\begin{vmatrix} 0 & -3 & 5 & 6 & -1 \\ 0 & 0 & 1 & 1 & 3 \\ 5 & -3 & 8 & -5 & 1 \\ 0 & 0 & 2 & 0 & 0 \\ 0 & 0 & 4 & -1 & 2 \end{vmatrix}$

26. $\begin{vmatrix} 3 & 1 & -3 & -2 & 5 \\ -1 & 5 & 1 & 0 & 1 \\ 5 & 0 & 3 & 0 & 0 \\ 4 & 0 & 0 & 0 & 0 \\ 2 & 2 & 4 & 0 & -1 \end{vmatrix}$

27. $\begin{vmatrix} -3 & 4 & 5 & 8 & -9 \\ 0 & 0 & 0 & 0 & 6 \\ 3 & 0 & 0 & -1 & 4 \\ 0 & 0 & 4 & 0 & 9 \\ -3 & 0 & 0 & 1 & 8 \end{vmatrix}$

28. $\begin{vmatrix} 3 & 4 & 1 & -2 & 0 \\ 1 & -2 & 3 & 0 & 1 \\ 0 & 4 & -2 & 1 & 0 \\ 5 & 0 & -3 & 0 & 0 \\ 2 & 1 & -2 & 2 & 0 \end{vmatrix}$

29. $\begin{vmatrix} 2 & -1 & 0 & 1 & -1 \\ 3 & 0 & 0 & 2 & 0 \\ 2 & 1 & 3 & 1 & 3 \\ 0 & 0 & 0 & -2 & 0 \\ 1 & 3 & -1 & 2 & 1 \end{vmatrix}$

30. $\begin{vmatrix} 2 & 1 & 3 & 0 & 0 \\ 6 & 1 & 5 & 2 & -1 \\ 1 & 4 & -5 & 9 & 3 \\ 3 & 2 & 5 & 7 & 2 \\ 2 & -3 & -2 & 4 & 6 \end{vmatrix}$

Use Cramer's Rule to solve the systems in Problems 31–40.

31. $\begin{cases} x + y + z = 6 \\ 2x - y + z = 3 \\ x - 2y - 3z = 6 \end{cases}$

32. $\begin{cases} 2x - y + z = 3 \\ x - 3y + 2z = 7 \\ x - y - z = -1 \end{cases}$

33. $\begin{cases} x + y + z = 4 \\ x + 3y + 2z = 4 \\ x - 2y + z = 7 \end{cases}$

34. $\begin{cases} x + y + z = 3 \\ 2x + z = -1 \\ y = 5 \end{cases}$

35. $\begin{cases} x + y + z = 3 \\ 3y - z = -11 \\ x = 4 \end{cases}$

36. $\begin{cases} 2x + y + z = -5 \\ 3x - y = -9 \\ z = -4 \end{cases}$

37. $\begin{cases} 2x + 2y + 3z = 1 \\ 2x - z = -11 \\ 3y + 2z = 6 \end{cases}$

38. $\begin{cases} x + 2y + z = 1 \\ x - 3y - 2z = 2 \\ 3x - 2y + z = 3 \end{cases}$

39. $\begin{cases} 2x - y + z = 4 \\ 3x - 2y + 2z = 3 \\ x - y + 3z = 2 \end{cases}$

40. $\begin{cases} 5x - 3y + 2z = 10 \\ 4x + 2y - 3z = 4 \\ 3x + y + 4z = -8 \end{cases}$

Use the determinant equation given in Problem 51 to find the equations of the lines passing through the points given in Problems 41–43.

41. $(-3, -2), (1, -3)$

42. $(4, -5), (7, -8)$

43. $(1, 5), (-2, -3)$

Find the absolute value of the determinant expression given in Problem 52 to find the areas of the triangles with vertices given in Problems 44–46.

44. $(1, 1), (-2, -3), (11, -3)$

45. $(-3, 12), (5, 6), (-3, -9)$

46. $(-8, 0), (12, 10), (4, -5)$

Solve the systems in Problems 47–50.

47. $\begin{cases} w + x - y + z = 7 \\ 2w + y - 3z = 1 \\ 2x - z + w = 4 \\ y - w + z = -4 \end{cases}$

48. $\begin{cases} 3t - u + x = 20 \\ x - 2t - u = 0 \\ 3u - 2x + 5t = 1 \end{cases}$

49. $\begin{cases} s + 3t - 2u = 4 \\ u + 2x - t = -5 \\ x - u - t = 0 \\ s - 2x = -1 \end{cases}$

50. $\begin{cases} 2x - y + w - v = -4 \\ 3x + 2w = 0 \\ x + y + 3z + w + 3v = 5 \\ -2w = -6 \\ x + 3y - z + 2w + v = 10 \end{cases}$

C

51. Prove that

$$\begin{vmatrix} x & y & 1 \\ x_1 & y_1 & 1 \\ x_2 & y_2 & 1 \end{vmatrix} = 0$$

is the equation of a line passing through (x_1, y_1) and (x_2, y_2).

52. Prove that the area of a triangle with vertices at (x_1, y_1), (x_2, y_2), and (x_3, y_3) is the absolute value of

$$\frac{1}{2}\begin{vmatrix} x_1 & y_1 & 1 \\ x_2 & y_2 & 1 \\ x_3 & y_3 & 1 \end{vmatrix}$$

53. If r_1, r_2, r_3, and r_4 are the fourth roots of 1, show that

$$\begin{vmatrix} r_1 & r_2 & r_3 & r_4 \\ r_2 & r_3 & r_4 & r_1 \\ r_3 & r_4 & r_1 & r_2 \\ r_4 & r_1 & r_2 & r_3 \end{vmatrix} = 0$$

For Problems 54–60, let

$$|A| = \begin{vmatrix} a_{11} & a_{12} & a_{13} \\ a_{21} & a_{22} & a_{23} \\ a_{31} & a_{32} & a_{33} \end{vmatrix}$$

54. Show that you obtain the same result if you expand along the first or third row.

55. Show that you obtain the same result if you expand along the second row or second column.

56. Prove the determinant reduction theorem for $|A|$. (Without loss of generality, prove it by multiplying the first row by k and adding it to the second row.)

57. Prove that $|A| = 0$ if two rows (say the first and second rows) are identical.

58. Prove that $|A| = 0$ if two columns (say the first and second columns) are identical.

59. Prove that if two rows of $|A|$ are interchanged (say the first and third), the resulting determinant is $-|A|$.

60. Prove that

$$k|A| = \begin{vmatrix} a_{11} & a_{12} & a_{13} \\ ka_{21} & ka_{22} & ka_{23} \\ a_{31} & a_{32} & a_{33} \end{vmatrix}$$

9.4 ALGEBRA OF MATRICES

Instead of considering the determinant of the coefficients, we will now arrange the coefficients of a system of linear equations into a rectangular **array** of numbers. Such an array of numbers is called a **matrix**. The number of columns and rows of a matrix need not be the same; but if they are, the matrix is called a **square matrix**. The **order** or **dimension** of a matrix is given by an expression $\boldsymbol{m \times n}$ (pronounced "m by n"), where m is the number of rows and n is the number of columns.

EXAMPLE 1

$$A = [5 \quad 2 \quad 1] \qquad B = [4 \quad 8 \quad -5]$$

$$C = \begin{bmatrix} 7 & 3 & 2 \\ 5 & -4 & -3 \end{bmatrix} \qquad D = \begin{bmatrix} 4 & -2 & 1 \\ -3 & 3 & -1 \\ 2 & 4 & -1 \end{bmatrix}$$

$$E = \begin{bmatrix} 3 & 4 & -1 \\ 2 & 0 & 5 \\ -4 & 2 & 3 \end{bmatrix} \qquad G = [g_{ij}]_{m,n}$$

Note that A and B have order 1×3; C has order 2×3; D and E have order 3×3—these are square **matrices** (plural form of *matrix*). Since m and n are the same for square matrices, we often refer to them as having order n: D and E have order 3.

Matrix G is an arbitrary m by n matrix; it is shorthand notation for

$$\begin{bmatrix} g_{11} & g_{12} & g_{13} & \cdots & g_{1n} \\ g_{21} & g_{22} & g_{23} & \cdots & g_{2n} \\ g_{31} & g_{32} & g_{33} & \cdots & g_{3n} \\ & & \vdots & & \\ g_{m1} & g_{m2} & g_{m3} & \cdots & g_{mn} \end{bmatrix}$$

■

Capital letters are generally used to denote matrices and the same lowercase letters for entries of that matrix. Also, do not confuse matrices, determinants, and their notation. A matrix M is an *array* of real numbers whereas the determinant $|M|$ is itself a single real number. The entries of a matrix are always shown enclosed in square brackets, whereas determinants are enclosed by vertical lines.

Following is a definition of matrix equality along with the fundamental matrix operations.

Matrix Operations

EQUALITY: $M = N$ if and only if matrices M and N have the same dimension and $m_{ij} = n_{ij}$ for all i and j.

ADDITION: $M + N = P$ if and only if M and N have the same dimension and $p_{ij} = m_{ij} + n_{ij}$ for all i and j.

MULTIPLICATION OF A MATRIX AND A REAL NUMBER: $cM = Mc = [cm_{ij}]$ for any real number c.

SUBTRACTION: $M - N = P$ if and only if matrices M and N have the same dimension and $p_{ij} = m_{ij} - n_{ij}$ for all i and j.

MULTIPLICATION: $MN = P$ if and only if the number of columns of matrix M is the same as the number of rows of matrix N and

$$p_{ij} = m_{i1}n_{1j} + m_{i2}n_{2j} + m_{i3}n_{3j} + \cdots + m_{in}n_{nj}$$

The matrix P has the same number of rows as M and the same number of columns as N.

All of these definitions, except multiplication, are straightforward, so we will consider multiplication separately after Example 2. If an addition or multiplication cannot be performed because of the order of the given matrices, the matrices are said to be **not conformable**.

EXAMPLE 2 Let A, B, C, D, and E be the matrices defined in Example 1.

a. $A + B = [5 \quad 2 \quad 1] + [4 \quad 8 \quad -5]$
$= [5 + 4 \quad 2 + 8 \quad 1 + (-5)]$
$= [9 \quad 10 \quad -4]$

b. $A + C$ is not defined because A and C are not conformable.

c. $E + D = \begin{bmatrix} 3 & 4 & -1 \\ 2 & 0 & 5 \\ -4 & 2 & 3 \end{bmatrix} + \begin{bmatrix} 4 & -2 & 1 \\ -3 & 3 & -1 \\ 2 & 4 & -1 \end{bmatrix}$

$= \begin{bmatrix} 7 & 2 & 0 \\ -1 & 3 & 4 \\ -2 & 6 & 2 \end{bmatrix}$ Add entry by entry.

d. $(-5)C = (-5)\begin{bmatrix} 7 & 3 & 2 \\ 5 & -4 & -3 \end{bmatrix}$

$= \begin{bmatrix} -35 & -15 & -10 \\ -25 & 20 & 15 \end{bmatrix}$ Multiply each entry by (-5).

e. $2A - 3B = 2[5 \quad 2 \quad 1] + (-3)[4 \quad 8 \quad -5]$
$= [10 \quad 4 \quad 2] + [-12 \quad -24 \quad 15]$
$= [-2 \quad -20 \quad 17]$ ■

You will need to study the definition of matrix multiplication carefully to understand the process. Consider the following specific example:

$$M = \begin{bmatrix} m_{11} & m_{12} & m_{13} & m_{14} \\ m_{21} & m_{22} & m_{23} & m_{24} \end{bmatrix} = \begin{bmatrix} 1 & 2 & 3 & 4 \\ 5 & 6 & 7 & 8 \end{bmatrix}$$

$$N = \begin{bmatrix} n_{11} & n_{12} & n_{13} \\ n_{21} & n_{22} & n_{23} \\ n_{31} & n_{32} & n_{33} \\ n_{41} & n_{42} & n_{43} \end{bmatrix} = \begin{bmatrix} -3 & 1 & -2 \\ 0 & -1 & 5 \\ -4 & 3 & -1 \\ 2 & 3 & -2 \end{bmatrix}$$

The matrix N and its entries are shown in color so you can keep track of them.

$$MN = \begin{bmatrix} m_{11} & m_{12} & m_{13} & m_{14} \\ m_{21} & m_{22} & m_{23} & m_{24} \end{bmatrix} \begin{bmatrix} n_{11} & n_{12} & n_{13} \\ n_{21} & n_{22} & n_{23} \\ n_{31} & n_{32} & n_{33} \\ n_{41} & n_{42} & n_{43} \end{bmatrix} = \begin{bmatrix} p_{11} & p_{12} & p_{13} \\ p_{21} & p_{22} & p_{23} \end{bmatrix}$$

These are the entries of the product.

These must be the same for multiplication to be conformable.

Order of M is 2×4 Order of N is 4×3

Order of P is 2×3; this is the order of the product.

Row 1 →, Row 2 →

$$MN = \begin{bmatrix} \mathbf{1} & \mathbf{2} & \mathbf{3} & \mathbf{4} \\ \mathbf{5} & \mathbf{6} & \mathbf{7} & \mathbf{8} \end{bmatrix} \begin{bmatrix} -3 & 1 & -2 \\ 0 & -1 & 5 \\ -4 & 3 & -1 \\ 2 & 3 & -2 \end{bmatrix} = \begin{bmatrix} p_{11} & p_{12} & p_{13} \\ p_{21} & p_{22} & p_{23} \end{bmatrix}$$

↑ Column 1 ↑ Column 2 ↑ Column 3

Entry in Product		
Row	**Column**	**Notation**
1	1	p_{11}
1	2	p_{12}
1	3	p_{13}
2	1	p_{21}
2	2	p_{22}
2	3	p_{23}

$$p_{11} = m_{11}n_{11} + m_{12}n_{21} + m_{13}n_{31} + m_{14}n_{41} = \mathbf{1}(-3) + \mathbf{2}(0) + \mathbf{3}(-4) + \mathbf{4}(2) = -7$$

$$p_{12} = m_{11}n_{12} + m_{12}n_{22} + m_{13}n_{32} + m_{14}n_{42} = \mathbf{1}(1) + \mathbf{2}(-1) + \mathbf{3}(3) + \mathbf{4}(3) = 20$$

$$p_{13} = m_{11}n_{13} + m_{12}n_{23} + m_{13}n_{33} + m_{14}n_{43} = \mathbf{1}(-2) + \mathbf{2}(5) + \mathbf{3}(-1) + \mathbf{4}(-2) = -3$$

$$p_{21} = m_{21}n_{11} + m_{22}n_{21} + m_{23}n_{31} + m_{24}n_{41} = \mathbf{5}(-3) + \mathbf{6}(0) + \mathbf{7}(-4) + \mathbf{8}(2) = -27$$

$$p_{22} = m_{21}n_{12} + m_{22}n_{22} + m_{23}n_{32} + m_{24}n_{42} = \mathbf{5}(1) + \mathbf{6}(-1) + \mathbf{7}(3) + \mathbf{8}(3) = 44$$

$$p_{23} = m_{21}n_{13} + m_{22}n_{23} + m_{23}n_{33} + m_{24}n_{43} = \mathbf{5}(-2) + \mathbf{6}(5) + \mathbf{7}(-1) + \mathbf{8}(-2) = -3$$

This example makes the process seem very lengthy, but it has been written out so that you could study the *process*. The way it would look in your work is shown in Example 3.

EXAMPLE 3 Find the products.

a. $\begin{bmatrix} \mathbf{1} & \mathbf{2} & \mathbf{3} & \mathbf{4} \\ \mathbf{5} & \mathbf{6} & \mathbf{7} & \mathbf{8} \end{bmatrix} \begin{bmatrix} -3 & 1 & -2 \\ 0 & -1 & 5 \\ -4 & 3 & -1 \\ 2 & 3 & -2 \end{bmatrix}$

$$= \begin{bmatrix} \mathbf{1}(-3) + \mathbf{2}(0) + \mathbf{3}(-4) + \mathbf{4}(2) & \mathbf{1}(1) + \mathbf{2}(-1) + \mathbf{3}(3) + \mathbf{4}(3) & \mathbf{1}(-2) + \mathbf{2}(5) + \mathbf{3}(-1) + \mathbf{4}(-2) \\ \mathbf{5}(-3) + \mathbf{6}(0) + \mathbf{7}(-4) + \mathbf{8}(2) & \mathbf{5}(1) + \mathbf{6}(-1) + \mathbf{7}(3) + \mathbf{8}(3) & \mathbf{5}(-2) + \mathbf{6}(5) + \mathbf{7}(-1) + \mathbf{8}(-2) \end{bmatrix}$$

$$= \begin{bmatrix} -7 & 20 & -3 \\ -27 & 44 & -3 \end{bmatrix}$$

b. $\begin{bmatrix} 3 & -1 & 4 \\ 2 & 1 & 0 \\ -1 & 3 & 2 \end{bmatrix} \begin{bmatrix} 5 & 1 & -1 \\ 2 & 3 & -2 \\ 0 & 3 & 4 \end{bmatrix}$

$$= \begin{bmatrix} 3(5) + (-1)(2) + 4(0) & 3(1) + (-1)(3) + 4(3) & 3(-1) + (-1)(-2) + 4(4) \\ 2(5) + (1)(2) + 0(0) & 2(1) + (1)(3) + 0(3) & 2(-1) + (1)(-2) + 0(4) \\ (-1)(5) + 3(2) + 2(0) & (-1)(1) + 3(3) + 2(3) & (-1)(-1) + 3(-2) + 2(4) \end{bmatrix} = \begin{bmatrix} 13 & 12 & 15 \\ 12 & 5 & -4 \\ 1 & 14 & 3 \end{bmatrix}$$

■

Systems of equations can be written in the form of matrix multiplication, as shown by Example 4.

EXAMPLE 4 Let

$$A = \begin{bmatrix} 1 & 2 & 3 \\ 4 & -1 & 5 \\ 3 & 2 & -1 \end{bmatrix} \qquad X = \begin{bmatrix} x \\ y \\ z \end{bmatrix} \qquad B = \begin{bmatrix} 3 \\ 16 \\ 5 \end{bmatrix}$$

What is $AX = B$?

Solution

$$AX = \begin{bmatrix} x + 2y + 3z \\ 4x - \ \ y + 5z \\ 3x + 2y - \ \ z \end{bmatrix}$$

so $AX = B$ is a matrix equation representing the system

$$\begin{cases} x + 2y + 3z = 3 \\ 4x - \ \ y + 5z = 16 \\ 3x - 2y - \ \ z = 5 \end{cases}$$

■

Properties for an algebra of matrices can also be developed. The $m \times n$ **zero matrix**, denoted by 0, is the matrix with m rows and n columns in which each entry is 0. The **identity matrix**, denoted by I_n, is the square matrix with n rows and n columns consisting of a 1 in each position on the **main diagonal** (entries $m_{11}, m_{22}, m_{33}, \ldots$) and zeros elsewhere:

$$I_2 = \begin{bmatrix} 1 & 0 \\ 0 & 1 \end{bmatrix} \qquad I_3 = \begin{bmatrix} 1 & 0 & 0 \\ 0 & 1 & 0 \\ 0 & 0 & 1 \end{bmatrix} \qquad I_4 = \begin{bmatrix} 1 & 0 & 0 & 0 \\ 0 & 1 & 0 & 0 \\ 0 & 0 & 1 & 0 \\ 0 & 0 & 0 & 1 \end{bmatrix}$$

The **additive inverse** of a matrix M is denoted by $-M$ and is defined by $(-1)M$; the **multiplicative inverse** of M is denoted by M^{-1} if it exists. (*Note:* $M^{-1} \neq 1/M$.) The following list summarizes the properties of matrices. Assume that matrices M, N, and P all have order n, which forces them to be conformable for the given operations.

TABLE 9.1
Properties of matrices

Property	Addition	Multiplication
Commutative	$M + N = N + M$	$MN \neq NM$ Matrix multiplication is not commutative.
Associative	$(M + N) + P = M + (N + P)$	$(MN)P = M(NP)$
Identity	$M + 0 = 0 + M = M$	$I_nM = MI_n = M$ For a square matrix M of order n.
Inverse	$M + (-M) = (-M) + M = 0$	$M(M^{-1}) = (M^{-1})M = I_n$ For a square matrix M for which M^{-1} exists.
Distributive	$M(N + P) = MN + MP$ and $(N + P)M = NM + PM$	

These properties are straightforward except for the inverse property. There seem to be two unanswered questions. Given a square matrix M, when does M^{-1} exist? And if it exists, how do you find it?

If a square matrix M has an inverse, it is called **nonsingular**. If

$$M = \begin{bmatrix} a_{11} & a_{12} \\ a_{21} & a_{22} \end{bmatrix} \quad \text{and it has an inverse} \quad M^{-1} = \begin{bmatrix} w & x \\ y & z \end{bmatrix}$$

then $MM^{-1} = I_2$ if and only if $|M| \neq 0$. To prove this, simply multiply MM^{-1} and solve the resulting systems for w, x, y, and z to find

$$w = \frac{a_{22}}{|M|} \qquad x = \frac{-a_{12}}{|M|} \qquad y = \frac{-a_{21}}{|M|} \qquad z = \frac{a_{11}}{|M|}$$

Thus we see that the existence of the inverse of a matrix depends on whether the value of the associated determinant is zero.

The procedure for finding an inverse depends on **elementary row operations** and on equivalent matrices. We say that a matrix M **is equivalent to** a matrix N, written $M \sim N$, if M can be changed into N by one or more of the following elementary row operations.

Elementary Row Operations

There are three elementary row operations:

1. Interchange any two rows.
2. Multiply all the elements of a row by the same nonzero real number.
3. Multiply all the elements of a row by a real number and add the product to the corresponding entry of another row.

If M has order n, write $[M \mid I_n]$, called the **augmented matrix**, and carry out the following procedure using only elementary row operations:

Procedure for Finding an Inverse Matrix

Step 1 Obtain a 1 in the first position on the main diagonal. If this is not possible (the first column has all zeros), the matrix does not have an inverse.

Step 2 Use the 1 in the first position on the main diagonal as a pivot to obtain zeros for all the entries in the first column under the 1.

Step 3 Obtain a 1 in the second position on the main diagonal (if possible).

Step 4 Use the 1 in the second position on the main diagonal as a pivot to obtain zeros for all the entries in the second column under the 1.

Step 5 Obtain a 1 in the third position on the main diagonal (if possible). Use this to obtain zeros for all entries below this 1. Continue this procedure until the main diagonal of M contains all 1's and all entries below those 1's are zeros.

Step 6 Use the 1 in the last position on the main diagonal to obtain zeros in all entries in the last column above the 1. Continue this process until all the entries above all the 1's on the diagonal of M are zero.

Step 7 The matrix in the position of I is the inverse of M.

EXAMPLE 5 Find the inverse, if possible, of the matrix

$$\begin{bmatrix} 1 & 2 \\ 1 & 4 \end{bmatrix}$$

Solution

$$\left[\begin{array}{cc|cc} 1 & 2 & 1 & 0 \\ 1 & 4 & 0 & 1 \end{array}\right] \sim \left[\begin{array}{cc|cc} 1 & 2 & 1 & 0 \\ 0 & 2 & -1 & 1 \end{array}\right]$$

Multiply row 1 by -1 and add to row 2.

$$\sim \left[\begin{array}{cc|cc} 1 & 2 & 1 & 0 \\ 0 & 1 & -\frac{1}{2} & \frac{1}{2} \end{array}\right]$$

Multiply row 2 by $\frac{1}{2}$.

$$\sim \left[\begin{array}{cc|cc} 1 & 0 & 2 & -1 \\ 0 & 1 & -\frac{1}{2} & \frac{1}{2} \end{array}\right]$$

Multiply row 2 by -2 and add to row 1.

The inverse is on the right side of the dotted line:

$$\begin{bmatrix} 2 & -1 \\ -\frac{1}{2} & \frac{1}{2} \end{bmatrix}$$

When a matrix has fractional entries, it is often rewritten with a fractional coefficient and integer entries in order to simplify subsequent arithmetic. For example, we would rewrite the above matrix as

$$\frac{1}{2}\begin{bmatrix} 4 & -2 \\ -1 & 1 \end{bmatrix}$$

The fractions obtained when finding inverses are often ugly. ■

EXAMPLE 6 Find the inverse of the matrix

$$A = \begin{bmatrix} 0 & 1 & 2 \\ 2 & -1 & 1 \\ -1 & 1 & 0 \end{bmatrix}$$

if it exists.

Solution

$$\left[\begin{array}{ccc|ccc} 0 & 1 & 2 & 1 & 0 & 0 \\ 2 & -1 & 1 & 0 & 1 & 0 \\ -1 & 1 & 0 & 0 & 0 & 1 \end{array}\right]$$

$$\sim \left[\begin{array}{ccc|ccc} -1 & 1 & 0 & 0 & 0 & 1 \\ 2 & -1 & 1 & 0 & 1 & 0 \\ 0 & 1 & 2 & 1 & 0 & 0 \end{array}\right]$$

Interchange row 1 and row 3.

$$\sim \left[\begin{array}{ccc|ccc} 1 & -1 & 0 & 0 & 0 & -1 \\ 2 & -1 & 1 & 0 & 1 & 0 \\ 0 & 1 & 2 & 1 & 0 & 0 \end{array}\right]$$

Multiply row 1 by -1.

$$\sim \left[\begin{array}{ccc|ccc} 1 & -1 & 0 & 0 & 0 & -1 \\ 0 & 1 & 1 & 0 & 1 & 2 \\ 0 & 1 & 2 & 1 & 0 & 0 \end{array}\right]$$

Multiply row 1 by -2 and add to row 2.

$$\sim\left[\begin{array}{rrr|rrr} 1 & -1 & 0 & 0 & 0 & -1 \\ 0 & 1 & 1 & 0 & 1 & 2 \\ 0 & 0 & 1 & 1 & -1 & -2 \end{array}\right]$$

Multiply row 2 by -1 and add to row 3.

$$\sim\left[\begin{array}{rrr|rrr} 1 & -1 & 0 & 0 & 0 & -1 \\ 0 & 1 & 0 & -1 & 2 & 4 \\ 0 & 0 & 1 & 1 & -1 & -2 \end{array}\right]$$

Multiply row 3 by -1 and add to row 2.

$$\sim\left[\begin{array}{rrr|rrr} 1 & 0 & 0 & -1 & 2 & 3 \\ 0 & 1 & 0 & -1 & 2 & 4 \\ 0 & 0 & 1 & 1 & -1 & -2 \end{array}\right]$$

Add row 2 to row 1.

$$A^{-1} = \begin{bmatrix} -1 & 2 & 3 \\ -1 & 2 & 4 \\ 1 & -1 & -2 \end{bmatrix}$$

■

HP-28S ACTIVITY

As you can see from Example 6, the procedure for finding the inverse of a matrix can be quite laborious. We will rework this example using the HP-28S. First, clear the stack and set the number display mode to two decimal places:

CLEAR MODE 2 FIX

Now you need to key in the elements of the matrix (a 3×3 matrix for this example); these are entered in one *column* at a time, starting with column 1. Separate each column by a bracket and make sure that a space or a comma separates each entry.

[[0 SPACE 2 SPACE 1 CHS
[1 SPACE 1 CHS SPACE 1
[2 SPACE 1 SPACE 0 ENTER

```
1: [[ 0.00 2.00 -1.00 ]
    [ 1.00 -1.00 1.00 ]
    [ 2.00 1.00 0.00 ]]
STO FIX SCI ENG DEG RAD
```

After entering the matrix, the inverse is found simply by pressing:

1/x

You can read the output to see:

$$A^{-1} = \begin{bmatrix} -1 & 2 & 3 \\ -1 & 2 & 4 \\ 1 & -1 & -2 \end{bmatrix}$$

Notice that the output gives the matrix entries ordered by *column*, just as the input was.

In Example 4 we saw how a system of equations can be written in matrix form. We can now see how to solve a system of linear equations by using the inverse. Consider a system of n linear equations with n unknowns whose matrix of coefficients A has an inverse A^{-1}:

$AX = B$	Given system
$(A^{-1})AX = A^{-1}B$	Multiply both sides by A^{-1}
$(A^{-1}A)X = A^{-1}B$	Associative property
$I_nX = A^{-1}B$	Inverse property
$X = A^{-1}B$	Identity property

This says that to solve a system, simply multiply A^{-1} and B to find X.

EXAMPLE 7 Solve the system

$$\begin{cases} y + 2z = 0 \\ 2x - y + z = -1 \\ y - x = 1 \end{cases}$$

Solution Write in matrix form:

$$A = \begin{bmatrix} 0 & 1 & 2 \\ 2 & -1 & 1 \\ -1 & 1 & 0 \end{bmatrix} \qquad X = \begin{bmatrix} x \\ y \\ z \end{bmatrix} \qquad B = \begin{bmatrix} 0 \\ -1 \\ 1 \end{bmatrix}$$

From Example 6,

$$A^{-1} = \begin{bmatrix} -1 & 2 & 3 \\ -1 & 2 & 4 \\ 1 & -1 & -2 \end{bmatrix}$$

Thus

$$X = A^{-1}B = \begin{bmatrix} -1 & 2 & 3 \\ -1 & 2 & 4 \\ 1 & -1 & -2 \end{bmatrix} \begin{bmatrix} 0 \\ -1 \\ 1 \end{bmatrix} = \begin{bmatrix} 0 - 2 + 3 \\ 0 - 2 + 4 \\ 0 + 1 - 2 \end{bmatrix} = \begin{bmatrix} 1 \\ 2 \\ -1 \end{bmatrix}$$

Therefore $\mathbf{x = 1, y = 2,}$ **and** $\mathbf{z = -1}$. ■

The method of solving a system by using the inverse matrix is very efficient if you know the inverse. Unfortunately, *finding* the inverse for one system is usually more work than using another method to solve the system. However, there are certain applications that yield the same system over and over, and the only thing to change is the constants. In this case the inverse method is worthwhile. And, finally, computers can often find approximations for inverse matrices quite easily, so this method might be appropriate if one is using a computer.

PROBLEM SET 9.4

A

In Problems 1–20, find the indicated matrices if possible.

$$A = \begin{bmatrix} 1 & 2 \\ 4 & 0 \\ -1 & 3 \\ 2 & 1 \end{bmatrix} \qquad B = \begin{bmatrix} 4 & 2 \\ -1 & 3 \end{bmatrix}$$

$$C = \begin{bmatrix} 1 & 0 & 0 & 0 \\ 0 & 1 & 0 & 0 \\ 0 & 0 & 1 & 0 \\ 0 & 0 & 0 & 1 \end{bmatrix}$$

$$D = \begin{bmatrix} 4 & 1 & 3 & 6 \\ -1 & 0 & -2 & 3 \end{bmatrix} \qquad E = \begin{bmatrix} 1 & 0 & 2 \\ 3 & -1 & 2 \\ 4 & 1 & 0 \end{bmatrix}$$

$$F = \begin{bmatrix} 1 & 4 & 0 \\ 3 & -1 & 2 \\ -2 & 1 & 5 \end{bmatrix} \qquad G = \begin{bmatrix} 8 & 1 & 6 \\ 3 & 5 & 7 \\ 4 & 9 & 2 \end{bmatrix}$$

1. $E + F$ **2.** EF **3.** EG **4.** $EF + EG$
5. $E(F + G)$ **6.** $2E - G$ **7.** FG **8.** GF
9. $(EF)G$ **10.** $E(FG)$ **11.** AB **12.** BA
13. B^2 **14.** CA **15.** BD **16.** DB
17. $(B + C)A$ **18.** $BA + CA$ **19.** C^3 **20.** CD

B

Find the inverse of each matrix in Problems 21–29, if it exists.

21. $\begin{bmatrix} 4 & -7 \\ -1 & 2 \end{bmatrix}$ **22.** $\begin{bmatrix} 8 & 6 \\ -2 & 4 \end{bmatrix}$

23. $\begin{bmatrix} 1 & 3 \\ 2 & 0 \end{bmatrix}$ **24.** $\begin{bmatrix} 1 & 0 & 2 \\ 2 & 1 & 0 \\ 0 & -2 & 9 \end{bmatrix}$

25. $\begin{bmatrix} 6 & 1 & 20 \\ 1 & -1 & 0 \\ 0 & 1 & 3 \end{bmatrix}$ **26.** $\begin{bmatrix} 4 & 1 & 0 \\ 2 & -1 & 4 \\ -3 & 2 & 1 \end{bmatrix}$

27. $\begin{bmatrix} 1 & 0 & 0 & 1 \\ 0 & 2 & 0 & 0 \\ 0 & 0 & 0 & 1 \\ 2 & 0 & 1 & 0 \end{bmatrix}$ **28.** $\begin{bmatrix} 0 & 1 & 2 & 0 \\ 0 & 0 & 0 & 1 \\ 1 & 1 & 3 & 0 \\ 2 & 4 & 0 & 0 \end{bmatrix}$

29. $\begin{bmatrix} 1 & 2 & 0 & 0 \\ 0 & 0 & 1 & 0 \\ 1 & 3 & 0 & 1 \\ 4 & 0 & 0 & 2 \end{bmatrix}$

Solve the systems in Problems 30–62 by solving the corresponding matrix equation with an inverse if possible. Problems 30–35 use the inverse found in Problem 21.

30. $\begin{cases} 4x - 7y = -2 \\ -x + 2y = 1 \end{cases}$ **31.** $\begin{cases} 4x - 7y = -65 \\ -x + 2y = 18 \end{cases}$

32. $\begin{cases} 4x - 7y = 48 \\ -x + 2y = -13 \end{cases}$ **33.** $\begin{cases} 4x - 7y = 2 \\ -x + 2y = 3 \end{cases}$

34. $\begin{cases} 4x - 7y = 5 \\ -x + 2y = 4 \end{cases}$ **35.** $\begin{cases} 4x - 7y = -3 \\ -x + 2y = 8 \end{cases}$

Problems 36–41 use the inverse found in Problem 22.

36. $\begin{cases} 8x + 6y = 12 \\ -2x + 4y = -14 \end{cases}$ **37.** $\begin{cases} 8x + 6y = 16 \\ -2x + 4y = 18 \end{cases}$

38. $\begin{cases} 8x + 6y = -6 \\ -2x + 4y = -26 \end{cases}$ **39.** $\begin{cases} 8x + 6y = -28 \\ -2x + 4y = 18 \end{cases}$

40. $\begin{cases} 8x + 6y = -26 \\ -2x + 4y = 12 \end{cases}$ **41.** $\begin{cases} 8x + 6y = -36 \\ -2x + 4y = -2 \end{cases}$

Problems 42–47 all use the same inverse.

42. $\begin{cases} 2x + 3y = 9 \\ x - 6y = -3 \end{cases}$ **43.** $\begin{cases} 2x + 3y = 2 \\ x - 6y = 16 \end{cases}$

44. $\begin{cases} 2x + 3y = 2 \\ x - 6y = -14 \end{cases}$ **45.** $\begin{cases} 2x + 3y = 9 \\ x - 6y = 42 \end{cases}$

46. $\begin{cases} 2x + 3y = -22 \\ x - 6y = 49 \end{cases}$ **47.** $\begin{cases} 2x + 3y = 12 \\ x - 6y = -24 \end{cases}$

Problems 48–53 use the inverse found in Problem 24.

48. $\begin{cases} x + 2z = 7 \\ 2x + y = 16 \\ -2y + 9z = -3 \end{cases}$ **49.** $\begin{cases} x + 2z = 4 \\ 2x + y = 0 \\ -2y + 9z = 19 \end{cases}$

50. $\begin{cases} x + 2z = 7 \\ 2x + y = 0 \\ -2y + 9z = 31 \end{cases}$ **51.** $\begin{cases} x + 2z = 7 \\ 2x + y = 1 \\ -2y + 9z = 28 \end{cases}$

52. $\begin{cases} x + 2z = 12 \\ 2x + y = 0 \\ -2y + 9z = 10 \end{cases}$ **53.** $\begin{cases} x + 2z = 5 \\ 2x + y = 8 \\ -2y + 9z = 9 \end{cases}$

Problems 54–56 use the inverse found in Problem 25.

54. $\begin{cases} 6x + y + 20z = 27 \\ x - y = 0 \\ y + 3z = 4 \end{cases}$ **55.** $\begin{cases} 6x + y + 20z = 14 \\ x - y = 1 \\ y + 3z = 1 \end{cases}$

56. $\begin{cases} 6x + y + 20z = 11 \\ x - y = 5 \\ y + 3z = -3 \end{cases}$

Problems 57–59 use the inverse found in Problem 26.

57. $\begin{cases} 4x + y = 6 \\ 2x - y + 4z = 12 \\ -3x + 2y + z = 4 \end{cases}$

58. $\begin{cases} 4x + y = 7 \\ 2x - y + 4z = -11 \\ -3x + 2y + z = -12 \end{cases}$

59. $\begin{cases} 4x + y = -10 \\ 2x - y + 4z = 20 \\ -3x + 2y + z = 20 \end{cases}$

Problems 60–62 use the inverse found in Problem 29.

60. $\begin{cases} x + 2y = 5 \\ z = 3 \\ x + 3y + w = 9 \\ 4x + 2w = 8 \end{cases}$

61. $\begin{cases} x + 2y = 0 \\ z = -4 \\ x + 3y + w = 4 \\ 4x + 2w = -2 \end{cases}$

62. $\begin{cases} x + 2y = 7 \\ z = -7 \\ x + 3y + w = 16 \\ 4x + 2w = -4 \end{cases}$

C

63. Let

$$M = \begin{bmatrix} a_{11} & a_{12} \\ a_{21} & a_{22} \end{bmatrix}$$

Show that if $|M| \neq 0$, then the inverse exists.

64. Let M be defined as in Problem 63. Show that if the inverse exists, then $|M| \neq 0$.

65. If M and N are nonsingular, then MN is nonsingular. Show that the inverse of MN is $(MN)^{-1} = N^{-1}M^{-1}$.

66. If M is nonsingular, show that if $MN = MP$ then $N = P$.

67. Prove that if M is nonsingular and $MN = M$, then $N = I_3$ for square matrices of order 3.

9.5 MATRIX SOLUTION OF SYSTEMS

Now that we have seen a variety of different methods for solving systems of linear equations, we will conclude our study of this topic by introducing the most general method yet: the **Gauss–Jordan method.** It is based on the method of solving a system by linear combinations but uses matrix notation to increase the efficiency of this method when working with more complicated systems. Instead of augmenting the matrix of the coefficients of the system as you did when finding the inverse, simply augment A_n with the constant terms. Next perform elementary row operations to transform A_n to I_n (if possible). The solution of the system will then be obvious, as shown by Examples 1 and 2.

EXAMPLE 1 Solve the system

$$\begin{cases} x + 2y - z = 0 \\ 2x + 3y - 2z = 3 \\ -x - 4y + 3z = -2 \end{cases}$$

Solution

$$\left[\begin{array}{ccc|c} 1 & 2 & -1 & 0 \\ 2 & 3 & -2 & 3 \\ -1 & -4 & 3 & -2 \end{array}\right] \sim \left[\begin{array}{ccc|c} 1 & 2 & -1 & 0 \\ 0 & -1 & 0 & 3 \\ 0 & -2 & 2 & -2 \end{array}\right]$$

Add -2 times the first row to the second row and add the first row to the third row.

$$\sim \left[\begin{array}{ccc|c} 1 & 2 & -1 & 0 \\ 0 & 1 & 0 & -3 \\ 0 & -2 & 2 & -2 \end{array}\right]$$

Multiply the second row by -1.

$$\sim\left[\begin{array}{ccc:c}1 & 2 & -1 & 0\\0 & 1 & 0 & -3\\0 & 0 & 2 & -8\end{array}\right]$$ Add two times the second row to the third row.

$$\sim\left[\begin{array}{ccc:c}1 & 2 & -1 & 0\\0 & 1 & 0 & -3\\0 & 0 & 1 & -4\end{array}\right]$$ Multiply the third row by $\frac{1}{2}$.

$$\sim\left[\begin{array}{ccc:c}1 & 2 & 0 & -4\\0 & 1 & 0 & -3\\0 & 0 & 1 & -4\end{array}\right]$$ Add the third row to the first row.

$$\sim\left[\begin{array}{ccc:c}1 & 0 & 0 & 2\\0 & 1 & 0 & -3\\0 & 0 & 1 & -4\end{array}\right]$$ Multiply the second row by -2 and add it to the first row.

This is equivalent to the system

$$\begin{aligned}1\cdot x+0\cdot y+0\cdot z&=2\\0\cdot x+1\cdot y+0\cdot z&=-3\\0\cdot x+0\cdot y+1\cdot z&=-4\end{aligned}\qquad\text{or}\qquad\begin{cases}x=2\\y=-3\\z=-4\end{cases}$$

The solution is now obvious: $\mathbf{(x, y, z) = (2, -3, -4)}$. ■

EXAMPLE 2 Solve the system

$$\begin{cases}w-x+2y+3z=-6\\2w+x-y+2z=-4\\w-3x+y-z=0\\w+2x+3y+4z=-1\end{cases}$$

Solution We will show the steps without mentioning which of the elementary row operations are being applied. Study this example carefully to make sure you can follow each step:

$$\left[\begin{array}{cccc:c}1 & -1 & 2 & 3 & -6\\2 & 1 & -1 & 2 & -4\\1 & -3 & 1 & -1 & 0\\1 & 2 & 3 & 4 & -1\end{array}\right]\sim\left[\begin{array}{cccc:c}1 & -1 & 2 & 3 & -6\\0 & 3 & -5 & -4 & 8\\0 & -2 & -1 & -4 & 6\\0 & 3 & 1 & 1 & 5\end{array}\right]$$

$$\sim\left[\begin{array}{cccc:c}1 & -1 & 2 & 3 & -6\\0 & 1 & -6 & -8 & 14\\0 & -2 & -1 & -4 & 6\\0 & 3 & 1 & 1 & 5\end{array}\right]\sim\left[\begin{array}{cccc:c}1 & -1 & 2 & 3 & -6\\0 & 1 & -6 & -8 & 14\\0 & 0 & -13 & -20 & 34\\0 & 0 & 19 & 25 & -37\end{array}\right]$$

$$\sim\left[\begin{array}{cccc:c}1 & -1 & 2 & 3 & -6\\0 & 1 & -6 & -8 & 14\\0 & 0 & -13 & -20 & 34\\0 & 0 & 6 & 5 & -3\end{array}\right]\sim\left[\begin{array}{cccc:c}1 & -1 & 2 & 3 & -6\\0 & 1 & -6 & -8 & 14\\0 & 0 & -1 & -10 & 28\\0 & 0 & 6 & 5 & -3\end{array}\right]$$

$$\sim\left[\begin{array}{rrrr:r}1&-1&2&3&-6\\0&1&-6&-8&14\\0&0&1&10&-28\\0&0&6&5&-3\end{array}\right]\sim\left[\begin{array}{rrrr:r}1&-1&2&3&-6\\0&1&-6&-8&14\\0&0&1&10&-28\\0&0&0&-55&165\end{array}\right]$$

$$\sim\left[\begin{array}{rrrr:r}1&-1&2&3&-6\\0&1&-6&-8&14\\0&0&1&10&-28\\0&0&0&1&-3\end{array}\right]\sim\left[\begin{array}{rrrr:r}1&-1&2&0&3\\0&1&-6&0&-10\\0&0&1&0&2\\0&0&0&1&-3\end{array}\right]$$

$$\sim\left[\begin{array}{rrrr:r}1&-1&0&0&-1\\0&1&0&0&2\\0&0&1&0&2\\0&0&0&1&-3\end{array}\right]\sim\left[\begin{array}{rrrr:r}1&0&0&0&1\\0&1&0&0&2\\0&0&1&0&2\\0&0&0&1&-3\end{array}\right]$$

Thus $\mathbf{w = 1, x = 2, y = 2}$, **and** $\mathbf{z = -3}$. ■

HP-28S ACTIVITY

You can solve systems of equations on the HP-28S by using matrix multiplication. In the last section we saw that a solution of the matrix equation $AX = B$ is

$$X = A^{-1}B$$

You can find the inverse of the matrix of the coefficients as described in the previous HP-28S activity box. Finally, multiply the inverse of A and B to find the answer. We show how this is done for Example 2.

First input A; but this time key in the coefficients by equation, not by column. Store this matrix in location A:

[[1 SPACE 1 CHS SPACE 2 SPACE 3
[2 SPACE 1 SPACE 1 CHS SPACE 2
[1 SPACE 3 CHS SPACE 1 SPACE 1 CHS
[1 SPACE 2 SPACE 3 SPACE 4 ENTER
' A STO

1: [[1.00 -1.00 2.00 ...
[2.00 1.00 -1.00 ...
[1.00 -3.00 1.00 ...
STO FIX SCI ENG DEG RAD

Notice from the display that you cannot see all the entries. If you wish to see the entries that are not shown, you can use the VIEW ↑ and VIEW ↓ keys. Next, input matrix B:

[[6 CHS [4 CHS [0 [1 CHS
ENTER ' B STO

3:
2:
1: [-6.00 -4.00 0.00 ...
STO FIX SCI ENG DEG RAD

Now compute the inverse of A and multiply by B:

A ENTER 1/X B ENTER ×

The result can now be seen by using the view keys: $w = 1$, $x = 2$, $y = 2$, and $z = -3$.

Our discussion concludes with two applied problems.

EXAMPLE 3 A rancher has to mix three types of feed for her cattle. The following analysis shows the amounts per bag (100 lb) of grain:

Grain	Protein	Carbohydrates	Sodium
A	7 lb	88 lb	1 lb
B	6 lb	90 lb	1 lb
C	10 lb	70 lb	2 lb

How many bags of each type of grain should she mix to provide 71 lb of protein, 854 lb of carbohydrates, and 12 lb of sodium?

Solution Let a, b, and c be the number of bags of grains A, B, and C respectively that are needed for the mixture. Then:

Grain	Protein	Carbohydrates	Sodium
A	$7a$	$88a$	a
B	$6b$	$90b$	b
C	$10c$	$70c$	$2c$
Total	71	854	12

Thus

$$\begin{cases} 7a + 6b + 10c = 71 \\ 88a + 90b + 70c = 854 \\ a + b + 2c = 12 \end{cases}$$

$$\left[\begin{array}{ccc|c} 7 & 6 & 10 & 71 \\ 88 & 90 & 70 & 854 \\ 1 & 1 & 2 & 12 \end{array}\right] \sim \left[\begin{array}{ccc|c} 1 & 1 & 2 & 12 \\ 88 & 90 & 70 & 854 \\ 7 & 6 & 10 & 71 \end{array}\right]$$

$$\sim \left[\begin{array}{ccc|c} 1 & 1 & 2 & 12 \\ 0 & 2 & -106 & -202 \\ 0 & -1 & -4 & -13 \end{array}\right] \sim \left[\begin{array}{ccc|c} 1 & 1 & 2 & 12 \\ 0 & 1 & -53 & -101 \\ 0 & -1 & -4 & -13 \end{array}\right]$$

$$\sim \left[\begin{array}{ccc|c} 1 & 1 & 2 & 12 \\ 0 & 1 & -53 & -101 \\ 0 & 0 & -57 & -114 \end{array}\right] \sim \left[\begin{array}{ccc|c} 1 & 1 & 2 & 12 \\ 0 & 1 & -53 & -101 \\ 0 & 0 & 1 & 2 \end{array}\right]$$

$$\sim \left[\begin{array}{ccc|c} 1 & 1 & 0 & 8 \\ 0 & 1 & 0 & 5 \\ 0 & 0 & 1 & 2 \end{array}\right] \sim \left[\begin{array}{ccc|c} 1 & 0 & 0 & 3 \\ 0 & 1 & 0 & 5 \\ 0 & 0 & 1 & 2 \end{array}\right]$$

Mix three bags of grain *A*, five bags of grain *B*, and two bags of grain *C*. ■

EXAMPLE 4 Suppose the equation of a certain parabola has the form

$$y = ax^2 + bx + c$$

Find the equation of the parabola passing through $(-1, -6)$, $(2, 9)$, and $(-2, -3)$.

Solution If a curve passes through a point, then the coordinates of that point must satisfy the equation

$$\begin{aligned} (-1,-6)&: \quad -6 = a(-1)^2 + b(-1) + c \\ (2,9)&: \quad 9 = a(2)^2 + b(2) + c \\ (-2,-3)&: \quad -3 = a(-2)^2 + b(-2) + c \end{aligned}$$

In standard form the system is

$$\begin{cases} a - b + c = -6 \\ 4a + 2b + c = 9 \\ 4a - 2b + c = -3 \end{cases}$$

$$\left[\begin{array}{ccc:c} 1 & -1 & 1 & -6 \\ 4 & 2 & 1 & 9 \\ 4 & -2 & 1 & -3 \end{array}\right] \sim \left[\begin{array}{ccc:c} 1 & -1 & 1 & -6 \\ 0 & 4 & 0 & 12 \\ 4 & -2 & 1 & -3 \end{array}\right] \sim \left[\begin{array}{ccc:c} 1 & -1 & 1 & -6 \\ 0 & 1 & 0 & 3 \\ 0 & 2 & -3 & 21 \end{array}\right]$$

$$\sim \left[\begin{array}{ccc:c} 1 & -1 & 1 & -6 \\ 0 & 1 & 0 & 3 \\ 0 & 0 & -3 & 15 \end{array}\right] \sim \left[\begin{array}{ccc:c} 1 & -1 & 1 & -6 \\ 0 & 1 & 0 & 3 \\ 0 & 0 & 1 & -5 \end{array}\right] \sim \left[\begin{array}{ccc:c} 1 & -1 & 0 & -1 \\ 0 & 1 & 0 & 3 \\ 0 & 0 & 1 & -5 \end{array}\right]$$

$$\sim \left[\begin{array}{ccc:c} 1 & 0 & 0 & 2 \\ 0 & 1 & 0 & 3 \\ 0 & 0 & 1 & -5 \end{array}\right]$$

Thus $a = 2$, $b = 3$, and $c = -5$, so the equation of the parabola is

$$\mathbf{y = 2x^2 + 3x - 5}$$

■

PROBLEM SET 9.5

A

Given the matrices in Problems 1–6, perform elementary row operations to obtain a 1 in the row 1, column 1 position.

1. $\left[\begin{array}{ccc:c} 3 & 1 & 2 & 1 \\ 0 & 2 & 4 & 5 \\ 1 & 3 & -4 & 9 \end{array}\right]$

2. $\left[\begin{array}{ccc:c} -2 & 3 & 5 & 9 \\ 1 & 0 & 2 & -8 \\ 0 & 1 & 0 & 5 \end{array}\right]$

3. $\left[\begin{array}{ccc:c} 2 & 4 & 10 & -12 \\ 6 & 3 & 4 & 6 \\ 10 & -1 & 0 & 1 \end{array}\right]$

4. $\left[\begin{array}{ccc:c} 5 & 20 & 15 & 6 \\ 7 & -5 & 3 & 2 \\ 12 & 0 & 1 & 4 \end{array}\right]$

5. $\left[\begin{array}{ccc:c} 5 & 6 & -3 & 4 \\ 4 & 1 & 9 & 2 \\ 7 & 6 & 1 & 3 \end{array}\right]$

6. $\left[\begin{array}{ccc:c} 4 & 8 & 5 & 9 \\ 3 & 2 & 1 & -4 \\ -2 & 5 & 0 & 1 \end{array}\right]$

Given the matrices in Problems 7–12, perform elementary row operations to obtain zeros under the 1 in the first column.

7. $\left[\begin{array}{ccc:c} 1 & 2 & -3 & 0 \\ 0 & 3 & 1 & 4 \\ 2 & 5 & 1 & 6 \end{array}\right]$

8. $\left[\begin{array}{ccc:c} 1 & 3 & -5 & 6 \\ -3 & 4 & 1 & 2 \\ 0 & 5 & 1 & 3 \end{array}\right]$

9. $\left[\begin{array}{rrr|r} 1 & 2 & 4 & 1 \\ -2 & 5 & 0 & 2 \\ -4 & 5 & 1 & 3 \end{array}\right]$ **10.** $\left[\begin{array}{rrr|r} 1 & 5 & 3 & 2 \\ 2 & 3 & -1 & 4 \\ 3 & 2 & 1 & 0 \end{array}\right]$

11. $\left[\begin{array}{rrrr|r} 1 & 4 & -1 & 3 & 3 \\ -3 & 4 & 6 & 4 & 0 \\ 5 & 1 & 9 & 1 & -2 \\ 1 & 0 & 2 & 0 & 0 \end{array}\right]$

12. $\left[\begin{array}{rrrr|r} 1 & 8 & 5 & 2 & -1 \\ 2 & 0 & 7 & 5 & -1 \\ 6 & -2 & 7 & 7 & 5 \\ 3 & 1 & 0 & 0 & 0 \end{array}\right]$

Given the matrices in Problems 13–18, perform elementary row operations to obtain a 1 in the second row, second column without changing the entries in the first column.

13. $\left[\begin{array}{rrr|r} 1 & 3 & 5 & 2 \\ 0 & 2 & 6 & -8 \\ 0 & 3 & 4 & 1 \end{array}\right]$ **14.** $\left[\begin{array}{rrr|r} 1 & 5 & -3 & 5 \\ 0 & 3 & 9 & -15 \\ 0 & 2 & 1 & 5 \end{array}\right]$

15. $\left[\begin{array}{rrr|r} 1 & 4 & -1 & 6 \\ 0 & 5 & 1 & 3 \\ 0 & 4 & 6 & 5 \end{array}\right]$ **16.** $\left[\begin{array}{rrr|r} 1 & 3 & -2 & 0 \\ 0 & 4 & 2 & 9 \\ 0 & 3 & 6 & 1 \end{array}\right]$

17. $\left[\begin{array}{rrr|r} 1 & 3 & -2 & 4 \\ 0 & 5 & 1 & 3 \\ 0 & 7 & 9 & 2 \end{array}\right]$ **18.** $\left[\begin{array}{rrr|r} 1 & 3 & -2 & 0 \\ 0 & 4 & 2 & 9 \\ 0 & 10 & -3 & 4 \end{array}\right]$

Given the matrices in Problems 19–21, perform elementary row operations to obtain a zero (or zeros) under the 1 in the second column without changing the entries in the first column.

19. $\left[\begin{array}{rrr|r} 1 & 5 & -3 & 2 \\ 0 & 1 & 4 & 5 \\ 0 & 3 & 4 & 2 \end{array}\right]$

20. $\left[\begin{array}{rrrr|r} 1 & 6 & -3 & 4 & 1 \\ 0 & 1 & 7 & 3 & 0 \\ 0 & 3 & 4 & 0 & -2 \\ 0 & -2 & 3 & 1 & 0 \end{array}\right]$

21. $\left[\begin{array}{rrrr|r} 1 & 7 & 6 & 6 & 2 \\ 0 & 1 & 9 & 2 & 1 \\ 0 & -4 & 6 & 1 & 1 \\ 0 & 5 & 8 & 10 & 3 \end{array}\right]$

Given the matrices in Problems 22–24, perform elementary row operations to obtain a 1 in the third row, third column without changing the entries in the first two columns.

22. $\left[\begin{array}{rrr|r} 1 & 3 & 4 & 5 \\ 0 & 1 & -3 & 6 \\ 0 & 0 & 5 & 10 \end{array}\right]$ **23.** $\left[\begin{array}{rrr|r} 1 & -3 & 4 & -5 \\ 0 & 1 & 3 & 6 \\ 0 & 0 & 8 & 12 \end{array}\right]$

24. $\left[\begin{array}{rrr|r} 1 & -2 & 5 & 6 \\ 0 & 1 & 9 & 1 \\ 0 & 0 & -3 & 5 \end{array}\right]$

Given the matrices in Problems 25–27, perform elementary row operations to obtain zeros above the 1 in the third column without changing the entries in the first or second columns.

25. $\left[\begin{array}{rrr|r} 1 & 3 & -1 & 5 \\ 0 & 1 & 2 & 6 \\ 0 & 0 & 1 & 4 \end{array}\right]$ **26.** $\left[\begin{array}{rrr|r} 1 & 6 & -3 & -2 \\ 0 & 1 & 4 & 5 \\ 0 & 0 & 1 & 3 \end{array}\right]$

27. $\left[\begin{array}{rrrr|r} 1 & 6 & 3 & 0 & -2 \\ 0 & 1 & -2 & 0 & 1 \\ 0 & 0 & 1 & 0 & 0 \\ 0 & 0 & 0 & 1 & -6 \end{array}\right]$

Given the matrices in Problems 28–30, perform elementary row operations to obtain a zero above the 1 in the second column and interpret the solution to the system if the variables are x, y, and z. For example,

$$\left[\begin{array}{rrr|r} 1 & 3 & 0 & 5 \\ 0 & 1 & 0 & 2 \\ 0 & 0 & 1 & 3 \end{array}\right] \sim \left[\begin{array}{rrr|r} 1 & 0 & 0 & -1 \\ 0 & 1 & 0 & 2 \\ 0 & 0 & 1 & 3 \end{array}\right]$$

so the solution is $(x, y, z) = (-1, 2, 3)$.

28. $\left[\begin{array}{rrr|r} 1 & 3 & 0 & 4 \\ 0 & 1 & 0 & -5 \\ 0 & 0 & 1 & 3 \end{array}\right]$ **29.** $\left[\begin{array}{rrr|r} 1 & -8 & 0 & 3 \\ 0 & 1 & 0 & 4 \\ 0 & 0 & 1 & -21 \end{array}\right]$

30. $\left[\begin{array}{rrr|r} 1 & -4 & 0 & 3 \\ 0 & 1 & 0 & -\frac{8}{3} \\ 0 & 0 & 1 & \frac{3}{5} \end{array}\right]$

Solve the systems given in Problems 31–45 by using the Gauss–Jordan method of solution.

31. $\begin{cases} 4x - 7y = -2 \\ -x + 2y = 1 \end{cases}$

32. $\begin{cases} 4x - 7y = -5 \\ 2y - x = 1 \end{cases}$

33. $\begin{cases} 8x + 6y = 17 \\ y - x = \frac{1}{2} \end{cases}$

34. $\begin{cases} 8x + 6y = 14 \\ 4y - 2x = 24 \end{cases}$

35. $\begin{cases} y = -\frac{2}{3}x + 3 \\ x - 6y = -3 \end{cases}$

36. $\begin{cases} y = \frac{1}{2}x + 1 \\ x + 3y = 18 \end{cases}$

37. $\begin{cases} x + y + z = 6 \\ 2x - y + z = 3 \\ x - 2y - 3z = -12 \end{cases}$

38. $\begin{cases} 2x - y + z = 3 \\ x - 3y + 2z = 7 \\ x - y - z = -1 \end{cases}$

39. $\begin{cases} x + y + z = 4 \\ x + 3y + 2z = 4 \\ x - 2y + z = 7 \end{cases}$

40. $\begin{cases} x + 2z = 13 \\ 2x + y = 8 \\ -2y + 9z = 41 \end{cases}$

41. $\begin{cases} x + 2z = 7 \\ x + y = 11 \\ -2y + 9z = -3 \end{cases}$

42. $\begin{cases} 4x + y + 2z = 7 \\ x + 2y = 0 \\ 3x - y - z = 7 \end{cases}$

43. $\begin{cases} 6x + y + 20z = 27 \\ x - y = 0 \\ y + z = 2 \end{cases}$

44. $\begin{cases} 2x - y + 4z = 13 \\ 3x + 6y = 0 \\ 2y - 3z = 3 + 3x \end{cases}$

45. $\begin{cases} 3x - 2y + z = 5 \\ 5x - 3y = 24 \\ 2y + z = -5 \end{cases}$

B

Solve the systems in Problems 46–54 by using the Gauss–Jordan method of solution.

46. $\begin{cases} 2x + 2y + 3z = 1 \\ 2x - z = -11 \\ 3y + 2z = 6 \end{cases}$

47. $\begin{cases} x + 2y + z = 1 \\ x - 3y - 2z = 2 \\ 3x - 2y + z = 3 \end{cases}$

48. $\begin{cases} 2x - y + z = 4 \\ 3x - 2y + 2z = 3 \\ x - y + 3z = 2 \end{cases}$

49. $\begin{cases} 2x + 10y - 6z = 28 \\ 5x - 3y + z = -14 \\ x + 5y - 3z = 14 \end{cases}$

50. $\begin{cases} x + 5y - 3z = 2 \\ 2x + 2y + z = -1 \\ 3x - y + 5z = 3 \end{cases}$

51. $\begin{cases} x + y + z = 1 \\ 2x - y + z = 2 \\ 3x + 2y - 3z = 3 \end{cases}$

52. $\begin{cases} x + 2y = 5 \\ z = 3 \\ w + x + 3y = 6 \\ 2w + 4x = 2 \end{cases}$

53. $\begin{cases} x + y + z + w = -2 \\ 2x + y + w = 2 \\ w + 3x + z = 5 \\ 2x + y + 3z = -5 \end{cases}$

54. $\begin{cases} 2x + y + z + w = 3 \\ x - y - z + 2w = -3 \\ x + 3y + 2z + 4w = -12 \\ x - y - z + w = 1 \end{cases}$

Suppose the equation of a certain parabola has the form $y = ax^2 + bx + c$. Find the equation of the parabola passing through the points given in Problems 55–57.

55. $(0, 5), (-1, 2), (3, 26)$

56. $(1, -2), (-2, -14), (3, -4)$

57. $(4, -4), (5, -5), (7, -1)$

C

58. ***Agriculture*** In order to control a certain type of crop disease, it is necessary to use 23 gal of chemical A and 34 gal of chemical B. The dealer can order commercial spray I, each container of which holds 5 gal of chemical A and 2 gal of chemical B, and commercial spray II, each container of which holds 2 gal of chemical A and 7 gal of chemical B. How many containers of each type of commercial spray should be used to attain exactly the right proportion of chemicals needed?

59. ***Business*** In order to manufacture a certain alloy, it is necessary to use 33 kg (kilograms) of metal A and 56 kg of metal B. It is cheaper for the manufacturer if she buys and mixes an alloy, each bar of which contains 3 kg of metal A and 5 kg of metal B, along with another alloy, each bar of which contains 4 kg of metal A and 7 kg of metal B. How much of the two alloys should she use to produce the alloy desired?

60. ***Business*** A candy maker mixes chocolate, milk, and almonds to produce three kinds of candy—I, II, and III—with the following proportions:

I: 7 lb chocolate, 5 gal milk, 1 oz almonds
II: 3 lb chocolate, 2 gal milk, 2 oz almonds
III: 4 lb chocolate, 3 gal milk, 3 oz almonds

If 67 lb of chocolate, 48 gal of milk, and 32 oz of almonds are available, how much of each kind of candy can be produced?

61. ***Business*** Using the data from Problem 60, how much of each type of candy can be produced with 62 lb of chocolate, 44 gal of milk, and 32 oz of almonds?

9.6 SYSTEMS OF INEQUALITIES

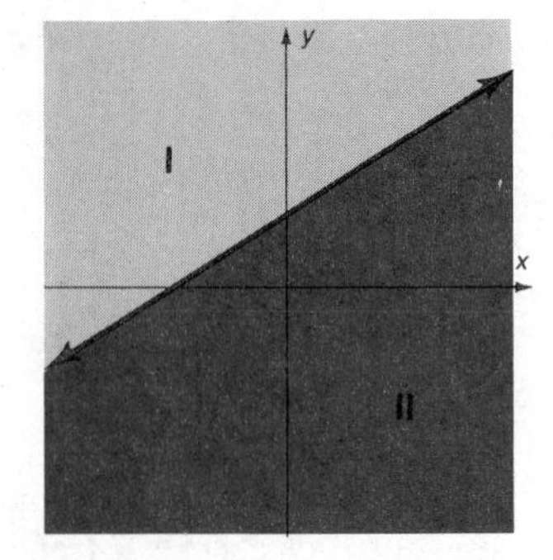

Figure 9.5 Half-planes

In previous sections we have discussed the simultaneous solution of a system of equations. In this section we will discuss the simultaneous solution of a **system of linear inequalities**. Let us begin by considering some preliminary ideas. The **solution** for an inequality in x and y is defined as an ordered pair that when substituted for (x, y) makes the inequality true. To **solve an inequality** means to find the set of all solutions, which is usually an infinite set. We therefore usually represent the solution graphically so that the **graph of the inequality** is the graph of all solutions of that inequality.

Graphing a linear inequality is similar to graphing a linear equation. A line divides the plane into three regions as shown in Figure 9.5. Regions I and II in Figure 9.5 are called **open half-planes**. Thus the three regions determined by the line are the open half-planes labeled I and II and the set of points on the line. The line is called the **boundary** of each open half-plane. An open half-plane, along with its boundary, is called a **closed half-plane**.

Every linear inequality with one or two variables determines an associated linear equation that is the boundary for the solution set of the inequality. For example, $2x + 3y \leq 12$ has a boundary line $2x + 3y = 12$. To graph the solution set of an inequality, begin by graphing the boundary line as shown in Figure 9.6. Next decide whether the solution is half-plane I or half-plane II. To do this, **choose *any* point not on the boundary**. For example, choose the point $(0, 0)$ in Figure 9.6—this choice, if not on the boundary, is usually the best because of the ease of the arithmetic involved. Notice from Figure 9.6 that the point $(0, 0)$ is in half-plane I. If $(0, 0)$ makes the inequality true, then the solution is the half-plane containing $(0, 0)$; if $(0, 0)$ makes the inequality false, then the solution set is the half-plane not containing $(0, 0)$. Checking by substituting $(0, 0)$ into $2x + 3y \leq 12$, you have

$$2(\mathbf{0}) + 3(\mathbf{0}) \leq 12 \qquad \text{True}$$

Therefore the solution set is the area shown as half-plane I. This is the shaded portion of Figure 9.7. Notice that the solution set is the closed half-plane I, since it includes the boundary. This is shown on the graph by using a solid line for the boundary. If the boundary is not included (when the inequality symbols are $<$ or $>$), a dashed line is used to indicate the boundary.

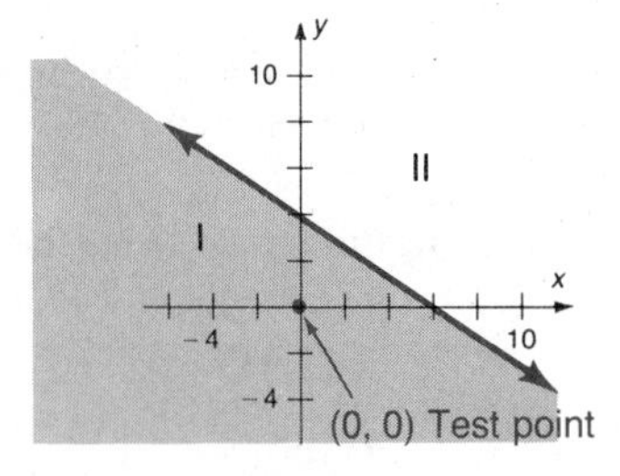

Figure 9.6

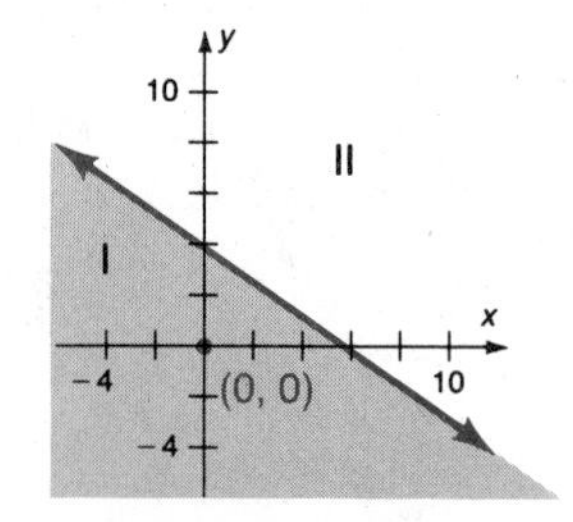

Figure 9.7 Graph of $2x + 3y \leq 12$

EXAMPLE 1 Graph $150x - 75y > 1875$.

Solution Graph the boundary line $150x - 75y = 1875$:

$$y = 2x - 25$$

Check some point, say $(0, 0)$, not on the boundary and test:

$$150(\mathbf{0}) - 75(\mathbf{0}) > 1875 \qquad \text{False}$$

The solution is the half-plane not including $(0, 0)$. It is the shaded portion of Figure 9.8.

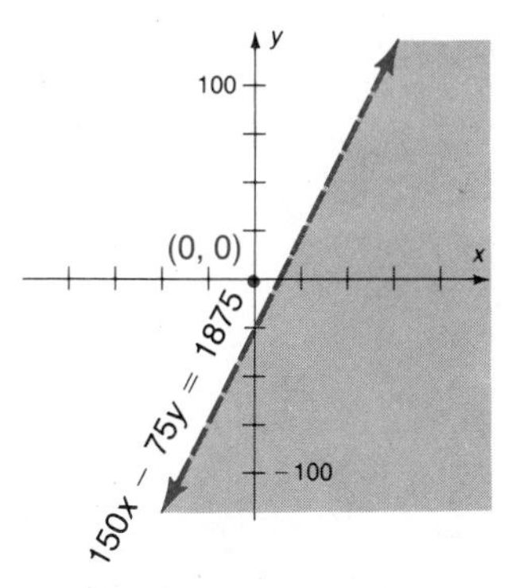

Figure 9.8 Graph of $150x - 75y > 1875$ ■

Many curves divide the plane into two regions with the curve serving as a boundary. The test point method illustrated for lines works efficiently in many different settings, as illustrated by Examples 2 and 3.

EXAMPLE 2 Graph $y \geq |x + 3| + 2$.

Solution Begin by graphing the boundary:

$$y = |x + 3| + 2$$

We recognize this as the *form* $y = |x|$ translated to the point $(-3, 2)$. This is drawn as a solid curve in Figure 9.9. Now this curve divides the plane into two regions, so plot a test point $(0, 0)$ and check the truth or falsity in the given inequality:

$$y \geq |x + 3| + 2$$
$$\mathbf{0} \geq |\mathbf{0} + 3| + 2$$
$$0 \geq 5 \qquad \text{False}$$

Since it is false, shade the region *not* containing the test point $(0, 0)$ as shown in Figure 9.9.

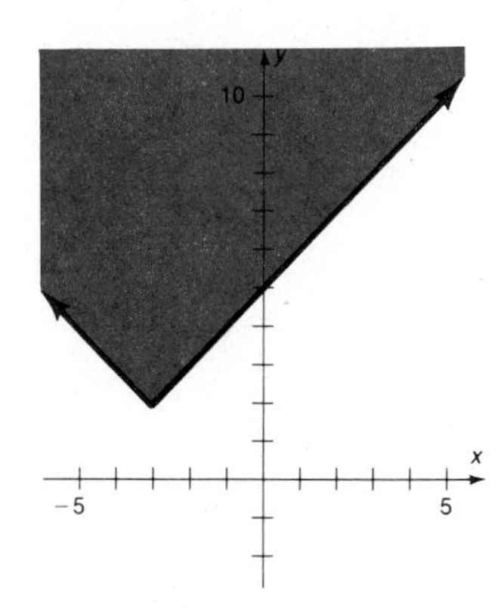

Figure 9.9 Graph of $y \geq |x + 3| + 2$ ■

EXAMPLE 3 Sketch the graph $y > x^2 + 2x + 3$.

Solution Consider the associated equation

$$y = x^2 + 2x + 3$$

or $y - 2 = (x + 1)^2$, which is a parabola that has vertex at $(-1, 2)$ and opens upward. Sketch this boundary equation as in Figure 9.10. In this case, the boundary is dashed because it is not included. Plot a test point $(0, 0)$ and check

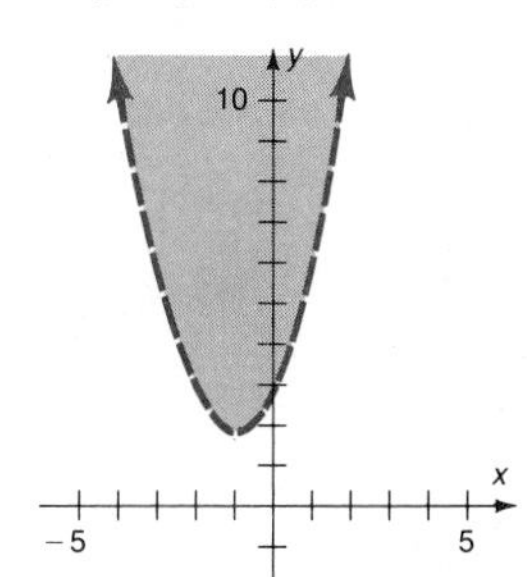

Figure 9.10 Graph of $y > x^2 + 2x + 3$

the inequality:

$$y > x^2 + 2x + 3$$
$$0 > 0^2 + 2 \cdot 0 + 3 \qquad \text{False}$$

Since it is false, (0, 0) is not in the solution set. Shade the appropriate region on the graph. ■

The solution of a **system of inequalities** refers to the intersection of the solutions of the individual inequalities in the system. The procedure for graphing the solution of a system of inequalities is to graph the solution of the individual inequalities and then shade in the intersection.

EXAMPLE 4 Graph the solution of the system

$$\begin{cases} y > -20x + 100 \\ x \geq 0 \\ y \geq 0 \\ x \leq 8 \end{cases}$$

Solution *Step 1:* Graph $y > -20x + 100$.

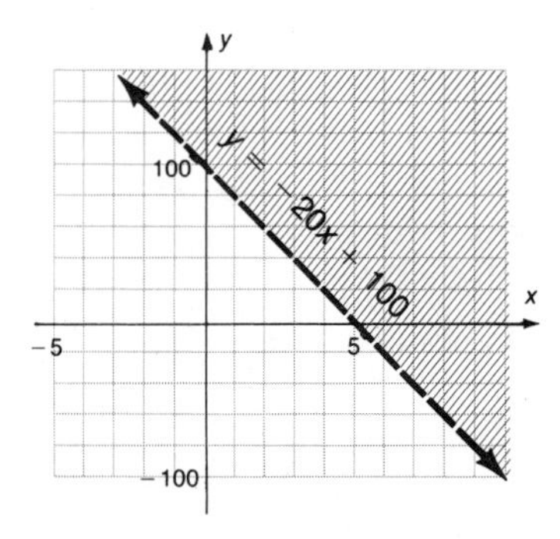

Step 2: Graph $x \geq 0$.

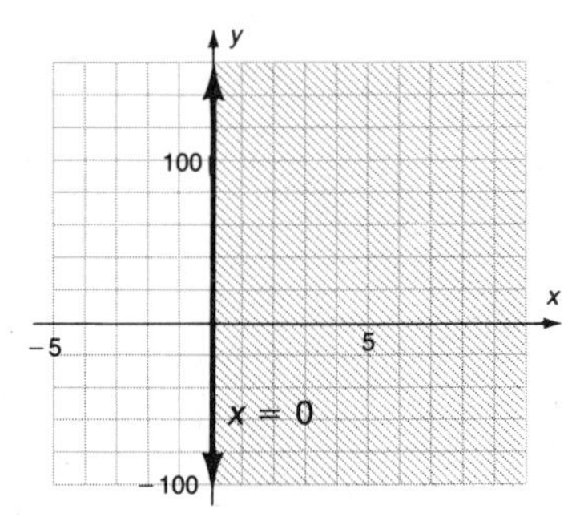

Step 3: Combine Steps 1 and 2. The intersection of the two graphs is the part that is shaded darker.

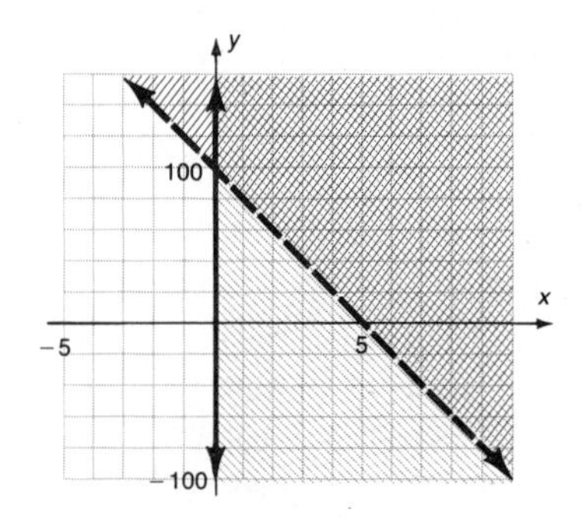

Step 4: Graph $y \geq 0$.

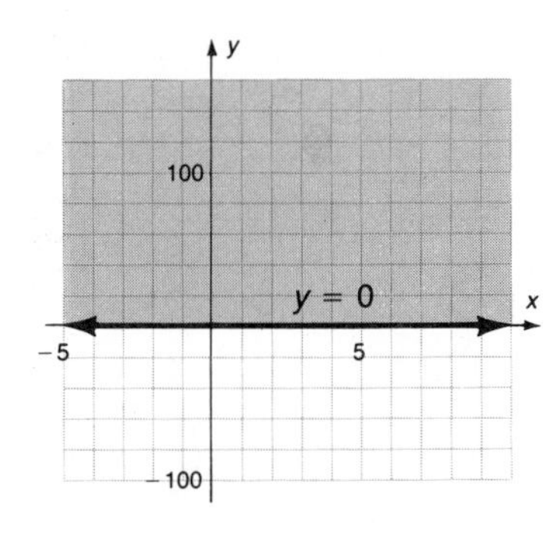

Step 5: Look at the intersection of Steps 3 and 4.

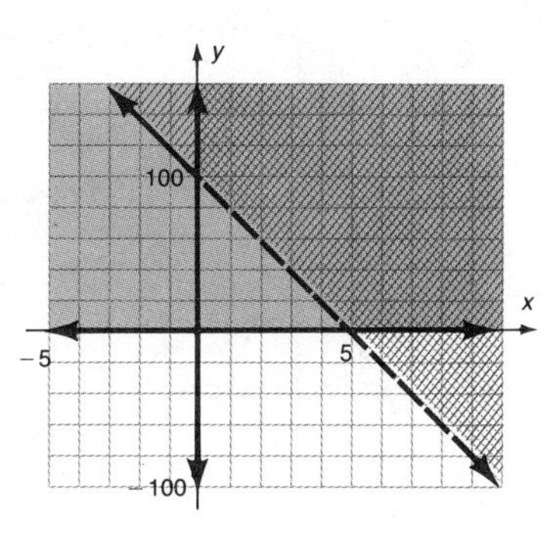

Step 6: Graph $x \leq 8$.

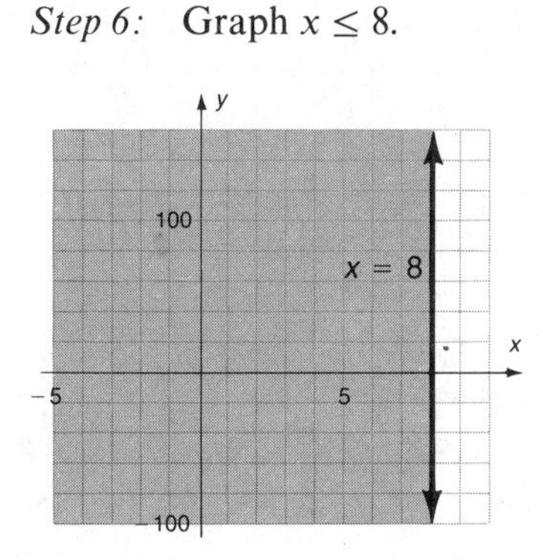

Step 7: Look at the intersection of Steps 5 and 6.

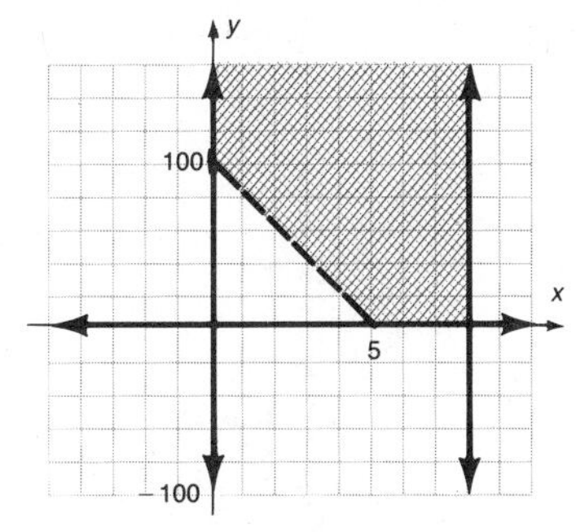

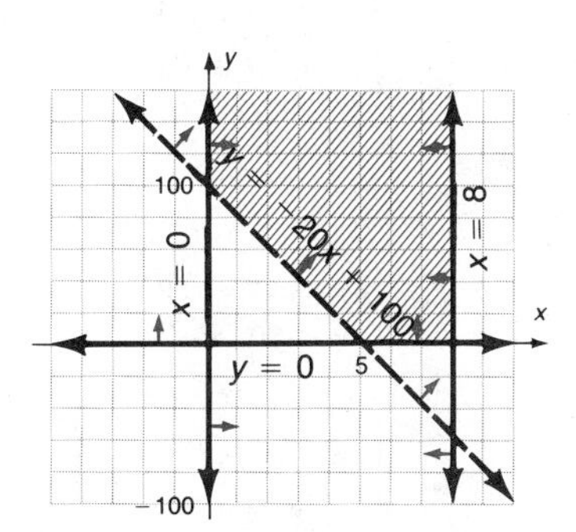

Figure 9.11 Graph of the solution of a system of inequalities

The way we have illustrated the steps of this solution was for explanation only; your work will not look at all like this. In practice, show the individual steps with little arrows on the boundaries and then shade in the intersection only after you have drawn in all of the boundaries. This device replaces the use of a lot of shading, which can be confusing if there are many inequalities in the system. The work for Example 4 is shown in Figure 9.11. ■

EXAMPLE 5 Graph the solution of the system

$$\begin{cases} 2x + y \leq 3 \\ x - y > 5 \\ x \geq 0 \\ y > -10 \end{cases}$$

Solution The graphs of the individual inequalities and their intersections are shown in Figure 9.12. Notice the use of arrows to show the solutions of the individual inequalities.

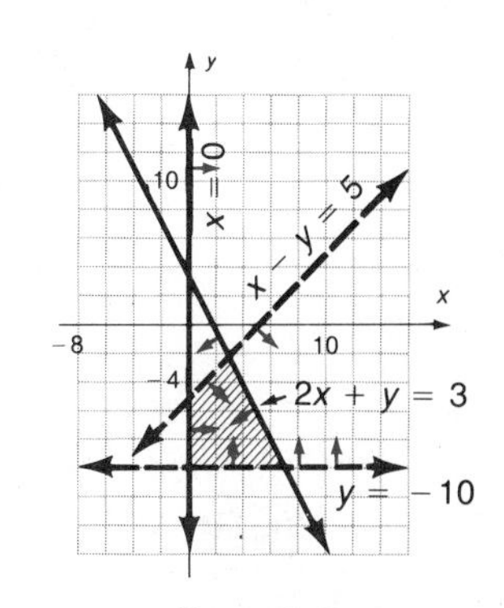

Figure 9.12 ■

PROBLEM SET 9.6

A

Graph the solution for each of the linear inequalities given in Problems 1–30.

1. $y \geq 2x - 3$
2. $y \geq 3x + 2$
3. $y \geq 4x + 5$
4. $y \leq \frac{1}{2}x + 2$
5. $y \leq \frac{2}{3}x - 4$
6. $y \leq \frac{4}{5}x - 3$
7. $x - y > 3$
8. $2x + y < -3$
9. $3x + y < 4$
10. $x + 2y > 4$
11. $x - 3y > 12$
12. $3x - 4y > 8$
13. $y \leq 6$
14. $x \geq 2$
15. $y < -2$
16. $y > 0$
17. $x < 0$
18. $x < 8$
19. $y \geq |x|$
20. $y < |x|$
21. $y < |x + 1|$
22. $y - 3 > |x + 1|$
23. $y + 1 \leq |x - 3|$
24. $y < |x - 1| + 2$
25. $y \leq x^2$
26. $y \geq (x - 3)^2$
27. $y - 1 \geq (x + 1)^2$
28. $y < x^2 - 6x + 7$
29. $x^2 - 4x + y + 5 < 0$
30. $x^2 + 6x + y + 7 > 0$

B

Graph the solution of each system given in Problems 31–57.

31. $\begin{cases} x \geq 0 \\ y \leq 0 \end{cases}$

32. $\begin{cases} x \geq 0 \\ y \geq 0 \end{cases}$

33. $\begin{cases} x \leq 0 \\ y \leq 0 \end{cases}$

34. $\begin{cases} x \geq 0 \\ y \geq 0 \\ x < 8 \\ y < 5 \end{cases}$

35. $\begin{cases} x \geq 0 \\ y \geq 0 \\ x < 5 \\ y < 6 \end{cases}$

36. $\begin{cases} x \geq 0 \\ y \geq 0 \\ x < 500 \\ y < 1000 \end{cases}$

37. $\begin{cases} 2x + y > 3 \\ 3x - y < 2 \end{cases}$

38. $\begin{cases} y \leq 3x - 4 \\ y \geq -2x + 5 \end{cases}$

39. $\begin{cases} 3x - 2y \geq 6 \\ 2x + 3y \leq 6 \end{cases}$

40. $\begin{cases} y - 5 \leq 0 \\ y \geq 0 \end{cases}$

41. $\begin{cases} x - 10 \leq 0 \\ x \geq 0 \end{cases}$

42. $\begin{cases} y - 25 \leq 0 \\ y \geq 0 \end{cases}$

43. $\begin{cases} -10 \leq x \\ x \leq 6 \\ 3 < y \\ y < 8 \end{cases}$

44. $\begin{cases} -5 < x \\ 3 \geq x \\ 5 > y \\ -2 \leq y \end{cases}$

45. $\begin{cases} -5 < x \\ x \leq 2 \\ -4 \leq y \\ y < 9 \end{cases}$

46. $\begin{cases} x \geq 0 \\ y \geq 0 \\ x + y \leq 9 \\ 2x - 3y \geq -6 \\ x - y \leq 3 \end{cases}$

47. $\begin{cases} x \geq 0 \\ y \geq 0 \\ x + y \leq 8 \\ y \leq 4 \\ x \leq 6 \end{cases}$

48. $\begin{cases} x \geq 0 \\ y \geq 0 \\ 2x + y \geq 8 \\ y \leq 5 \\ x - y \leq 2 \\ 3x - y \geq 5 \end{cases}$

49. $\begin{cases} 2x - 3y + 30 \geq 0 \\ 3x - 2y + 20 \leq 0 \\ x \leq 0 \\ y \geq 0 \end{cases}$

50. $\begin{cases} 2x + 3y \leq 30 \\ 3x + 2y \geq 20 \\ x \geq 0 \\ y \geq 0 \end{cases}$

51. $\begin{cases} x + y - 10 \leq 0 \\ x + y + 4 \geq 0 \\ x - y \leq 6 \\ y - x \leq 4 \end{cases}$

52. $\begin{cases} 3y = x^2 + 21 \\ y \leq 10 \end{cases}$

53. $\begin{cases} 3y + 16x = x^2 + 85 \\ y \leq 10 \end{cases}$

54. $\begin{cases} 3y + x^2 = 2x + 25 \\ -1 \leq x \leq 2 \end{cases}$

55. $\begin{cases} 4y + 6x + 3 = x^2 \\ |x - 3| \leq 4 \end{cases}$

56. $\begin{cases} 3y + x^2 + 23 = 14x \\ |x - 7| \leq 2 \end{cases}$

57. $\begin{cases} 3x^2 + 14 = 2y + 12x \\ |x - 2| \leq 2 \end{cases}$

C

In Problems 58–63 sketch the systems of inequalities.

58. $\begin{cases} y \geq x^2 \\ 5x - 4y + 26 \geq 0 \end{cases}$

59. $\begin{cases} y < 4 - x^2 \\ y > \frac{1}{2}x \end{cases}$

60. $\begin{cases} y \geq \frac{1}{2}(x - 2)^2 \\ y - 4 < \frac{1}{4}(x - 2)^2 \end{cases}$

61. $\begin{cases} y - 2 \geq \frac{1}{3}(x - 3)^2 \\ y - 4 \leq -\frac{1}{3}(x - 3)^2 \end{cases}$

62. $\begin{cases} y + 2 \geq \frac{1}{8}x^2 \\ x - \frac{3}{2} \geq \frac{1}{2}(y + 1)^2 \\ y + 1 \geq \frac{1}{2}(x - \frac{3}{2})^2 \end{cases}$

63. $\begin{cases} x + 1 \geq (y - 5)^2 \\ y - 3 \geq (x + 1)^2 \end{cases}$

9.7 LINEAR PROGRAMMING*

This section introduces you to the topic of linear programming. Linear programming is a branch of mathematics that can be applied when you are interested in maximizing or minimizing a linear function. It was developed during World War II, and today it is used in a variety of applications, such as maximizing profits, minimizing costs, finding the most efficient shipping schedules, minimizing waste, securing the proper mix of ingredients, controlling inventories, and finding the most efficient assignment of personnel. In this section we apply the ideas of systems of inequalities to solve certain types of linear programming problems graphically. In later courses you will learn how to apply the processes of linear programming to more advanced applications.

We begin with some terminology. A set of points S is called a **convex set** if, for *any* two points P and Q in S, the entire segment $\overline{PQ}$ is in S. Some examples are shown in Figure 9.13.

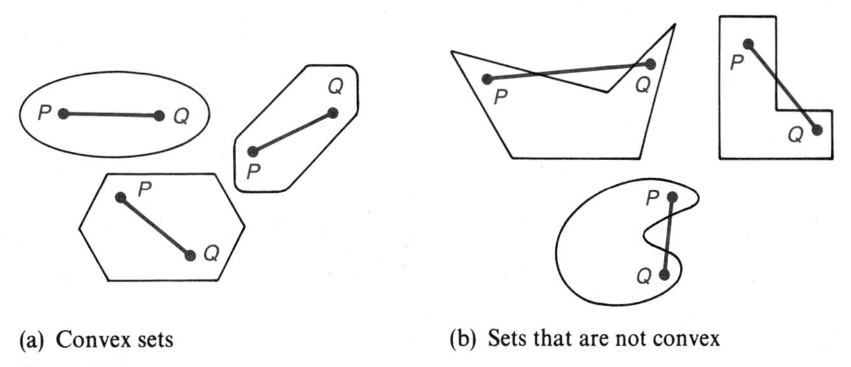

(a) Convex sets (b) Sets that are not convex

Figure 9.13

The applications of linear programming can be generalized as follows. Suppose you are given a function $c_1x + c_2y$ in two variables x and y with constants c_1 and c_2, and the problem is to find the maximum or minimum value of this function. This function is called the **objective function.** This objective function is subject to certain limitations called **constraints.** In linear programming these constraints are specified by a system of linear inequalities whose solution forms a convex set S. Each point of the set S is called a **feasible solution,** and a point at which the objective function takes on a maximum or a minimum value is called an **optimum solution.** The proof of the following **Linear Programming Theorem** is beyond the scope of this course, so it is stated without proof.

---

* Optional section

Linear Programming Theorem

> A linear function in two variables,
>
> $$c_1x + c_2y$$
>
> defined over a convex set S whose sides are line segments, takes on its maximum value at a corner point of S and its minimum value at a corner point of S.*

In summary, to solve a linear programming problem:

1. Find the objective function (the quantity to be maximized or minimized).
2. Graph the constraints defined by a system of linear inequalities; the simultaneous solution is called the set S.
3. Find the corners of S; this may require the solution of a system of two equations with two unknowns, one for each corner.
4. Find the value of the objective function for the coordinates of each corner point. The largest value is the maximum; the smallest value is the minimum.

EXAMPLE 1 Maximize $C = 4x + 5y$, subject to:

$$\begin{cases} 2x + 5y \leq 25 \\ 6x + 5y \leq 45 \\ x \geq 0 \\ y \geq 0 \end{cases}$$

Solution The constraints give a set of feasible solutions; graph this system of inequalities as was illustrated in Section 9.6. This region is shown in Figure 9.14. To solve the linear

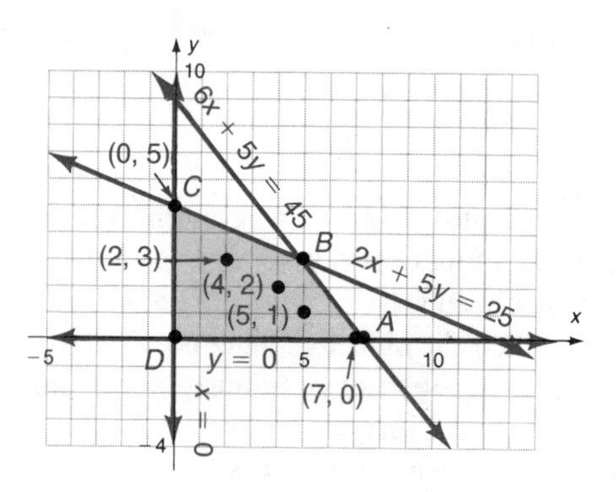

Figure 9.14

* If S is unbounded, there may or may not be an optimum value, but if there is, then it must occur at a corner point. Consideration of this type of problem is beyond the scope of this book.

programming problem we must now find the feasible solution which makes the objective function as large as possible. Consider the following table of possible solutions:

Feasible Solution (A point in the solution set of the system)	Objective Function $C = 4x + 5y$
$(2, 3)$	$4(2) + 5(3) = 8 + 15 = 23$
$(4, 2)$	$4(4) + 5(2) = 16 + 10 = 26$
$(5, 1)$	$4(5) + 5(1) = 20 + 5 = 25$
$(7, 0)$	$4(7) + 5(0) = 28 + 0 = 28$
$(0, 5)$	$4(0) + 5(5) = 0 + 25 = 25$

The point from the table that makes the objective function the largest is $(7, 0)$. But is this the largest for all feasible solutions? How about $(6, 1)$? or $(5, 3)$? The Linear Programming Theorem tells us that the maximum value will occur at a corner point. The corner points are labeled A, B, C, and D on Figure 9.14. Some corner points can usually be found by inspection. In this example we see $\boldsymbol{D = (0, 0)}$ and $\boldsymbol{C = (0, 5)}$. Other corner points may require some work with the boundary *lines*. Use the equations of the boundaries, not the inequalities giving the regions, to find the points of intersection.

Point A: System $\begin{cases} y = 0 \\ 6x + 5y = 45 \end{cases}$

Solve by substitution: $6x + 5(0) = 45$

$$x = \frac{45}{6} = \frac{15}{2} \qquad \boldsymbol{A = \left(\frac{15}{2}, 0\right)}$$

Point B: System $\begin{cases} 2x + 5y = 25 \\ 6x + 5y = 45 \end{cases}$

Solve by adding: $\begin{cases} -2x - 5y = -25 \\ 6x + 5y = 45 \end{cases}$

$$4x = 20$$
$$x = 5$$

If $x = 5$, then $2(5) + 5y = 25$

$$5y = 15$$
$$y = 3 \qquad \boldsymbol{B = (5, 3)}$$

The corner points are: $(0, 0)$, $(0, 5)$, $(\frac{15}{2}, 0)$, and $(5, 3)$. The maximum value of $C = 4x + 5y$ will occur at one of these points. Compare the following table with the previous one:

Corner Points	Objective Function $C = 4x + 5y$	
$(0, 0)$	$4(0) + 5(0) = 0$	← Minimum
$(0, 5)$	$4(0) + 5(5) = 25$	
$(\frac{15}{2}, 0)$	$4(\frac{15}{2}) + 5(0) = 30$	
$(5, 3)$	$4(5) + 5(3) = 35$	← Maximum

Look for the maximum value of C on *this* list; it is 35 so **the maximum value of C is 35.** ■

EXAMPLE 2 A farmer has 100 acres on which to plant two crops: corn or wheat. To produce these crops, there are certain expenses:

	Item	Cost per Acre
Corn:	seed	\$ 12
	fertilizer	\$ 58
	planting/care/harvesting	\$ 50
	TOTAL	\$120
Wheat:	seed	\$ 40
	fertilizer	\$ 80
	planting/care/harvesting	\$ 90
	TOTAL	\$210

After the harvest, the farmer must store the crops awaiting proper market conditions. Each acre yields an average of 110 bushels of corn or 30 bushels of wheat. The limitations of resources are as follows:

Available capital:	\$15,000
Available storage facilities:	4000 bushels

If the net profit (after all expenses have been subtracted) per bushel of corn is \$1.30 and for wheat is \$2.00, how should the farmer plant the 100 acres to maximize the profits?

Solution First, you might try to solve this problem using your intuition. If you plant all 100 acres with wheat the production is $30 \times 100 = 3000$ bushels for a net profit of $3000 \times 2 = \$6000$. But the farmer could not plant the entire 100 acres because wheat costs \$210 per acre for a total of \$21,000 (210×100) and there is only \$15,000 available. On the other hand, if the farmer plants 100 acres of corn the total cost is $\$120 \times 100 = \$12{,}000$. For this option, the net profit is $110 \times \$1.30 \times 100 = \$14{,}300$, but the production is $110 \times 100 = 11{,}000$ bushels, which exceeds the available storage capacity of 4000 bushels. Clearly a mix is necessary.

To formulate a mathematical model, begin by letting

$x =$ number of acres to be planted in corn
$y =$ number of acres to be planted in wheat

There are certain limitations or constraints.

$\mathbf{x \geq 0}$ The number of acres planted cannot be negative. *These constraints will apply in almost every model even though they are not explicitly stated as*

$\mathbf{y \geq 0}$ *part of the given problem.*

$\mathbf{x + y \leq 100}$ The amount of available land is 100 acres. Why not $x + y = 100$? It might be more profitable for the farmer to leave some land out of production. That is, it is not *necessary* to plant all the land.

$120x =$ expenses for planting the corn
$210y =$ expenses for planting the wheat

$\mathbf{120x + 210y \leq 15{,}000}$ The total expenses cannot exceed \$15,000; this is the *available capital.*

$110x =$ yield of acreage planted in corn
$30y =$ yield of acreage planted in wheat

$\mathbf{110x + 30y \leq 4000}$ The total yield cannot exceed the storage capacity of 4000 bushels.

Summary of constraints (in boldface above):

$$\begin{cases} x \geq 0 \\ y \geq 0 \\ x + y \leq 100 \\ 120x + 210y \leq 15{,}000 \\ 110x + 30y \leq 4000 \end{cases}$$

Now, let $P =$ total profit. The farmer wants to maximize the profit, P.

$$\begin{aligned} \text{profit from the corn} &= \text{value} \times \text{amount} \\ &= 1.30 \times 110x \\ &= 143x \end{aligned}$$

$$\begin{aligned} \text{profit from the wheat} &= \text{value} \times \text{amount} \\ &= 2.00 \times 30y \\ &= 60y \end{aligned}$$

$$\begin{aligned} P &= \text{profit from the corn} + \text{profit from the wheat} \\ &= 143x + 60y \end{aligned}$$

The linear programming model is stated as follows:

Maximize: $P = 143x + 60y$
Subject to: $\begin{cases} x \geq 0 \\ y \geq 0 \\ x + y \leq 100 \\ 120x + 210y \leq 15{,}000 \\ 110x + 30y \leq 4000 \end{cases}$

First graph the set of feasible solutions by graphing the system of inequalities (see Figure 9.15).

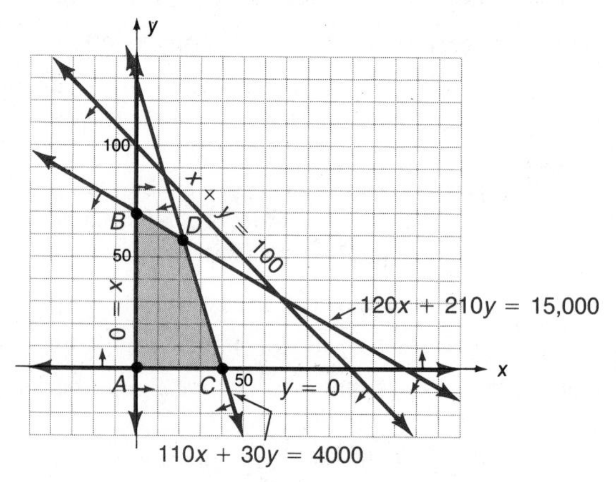

Figure 9.15

Next, find the corner points:

$\boldsymbol{A = (0, 0)}$ By inspection

B: $\begin{cases} 120x + 210y = 15{,}000 \\ x = 0 \end{cases}$

$$120(0) + 210y = 15{,}000$$
$$y = \frac{15{,}000}{210}$$
$$= \frac{500}{7}$$

$$\boldsymbol{B = \left(0, \frac{500}{7}\right)}$$

C: $\begin{cases} 110x + 30y = 4000 \\ y = 0 \end{cases}$

$$110x + 30(0) = 4000$$
$$110x = 4000$$
$$x = \frac{400}{11}$$

$$\boldsymbol{C = \left(\frac{400}{11}, 0\right)}$$

D: $-7\begin{cases} 110x + 30y = 4000 \\ 120x + 210y = 15{,}000 \end{cases}$ $\begin{cases} -770x - 210y = -28{,}000 \\ 120x + 210y = 15{,}000 \end{cases}$

$$-650x = -13{,}000$$
$$x = 20$$
$$110(20) + 30y = 4000$$
$$30y = 1800$$
$$y = 60$$

$$\boldsymbol{D = (20, 60)}$$

Use the Linear Programming Theorem and check the corner points.

Corner Point	Objective Function $P = 143x + 60y$	
$(0, 0)$	$143(0) + 60(0) = 0$	← Minimum value
$\left(0, \frac{500}{7}\right)$	$143(0) + 60\left(\frac{500}{7}\right) \approx 4286$	
$\left(\frac{400}{11}, 0\right)$	$143\left(\frac{400}{11}\right) + 60(0) = 5200$	
$(20, 60)$	$143(20) + 60(60) = 6460$	← Maximum value

The maximum value of P is at $(20, 60)$. **This means to maximize profit the farmer should plant 20 acres in corn, plant 60 acres in wheat, and leave 20 acres unplanted.** ■

Notice from the graph in Example 2 that some of the constraints could be eliminated from the problem and everything else would remain unchanged. For example, the boundary $x + y = 100$ was not necessary in finding the maximum value of P. Such a condition is said to be a **superfluous constraint**. It is not uncommon to have superfluous constraints in a linear programming problem. Suppose, however, that the farmer in Example 2 contracted to have the grain stored at a neighboring farm and now the contract calls for *at least* 4000 bushels to be stored. This change from $110x + 30y \leq 4000$ to $110x + 30y \geq 4000$ *now* makes the condition $x + y \leq 100$ important to the solution of the problem. Therefore, you must be careful about superfluous constraints even though they do not affect the solution at the present time.

The next example is solved more succinctly to show you the way your work will probably look.

EXAMPLE 3 The Sticky Widget Company makes two types of widgets: regular and deluxe. Each widget is produced at a station consisting of a machine and a person who finishes the widgets by hand. The regular widget requires 2 hr of machine time and 1 hr of finishing time. The deluxe widget requires 3 hr of machine time and 5 hr of finishing time. The profit on the regular widget is \$25; on the deluxe widget it is \$30. If the workday is 8 hr, how many of each type of widget should be produced at each station to maximize the profit?

Solution Let x = number of regular widgets produced

y = number of deluxe widgets produced

$$\begin{cases} \left.\begin{aligned} x &\geq 0 \\ y &\geq 0 \end{aligned}\right\} & \text{The number of widgets must be nonnegative.} \\ 2x + 3y \leq 8 & \text{The machine workday is not more than 8 hr.} \\ x + 5y \leq 8 & \text{The person's workday is not more than 8 hr.} \end{cases}$$

The set S is shown in color in Figure 9.16. The corner points are found by considering the piecewise simultaneous solution of the equations of the boundary lines. You can find these intersection points by inspection or by following the methods of this chapter.

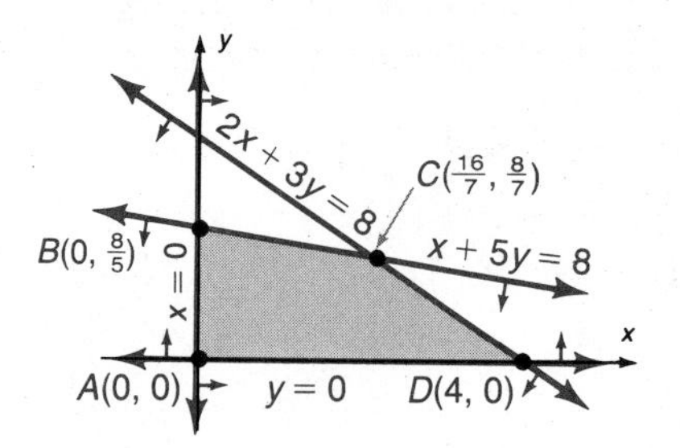

Figure 9.16

Corner Point	System to Solve	Solution	Objective Function $25x+30y$	Maximum or Minimum
A	$\begin{cases} x=0 \\ y=0 \end{cases}$	$(0, 0)$	$25(0)+30(0)=0$	minimum
B	$\begin{cases} x=0 \\ x+5y=8 \end{cases}$	$(0, \frac{8}{5})$	$25(0)+30(\frac{8}{5})=48$	
C	$\begin{cases} 2x+3y=8 \\ x+5y=8 \end{cases}$	$(\frac{16}{7}, \frac{8}{7})$	$25(\frac{16}{7})+30(\frac{8}{7})\approx 91.43$	
D	$\begin{cases} y=0 \\ 2x+3y=8 \end{cases}$	$(4, 0)$	$25(4)+30(0)=100$	maximum

Profits are maximized if only regular widgets are produced. **The company should produce 4 regular widgets per day at each station.** ■

EXAMPLE 4 The Sticky Widget Company of the preceding example changes its operating procedure so that the profit on a regular widget is \$20 and that on a deluxe widget is \$35. All the constraints of the previous example are the same. How many of each type of widget should be produced to maximize the profits?

Solution The objective function to be maximized is $20x + 35y$. Check the values at each of the corner points:

A: $20(0) + 35(0) = 0$ minimum

B: $20(0) + 35\left(\frac{8}{5}\right) = 56$

C: $20\left(\frac{16}{7}\right) + 35\left(\frac{8}{7}\right) \approx 85.71$ maximum

D: $20(4) + 35(0) = 80$

The profits are maximized if the company produces 16 regular widgets and 8 deluxe widgets at each station every 7 working days. ■

The method discussed here can be generalized to higher dimensions by means of more sophisticated methods for solving these linear programming problems. The general method for solving a linear programming problem is called the *simplex method* and is usually discussed in a course on linear programming.

PROBLEM SET 9.7

A

For Problems 1–6, decide whether the given point is a feasible solution for the constraints

$$\begin{cases} x \geq 0 \\ y \geq 0 \\ 3x + 2y \leq 10 \\ 2x + 4y \leq 8 \end{cases}$$

1. $(1, 3)$ **2.** $(2, 1)$ **3.** $(1, 2)$ **4.** $(-1, 4)$

5. $(2, 2)$ **6.** $(0, 4)$

For Problems 7–12, decide whether the given point is a corner point for the constraints

$$\begin{cases} x \geq 0 \\ y \geq 0 \\ 2x + 3y \geq 120 \\ 2x + y \geq 80 \end{cases}$$

7. $(0, 0)$ **8.** $(0, 80)$ **9.** $(80, 0)$ **10.** $(60, 0)$

11. $(30, 20)$ **12.** $(20, 30)$

Find the corner points for each set of feasible solutions given in Problems 13–24.

13. $\begin{cases} x \geq 0 \\ y \geq 0 \\ 2x + y \leq 12 \\ x + 2y \leq 9 \end{cases}$

14. $\begin{cases} x \geq 0 \\ y \geq 0 \\ 2x + 5y \leq 20 \\ 2x + y \leq 12 \end{cases}$

15. $\begin{cases} x \geq 0 \\ y \geq 0 \\ 3x + 2y \leq 12 \\ x + 2y \leq 8 \end{cases}$

16. $\begin{cases} x \geq 0 \\ y \geq 0 \\ x \leq 10 \\ y \leq 8 \\ 3x + 2y \geq 12 \end{cases}$

17. $\begin{cases} x \geq 0 \\ y \geq 0 \\ x + y \leq 8 \\ y \leq 4 \\ x \leq 6 \end{cases}$

18. $\begin{cases} x \geq 0 \\ y \geq 0 \\ x + y \geq 6 \\ -2x + y \geq -16 \\ y \leq 9 \end{cases}$

19. $\begin{cases} x \geq 0 \\ y \geq 0 \\ 3x + 2y \leq 8 \\ x + 5y \leq 8 \end{cases}$

20. $\begin{cases} x \geq 0 \\ y \geq 0 \\ x \leq 8 \\ y \geq 2 \\ x + y \leq 10 \\ x \leq 3y \end{cases}$

21. $\begin{cases} x \geq 0 \\ y \geq 0 \\ 10x + 5y \geq 200 \\ 2x + 5y \geq 100 \\ 3x + 4y \geq 120 \end{cases}$

22. $\begin{cases} x \geq 0 \\ y \geq 0 \\ x + y \leq 9 \\ 2x - 3y \geq -6 \\ x - y \leq 3 \end{cases}$

23. $\begin{cases} x \geq 0 \\ y \geq 0 \\ 2x + y \geq 8 \\ y \leq 5 \\ x - y \leq 2 \\ 3x - 2y \geq 5 \end{cases}$

24. $\begin{cases} x \geq 0 \\ y \geq 0 \\ 2x + y \geq 8 \\ x - 2y \leq 7 \\ x - y \geq -3 \\ x \leq 9 \end{cases}$

B

Find the optimum value for each objective function given in Problems 25–44.

25. Maximize $W = 30x + 20y$ subject to the constraints of Problem 13.

26. Maximize $P = 40x + 10y$ subject to the constraints of Problem 13.

27. Maximize $T = 100x + 10y$ subject to the constraints of Problem 14.

28. Maximize $V = 20x + 30y$ subject to the constraints of Problem 14.

29. Maximize $P = 100x + 100y$ subject to the constraints of Problem 15.

30. Maximize $T = 30x + 10y$ subject to the constraints of Problem 15.

31. Minimize $C = 24x + 12y$ subject to the constraints of Problem 16.

32. Minimize $K = 6x + 18y$ subject to the constraints of Problem 16.

33. Maximize $F = 2x - 3y$ subject to the constraints of Problem 17.

34. Maximize $P = 5x + y$ subject to the constraints of Problem 17.

35. Minimize $I = 90x + 20y$ subject to the constraints of Problem 18.

36. Minimize $T = 400x + 100y$ subject to the constraints of Problem 18.

37. Maximize $P = 23x + 46y$ subject to the constraints of Problem 18.

38. Minimize $C = 12x + 15y$ subject to the constraints of Problem 18.

39. Maximize $P = 6x + 3y$ subject to the constraints of Problem 22.

40. Maximize $T = x + 6y$ subject to the constraints of Problem 22.

41. Minimize $X = 5x + 3y$ subject to the constraints of Problem 23.

42. Minimize $A = 2x - 3y$ subject to the constraints of Problem 23.

43. Minimize $K = 140x + 250y$ subject to the constraints of Problem 24.

44. Minimize $C = 640x - 130y$ subject to the constraints of Problem 24.

Write a linear programming model including the objective function and the set of constraints for Problems 45–51. DO NOT SOLVE, but be sure to define all your variables.

45. ***Allocation of Resources in Manufacturing*** The Wadsworth Widget Company manufactures two types of widgets: regular and deluxe. Each widget is produced at a station consisting of a machine and a person who finishes the widgets by hand. The regular widget requires 3 hr of machine time and 2 hr of finishing time. The deluxe widget requires 2 hr of machine time and 4 hr of finishing time. The profit on the regular widget is \$25; on the deluxe widget it is \$30. If the workday is 8 hours, how many of each type of widget should be produced at each station per day in order to maximize the profit?

46. ***Diet Problem*** A convalescent hospital wishes to provide, at a minimum cost, a diet that has a minimum of 200 g of carbohydrates, 100 g of protein, and 120 g of fats per day. These requirements can be met with two foods:

Food	Carbohydrates	Protein	Fats
A	10 g	2 g	3 g
B	5 g	5 g	4 g

If food A costs \$0.29 per gram and food B, \$0.15 per gram, how many grams of each food should be purchased for each patient per day in order to meet the minimum requirements at the lowest cost?

47. ***Investment Application*** Brown Bros., Inc. is an investment company doing an analysis of the pension fund for a certain company. The fund has a maximum of \$10 million to invest in two places: no more than \$8 million in stocks yielding 12%, and at least \$2 million in long-term bonds yielding 8%. The bond-to-stock investment ratio cannot be more than 3 to 1. How should Brown Bros. advise their client so that the investments yield the maximum yearly return?

48. ***Office Management*** An office manager decides to purchase computers for each office worker. MAC computers cost \$900 each, will provide 3MB storage capability, and will take up 5 sq ft of desk space. IBM PC computers will cost \$1200 each, will provide 4MB storage, and will take up 3 sq ft of desk space. Proper utilization of space requires that a total of not more than 165 sq ft be used for this new equipment. If there is \$36,000 available to spend on computers, how many of each type of computer should the office manager purchase if she wishes to maximize storage capacity?

49. ***Allocation of Resources*** Foley's Motel has 200 rooms and a restaurant that seats 50 people. Experience shows that 40% of the commercial guests and 20% of the other guests eat in the restaurant. Suppose there is a net profit of \$20 per day from each commercial guest and \$24 per day from each other guest. Find the number of commercial and other guests needed in order to maximize the net profits assuming exactly one guest per room and that the capacity of the restaurant will not be exceeded.

50. ***Operating Costs*** Karlin Enterprises manufactures two electronic games. Standing orders require that at least 24,000 space battle and 5000 football games be produced. The company has two factories: The Gainesville plant can produce 600 space battle games and 100 football games per day; the Sacramento plant can produce 300 space battle games and 100 football games per day. If the Gainesville plant costs \$20,000 per day to operate and the Sacramento factory costs \$15,000 per day, find the number of days per month each factory should operate to minimize the cost. (Assume that each month has 30 days.)

51. ***Production Problem*** The Alco Company manufactures two products, Alpha and Beta. Each product must pass through two processing operations, and all materials are introduced at the first operation. Alco may produce either one product exclusively or various combinations of both products subject to the following constraints:

	First Process	Second Process	Profit per Unit
Hours required to produce one unit of:			
Alpha	1 hr	1 hr	$5.00
Beta	3 hr	2 hr	$8.00
Total capacity in hours per day:	1200 hr	1000 hr	

A shortage of technical labor has limited Alpha production to no more than 700 units per day. There are no constraints on the production of Beta other than the hour constraints in the schedule shown above. How many of each product should be manufactured in order to maximize the profit?

Solve the linear programming problems in Problems 52–60.

52. ***Allocation of Resources in Production*** Suppose the net profit per bushel of corn in Example 2 increased to $2.00 and the net profit per bushel of wheat dropped to $1.50. Maximize the profit if the other conditions in the example remain the same.

53. ***Allocation of Resources in Production*** Suppose the farmer in Example 2 contracted to have the grain stored at a neighboring farm and the contract calls for at least 4000 bushels to be stored. How many acres should be planted in corn and how many in wheat to maximize profits if the other conditions in Example 2 remain the same?

54. ***Allocation of Resources in Production*** A farmer has 500 acres on which to plant two crops, corn and wheat. It costs $120 per acre to produce corn and $60 per acre to produce wheat and there is $24,000 available to pay for this year's production. If the yields per acre are 100 bushels of corn and 40 bushels of wheat and the farmer has contracted to store at least 18,000 bushels, how much should the farmer plant to maximize profits when the net profit is $1.20 per bushel for corn and $2.50 per bushel for wheat?

55. ***Allocation of Resources in Manufacturing*** The Wadsworth Widget Company manufactures two types of widgets: regular and deluxe. Each widget is produced at a station consisting of a machine and a person who finishes the widgets by hand. The regular widget requires 2 hr of machine time and 1 hr of finishing time. The deluxe widget requires 3 hr of machine time and 5 hr of finishing time. The profit on the regular widget is $25; on the deluxe widget it is $30. If the workday is 8 hours, how many of each type of widget would be produced at each station per day in order to maximize the profit?

C

56. ***Allocation of Resources in Manufacturing*** The Thompson Company manufactures two industrial products, standard ($45 profit per item) and economy ($30 profit per item). These items are built using machine time and manual labor. The standard product requires 3 hr of machine time and 2 hr of manual labor. The economy model requires 3 hr of machine time and no manual labor. If the week's supply of manual labor is limited to 800 hr and machine time to 15,000 hr, how much of each type of product should be produced each week in order to maximize the profit?

57. ***Diet Problem*** A convalescent hospital wishes to provide at a minimum cost, a diet that has at least 200 g of carbohydrates, 100 g of protein, and 120 g of fats per day. These requirements can be met with two foods:

Food	Carbohydrates	Protein	Fats
A	10 g	2 g	3 g
B	5 g	5 g	4 g

If food *A* costs $0.29 per gram and food *B* is $0.15 per gram how many grams of each food should be purchased for each patient per day in order to meet the minimum requirements at the lowest cost?

58. ***Diet Problem*** The following carbohydrate information is given on the sides of two cereal boxes (for 1 oz of cereal with 1/2 cup of whole milk):

	Starch and Related Carbohydrates	Sucrose and Other Sugars
Kellogg's Corn Flakes	23 g	7 g
Post Honeycombs	14 g	17 g

What is the minimum cost in order to receive at least 322 g starch and 119 g sucrose by consuming these two cereals if Corn Flakes cost $0.07 per ounce and Honeycombs cost $0.19 per ounce?

59. ***Investment Application*** Brown Bros., Inc. is an investment company doing an analysis of the pension fund for a certain company. The fund has a maximum of $10 million to invest in two places: no more than $8 million in stocks yielding 12%, and at least \$2 million in long-term bonds yielding 8%. The bond-to-stock investment ratio cannot be more than 3 to 1. How should Brown Bros. advise their client so that the investments yield the maximum yearly return?

60. ***Investment Application*** Your broker tells you of two investments she thinks are worthwhile. She advises a new issue of Pertec stock which should yield 20% over the next year, and then to balance your account she advises Campbell Municipal Bonds with a 10% yearly yield. The stock-to-bond ratio should be no less than 3 to 1. If you have no more than \$100,000 to invest and do not want to invest more than \$70,000 in Pertec or less than \$20,000 in bonds, how much should be invested in each to maximize your return?

9.8 CHAPTER 9 SUMMARY

The material of this chapter is reviewed in the following list of objectives. After each objective there are some practice questions. For a sample test, select the first question of each set and check your answers with the answer section. For a sample test without answers, use the second question of each set. Additional practice is given by the other questions in each set. If you are having trouble with a particular type of problem, look back to that section for extra help.

9.1 SYSTEMS OF EQUATIONS

Objective 1 *Solve a system of equations by graphing.*

1. $\begin{cases} 2x - y = -8 \\ y = \frac{3}{5}x + 1 \end{cases}$

2. $\begin{cases} y = \frac{1}{3}x - 5 \\ 2x - 4y - 12 = 0 \end{cases}$

3. $\begin{cases} 2x - 3y + 21 = 0 \\ 5x + 4y - 5 = 0 \end{cases}$

4. $\begin{cases} y = x^2 - 6x + 5 \\ x + y - 5 = 0 \end{cases}$

Objective 2 *Solve a system of equations by substitution.*

5. $\begin{cases} 4x + 3y = -18 \\ y = -\frac{2}{3}x - 2 \end{cases}$

6. $\begin{cases} 5y - 3x = 0 \\ y = x + 2 \end{cases}$

7. $\begin{cases} x + 2y = 26 \\ 5x - 2y = -122 \end{cases}$

8. $\begin{cases} y = x^2 + 4x - 3 \\ x - y + 1 = 0 \end{cases}$

Objective 3 *Solve a system of equations by linear combinations.*

9. $\begin{cases} 2x + 3y = 6 \\ 3x + 2y = -1 \end{cases}$

10. $\begin{cases} 5x - 2y = 30 \\ 3x - 2y = -2 \end{cases}$

11. $\begin{cases} 4x + 3y = -7 \\ 2x - 5y = 55 \end{cases}$

12. $\begin{cases} 3x^2 - 2y^2 = 9 \\ x^2 + y^2 = 8 \end{cases}$

9.2 CRAMER'S RULE

Objective 4 *Solve a system of two equations in two unknowns by using Cramer's Rule.*

13. $\begin{cases} 3x - y = 2 \\ x + 5y = -3 \end{cases}$

14. $\begin{cases} 5x - 3y = 2 \\ 2x + 4y = 5 \end{cases}$

15. $\begin{cases} 3x + 4y = 0 \\ 6x + 8y = 0 \end{cases}$

16. $\begin{cases} y = m_1x + b_1 \\ y = m_2x + b_2 \end{cases}$

9.3 PROPERTIES OF DETERMINANTS

Objective 5 *Evaluate determinants.*

17. $\begin{vmatrix} 1 & 3 & -2 \\ 4 & 5 & 1 \\ 3 & -2 & 4 \end{vmatrix}$

18. $\begin{vmatrix} 3 & 1 & -2 \\ 4 & 4 & 0 \\ 2 & -3 & 1 \end{vmatrix}$

19. $\begin{vmatrix} 3 & -2 & 0 \\ 2 & 5 & 8 \\ 5 & 3 & 5 \end{vmatrix}$

20. $\begin{vmatrix} 0 & 3 & 0 & 0 \\ 3 & 4 & 0 & 2 \\ 3 & 8 & -4 & 1 \\ 1 & -5 & 0 & -1 \end{vmatrix}$

Objective 6 *Solve a system of n unknowns by using Cramer's Rule.*

21. $\begin{cases} 3x - 2y + z = 9 \\ 2x + 5y - 3z = 17 \\ x - 3y + 2z = -2 \end{cases}$

22. $\begin{cases} 3x + y - 2z = 2 \\ 4x + 4y = 8 \\ 2x - 3y + z = 0 \end{cases}$

23. $\begin{cases} 2x + z = 6 + y \\ 3x + z = 8 + 2y \\ x + y = 11 - 3z \end{cases}$

24. $\begin{cases} w + 2z = 7 \\ x - 3y = 5 \\ z - 4w = -1 \\ 2w + y = 5 \end{cases}$

9.4 ALGEBRA OF MATRICES

Objective 7 *Perform matrix operations.* Let

$$A = \begin{bmatrix} 3 & -2 & 1 \\ -2 & 1 & 4 \\ 1 & -3 & 1 \end{bmatrix} \quad B = \begin{bmatrix} 2 & 1 & 0 \\ -1 & 7 & 3 \\ 2 & -3 & 5 \end{bmatrix} \quad \text{and } C = \begin{bmatrix} 2 & 4 & -2 \\ 1 & -1 & 2 \end{bmatrix}$$

Find, if possible:

25. $A + B$ **26.** $A(B + C)$ **27.** AB **28.** $C(BA)$

Objective 8 *Find the inverse of a matrix (if it exists).*

29. $\begin{bmatrix} 1 & -2 \\ -3 & 7 \end{bmatrix}$ **30.** $\begin{bmatrix} 4 & -3 \\ 5 & 4 \end{bmatrix}$ **31.** $\begin{bmatrix} 1 & 1 \\ 1 & 1 \end{bmatrix}$ **32.** $\begin{bmatrix} 1 & 2 & 2 \\ 1 & 3 & 2 \\ 1 & 2 & 3 \end{bmatrix}$

Objective 9 *Solve a system of equations by using an inverse matrix.* The matrices

$$\begin{bmatrix} -1 & 1 & -1 \\ 2 & -1 & 2 \\ 2 & -1 & 1 \end{bmatrix} \quad \text{and} \quad \begin{bmatrix} 1 & 0 & 1 \\ 2 & 1 & 0 \\ 0 & 1 & -1 \end{bmatrix}$$

are inverses. Use this information to solve the systems in Problems 33–36.

33. $\begin{cases} -x + y - z = 6 \\ 2x - y + 2z = -5 \\ 2x - y + z = -5 \end{cases}$

34. $\begin{cases} y = x + z \\ 2x + 2z = y + 1 \\ 2x + z = y + 3 \end{cases}$

35. $\begin{cases} x + z = 11 \\ 2x + y = 14 \\ y - z = -5 \end{cases}$

36. $\begin{cases} x + z = -4 \\ 2x + y = -4 \\ y - z = 1 \end{cases}$

9.5 MATRIX SOLUTION OF SYSTEMS

Objective 10 *Solve a system of equations by using the Gauss–Jordan method.*

37. $\begin{cases} 2x + 3y - z = -2 \\ x + y - 5z = -3 \\ 5x - 7y - 10z = 60 \end{cases}$

38. $\begin{cases} x - 3y + z = 11 \\ 2x + y - z = 0 \\ x - 2y + 4z = 25 \end{cases}$

39. $\begin{cases} 2x - y + z = -3 \\ 3x + y - 2z = 11 \\ 5x - 2y + 3z = -8 \end{cases}$

40. $\begin{cases} w + 3x = -5 \\ x + y = -4 \\ 2x - 3z = -7 \\ z + y = -1 \end{cases}$

9.6 SYSTEMS OF INEQUALITIES

Objective 11 Graph a linear inequality in two variables.

41. $3x - 2y + 16 \geq 9$

42. $y < x - 1$

43. $25x - 5y + 120 \geq 0$

44. $y \geq x^2 - 8x + 19$

Objective 12 Graph the solutions of a system of inequalities.

45. $\begin{cases} 2x + 7y \geq 420 \\ 2x + 2y \leq 500 \\ x > 50 \\ y < 100 \end{cases}$

46. $\begin{cases} y \geq 0 \\ 3x + 2y > -3 \\ x - y < 0 \end{cases}$

47. $\begin{cases} x - y - 8 \leq 0 \\ x - y + 4 \geq 0 \\ x + y \geq -5 \\ y + x \leq 4 \end{cases}$

48. $\begin{cases} 2x + 3y \leq 600 \\ x + 2y \leq 360 \\ 3x - 2y \leq 480 \\ x \geq 0 \\ y \geq 0 \end{cases}$

*9.7 LINEAR PROGRAMMING

Objective 13 Solve a linear programming problem.

49. Maximize $P = 5x + 7y$ subject to:

$$\begin{cases} x \geq 0 \\ y \geq 0 \\ 4x + y \geq 7 \\ x + 2y \leq 12 \\ x \leq 7 \end{cases}$$

50. Minimize $C = 8x + 5y$ subject to:

$$\begin{cases} x \geq 0 \\ y \geq 0 \\ 2x + y \geq 8 \\ x - y \leq 2 \end{cases}$$

51. Maximize $M = 5x + 4y$ subject to:

$$\begin{cases} 2x + y \leq 420 \\ 2x + 2y \leq 500 \\ 2x + 3y \leq 600 \\ x \geq 0 \\ y \geq 0 \end{cases}$$

52. Maximize $S = x + y$ subject to:

$$\begin{cases} x \geq 0 \\ y \geq 0 \\ 12x + 150y \leq 1200 \\ 6x + 200y \leq 1200 \\ 16x + 50y \leq 800 \end{cases}$$

Objective 14 Solve applied problems involving systems of equations and linear programming.

53. ***Chemistry*** Two commercial preparations, I and II, contain two ingredients, *A* and *B*, in the following proportions:

	A	*B*
I	70%	30%
II	40%	60%

How many grams of each preparation must be mixed to obtain 60 g of a preparation that contains the ingredients in equal parts?

* Optional section

54. ***Business*** A manufacturer of auto accessories uses three basic parts, A, B, and C, in two products, I and II, in the following proportions:

	A	B	C
I	2	1	4
II	2	2	1

The inventory shows 1250 of A, 900 of B, and 2000 of C on hand. Maximize $5x + 7y$, where x is the number of units of product I and y the number of units of product II.

55. ***Medicine*** A hospital wishes to provide for its patients a diet that has a minimum of 100 g of carbohydrates, 60 g of proteins, and 40 g of fats per day. These requirements can be met with two foods:

Food	Carbohydrates	Proteins	Fats
A	6 g	3 g	1 g
B	2 g	2 g	2 g

It is also important to minimize costs, and food A costs \$0.14 per gram and food B costs \$0.06 per gram. How many grams of each food should be bought for each patient per day in order to meet the minimum daily requirements at the lowest cost?

56. ***Utilization of Sales Force*** Bradbury Bros. Realty plans to open several new branch offices, which will employ either four or six people, and has \$1,275,000 in capital for this expansion. A four-person branch requires an initial cash outlay of \$175,000, and a six-person branch, \$200,000. Bradbury Bros. has also decided not to hire more than 32 new people and will not open more than 10 branches. How many of each type of branch should be opened to maximize cash inflow if it is expected to be \$50,000 for four-person branches and \$65,000 for six-person branches?

Sir Isaac Newton (1642–1727)

Analysis and natural philosophy owe their most important discoveries to this fruitful means, which is called "induction." Newton was indebted to it for his theorem of the binomial and the principle of universal gravity.

Pierre-Simon Laplace
A Philosophical Essay on Probabilities

I don't know what I may seem to the world, but, as to myself, I seem to have been only as a boy playing on the seashore, and diverting myself in now and then finding a smoother pebble or a prettier shell than ordinary, whilst the great ocean of truth lay all undiscovered before me.

Sir Isaac Newton
Quoted by Rev. J. Spence, *Anecdotes, Observations, and Characters of Books and Men* (1858)

HISTORICAL NOTE

Isaac Newton was one of the greatest mathematicians of all time. However, when he was a schoolboy at Grantham he was last in his class until he wanted to beat a bully both physically and mentally. He succeeded on both counts. At 18, he entered Trinity College, Cambridge and remained there until 1696, first as a student and later as a professor. In 1665, the university closed for a year because of bubonic plague, and Newton found himself at home for a year. During this year he interested himself in various physical questions, and among other things he formulated the basis of his theory of gravitation and invented calculus. Leibniz, who was featured at the beginning of Chapter 9, invented calculus at about the same time.

The Binomial Theorem of this chapter was proved for real exponents by Newton in 1665. His statement of this theorem is awkward by today's standards, but remember that he found it by a laborious trial-and-error procedure. The following statement of the Binomial Theorem is attributed to Newton:*

$$\overline{P + PQ}\Big|^{\frac{m}{n}} = P^{\frac{m}{n}} + \frac{m}{n}AQ + \frac{m-n}{2n}BQ + \frac{m-2n}{3n}CQ + \frac{m-3n}{4n}DQ + \frac{m-4n}{5n}EQ + \cdots$$

* From D. E. Smith, *Source Book in Mathematics* (New York: McGraw-Hill), p. 224.

10

Additional Topics in Algebra

CONTENTS

PREVIEW

Two of the most far-reaching results in algebra are presented in this chapter, *mathematical induction* and the *Binomial Theorem.* The three sections of this chapter that deal with *sequences* and *series* should be considered together as a unit. There are 9 objectives in this chapter, which are listed on pages 419–420.

PERSPECTIVE

Mathematical induction is used in calculus in the introduction to the idea of integration. It forms the basis of a number of the results that are needed to formulate some integral theorems. In addition, series are a major topic of consideration in calculus. The material introduced in this chapter forms the starting point for the ideas of convergence and divergence of series in calculus. In the following example taken from a calculus book, we see the so-called *p-series,* which you can find in Section 10.5:

$$\frac{1}{1} + \frac{1}{2^p} + \frac{1}{3^p} + \frac{1}{4^p} + \frac{1}{5^p} + \cdots$$

$$S_{10} = \frac{1}{e} + \frac{2}{e^2} + \frac{3}{e^3} + \cdots + \frac{10}{e^{10}} \approx 0.92037$$

Now, since the terms of the series are positive, we know that

$$S_{10} < S \le S_{10} + 0.00050$$

from which we conclude that

$$0.92037 < S \le 0.92087$$

In the remainder of this section, we investigate a second type of series that has a simple arithmetic test for convergence or divergence.

DEFINITION OF p-SERIES A series of the form

$$\sum_{n=1}^{\infty} \frac{1}{n^p} = \frac{1}{1^p} + \frac{1}{2^p} + \frac{1}{3^p} + \cdots$$

is called a **p-series,** where p is a positive constant. For $p = 1$, the series

$$\sum_{n=1}^{\infty} \frac{1}{n} = 1 + \frac{1}{2} + \frac{1}{3} + \cdots$$

is called the **harmonic series.**

10.1 MATHEMATICAL INDUCTION

Mathematical induction is an important method of proof in mathematics. Let us begin with a simple example. Suppose you want to know the sum of the first n odd integers. You could begin by looking for a pattern:

$$
\begin{aligned}
&1 &&= 1\\
&1 + 3 &&= 4\\
&1 + 3 + 5 &&= 9\\
&1 + 3 + 5 + 7 &&= 16\\
&1 + 3 + 5 + 7 + 9 &&= 25
\end{aligned}
$$

Do you see a pattern here? It appears that the sum of the first n odd numbers is n^2 since $1 = 1^2, 4 = 2^2, 9 = 3^2, 16 = 4^2$, and so on. Now *prove deductively* that

$$1 + 3 + 5 + \cdots + \underbrace{(2n - 1)}_{n\text{th odd number}} = n^2$$

is true for all positive integers n. How can you proceed? Use a method called **mathematical induction**, which is used to prove certain propositions about the positive integers. The proposition is denoted by $P(n)$. In this case, $P(n)$ is the proposition

$$P(n):\quad 1 + 3 + 5 + \cdots + (2n - 1) = n^2$$

Then

$$
\begin{aligned}
&P(1): && 1 = 1^2\\
&P(2): && 1 + 3 = 2^2\\
&P(3): && 1 + 3 + 5 = 3^2\\
&P(4): && 1 + 3 + 5 + 7 = 4^2\\
&\quad\vdots && \quad\vdots\\
&P(100): && 1 + 3 + 5 + \cdots + 199 = 100^2\\
&P(x - 1): && 1 + 3 + 5 + \cdots + (2x - 3) = (x - 1)^2\\
&P(x): && 1 + 3 + 5 + \cdots + (2x - 1) = x^2\\
&P(x + 1): && 1 + 3 + 5 + \cdots + (2x + 1) = (x + 1)^2
\end{aligned}
$$

We want to show that $P(n)$ is true for *all* n when $n = 1, n = 2, \ldots$ (n a positive integer). Let S denote the set of positive integers for which $P(n)$ is true. If we can show that 1 is in S and that if k is in S then $k + 1$ is in S, we know that all positive integers are in S. This leads to a statement of the **Principle of Mathematical Induction**.

Principle of Mathematical Induction (PMI)

> If a given proposition $P(n)$ is true for $P(1)$ and if the truth of $P(k)$ implies the truth of $P(k + 1)$, then $P(n)$ is true for all positive integers.

This sets up the following procedure for proof by mathematical induction.

1. **Prove $P(1)$ is true.**
2. **Assume $P(k)$ is true.**
3. **Prove $P(k + 1)$ is true.**
4. **Conclude that $P(n)$ is true for all positive integers n.**

Drawing by Levin; © 1976 The New Yorker Magazine, Inc.

Students often have a certain uneasiness when they first use the Principle of Mathematical Induction as a method of proof. Suppose this principle is used with a stack of large dominoes set up as shown in the cartoon. How can the man in the cartoon be certain of knocking over all the dominoes?

1. He would have to be able to knock over the first one.
2. He would have to have the dominoes arranged so that if the kth domino falls, then the next one, the $(k + 1)$st, will also fall. That is, each domino is set up so that if it falls it causes the next one to fall. We have set up a kind of chain reaction here. The first domino falls; this knocks over the next one (the second domino); the second one knocks over the next one (the third domino); the third one knocks over the next one; and so on. This continues until all the dominoes are knocked over.

EXAMPLE 1 Prove that $1 + 3 + 5 + \cdots + (2n - 1) = n^2$ is true for all positive integers n.

Solution Proof:

Step 1 Prove $P(1)$ true: $1 = 1^2$ is true.

Step 2 Assume $P(k)$ true: $1 + 3 + 5 + \cdots + (2k - 1) = k^2$.

Step 3 Prove $P(k + 1)$ true.
To prove:

$$1 + 3 + 5 + \cdots + [2(k + 1) - 1] = (k + 1)^2$$

or

$$1 + 3 + 5 + \cdots + (2k + 1) = (k + 1)^2$$

Since $2(k + 1) - 1 = 2k + 2 - 1$
$= 2k + 1$

Statements	Reasons
1. $1 + 3 + 5 + \cdots + (2k - 1) = k^2$	1. By hypothesis (step 2)
2. $1 + 3 + 5 + \cdots + (2k - 1) + (2k + 1) = k^2 + (2k + 1)$	2. Add $(2k + 1)$ to both sides
3. $1 + 3 + 5 + \cdots + (2k - 1) + (2k + 1) = k^2 + 2k + 1$	3. Associative
4. $1 + 3 + 5 + \cdots + (2k - 1) + (2k + 1) = (k + 1)^2$	4. Factoring (distributive)
5. $1 + 3 + 5 + \cdots + (2k - 1) + [2(k + 1) - 1] = (k + 1)^2$	5. $2k + 1 = 2k + 2 - 1$ $= 2(k + 1) - 1$

Step 4 The proposition $P(n)$ is true for all positive integers n by PMI (the Principle of Mathematical Induction). ■

Remember: A single example showing that a proposition is false serves as a counterexample to disprove the proposition.

EXAMPLE 2 Prove or disprove that $2 + 4 + 6 + \cdots + 2n = n(n + 1)$ is true for all positive integers n.

Solution *Step 1* Prove $P(1)$ true: $2 \stackrel{?}{=} 1(1 + 1)$
$2 = 2$; it is true

Step 2 Assume $P(k)$.
Hypothesis: $2 + 4 + 6 + \cdots + 2k = k(k + 1)$

Step 3 Prove $P(k + 1)$.
To prove: $2 + 4 + 6 + \cdots + 2(k + 1) = (k + 1)(k + 2)$

Statements	Reasons
1. $2 + 4 + 6 + \cdots + 2k = k(k + 1)$	1. Hypothesis (step 2)
2. $2 + 4 + 6 + \cdots + 2k + 2(k + 1)$ $= k(k + 1) + 2(k + 1)$	2. Add $2(k + 1)$ to both sides
3. $= (k + 1)(k + 2)$	3. Factoring

Step 4 The proposition $P(n)$ is true for all positive integers n by PMI. ■

Mathematical induction does not apply only to propositions involving sums of terms, as shown by the next examples.

EXAMPLE 3 Prove or disprove that $n^3 + 2n$ is divisible by 3 for all positive integers n.

Solution Proof: An integer is divisible by 3 if it has a factor of 3.

Step 1 Prove $P(1)$: $1^3 + 2 \cdot 1 = 3$, which is divisible by 3; thus, $P(1)$ is true.

Step 2 Assume $P(k)$.
Hypothesis: $k^3 + 2k$ is divisible by 3

Step 3 Prove $P(k + 1)$.
To prove: $(k + 1)^3 + 2(k + 1)$ is divisible by 3

Statements	Reasons
1. $(k + 1)^3 + 2(k + 1)$ $= k^3 + 3k^2 + 3k + 1 + 2k + 2$	1. Distributive, associative, and commutative axioms
2. $= (3k^2 + 3k + 3) + (k^3 + 2k)$	2. Commutative and associative
3. $= 3(k^2 + k + 1) + (k^3 + 2k)$	3. Distributive
4. $3(k^2 + k + 1)$ is divisible by 3	4. Definition of divisibility by 3
5. $k^3 + 2k$ is divisible by 3	5. Hypothesis
6. $(k + 1)^3 + 2(k + 1)$ is divisible by 3	6. Both terms are divisible by 3; therefore the sum is divisible by 3

Step 4 The proposition $P(n)$ is true for all positive integers n by PMI. ■

EXAMPLE 4 Prove or disprove that $n + 1$ is prime for all positive integers n.

Solution Proof:

Step 1 Prove $P(1)$: $1 + 1 = 2$ is a prime

Step 2 Assume $P(k)$.
Hypothesis: $k + 1$ is a prime

Step 3 Prove $P(k + 1)$.
To prove: $(k + 1) + 1$ is a prime; it is not possible since $(k + 1) + 1 = k + 2$, which is not prime whenever k is an even positive integer.

Step 4 Any conclusion? You cannot conclude that it is false—only that induction does not work. But it is in fact false, and a counterexample is found by letting $n = 3$; then $n + 1 = 4$ is not prime. ■

Example 4 shows that even though you made an assumption in Step 2, it is not going to change the conclusion. If the proposition you are trying to prove is not true, making the assumption in Step 2 that it is true will *not* enable you to prove it true. Another common mistake of students working with mathematical induction for the first time is illustrated by Example 5.

EXAMPLE 5 Prove that $1 \cdot 2 \cdot 3 \cdot 4 \cdots n < 0$ for all positive integers n.

Solution Students often slip into the habit of skipping either the first or second step in a proof by mathematical induction. This is dangerous. It is important to check every step. Suppose a careless person did not verify the first step. Then the results could be as follows:

Step 2 Assume $P(k)$.
Hypothesis: $1 \cdot 2 \cdot 3 \cdots k < 0$

Step 3 Prove $P(k + 1)$.
To prove: $1 \cdot 2 \cdot 3 \cdots k \cdot (k + 1) < 0$
Proof: $1 \cdot 2 \cdot 3 \cdots k < 0$ by hypothesis; $k + 1$ is positive since k is a positive integer; therefore

$$\underbrace{1 \cdot 2 \cdot 3 \cdots k}_{\substack{\uparrow \\ \text{negative}}} \cdot \underbrace{(k + 1)}_{\substack{\uparrow \\ \text{positive}}} < 0$$

Step 3 is proved.

Step 4 The proposition is not true for all positive integers, since the first step, $1 < 0$, does not hold. ■

PROBLEM SET 10.1

A

If $P(n)$ represents each statement given in Problems 1–10, state and prove or disprove $P(1)$.

1. $5 + 9 + 13 + \cdots + (4n + 1) = n(2n + 3)$
2. $3 + 9 + 15 + \cdots + (6n - 3) = 3n^2$
3. $2^2 + 4^2 + 6^2 + \cdots + (2n)^2 = \dfrac{2n(n + 1)(2n + 1)}{3}$
4. $1^3 + 2^3 + 3^3 + \cdots + n^3 = \dfrac{n^2(n + 1)^2}{4}$
5. $\cos(\theta + n\pi) = (-1)^n \cos\theta$
6. $\sin(\frac{\pi}{4} + n\pi) = (-1)^n(\frac{\sqrt{2}}{2})$
7. $n^2 + n$ is even
8. $n^3 - n + 3$ is divisible by 3 (Steps in proof may vary.)
9. $\left(\dfrac{2}{3}\right)^{n+1} < \left(\dfrac{2}{3}\right)^n$
10. $1 + 2n \leq 3^n$

If $P(n)$ represents each statement given in Problems 11–20, state $P(k)$ and $P(k + 1)$.

11. $5 + 9 + 13 + \cdots + (4n + 1) = n(2n + 3)$
12. $3 + 9 + 15 + \cdots + (6n - 3) = 3n^2$
13. $2^2 + 4^2 + 6^2 + \cdots + (2n)^2 = \dfrac{2n(n + 1)(2n + 1)}{3}$
14. $1^3 + 2^3 + 3^3 + \cdots + n^3 = \dfrac{n^2(n + 1)^2}{4}$
15. $\cos(\theta + n\pi) = (-1)^n \cos\theta$
16. $\sin(\frac{\pi}{4} + n\pi) = (-1)^n(\frac{\sqrt{2}}{2})$
17. $n^2 + n$ is even
18. $n^3 - n + 3$ is divisible by 3
19. $\left(\dfrac{2}{3}\right)^{n+1} < \left(\dfrac{2}{3}\right)^n$
20. $1 + 2n \leq 3^n$

In each of Problems 21–35, prove that the given formula is true for all positive integers n.

21. $1 + 2 + 3 + \cdots + n = \dfrac{n(n + 1)}{2}$
22. $1 + 4 + 7 + \cdots + (3n - 2) = \dfrac{n(3n - 1)}{2}$
23. $5 + 9 + 13 + \cdots + (4n + 1) = n(2n + 3)$
24. $3 + 9 + 15 + \cdots + (6n - 3) = 3n^2$

25. $2 + 7 + 12 + \cdots + (5n - 3) = \frac{n(5n - 1)}{2}$

26. $1^2 + 2^2 + 3^2 + \cdots + n^2 = \frac{n(n + 1)(2n + 1)}{6}$

27. $1^2 + 3^2 + 5^2 + \cdots + (2n - 1)^2 = \frac{n(2n - 1)(2n + 1)}{3}$

28. $1^3 + 2^3 + 3^3 + \cdots + n^3 = \frac{n^2(n + 1)^2}{4}$

29. $2^2 + 4^2 + 6^2 + \cdots + (2n)^2 = \frac{2n(n + 1)(2n + 1)}{3}$

30. $1 \cdot 2 + 2 \cdot 3 + 3 \cdot 4 + \cdots + n(n + 1) = \frac{n(n + 1)(n + 2)}{3}$

31. $1 \cdot 3 + 2 \cdot 4 + 3 \cdot 5 + \cdots + n(n + 2) = \frac{n(n + 1)(2n + 7)}{6}$

32. $3 + 3^2 + \cdots + 3^n = \frac{3^{n+1} - 3}{2}$

33. $1 + 5 + 5^2 + \cdots + 5^n = \frac{5^{n+1} - 1}{4}$

34. $1 + r + r^2 + \cdots + r^n = \frac{r^{n+1} - 1}{r - 1}$

35. $\log(a_1 a_2 \cdots a_n) = \log a_1 + \log a_2 + \cdots + \log a_n$ for $n \geq 2$ (Assume all the a_i are positive real numbers.)

B

Define $b^{n+1} = b^n \cdot b$ and $b^0 = 1$. Use this definition to prove the properties of exponents in Problems 36–39 for all positive integers n.

36. $b^m \cdot b^n = b^{m+n}$

37. $(b^m)^n = b^{mn}$

38. $(ab)^n = a^n b^n$

39. $\left(\frac{a}{b}\right)^n = \frac{a^n}{b^n}$

Prove that the statements in Problems 40–53 are true for every positive integer n.

40. $\cos(\theta + n\pi) = (-1)^n(\cos\theta)$

41. $\sin(\frac{\pi}{4} + n\pi) = (-1)^n(\frac{\sqrt{2}}{2})$

42. $\cos(\frac{\pi}{3} + n\pi) = \frac{(-1)^n}{2}$

43. $n^2 + n$ is even

44. $n^5 - n$ is divisible by 5

45. $n(n + 1)(n + 2)$ is divisible by 6

46. $n^3 - n + 3$ is divisible by 3

47. $10^{n+1} + 3 \cdot 10^n + 5$ is divisible by 9

48. $(1 + n)^2 \geq 1 + n^2$

49. $2^n > n$

50. $\left(\frac{2}{3}\right)^{n+1} < \left(\frac{2}{3}\right)^n$

51. $1 + 2n \leq 3^n$

52. $\frac{1}{2} + \frac{1}{3} + \frac{1}{4} + \frac{1}{5} + \cdots + \frac{1}{n + 1} < n$

53. $1 + 2 + 3 + \cdots + n < \frac{(2n + 1)^2}{8}$

54. Prove the generalized distributive property: $a(b_1 + b_2 + \cdots + b_n) = ab_1 + ab_2 + \cdots + ab_n$.

C

55. Notice the following:

$$1^3 = 1^2$$
$$1^3 + 2^3 = 3^2$$
$$1^3 + 2^3 + 3^3 = 6^2$$
$$1^3 + 2^3 + 3^3 + 4^3 = 10^2$$

Make a conjecture based on this pattern and then prove your conjecture.

56. Notice the following:

$$1 = 1$$
$$1 + 4 = 5$$
$$1 + 4 + 7 = 12$$
$$1 + 4 + 7 + 10 = 22$$

Make a conjecture based on this pattern and then prove your conjecture.

57. Notice the following:

$$2 = 2$$
$$2 + 2 \cdot 3 = 8$$
$$2 + (2 \cdot 3) + (2 \cdot 3^2) = 26$$
$$2 + (2 \cdot 3) + (2 \cdot 3^2) + (2 \cdot 3^3) = 80$$

Make a conjecture based on this pattern and then prove your conjecture.

58. Notice the following:

$$(-1)^1 = -1$$
$$(-1)^1 + (-1)^2 = 0$$
$$(-1)^1 + (-1)^2 + (-1)^3 = -1$$
$$(-1)^1 + (-1)^2 + (-1)^3 + (-1)^4 = 0$$

Make a conjecture based on this pattern and then prove your conjecture.

59. Prove $\sin x + \sin^2 x + \cdots + \sin^n x = \frac{\sin^{n+1} x - \sin x}{\sin x - 1}$ for all positive integers n.

60. Prove $e^x + e^{2x} + \cdots + e^{nx} = \frac{e^{(n+1)x} - e^x}{e^x - 1}$ for all positive integers n.

61. Use mathematical induction to prove De Moivre's Theorem: $(r \operatorname{cis} \theta)^n = r^n \operatorname{cis}(n\theta)$ for every positive integer n.

10.2

BINOMIAL THEOREM

In mathematics it is frequently necessary to expand $(a + b)^n$. If n is very large, direct calculation is rather tedious so we try to find an easy pattern that will not only help us find $(a + b)^n$ but will also allow us to find any given term in that expansion.

Consider the powers of $(a + b)$, which are found by direct multiplication:

$$\begin{aligned}
(a+b)^0 &= 1 \\
(a+b)^1 &= \mathbf{1}\cdot a + \mathbf{1}\cdot b \\
(a+b)^2 &= \mathbf{1}\cdot a^2 + \mathbf{2}\cdot ab + \mathbf{1}\cdot b^2 \\
(a+b)^3 &= \mathbf{1}\cdot a^3 + \mathbf{3}\cdot a^2b + \mathbf{3}\cdot ab^2 + \mathbf{1}\cdot b^3 \\
(a+b)^4 &= \mathbf{1}\cdot a^4 + \mathbf{4}\cdot a^3b + \mathbf{6}\cdot a^2b^2 + \mathbf{4}\cdot ab^3 + \mathbf{1}\cdot b^4 \\
(a+b)^5 &= \mathbf{1}\cdot a^5 + \mathbf{5}\cdot a^4b + \mathbf{10}\cdot a^3b^2 + \mathbf{10}\cdot a^2b^3 + \mathbf{5}\cdot ab^4 + \mathbf{1}\cdot b^5 \\
&\vdots
\end{aligned}$$

Ignore the coefficients and focus your attention only on the variables:

$$\begin{array}{llllll}
(a+b)^1: & a & b & & & \\
(a+b)^2: & a^2 & ab & b^2 & & \\
(a+b)^3: & a^3 & a^2b & ab^2 & b^3 & \\
(a+b)^4: & a^4 & a^3b & a^2b^2 & ab^3 & b^4 \\
& & & \vdots & &
\end{array}$$

Do you see a pattern? As you read from left to right, the powers of a decrease and the powers of b increase. Notice that the sum of the exponents for each term is the same as the original exponent:

$$(a+b)^n: \quad a^nb^0 \quad a^{n-1}b^1 \quad a^{n-2}b^2 \cdots a^{n-r}b^r \cdots a^2b^{n-2} \quad a^1b^{n-1} \quad a^0b^n$$

Next consider the coefficients:

$$\begin{array}{lccccccccccc}
(a+b)^0: & & & & & & 1 & & & & & \\
(a+b)^1: & & & & & 1 & & 1 & & & & \\
(a+b)^2: & & & & 1 & & 2 & & 1 & & & \\
(a+b)^3: & & & 1 & & 3 & & 3 & & 1 & & \\
(a+b)^4: & & 1 & & 4 & & 6 & & 4 & & 1 & \\
(a+b)^5: & 1 & & 5 & & 10 & & 10 & & 5 & & 1 \\
& & & & & & \vdots & & & & &
\end{array}$$

Do you see this pattern? This arrangement of numbers is called **Pascal's triangle**. The rows and columns of Pascal's triangle are usually numbered as shown in Figure 10.1.

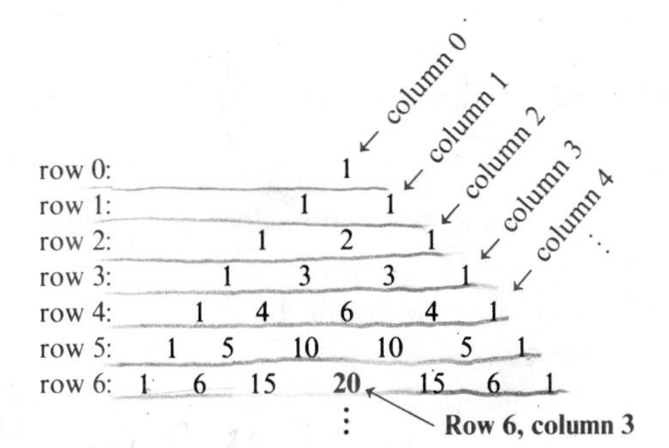

Figure 10.1 Pascal's triangle

Notice that the rows of Pascal's triangle are numbered to correspond to the exponent n on $(a + b)^n$. There are many relationships associated with this pattern, but we are concerned with an expression representing the entries in the pattern. Do you see how to generate additional rows of the triangle?

1. Each row begins and ends with a 1.
2. Notice that we began counting the rows with row 0. This is because after row 0, the second entry in the row is the same as the row number. Thus row 7 begins 1 7 ....
3. The triangle is symmetric about the middle. This means that the entries of each row are the same at the beginning and the end. Thus row 7 ends with ... 7 1. (This property is proved in Problem 58.)
4. To find new entries, we can simply add the two entries just above in the preceding row. Thus row 7 is found by looking at row 6:

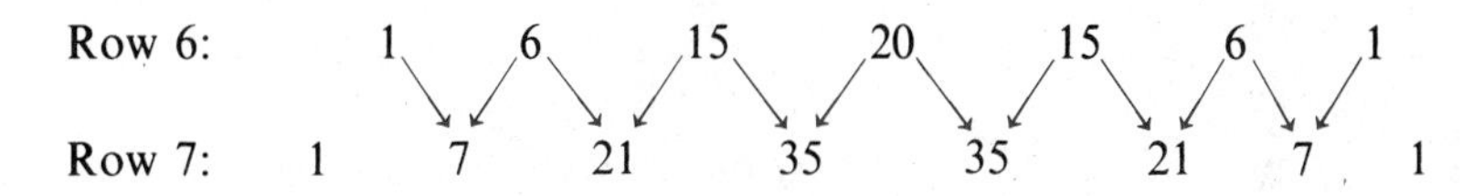

(This property is proved in Problem 57.)

Write $\binom{n}{r}$ to represent the element in the nth row and rth column of the triangle. Therefore:

$$\binom{0}{0} = 1$$

$$\binom{1}{0} = 1 \quad \binom{1}{1} = 1$$

$$\binom{2}{0} = 1 \quad \binom{2}{1} = 2 \quad \binom{2}{2} = 1$$

$$\binom{3}{0} = 1 \quad \binom{3}{1} = 3 \quad \binom{3}{2} = 3 \quad \binom{3}{3} = 1$$

$$\vdots$$

Using this notation we can now state a very important theorem in mathematics called the **Binomial Theorem**:

Binomial Theorem

For any positive integer n,

$$(a+b)^n = \binom{n}{0}a^n + \binom{n}{1}a^{n-1}b + \binom{n}{2}a^{n-2}b^2 + \cdots + \binom{n}{r}a^{n-r}b^r + \cdots + \binom{n}{n-2}a^2b^{n-2} + \binom{n}{n-1}ab^{n-1} + \binom{n}{n}b^n$$

The proof of the Binomial Theorem is by mathematical induction and the procedure is lengthy, so we leave the proof for the problem set. You are led through the steps and are asked to fill in the details in Problems 60–63.

EXAMPLE 1 Find $(x+y)^8$.

Solution Use Pascal's triangle to obtain the coefficients in the expansion. Thus

$$(x+y)^8 = x^8 + 8x^7y + 28x^6y^2 + 56x^5y^3 + 70x^4y^4 + 56x^3y^5 + 28x^2y^6 + 8xy^7 + y^8$$ ■

EXAMPLE 2 Find $(x-2y)^4$.

Solution In this example, $a = x$ and $b = -2y$ and the coefficients are in Pascal's triangle.

$$(x-2y)^4 = 1\cdot x^4 + 4\cdot x^3(-2y)^1 + 6\cdot x^2(-2y)^2 + 4\cdot x(-2y)^3 + 1\cdot(-2y)^4$$
$$= x^4 - 8x^3y + 24x^2y^2 - 32xy^3 + 16y^4$$ ■

If the power of the binomial is very large, then Pascal's triangle is not efficient, so the next step is to find a formula for $\binom{n}{r}$. This formula is found by using a notation called **factorial notation**.

Factorial Notation

The symbol

$$n! = 1\cdot 2\cdot 3\cdots(n-1)\cdot n$$

is called ***n* factorial** (n a natural number). Also, we define $0! = 1$ and $1! = 1$.

EXAMPLE 3

$0! = 1$
$1! = 1$
$2! = 1\cdot 2 = 2$
$3! = 1\cdot 2\cdot 3 = 6$
$4! = 1\cdot 2\cdot 3\cdot 4 = 24$
$5! = 1\cdot 2\cdot 3\cdot 4\cdot 5 = 120$

$$6! = 1 \cdot 2 \cdot 3 \cdot 4 \cdot 5 \cdot 6 = 720$$
$$7! = 1 \cdot 2 \cdot 3 \cdot 4 \cdot 5 \cdot 6 \cdot 7 = 5040$$
$$8! = 1 \cdot 2 \cdot 3 \cdot \cdots \cdot 7 \cdot 8 = 40{,}320$$
$$9! = 1 \cdot 2 \cdot 3 \cdot \cdots \cdot 8 \cdot 9 = 362{,}880$$
$$10! = 1 \cdot 2 \cdot 3 \cdot \cdots \cdot 9 \cdot 10 = 3{,}628{,}800$$

■

In finding the values of Example 3, notice that $3! = 3 \cdot 2!$, $4! = 4 \cdot 3!$, $5! = 5 \cdot 4!, \ldots$; in general $n! = n(n-1)!$

EXAMPLE 4 Evaluate the given expressions.

a. $5! - 4! = 120 - 24$
$= \mathbf{96}$

b. $(5-4)! = 1!$
$= \mathbf{1}$

c. $\dfrac{8!}{4!} = \dfrac{8 \cdot 7 \cdot 6 \cdot 5 \cdot \not{4} \cdot \not{3} \cdot \not{2} \cdot \not{1}}{\not{4} \cdot \not{3} \cdot \not{2} \cdot \not{1}}$
$= 8 \cdot 7 \cdot 6 \cdot 5$
$= \mathbf{1680}$

d. $\left(\dfrac{8}{4}\right)! = 2!$
$= \mathbf{2}$

e. $\dfrac{10!}{8!} = \dfrac{10 \cdot 9 \cdot \not{8!}}{\not{8!}}$
$= \mathbf{90}$

Notice that

$$10! = 10 \cdot 9!$$
$$= 10 \cdot 9 \cdot 8!$$
$$= 10 \cdot 9 \cdot 8 \cdot 7!$$

and so on.

■

Now we can state a formula for $\binom{n}{r}$ using this notation.

Binomial Coefficient

The symbol $\binom{n}{r}$ is defined for integers r and n such that $0 \le r \le n$.

$$\binom{n}{r} = \frac{n!}{r!(n-r)!}$$

is called the **binomial coefficient n, r**.

EXAMPLE 5 Find $\binom{3}{2}$ both by Pascal's triangle and by formula.

Solution Row 3, column 2; entry in Pascal's triangle is 3. By formula,

$$\binom{3}{2} = \frac{3!}{2!(3-2)!}$$
$$= \frac{3!}{2!1!}$$
$$= \mathbf{3}$$

■

EXAMPLE 6 Find $\binom{6}{4}$ both by Pascal's triangle and by formula.

Solution Row 6, column 4; entry is 15. By formula,

$$\binom{6}{4} = \frac{6!}{4!(6-4)!}$$
$$= \frac{6!}{4!2!}$$
$$= \frac{6 \cdot 5 \cdot 4!}{2 \cdot 1 \cdot 4!}$$
$$= \mathbf{15}$$

■

EXAMPLE 7 Evaluate the given expressions. (You would not use Pascal's triangle for these.)

a. $$\binom{52}{2} = \frac{52!}{2!(52-2)!}$$
$$= \frac{52!}{2!50!}$$
$$= \frac{52 \cdot 51 \cdot 50!}{2 \cdot 1 \cdot 50!}$$
$$= \mathbf{1326}$$

b. $$\binom{n}{n} = \frac{n!}{n!(n-n)!}$$
$$= \frac{n!}{n!0!}$$
$$= \mathbf{1}$$

c. $$\binom{n}{n-1} = \frac{n!}{(n-1)![n-(n-1)]!}$$
$$= \frac{n!}{(n-1)!1!}$$
$$= \frac{n(n-1)!}{(n-1)!}$$
$$= \boldsymbol{n}$$

■

EXAMPLE 8 Find $(a+b)^{15}$.

Solution The power is rather large, so use the formula to find the coefficients.

$$(a+b)^{15} = \binom{15}{0}a^{15} + \binom{15}{1}a^{14}b + \binom{15}{2}a^{13}b^2 + \cdots + \binom{15}{14}ab^{14} + \binom{15}{15}b^{15}$$
$$= \frac{15!}{0!15!}a^{15} + \frac{15!}{1!14!}a^{14}b + \frac{15!}{2!13!}a^{13}b^2 + \cdots + \frac{15!}{14!1!}ab^{14} + \frac{15!}{15!0!}b^{15}$$
$$= a^{15} + 15a^{14}b + 105a^{13}b^2 + \cdots + 15ab^{14} + b^{15}$$

■

EXAMPLE 9 Find the coefficient of the term x^2y^{10} in the expansion of $(x + 2y)^{12}$.

Solution $n = 12$, $r = 10$, $a = x$, and $b = 2y$; thus

$$\binom{12}{10}x^2(2y)^{10} = \frac{12!}{10!2!}(2)^{10}x^2y^{10}$$

The coefficient is 66(1024) = **67,584**. ■

PROBLEM SET 10.2

A

Evaluate the expressions in Problems 1–24.

1. $4! - 2!$ **2.** $5! - 3!$

3. $(4 - 2)!$ **4.** $(6 - 3)!$

5. $\frac{9!}{7!}$ **6.** $\frac{10!}{6!}$

7. $\frac{12!}{10!}$ **8.** $\frac{10!}{4!6!}$

9. $\frac{12!}{3!(12-3)!}$ **10.** $\frac{15!}{5!(15-5)!}$

11. $\frac{20!}{3!(20-3)!}$ **12.** $\frac{52!}{3!(52-3)!}$

13. $\binom{8}{1}$ **14.** $\binom{5}{4}$

15. $\binom{8}{2}$ **16.** $\binom{52}{3}$

17. $\binom{8}{3}$ **18.** $\binom{7}{4}$

19. $\binom{8}{4}$ **20.** $\binom{5}{5}$

21. $\binom{52}{2}$ **22.** $\binom{10}{1}$

23. $\binom{1000}{1}$ **24.** $\binom{574}{2}$

B

In Problems 25–40, expand by using Pascal's triangle or by formula.

25. $(a + b)^6$ **26.** $(a + b)^7$

27. $(2x + 3)^3$ **28.** $(2x - 3)^4$

29. $(x + y)^5$ **30.** $(x - y)^6$

31. $(3x + 2)^5$ **32.** $(3x - 2)^4$

33. $(x + y)^4$ **34.** $(2x + 3y)^4$

35. $(\frac{1}{2}x + y^3)^3$ **36.** $(x^{-2} + y^{-2})^4$

37. $(x^{1/2} + y^{1/2})^4$ **38.** $(1 + x)^{10}$

39. $(1 - x)^8$ **40.** $(1 - 2y)^6$

Find the first four terms in the expressions given in Problems 41–49.

41. $(a + b)^{10}$ **42.** $(a + b)^{12}$

43. $(a + b)^{14}$ **44.** $(x - y)^{15}$

45. $(x + 2y)^{16}$ **46.** $(x + \sqrt{2})^8$

47. $(x - 2y)^{12}$ **48.** $(1 - 0.03)^{13}$

49. $(1 - 0.02)^{12}$

50. Find the coefficient of a^5b^6 in $(a - b)^{11}$.

51. Find the coefficient of a^4b in $(a^2 - 2b)^8$.

52. Find the coefficient of $a^{10}b^4$ in $(a + b)^{14}$.

53. Find the coefficient of $x^{10}y^2$ in $(2x^2 + \sqrt{y})^9$.

C

54. What is the constant term in the expansion of $(9x^{-1} + x^2/3)^6$?

55. What is the constant term in the expansion of $(4y^{-2} + y^3/2)^5$?

56. Show that

$$\binom{n}{0} + \binom{n}{1} + \binom{n}{2} + \cdots + \binom{n}{n-1} + \binom{n}{n} = 2^n$$

This says that the sum of the entries of the nth row of Pascal's triangle is 2^n.

57. Show that

$$\binom{n-1}{r-1} + \binom{n-1}{r} = \binom{n}{r}$$

This says that to find any entry in Pascal's triangle (except the first and last), simply add the entries above.

58. Show that

$$\binom{n}{r} = \binom{n}{n-r}$$

This says that Pascal's triangle is symmetric.

59. Use the formula $\binom{n}{k} = \dfrac{n!}{k!(n-k)!}$ to prove

$$\binom{k}{r} + \binom{k}{r-1} = \binom{k+1}{r}$$

Problems 60–63 will lead you through the induction proof of the Binomial Theorem. Let n be any positive integer. To prove:

$$(a+b)^n = \binom{n}{0}a^n + \binom{n}{1}a^{n-1}b + \binom{n}{2}a^{n-2}b^2 + \cdots + \binom{n}{r}a^{n-r}b^r + \cdots + \binom{n}{n-2}a^2b^{n-2} + \binom{n}{n-1}ab^{n-1} + \binom{n}{n}b^n$$

60. Prove it true for $n = 1$.

61. Assume it true for $n = k$. Fill in the statement of the hypothesis.

62. Prove it true for $n = k + 1$.
To prove:

$$(a+b)^{k+1} = a^{k+1} + \cdots + \left[\binom{k}{r} + \binom{k}{r-1}\right]a^{k-r+1}b^r + \cdots + b^{k+1}$$

Hint: To prove this you will need to use the results of Problem 59.

63. Tie together Problems 60–62 to complete the proof of the Binomial Theorem.

64. What is wrong with the following "proof," which results in $1 = 2$?

$$(a+b)^n = a^n + na^{n-1}b + \frac{n(n-1)}{2!}a^{n-2}b^2 + \cdots + nab^{n-1} + b^n \quad \text{From the Binomial Theorem}$$

Let $n = 0$. Then

$$(a+b)^0 = a^0 + 0 + 0 + \cdots + 0 + b^0$$

By substitution and zero multiplication

$$1 = 1 + 0 + 0 + \cdots + 0 + 1$$

By definition of zero exponent

$$1 = 2$$

10.3 SEQUENCES, SERIES, AND SUMMATION NOTATION

Patterns and proof are two of the cornerstones upon which mathematics is founded. In this chapter these two ideas are tied together. Consider a function whose domain is the set of counting numbers. Remember that the set of counting numbers is $N = \{1, 2, 3, 4, \ldots, n, \ldots\}$.

Infinite Sequence

An **infinite sequence** is a function s with a domain that consists of the set of counting numbers. The number $s(1)$ is called the *first term* of the sequence, $s(2)$ the *second term*, and $s(n)$ the *nth term* or **general term** of the sequence.

For convenience we sometimes refer to an infinite sequence simply as a **sequence** or **progression**. Sometimes we talk of a **finite sequence**, which means that the domain is the finite set $\{1, 2, 3, \ldots, n-1, n\}$ for some natural number n.

Two special types of sequences are studied in this chapter. The first is an **arithmetic sequence**. In an arithmetic sequence, there is a common difference between successive terms. That is, if any term is subtracted from the next term, the result is always the same, and this number is called the **common difference**.

EXAMPLE 1 Show that 1, 4, 7, 10, 13, ____, ... is an arithmetic sequence and find the missing term.

Solution Look for a common difference by subtracting each term from the succeeding term:

$$4 - 1 = 3$$
$$7 - 4 = 3$$
$$10 - 7 = 3$$
$$13 - 10 = 3$$

The common difference is 3 (be sure to find this for *each* of the *given* terms).

$$x - 13 = 3$$ The common difference is 3; x is the missing term.

Thus $\boldsymbol{x = 16}$ is the next term (which is found by solving the equation). In order to view this as a function whose domain is the set of counting numbers, you will need to wait until we discuss the general term of the sequence in the next section. ■

The second type of sequence is called a **geometric sequence**. In a geometric sequence, there is a common ratio between successive terms. If any term is divided into the next term, the result is always the same, and this number is called the **common ratio**.

EXAMPLE 2 Show that 2, 4, 8, 16, 32, ____, ... is a geometric sequence and find the missing term.

Solution

$$\frac{4}{2} = 2$$
$$\frac{8}{4} = 2$$
$$\frac{16}{8} = 2$$
$$\frac{32}{16} = 2$$

The common ratio is 2.

$$\frac{x}{32} = 2$$ (where x is the next term)

Thus

$\boldsymbol{x = 64}$ ← The next term ■

There are other sequences, as shown by Example 3, that are neither arithmetic nor geometric.

EXAMPLE 3 Show that 1, 1, 2, 3, 5, 8, 13, ____, ____, ____, ... is neither arithmetic nor geometric. Find the missing terms.

Solution First check to see if it is arithmetic: $1 - 1 = 0$; $2 - 1 = 1$; there is no *common* difference. Next check to see if it is geometric: $\frac{1}{1} = 1$; $\frac{2}{1} = 2$; there is no *common* ratio. Look for another pattern:

$$1 + 1 = 2$$
$$1 + 2 = 3$$
$$2 + 3 = 5$$
$$3 + 5 = 8$$
$$5 + 8 = 13$$

It looks like the pattern is obtained by adding the two preceding terms:

$$8 + 13 = \mathbf{21}$$
$$13 + 21 = \mathbf{34}$$
$$21 + 34 = \mathbf{55}$$

The missing terms are 21, 34, and 55. ■

A new notation is generally used when working with sequences. Remember that the domain is the set of counting numbers, so a sequence could be defined by

$$s(n) = 3n - 2 \quad \text{where } n \in \{1, 2, 3, \ldots\}$$

Thus

$$s(1) = 3(1) - 2 = 1$$
$$s(2) = 3(2) - 2 = 4$$
$$s(3) = 3(3) - 2 = 7$$
$$\vdots$$

Instead of writing $s(1)$, however, the notation s_1 is used; in place of $s(2)$, s_2; in place of $s(n)$, s_n. Thus s_{15} means the fifteenth term of the sequence. It is found in the same fashion as though the notation $s(15)$ were used:

$$\begin{aligned} s_{15} &= 3(15) - 2 \\ &= 43 \end{aligned}$$

EXAMPLE 4 Find the first four terms of $s_n = 26 - 6n$.

Solution

$$s_1 = 26 - 6(1) = 20$$
$$s_2 = 26 - 6(2) = 14$$
$$s_3 = 26 - 6(3) = 8$$
$$s_4 = 26 - 6(4) = 2$$

The sequence is **20, 14, 8, 2, ...** . It is an arithmetic sequence. ■

EXAMPLE 5 Find the first four terms of $s_n = (-2)^n$.

Solution

$$s_1 = (-2)^1 = -2$$
$$s_2 = (-2)^2 = 4$$
$$s_3 = (-2)^3 = -8$$
$$s_4 = (-2)^4 = 16$$

The sequence is **−2, 4, −8, 16, ...** It is a geometric sequence. ■

EXAMPLE 6 Find the first four terms of $s_n = s_{n-1} + s_{n-2}$, $n \geq 3$, where $s_1 = 1$ and $s_2 = 2$.

Solution

$s_1 = 1$ Given

$s_2 = 2$ Given

$$\begin{aligned} s_3 &= s_2 + s_1 \\ &= 2 + 1 \quad \text{By substitution} \\ &= 3 \end{aligned}$$

$$\begin{aligned} s_4 &= s_3 + s_2 \\ &= 3 + 2 \\ &= 5 \end{aligned}$$

The sequence is **1, 2, 3, 5, ...** This sequence is neither arithmetic nor geometric. ■

EXAMPLE 7 Find the first four terms of $s_n = 2n$.

Solution $s_1 = 2, s_2 = 4, s_3 = 6, s_4 = 8$

The sequence is **2, 4, 6, 8, ...** ■

EXAMPLE 8 Find the first four terms of $s_n = 2n + (n - 1)(n - 2)(n - 3)(n - 4)$.

Solution

$$s_1 = 2(1) + 0 = 2$$
$$s_2 = 2(2) + 0 = 4$$
$$s_3 = 2(3) + 0 = 6$$
$$s_4 = 2(4) + 0 = 8$$

The sequence is **2, 4, 6, 8, ...** ■

If you are given a general term, you can find a unique sequence. However, Examples 7 and 8 show that if only a finite number of successive terms is known and no general term is given, then a *unique* general term cannot be given. That is, if we are given the sequence

2, 4, 6, 8, ____

the next term is probably 10 (if we are thinking of the general term of Example 7), but it *may* be something different. In Example 8, $s_1 = 2, s_2 = 4, s_3 = 6, s_4 = 8$, and

$$\begin{aligned} s_5 &= 2(5) + (5 - 1)(5 - 2)(5 - 3)(5 - 4) \\ &= 10 + (4)(3)(2)(1) \\ &= 34 \end{aligned}$$

In general, you are looking for the simplest general term; nevertheless, you must remember that answers are not unique *unless the general term is given.*

It is sometimes necessary to consider the sum of the terms of a sequence. This sum is called a **series**.

Finite Series

> The indicated sum of the terms of a finite sequence $s_1, s_2, s_3, \ldots, s_n$ is called a **finite series** and is denoted by
>
> $$S_n = s_1 + s_2 + s_3 + \cdots + s_n$$

EXAMPLE 9 Let $s_n = 26 - 6n$ from Example 4. Find S_4.

Solution

$$\begin{aligned} S_4 &= s_1 + s_2 + s_3 + s_4 \\ &= 20 + 14 + 8 + 2 \qquad \text{From Example 4} \\ &= \mathbf{44} \end{aligned}$$

■

EXAMPLE 10 Let $s_n = (-1)^n n^2$. Find S_3.

Solution $s_1 = (-1)^1(1)^2 = -1$; $s_2 = (-1)^2(2)^2 = 4$; $s_3 = (-1)^3(3)^2 = -9$. Now find S_3:

$$\begin{aligned} S_3 &= s_1 + s_2 + s_3 \\ &= (-1) + 4 + (-9) \\ &= \mathbf{-6} \end{aligned}$$

■

The terms of the sequence in Example 10 alternate in sign:

$-1, 4, -9, 16, \ldots$

A factor of $(-1)^n$ or $(-1)^{n+1}$ in the general term will cause the sign of the terms to alternate, creating a series called an **alternating series**.

The next two sections discuss more efficient ways of finding the general term of a sequence, as well as more efficient ways of finding the sum of n terms of these sequences. We will use the following **summation notation** to simplify the way we express sums.

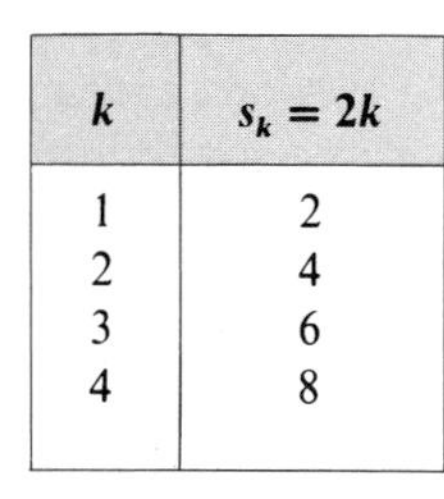

k	$s_k = 2k$
1	2
2	4
3	6
4	8

Consider the function $s_k = 2k$ with the domain $N = \{1, 2, 3, 4\}$. The sum of the terms of this finite arithmetic sequence is the sum of the series

$2 + 4 + 6 + 8$

as shown in the table in the margin. Now denote this sum by using the symbol Σ (called **sigma**) as follows:

This is the last natural number in the domain.

$$\sum_{k=1}^{4} 2k = 2 + 4 + 6 + 8$$

This is the function being evaluated; it is the general term of the sequence.

This is the first natural number in the domain.

This variable is called the **index of summation**.

This symbol means to evaluate the function for each number in the domain and *add* the resulting terms.

Thus

$$\sum_{k=1}^{4} 2k = 2 + 4 + 6 + 8 = 20$$

EXAMPLE 11 Let $s_k = 2k + 1$ and $N = \{3, 4, 5, 6\}$. Evaluate $\sum_{k=3}^{6} (2k + 1)$.

Solution

	k	$s_k = 2k + 1$
First natural number in domain; $k = 3$ →	3	7
	4	9
	5	11
Last natural number in domain; $k = 6$ →	6	13

↑ Σ means to add these values

Thus

$$\sum_{k=3}^{6} (2k + 1) = 7 + 9 + 11 + 13 = \mathbf{40}$$

(9: $k = 4$; 11: $k = 5$; 13: $k = 6$)

This is obtained by letting $k = 3$ and evaluating $(2k + 1)$. ■

EXAMPLE 12 Expand $\sum_{k=3}^{n} \frac{1}{2^k}$

Solution

$$\sum_{k=3}^{n} \frac{1}{2^k} = \underset{k=3}{\frac{1}{8}} + \underset{k=4}{\frac{1}{16}} + \underset{k=5}{\frac{1}{32}} + \cdots + \underset{k=n}{\frac{1}{2^n}}$$

■

EXAMPLE 13 Write the sum of the arithmetic series $S_n = a + (a + d) + (a + 2d) + \cdots + (a + nd)$ using sigma notation.

Solution

$$S_n = \sum_{k=1}^{n} [a + (k - 1)d]$$

■

EXAMPLE 14 Write the sum of the finite geometric series $S_n = a + ar + ar^2 + \cdots + ar^{n-1}$ using sigma notation.

Solution

$$S_n = \sum_{k=1}^{n} ar^{k-1}$$

■

EXAMPLE 15 Write the Binomial Theorem using summation notation.

Solution

$$(a + b)^n = \sum_{k=0}^{n} \binom{n}{k} a^{n-k} b^k$$

■

PROBLEM SET 10.3

A

For the sequences in Problems 1–16, answer the following questions. The answers are not necessarily unique.

a. *Classify each as arithmetic, geometric, or neither.*

b. *If arithmetic, state the common difference d; if geometric, state the common ratio r; if neither, state a pattern in your own words.*

c. *Supply the missing term.*

1. 2, 5, 8, 11, 14, ____

2. 1, 2, 1, 1, 2, 1, 1, 1, 2, 1, 1, 1, 1, ____

3. 3, 6, 12, 24, 48, ____

4. 5, −15, 45, −135, 405, ____

5. 100, 99, 97, 94, 90, ____

6. 1, 1, 2, 3, 5, 8, 13, ____

7. p, pq, pq^2, pq^3, pq^4, ____

8. 97, 86, 75, 64, ____

9. 8, 12, 18, 26, ____

10. $5^5, 5^4, 5^3, 5^2$, ____

11. 2, 5, 2, 5, 5, 2, 5, 5, 5, ____

12. 5, −5, −15, −25, −35, ____

13. $1, \frac{1}{2}, \frac{1}{3}, \frac{2}{3}, \frac{1}{4}, \frac{3}{4}, \frac{1}{5}, \frac{2}{5}, \frac{3}{5}, \frac{4}{5}, \frac{1}{6}$, ____

14. $\frac{4}{3}, 2, 3, 4\frac{1}{2}$, ____

15. 1, 8, 27, 64, 125, ____

16. 2, 8, 18, 32, ____

Find the first three terms of the sequence with the nth term given in Problems 17–28.

17. $s_n = 4n - 3$

18. $s_n = a + nd$

19. $s_n = ar^{n-1}$

20. $s_n = \dfrac{n-1}{n+1}$

21. $s_n = (-1)^n$

22. $s_n = (-1)^n(n+1)$

23. $s_n = 1 + \dfrac{1}{n}$

24. $s_n = \dfrac{n+1}{n}$

25. $s_n = 2$

26. $s_n = -5$

27. $\cos nx$

28. $\dfrac{\sin n}{n^2}$

Evaluate the expressions in Problems 29–40.

29. $\sum_{k=2}^{6} k$

30. $\sum_{m=1}^{4} m^2$

31. $\sum_{n=0}^{6} (2n+1)$

32. $\sum_{p=1}^{6} 2p$

33. $\sum_{k=2}^{5} (10 - 2k)$

34. $\sum_{k=2}^{5} (100 - 5k)$

35. $\sum_{k=1}^{5} (-2)^{k-1}$

36. $\sum_{k=0}^{4} 3(-2)^k$

37. $\sum_{k=0}^{3} 2(3^k)$

38. $\sum_{k=1}^{3} (-1)^k(k^2+1)$

39. $\sum_{k=1}^{10} [1^k + (-1)^k]$

40. $\sum_{k=0}^{5} [2^k + (-2)^k]$

B

41. Find the fifteenth term of the sequence in Problem 17.

42. Find the 102nd term of the sequence in Problem 20.

43. Find the tenth term of the sequence in Problem 21.

44. Find the twentieth term of the sequence in Problem 22.

45. Find the third term of the sequence $(-1)^{n+1}5^{n+1}$.

46. Find the second term of the sequence $(-1)^{n-1}7^{n-1}$.

47. Find the first five terms of the sequence where $s_1 = 2$ and $s_n = 3s_{n-1}, n \geq 2$.

48. Find the first five terms of the sequence where $s_1 = 3$ and $s_n = \frac{1}{3}s_{n-1}, n \geq 2$.

49. Find the first five terms of the sequence where $s_1 = 1$, $s_2 = 1$, and $s_n = s_{n-1} + s_{n-2}, n \geq 3$.

50. Find the first five terms of the sequence where $s_1 = 1$, $s_2 = 2$, and $s_n = s_{n-1} + s_{n-2}, n \geq 3$.

51. Write $\dfrac{1}{2} + \dfrac{1}{4} + \dfrac{1}{8} + \cdots + \dfrac{1}{2^r} + \cdots + \dfrac{1}{128}$ using summation notation.

52. Write $2 + 4 + 6 + \cdots + 2n + \cdots + 100$ using summation notation.

53. Write $1 + 6 + 36 + 216 + 1296$ using summation notation. The rth term is 6^{r-1}.

54. Write $5 + 15 + 45 + 135 + 405$ using summation notation. The rth term is $5 \cdot 3^{r-1}$.

Classify the situations in Problems 55–59 as arithmetic or geometric. Do NOT answer the questions.

55. Suppose that a teacher obtains a job with a starting salary of $15,000 and receives a $500 raise every year thereafter. What will the teacher's salary be in 10 years?

56. Suppose that an autoworker obtains a job with a starting salary of $20,000 and receives an 8% raise every year thereafter. What will the autoworker's salary be in 10 years?

57. Suppose that a teacher obtains a job with a starting salary of $15,000 and receives a $3\frac{1}{3}$% raise every year thereafter. What will the teacher's salary be in 10 years?

58. A grocery clerk must stack 30 cases of canned fruit, each containing 24 cans. He decides to display the cans by stacking them in a pyramid where each row after the bottom row contains one less can. Is it possible to use all the cans and end up with a top row of only one can?

59. Suppose that a chain letter asks you to send copies to 10 of your friends. If everyone carries out the directions and sends the chain letter, and if nobody receives more than one chain letter, how many people will be involved in 10 mailings?

60. Compare Problems 55 and 57. Since $3\frac{1}{3}\%$ of \$15,000 is \$500 do you think that the total amount earned in 10 years will be the same for both of these problems? If not, which one do you feel will be the larger, and why?

C

61. **a.** Write out $\sum_{j=1}^{r} a_j b_j$ without summation notation.

b. Let $b_j = k$ and show that $\sum_{j=1}^{r} ka_j = k \sum_{j=1}^{r} a_j$.

c. Let $a_j = 1$ and show that $\sum_{j=1}^{r} k = kr$.

62. Show that $\sum_{k=1}^{n} (a_k + b_k) = \sum_{k=1}^{n} a_k + \sum_{k=1}^{n} b_k$.

Find the next term for the sequences in Problems 63–65. Use any rule you can defend.

63. 1, 3, 4, 7, 11, 18, 29, ____

64. 225, 625, 1225, 2025, ____

65. 8, 5, 4, 9, 1, ____

10.4

ARITHMETIC SEQUENCES AND SERIES

This section focuses on arithmetic sequences and series.

General Term of an Arithmetic Sequence

> The **general term of an arithmetic sequence** $a_1, a_2, a_3, \ldots, a_n, \ldots$ with common difference d is
>
> $$a_n = a_1 + (n - 1)d$$
>
> for $n \geq 1$.

Even though this formula can be proved using mathematical induction, let us consider the first few terms to see where it came from.

Remember that an arithmetic sequence is one in which each succeeding term follows from the previous term by adding the common difference d. Thus,

for $n = 2$, then $a_2 = a_1 + d$;
for $n = 3$, then $a_3 = a_2 + d = a_1 + d + d = a_1 + 2d$;
for $n = 4$, then $a_4 = a_3 + d = a_1 + 2d + d = a_1 + 3d$;
⋮

Do you see how each step is substituted into the next step? Look for a pattern:

$$a_5 = a_1 + 4d$$

↑ n (under a_5); ↑ 1 less than n (under 4)

Thus, it is reasonable to write

$$a_n = a_1 + (n - 1)d$$

EXAMPLE 1 Find the general term of the arithmetic sequence 18, 14, 10, 6, ...

Solution $a_1 = 18$; $d = 14 - 18 = -4$; thus

$$\begin{aligned} a_n &= a_1 + (n - 1)d \\ &= 18 + (n - 1)(-4) \\ &= 18 - 4n + 4 \\ &= \mathbf{22 - 4n} \end{aligned}$$

■

EXAMPLE 2 If $a_5 = 14$, $a_{10} = 34$, find d.

Solution Use the formula $a_n = a_1 + (n - 1)d$:

$$a_5 = a_1 + 4d \quad \text{or, since } a_5 = 14, \quad 14 = a_1 + 4d$$
$$a_{10} = a_1 + 9d \quad \text{or, since } a_{10} = 34, \quad 34 = a_1 + 9d$$

This can be written as the system

$$\begin{cases} a_1 + 4d = 14 \\ a_1 + 9d = 34 \end{cases}$$

Multiply the first equation by -1 and add:

$$5d = 20$$
$$\mathbf{d = 4}$$

■

What is the total number of blocks shown in Figure 10.2?

Figure 10.2 How many blocks?

The number of blocks in the successive rows form an arithmetic sequence: 1, 6, 11, 16. The total number of blocks is the sum

$$\sum_{j=1}^{4} (-4 + 5j)$$

The indicated sum of an arithmetic sequence is called an **arithmetic series**. Let A_n denote the arithmetic series

$$\sum_{k=1}^{n} a_k = a_1 + a_2 + a_3 + \cdots + a_n$$

EXAMPLE 3 **a.** How many blocks are in the stack shown in Figure 10.2?
b. How many blocks are in 10 rows of a stack of blocks similar to the one shown in Figure 10.2?

Solution **a.** $A_1 = a_1 = 1$

$A_2 = a_1 + a_2 = 1 + 6 = 7$

$A_3 = a_1 + a_2 + a_3 = 1 + 6 + 11 = 18$

$A_4 = a_1 + a_2 + a_3 + a_4 = 1 + 6 + 11 + 16 = \mathbf{34}$

This is the number of blocks in Figure 10.2.

b. $A_{10} = a_1 + a_2 + \cdots + a_9 + a_{10}$

$= 1 + 6 + 11 + 16 + 21 + 26 + 31 + 36 + 41 + 46$

$=$ **235 blocks in a stack 10 rows high** ■

In Example 3, we found A_{10} by brute force addition. This method would not be practical for large n, however, so it is desirable to find a general formula for A_n as we did for a_n. Consider the following method for finding A_{10} of Example 3:

$$A_{10} = 1 + 6 + 11 + 16 + 21 + 26 + 31 + 36 + 41 + 46$$

and

$$A_{10} = 46 + 41 + 36 + 31 + 26 + 21 + 16 + 11 + 6 + 1$$

Add these equations term by term:

$$A_{10} + A_{10} = (1 + 46) + (6 + 41) + (11 + 36) + \cdots + (41 + 6) + (46 + 1)$$
$$2A_{10} = 47 + 47 + 47 + \cdots + 47 + 47$$

Notice that the sums of all the numbers within parentheses are equal to the sum of the first and last terms. Instead of doing a lot of addition, the result can be found by multiplication. In this example $n = 10$, so the number of terms (without directly counting them) is 10. Thus

Number of terms

$$2A_{10} = 10(47)$$

Sum of first and last terms

$$A_{10} = \frac{10(47)}{2}$$
$$= 10\left(\frac{47}{2}\right)$$

Average of first and last terms

$$= 235$$

Arithmetic Series

For an arithmetic sequence $a_1, a_2, a_3, \ldots, a_n$, with common difference d, the sum of the arithmetic series

$$\sum_{k=1}^{n} a_k = a_1 + a_2 + a_3 + \cdots + a_n$$

is

$$A_n = n\left(\frac{a_1 + a_n}{2}\right) \quad \text{or, equivalently,} \quad A_n = \frac{n}{2}[2a_1 + (n - 1)d]$$

Proof

$$A_n = a_1 + a_2 + a_3 + \cdots + a_{n-1} + a_n$$
$$A_n = a_n + a_{n-1} + \cdots + a_3 + a_2 + a_1$$

Add these equations term by term:

$$2A_n = (a_1 + a_n) + (a_2 + a_{n-1}) + \cdots + (a_{n-1} + a_2) + (a_n + a_1)$$

All the quantities enclosed by parentheses are equal because

$$\begin{aligned}
a_1 + a_n &= a_1 + [a_1 + (n-1)d] \\
&= \mathbf{2a_1 + (n-1)d} \\
a_2 + a_{n-1} &= (a_1 + d) + [a_1 + (n-2)d] \\
&= \mathbf{2a_1 + (n-1)d} \\
a_3 + a_{n-2} &= (a_1 + 2d) + [a_1 + (n-3)d] \\
&= \mathbf{2a_1 + (n-1)d} \\
&\vdots
\end{aligned}$$

Since there is a total of n such sums,

$$2A_n = n[2a_1 + (n-1)d]$$

$$A_n = \frac{n}{2}[2a_1 + (n-1)d]$$

For the other part of the formula, replace $2a_1 + (n-1)d$ with $a_1 + a_n$ because

$$\begin{aligned}
a_1 + a_n &= a_1 + [a_1 + (n-1)d] \\
&= 2a_1 + (n-1)d
\end{aligned}$$

Therefore

$$A_n = n\left(\frac{a_1 + a_n}{2}\right)$$

□

EXAMPLE 4 Find A_{100} for the sequence of Example 3.

Solution $a_1 = 1$; $n = 100$; $d = 6 - 1 = 5$; thus

$$\begin{aligned}
A_{100} &= \frac{100}{2}[2(1) + 99(5)] \\
&= 50[2 + 495] \\
&= \mathbf{24{,}850}
\end{aligned}$$

■

EXAMPLE 5 Find A_{10} where $a_1 = 6$ and $a_7 = -18$.

Solution First find d:

$a_n = a_1 + (n-1)d$

$a_7 = a_1 + (7-1)d$ Replace n by 7.

$-18 = 6 + 6d$ Substitute the given values of a_7 and a_1.

$-24 = 6d$

$-4 = d$

Then find A_{10}:

$$A_n = \frac{n}{2}[2a_1 + (n-1)d]$$

$$A_{10} = \frac{10}{2}[2(6) + (10-1)(-4)]$$
$$= 5[12 - 36]$$
$$= \mathbf{-120}$$

■

PROBLEM SET 10.4

A

Write out the first four terms of the arithmetic sequences in Problems 1–12.

1. $a_1 = 5, d = 4$
2. $a_1 = -5, d = 3$
3. $a_1 = 85, d = 3$
4. $a_1 = 50, d = -10$
5. $a_1 = 100, d = -5$
6. $a_1 = 20, d = -4$
7. $a_1 = -\frac{5}{2}, d = \frac{1}{2}$
8. $a_1 = \frac{2}{3}, d = -\frac{5}{3}$
9. $a_1 = \sqrt{12}, d = \sqrt{3}$
10. $a_1 = 5, d = x$
11. $a_1 = x, d = y$
12. $a_1 = m, d = 2m$

Find a_1 and d for the arithmetic sequences in Problems 13–21.

13. $5, 8, 11, \ldots$
14. $5, 5, 5, \ldots$
15. $6, 11, 16, \ldots$
16. $35, 46, 57, \ldots$
17. $-8, -1, 6, \ldots$
18. $-1, 1, 3, \ldots$
19. $x, 2x, 3x, \ldots$
20. $x + \sqrt{3}, x + \sqrt{12}, x + \sqrt{27}, \ldots$
21. $x - 5b, x - 3b, x - b, \ldots$

B

Find an expression for the general term of each arithmetic sequence in Problems 22–30.

22. $5, 8, 11, \ldots$
23. $5, 5, 5, \ldots$
24. $6, 11, 16, \ldots$
25. $35, 46, 57, \ldots$
26. $-8, -1, 6, \ldots$
27. $-1, 1, 3, \ldots$
28. $x, 2x, 3x, \ldots$
29. $x + \sqrt{3}, x + \sqrt{12}, x + \sqrt{27}, \ldots$
30. $x - 5b, x - 3b, x - b, \ldots$

Find the indicated quantity for each arithmetic sequence in Problems 31–45.

31. $a_1 = 6, d = 5; a_{20}$
32. $a_1 = 35, d = 11; a_{10}$
33. $a_1 = -20, d = 5; a_{10}$
34. $a_1 = 35, d = 11; A_{10}$
35. $a_1 = -7, d = -2; A_{100}$
36. $a_1 = 15, d = -4; A_{50}$
37. $a_1 = -5, a_{30} = -63; d$
38. $a_1 = 4, a_6 = 24; d$
39. $a_1 = -13, a_{10} = 5; d$
40. $a_1 = 4, a_6 = 24; A_{15}$
41. $a_1 = -5, a_{30} = -63; A_{10}$
42. $a_1 = 110, a_{11} = 0; d$
43. $a_5 = 27, a_{10} = 47; d$
44. $a_4 = 36, a_5 = 60; d$
45. $a_3 = 36, a_5 = 60; d$

46. a. How many blocks are shown in Figure 10.3?
b. How many blocks would there be in 100 rows of a stack of blocks like the one shown in Figure 10.3?

Figure 10.3 How many blocks?

47. Find the sum of the first 20 terms of the arithmetic sequence with first term 100 and common difference 50.

48. Find the sum of the first 50 terms of the arithmetic sequence with first term -15 and common difference 5.

49. Find the sum of the even integers between 41 and 99.

50. Find the sum of the odd integers between 100 and 80.

51. Find the sum of the first n odd integers.

52. Find the sum of the first n even integers.

53. A sequence $s_1, s_2, \ldots, s_n$ is a **harmonic sequence** if its reciprocals form an arithmetic sequence. Which of the following are harmonic sequences?
a. $1, \frac{1}{2}, \frac{1}{3}, \frac{1}{4}, \frac{1}{5}, \ldots$
b. $\frac{1}{2}, \frac{1}{5}, \frac{1}{8}, \frac{1}{11}, \frac{1}{14}, \ldots$

c. $2, \frac{2}{3}, \frac{2}{5}, \frac{2}{7}, \ldots$
d. $\frac{1}{5}, -\frac{1}{5}, -\frac{1}{15}, -\frac{1}{25}, \ldots$
e. $\frac{3}{4}, \frac{1}{2}, \frac{1}{3}, \frac{2}{9}, \ldots$

54. Consider the arithmetic sequence a_1, x, a_3. The number x can be found as follows:

$$+\begin{cases} x = a_1 + d \\ x = a_3 - d \end{cases}$$
$$2x = a_1 + a_3 \qquad \text{By adding}$$
$$x = \frac{a_1 + a_3}{2}$$

Here x is called the **arithmetic mean** between a_1 and a_3. Find the arithmetic mean between each of the given pairs of numbers.

a. 1, 8 **b.** 1, 7 **c.** $-5, 3$ **d.** 80, 88 **e.** 40, 56

55. Find the arithmetic mean (see Problem 54) between each of the given pairs of numbers.

a. 4, 20 **b.** 4, 15 **c.** $\frac{1}{2}, \frac{1}{3}$ **d.** $-10, -2$ **e.** $-\frac{2}{3}, \frac{4}{5}$

C

56. Suppose you were hired for a job paying \$21,000 per year and were given the following options:

Option A: annual salary increase of \$1440
Option B: semiannual salary increase of \$360

Which is the better option?

57. Repeat Problem 56 for the following options:

Option C: quarterly salary increase of \$90
Option D: monthly salary increase of \$10

58. What are the differences in the amounts earned in the first year from options A to D in Problems 56 and 57?

59. Repeat Problem 58 for the first 2 years.

60. Write the arithmetic series for the total amount of money earned in 10 years under the following options described in Problem 56.

a. option A **b.** option B

10.5

GEOMETRIC SEQUENCES AND SERIES

As with arithmetic sequences, we will denote the terms of a geometric sequence by using a special notation. Let $g_1, g_2, g_3, \ldots, g_n$ be the terms of a geometric sequence. To find the general term of a geometric sequence, use r to represent the common ratio. Thus

$$g_2 = rg_1$$
$$g_3 = rg_2 = r(rg_1) = r^2g_1$$
$$g_4 = rg_3 = r(r^2g_1) = r^3g_1$$
$$\vdots$$
$$g_n = g_1r^{n-1}$$

General Term of a Geometric Sequence

For a geometric sequence $g_1, g_2, g_3, \ldots, g_n$ with a common ratio r,

$$g_n = g_1r^{n-1}$$

for every $n \geq 1$.

EXAMPLE 1 Find the general term of the geometric sequence 50, 100, 200, ...

Solution

$$g_1 = 50 \qquad r = \frac{100}{50} = 2$$
$$g_n = 50(2)^{n-1}$$
$$= 2 \cdot 5^2 \cdot 2^{n-1}$$
$$= \mathbf{2^n \cdot 5^2}$$

■

EXAMPLE 2 Find r when $g_1 = 20$ and $g_5 = 200$.

Solution Use $g_n = g_1 r^{n-1}$:

$$g_5 = 20r^4$$
$$200 = 20r^4$$
$$10 = r^4$$
$$r = \sqrt[4]{10}$$

■

Suppose you receive the following letter:

> Dear Friend,
>
> This is a chain letter...
>
> Copy this letter six times and send it to six of your friends. In twenty days, you will have good luck.
>
> If you break this chain, you will have bad luck!...

Consider the number of people that could become involved with this chain letter if we assume that everyone carries out their task and does not break the chain. The first mailing would consist of six letters with seven people involved.

$$1 + 6 = 7$$

The second mailing would involve 43 people since the second mailing of 36 letters (each of the six people receiving a letter sends out six more letters) is added to the total:

$$1 + 6 + 36 = 43$$

The number of letters in each successive mailing is a number of a geometric sequence:

1st mailing: $6 = 6$
2nd mailing: $6^2 = 36$
3rd mailing: $6^3 = 216$
4th mailing: $6^4 = 1{,}296$
⋮
10th mailing: $6^{10} = 60{,}466{,}176$
11th mailing: $6^{11} = 362{,}797{,}056$

By the eleventh mailing, more letters would have to be sent than there are people in the United States! The number of letters in only two more mailings would exceed the number of men, women, and children in the whole world.

How many people are involved in 11 mailings assuming that no person receives a letter more than once? To answer this question, consider the series associated with the geometric sequence. Then G_n represents the sum of the first n terms of the geometric series. In this problem, then, you need to find G_{11}:

$$G_{11} = 1 + 6 + 36 + 216 + \cdots + 60{,}466{,}176 + 362{,}797{,}056$$

You could add these on your calculator, but that would take a long time; instead write these numbers by using exponents:

$$G_{11} = 1 + 6 + 6^2 + 6^3 + \cdots + 6^{10} + 6^{11}$$

Next multiply both sides by 6:

$$6G_{11} = 6 + 6^2 + 6^3 + 6^4 + \cdots + 6^{11} + 6^{12}$$

Finally, consider $G_{11} - 6G_{11}$:

$$G_{11} - 6G_{11} = 1 - 6^{12}$$

$$-5G_{11} = 1 - 6^{12}$$

$$G_{11} = \frac{1 - 6^{12}}{-5} \quad \text{or} \quad \frac{1}{5}(6^{12} - 1)$$

This number is easy to find on your calculator. But, more important, it leads to a procedure for finding G_n in general, where G_n represents the geometric series

$$\sum_{k=1}^{n} g_k = g_1 + g_2 + g_3 + \cdots + g_n$$

The next step is to find a formula for G_n:

$$G_n = g_1 + g_1 r + g_1 r^2 + g_1 r^3 + \cdots + g_1 r^{n-1}$$

Multiply both sides by r:

$$rG_n = g_1 r + g_1 r^2 + g_1 r^3 + g_1 r^4 + \cdots + g_1 r^n$$

Notice that, except for the first and last terms, all the terms in the expressions for G_n and rG_n are the same, so that

$$G_n - rG_n = g_1 - g_1 r^n$$

Now solve for G_n:

$$(1 - r)G_n = g_1(1 - r^n)$$

$$G_n = \frac{g_1(1 - r^n)}{1 - r} \quad (r \neq 1)$$

Geometric Series

For a geometric series $\sum_{k=1}^{n} g_k = g_1 + g_2 + g_3 + \cdots + g_n$ with common ratio $r \neq 1$,

$$G_n = \frac{g_1(1 - r^n)}{1 - r}$$

EXAMPLE 3 Find the sum of the first five terms of a geometric series with $g_1 = -15$ and $r = 2$.

Solution

$$G_n = \frac{g_1(1-r^n)}{1-r} = \frac{-15(1-2^n)}{1-2} = \frac{-15}{-1}(1-2^n) = 15(1-2^n)$$

$$G_5 = 15(1-2^5) = 15(1-32) = \mathbf{-465}$$

■

EXAMPLE 4 Find

$$\sum_{k=1}^{10}\left(\frac{1}{2}\right)^k = \left(\frac{1}{2}\right)^1 + \left(\frac{1}{2}\right)^2 + \left(\frac{1}{2}\right)^3 + \left(\frac{1}{2}\right)^4 + \cdots + \left(\frac{1}{2}\right)^{10}$$

$$= \frac{1}{2} + \frac{1}{4} + \frac{1}{8} + \frac{1}{16} + \cdots + \frac{1}{1024}$$

This is a geometric series with $g_1 = \frac{1}{2}$ and $r = \frac{1}{2}$.

$$\sum_{k=1}^{10}\left(\frac{1}{2}\right)^k = G_{10} = \frac{\frac{1}{2}[1-(\frac{1}{2})^{10}]}{1-\frac{1}{2}}$$

$$= 1 - \left(\frac{1}{2}\right)^{10}$$

$$= 1 - \frac{1}{1024}$$

$$= \mathbf{\frac{1023}{1024}}$$

■

The geometric series presented above is finite. Consider now an infinite geometric series. Remember:

1. $s_1, s_2, s_3, \ldots$ is a ***sequence***.
2. $s_1 + s_2 + s_3 + \cdots + s_n$ is a ***series*** denoted by S_n.
3. If we consider $s_1 + s_2 + s_3 + \cdots + s_n$, then S_n is called the ***nth partial sum***.
4. Now consider $S_1, S_2, S_3, \ldots$. This is a ***sequence of partial sums***.

Consider the sequence of partial sums for Example 4.

$$G_1 = \frac{1}{2}$$

$$G_2 = \frac{1}{2} + \frac{1}{4} = \frac{3}{4}$$

$$G_3 = \frac{1}{2} + \frac{1}{4} + \frac{1}{8} = \frac{7}{8}$$

$$\vdots$$

$$G_{10} = \frac{1023}{1024}$$ This result was found in Example 4.

It appears that the partial sums are getting closer to 1 as n becomes large. We *can* find the sum of an infinite geometric sequence. Consider

$$\begin{aligned} G_n &= \frac{g_1(1 - r^n)}{1 - r} \\ &= \frac{g_1 - g_1 r^n}{1 - r} \\ &= \frac{g_1}{1 - r} - \frac{g_1}{1 - r} r^n \end{aligned}$$

Now g_1, r, and $1 - r$ are fixed numbers. If $|r| < 1$, then $r^n \to 0$ as $n \to \infty$ and thus

$$G_n \to \frac{g_1}{1 - r} \text{ as } n \to \infty$$

because the second term is approaching zero. This is not a proof, of course, but it does lead to the following result, which can be proved in a calculus course.

Sum of an Infinite Geometric Series

If $g_1, g_2, g_3, \ldots, g_n$ is an infinite geometric sequence with a common ratio r such that $|r| < 1$, then its sum is denoted by G and found by

$$G = \frac{g_1}{1 - r}$$

If $|r| \geq 1$, the infinite geometric series has no sum.

EXAMPLE 5 Find the sum of the series $100 + 50 + 25 + \cdots$ if possible.

Solution Since $g_1 = 100$ and $r = \frac{1}{2}$, then $G = \dfrac{100}{(1 - \frac{1}{2})} = \mathbf{200}$. ■

EXAMPLE 6 Find the sum of the series $-5 + 10 - 20 + \cdots$ if possible.

Solution $g_1 = -5$ and $r = -2$; since $|r| \geq 1$, this infinite series does not have a sum. ■

EXAMPLE 7 The repeating decimal $0.\overline{72} = 0.72727272\ldots$ is a rational number and can therefore be written as the quotient of two integers. Find this representation by using a geometric series.

Solution $0.727272\ldots = 0.72 + 0.0072 + 0.000072 + \cdots$

Notice that two digits are repeating, so form a series using the repeating digits and leading zeros as placeholders.

$= 0.72 + 0.72(0.01) + 0.72(0.0001) + \cdots$

Factor out the common (repeating) digits.

$= 0.72 + 0.72(0.01) + 0.72(0.01)^2 + \cdots$

Now you can see the common ratio.

Now $g_1 = 0.72$ and $r = 0.01$; then

$$G = \frac{0.72}{1 - 0.01} \quad \text{Since } |r| < 1$$
$$= \frac{0.72}{0.99}$$

Remember: You want to write this out as a fraction, not as a decimal.

$$= \frac{72/100}{99/100}$$
$$= \frac{72}{99}$$
$$= \frac{\mathbf{8}}{\mathbf{11}}$$

■

PROBLEM SET 10.5

A

Write out the first three terms of the geometric sequences in Problems 1–9 with first term g_1 and common ratio r.

1. $g_1 = 5, r = 3$
2. $g_1 = -12, r = 3$
3. $g_1 = 1, r = -2$
4. $g_1 = 1, r = 2$
5. $g_1 = -15, r = \frac{1}{5}$
6. $g_1 = 625, r = -\frac{1}{5}$
7. $g_1 = 8, r = x$
8. $g_1 = a, r = \frac{1}{2}$
9. $g_1 = x, r = y$

Find g_1 and r for the geometric sequences in Problems 10–15.

10. $3, 6, 12, \ldots$
11. $7, 14, 28, \ldots$
12. $1, \frac{1}{2}, \frac{1}{4}, \ldots$
13. $100, 50, 25, \ldots$
14. $x, x^2, x^3, \ldots$
15. $xyz, xy, \ldots$

Find the general term for each of the geometric sequences in Problems 16–21.

16. $3, 6, 12, \ldots$
17. $7, 14, 28, \ldots$
18. $1, \frac{1}{2}, \frac{1}{4}, \ldots$
19. $100, 50, 25, \ldots$
20. $x, x^2, x^3, \ldots$
21. $xyz, xy, \ldots$

Find the sum, if possible, of the infinite geometric series in Problems 22–27.

22. $1 + \frac{1}{2} + \frac{1}{4} + \cdots$
23. $1000 + 500 + 250 + \cdots$
24. $100 + 50 + 25 + \cdots$
25. $-20 + 10 - 5 + \cdots$
26. $-45 - 15 - 5 - \cdots$
27. $-216 - 36 - 6 - \cdots$

B

Find the indicated quantities in Problems 28–39 for the given geometric sequences.

28. $g_1 = 6, r = 3; g_5$
29. $g_1 = 100, r = \frac{1}{10}; g_{10}$
30. $g_1 = 6, r = 3; G_5$
31. $g_1 = 7, g_8 = 896; r$
32. $g_1 = 1, r = 10; G_{10}$
33. $g_1 = 3, r = \frac{1}{5}; G_3$
34. $g_1 = \frac{1}{3}, r = \frac{1}{3}; G$
35. $g_1 = \frac{1}{4}, r = \frac{1}{4}; G$
36. $g_1 = 1, r = 0.08; G$
37. $\sum_{k=1}^{4} \left(\frac{1}{10}\right)^k$
38. $\sum_{k=1}^{4} \left(\frac{1}{3}\right)^k$
39. $\sum_{k=1}^{4} \left(\frac{1}{4}\right)^k$

Represent each repeating decimal in Problems 40–51 as the quotient of two integers by considering an infinite geometric series.

40. $0.\overline{4}$
41. $0.\overline{5}$
42. $0.\overline{9}$
43. $0.\overline{27}$
44. $0.\overline{18}$
45. $0.\overline{45}$
46. $0.\overline{418}$
47. $0.\overline{218}$
48. $0.\overline{123}$
49. $2.\overline{45}$
50. $5.03\overline{1}$
51. $2.25\overline{34}$

52. ***Social Science*** Suppose that a chain letter asks you to send copies to 10 of your friends. If everyone carries out the directions and sends the chain letter, and if nobody receives more than one chain letter, how many people will be involved in five mailings? Count yourself, the 10 letters you mail, the letters they mail, and so forth.

53. ***Social Science*** According to the 1980 census, the U.S. population is 226,504,825. If everyone follows the directions in the chain letter mentioned in Problem 52, how many mailings would be necessary to include the *entire* U.S. population?

54. A new type of Superball advertises that it will rebound to nine-tenths of its original height. If it is dropped from a height of 10 ft, how far will the ball travel before coming to rest?

55. Repeat Problem 54 for a ball that rebounds to two-thirds of its original height.

56. ***Business*** Suppose that a piece of machinery costing \$10,000 depreciates 20% of its present value each year. That is, the first year \$10,000(0.20) = \$2000 is depreciated. The second year's depreciation is

$$\$8000(0.20) = \$1600$$

since the value for the second year is \$10,000 − \$2000 = \$8000. The third year's depreciation is

$$\$6400(0.20) = \$1280$$

If the depreciation is calculated this way indefinitely, what is the total depreciation?

57. ***Business*** Winnie Winner wins \$100 in a pie-baking contest run by the Hi-Do Pie Co. The company gives Winnie the \$100. However, the tax collector wants 20% of the \$100. Winnie pays the tax. But then she realizes that she didn't really win a \$100 prize and tells her story to the Hi-Do Co. The friendly Hi-Do Co. gives Winnie the \$20 she paid in taxes. Unfortunately, the tax collector now wants 20% of the \$20. She pays the tax again and then goes back to the Hi-Do Co. with her story. Assume that this can go on indefinitely. How much money does the Hi-Do Co. have to give Winnie so that she will really win \$100? How much does she pay in taxes?

58. Consider the geometric sequence g_1, x, g_3. The number x can be found by considering

$$\frac{x}{g_1} = r \quad \text{and} \quad \frac{g_3}{x} = r$$

Thus

$$\frac{x}{g_1} = \frac{g_3}{x} \quad \text{so} \quad x^2 = g_1 g_3$$

This equation has two solutions:

$$x = \sqrt{g_1 g_3} \quad \text{and} \quad x = -\sqrt{g_1 g_3}$$

If g_1 and g_3 are both positive, then $\sqrt{g_1 g_3}$ is called the **geometric mean** of g_1 and g_3. If g_1 and g_3 are both negative, then $-\sqrt{g_1 g_3}$ is called the **geometric mean.** Find the geometric mean of each of the given pairs of numbers:

a. 1, 8 **b.** 2, 8 **c.** −5, −3
d. −10, −2 **e.** 4, 20

C

Find the sum of the infinite geometric series in Problems 59–62.

59. $2 + \sqrt{2} + 1 + \cdots$

60. $3 + \sqrt{3} + 1 + \cdots$

61. $(1 + \sqrt{2}) + 1 + (-1 + \sqrt{2}) + \cdots$

62. $(\sqrt{2} - 1) + 1 + (\sqrt{2} + 1) + \cdots$

63. Find three distinct numbers with a sum equal to 9 so that these numbers form an arithmetic sequence and their squares form a geometric sequence.

64. The infinite series

$$\lim_{n\to\infty} \sum_{k=1}^{n} \frac{1}{k^p} = \frac{1}{1^p} + \frac{1}{2^p} + \frac{1}{3^p} + \cdots$$

is called the ***p*-series.** Look at the sequence of partial sums $s_1, s_2, s_3, s_4, \ldots$, and state whether you think the given infinite p-series has a limit.

a. $p = 1$ **b.** $p = 2$ **c.** $p = 3$ **d.** $p = 0.5$

65. Can you make a conjecture (guess) about the values of p in Problem 64 for which the infinite p-series has a limit?

66. Square $ABCD$ has sides of length 1. Square $EFGH$ is formed by connecting the midpoints of the sides of the first square, as shown in Figure 10.4. Assume that the pattern of shaded regions in the square is continued indefinitely. What is the total area of the shaded regions?

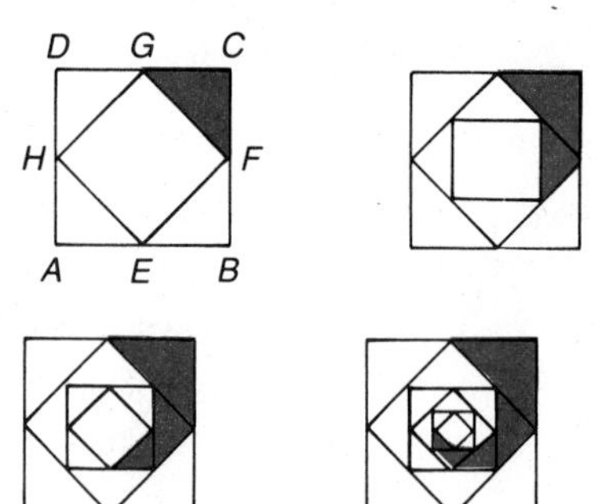

Figure 10.4 Find the area of the shaded regions.

67. Repeat Problem 66 for a square whose sides have length a.

10.6 CHAPTER 10 SUMMARY

The material of this chapter is reviewed in the following list of objectives. After each objective there are some practice questions. For a sample test, select the first question of each set and check your answers with the answer section. For a sample test without answers, use the second question of each set. Additional practice is given by the other questions in each set. If you are having trouble with a particular type of problem, look back to that section for extra help.

10.1 MATHEMATICAL INDUCTION

Objective 1 State the Principle of Mathematical Induction and prove propositions using this principle.

1. State the Principle of Mathematical Induction.
2. If $4 + 8 + 12 + \cdots + 4n = 2n(n + 1)$, then state the proposition $P(1)$ and prove it is true.
3. State the propositions $P(k)$ and $P(k + 1)$ for the proposition given in Problem 2.
4. Prove that $4 + 8 + 12 + \cdots + 4n = 2n(n + 1)$ for all positive integers n.

10.2 BINOMIAL THEOREM

Objective 2 Use Pascal's triangle to expand binomials. Expand:

5. $(a + b)^5$ **6.** $(x - y)^5$ **7.** $(2x + y)^5$ **8.** $(3x^2 - y^3)^5$

Objective 3 Simplify expressions containing factorial and binomial coefficient notations.

9. $\dfrac{52!}{5!\,47!}$ **10.** $\left(\dfrac{8}{4}\right)!$ **11.** $\dbinom{8}{4}$ **12.** $\dbinom{p}{q}$

Objective 4 State the Binomial Theorem, expand binomials by formula, and find particular terms in a binomial expansion.

13. State the binomial formula with and without summation notation.
14. Write out the first four terms in the expansion of $(a + b)^{18}$.
15. Write the term that includes y^r in the expansion of $(x - y)^{15}$.
16. What is the coefficient of x^8y^4 in the expansion of $(x + 2y)^{12}$?

10.3 SEQUENCES, SERIES, AND SUMMATION NOTATION

Objective 5 Classify a sequence as arithmetic, geometric, or neither. If it is arithmetic or geometric find d or r, and give the general term; if it is neither, find the pattern and give the next two terms.

17. 1, 11, 21, 31, ... **18.** 1, 11, 121, 1331, ...
19. 1, 11, 111, 1111, ... **20.** 54, 18, 6, 2, ...

10.4, 10.5 ARITHMETIC AND GEOMETRIC SEQUENCES AND SERIES

Objective 6 Use summation notation with series. Evaluate:

21. $\sum_{k=0}^{3} 3^k$ **22.** $\sum_{k=1}^{4} 5$
23. $\sum_{k=1}^{10} 2(3)^{k-1}$ **24.** $\sum_{k=1}^{100} [5 + (k - 1)4]$

Objective 7 Find the sum of n terms of an arithmetic or geometric sequence. Find the sum of the first 10 terms of the following sequences.

25. 1, 11, 21, 31, ... **26.** 1, 11, 121, 1331, ...
27. 54, 18, 6, 2, ... **28.** 5, 5, 5, 5, ...

Objective 8 *Work with the following formulas for sequences and series:*

Arithmetic Sequence

$$a_n = a_1 + (n - 1)d$$

Geometric Sequence

$$g_n = g_1 r^{n-1}$$

Arithmetic Series

$$A_n = n\left(\frac{a_1 + a_n}{2}\right)$$

$$A_n = \frac{n}{2}[2a_1 + (n - 1)d]$$

Geometric Series

$$G_n = \frac{g_1(1 - r^n)}{1 - r}$$

$$G = \frac{g_1}{1 - r}, |r| < 1$$

Find the indicated quantities for the given sequences:

29. $a_1 = 2, a_{10} = 20; d, A_{10}$

30. $a_2 = 5, d = 13; a_n$

31. $g_1 = 5, r = 2; g_{10}, G_5$

32. $a_1 = 50, d = -5; a_{10}, A_5$

Objective 9 *Find the sum of an infinite geometric series including the fractional representation of a repeating decimal.*

33. Find the sum of the infinite series $1000 + 500 + 250 + \cdots$.

34. Find the sum of the infinite series $\frac{1}{8} + \frac{1}{16} + \frac{1}{32} + \cdots$.

35. Find $2.\overline{18}$ as the quotient of two integers by considering an infinite geometric series.

36. Find $3.1\overline{6}$ as the quotient of two integers by considering an infinite geometric series.

Rene Descartes (1596–1650)

It is impossible not to feel stirred at the thought of the emotions of men at certain historic moments of adventure and discovery — Columbus when he first saw the Western shore, Pizarro when he stared at the Pacific Ocean, Franklin when the electric spark came from the string of his kite, Galileo when he first turned his telescope to the heavens. Such moments are also granted to students in the abstract regions of thought, and high among them must be placed the morning when Descartes lay in bed and invented the method of coordinate geometry.

A. N. Whitehead
An Introduction to Mathematics

HISTORICAL NOTE

In this text we have seen many ideas that revolutionized mathematics and the nature of thought. None is more profound than the powerful idea of associating ordered pairs of real numbers with points in a plane, thereby making possible a correspondence between curves in the plane and equations in two variables. This idea came from the French mathematician René Descartes.

Descartes was born in France on March 31, 1596. He came from a wealthy family, but was a weakling with frail health. Because of his delicate health he formed a lifelong habit of spending the mornings in bed, and he did much of his work during those morning hours.

According to legend, Descartes thought of his coordinate system while he was lying in bed watching a fly crawl around on the ceiling of the room. He noticed that the path of the fly could be described if he knew the relation connecting the fly's distances from the walls.

In 1649, Descartes was invited to tutor Queen Christina of Sweden. Since she was from a royal family Descartes thought the job would be fun, and he accepted the invitation. However, when he arrived, he found that she wanted her lessons at daybreak. Since he was not used to getting up early, he soon became ill and died of pneumonia within a year.

It should be pointed out that 1800 years earlier the Greek mathematician Apollonius (*c.* 262–190 B.C.) made a thorough investigation of the *conic sections* introduced in this chapter. He considered himself a rival of Archimedes and was the first person to use the words *parabola*, *ellipse*, and *hyperbola*. His methods were expounded in an eight-volume work called *Conics*, and his work was so modern it is sometimes judged to be an analytic geometry preceding Descartes.

11

Analytic Geometry—Conic Sections

CONTENTS

* Optional section

PREVIEW

The conic sections consist of the circles (already considered), parabolas, ellipses, and hyperbolas. In this chapter we consider the standard and general forms of each of these curves as well as the translated and rotated forms. (Section 11.4 on rotations may easily be omitted.) There are 10 objectives for this chapter, which are listed on page 459.

PERSPECTIVE

Although many calculus books include analytic geometry, they often assume that students have a rudimentary knowledge of the fundamentals, in particular of conic sections. In the following page taken from a calculus book, we see hyperbolas being used in a navigational system.

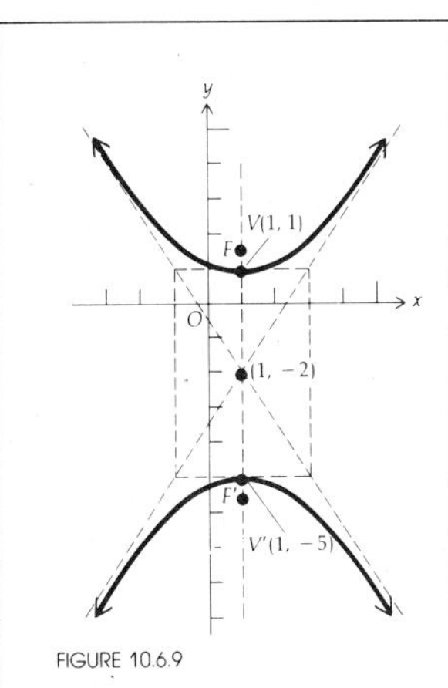

FIGURE 10.6.9

$$9(x^2 - 2x + 1) - 4(y^2 + 4y + 4) = -29 + 9 - 16$$

$$9(x - 1)^2 - 4(y + 2)^2 = -36$$

$$\frac{(y+2)^2}{9} - \frac{(x-1)^2}{4} = 1 \qquad (14)$$

Equation (14) has the form of (11); so the graph is a hyperbola whose principal axis is parallel to the y axis and whose center is at $(1, -2)$.

EXAMPLE 5

Find the vertices and foci of the hyperbola of Example 4. Draw a sketch showing the hyperbola, its asymptotes, and the foci.

SOLUTION

From (14) we observe that $a = 3$ and $b = 2$. Because the principal axis is vertical, the center is at $(1, -2)$, and $a = 3$, it follows that the vertices are at the points $V(1, 1)$ and $V'(1, -5)$. For a hyperbola, $c^2 = a^2 + b^2$; therefore $c^2 = 9 + 4$ and $c = \sqrt{13}$. Therefore the foci are at $F(1, -2 + \sqrt{13})$ and $F'(1, -2 - \sqrt{13})$. Figure 10.6.9 shows the hyperbola, its asymptotes, and the foci.

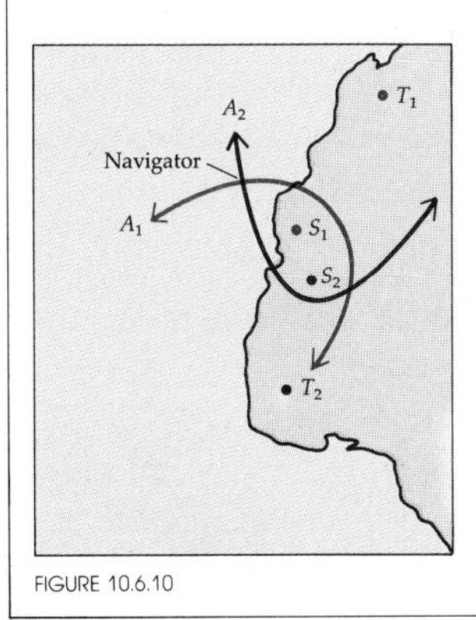

FIGURE 10.6.10

The property of the hyperbola given in Definition 10.6.1 forms the basis of several important navigational systems. These systems involve a network of pairs of radio transmitters at fixed positions at a known distance from one another. The transmitters send out radio signals that are received by a navigator. The difference in arrival time of the two signals determines the difference $2a$ of the distances from the navigator. Thus the navigator's position is known to be somewhere along one arc of a hyperbola having foci at the locations of the two transmitters. One arc, rather than both, is determined because of the signal delay between the two transmitters that is built into the system. The procedure is then repeated for a different pair of radio transmitters, and another arc of a hyperbola that contains the navigator's position is determined. The point of intersection of the two hyperbolic arcs is the actual position. For example, in Fig. 10.6.10 suppose a pair of transmitters is located at points T_1 and S_1 and the signals from this pair determine the hyperbolic arc A_1. Another pair of transmitters is located at points T_2 and S_2 and hyperbolic arc A_2 is determined from their signals. Then the intersection of A_1 and A_2 is the position of the navigator.

11.1 PARABOLAS

In Chapter 2 we looked at quadratic functions of the form

$$y = ax^2 + bx + c \qquad (a \neq 0)$$

The graph of this quadratic equation is a parabola, but not all parabolas can be represented by this equation, because not all parabolas are graphs of functions. Consider the general second-degree equation

$$Ax^2 + Bxy + Cy^2 + Dx + Ey + F = 0$$

for any constants A, B, C, D, E, and F. If $A = B = C = 0$, then the equation is not quadratic but linear (first degree); and if at least one of A, B, or C is not zero, then the equation is quadratic.

Historically, second-degree equations in two variables were first considered in a geometric context and were called **conic sections** because the curves they represent can be described as the intersections of a double-napped right circular cone and a plane. There are three general ways a plane can intersect a cone, as shown in Figure 11.1. (Several special cases are discussed later.)

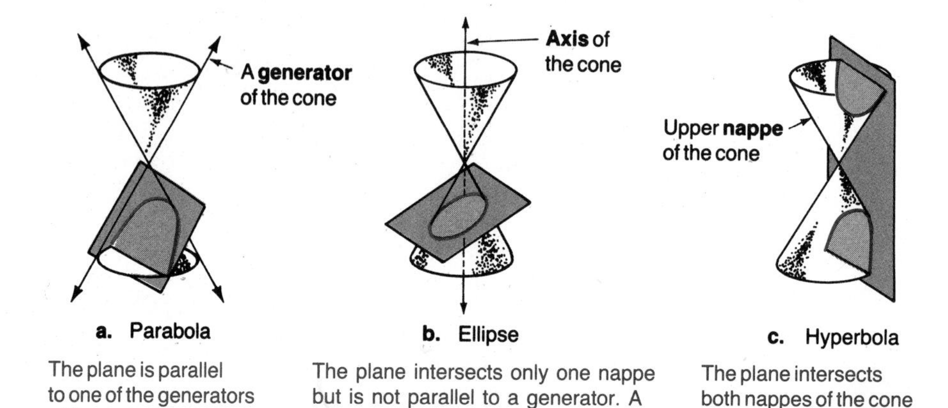

Figure 11.1 Conic sections

Reconsider the parabola; this time take a different geometric viewpoint:

Parabola

> A **parabola** is a set of all points in the plane equidistant from a given point (called the **focus**) and a given line (called the **directrix**).

To obtain the graph of a parabola from this definition, you can use the special type of graph paper shown in Figure 11.2, where F is the focus and L is the directrix.

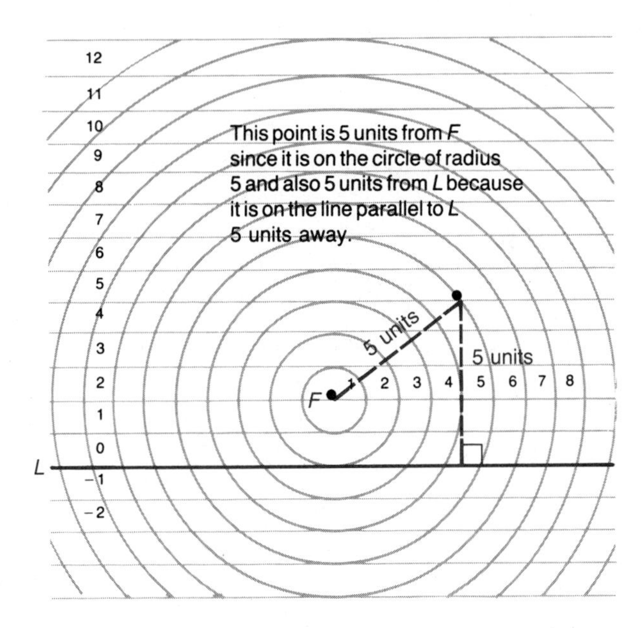

Figure 11.2 Parabola graph paper

To sketch a parabola using the definition, let F be any point and let L be any line, as shown in Figure 11.2. Plot points in the plane equidistant from the focus and the directrix. Draw a line through the focus and perpendicular to the directrix. This line is called the **axis** of the parabola. Let V be the point on this line halfway between the focus and the directrix. This is the point of the parabola nearest to both the focus and the directrix. It is called the **vertex** of the parabola. Plot other points equidistant from F and L as shown in Figure 11.3.

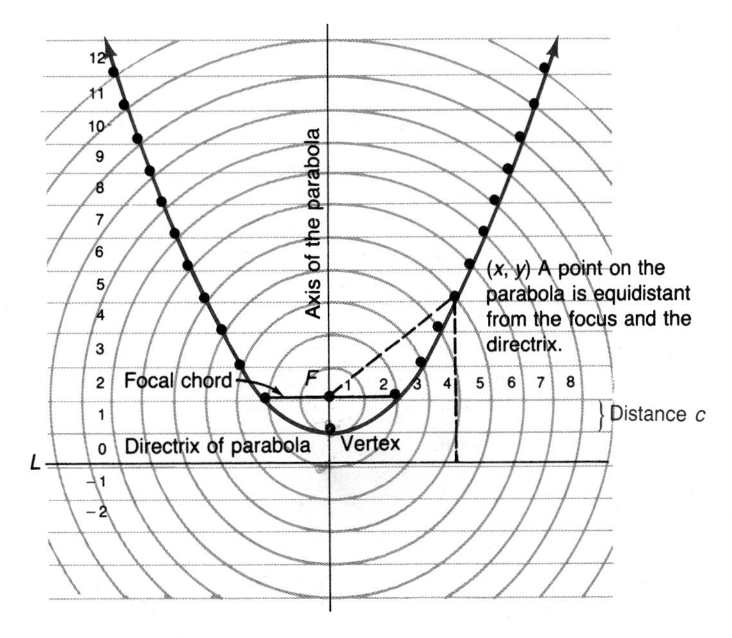

Figure 11.3 Parabola graphed from definition

In Figure 11.3, let c be the distance from the vertex to the focus. Notice that the distance from the vertex to the directrix is also c. Consider the segment that passes through the focus perpendicular to the axis and with endpoints on the parabola. This segment has length $4c$ and is called the **focal chord**.

To obtain the equation of a parabola, first consider a special case—a parabola with focus $F(0, c)$ and directrix $y = -c$, where c is any positive number. This parabola must have its vertex at the origin (remember that the vertex is halfway between the focus and the directrix) and must open upward as shown in Figure 11.4.

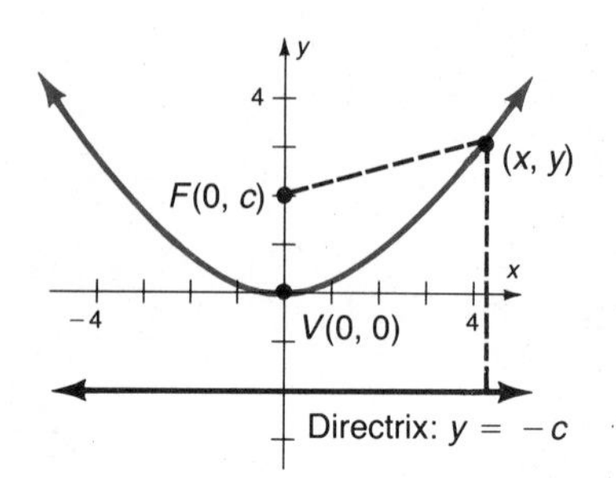

Figure 11.4 Graph of the parabola $x^2 = 4cy$

Let (x, y) be any point on the parabola. Then, from the definition of a parabola,

Distance from (x, y) to $(0, c)$ = distance from (x, y) to directrix

or

$$\sqrt{(x-0)^2 + (y-c)^2} = y + c$$

Square both sides:

$$x^2 + (y-c)^2 = (y+c)^2$$
$$x^2 + y^2 - 2cy + c^2 = y^2 + 2cy + c^2$$
$$x^2 = 4cy$$

This is the equation of the parabola with vertex $(0, 0)$ and directrix $y = -c$.

You can repeat this argument for parabolas that have vertex at the origin and open downward, to the left, and to the right to obtain the results summarized below. A positive number c, the distance from the focus to the vertex, is assumed given. These are called the **standard-form parabola equations** with vertex $(0, 0)$.

Standard-Form Equations for Parabolas with Vertex (0, 0)

Parabola	Focus	Directrix	Vertex	Equation
Opens *upward*	$(0, c)$	$y = -c$	$(0, 0)$	$x^2 = 4cy$
Opens *downward*	$(0, -c)$	$y = c$	$(0, 0)$	$x^2 = -4cy$
Opens *right*	$(c, 0)$	$x = -c$	$(0, 0)$	$y^2 = 4cx$
Opens *left*	$(-c, 0)$	$x = c$	$(0, 0)$	$y^2 = -4cx$

EXAMPLE 1 Graph $x^2 = 8y$.

Solution This equation represents a parabola that opens upward. The vertex is $(0, 0)$, and notice by inspection that

$$4c = 8$$
$$c = 2$$

Thus the focus is $(0, 2)$. After plotting the vertex $V(0, 0)$ and the focus $F(0, 2)$, the only question is the width of the parabola. In Chapter 2 you plotted a couple of points or found the y intercept. The method in these examples is more useful and efficient for determining the graph of a parabola. In this chapter we will determine the width of the parabola by using the focal chord. Remember that the **length of the focal chord is 4*c***, so that in this case it is 8. Do you see that in each case $4c$ is the absolute value of the coefficient of the first-degree term in the standard-form equations? Since a parabola is symmetric with respect to its axis, draw a segment of length 8 with the midpoint at F. Using these three points (the vertex and the endpoints of the focal chord), sketch the parabola as shown in Figure 11.5. ■

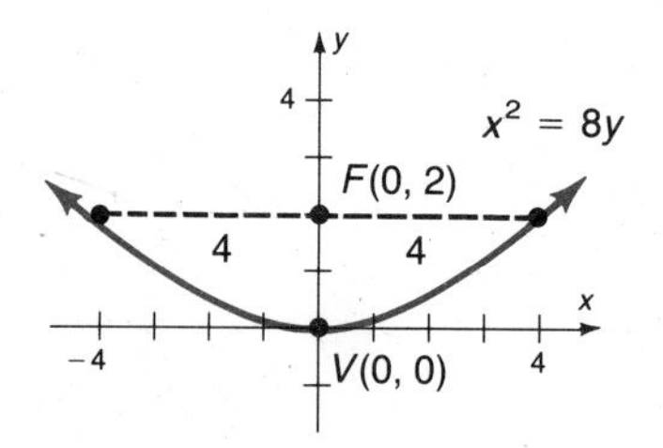

Figure 11.5 Graph of the parabola $x^2 = 8y$

EXAMPLE 2 Graph $y^2 = -12x$.

Solution This equation represents a parabola that opens left. The vertex is $(0,0)$ and

$$4c = 12$$
$$c = 3$$

(recall that c is positive), so the focus is $(-3,0)$. The length of the focal chord is 12 and the parabola is drawn as in Figure 11.6.

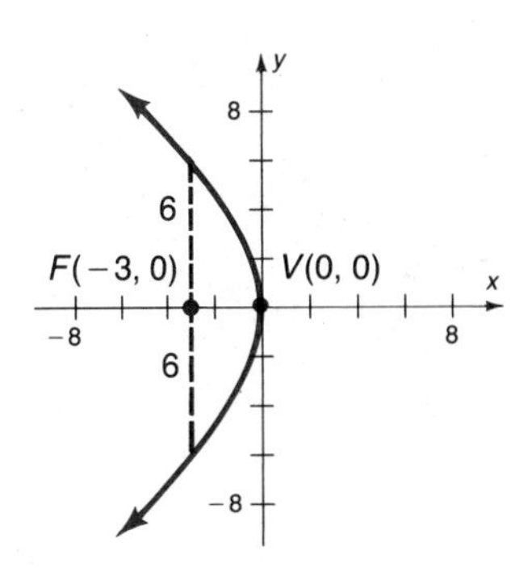

Figure 11.6 Graph of the parabola $y^2 = -12x$ ■

EXAMPLE 3 Graph $2y^2 - 5x = 0$.

Solution You might first put the equation into standard form by solving for the second-degree term:

$$y^2 = \frac{5}{2}x$$

The vertex is $(0,0)$, and

$$4c = \frac{5}{2}$$

$$c = \frac{5}{8}$$

Thus the parabola opens to the right, the focus is $(\frac{5}{8},0)$, and the length of the focal chord is $\frac{5}{2}$, as shown in Figure 11.7.

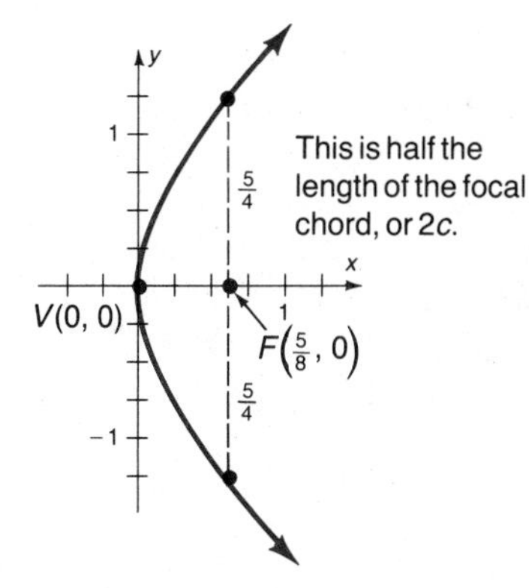

Figure 11.7 Graph of the parabola $2y^2 - 5x = 0$ ■

There are two basic types of problems you will need to solve concerning each curve in analytic geometry:

1. Given the equation, draw the graph; this is what you did in Examples 1 to 3.
2. Given the graph (or information about the graph), write the equation. Example 4 is a problem of this type.

EXAMPLE 4 Find the equation of the parabola with directrix $y = 4$ and focus at $F(0, -4)$.

Solution This curve is a parabola that opens downward with vertex at the origin, as shown in Figure 11.8, page 428. The value for c is found by inspection: $c = 4$. Thus $4c = 16$. Since the equation is of the form $x^2 = -4cy$, the desired equation is (by substitution)

$$\mathbf{x^2 = -16y}$$

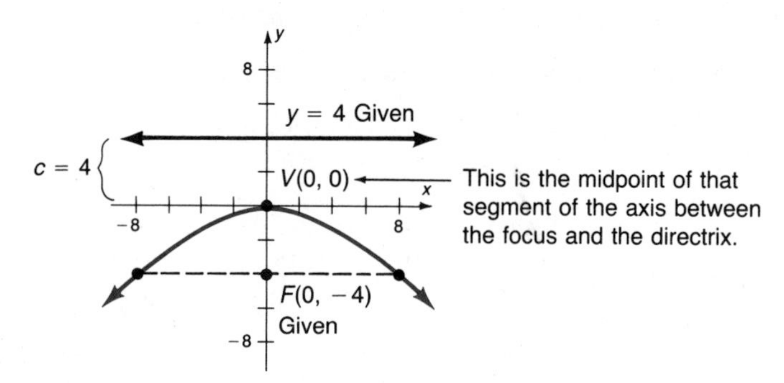

Figure 11.8 Graph of the parabola with focus at $(0, -4)$ and directrix $y = 4$

■

The types of parabolas we have been considering are quite limited, since we have assumed that the vertex is at the origin and the directrix is parallel to one of the coordinate axes. Suppose, however, that you are given a parabola with vertex at (h, k) and a directrix parallel to one of the coordinate axes. In Section 2.4 we showed that you can translate the axes to (h, k) by a substitution:

$$x' = x - h$$
$$y' = y - k$$

Therefore the **standard-form parabola equations** with vertex (h, k) can be summarized by the following table.

Standard-Form Equations for Translated Parabolas

Parabola	Focus	Directrix	Vertex	Equation
Opens upward	$(h, k+c)$	$y=k-c$	(h, k)	$(x-h)^2=4c(y-k)$
Opens downward	$(h, k-c)$	$y=k+c$	(h, k)	$(x-h)^2=-4c(y-k)$
Opens right	$(h+c, k)$	$x=h-c$	(h, k)	$(y-k)^2=4c(x-h)$
Opens left	$(h-c, k)$	$x=h+c$	(h, k)	$(y-k)^2=-4c(x-h)$

For the rest of the material in this chapter you will need to remember the procedure for **completing the square**. Examples 5 and 6 should provide enough detail to refresh your memory; if you need additional review, see Section 2.5, pages 74–75.

EXAMPLE 5 Sketch $x^2 + 4y + 8x + 4 = 0$.

Solution *Step 1* Associate together the terms involving the variable that is squared:

$$x^2 + 8x = -4y - 4$$

Step 2 Complete the square for the variable that is squared.

Coefficient is 1; if it is not, divide both sides by this coefficient.

$$x^2 + 8x + \left(\frac{1}{2} \cdot 8\right)^2 = -4y - 4 + \left(\frac{1}{2} \cdot 8\right)^2$$

Take one-half of this coefficient, square it, and add it to both sides.

$$x^2 + 8x + \mathbf{16} = -4y - 4 + \mathbf{16}$$
$$(x + 4)^2 = -4y + 12$$

Step 3 Factor out the coefficient of the first-degree term:

$$(x + 4)^2 = -4(y - 3)$$

Step 4 Determine the vertex by inspection. Plot (h, k); in this example, the vertex is $(-4, 3)$. (See Figure 11.9.)

Step 5 Determine the focus. By inspection, $4c = 4$, $c = 1$, and the parabola opens downward from the vertex as shown in Figure 11.9.

Step 6 Plot the endpoints of the focal chord; $4c = 4$. Draw the parabola as shown in Figure 11.9.

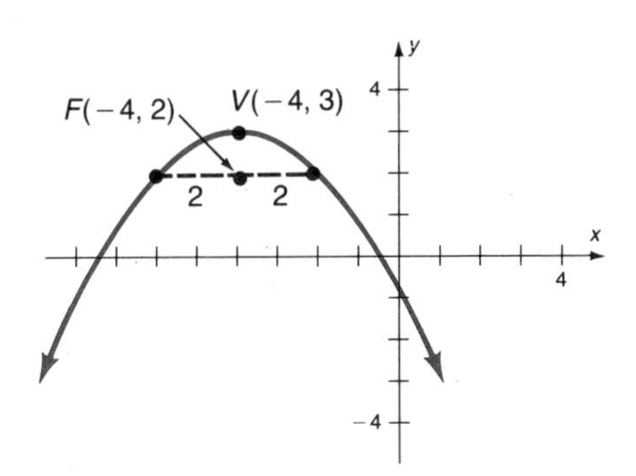

Figure 11.9 Graph of $x^2 + 4y + 8x + 4 = 0$

■

EXAMPLE 6 Sketch $2y^2 + 6y + 5x + 10 = 0$.

Solution

$$2y^2 + 6y = -5x - 10$$

$$y^2 + 3y = -\frac{5}{2}x - 5$$

Divide both sides by 2 so the leading coefficient is 1; then complete the square.

$$y^2 + 3y + \frac{9}{4} = -\frac{5}{2}x - 5 + \frac{9}{4}$$

$$\left(y + \frac{3}{2}\right)^2 = -\frac{5}{2}x - \frac{11}{4}$$

$$\left(y + \frac{3}{2}\right)^2 = -\frac{5}{2}\left(x + \frac{11}{10}\right)$$

The last step is to factor out the coefficient of the first-degree term.

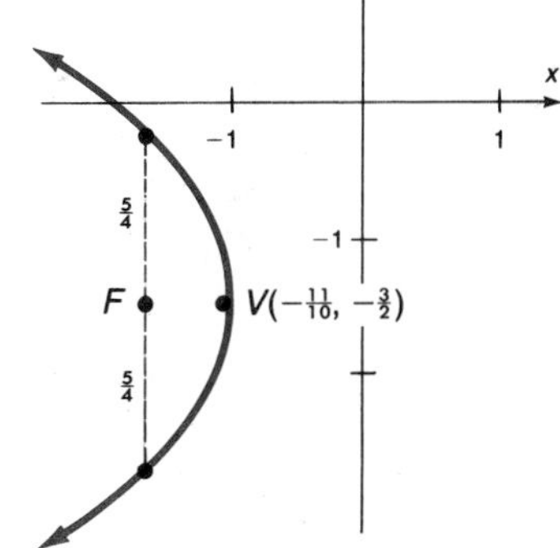

Figure 11.10 Graph of $2y^2 + 6y + 5x + 10 = 0$

The vertex is $(-\frac{11}{10}, -\frac{3}{2})$ and $4c = \frac{5}{2}$, $c = \frac{5}{8}$. Sketch the curve as shown in Figure 11.10. ■

EXAMPLE 7 Find the equation of the parabola with focus at $(4, -3)$ and directrix the line $x + 2 = 0$.

Solution Sketch the given information as shown in Figure 11.11, page 430. The vertex is $(1, -3)$ since it must be equidistant from F and the directrix. Note that $c = 3$. Thus substitute into the equation

$$(y-k)^2 = 4c(x-h)$$

since the parabola opens to the right. The desired equation is

$$(\mathbf{y}+\mathbf{3})^2 = \mathbf{12}(\mathbf{x}-\mathbf{1})$$

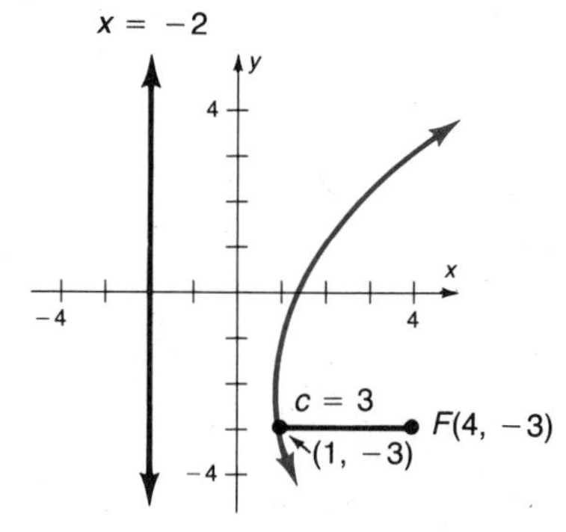

Figure 11.11 ■

By sketching the focus and directrix, it is easy to find c and the vertex by inspection. The sketch of the curve is not necessary in order to find this equation.

EXAMPLE 8 Find the equation of the parabola with vertex at $(2, -1)$, axis parallel to the y axis, and passing through $(3, 2)$.

Solution Sketch the given information as shown in Figure 11.12. The parabola opens upward and thus has the form

$$(x-h)^2 = 4c(y-k)$$

Since the vertex is $(2, -1)$,

$$(x-2)^2 = 4c(y+1)$$

Also, since it passes through $(3, 2)$, this point must satisfy the equation

$$(3-2)^2 = 4c(2+1)$$

Solving for c,

$$c = \frac{1}{12}$$

Therefore the desired equation is

$$(\mathbf{x}-\mathbf{2})^2 = \frac{\mathbf{1}}{\mathbf{3}}(\mathbf{y}+\mathbf{1})$$

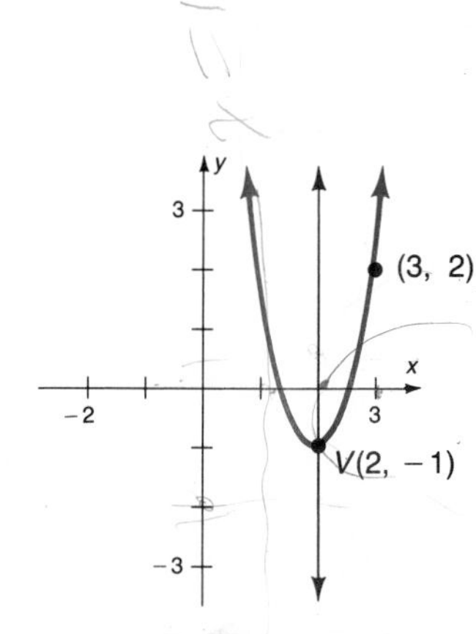

Figure 11.12 ■

PROBLEM SET 11.1

A

In Problems 1–6, use the definition of a parabola to sketch a parabola such that the distance between the focus and the directrix is the given number. (See Figure 11.2.)

1. 4 **2.** 6 **3.** 8 **4.** 10

5. 3 **6.** 7

Sketch the parabola satisfying the conditions given in Problems 7–16.

7. Directrix $x = 0$; focus at $(5, 0)$

8. Directrix $y = 0$; focus at $(0, -3)$

9. Directrix $x - 3 = 0$; vertex at $(-1, 2)$

10. Directrix $y + 4 = 0$; vertex at $(4, -1)$
11. Vertex at $(-2, -3)$; focus at $(-2, 3)$
12. Vertex at $(-3, 4)$; focus at $(1, 4)$
13. Vertex at $(-3, 2)$ and passing through $(-2, -1)$; axis parallel to the y axis
14. Vertex at $(4, 2)$ and passing through $(-3, -4)$; axis parallel to the x axis
15. The set of all points with distances from $(4, 3)$ that equal their distances from $(0, 3)$
16. The set of all points with distances from $(4, 3)$ that equal their distances from $(-2, 1)$

Sketch the curves given by the equations in Problems 17–40. Label the focus F and the vertex V in Problems 17–30.

17. $y^2 = 8x$
18. $y^2 = -12x$
19. $y^2 = -20x$
20. $4x^2 = 10y$
21. $3x^2 = -12y$
22. $2x^2 = -4y$
23. $2x^2 + 5y = 0$
24. $5y^2 + 15x = 0$
25. $3y^2 - 15x = 0$
26. $4y^2 + 3x = 12$
27. $5x^2 + 4y = 20$
28. $4x^2 + 3y = 12$
29. $(y - 1)^2 = 2(x + 2)$
30. $(y + 3)^2 = 3(x - 1)$
31. $(x + 2)^2 = 2(y - 1)$
32. $(x - 1)^2 = 3(y + 3)$
33. $(x - 1) = -2(y + 2)$
34. $(x + 3) = -3(y - 1)$
35. $y^2 + 4x - 3y + 1 = 0$
36. $y^2 - 4x + 10y + 13 = 0$
37. $2y^2 + 8y - 20x + 148 = 0$
38. $x^2 + 9y - 6x + 18 = 0$
39. $9x^2 + 6x + 18y - 23 = 0$
40. $9x^2 + 6y + 18x - 23 = 0$

B

Find the equation of each curve in Problems 41–50. You sketched these curves in Problems 7–16.

41. Directrix $x = 0$; focus at $(5, 0)$
42. Directrix $y = 0$; focus at $(0, -3)$
43. Directrix $x - 3 = 0$; vertex at $(-1, 2)$
44. Directrix $y + 4 = 0$; vertex at $(4, -1)$
45. Vertex at $(-2, -3)$; focus at $(-2, 3)$
46. Vertex at $(-3, 4)$; focus at $(1, 4)$
47. Vertex at $(-3, 2)$ and passing through $(-2, -1)$; axis parallel to the y axis
48. Vertex at $(4, 2)$ and passing through $(-3, -4)$; axis parallel to the x axis
49. The set of all points with distances from $(4, 3)$ that equal their distances from $(0, 3)$
50. The set of all points with distances from $(4, 3)$ that equal their distances from $(-2, 1)$

51. ***Physics*** If the path of a baseball is parabolic and is 200 ft wide at the base and 50 ft high in the vertex, write the equation that gives the path of the baseball if the origin is the point of departure for the ball.

52. ***Engineering*** A parabolic archway has the dimensions shown in Figure 11.13. Find the equation of the parabolic portion.

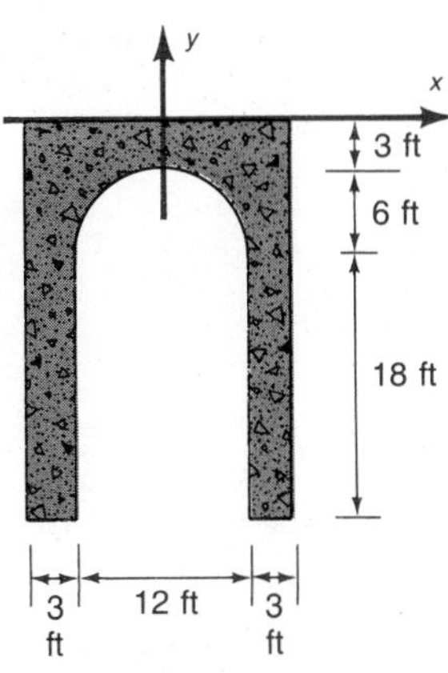

Figure 11.13 A parabolic archway

53. ***Engineering*** A radar antenna is constructed so that a cross section along its axis is a parabola with the receiver at the focus. Find the focus if the antenna is 12 m across and its depth is 4 m. See Figure 11.14.

a. Radar antenna

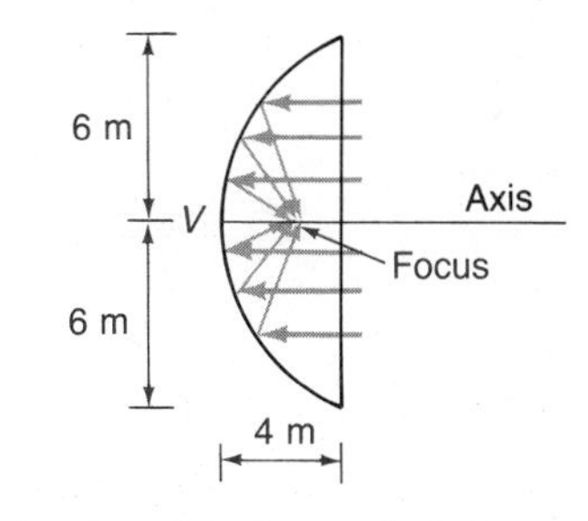

b. Dimensions for radar antenna in Problem 53

Figure 11.14 A three-dimensional model of a parabola is a **parabolic reflector** (or **parabolic mirror**). A radar antenna serves as an example. If a source of light is placed at the focus of the mirror, the light rays will reflect from the mirror as rays parallel to the axis (as in an automobile headlamp). The radar antenna works in reverse—parallel incoming rays are focused at a single location. (Photo courtesy of Wide World Photos, Inc.)

54. ***Engineering*** If the diameter of the parabolic reflector in Problem 53 is 16 cm and the depth is 8 cm, find the focus. (See Figure 11.14.)

C

Find the points of intersection (if any) for each line and parabola in Problems 55–58. Show the result both algebraically and geometrically.

55. $\begin{cases} y = 2x + 10 \\ y = x^2 + 4x + 7 \end{cases}$

56. $\begin{cases} y = -2x + 4 \\ y = x^2 - 12x + 25 \end{cases}$

57. $\begin{cases} 2x + y - 7 = 0 \\ y^2 - 6y - 4x + 17 = 0 \end{cases}$

58. $\begin{cases} 3x + 2y - 5 = 0 \\ y^2 + 4y + 3x - 4 = 0 \end{cases}$

59. ***Sports*** Phil Lee, a physical education expert, has made a study and determined that a woman who runs the 100-m dash reaches her peak at age 20. Her time T, for a 100-m dash at a particular age A, is

$$T = c_1(A - 20)^2 + c_2$$

for constants c_1 and c_2. If Wyomia Tyus ran the race when she was 16 years old in 11.4 sec and when she was 20 in 11.0 sec, predict her time for running the race when she is 40 years old. Graph this function.

60. ***Sports*** According to another physical education expert, Jim Kintzi, the peak age for a woman runner is 24. Using this information, answer the questions posed in Problem 59.

61. Derive the equation of a parabola with $F(0, -c)$, where c is a positive number and the directrix is the line $y = c$.

62. Derive the equation of a parabola with $F(c, 0)$, where c is a positive number and the directrix is the line $x = -c$.

63. Derive the equation of a parabola with $F(-c, 0)$, where c is a positive number and the directrix is the line $x = c$.

64. Show that the length of the focal chord for the parabola $y^2 = 4cx$ is $4c$ $(c > 0)$.

65. Let L: $Ax + By + C = 0$ be any nonvertical line and let $P(x_0, y_0)$ be any point not on the line.
    a. Find the slope of L.
    b. Let L' be a line through P perpendicular to L. Find the slope of L'.
    c. Find the equation of L'.
    d. Let Q be the point of intersection of L and L'. Find Q.
    e. Find the distance between P and Q to show that the distance from a point to a line is

$$d = \frac{|Ax_0 + By_0 + C|}{\sqrt{A^2 + B^2}}$$

66. Find the equation of the parabola with focus at $(4, -3)$ and directrix $x - y + 3 = 0$. (*Hint:* Use the definition of a parabola and the formula derived in Problem 65.)

67. Find the equation of the parabola with focus at $(3, -5)$ and directrix $12x - 5y + 4 = 0$. (*Hint:* Use the definition of a parabola and the formula derived in Problem 65.)

11.2 ELLIPSES

The second conic considered in this chapter is called an *ellipse*.

Ellipse

> An **ellipse** is the set of all points in the plane such that, for each point on the ellipse, the sum of its distances from two fixed points is a constant.

The fixed points are called the **foci** (plural of **focus**). To see what an ellipse looks like, we will use the special type of graph paper shown in Figure 11.15a, where F_1 and F_2 are the foci.

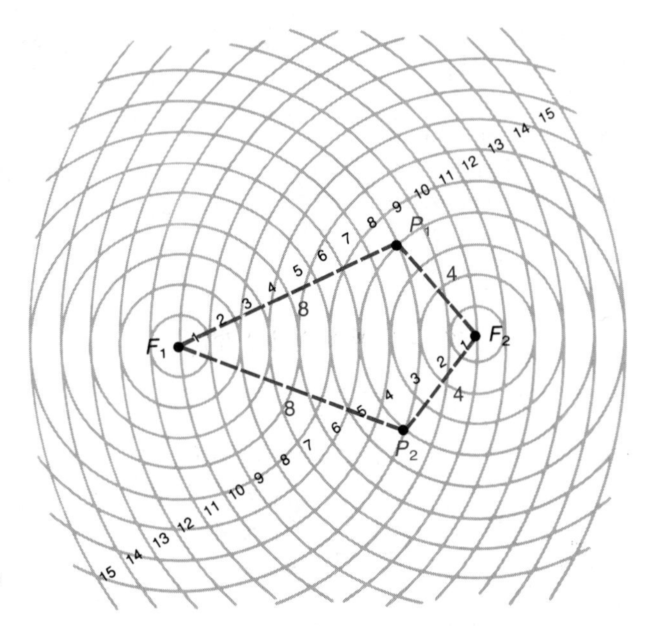

Figure 11.15a Ellipse graph paper

Let the given constant be 12. Plot all the points in the plane so that the sum of their distances from the foci is 12. If a point is 8 units from F_1, for example, then it is 4 units from F_2 and you can plot the points P_1 and P_2. The completed graph of this ellipse is shown in Figure 11.15b.

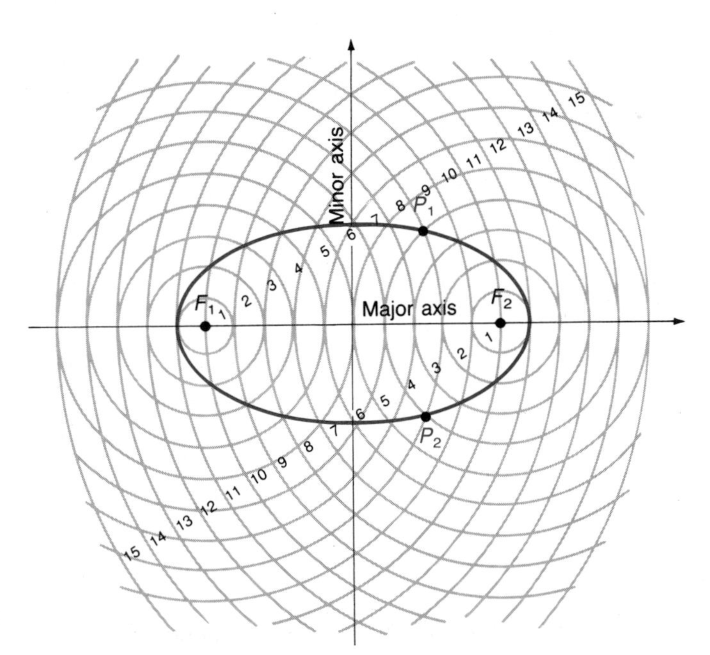

Figure 11.15b Graphing an ellipse

The line passing through F_1 and F_2 is called the **major axis**. The **center** is the midpoint of the segment $\overline{F_1F_2}$. The line passing through the center perpendicular to the major axis is called the **minor axis**. The ellipse is symmetric with respect to both the major and minor axes.

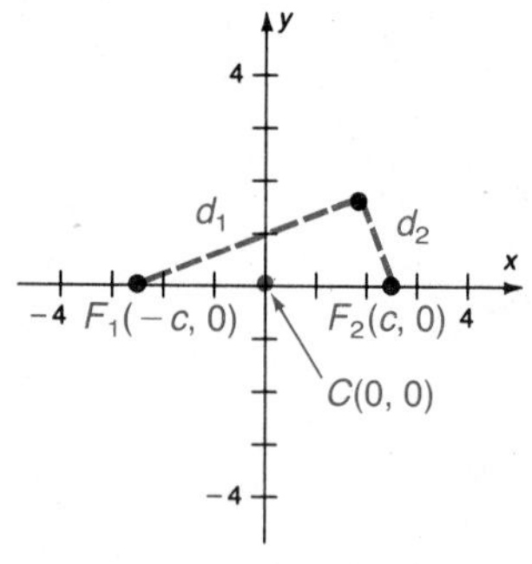

Figure 11.16 Developing the equation of an ellipse by using the definition

To find the equation of an ellipse, first consider a special case where the center is at the origin. Let the distance from the center to a focus be the positive number c; that is, let $F_1(-c, 0)$ and $F_2(c, 0)$ be the foci and let the constant distance be $2a$, as shown in Figure 11.16.

Notice that the center of the ellipse in Figure 11.16 is $(0, 0)$. Let $P(x, y)$ be any point on the ellipse, and use the distance formula and the definition of an ellipse to derive the equation of this ellipse:

$$d_1 + d_2 = 2a$$

or

$$\sqrt{(x+c)^2 + (y-0)^2} + \sqrt{(x-c)^2 + (y-0)^2} = 2a$$

Simplifying,

$$\sqrt{(x+c)^2 + y^2} = 2a - \sqrt{(x-c)^2 + y^2} \qquad \text{Isolate one radical.}$$

$$(x+c)^2 + y^2 = 4a^2 - 4a\sqrt{(x-c)^2 + y^2} + (x-c)^2 + y^2 \qquad \text{Square both sides.}$$

$$x^2 + 2cx + c^2 + y^2 = 4a^2 - 4a\sqrt{(x-c)^2 + y^2} + x^2 - 2cx + c^2 + y^2$$

$$4a\sqrt{(x-c)^2 + y^2} = 4a^2 - 4cx$$

$$\sqrt{(x-c)^2 + y^2} = a - \frac{c}{a}x \qquad \text{Since } a \neq 0 \text{, divide by } 4a.$$

$$(x-c)^2 + y^2 = \left(a - \frac{c}{a}x\right)^2 \qquad \text{Square both sides again.}$$

$$x^2 - 2cx + c^2 + y^2 = a^2 - 2cx + \frac{c^2}{a^2}x^2$$

$$x^2 + y^2 = a^2 - c^2 + \frac{c^2}{a^2}x^2$$

$$x^2 - \frac{c^2}{a^2}x^2 + y^2 = a^2 - c^2$$

$$\left(1 - \frac{c^2}{a^2}\right)x^2 + y^2 = a^2 - c^2$$

$$\frac{a^2 - c^2}{a^2}x^2 + y^2 = a^2 - c^2$$

$$\frac{x^2}{a^2} + \frac{y^2}{a^2 - c^2} = 1 \qquad \text{Divide both sides by } a^2 - c^2.$$

Let $b^2 = a^2 - c^2$; then

$$\frac{x^2}{a^2} + \frac{y^2}{b^2} = 1$$

This substitution is a notational change that will simplify the equation.

If $x = 0$, the y intercepts are obtained:

$$\frac{y^2}{b^2} = 1$$

$$y = \pm b$$

If $y = 0$, the x intercepts are obtained: $x = \pm a$. The intercepts on the major axis are called the **vertices** of the ellipse.

The equation of the ellipse with major axis vertical, $F_1(0, c)$, $F_2(0, -c)$, and constant distance $2a$ is found in a similar fashion. Simplifying the equation as before,

$$\frac{y^2}{a^2} + \frac{x^2}{b^2} = 1$$

where $b^2 = a^2 - c^2$.

Notice that in both cases a^2 must be larger than both c^2 and b^2. If it were not, a square number would be equal to a negative number, which is a contradiction in the set of real numbers.

Standard-Form Equations for Ellipses with Center (0, 0)

Ellipse	Foci	Constant distance	Center	Equation
Horizontal	$(-c, 0), (c, 0)$	$2a$	$(0, 0)$	$\frac{x^2}{a^2} + \frac{y^2}{b^2} = 1$
Vertical	$(0, c), (0, -c)$	$2a$	$(0, 0)$	$\frac{y^2}{a^2} + \frac{x^2}{b^2} = 1$
where $b^2 = a^2 - c^2$ or $c^2 = a^2 - b^2$				

EXAMPLE 1 Sketch $\frac{x^2}{9} + \frac{y^2}{4} = 1$.

Solution The center of the ellipse is $(0, 0)$. The x intercepts are ± 3 (these are the vertices) and the y intercepts are ± 2. Sketch the ellipse as shown in Figure 11.17.

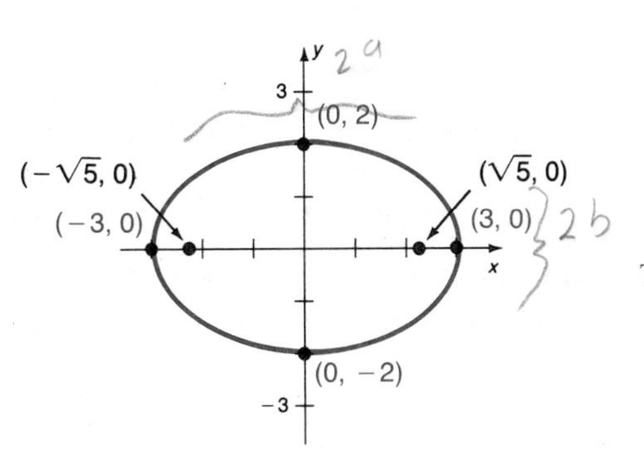

Figure 11.17 Graph of $\frac{x^2}{9} + \frac{y^2}{4} = 1$

The foci can also be found, since

$$c^2 = a^2 - b^2$$
$$c^2 = 9 - 4$$
$$c = \pm\sqrt{5}$$

■

EXAMPLE 2 Sketch $\frac{x^2}{4} + \frac{y^2}{9} = 1$.

Solution Here $a^2 = 9$ and $b^2 = 4$, which is an ellipse with major axis vertical. The x intercepts are ± 2 and the y intercepts are ± 3 (these are the vertices). The sketch is shown in Figure 11.18, at the top of the next page.

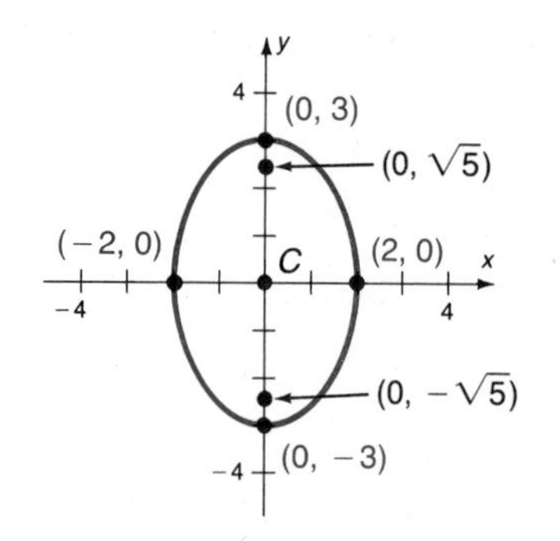

Figure 11.18 Graph of $\frac{x^2}{4} + \frac{y^2}{9} = 1$

The foci are found by

$$c^2 = a^2 - b^2$$
$$c^2 = 9 - 4$$
$$c = \pm\sqrt{5}$$

Remember: The foci are always on the major axis, so plot $(0, \sqrt{5})$ and $(0, -\sqrt{5})$ for the foci. ■

If the center of the ellipse is at (h, k), the equations can be written in terms of a translation as shown by the following table.

Standard-Form Equations for Translated Ellipses

Ellipse	Center	Foci	Intercepts: Major axis	Intercepts: Minor axis	Equation
Horizontal	(h, k)	$(h + c, k)$ $(h - c, k)$	$(h - a, k)$ $(h + a, k)$	$(h, k + b)$ $(h, k - b)$	$\frac{(x - h)^2}{a^2} + \frac{(y - k)^2}{b^2} = 1$
Vertical	(h, k)	$(h, k + c)$ $(h, k - c)$	$(h, k + a)$ $(h, k - a)$	$(h - b, k)$ $(h + b, k)$	$\frac{(y - k)^2}{a^2} + \frac{(x - h)^2}{b^2} = 1$

The segment from the center to a vertex on the major axis is called a **semimajor axis** and has length a; the segment from the center to an intercept on the minor axis is called a **semiminor axis** and has length b. Use a and b to plot four points, which are in turn used to draw the curve.

EXAMPLE 3 Graph $\frac{(x - 3)^2}{25} + \frac{(y - 1)^2}{16} = 1$.

Solution *Step 1* Plot the center (h, k). By inspection, the center of this ellipse is $(3, 1)$. This becomes the center of a new translated coordinate system. The vertices and foci are now measured with reference to the new origin at $(3, 1)$.

Step 2 Plot the x' and y' intercepts. These are ± 5 and ± 4, respectively. Remember to measure these distances from $(3, 1)$ as shown in Figure 11.19.

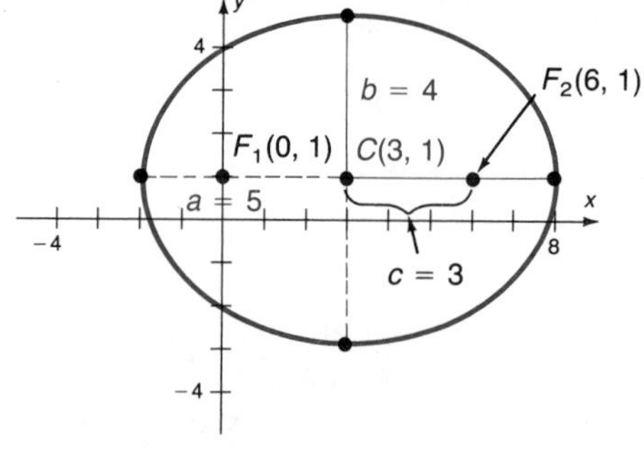

Figure 11.19 Graph of $\frac{(x - 3)^2}{25} + \frac{(y - 1)^2}{16} = 1$

The foci are found by

$$c^2 = a^2 - b^2$$
$$= 25 - 16$$
$$= \pm 9$$

The distance from the center to either focus is 3, so the coordinates of the foci are $(6, 1)$ and $(0, 1)$. ■

EXAMPLE 4 Graph $3x^2 + 4y^2 + 24x - 16y + 52 = 0$.

Solution *Step 1* Associate together the x and the y terms:

$$(3x^2 + 24x) + (4y^2 - 16y) = -52$$

Step 2 Complete the squares in both x and y. This requires that the coefficients of the squared terms be 1. You can accomplish this by factoring:

$$3(x^2 + 8x \qquad) + 4(y^2 - 4y \qquad) = -52$$

Next complete the square for both x and y, being sure to add the same number to both sides:

Add 16 to both sides.

$$3(x^2 + 8x + 16) + 4(y^2 - 4y + 4) = -52 + 48 + 16$$

Add 48 to both sides.

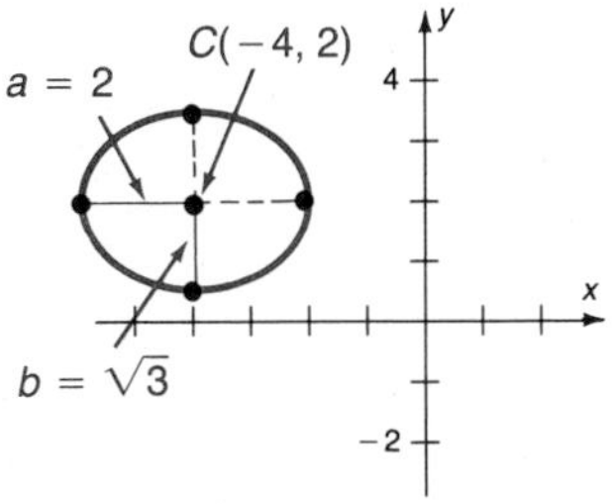

Step 3 Factor:

$$3(x + 4)^2 + 4(y - 2)^2 = 12$$

Step 4 Divide both sides by 12:

$$\frac{(x + 4)^2}{4} + \frac{(y - 2)^2}{3} = 1$$

Step 5 Plot the center (h, k). By inspection, you can see the center is $(-4, 2)$. The vertices are at ± 2, and the length of the semiminor axis is $\sqrt{3}$, as shown in Figure 11.20. ■

Figure 11.20 Graph of $3x^2 + 4y^2 + 24x - 16y + 52 = 0$

We have seen that some ellipses are more circular and some more flat than others. A measure of the amount of flatness of an ellipse is called its **eccentricity**, which is defined as

$$\epsilon = \frac{c}{a}$$

Notice that

$$\epsilon = \frac{c}{a} = \frac{\sqrt{a^2 - b^2}}{a} = \sqrt{\frac{a^2 - b^2}{a^2}} = \sqrt{1 - \left(\frac{b}{a}\right)^2}$$

Since $c < a$, ϵ is between 0 and 1. If $a = b$, then $\epsilon = 0$ and the conic is a **circle**. If the ratio b/a is small, then the ellipse is very flat. Thus, for an ellipse,

$$0 \leq \epsilon < 1$$

and ϵ measures the amount of roundness of the ellipse.

Consider a circle; that is, suppose $a = b$. In this case, let $r = a = b$ and call this distance the **radius**. You can see that a circle is a special case of an ellipse, as given in the following box:

Standard-Form Equation of a Circle

The equation of a **circle** with center at (h, k) and radius r is

$$(x-h)^2 + (y-k)^2 = r^2$$

EXAMPLE 5 Graph $x^2 + y^2 + 6x - 14y + 22 = 0$.

Solution Complete the square in x and y:

$$(x^2 + 6x \qquad) + (y^2 - 14y \qquad) = -22$$
$$(x^2 + 6x + \mathbf{9}) + (y^2 - 14y + \mathbf{49}) = -22 + \mathbf{9} + \mathbf{49}$$
$$(x+3)^2 + (y-7)^2 = 36$$

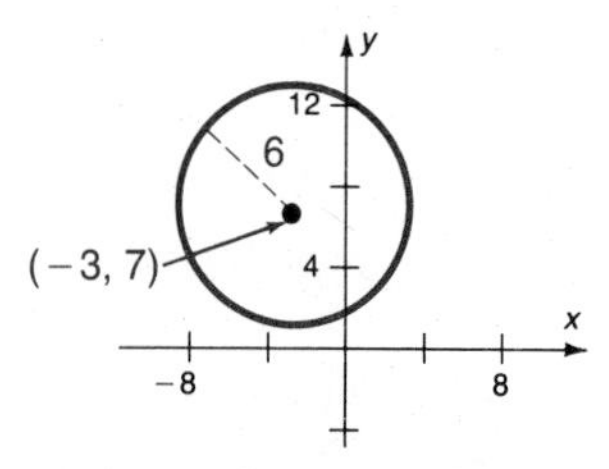

Figure 11.21 Graph of $x^2 + y^2 + 6x - 14y + 22 = 0$

This is a circle with center at $(-3, 7)$ and radius 6, as shown in Figure 11.21. ∎

EXAMPLE 6 Find the equation of the circle passing through the points $(-3, 4)$, $(4, 5)$, and $(1, -4)$. Sketch the graph.

Solution Now $(x-h)^2 + (y-k)^2 = r^2$ can be written as $x^2 + y^2 + C_1x + C_2y + C_3 = 0$ for some constants C_1, C_2, and C_3. Since the points lie on the circle, they satisfy the equation. Thus

$$\begin{aligned}(-3,4)&: & -3C_1 + 4C_2 + C_3 &= -25\\ (4,5)&: & 4C_1 + 5C_2 + C_3 &= -41\\ (1,-4)&: & C_1 - 4C_2 + C_3 &= -17\end{aligned}$$

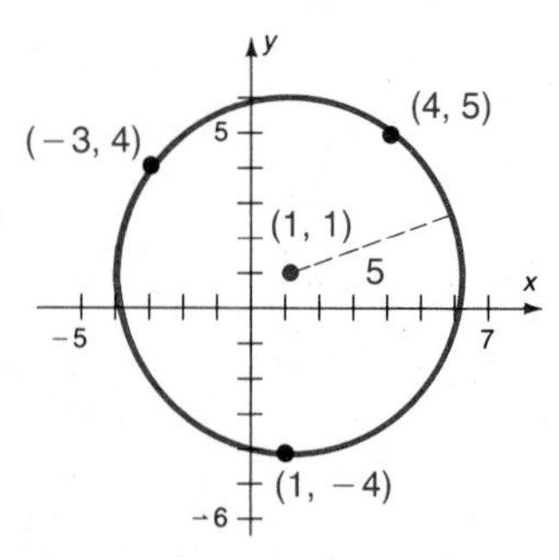

Figure 11.22 Graph of the circle passing through $(-3, 4)$, $(4, 5)$, and $(1, -4)$

This is a system of three linear equations in three unknowns that can be solved simultaneously by using the methods given in Chapter 9. (We will not show the details of this solution here.) After several steps you will find that $C_1 = -2$, $C_2 = -2$, and $C_3 = -23$. Now substitute these values into the equation of the circle:

$$x^2 + y^2 - 2x - 2y - 23 = 0$$
$$(x^2 - 2x + 1) + (y^2 - 2y + 1) = 23 + 1 + 1$$
$$\mathbf{(x-1)^2 + (y-1)^2 = 25}$$

The graph is shown in Figure 11.22. ∎

EXAMPLE 7 Find the equation of the ellipse with vertices at $(3, 2)$ and $(3, -4)$ and foci at $(3, \sqrt{5} - 1)$ and $(3, -\sqrt{5} - 1)$.

Solution By inspection, the ellipse is vertical, and it is centered at $(3, -1)$, where $a = 3$ and $c = \sqrt{5}$ (see Figure 11.23). Thus

$$c^2 = a^2 - b^2$$
$$5 = 9 - b^2$$
$$b^2 = 4$$

The equation is

$$\frac{(y+1)^2}{9} + \frac{(x-3)^2}{4} = 1$$

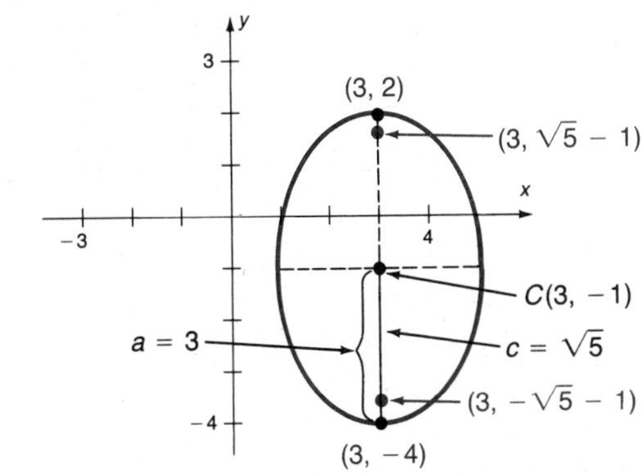

Figure 11.23

■

EXAMPLE 8 Find the equation of the set of all points with the sum of distances from $(-3, 2)$ and $(5, 2)$ equal to 16.

Solution By inspection, the ellipse is horizontal and is centered at $(1, 2)$. You are given

$$2a = 16 \quad \text{This is the sum of the distances.}$$
$$a = 8$$

and $c = 4$—the distance from the center $(1, 2)$ to a focus $(5, 2)$. Thus

$$c^2 = a^2 - b^2$$
$$16 = 64 - b^2$$
$$b^2 = 48$$

The equation is

$$\frac{(x-1)^2}{64} + \frac{(y-2)^2}{48} = 1$$

■

EXAMPLE 9 Find the equation of the ellipse with foci at $(-3, 6)$ and $(-3, 2)$ with $\epsilon = \frac{1}{5}$.

Solution By inspection, the ellipse is vertical and is centered at $(-3, 4)$ with $c = 2$. Since

$$\epsilon = \frac{c}{a} = \frac{1}{5}$$

and $c = 2$, then

$$\frac{2}{a} = \frac{1}{5}$$

which implies $a = 10$. Just because

$$\frac{c}{a} = \frac{1}{5}$$

you cannot assume that $c = 1$ and $a = 5$; all you know is that the reduced *ratio* of c to a is $\frac{1}{5}$. Also,

$$c^2 = a^2 - b^2$$
$$4 = 100 - b^2$$
$$b^2 = 96$$

Thus the equation is

$$\frac{(y-4)^2}{100} + \frac{(x+3)^2}{96} = 1$$

■

PROBLEM SET 11.2

A

Sketch the curves in Problems 1–18.

1. $\frac{x^2}{4} + \frac{y^2}{9} = 1$
2. $\frac{x^2}{25} + \frac{y^2}{36} = 1$
3. $x^2 + \frac{y^2}{9} = 1$
4. $4x^2 + 9y^2 = 36$
5. $25x^2 + 16y^2 = 400$
6. $36x^2 + 25y^2 = 900$
7. $3x^2 + 2y^2 = 6$
8. $4x^2 + 3y^2 = 12$
9. $5x^2 + 10y^2 = 7$
10. $(x-2)^2 + (y+3)^2 = 25$
11. $(x+4)^2 + (y-2)^2 = 49$
12. $(x-1)^2 + (y-1)^2 = \frac{1}{4}$
13. $\frac{(x+3)^2}{81} + \frac{(y-1)^2}{49} = 1$
14. $\frac{(x-3)^2}{16} + \frac{(y-2)^2}{9} = 1$
15. $\frac{(x+2)^2}{25} + \frac{(y+4)^2}{9} = 1$
16. $3(x+1)^2 + 4(y-1)^2 = 12$
17. $10(x-5)^2 + 6(y+2)^2 = 60$
18. $5(x+2)^2 + 3(y+4)^2 = 60$

Sketch the curves in Problems 19–30.

19. The set of points 6 units from the point $(4, 5)$
20. The set of points 3 units from the point $(-2, 3)$
21. The set of points 6 units from $(-1, -4)$
22. The set of points such that the sum of the distances from $(-6, 0)$ and $(6, 0)$ is 20
23. The set of points such that the sum of the distances from $(4, 0)$ and $(-4, 0)$ is 10
24. The set of points such that the sum of the distances from $(-4, 1)$ and $(2, 1)$ is 10
25. The ellipse with vertices at $(0, 7)$ and $(0, -7)$ and foci at $(0, 5)$ and $(0, -5)$
26. The ellipse with vertices at $(4, 3)$ and $(4, -5)$ and foci at $(4, 2)$ and $(4, -4)$
27. The ellipse with vertices $(-6, 3)$ and $(4, 3)$ and foci at $(-4, 3)$ and $(2, 3)$
28. The ellipse with foci at $(-4, -3)$ and $(2, -3)$ with eccentricity $\frac{4}{5}$
29. The circle passing through $(2, 2)$, $(-2, -6)$, and $(5, 1)$
30. The ellipse passing through $(5, 2)$ and $(3, \sqrt{5})$ with axes along the coordinate axes

B

Find the equations of the curves in Problems 31–42. These are the curves you sketched in Problems 19–30.

31. The set of points 6 units from the point $(4, 5)$
32. The set of points 3 units from the point $(-2, 3)$
33. The set of points 6 units from $(-1, -4)$
34. The set of points such that the sum of the distances from $(-6, 0)$ and $(6, 0)$ is 20
35. The set of points such that the sum of the distances from $(4, 0)$ and $(-4, 0)$ is 10
36. The set of points such that the sum of the distances from $(-4, 1)$ and $(2, 1)$ is 10
37. The ellipse with vertices at $(0, 7)$ and $(0, -7)$ and foci at $(0, 5)$ and $(0, -5)$
38. The ellipse with vertices at $(4, 3)$ and $(4, -5)$ and foci at $(4, 2)$ and $(4, -4)$
39. The ellipse with vertices $(-6, 3)$ and $(4, 3)$ and foci at $(-4, 3)$ and $(2, 3)$
40. The ellipse with foci at $(-4, -3)$ and $(2, -3)$ and eccentricity $\frac{4}{5}$
41. The circle passing through $(2, 2)$, $(-2, -6)$, and $(5, 1)$
42. The ellipse passing through $(5, 2)$ and $(3, \sqrt{5})$ with axes along the coordinate axes

Sketch the curves in Problems 43–54.

43. $x^2 + 4x + y^2 + 6y - 12 = 0$

44. $9x^2 + 4y^2 - 18x + 16y - 11 = 0$

45. $16x^2 + 9y^2 + 96x - 36y + 36 = 0$

46. $y^2 + 6y + 25x + 159 = 0$

47. $3x^2 + 4y^2 + 2x - 8y + 4 = 0$

48. $144x^2 + 72y^2 - 72x + 48y - 7 = 0$

49. $y^2 + 4x^2 + 2y - 8x + 1 = 0$

50. $4y^2 + x^2 - 16y + 4x - 8 = 0$

51. $x^2 + y^2 - 10x - 14y - 70 = 0$

52. $x^2 + y^2 - 4x + 10y + 15 = 0$

53. $4x^2 + y^2 + 24x + 4y + 16 = 0$

54. $x^2 + 9y^2 - 4x - 18y - 14 = 0$

55. Derive the equation of the ellipse with foci at $(0, c)$ and $(0, -c)$ and constant distance $2a$. Let $b^2 = a^2 - c^2$. Show all your work.

56. Derive the equation of a circle with center (h, k) and radius r by using the distance formula.

The following is needed for Problems 57–59.
The orbit of a planet can be described by

$$\frac{x^2}{a^2} + \frac{y^2}{b^2} = 1$$

*with the sun at one focus. The orbit is commonly identified by its **major axis** $2a$ and its eccentricity ϵ.*

$$\epsilon = \frac{c}{a} = \sqrt{1 - \frac{b^2}{a^2}} \qquad (a \geq b > 0)$$

where c is the distance between the center and a focus.

57. The orbits of the planets can be described as standard-form ellipses with the sun at one focus (see Figure 11.24). The point at which a planet is farthest from the sun is called **aphelion**, and the point at which it is closest is called **perihelion**. (For a more detailed discussion of planetary orbits, see the Extended Application at the end of Chapter 12.) If the major axis of the earth's orbit is 186,000,000 miles and its eccentricity is $\frac{1}{62}$, how far is the earth from the sun at aphelion and at perihelion?

58. If the planet Mercury is 28 million miles from the sun at perihelion, and the eccentricity of its orbit is $\frac{1}{5}$, how long is the major axis of Mercury's orbit? (See Problem 57.)

59. The moon's orbit is elliptical with the earth at one focus. The point at which the moon is farthest from the earth is called **apogee**, and the point at which it is closest is called **perigee**. If the moon is 199,000 miles from the earth at apogee and the major axis of its orbit is 378,000 miles, what is the eccentricity of the moon's orbit?

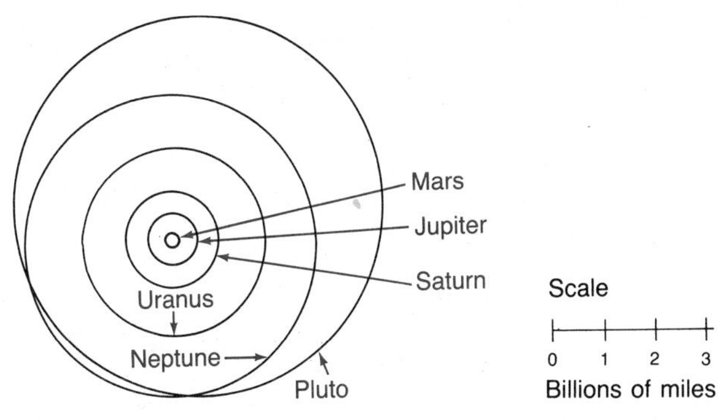

Figure 11.24 Planetary orbits: The orbits of planets or satellites serve as an example of ellipses. (See the Extended Application following Chapter 12 for a discussion of this application.)

C

60. ***Engineering*** A stone tunnel is to be constructed such that the opening is a semielliptic arch as shown in Figure 11.25. It is necessary to know the height at 4-ft intervals from the center. That is, how high is the tunnel at 4, 8, 12, 16, and 20 ft from the center? (Answer to the nearest tenth of a foot.)

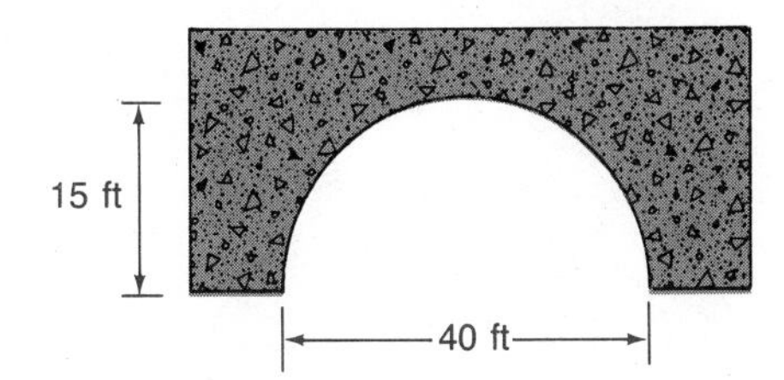

Figure 11.25 Semielliptic arch

61. Find the points of intersection (if any) of the ellipse given by $16(x - 2)^2 + 9(y + 1)^2 = 144$ and the line $4x - 3y - 23 = 0$. Graph both equations.

62. Find the points of intersection (if any) of the circle given by $x^2 + y^2 + 4x + 6y + 4 = 0$ and the parabola $x^2 + 4x + 8y + 4 = 0$. Graph both equations.

63. If we are given an ellipse with foci at $(-c, 0)$ and $(c, 0)$ and vertices at $(-a, 0)$ and $(a, 0)$, we define the *directrices* of the ellipse as the lines $x = a/\epsilon$ and $x = -a/\epsilon$. Show that an ellipse is the set of all points with distances from $F(c, 0)$ equal to ϵ times their distances from the line $x = a/\epsilon$ $(a > 0, c > 0)$.

64. A line segment through a focus parallel to a directrix (see Problem 63) and cut off by the ellipse is called the *focal chord.* Show that the length of the focal chord of the following ellipse is $2b^2/a$: $\dfrac{x^2}{a^2} + \dfrac{y^2}{b^2} = 1$

11.3 HYPERBOLAS

The last of the conic sections to be considered has a definition similar to that of the ellipse.

Hyperbola

> A **hyperbola** is the set of all points in the plane such that, for each point on the hyperbola, the difference of its distances from two fixed points is a constant.

The fixed points are called the **foci**. A hyperbola with foci at F_1 and F_2, where the given constant is 8, is shown in Figure 11.26.

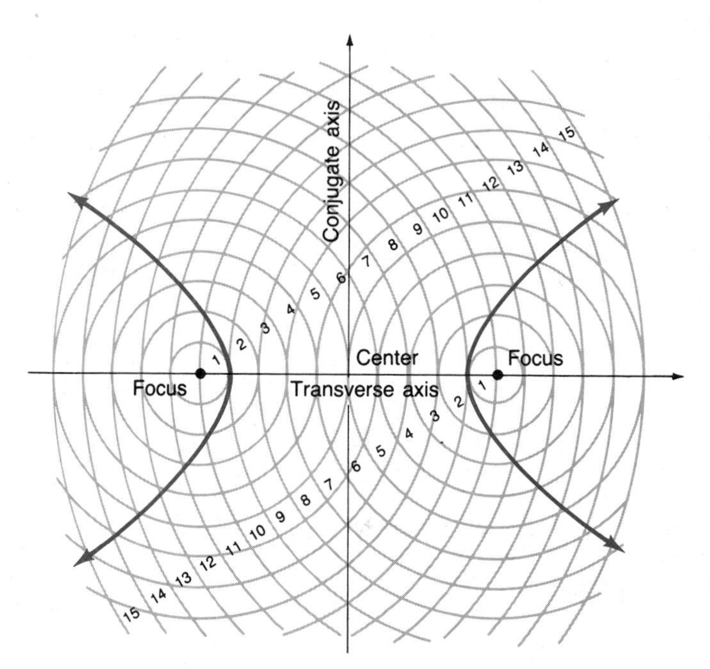

Figure 11.26 Graph of a hyperbola from the definition

The line passing through the foci is called the **transverse axis**. The **center** is the midpoint of the segment connecting the foci. The line passing through the center perpendicular to the transverse axis is called the **conjugate axis**. The hyperbola is symmetric with respect to both the transverse and the conjugate axes.

If you use the definition, you can derive the equation for a hyperbola with foci at $(-c, 0)$ and $(c, 0)$ with constant distance $2a$. If (x, y) is any point on the curve, then

$$\left|\sqrt{(x+c)^2+(y-0)^2}-\sqrt{(x-c)^2+(y-0)^2}\right|=2a$$

The procedure for simplifying this expression is the same as that shown for the ellipse, so the details are left as a problem (see Problems 59 and 60). After several steps, you should obtain

$$\frac{x^2}{a^2}-\frac{y^2}{c^2-a^2}=1$$

If $b^2=c^2-a^2$, then

$$\frac{x^2}{a^2}-\frac{y^2}{b^2}=1$$

which is the standard-form equation. Notice that $c^2=a^2-b^2$ for the ellipse and that $c^2=a^2+b^2$ for the hyperbola. For the ellipse it is necessary that $a^2>b^2$, but for the hyperbola there is no restriction on the relative sizes for a and b.

Repeat the argument for a hyperbola with foci $(0,c)$ and $(0,-c)$, and you will obtain the other standard-form equation for a hyperbola with a vertical transverse axis.

Standard-Form Equations for the Hyperbola with Center (0, 0)

Hyperbola	Foci	Constant distance	Center	Equation
Horizontal	$(-c,0),(c,0)$	$2a$	$(0,0)$	$\frac{x^2}{a^2}-\frac{y^2}{b^2}=1$
Vertical	$(0,c),(0,-c)$	$2a$	$(0,0)$	$\frac{y^2}{a^2}-\frac{x^2}{b^2}=1$
where $b^2=c^2-a^2$ or $c^2=a^2+b^2$				

As with the other conics, we will sketch a hyperbola by determining some information about the curve directly from the equation by inspection. The points of intersection of the hyperbola with the transverse axis are called the **vertices**. For

$$\frac{x^2}{a^2}-\frac{y^2}{b^2}=1$$

and

$$\frac{y^2}{a^2}-\frac{x^2}{b^2}=1$$

notice that the vertices occur at $(a,0)$, $(-a,0)$ and $(0,a)$, $(0,-a)$, respectively. The number $2a$ is the **length of the transverse axis**. The hyperbola does not intersect the conjugate axis, but if you plot the points $(0,b)$, $(0,-b)$ and $(-b,0)$, $(b,0)$, respectively, you determine a segment on the conjugate axis called the **length of the conjugate axis**.

EXAMPLE 1 Sketch $\dfrac{x^2}{4} - \dfrac{y^2}{9} = 1.$

Solution The center of the hyperbola is $(0, 0)$, $a = 2$, and $b = 3$. Plot the vertices at ± 2, as shown in Figure 11.27. The transverse axis is along the x axis and the conjugate axis is along the y axis. Plot the length of the conjugate axis at ± 3. We call the points the **pseudovertices**, since the curve does not actually pass through these points.

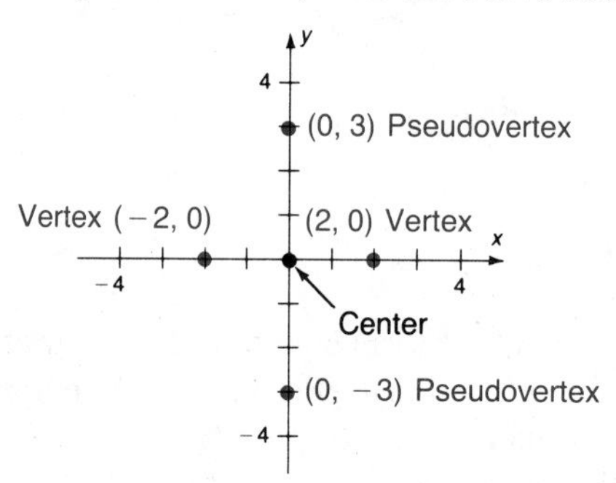

Figure 11.27

Next draw lines through the vertices and pseudovertices parallel to the axes of the hyperbola. These lines form what we will call the **central rectangle**. The diagonal lines passing through the corners of the central rectangle are **slant asymptotes** for the hyperbola, as shown in Figure 11.28; they aid in sketching the hyperbola.

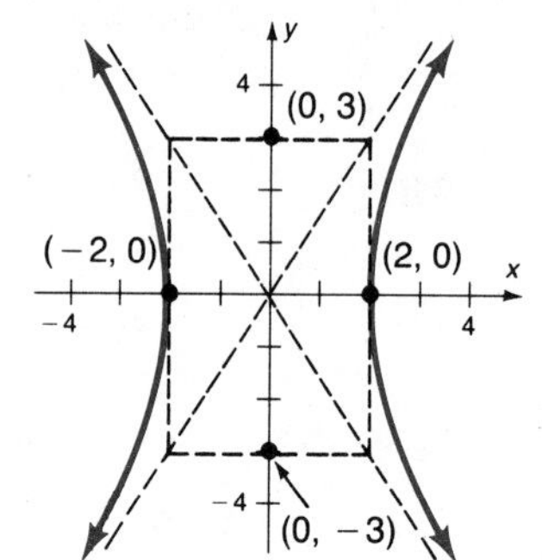

Figure 11.28 Graph of $\dfrac{x^2}{4} - \dfrac{y^2}{9} = 1$

■

For the general hyperbola given by the equation

$$\frac{x^2}{a^2} - \frac{y^2}{b^2} = 1$$

the equations of the slant asymptotes described in Example 1 are

$$y = \frac{b}{a}x \quad \text{and} \quad y = -\frac{b}{a}x$$

To justify this result, solve the equation of the hyperbola for y:

$$\frac{y^2}{b^2} = \frac{x^2}{a^2} - 1$$

$$y^2 = b^2\left(\frac{x^2 - a^2}{a^2}\right)$$

$$y = \pm\frac{b}{a}\sqrt{x^2 - a^2}$$

Now write

$$y = \pm\frac{b}{a}\sqrt{x^2\left(1 - \frac{a^2}{x^2}\right)}$$

$$= \pm\frac{bx}{a}\sqrt{1 - \frac{a^2}{x^2}}$$

in order to see that, as $|x| \to \infty$,

$$\sqrt{1 - \frac{a^2}{x^2}} \to 1$$

So, as $|x|$ becomes large,

$$y \to \pm\frac{b}{a}x$$

The same result is obtained for the hyperbola

$$\frac{y^2}{a^2} - \frac{x^2}{b^2} = 1$$

If the center of the hyperbola is (h, k), the following equations for a hyperbola are obtained:

Standard-Form Equations for Translated Hyperbolas

Hyperbola	Center	Foci	Vertices	Pseudo-vertices	Equations
Horizontal	(h, k)	$(h + c, k)$ $(h - c, k)$	$(h - a, k)$ $(h + a, k)$	$(h, k + b)$ $(h, k - b)$	$\frac{(x-h)^2}{a^2} - \frac{(y-k)^2}{b^2} = 1$
Vertical	(h, k)	$(h, k + c)$ $(h, k - c)$	$(h, k + a)$ $(h, k - a)$	$(h - b, k)$ $(h + b, k)$	$\frac{(y-k)^2}{a^2} - \frac{(x-h)^2}{b^2} = 1$

EXAMPLE 2 Sketch $16x^2 - 9y^2 - 128x - 18y + 103 = 0$.

Solution Complete the square in both x and y:

$$(16x^2 - 128x) + (-9y^2 - 18y) = -103$$
$$16(x^2 - 8x \qquad) - 9(y^2 + 2y \qquad) = -103$$
$$16(x^2 - 8x + 16) - 9(y^2 + 2y + 1) = -103 + 256 - 9$$
$$16(x - 4)^2 - 9(y + 1)^2 = 144$$
$$\mathbf{\frac{(x-4)^2}{9} - \frac{(y+1)^2}{16} = 1}$$

The graph is shown in Figure 11.29. ■

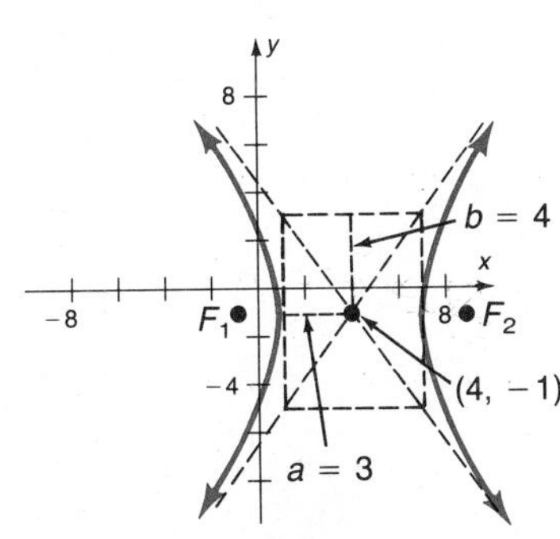

Figure 11.29 Sketch of $16x^2 - 9y^2 - 128x - 18y + 103 = 0$

EXAMPLE 3 Find the equation of the hyperbola with vertices at $(2, 4)$ and $(2, -2)$ and foci at $(2, 6)$ and $(2, -4)$.

Solution Plot the given points as shown in Figure 11.30. Notice that the center of the hyperbola is $(2, 1)$ since it is the midpoint of the segment connecting the foci. Also, $c = 5$ and $a = 3$. Since

$$c^2 = a^2 + b^2$$

you have

$$25 = 9 + b^2$$

$$b^2 = 16$$

and the equation is

$$\frac{(y-1)^2}{9} - \frac{(x-2)^2}{16} = 1$$

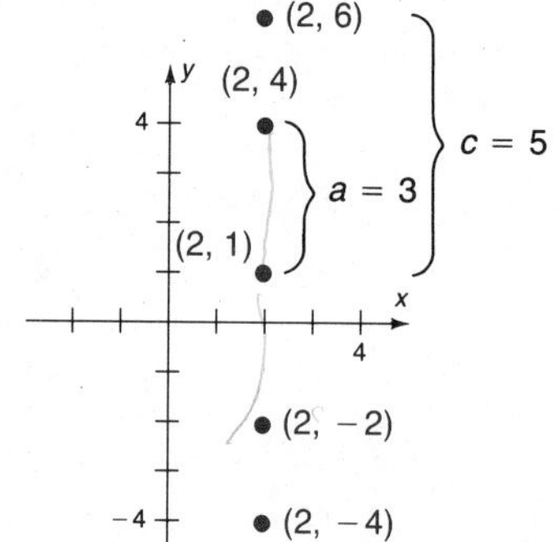

Figure 11.30

■

The eccentricity of the hyperbola and parabola is defined by the same equation that was used for the ellipse, namely

$$\epsilon = \frac{c}{a}$$

Remember that for the ellipse, $0 \leq \epsilon < 1$; however, for the hyperbola $c > a$ so $\epsilon > 1$, and for the parabola $c = a$ so $\epsilon = 1$.

EXAMPLE 4 Find the equation of the hyperbola with foci at $(-3, 2)$ and $(5, 2)$ and with eccentricity $\frac{3}{2}$.

Solution The center of the hyperbola is $(1, 2)$ and $c = 4$. Also, since

$$\epsilon = \frac{c}{a} = \frac{3}{2}$$

you have

$$\frac{4}{a} = \frac{3}{2}$$

$$a = \frac{8}{3}$$

Since $c^2 = a^2 + b^2$,

$$16 = \frac{64}{9} + b^2$$

$$b^2 = \frac{80}{9}$$

Thus the equation is

$$\frac{(x-1)^2}{\frac{64}{9}} - \frac{(y-2)^2}{\frac{80}{9}} = 1 \quad \text{or} \quad \frac{9(x-1)^2}{64} - \frac{9(y-2)^2}{80} = 1$$

■

EXAMPLE 5 Find the set of points such that the difference of their distances from (6, 2) and (6, −5) is always 3.

Solution This is a hyperbola with center $(6, -\frac{3}{2})$ and $c = \frac{7}{2}$. Also $2a = 3$, so $a = \frac{3}{2}$. Since

$$c^2 = a^2 + b^2$$

you have

$$\frac{49}{4} = \frac{9}{4} + b^2$$

$$b^2 = 10$$

The equation is

$$\frac{\left(y+\frac{3}{2}\right)^2}{\frac{9}{4}} - \frac{(x-6)^2}{10} = 1 \quad \text{or} \quad \frac{4\left(y+\frac{3}{2}\right)^2}{9} - \frac{(x-6)^2}{10} = 1$$

■

Conic Section Summary

We have now considered the graphs of equations of the form

$$Ax^2 + Bxy + Cy^2 + Dx + Ey + F = 0$$

Geometrically they represent the intersection of a plane and a cone, usually resulting in a line, parabola, ellipse, or hyperbola. However, there are certain positions of the plane that result in what are called **degenerate conics**. To visualize some of these degenerate conics, first consider Figure 11.31.

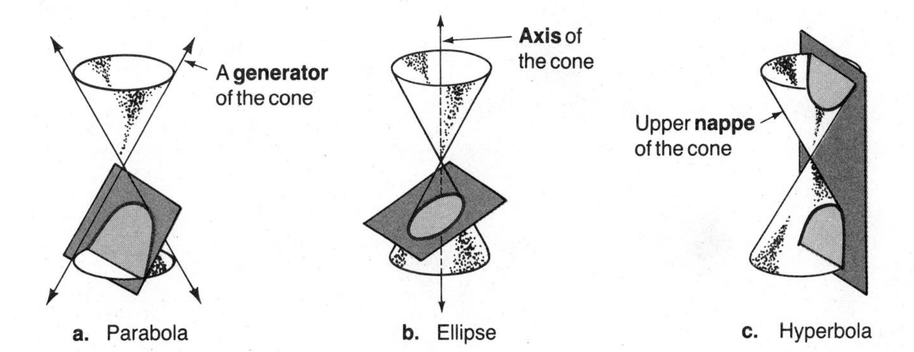

Figure 11.31 Conic sections

For a **degenerate parabola**, visualize the cone (see Figure 11.31a) situated so that one of its generators lies in the plane; a line results. For a **degenerate ellipse**, visualize the plane intersecting at the vertex of the upper and lower nappes (see Figure 11.31b); a point results. And finally, for a **degenerate hyperbola**, visualize the plane situated so that the axis of the cone lies in the plane (see Figure 11.31c); a pair of intersecting lines results.

EXAMPLE 6 Sketch $\frac{(x-2)^2}{4} + \frac{(y+3)^2}{9} = 0$

Solution There is only one point that satisfies this equation—namely $(2, -3)$. This is an example of a *degenerate ellipse*. Notice that except for the zero the equation has the "form of an ellipse." See Figure 11.32.

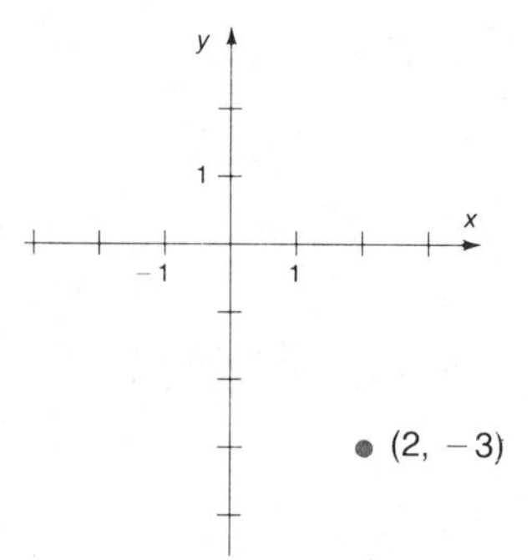

Figure 11.32 Graph of $\frac{(x-2)^2}{4} + \frac{(y+3)^2}{9} = 0$ ■

EXAMPLE 7 Sketch $\frac{x^2}{4} - \frac{y^2}{9} = 0$.

Solution This equation has the "form of a hyperbola," but because of the zero it cannot be put into standard form. You can, however, treat this as a factored form as described in Section 4.1:

$$\left(\frac{x}{2} - \frac{y}{3}\right)\left(\frac{x}{2} + \frac{y}{3}\right) = 0$$

$$\frac{x}{2} - \frac{y}{3} = 0 \quad \text{or} \quad \frac{x}{2} + \frac{y}{3} = 0$$

The graph (Figure 11.33) is a *degenerate hyperbola*.

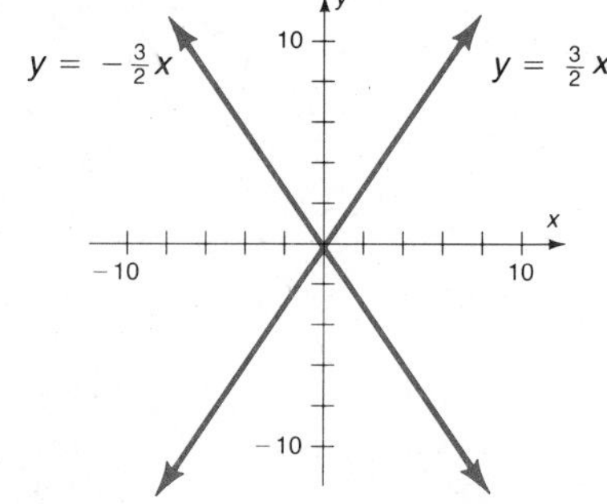

Figure 11.33 Graph of $\frac{x^2}{4} - \frac{y^2}{9} = 0$ ■

It is important to be able to recognize the curve by inspection of the equation before you begin. The first thing to notice is whether or not there is an xy term.

Up to now we have considered conics for which $B = 0$ (that is, no xy term). If $B = 0$, then:

B = 0

Type of curve	Degree of equation	Degree in x	Degree in y	Relationship to the general equation $Ax^2+Bxy+Cy^2+Dx+Ey+F=0$
1. Line	First	First	First	$A = C = 0$
2. Parabola	Second	First	Second	$A = 0$ and $C \neq 0$
		Second	First	$A \neq 0$ and $C = 0$
3. Ellipse	Second	Second	Second	A and C have same sign
4. Circle	Second	Second	Second	$A = C$
5. Hyperbola	Second	Second	Second	A and C have opposite signs

In the next section, we will graph conics for which $B \neq 0$ (that is, there is an xy term). However, we will begin in this section by identifying this type of conic. This identification consists of calculating $B^2 - 4AC$. (Remember, $b^2 - 4ac$ in the quadratic formula was called the discriminant.)

Type of curve	Relationship to general equation
1. Ellipse	$B^2 - 4AC < 0$
2. Parabola	$B^2 - 4AC = 0$
3. Hyperbola	$B^2 - 4AC > 0$

These tests do not distinguish degenerate cases. This means that the test may tell you the curve is an ellipse, but it may turn out to be a single point. Remember too that a circle is a special case of the ellipse. The expression $B^2 - 4AC$ is called the **discriminant**.

EXAMPLE 8 Identify each curve.

Solution

a. $x^2 + 4xy + 4y^2 = 9$
$B^2 - 4AC = 16 - 4(1)(4) = 0$; parabola

b. $2x^2 + 3xy + y^2 = 25$
$B^2 - 4AC = 9 - 4(2)(1) > 0$; hyperbola

c. $x^2 + xy + y^2 - 8x + 8y = 0$
$B^2 - 4AC = 1 - 4(1)(1) < 0$; ellipse

d. $xy = 5$
$B^2 - 4AC = 1 - 4(0)(0) > 0$; hyperbola ■

It is important to remember the standard-form equations of the conics and some basic information about these curves. The important ideas are summarized in the following conic summary.

Summary of Standard-Position Conics

	Parabola	Ellipse	Hyperbola
Definition	All points equidistant from a given point and a given line	All points with the sum of distances from two fixed points constant	All points with the difference of distances from two fixed points constant
Equations	Up: $x^2 = 4cy$ Down: $x^2 = -4cy$ Right: $y^2 = 4cx$ Left: $y^2 = -4cx$ c is a positive number and the distance from the vertex to the focus; $4c$ is the length of the focal chord	$c^2 = a^2 - b^2$ Horizontal axis: $\frac{x^2}{a^2} + \frac{y^2}{b^2} = 1$ Vertical axis: $\frac{y^2}{a^2} + \frac{x^2}{b^2} = 1$	$c^2 = a^2 + b^2$ Horizontal axis: $\frac{x^2}{a^2} - \frac{y^2}{b^2} = 1$ Vertical axis: $\frac{y^2}{a^2} - \frac{x^2}{b^2} = 1$

	Parabola	Ellipse	Hyperbola
Recognition	Second-degree equation; linear in one variable, quadratic in the other variable	Second-degree equation; coefficients of x^2 and y^2 have same sign	Second-degree equation; coefficients of x^2 and y^2 have different signs
Eccentricity	$\epsilon = 1$	$0 \leq \epsilon < 1$	$\epsilon > 1$
Directrix	Perpendicular to axis c units from the vertex (one directrix)	Perpendicular to major axis $\pm a/\epsilon$ units from center (two directrices); see Problem Set 11.2, Problem 63	Perpendicular to transverse axis $\pm a/\epsilon$ units from center (two directrices); see Problem 64 below
Translations to the point (h, k): $\mathbf{x' = x - h}$ and $\mathbf{y' = y - k}$			

PROBLEM SET 11.3

A

Sketch the curves in Problems 1–21.

1. $x^2 - y^2 = 1$

2. $x^2 - y^2 = 4$

3. $y^2 - x^2 = 1$

4. $\dfrac{x^2}{4} - \dfrac{y^2}{9} = 1$

5. $\dfrac{x^2}{9} - \dfrac{y^2}{4} = 1$

6. $\dfrac{y^2}{9} - \dfrac{x^2}{4} = 1$

7. $\dfrac{x^2}{16} - \dfrac{y^2}{25} = 1$

8. $\dfrac{y^2}{16} - \dfrac{x^2}{25} = 1$

9. $\dfrac{x^2}{36} - \dfrac{y^2}{9} = 1$

10. $36y^2 - 25x^2 = 900$

11. $3x^2 - 4y^2 = 12$

12. $3y^2 = 4x^2 + 12$

13. $3x^2 - 4y^2 = 5$

14. $4y^2 - 4x^2 = 5$

15. $4y^2 - x^2 = 9$

16. $\dfrac{(x-2)^2}{4} - \dfrac{(y+3)^2}{16} = 1$

17. $\dfrac{(x+3)^2}{8} - \dfrac{(y-1)^2}{5} = 1$

18. $\dfrac{(y-1)^2}{6} - \dfrac{(x+2)^2}{8} = 1$

19. $\dfrac{(x-2)^2}{16} - \dfrac{(y+1)^2}{9} = 1$

20. $\dfrac{(y+2)^2}{25} - \dfrac{(x+1)^2}{16} = 1$

21. $\dfrac{(y-1)^2}{\frac{1}{9}} - \dfrac{(x+2)^2}{\frac{1}{4}} = 1$

In Problems 22–33, identify and sketch the curve.

22. $2x - y - 8 = 0$

23. $2x + y - 10 = 0$

24. $4x^2 - 16y = 0$

25. $\dfrac{(x-3)^2}{4} - \dfrac{(y+2)}{6} = 1$

26. $\dfrac{(x-3)^2}{9} - \dfrac{(y+2)^2}{25} = 1$

27. $\dfrac{x-3}{9} + \dfrac{y-2}{25} = 1$

28. $(x+3)^2 + (y-2)^2 = 0$

29. $9(x+3)^2 + 4(y-2) = 0$

30. $9(x+3)^2 - 4(y-2)^2 = 0$

31. $x^2 + 8(y-12)^2 = 16$

32. $x^2 + 64(y+4)^2 = 16$

33. $x^2 + y^2 - 3y = 0$

B

Find the equations of the curves in Problems 34–39.

34. The hyperbola with vertices at $(0, 5)$ and $(0, -5)$ and foci at $(0, 7)$ and $(0, -7)$

35. The set of points such that the difference of their distances from $(-6, 0)$ and $(6, 0)$ is 10

36. The hyperbola with foci at $(5, 0)$ and $(-5, 0)$ and eccentricity 5

37. The hyperbola with vertices at $(4, 4)$ and $(4, 8)$ and foci at $(4, 3)$ and $(4, 9)$

38. The set of points such that the difference of their distances from $(4, -3)$ and $(-4, -3)$ is 6

39. The hyperbola with vertices at $(-2, 0)$ and $(6, 0)$ passing through $(10, 3)$

Sketch the curves in Problems 40–47.

40. $5(x-2)^2 - 2(y+3)^2 = 10$

41. $4(x+4)^2 - 3(y+3)^2 = -12$

42. $3x^2 - 4y^2 = 12x + 80y + 88$

43. $9x^2 - 18x - 11 = 4y^2 + 16y$

44. $4y^2 - 8y + 4 = 3x^2 - 2x$

45. $x^2 - 4x + y^2 + 6y - 12 = 0$

46. $x^2 - y^2 = 2x + 4y - 3$

47. $3x^2 - 5y^2 + 18x + 10y - 8 = 0$

In Problems 48–57, identify and sketch the curve.

48. $4x^2 - 3y^2 - 24y - 112 = 0$

49. $y^2 - 4x + 2y + 21 = 0$

50. $9x^2 + 2y^2 - 48y + 270 = 0$

51. $x^2 + 4x + 12y + 64 = 0$

52. $y^2 - 6y - 4x + 5 = 0$

53. $100x^2 - 7y^2 + 98y - 368 = 0$

54. $x^2 + y^2 + 2x - 4y - 20 = 0$

55. $4x^2 + 12x + 4y^2 + 4y + 1 = 0$

56. $x^2 - 4y^2 - 6x - 8y - 11 = 0$

57. $9x^2 + 25y^2 - 54x - 200y + 256 = 0$

C

58. Consider a person A who fires a rifle at a distant gong B. Assuming that the ground is flat, where must you stand to hear the sound of the gun and the sound of the gong simultaneously? *Hint:* To answer this question, let x be the distance sound travels in the length of time it takes the bullet to travel from the gun to the gong. Show that the person who hears the sounds simultaneously must stand on a branch of a hyperbola (the one nearest the target) so that the difference of the distances from A to B is x.

59. Derive the equation of the hyperbola with foci at $(-c, 0)$ and $(c, 0)$ and constant distance $2a$. Let $b^2 = c^2 - a^2$. Show all your work.

60. Derive the equation of the hyperbola with foci at $(0, c)$ and $(0, -c)$ and constant distance $2a$. Let $b^2 = c^2 - a^2$. Show all your work.

61. Let d represent the vertical distance between the hyperbola in the first quadrant

$$y = \frac{b}{a}\sqrt{x^2 - a^2}$$

and the line

$$y = \frac{b}{a}x$$

in the first quadrant. Show that as $|x| \to \infty$, then $d \to 0$.

62. Refer to Problem 61. Show what happens to d as $|x| \to \infty$ if (x, y) is in Quadrant IV.

63. Given the hyperbola

$$\frac{x^2}{a^2} - \frac{y^2}{b^2} = 1$$

show that the length of the diagonal of the central rectangle of the hyperbola is $2c$.

64. Given the hyperbola

$$\frac{x^2}{a^2} - \frac{y^2}{b^2} = 1$$

we define the *directrices* of the hyperbola as the lines

$$x = \frac{a}{\epsilon} \quad \text{and} \quad x = -\frac{a}{\epsilon}$$

Show that the hyperbola is the set of all points with distances from $F(c, 0)$ that are equal to ϵ times their distances from the line $x = a/\epsilon$.

65. A line through a focus parallel to a directrix and cut off by the hyperbola is called the *focal chord*. Show that the length of the focal chord of the following hyperbola is $2b^2/a$:

$$\frac{x^2}{a^2} - \frac{y^2}{b^2} = 1$$

11.4 ROTATIONS*

In Section 2.4 we introduced the idea of a **translation** in order to simplify the equation of a curve by writing it relative to a new translated coordinate system. In this section we will consider the idea of a **rotation** in which the equation of a curve

* Optional section

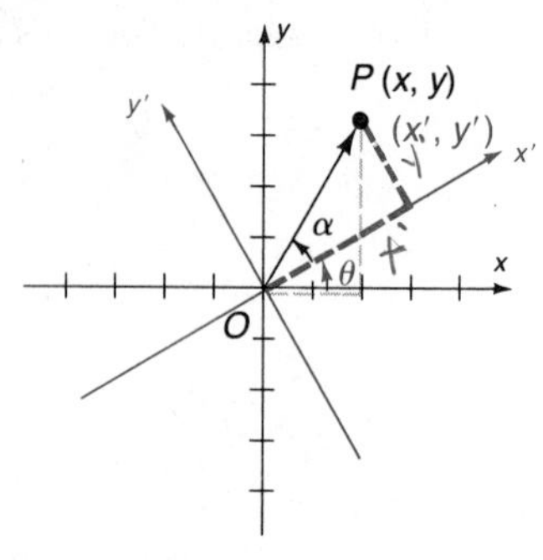

Figure 11.34 Rotation of axes

can be simplified by writing it in terms of a rotated coordinate system. Suppose the coordinate axes are rotated through an angle θ ($0 < \theta < 90°$). The relationship between the old coordinates (x, y) and the new coordinates (x', y') can be found by considering Figure 11.34.

Let O be the origin and P be a point with coordinates (x, y) relative to the old coordinate system and (x', y') relative to the new rotated coordinate system. Let θ be the amount of rotation and let α be the angle between the x' axis and $|OP|$. Then, using the definition of sine and cosine,

$$x = |OP|\cos(\theta + \alpha) \qquad x' = |OP|\cos\alpha$$
$$y = |OP|\sin(\theta + \alpha) \qquad y' = |OP|\sin\alpha$$

Now

$$\begin{aligned} x = |OP|\cos(\theta + \alpha) &= |OP|[\cos\theta\cos\alpha - \sin\theta\sin\alpha] \\ &= |OP|\cos\theta\cos\alpha - |OP|\sin\theta\sin\alpha \\ &= (|OP|\cos\alpha)\cos\theta - (|OP|\sin\alpha)\sin\theta \\ &= x'\cos\theta - y'\sin\theta \end{aligned}$$

Also,

$$\begin{aligned} y = |OP|\sin(\theta + \alpha) &= |OP|\sin\theta\cos\alpha + |OP|\cos\theta\sin\alpha \\ &= x'\sin\theta + y'\cos\theta \end{aligned}$$

Therefore:

Rotation of Axes Formulas

$$x = x'\cos\theta - y'\sin\theta \qquad y = x'\sin\theta + y'\cos\theta$$

All the curves considered in this chapter can be characterized by the general second-degree equation

$$Ax^2 + Bxy + Cy^2 + Dx + Ey + F = 0$$

Notice that the xy term has not appeared before. The presence of this term indicates that the conic has been rotated. Thus in this section we assume that $B \neq 0$ and now need to determine the amount this conic has been rotated from standard position. That is, the new axes should be rotated the same amount as the given conic so that it will be in standard position after the rotation. To find out how much to rotate the axes, substitute

$$x = x'\cos\theta - y'\sin\theta \qquad y = x'\sin\theta + y'\cos\theta$$

into

$$Ax^2 + Bxy + Cy^2 + Dx + Ey + F = 0 \qquad (B \neq 0)$$

After a lot of simplifying you will obtain

$$\begin{aligned} &(A\cos^2\theta + B\cos\theta\sin\theta + C\sin^2\theta)x'^2 \\ &\quad + [B(\cos^2\theta - \sin^2\theta) + 2(C - A)\sin\theta\cos\theta]x'y' \\ &\quad + (A\sin^2\theta - B\sin\theta\cos\theta + C\cos^2\theta)y'^2 + (D\cos\theta + E\sin\theta)x' \\ &\quad + (-D\sin\theta + E\cos\theta)y' + F = 0 \end{aligned}$$

This looks terrible, but it is still in the form

$$A'x'^2 + B'x'y' + C'y'^2 + D'x' + E'y' + F = 0$$

You want to choose θ so that $B' = 0$. This will give you a standard position relative to the new coordinate axes. That is,

$$B(\cos^2\theta - \sin^2\theta) + 2(C - A)\sin\theta\cos\theta = 0$$

$$B\cos 2\theta + (C - A)\sin 2\theta = 0$$

$$B\cos 2\theta = (A - C)\sin 2\theta$$

$$\frac{\cos 2\theta}{\sin 2\theta} = \frac{A - C}{B}$$

Using double-angle identities, $B \neq 0, \theta \neq 0$.

Simplifying, you obtain the following result:

Amount of Rotation Formula

$$\cot 2\theta = \frac{A - C}{B}$$

$\tan 2\theta = \frac{B}{A-C}$

Notice that we required $0 < \theta < 90°$, so 2θ is in Quadrant I or Quadrant II. This means that if $\cot 2\theta$ is positive, then 2θ must be in Quadrant I; if $\cot 2\theta$ is negative, then 2θ is in Quadrant II.

In Examples 1 to 4, find the appropriate rotation so that the given curve will be in standard position relative to the rotated axes. Also find the x and y values in the new coordinate system.

EXAMPLE 1 $xy = 6$

Solution

$$\cot 2\theta = \frac{A - C}{B} = \frac{0 - 0}{1} = 0$$

Thus $2\theta = 90°$ and $\theta = 45°$

$$x = x'\cos\theta - y'\sin\theta = x'\cos 45° - y'\sin 45° = x'\left(\frac{1}{\sqrt{2}}\right) - y'\left(\frac{1}{\sqrt{2}}\right) = \frac{1}{\sqrt{2}}(x' - y')$$

$$y = x'\sin\theta + y'\cos\theta = x'\sin 45° + y'\cos 45° = x'\left(\frac{1}{\sqrt{2}}\right) + y'\left(\frac{1}{\sqrt{2}}\right) = \frac{1}{\sqrt{2}}(x' + y')$$

■

EXAMPLE 2 $7x^2 - 6\sqrt{3}xy + 13y^2 - 16 = 0$

Solution

$$\cot 2\theta = \frac{A - C}{B} = \frac{7 - 13}{-6\sqrt{3}} = \frac{1}{\sqrt{3}}$$

Thus $2\theta = 60°$ and $\theta = 30°$

$$x = x'\cos\theta - y'\sin\theta = x'\cos 30° - y'\sin 30° = x'\left(\frac{\sqrt{3}}{2}\right) - y'\left(\frac{1}{2}\right) = \frac{1}{2}(\sqrt{3}x' - y')$$

$$y = x'\sin\theta + y'\cos\theta = x'\sin 30° + y'\cos 30° = x'\left(\frac{1}{2}\right) + y'\left(\frac{\sqrt{3}}{2}\right) = \frac{1}{2}(x' + \sqrt{3}y')$$

■

EXAMPLE 3 $x^2 - 4xy + 4y^2 + 5\sqrt{5}y - 10 = 0$

Solution

$$\cot 2\theta = \frac{1-4}{-4} = \frac{3}{4}$$

Since this is not an exact value for θ (as it was in Examples 1 and 2), you will need to use some trigonometric identities to find $\cos\theta$ and $\sin\theta$. If $\cot 2\theta = \frac{3}{4}$, then $\tan 2\theta = \frac{4}{3}$ and $\sec 2\theta = \sqrt{1 + (\frac{4}{3})^2} = \frac{5}{3}$. Then $\cos 2\theta = \frac{3}{5}$ and you can now apply the half-angle identities:

$$\cos\theta = \pm\sqrt{\frac{1+\cos 2\theta}{2}} \qquad \sin\theta = \pm\sqrt{\frac{1-\cos 2\theta}{2}}$$

$$\cos\theta = \sqrt{\frac{1+(\frac{3}{5})}{2}} \qquad \sin\theta = \sqrt{\frac{1-(\frac{3}{5})}{2}}$$

Positive because θ is in Quadrant I

$$= \frac{2}{\sqrt{5}} \qquad = \frac{1}{\sqrt{5}}$$

To find the amount of rotation, use a calculator and one of the preceding equations to find $\theta \approx 26.6°$ Finally, the rotation of axes formulas provide

$$x = x'\cos\theta - y'\sin\theta \qquad y = x'\sin\theta + y'\cos\theta$$

$$= x'\left(\frac{2}{\sqrt{5}}\right) - y'\left(\frac{1}{\sqrt{5}}\right) \qquad = x'\left(\frac{1}{\sqrt{5}}\right) + y'\left(\frac{2}{\sqrt{5}}\right)$$

$$= \frac{1}{\sqrt{5}}(2x' - y') \qquad = \frac{1}{\sqrt{5}}(x' + 2y')$$

■

EXAMPLE 4 $10x^2 + 24xy + 17y^2 - 9 = 0$

Solution

$$\cot 2\theta = \frac{10-17}{24}$$

$$= -\frac{7}{24}$$

Since this is negative, 2θ must be in Quadrant II; this means that $\sec 2\theta$ is negative in the following sequence of identities: Since $\cot 2\theta = -\frac{7}{24}$, then $\tan 2\theta = -\frac{24}{7}$ and $\sec 2\theta = -\sqrt{1 + (-\frac{24}{7})^2} = -\frac{25}{7}$. Thus $\cos 2\theta = -\frac{7}{25}$, which gives

$$\cos\theta = \sqrt{\frac{1+(-\frac{7}{25})}{2}} \qquad \sin\theta = \sqrt{\frac{1-(-\frac{7}{25})}{2}}$$

$$= \frac{3}{5} \qquad = \frac{4}{5}$$

Using either of these equations and a calculator, you find that the rotation is

$\theta \approx 53.1^\circ$. The rotation of axes formulas provide

$$x = x'\cos\theta - y'\sin\theta \qquad y = x'\sin\theta + y'\cos\theta$$
$$= x'\left(\frac{3}{5}\right) - y'\left(\frac{4}{5}\right) \qquad = x'\left(\frac{4}{5}\right) + y'\left(\frac{3}{5}\right)$$
$$= \frac{1}{5}(3x' - 4y') \qquad = \frac{1}{5}(4x' + 3y')$$

■

Procedure for Sketching a Rotated Conic

1. Identify curve.
2. a. Find the angle of rotation.
   b. Find x and y in the new coordinate system.
   c. Substitute the values found in Step 2b into the given equation and simplify.
3. Translate axes, if necessary.
4. Sketch the resulting equation relative to the new x' and y' axes.

EXAMPLE 5 Sketch $xy = 6$.

Solution From Example 1, the rotation is $\theta = 45^\circ$ and

$$x = \frac{1}{\sqrt{2}}(x' - y') \qquad y = \frac{1}{\sqrt{2}}(x' + y')$$

Substitute these values into the original equation $xy = 6$:

$$\left[\frac{1}{\sqrt{2}}(x' - y')\right]\left[\frac{1}{\sqrt{2}}(x' + y')\right] = 6$$

Simplify (see Problem 1 for the details) to obtain

$$x'^2 - y'^2 = 12$$
$$\frac{x'^2}{12} - \frac{y'^2}{12} = 1$$

This curve is a hyperbola that has been rotated 45°. Draw the rotated axis and sketch this equation relative to the rotated axes. The result is shown in Figure 11.35.

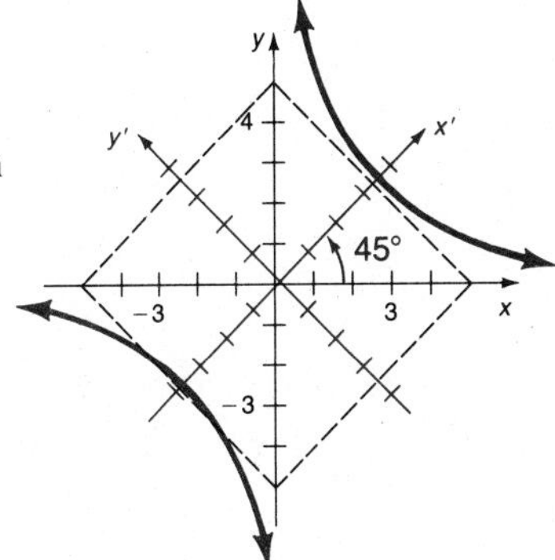

Figure 11.35 Graph of $xy = 6$ ■

EXAMPLE 6 Sketch $7x^2 - 6\sqrt{3}xy + 13y^2 - 16 = 0$.

Solution From Example 2, $\theta = 30^\circ$ and

$$x = \frac{1}{2}(\sqrt{3}x' - y') \qquad y = \frac{1}{2}(x' + \sqrt{3}y')$$

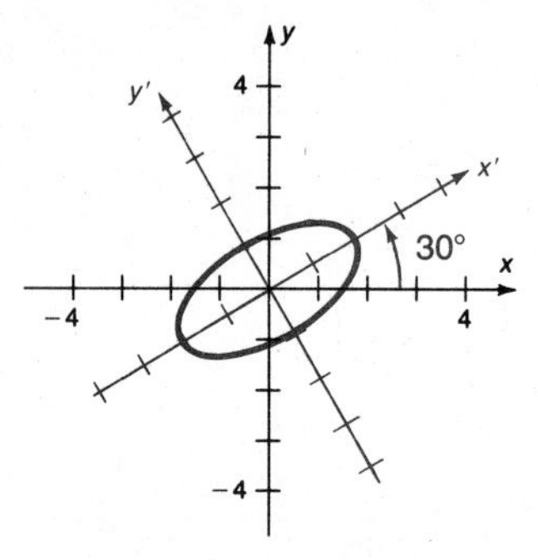

Figure 11.36 Graph of $7x^2 - 6\sqrt{3}xy + 13y^2 - 16 = 0$

Substitute into the original equation:

$$7\left(\frac{1}{2}\right)^2(\sqrt{3}x' - y')^2 - 6\sqrt{3}\left(\frac{1}{2}\right)(\sqrt{3}x' - y')\left(\frac{1}{2}\right)(x' + \sqrt{3}y') + 13\left(\frac{1}{2}\right)^2(x' + \sqrt{3}y')^2 - 16 = 0$$

Simplify (see Problem 2 for the details) to obtain

$$\frac{x'^2}{4} + \frac{y'^2}{1} = 1$$

This curve is an ellipse with a 30° rotation. The sketch is shown in Figure 11.36. ■

EXAMPLE 7 Sketch $x^2 - 4xy + 4y^2 + 5\sqrt{5}y - 10 = 0$.

Solution From Example 3, the rotation is $\theta \approx 26.6°$ and

$$x = \frac{1}{\sqrt{5}}(2x' - y') \qquad y = \frac{1}{\sqrt{5}}(x' + 2y')$$

Substitute

$$\frac{1}{5}(2x' - y')^2 - 4\left(\frac{1}{5}\right)(2x' - y')(x' + 2y') + 4\left(\frac{1}{5}\right)(x' + 2y')^2 + 5\sqrt{5}\left(\frac{1}{\sqrt{5}}\right)(x' + 2y') - 10 = 0$$

Simplify (see Problem 3 for the details) to obtain

$$y'^2 + 2y' = -x' + 2$$

This curve is a parabola with a rotation of about 26.6°. Next complete the square to obtain

$$(y' + 1)^2 = -(x' - 3)$$

This sketch is shown in Figure 11.37.

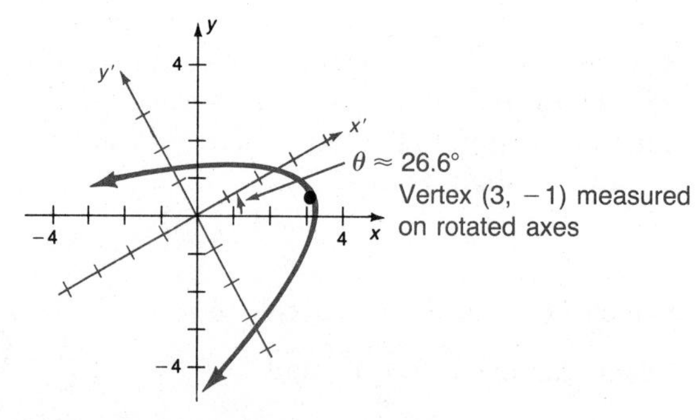

Figure 11.37 Graph of $x^2 - 4xy + 4y^2 + 5\sqrt{5}y - 10 = 0$

■

To graph the general second-degree equation

$$Ax^2 + Bxy + Cy^2 + Dx + Ey + F = 0$$

follow the steps shown in Figure 11.38.

Step 1 Identify curve (this chart does not identify degenerate cases).

Step 2 Rotate axis; if $B = 0$, go to step 3.

Step 3 Translate axis.

Step 4 Graph the standard-form equation.

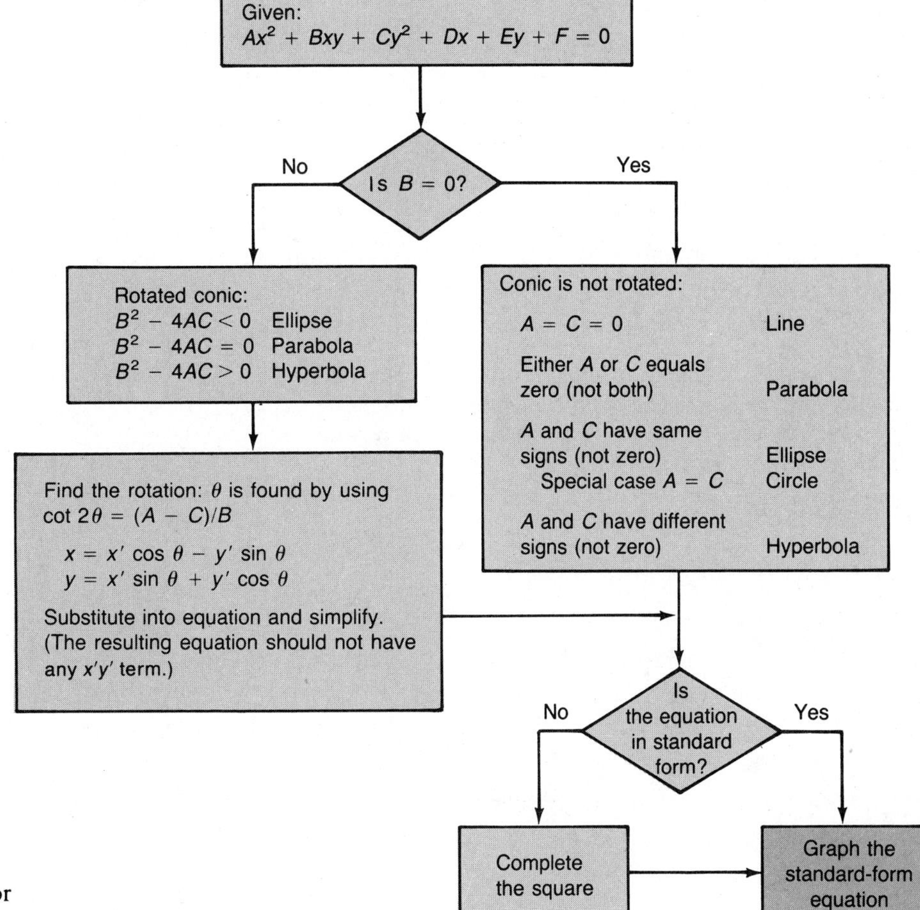

Figure 11.38 Procedure for graphing conics

PROBLEM SET 11.4

A

Because of the amount of arithmetic and algebra involved, many algebraic steps were left out of the examples of this section. In Problems 1–3, fill in the details left out of the indicated example.

1. Example 5

2. Example 6

3. Example 7

Identify the curves whose equations are given in Problems 4–21.

4. $xy = 10$

5. $xy = -1$

6. $xy = -4$

7. $13x^2 - 10xy + 13y^2 - 72 = 0$

8. $5x^2 - 26xy + 5y^2 + 72 = 0$

9. $x^2 + 4xy + 4y^2 + 10\sqrt{5}x = 9$
10. $5x^2 - 4xy + 8y^2 = 36$
11. $23x^2 + 26\sqrt{3}xy - 3y^2 - 144 = 0$
12. $3x^2 + 2\sqrt{3}xy + y^2 + 16x - 16\sqrt{3}y = 0$
13. $24x^2 + 16\sqrt{3}xy + 8y^2 - x + \sqrt{3}y - 8 = 0$
14. $3x^2 - 2\sqrt{3}xy + y^2 + 24x + 24\sqrt{3}y = 0$
15. $13x^2 - 6\sqrt{3}xy + 7y^2 + (16\sqrt{3} - 8)x + (-16 - 8\sqrt{3})y + 16 = 0$
16. $3x^2 - 10xy + 3y^2 - 32 = 0$
17. $5x^2 - 3xy + y^2 + 65x - 25y + 203 = 0$
18. $5x^2 - 6xy + y^2 + 4x - 3y + 10 = 0$
19. $4x^2 + 3xy + 2y^2 - 5x - 4y - 18 = 0$
20. $4x^2 + 2\sqrt{3}xy + 3y^2 - 6x + 12y - 15 = 0$
21. $2x^2 - \sqrt{56}xy + 7y^2 + 9x - 7y + 112 = 0$

B

Find the appropriate rotation in Problems 22–35 so that the given curve will be in standard position relative to the rotated axes. Also find the x and y values in the new coordinate system by using the rotation of axes formulas.

22. $xy = 10$
23. $xy = -1$
24. $xy = -4$
25. $13x^2 - 10xy + 13y^2 - 72 = 0$
26. $5x^2 - 26xy + 5y^2 + 72 = 0$
27. $x^2 + 4xy + 4y^2 + 10\sqrt{5}x = 9$
28. $5x^2 - 4xy + 8y^2 = 36$
29. $23x^2 + 26\sqrt{3}xy - 3y^2 - 144 = 0$
30. $3x^2 + 2\sqrt{3}xy + y^2 + 16x - 16\sqrt{3}y = 0$
31. $24x^2 + 16\sqrt{3}xy + 8y^2 - x + \sqrt{3}y - 8 = 0$
32. $3x^2 - 2\sqrt{3}xy + y^2 + 24x + 24\sqrt{3}y = 0$
33. $13x^2 - 6\sqrt{3}xy + 7y^2 + (16\sqrt{3} - 8)x + (-16 - 8\sqrt{3})y + 16 = 0$
34. $3x^2 - 10xy + 3y^2 - 32 = 0$
35. $5x^2 - 3xy + y^2 + 65x - 25y + 203 = 0$

C

Sketch the curves in Problems 36–59.

36. $xy = 10$ 37. $xy = -1$ 38. $xy = -4$ 39. $xy = 8$
40. $8x^2 - 4xy + 5y^2 = 36$
41. $13x^2 - 10xy + 13y^2 - 72 = 0$
42. $5x^2 - 26xy + 5y^2 + 72 = 0$
43. $x^2 + 4xy + 4y^2 + 10\sqrt{5}x = 9$
44. $5x^2 - 4xy + 8y^2 = 36$
45. $23x^2 + 26\sqrt{3}xy - 3y^2 - 144 = 0$
46. $3x^2 + 2\sqrt{3}xy + y^2 + 16x - 16\sqrt{3}y = 0$
47. $24x^2 + 16\sqrt{3}xy + 8y^2 - x + \sqrt{3}y - 8 = 0$
48. $3x^2 - 2\sqrt{3}xy + y^2 + 24x + 24\sqrt{3}y = 0$
49. $3x^2 - 10xy + 3y^2 - 32 = 0$
50. $x^2 + 2xy + y^2 + 12\sqrt{2}x - 6 = 0$
51. $10x^2 + 24xy + 17y^2 - 9 = 0$
52. $5x^2 - 3xy + y^2 + 65x - 25y + 203 = 0$
53. $3xy - 4y^2 + 18 = 0$
54. $17x^2 - 12xy + 8y^2 - 80 = 0$
55. $5x^2 - 8xy + 5y^2 - 9 = 0$
56. $16x^2 - 24xy + 9y^2 - 60x - 80y + 100 = 0$
57. $13x^2 - 6\sqrt{3}xy + 7y^2 + (16\sqrt{3} - 8)x + (-16 - 8\sqrt{3})y + 16 = 0$
58. $x^2 + 2\sqrt{3}xy + 3y^2 + 2\sqrt{3}x + 2y - 16 = 0$
59. $21x^2 + 10\sqrt{3}xy + 31y^2 - 72x - 16\sqrt{3}x - 72\sqrt{3}y + 16y + 16 = 0$
60. Let $Ax^2 + Bxy + Cy^2 + Dx + Ey + F = 0$. Show that $B^2 - 4AC = B'^2 - 4A'C'$ for any angle θ through which the axes may be rotated and that A', B', and C' are the values given on pages 452–453. Use this fact to prove that (if the graph exists)

    If $B^2 - 4AC = 0$, the graph is a parabola
    If $B^2 - 4AC < 0$, the graph is an ellipse
    If $B^2 - 4AC > 0$, the graph is a hyperbola

11.5

CHAPTER 11 SUMMARY

The material of this chapter is reviewed in the following list of objectives. After each objective there are some practice questions. For a sample test, select the first question of each set and check your answers with the answer section. For a sample test without answers, use the second question of each set. Additional practice is given by the other questions in each set. If you are having trouble with a particular type of problem, look back to that section for extra help.

11.1 PARABOLAS

Objective 1 *Graph parabolas.*

1. $x^2 = y$
2. $(y - 1)^2 = 8(x + 2)$
3. $8y^2 - x - 32y + 31 = 0$
4. $y^2 + 4x + 4y = 0$

Objective 2 *Find the equations of parabolas given certain information about the graph.*

5. Vertex at $(6, 3)$; directrix $x = 1$
6. Directrix $y - 3 = 0$; focus $(-3, -2)$
7. Vertex $(-3, 5)$; focus $(-3, -1)$
8. Vertex at $(4, 2)$ and passing through $(-3, -4)$; axis parallel to the y axis

11.2 ELLIPSES

Objective 3 *Graph ellipses.*

9. $25x^2 + 16y^2 = 400$
10. $5(x + 3)^2 + 9(y - 2)^2 = 45$
11. $x^2 + y^2 = 4x + 2y - 3$
12. $9x^2 + 16y^2 - 90x - 32y + 97 = 0$

Objective 4 *Find the equations of ellipses given certain information about the graph.*

13. The ellipse with the center at $(4, 1)$, a focus at $(5, 1)$, and a semimajor axis 2
14. The set of points such that the sum of the distances from $(-3, 4)$ and $(-7, 4)$ is 12
15. The set of points 8 units from the point $(-1, -2)$
16. The ellipse with foci at $(2, 3)$ and $(-1, 3)$ with eccentricity $\frac{3}{5}$.

11.3 HYPERBOLAS

Objective 5 *Graph hyperbolas.*

17. $x^2 - y^2 + x - y = 3$
18. $x(x - y) = y(y - x) - 1$
19. $5(x + 2)^2 - 3(y + 4)^2 = 60$
20. $12x^2 - 4y^2 + 24x - 8y + 4 = 0$

Objective 6 *Find the equations of hyperbolas given certain information about the graph.*

21. The set of points with the difference of distances from $(-3, 4)$ and $(-7, 4)$ equal to 2
22. The hyperbola with vertices at $(-3, 0)$ and $(3, 0)$ and foci at $(5, 0)$ and $(-5, 0)$
23. The hyperbola with vertices at $(0, -3)$ and $(0, 3)$ and eccentricity $\frac{5}{3}$
24. The hyperbola with vertices at $(-3, 1)$ and $(-5, 1)$ and foci at $(-4 - \sqrt{6}, 1)$ and $(-4 + \sqrt{6}, 1)$

Objective 7 *Know the definition and standard-form equations for the conic section.* State the appropriate standard-form equation.

25. Horizontal ellipse
26. Vertical hyperbola
27. Parabola opening right
28. Circle

Objective 8 *Graph conic sections. Name the type of curve (by inspection) and then graph the curve.*

29. $3x - 2y^2 - 4y + 7 = 0$
30. $\dfrac{x}{16} + \dfrac{y}{4} = 1$
31. $25x^2 + 9y^2 = 225$
32. $25(x - 2)^2 + 25(y + 1)^2 = 400$

*11.4 ROTATIONS

Objective 9 *Use the rotation of axes formulas and the amount of rotation formula.*

33. What is the rotation for the curve whose equation is $xy - 7 = 0$?
34. What is the rotation for the curve whose equation is $5x^2 + 4xy + 5y^2 + 3x - 2y + 5 = 0$?
35. What is the rotation for the curve whose equation is $4x^2 + 4xy + y^2 + 3x - 2y + 7 = 0$?
36. Use the rotation formulas to rewrite the equation in Problem 33 so that there is no xy term.

Objective 10 *Identify the conic by looking at its equation.*

37. $xy + x^2 - 3x = 5$
38. $x^2 + y^2 + xy + 3x - y = 3$
39. $x^2 + 2xy + y^2 = 10$
40. $(x - 1)(y + 1) = 7$

---

* Optional section

Charlotte Angas Scott (1858–1931)

Indeed, mathematics, the indispensable tool of the sciences, defying the senses to follow its splendid flights, is demonstrating today, as it never has been demonstrated before, the supremacy of the pure reason.

Nicholas Butler
The Meaning of Education and Other Essays and Addresses

HISTORICAL NOTE

The first prominent woman mathematician in America was Charlotte Angas Scott. She is best known because of her lifelong work as an educator. She studied in England, but soon after receiving her Ph.D., she came to America and developed the mathematics program at Bryn Mawr College. Today, it is difficult for us to understand the problems encountered by a woman in Scott's day who wished to receive a college education. Scott, for example, was permitted to take her final undergraduate examinations at Cambridge only informally. According to Karen Rappaport,*

"Scott tied for eighth place in the mathematics exam. Mathematics was an unprecedented area for a woman to excel in and the achievement attracted public attention, especially when her name was not mentioned at the official ceremony. When the official name was read, loud shouts of 'Scott of Girton' could be heard in the gallery. The public also responded to the slight. The February 7, 1880 issue of Punch stated: 'But when the academy doors are reopened to the Ladies let them be opened to their full worth. Let us not hear of any restrictions or exclusions from this or that function or privilege....' As a result of this support and a public petition, women were formally admitted to the Tripos Exams the following year but they were still not permitted Cambridge degrees. (That event did not occur until 1948.)"

Mathematically, Scott was concerned with the study of specific algebraic curves of degree higher than 2. She published thirty papers in the field of algebraic geometry, as well as a text, *Modern Analytic Geometry*, in 1894. In 1922, members of the American Mathematical Society and many former students organized a dinner to honor Scott. Alfred North Whitehead came from England to give the main address, in which he stated: "A life's work such as that of Professor Charlotte Angas Scott is worth more to the world than many anxious efforts of diplomatists."

* Karen D. Rappaport, "Two American Women Mathematicians," *MATYC Journal*, Fall (1980), p. 203.

12

Additional Topics in Analytic Geometry

CONTENTS

PREVIEW

In the first four sections of this chapter we introduce vectors, along with the basic vector operations. The remainder of this chapter is concerned with polar coordinates, polar-form curves, and parametric equations. The 23 objectives for this chapter are listed on pages 511–514.

PERSPECTIVE

Vectors will be an important tool in your more advanced work with mathematics. In addition, polar coordinates open up avenues of investigation that would be very difficult if we were limited to rectangular coordinates. The page below, from a leading calculus book, shows a curve called a *cardioid* (Example 1), a circle, and a curve called a limaçon (Example 2). Their respective equations are given in polar coordinates as

$$r = 1 + \sin\theta \qquad r = 5\sin\theta \qquad r = 2 + \sin\theta$$

We will introduce these curves in Section 12.5.

EXAMPLE 1

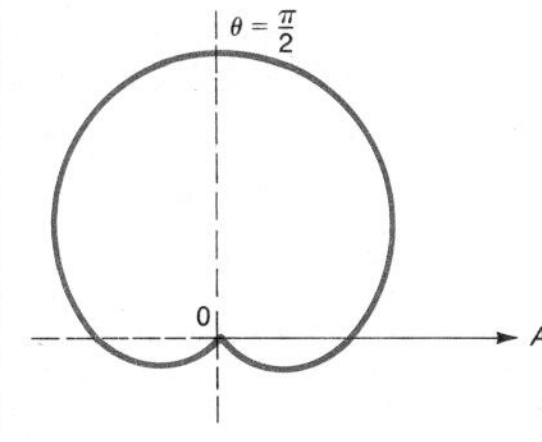

Figure 3
The cardioid $r = 1 + \sin\theta$

Calculate the area of the region enclosed by the graph of the cardioid $r = 1 + \sin\theta$.

Solution. The curve is symmetric about the line $\theta = \pi/2$ and is sketched in Figure 3. (See Example 8.6.4.) Thus we need only calculate the area for θ in $[-\pi/2, \pi/2]$ and then multiply it by 2. We have

$$\text{half the area} = \frac{1}{2}\int_{-\pi/2}^{\pi/2} (1+\sin\theta)^2\,d\theta = \frac{1}{2}\int_{-\pi/2}^{\pi/2} (1 + 2\sin\theta + \sin^2\theta)\,d\theta$$

$$= \frac{1}{2}\theta\Big|_{-\pi/2}^{\pi/2} + \int_{-\pi/2}^{\pi/2} \sin\theta\,d\theta + \frac{1}{4}\int_{-\pi/2}^{\pi/2} (1 - \cos 2\theta)\,d\theta$$

$$= \frac{\pi}{2} + 0 + \frac{\pi}{4} = \frac{3\pi}{4},$$

and the total area $A = 2 \cdot 3\pi/4 = 3\pi/2$.

EXAMPLE 2

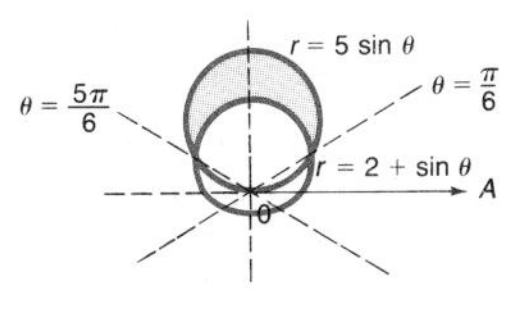

Figure 4
Region between a circle and a limaçon

Find the area inside the circle $r = 5\sin\theta$ and outside the limaçon $r = 2 + \sin\theta$.

Solution. The region is sketched in Figure 4. The two curves intersect whenever $2 + \sin\theta = 5\sin\theta$, or $\sin\theta = \frac{1}{2}$, so that $\theta = \pi/6$ and $\theta = 5\pi/6$. The area of the region can be calculated by calculating the area of the limaçon between $\pi/6$ and $5\pi/6$ and subtracting it from the area of the circle for θ in that interval. By symmetry, we need only calculate the area between $\theta = \pi/6$ and $\theta = \pi/2$ and then multiply it by 2. We therefore have

$$A = \int_{\pi/6}^{\pi/2} (5\sin\theta)^2\,d\theta - \int_{\pi/6}^{\pi/2} (2+\sin\theta)^2\,d\theta$$

(we have already multiplied by 2)

12.1 VECTORS

Many applications of mathematics involve quantities that have *both* magnitude and direction, such as forces, velocities, accelerations, and displacements. Vectors are used to describe such quantities. A **vector** is a directed line segment specifying both a magnitude and a direction. The length of the vector represents the **magnitude** of the quantity being represented; the **direction** of the vector represents the direction of the quantity. Two vectors are **equal** if they have the same magnitude and direction.

Suppose we choose a point O in the plane and call it the origin. A vector is a directed line segment from O to a point $P(x, y)$ in the plane. This vector is denoted by $\overrightarrow{OP}$ or **v**. In the text we use **v**; in your work you will write $\vec{v}$. The magnitude of $\overrightarrow{OP}$ is denoted by $|\overrightarrow{OP}|$ or $|\mathbf{v}|$. The vector from O to O is called the zero vector **0**.

If **v** and **w** represent any two vectors having different (but not opposite) directions, then the **sum** or **resultant** is the vector drawn as the diagonal of the parallelogram having **v** and **w** as the adjacent sides, as shown in Figure 12.1. The vectors **v** and **w** are called **components**.

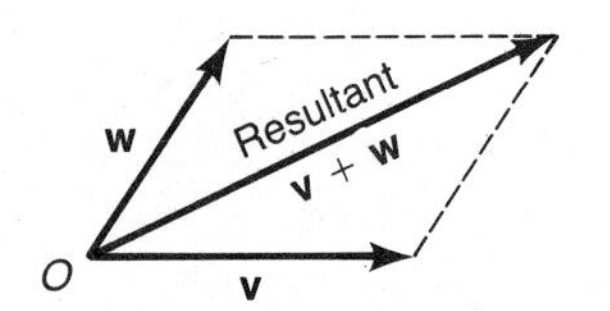

Figure 12.1 Resultant of vectors **v** and **w**

There are basically two types of vector problems dealing with addition of vectors. The first is to find the *resultant vector*. To do this you use a right triangle, the Law of Sines, or the Law of Cosines. The second problem is to *resolve* a vector into two component vectors. If the two vectors form a right angle, they are called **rectangular components**. You will usually resolve a vector into rectangular components.

EXAMPLE 1 Consider two forces, one with magnitude 3.0 in a N20°W direction and the other with magnitude 7.0 in a S50°W direction. Find the resultant vector.

Solution Sketch the given vectors and draw the parallelogram formed by these vectors. The diagonal is the resultant vector as shown in Figure 12.2. You can easily find $\theta = 110°$, but you really need an angle inside the shaded triangle. Use the property from geometry that adjacent angles in a parallelogram are supplementary (they add up to 180°). This tells you that $\phi = 70°$. Thus you know SAS, so you should use the Law of Cosines to find the magnitude $|\mathbf{v}|$:

$$|\mathbf{v}|^2 = 3^2 + 7^2 - 2(3)(7)\cos 70°$$
$$|\mathbf{v}| \approx 6.60569103$$

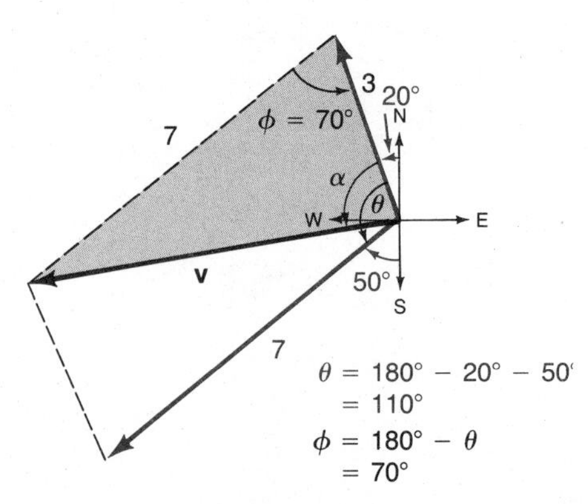

Figure 12.2

The direction of **v** can be found by using the Law of Sines to derive α:

$$\frac{\sin 70^\circ}{|\mathbf{v}|} = \frac{\sin \alpha}{7}$$

$$\sin \alpha = \frac{7}{|\mathbf{v}|} \sin 70^\circ$$

$$\approx 0.99578505$$

Thus $\alpha \approx 84.737565^\circ$. (Find $\alpha = \text{Sin}^{-1} 0.99578505$; do not forget significant digits—see Appendix B.)

Since $20^\circ + 85^\circ = 105^\circ$, you can see that the direction of **v** should be measured from the south. Thus the magnitude of **v** is 6.6 and the direction is S75°W. ■

EXAMPLE 2 Suppose a vector **v** has a magnitude of 5.00 and a direction given by $\theta = 30.0^\circ$, where θ is the angle the vector makes with the positive x axis. Resolve this vector into horizontal and vertical components. (Do not forget significant digits—see Appendix B.)

Solution Let $\mathbf{v}_x$ be the horizontal component and $\mathbf{v}_y$ be the vertical component as shown in Figure 12.3. Then

$$\cos \theta = \frac{|\mathbf{v}_x|}{|\mathbf{v}|} \qquad \sin \theta = \frac{|\mathbf{v}_y|}{|\mathbf{v}|}$$

$$|\mathbf{v}_x| = |\mathbf{v}| \cos \theta = 5 \cos 30^\circ = \frac{5}{2}\sqrt{3} \approx 4.33$$

$$\mathbf{v}_y = |\mathbf{v}| \sin \theta = 5 \sin 30^\circ = \frac{5}{2} = 2.50$$

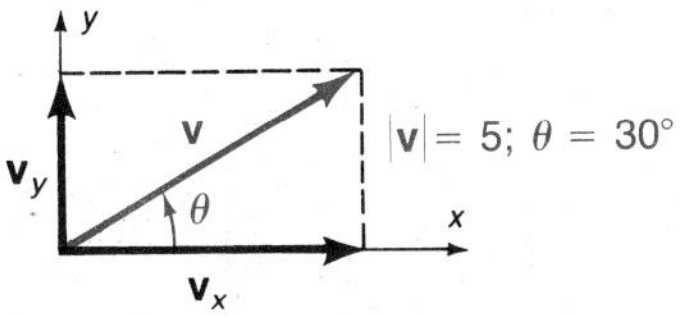

Figure 12.3 Resolving a vector

Thus $\mathbf{v}_x$ is a horizontal vector with magnitude 4.33 and $\mathbf{v}_y$ is a vertical vector with magnitude 2.50. ■

The process of resolving a vector can be simplified by using scalar multiplication and two special vectors. **Scalar multiplication** is the multiplication of a vector by a real number. It is called scalar multiplication because real numbers are sometimes called scalars. If c is a positive real number and **v** is a vector, then the vector $c\mathbf{v}$ is a vector representing the scalar multiplication of c and **v**. It is defined geometrically as a vector in the same direction as **v** but with a magnitude c times the original magnitude of **v**, as shown in Figure 12.4. If c is a negative real number, then the scalar multiplication results in a vector in exactly the opposite direction as **v** with length c times as long, as shown in Figure 12.4. If $c = 0$, then $c\mathbf{v}$ is the zero vector.

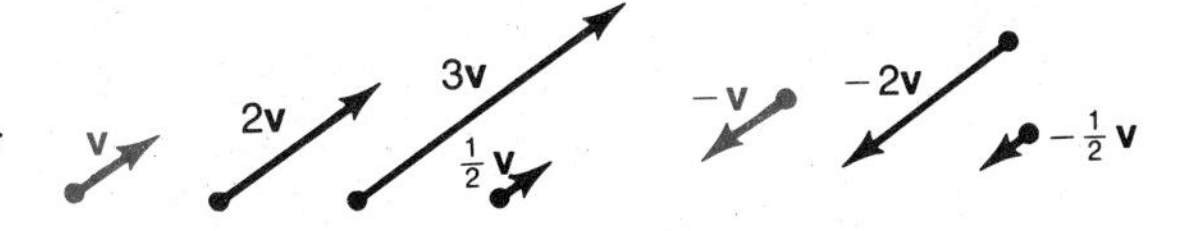

Figure 12.4 Examples of scalar multiplication

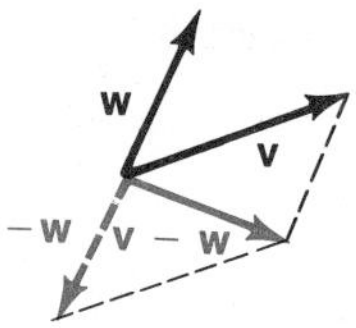

Scalar multiplication allows us to define **subtraction** for vectors:

$$\mathbf{v} - \mathbf{w} = \mathbf{v} + (-\mathbf{w})$$

Geometrically, subtraction is shown in Figure 12.5.

Figure 12.5 Vector subtraction

Two special vectors help us to treat vectors algebraically as well as geometrically:

Definition of the i and j Vectors

> **i** is the vector of unit length in the direction of the positive x axis.
> **j** is the vector of unit length in the direction of the positive y axis.

Example 2 shows how the vector **v** with magnitude 5.00 and $\theta = 30.0°$ can be resolved into rectangular components:

1. $\mathbf{v}_x$ (horizontal component) with magnitude 4.33
2. $\mathbf{v}_y$ (vertical component) with magnitude 2.50

However, since **i** has unit length and is horizontal and **j** has unit length and is vertical, we can write

$$\mathbf{v}_x = 4.33\mathbf{i} \quad \text{and} \quad \mathbf{v}_y = 2.50\mathbf{j}$$

Consequently,

$$\begin{aligned}\mathbf{v} &= \mathbf{v}_x + \mathbf{v}_y \\ &= 4.33\mathbf{i} + 2.50\mathbf{j}\end{aligned}$$

In general, any vector **v** can be written as

$$\mathbf{v} = a\mathbf{i} + b\mathbf{j}$$

where a and b are the magnitude of the horizontal and vertical components, respectively. This is called the **algebraic representation of a vector** and is shown in Figure 12.6.

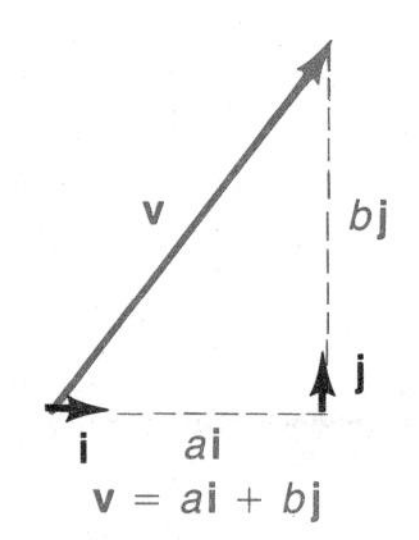

Figure 12.6 Algebraic representation of a vector

EXAMPLE 3 Find the algebraic representation for a vector **v** with magnitude 10 making an angle of 60° with the positive x axis.

Solution Figure 12.7 shows the general procedure for writing the algebraic representation of a vector when given the magnitude and direction of that vector.

$$|b\mathbf{j}| = b$$

$$|a\mathbf{i}| = a$$

$$\cos\theta = \frac{a}{|\mathbf{v}|} \qquad a = |\mathbf{v}|\cos\theta$$

$$\sin\theta = \frac{b}{|\mathbf{v}|} \qquad b = |\mathbf{v}|\sin\theta$$

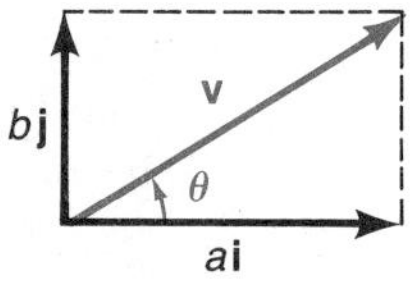

Figure 12.7
$\mathbf{v} = |\mathbf{v}|\cos\theta\mathbf{i} + |\mathbf{v}|\sin\theta\mathbf{j}$

From Figure 12.7,

$$a = |\mathbf{v}|\cos\theta \qquad b = |\mathbf{v}|\sin\theta$$
$$= 10\cos 60^\circ \qquad = 10\sin 60^\circ$$
$$= 5.0 \qquad \approx 8.7$$

Therefore

$$\mathbf{v} = 5.0\mathbf{i} + 8.7\mathbf{j}$$

■

EXAMPLE 4 Find the algebraic representation for a vector $\mathbf{v}$ with initial point $(4, -3)$ and endpoint $(-2, 4)$.

Solution Figure 12.8 shows the general procedure for writing the algebraic representation of a vector when given the endpoints of that vector.

$$a = x_2 - x_1 \quad \text{and} \quad b = y_2 - y_1$$

From Figure 12.8,

$$a = x_2 - x_1 \qquad b = y_2 - y_1$$
$$= -2 - 4 \qquad = 4 - (-3)$$
$$= -6 \qquad = 7$$

Thus $\mathbf{v} = -6\mathbf{i} + 7\mathbf{j}$

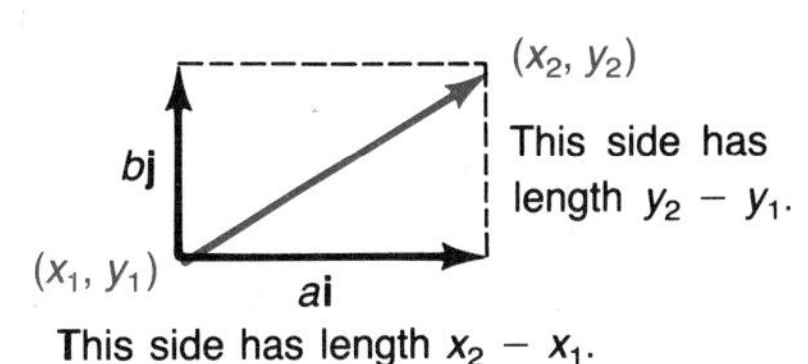

Figure 12.8
$\mathbf{v} = (x_2 - x_1)\mathbf{i} + (y_2 - y_1)\mathbf{j}$

■

EXAMPLE 5 Find the magnitude of the vector in Example 4.

Solution $\mathbf{v} = -6\mathbf{i} + 7\mathbf{j}$. Thus

$$|\mathbf{v}| = \sqrt{(-6)^2 + (7)^2}$$
$$= \sqrt{36 + 49}$$
$$= \sqrt{85}$$

■

Example 5 leads to the following general result.

Magnitude of a Vector

The **magnitude** of a vector $\mathbf{v} = a\mathbf{i} + b\mathbf{j}$ is given by

$$|\mathbf{v}| = \sqrt{a^2 + b^2}$$

The operations of addition, subtraction, and scalar multiplication can also be stated algebraically. Let $\mathbf{v} = a\mathbf{i} + b\mathbf{j}$ and $\mathbf{w} = c\mathbf{i} + d\mathbf{j}$. Then

$$\mathbf{v} + \mathbf{w} = (a + c)\mathbf{i} + (b + d)\mathbf{j}$$
$$\mathbf{v} - \mathbf{w} = (a - c)\mathbf{i} + (b - d)\mathbf{j}$$
$$c\mathbf{v} = ca\mathbf{i} + cb\mathbf{j}$$

EXAMPLE 6 Let $\mathbf{v} = 6\mathbf{i} + 4\mathbf{j}$ and $\mathbf{w} = -2\mathbf{i} + 3\mathbf{j}$.

Solution

a. $|\mathbf{v}| = \sqrt{6^2 + 4^2} = \sqrt{36 + 16} = 2\sqrt{13}$

b. $|\mathbf{w}| = \sqrt{(-2)^2 + 3^2} = \sqrt{4 + 9} = \sqrt{13}$

c. $\mathbf{v} + \mathbf{w} = (6 - 2)\mathbf{i} + (4 + 3)\mathbf{j} = 4\mathbf{i} + 7\mathbf{j}$

d. $\mathbf{v} - \mathbf{w} = (6 + 2)\mathbf{i} + (4 - 3)\mathbf{j} = 8\mathbf{i} + \mathbf{j}$

e. $-\mathbf{v} = (-1)6\mathbf{i} + (-1)4\mathbf{j} = -6\mathbf{i} - 4\mathbf{j}$

f. $-2\mathbf{w} = (-2)(-2)\mathbf{i} + (-2)(3)\mathbf{j} = 4\mathbf{i} - 6\mathbf{j}$ ■

You can see from Example 6 that the algebraic representation of a vector makes it easy to handle vectors and their operations. Another advantage of the algebraic representation is that it specifies a direction and a magnitude, but not a particular location. The directed line segment in Example 4 defined a given vector. However, there are infinitely many other vectors represented by the form $-6\mathbf{i} + 7\mathbf{j}$. Two of these are shown in Figure 12.9.

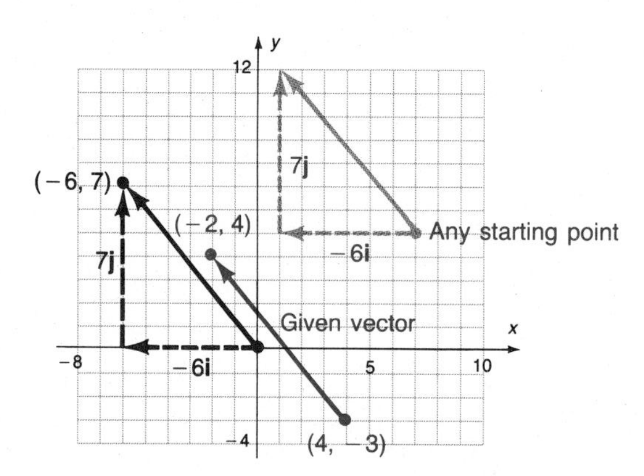

Figure 12.9 Some vectors represented by $\mathbf{v} = -6\mathbf{i} + 7\mathbf{j}$

WARNING: Note the usage of the words *product* and *multiplication.*

In arithmetic the words *multiplication* and *product* are used to mean the same thing. When working with vectors, however, these words are used to denote different ideas. Recall that scalar multiplication does not tell how to multiply two vectors; instead it tells how to multiply a scalar and a vector. Now we define an operation called **scalar product** in order to multiply two vectors to obtain a number. It is called *scalar* product because a number or scalar is obtained as an answer. In Section 12.4 we will define another vector multiplication, called **vector product**, in which a vector is obtained as an answer.

Definition of Scalar Product

> Let $\mathbf{v} = a\mathbf{i} + b\mathbf{j}$ and $\mathbf{w} = c\mathbf{i} + d\mathbf{j}$. Then the **scalar product**, written $\mathbf{v} \cdot \mathbf{w}$, is defined by
>
> $$\mathbf{v} \cdot \mathbf{w} = ac + bd$$

Sometimes this product is called the **dot product**, from the form in which it is written.

EXAMPLE 7 Find the scalar product of the given vectors.

Solution **a.** If $\mathbf{v} = 2\mathbf{i} + 5\mathbf{j}$ and $\mathbf{w} = 6\mathbf{i} - 3\mathbf{j}$, then

$$\begin{aligned}\mathbf{v} \cdot \mathbf{w} &= 2(6) + 5(-3)\\ &= 12 - 15\\ &= -3\end{aligned}$$

b. If $\mathbf{v} = \cos 30°\mathbf{i} + \sin 30°\mathbf{j}$ and $\mathbf{w} = \cos 60°\mathbf{i} - \sin 60°\mathbf{j}$, then

$$\begin{aligned}\mathbf{v} \cdot \mathbf{w} &= \cos 30° \cos 60° - \sin 30° \sin 60°\\ &= \cos(30° + 60°)\\ &= 0\end{aligned}$$

c. If $\mathbf{v} = -\sqrt{3}\mathbf{i} + \sqrt{2}\mathbf{j}$ and $\mathbf{w} = 3\sqrt{3}\mathbf{i} + 5\sqrt{2}\mathbf{j}$, then

$$\begin{aligned}\mathbf{v} \cdot \mathbf{w} &= (-\sqrt{3})(3\sqrt{3}) + \sqrt{2}(5\sqrt{2})\\ &= -9 + 10\\ &= 1\end{aligned}$$

d. If $\mathbf{v} = 2\mathbf{i} - 3\mathbf{j}$ and $\mathbf{w} = 4\mathbf{i} + a\mathbf{j}$, then

$$\mathbf{v} \cdot \mathbf{w} = 8 - 3a$$

■

There is a very useful geometric property for scalar product that is apparent if we find an expression for the angle between two vectors:

Angle between Vectors

> The angle θ between vectors $\mathbf{v}$ and $\mathbf{w}$ is found by
>
> $$\cos\theta = \frac{\mathbf{v} \cdot \mathbf{w}}{|\mathbf{v}||\mathbf{w}|}$$

Proof Let $\mathbf{v} = a\mathbf{i} + b\mathbf{j}$ and $\mathbf{w} = c\mathbf{i} + d\mathbf{j}$ be drawn with their bases at the origin, as shown in Figure 12.10. Let x be the distance between the endpoints of the vectors. Then, by the Law of Cosines,

$$\begin{aligned}\cos\theta &= \frac{|\mathbf{v}|^2 + |\mathbf{w}|^2 - x^2}{2|\mathbf{v}||\mathbf{w}|}\\ &= \frac{(\sqrt{a^2 + b^2})^2 + (\sqrt{c^2 + d^2})^2 - (\sqrt{(a - c)^2 + (b - d)^2})^2}{2|\mathbf{v}||\mathbf{w}|}\\ &= \frac{a^2 + b^2 + c^2 + d^2 - (a^2 - 2ac + c^2 + b^2 - 2bd + d^2)}{2|\mathbf{v}||\mathbf{w}|}\\ &= \frac{2ac + 2bd}{2|\mathbf{v}||\mathbf{w}|}\\ &= \frac{ac + bd}{|\mathbf{v}||\mathbf{w}|}\\ &= \frac{\mathbf{v} \cdot \mathbf{w}}{|\mathbf{v}||\mathbf{w}|}\end{aligned}$$

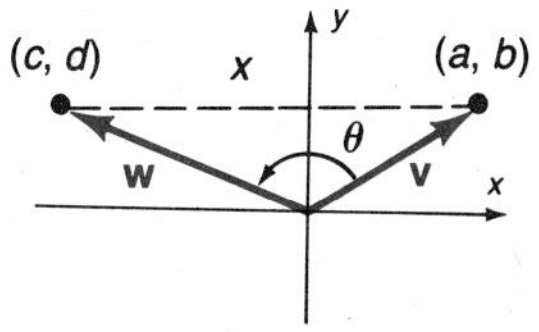

Figure 12.10 Finding the angle between two vectors

□

There is a useful geometric property of vectors whose directions differ by 90°. If you are dealing with lines that meet at a 90° angle, they are called **perpendicular lines**, but if vectors form a 90° angle, they are called **orthogonal vectors**. Notice that if $\theta = 90°$, then $\cos 90° = 0$; therefore, from the angle between vectors formula,

$$0 = \frac{\mathbf{v} \cdot \mathbf{w}}{|\mathbf{v}||\mathbf{w}|}$$

If you multiply both sides by $|\mathbf{v}||\mathbf{w}|$, the result is the following important condition.

Orthogonal Vectors

> Vectors $\mathbf{v}$ and $\mathbf{w}$ are orthogonal if and only if $\mathbf{v} \cdot \mathbf{w} = 0$.

EXAMPLE 8 Show that $\mathbf{v} = 3\mathbf{i} - 2\mathbf{j}$ and $\mathbf{w} = 6\mathbf{i} + 9\mathbf{j}$ are orthogonal.

Solution

$$\begin{aligned} \mathbf{v} \cdot \mathbf{w} &= 3(6) + (-2)(9) \\ &= 18 - 18 \\ &= 0 \end{aligned}$$

Since the scalar product is zero, the vectors are orthogonal. ■

EXAMPLE 9 Find a so that $\mathbf{v} = 3\mathbf{i} + a\mathbf{j}$ and $\mathbf{w} = \mathbf{i} - 2\mathbf{j}$ are orthogonal.

Solution

$$\mathbf{v} \cdot \mathbf{w} = 3 - 2a$$

If they are orthogonal, then

$$3 - 2a = 0$$

$$a = \frac{3}{2}$$ ■

PROBLEM SET 12.1

A

Find the algebraic representation for each vector given in Problems 1–18. $|\mathbf{v}|$ is the magnitude of the vector $\mathbf{v}$, and θ is the angle the vector makes with the positive x axis. A and B are the endpoints of the vector $\mathbf{v}$, and A is the base point. Draw each vector.

1. $|\mathbf{v}| = 12, \theta = 60°$
2. $|\mathbf{v}| = 8, \theta = 30°$
3. $|\mathbf{v}| = \sqrt{2}, \theta = 45°$
4. $|\mathbf{v}| = 9, \theta = 45°$
5. $|\mathbf{v}| = 7, \theta = 23°$
6. $|\mathbf{v}| = 5, \theta = 72°$
7. $|\mathbf{v}| = 4, \theta = 112°$
8. $|\mathbf{v}| = 10, \theta = 214°$
9. $A(4, 1), B(2, 3)$
10. $A(-1, -3), B(4, 5)$
11. $A(1, -2), B(-5, -7)$
12. $A(6, -8), B(5, -2)$
13. $A(-3, 2), B(5, -8)$
14. $A(0, 0), B(-3, -4)$
15. $A(7, 1), B(0, 0)$
16. $A(2, 9), B(-5, 8)$
17. $A(6, -1); B(-3, -7)$
18. $A(-2, -8), B(-4, 7)$

Find the magnitude of each of the vectors given in Problems 19–27.

19. $\mathbf{v} = 3\mathbf{i} + 4\mathbf{j}$
20. $\mathbf{v} = 5\mathbf{i} - 12\mathbf{j}$
21. $\mathbf{v} = 6\mathbf{i} - 7\mathbf{j}$
22. $\mathbf{v} = -3\mathbf{i} + 5\mathbf{j}$
23. $\mathbf{v} = -2\mathbf{i} + 2\mathbf{j}$
24. $\mathbf{v} = 5\mathbf{i} - 8\mathbf{j}$
25. $\mathbf{v} = \mathbf{i} - 3\mathbf{j}$
26. $\mathbf{v} = 2\mathbf{i} - \mathbf{j}$
27. $\mathbf{v} = 4\mathbf{i} + 5\mathbf{j}$

State whether the given pairs of vectors in Problems 28–33 are orthogonal.

28. $\mathbf{v} = 3\mathbf{i} - 2\mathbf{j}; \mathbf{w} = 6\mathbf{i} + 9\mathbf{j}$
29. $\mathbf{v} = 2\mathbf{i} + 3\mathbf{j}; \mathbf{w} = 6\mathbf{i} - 9\mathbf{j}$
30. $\mathbf{v} = 4\mathbf{i} - 5\mathbf{j}; \mathbf{w} = 8\mathbf{i} + 10\mathbf{j}$
31. $\mathbf{v} = 5\mathbf{i} + 4\mathbf{j}; \mathbf{w} = 8\mathbf{i} - 10\mathbf{j}$
32. $\mathbf{v} = 2\mathbf{i} - 3\mathbf{j}; \mathbf{w} = 3\mathbf{i} + 2\mathbf{j}$
33. $\mathbf{v} = \mathbf{i}; \mathbf{w} = \mathbf{j}$

B

34. ***Navigation*** A woman sets out in a rowboat heading due west and rows at 4.8 mph. The current is carrying the boat due south at 12 mph. What is the true course of the rowboat, and how fast is the boat traveling relative to the ground?

35. ***Aviation*** An airplane is headed due west at 240 mph. The wind is blowing due south at 43 mph. What is the true course of the plane, and how fast is it traveling across the ground?

36. ***Avivation*** An airplane is heading 535°W with a velocity of 723 mph. How far south has it traveled in 1 hr?

37. ***Aviation*** An airplane is heading N43.0°E with a velocity of 248 mph. How far east has it traveled in 2 hr?

In Problems 38–49, find $\mathbf{v} \cdot \mathbf{w}$, $|\mathbf{v}|$, $|\mathbf{w}|$, *and* $\cos\theta$, *where* θ *is the angle between* $\mathbf{v}$ *and* $\mathbf{w}$.

38. $\mathbf{v} = 3\mathbf{i} + 4\mathbf{j}$, $\mathbf{w} = 5\mathbf{i} + 12\mathbf{j}$

39. $\mathbf{v} = 8\mathbf{i} - 6\mathbf{j}$, $\mathbf{w} = -5\mathbf{i} + 12\mathbf{j}$

40. $\mathbf{v} = 2\mathbf{i} + \sqrt{5}\mathbf{j}$, $\mathbf{w} = 3\sqrt{5}\mathbf{i} - 3\mathbf{j}$

41. $\mathbf{v} = 7\mathbf{i} - \sqrt{15}\mathbf{j}$, $\mathbf{w} = 2\sqrt{15}\mathbf{i} + 14\mathbf{j}$

42. $\mathbf{v} = -2\mathbf{i} + 3\mathbf{j}$, $\mathbf{w} = 6\mathbf{i} + 5\mathbf{j}$

43. $\mathbf{v} = 3\mathbf{i} + 9\mathbf{j}$, $\mathbf{w} = 2\mathbf{i} - 5\mathbf{j}$

44. $\mathbf{v} = \mathbf{i}$, $\mathbf{w} = \mathbf{i}$

45. $\mathbf{v} = \mathbf{j}$, $\mathbf{w} = \mathbf{j}$

46. $\mathbf{v} = \mathbf{i}$, $\mathbf{w} = -\mathbf{j}$

47. $\mathbf{v} = 5\mathbf{i} - \mathbf{j}$, $\mathbf{w} = 2\mathbf{i} + 3\mathbf{j}$

48. $\mathbf{v} = 4\mathbf{i} + 2\mathbf{j}$, $\mathbf{w} = 3\mathbf{i} - \mathbf{j}$

49. $\mathbf{v} = \mathbf{i} + \mathbf{j}$, $\mathbf{w} = \mathbf{i}$

In Problems 50–55, find the angle θ *to the nearest degree,* $0° \le \theta \le 180°$, *between the vectors* $\mathbf{v}$ *and* $\mathbf{w}$.

50. $\mathbf{v} = \frac{1}{2}\mathbf{i} + \frac{\sqrt{3}}{2}\mathbf{j}$, $\mathbf{w} = \frac{1}{2}\mathbf{i} + \frac{1}{2}\mathbf{j}$

51. $\mathbf{v} = \sqrt{2}\mathbf{i} - \sqrt{2}\mathbf{j}$, $\mathbf{w} = \frac{\sqrt{3}}{2}\mathbf{i} + \frac{1}{2}\mathbf{j}$

52. $\mathbf{v} = \mathbf{j}$, $\mathbf{w} = \frac{1}{2}\mathbf{i} - \frac{\sqrt{3}}{2}\mathbf{j}$

53. $\mathbf{v} = -\mathbf{i}$, $\mathbf{w} = -2\sqrt{2}\mathbf{i} + 2\sqrt{2}\mathbf{j}$

54. $\mathbf{v} = 2\mathbf{i} + 3\mathbf{j}$, $\mathbf{w} = -\mathbf{i} + 4\mathbf{j}$

55. $\mathbf{v} = -3\mathbf{i} + 2\mathbf{j}$, $\mathbf{w} = 6\mathbf{i} + 9\mathbf{j}$

In Problems 56–58, find a number a so that the given vectors are orthogonal.

56. $\mathbf{v} = 2\mathbf{i} + 3\mathbf{j}$, $\mathbf{w} = 5\mathbf{i} + a\mathbf{j}$

57. $\mathbf{v} = 4\mathbf{i} - a\mathbf{j}$, $\mathbf{w} = -2\mathbf{i} + 5\mathbf{j}$

58. $\mathbf{v} = a\mathbf{i} + 5\mathbf{j}$, $\mathbf{w} = a\mathbf{i} - 15\mathbf{j}$

C

59. ***Aviation*** A pilot is flying at an airspeed of 241 mph in a wind blowing 20.4 mph from the east. In what direction must the pilot head in order to fly due north? What is the pilot's speed relative to the ground?

60. ***Aviation*** Answer the questions posed in Problem 59 with the pilot wishing to fly due south.

61. ***Space Science*** The weight of astronauts on the moon is about one-sixth of their weight on earth. This fact has a marked effect on such simple acts as walking, running, and jumping. To study these effects and to train astronauts for working under lunar-gravity conditions, scientists at NASA's Langley Research Center have designed an inclined-plane apparatus to simulate reduced gravity. The apparatus consists of a sling that holds the astronaut in a position perpendicular to the inclined plane (see Figure 12.11). The sling is attached to one end of a long cable that runs parallel to the inclined plane. The other end of the cable is attached to a trolley that runs along an overhead track. This device allows the astronaut to move freely in a plane perpendicular to the inclined plane. Let W be the astronaut's mass and θ be the angle between the inclined plane and the ground. Make a vector diagram showing the tension in the cable and the force exerted by the inclined plane against the feet of the astronaut.

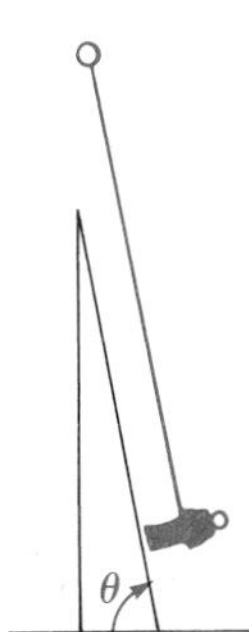

Figure 12.11

62. ***Space Science*** From the point of view of the astronaut in Problem 61, the inclined plane is the ground and the astronaut's simulated mass (that is, the downward force against the inclined plane) is $\mathbf{W}\cos\theta$. What value of θ is required in order to simulate lunar gravity?

12.2 PROPERTIES OF VECTORS

In this section we will look at several properties of vectors that will be useful in more advanced mathematics courses. Suppose you are given a line $ax + by + c = 0$. A vector perpendicular to a line is called a **normal** to the line. We will show that

$$\mathbf{N} = a\mathbf{i} + b\mathbf{j}$$

is normal to

$$L: ax + by + c = 0$$

Notice that, once L is given, **N** can be found by inspection. For example, if

$$L: 3x + 4y + 5 = 0 \text{ then}$$

$$\mathbf{N} = 3\mathbf{i} + 4\mathbf{j}$$

If

$$L: 5x - 3y + 7 = 0 \text{ then}$$

$$\mathbf{N} = 5\mathbf{i} - 3\mathbf{j}$$

In general, if $P_1(x_1, y_1)$ and $P_2(x_2, y_2)$ are any two points on L, then $\overrightarrow{P_1P_2}$ is a representative of the vector determined by the line. That is,

$$\overrightarrow{P_1P_2} = (x_2 - x_1)\mathbf{i} + (y_2 - y_1)\mathbf{j}$$

We must show that $\overrightarrow{P_1P_2}$ and $\mathbf{N} = a\mathbf{i} + b\mathbf{j}$ are orthogonal. That is, we must show that

$$\overrightarrow{P_1P_2} \cdot \mathbf{N} = 0$$

Now,

$$\overrightarrow{P_1P_2} \cdot \mathbf{N} = a(x_2 - x_1) + b(y_2 - y_1)$$

Since $P_1(x_1, y_1)$ and $P_2(x_2, y_2)$ are on the line, they make the equation true:

$$ax_1 + by_1 + c = 0$$

$$ax_2 + by_2 + c = 0$$

Subtracting,

$$a(x_2 - x_1) + b(y_2 - y_1) = 0$$

Thus,

$$\begin{aligned}\overrightarrow{P_1P_2} \cdot \mathbf{N} &= a(x_2 - x_1) + b(y_2 - y_1)\\ &= 0\end{aligned}$$

EXAMPLE 1 Find a vector determined by the line $5x - 3y - 15 = 0$. Also find a normal vector. Show that they are orthogonal.

Solution A normal vector is $5\mathbf{i} - 3\mathbf{j}$.

A vector determined by the line is found by first obtaining two points on the line. If $x = 0$, then $y = -5$; if $y = 0$, then $x = 3$. Thus, two points on the line are $(0, -5)$ and $(3, 0)$. Hence $3\mathbf{i} + 5\mathbf{j}$ is a vector determined by the line.

Finally, $(5\mathbf{i} - 3\mathbf{j}) \cdot (3\mathbf{i} + 5\mathbf{j}) = 15 - 15 = 0$, so they are orthogonal (see Figure 12.12).

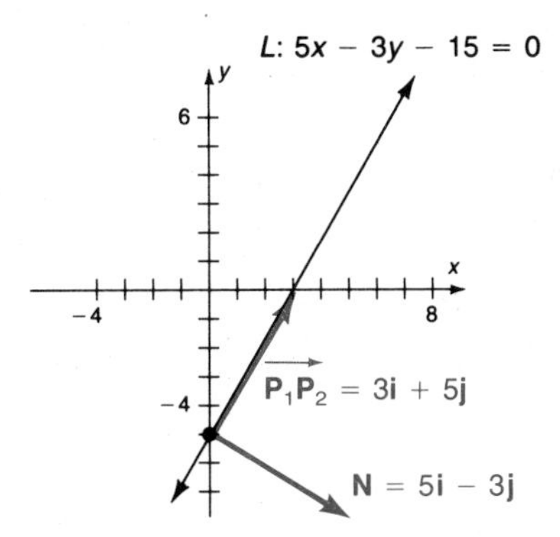

Figure 12.12

Let **v** and **w** be two vectors whose representatives have a common base point. If we drop a perpendicular from the head of **v** to the line determined by **w**, we determine a vector called *the vector projection of* **v** *onto* **w**, denoted by **u**, as shown in Figure 12.13.

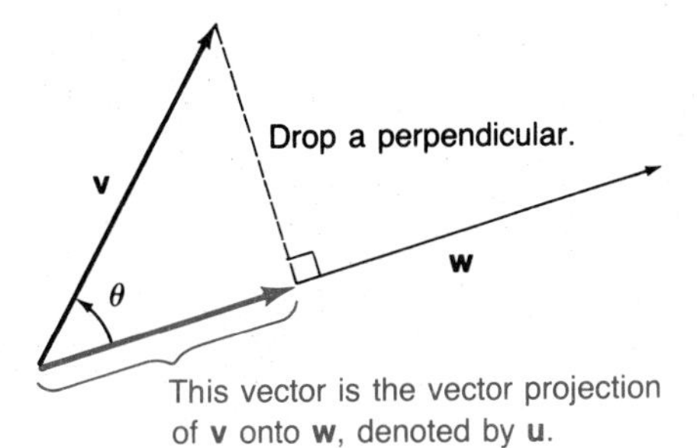

Figure 12.13 Projection of **v** onto **w**

The *scalar projection* is the *length* of the vector projection. Let θ be an acute angle between **v** and **w**. Then

$$\cos\theta = \frac{|\mathbf{u}|}{|\mathbf{v}|} \qquad \text{By definition of cosine}$$

Thus,

$$\begin{aligned} |\mathbf{u}| &= |\mathbf{v}|\cos\theta \\ &= |\mathbf{v}|\frac{\mathbf{v}\cdot\mathbf{w}}{|\mathbf{v}|\,|\mathbf{w}|} \\ &= \frac{\mathbf{v}\cdot\mathbf{w}}{|\mathbf{w}|} \end{aligned}$$

If $90^\circ < \theta < 180^\circ$, then $\cos\theta \le 0$ so $|\mathbf{v}|\cos\theta$ is a negative number. Since we want $|\mathbf{u}|$ to be nonnegative (since it is a length), we introduce an absolute value:

$$|\mathbf{u}| = \left|\frac{\mathbf{v}\cdot\mathbf{w}}{|\mathbf{w}|}\right|$$

In order to find the vector projection, we notice

$$\mathbf{u} = s\mathbf{w}$$

for some scalar s. Since we do not know $\mathbf{u}$ and do not know s, this equation is not much help. Therefore we use

$$|\mathbf{u}| = s|\mathbf{w}|$$

since we know $|\mathbf{u}|$ (it is the scalar projection we found above). Thus,

$$\frac{\mathbf{v} \cdot \mathbf{w}}{|\mathbf{w}|} = s|\mathbf{w}|$$

$$s = \frac{\mathbf{v} \cdot \mathbf{w}}{|\mathbf{w}|^2}$$

$$= \frac{\mathbf{v} \cdot \mathbf{w}}{\mathbf{w} \cdot \mathbf{w}}$$

We can now find $\mathbf{u}$:

$$\mathbf{u} = s\mathbf{w}$$

$$= \left(\frac{\mathbf{v} \cdot \mathbf{w}}{\mathbf{w} \cdot \mathbf{w}}\right)\mathbf{w}$$

In summary:

Projections of v onto w

Scalar Projection (a number): $\left|\dfrac{\mathbf{v} \cdot \mathbf{w}}{|\mathbf{w}|}\right|$

Vector Projection (a vector): $\left(\dfrac{\mathbf{v} \cdot \mathbf{w}}{\mathbf{w} \cdot \mathbf{w}}\right)\mathbf{w}$

EXAMPLE 2 Find the scalar and vector projections of $\mathbf{v} = 5\mathbf{i} - 3\mathbf{j}$ onto $\mathbf{w} = 7\mathbf{i} + 4\mathbf{j}$.

Solution Vector projection:

$$\left(\frac{\mathbf{v} \cdot \mathbf{w}}{\mathbf{w} \cdot \mathbf{w}}\right)\mathbf{w} = \frac{35 - 12}{49 + 16}\mathbf{w}$$

$$= \frac{23}{65}(7\mathbf{i} + 4\mathbf{j})$$

$$= \frac{161}{65}\mathbf{i} + \frac{92}{65}\mathbf{j}$$

Scalar projection: We could find the length of the vector projection by calculating

$$\sqrt{\left(\frac{161}{65}\right)^2 + \left(\frac{92}{65}\right)^2}$$

or we can use

$$\left|\frac{\mathbf{v} \cdot \mathbf{w}}{|\mathbf{w}|}\right|$$

Since the latter is easier to calculate, we find

$$\left|\frac{23}{\sqrt{49+16}}\right| = \frac{23}{\sqrt{65}}$$

■

EXAMPLE 3 Find the scalar projection of $\mathbf{v} = 3\mathbf{i} - 2\mathbf{j}$ onto $\mathbf{w} = 2\mathbf{i} + 4\mathbf{j}$.

Solution The scalar projection of $\mathbf{v}$ onto $\mathbf{w}$ is

$$\begin{aligned}\left|\frac{\mathbf{v}\cdot\mathbf{w}}{|\mathbf{w}|}\right| &= \left|\frac{6-8}{\sqrt{4+16}}\right| \\ &= \left|\frac{-2}{2\sqrt{5}}\right| = \frac{1}{\sqrt{5}}\end{aligned}$$

We can check this result by finding the length of the vector projection of $\mathbf{v}$ onto $\mathbf{w}$:

$$\begin{aligned}\mathbf{u} &= \left(\frac{\mathbf{v}\cdot\mathbf{w}}{\mathbf{w}\cdot\mathbf{w}}\right)\mathbf{w} \qquad & |\mathbf{u}| &= \sqrt{\left(-\frac{1}{5}\right)^2 + \left(-\frac{2}{5}\right)^2} \\ &= \left(\frac{-2}{4+16}\right)\mathbf{w} & &= \sqrt{\frac{1+4}{25}} = \frac{1}{5}\sqrt{5} \\ &= \frac{-1}{10}(2\mathbf{i}+4\mathbf{j}) \\ &= -\frac{1}{5}\mathbf{i} - \frac{2}{5}\mathbf{j}\end{aligned}$$

■

The final property of vectors we will consider in this section enables us to derive a formula for finding the distance from a point to a line. In Section 11.1 (Problem 65) this formula was derived with a great deal of effort without using vectors. We can now derive the same formula quite easily by using vector ideas.

Let L be any given line and P any given point not on L. By the distance from P to L we mean the perpendicular distance d as shown in Figure 12.14a. We wish to find this distance.

If L is a vertical line, then the distance from P to L is easy to find. (Why?) If L is not vertical, then we let B be the y intercept and $\mathbf{N}$ a normal to L. Let the base of $\mathbf{N}$ be drawn at B, as shown in Figure 12.14b.

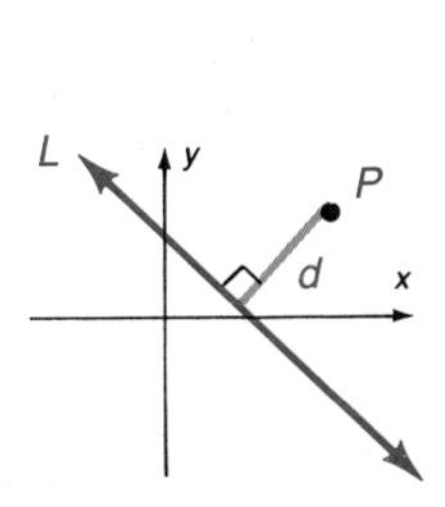

Figure 12.14a Distance from P to L

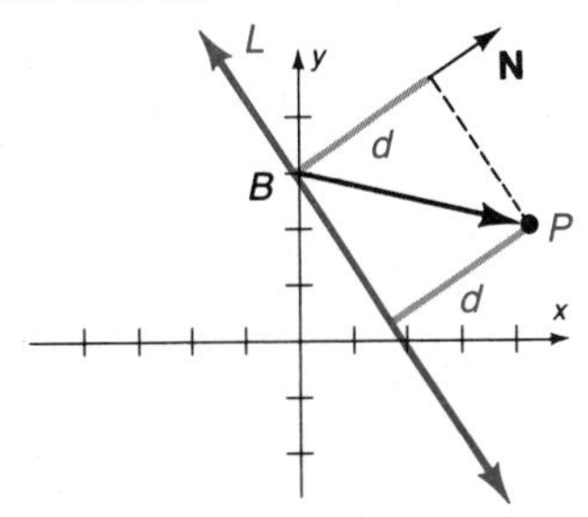

Figure 12.14b Procedure for finding the distance from a point to a line

The distance we seek is seen to be the scalar projection of $\overrightarrow{BP}$ onto **N**. Thus

Vector Formula for the Distance from a Point to a Line

$$d = \left|\frac{\overrightarrow{BP} \cdot \mathbf{N}}{|\mathbf{N}|}\right|$$

EXAMPLE 4 Find the distance from the point $(5, -3)$ to the line $4x + 3y - 15 = 0$.

Solution P is $(5, -3)$ and B is $(0, 5)$. Then

$$\overrightarrow{BP} = 5\mathbf{i} - 8\mathbf{j}$$
$$\mathbf{N} = 4\mathbf{i} + 3\mathbf{j}$$
$$|\mathbf{N}| = \sqrt{4^2 + 3^2} = 5$$
$$\overrightarrow{BP} \cdot \mathbf{N} = 20 - 24 = -4$$

Thus,

$$d = \left|\frac{\overrightarrow{BP} \cdot \mathbf{N}}{|\mathbf{N}|}\right| = \left|\frac{-4}{5}\right| = \frac{4}{5}$$

■

PROBLEM SET 12.2

A

Find a vector normal to each line given in Problems 1–12.

1. $2x - 3y + 4 = 0$
2. $x + y - 1 = 0$
3. $x - y + 3 = 0$
4. $4x + 5y - 3 = 0$
5. $3x - 2y + 1 = 0$
6. $5x + y - 3 = 0$
7. $9x + 7y - 5 = 0$
8. $6x - 3y + 2 = 0$
9. $4x - y - 12 = 0$
10. $y = \frac{2}{3}x - 5$
11. $y = -\frac{1}{2}x - 10$
12. $y = -\frac{5}{8}x + 4$

Find a vector determined by each line given in Problems 13–24.

13. $2x - 3y + 4 = 0$
14. $x + y - 1 = 0$
15. $x - y + 3 = 0$
16. $4x + 5y - 3 = 0$
17. $3x - 2y + 1 = 0$
18. $5x + y - 3 = 0$
19. $9x + 7y - 5 = 0$
20. $6x - 3y + 2 = 0$
21. $4x - y - 12 = 0$
22. $y = \frac{2}{3}x - 5$
23. $y = -\frac{1}{2}x - 10$
24. $y = -\frac{5}{8}x + 4$

In Problems 25–30, find the scalar projection of **v** *onto* **w**.

25. $\mathbf{v} = 3\mathbf{i} + 4\mathbf{j}$, $\mathbf{w} = 5\mathbf{i} + 12\mathbf{j}$
26. $\mathbf{v} = 8\mathbf{i} - 6\mathbf{j}$, $\mathbf{w} = -5\mathbf{i} + 12\mathbf{j}$
27. $\mathbf{v} = 7\mathbf{i} - \sqrt{15}\mathbf{j}$, $\mathbf{w} = 2\sqrt{15}\mathbf{i} + 14\mathbf{j}$
28. $\mathbf{v} = 2\mathbf{i} + \sqrt{5}\mathbf{j}$, $\mathbf{w} = 3\sqrt{5}\mathbf{i} - 3\mathbf{j}$
29. $\mathbf{v} = -2\mathbf{i} + 3\mathbf{j}$, $\mathbf{w} = 6\mathbf{i} + 5\mathbf{j}$
30. $\mathbf{v} = 3\mathbf{i} + 9\mathbf{j}$, $\mathbf{w} = 2\mathbf{i} - 5\mathbf{j}$

In Problems 31–36, find the vector projection of **v** *onto* **w**.

31. $\mathbf{v} = 3\mathbf{i} + 4\mathbf{j}$, $\mathbf{w} = 5\mathbf{i} + 12\mathbf{j}$
32. $\mathbf{v} = 8\mathbf{i} - 6\mathbf{j}$, $\mathbf{w} = -5\mathbf{i} + 12\mathbf{j}$
33. $\mathbf{v} = 7\mathbf{i} - \sqrt{15}\mathbf{j}$, $\mathbf{w} = 2\sqrt{15}\mathbf{i} + 14\mathbf{j}$
34. $\mathbf{v} = 2\mathbf{i} + \sqrt{5}\mathbf{j}$, $\mathbf{w} = 3\sqrt{5}\mathbf{i} - 3\mathbf{j}$
35. $\mathbf{v} = -2\mathbf{i} + 3\mathbf{j}$, $\mathbf{w} = 6\mathbf{i} + 5\mathbf{j}$
36. $\mathbf{v} = 3\mathbf{i} + 9\mathbf{j}$, $\mathbf{w} = 2\mathbf{i} - 5\mathbf{j}$

B

Find the distance from the given point to the given line in Problems 37–50.

37. $3x - 4y + 8 = 0$; $(4, 5)$
38. $5x - 12y + 15 = 0$; $(6, -3)$
39. $3x - 4y + 8 = 0$; $(9, -3)$
40. $5x - 12y + 15 = 0$; $(-2, 6)$
41. $4x + 3y - 5 = 0$; $(-1, -1)$

42. $12x + 5y - 2 = 0; (3, 5)$ **43.** $4x + 3y - 5 = 0; (6, 1)$

44. $12x + 5y - 2 = 0; (4, -3)$

45. $x - 3y + 15 = 0; (1, -6)$ **46.** $6x - y - 10 = 0; (-5, 6)$

47. $x - 3y + 15 = 0; (8, 14)$ **48.** $6x - y - 10 = 0; (8, 10)$

49. $2x - 5y = 0; (4, 5)$ **50.** $4x + 7y = 0; (5, 10)$

Find the area of the triangle determined by the given points in Problems 51–56.

51. $(1, 2), (4, 5), (-5, 3)$ **52.** $(-1, 1), (4, 3), (1, -1)$

53. $(0, 0), (5, -3), (-2, -7)$ **54.** $(5, 6), (-3, 5), (0, 0)$

55. $(3, 0), (0, 8), (-4, 6)$ **56.** $(0, 6), (-5, 2), (-3, -6)$

C

57. Prove the commutative law $\mathbf{u} \cdot \mathbf{v} = \mathbf{v} \cdot \mathbf{u}$ for any vectors $\mathbf{u}$ and $\mathbf{v}$.

58. Prove the distributive law $\mathbf{u} \cdot (\mathbf{v} + \mathbf{w}) = \mathbf{u} \cdot \mathbf{v} + \mathbf{u} \cdot \mathbf{w}$ for any vectors $\mathbf{u}$, $\mathbf{v}$, and $\mathbf{w}$.

59. Let $\mathbf{u}$ be the projection of $\mathbf{v}$ onto $\mathbf{w}$. Show that $\mathbf{v} - \mathbf{u}$ is orthogonal to $\mathbf{w}$.

60. Let $\mathbf{v} = a\mathbf{i} + b\mathbf{j}$ and $\mathbf{w} = c\mathbf{i} + d\mathbf{j}$ be two vectors. Use the Law of Cosines to show that

$$\cos\theta = \frac{ac + bd}{|\mathbf{v}||\mathbf{w}|}$$

where θ is the angle between the vectors. *Hint:* The Law of Cosines states that $c^2 = a^2 + b^2 - 2ab\cos\gamma$, where a, b, and c are sides of a triangle and γ is the angle opposite side c.

12.3 THREE-DIMENSIONAL COORDINATE SYSTEM*

Many real-life models involve more than one variable. Suppose, for example, we consider one of the most fundamental applications, that of the cost of producing an item. If Ballad Corporation produces a single record with fixed costs of $2000 and a unit cost of $0.35, then

$$C(x) = 2000 + 0.35x$$

for x records produced. However, if a second record is produced with additional fixed costs of $500 and a unit cost of $0.30, then the total cost of producing x records of the first type and y records of the second type requires what we call a **function of two independent variables** x and y:

$$C(x, y) = 2500 + 0.35x + 0.30y$$

Function of Two or More Variables

Suppose D is a collection of ordered n-tuples of real numbers $(x_1, x_2, \ldots, x_n)$. Then a function f with **domain** D is a rule that assigns a number

$$z = f(x_1, x_2, \ldots, x_n)$$

to each n-tuple in D. The **range** of the function is the set of z values the function assumes. Then z is called the **dependent variable** of f, and f is said to be a **function of the n independent variables** $x_1, x_2, \ldots, x_n$.

You have already considered many examples of functions of several variables, as Example 1 shows.

* Optional section

EXAMPLE 1

Area of a rectangle:	$K(l, w) = lw$
Volume of a box:	$V(l, w, h) = lwh$
Simple interest:	$I(P, r, t) = P(1 + rt)$
Compound interest:	$A(P, r, t, n) = P\left(1 + \frac{r}{n}\right)^{nt}$

Find each of the values requested below and interpret your results in terms of what you know about each of the above formulas.

a. $K(25, 15)$ **b.** $V(5, 20, 30)$
c. $I(100{,}000, 0.08, 15)$ **d.** $A(450{,}000, 0.09, 30, 12)$

Solution

a. $K(25, 15) = 25(15) = 375$; the area of a 25 by 15 rectangle is 375 square units.

b. $V(5, 20, 30) = 5(20)(30) = 3000$; the volume of a 5 by 20 by 30 box is 3000 cubic units.

c. $I(100{,}000, 0.08, 15) = 100{,}000[1 + 0.08(15)] = 100{,}000[2.2] = 220{,}000$; the future value of a \$100,000 investment at 8% simple interest for 15 years is \$220,000.

d. $A(450{,}000, 0.09, 30, 12) = 450{,}000(1 + \frac{0.09}{12})^{30(12)} = 450{,}000(1.0075)^{360} \approx 6{,}628{,}759.26$; the future value of a \$450,000 investment at 9% compounded monthly is approximately \$6,628,759.26. ■

Even though a function of several variables has been defined for the general case and Example 1 shows functions of several variables, this chapter focuses primarily on functions of two variables. That is, $z = f(x, y)$ is the notation used for z, a function of two independent variables x and y. In order to graph such a function we need to consider **ordered triplets** (x, y, z) and a **three-dimensional coordinate system**, just as we have already considered ordered pairs (x, y) and a two-dimensional coordinate system. We draw a coordinate system with three mutually perpendicular axes, as shown in Figure 12.15a. Notice in Figure 12.15a that only the octant for which x, y, and z are all positive is shown, but in Figure 12.15b you can see all eight octants.

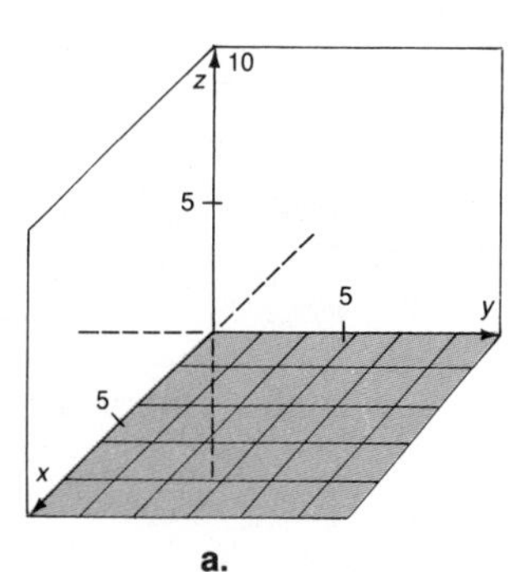

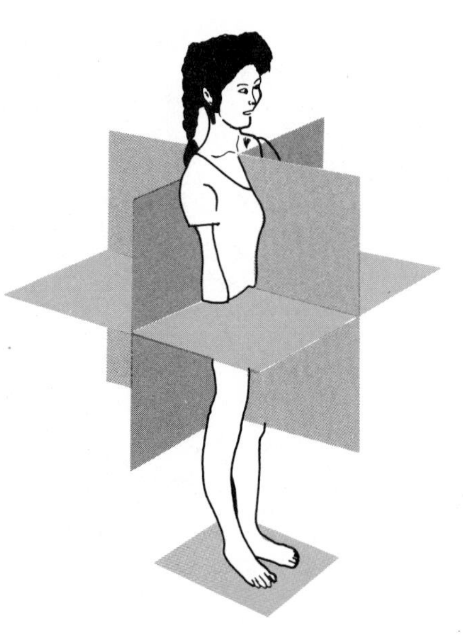

Figure 12.15 Three-dimensional coordinate system

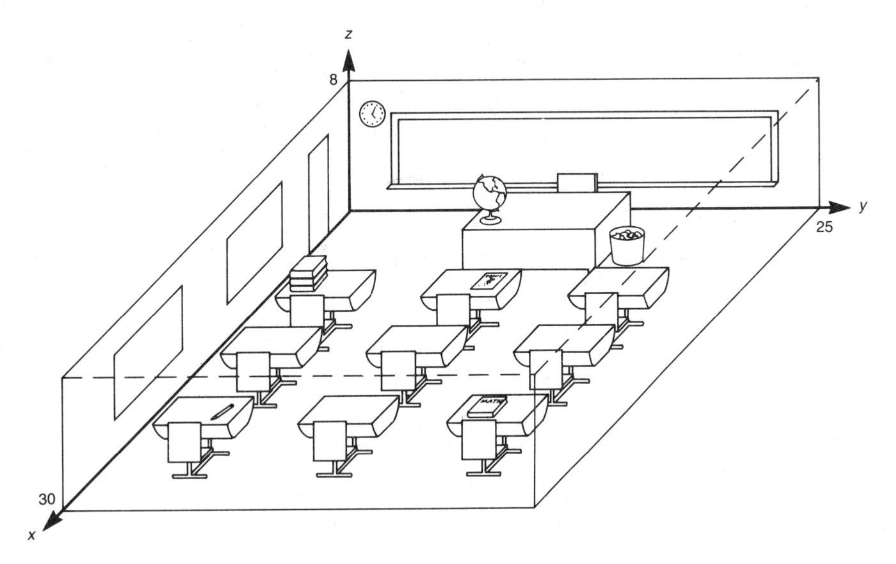

Figure 12.16 A typical classroom; assume the dimensions are 25 by 30 feet with an 8-foot ceiling

Think of the x axis and the y axis as lying in the plane of the floor and the z axis as a line perpendicular to the floor. All the graphs we have drawn in this book would now be drawn on the floor.

If you orient yourself in a room (your classroom, for example), as shown in Figure 12.16, you will notice certain important planes:

Floor:	$\boldsymbol{xy}$ **plane**; equation is $z = 0$
Ceiling:	plane parallel to the xy plane; equation is $z = 8$
Front wall:	$\boldsymbol{yz}$ **plane**; equation is $x = 0$
Back wall:	plane parallel to the yz plane; equation is $x = 30$
Left side wall:	$\boldsymbol{xz}$ **plane**; equation is $y = 0$
Right side wall:	plane parallel to the xz plane; equation is $y = 25$

The xy, xz, and yz planes are called the **coordinate planes**. Name the coordinates of several objects in the figure.

EXAMPLE 2 Graph the following ordered triplets:

a. $(10, 20, 10)$ **b.** $(-12, 6, 12)$
c. $(-12, -18, 6)$ **d.** $(20, -10, 18)$

Solution

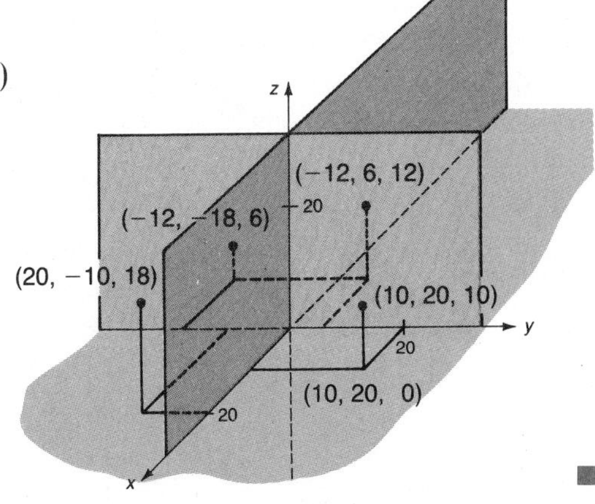

■

Just as points in the plane are associated with ordered pairs satisfying an equation in two variables, points in space are associated with ordered triplets satisfying an equation. The graph of any function of the form $z = f(x, y)$ is called a **surface**. It is beyond the scope of this course to have you spend a great deal of time graphing three-dimensional surfaces, but computer programs are available that have simplified the task of graphing surfaces such as those shown in Figure 12.17.

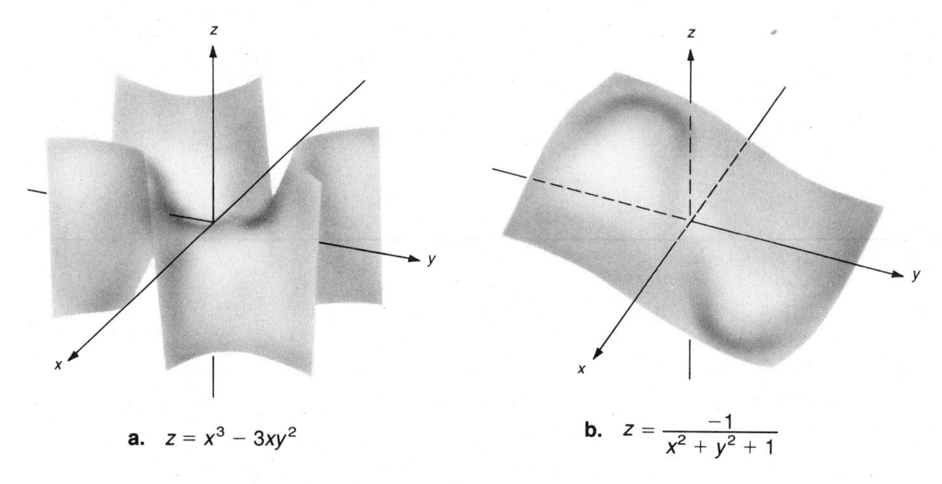

Figure 12.17 Graphs of surfaces in three dimensions

The remainder of this section is a brief introduction to some of the more common three-dimensional surfaces.

Planes

The graph of $ax + by + cz = d$ is a **plane** if a, b, c, and d are real numbers (not all zero).

EXAMPLE 3 Graph the planes defined by the given equations.

a. $x + 3y + 2z = 6$ **b.** $y + z = 5$ **c.** $x = 4$

Solution It is customary to show only the portion of the graph that lies in the **first octant** (that is, where x, y, and z are all positive). To graph a plane, find some ordered triplets satisfying the equation. The best ones to use are often those that fall on the coordinate axes.

a. Let $x = 0$ and $y = 0$; then $z = 3$; point is $(0, 0, 3)$. Let $x = 0$ and $z = 0$; then $y = 2$; point is $(0, 2, 0)$. Let $y = 0$ and $z = 0$; then $x = 6$; point is $(6, 0, 0)$. Plot these points as shown in Figure 12.18a and use them to draw the plane.

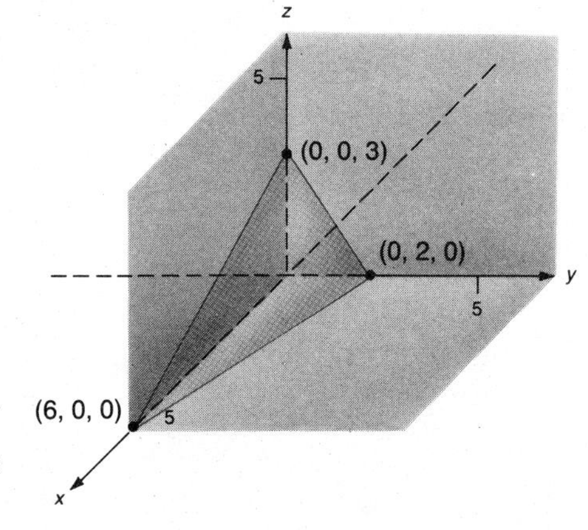

Figure 12.18a Graph of plane $x + 3y + 2z = 6$

b. When one of the variables is missing from the equation of a plane, then that plane is parallel to the axis corresponding to the missing variable; in this case it is parallel to the x axis. Draw the line $y + z = 5$ on the yz plane, and then complete the plane as shown in Figure 12.18b.

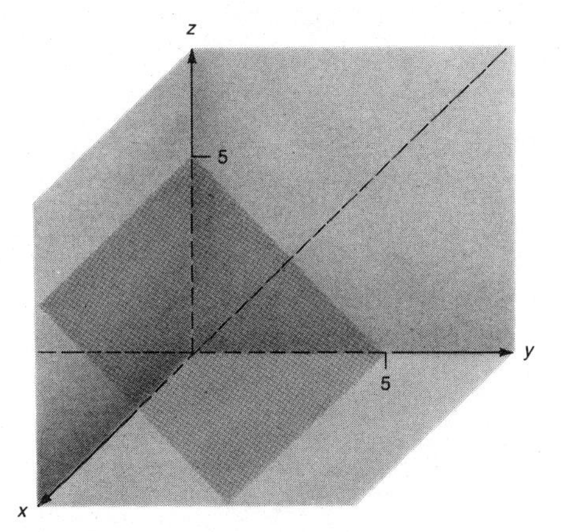

Figure 12.18b Graph of $y + z = 5$

c. When two variables are missing, then the plane is parallel to one of the coordinate planes, as shown in Figure 12.18c.

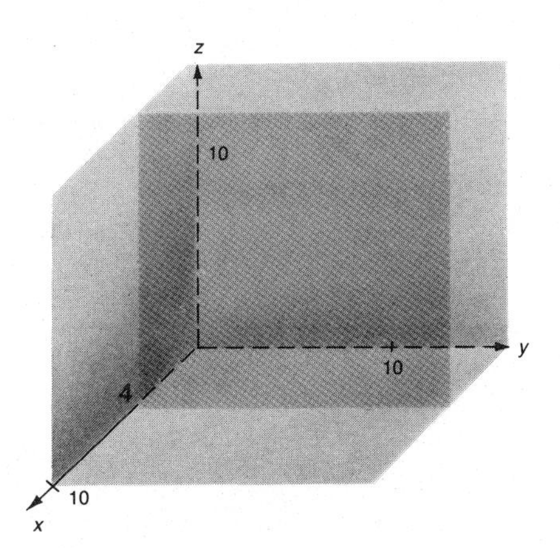

Figure 12.18c Graph of $x = 4$

■

Quadric Surfaces

The graph of the equation

$$Ax^2 + By^2 + Cz^2 + Dxy + Exz + Fyz + Gx + Hy + Iz + J = 0$$

is called a **quadric surface**. The **trace** of a curve is found by setting one of the variables equal to a constant and then graphing the resulting curve. If $x = k$ (k a constant), then the resulting curve is drawn in the plane $x = k$, which is parallel to the yz plane. Similarly, if $y = k$, then the curve is drawn in the plane $y = k$, which is parallel to the xz plane; and if $z = k$, then the curve is drawn in the plane $z = k$, which is parallel to the xy plane. Table 12.1 (page 480) shows the quadric surfaces.

TABLE 12.1 Quadric surfaces

Surface	Description	Surface	Description	Surface	Description
Elliptic cone 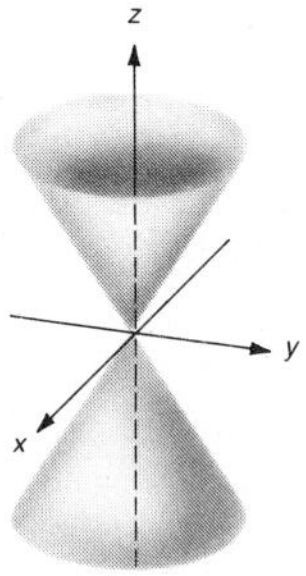	The trace in the xy plane is a point; in planes parallel to the xy plane it is an ellipse. Traces in the xz and yz planes are intersecting lines; in planes parallel to these they are hyperbolas. $z^2 = \frac{x^2}{a^2} + \frac{y^2}{b^2}$	Elliptic paraboloid 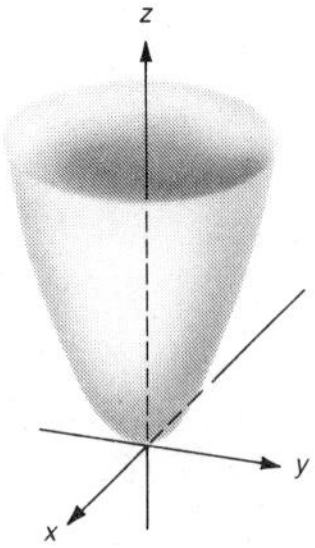	The trace in the xy plane is a point; in planes parallel to the xy plane it is an ellipse. Traces in the xz and yz planes are parabolas. $z = \frac{x^2}{a^2} + \frac{y^2}{b^2}$	Ellipsoid	The traces in the coordinate planes are ellipses. $\frac{x^2}{a^2} + \frac{y^2}{b^2} + \frac{z^2}{c^2} = 1$ 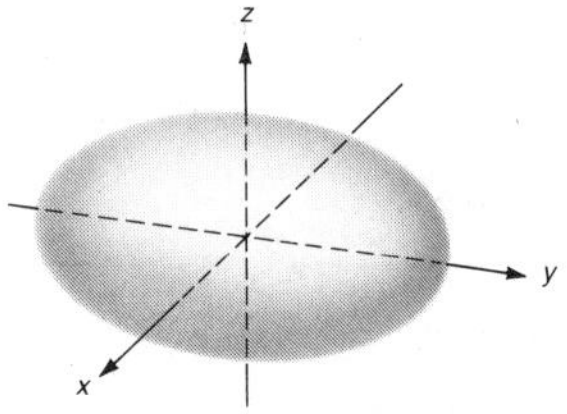
Hyperboloid of one sheet 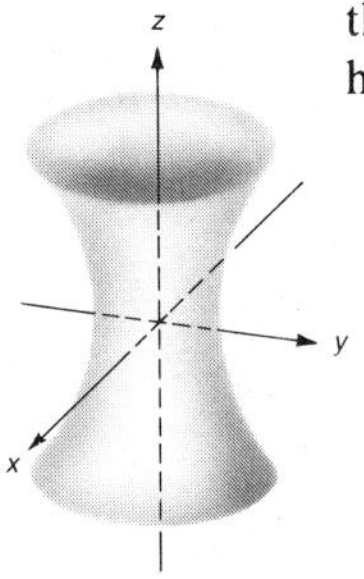	The trace in the xy plane is an ellipse; in the xz and yz planes the traces are hyperbolas. $\frac{x^2}{a^2} + \frac{y^2}{b^2} - \frac{z^2}{c^2} = 1$	Hyperboloid of two sheets 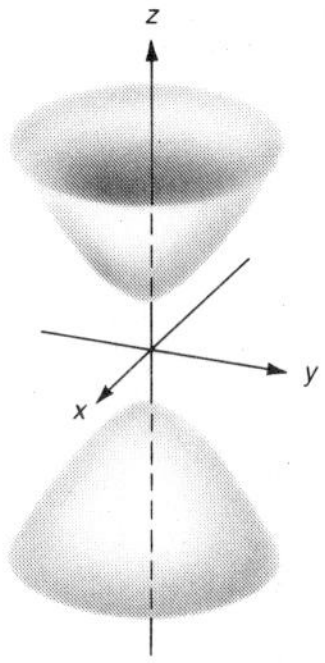	There is no trace in the xy plane. In planes parallel to the xy plane, which intersect the surface, the traces are ellipses. Traces in the xz and yz planes are the hyperbolas. $\frac{x^2}{a^2} + \frac{y^2}{b^2} - \frac{z^2}{c^2} = -1$	Sphere	If $a^2 = b^2 = c^2 = r^2$, then the graph is a sphere. $x^2 + y^2 + z^2 = r^2$ 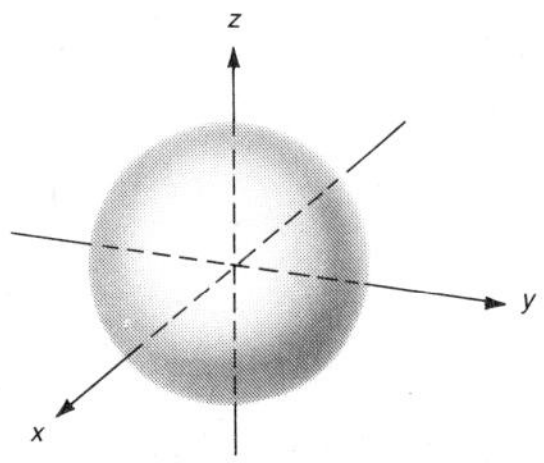
Hyperbolic paraboloid	The trace in the xy plane is two intersecting lines; in planes parallel to the xy plane the traces are hyperbolas. Traces in the xz and yz planes are parabolas. $z = \frac{y^2}{b^2} - \frac{x^2}{a^2}$	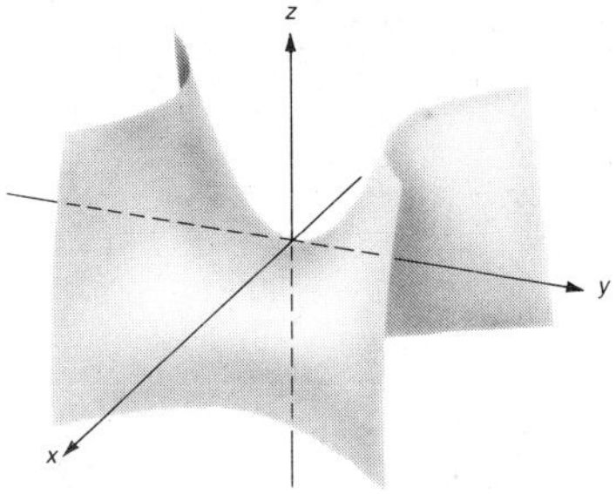			

Circular Cylinders

The graphs of

$$y^2 + z^2 = r^2 \qquad x^2 + z^2 = r^2 \qquad x^2 + y^2 = r^2$$

are **right circular cylinders** of radius r, parallel to the x axis, y axis, and z axis, respectively.

EXAMPLE 4 Graph the following equations:

a. $x^2 + y^2 = 9$ **b.** $y^2 + z^2 = 16$ **c.** $x^2 + z^2 = 25$

Solution **a.** This is a cylinder parallel to the z axis (the z variable is missing), as shown in Figure 12.19a.

b. This is a cylinder parallel to the x axis, as shown in Figure 12.19b.

c. This is a cylinder parallel to the y axis, as shown in Figure 12.19c.

Figure 12.19 Graphs of right circular cylinders

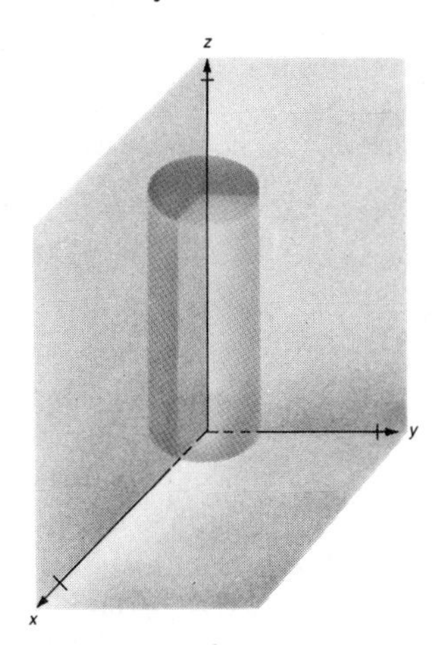

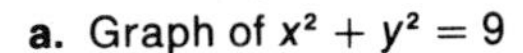

a. Graph of $x^2 + y^2 = 9$

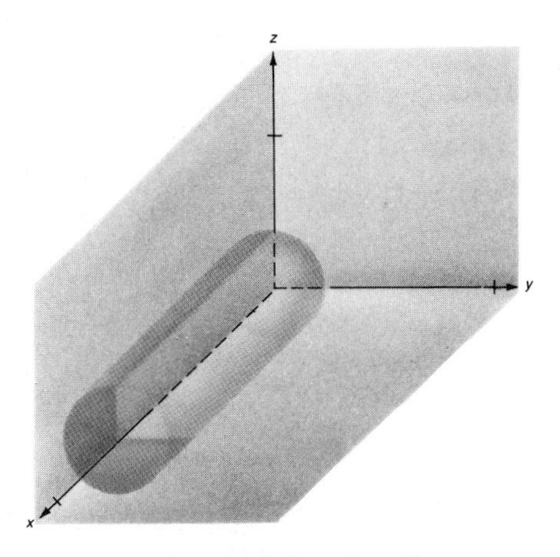

b. Graph of $y^2 + z^2 = 16$

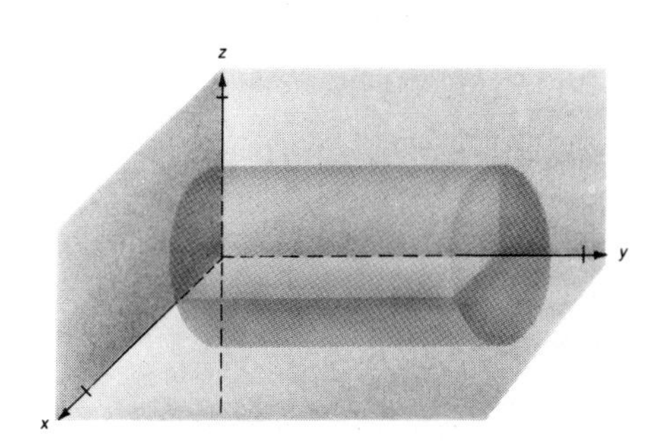

c. Graph of $x^2 + z^2 = 25$ ■

PROBLEM SET 12.3

A

Evaluate the functions K, V, I, and A from Example 1 in Problems 1–16, and interpret your results.

1. $K(15, 35)$ **2.** $K(45, 90)$

3. $K(\sqrt{2}, \sqrt{3})$ **4.** $K(\frac{\pi}{2}, \frac{\pi}{3})$

5. $V(3, 5, 8)$ **6.** $V(15, 25, 8)$

7. $V(\sqrt{2}, 2\sqrt{5}, 4)$ **8.** $V(3\pi, 2\pi, 1)$

9. $I(500, 0.05, 3)$ **10.** $I(1250, 0.08, 12)$

11. $I(1000, 0.11, 4)$ **12.** $I(850, 0.18, 2)$

13. $A(2500, 0.12, 6, 4)$ **14.** $A(5500, 0.18, 5, 6)$

15. $A(110{,}000, 0.09, 30, 12)$ **16.** $A(250{,}000, 0.15, 15, 12)$

Evaluate $f(x, y) = x^2 - 2xy + y^2$ *for the values in Problems 17–22.*

17. $f(2, 3)$ **18.** $f(-1, 4)$

19. $f(-2, 5)$ **20.** $f(0, 6)$

21. $f(a, 6)$ **22.** $f(3, b)$

Evaluate

$$g(x, y) = \frac{2x - 4y}{x^2 + y^2}$$

for the values in Problems 23–28.

23. $g(2, 1)$ **24.** $g(-3, 2)$

25. $g(5, -3)$ **26.** $g(-3, -4)$

27. $g(s, 0)$ **28.** $g(4, t)$

Evaluate

$$h(x, y) = \frac{e^{xy}}{\sqrt{x^2 + y^2}}$$

for the values in Problems 29–34.

29. $h(0, 5)$ **30.** $h(-2, 3)$

31. $h(-2, -3)$ **32.** $h(1, 1)$

33. $h(\pi, 0)$ **34.** $h(\sqrt{2}, 0)$

Graph the ordered triplets in Problems 35–41.

35. a. $(1, 2, 3)$ **b.** $(-3, 2, 4)$

36. a. $(1, -4, 3)$ **b.** $(-5, -9, 4)$

37. a. $(2, 4, 3)$ **b.** $(-3, 2, -4)$

38. a. $(-6, 8, -10)$ **b.** $(-1, -2, -3)$

39. a. $(10, 5, 10)$ **b.** $(5, -5, -5)$

40. a. $(3, -2, -4)$ **b.** $(-5, -1, 3)$

41. a. $(\pi, \frac{\pi}{2}, \frac{3\pi}{2})$ **b.** $(\sqrt{2}, \frac{1}{2}\sqrt{2}, 2\sqrt{2})$

B

Graph the surfaces in Problems 42–57.

42. $2x + y + 3z = 6$ **43.** $x + 2y + 5z = 10$

44. $x + y + z = 1$ **45.** $3x - 2y - z = 12$

46. $z^2 = \frac{x^2}{4} + \frac{y^2}{9}$ **47.** $z = \frac{x^2}{4} + \frac{y^2}{9}$

48. $\frac{x^2}{1} + \frac{y^2}{4} + \frac{z^2}{9} = 1$ **49.** $\frac{x^2}{9} + \frac{y^2}{4} + \frac{z^2}{25} = 1$

50. $x^2 + y^2 + z^2 = 9$ **51.** $z = x^2 + y^2$

52. $\frac{x^2}{9} - \frac{y^2}{1} + \frac{z^2}{4} = 1$ **53.** $\frac{x^2}{9} + \frac{y^2}{1} - \frac{z^2}{4} = -1$

54. $y^2 + z^2 = 25$ **55.** $x^2 + y^2 = 36$

56. $x^2 + z^2 = 4$ **57.** $y^2 + z^2 = 20$

58. ***Psychology*** The intelligence quotient (IQ) is defined as

$$Q(x, y) = \frac{100x}{y}$$

where Q is the IQ, x is a person's mental age as measured on a standardized test, and y is a person's chronological age measured in years. Find and interpret:
a. $Q(15, 13)$ **b.** $Q(6, 9)$
c. $Q(15, 15)$ **d.** $Q(10.5, 9.8)$

59. ***Business*** A company manufactures two types of golf carts. The first has a fixed cost of \$2500, a variable cost of \$800, and x are produced. The second has a fixed cost of \$1200, a variable cost of \$550, and y are produced. Write a cost function $C(x, y)$ and find:
a. $C(10, 15)$ **b.** $C(5, 25)$
c. $C(15, 10)$ **d.** $C(0, 30)$

60. ***Business*** If the revenue function for the golf carts in Problem 59 is $R(x, y) = 1500x + 900y$, find:
a. $R(10, 15)$ **b.** $R(5, 25)$
c. $R(15, 10)$ **d.** $R(0, 30)$

61. If the dimensions of the room in Figure 12.16 are x feet wide, y feet long, and z feet high, and if ceiling material is \$2 per square foot, wall material is \$0.75 per square foot, and floor material is \$1.25 per square foot, write a cost function for the ceiling, floor, and wall (assuming no doors or windows).

62. If $C(x, y, z)$ is the cost function for Problem 61, find:
a. $C(25, 30, 8)$ **b.** $C(12, 14, 8)$ **c.** $C(15, 20, 10)$

63. ***Medicine*** The amount of blood flowing in a blood vessel measured in milliliters is given by $F(l, r) = 0.002l/r^4$, where l is the length of the blood vessel and r is the radius. Find:
a. $F(3.1, 0.002)$ **b.** $F(15.3, 0.001)$ **c.** $F(6, 0.005)$

12.4 VECTORS IN THREE DIMENSIONS*

In this section we will consider vectors in three dimensions. Define a unit vector (a vector of length 1) called **k** as the vector in the direction of the positive z axis. The **k** vector is used in conjunction with the **i** and **j** unit vectors to define a three-dimensional vector as illustrated by Example 1.

EXAMPLE 1 Let O be the point $(0, 0, 0)$ and P be $(5, 4, 3)$. Find the vector $\overrightarrow{OP}$.

---

* Optional section

Solution

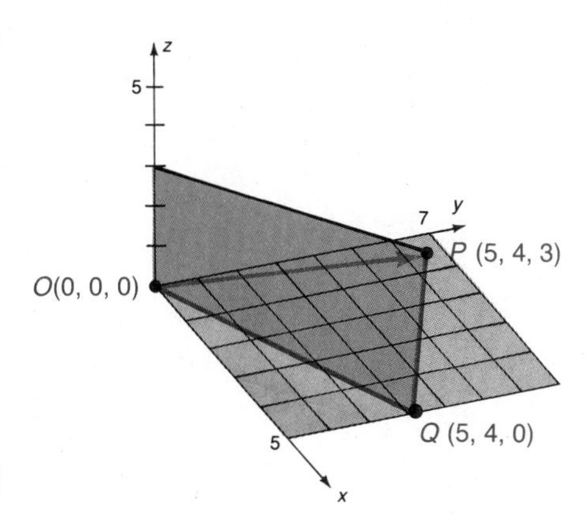

Figure 12.20 Vector $\overrightarrow{OP}$

Let Q be the point $(5, 4, 0)$. Then $\overrightarrow{OQ} = 5\mathbf{i} + 4\mathbf{j}$. Also, we note that $\overrightarrow{OP}$ is the diagonal of a triangle in the plane determined by $\overrightarrow{OP}$ and $\overrightarrow{OQ}$ (see Figure 12.20). Then $\overrightarrow{OP} = (5\mathbf{i} + 4\mathbf{j}) + 3\mathbf{k} = 5\mathbf{i} + 4\mathbf{j} + 3\mathbf{k}$. ■

Representation of a Vector Determined by Two Points

If A is the point (x_A, y_A, z_A) and B is (x_B, y_B, z_B), then the vector $\mathbf{v}$ determined by $\overrightarrow{AB}$ is

$$\mathbf{v} = (x_B - x_A)\mathbf{i} + (y_B - y_A)\mathbf{j} + (z_B - z_A)\mathbf{k}$$

Let $a_1 = (x_B - x_A)$, $a_2 = (y_B - y_A)$, and $a_3 = (z_B - z_A)$. Then

$$\mathbf{v} = a_1\mathbf{i} + a_2\mathbf{j} + a_3\mathbf{k}$$

The terminology and processes in three dimensions are identical or similar to those used in two dimensions.

Magnitude of a Vector

The **magnitude** of a vector is the length of the vector. If $\mathbf{v} = a_1\mathbf{i} + a_2\mathbf{j} + a_3\mathbf{k}$, then

$$|\mathbf{v}| = \sqrt{a_1^2 + a_2^2 + a_3^2}$$

EXAMPLE 2 Given $A(4, -2, 3)$ and $B(-1, 3, 5)$, find a vector whose representative is $\overrightarrow{AB}$ and also find its magnitude.

Solution

$$\begin{aligned} \mathbf{v} &= (-1 - 4)\mathbf{i} + [3 - (-2)]\mathbf{j} + (5 - 3)\mathbf{k} \\ &= -5\mathbf{i} + 5\mathbf{j} + 2\mathbf{k} \\ |\mathbf{v}| &= \sqrt{(-5)^2 + (5)^2 + (2)^2} \\ &= \sqrt{25 + 25 + 4} \\ &= \sqrt{54} \end{aligned}$$

■

In two dimensions vectors are defined geometrically and then the algebraic characterization of those operations is considered. In three dimensions, the procedure is exactly reversed. We define the operations algebraically and then interpret the results geometrically.

Addition, Subtraction, and Scalar Multiplication of Vectors

Let $\mathbf{v} = a_1\mathbf{i} + a_2\mathbf{j} + a_3\mathbf{k}$ and $\mathbf{w} = b_1\mathbf{i} + b_2\mathbf{j} + b_3\mathbf{k}$ and let s be any scalar. Then,

ADDITION: $\mathbf{v} + \mathbf{w} = (a_1 + b_1)\mathbf{i} + (a_2 + b_2)\mathbf{j} + (a_3 + b_3)\mathbf{k}$

SUBTRACTION: $\mathbf{v} - \mathbf{w} = (a_1 - b_1)\mathbf{i} + (a_2 - b_2)\mathbf{j} + (a_3 - b_3)\mathbf{k}$

SCALAR MULTIPLICATION: $s\mathbf{v} = sa_1\mathbf{i} + sa_2\mathbf{j} + sa_3\mathbf{k}$

EXAMPLE 3 Let $\mathbf{v} = 3\mathbf{i} + 2\mathbf{j} - \mathbf{k}$ and $\mathbf{w} = 2\mathbf{i} - 5\mathbf{j} + 2\mathbf{k}$. Find $|\mathbf{v}|$, $\mathbf{v} + \mathbf{w}$, $\mathbf{v} - \mathbf{w}$, $3\mathbf{v}$, and $3\mathbf{v} - 2\mathbf{w}$.

Solution

$$\begin{aligned} |\mathbf{v}| &= \sqrt{3^2 + 2^2 + (-1)^2} \\ &= \sqrt{9 + 4 + 1} \\ &= \sqrt{14} \end{aligned}$$

$$\begin{aligned} \mathbf{v} + \mathbf{w} &= (3 + 2)\mathbf{i} + (2 - 5)\mathbf{j} + (-1 + 2)\mathbf{k} \\ &= 5\mathbf{i} - 3\mathbf{j} + \mathbf{k} \end{aligned}$$

$$\begin{aligned} \mathbf{v} - \mathbf{w} &= (3 - 2)\mathbf{i} + (2 + 5)\mathbf{j} + (-1 - 2)\mathbf{k} \\ &= \mathbf{i} + 7\mathbf{j} - 3\mathbf{k} \end{aligned}$$

$$\begin{aligned} 3\mathbf{v} &= (3)3\mathbf{i} + (3)2\mathbf{j} + (3)(-1)\mathbf{k} \\ &= 9\mathbf{i} + 6\mathbf{j} - 3\mathbf{k} \end{aligned}$$

$$\begin{aligned} 3\mathbf{v} - 2\mathbf{w} &= (9\mathbf{i} + 6\mathbf{j} - 3\mathbf{k}) + (-4\mathbf{i} + 10\mathbf{j} - 4\mathbf{k}) \\ &= 5\mathbf{i} + 16\mathbf{j} - 7\mathbf{k} \end{aligned}$$

■

Geometrically, multiplying by a positive scalar changes the magnitude of a vector but does not change its direction. Multiplication by a negative scalar reverses its direction as well as changing its magnitude.

EXAMPLE 4 Find a vector in the direction of $\mathbf{v} = 3\mathbf{i} - 4\mathbf{j} + 12\mathbf{k}$ with unit length.

Solution

$$|\mathbf{v}| = \sqrt{9 + 16 + 144} = 13$$

Thus, $\frac{1}{13}\mathbf{v} = \frac{3}{13}\mathbf{i} - \frac{4}{13}\mathbf{j} + \frac{12}{13}\mathbf{k}$ is a vector in the same direction (since we multiplied by a scalar) but with unit length. ■

Unit Vector in a Given Direction

If $\mathbf{v}$ is a nonzero vector, a unit vector in the direction of $\mathbf{v}$ is given by

$$\frac{\mathbf{v}}{|\mathbf{v}|}$$

The definition of **scalar product** (also called *dot product* or *inner product*) is the same regardless of the dimension of the vectors **v** and **w**. Remember:

Scalar Product

The scalar product, denoted by $\mathbf{v} \cdot \mathbf{w}$, is defined by

$$\mathbf{v} \cdot \mathbf{w} = |\mathbf{v}||\mathbf{w}| \cos \theta$$

where θ is the angle between **v** and **w**.

The main properties of scalar product are:

Parallel and Orthogonal Vectors

Two vectors are parallel if and only if

$$\mathbf{v} \cdot \mathbf{w} = \pm |\mathbf{v}| |\mathbf{w}|$$

Two vectors are orthogonal if and only if

$$\mathbf{v} \cdot \mathbf{w} = 0$$

To calculate the scalar product, we first need to find $\cos \theta$ in three dimensions. Let

$$\mathbf{v} = a_1\mathbf{i} + a_2\mathbf{j} + a_3\mathbf{k} \quad \text{and} \quad \mathbf{w} = b_1\mathbf{i} + b_2\mathbf{j} + b_3\mathbf{k}$$

Using the Law of Cosines, it can be shown that

$$\cos \theta = \frac{a_1b_1 + a_2b_2 + a_3b_3}{|\mathbf{v}| |\mathbf{w}|}$$

Thus we see that

Procedure for Finding Scalar Product

$$\mathbf{v} \cdot \mathbf{w} = a_1b_1 + a_2b_2 + a_3b_3$$

For Examples 5–10, let $\mathbf{v} = 3\mathbf{i} + 2\mathbf{j} - \mathbf{k}$ and $\mathbf{w} = 4\mathbf{i} - 3\mathbf{j} + 2\mathbf{k}$.

EXAMPLE 5 Find $\mathbf{v} \cdot \mathbf{w}$.

Solution

$$\begin{aligned} \mathbf{v} \cdot \mathbf{w} &= 3(4) + 2(-3) + (-1)(2) \\ &= 12 - 6 - 2 \\ &= 4 \end{aligned}$$

■

EXAMPLE 6 Find $|\mathbf{v}|$ and $|\mathbf{w}|$.

Solution

$$\begin{aligned} |\mathbf{v}| &= \sqrt{9 + 4 + 1} \\ &= \sqrt{14} \end{aligned}$$

$$\begin{aligned} |\mathbf{w}| &= \sqrt{16 + 9 + 4} \\ &= \sqrt{29} \end{aligned}$$

■

EXAMPLE 7 Find $\cos\theta$, where θ is the angle between $\mathbf{v}$ and $\mathbf{w}$.

Solution

$$\cos\theta = \frac{\mathbf{v}\cdot\mathbf{w}}{|\mathbf{v}|\,|\mathbf{w}|}$$

$$= \frac{4}{\sqrt{14}\sqrt{29}} \quad \text{From Examples 5 and 6}$$

$$= \frac{4}{\sqrt{406}}$$

$$= \frac{2}{203}\sqrt{406}$$

■

EXAMPLE 8 Find $\mathbf{v} - s\mathbf{w}$.

Solution

$$\mathbf{v} - s\mathbf{w} = (3\mathbf{i} + 2\mathbf{j} - \mathbf{k}) - (4s\mathbf{i} - 3s\mathbf{j} + 2s\mathbf{k})$$
$$= (3 - 4s)\mathbf{i} + (2 + 3s)\mathbf{j} + (-1 - 2s)\mathbf{k}$$

■

EXAMPLE 9 Find s so that $\mathbf{v}$ and $\mathbf{v} - s\mathbf{w}$ are orthogonal.

Solution If $\mathbf{v}$ and $\mathbf{v} - s\mathbf{w}$ are orthogonal, then

$$\mathbf{v}\cdot(\mathbf{v} - s\mathbf{w}) = 0$$
$$3(3 - 4s) + 2(2 + 3s) + (-1)(-1 - 2s) = 0$$
$$9 - 12s + 4 + 6s + 1 + 2s = 0$$
$$-4s = -14$$
$$s = \frac{7}{2}$$

■

EXAMPLE 10 Find $\mathbf{v}\cdot\mathbf{v}$ and $|\mathbf{v}|^2$

Solution

$$\mathbf{v}\cdot\mathbf{v} = 3(3) + 2(2) + (-1)(-1) = 9 + 4 + 1 = 14$$

$$|\mathbf{v}|^2 = (\sqrt{14})^2 = 14$$

■

Notice in Example 10 that $\mathbf{v}\cdot\mathbf{v} = |\mathbf{v}|^2$. In general, for any vector $\mathbf{v}$,

$$\mathbf{v}\cdot\mathbf{v} = |\mathbf{v}|^2$$

The formulas for both vector and scalar projections are the same in three dimensions as we found them to be in two dimensions.

Vector and Scalar Projections

VECTOR PROJECTION OF V ONTO W:

$$\left(\frac{\mathbf{v}\cdot\mathbf{w}}{\mathbf{w}\cdot\mathbf{w}}\right)\mathbf{w}$$

SCALAR PROJECTION OF V ONTO W:

$$|\mathbf{v}|\cos\theta = \left|\frac{\mathbf{v}\cdot\mathbf{w}}{|\mathbf{w}|}\right|$$

where θ is the acute angle between $\mathbf{v}$ and $\mathbf{w}$.

EXAMPLE 11 Let $\mathbf{v} = 2\mathbf{i} - 6\mathbf{j} + 3\mathbf{k}$ and $\mathbf{w} = \mathbf{i} + 2\mathbf{j} - 2\mathbf{k}$. Find the vector and scalar projections of $\mathbf{v}$ onto $\mathbf{w}$.

Solution The vector projection of $\mathbf{v}$ onto $\mathbf{w}$ is

$$\begin{aligned}\left(\frac{\mathbf{v}\cdot\mathbf{w}}{\mathbf{w}\cdot\mathbf{w}}\right)\mathbf{w} &= \left(\frac{2-12-6}{1+4+4}\right)(\mathbf{i}+2\mathbf{j}-2\mathbf{k})\\ &= \frac{-16}{9}(\mathbf{i}+2\mathbf{j}-2\mathbf{k})\\ &= \frac{-16}{9}\mathbf{i} - \frac{32}{9}\mathbf{j} + \frac{32}{9}\mathbf{k}\end{aligned}$$

For the scalar projection we could find the length of this vector directly (heaven forbid), or use the formula:

$$\begin{aligned}|\mathbf{v}|\cos\theta &= \left|\frac{\mathbf{v}\cdot\mathbf{w}}{|\mathbf{w}|}\right|\\ &= \left|\frac{-16}{\sqrt{1+4+4}}\right| = \left|\frac{-16}{\sqrt{9}}\right| = \frac{16}{3}\end{aligned}$$

■

We have seen that scalar product is an important vector operation which we have defined for vectors in both two and three dimensions. The result of multiplying two vectors using scalar multiplication is a real number, or scalar. There is another operation, called **vector product**, which is defined only for three-dimensional vectors and gives a vector as the result.

Vector Product

If $\mathbf{v} = a_1\mathbf{i} + a_2\mathbf{j} + a_3\mathbf{k}$ and $\mathbf{w} = b_1\mathbf{i} + b_2\mathbf{j} + b_3\mathbf{k}$, the vector product, written $\mathbf{v} \times \mathbf{w}$, is the vector

$$(a_2b_3 - a_3b_2)\mathbf{i} + (a_3b_1 - a_1b_3)\mathbf{j} + (a_1b_2 - a_2b_1)\mathbf{k}$$

These terms can be obtained by using a determinant:

$$\mathbf{v}\times\mathbf{w} = \begin{vmatrix}\mathbf{i} & \mathbf{j} & \mathbf{k}\\ a_1 & a_2 & a_3\\ b_1 & b_2 & b_3\end{vmatrix}$$

Sometimes this product is called the *cross product* or *outer product*.

WARNING: $\mathbf{v}\times\mathbf{w} \neq \mathbf{w}\times\mathbf{v}$

Vector product is *not* commutative, since $\mathbf{v}\times\mathbf{w} \neq \mathbf{w}\times\mathbf{v}$. Using properties of determinants we see that $\mathbf{v}\times\mathbf{w} = -(\mathbf{w}\times\mathbf{v})$: Now,

$$\mathbf{v}\times\mathbf{w} = \begin{vmatrix}\mathbf{i} & \mathbf{j} & \mathbf{k}\\ a_1 & a_2 & a_3\\ b_1 & b_2 & b_3\end{vmatrix} = -\begin{vmatrix}\mathbf{i} & \mathbf{j} & \mathbf{k}\\ b_1 & b_2 & b_3\\ a_1 & a_2 & a_3\end{vmatrix} = -(\mathbf{w}\times\mathbf{v})$$

The vector $(\mathbf{v}\times\mathbf{w})$ is orthogonal with both the vectors $\mathbf{v}$ and $\mathbf{w}$. To see this we consider the dot product of the vectors $\mathbf{v}$ and $(\mathbf{v}\times\mathbf{w})$. If this product is zero, then the vectors are orthogonal.

$$\begin{aligned}\mathbf{v}\cdot(\mathbf{v}\times\mathbf{w}) &= (a_1\mathbf{i}+a_2\mathbf{j}+a_3\mathbf{k})\cdot[(a_2b_3-a_3b_2)\mathbf{i}+(a_3b_1-a_1b_3)\mathbf{j}+(a_1b_2-a_2b_1)\mathbf{k}]\\ &= a_1a_2b_3-a_1a_3b_2+a_2a_3b_1-a_1a_2b_3+a_1a_3b_2-a_2a_3b_1\\ &= 0\end{aligned}$$

Similarly, we can show that $\mathbf{w}\cdot(\mathbf{v}\times\mathbf{w})=0$

We have just proved a geometric property of vectors by using algebra. The vector product of two vectors is a vector orthogonal to the two given vectors. The only way this can occur is in a three-dimensional setting. Any two distinct nonzero three-dimensional vectors that are not parallel (that is, not scalar multiples of one another) can be arranged so that they have a common base and will therefore determine a plane. Then the vector product *must* be orthogonal to this plane as shown in Figure 12.21.

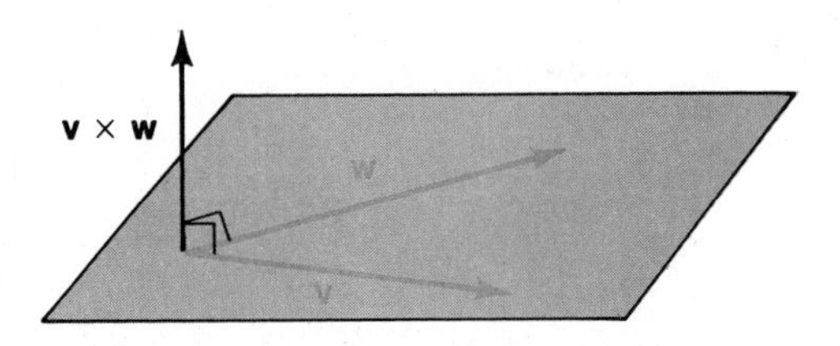

Figure 12.21 Vector product of two vectors

Geometric Property of Vector Product

> If **v** and **w** are not parallel, then they determine a plane and **v** × **w** is a *vector orthogonal to this plane. That is, it is orthogonal to both* **v** *and* **w**.

The magnitude, or length, of this vector is denoted by $|\mathbf{v}\times\mathbf{w}|$. Using the formula for magnitude, we can show that

$$|\mathbf{v}\times\mathbf{w}| = |\mathbf{v}||\mathbf{w}|\sin\theta$$

The distributive property holds for vector product:

$$\mathbf{u}\times(\mathbf{v}+\mathbf{w}) = (\mathbf{u}\times\mathbf{v})+(\mathbf{u}\times\mathbf{w})$$
$$(\mathbf{v}+\mathbf{w})\times\mathbf{u} = (\mathbf{v}\times\mathbf{u})+(\mathbf{w}\times\mathbf{u})$$

EXAMPLE 12 Find $\mathbf{i}\times\mathbf{j}$.

Solution

$$\mathbf{i}\times\mathbf{j} = \begin{vmatrix}\mathbf{i} & \mathbf{j} & \mathbf{k}\\ 1 & 0 & 0\\ 0 & 1 & 0\end{vmatrix} = \mathbf{k}$$ ■

Similarly:

$$\mathbf{j}\times\mathbf{k}=\mathbf{i} \qquad \mathbf{k}\times\mathbf{i}=\mathbf{j} \qquad \mathbf{i}\times\mathbf{j}=\mathbf{k}$$
$$\mathbf{i}\times\mathbf{i}=\mathbf{j}\times\mathbf{j}=\mathbf{k}\times\mathbf{k}=0$$

EXAMPLE 13 Find $\mathbf{v} \times \mathbf{w}$, where $\mathbf{v} = 2\mathbf{i} + \mathbf{j} + \mathbf{k}$ and $\mathbf{w} = -\mathbf{i} + 2\mathbf{j} + 3\mathbf{k}$.

Solution

$$\mathbf{v} \times \mathbf{w} = \begin{vmatrix} \mathbf{i} & \mathbf{j} & \mathbf{k} \\ 2 & 1 & 1 \\ -1 & 2 & 3 \end{vmatrix}$$

$$= \begin{vmatrix} 1 & 1 \\ 2 & 3 \end{vmatrix}\mathbf{i} - \begin{vmatrix} 2 & 1 \\ -1 & 3 \end{vmatrix}\mathbf{j} + \begin{vmatrix} 2 & 1 \\ -1 & 2 \end{vmatrix}\mathbf{k}$$

$$= \mathbf{i} - 7\mathbf{j} + 5\mathbf{k}$$

■

PROBLEM SET 12.4

A

In Problems 1–6, find the vector $\overrightarrow{AB}$ determined by the given points. Also find its magnitude.

1. $A(2, -1, 3), B(4, 5, -3)$ **2.** $A(2, -1, 3), B(4, 5, 3)$

3. $A(4, -3, 1), B(-1, 3, -2)$ **4.** $A(4, 1, 5), B(-3, 0, 2)$

5. $A(7, 1, 0), B(3, -2, 5)$ **6.** $A(2, -1, 2), B(4, -1, -1)$

Let $\mathbf{v} = 3\mathbf{i} - 2\mathbf{j} + \mathbf{k}$ and $\mathbf{w} = 4\mathbf{i} + \mathbf{j} - 3\mathbf{k}$. Find the scalars or vectors requested in Problems 7–9.

7. $|\mathbf{v}|$ **8.** $\mathbf{v} + \mathbf{w}$ **9.** $2\mathbf{v} + 3\mathbf{w}$

Let $\mathbf{v} = 5\mathbf{i} - 3\mathbf{j} + 2\mathbf{k}$ and $\mathbf{w} = -\mathbf{i} + 2\mathbf{j} - 3\mathbf{k}$. Find the scalars or vectors requested in Problems 10–12.

10. $|\mathbf{w}|$ **11.** $\mathbf{v} - \mathbf{w}$ **12.** $3\mathbf{v} - \mathbf{w}$

Let $\mathbf{v} = 5\mathbf{i} + 4\mathbf{k}$ and $\mathbf{w} = \mathbf{j} + 3\mathbf{k}$. Find the vectors requested in Problems 13–15.

13. $\mathbf{v} + \mathbf{w}$ **14.** $\mathbf{v} - \mathbf{w}$ **15.** $5\mathbf{v} - 3\mathbf{w}$

In Problems 16–24, find $\mathbf{v} \cdot \mathbf{w}$ for the given vectors.

16. $\mathbf{v} = \mathbf{i}$, $\mathbf{w} = \mathbf{j}$ **17.** $\mathbf{v} = \mathbf{k}$, $\mathbf{w} = \mathbf{k}$ **18.** $\mathbf{v} = 3\mathbf{i} + 2\mathbf{k}$, $\mathbf{w} = 2\mathbf{i} + \mathbf{j}$

19. $\mathbf{v} = \mathbf{i} - 3\mathbf{j}$, $\mathbf{w} = \mathbf{i} + 5\mathbf{k}$ **20.** $\mathbf{v} = 3\mathbf{i} - 2\mathbf{j} + 4\mathbf{k}$, $\mathbf{w} = \mathbf{i} + 4\mathbf{j} - 7\mathbf{k}$

21. $\mathbf{v} = 5\mathbf{i} - \mathbf{j} + 2\mathbf{k}$, $\mathbf{w} = 2\mathbf{i} + \mathbf{j} - 3\mathbf{k}$ **22.** $\mathbf{v} = 3\mathbf{i} - \mathbf{j} + 2\mathbf{k}$, $\mathbf{w} = 2\mathbf{i} + 3\mathbf{j} - 4\mathbf{k}$

23. $\mathbf{v} = -\mathbf{j} + 4\mathbf{k}$, $\mathbf{w} = 5\mathbf{i} + 6\mathbf{k}$ **24.** $\mathbf{v} = \mathbf{i} - 6\mathbf{j} + 10\mathbf{k}$, $\mathbf{w} = -\mathbf{i} + 5\mathbf{j} - 6\mathbf{k}$

In Problems 25–33, find $\mathbf{v} \times \mathbf{w}$ for the given vectors.

25. $\mathbf{v} = \mathbf{i}$, $\mathbf{w} = \mathbf{j}$ **26.** $\mathbf{v} = \mathbf{k}$, $\mathbf{w} = \mathbf{k}$ **27.** $\mathbf{v} = 3\mathbf{i} + 2\mathbf{k}$, $\mathbf{w} = 2\mathbf{i} + \mathbf{j}$

28. $\mathbf{v} = \mathbf{i} - 3\mathbf{j}$, $\mathbf{w} = \mathbf{i} + 5\mathbf{k}$ **29.** $\mathbf{v} = 3\mathbf{i} - 2\mathbf{j} + 4\mathbf{k}$, $\mathbf{w} = \mathbf{i} + 4\mathbf{j} - 7\mathbf{k}$

30. $\mathbf{v} = 5\mathbf{i} - \mathbf{j} + 2\mathbf{k}$, $\mathbf{w} = 2\mathbf{i} + \mathbf{j} - 3\mathbf{k}$ **31.** $\mathbf{v} = 3\mathbf{i} - \mathbf{j} + 2\mathbf{k}$, $\mathbf{w} = 2\mathbf{i} + 3\mathbf{j} - 4\mathbf{k}$

32. $\mathbf{v} = -\mathbf{j} + 4\mathbf{k}$, $\mathbf{w} = 5\mathbf{i} + 6\mathbf{k}$ **33.** $\mathbf{v} = \mathbf{i} - 6\mathbf{j} + 10\mathbf{k}$, $\mathbf{w} = -\mathbf{i} + 5\mathbf{j} - 6\mathbf{k}$

B

In Problems 34–39, find a unit vector in the direction of the given vector.

34. $3\mathbf{i} + 4\mathbf{j}$ **35.** $5\mathbf{i} + 12\mathbf{k}$

36. $\mathbf{i} + \mathbf{j} + \mathbf{k}$ **37.** $3\mathbf{i} + 12\mathbf{j} - 4\mathbf{k}$

38. $2\mathbf{i} - 2\mathbf{j} + \mathbf{k}$ **39.** $4\mathbf{i} + 2\mathbf{j} - 3\mathbf{k}$

In Problems 40–43, let $\mathbf{v} = 4\mathbf{i} - \mathbf{j} + \mathbf{k}$ and $\mathbf{w} = 2\mathbf{i} + 3\mathbf{j} - \mathbf{k}$. Find:

40. $\mathbf{v} \cdot \mathbf{w}$

41. $\cos\theta$, where θ is the angle between $\mathbf{v}$ and $\mathbf{w}$

42. a scalar s such that $\mathbf{v}$ is orthogonal to $\mathbf{v} - s\mathbf{w}$

43. scalars s and t such that $s\mathbf{v} + t\mathbf{w}$ is orthogonal to $\mathbf{w}$

In Problems 44–47, let $\mathbf{v} = 2\mathbf{i} + 3\mathbf{k}$ and $\mathbf{w} = 2\mathbf{j} - 3\mathbf{k}$. Find:

44. $\mathbf{v} \cdot \mathbf{w}$

45. $\cos\theta$, where θ is the angle between $\mathbf{v}$ and $\mathbf{w}$

46. a scalar s such that $\mathbf{v}$ is orthogonal to $\mathbf{v} - s\mathbf{w}$

47. scalars s and t such that $s\mathbf{v} + t\mathbf{w}$ is orthogonal to $\mathbf{w}$

In Problems 48–51, let $\mathbf{v} = \mathbf{i} - 3\mathbf{j} + 2\mathbf{k}$ and $\mathbf{w} = \mathbf{i} + \mathbf{j} + 5\mathbf{k}$. Find:

48. $\mathbf{v} \times \mathbf{w}$

49. $\cos\theta$, where θ is the angle between $\mathbf{v}$ and $\mathbf{w}$

50. a scalar s such that $\mathbf{v}$ is orthogonal to $\mathbf{v} - s\mathbf{w}$

51. scalars s and t such that $s\mathbf{v} + t\mathbf{w}$ is orthogonal to $\mathbf{w}$

In Problems 52–55, let $\mathbf{v} = 2\mathbf{i} - 3\mathbf{j} + 6\mathbf{k}$ and $\mathbf{w} = 4\mathbf{i} + 3\mathbf{k}$. Find:

52. $\mathbf{v} \times \mathbf{w}$

53. $\cos\theta$, where θ is the angle between $\mathbf{v}$ and $\mathbf{w}$

54. a scalar s such that $\mathbf{v}$ is orthogonal to $\mathbf{v} - s\mathbf{w}$

55. scalars s and t such that $s\mathbf{v} + t\mathbf{w}$ is orthogonal to $\mathbf{w}$

C

In Problems 56–61, find the vector and scalar projections of $\mathbf{v}$ *onto* $\mathbf{w}$.

56. $\mathbf{v} = 3\mathbf{i} + 4\mathbf{j} + 12\mathbf{k}$
$\mathbf{w} = 2\mathbf{i} + \mathbf{j} + \mathbf{k}$

57. $\mathbf{v} = 5\mathbf{i} + 2\mathbf{j} + \mathbf{k}$
$\mathbf{w} = -\mathbf{i} + \mathbf{j} + \mathbf{k}$

58. $\mathbf{v} = \mathbf{i} - 2\mathbf{j} + \mathbf{k}$
$\mathbf{w} = \mathbf{i} + 2\mathbf{j} + 5\mathbf{k}$

59. $\mathbf{v} = 2\mathbf{i} + \mathbf{j} - 3\mathbf{k}$
$\mathbf{w} = 5\mathbf{i} + \mathbf{j} - 3\mathbf{k}$

60. $\mathbf{v} = -\mathbf{i} - 2\mathbf{k}$
$\mathbf{w} = 3\mathbf{i} - 4\mathbf{j} + 2\mathbf{k}$

61. $\mathbf{v} = 2\mathbf{i} + 3\mathbf{j} - \mathbf{k}$
$\mathbf{w} = 3\mathbf{i} - 4\mathbf{j} + 2\mathbf{k}$

62. Explain whether each of the following products is a scalar or a vector, or does not exist.
a. $\mathbf{u} \times (\mathbf{v} \cdot \mathbf{w})$ **b.** $\mathbf{u} \cdot (\mathbf{v} \cdot \mathbf{w})$ **c.** $\mathbf{u} \times (\mathbf{v} \times \mathbf{w})$

63. Explain whether each of the following products is a scalar or a vector, or does not exist.
a. $(\mathbf{u} \times \mathbf{v}) \cdot (\mathbf{u} \times \mathbf{w})$ **b.** $(\mathbf{u} \times \mathbf{v}) \times (\mathbf{u} \times \mathbf{w})$
c. $\mathbf{u} \cdot (\mathbf{v} \times \mathbf{w})$

12.5 POLAR COORDINATES*

Up to this point in the book, we have plotted points using rectangular coordinates—that is, measuring out a distance on the x axis and then a distance on the y axis to plot the point (x, y). Now we consider a different method of locating points in the Cartesian plane. In this system, we measure a rotation of the x axis (a positive or a negative rotation θ) and then measure a (positive or negative) distance r along the rotated x axis. A point fixed in this fashion is denoted by $P(r, \theta)$ and is called a **polar coordinate**. The point $(0, 0)$ is called the **origin** when using rectangular coordinates and the **pole** when using polar coordinates.

A point (r, θ) is easy to visualize if you think of the x axis with its positive values to the right of the origin and negative values to the left. First, rotate the x axis through the angle whose measure is θ (positive, counterclockwise; negative, clockwise). After this rotation, plot either the positive or negative value of r on this number line.

EXAMPLE 1 Plot the following points: $A(4, \frac{\pi}{3})$, $B(3, 3)$, $C(-4, \frac{\pi}{3})$, $D(-3, 3)$, $E(8, -\frac{\pi}{6})$, $F(-8, -\frac{\pi}{6})$, $G(6, \frac{5\pi}{6})$, $H(6, \frac{3\pi}{2}) = I(-6, \frac{\pi}{2}) = J(6, -\frac{\pi}{2})$

Solution WARNING: Make sure you understand how each one of these points is plotted.

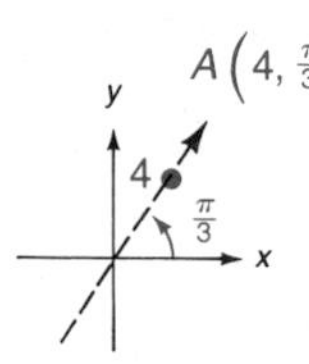

First, angle $\frac{\pi}{3}$
Next, 4 units in the positive direction on the rotated x axis

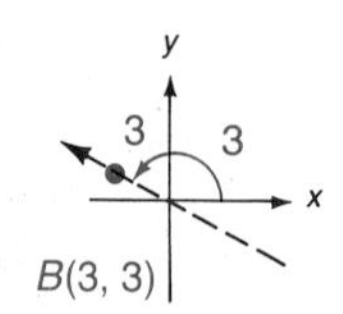

First, angle 3
Next, 3 units in the positive direction

* This section does not require Sections 12.1–12.4.

$C\left(-4, \frac{\pi}{3}\right)$

First, angle $\frac{\pi}{3}$
Next, 4 units in the *negative* direction

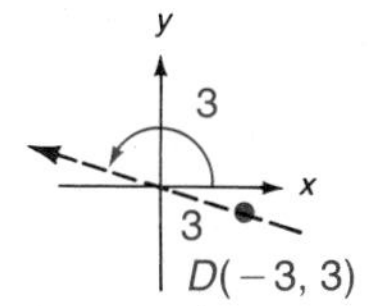

First, angle 3
Next, 3 units in the negative direction

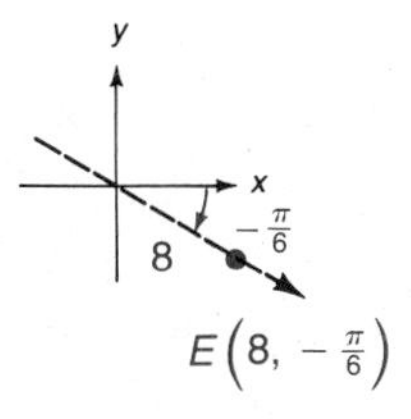

First, angle $-\frac{\pi}{6}$
Next, 8 units in the positive direction

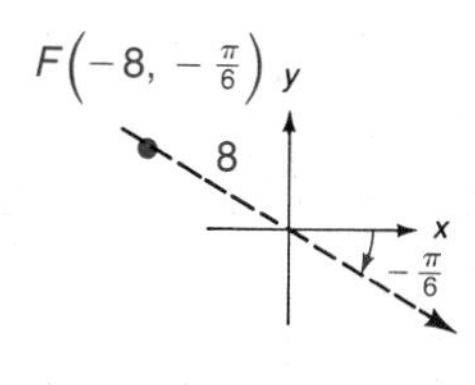

First, angle $-\frac{\pi}{6}$
Next, 8 units in the negative direction

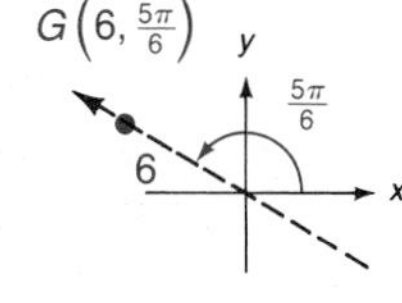

First, angle $\frac{5\pi}{6}$
Next, 6 units in the positive direction

Notice the direction of the arrow; do you see why it is pointing in the indicated direction for each example?

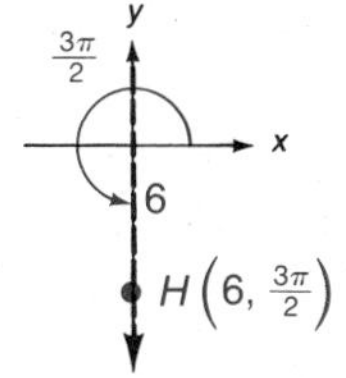

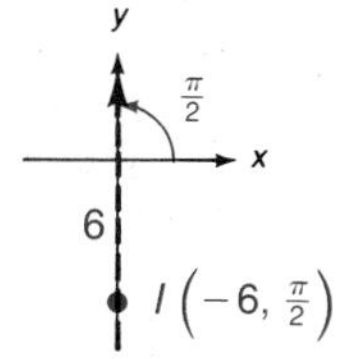

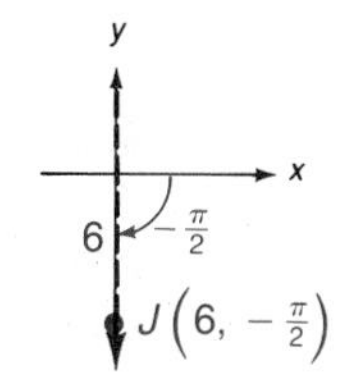

Notice that the points H, I, and J are all the same point, but with different coordinates ■

One thing you will notice from Example 1 is that ordered pairs in polar form are not associated in a one-to-one fashion with points in the plane. Indeed, given any point in the plane, there are infinitely many ordered pairs of polar coordinates associated with that point in polar form. If you are given a point (r, θ) other than the pole in polar coordinates, then $(-r, \theta + \pi)$ also represents the same point. In addition, there are also infinitely many others, all of which have the same first component as one of these, and have second components that are multiples of 2π added to these angles. We call (r, θ) and $(-r, \theta + \pi)$ the **primary representations of the point** if the angles θ and $\theta + \pi$ are between zero and 2π.

Primary Representations of a Point in Polar Form

Every point in polar form has two primary representations:

$$(r, \theta), \text{ where } 0 \leq \theta < 2\pi \quad \text{and} \quad (-r, \pi + \theta), \text{ where } 0 \leq \pi + \theta < 2\pi$$

EXAMPLE 2 Give both primary representations for each of the given points:

a. $(3, \frac{\pi}{4})$ has primary representations $\mathbf{(3, \frac{\pi}{4})}$ and $\mathbf{(-3, \frac{5\pi}{4})}$.

$\frac{\pi}{4} + \pi = \frac{5\pi}{4}$

b. $(5, \frac{5\pi}{4})$ has primary representations $\mathbf{(5, \frac{5\pi}{4})}$ and $\mathbf{(-5, \frac{\pi}{4})}$.

$\frac{5\pi}{4} + \pi = \frac{9\pi}{4}$, but $(-5, \frac{9\pi}{4})$ is not a primary representation of the point $(5, \frac{5\pi}{4})$ since $\frac{9\pi}{4} > 2\pi$. Use $\frac{\pi}{4}$ since it is coterminal with $\frac{9\pi}{4}$ and satisfies $0 \le \frac{\pi}{4} < 2\pi$.

c. $(-6, -\frac{2\pi}{3})$ has primary representations $\mathbf{(-6, \frac{4\pi}{3})}$ and $\mathbf{(6, \frac{\pi}{3})}$.

d. $(9, 5)$ has primary representations $\mathbf{(9, 5)}$ and $\mathbf{(-9, 5 - \pi)}$; a point like $(-9, 5 - \pi)$ is usually approximated by writing $(-9, 1.86)$.

Notice that $(-9, 5 + \pi)$ is not a primary representation, since $5 + \pi > 2\pi$.

e. $(9, 7)$ has primary representations $(9, 7 - 2\pi)$ or $\mathbf{(9, 0.72)}$ and $(-9, 7 - \pi)$ or $\mathbf{(-9, 3.86)}$. ■

The relationship between rectangular and polar coordinates can easily be found by using the definition of the trigonometric functions (see Figure 12.22).

Relationship between Rectangular and Polar Coordinates

1. To change **from polar to rectangular coordinates:**

$$x = r\cos\theta$$
$$y = r\sin\theta$$

2. To change **from rectangular to polar coordinates:**

$$r = \sqrt{x^2 + y^2} \qquad \theta' = \tan^{-1}\left|\frac{y}{x}\right|, x \neq 0$$

where θ' is the reference angle for θ. Place θ in the proper quadrant by noting the signs of x and y. If $x = 0$, then $\theta' = \frac{\pi}{2}$.

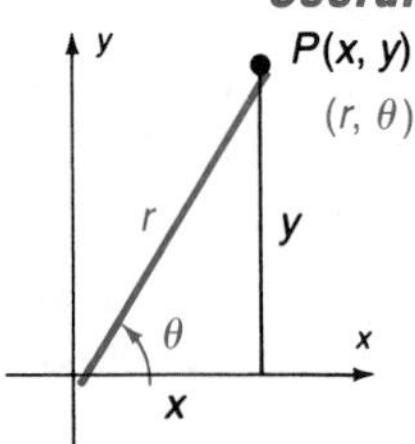

Figure 12.22 Relationship between rectangular and polar coordinates

EXAMPLE 3 Change the polar coordinates $(-3, \frac{5\pi}{4})$ to rectangular coordinates.

Solution

$$x = -3\cos\frac{5\pi}{4} = -3\left(-\frac{\sqrt{2}}{2}\right) = \frac{3\sqrt{2}}{2}$$

$$y = -3\sin\frac{5\pi}{4} = -3\left(-\frac{\sqrt{2}}{2}\right) = \frac{3\sqrt{2}}{2}$$

$$\underbrace{\left(-3, \frac{5\pi}{4}\right)}_{\text{polar form}} = \underbrace{\left(\frac{3\sqrt{2}}{2}, \frac{3\sqrt{2}}{2}\right)}_{\text{rectangular form}}$$

■

EXAMPLE 4 Write both primary representations of the polar-form coordinates for the point whose rectangular coordinates are

$$\left(\frac{5\sqrt{3}}{2}, -\frac{5}{2}\right)$$

Solution

$$r = \sqrt{\left(\frac{5\sqrt{3}}{2}\right)^2 + \left(-\frac{5}{2}\right)^2} \qquad \theta' = \tan^{-1}\left|\frac{-\frac{5}{2}}{\frac{5\sqrt{3}}{2}}\right|$$

$$= \sqrt{\frac{75}{4} + \frac{25}{4}}$$

$$= 5 \qquad\qquad = \tan^{-1}\left(\frac{1}{\sqrt{3}}\right)$$

$$= \frac{\pi}{6} \qquad \theta = \frac{11\pi}{6} \text{ (Quadrant IV)}$$

$$\underbrace{\left(\frac{5\sqrt{3}}{2}, -\frac{5}{2}\right)}_{\text{rectangular form}} = \underbrace{\left(5, \frac{11\pi}{6}\right) = \left(-5, \frac{5\pi}{6}\right)}_{\text{polar form}}$$

$\frac{11\pi}{6} + \pi = \frac{17\pi}{6}$, and $\frac{17\pi}{6}$ is coterminal with $\frac{5\pi}{6}$

■

We now turn our attention to polar-form graphing. The basic procedure is to plot points to determine a curve's general shape and then to make some generalizations that will simplify the graphing of similar curves. Because the representation of points in polar form by ordered pairs of real numbers is not unique, we need the following definition of what it means for a polar-form point to **satisfy** an equation.

A Point Satisfying an Equation

An ordered pair representing a polar-form point (other than the pole) **satisfies an equation** involving a trigonometric function of $n\theta$ (n an integer) if and only if at least one of its primary representations satisfies the given equation.

The easiest graphs to consider are those for which either r or θ is a constant. For example, $r = 5$ is the equation of the set of all ordered pairs (r, θ) for which the first component (r) is 5. This means that points $(5, 0), (5, 1), (5, 2), (5, 3), \ldots$ all satisfy this equation. The graph shown in Figure 12.23a is seen to be a circle with center at the pole and radius 5. Similarly, $\theta = 1$ is the set of all ordered pairs for which the second component (θ) is 1. This means that the points $(0, 1), (1, 1), (2, 1), (3, 1), (-1, 1), \ldots$ all satisfy the equation $\theta = 1$. The graph shown in Figure 12.23b is seen to be a line.

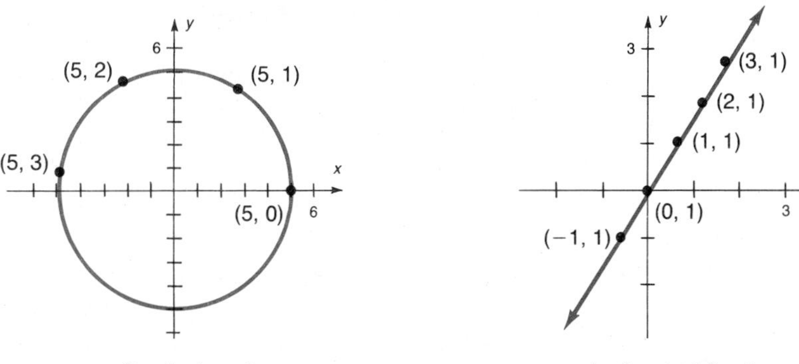

Figure 12.23 Polar-form graphs for which one of the components is a constant

The following examples show some nontrivial polar graphs. Remember, the basic ideas of polar-form graphing are identical to those of rectangular-form graphing. You are looking for ordered pairs satisfying the given equation. Each of these ordered pairs is then plotted on a Cartesian coordinate system; the only difference with polar-form graphing is the first component is a distance from the pole and the second component represents an angle of rotation.

EXAMPLE 5 Graph $r = 2(1 - \cos\theta)$.

Solution First construct a table of values by choosing values for θ and approximating the corresponding values for r:

θ	0	$\frac{\pi}{6}$	$\frac{\pi}{3}$	$\frac{\pi}{2}$	$\frac{2\pi}{3}$	$\frac{5\pi}{6}$	π	$\frac{7\pi}{6}$	$\frac{4\pi}{3}$	$\frac{3\pi}{2}$	$\frac{5\pi}{3}$	$\frac{11\pi}{6}$
r (approx. value)	0	0.27	1	2	3	3.7	4	3.7	3	2	1	0.27

These points are plotted and then connected as in Figure 12.24.

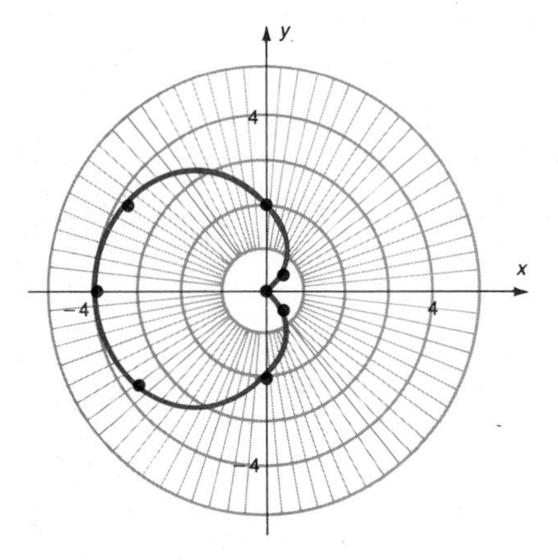

Figure 12.24 Graph of $r = 2(1 - \cos\theta)$

■

The curve in Example 5 is called a **cardioid** because it is heart-shaped. Compare the curve graphed in Example 5 with the general curve $r = a(1 - \cos\theta)$, which is called the **standard-position cardioid**. Consider the following table of values:

θ	0	$\frac{\pi}{2}$	π	$\frac{3\pi}{2}$
r	0	a	$2a$	a

These values for θ should be included whenever you are making a graph in polar coordinates and using the method of plotting points.

These reference points are all that is necessary to plot for future standard-position cardioids, because they will all have the same shape as the one shown in Figure 12.24.

What about cardioids that are not in standard position? In Chapter 2 translations were considered, but the translation of a polar-form curve is rather difficult since points are not labeled in a rectangular fashion. You can, however, easily rotate a polar-form curve.

Rotation of Polar-Form Graphs

> The polar graph $r = f(\theta - \alpha)$ is the same as the polar graph of $r = f(\theta)$ that has been rotated through an angle α.

EXAMPLE 6 Graph $r = 3 - 3\cos(\theta - \frac{\pi}{6})$.

Solution Recognize this as a cardioid with $a = 3$ and a rotation of $\frac{\pi}{6}$. Plot the four points shown in Figure 12.25 and draw the cardioid.

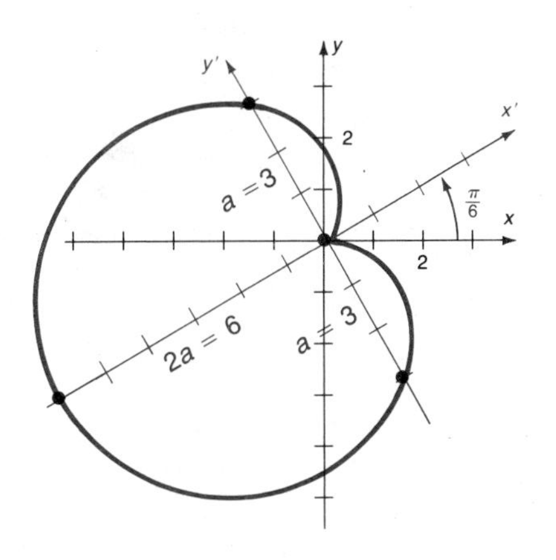

Notice that the $x'y'$ axis is drawn by rotating the xy axis through an angle of $\frac{\pi}{6}$. The standard cardioid is then drawn on the rotated axis.

Figure 12.25 Graph of $r = 3 - 3\cos(\theta - \frac{\pi}{6})$

■

EXAMPLE 7 Graph $r = 4(1 - \sin\theta)$.

Solution Notice that

$$\sin\theta = \cos\left(\frac{\pi}{2} - \theta\right)$$
$$= \cos\left(\theta - \frac{\pi}{2}\right)$$

Thus, $r = 4[1 - \cos(\theta - \frac{\pi}{2})]$. This is a cardioid with $a = 4$ that has been rotated $\frac{\pi}{2}$. The graph is shown in Figure 12.26. A curve of the form $r = a(1 - \sin\theta)$ is a cardioid that has been rotated $\frac{\pi}{2}$.

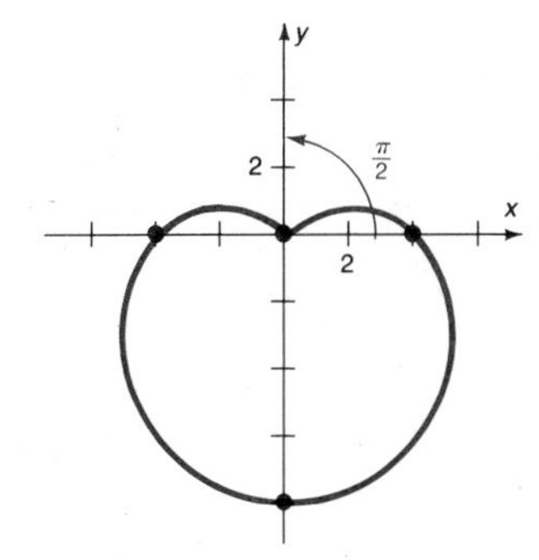

Figure 12.26 Graph of $r = 4(1 - \sin\theta)$ ■

The cardioid is only one of the interesting polar-form curves. It is a special case of a curve called a **limaçon**, which is developed in the exercises (see Problem 55). The next example illustrates a curve called a **rose curve**.

EXAMPLE 8 Graph $r = 4\cos 2\theta$.

Solution When presented with a polar-form curve that you do not recognize, graph the curve by plotting points. You can use tables, exact values, or a calculator. A calculator was used to find the values in the table at the top of the next page.

θ	0	$\frac{\pi}{12}$	$\frac{\pi}{6}$	$\frac{\pi}{4}$	$\frac{\pi}{3}$	$\frac{5\pi}{12}$	$\frac{\pi}{2}$	$\frac{7\pi}{12}$	$\frac{2\pi}{3}$	$\frac{3\pi}{4}$	$\frac{5\pi}{6}$	$\frac{11\pi}{12}$
r (approx. value)	4	3.5	2	0	-2	-3.5	-4	-3.5	-2	0	2	3.5

θ	π	$\frac{13\pi}{12}$	$\frac{7\pi}{6}$	$\frac{5\pi}{4}$	$\frac{4\pi}{3}$	$\frac{17\pi}{12}$	$\frac{3\pi}{2}$	$\frac{19\pi}{12}$	$\frac{5\pi}{3}$	$\frac{7\pi}{4}$	$\frac{11\pi}{6}$	$\frac{23\pi}{12}$
r (approx. value)	4	3.5	2	0	-2	-3.5	-4	-3.5	-2	0	2	3.5

The graph is shown in Figure 12.27.

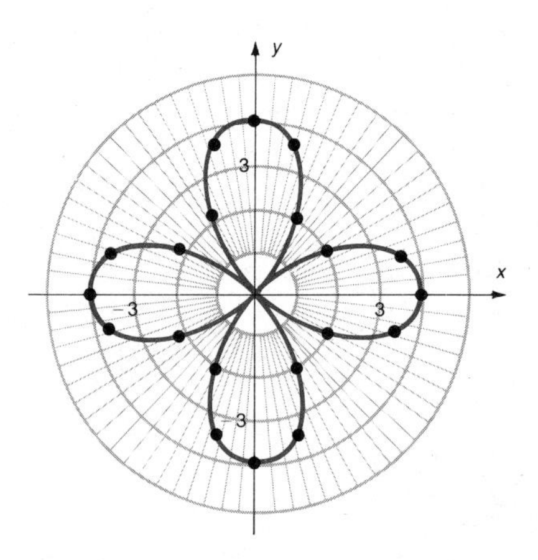

Figure 12.27 Graph of $r = 4\cos 2\theta$

■

In general,

$$r = a\cos n\theta$$

is a four-leaved rose if $n = 2$ and the length of the leaves is a. If n is an even number, the curve has $2n$ leaves; if n is odd, the number of leaves is n. These leaves are equally spaced on a circle of radius a.

EXAMPLE 9 Graph $r = 4\cos 2(\theta - \frac{\pi}{4})$.

Solution This is a rose curve with four leaves of length 4 equally spaced on a circle. However, this curve has been rotated $\frac{\pi}{4}$, as shown in Figure 12.28.

Figure 12.28 Graph of $r = 4\cos 2(\theta - \frac{\pi}{4})$.

■

Notice that Example 9 can be rewritten as a sine curve:

$$\begin{aligned} r &= 4\cos 2\left(\theta - \frac{\pi}{4}\right) \\ &= 4\cos\left(2\theta - \frac{\pi}{2}\right) \\ &= 4\cos\left(\frac{\pi}{2} - 2\theta\right) \\ &= 4\sin 2\theta \end{aligned}$$

These steps can be reversed to graph a rose curve written in terms of a sine function.

EXAMPLE 10 Graph $r = 5\sin 4\theta$.

Solution

$$\begin{aligned} r &= 5\sin 4\theta \\ &= 5\cos\left(\frac{\pi}{2} - 4\theta\right) \\ &= 5\cos\left(4\theta - \frac{\pi}{2}\right) \\ &= 5\cos 4\left(\theta - \frac{\pi}{8}\right) \end{aligned}$$

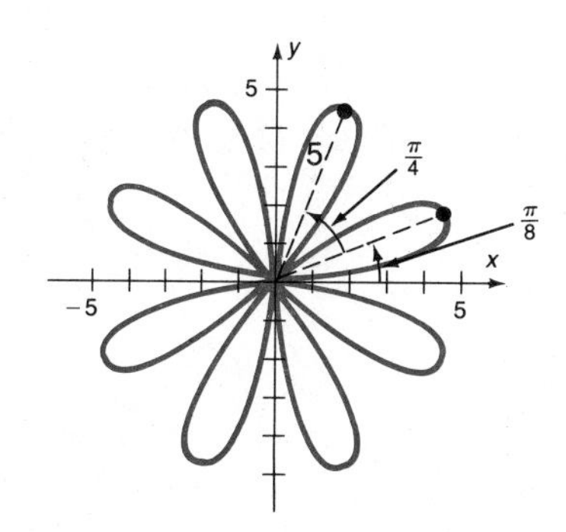

Figure 12.29 Graph of $r = 5\sin 4\theta$

Recognize this as a rose curve rotated $\pi/8$. There are eight leaves of length 5. The leaves are a distance $\frac{\pi}{4}$ apart, as shown in Figure 12.29. ■

The third and last general type of polar-form curve we will consider is called a **lemniscate** and has the general form

$$r^2 = a^2\cos 2\theta$$

EXAMPLE 11 Graph $r^2 = 16\cos 2\theta$.

Solution As before, when graphing a curve for the first time, begin by plotting points. For this example be sure to obtain two values for r when solving this quadratic equation. For example, if $\theta = 0$, then $\cos 2\theta = 1$ and $r^2 = 16$, so $r = 4$ or -4.

θ	0	$\frac{\pi}{12}$	$\frac{\pi}{6}$	$\frac{\pi}{4}$	$\frac{\pi}{4}$ to $\frac{3\pi}{4}$	$\frac{5\pi}{6}$	$\frac{11\pi}{12}$	π
r (approx. value)	± 4	± 3.7	± 2.8	0	undefined	± 2.8	± 3.7	± 4

Notice that for $\frac{\pi}{4} < \theta < \frac{3\pi}{4}$ there are no values for r, since $\cos 2\theta$ is negative. For $\pi \le \theta \le 2\pi$, the values repeat the sequence given above, so these points are plotted and then connected, as shown in Figure 12.30.

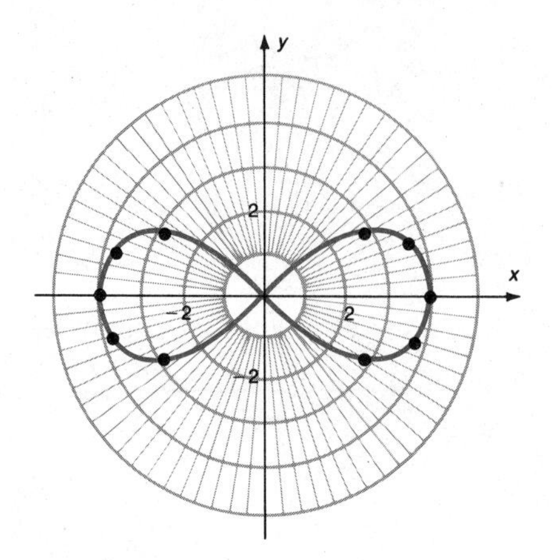

Figure 12.30 Graph of $r^2 = 9\cos 2\theta$

■

The graph of $r^2 = a^2 \sin 2\theta$ is also a lemniscate. There are always two leaves to a lemniscate and the length of the leaves is a. The sine function can be considered as a rotation of the cosine function.

EXAMPLE 12 Graph $r^2 = 16\sin 2\theta$.

Solution

$$\begin{aligned} r^2 &= 16\sin 2\theta \\ &= 16\cos\left(\frac{\pi}{2} - 2\theta\right) \\ &= 16\cos\left(2\theta - \frac{\pi}{2}\right) \\ &= 16\cos 2\left(\theta - \frac{\pi}{4}\right) \end{aligned}$$

This is a lemniscate whose leaf has length $\sqrt{16} = 4$ and is rotated $\frac{\pi}{4}$ as shown in Figure 12.31.

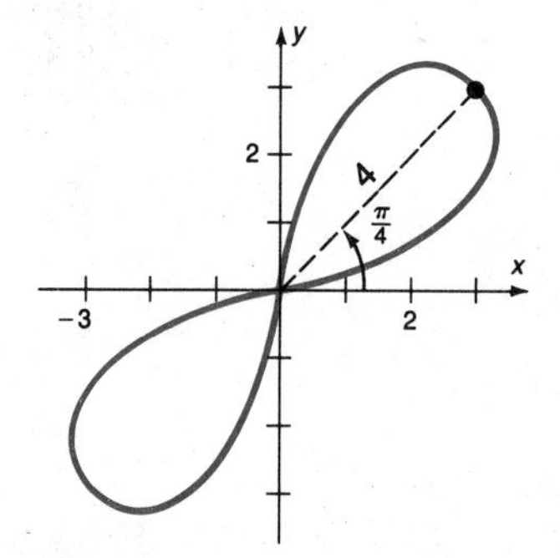

Figure 12.31 Graph of $r^2 = 16\sin 2\theta$

■

We conclude this section by summarizing the special types of polar-form curves we have developed. There are many others, some of which are presented in the problems, but these three special types are the most common.

Cardioid:

$r = a(1 \pm \cos\theta)$ **or** $r = a(1 \pm \sin\theta)$

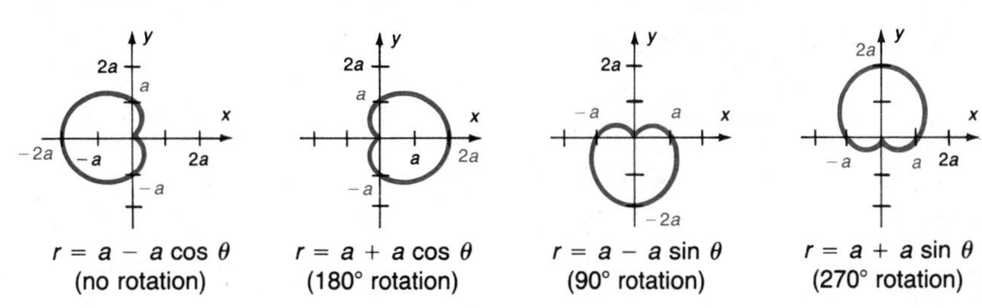

$r = a - a\cos\theta$ (no rotation) | $r = a + a\cos\theta$ (180° rotation) | $r = a - a\sin\theta$ (90° rotation) | $r = a + a\sin\theta$ (270° rotation)

Rose Curve:

$r = a\cos n\theta$ **or** $r = a\sin n\theta$ (n is a positive integer). Length of each leaf is a.

1. If n is odd, the rose is n-leaved.

   One leaf: If $n = 1$, the rose is a curve with one petal and is circular.

2. If n is even, the rose is $2n$-leaved.

   Two leaves: See the lemniscate described below.

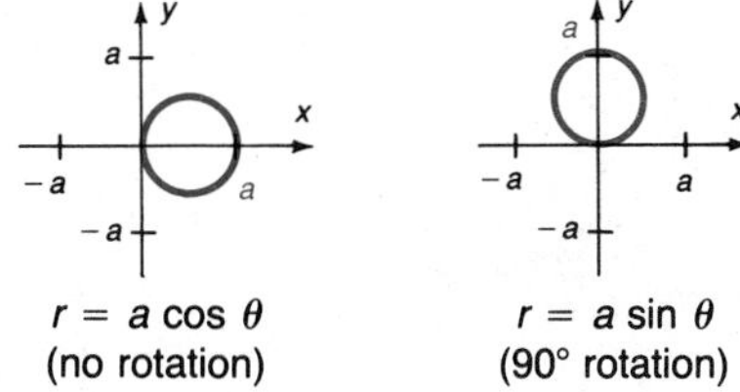

$r = a\cos\theta$ (no rotation) | $r = a\sin\theta$ (90° rotation)

Three leaves: $n = 3$ | *Four leaves:* $n = 2$

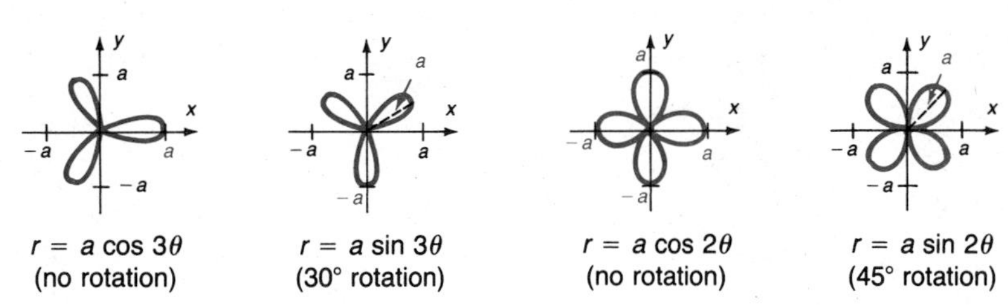

$r = a\cos 3\theta$ (no rotation) | $r = a\sin 3\theta$ (30° rotation) | $r = a\cos 2\theta$ (no rotation) | $r = a\sin 2\theta$ (45° rotation)

Lemniscate:

$r^2 = a^2\cos 2\theta$ **or** $r^2 = a^2\sin 2\theta$

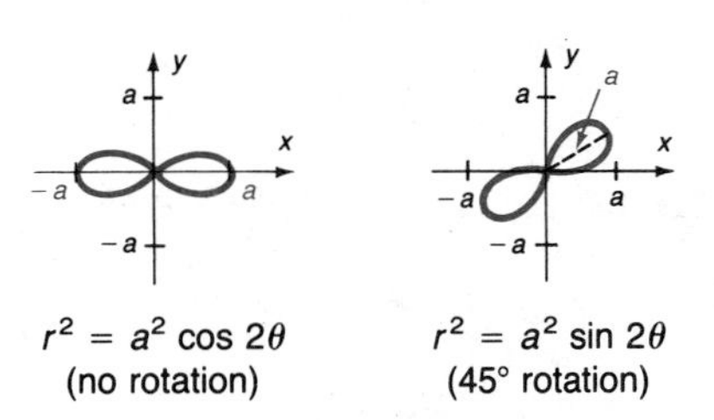

$r^2 = a^2\cos 2\theta$ (no rotation) | $r^2 = a^2\sin 2\theta$ (45° rotation)

PROBLEM SET 12.5

A

In Problems 1–10, plot each of the given polar-form points. Give both primary representations, and give the rectangular coordinates of the point. In Problems 7–9 approximate values to two decimal places. Otherwise use exact values.

1. $(4, \frac{\pi}{4})$
2. $(6, \frac{\pi}{3})$
3. $(5, \frac{2\pi}{3})$
4. $(3, -\frac{\pi}{6})$
5. $(\frac{3}{2}, -\frac{5\pi}{6})$
6. $(5, -\frac{\pi}{2})$
7. $(-4, 4)$
8. $(4, 10)$
9. $(-4, 5\pi)$
10. $(-3, -\frac{7\pi}{3})$

In Problems 11–20, plot the given rectangular-form points and give both primary representations in polar form. In Problems 17–20 approximate values to the nearest hundredth.

11. $(5, 5)$
12. $(-1, \sqrt{3})$
13. $(2, -2\sqrt{3})$
14. $(-2, -2)$
15. $(3, -3)$
16. $(-6, 6)$
17. $(-\sqrt{3}, 1)$
18. $(4, 3)$
19. $(-12, 5)$
20. $(3, 7)$

Identify each of the curves in Problems 21–35 as a cardioid, a rose curve (state number of leaves), a lemniscate, or none of the above.

21. $r^2 = 9\cos 2\theta$
22. $r = 2\sin 2\theta$
23. $r = 3\sin 3\theta$
24. $r^2 = 2\cos 2\theta$
25. $r = 2 - 2\cos\theta$
26. $r = 3 + 3\sin\theta$
27. $r^2 = \sin 3\theta$
28. $r = 4\sin 30^\circ$
29. $r = 5\cos 60^\circ$
30. $r = 5\sin 8\theta$
31. $r = 3\theta$
32. $r\theta = 3$
33. $\theta = \tan\frac{\pi}{4}$
34. $r = 5(1 - \sin\theta)$
35. $\cos\theta = 1 - r$

B

Sketch each of the curves given in Problems 36–50. If it is not one of the types discussed in the text, graph it by plotting points.

36. $r = 2(1 + \cos\theta)$
37. $r = 3(1 - \sin\theta)$
38. $r = 4(1 + \sin\theta)$
39. $r = 4\cos 2\theta$
40. $r = 5\sin 3\theta$
41. $r = 3\cos 3\theta$
42. $r = 2\cos\theta$
43. $r^2 = 9\cos 2\theta$
44. $r^2 = 16\cos 2\theta$
45. $r^2 = 16\sin 2\theta$
46. $r = 5\sin\frac{\pi}{6}$
47. $r = 9\cos\frac{\pi}{3}$
48. $r = \theta$
49. $r = 3\theta$
50. $r = \tan\theta$

51. Derive the equations for changing from polar coordinates to rectangular coordinates.

52. Derive the equations for changing from rectangular coordinates to polar coordinates.

C

53. What is the distance between the polar-form points $(3, \frac{\pi}{3})$ and $(7, \frac{\pi}{4})$?

54. What is the distance between the polar-form points (r, θ) and (c, α)?

55. The **limaçon** is a curve of the form $r = b \pm a\cos\theta$ or $r = b \pm a\sin\theta$ where $a > 0$, $b > 0$. (This problem is required for Section 12.6.) There are four types of limaçons.
a. $b/a < 1$ (limaçon with inner loop): graph $r = 2 - 3\cos\theta$ by plotting points.
b. $b/a = 1$ (cardioid): graph $r = 2 - 2\cos\theta$.
c. $1 < b/a < 2$ (limaçon with a dimple): graph $r = 3 - 2\cos\theta$ by plotting points.
d. $b/a \geq 2$ (convex limaçon): graph $r = 3 - \cos\theta$ by plotting points.

56. Graph the following limaçons (see Problem 55):
a. $r = 1 - 2\cos\theta$ **b.** $r = 2 - \cos\theta$
c. $r = 2 + 3\cos\theta$ **d.** $r = 2 - 3\sin\theta$
e. $r = 2 + 3\sin\theta$ **f.** $r = 3 - 2\sin\theta$

57. *Spirals* are interesting mathematical curves. There are three general types of spirals:
a. A spiral of Archimedes has the form $r = a\theta$; graph $r = 2\theta$ $(\theta > 0)$ by plotting points.
b. A hyperbolic spiral has the form $r\theta = a$; graph $r\theta = 2$ $(\theta > 0)$ by plotting points.
c. A logarithmic spiral has the form $r = a^{k\theta}$; graph $r = 2^\theta$ $(\theta > 0)$ by plotting points.

58. Identify and graph the following spirals (see Problem 57). Assume $\theta > 0$.
a. $r = \theta$ **b.** $r = -\theta$ **c.** $r\theta = 1$
d. $r\theta = -1$ **e.** $r = 2^{2\theta}$ **f.** $r = 3^\theta$

59. The *strophoid* is a curve of the form $r = a\cos 2\theta\sec\theta$; graph this curve where $a = 2$ by plotting points.

60. The *bifolium* has the form $r = a\sin\theta\cos^2\theta$; graph this curve where $a = 1$ by plotting points.

61. The *folium of Descartes* has the form

$$r = \frac{3a\sin\theta\cos\theta}{\sin^3\theta + \cos^3\theta}$$

Graph this curve where $a = 2$ by plotting points.

12.6 INTERSECTION OF POLAR-FORM CURVES*

In order to find the points of intersection of graphs in rectangular form, you need only find the simultaneous solution of the equations that define those graphs. It is not always necessary to draw the graphs. This follows because in a rectangular coordinate system there is a one-to-one correspondence between ordered pairs satisfying an equation and points on its graph.

However, as you saw in the last section, this one-to-one property is lost when you are working with polar coordinates. This means that the simultaneous solution of two equations in polar form may introduce extraneous points of intersection or may even fail to yield all points of intersection. For this reason, our method for finding the intersection of polar-form curves will include sketching the graphs.

EXAMPLE 1 Find the points of intersection of the curves $r = 2\cos\theta$ and $r = 2\sin\theta$.

Solution First consider the simultaneous solution of the system of equations:

$$\begin{cases} r = 2\cos\theta \\ r = 2\sin\theta \end{cases}$$

By substitution,

$$2\sin\theta = 2\cos\theta$$
$$\sin\theta = \cos\theta$$
$$\theta = \frac{\pi}{4} + 2n\pi, \frac{5\pi}{4} + 2n\pi \qquad (n \text{ an integer})$$

Then find r using the primary representations for θ:

$$r = 2\cos\frac{\pi}{4} \qquad r = 2\cos\frac{5\pi}{4}$$
$$= 2\left(\frac{\sqrt{2}}{2}\right) \qquad = 2\left(\frac{-\sqrt{2}}{2}\right)$$
$$= \sqrt{2} \qquad = -\sqrt{2}$$

This gives the points $(\sqrt{2}, \frac{\pi}{4})$ and $(-\sqrt{2}, \frac{5\pi}{4})$. Writing the primary representations for these points, we see that they represent the same point. Thus the simultaneous solution yields one point of intersection. Next consider the graphs of these curves as shown in Figure 12.32.

---

* Optional section

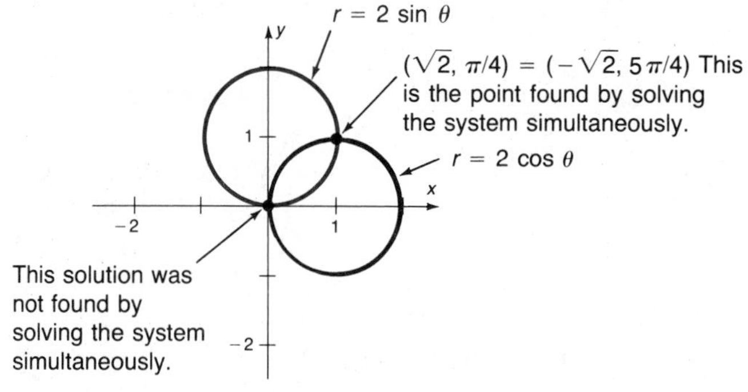

Figure 12.32 Graphs of $r = 2\cos\theta$ and $r = 2\sin\theta$

It looks like $(0, 0)$ is also a point of intersection. Check this point in each of the given equations:

$$r = 2\cos\theta: \quad \text{If } r = 0, \text{ then } \theta = \frac{\pi}{2}$$

$$r = 2\sin\theta: \quad \text{If } r = 0, \text{ then } \theta = \pi$$

At first it does not seem that $(0, 0)$ satisfies the equations since $r = 0$ gives $(0, \frac{\pi}{2})$ and $(0, \pi)$, respectively. Notice that these coordinates are different and do not satisfy the equations simultaneously. But if you plot these coordinates you will see that $(0, 0)$, $(0, \frac{\pi}{2})$, and $(0, \pi)$ are all the same point. Points of intersection for the given curves are $\mathbf{(0, 0)}$, $\mathbf{(\sqrt{2}, \frac{\pi}{4})}$. ■

The pole is often a solution for a system of equations even though it may not satisfy the equations simultaneously. This is because when $r = 0$, all values of θ will yield the same point—namely the pole. For this reason it is necessary to check separately to see if the pole lies on the given graph.

Graphical Solution of the Intersection of Polar Curves

1. Find the simultaneous solution of the given system of equations.
2. Determine whether the pole lies on the two graphs.
3. Graph the curves and look for other points of intersection.

EXAMPLE 2 Find the points of intersection of the curves $r = 2 + 4\cos\theta$ and $r = 6\cos\theta$.

Solution *Step 1* Solve the equations simultaneously:

$$\begin{cases} r = 2 + 4\cos\theta \\ r = 6\cos\theta \end{cases}$$

By substitution,

$$\begin{aligned} 6\cos\theta &= 2 + 4\cos\theta \\ 2\cos\theta &= 2 \\ \cos\theta &= 1 \\ \theta &= 0 \end{aligned}$$

If $\theta = 0$ then $r = 6$, so an intersection point is $(6, 0)$.

Step 2 Determine whether the pole lies on the graphs.

1. If $r = 0$, then

$$0 = 2 + 4\cos\theta$$

$$\cos\theta = -\frac{1}{2}$$

$$\theta = \frac{2\pi}{3}, \frac{4\pi}{3}$$

Thus $(0, \frac{2\pi}{3})$ and $(0, \frac{4\pi}{3})$ satisfy the first equation and the pole lies on this graph.

2. If $r = 0$, then

$$0 = 6\cos\theta$$

$$\theta = \frac{\pi}{2}, \frac{3\pi}{2}$$

Thus $(0, \frac{\pi}{2})$ and $(0, \frac{3\pi}{2})$ satisfy the second equation so the pole also lies on this graph.

Step 3 Graph the curves and look for other points of intersection. The first curve is a limaçon; the second is a circle as shown in Figure 12.33.

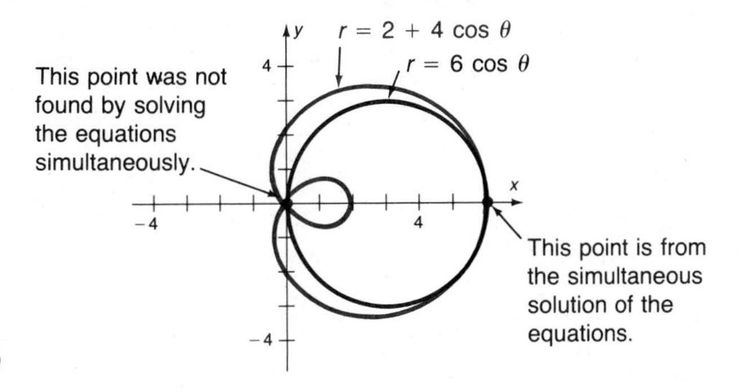

Figure 12.33 Graphs of $r = 2 + 4\cos\theta$ and $r = 6\cos\theta$

Notice that there are no other points of intersection, so the intersection points are **(6, 0)** and **(0, 0)**. ■

EXAMPLE 3 Find the points of intersection of the curves $r = \frac{3}{2} - \cos\theta$ and $\theta = \frac{2\pi}{3}$.

Solution Solve

$$\begin{cases} r = \dfrac{3}{2} - \cos\theta \\ \theta = \dfrac{2\pi}{3} \end{cases}$$

By substitution,

$$\begin{aligned} r &= \frac{3}{2} - \cos\frac{2\pi}{3} \\ &= \frac{3}{2} - \left(-\frac{1}{2}\right) \\ &= 2 \end{aligned}$$

Solution: $(2, \frac{2\pi}{3})$.

If $r = 0$, the first equation has no solution since

$$0 = \frac{3}{2} - \cos\theta$$

$$\cos\theta = \frac{3}{2}$$

and $\cos\theta$ cannot be larger than 1. Now look at the graph (Figure 12.34).

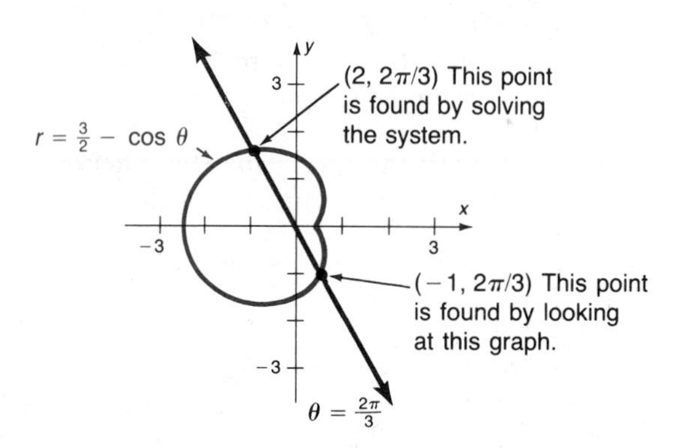

Figure 12.34 Graphs of $r = \frac{3}{2} - \cos\theta$ and $\theta = \frac{2\pi}{3}$

From the graph we see that $(-1, \frac{2\pi}{3})$ looks like a point of intersection. It satisfies the equation $\theta = \frac{2\pi}{3}$, but what about $r = \frac{3}{2} - \cos\theta$?

$$\text{Check}\left(-1, \frac{2\pi}{3}\right): \quad -1 \stackrel{?}{=} \frac{3}{2} - \cos\left(\frac{2\pi}{3}\right)$$

$$= \frac{3}{2} - \left(-\frac{1}{2}\right)$$

$$= 2 \qquad \text{(not satisfied)}$$

But check the other representation:

$$\text{Check}\left(-1, \frac{2\pi}{3}\right) = \left(1, \frac{5\pi}{3}\right): \quad 1 \stackrel{?}{=} \frac{3}{2} - \cos\left(\frac{5\pi}{3}\right)$$

$$= \frac{3}{2} - \left(\frac{1}{2}\right)$$

$$= 1 \quad \text{(satisfied)}$$

You would not have found this other point without checking the graph. The points of intersection are $\mathbf{(2, \frac{2\pi}{3})}$ and $\mathbf{(-1, \frac{2\pi}{3})}$. ■

Now you might ask if there is a procedure that will yield all the points of intersection without relying on the graph. The answer is yes—if you realize that polar-form equations can have different representations just as we found for polar-form points. Recall that a polar-form point has two primary representations:

$$(r, \theta) \quad \text{and} \quad (-r, \theta + \pi)$$

where the angle (θ or $\theta + \pi$, respectively) is between zero and 2π. For every polar-form equation $r = f(\theta)$, there are equations

$$r = (-1)^n f(\theta + n\pi)$$

for n any integer that yield exactly the same curve.

Analytic Solution for the Intersection of Polar Curves

1. Solve each equation of one graph simultaneously with each equation of the other graph. If $r = f(\theta)$ is the given equation, the other equations of the same graph are

$$r = (-1)^n f(\theta + n\pi)$$

2. Determine if the pole lies on the two graphs.

EXAMPLE 4 Find the points of intersection of the curves

$$r = 1 - 2\cos\theta \quad \text{and} \quad r = 1$$

Solution Check pole: If $r = 0$, then the second equation ($r = 1$) is not satisfied. That is, the graph of the second equation does not pass through the origin. Next solve the equations simultaneously; write out the alternative forms of the equations.

$\mathbf{r = 1 - 2\cos\theta}$: $\mathbf{r = (-1)^n[1 - 2\cos(\theta + n\pi)]}$

If $n = 0$: $r = 1 - 2\cos\theta$

If $n = 1$: $r = -[1 - 2\cos(\theta + \pi)]$
$= -[1 + 2\cos\theta]$
$= -1 - 2\cos\theta$

For other integral values one of these two equations is repeated.

$\mathbf{r = 1}$: $\mathbf{r = (-1)^n \cdot 1}$

If $n = 0$: $r = 1$

If $n = 1$: $r = -1$

For other integral values one of these two equations is repeated.

Solve the systems

$$\begin{cases} r = 1 - 2\cos\theta \\ r = 1 \end{cases} \quad \begin{cases} r = 1 - 2\cos\theta \\ r = -1 \end{cases} \quad \begin{cases} r = -1 - 2\cos\theta \\ r = 1 \end{cases} \quad \begin{cases} r = -1 - 2\cos\theta \\ r = -1 \end{cases}$$

$1 = 1 - 2\cos\theta$	$-1 = 1 - 2\cos\theta$	$1 = -1 - 2\cos\theta$	$-1 = -1 - 2\cos\theta$
$0 = \cos\theta$	$1 = \cos\theta$	$-1 = \cos\theta$	$0 = \cos\theta$
$\theta = \frac{\pi}{2}, \frac{3\pi}{2}$	$\theta = 0$	$\theta = \pi$	Same as first equation

Points: $(1, \frac{\pi}{2}), (1, \frac{3\pi}{2}), (-1, 0), (1, \pi)$

Notice that $(-1, 0)$ and $(1, \pi)$ are the same point, so the solution is $\mathbf{(1, \frac{\pi}{2}), (1, \frac{3\pi}{2}), (1, \pi)}$.

The graphs of these curves are shown in Figure 12.35. ■

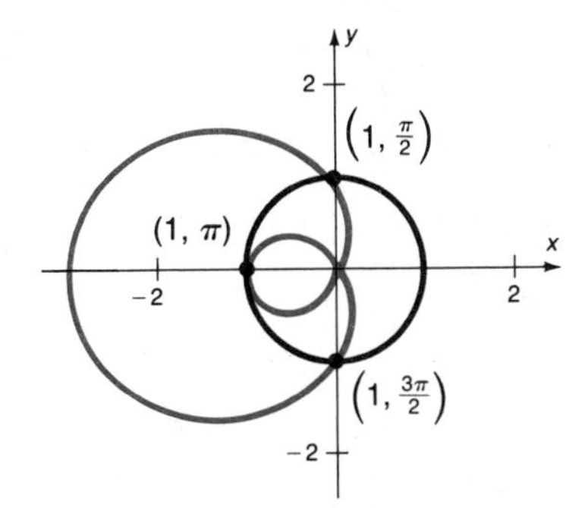

Figure 12.35 Graphs of $r = 1 - 2\cos\theta$ and $r = 1$

PROBLEM SET 12.6

A

Sketch each pair of equations in Problems 1–15 on the same coordinate axes.

1. $\begin{cases} r = 4\cos\theta \\ r = 4\sin\theta \end{cases}$

2. $\begin{cases} r = 8\cos\theta \\ r = 8\sin\theta \end{cases}$

3. $\begin{cases} r = 2\cos\theta \\ r = 1 \end{cases}$

4. $\begin{cases} r = 4\cos\theta \\ r = 2 \end{cases}$

5. $\begin{cases} r^2 = 9\cos\theta \\ r = 3 \end{cases}$

6. $\begin{cases} r^2 = 4\sin\theta \\ r = 2 \end{cases}$

7. $\begin{cases} r = 2(1 + \cos\theta) \\ r = 2(1 - \cos\theta) \end{cases}$

8. $\begin{cases} r = 2(1 + \sin\theta) \\ r = 2(1 - \sin\theta) \end{cases}$

9. $\begin{cases} r^2 = 4\cos 2\theta \\ r = 2 \end{cases}$

10. $\begin{cases} r^2 = 9\sin 2\theta \\ r = 3 \end{cases}$

11. $\begin{cases} r^2 = 4\sin 2\theta \\ r = 2\sqrt{2}\cos\theta \end{cases}$

12. $\begin{cases} r^2 = 9\cos 2\theta \\ r = 3\sqrt{2}\sin\theta \end{cases}$

13. $\begin{cases} r^2 = \cos 2\theta \\ r^2 = -\cos 2\theta \end{cases}$

14. $\begin{cases} r = 1 + \sin\theta \\ r = 1 + \cos\theta \end{cases}$

15. $\begin{cases} r = a(1 + \sin\theta) \\ r = a(1 - \sin\theta) \end{cases}$

Find the points of intersection of the curves given by the equations in Problems 16–30. You need to give only one primary representation for each point of intersection; give the one with a positive r. You graphed each of these systems in Problems 1–15.

16. $\begin{cases} r = 4\cos\theta \\ r = 4\sin\theta \end{cases}$

17. $\begin{cases} r = 8\cos\theta \\ r = 8\sin\theta \end{cases}$

18. $\begin{cases} r = 2\cos\theta \\ r = 1 \end{cases}$

19. $\begin{cases} r = 4\cos\theta \\ r = 2 \end{cases}$

20. $\begin{cases} r^2 = 9\cos\theta \\ r = 3 \end{cases}$

21. $\begin{cases} r^2 = 4\sin\theta \\ r = 2 \end{cases}$

22. $\begin{cases} r = 2(1 + \cos\theta) \\ r = 2(1 - \cos\theta) \end{cases}$

23. $\begin{cases} r = 2(1 + \sin\theta) \\ r = 2(1 - \sin\theta) \end{cases}$

24. $\begin{cases} r^2 = 4\cos 2\theta \\ r = 2 \end{cases}$

25. $\begin{cases} r^2 = 9\sin 2\theta \\ r = 3 \end{cases}$

26. $\begin{cases} r^2 = 4\sin 2\theta \\ r = 2\sqrt{2}\cos\theta \end{cases}$

27. $\begin{cases} r^2 = 9\cos 2\theta \\ r = 3\sqrt{2}\sin\theta \end{cases}$

28. $\begin{cases} r^2 = \cos 2\theta \\ r^2 = -\cos 2\theta \end{cases}$

29. $\begin{cases} r = 1 + \sin\theta \\ r = 1 + \cos\theta \end{cases}$

30. $\begin{cases} r = a(1 + \sin\theta) \\ r = a(1 - \sin\theta) \end{cases}$

B

Sketch each pair of equations in Problems 31–45 on the same coordinate axes.

31. $\begin{cases} r = 2\theta \\ \theta = \frac{\pi}{6} \end{cases}$

32. $\begin{cases} r = 3\theta \\ \theta = \frac{\pi}{3} \end{cases}$

33. $\begin{cases} r^2 = \cos 2\theta \\ r = \sqrt{2}\sin\theta \end{cases}$

34. $\begin{cases} r = 2(1 - \cos\theta) \\ r = 4\sin\theta \end{cases}$

35. $\begin{cases} r = 2(1 - \cos\theta) \\ r = 4\sin\theta \end{cases}$

36. $\begin{cases} r = 2(1 - \sin\theta) \\ r = 4\cos\theta \end{cases}$

37. $\begin{cases} r = 2\cos\theta + 1 \\ r = \sin\theta \end{cases}$

38. $\begin{cases} r = 2\sin\theta + 1 \\ r = \cos\theta \end{cases}$

39. $\begin{cases} r = \dfrac{5}{3 - \cos\theta} \\ r = 2 \end{cases}$

40. $\begin{cases} r = \dfrac{2}{1 + \cos\theta} \\ r = 2 \end{cases}$

41. $\begin{cases} r = \dfrac{4}{1 - \cos\theta} \\ r = 2\cos\theta \end{cases}$

42. $\begin{cases} r = \dfrac{1}{1 + \cos\theta} \\ r = 2(1 - \cos\theta) \end{cases}$

43. $\begin{cases} r = a\cos\theta \\ r = a\sec\theta \end{cases}$

44. $\begin{cases} r = a\sin\theta \\ r = a\csc\theta \end{cases}$

45. $\begin{cases} r\sin\theta = 1 \\ r = 4\sin\theta \end{cases}$

Find the points of intersection of the curves given by the equations in Problems 46–60. You need to give only one primary representation for each point of intersection ($0 \le \theta < 2\pi$); give the one with a positive r. You graphed each of these systems in Problems 31–45.

46. $\begin{cases} r = 2\theta \\ \theta = \frac{\pi}{6} \end{cases}$

47. $\begin{cases} r = 3\theta \\ \theta = \frac{\pi}{3} \end{cases}$

48. $\begin{cases} r^2 = \cos 2\theta \\ r = \sqrt{2}\sin\theta \end{cases}$

49. $\begin{cases} r = 2(1 - \cos\theta) \\ r = 4\sin\theta \end{cases}$

50. $\begin{cases} r = 2(1 - \cos\theta) \\ r = 4\sin\theta \end{cases}$

51. $\begin{cases} r = 2(1 - \sin\theta) \\ r = 4\cos\theta \end{cases}$

52. $\begin{cases} r = 2\cos\theta + 1 \\ r = \sin\theta \end{cases}$

53. $\begin{cases} r = 2\sin\theta + 1 \\ r = \cos\theta \end{cases}$

54. $\begin{cases} r = \dfrac{5}{3 - \cos\theta} \\ r = 2 \end{cases}$

55. $\begin{cases} r = \dfrac{2}{1 + \cos\theta} \\ r = 2 \end{cases}$

56. $\begin{cases} r = \dfrac{4}{1 - \cos\theta} \\ r = 2\cos\theta \end{cases}$

57. $\begin{cases} r = \dfrac{1}{1 + \cos\theta} \\ r = 2(1 - \cos\theta) \end{cases}$

58. $\begin{cases} r = a\cos\theta \\ r = a\sec\theta \end{cases}$

59. $\begin{cases} r = a\sin\theta \\ r = a\csc\theta \end{cases}$

60. $\begin{cases} r\sin\theta = 1 \\ r = 4\sin\theta \end{cases}$

12.7 PARAMETRIC EQUATIONS

Up to now, the curves we have discussed have been represented by a single equation. However, there is another way of representing curves which is often useful. This new representation defines the x and y in (x, y) so that they are *each* functions of some other variable, say t. That is, let

$$x = g(t) \quad \text{and} \quad y = h(t)$$

for functions g and h, where the domain of these functions is some interval I. For example, let

$$g(t) = 1 + 3t \quad \text{and} \quad h(t) = 2t \qquad (\text{for } 0 \le t \le 5)$$

Then, if $t = 1$,

$$x = g(1) = 1 + 3(1) = (4)$$
$$y = h(1) = 2(1) = 2$$

Then the point $(x, y) = (4, 2)$ when $t = 1$. Other values are shown in the following table and plotted in Figure 12.36.

t	0	1	2	3	4	5
x	1	4	7	10	13	16
y	0	2	4	6	8	10

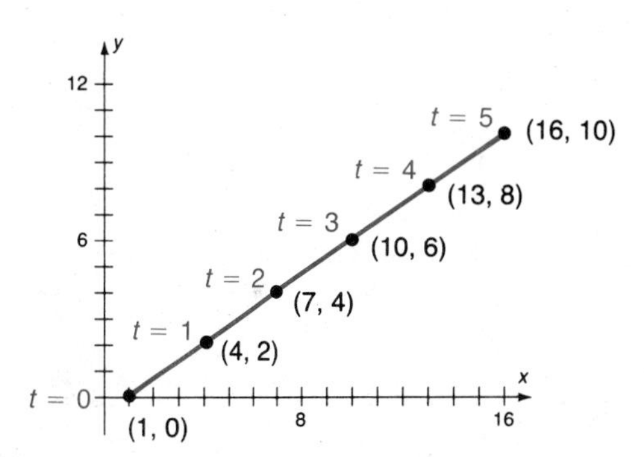

Figure 12.36 Graphs of of $x = 1 + 3t$, $y = 2t$

The variable t is called a **parameter** and the equations $x = 1 + 3t$ and $y = 2t$ are called **parametric equations** for the line segment shown in Figure 12.36.

Parameter and Parametric Equations

Let t be a number in an interval I. Consider the curve defined by the set of ordered pairs (x, y), where

$$x = f(t) \quad \text{and} \quad y = g(t)$$

for functions f and g defined on I. Then the variable t is called a **parameter** and the equations $x = f(t)$ and $y = g(t)$ are called **parametric equations** for the curve defined by (x, y).

EXAMPLE 1 Plot the curve represented by the parametric equations

$$x = \cos\theta$$
$$y = \sin\theta$$

Solution The parameter is θ and you can generate a table of values by using Table C.III, exact values, or a calculator:

θ	0°	15°	30°	45°	60°	75°	90°	120°	...
x	1.00	0.97	0.87	0.71	0.50	0.26	0.00	−0.50	...
y	0.00	0.26	0.50	0.71	0.87	0.97	1.00	0.87	...

These points are plotted in Figure 12.37. If the plotted points are connected, you can see that the curve is a circle.

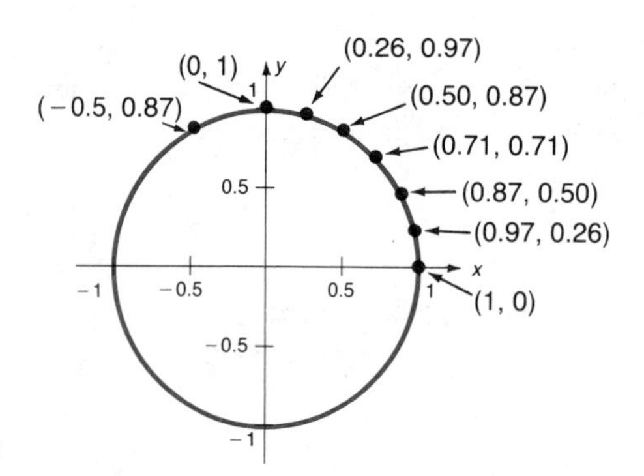

Figure 12.37 Graph of $x = \cos\theta$, $y = \sin\theta$

■

It is possible to recognize the parametric equations in Example 1 as a unit circle if you square both sides of the equation and add:

$$x^2 = \cos^2\theta$$
$$y^2 = \sin^2\theta$$
$$x^2 + y^2 = \cos^2\theta + \sin^2\theta$$

So

$$x^2 + y^2 = 1$$

This process is called **eliminating the parameter**.

EXAMPLE 2 Eliminate the parameter for the parametric equations $x = t + 2$, $y = t^2 + 2t - 1$.

Solution Solve the first equation for t: $t = x - 2$. Substitute into the second equation:

$$\begin{aligned} y &= (x-2)^2 + 2(x-2) - 1 \\ &= x^2 - 4x + 4 + 2x - 4 - 1 \\ &= x^2 - 2x - 1 \end{aligned}$$

You can recognize $y = x^2 - 2x - 1$ as a parabola. It can now be graphed by completing the square or by using the parametric equations and plotting points as illustrated by Example 1. ■

EXAMPLE 3 Eliminate the parameter for the parametric equations

$$x = t^2 - 3t + 1, \; y = -t^2 + 2t + 3$$

Solution It is not as easy to solve one of these equations for t as it was in Example 2. You can, however, add one equation to the other:

$$x + y = -t + 4 \quad \text{or} \quad t = 4 - x - y$$

This can be substituted into either equation to give

$$\begin{aligned} x &= (4 - x - y)^2 - 3(4 - x - y) + 1 \\ &= 16 - 4x - 4y - 4x + x^2 + xy - 4y + xy + y^2 - 12 + 3x + 3y + 1 \\ &= x^2 + 2xy + y^2 - 5x - 5y + 5 \end{aligned}$$

The curve whose equation is $x^2 + 2xy + y^2 - 6x - 5y + 5 = 0$ is a rotated parabola since $B^2 - 4AC = 4 - 4(1)(1) = 0$. ■

EXAMPLE 4 Eliminate the parameter for the parametric equations $x = 2^t$, $y = 2^{t+1}$

Solution

$$\begin{aligned} \frac{y}{x} &= \frac{2^{t+1}}{2^t} \\ &= 2^{t+1-t} \\ &= 2 \\ y &= 2x \end{aligned}$$

WARNING: This answer is not complete; see discussion below.

Consider the parametric equations given in this example a little more closely. You can plot a curve by using the parametric equations or by eliminating the parameter. Plot the equations both ways:

1. Parametric form:

t	0	1	2	3
x	1	2	4	8
y	2	4	8	16

The values on this table are found by substitution. For example, if $t = 0$ then

$$x = 2^0 = 1 \quad \text{and} \quad y = 2^{0+1} = 2$$

Can $x = 3$? Solve

$$\begin{aligned} 3 &= 2^t \\ \log 3 &= t \log 2 \\ t &= \frac{\log 3}{\log 2} \\ &\approx 1.5850 \end{aligned}$$

and then

$$y = 2^{1.5850+1} \approx 6$$

Can $x = 0$? No, since $0 = 2^t$ has no solution.
Can $x = -1$? No, since $2^t > 0$
The graph is shown in Figure 12.38.

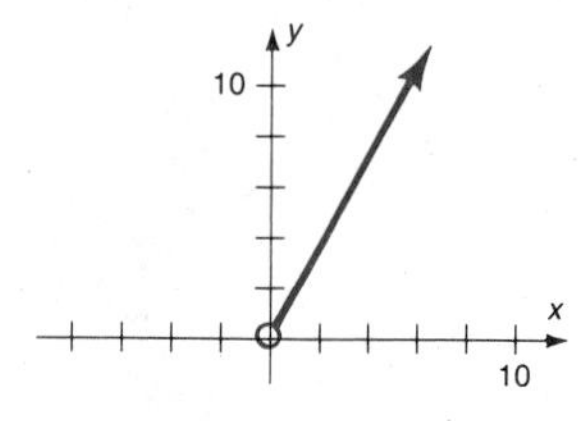

Figure 12.38 Graph of $x = 2^t$, $y = 2^{t+1}$

2. **Eliminate the parameter as shown above:**

$$y = 2x$$

Can $x = 3$? Solve $y = 2(3) = 6$
Can $x = 0$? Solve $y = 2(0) = 0$
Can $x = -1$? Solve $y = 2(-1) = -2$
The graph is shown in Figure 12.39.

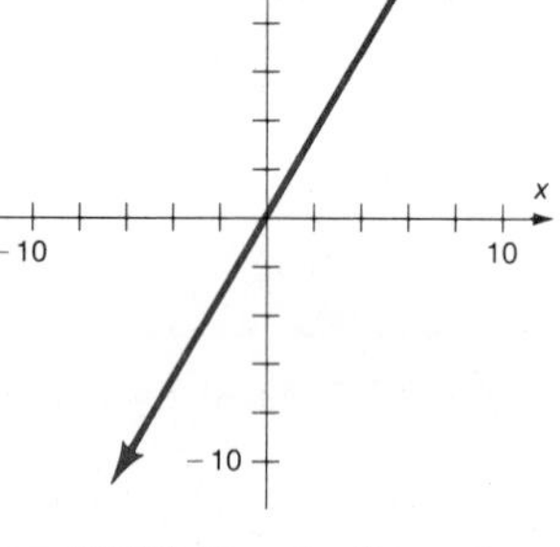

Figure 12.39 Graph of $y = 2x$

Of course, what you notice by comparing Figures 12.38 and 12.39 is that **you must be very careful about the domain for x when eliminating the parameter**. Thus when you eliminate the parameter t and write $\mathbf{y = 2x}$, you must include the condition $\mathbf{x > 0}$ (from $x = 2^t$, x is nonnegative). ■

Note also that sometimes it is impossible to eliminate the parameter in any simple way, as in the equations

$$x = 3t^5 - 4t^2 + 7t + 11$$
$$y = 4t^{13} + 12t^5 + 6t^4 + 16$$

You must therefore be able to graph parametric equations using *both* methods since one or the other might not be appropriate for a particular graph.

PROBLEM SET 12.7

A

Plot the curves in Problems 1–20 by plotting points.

1. $x = 4t$, $y = -2t$
2. $x = t + 1$, $y = 2t$
3. $x = t$, $y = 2 + \frac{2}{3}(t - 1)$
4. $x = t$, $y = 3 - \frac{3}{5}(t + 2)$
5. $x = 2t$, $y = t^2 + t + 1$
6. $x = 3t$, $y = t^2 - t + 6$
7. $x = t$, $y = t^2 + 2t + 3$
8. $x = t$, $y = 2t^2 - 5t + 6$
9. $x = 3\cos\theta$, $y = 3\sin\theta$
10. $x = 2\cos\theta$, $y = 2\sin\theta$
11. $x = 4\cos\theta$, $y = 3\sin\theta$
12. $x = 5\cos\theta$, $y = 2\sin\theta$
13. $x = t^2 + 2t + 3$, $y = t^2 + t - 4$
14. $x = t^2 - 2t + 3$, $y = t^2 - t + 4$
15. $x = t^2 + 3t - 4$, $y = 2t^2 + 4t - 1$
16. $x = 2t^2 + t + 6$, $y = t^2 + t + 6$
17. $x = 3^t$, $y = 3^{t+1}$
18. $x = 2^t$, $y = 2^{1-t}$
19. $x = e^t$, $y = e^{t+1}$
20. $x = e^t$, $y = e^{1-t}$

Eliminate the parameter in Problems 21–40 and plot the resulting equations.

21. $x = 4t$, $y = -2t$
22. $x = t + 1$, $y = 2t$
23. $x = t$, $y = 2 + \frac{2}{3}(t - 1)$
24. $x = t$, $y = 3 - \frac{3}{5}(t + 2)$
25. $x = 2t$, $y = t^2 + t + 1$
26. $x = 3t$, $y = t^2 - t + 6$
27. $x = t$, $y = t^2 + 2t + 3$
28. $x = t$, $y = 2t^2 - 5t + 6$
29. $x = 3\cos\theta$, $y = 3\sin\theta$
30. $x = 2\cos\theta$, $y = 2\sin\theta$
31. $x = 4\cos\theta$, $y = 3\sin\theta$
32. $x = 5\cos\theta$, $y = 2\sin\theta$
33. $x = t^2 + 2t + 3$, $y = t^2 + t - 4$
34. $x = t^2 - 2t + 3$, $y = t^2 - t + 4$
35. $x = t^2 + 3t - 4$, $y = 2t^2 + 4t - 1$
36. $x = 2t^2 + t + 6$, $y = t^2 + t + 6$
37. $x = 3^t$, $y = 3^{t+1}$
38. $x = 2^t$, $y = 2^{1-t}$
39. $x = e^t$, $y = e^{t+1}$
40. $x = e^t$, $y = e^{1-t}$

B

Plot the curves in Problems 41–60 by any convenient method.

41. $x = 60t, y = 80t - 16t^2$ **42.** $x = 30t, y = 60t - 9t^2$

43. $x = 10\cos t, y = 10\sin t$ **44.** $x = 8\sin t, y = 8\cos t$

45. $x = 5\cos\theta, y = 3\sin\theta$ **46.** $x = 4\cos\theta, y = 2\sin\theta$

47. $x = t^2, y = t^3$ **48.** $x = t^3 + 1, y = t^3 - 1$

49. $x = e^t, y = e^{t+2}$ **50.** $x = e^t, y = e^{t-2}$

51. $x = \theta - \sin\theta, y = 1 - \cos\theta$

52. $x = \theta + \sin\theta, y = 1 - \cos\theta$

53. $x = 4\tan 2t, y = 3\sec 2t$

54. $x = 2\tan 2t, y = 4\sec 2t$

55. $x = 1 + \cos t, y = 3 - \sin t$

56. $x = 2 - \sin t, y = -3 + \cos t$

57. $x = 3\cos\theta + \cos 3\theta, y = 3\sin\theta - \sin 3\theta$

58. $x = \cos t + t\sin t, y = \sin t - t\cos t$

59. $x = \tan t, y = \cot t$ **60.** $x = \cos\theta, y = \sec\theta$

C

61. ***Physics*** Suppose a light is attached to the edge of a bike wheel. The path of the light is shown in Figure 12.40. If the

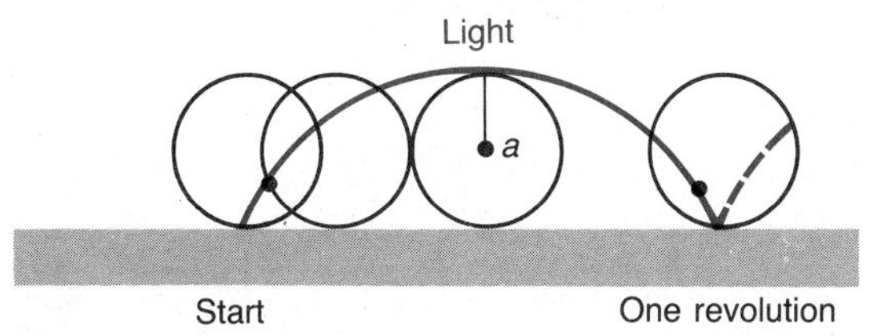

Figure 12.40 Graph of a cycloid

radius of the wheel is a, find the equation for the path of the light. Such a curve is called a *cycloid. Hint:* Consider Figure 12.41 and find the coordinates of $P(x, y)$. Notice that

$$x = |OA|$$
$$y = |PA|$$

Find x and y in terms of θ, the amount of rotation in radians.

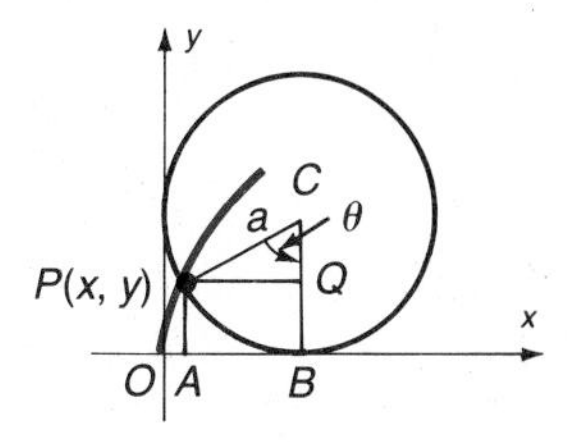

Figure 12.41

62. Suppose a string is wound around a circle of radius a. The string is then unwound in the plane of the circle while it is held tight, as shown in Figure 12.42. Find the equation for this curve, called the *involute of a circle. Hint:* Consider Figure 12.43 and find the coordinates of $P(x, y)$. Notice that

$$x = |OB|$$
$$y = |PB|$$

Find x and y in terms of θ, the amount of rotation in radians.

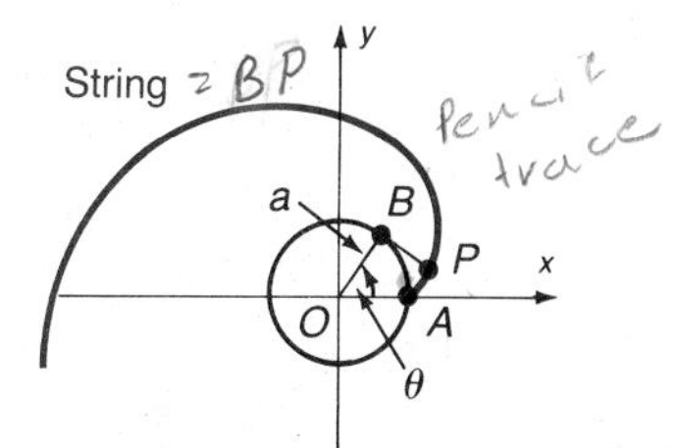

Figure 12.42 Graph of the involute of a circle

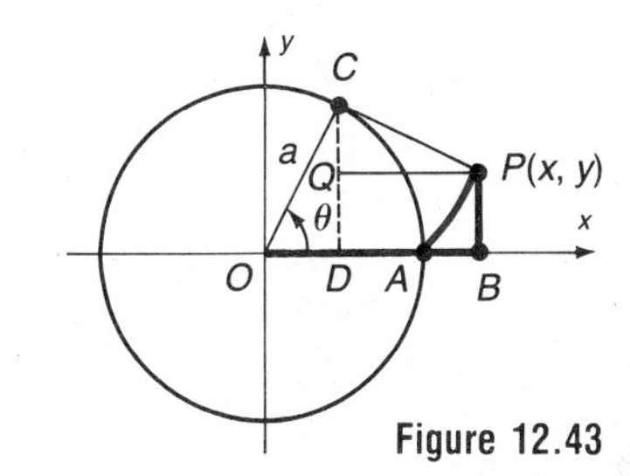

Figure 12.43

12.8 CHAPTER 12 SUMMARY

The material of this chapter is reviewed in the following list of objectives. After each objective there are some practice questions. For a sample test, select the first question of each set and check your answers with the answer section. For a sample test without answers, use the second question of each set. Additional practice is given by the other questions in each set. If you are having trouble with a particular type of problem, look back to that section for extra help.

12.1 VECTORS *Objective 1 Find the resultant vector of two given forces.*

1. If $\mathbf{v} = 5\mathbf{i} - 2\mathbf{j}$ and $\mathbf{w} = -7\mathbf{i} + 3\mathbf{j}$, find the resultant vector.
2. If $\mathbf{v} = -3\mathbf{i} + 2\mathbf{j}$ and $\mathbf{w} = 8\mathbf{i} - 3\mathbf{j}$, find the resultant vector.
3. Consider two forces, one with magnitude 8.0 in a N20°E direction and the other with a magnitude of 9.0 in a S50°E direction. What is the resultant vector?

4. An object is hurled from a catapult due east with a velocity of 38 feet per second (fps). If the wind is blowing due south at 15 mph (22 fps), what are the true bearing and the velocity of the object?

Objective 2 *Resolve a given vector and find its algebraic representation.*

5. Resolve the vector with magnitude 4.5 and $\theta = 51°$ into horizontal and vertical components.
6. Resolve the vector with magnitude 12 and $\theta = 30°$ into horizontal and vertical components.
7. Find the algebraic representation of a vector determined by $\overrightarrow{AB}$ with $A(0,0)$ and $B(5,12)$.
8. Find the algebraic representation of a vector determined by $\overrightarrow{AB}$ with $A(-3,-5)$, $B(-6,-10)$.

Objective 3 *Find the magnitude of a vector.*

9. $\mathbf{v} = 5\mathbf{i} - 2\mathbf{j}$
10. $\mathbf{w} = -7\mathbf{i} + 3\mathbf{j}$
11. $\mathbf{v} = 6\mathbf{i}$
12. $\mathbf{w} = \cos 42°\mathbf{i} + \sin 42°\mathbf{j}$

Objective 4 *Know the definition of scalar product and be able to find the scalar product of two vectors.*

13. Define scalar product.
14. Find the scalar product of $\mathbf{v}$ and $\mathbf{w}$ where $\mathbf{v} = 4\mathbf{i} - 2\mathbf{j}$ and $\mathbf{w} = -7\mathbf{i} + 3\mathbf{j}$.
15. Find $(3\mathbf{i} + 5\mathbf{j}) \cdot (-2\mathbf{i} + 3\mathbf{j})$
16. Find $(e\mathbf{i} - e^{-1}\mathbf{j}) \cdot (e^2\mathbf{i} - e\mathbf{j})$

Objective 5 *Find the cosine of the angle between two vectors, or use the inverse cosine function to find that angle.*

17. Find the cosine of the angle between the vectors $3\mathbf{i} - 2\mathbf{j}$ and $5\mathbf{i} + 12\mathbf{j}$.
18. Find the angle between the vectors $5\mathbf{i} - 2\mathbf{j}$ and $-7\mathbf{i} + 3\mathbf{j}$.
19. Find the angle between the vectors $\cos 15°\mathbf{i} + \sin 15°\mathbf{j}$ and $\cos 20°\mathbf{i} + \sin 20°\mathbf{j}$.
20. Find the cosine of the angle between the vectors $5\mathbf{i}$ and $-3\mathbf{i} - \mathbf{j}$.

12.2 PROPERTIES OF VECTORS

Objective 6 *Find a vector normal to a given line and a vector determined by a line.* After you have found the vector normal to the given line and the vector determined by the line, check your answer by showing that the dot product of these vectors is 0.

21. $5x - 12y + 3 = 0$
22. $2x + 3y + 12 = 0$
23. $x - 5y + 4 = 0$
24. $x + y = 0$

Objective 7 *Find scalar and vector projections.* For the given vectors, find the vector and scalar projections of $\mathbf{v}$ onto $\mathbf{w}$.

25. $\mathbf{v} = 2\mathbf{i} + \sqrt{5}\mathbf{j}$, $\mathbf{w} = 3\sqrt{5}\mathbf{i} + 3\mathbf{j}$
26. $\mathbf{v} = 5\mathbf{i} - 12\mathbf{j}$, $\mathbf{w} = 3\mathbf{i} + 4\mathbf{j}$
27. $\mathbf{v} = 9\mathbf{i} + \mathbf{j}$, $\mathbf{w} = \mathbf{i} - 9\mathbf{j}$
28. $\mathbf{v} = -2\mathbf{i} - 7\mathbf{j}$, $\mathbf{w} = 4\mathbf{i} + 3\mathbf{j}$

Objective 8 *Find the distance from a given point to a given line.*

29. $5x + 12y + 8 = 0$; $(1, 5)$
30. $6x - 2y + 5 = 0$; $(-5, -10)$
31. $2x - 3y - 5 = 0$; $(-10, 2)$
32. $5x + 3y = 0$; $(8, 10)$

Objective 9 *Find a vector $\overrightarrow{AB}$ given the points A and B.**

33. $A(0,0,0)$, $B(4,8,-2)$
34. $A(-2,3,1)$, $B(-8,5,-4)$
35. $A(6,1,2)$, $B(-2,0,5)$
36. $A(0,0,5)$, $B(8,1,0)$

---

* For two dimensions (Section 12.2), use the first two components only. If you studied Section 12.4, find these three-dimensional vectors.

*12.3 THREE-DIMENSIONAL COORDINATE SYSTEM

Objective 10 Evaluate functions of several variables. Evaluate the functions in Problems 37–40 for the point $(5, -12)$.

37. $f(x, y) = 3x^2 - 2xy + y^2$
38. $g(x, y) = e^{x/y}$
39. $k(l, w) = lw$
40. $P(L, K) = L^{0.5}K^3$

Objective 11 Plot the points in three dimensions.

41. a. $(5, 0, 0)$ **b.** $(3, -5, 5)$
42. a. $(0, 1, 0)$ **b.** $(-1, 2, 1)$
43. a. $(0, 0, 4)$ **b.** $(5, 10, 8)$
44. a. $(2, 5, 7)$ **b.** $(-2, -5, -4)$

Objective 12 Graph surfaces in space.

45. $x + y + 2z = 10$
46. $z = x^2 + y^2$
47. $x^2 + y^2 - z^2 = 0$
48. $x^2 + y^2 = 9$

*12.4 VECTORS IN THREE DIMENSIONS

Objective 13 Carry out vector operations, including magnitude, for three-dimensional vectors. Let $\mathbf{v} = 3\mathbf{i} - 4\mathbf{j} + \mathbf{k}$ and $\mathbf{w} = -\mathbf{i} + 3\mathbf{j} - \mathbf{k}$.

49. $\mathbf{v} - \mathbf{w}$
50. $|\mathbf{v}| - |\mathbf{w}|$
51. $2\mathbf{v} + 3\mathbf{w}$
52. $\mathbf{v} - 2\mathbf{w}$

Objective 14 Find the scalar product for three-dimensional vectors.

53. $\mathbf{i} \cdot 2\mathbf{j}$
54. $(\mathbf{i} + \mathbf{j} + \mathbf{k}) \cdot (-2\mathbf{i} + 5\mathbf{k})$
55. $(2\mathbf{i} - \mathbf{j} - \mathbf{k}) \cdot (3\mathbf{i} + \mathbf{j} + \mathbf{k})$
56. $(\mathbf{i} + \mathbf{j} - 5\mathbf{k}) \cdot (4\mathbf{j} - \mathbf{k})$

Objective 15 Find the vector product for three-dimensional vectors.

57. $\mathbf{i} \times 2\mathbf{j}$
58. $(\mathbf{i} + \mathbf{j} + \mathbf{k}) \times (-2\mathbf{i} + 5\mathbf{k})$
59. $(2\mathbf{i} - \mathbf{j} - \mathbf{k}) \times (3\mathbf{i} + \mathbf{j} + \mathbf{k})$
60. $(\mathbf{i} + \mathbf{j} - 5\mathbf{k}) \times (4\mathbf{j} - \mathbf{k})$

Objective 16 Find a unit vector in a given direction; find the angle between two three-dimensional vectors; apply vector operations to find scalars; find vector and scalar projections. Let $\mathbf{v} = \mathbf{i} + \mathbf{j} - \sqrt{7}\mathbf{k}$ and $\mathbf{w} = -\mathbf{i} - 2\mathbf{j} + \mathbf{k}$.

61. Find a unit vector in the direction of $\mathbf{v}$.
62. Find the cosine of the angle between $\mathbf{v}$ and $\mathbf{w}$.
63. Find a scalar s such that $\mathbf{v}$ is orthogonal to $s\mathbf{v} + \mathbf{w}$.
64. Find the vector and scalar projections of $\mathbf{v}$ onto $\mathbf{w}$.

12.5 POLAR COORDINATES

Objective 17 Plot points in polar form and give both primary representations. Approximate to four decimal places.

65. $(5, \sqrt{75})$
66. $(3, -\frac{2\pi}{3})$
67. $(-2, 2)$
68. $(-5, 9.4248)$

Objective 18 Change from rectangular to polar form and from polar form to rectangular form. Use exact values if possible; otherwise approximate to four decimal places.

69. Change the polar-form point $(3, -\frac{2\pi}{3})$ to rectangular form.
70. Change the polar-form point $(5, \sqrt{75})$ to rectangular form.
71. Change the rectangular-form point $(3, -3)$ to polar form.
72. Change the rectangular-form point $(3, \sqrt{3})$ to polar form.

Objective 19 Sketch polar-form curves.

73. $r = 2\cos 2\theta$
74. $r = 2 + 2\cos\theta$
75. $r^2 = 25\cos 2\theta$
76. $r = \tan 45^\circ$

Objective 20 Identify cardioids, rose curves, and lemniscates by looking at the equation.

77. a. $r^2 = 5\cos 2\theta$ **b.** $r = 5\cos 2\theta$

---

* Optional sections

78. **a.** $r = 5\theta$ **b.** $r = 5\cos 3\theta$
79. **a.** $r = 3 + 5\cos\theta$ **b.** $r = 5 + 5\cos\theta$
80. **a.** $\theta = \frac{\pi}{2}$ **b.** $\frac{\pi}{2}\cos\theta = \frac{\pi}{2} - r$

*12.6 INTERSECTION OF POLAR-FORM CURVES

Objective 21 *Find the intersection of polar-form curves using both the graphing and the analytic methods.*

81. Show the intersection of the curves $r = 4 - 4\sin\theta$ and $4r = 4 - 8\sin\theta$ using the graphing method.
82. Show the intersection of the curves $r = 5\sin\theta$ and $r = 6\cos\theta$ using the graphing method.
83. Find the intersection of the curves in Problem 81 using the analytic method.
84. Find the intersection of the curves in Problem 82 using the analytic method. (Give answer correct to two significant figures.)

12.7 PARAMETRIC EQUATIONS

Objective 22 *Sketch a curve represented by parametric equations by plotting points.*

85. $x = 2 + 5t,\ y = -1 - 3t$
86. $x = 3\cos\theta,\ y = 5\sin\theta$
87. $x = e^{4t},\ y = e^{4t-2}$
88. $x = t^2 + 3t - 1,\ y = t^2 + 2t + 5$

Objective 23 *Eliminate the parameter, if possible, to sketch a curve defined by parametric equations.*

89. $x = 2 + 5t,\ y = -1 - 3t$
90. $x = 3\cos\theta,\ y = 5\sin\theta$
91. $x = e^{4t},\ y = e^{4t-2}$
92. $x = t^2 + 3t - 1,\ y = t^2 + 2t + 5$

---

* Optional section

CUMULATIVE REVIEW III

Suggestions for study of Chapters 9–12:

Make a list of important ideas from Chapters 9–12. Study this list. Use the objectives at the end of each chapter to help you make up this list.

Work some practice problems. A good source of problems is the set of chapter objectives at the end of each chapter. You should try to work at least one problem from each objective:

Chapter 9, pp. 382–384; work 14 problems
Chapter 10, pp. 419–420; work 9 problems
Chapter 11, p. 459; work 10 problems
Chapter 12, pp. 511–514; work 23 problems

Check the answers for the practice problems you worked. The first and third problems of each objective have their answers listed in the back of the book.

Additional problems. Work additional odd-numbered problems (answers in the back of the book) from the problem sets as needed. Focus on the problems you missed in the chapter summaries.

Work the problems in the following cumulative review. These problems should be done after you have studied the material. They should take you about 1 hour and will serve as a sample test for Chapters 9–12. Assume that all variables are restricted so that each expression is defined. All the answers for these questions are provided in the back of the book for self-checking.

PRACTICE TEST FOR CHAPTERS 9–12

1. Find the next term of each given sequence, classify it as arithmetic, geometric, or neither. If arithmetic, give the common difference; if geometric, give the common ratio; and if neither, explain the pattern.
a. 45, 30, 20 **b.** 45, 30, 15 **c.** 45, 30, 25

2. Solve each system by the indicated method.
a. By addition:

$$\begin{cases} 5x + 3y = 5 \\ 3x + 2y = 4 \end{cases}$$

b. By substitution:

$$\begin{cases} x + y = 1 \\ y = x^2 + 4x + 5 \end{cases}$$

c. By graphing

$$\begin{cases} 2x - y - 5 = 0 \\ 2x + y + 1 = 0 \end{cases}$$

d. Represent the system

$$\begin{cases} x + z = 1 \\ x + y + z = 0 \\ 2x + y = 3 \end{cases}$$

as an augmented matrix and solve using the Gauss–Jordan method.

3. Find the indicated matrices, if possible, where:

$$A = \begin{bmatrix} 1 & 0 \\ 2 & -1 \end{bmatrix} \quad B = \begin{bmatrix} 0 & 1 & 0 \\ 1 & 0 & -1 \end{bmatrix}$$

$$C = \begin{bmatrix} 1 & 0 \\ -2 & 1 \\ 0 & 1 \end{bmatrix}$$

a. $A + BC$ **b.** $AB - AC$ **c.** BCA

4. Identify the curve represented by the given equation.
a. $3x - 2y^2 - 4y + 7 = 0$ **b.** $y^2 = x^2 - 1$
c. $xy + y^2 - 3x = 5$ **d.** $x^2 + y^2 + x - y = 3$
e. $\dfrac{x}{4} + \dfrac{y}{9} = 1$ **f.** $(x - 1)(y + 1) = 7$
g. $r = 3 \sin 2\theta$ **h.** $r^2 = 3 \sin 2\theta$
i. $r - 5 = 5 \cos \theta$ **j.** $r = 2 \sin \frac{\pi}{2}$

5. Write the equations for the given curves.
a. The ellipse with center (2, 1), focus (1, 1), and semimajor axis 2
b. The set of points so that the difference of the distance from (3, 0) and (9, 0) is always 4
c. The parabola whose vertex is (4, 3) and directrix is $x = 1$

6. Sketch the given curves.
a. $3x - 2y^2 - 4y + 7 = 0$
b. $25x^2 - 16y^2 = 400$
c. $x = 4 \cos \theta$, $y = 3 \sin \theta$
d. $r = 3 \sin 3\theta$
e. $r = 5 - 5 \cos \theta$

*7. Sketch the given surfaces.
   a. $2x + y + z = 8$
   b. $y^2 + z^2 = 0$

8. a. State the Binomial Theorem using summation notation.
   b. State the Binomial Theorem without using summation notation.
   c. Use the Binomial Theorem to find $(x - 3y)^4$.

9. Prove that

$$1^3 + 2^3 + 3^3 + \cdots + n^3 = \frac{n^2(n + 1)^2}{4}$$

****10.** Let $\mathbf{v} = 5\mathbf{i} - \sqrt{5}\mathbf{j} + \mathbf{k}$ and $\mathbf{w} = -2\mathbf{i} + 3\sqrt{5}\mathbf{j} - 2\mathbf{k}$. Find the requested numbers or vectors, if possible.
   a. $\mathbf{v} - 2\mathbf{w}$ **b.** $\mathbf{v} \cdot \mathbf{w}$ **c.** $|\mathbf{w}|$
   d. A unit vector in the direction of $\mathbf{v}$
   e. $\mathbf{v} \times \mathbf{w}$

* From optional section
** From optional section; you can work parts **a–d** if you did not cover Section 12.4 by using $\mathbf{v} = 5\mathbf{i} - \sqrt{5}\mathbf{j}$ and $\mathbf{w} = -2\mathbf{i} + 3\sqrt{5}\mathbf{j}$.

Extended Application: PLANETARY ORBITS

A 'Planet X' Way Out There?

LIVERMORE, Calif. (UPI)—A scientist at the Lawrence Livermore Laboratory suggested Friday the existence of a "Planet X" three times as massive as Saturn and nearly six billion miles from earth.

The planet, far beyond Pluto which is currently the outermost of the nine known planets of the solar system, was predicted on sophisticated mathematical computations of the movements of Halley's Comet.

Joseph L. Brady, a Lawrence mathematician and an authority on the comet, reported the calculations in the Journal of the Astronomical Society of the Pacific.

Brady said he and his colleagues, Edna M. Carpenter and Francis H. McMahon, used a computer to process mathematical observations of the strange deviations in Halley's Comet going back to before Christ.

Lawrence officials said the existence of a 10th planet has been predicted before, but Brady is the first to predict its orbit, mass and position.

Brady said the planet was about 65 times as far from the sun as earth, which is about 93 million miles from the sun. From earth "Planet X" would be located in the constellation Casseiopeia on the border of the Milky Way.

The size and location of "Planet X" were proposed to account for mysterious deviations in the orbit of Halley's Comet. But the calculations subsequently were found to account for deviations in the orbits of two other reappearing comets, Olbers and Pons-Brooks, Brady said.

No contradiction between the proposed planet and the known orbits of comets and other planets has been found.

The prediction of unseen planets is not new. The location of Neptune was predicted in 1846 on the basis of deviations in the orbit of Uranus. Deviations in Neptune's orbit led to a prediction of Pluto's location in 1915.

Although no such deviations of Pluto have been found, Brady pointed out that since its discovery in 1930, Pluto has been observed through less than one-fourth of its revolution around the sun and a complete picture of its orbit is not available.

Brady said "Planet X" may be as elusive as Pluto was a half century ago. It took 15 years to find it from the time of its prediction.

"The proposed planet is located in the densely populated Milky Way where even a tiny area encompasses thousands of stars, many of which are brighter than we expect this planet to be," he said. "If it exists, it will be extremely difficult to find."

"Planet X," in its huge orbit, takes 600 years to complete a revolution around the sun, he said.

In 1986 Halley's Comet returned for our once-every-76-year view. By using mathematics and a knowledge of the conic sections, we can predict the comet's path as well as its exact arrival time (closest to earth on April 10, 1986).

After the planet Uranus was discovered in 1781, its motion revealed gravitational perturbations caused by an unknown planet. Independent mathematical calculations by Urbain Leverrier and John Couch Adams predicted the position of this unknown planet—the discovery of Neptune in 1846 is one of the greatest triumphs of celestial mechanics and mathematics in the history of astronomy. A similar search lead to the discovery of Pluto in 1930. The article reproduced here is dated April 30, 1972, but at the time of this printing no planet has yet been found. Remember, though, that it took 15 years to find Pluto after the time of its prediction.

The orbits of the planets are elliptical in shape. If the sun is placed at one of the foci of a giant ellipse, the orbit of the earth is elliptical. The **perihelion** is the point where the planet comes closest to the sun; the **aphelion** is the farthest distance the planet travels from the sun. The eccentricity of a planet tells us the amount of roundness of that planet's orbit. The eccentricity of a circle is 0 and that of a parabola is 1. The eccentricity for each planet in our solar system is given here:

Planet	Eccentricity
Mercury	0.194
Venus	0.007
Earth	0.017
Mars	0.093
Jupiter	0.048
Saturn	0.056
Uranus	0.047
Neptune	0.009
Pluto	0.249

Planet
A Sun P
Aphelion Perihelion

EXAMPLE 1 The orbit of the earth around the sun is elliptical with the sun at one focus. If the semimajor axis of this orbit is 9.3×10^7 mi and the eccentricity is about 0.017, determine the greatest and least distance of the earth from the sun (correct to two significant digits).

Solution We are given $a = 9.3 \times 10^7$ and $\epsilon = 0.17$. Now

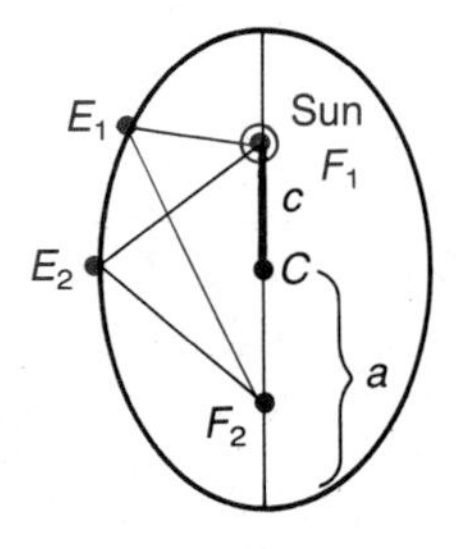

$$\epsilon = \frac{c}{a}$$

$$0.017 = \frac{c}{9.3 \times 10^7}$$

$$c \approx 1.581 \times 10^6$$

The greatest distance is $a + c \approx \mathbf{9.5 \times 10^7}$.
The least distance is $a - c \approx \mathbf{9.1 \times 10^7}$. ■

The orbit of a satellite can be calculated from Kepler's third law:

$$\frac{\text{Mass of (planet + satellite)}}{\text{Mass of (sun + planet + satellite)}} = \frac{(\text{semimajor axis of satellite orbit})^3}{(\text{semimajor axis of planet orbit})^3} \times \frac{(\text{period of planet})^2}{(\text{period of satellite})^2}$$

EXAMPLE 2 If the mass of the earth is 6.58×10^{21} tons, the sun 2.2×10^{27} tons, and the moon 8.1×10^{19} tons, calculate the orbit of the moon. Assume that the semimajor axis of the earth is 9.3×10^7 mi, the period of the earth 365.25 days, and the period of the moon 27.3 days.

Solution First solve the equation given as Kepler's third law for the unknown—the semimajor axis of the satellite orbit:

$$(\text{Semimajor axis of satellite orbit})^3$$
$$= \frac{\text{mass of (planet + satellite)}}{\text{mass of (sun + planet + satellite)}} \times \frac{(\text{period of satellite})^2}{(\text{period of planet})^2} \times (\text{semimajor axis of planet orbit})^3$$

Thus

$$\begin{aligned}&(\text{Semimajor axis of satellite orbit})^3\\&= \frac{(6.58 \cdot 10^{21}) + (8.1 \cdot 10^{19})}{(2.2 \cdot 10^{27}) + (6.58 \cdot 10^{21}) + (8.1 \cdot 10^{19})} \times \frac{(27.3)^2}{(365.25)^2} \times (9.3 \cdot 10^7)^3\\&\approx \frac{3.993131141 \cdot 10^{48}}{2.934975261 \cdot 10^{32}}\\&\approx 1.360533151 \cdot 10^{16}\end{aligned}$$

$$\text{Semimajor axis of satellite orbit} \approx 2.387278258 \cdot 10^5$$

The moon's orbit has a semimajor axis of about 239,000 mi. ■

Extended Application Problems—Planetary Orbits

1. The orbit of Mars about the sun is elliptical with the sun at one focus. If the semimajor axis of this orbit is 1.4×10^8 mi and the eccentricity is about 0.093, determine the greatest and least distance of Mars from the sun, correct to two significant digits.
2. The orbit of Venus about the sun is elliptical with the sun at one focus. If the semimajor axis of this orbit is 6.7×10^7 mi and the eccentricity is about 0.007, determine the greatest and least distance of Venus from the sun, correct to two significant digits.
3. The orbit of Neptune about the sun is elliptical with the sun at one focus. If the semimajor axis of this orbit is 3.66×10^9 mi and the eccentricity is about 0.009, determine the greatest and least distance of Neptune from the sun, correct to two significant digits.
4. If an unknown Planet X has an elliptical orbit about the sun with a semimajor axis of 6.89×10^{10} mi and eccentricity of about 0.35, what is the least distance between Planet X and the sun?
5. If the mass of Mars is 7.05×10^{20} tons, its satellite Phobos 4.16×10^{12} tons, and the sun 2.2×10^{27} tons, calculate the length of the semimajor axis of Phobos. Assume that the semimajor axis of Mars is 1.4×10^8 mi, the period of Mars 693.5 days, and the period of Phobos 0.3 day.
6. Use the information in Problem 5 to find the approximate mass of the Martian satellite Deimos if its elliptical orbit has a semimajor axis of 15,000 mi and a period of 1.26 days.
7. If the perihelion distance of Mercury is 2.9×10^7 mi and the aphelion distance is 4.3×10^7 mi, write the equation for the orbit of Mercury.
8. If the perihelion distance of Venus is 6.7234×10^7 mi and the aphelion distance is 6.8174×10^7 mi, write the equation for the orbit of Venus.
9. If the perihelion distance of earth is 9.225×10^7 mi and the aphelion distance is 9.542×10^7 mi, write the equation for the orbit of earth.
10. If the perihelion distance of Mars is 1.29×10^8 mi and the aphelion distance is 1.55×10^8 mi, write the equation for the orbit of Mars.

Algebraic (Symbolic) Manipulation and Graphing by Calculator

This appendix addresses itself to the task of comparing three different graphic calculators: the Casio fx-7000G,* the Sharp EL-5200, and the Hewlett-Packard HP-28S. In 1972, Hewlett-Packard announced and began selling what they called the HP-35 electronic calculator, the first such device available to the scientific community. The introductory price was $385. In the few years since its introduction we have seen its use become widespread and the price drop dramatically (to around $10). The new *NCTM Standards* for school mathematics makes the following assumption: "All students will have a calculator with functions consistent with the tasks envisioned in this curriculum. Calculators should include the following features: algebraic logic including order of operations; computation in decimal and common fraction form; constant function for addition, subtraction, multiplication, and division; and memory, percent, square root, exponent, reciprocal, and +/− keys." This first generation of calculators is now a permanent part of our curriculum, and the use of these calculators has been integrated into the writing and development of this book.

Today, we stand at the threshold of the next generation of calculators. These calculators are capable of graphing curves and doing algebraic manipulation. They will revolutionize the way mathematics is taught in the future. However, this new type of calculator is still too expensive, and too difficult to learn to use, for widespread use at the present time. When I was preparing this edition, I asked many of you if you wanted these calculators integrated into the text. I found that most of you want to know something about the new calculators, but you do not want them used or assumed in the text's development. Therefore, these calculators are not assumed in this book. However, I have included special boxes titled "HP-28S Activity" sprinkled throughout the book in order to give you some idea of how they might be used in a precalculus course. If you do not own one of these graphic calculators, you may be interested in knowing about the state of our present calculator technology. If you do own this type of calculator, you can use these boxes to help you work some of the problems in the book. These special boxes are, of course, optional. The remainder of this appendix will give you some insight into the types of problems you can solve with these calculators.

---

* The fx-7000G and fx-8000G operate in the same fashion. The 8000 has a printer interface and a little more memory.

Comparisons Among the Calculators

As you might imagine, there are basic and quite specific differences among the available calculators. The following table compares the salient features of each of the three graphic calculators discussed here. Features that are germane to this textbook are discussed (and compared) in more detail in the remainder of this appendix.

Feature	Hewlett-Packard HP-28S	Casio fx-8000G	Sharp EL-5200
Memory	32K	1.466K*	8K*
Notation used			
RPN	Yes	No	No
Algebraic	Yes	Yes	Yes
Built-in commands	250**	255**	194**
Graphics	Yes	Yes	Yes
Matrices and vectors	Yes	No	Yes
Polynomial solutions	Yes	No	Yes
Complex numbers	Yes	No	Yes
Statistics	Yes	Yes	Yes
Computer arithmetic	Yes	Yes	Yes
Calculus			
Differentiation	Yes	No	Yes
Integration	Yes	No	Yes
Series	Yes	No	Yes
Languages	Several	Casio	Sharp
Printer available	Yes	Yes	Yes
Programmable	Yes	Yes	Yes

* Although Casio and Sharp list their memory in Kbytes, this memory is based entirely upon the number of program steps, which they equate to bytes, and cannot be related to computer bytes.

** It is impossible to know what is really meant by this number. However, it usually, but not always, is a good indicator of calculator versatility.

Feature Comparison: Algebraic Notation

Most calculators, including the simple four-function calculators that perform the basic operations ($+, -, \times, \div$), use variations of what can be called the *algebraic calculator interface*. What is meant by this terminology? To evaluate the arithmetic expression $2 + 5 - 6$, you would press the keys labeled [2] [+] [5] [−] [6] [=] in the sequence just shown, thereby closely approximating the manner in which one would write an expression on a piece of paper.

This particular interface between the user and the calculator works well if we limit consideration to the four operators $+, -, \times, \div$. But most of us in technical fields need (or, at least, desire) calculators with more sophistication than the basic four-function calculator. The next higher level of sophistication allows an operator to use parentheses to make use of the principle we call the *hierarchy of enclosures and operations*. A further level of sophistication introduces *prefix functions* (sin, cos, log, etc.) and leads to two variations of algebraic calculators—ordinary and direct formula entry.

Ordinary algebraic calculators use a combination of styles. *Infix* operators (+, −, ×, ÷ that appear *between* factors in an expression) and *prefix* operators (sin, cos, log, etc., which appear as prefixes). For example,

$$5 + \sin 45°$$

contains the infix operator + and the prefix function sin. To evaluate this expression using an ordinary algebraic calculator, you would press

[5] [+] [45] [sin] [=]

Direct formula entry calculators allow the user to key in an entire expression in its algebraic form and then evaluate the expression by pressing some kind of termination key, such as [ENTER], [ENDLINE], [EVAL], or [EXE]. For the example shown above, you would press

[5] [+] [sin] [45] [EXE]

RPN Notation

A Polish logician, Jan Lukasiewicz, developed a mathematical logic now known as **Polish notation**. Whereas conventional algebraic notation places arithmetic operators *between* factors in the expression, Polish notation places the arithmetic operators *before* the factors. A variation of the logic developed by Lukasiewicz places the operators *after* the factors. This variation is called **Reverse Polish Notation**, or **RPN**.

The basic idea in RPN is to enter numbers or expressions in the calculator *first*, and then execute commands that act on the entries, which can now be called *arguments*. A stack-type calculator (such as the HP-28S) holds these entries in the sequence they were keyed into the device.

Given the expression 5 + sin 45, let us evaluate the expression on each calculator discussed here and see if we can determine the notation used by that calculator.

The Casio fx-7000G keystrokes are

[5] [+] [sin] [45] [EXE]

The expression has a value of 5.707106781.* Notation used is direct formula entry.

The Sharp EL-5200 keystrokes are

[5] [+] [sin] [45] [=]

yielding the result 5.70710678.*As with the Casio, notation is direct formula entry.

The HP-28S keystrokes can be made two ways: In RPN notation, the keystrokes are:

[5] [ENTER]	Stores 5 in stack register 1
[TRIG]	Selects trigonometric menu

* In reading the display of the three calculators, you will find that the Casio provides for 10-decimal numbers; the Sharp provides for 9-decimal numbers; and the HP provides for 12-decimal numbers.

[45] [sin]	Stores the value of sin 45 in stack register 1; the number 5 previously stored in stack register 1 moves to stack register 2
[+]	Adds the contents of stack registers 1 and 2

In algebraic notation, the keystrokes are:

['] [5] [+] [TRIG] [sin] [45] [EVAL]

which yields a value of 5.70710678119 in direct formula entry notation. The single quote symbol is required to indicate to the calculator that algebraic notation mode has been selected. The keystroke shown in color indicates a menu key.

When we compare the direct formula entry features of these three calculators, we discover that the HP-28S retains each result computed in a stack register, whereas each result obtained from the Casio or Sharp calculator must be recorded if we wish to refer to that result later. With a dynamic stack (limited in registers only by memory), a user can break up a large calculation into smaller subcalculations and retain the intermediate results—the essence of RPN arithmetic.

Graphics Capabilities: An Overview

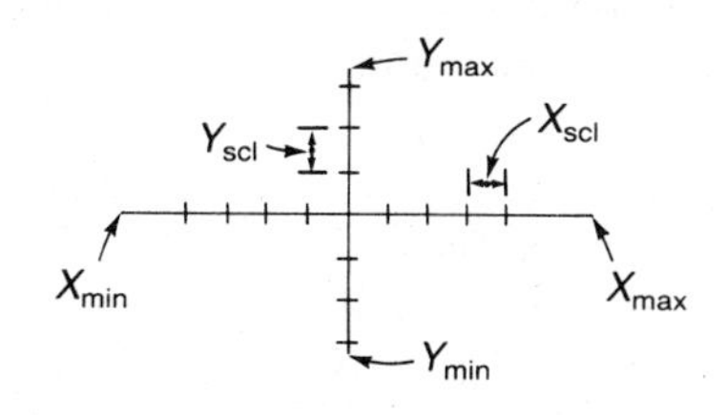

Each of the three calculators has default values for what we will call X_{min}, X_{max}, Y_{min}, Y_{max}, X_{scl}, and Y_{scl}. See the diagram in the margin for an explanation of each term listed. Close examination of the owner's manual accompanying each calculator will lend some reason to chosen values for the default ranges. As a prelude to reading the manual, we will look at the Casio default range. On this calculator the screen (a liquid crystal device, LCD) is comprised of a series of square pixels (dots). The graphing area is 95 pixels wide and 63 pixels high. The default values are $X_{min} = -4.7$, $X_{max} = 4.7$, $Y_{min} = -3.1$, $Y_{max} = 3.1$, $X_{scl} = 1$, and $Y_{scl} = 1$. Simply counting the pixels from one end of the graph axes to the other results in finding 94 pixels used along the horizontal axis and 62 pixels used on the vertical axis. Using the above default parameters, each tick mark interval along each axis is comprised of 10 pixels. Similar default values are used on the Sharp and Hewlett-Packard calculators. It is important to be aware that each calculator has resolution restrictions, but the important uses of graphing far outstrip the lack of resolution.

Each calculator can produce graphs of many mathematical expressions and statistical data. The list below includes those graphs that all three calculators can draw:

Straight lines
Trigonometric functions and their inverses
Hyperbolic functions and their inverses
Logarithmic expressions
Polynomial expressions
Single-variable statistical graphs
- Bar graphs
- Line graphs
- Normal distribution curves

Paired-variable statistical graphs
- Linear regression
- Data distribution
- Standard function

An example of graphing a polynomial function is shown on page 136.

Using the Graphs

In addition to providing a graphical representation of the algebraic expression, a user will find other uses for the graphics capabilities of these three calculators. The Casio and Sharp instruments (calculators *are* instruments) provide a TRACE procedure. With cursor keys, the user can literally trace the graph generated and determine, at any point along the graph, the (x, y) coordinates of that point. The Casio and Sharp calculators accomplish with TRACE what the HP-28S accomplishes in a bit more sophisticated manner—they *digitize* a point on the curve. The HP-28S provides a real digitizing function. With this function, the user may digitize points, redefine the corners of a graph, and redraw the graph to a larger or smaller scale. Sharp provides for enlarging and reducing the entire graph with a ZOOM command. A complete replot is necessary to enlarge or reduce a graph when using the Casio. By digitizing the points at which a graph crosses the x axis, we can determine solutions to (roots of) polynomials, maximums and minimums of graphs, and slopes of lines.

Another way to use the graphics capabilities of two of these calculators (the Sharp and HP) is to use their equation solving capabilities after a graph has been drawn. In effect, the user has only to push the key that invokes the equation solver routine to find zeros (Sharp and HP) or extrema (HP only). The Casio provides no such function.

EXAMPLE Find the roots for $y = 2x^2 - 4x + 3$. The HP-28S graphs the equation. Because the graph does not intersect the x axis, there are no real roots; that is, the roots are complex. The HP-28S returns the two roots $1 + 0.707106781188i$ and $1 - 0.707106781188i$. The Sharp will also graph the equation, but activating the equation solver yields the message NO SOLUTION. The Casio will graph the equation, but will not solve it. ■

An example showing the solution of a quadratic equation is shown on page 25.

Matrices and Vectors

The Sharp EL-5200 and the HP-28S are capable of performing matrix operations, although the EL-5200 is quite limited in scope. The Casio fx-7000G will perform no matrix operations. The table lists the operations available on the Sharp and Hewlett-Packard instruments. An asterisk to the right of an operation listed indicates that the calculator is capable of performing that operation.

Operation	HP-28S	EL-5200
Sum of matrices	*	*
Matrix multiplication	*	*
Inverse of a matrix	*	*
Transposition of a matrix	*	*
Determinant of a matrix	*	
Conjugate of a complex matrix	*	
Minor of a matrix	*	
Compute rank of a matrix	*	
Hermitian matrices	*	
Systems of linear equations	*	*
Vector orthogonality	*	
Vector length	*	
Normalization	*	
Eigenvalues from expansion	*	
Eigenvectors	*	
Eigenvalues from $\|\lambda I - A\|$	*	

Examples showing how to find the inverse of a matrix and how to solve systems of linear equations with matrices are shown on pages 356 and 361.

Polynomial Solutions

Earlier, we mentioned that by using the TRACE function on the Casio and Sharp calculators, we are able to find real roots of polynomial equations. Further, the EL-5200 performs an equation SOLVE routine after first graphing the equation in question. Neither of these calculators can solve equations directly. The HP-28S is *by far* the most versatile of the three when it comes to solving polynomial equations. The graphing, root-finding, and extrema-finding algorithms are very powerful. Further, it is not necessary to graph a function to find either roots or extrema.

An example showing the solution of a polynomial equation is shown on page 146.

Algebra, Geometry, and Trigonometry

In the area of algebra, the HP-28S is unique. It can be made to factor expressions, collect terms, solve for roots, solve for extrema (maxima and minima), do polynomial long division, operate on trigonometric functions of one or two angles, plot sums of trigonometric functions, solve trigonometric equations, prove trigonometric identities, plot equations in polar coordinates, evaluate and plot conic section curves (circle, ellipse, parabola, hyperbola), perform complex arithmetic, solve systems of linear equations (homogeneous or nonhomogeneous), and perform other myriad algebraic chores. The capabilities of the EL-5200 and fx-7000G, while quite extensive, are no match for the HP-28S.

An example showing how to convert forms of a complex number is shown on page 282.

B

Significant Digits

When you work with numbers that arise from counting, the numbers are **exact**. If you work with measurements, however, the quantities are necessarily **approximations**. The digits known to be correct in a number obtained by a measurement are called **significant digits**. The digits 1, 2, 3, 4, 5, 6, 7, 8, and 9 are always significant, whereas the digit 0 may or may not be significant, as described by the following two rules:

1. Zeros that come between two other digits are significant, as in 203 or 10.04.
2. If the only function of the zero is to place the decimal point, it is not significant, as in

$$\underset{\substack{\uparrow \\ \text{placeholders}}}{0.\underbrace{000}023} \quad \text{or} \quad \underset{\substack{\uparrow \\ \text{placeholders}}}{23,\underbrace{000}}$$

   If it does more than fix the decimal point, it is significant, as in

$$0.0023\underset{\substack{\uparrow \\ \text{This digit is significant.}}}{0} \quad \text{or} \quad 23,\underset{\substack{\uparrow \\ \text{These are significant since they come between two other digits.}}}{\underbrace{000.0}}1$$

This second rule can, of course, result in certain ambiguities, as in 23,000 (measured to the *exact* unit). To avoid confusion, we use scientific notation in this case:

2.3×10^4 two significant digits
2.3000×10^4 five significant digits

EXAMPLE 1 Two significant digits: 46, 0.00083, 4.0×10^1, 0.050. ■

EXAMPLE 2 Three significant digits: 523, 403, 4.00×10^2, 0.000800. ■

EXAMPLE 3 Four significant digits: 600.1, 4.000×10^1, 0.0002345. ■

When we are doing calculations with approximate numbers (particularly when using a calculator), it is often necessary to round off results.

Procedure for Rounding

To round off numbers:

1. Increase the last retained digit by 1 if the residue is 5 or greater.
2. Retain the last digit unchanged if the residue is less than 5.

Elaborate rules for calculating approximate data can be developed (when it is necessary for some applications, as in chemistry), but there are two simple rules that will work satisfactorily for the material in this book.

Significant-Digit Operations

ADDITION–SUBTRACTION: Add or subtract in the usual fashion, and then round off the result so that the last digit retained is in the column farthest to the right in which both given numbers have significant digits.

MULTIPLICATION–DIVISION: Multiply or divide in the usual fashion, and then round off the result to the smaller number of significant digits found in either of the given numbers.

These rules are particularly important when we are using a calculator, since the results obtained will look much more accurate on the calculator than they actually are. Suppose we want to calculate

$$b = \frac{50}{\tan 35^\circ}$$

1. By division: From Table C.III, $\tan 35^\circ \approx 0.7002$. Thus

$$b \approx \frac{50}{0.7002} = 71.408169090\ldots$$

2. By multiplication:

$$\begin{aligned} b = \frac{50}{\tan 35^\circ} &= 50 \cot 35^\circ \\ &\approx 50(1.4281) \qquad \text{from Table C.III} \\ &= 71.405 \end{aligned}$$

3. By calculator with algebraic logic (set to degrees): [50] [÷] [35] [tan] [=]

   By calculator with RPN logic: [50] [ENTER] [35] [tan] [÷]
   The answer is now displayed: `71.40740034`

$$b = \frac{50}{\tan 35^\circ} \approx 71.40740034$$

Notice that all these answers differ. Now use the multiplication–division rule.

50: This number has one or two significant digits; there is no ambiguity if we write 5×10^1 or 5.0×10^1. If the given data include a number with a doubtful degree of accuracy, *in this book we will assume the maximum degree of accuracy*. Thus 50 has two significant digits.

tan 35°: From Table C.III we find this number to four significant digits; on a calculator you may have 8, 10, or 12 significant digits (depending on the calculator).

The result of this division is correct to two significant digits:

$$b = \frac{50}{\tan 35^\circ} = 71$$

which agrees with all the preceding methods of solution.

In solving triangles in this text, we will assume a certain relationship in the accuracy of the measurement between the sides and the angles.

Significant Digits in Solving Triangles

Accuracy in Sides	Accuracy in Angles
Two significant digits	Nearest degree
Three significant digits	Nearest tenth of a degree
Four significant digits	Nearest hundredth of a degree

This chart means that if the data include one side given with two significant digits and another with three significant digits, the angle would be computed to the nearest degree. If one side is given to four significant digits and an angle to the nearest tenth of a degree, the other sides would be given to three significant digits and the angles computed to the nearest tenth of a degree. In general, results should not be more accurate than the least accurate item of the given data.

If you have access only to a four-function calculator, you can use Table C.III in conjunction with your calculator. For example, to find b, you first find

$$\tan 35^\circ \approx 0.7002$$

and then calculate

$$b \approx \frac{50}{0.7002}$$ Algebraic logic: [50] [÷] [0.7002] [=]

$$\approx 71.41$$ RPN logic: [50] [ENTER] [0.7002] [÷]

or, to two significant digits, $b = 71$. Of course, one of the great advantages of calculators is that they enable us to work with much greater accuracy without having to use interpolation or tables. See Problems 16–22 of Problem Set 5.3 for some practice problems with significant digits.

Tables

TABLE C.1
Powers of e

x	e^x	e^{-x}	x	e^x	e^{-x}	x	e^x	e^{-x}
0.00	1.000	1.000	0.50	1.649	0.607	1.00	2.718	0.368
0.01	1.010	0.990	0.51	1.665	0.600	1.01	2.746	0.364
0.02	1.020	0.980	0.52	1.682	0.595	1.02	2.773	0.361
0.03	1.031	0.970	0.53	1.699	0.589	1.03	2.801	0.357
0.04	1.041	0.961	0.54	1.716	0.583	1.04	2.829	0.353
0.05	1.051	0.951	0.55	1.733	0.577	1.05	2.858	0.350
0.06	1.062	0.942	0.56	1.751	0.571	1.06	2.886	0.346
0.07	1.073	0.932	0.57	1.768	0.566	1.07	2.915	0.343
0.08	1.083	0.923	0.58	1.786	0.560	1.08	2.945	0.340
0.09	1.094	0.914	0.59	1.804	0.554	1.09	2.974	0.336
0.10	1.105	0.905	0.60	1.822	0.549	1.10	3.004	0.333
0.11	1.116	0.896	0.61	1.840	0.543	1.11	3.034	0.330
0.12	1.127	0.887	0.62	1.859	0.538	1.12	3.065	0.326
0.13	1.139	0.878	0.63	1.878	0.533	1.13	3.096	0.323
0.14	1.150	0.869	0.64	1.896	0.527	1.14	3.127	0.320
0.15	1.162	0.861	0.65	1.916	0.522	1.15	3.158	0.317
0.16	1.174	0.852	0.66	1.935	0.517	1.16	3.190	0.313
0.17	1.185	0.844	0.67	1.954	0.512	1.17	3.222	0.310
0.18	1.197	0.835	0.68	1.974	0.507	1.18	3.254	0.307
0.19	1.209	0.827	0.69	1.994	0.502	1.19	3.287	0.304
0.20	1.221	0.819	0.70	2.014	0.497	1.20	3.320	0.301
0.21	1.234	0.811	0.71	2.034	0.492	1.21	3.353	0.298
0.22	1.246	0.803	0.72	2.054	0.487	1.22	3.387	0.295
0.23	1.259	0.795	0.73	2.075	0.482	1.23	3.421	0.292
0.24	1.271	0.787	0.74	2.096	0.477	1.24	3.456	0.289
0.25	1.284	0.779	0.75	2.117	0.472	1.25	3.490	0.287
0.26	1.297	0.771	0.76	2.138	0.468	1.26	3.525	0.284
0.27	1.310	0.763	0.77	2.160	0.463	1.27	3.561	0.281
0.28	1.323	0.756	0.78	2.182	0.458	1.28	3.597	0.278
0.29	1.336	0.748	0.79	2.203	0.454	1.29	3.633	0.275
0.30	1.350	0.741	0.80	2.226	0.449	1.30	3.669	0.273
0.31	1.363	0.733	0.81	2.248	0.445	1.31	3.706	0.270
0.32	1.377	0.726	0.82	2.270	0.440	1.32	3.743	0.267
0.33	1.391	0.719	0.83	2.293	0.436	1.33	3.781	0.264
0.34	1.405	0.712	0.84	2.316	0.432	1.34	3.819	0.262
0.35	1.419	0.705	0.85	2.340	0.427	1.35	3.857	0.259
0.36	1.433	0.698	0.86	2.363	0.423	1.36	3.896	0.257
0.37	1.448	0.691	0.87	2.387	0.419	1.37	3.935	0.254
0.38	1.462	0.684	0.88	2.441	0.415	1.38	3.975	0.252
0.39	1.477	0.677	0.89	2.435	0.411	1.39	4.015	0.249
0.40	1.492	0.670	0.90	2.460	0.407	1.40	4.055	0.247
0.41	1.507	0.664	0.91	2.484	0.403	1.41	4.096	0.244
0.42	1.522	0.657	0.92	2.509	0.399	1.42	4.137	0.242
0.43	1.537	0.651	0.93	2.535	0.395	1.43	4.179	0.239
0.44	1.553	0.644	0.94	2.560	0.391	1.44	4.221	0.237
0.45	1.568	0.638	0.95	2.586	0.387	1.45	4.263	0.235
0.46	1.584	0.631	0.96	2.612	0.383	1.46	4.306	0.232
0.47	1.600	0.625	0.97	2.638	0.379	1.47	4.349	0.230
0.48	1.616	0.619	0.98	2.664	0.375	1.48	4.393	0.228
0.49	1.632	0.613	0.99	2.691	0.372	1.49	4.437	0.225

TABLE C.I
Powers of e
(continued)

x	e^x	e^{-x}	x	e^x	e^{-x}	x	e^x	e^{-x}
1.50	4.482	0.223	2.00	7.389	0.135	2.50	12.182	0.082
1.51	4.527	0.221	2.01	7.463	0.134	2.51	12.305	0.081
1.52	4.572	0.219	2.02	7.538	0.133	2.52	12.429	0.080
1.53	4.618	0.217	2.03	7.614	0.131	2.53	12.554	0.080
1.54	4.665	0.214	2.04	7.691	0.130	2.54	12.680	0.079
1.55	4.712	0.212	2.05	7.768	0.129	2.55	12.807	0.078
1.56	4.759	0.210	2.06	7.846	0.127	2.56	12.936	0.077
1.57	4.807	0.208	2.07	7.925	0.126	2.57	13.066	0.077
1.58	4.855	0.206	2.08	8.004	0.125	2.58	13.197	0.076
1.59	4.904	0.204	2.09	8.085	0.124	2.59	13.330	0.075
1.60	4.953	0.202	2.10	8.166	0.122	2.60	13.464	0.074
1.61	5.003	0.200	2.11	8.248	0.121	2.61	13.599	0.074
1.62	5.053	0.198	2.12	8.331	0.120	2.62	13.736	0.073
1.63	5.104	0.196	2.13	8.415	0.119	2.63	13.874	0.072
1.64	5.155	0.194	2.14	8.499	0.118	2.64	14.013	0.071
1.65	5.207	0.192	2.15	8.585	0.116	2.65	14.154	0.071
1.66	5.259	0.190	2.16	8.671	0.115	2.66	14.296	0.070
1.67	5.312	0.188	2.17	8.758	0.114	2.67	14.440	0.069
1.68	5.366	0.186	2.18	8.846	0.113	2.68	14.585	0.069
1.69	5.420	0.185	2.19	8.935	0.112	2.69	14.732	0.068
1.70	5.474	0.183	2.20	9.025	0.111	2.70	14.880	0.067
1.71	5.529	0.181	2.21	9.116	0.110	2.71	15.029	0.067
1.72	5.585	0.179	2.22	9.207	0.109	2.72	15.180	0.066
1.73	5.641	0.177	2.23	9.300	0.108	2.73	15.333	0.065
1.74	5.697	0.176	2.24	9.393	0.106	2.74	15.487	0.065
1.75	5.755	0.174	2.25	9.488	0.105	2.75	15.643	0.064
1.76	5.812	0.172	2.26	9.583	0.104	2.76	15.800	0.063
1.77	5.871	0.170	2.27	9.679	0.103	2.77	15.959	0.063
1.78	5.930	0.169	2.28	9.777	0.102	2.78	16.119	0.062
1.79	5.989	0.167	2.29	9.875	0.101	2.79	16.281	0.061
1.80	6.050	0.165	2.30	9.974	0.100	2.80	16.445	0.061
1.81	6.110	0.164	2.31	10.074	0.099	2.81	16.610	0.060
1.82	6.172	0.162	2.32	10.176	0.098	2.82	16.777	0.060
1.83	6.234	0.160	2.33	10.278	0.097	2.83	16.945	0.059
1.84	6.297	0.159	2.34	10.381	0.096	2.84	17.116	0.058
1.85	6.360	0.157	2.35	10.486	0.095	2.85	17.288	0.058
1.86	6.424	0.156	2.36	10.591	0.094	2.86	17.462	0.057
1.87	6.488	0.154	2.37	10.697	0.093	2.87	17.637	0.057
1.88	6.553	0.153	2.38	10.805	0.093	2.88	17.814	0.056
1.89	6.619	0.151	2.39	10.913	0.092	2.89	17.993	0.056
1.90	6.686	0.150	2.40	11.023	0.091	2.90	18.174	0.055
1.91	6.753	0.148	2.41	11.134	0.090	2.91	18.357	0.054
1.92	6.821	0.147	2.42	11.246	0.089	2.92	18.541	0.054
1.93	6.890	0.145	2.43	11.359	0.088	2.93	18.728	0.053
1.94	6.959	0.144	2.44	11.473	0.087	2.94	18.916	0.053
1.95	7.029	0.142	2.45	11.588	0.086	2.95	19.016	0.052
1.96	7.099	0.141	2.46	11.705	0.085	2.96	19.298	0.052
1.97	7.171	0.139	2.47	11.822	0.085	2.97	19.492	0.051
1.98	7.243	0.138	2.48	11.941	0.084	2.98	19.688	0.051
1.99	7.316	0.137	2.49	12.061	0.083	2.99	19.886	0.050
						3.00	20.086	0.050

TABLE C.II
Common logarithms

N	0	1	2	3	4	5	6	7	8	9
1.0	.0000	.0043	.0086	.0128	.0170	.0212	.0253	.0294	.0334	.0374
1.1	.0414	.0453	.0492	.0531	.0569	.0607	.0645	.0682	.0719	.0755
1.2	.0792	.0828	.0864	.0899	.0934	.0969	.1004	.1038	.1072	.1106
1.3	.1139	.1173	.1206	.1239	.1271	.1303	.1335	.1367	.1399	.1430
1.4	.1461	.1492	.1523	.1553	.1584	.1614	.1644	.1673	.1703	.1732
1.5	.1761	.1790	.1818	.1847	.1875	.1903	.1931	.1959	.1987	.2014
1.6	.2041	.2068	.2095	.2122	.2148	.2175	.2201	.2227	.2253	.2279
1.7	.2304	.2330	.2355	.2380	.2405	.2430	.2455	.2480	.2504	.2529
1.8	.2553	.2577	.2601	.2625	.2648	.2672	.2695	.2718	.2742	.2765
1.9	.2788	.2810	.2833	.2856	.2878	.2900	.2923	.2945	.2967	.2989
2.0	.3010	.3032	.3054	.3075	.3096	.3118	.3139	.3160	.3181	.3201
2.1	.3222	.3243	.3263	.3284	.3304	.3324	.3345	.3365	.3385	.3404
2.2	.3424	.3444	.3464	.3483	.3502	.3522	.3541	.3560	.3579	.3598
2.3	.3617	.3636	.3655	.3674	.3692	.3711	.3729	.3747	.3766	.3784
2.4	.3802	.3820	.3838	.3856	.3874	.3892	.3909	.3927	.3945	.3962
2.5	.3979	.3997	.4014	.4031	.4048	.4065	.4082	.4099	.4116	.4133
2.6	.4150	.4166	.4183	.4200	.4216	.4232	.4249	.4265	.4281	.4298
2.7	.4314	.4330	.4346	.4362	.4378	.4393	.4409	.4425	.4440	.4456
2.8	.4472	.4487	.4502	.4518	.4533	.4548	.4564	.4579	.4594	.4609
2.9	.4624	.4639	.4654	.4669	.4683	.4698	.4713	.4728	.4742	.4757
3.0	.4771	.4786	.4800	.4814	.4829	.4843	.4857	.4871	.4886	.4900
3.1	.4914	.4928	.4942	.4955	.4969	.4983	.4997	.5011	.5024	.5038
3.2	.5051	.5065	.5079	.5092	.5105	.5119	.5132	.5145	.5159	.5172
3.3	.5185	.5198	.5211	.5224	.5237	.5250	.5263	.5276	.5289	.5302
3.4	.5315	.5328	.5340	.5353	.5366	.5378	.5391	.5403	.5416	.5428
3.5	.5441	.5453	.5465	.5478	.5490	.5502	.5514	.5527	.5539	.5551
3.6	.5563	.5575	.5587	.5599	.5611	.5623	.5635	.5647	.5658	.5670
3.7	.5682	.5694	.5705	.5717	.5729	.5740	.5752	.5763	.5775	.5786
3.8	.5798	.5809	.5821	.5832	.5843	.5855	.5866	.5877	.5888	.5899
3.9	.5911	.5922	.5933	.5944	.5955	.5966	.5977	.5988	.5999	.6010
4.0	.6021	.6031	.6042	.6053	.6064	.6075	.6085	.6096	.6107	.6117
4.1	.6128	.6138	.6149	.6160	.6170	.6180	.6191	.6201	.6212	.6222
4.2	.6232	.6243	.6253	.6263	.6274	.6284	.6294	.6304	.6314	.6325
4.3	.6335	.6345	.6355	.6365	.6375	.6385	.6395	.6405	.6415	.6425
4.4	.6435	.6444	.6454	.6464	.6474	.6484	.6493	.6503	.6513	.6522
4.5	.6532	.6542	.6551	.6561	.6571	.6580	.6590	.6599	.6609	.6618
4.6	.6628	.6637	.6646	.6656	.6665	.6675	.6684	.6693	.6702	.6712
4.7	.6721	.6730	.6739	.6749	.6758	.6767	.6776	.6785	.6794	.6803
4.8	.6812	.6821	.6830	.6839	.6848	.6857	.6866	.6875	.6884	.6893
4.9	.6902	.6911	.6920	.6928	.6937	.6946	.6955	.6964	.6972	.6981
5.0	.6990	.6998	.7007	.7016	.7024	.7033	.7042	.7050	.7059	.7067
5.1	.7076	.7084	.7093	.7101	.7110	.7118	.7126	.7135	.7143	.7152
5.2	.7160	.7168	.7177	.7185	.7193	.7202	7210	.7218	.7226	.7235
5.3	.7243	.7251	.7259	.7267	.7275	.7284	.7292	.7300	.7308	.7316
5.4	.7324	.7332	.7340	.7348	.7356	.7364	.7372	.7380	.7388	.7396
N	0	1	2	3	4	5	6	7	8	9

TABLE C.II
Common logarithms
(continued)

N	0	1	2	3	4	5	6	7	8	9
5.5	.7404	.7412	.7419	.7427	.7435	.7443	.7451	.7459	.7466	.7474
5.6	.7482	.7490	.7497	.7505	.7513	.7520	.7528	.7536	.7543	.7551
5.7	.7559	.7566	.7574	.7582	.7589	.7597	.7604	.7612	.7619	.7627
5.8	.7634	.7642	.7649	.7657	.7664	.7672	.7679	.7686	.7694	.7701
5.9	.7709	.7716	.7723	.7731	.7738	.7745	.7752	.7760	.7767	.7774
6.0	.7782	.7789	.7796	.7803	.7810	.7818	.7825	.7832	.7839	.7846
6.1	.7853	.7860	.7868	.7875	.7882	.7889	.7896	.7903	.7910	.7917
6.2	.7924	.7931	.7938	.7945	.7952	.7959	.7966	.7973	.7980	.7987
6.3	.7993	.8000	.8007	.8014	.8021	.8028	.8035	.8041	.8048	.8055
6.4	.8062	.8069	.8075	.8082	.8089	.8096	.8102	.8109	.8116	.8122
6.5	.8129	.8136	.8142	.8149	.8156	.8162	.8169	.8176	.8182	.8189
6.6	.8195	.8202	.8209	.8215	.8222	.8228	.8235	.8241	.8248	.8254
6.7	.8261	.8267	.8274	.8280	.8287	.8293	.8299	.8306	.8312	.8319
6.8	.8325	.8331	.8338	.8344	.8351	.8357	.8363	.8370	.8376	.8382
6.9	.8388	.8395	.8401	.8407	.8414	.8420	.8426	.8432	.8439	.8445
7.0	.8451	.8457	.8463	.8470	.8476	.8482	.8488	.8494	.8500	.8506
7.1	.8513	.8519	.8525	.8531	.8537	.8543	.8549	.8555	.8561	.8567
7.2	.8573	.8579	.8585	.8591	.8597	.8603	.8609	.8615	.8621	.8627
7.3	.8633	.8639	.8645	.8651	.8657	.8663	.8669	.8675	.8681	.8686
7.4	.8692	.8698	.8704	.8710	.8716	.8722	.8727	.8733	.8739	.8745
7.5	.8751	.8756	.8762	.8768	.8774	.8779	.8785	.8791	.8797	.8802
7.6	.8808	.8814	.8820	.8825	.8831	.8837	.8842	.8848	.8854	.8859
7.7	.8865	.8871	.8876	.8882	.8887	.8893	.8899	.8904	.8910	.8915
7.8	.8921	.8927	.8932	.8938	.8943	.8949	.8954	.8960	.8965	.8971
7.9	.8976	.8982	.8987	.8993	.8998	.9004	.9009	.9015	.9020	.9025
8.0	.9031	.9036	.9042	.9047	.9053	.9058	.9063	.9069	.9074	.9079
8.1	.9085	.9090	.9096	.9101	.9106	.9112	.9117	.9122	.9128	.9133
8.2	.9138	.9143	.9149	.9154	.9159	.9165	.9170	.9175	.9180	.9186
8.3	.9191	.9196	.9201	.9206	.9212	.9217	.9222	.9227	.9232	.9238
8.4	.9243	.9248	.9253	.9258	.9263	.9269	.9274	.9279	.9284	.9289
8.5	.9294	.9299	.9304	.9309	.9315	.9320	.9325	.9330	.9335	.9340
8.6	.9345	.9350	.9355	.9360	.9365	.9370	.9375	.9380	.9385	.9390
8.7	.9395	.9400	.9405	.9410	.9415	.9420	.9425	.9430	.9435	.9440
8.8	.9445	.9450	.9455	.9460	.9465	.9469	.9474	.9479	.9484	.9489
8.9	.9494	.9499	.9504	.9509	.9513	.9518	.9523	.9528	.9533	.9538
9.0	.9542	.9547	.9552	.9557	.9562	.9566	.9571	.9576	.9581	.9586
9.1	.9590	.9595	.9600	.9605	.9609	.9614	.9619	.9624	.9628	.9633
9.2	.9638	.9643	.9647	.9652	.9657	.9661	.9666	.9671	.9675	.9680
9.3	.9685	.9689	.9694	.9699	.9703	.9708	.9713	.9717	.9722	.9727
9.4	.9731	.9736	.9741	.9745	.9750	.9754	.9759	.9763	.9768	.9773
9.5	.9777	.9782	.9786	.9791	.9795	.9800	.9805	.9809	.9814	.9818
9.6	.9823	.9827	.9832	.9836	.9841	.9845	.9850	.9854	.9859	.9863
9.7	.9868	.9872	.9877	.9881	.9886	.9890	.9894	.9899	.9903	.9908
9.8	.9912	.9917	.9921	.9926	.9930	.9934	.9939	.9943	.9948	.9952
9.9	.9956	.9961	.9965	.9969	.9974	.9978	.9983	.9987	.9991	.9996
N	0	1	2	3	4	5	6	7	8	9

TABLE C.III Trigonometric functions

Rad	Deg	cos	sin	tan		
.00	**.0**	1.0000	.0000	.0000	**90.0**	1.57
.00	.1	1.0000	.0017	.0017	89.9	1.57
.00	.2	1.0000	.0035	.0035	89.8	1.57
.01	.3	1.0000	.0052	.0052	89.7	1.57
.01	.4	1.0000	.0070	.0070	89.6	1.56
.01	**.5**	1.0000	.0087	.0087	**89.5**	1.56
.01	.6	.9999	.0105	.0105	89.4	1.56
.01	.7	.9999	.0122	.0122	89.3	1.56
.01	.8	.9999	.0140	.0140	89.2	1.56
.02	.9	.9999	.0157	.0157	89.1	1.56
.02	**1.0**	.9998	.0175	.0175	**89.0**	1.55
.02	1.1	.9998	.0192	.0192	88.9	1.55
.02	1.2	.9998	.0209	.0209	88.8	1.55
.02	1.3	.9997	.0227	.0227	88.7	1.55
.02	1.4	.9997	.0244	.0244	88.6	1.55
.03	**1.5**	.9997	.0262	.0262	**88.5**	1.54
.03	1.6	.9996	.0279	.0279	88.4	1.54
.03	1.7	.9996	.0297	.0297	88.3	1.54
.03	1.8	.9995	.0314	.0314	88.2	1.54
.03	1.9	.9995	.0332	.0332	88.1	1.54
.03	**2.0**	.9994	.0349	.0349	**88.0**	1.54
.04	2.1	.9993	.0366	.0367	87.9	1.53
.04	2.2	.9993	.0384	.0384	87.8	1.53
.04	2.3	.9992	.0401	.0402	87.7	1.53
.04	2.4	.9991	.0419	.0419	87.6	1.53
.04	**2.5**	.9990	.0436	.0437	**87.5**	1.53
.05	2.6	.9990	.0454	.0454	87.4	1.53
.05	2.7	.9989	.0471	.0472	87.3	1.52
.05	2.8	.9988	.0488	.0489	87.2	1.52
.05	2.9	.9987	.0506	.0507	87.1	1.52
.05	**3.0**	.9986	.0523	.0524	**87.0**	1.52
.05	3.1	.9985	.0541	.0542	86.9	1.52
.06	3.2	.9984	.0558	.0559	86.8	1.51
.06	3.3	.9983	.0576	.0577	86.7	1.51
.06	3.4	.9982	.0593	.0594	86.6	1.51
.06	**3.5**	.9981	.0610	.0612	**86.5**	1.51
.06	3.6	.9980	.0628	.0629	86.4	1.51
.06	3.7	.9979	.0645	.0647	86.3	1.51
.07	3.8	.9978	.0663	.0664	86.2	1.50
.07	3.9	.9977	.0680	.0682	86.1	1.50
.07	**4.0**	.9976	.0698	.0699	**86.0**	1.50
.07	4.1	.9974	.0715	.0717	85.9	1.50
.07	4.2	.9973	.0732	.0734	85.8	1.50
.08	4.3	.9972	.0750	.0752	85.7	1.50
.08	4.4	.9971	.0767	.0769	85.6	1.49
.08	**4.5**	.9969	.0785	.0787	**85.5**	1.49
.08	4.6	.9968	.0802	.0805	85.4	1.49
.08	4.7	.9966	.0819	.0822	85.3	1.49
.08	4.8	.9965	.0837	.0840	85.2	1.49
.09	4.9	.9963	.0854	.0857	85.1	1.49
.09	**5.0**	.9962	.0872	.0875	**85.0**	1.49
		sin	cos	cot	Deg	Rad

Rad	Deg	cos	sin	tan		
.09	**5.0**	.9962	.0872	.0875	**85.0**	1.48
.09	5.1	.9960	.0889	.0892	84.9	1.48
.09	5.2	.9959	.0906	.0910	84.8	1.48
.09	5.3	.9957	.0924	.0928	84.7	1.48
.09	5.4	.9956	.0941	.0945	84.6	1.48
.10	**5.5**	.9954	.0958	.0963	**84.5**	1.47
.10	5.6	.9952	.0976	.0981	84.4	1.47
.10	5.7	.9951	.0993	.0998	84.3	1.47
.10	5.8	.9949	.1011	.1016	84.2	1.47
.10	5.9	.9947	.1028	.1033	84.1	1.47
.10	**6.0**	.9945	.1045	.1051	**84.0**	1.47
.11	6.1	.9943	.1063	.1069	83.9	1.46
.11	6.2	.9942	.1080	.1086	83.8	1.46
.11	6.3	.9940	.1097	.1104	83.7	1.46
.11	6.4	.9938	.1115	.1122	83.6	1.46
.11	**6.5**	.9936	.1132	.1139	**83.5**	1.46
.12	6.6	.9934	.1149	.1157	83.4	1.46
.12	6.7	.9932	.1167	.1175	83.3	1.45
.12	6.8	.9930	.1184	.1192	83.2	1.45
.12	6.9	.9928	.1201	.1210	83.1	1.45
.12	**7.0**	.9925	.1219	.1228	**83.0**	1.45
.12	7.1	.9923	.1236	.1246	82.9	1.45
.13	7.2	.9921	.1253	.1263	82.8	1.45
.13	7.3	.9919	.1271	.1281	82.7	1.44
.13	7.4	.9917	.1288	.1299	82.6	1.44
.13	**7.5**	.9914	.1305	.1317	**82.5**	1.44
.13	7.6	.9912	.1323	.1334	82.4	1.44
.13	7.7	.9910	.1340	.1352	82.3	1.44
.14	7.8	.9907	.1357	.1370	82.2	1.43
.14	7.9	.9905	.1374	.1388	82.1	1.43
.14	**8.0**	.9903	.1392	.1405	**82.0**	1.43
.14	8.1	.9900	.1409	.1423	81.9	1.43
.14	8.2	.9898	.1426	.1441	81.8	1.43
.14	8.3	.9895	.1444	.1459	81.7	1.43
.15	8.4	.9893	.1461	.1477	81.6	1.42
.15	**8.5**	.9890	.1478	.1495	**81.5**	1.42
.15	8.6	.9888	.1495	.1512	81.4	1.42
.15	8.7	.9885	.1513	.1530	81.3	1.42
.15	8.8	.9882	.1530	.1548	81.2	1.42
.16	8.9	.9880	.1547	.1566	81.1	1.42
.16	**9.0**	.9877	.1564	.1584	**81.0**	1.41
.16	9.1	.9874	.1582	.1602	80.9	1.41
.16	9.2	.9871	.1599	.1620	80.8	1.41
.16	9.3	.9869	.1616	.1638	80.7	1.41
.16	9.4	.9866	.1633	.1655	80.6	1.41
.17	**9.5**	.9863	.1650	.1673	**80.5**	1.40
.17	9.6	.9860	.1668	.1691	80.4	1.40
.17	9.7	.9857	.1685	.1709	80.3	1.40
.17	9.8	.9854	.1702	.1727	80.2	1.40
.17	9.9	.9851	.1719	.1745	80.1	1.40
.17	**10.0**	.9848	.1736	.1763	**80.0**	1.40
		sin	cos	cot	Deg	Rad

TABLE C.III Trigonometric functions (continued)

Rad	Deg	cos	sin	tan		
.17	**10.0**	.9848	.1736	.1763	**80.0**	1.40
.18	10.1	.9845	.1754	.1781	79.9	1.39
.18	10.2	.9842	.1771	.1799	79.8	1.39
.18	10.3	.9839	.1788	.1817	79.7	1.39
.18	10.4	.9836	.1805	.1835	79.6	1.39
.18	**10.5**	.9833	.1822	.1853	**79.5**	1.39
.19	10.6	.9829	.1840	.1871	79.4	1.39
.19	10.7	.9826	.1857	.1890	79.3	1.38
.19	10.8	.9823	.1874	.1908	79.2	1.38
.19	10.9	.9820	.1891	.1926	79.1	1.38
.19	**11.0**	.9816	.1908	.1944	**79.0**	1.38
.19	11.1	.9813	.1925	.1962	78.9	1.38
.20	11.2	.9810	.1942	.1980	78.8	1.38
.20	11.3	.9806	.1959	.1998	78.7	1.37
.20	11.4	.9803	.1977	.2016	78.6	1.37
.20	**11.5**	.9799	.1994	.2035	**78.5**	1.37
.20	11.6	.9796	.2011	.2053	78.4	1.37
.20	11.7	.9792	.2028	.2071	78.3	1.37
.21	11.8	.9789	.2045	.2089	78.2	1.36
.21	11.9	.9785	.2062	.2107	78.1	1.36
.21	**12.0**	.9871	.2079	.2126	**78.0**	1.36
.21	12.1	.9778	.2096	.2144	77.9	1.36
.21	12.2	.9774	.2113	.2162	77.8	1.36
.21	12.3	.9770	.2130	.2180	77.7	1.36
.22	12.4	.9767	.2147	.2199	77.6	1.35
.22	**12.5**	.9763	.2164	.2217	**77.5**	1.35
.22	12.6	.9759	.2181	.2235	77.4	1.35
.22	12.7	.9755	.2198	.2254	77.3	1.35
.22	12.8	.9751	.2215	.2272	77.2	1.35
.23	12.9	.9748	.2233	.2290	77.1	1.35
.23	**13.0**	.9744	.2250	.2309	**77.0**	1.34
.23	13.1	.9740	.2267	.2327	76.9	1.34
.23	13.2	.9736	.2284	.2345	76.8	1.34
.23	13.3	.9732	.2300	.2364	76.7	1.34
.23	13.4	.9728	.2317	.2382	76.6	1.34
.24	**13.5**	.9724	.2334	.2401	**76.5**	1.34
.24	13.6	.9720	.2351	.2419	76.4	1.33
.24	13.7	.9715	.2368	.2438	76.3	1.33
.24	13.8	.9711	.2385	.2456	76.2	1.33
.24	13.9	.9707	.2402	.2475	76.1	1.33
.24	**14.0**	.9703	.2419	.2493	**76.0**	1.33
.25	14.1	.9699	.2436	.2512	75.9	1.32
.25	14.2	.9694	.2453	.2530	75.8	1.32
.25	14.3	.9690	.2470	.2549	75.7	1.32
.25	14.4	.9686	.2487	.2568	75.6	1.32
.25	**14.5**	.9681	.2504	.2586	**75.5**	1.32
.25	14.6	.9677	.2521	.2605	75.4	1.32
.26	14.7	.9673	.2538	.2623	75.3	1.31
.26	14.8	.9668	.2554	.2642	75.2	1.31
.26	14.9	.9664	.2571	.2661	75.1	1.31
.26	**15.0**	.9659	.2588	.2679	**75.0**	1.31
		sin	cos	cot	Deg	Rad

Rad	Deg	cos	sin	tan		
.26	**15.0**	.9659	.2588	.2679	**75.0**	1.31
.26	15.1	.9655	.2605	.2698	74.9	1.31
.27	15.2	.9650	.2622	.2717	74.8	1.31
.27	15.3	.9646	.2639	.2736	74.7	1.30
.27	15.4	.9641	.2656	.2754	74.6	1.30
.27	**15.5**	.9636	.2672	.2773	**74.5**	1.30
.27	15.6	.9632	.2689	.2792	74.4	1.30
.27	15.7	.9627	.2706	.2811	74.3	1.30
.28	15.8	.9622	.2723	.2830	74.2	1.30
.28	15.9	.9617	.2740	.2849	74.1	1.29
.28	**16.0**	.9613	.2756	.2867	**74.0**	1.29
.28	16.1	.9608	.2773	.2886	73.9	1.29
.28	16.2	.9603	.2790	.2905	73.8	1.29
.28	16.3	.9598	.2807	.2924	73.7	1.29
.29	16.4	.9593	.2823	.2943	73.6	1.28
.29	**16.5**	.9588	.2840	.2962	**73.5**	1.28
.29	16.6	.9583	.2857	.2981	73.4	1.28
.29	16.7	.9578	.2874	.3000	73.3	1.28
.29	16.8	.9573	.2890	.3019	73.2	1.28
.29	16.9	.9568	.2907	.3038	73.1	1.28
.30	**17.0**	.9563	.2924	.3057	**73.0**	1.27
.30	17.1	.9558	.2940	.3076	72.9	1.27
.30	17.2	.9553	.2957	.3096	72.8	1.27
.30	17.3	.9548	.2974	.3115	72.7	1.27
.30	17.4	.9542	.2990	.3134	72.6	1.27
.31	**17.5**	.9537	.3007	.3153	**72.5**	1.27
.31	17.6	.9532	.3024	.3172	72.4	1.26
.31	17.7	.9527	.3040	.3191	72.3	1.26
.31	17.8	.9521	.3057	.3211	72.2	1.26
.31	17.9	.9516	.3074	.3230	72.1	1.26
.31	**18.0**	.9511	.3090	.3249	**72.0**	1.26
.32	18.1	.9505	.3107	.3269	71.9	1.25
.32	18.2	.9500	.3123	.3288	71.8	1.25
.32	18.3	.9494	.3140	.3307	71.7	1.25
.32	18.4	.9489	.3156	.3327	71.6	1.25
.32	**18.5**	.9483	.3173	.3346	**71.5**	1.25
.32	18.6	.9478	.3190	.3365	71.4	1.25
.33	18.7	.9472	.3206	.3385	71.3	1.24
.33	18.8	.9466	.3223	.3404	71.2	1.24
.33	18.9	.9461	.3239	.3424	71.1	1.24
.33	**19.0**	.9455	.3256	.3443	**71.0**	1.24
.33	19.1	.9449	.3272	.3463	70.9	1.24
.34	19.2	.9444	.3289	.3482	70.8	1.24
.34	19.3	.9438	.3305	.3502	70.7	1.23
.34	19.4	.9432	.3322	.3522	70.6	1.23
.34	**19.5**	.9426	.3338	.3541	**70.5**	1.23
.34	19.6	.9421	.3355	.3561	70.4	1.23
.34	19.7	.9415	.3371	.3581	70.3	1.23
.35	19.8	.9409	.3387	.3600	70.2	1.23
.35	19.9	.9403	.3404	.3620	70.1	1.22
.35	**20.0**	.9397	.3420	.3640	**70.0**	1.22
		sin	cos	cot	Deg	Rad

TABLE C.III Trigonometric functions (continued)

Rad	Deg	cos	sin	tan		
.35	**20.0**	.9397	.3420	.3640	**70.0**	1.22
.35	20.1	.9391	.3437	.3659	69.9	1.22
.35	20.2	.9385	.3453	.3679	69.8	1.22
.35	20.3	.9379	.3469	.3699	69.7	1.22
.36	20.4	.9373	.3486	.3719	69.6	1.21
.36	**20.5**	.9367	.3502	.3739	**69.5**	1.21
.36	20.6	.9361	.3518	.3759	69.4	1.21
.36	20.7	.9354	.3535	.3779	69.3	1.21
.36	20.8	.9348	.3551	.3799	69.2	1.21
.36	20.9	.9342	.3567	.3819	69.1	1.21
.37	**21.0**	.9336	.3584	.3839	**69.0**	1.20
.37	21.1	.9330	.3600	.3859	68.9	1.20
.37	21.2	.9323	.3616	.3879	68.8	1.20
.37	21.3	.9317	.3633	.3899	68.7	1.20
.37	21.4	.9311	.3649	.3919	68.6	1.20
.38	**21.5**	.9304	.3665	.3939	**68.5**	1.20
.38	21.6	.9298	.3681	.3959	68.4	1.19
.38	21.7	.9291	.3697	.3979	68.3	1.19
.38	21.8	.9285	.3714	.4000	68.2	1.19
.38	21.9	.9278	.3730	.4020	68.1	1.19
.38	**22.0**	.9272	.3746	.4040	**68.0**	1.19
.39	22.1	.9265	.3762	.4061	67.9	1.19
.39	22.2	.9259	.3778	.4081	67.8	1.18
.39	22.3	.9252	.3795	.4101	67.7	1.18
.39	22.4	.9245	.3811	.4122	67.6	1.18
.39	**22.5**	.9239	.3827	.4142	**67.5**	1.18
.39	22.6	.9232	.3843	.4163	67.4	1.18
.40	22.7	.9225	.3859	.4183	67.3	1.17
.40	22.8	.9219	.3875	.4204	67.2	1.17
.40	22.9	.9212	.3891	.4224	67.1	1.17
.40	**23.0**	.9205	.3907	.4245	**67.0**	1.17
.40	23.1	.9198	.3923	.4265	66.9	1.17
.40	23.2	.9191	.3939	.4286	66.8	1.17
.41	23.3	.9184	.3955	.4307	66.7	1.16
.41	23.4	.9178	.3971	.4327	66.6	1.16
.41	**23.5**	.9171	.3987	.4348	**66.5**	1.16
.41	23.6	.9164	.4003	.4369	66.4	1.16
.41	23.7	.9157	.4019	.4390	66.3	1.16
.42	23.8	.9150	.4035	.4411	66.2	1.16
.42	23.9	.9143	.4051	.4431	66.1	1.15
.42	**24.0**	.9135	.4067	.4452	**66.0**	1.15
.42	24.1	.9128	.4083	.4473	65.9	1.15
.42	24.2	.9121	.4099	.4494	65.8	1.15
.42	24.3	.9114	.4115	.4515	65.7	1.15
.43	24.4	.9107	.4131	.4536	65.6	1.14
.43	**24.5**	.9100	.4147	.4557	**65.5**	1.14
.43	24.6	.9092	.4163	.4578	65.4	1.14
.43	24.7	.9085	.4179	.4599	65.3	1.14
.43	24.8	.9078	.4195	.4621	65.2	1.14
.43	24.9	.9070	.4210	.4642	65.1	1.14
.44	**25.0**	.9063	.4226	.4663	**65.0**	1.14
		sin	cos	cot	Deg	Rad

Rad	Deg	cos	sin	tan		
.44	**25.0**	.9063	.4226	.4663	**65.0**	1.13
.44	25.1	.9056	.4242	.4684	64.9	1.13
.44	25.2	.9048	.4258	.4706	64.8	1.13
.45	25.3	.9041	.4274	.4727	64.7	1.13
.45	25.4	.9033	.4289	.4748	64.6	1.13
.45	**25.5**	.9026	.4305	.4770	**64.5**	1.13
.45	25.6	.9018	.4321	.4791	64.4	1.12
.45	25.7	.9011	.4337	.4813	64.3	1.12
.45	25.8	.9003	.4352	.4834	64.2	1.12
.45	25.9	.8996	.4368	.4856	64.1	1.12
.45	**26.0**	.8988	.4384	.4877	**64.0**	1.12
.46	26.1	.8980	.4399	.4899	63.9	1.12
.46	26.2	.8973	.4415	.4921	63.8	1.11
.46	26.3	.8965	.4431	.4942	63.7	1.11
.46	26.4	.8957	.4446	.4964	63.6	1.11
.46	**26.5**	.8949	.4462	.4986	**63.5**	1.11
.46	26.6	.8942	.4478	.5008	63.4	1.11
.47	26.7	.8934	.4493	.5029	63.3	1.10
.47	26.8	.8926	.4509	.5051	63.2	1.10
.47	26.9	.8918	.4524	.5073	63.1	1.10
.47	**27.0**	.8910	.4540	.5095	**63.0**	1.10
.47	27.1	.8902	.4555	.5117	62.9	1.10
.47	27.2	.8894	.4571	.5139	62.8	1.10
.48	27.3	.8886	.4586	.5161	62.7	1.09
.48	27.4	.8878	.4602	.5184	62.6	1.09
.48	**27.5**	.8870	.4617	.5206	**62.5**	1.09
.48	27.6	.8862	.4633	.5228	62.4	1.09
.48	27.7	.8854	.4648	.5250	62.3	1.09
.49	27.8	.8846	.4664	.5272	62.2	1.09
.49	27.9	.8838	.4679	.5295	62.1	1.08
.49	**28.0**	.8829	.4695	.5317	**62.0**	1.08
.49	28.1	.8821	.4710	.5340	61.9	1.08
.49	28.2	.8813	.4726	.5362	61.8	1.08
.49	28.3	.8805	.4741	.5384	61.7	1.08
.50	28.4	.8796	.4756	.5407	61.6	1.08
.50	**28.5**	.8788	.4772	.5430	**61.5**	1.07
.50	28.6	.8780	.4787	.5452	61.4	1.07
.50	28.7	.8771	.4802	.5475	61.3	1.07
.50	28.8	.8763	.4818	.5498	61.2	1.07
.50	28.9	.8755	.4833	.5520	61.1	1.07
.51	**29.0**	.8746	.4848	.5543	**61.0**	1.06
.51	29.1	.8738	.4863	.5566	60.9	1.06
.51	29.2	.8729	.4879	.5589	60.8	1.06
.51	29.3	.8721	.4894	.5612	60.7	1.06
.51	29.4	.8712	.4909	.5635	60.6	1.06
.51	**29.5**	.8704	.4924	.5658	**60.5**	1.06
.52	29.6	.8695	.4939	.5681	60.4	1.05
.52	29.7	.8686	.4955	.5704	60.3	1.05
.52	29.8	.8678	.4970	.5727	60.2	1.05
.52	29.9	.8669	.4985	.5750	60.1	1.05
.52	**30.0**	.8660	.5000	.5774	**60.0**	1.05
		sin	cos	cot	Deg	Rad

TABLE C.III Trigonometric functions (continued)

Rad	Deg	cos	sin	tan		
.52	**30.0**	.8660	.5000	.5774	**60.0**	1.05
.53	30.1	.8652	.5015	.5797	59.9	1.05
.53	30.2	.8643	.5030	.5820	59.8	1.04
.53	30.3	.8634	.5045	.5844	59.7	1.04
.53	30.4	.8625	.5060	.5867	59.6	1.04
.53	**30.5**	.8616	.5075	.5890	**59.5**	1.04
.53	30.6	.8607	.5090	.5914	59.4	1.04
.54	30.7	.8599	.5105	.5938	59.3	1.03
.54	30.8	.8590	.5120	.5961	59.2	1.03
.54	30.9	.8581	.5135	.5985	59.1	1.03
.54	**31.0**	.8572	.5150	.6009	**59.0**	1.03
.54	31.1	.8563	.5165	.6032	58.9	1.03
.54	31.2	.8554	.5180	.6056	58.8	1.03
.55	31.3	.8545	.5195	.6080	58.7	1.02
.55	31.4	.8536	.5210	.6104	58.6	1.02
.55	**31.5**	.8526	.5225	.6128	**58.5**	1.02
.55	31.6	.8517	.5240	.6152	58.4	1.02
.55	31.7	.8508	.5255	.6176	58.3	1.02
.56	31.8	.8499	.5270	.6200	58.2	1.02
.56	31.9	.8490	.5284	.6224	58.1	1.01
.56	**32.0**	.8480	.5299	.6249	**58.0**	1.01
.56	32.1	.8471	.5314	.6273	57.9	1.01
.56	32.2	.8462	.5329	.6297	57.8	1.01
.56	32.3	.8453	.5344	.6322	57.7	1.01
.57	32.4	.8443	.5358	.6346	57.6	1.01
.57	**32.5**	.8434	.5373	.6371	**57.5**	1.00
.57	32.6	.8425	.5388	.6395	57.4	1.00
.57	32.7	.8415	.5402	.6420	57.3	1.00
.57	32.8	.8406	.5417	.6445	57.2	1.00
.57	32.9	.8396	.5432	.6469	57.1	1.00
.58	**33.0**	.8387	.5446	.6494	**57.0**	.99
.58	33.1	.8377	.5461	.6519	56.9	.99
.58	33.2	.8368	.5476	.6544	56.8	.99
.58	33.3	.8358	.5490	.6569	56.7	.99
.58	33.4	.8348	.5505	.6594	56.6	.99
.58	**33.5**	.8339	.5519	.6619	**56.5**	.99
.59	33.6	.8329	.5534	.6644	56.4	.98
.59	33.7	.8320	.5548	.6669	56.3	.98
.59	33.8	.8310	.5563	.6694	56.2	.98
.59	33.9	.8300	.5577	.6720	56.1	.98
.59	**34.0**	.8290	.5592	.6745	**56.0**	.98
.60	34.1	.8281	.5606	.6771	55.9	.98
.60	34.2	.8271	.5621	.6796	55.8	.97
.60	34.3	.8261	.5635	.6822	55.7	.97
.60	34.4	.8251	.5650	.6847	55.6	.97
.60	**34.5**	.8241	.5664	.6873	**55.5**	.97
.60	34.6	.8231	.5678	.6899	55.4	.97
.61	34.7	.8221	.5693	.6924	55.3	.97
.61	34.8	.8211	.5707	.6950	55.2	.96
.61	34.9	.8202	.5721	.6976	55.1	.96
.61	**35.0**	.8192	.5736	.7002	**55.0**	.96
		sin	cos	cot	Deg	Rad

Rad	Deg	cos	sin	tan		
.61	**35.0**	.8192	.5736	.7002	**55.0**	.96
.61	35.1	.8181	.5750	.7028	54.9	.96
.61	35.2	.8171	.5764	.7054	54.8	.96
.62	35.3	.8161	.5779	.7080	54.7	.95
.62	35.4	.8151	.5793	.7107	54.6	.95
.62	**35.5**	.8141	.5807	.7133	**54.5**	.95
.62	35.6	.8131	.5821	.7159	54.4	.95
.62	35.7	.8121	.5835	.7186	54.3	.95
.62	35.8	.8111	.5850	.7212	54.2	.95
.63	35.9	.8100	.5864	.7239	54.1	.94
.63	**36.0**	.8090	.5878	.7265	**54.0**	.94
.63	36.1	.8080	.5892	.7292	53.9	.94
.63	36.2	.8070	.5906	.7319	53.8	.94
.63	36.3	.8059	.5920	.7346	53.7	.94
.64	36.4	.8049	.5934	.7373	53.6	.94
.64	**36.5**	.8039	.5948	.7400	**53.5**	.93
.64	36.6	.8028	.5962	.7427	53.4	.93
.64	36.7	.8018	.5976	.7454	53.3	.93
.64	36.8	.8007	.5990	.7481	53.2	.93
.64	36.9	.7997	.6004	.7508	53.1	.93
.65	**37.0**	.7986	.6018	.7536	**53.0**	.93
.65	37.1	.7976	.6032	.7563	52.9	.92
.65	37.2	.7965	.6046	.7590	52.8	.92
.65	37.3	.7955	.6060	.7618	52.7	.92
.65	37.4	.7944	.6074	.7646	52.6	.92
.65	**37.5**	.7934	.6088	.7673	**52.5**	.92
.66	37.6	.7923	.6101	.7701	52.4	.91
.66	37.7	.7912	.6115	.7729	52.3	.91
.66	37.8	.7902	.6129	.7757	52.2	.91
.66	37.9	7891	.6143	.7785	52.1	.91
.66	**38.0**	.7880	.6157	.7813	**52.0**	.91
.66	38.1	.7869	.6170	.7841	51.9	.91
.67	38.2	.7859	.6184	.7869	51.8	.90
.67	38.3	.7848	.6198	.7898	51.7	.90
.67	38.4	.7837	.6211	.7926	51.6	.90
.67	**38.5**	.7826	.6225	.7954	**51.5**	.90
.67	38.6	.7815	.6239	.7983	51.4	.90
.68	38.7	.7804	.6252	.8012	51.3	.90
.68	38.8	.7793	.6266	.8040	51.2	.89
.68	38.9	.7782	.6280	.8069	51.1	.89
.68	**39.0**	.7771	.6293	.8098	**51.0**	.89
.68	39.1	.7760	.6307	.8127	50.9	.89
.68	39.2	.7749	.6320	.8156	50.8	.89
.69	39.3	.7738	.6334	.8185	50.7	.88
.69	39.4	.7727	.6347	.8214	50.6	.88
.69	**39.5**	.7716	.6361	.8243	**50.5**	.88
.69	39.6	.7705	.6374	.8273	50.4	.88
.69	39.7	.7694	.6388	.8302	50.3	.88
.69	39.8	.7683	.6401	.8332	50.2	.88
.70	39.9	.7672	.6414	.8361	50.1	.87
.70	**40.0**	.7660	.6428	.8391	**50.0**	.87
		sin	cos	cot	Deg	Rad

TABLE C.III Trigonometric functions (continued)

Rad	Deg	cos	sin	tan		
.70	**40.0**	.7660	.6428	.8391	**50.0**	.87
.70	40.1	.7649	.6441	.8421	49.9	.87
.70	40.2	.7638	.6455	.8451	49.8	.87
.70	40.3	.7627	.6468	.8481	49.7	.87
.71	40.4	.7615	.6481	.8511	49.6	.87
.71	**40.5**	.7604	.6494	.8541	**49.5**	.86
.71	40.6	.7593	.6508	.8571	49.4	.86
.71	40.7	.7581	.6521	.8601	49.3	.86
.71	40.8	.7570	.6534	.8632	49.2	.86
.71	40.9	.7559	.6547	.8662	49.1	.86
.72	**41.0**	.7547	.6561	.8693	**49.0**	.86
.72	41.1	.7536	.6574	.8724	48.9	.85
.72	41.2	.7524	.6587	.8754	48.8	.85
.72	41.3	.7513	.6600	.8785	48.7	.85
.72	41.4	.7501	.6613	.8816	48.6	.85
.72	**41.5**	.7490	.6626	.8847	**48.5**	.85
.73	41.6	.7478	.6639	.8878	48.4	.84
.73	41.7	.7466	.6652	.8910	48.3	.84
.73	41.8	.7455	.6665	.8941	48.2	.84
.73	41.9	.7443	.6678	8972	48.1	.84
.73	**42.0**	.7431	.6691	.9004	**48.0**	.84
.73	42.1	.7420	.6704	.9036	47.9	.84
.74	42.2	.7408	.6717	.9067	47.8	.83
.74	42.3	.7396	.6730	.9099	47.7	.83
.74	42.4	.7385	.6743	.9131	47.6	.83
.74	**42.5**	.7373	.6756	.9163	**47.5**	.83
		sin	cos	cot	Deg	Rad

Rad	Deg	cos	sin	tan		
.74	**42.5**	.7373	.6756	.9163	**47.5**	.83
.74	42.6	.7361	.6769	.9195	47.4	.83
.75	42.7	.7349	.6782	.9228	47.3	.83
.75	42.8	.7337	.6794	.9260	47.2	.82
.75	42.9	.7325	.6807	.9293	47.1	.82
.75	**43.0**	.7314	.6820	.9325	**47.0**	.82
.75	43.1	.7302	.6833	.9358	46.9	.82
.75	43.2	.7290	.6845	.9391	46.8	.82
.76	43.3	.7278	.6858	.9424	46.7	.82
.76	43.4	.7266	.6871	.9457	46.6	.81
.76	**43.5**	.7254	.6884	.9490	**46.5**	.81
.76	43.6	.7242	.6896	.9523	46.4	.81
.76	43.7	.7230	.6909	.9556	46.3	.81
.76	43.8	.7218	.6921	.9590	46.2	.81
.77	43.9	.7206	.6934	.9623	46.1	.80
.77	**44.0**	.7193	.6947	.9657	**46.0**	.80
.77	44.1	.7181	.6959	.9691	45.9	.80
.77	44.2	.7169	.6972	.9725	45.8	.80
.77	44.3	.7157	.6984	.9759	45.7	.80
.77	44.4	.7145	.6997	.9793	45.6	.80
.78	**44.5**	.7133	.7009	.9827	**45.5**	.79
.78	44.6	.7120	.7022	.9861	45.4	.79
.78	44.7	.7108	.7034	.9896	45.3	.79
.78	44.8	.7096	.7046	.9930	45.2	.79
.78	44.9	.7083	.7059	.9965	45.1	.79
.79	**45.0**	.7071	.7071	1.0000	**45.0**	.79
		sin	cos	cot	Deg	Rad

D

Answers

PROBLEM SET 1.1, PAGES 8–9

1. a. $\mathbb{Z}$, $\mathbb{Q}$, and $\mathbb{R}$ **b.** $\mathbb{Q}$ and $\mathbb{R}$ **c.** $\mathbb{N}$, $\mathbb{W}$, $\mathbb{Z}$, $\mathbb{Q}$, and $\mathbb{R}$ **d.** $\mathbb{Q}'$ and $\mathbb{R}$ **e.** $\mathbb{Q}'$ and $\mathbb{R}$ **3. a.** $\mathbb{Q}$ and $\mathbb{R}$ **b.** $\mathbb{Q}$ and $\mathbb{R}$ **c.** $\mathbb{Q}'$ and $\mathbb{R}$ **d.** $\mathbb{Q}$ and $\mathbb{R}$ **e.** $\mathbb{Q}'$ and $\mathbb{R}$ **5. a.** $\mathbb{N}$, $\mathbb{W}$, $\mathbb{Z}$, $\mathbb{Q}$, and $\mathbb{R}$ **b.** $\mathbb{Q}'$ and $\mathbb{R}$ **c.** $\mathbb{Q}'$ and $\mathbb{R}$ **d.** $\mathbb{Q}$ and $\mathbb{R}$ **e.** $\mathbb{Q}$ and $\mathbb{R}$ **7.** 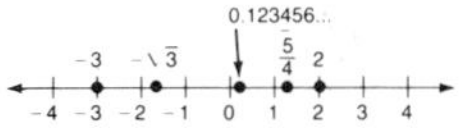 **9.** 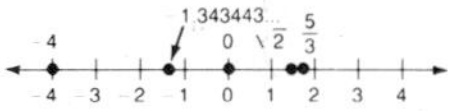 **11.**

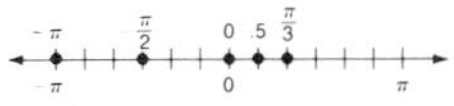

13. a. $<$ **b.** $<$ **15. a.** $<$ **b.** $=$ **17. a.** $=$ **b.** $<$ **19.** reflexive **21.** distributive **23.** closure **25.** transitive **27.** substitution **29.** closure **31.** multiplicative inverse **33.** distributive **35.** identity **37.** no **39.** no **41.** yes **43.** no **45.** Answers vary; $\frac{12}{4} \neq \frac{4}{12}$ **47.** no **49.** $-\dfrac{\pi}{3} - 1$ **51.** $\dfrac{3}{\pi + 3}$

53. addition properties: closure, commutative, associative; multiplication properties: closure, commutative, associative, identity; distributive for multiplication over addition
55. addition properties: closure, commutative, associative, identity, inverse; multiplication properties: closure, commutative, associative, identity; distributive for multiplication over addition
57. not closed for either addition or multiplication; commutative and associative for those sums and products in the set; distributive for multiplication over addition for those sums and products in the set
59. multiplication properties: closure, commutative, associative, identity and inverse

PROBLEM SET 1.2, PAGES 17–18

1. a. $(3, 7)$  **b.** $(-4, -1)$  **c.** $[-2, 6]$  **d.** $(-3, 0]$ 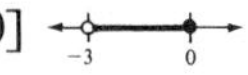 **3. a.** $(-\infty, -3]$  **b.** $[-2, \infty)$ **c.** $(-\infty, 0)$ **d.** $(2, \infty)$ **5. a.** **b.** **c.** **d.** **7. a.** **b.** **c.** **d.** **9. a.** $-4 \leq x \leq 2$ **b.** $-1 \leq x \leq 2$ **c.** $0 < x < 8$ **d.** $-5 < x \leq 3$ **11. a.** $x < 2$ **b.** $x > 6$ **c.** $x > -1$ **d.** $x \leq 3$ **13. a.** $\pi - 2$ **b.** $5 - \pi$ **c.** $2\pi - 6$ **d.** $7 - 2\pi$ **15. a.** $\sqrt{2} - 1$ **b.** $2 - \sqrt{2}$ **c.** $1 - \dfrac{\pi}{6}$ **d.** $\dfrac{2\pi}{3} - 1$ **17. a.** $\pi - 3$ **b.** 119 **c.** $3 - \sqrt{5}$
19. $\{5, -5\}$ **21.** $\{\ \}$ or $\varnothing$ **23.** $\{7, -1\}$ **25.** $\{24, -6\}$ **27.** $\{\ \}$ or $\varnothing$ **29.** $\{\frac{2}{5}, -2\}$ **31.** $[7, \infty)$ **33.** $(-\infty, -41]$ **35.** $(-\infty, 2)$ **37.** $(-\frac{8}{5}, 0)$ **39.** $(1, 3]$ **41.** $[-4, 2)$ **43.** $(-\frac{15}{2}, 4]$ **45.** $[-2, 8]$ **47.** $(2.999, 3.001)$ **49.** $\{\ \}$ or $\varnothing$ **51.** $(\frac{1}{3}, \frac{7}{3})$ **53.** $(-\infty, -6) \cup (-1, \infty)$
55. You must drive less than 200 miles per day. **57.** $68 < F < 86$

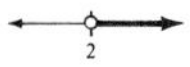

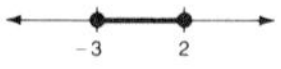

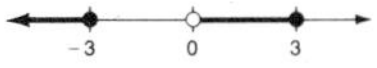

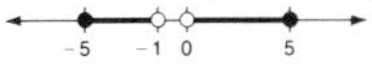

59. values in thousands of dollars **a.** $s < 10$ or $s > 20$ **b.** $|s - 15| > 5$ **61.** $(-\infty, -3) \cup (9, \infty)$ **63.** $[-2, 0]$
65. $\{1, -\frac{3}{2}\}$ **67.** Answers vary. **69.** Answers vary.

PROBLEM SET 1.3, PAGE 21

1. $6i$ **3.** $7i$ **5.** $2i\sqrt{5}$ **7.** $8 + 7i$ **9.** $-5i$ **11.** $1 - 6i$ **13.** $3 + 3i$ **15.** 10 **17.** $2 + 5i$ **19.** $7 + i$
21. 29 **23.** 34 **25.** 1 **27.** $-i$ **29.** -1 **31.** 1 **33.** $-i$ **35.** -1 **37.** $32 - 24i$ **39.** $-9 + 40i$
41. $-198 - 10i$ **43.** $-\frac{3}{2} + \frac{3}{2}i$ **45.** $1 + i$ **47.** $-2i$ **49.** $-\frac{1}{5} - \frac{3}{5}i$ **51.** $\frac{1}{5} - \frac{2}{5}i$ **53.** $1 - i$ **55.** $-1 + i$
57. $\frac{-45}{53} + \frac{28}{53}i$ **59.** $0.3131 + 2.2281i$ **61.** $(-1 - \sqrt{3}) + 2i$ **63.** $\left(\frac{5 + 2\sqrt{3}}{13}\right) + \left(\frac{1 + 3\sqrt{3}}{13}\right)i$
65. $(13 + 8\sqrt{3}) + (12 + 4\sqrt{3})i$ **67.** Answers vary.

PROBLEM SET 1.4, PAGES 28–29

1. $\{3, -5\}$ **3.** $\{2, -9\}$ **5.** $\{\frac{4}{5}, -\frac{1}{2}\}$ **7.** $\{4, -\frac{2}{9}\}$ **9.** $\{\frac{4}{3}\}$ **11.** $\{0, 2\}$ **13.** $\{1, -5\}$ **15.** $\{2, -4\}$
17. $\{-3, -4\}$ **19.** $\{5 \pm 3\sqrt{3}\}$ **21.** $\left\{\frac{3 \pm \sqrt{13}}{2}\right\}$ **23.** $\{\frac{2}{3}, -\frac{1}{2}\}$ **25.** $\{1, -6\}$ **27.** $\{5\}$ **29.** $\{\frac{1}{4}, -\frac{2}{3}\}$
31. $\varnothing$ over $\mathbb{R}$; $\{2 \pm i\}$ over the complex numbers **33.** $\left\{\pm\frac{\sqrt{5}}{2}\right\}$ **35.** $\{0, \frac{7}{3}\}$ **37.** $\{2, -\frac{1}{3}\}$ **39.** $\left\{\frac{-5 \pm \sqrt{73}}{8}\right\}$
41. $\left\{\frac{-3 \pm \sqrt{17}}{4}\right\}$ **43.** $\left\{\frac{-3 \pm \sqrt{9 + 16\sqrt{5}}}{8}\right\}$ **45.** $\left\{\frac{2 \pm \sqrt{4 + 3\sqrt{5}}}{3}\right\}$ **47.** $\left\{\frac{-w \pm \sqrt{w^2 - 40}}{4}\right\}$
49. $\left\{\frac{-5 \pm \sqrt{12y - 23}}{6}\right\}$ **51.** $\left\{2, \frac{3t + 2}{4}\right\}$ **53.** $\left\{\frac{-1 \pm \sqrt{8y - 47}}{4}\right\}$ **55.** $\{3 \pm \sqrt{4y - y^2}\}$
57. a. 1454 ft (let $t = 0$) **b.** about 11.2 sec **59.** It will be in the air for 8 sec. **61.** It will take 7.9 sec.

PROBLEM SET 1.5, PAGES 32–33

1. $(-3, 0)$  **3.** $(-\infty, 2] \cup [6, \infty)$ **5.** $(-7, 8)$
7. $[-2, \frac{1}{2}]$ 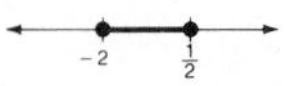 **9.** $(-\infty, -\frac{2}{3}) \cup (3, \infty)$ 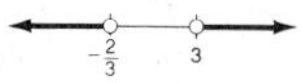 **11.** $(-\infty, -2] \cup [8, \infty)$
13. $(-\infty, \frac{1}{3}) \cup (4, \infty)$ 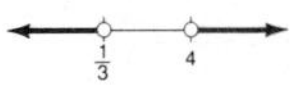 **15.** $(-\infty, -4] \cup [0, 3]$ 
17. $[-3, 2] \cup [4, \infty]$  **19.** $(-\infty, -\frac{5}{2}) \cup (-1, \frac{7}{3})$ 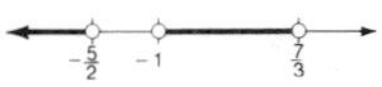 **21.** $(-2, 0)$ 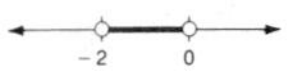
23. $(-\infty, 0) \cup (8, \infty)$ 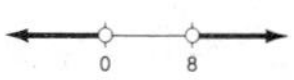 **25.** $(-5, 2]$  **27.** $(-\infty, 0) \cup (\frac{1}{2}, 5)$ 
29. $(-\infty, -2) \cup (0, 3)$ 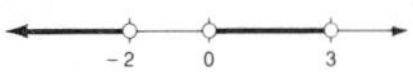 **31.** $(-\infty, -3] \cup [3, \infty)$ **33.** $(-\infty, \infty)$
35. $(-\infty, -2) \cup (3, \infty)$ 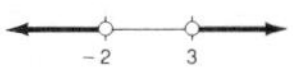 **37.** $[2, 3]$  **39.** $[-5, 1]$ 
41. $(1 - \sqrt{3}, 1 + \sqrt{3})$ 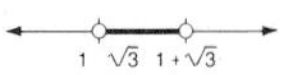 **43.** $(-\infty, \infty)$ 
45. $\left(-\infty, \frac{-3 - \sqrt{37}}{2}\right] \cup \left[\frac{-3 + \sqrt{37}}{2}, \infty\right)$ **47.** $[-5, -3) \cup [0, 3] \cup (4, \infty)$ 
49. $(-3, 2) \cup [12, \infty)$ **51.** $[-13, -\frac{1}{2}) \cup [0, \frac{1}{3})$
53. $(-\infty, 2) \cup (3, 4]$ 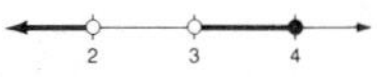 **55.** all numbers except those between -17 and 20 **57.** all numbers except those between -3 and 0, inclusive **59.** The width must be greater than 3 and the length greater than 6.

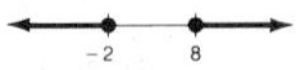

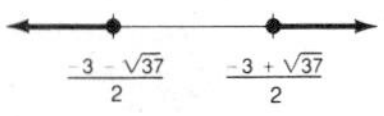

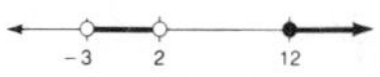

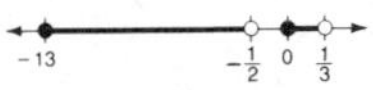

PROBLEM SET 1.6, PAGES 39–41

1. 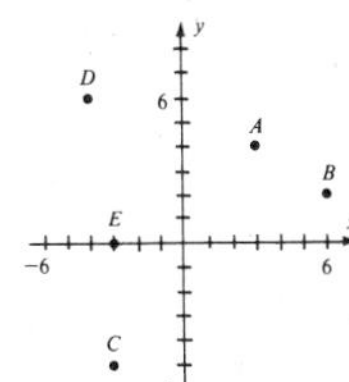

3. 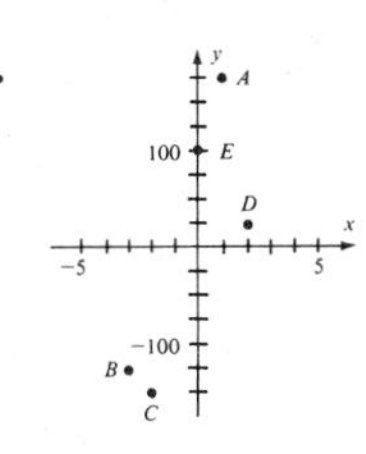

5. 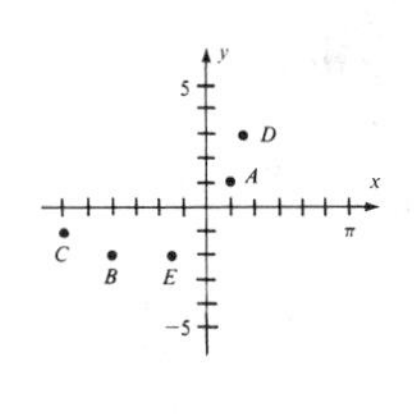

7. Answers vary. 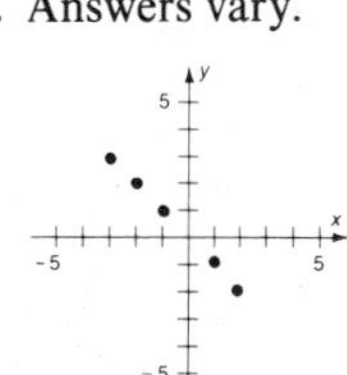

9. Answers vary. 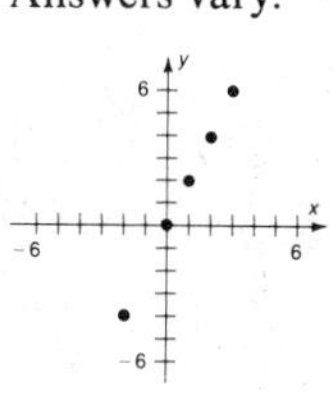

11. Answers vary. 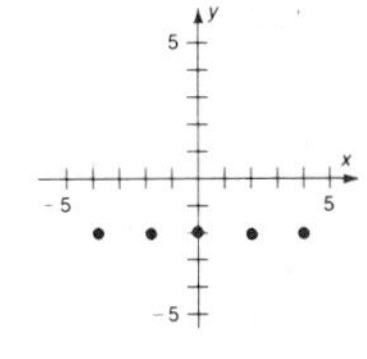

13. 5 **15.** $2\sqrt{5}$ **17.** $\sqrt{37}$ **19.** $-5x$ **21.** $5x$ **23.** $(7, \frac{13}{2})$ **25.** $(\frac{5}{2}, -1)$

27. $(-\frac{3}{2}, -2)$ **29.** $(-x, \frac{7}{2}x)$

31. 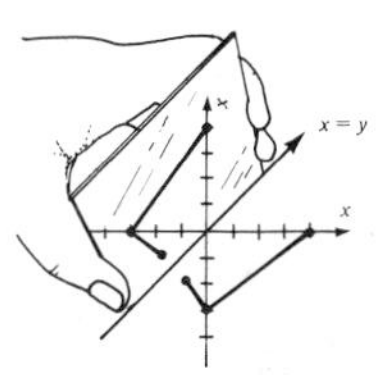

33. 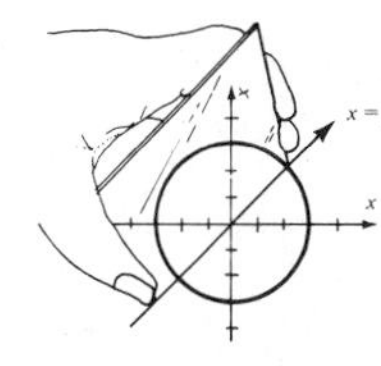

35. 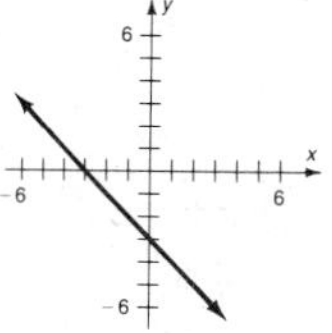

37. 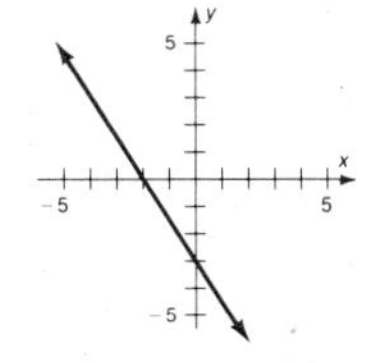

39. 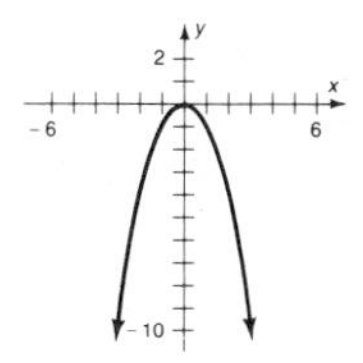

41. 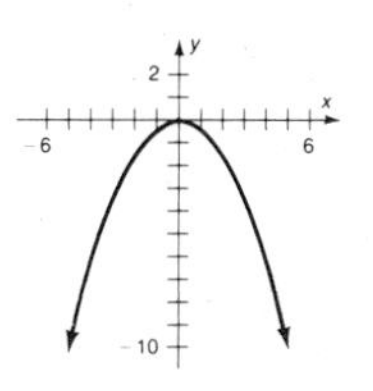

43. 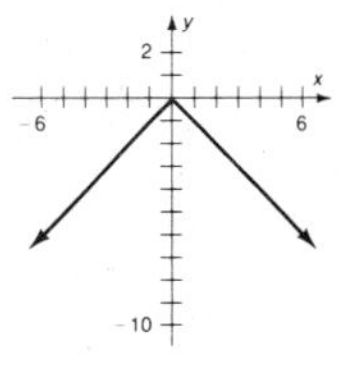

45. 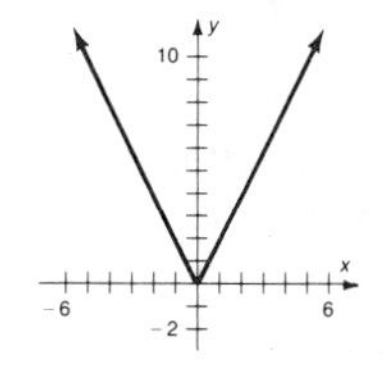

47. 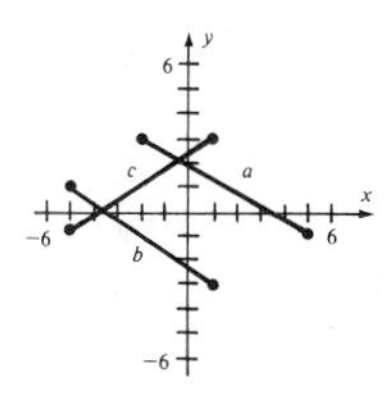

49. 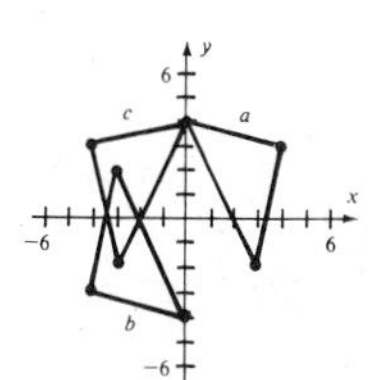

51. 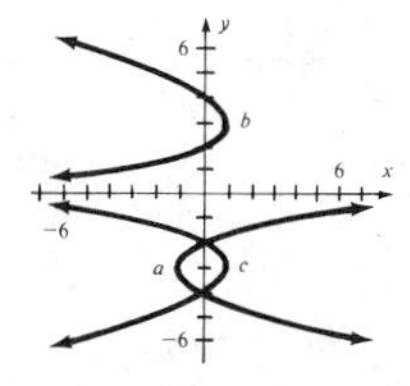

53. $d_1 = \sqrt{40}, d_2 = \sqrt{85}, d_3 = 9$; not a right triangle **55.** $d_1 = 5, d_2 = \sqrt{73}, d_3 = \sqrt{74}$; not a right triangle

57. $d_1 = \sqrt{52}, d_2 = \sqrt{65}, d_3 = \sqrt{13}$; it is a right triangle. **59.** $(x-2)^2 + (y-3)^2 = 49$ **61.** $(x+4)^2 + (y-1)^2 = 9$

63. $(0, 4 \pm 2\sqrt{15})$ **65.**

67. 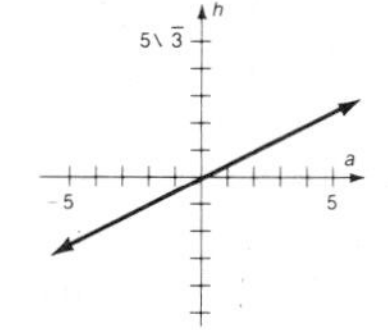

69. Answers vary. **71.** $9x^2 + 25y^2 = 225$

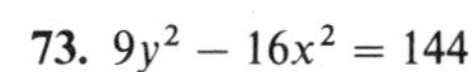

73. $9y^2 - 16x^2 = 144$

CHAPTER 1 SUMMARY, PAGES 41–43

Let $\mathbb{N}$ = {counting numbers}, $\mathbb{W}$ = {whole numbers}, $\mathbb{Z}$ = {integers}, $\mathbb{Q}$ = {rationals}, $\mathbb{Q}'$ = {irrationals} and $\mathbb{R}$ = {reals}.

1. $\frac{14}{7}$: $\mathbb{N}$, $\mathbb{W}$, $\mathbb{Z}$, $\mathbb{Q}$, and $\mathbb{R}$
$\sqrt{144}$: $\mathbb{N}$, $\mathbb{W}$, $\mathbb{Z}$, $\mathbb{Q}$, and $\mathbb{R}$
$6.\bar{2}$: $\mathbb{Q}$ and $\mathbb{R}$
π: $\mathbb{Q}'$ and $\mathbb{R}$
3. $3.\bar{1}$: $\mathbb{Q}$ and $\mathbb{R}$
$\frac{5\pi}{6}$: $\mathbb{Q}'$ and $\mathbb{R}$
$\frac{22}{7}$: $\mathbb{Q}$ and $\mathbb{R}$
$\sqrt{10}$: $\mathbb{Q}'$ and $\mathbb{R}$
5. **7.**

9. = **11.** < **13.** $a(b + c)$ **15.** $a(b + c) = 5$ **17.** $(b + c)a$ **19.** $ab + ac$ **21.** $(-4, 2)$

23. $(-3, \infty)$ **25.** **27.** **29.** $-8 \le x < -5$ **31.** $x < 3$ **33.** $(-\infty, 4]$

35. $(-8, -4]$ **37.** $\sqrt{11}$ **39.** $\sqrt{11} - 3$ **41.** 8 **43.** $\pi + 2$ **45.** $\{-8, 8\}$ **47.** $\{-\frac{11}{2}, \frac{5}{2}\}$ **49.** $(-1, 9)$

51. $[-10, 15]$ **53.** i **55.** 29 **57.** $\{-3, 4\}$ **59.** $\{3, -\frac{5}{2}\}$ **61.** $\{-3, 5\}$ **63.** $\{-4, -5\}$ **65.** $\left\{\frac{5 \pm \sqrt{13}}{2}\right\}$

67. $\{-1 \pm \sqrt{6}\}$ **69.** $(-\frac{1}{3}, 1)$ **71.** $(-\infty, \infty)$ **73.** $(-1, 0) \cup (3, \infty)$ **75.** $(-\infty, -1) \cup (0, 2) \cup (2, \infty)$

77–79.

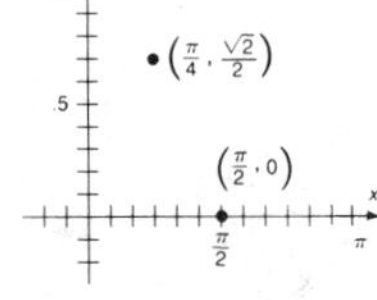

81. $d = \sqrt{(\gamma - \alpha)^2 + (\delta - \beta)^2}$ **83.** $-5x$ **85.** $\left(\frac{\alpha+\gamma}{2}, \frac{\beta+\delta}{2}\right)$ **87.** $(3x, \frac{5x}{2})$

89. a set of ordered pairs **91.** satisfies **93.** **95.** **97.**

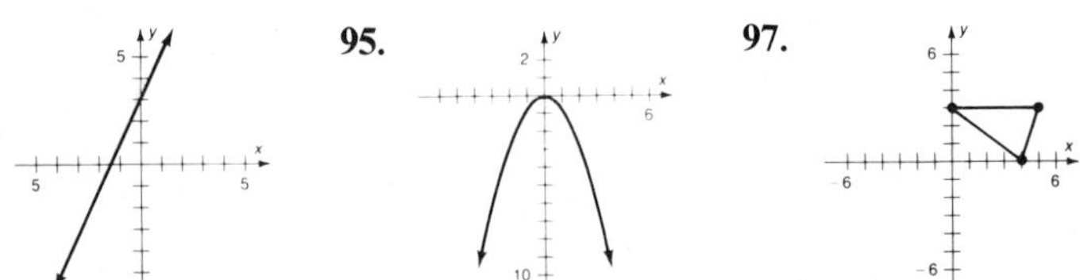

99.

PROBLEM SET 2.1, PAGES 50–52

1. onto, one-to-one function **3.** not a function **5.** function **7.** function **9.** function **11.** not a function **13.** function **15.** not a function **17.** function **19.** function **21.** not a function **23.** function **25.** function **27.** not a function **29.** function **31.** $R(9, g(9))$; $S(a, g(a))$ **33.** $P(x_0, F(x_0))$; $Q(x_0 + h, F(x_0 + h))$
35. $T(x_0, K(x_0))$; $U(x_0 + h, K(x_0 + h))$
37. D: $[-5, 6]$; R: $[-6, 3]$; intercepts: $(-4\frac{1}{2}, 0)$, $(4, 0)$, $(0, 3)$; increasing on $(-5, -3)$; constant on $(-3, 3)$; decreasing on $(3, 6)$
39. D: $[-4, 7]$; R: $[-4, 5]$; intercepts: $(0, 5)$, $(3, 0)$, $(6, 0)$; constant on $(-4, 1)$; decreasing on $(1, 3)$ and $(5, 7)$; increasing on $(3, 5)$
41. D: $[-5, 3) \cup (3, \infty)$; R: $[-3, 6) \cup (6, \infty)$; intercepts: $(0, -3)$, $(-\frac{7}{4}, 0)$, $(\frac{7}{4}, 0)$; constant on $(-5, -2)$; decreasing on $(-2, 0)$; increasing on $(0, \infty)$
43. D: $[-6, 6]$; R: $[-5, 5]$; intercepts: $(-3, 0)$, $(0, 5)$, $(3, 0)$; increasing on $(-6, 0)$; decreasing on $(0, 6)$
45. a. \$0.92 **b.** \$0.45 **47.** \$1.15 **49. a.** \$0.51 **b.** $e(1984) - e(1944)$
51. a. \$0.02225 **b.** average annual change of price of gasoline from 1944 to 1984 **53. a.** \$.018, $\frac{s(1954) - s(1944)}{10}$
b. \$0.0125; $\frac{s(1964) - s(1944)}{20}$ **c.** \$0.058; $\frac{s(1974) - s(1944)}{30}$ **d.** \$0.02875; $\frac{s(1984) - s(1944)}{40}$ **e.** $\frac{s(1944 + h) - s(1944)}{h}$
55. a. 63,800 **b.** the average annual change in the number of marriages from 1977 to $(1977 + h)$
57. It also has 5 elements. **59.** Answers vary.

PROBLEM SET 2.2, PAGES 58–59

1. a. 1 **b.** 5 **c.** -5 **d.** $2\sqrt{5} + 1$ **e.** $2\pi + 1$ **3. a.** $2w + 1$ **b.** $2w^2 - 1$ **c.** $2t^2 - 1$ **d.** $2v^2 - 1$ **e.** $2m + 1$ **5. a.** $3 + 2\sqrt{2}$ **b.** $5 + 4\sqrt{2}$ **c.** $2t^2 + 12t + 17$ **d.** $2t^2 + 4t + 3$ **e.** $2m^2 - 4m + 1$ **7.** 2 **9.** 2

11. $4t + 2h$ **13. a.** $w^2 - 1$ **b.** $h^2 - 1$ **c.** $w^2 + 2wh + h^2 - 1$ **d.** $w^2 + h^2 - 2$ **15. a.** $x^4 - 1$ **b.** $x - 1$ **c.** $x^2 + 2xh + h^2 - 1$ **d.** $x^2 - 1$ **17.** 2 **19.** $2x + h$ **21.** -2 **23.** $\dfrac{|2x + 2h + 1| - |2x + 1|}{h}$

25. $6x + 2 + 3h$ **27.** $\dfrac{-1}{x(x + h)}$ **29.** $(-\infty, \infty)$ **31.** $(-\infty, -2) \cup (-2, \infty)$ **33.** $[-\frac{1}{2}, \infty)$ **35.** $[-2, 1]$

37. $(-\infty, -2) \cup (-2, 2) \cup (2, \infty)$ **39.** not equal **41.** not equal **43.** not equal **45.** even **47.** neither **49.** neither **51.** even **53.** even **55.** For 50 units the average cost is \$13; for 100 units the average cost is \$7. $\dfrac{C(100) - C(50)}{50} = \$1$ **57.** $\dfrac{C(x + h) - C(x)}{h} = -0.040x + 4 - 0.02h$

59. Answers vary. It is the average rate of change for an object falling between time x and $x + h$.
61. Answers vary. **63.** Answers vary.

PROBLEM SET 2.3, PAGES 65–66

The graphs for Problems 1–9 should also be shown. **1.** 1 **3.** $\frac{13}{5}$ **5.** $-\frac{7}{3}$ **7.** Slope is undefined. **9.** 0
11. $m = -4; b = -1$ **13.** $m = \frac{1}{5}; b = -\frac{6}{5}$ **15.** $m = 300; b = 0$ **17.** $m = 0; b = -2$ **19.** $m = \frac{1}{3}; b = \frac{2}{3}$

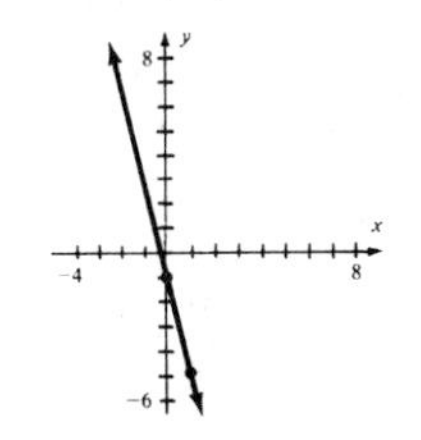
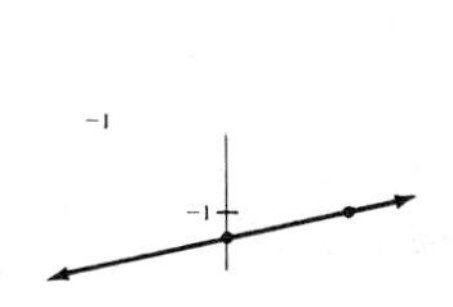
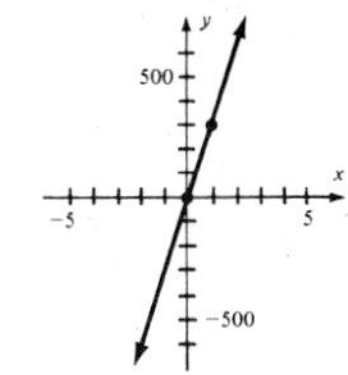
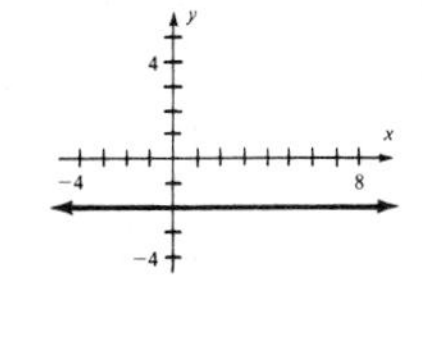
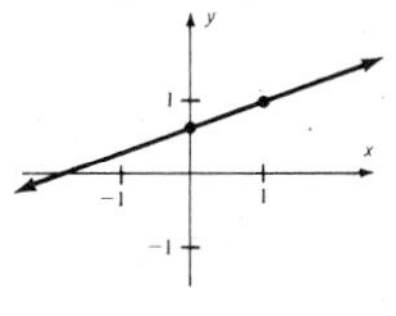

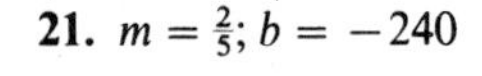
21. $m = \frac{2}{5}; b = -240$ **23.**
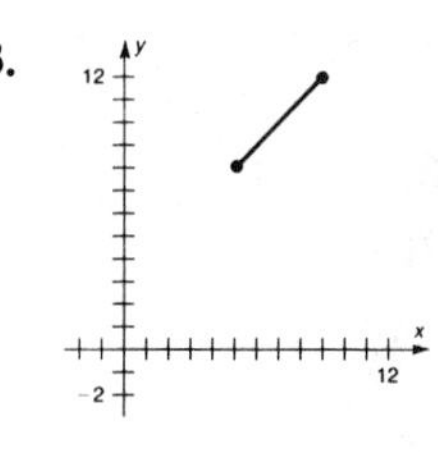

25.
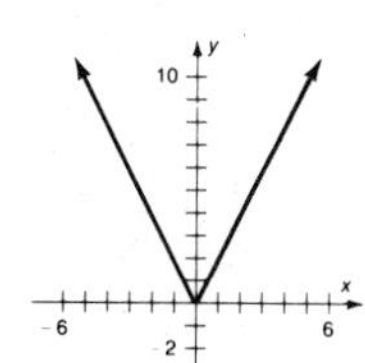

27.
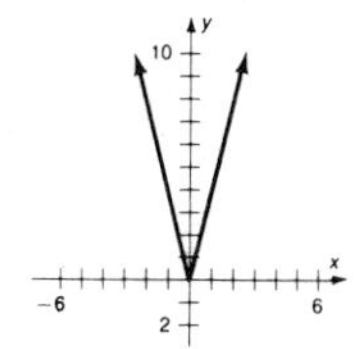

29. $m_{AN} = \frac{2}{3}; m_{AG} = -8; m_{NG} = -\frac{3}{2}$; thus, $m_{AN} \cdot m_{NG} = -1$. Thus AN and NG are perpendicular and it is a right triangle.
31. $m_{RE} = \frac{7}{3}; m_{EC} = -\frac{1}{2}; m_{CT} = \frac{6}{5}$; not a parallelogram **33.** $m_{PA} = 1; m_{AR} = -7; m_{RL} = 1; m_{LP} = -7$; it is a parallelogram.
35. $m_{PR} = 3; m_{AL} = -\frac{1}{3}$; yes **37.** $5x - y + 6 = 0$ **39.** $y = 0$ **41.** $3x - y - 3 = 0$ **43.** $x - 2y + 3 = 0$
45. $x - 2y + 2 = 0$ **47.** $2x - y - 4 = 0$ **49.** $2x + 3y - 16 = 0$ **51.** $2x + y + 4 = 0$

53. a. $A(x_0, f(x_0)); B(x_0 + \Delta x, f(x_0 + \Delta x))$ **b.** $\dfrac{f(x_0 + \Delta x) - f(x_0)}{\Delta x}$

55. a. $A(x_0, H(x_0)); B(x_0 + h, H(x_0 + h))$ **b.** $\dfrac{H(x_0 + h) - H(x_0)}{h}$ **57.** $(1, 10), (2, 20); 10x - y = 0$

59. $(20, 11.2)$ and $(30, 14.2)$; $3x - 10y + 52 = 0$; if $x = 40$, then $y = 17.2$, so the projected population in 1990 is 17.2 million.
61. $y - 60 = 0$; the cost is \$60. **63–67.** Answers vary.

PROBLEM SET 2.4, PAGE 71

1. $(6, 3)$ **3.** $(6, -1)$ **5.** $(0, \sqrt{2})$ **7.** $(3, -4)$ **9.** $(0, 0)$ **11.** $(-3, 0)$ **13.** $y - 3 = (x - 2)^2$ **15.** $y + 1 = x^2$
17. $y + 2 = (x + \sqrt{3})^2$ **19.** $y + 6 = |x|$ **21.** $y + \sqrt{3} = |x - \sqrt{2}|$ **23.** $y' = x'^2$ **25.** $y' = -5x'^2$
27. $y' = -2x'^2$ **29.** $y' = |x'|$ **31.** $y' = -2|x'|$

33. 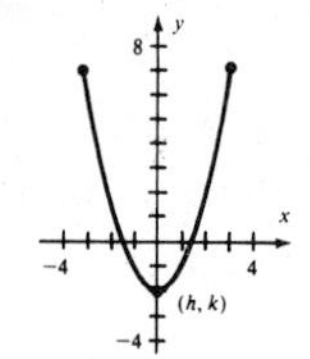

35. 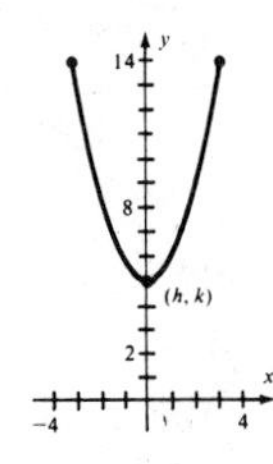

37. 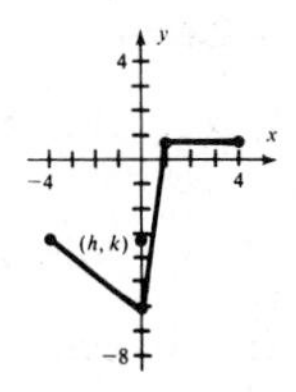

39. 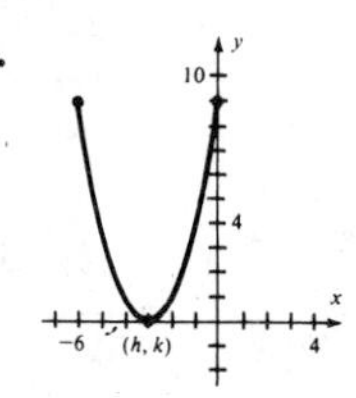

41. 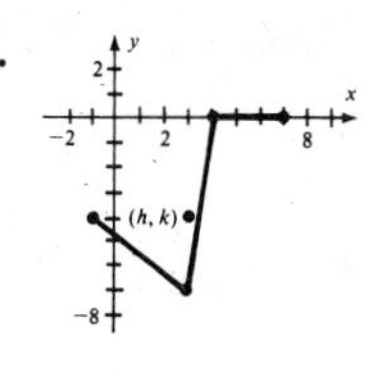

43. 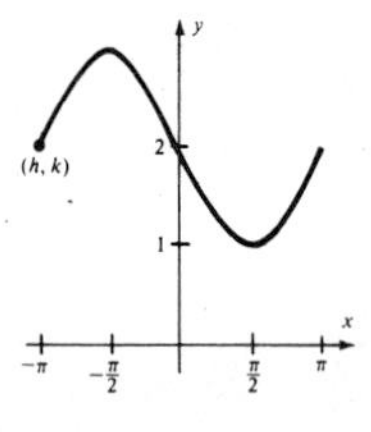

45. 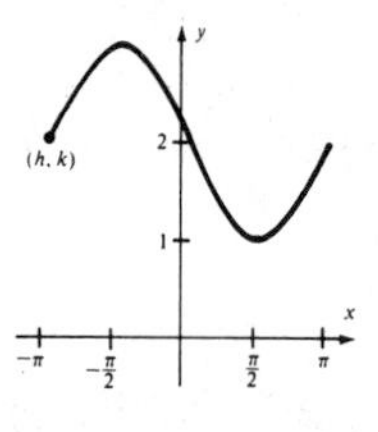

47. 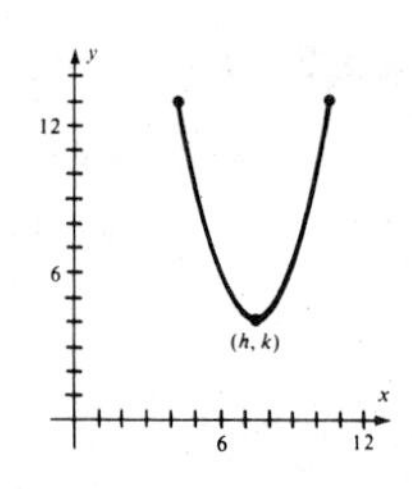

49. 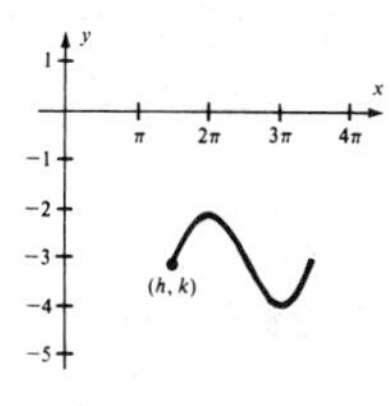

51. 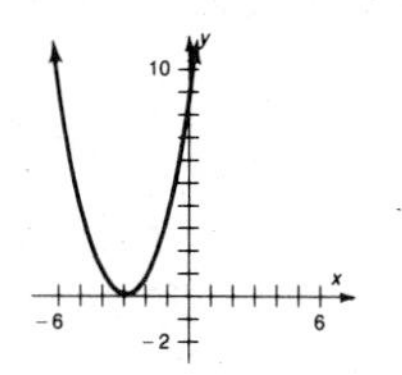

53. 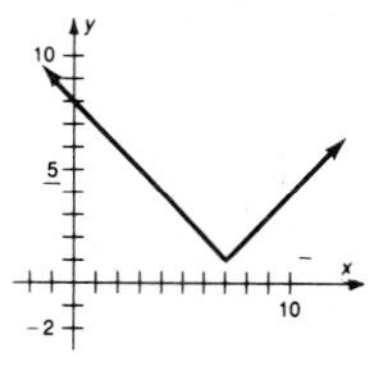

55. 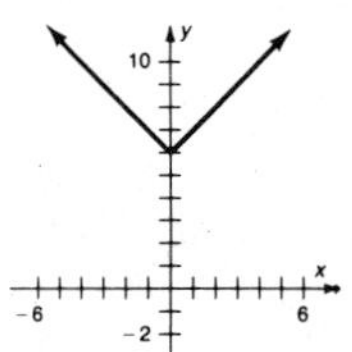

57. 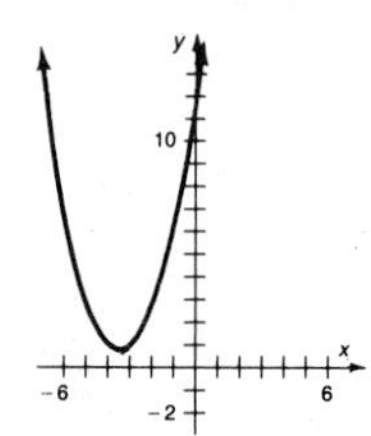

59. 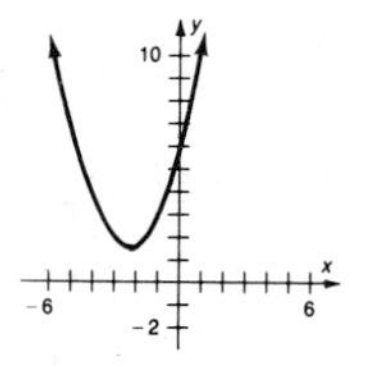

61. 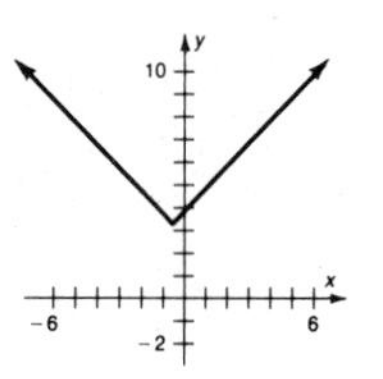

63. 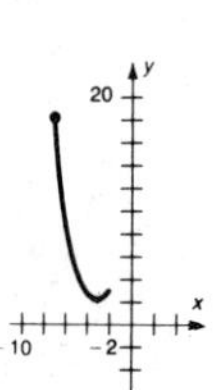

65. 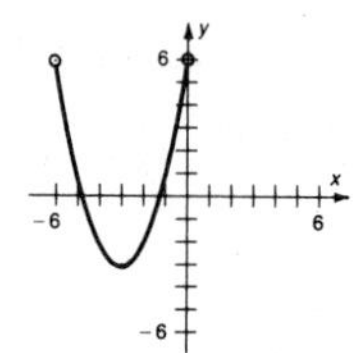

PROBLEM SET 2.5, PAGES 76–78

1. 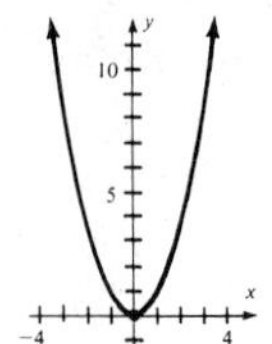

3. 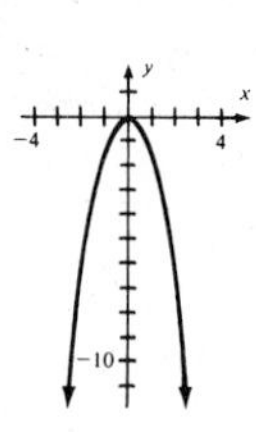

5. 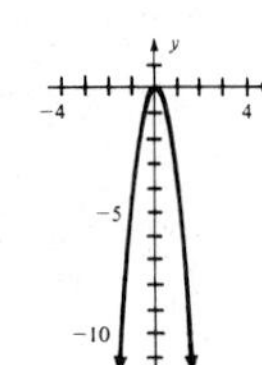

7. 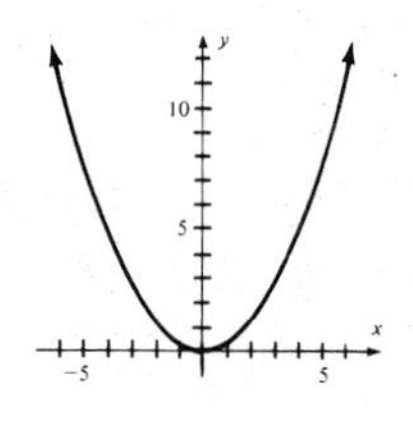

9. 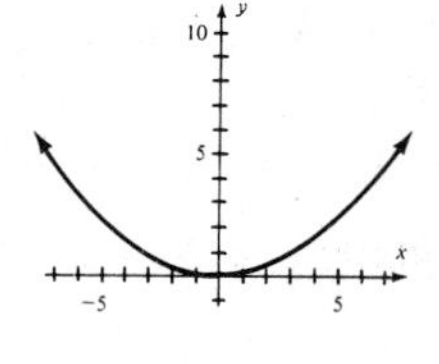

11. 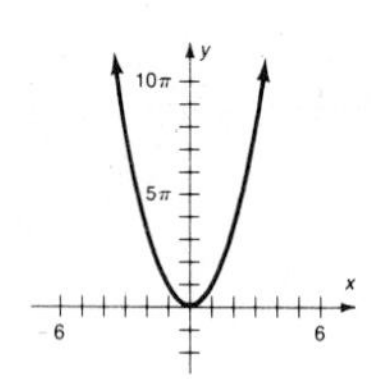

13. 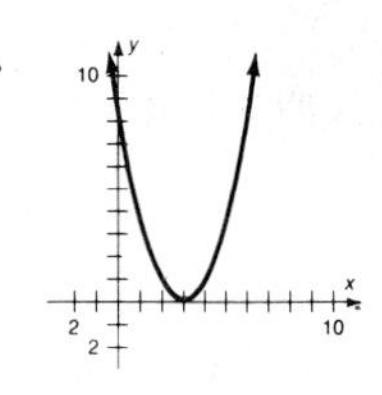

15. 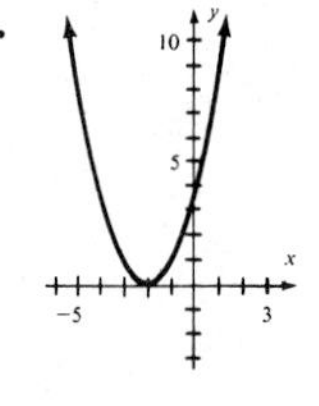

17. 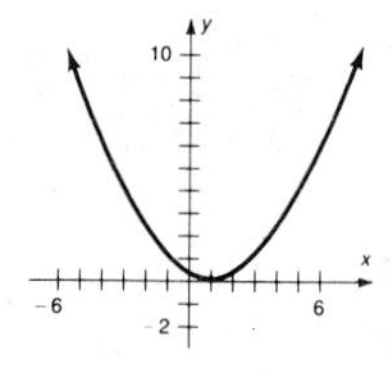

19. 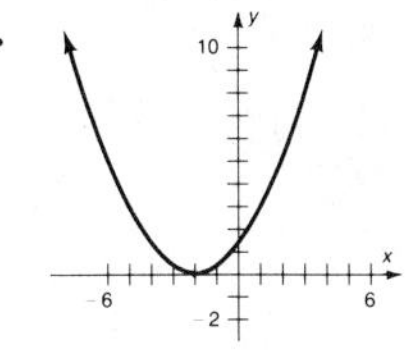

21. 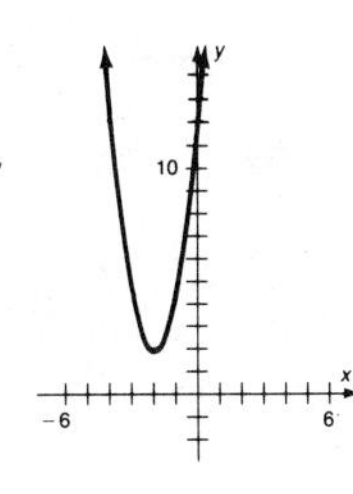

23. 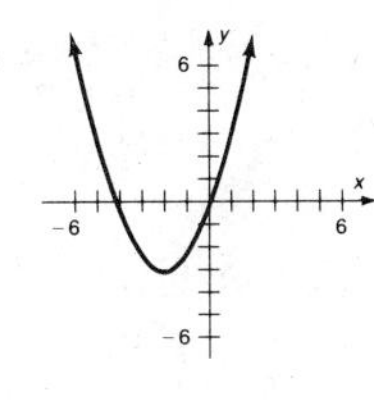

25. 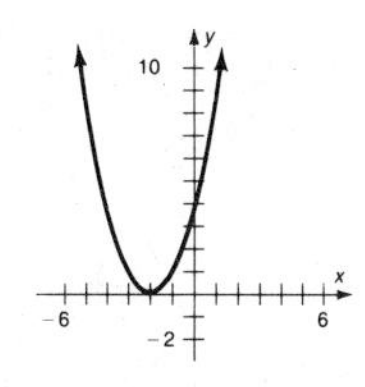

27. 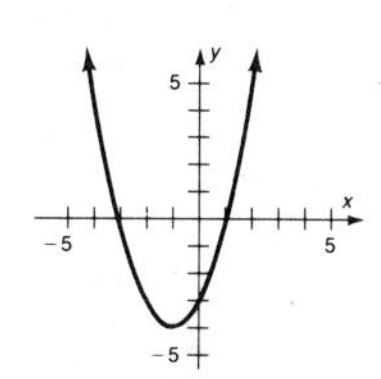

29. 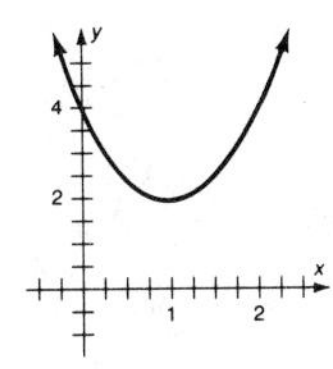

31. 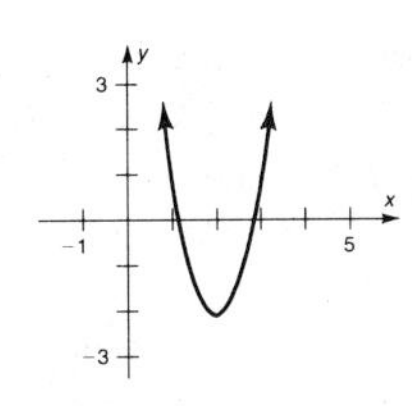

33. 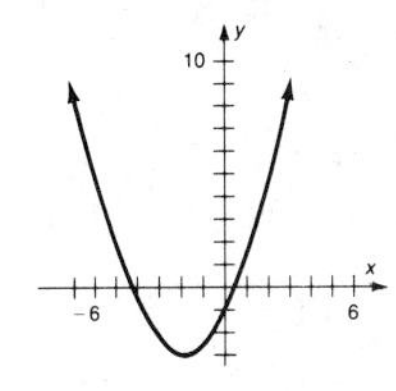

35. 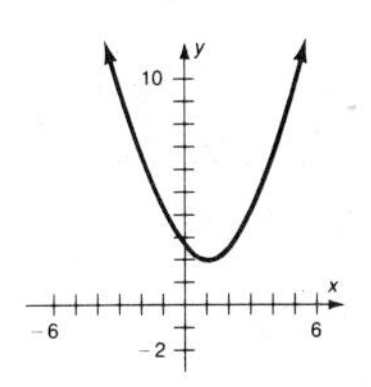

37. 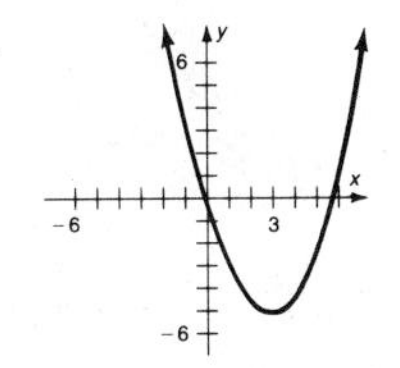

39. 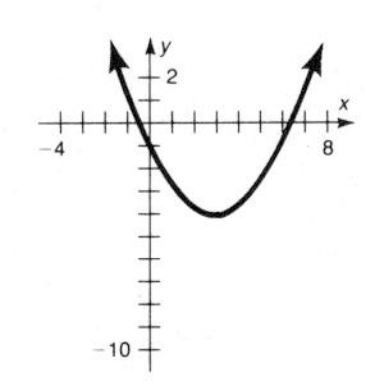

41. 3 **43.** -15 **45.** $\frac{2}{3}$

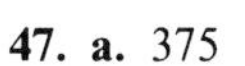

47. a. 375 **b.** \$250,000 loss **c.** \$1,156,250 **49.** \$650
51. a. If $x = 9$, then $y - 18 = -2$, so $y = 16$; height is 16 ft. **b.** If $x = 18$, then $y - 18 = -\frac{2}{81}(18)^2$ so $y = 10$; height is 10 ft.
53. Dimensions are 25 ft by 25 ft; area $= 625$ ft^2. **55.** The maximum height is about 3456 ft (to the nearest foot).

57. 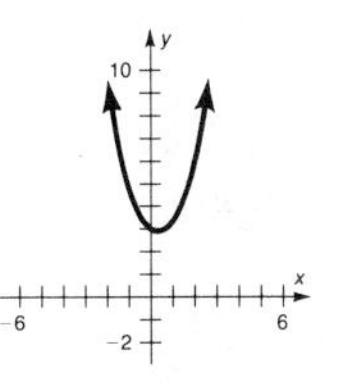

59. 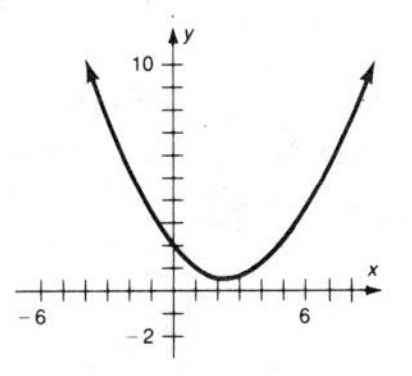

61. 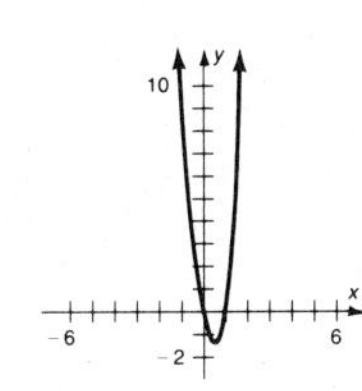

63. a. $y = 2x$ **b.**

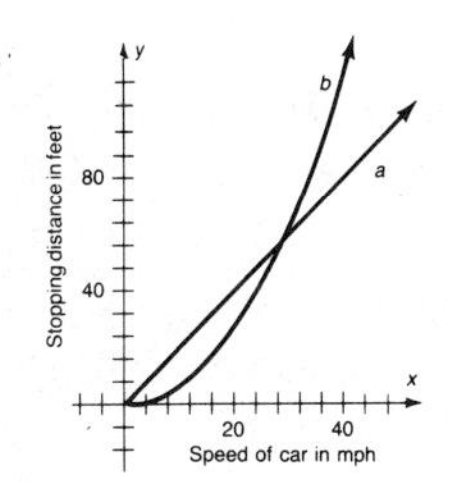

c. Answers vary; at 0 mph and at about 28 mph.

PROBLEM SET 2.6, PAGES 83–85

1. $(f \circ g)(x) = 2x + 9$; $(g \circ f)(x) = 2x + 3$ **3.** $(f \circ g)(x) = x$; $(g \circ f)(x) = x$
5. $(f \circ g)(x) = \dfrac{x^2 - x - 2}{x^2 - x + 1}$; $(g \circ f)(x) = \dfrac{-3x + 6}{(x + 1)^2}$ **7.** $(f \circ g)(x) = \dfrac{1}{x^2 - 2}$; $(g \circ f)(x) = \dfrac{-x^2 + 2x}{(x - 1)^2}$
9. $(f \circ g)(x) = 3$; $(g \circ f)(x) = 45$ **11.** $f \circ g = \{(5, 12), (6, 3)\}$; $g \circ f = \varnothing$ **13.** inverses **15.** not inverses
17. not inverses **19.** $f^{-1} = \{(5, 4), (3, 6), (1, 7), (4, 2)\}$ **21.** $f^{-1}(x) = x - 3$ **23.** $g^{-1}(x) = \frac{x}{5}$
25. The inverse function does not exist because h is not a one-to-one function. **27.** $f^{-1}(x) = x$
29. The inverse function does not exist because f is not a one-to-one function. **31.** $f^{-1}(x) = \dfrac{1 + 2x}{x}$; $x \neq 0$
33. $f^{-1}(x) = \dfrac{-3(x + 2)}{3x - 2}$; $x \neq \dfrac{2}{3}$ **35. a.** $4x^2 - 4x + 1$ **b.** $6x + 3$ **c.** $36x^2 + 36x + 9$ **d.** $36x^2 + 36x + 9$
37. a. x **b.** $\frac{1}{2}x^2 - \frac{3}{2}$ **c.** $x^2 + 1$ **d.** $x^2 + 1$ **39. a.** $6x - 13$ **b.** $2x - 3$ **c.** $6x - 7$ **d.** $6x - 7$
41. a. $4x$ **b.** $4x$ **c.** $8x$ **d.** $8x$ **43. a.** -4 **b.** -2 **c.** 3 **d.** 5 **e.** 8
45. a. 5.5 **b.** 6 **c.** 12 **d.** 4 **e.** 0 **47. a.** 1 **b.** -1 **c.** 7 **d.** 14 **e.** 7

49. a. $D: [0, 15]$; $R: [-6, 10]$ **b.** $D: [-6, 10]$; $R: [0, 15]$ **51.** inverses **53.** not inverses **55. a.** \$300 **b.** \$84 **c.** $(p \circ c)(n) = \dfrac{60n + 240}{n}$ **57. a.** 144π **b.** $(S \circ r)(t) = 36\pi t^2$ **c.** $(0, \frac{8}{3}]$ **59.** $f^{-1}(x) = -\sqrt{x}$ on $[0, \infty)$ **61.** $f^{-1}(x) = \sqrt{x - 1}$ on $[1, \infty)$ **63.** $f^{-1}(x) = -\frac{1}{2}\sqrt{2x}$ on $[2, 200]$ **65.** $f^{-1}(x) = \dfrac{1 + x}{2x}$ on $(0, \infty)$ **67.** $f^{-1}(x) = x - 1$ on $[0, \infty)$ **69. a.** 1 **b.** 4 **c.** 9 **d.** k^2

CHAPTER 2 SUMMARY, PAGES 85–87

1. onto, one-to-one function **3.** not a function **5.** $D: [-5, 11]$; $R: [-3, 7]$ **7.** $(-3.4, 0), (9, 0), (0, 2), (3.4, 0)$ **9. a.** 11 **b.** -11 **11. a.** $3w - 1$ **b.** $5 - w^2 - 2wh - h^2$ **13.** $-2x - h$ **15.** 0 **17.** $(-\infty, 1] \cup [6, \infty)$ **19.** $(-\infty, \infty)$ **21.** not equal **23.** not equal **25.** neither **27.** neither

29.

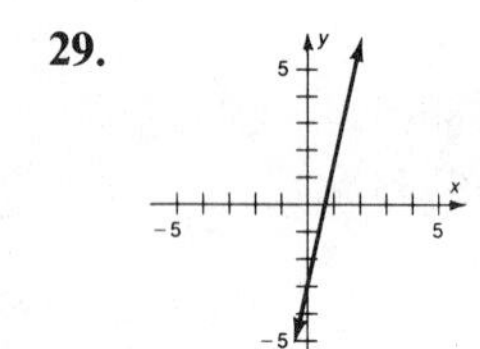

31.

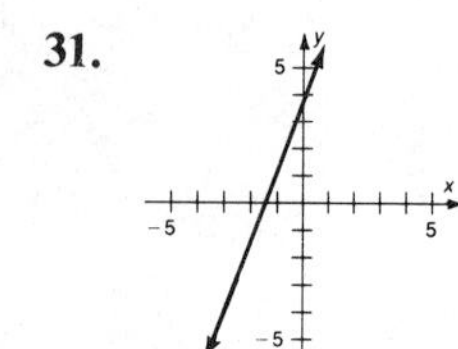

33. $-\frac{5}{4}$ **35.** $\frac{7}{5}$ **37.**

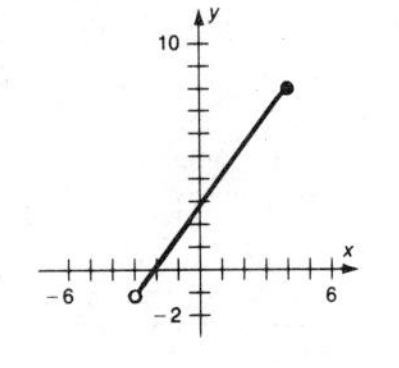

39.

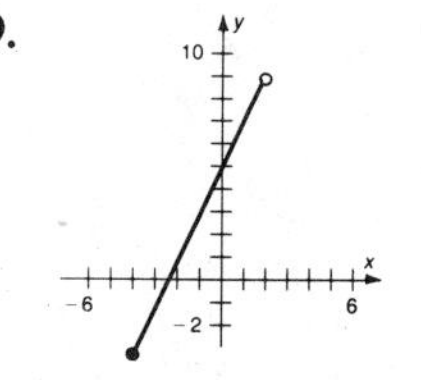

41.

43.

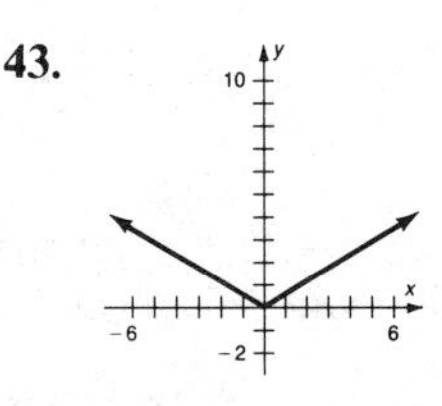

45. $m_{AB} = -\frac{5}{4}$; $m_{BC} = \frac{5}{7}$; $m_{CA} = -\frac{1}{3}$; not a right triangle **47.** $-\frac{2}{3}$

49. $Ax + By + C = 0$; (x, y) is any point on the line and A, B, C are any constants (not all zero). **51.** $y = mx + b$; (x, y) is any point on the line, m is the slope and b is the y intercept. **53.** $2x - 3y - 27 = 0$ **55.** $5x + 8y + 23 = 0$ **57.** $(-\pi, 6)$ **59.** $(4, 0)$ **61.** $y - 3 = 9(x + \sqrt{2})^2$ **63.** $y = -2(x - \pi)^2$ **65.** $y' = 3x'^2$ **67.** $y' = 5x'^2$ **69.**

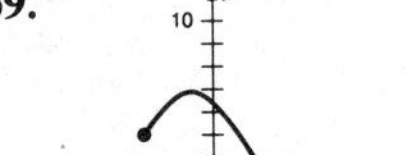

71. **73.**

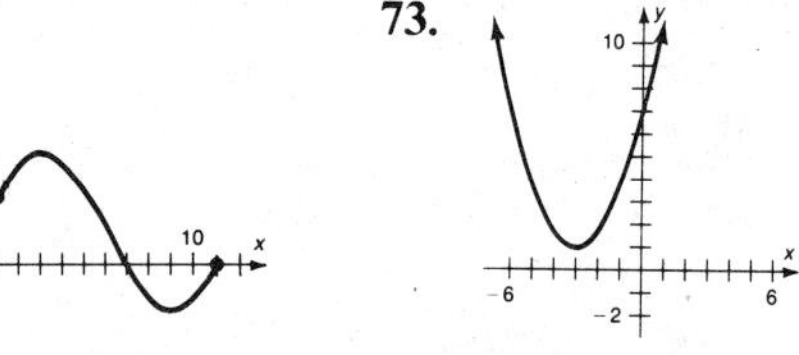

75.

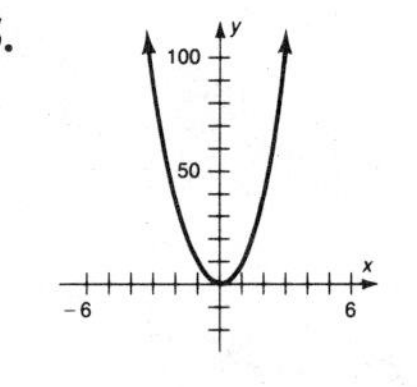

77. $y + 2 = (x + 1)^2$

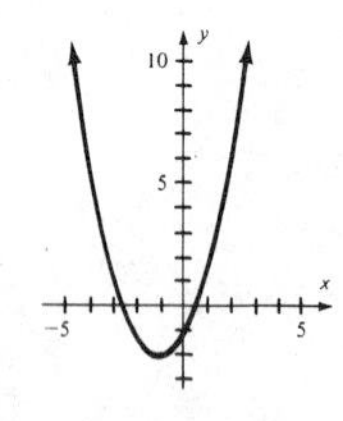

79. $y - 1 = (x - 3)^2$

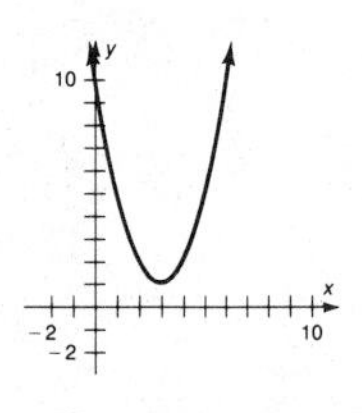

81. maximum value 250 at $x = -6$ **83.** minimum value -850 at $x = 5$

85. $14 - 3x^2$ **87.** 17 **89.** no **91.** no **93.** $f^{-1}(x) = \dfrac{x + 1}{3}$ **95.** $f^{-1}(x) = 2x - 10$

PROBLEM SET 3.1, PAGES 93–94

1. $(0, 5), (-5, 0)$ **3.** $(0, -4), (2, 0), (-2, 0)$ **5.** $(0, 0)$ **7.** $(0, 3)$ **9.** $(0, \frac{9}{2}), (3, 0), (-3, 0)$ **11.** $(0, -2), (2, 0)$ **13.** $(0, 1), (1, 0)$ **15.** $(0, \sqrt{3}), (-3, 0)$ **17.** $(\frac{5}{2}, 0), (-\frac{5}{2}, 0)$ **19.** none **21.** $(0, 2), (0, -2), (2, 0), (-2, 0)$ **23.** $(0, -3)$

25. none **27.** y axis **29.** origin **31.** none **33.** none **35.** none **37.** none **39.** none
41. x axis, y axis, origin **43.** origin **45.** none **47.** $(-\frac{1}{5}, 3), (3, 3)$ **49.** $(-4, -16)$ **51.** none
53. $(0, -1)$; no symmetry **55.** $(2, 0), (-2, 0), (0, 3), (0, -3)$; x axis, y axis, origin **57.** $(0, 1), (-1, 0)$; no symmetry
59. $(2, 0), (-2, 0)$; x axis, y axis, origin **61.** $(16, 0)$; x axis **63.** Answers vary. **65.** Answers vary.

PROBLEM SET 3.2, PAGES 99–100

1. $y = x + 1; x \neq -2$

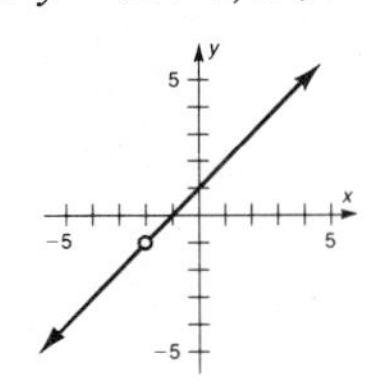

3. $y = 2x + 1; x \neq -1$

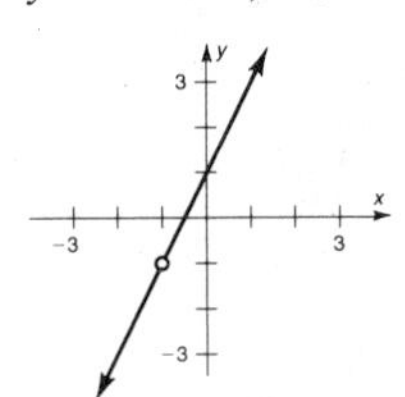

5. $y = x + 3; x \neq -1, 2$

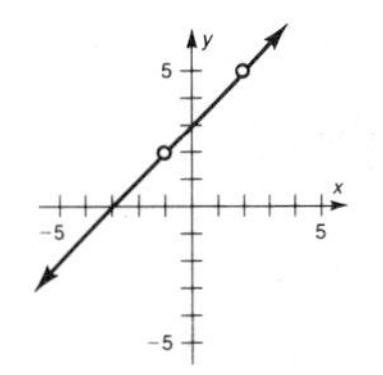

7. $y = 3x + 2; x \neq -2, 1$

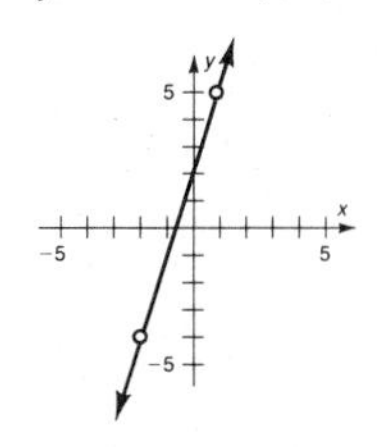

9. $y = (x + 1)(x - 1); x \neq -2$

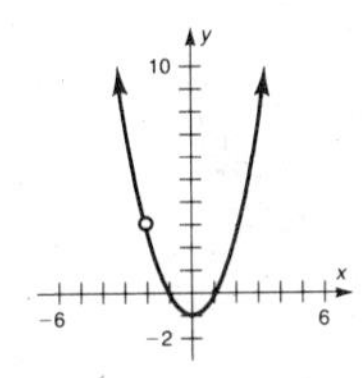

11. $y = (x - 3)(x + 4); x \neq -2, 1$

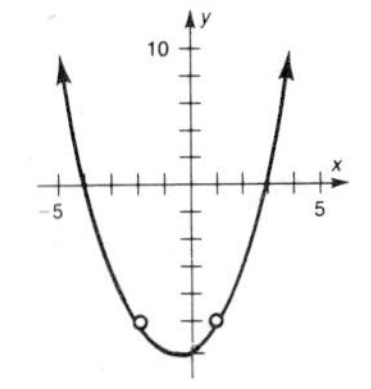

13. $y = x - 4; x \neq -3$

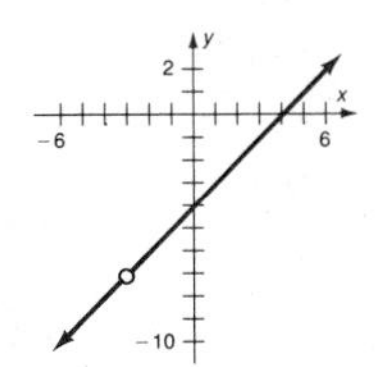

15. $y = x - 3; x \neq -2$

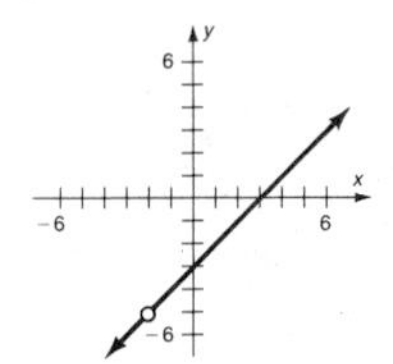

17. $y = 3x - 4; x \neq -\frac{1}{2}$

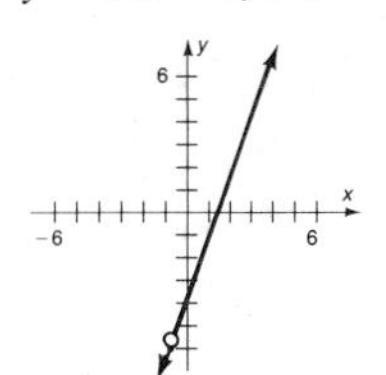

19. $y = x^2 + 4x + 2; x \neq -2$

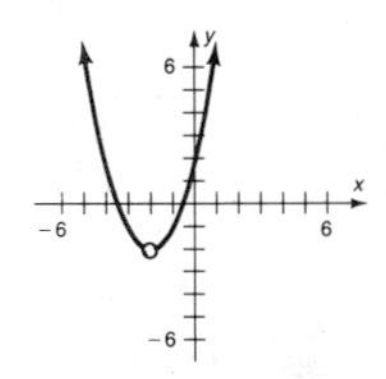

21. $y = x^2 + 10x + 20; x \neq -2$

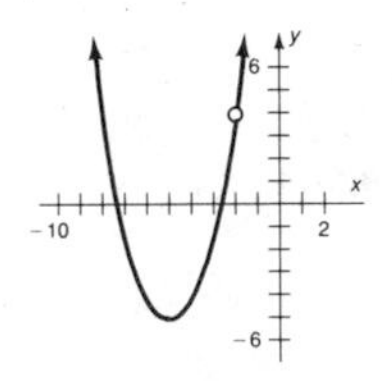

23. $D = [0, \infty); R = [0, \infty)$

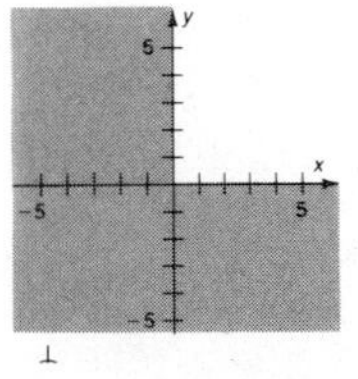

25. $D = [2, \infty); R = [0, \infty)$

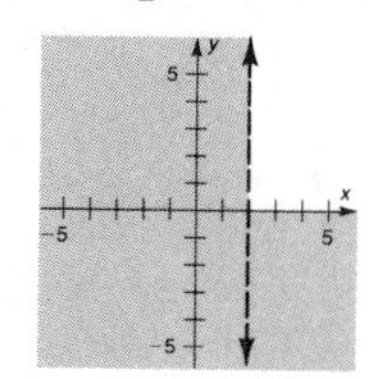

27. $D = [\frac{1}{2}, \infty); R = [0, \infty)$

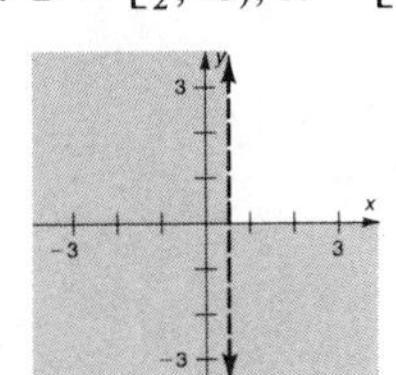

29. $D = [0, \infty); R = [0, \infty)$

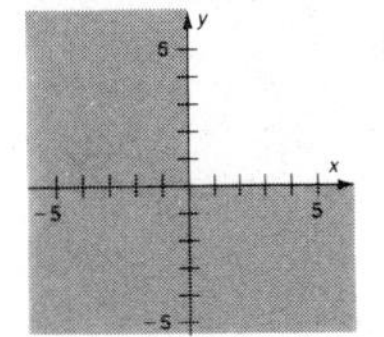

31. $D = [0, \infty)$; $R = [0, \infty)$

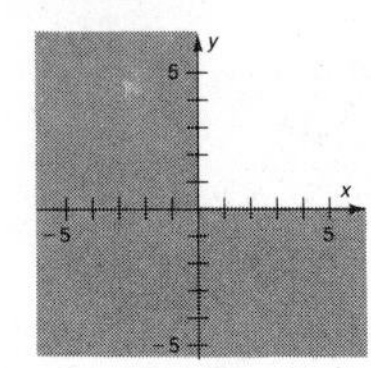

33. $D = [0, \infty)$; $R = (-\infty, \infty)$

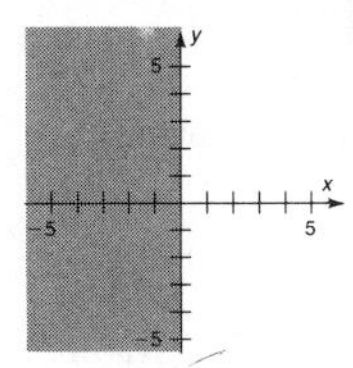

35. $D = (-\infty, -3] \cup [3, \infty)$;
$R = (-\infty, \infty)$

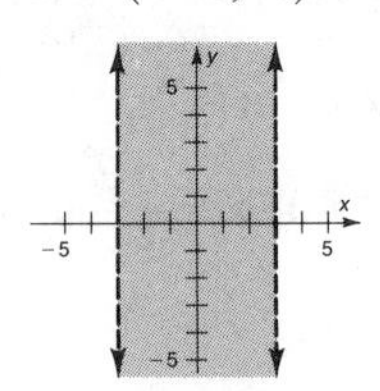

37. $D = (-\infty, -2] \cup [2, \infty)$;
$R = [0, \infty)$

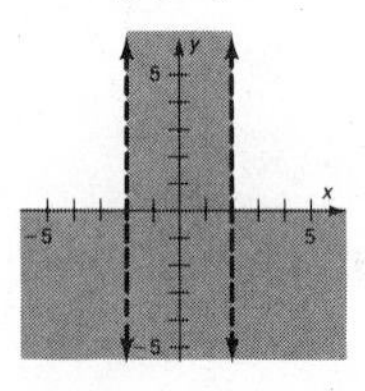

39. $D = (-\infty, -4] \cup [3, \infty)$;
$R = [0, \infty)$

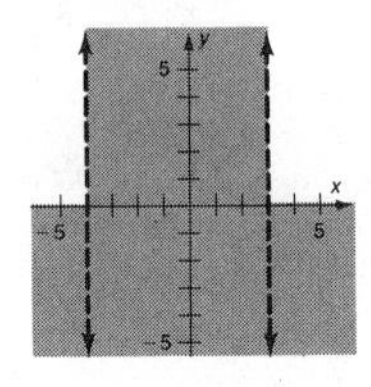

41. $D = [-1, 1] \cup (2, \infty)$;
$R = [0, \infty)$

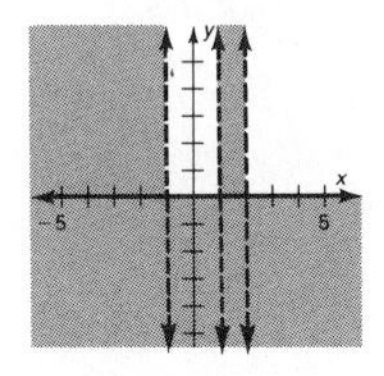

43. $D = (-1, 1)$; $R = (-\infty, \infty)$

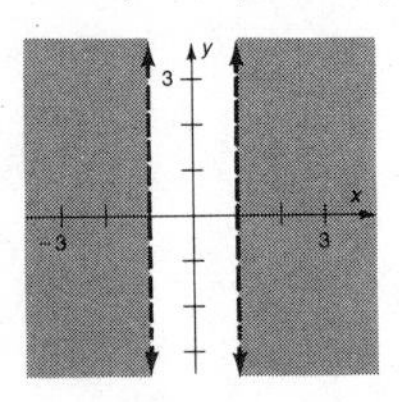

45. $D = [-3, 3]$; $R = [-4, 4]$

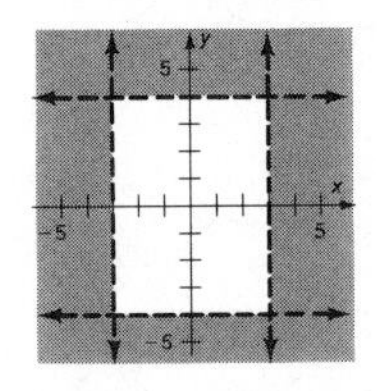

47. $D = [-2, 4]$; $R = [-6, 0]$

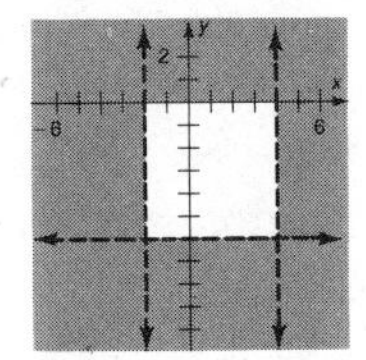

49. $D = (-\infty, -1) \cup (1, \infty)$;
$R = (-\infty, 0) \cup (0, \infty)$

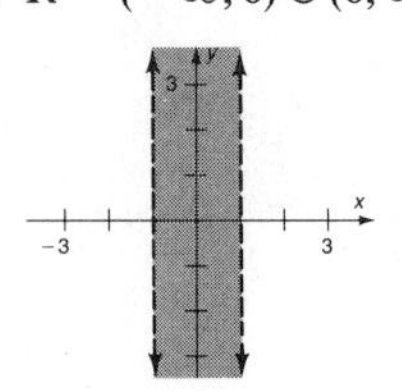

51. $D = [0, 4]$; $R = [-3, 1]$

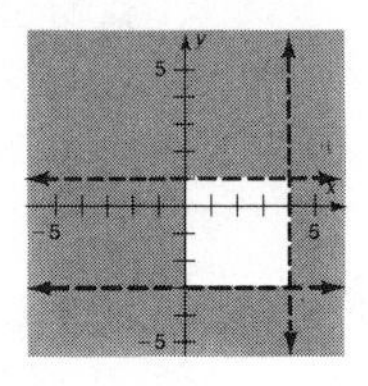

53. $D = (0, \infty)$;
$R = (-\infty, -\sqrt{2}] \cup [\sqrt{2}, \infty)$

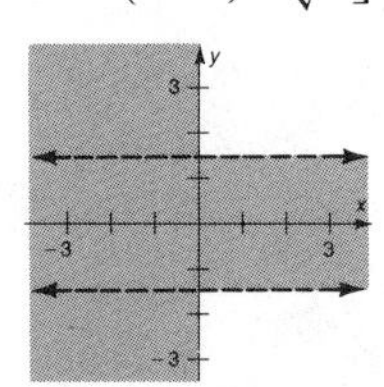

55. $D = (-\infty, \infty)$; $R = [0, 3)$

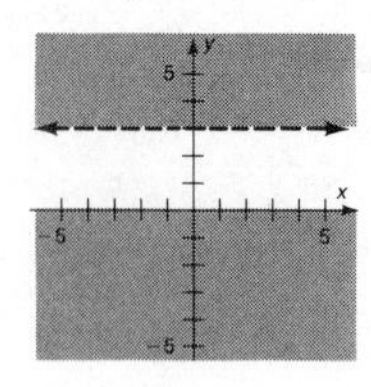

57. $D = (-\infty, 0] \cup (4, \infty)$;
$R = (-\infty, -2) \cup (2, \infty)$

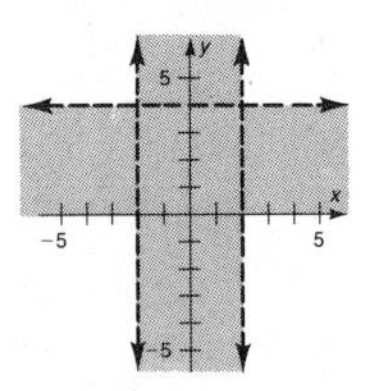

59. $D = (-\infty, -1) \cup (-1, 1) \cup (1, \infty)$;
$R = (-\infty, \infty)$

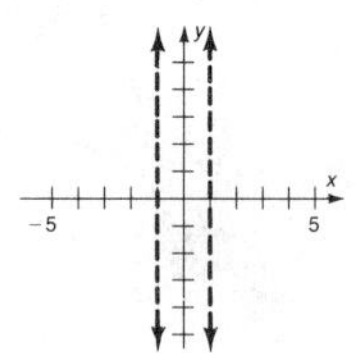

61. $D = (-\infty, -2) \cup [0, 1] \cup (2, \infty)$;

$$R = \left(-\infty, -\sqrt{\frac{2+\sqrt{3}}{4}}\right] \cup \left[-\sqrt{\frac{2-\sqrt{3}}{4}}, \sqrt{\frac{2-\sqrt{3}}{4}}\right] \cup \left[\sqrt{\frac{2+\sqrt{3}}{4}}, \infty\right)$$

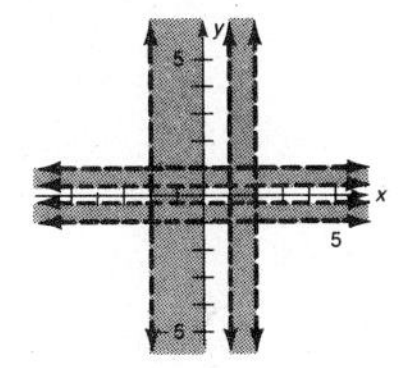

PROBLEM SET 3.3, PAGE 107

1. doesn't exist **3.** 0 **5.** $\frac{1}{2}$ **7.** doesn't exist **9.** 12 **11.** 1 **13.** doesn't exist **15.** 0 **17.** $\frac{1}{4}$ **19.** 2
21. doesn't exist **23.** 5 **25.** doesn't exist **27.** $\frac{4}{3}$ **29.** $\frac{3}{5}$ **31.** $x = 0$; $y = 0$ **33.** $x = 0$; $y = 1$
35. $x = 0$; $y = 2$ **37.** $x = -3$; $y = 0$ **39.** $x = 4$; $y = x + 4$ **41.** $x = 1$; $y = -x - 1$ **43.** $x = 2$; $x = -3$; $y = 0$
45. $x = 5$; $x = -4$; $y = 0$ **47.** none **49.** none **51.** none **53.** none **55.** none **57.** none
59. $x = 2$; $x = -4$; $y = 1$

PROBLEM SET 3.4, PAGES 115–116

1. A function; $D = (-\infty, 0) \cup (0, \infty)$; $R = (-\infty, 0) \cup (0, \infty)$; symmetric with respect to the origin; asymptotes are $x = 0$ and $y = 0$; there are no intercepts
3. A function; $D = (-\infty, 0) \cup (0, \infty)$; $R = (-\infty, 1) \cup (1, \infty)$; asymptotes are $x = 0$ and $y = 1$; the intercept is $(-1, 0)$
5. A function; $D = (-\infty, 4]$; $R = [0, \infty)$; intercepts are $(0, 2)$ and $(4, 0)$
7. A function; $D = [6, \infty)$; $R = (-\infty, 0]$; the intercept is $(6, 0)$
9. A function; $D = (-\infty, -2) \cup (-2, 2) \cup (2, \infty)$; $R = (-\infty, -\frac{1}{4}] \cup (0, \infty)$; symmetric with respect to the y axis; asymptotes are $x = 2$, $x = -2$, and $y = 0$; the intercept is $(0, -\frac{1}{4})$
11. A function; $D = (-\infty, -2) \cup (-2, \infty)$; $R = (-\infty, \infty)$; asymptotes are $x = -2$ and $y = 2x - 3$; the intercepts are $(0, -5)$, $(-\frac{5}{2}, 0)$, $(2, 0)$
13. A function; $D = (-\infty, -\frac{1}{2}) \cup (-\frac{1}{2}, \infty)$; $R = [-1, \frac{5}{4}) \cup (\frac{5}{4}, \infty)$; the intercepts are $(0, 0)$, $(2, 0)$
15. A function; $D = [0, \infty)$; $R = (-\infty, \frac{1}{4}]$; the intercepts are $(0, 0)$ and $(1, 0)$
17. A function; $D = (-\infty, -3] \cup [1, \infty)$; $R = [0, \infty)$; the intercepts are $(-3, 0)$ and $(1, 0)$
19. A function; $D = (-2, 2)$; $R = (-\infty, \infty)$; asymptotes are $x = 2$ and $x = -2$; the intercept is $(0, 0)$
21. Not a function; $D = [-5, 5]$; $R = [-5, 5]$; symmetric with respect to the x axis, y axis, and origin; the intercepts are $(5, 0)$, $(-5, 0)$, $(0, 5)$, and $(0, -5)$
23. Not a function; $D = [-4, 4]$; $R = [-3, 3]$; symmetric with respect to the x axis, y axis, and origin; the intercepts are $(0, -3)$, $(0, 3)$, $(4, 0)$, and $(-4, 0)$
25. A function; $D = (-\infty, 0) \cup (0, \infty)$; $R = (-\infty, -2] \cup [2, \infty)$; symmetric with respect to the origin; asymptotes are $x = 0$ and $y = x$; there are no intercepts
27. Not a function; $D = [-1, 1] \cup (4, \infty)$; $R = (-\infty, -\sqrt{8 + 2\sqrt{15}}] \cup (-\sqrt{8 - 2\sqrt{15}}, \sqrt{8 - 2\sqrt{15}}) \cup [\sqrt{8 + 2\sqrt{15}}, \infty)$; asymptote is $x = 4$; the intercepts are $(0, \frac{1}{2})$, $(0, -\frac{1}{2})$, $(1, 0)$, and $(-1, 0)$
29. Not a function; $D = (-\infty, -3) \cup (-3, 3] \cup (5, \infty)$; $R = (-\infty, -1) \cup (-1, 1) \cup (1, \infty)$; symmetric with respect to the x axis; asymptotes are $x = 5$, $y = 1$, and $y = -1$; the intercepts are $(0, \sqrt{15}/5)$, $(0, -\sqrt{15}/5)$, and $(3, 0)$

31.

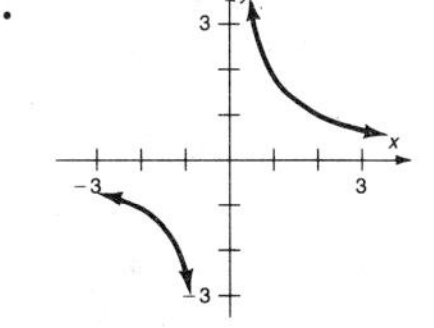

33.

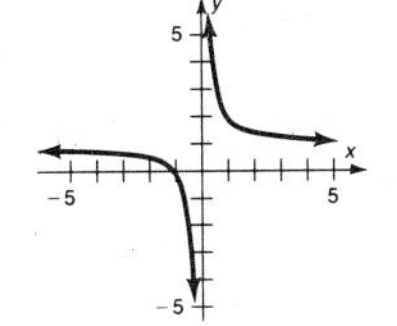

35.

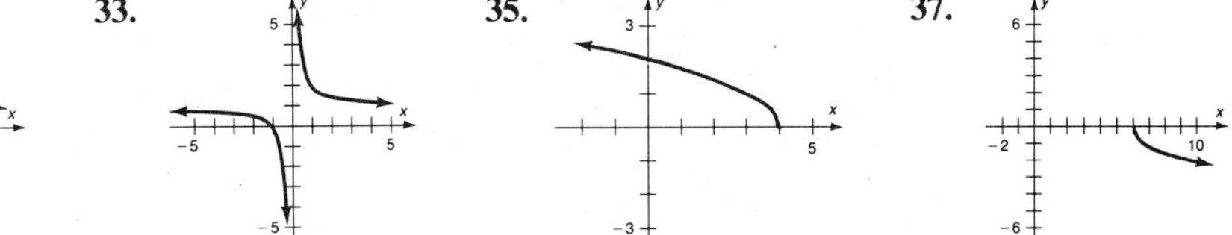

37.

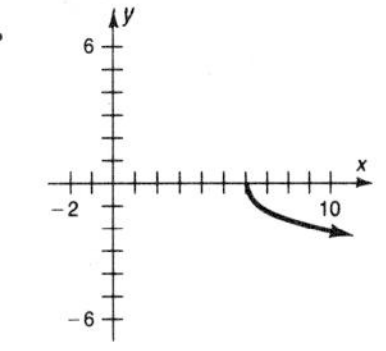

39.

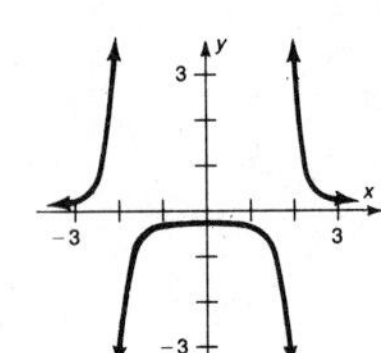

41.

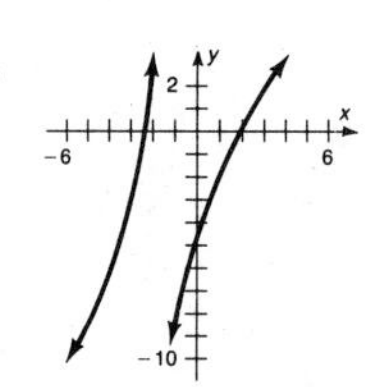

43.

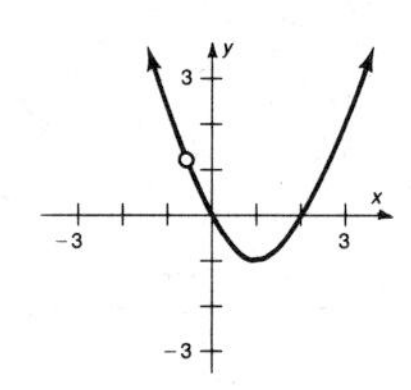

45.

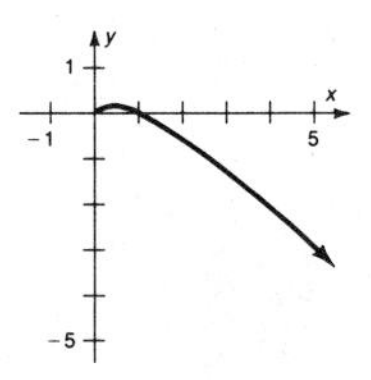

47.

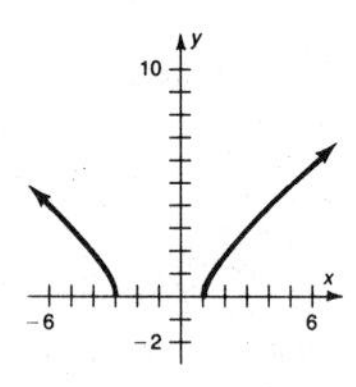

49.

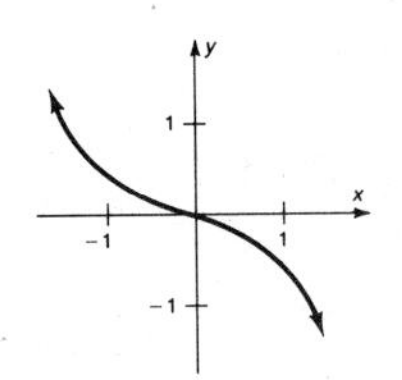

51.

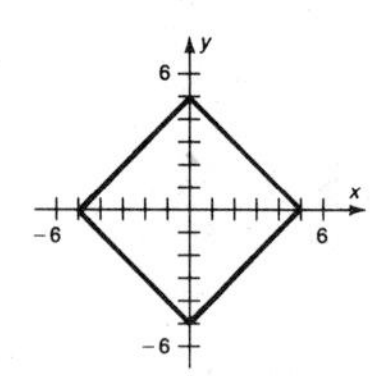

53.

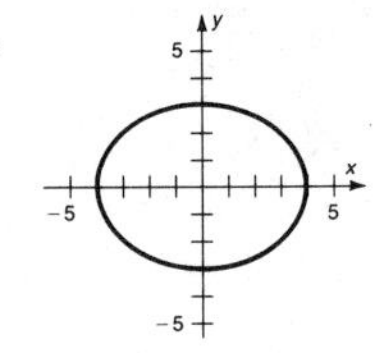

55.

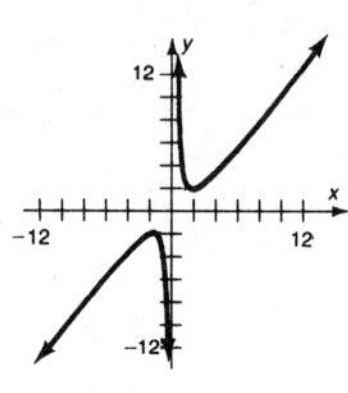

57.

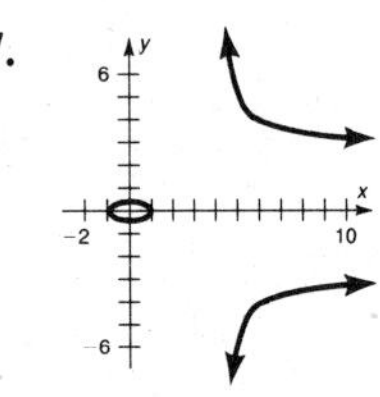

59.

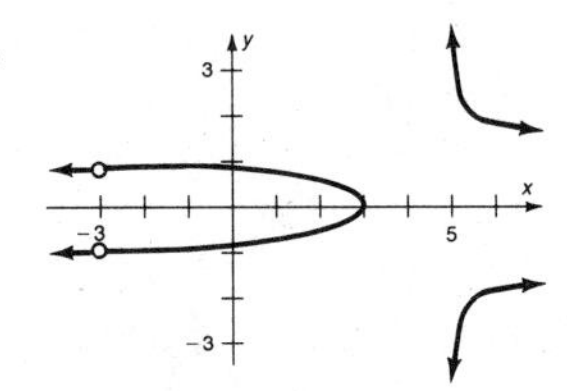

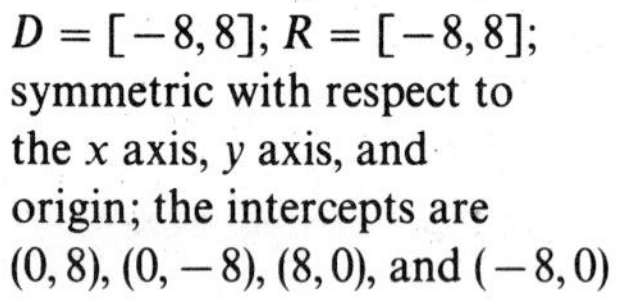

61. Not a function; $D = [-8, 8]$; $R = [-8, 8]$; symmetric with respect to the x axis, y axis, and origin; the intercepts are $(0, 8)$, $(0, -8)$, $(8, 0)$, and $(-8, 0)$

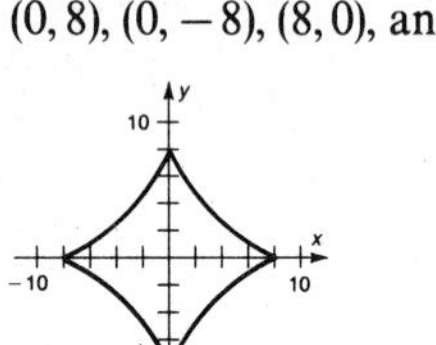

63. Not a function; $D = (-\infty, -3) \cup [-2, 2] \cup [3, \infty)$; $R = (-\infty, -1) \cup [-\frac{2}{3}, \frac{2}{3}] \cup (1, \infty)$; symmetric with respect to the x axis, y axis, and origin; the asymptotes are $x = 3$, $x = -3$, $y = 1$, $y = -1$; the intercepts are $(0, \frac{2}{3})$, $(0, -\frac{2}{3})$, $(2, 0)$, and $(-2, 0)$

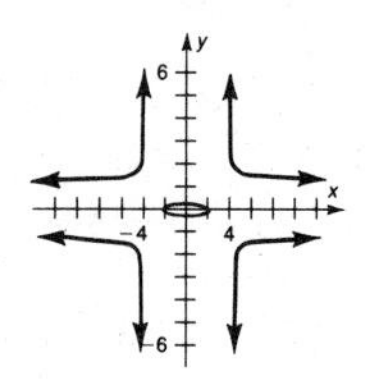

65. Not a function; $D = [-2, 2]$; $R = [-2, 2]$; symmetric with respect to the x axis, y axis, and origin; the intercepts are $(0, 0)$, $(2, 0)$, and $(-2, 0)$

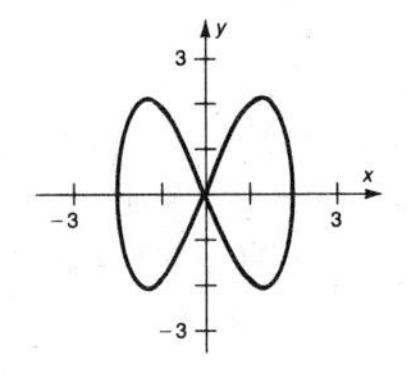

CHAPTER 3 SUMMARY, PAGES 116–117

1. $(0, -20)$, $(2, 0)$, $(-\frac{5}{3}, 0)$, $(-2, 0)$ **3.** $(\frac{1}{2}\sqrt{2}, 0)$, $(-\frac{1}{2}\sqrt{2}, 0)$ **5.** x axis **7.** origin

9. $y = x^2 - 4x + 4, x \neq 4$ **11.** $y = x - 2, x \neq 2, 3$ **13.** $D = (-\infty, \infty)$; $R = [5, \infty)$

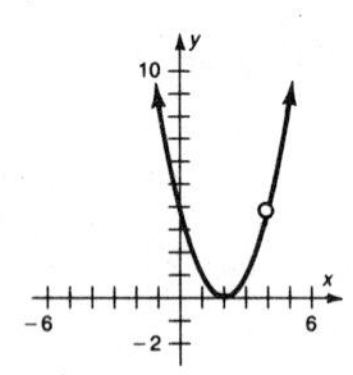

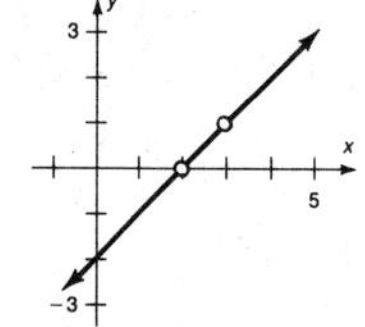

15. $D = [-5, 1)$; $R = [0, \infty)$ **17.** $\frac{3}{4}$ **19.** 0 **21.** $x = \frac{1}{2}$; $y = 3x - 4$ **23.** $x = 3, x = -3$

25. Not a function;
$y = \pm\sqrt{\dfrac{x^2+25}{x^2+1}}$; $x = \pm\sqrt{\dfrac{y^2-25}{1-y^2}}$;
$D = (-\infty, \infty)$; $R = [-5, -1) \cup (1, 5]$;
symmetric with respect to the x axis, y axis, and origin;
the intercepts are $(0, 5)$ and $(0, -5)$

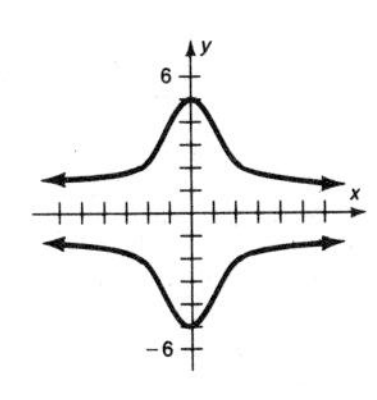

27. Not a function;
$D = [-1, 1]$; $R = [-1, 1]$;
symmetric with respect to the x axis, y axis and origin;
the intercepts are $(1, 0)$, $(-1, 0)$, $(0, -1)$ and $(0, 1)$

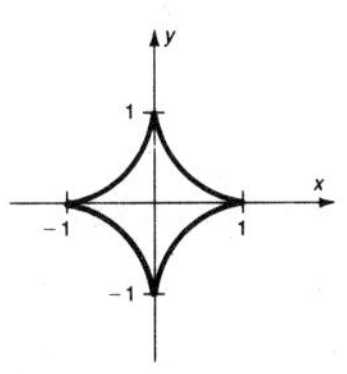

PROBLEM SET 4.1, PAGES 125–126

1. a. 11 **b.** 4 **c.** -10 **3. a.** 26 **b.** 0 **c.** -4 **5. a.** $x^3 - 7x^2 + 6x - 6$ **b.** $15x^3 - 22x^2 + 5x + 2$
7. a. $x^2 + 3x + 2$ **b.** $y^2 + y - 6$ **c.** $x^2 - x - 2$ **d.** $y^2 - y - 6$ **9. a.** $2x^2 - x - 1$ **b.** $2x^2 - 5x + 3$
c. $3x^2 + 4x + 1$ **d.** $3x^2 + 5x + 2$ **11. a.** $a^2 + 4a + 4$ **b.** $b^2 - 4b + 4$ **c.** $x^2 + 8x + 16$ **d.** $y^2 - 6y + 9$
13. a. $m(e + i + y)$ **b.** $(a - b)(a + b)$ **c.** irreducible **d.** $(a - b)(a^2 + ab + b^2)$ **15. a.** $(a + b)^3$ **b.** $(p - q)^3$
c. $(d - c)^3$ **d.** $xy(x + y)$ **17. a.** $(3x + 1)(x - 2)$ **b.** $(3y - 2)(2y - 1)$ **c.** $b(4a - 1)(2a + 3)$ **d.** $2(s - 8)(s + 3)$
19. a. $3x^3 + 8x^2 - 9x + 2$ **b.** $2x^3 + 5x^2 - 8x - 5$ **21. a.** $2x^3 - 3x^2 - 8x - 3$ **b.** $6x^3 + 17x^2 - 4x - 3$
23. a. $x^3 - 3x^2 + 4$ **b.** $x^3 - 3x - 2$ **25.** $(x - y - 1)(x - y + 1)$ **27.** $5(a - 1)(5a + 1)$
29. $-\frac{1}{25}(3x + 10)(7x + 10)$ **31.** $\dfrac{1}{y^8}(x^3 - 13y^4)(x^3 + 13y^4)$ **33.** $(a + b - x - y)(a + b + x + y)$
35. $(2x - 3)(x + 2)$ **37.** $(6x - 1)(x + 8)$ **39.** $(6x + 1)(x + 8)$ **41.** $(4x - 3)(x + 4)$ **43.** $(9x - 2)(x - 6)$
45. $(2x - 1)(2x + 1)(x + 2)(x - 2)$ **47.** $(x + 2)(x + 1)(x^2 - 2x + 4)(x^2 - x + 1)$
49. $\frac{1}{36}(2x + 1)(2x - 1)(3x - 1)(3x + 1)$ **51.** $\frac{1}{32}(2x + 1)^2(2x - 1)(4x^2 - 2x + 1)$

53.

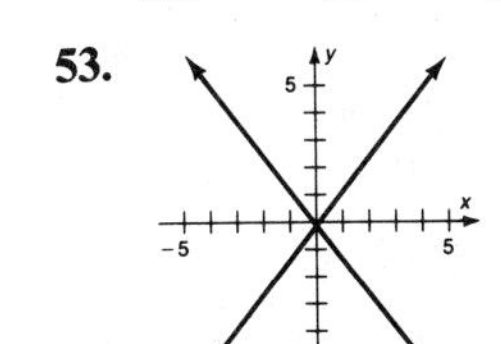

55.

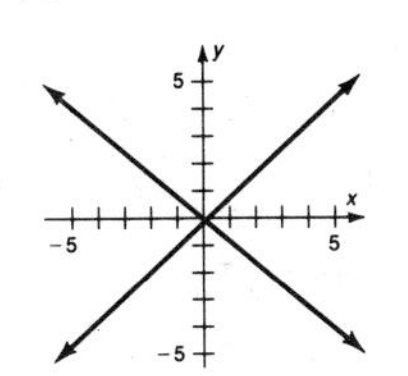

57.

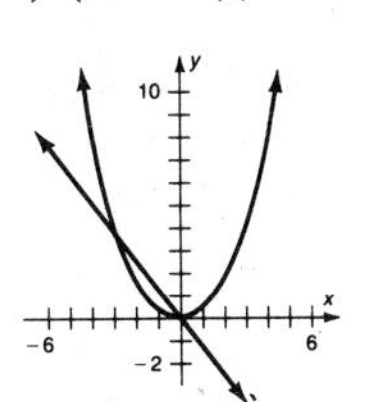

59.

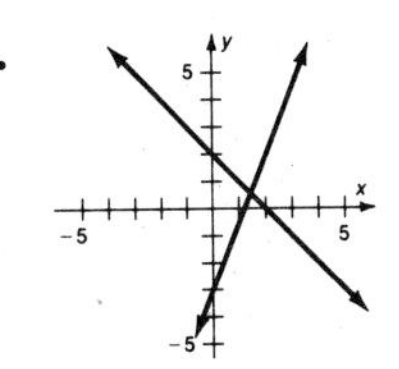

61.

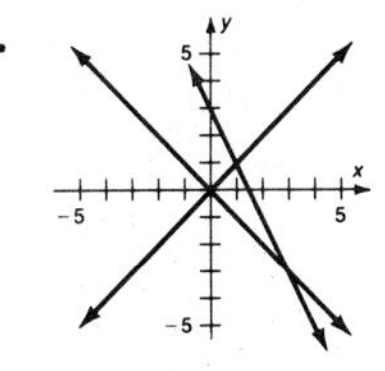

63. $2x^6 - 23x^5 + 103x^4 - 232x^3 + 306x^2 - 297x + 189$ **65.** $6x^4 - x^3 - 6x^2 + 25x - 12$
67. $x^9 - 6x^8 + 15x^7 - 35x^6 + 75x^5 - 96x^4 + 136x^3 - 165x^2 + 75x - 125$ **69.** $(x^n - y^n)(x^{2n} + x^ny^n + y^{2n})$
71. $(x^n - y^n)^2$ **73.** $x(x + 5)$ **75.** $(x - 1)(x + 1)(x + 2)(x^2 - 2x + 4)$ **77.** $(x + y - a - b)(x + y + a + b)$
79. $(x + y + a + b)(x^2 + 2xy + y^2 - ax - by - ay - bx + a^2 + 2ab + b^2)$ **81.** $(x + 1)(x - 4)(x^2 - 3x - 8)$
83. $s(3s + 5t)$ **85.** $2(x - 5)^3(x + 3)(3x + 1)$ **87.** $3(2x - 3)^2(1 - x)^2(5 - 4x)$
89. $(x - 2)^4(x^2 + 1)^2(11x^2 - 12x + 5)$

PROBLEM SET 4.2, PAGES 131–133

1. a. 3 **b.** -9 **c.** -6 **d.** $3x^2 - 3x + 5$ **3. a.** -2 **b.** -4 **c.** -10 **d.** 16 **e.** 12 **f.** -12
g. $2x^2 - 8x + 6$ **5. a.** -4 **b.** 1 **c.** 3 **d.** 8 **e.** -4 **f.** -20 **g.** -4 **h.** $x^2 + 3x - 4 + \dfrac{-4}{x + 4}$
7. a. 1 **b.** 1 **c.** 2 **d.** 0 **e.** -5 **f.** -2 **g.** 2 **h.** 0 **i.** $x^3 + 3x^2 + 3x - 2$ **9. a.** -2 **b.** 1
c. 0 **d.** 0 **e.** 0 **f.** -8 **g.** 0 **h.** 16 **i.** 16 **j.** -32 **k.** -32 **l.** -64
m. $x^4 - 2x^3 + 4x^2 - 8x + 16 + \dfrac{-64}{x + 2}$ **11.** $3x^2 + 7x + 25$ **13.** $x^3 - 7x^2 + 8x - 8$ **15.** $2x^3 - 6x^2 + 3x - 1$

17. $4x^4 + x^3 - 4x^2 - 4x - 4$ **19.** $3x^2 + 4x + 12$ **21.** $x^2 + x - 6$ **23.** $x^2 + x - 12$
25. $4x^3 + 8x^2 - 7x - 7$ **27.** $x^3 - 2x^2 + 3x - 4$ **29.** $3x^3 - 2x^2 - 5$ **31.** $Q(x) = 2x^3 - 6x^2 + 2x - 2$; $R(x) = 0$
33. $Q(x) = x^3 - x^2 + 1$; $R(x) = 1 - x$ **35.** $Q(x) = 3x^2 - 7x + 5$; $R(x) = 0$ **37.** $3x^2 + 2x + 1$ **39.** $x^3 + x^2 + 2$
41. $5x^3 - 5x^2 - 5x + 3$ **43.** $x^4 - 5x^3 + 10x^2 - 18x + 36$ **45.** $4x^2 - 5x + 7$ **47.** $5x^4 - 10x^3 + 20x^2 - 40x + 78$
49. $x^4 + 3x^2 - 7x$ **51.** $x^2 + 2x - 5$ **53.** $2x - 1$ **55.** $2x^2 + 3x - 4$
57. $x^2 + 4x - 5 = (x + 5)(x - 1)$ **59.** $K = 7$

$$\begin{array}{r|rrrrr}
1 & 1 & 7 & 5 & -23 & 10 \\
 & & 1 & 8 & 13 & -10 \\
\hline
 & 1 & 8 & 13 & -10 & 0 \\
-5 & & -5 & -15 & 10 & \\
\hline
 & 1 & 3 & -2 & 0 &
\end{array}$$

$Q(x) = x^2 + 3x - 2$

61. a.
$$\begin{array}{r}
x^2 - x + 3 \\
2x + 1\overline{)2x^3 - x^2 + 5x + 3} \\
\underline{2x^3 + x^2} \\
-2x^2 \\
\underline{-2x^2 - x} \\
6x \\
\underline{6x + 3} \\
0
\end{array}$$

b.
$$\begin{array}{r|rrrr}
-\frac{1}{2} & 2 & -1 & 5 & 3 \\
 & & -1 & 1 & -3 \\
\hline
 & 2 & -2 & 6 & 0
\end{array}$$

$2x^2 - 2x + 6$

c. The answer to part **b** is double that of part **a.** **d.** It will be doubled.

e. Divide by $x + \frac{b}{a}$ synthetically ($a \neq 0$) and then divide the quotient by a and keep the remainder.

PROBLEM SET 4.3, PAGES 138–139

1. a. -3 **b.** -19 **c.** -4 **d.** 842 **e.** -448 **3. a.** -824 **b.** -758 **c.** -812 **d.** -890
e. -1022 **5. a.** -10 **b.** -8 **c.** 68 **d.** 2140 **e.** 380 **7. a.** -3 **b.** 8 **c.** 0 **d.** $-\frac{5}{2}$ **e.** 840
9. a. 0 **b.** 42 **c.** 0 **d.** 0 **e.** 0 **11. a.** 0 **b.** 0 **c.** 0 **d.** -54 **e.** -10

13. **15.** **17.** **19.** **21.**

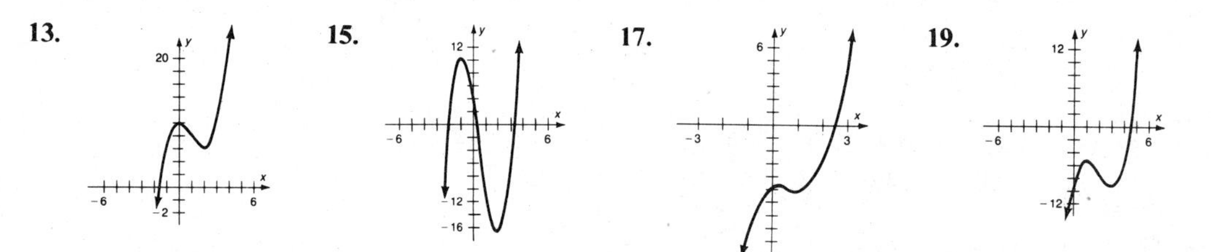

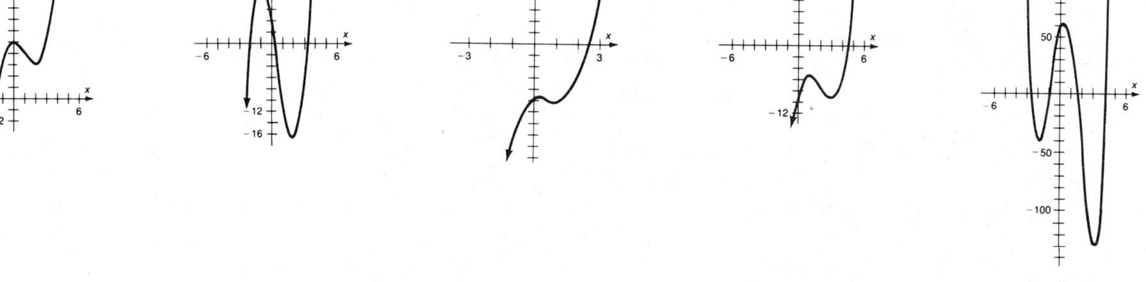

23. **25.** **27.** **29.** **31.**

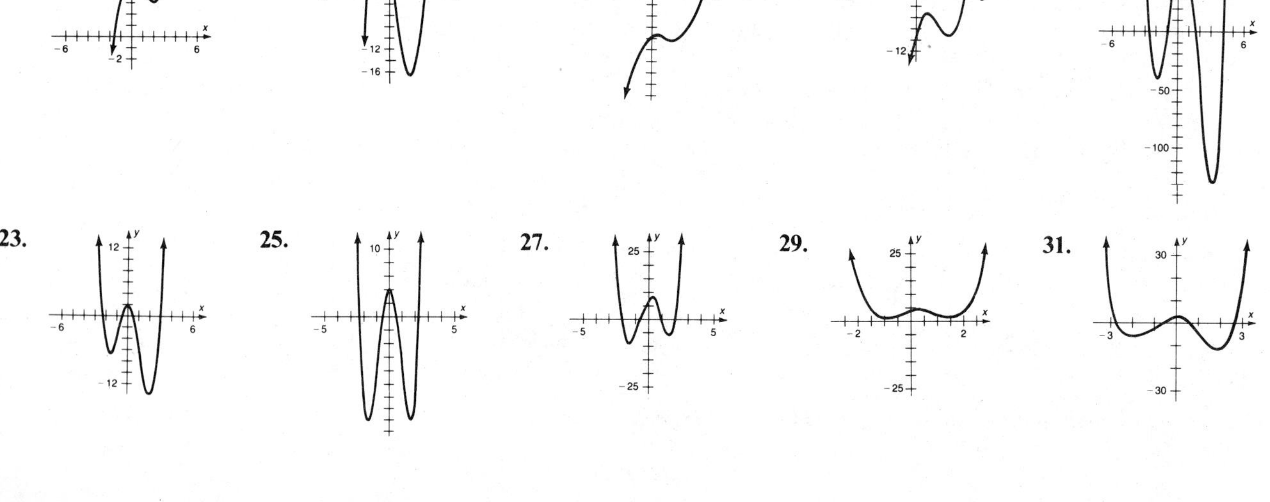

33. 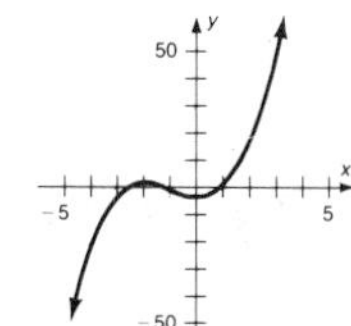

35. 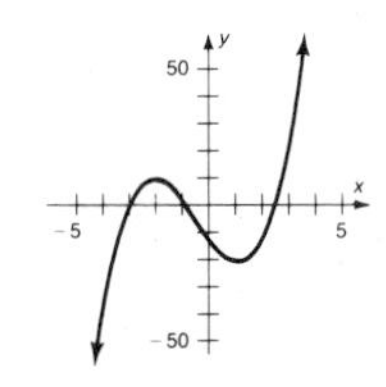

37. 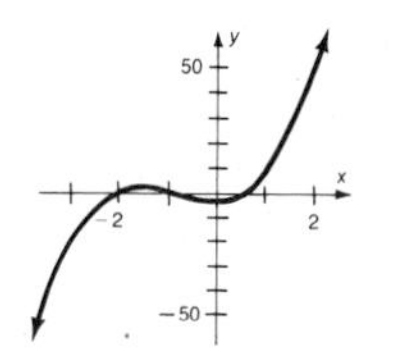

39. 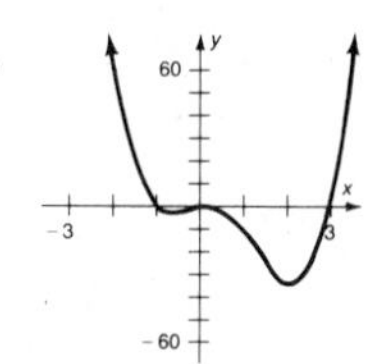

41. 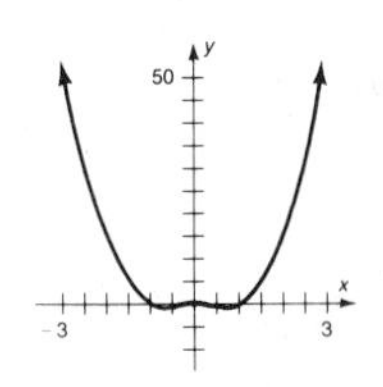

43. 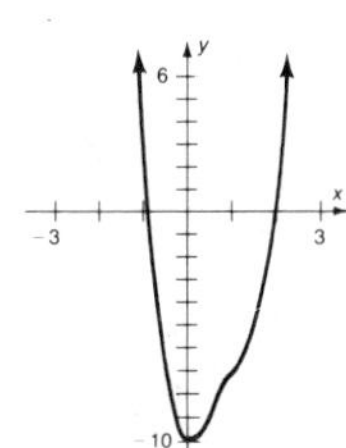

45. 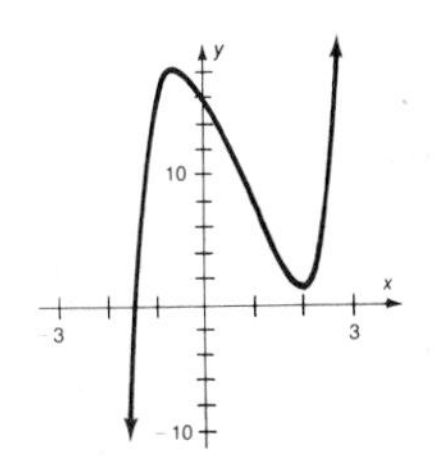

47. 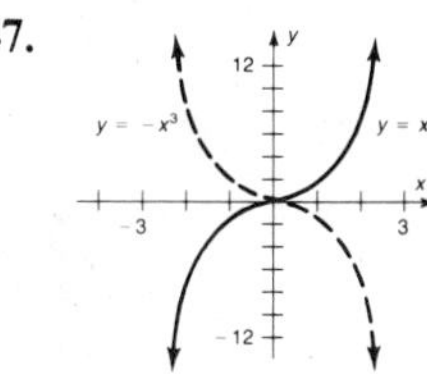

49. 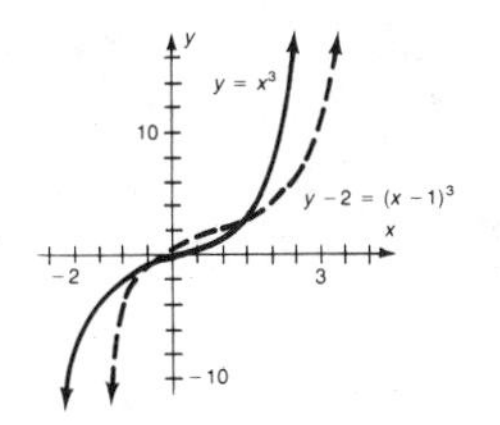

51. 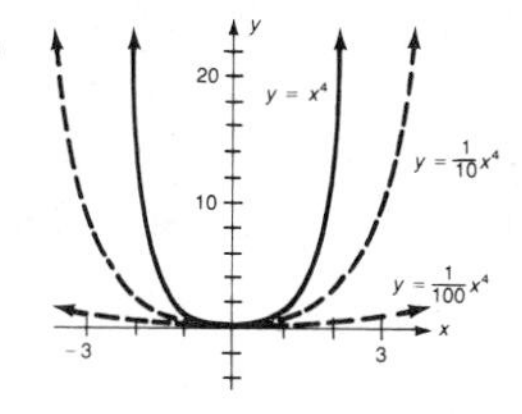

53. Answers vary.

55. 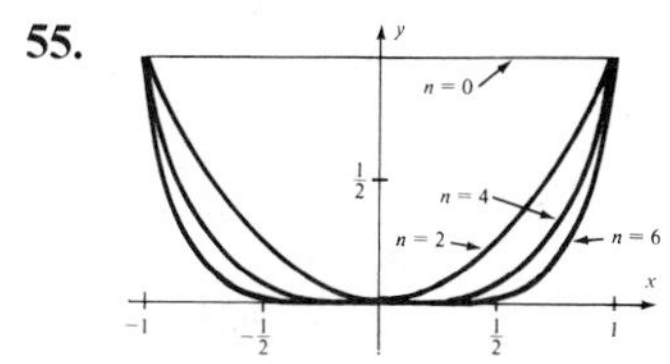

57. 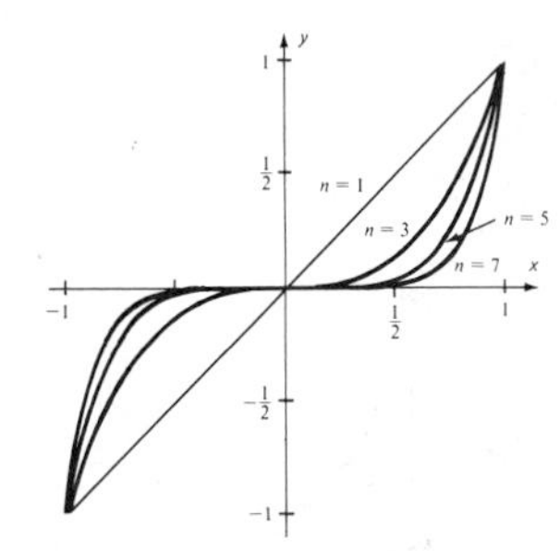

59. Answers vary.

PROBLEM SET 4.4, PAGES 147–148

1. 2, -3 (mult. 2) **3.** 0 (mult. 3), $\frac{3}{2}$ (mult. 2) **5.** -2, 1 (mult. 2), -1 (mult. 2) **7.** -5 (mult. 2), 3 (mult. 2)
9. 0 (mult. 2), 4 (mult. 2) **11.** 0 (mult. 2), 3 (mult. 2), -3 (mult. 2) **13.** 4, 2, or 0 pos; 0 neg **15.** 2 or 0 pos; 1 neg
17. 1 pos; 2 or 0 neg **19.** 1 pos; 2 or 0 neg **21.** 2 or 0 pos; 2 or 0 neg **23.** 2 or 0 pos; 1 neg **25.** $\pm 1, \pm 3, \pm 5, \pm 15$
27. $\pm 1, \pm 2, \pm 3, \pm 4, \pm 6, \pm 12, \pm\frac{1}{2}, \pm\frac{3}{2}$ **29.** $\pm 1, \pm 2, \pm 3, \pm 4, \pm 6, \pm 12$
31. $\pm 1, \pm 3, \pm 5, \pm 15, \pm\frac{1}{2}, \pm\frac{3}{2}, \pm\frac{5}{2}, \pm\frac{15}{2}$ **33.** $\pm 1, \pm 2, \pm 3, \pm 6, \pm 9, \pm 18$ **35.** $\pm 1, \pm\frac{1}{2}, \pm\frac{1}{3}, \pm\frac{1}{6}$
37. $\{1, 2, -2\}$ **39.** $\{2, -3, 3\}$ **41.** $\{2, -2, -3\}$ **43.** $\{-3, -\frac{1}{2}, 5\}$ **45.** $\{1, -1, -6, 3\}$ **47.** $\{-3, -5, -7\}$
49. $\{\frac{1}{2}, -\frac{5}{2}, \frac{7}{2}\}$ **51.** $\{-1, 3, -3, -4\}$ **53.** $\{0, 1, -1, -3\}$ **55.** $\{2, -2\}$ **57.** yes
59. The dimensions of the original cube are 8 cm on each side. **61.** 1.76224... or 1.8 to the nearest tenth
63. The roots, to the nearest tenth, are -0.2, 2.6, 0.4, and -4.8.

PROBLEM SET 4.5, PAGES 152–153

1. $-24 + 4i$ **3.** $24 - 14\sqrt{2}$ **5.** $12 + 16i$ **7.** $-30 + 6i$ **9.** $62 - 38\sqrt{3}$ **11.** 0 **13.** $-48i$ **15.** 0 **17.** 0
19. 5 **21.** 0 **23.** 0 **25.** -3 **27.** 0 **29.** $30 + 6i$ **31.** not a root **33.** yes; $1 + 2i$ **35.** not a root

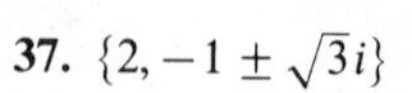

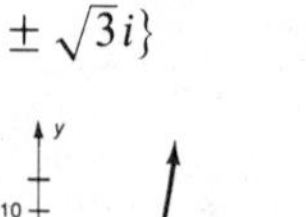

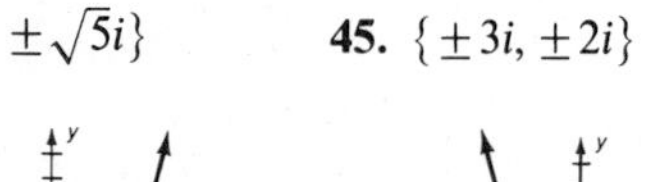

37. $\{2, -1 \pm \sqrt{3}i\}$ **39.** $\left\{5, \dfrac{-5 \pm 5\sqrt{3}i}{2}\right\}$ **41.** $\{\pm 3, \pm 3i\}$ **43.** $\{\pm 2i, \pm\sqrt{5}i\}$ **45.** $\{\pm 3i, \pm 2i\}$

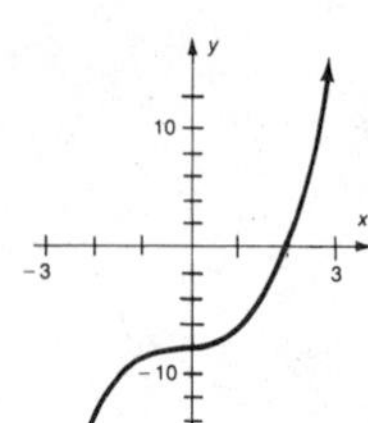

47. $\{3 \pm i, 4 \pm i\}$ **49.** $\left\{2 \pm \sqrt{5}, \dfrac{3 \pm \sqrt{11}i}{2}\right\}$ **51.** $\{-\frac{1}{2}, 1 \pm \sqrt{2}i\}$ **53.** $\{\pm 1, 1 \pm 2i\}$ **55.** $\left\{5, -\dfrac{1}{3}, \dfrac{3 \pm \sqrt{7}}{3}\right\}$

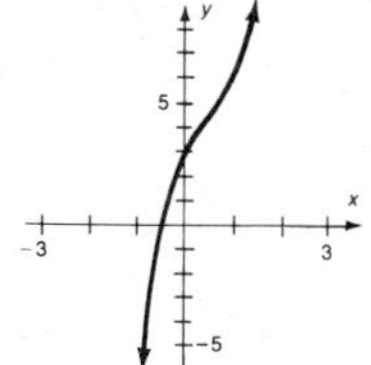

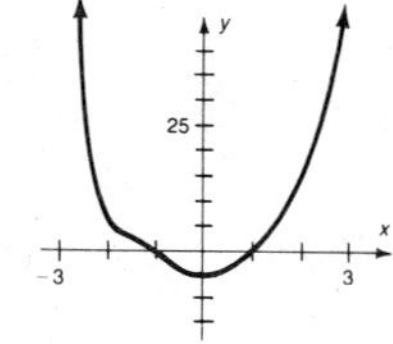

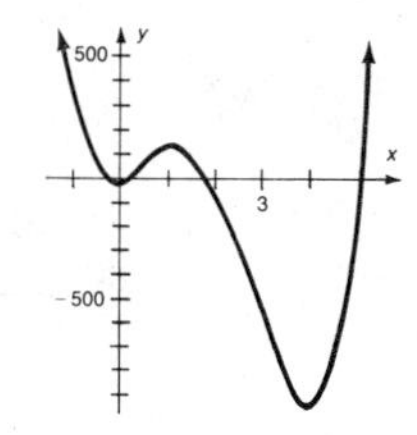

57. $\{\pm 3i, \pm 2i\}$ **59.** $\{2 \pm i, \pm\sqrt{2}\}$ **61.** $\{-1, \frac{3}{2}, \pm\sqrt{5}\}$ **63.** $\{\frac{3}{2}, -3 \pm \sqrt{2}, \pm i\sqrt{2}\}$ **65.** $\left\{\pm\sqrt{2}, \dfrac{-1 \pm \sqrt{3}i}{2}\right\}$

PROBLEM SET 4.6, PAGE 156

1.

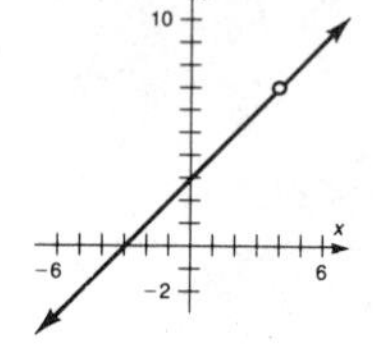

3.

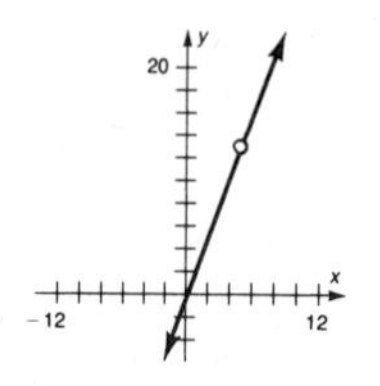

5.

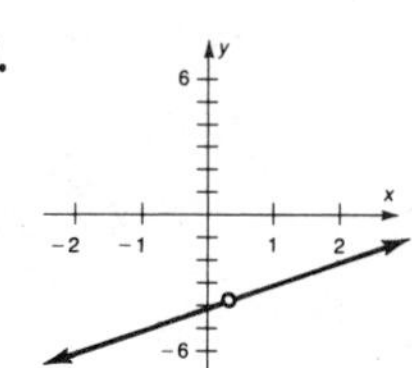

7.

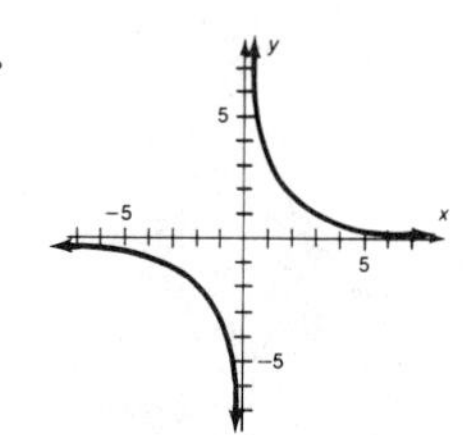

9.

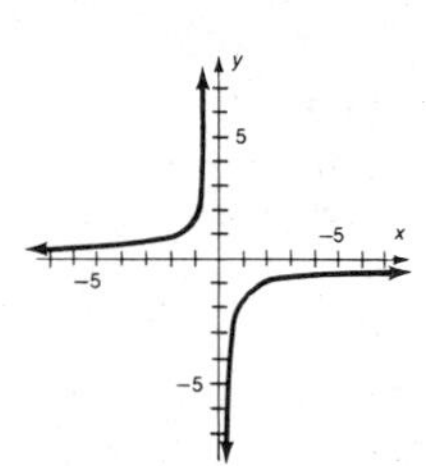

11.

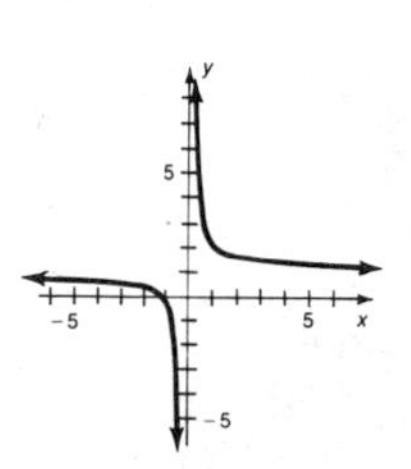

13.

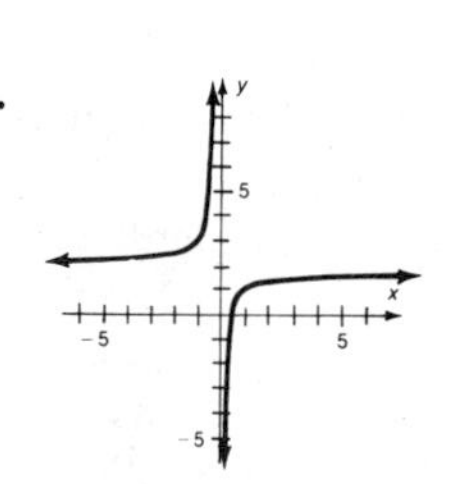

15.

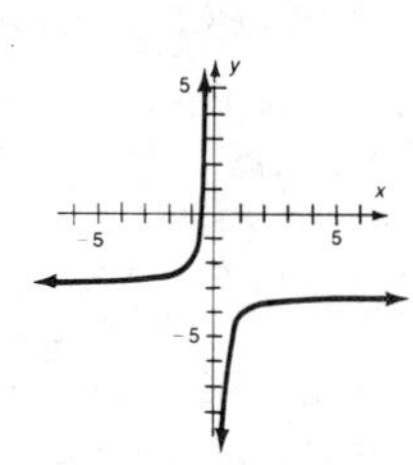

17.

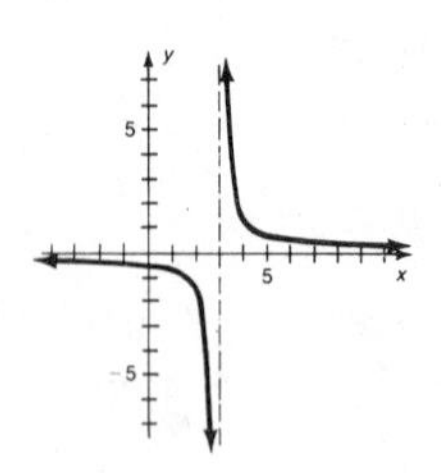

19.

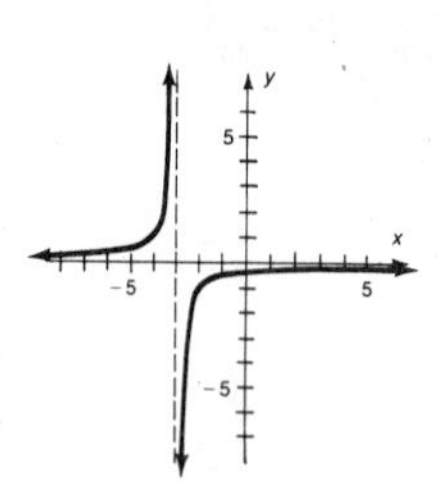

21.

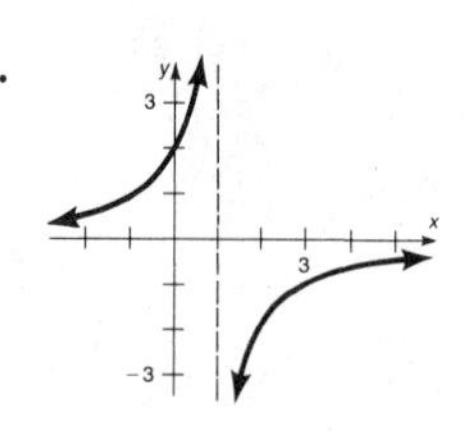

23.

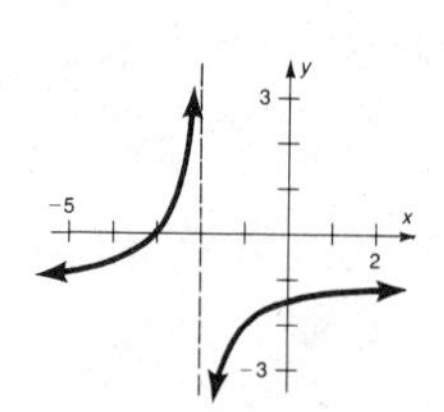

25. 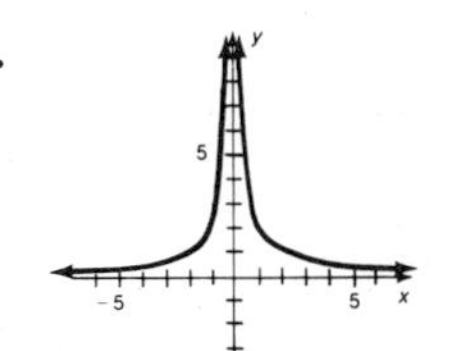

27. 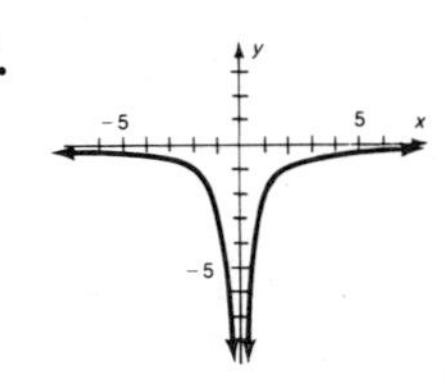

29. 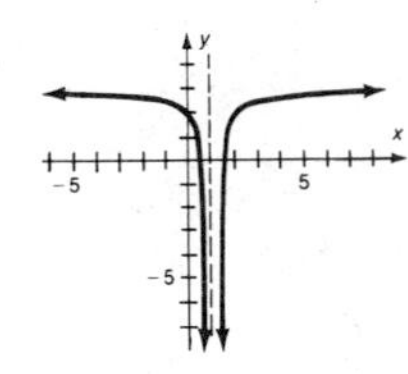

31. 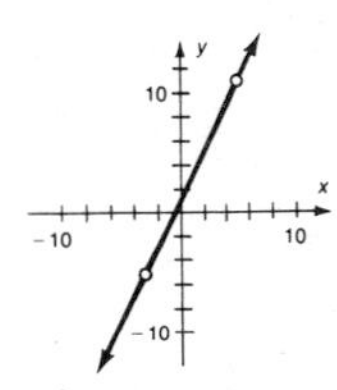

33. 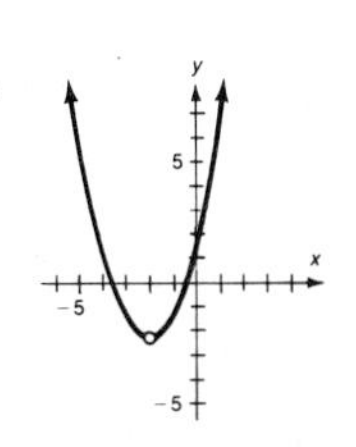

35. 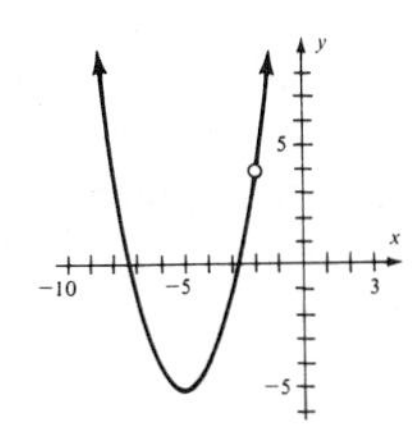

37. 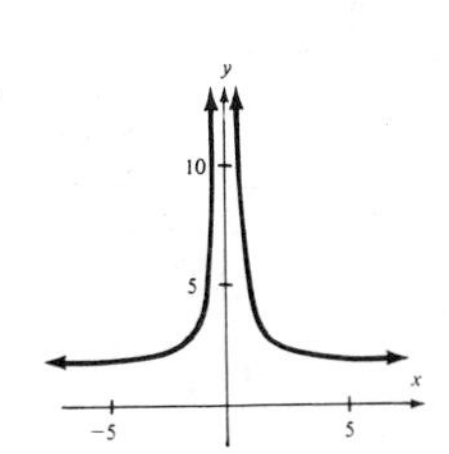

39. 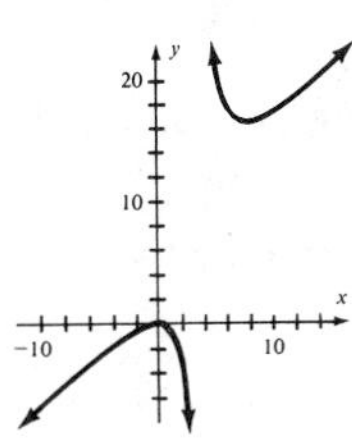

41. 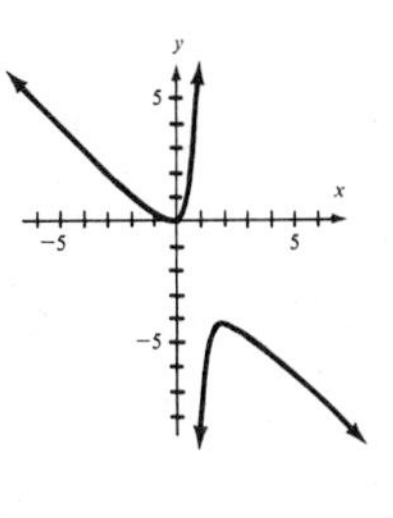

43. 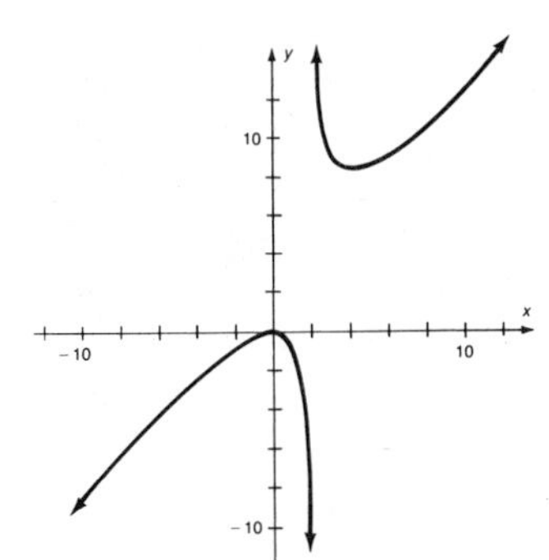

45. 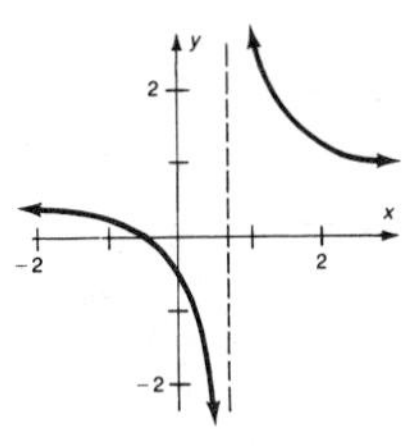

47. 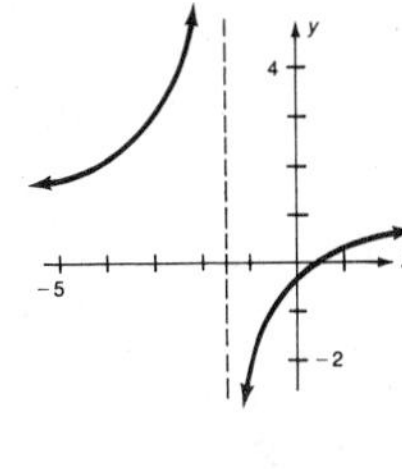

49. 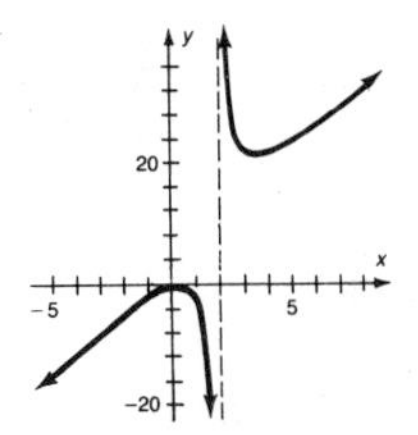

51. 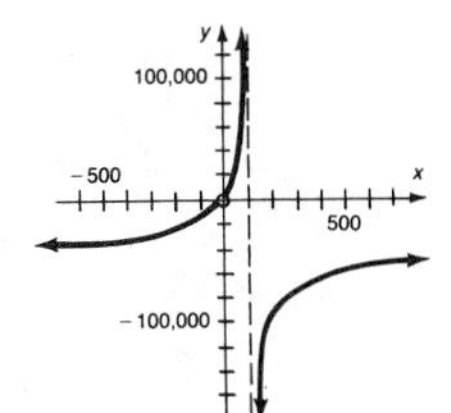

53. 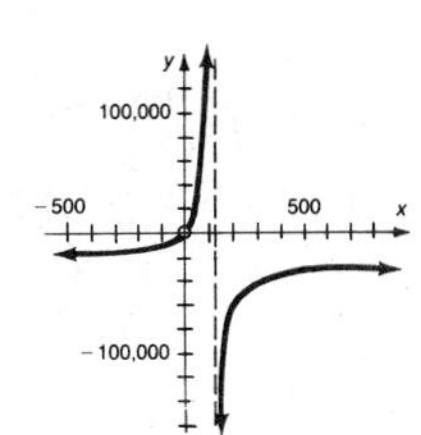

55. 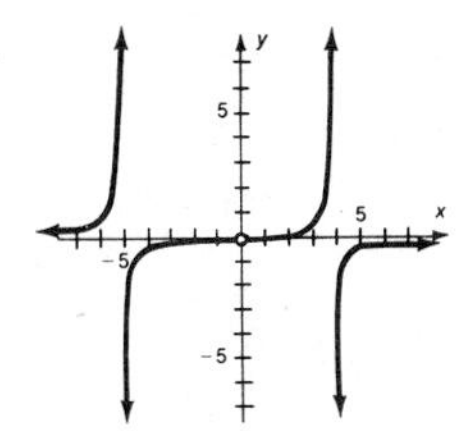

57. 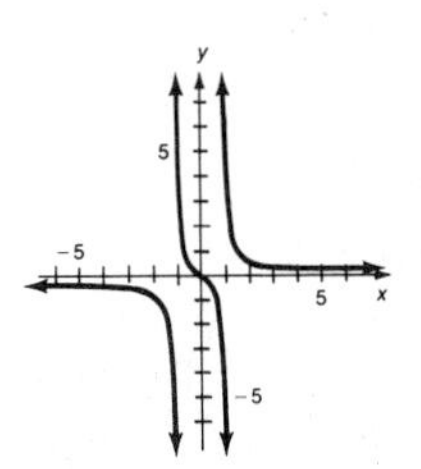

59. 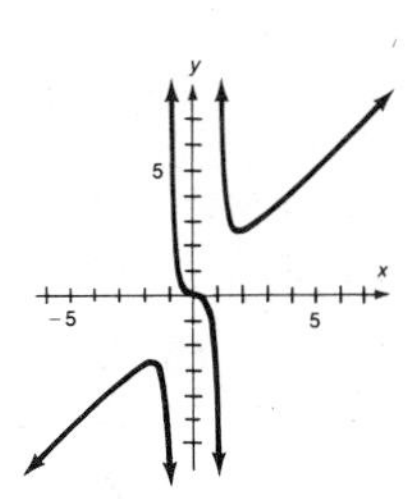

61. Answers vary.

PROBLEM SET 4.7, PAGES 160–161

1. $(x + 3); (x + 4)$ **3.** $(x - 3); (2x + 1)$ **5.** $(x - 8); (x - 6)$ **7.** $x; (x - 1); (x + 1)$ **9.** $x; (x - 2); (x + 2)$ **11.** $(x - 5); (x - 5)^2$ **13.** $x; (x - 1); (x - 1)^2$ **15.** $(x - 2); (x - 2)^2; (x - 2)^3$ **17.** $(x - 3); (x + 1)$ **19.** $x; (x + 1); (x - 2)$ **21.** $x; (x - 4); (x - 1)$ **23.** $(1 - x); (1 + x); (1 + x^2)$ **25.** $\frac{1}{x} + \frac{2}{x^2} + \frac{5}{x^3}$ **27.** $\frac{2}{x} - \frac{5}{x^2} + \frac{4}{x^3}$ **29.** $\frac{1}{x + 4} - \frac{1}{x + 5}$ **31.** $\frac{4}{x - 2} + \frac{3}{x - 1}$ **33.** $\frac{2}{x + 2} + \frac{5}{x - 4}$ **35.** $\frac{4}{x + 3} - \frac{2}{x - 2}$

37. $\frac{1}{x-2}+\frac{3}{x+2}$ **39.** $\frac{3}{x-4}-\frac{2}{x-5}$ **41.** $\frac{3}{x}-\frac{3}{x-1}+\frac{4}{x+1}$ **43.** $\frac{2}{x-2}+\frac{3}{(x-2)^2}$ **45.** $\frac{1}{x}+\frac{3}{(x+1)^2}$ **47.** $\frac{2}{x+1}+\frac{4}{(x+1)^2}-\frac{3}{(x+1)^3}$ **49.** $\frac{5}{6(x+5)}+\frac{1}{6(x-1)}$ **51.** $\frac{13}{3(x-2)}+\frac{8}{3(x+1)}$ **53.** $\frac{1}{x}+\frac{3}{x+3}-\frac{4}{x-2}$ **55.** $\frac{3}{x-1}+\frac{2x-4}{x^2+1}$ **57.** $x+2+\frac{3}{x-1}+\frac{1}{(x-1)^2}$ **59.** $\frac{1}{1-x}-\frac{3}{1+x}+\frac{2x+1}{1+x^2}$

CHAPTER 4 SUMMARY, PAGES 161–163

1. $P(x)=a_nx^n+a_{n-1}x^{n-1}+a_{n-2}x^{n-2}+\cdots+a_1x+a_0$, where n is an integer greater than or equal to zero and the coefficients are real numbers. **3.** b^{m+n} **5.** $9x^4+15x^3-101x^2-71x-12$ **7.** $-3w^3-4w^2+38w+13$ **9.** $\frac{1}{y^2}(2x-2xy-y^2)(2x+2xy+y^2)$ **11.** $(2x-1)(x+1)(4x^2+2x+1)(x^2-x+1)$ **13.**

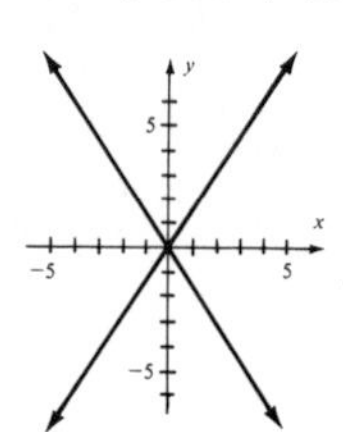

15.

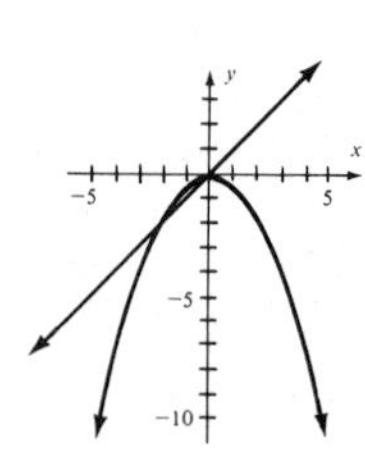

17. $2x+3$ **19.** $3x^3-2x^2+3x+\frac{-2}{2x+1}$ **21.** $x^3-2x^2+6x-13$ **23.** $3x^2+8x+4$ **25.** -42 **27.** 0 **29.**

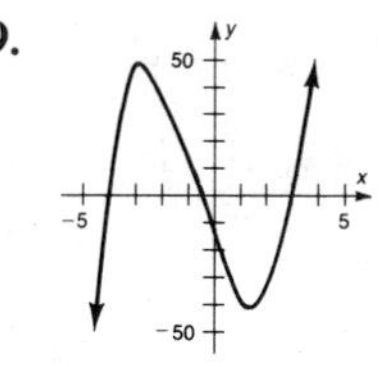

31.

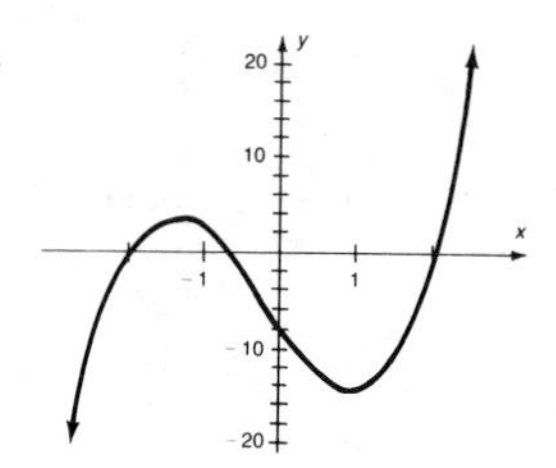

33. 1 (multiplicity 3); -2 (multiplicity 2) **35.** 0 (multiplicity 2); 4 (multiplicity 2); -4 (multiplicity 2) **37.** $\pm1, \pm2, \pm3, \pm4, \pm6, \pm12, \pm\frac{1}{3}, \pm\frac{2}{3}, \pm\frac{4}{3}$ **39.** $\pm1, \pm2, \pm4, \pm8, \pm\frac{1}{3}, \pm\frac{2}{3}, \pm\frac{4}{3}, \pm\frac{8}{3}$ **41.** 1 pos real root; 2 or 0 neg real roots **43.** 1 pos real root; 2 or 0 neg real roots **45.** $\{-\frac{1}{3}, -4, 3\}$ **47.** $\{2, -2, -\frac{2}{3}\}$ **49.** 0 **51.** 0 **53.** $\{2, -3, 2\pm\sqrt{3}\}$ **55.** $\{0, -1, \pm\sqrt{14}\}$ **57.** $\{1\pm i, \pm\sqrt{3}i\}$ **59.** $\{1\pm\sqrt{3}i, \pm\sqrt{5}\}$ **61.**

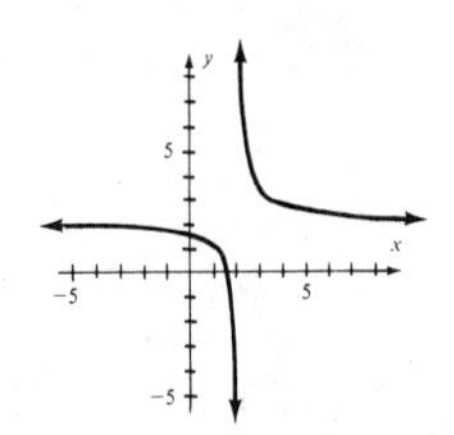

63.

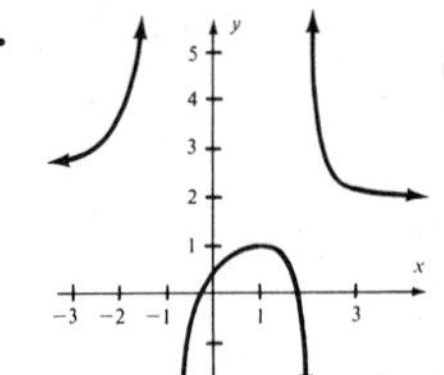

65. $\frac{3}{x-1}+\frac{2}{x-2}+\frac{-1}{(x-2)^2}$ **67.** $2+\frac{3}{x+5}+\frac{6}{x-3}$

PROBLEM SET 5.1, PAGES 171–172

1. 5 **3.** -3 **5.** 6 **7.** 16 **9.** -8^3 or -512 **11.** not defined **13.** 8 **15.** 108 **17.** $\frac{1}{100}$ **19.** $\frac{1}{1000}$ **21.** $x+1$ **23.** $1+x$ **25.** $\frac{x^2}{y}$ **27.** $x+2x^{1/2}y^{1/2}+y$ **29.** $x+y$

31. 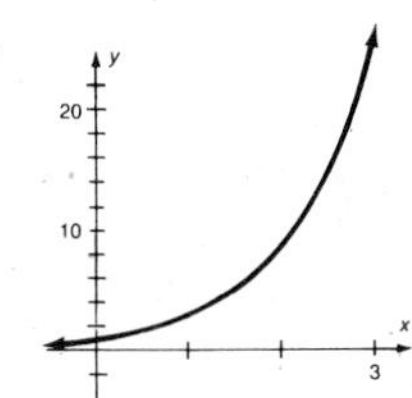

33. 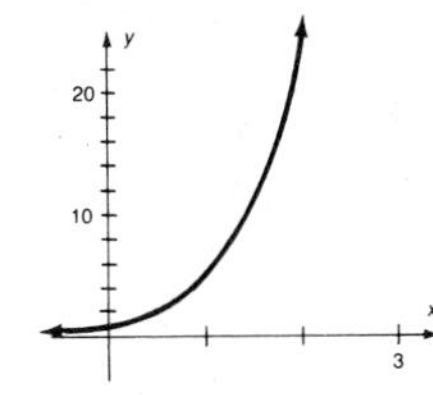

35. 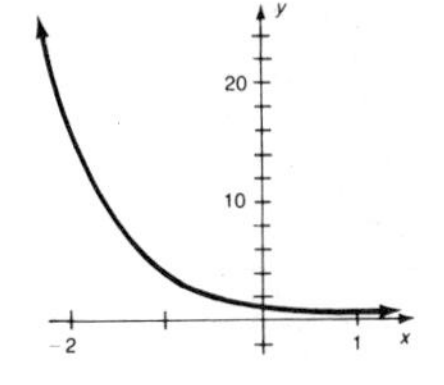

37. 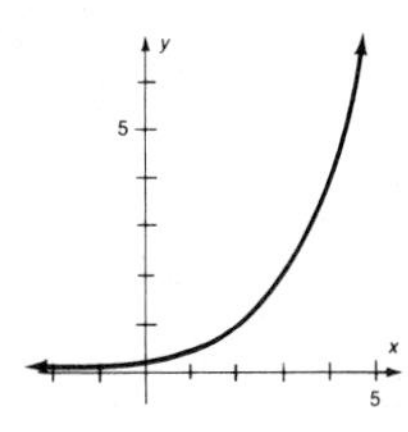

39. 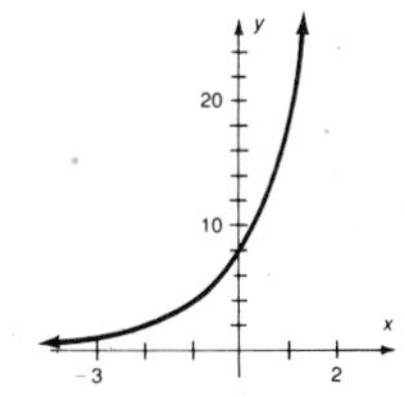

41. 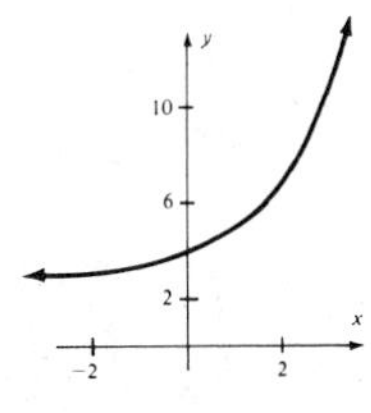

43. 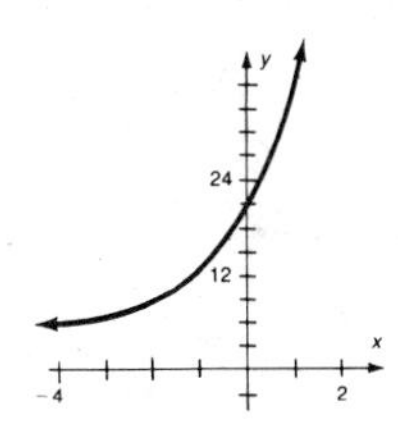

45. 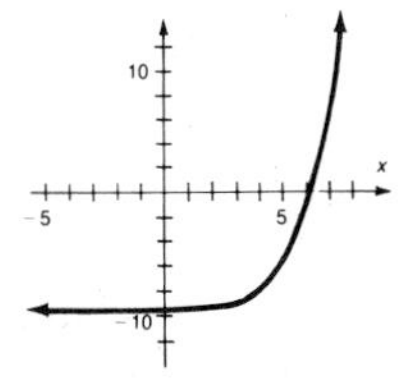

47. 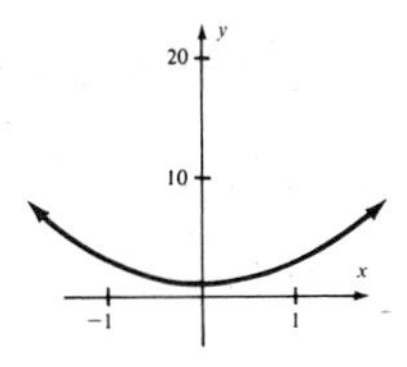

49. 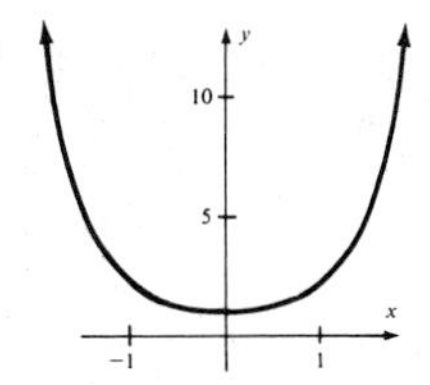

51. 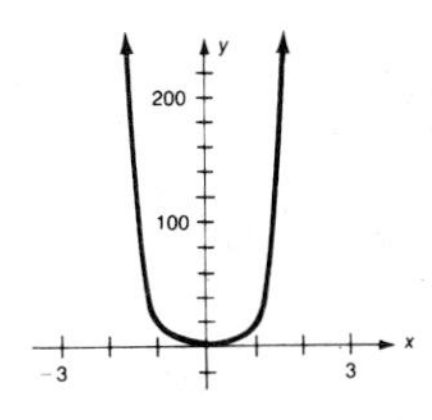

53. 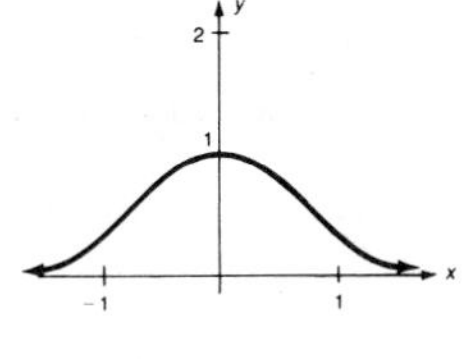

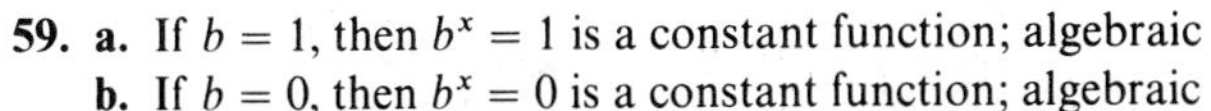

55. $2^{\sqrt{2}} \approx 2.7$ **57.** $10^{\sqrt{2}} \approx 26.0$

59. a. If $b = 1$, then $b^x = 1$ is a constant function; algebraic
b. If $b = 0$, then $b^x = 0$ is a constant function; algebraic

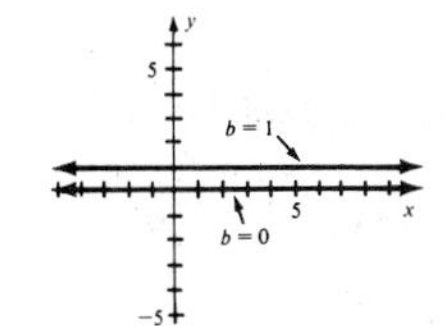

61. 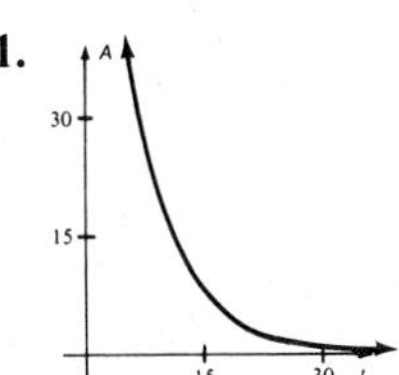

63. 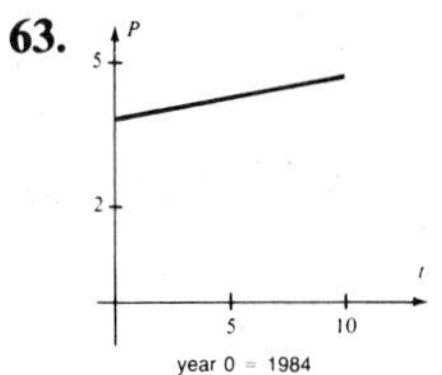

PROBLEM SET 5.2, PAGES 182–183

1. $\log_2 64 = 6$ **3.** $\log 1000 = 3$ **5.** $\log_5 125 = 3$ **7.** $\log_n m = p$ **9.** $\log_{1/3} 9 = -2$ **11.** $\log_9 \frac{1}{3} = -\frac{1}{2}$
13. $10^4 = 10{,}000$ **15.** $10^0 = 1$ **17.** $e^2 = e^2$ **19.** $e^5 = x$ **21.** $2^{-3} = \frac{1}{8}$ **23.** $4^{\frac{1}{2}} = 2$ **25.** $m^P = n$
27. $x^3 = 8$ **29.** 2 **31.** 4 **33.** -1 **35.** 2 **37.** 0.63042788 **39.** 0.92582757 **41.** 4.85491302
43. -0.49349497 or $9.50650503 - 10$ **45.** 0.81977983 **47.** 0.69314718 **49.** 2.56494936 **51.** 1.98787435
53. Answers vary. **55.** \$3,207.14 **57.** \$4,055.20 **59.** \$10,285.33 **61. a.** 3572 **b.** 4746 **c.** \$116,619
63. a. $M \approx 3.89$; about 3.9 or 4.0 on the Richter scale **b.** $M \approx 8.8$; about 8.8 or 9 on the Richter scale

PROBLEM SET 5.3, PAGES 188–190

1. a. 2 **b.** 7 **3. a.** -1 **b.** 4 **5. a.** $2\sqrt{7}$ **b.** 3 **7. a.** $\sqrt{e}$ **b.** e **9. a.** e^4 **b.** 14 **11. a.** $\pm 5\sqrt{5}$
b. $\pm 2\sqrt{3}$ **13. a.** $\frac{7}{2}$ **b.** all nonzero x **15. a.** 32 **b.** 10^5 **17. a.** $5{,}517{,}840 \approx 5{,}520{,}000$ **b.** $4914 \approx 4910$
19. a. $378{,}193{,}771 \approx 380{,}000{,}000$ **b.** $93{,}877.1 \approx 94{,}000$ **21. a.** $40 \approx 40.0$ **b.** $0.0938086 \approx 0.0938$

23. 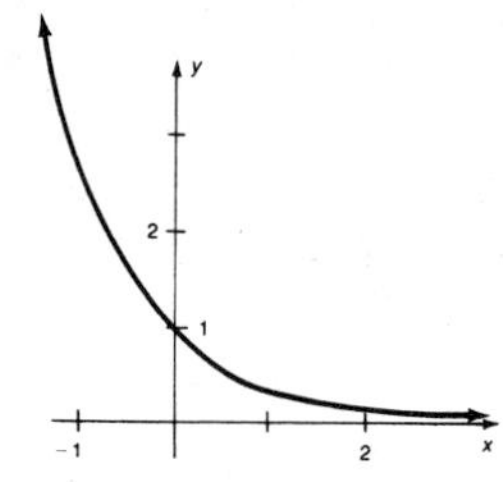

25. 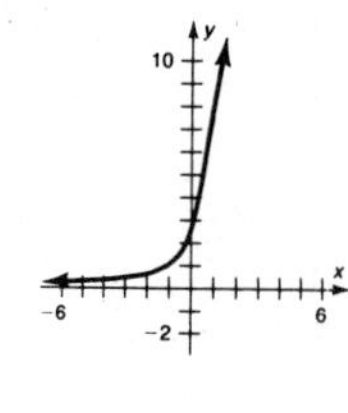

27. 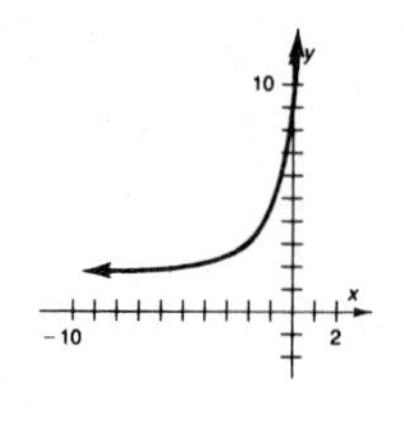

29. 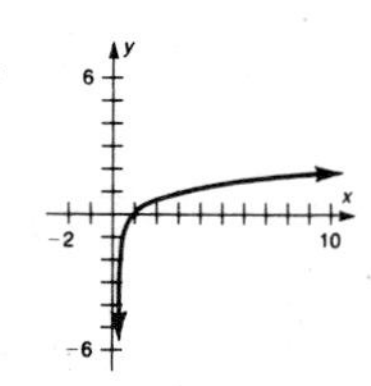

31. 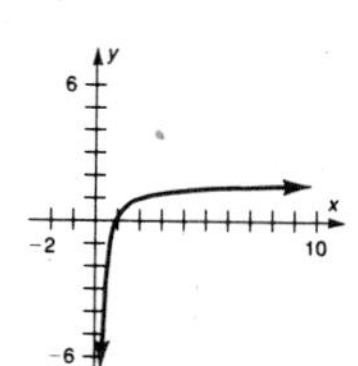

33. 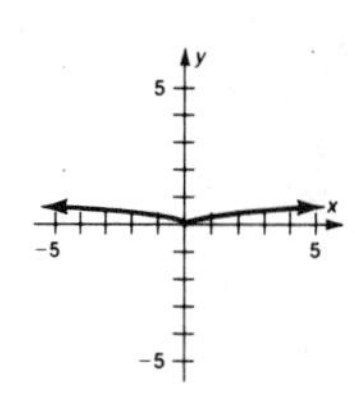

35. 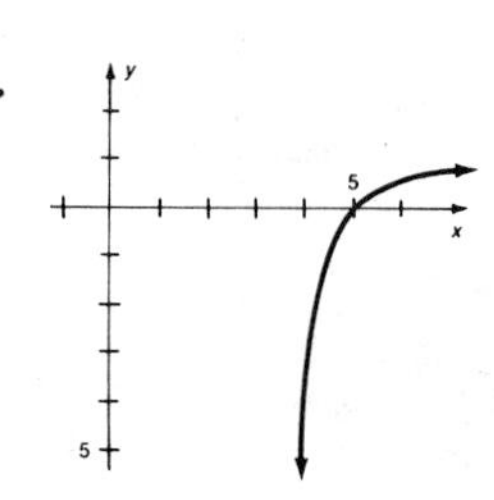

37. 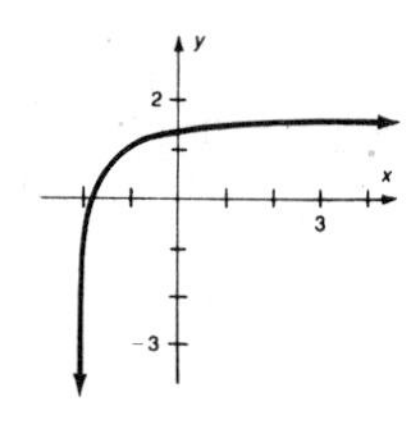

39. 0 **41.** $1250\sqrt{3}$ **43.** 100 **45.** 2, 6 **47.** 25 **49.** 3, 6 **51.** 243 **53.** $5e^{-2}$ **55.** 3 **57.** \$1131.47
59. \$1530.69 **61. a.** about 29 days **b.** No; if $N = 80$, then t is not defined. **c.** $N = 80(1 - e^{-t/62.5})$

63. a. $E = 10^{1.5M + 11.8}$ **b.** $1.12201 \cdot 10^{23}$ ergs **65.** 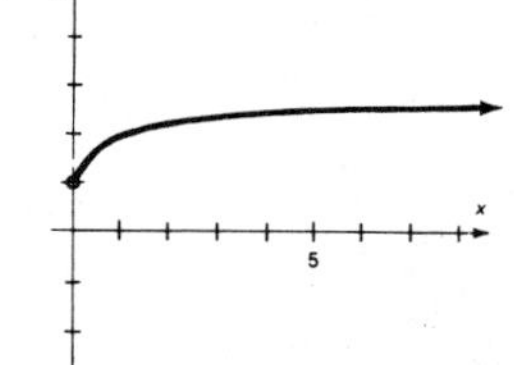 **67.** Answers vary.

PROBLEM SET 5.4, PAGES 196–197

1. 7 **3.** $\frac{5}{3}$ **5.** $\frac{2}{3}$ **7.** $-\frac{4}{3}$ **9.** 3 **11.** -2 **13.** $\frac{1}{4}$ **15.** $-\frac{2}{15}$ **17.** 2.1132828 **19.** -1.2850972
21. 2.011465868 **23.** 1.62324929 **25.** -2.48811664 **27.** -0.66958623 **29.** 0.78746047 **31.** 2.12907205
33. 1.15129255 **35.** -0.57564627 **37.** -0.049306144 **39.** 2.7429396 **41.** 1.4306766 **43.** -2.321928
45. -1.2920297 **47.** 11.89566 periods; about 6 years **49.** 31 quarters; about $7\frac{3}{4}$ years
51. $k \approx 0.023104906$ **53.** $t \approx 10.813529$ years **55.** $t \approx 11{,}400$ years **57.** about 17,400 years

59. It is about $\frac{1}{2}$ mile above sea level. **61.** $t = -250\ln\left(\frac{P}{50}\right)$ **63. a.** $\frac{x^5}{5!} + \frac{x^6}{6!}$ **b.** $\frac{x^{r-1}}{(r-1)!}$

c.
$$e = 1 + 1 + \frac{1}{2} + \frac{1}{2 \cdot 3} + \frac{1}{2 \cdot 3 \cdot 4} + \frac{1}{5!} + \frac{1}{6!} + \cdots$$
$$= 2 + 0.5 + 0.1666\ldots + 0.041666\ldots + 0.008333\ldots + 0.0013888\ldots + 0.0001984127\ldots + \cdots$$
$$\approx 2.718$$

d. $e^{0.5} = 1 + 0.5 + \frac{(0.5)^2}{2} + \frac{(0.5)^3}{2 \cdot 3} + \frac{(0.5)^4}{2 \cdot 3 \cdot 4} + \frac{(0.5)^5}{5!} + \cdots \approx 1.649$ **65.** -0.0968291 **67.** 4.6052702

CHAPTER 5 SUMMARY, PAGES 197–198

1. 25 **3.** $\sqrt{3}$ **5.**

7. 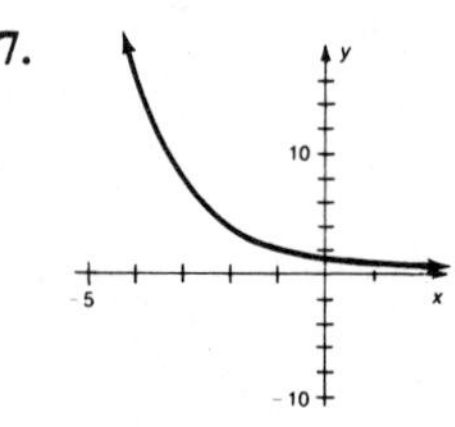

9. $\log\sqrt{10} = 0.5$ **11.** $\log_9 729 = 3$

13. $10^0 = 1$ **15.** $2^6 = 64$ **17.** 0.4771212547 **19.** -2.677780705 **21.** 1.098612289 **23.** -4.342805922 **25.** 3
27. 5 **29.** $\log_b A + \log_b B$ **31.** $p \log_b A$ **33.** **35.** **37.** 2 **39.** 243

41. $831{,}450 \approx 831{,}000$ **43.** 47,045,881 **45.** 0.096910013 **47.** ± 0.5888866497 **49.** 1.304007668
51. -2.768345799 **53.** 3 **55.** 1.771243749 **57.** 9.67957033 **59.** 4.191806549 **61.** $k \approx 0.026$
63. \$131,894.24 **65.** \$36,018.87

CUMULATIVE REVIEW I, PAGE 199

1. a. 14 **b.** $2t - 1$ **c.** $3t^2 + 6th + 3h^2 + 2$ **d.** $12x^2 - 12x + 5$ **e.** $6x^2 + 3$ **f.** $6x + 3h$
2. a. no inverse (it is not one-to-one) **b.** $y = \frac{1}{2}(x + 1)$

3. a.

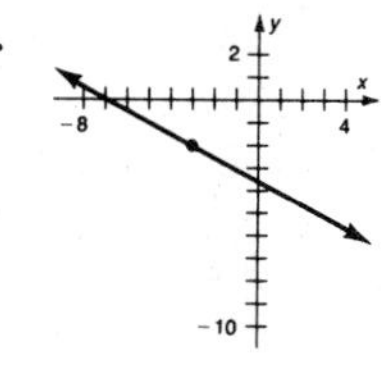

b.

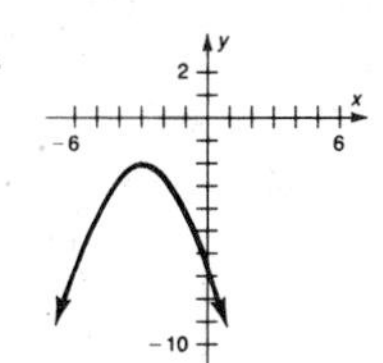

4. a. 0 **b.** 27 **c.** -36 **d.** 0 **e.** $\{-1, \frac{3}{2}, \frac{9}{2}\}$

5.

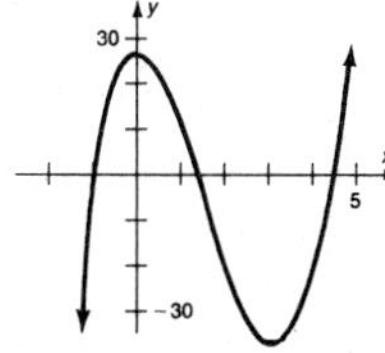

6. $\{3 \pm \sqrt{7}, \frac{1}{2}, -\frac{3}{2}\}$ **7. a.** $D = (-\infty, -1) \cup (-1, 2) \cup (2, \infty)$ **b.** no symmetry

c. yes; $x = 2$, $x = -1$, and $y = 2$; the curve passes through the horizontal asymptote at $(5, 2)$. **d.** $(\frac{1}{2}, 0)$, $(1, 0)$, $(0, -\frac{1}{2})$

e.

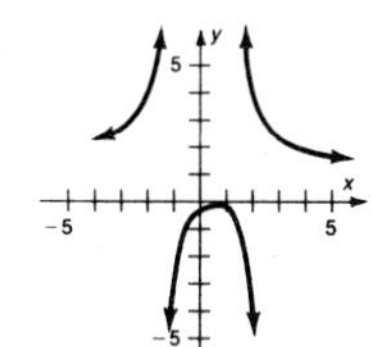

8. a.

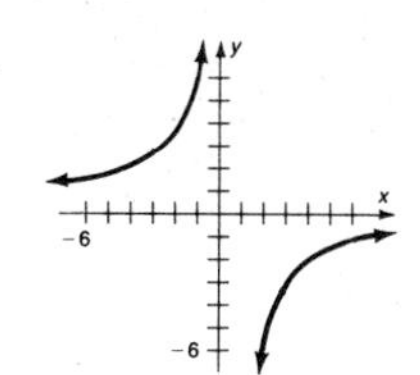

b.

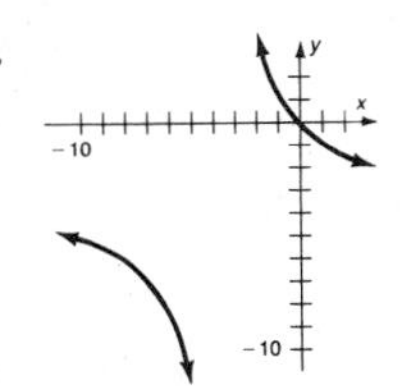

c.

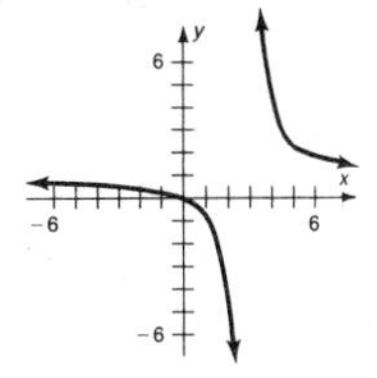

9. a. $\frac{5}{3}$ **b.** 92.4 **c.** 1 **d.** 81 **10. a.** $k = \dfrac{1}{5700}$ **b.** about 2700 years **c.** $t = -\dfrac{1}{k}\log_2 \dfrac{A}{A_0}$

EXTENDED APPLICATION: POPULATION GROWTH, PAGES 202–203

1. 0.9% **3.** 1.9% **5.** 8 yr, 105 days (Oct. 18, 1994 if you account for leap years)
7. 21 yr, 133 days (Nov. 12, 2007 if you account for leap years) **9. a.** 1.3% decline **b.** 493,000

11.

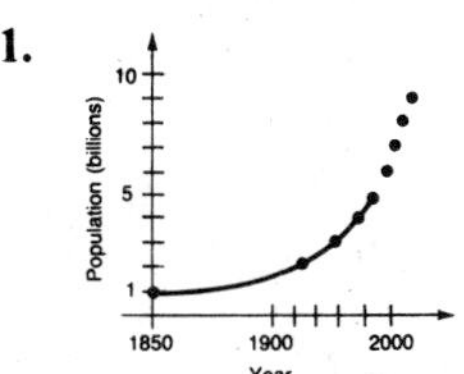

13. a.

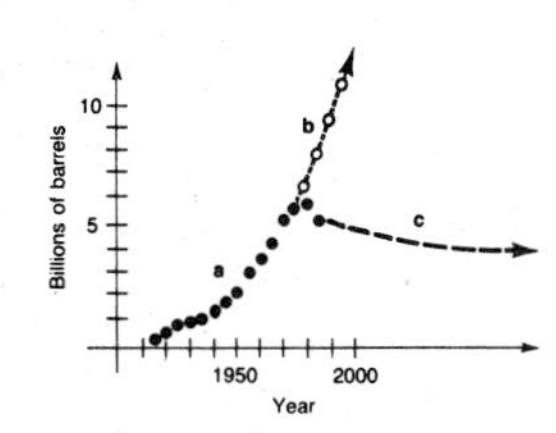

b. Answers may vary; 1980, 6.5 billion; 1985, 7.9 billion; 1990, 9.2 billion; 1995, 10 billion.

c. Answers may vary: 1990, 5.1 billion; 1995, 5.0 billion. **d.** Answers vary. **15.** Answers vary.

PROBLEM SET 6.1, PAGES 213–214

1. a. $\frac{\pi}{6}$ **b.** $\frac{\pi}{2}$ **c.** $\frac{3\pi}{2}$ **d.** $\frac{\pi}{4}$ **3. a.** 180° **b.** 45° **c.** 60° **d.** 360°

5. a. **b.** 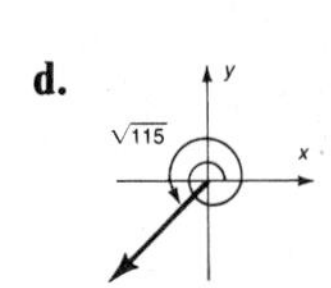 **c.** **d.** **7. a.** **b.**

c. **d.** **9. a.** **b.** **c.** **d.**

11. a. 40° **b.** 180° **c.** 30° **d.** 330° **13. a.** 240° **b.** 140° **c.** 180° **d.** 280° **15. a.** $\frac{7\pi}{4}$ **b.** $\frac{\pi}{4}$ **c.** $\frac{5\pi}{3}$ **d.** $2\pi - 2$ **17. a.** $9 - 2\pi \approx 2.7168$ **b.** $2\pi - 5 \approx 1.2832$ **c.** $\sqrt{50} - 2\pi \approx 0.7879$ **d.** $2\pi - 6 \approx 0.2832$ **19. a.** 0.5236 **b.** 5.4978 **c.** 5.4867 **d.** 3.1415 **21. a.** 60° **b.** 60° **c.** 60° **d.** 45° **23. a.** $\frac{\pi}{12}$ **b.** $\frac{\pi}{3}$ **c.** $\frac{\pi}{6}$ **d.** $\frac{\pi}{4}$ **25. a.** $\frac{\pi}{4}$ **b.** 0 **c.** $\frac{\pi}{6}$ **d.** $\frac{\pi}{3}$ **27. a.** 0.7879 **b.** 0.4250 **c.** 0.5236 **d.** 0.7854 **29.** 18.00° **31.** 300.00° **33.** 30.00° **35.** −171.89° **37.** −143.24° **39.** 29.22° **41.** $\frac{\pi}{9}$ **43.** $\frac{-11\pi}{9}$ **45.** $\frac{17\pi}{36}$ **47.** 5.48 **49.** 6.11 **51.** 17.19 **53.** 14.04 cm **55.** 4.19 m **57.** 4.89 ft **59.** 14.07 cm **61.** 87.27 cm **63.** about 440 km **65.** The diameter of the moon is about 5200 km.

PROBLEM SET 6.2, PAGES 220–221

1. 0.6 **3.** −0.6 **5.** −0.4 **7.** 2.9 **9. a.** 0.6427876 **b.** 0.3420201 **11. a.** 0.3639702 **b.** −1.4142136 **13. a.** −3.0776835 **b.** −1.1923633 **15. a.** 0.8414709 **b.** 0.0174524 **17. a.** −1.3386481 **b.** −14.136833 **19. a.** −1.2281621 **b.** −0.4999664 **21. a.** 0.2876553 **b.** 0.5455784 **23. a.** 2.1164985 **b.** 1.8899139 **25. a.** 2.6327477 **b.** −3.4184818 **27. a.** −5.4914632 **b.** −1.6389075 **29.** + **31.** − **33.** − **35.** −

37. − **39.** − **41.** II, III **43.** III **45.** III **47.** $\cos\theta = -\frac{3}{5}$ $\sec\theta = -\frac{5}{3}$ $\sin\theta = \frac{4}{5}$ $\csc\theta = \frac{5}{4}$ $\tan\theta = -\frac{4}{3}$ $\cot\theta = -\frac{3}{4}$

49. $\cos\theta = \frac{5}{13}$ $\sec\theta = \frac{13}{5}$ $\sin\theta = \frac{12}{13}$ $\csc\theta = \frac{13}{12}$ $\tan\theta = \frac{12}{5}$ $\cot\theta = \frac{5}{12}$

51. $\cos\theta = \frac{5}{13}$ $\sec\theta = \frac{13}{5}$ $\sin\theta = -\frac{12}{13}$ $\csc\theta = -\frac{13}{12}$ $\tan\theta = -\frac{12}{5}$ $\cot\theta = -\frac{5}{12}$

53. $\cos\theta = -\frac{6}{37}\sqrt{37}$ $\sec\theta = -\frac{1}{6}\sqrt{37}$ $\sin\theta = \frac{1}{37}\sqrt{37}$ $\csc\theta = \sqrt{37}$ $\tan\theta = -\frac{1}{6}$ $\cot\theta = -6$ **55.** $|x|\sqrt{2}$ **57.** $-x\sqrt{2}$ **59.** $3x$ **61.** $\frac{1}{2}$

63. −4 **65.** 0.8415

PROBLEM SET 6.3, PAGES 225–226

1. a. 1 **b.** 1 **c.** $\frac{\sqrt{3}}{2}$ **d.** $\frac{\sqrt{3}}{2}$ **3. a.** 0 **b.** 1 **c.** 0 **d.** $\frac{\sqrt{2}}{2}$ **5. a.** $\sqrt{2}$ **b.** undefined **c.** 2 **d.** $\sqrt{3}$ **7. a.** 1 **b.** −1 **c.** −1 **d.** 0 **9. a.** $\frac{1}{2}$ **b.** $\frac{1}{2}$ **c.** $\frac{\sqrt{2}}{2}$ **d.** 1 **11. a.** 0.5650 **b.** 0.5850 **13. a.** 0.6428 **b.** 0.1763 **15. a.** −0.3640 **b.** −0.1736 **17. a.** 0.7337 **b.** −0.4791 **19. a.** 0.9320 **b.** 0.7961 **21. a.** 1.5574 **b.** 0.0709 **23. a.** −0.7470 **b.** 0.1411 **25. a.** 0.9004 **b.** 0.3555 **27.** Answers vary.

29. $\cos\frac{5\pi}{4} = \frac{-\sqrt{2}}{2}$ **31.** $\sin\frac{-\pi}{4} = \frac{-\sqrt{2}}{2}$ **33.** $\cos 210° = \frac{-\sqrt{3}}{2}$ **35.** 1 **37.** -1 **39.** $\sqrt{2}$ **41.** $\frac{1}{2}$ **43.** 1 **45.** $\frac{1}{2}$ **47.** 1 **49.** $\frac{\sqrt{2}}{2}$ **51.** $\frac{2}{3}\sqrt{3}$ **53.** $\frac{\sqrt{3}}{3}$ **55.** $\frac{\sqrt{3}}{2}$ **57.** $-\sqrt{3}$ **59.** $\frac{1}{2}$ **61–63.** Answers vary.

PROBLEM SET 6.4, PAGES 234–235

1.

x = angle	$\frac{2\pi}{3}$	$\frac{3\pi}{4}$	$\frac{5\pi}{6}$	$\frac{7\pi}{6}$	$\frac{5\pi}{4}$	$\frac{4\pi}{3}$	$\frac{7\pi}{4}$	$\frac{11\pi}{6}$
Quadrant	II; −	II; −	II; −	III; −	III; −	III; −	IV; +	IV; +
$y = \cos x$	$-\frac{1}{2}$	$\frac{-\sqrt{2}}{2}$	$\frac{-\sqrt{3}}{2}$	$\frac{-\sqrt{3}}{2}$	$\frac{-\sqrt{2}}{2}$	$-\frac{1}{2}$	$\frac{\sqrt{2}}{2}$	$\frac{\sqrt{3}}{2}$
y (approximate)	−0.5	−0.71	−0.87	−0.87	−0.71	−0.5	0.71	0.87

See Figure 6.17 in text for graph.

3. See Figure 6.14 in text for graph.

5.

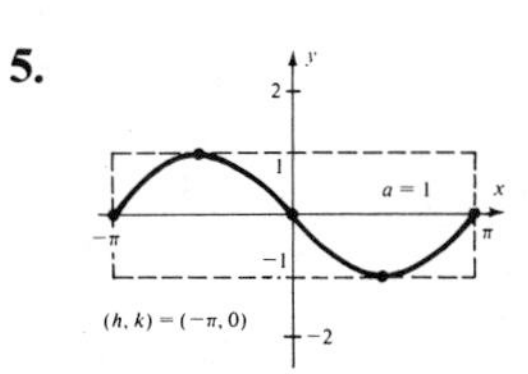

7.

9.

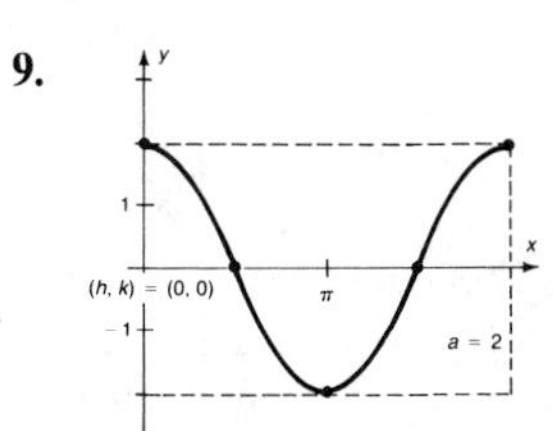

11.

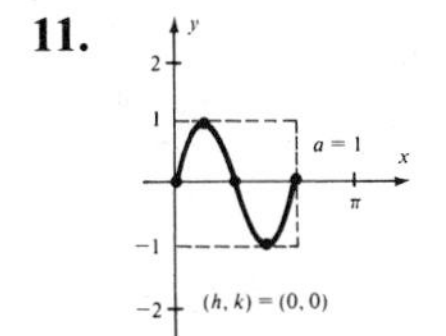

13.

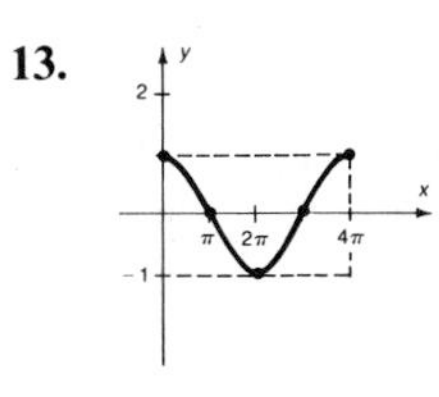

15.

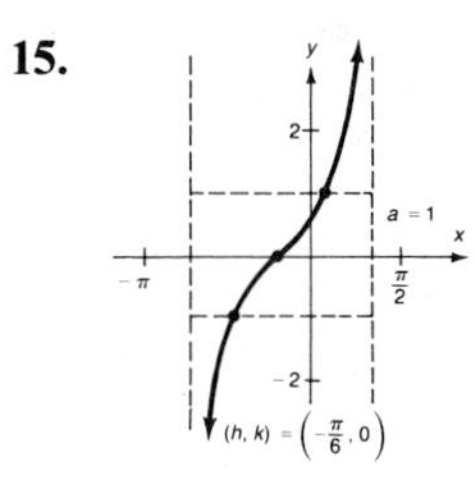

17.

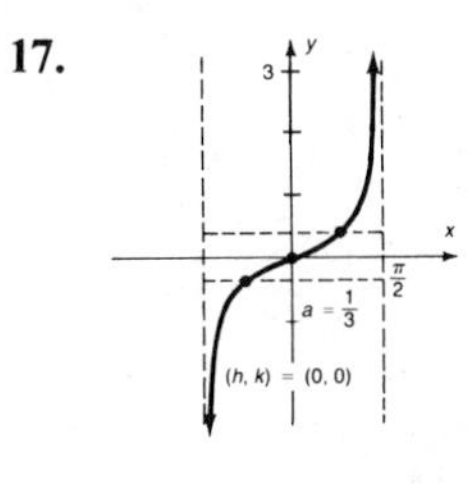

19.

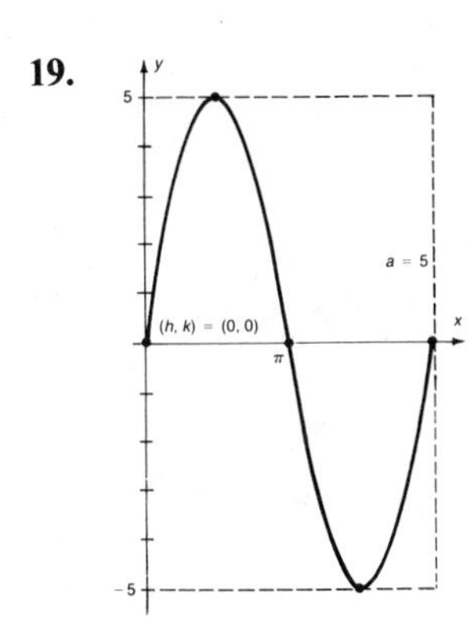

21.

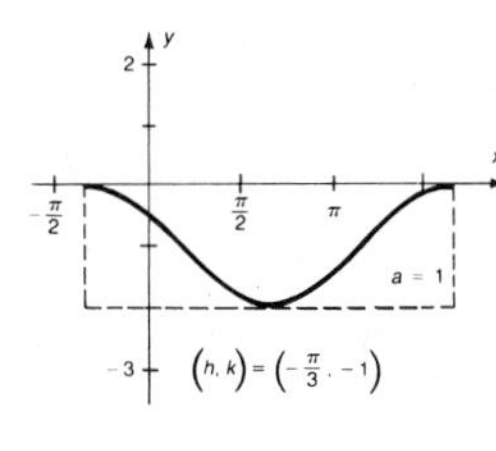

23.

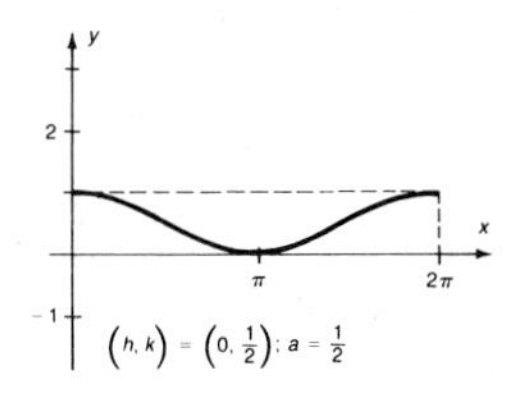

25.

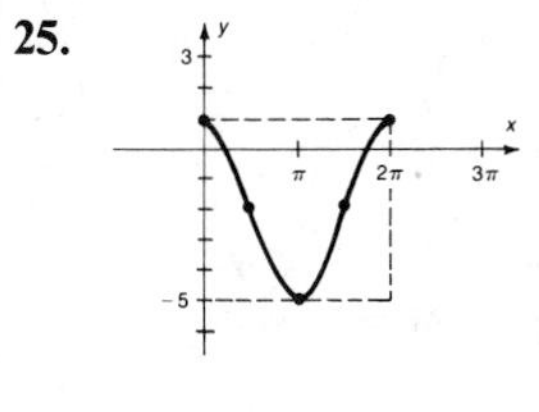

27.

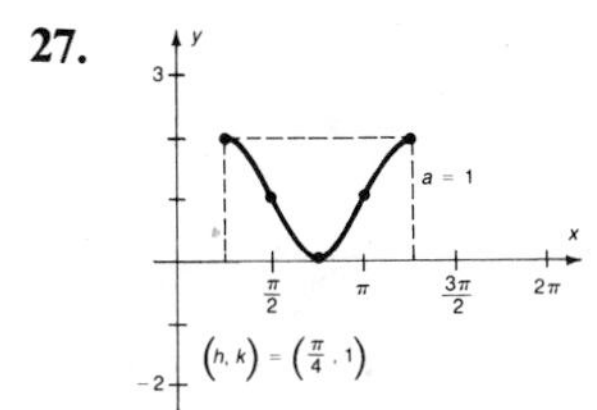

29.

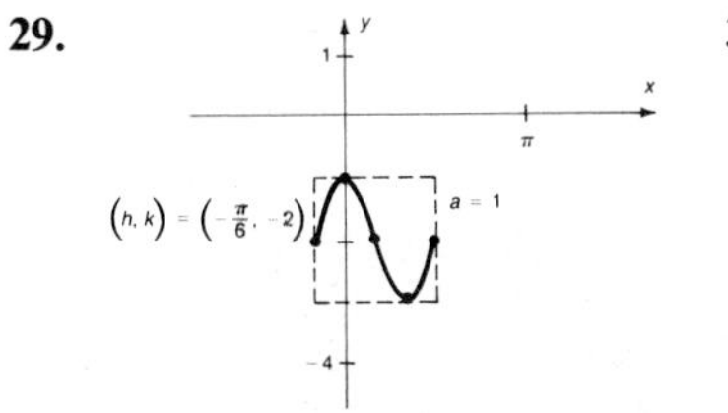

31.

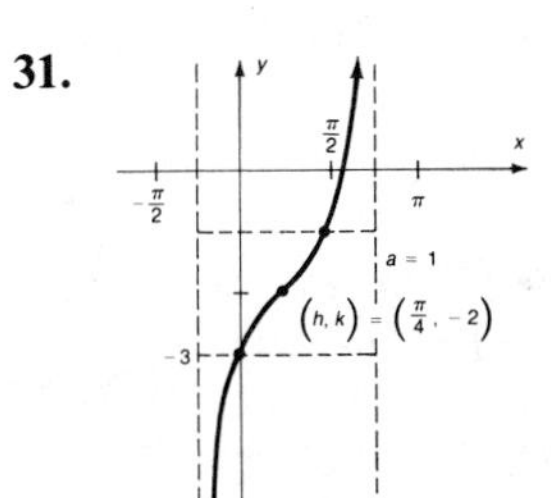

33.

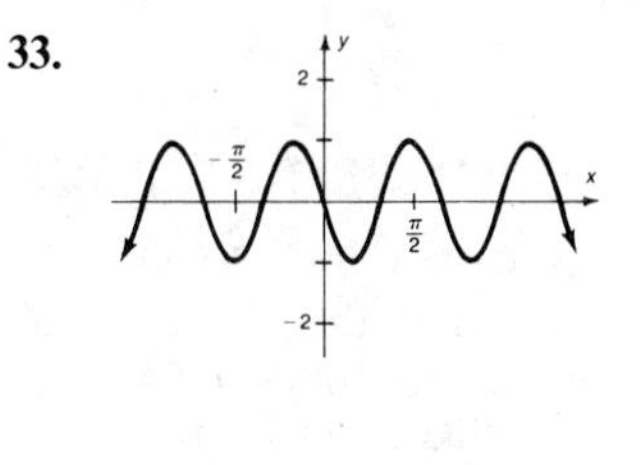

35.
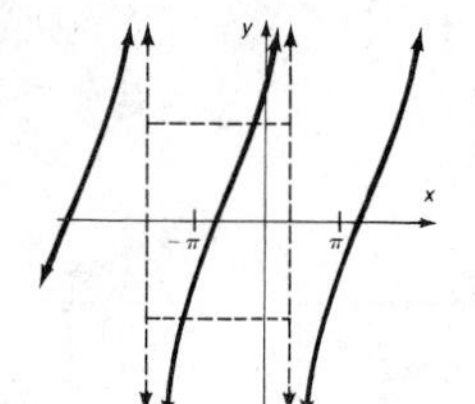

37.
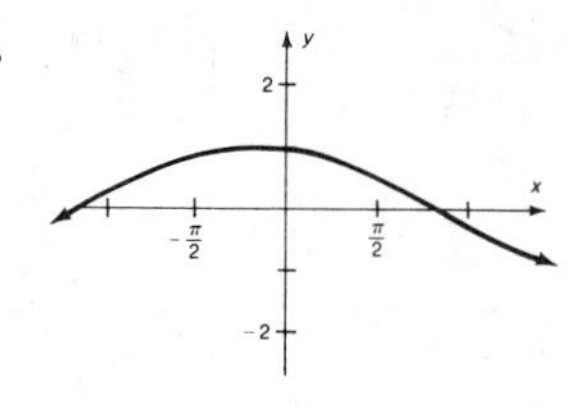

39.
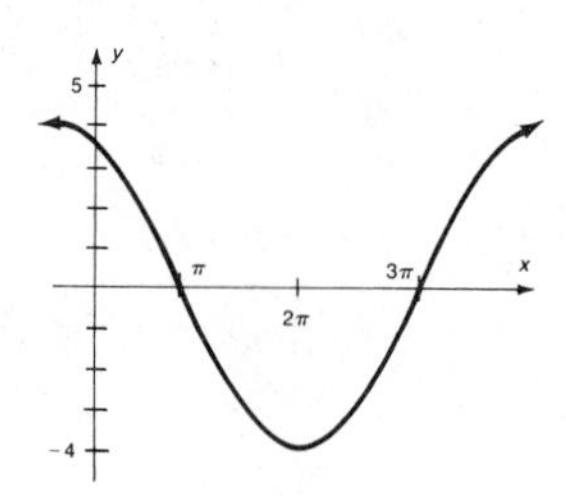

41.
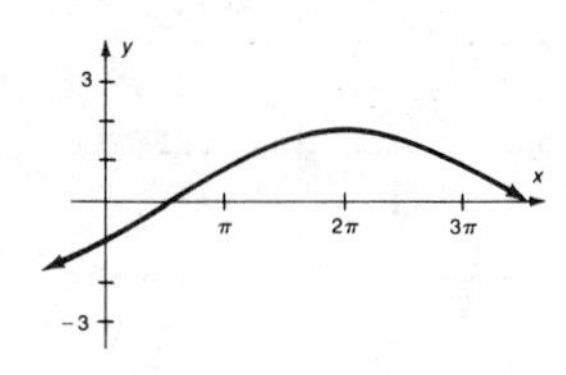

43.
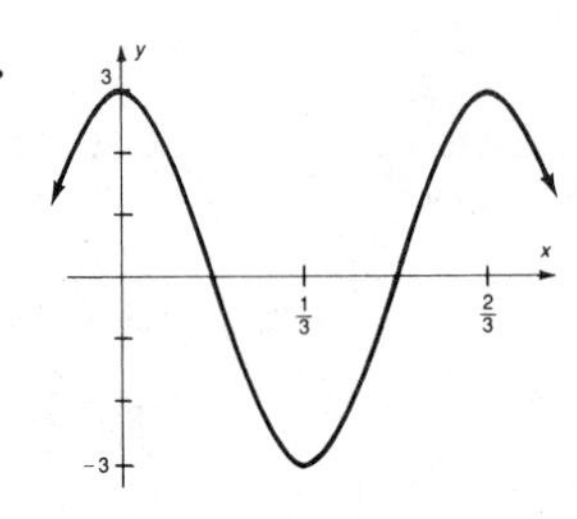

45.
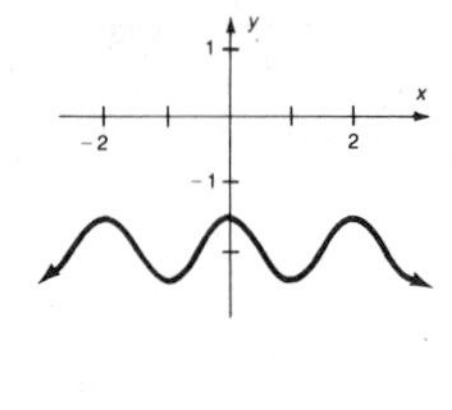

47.
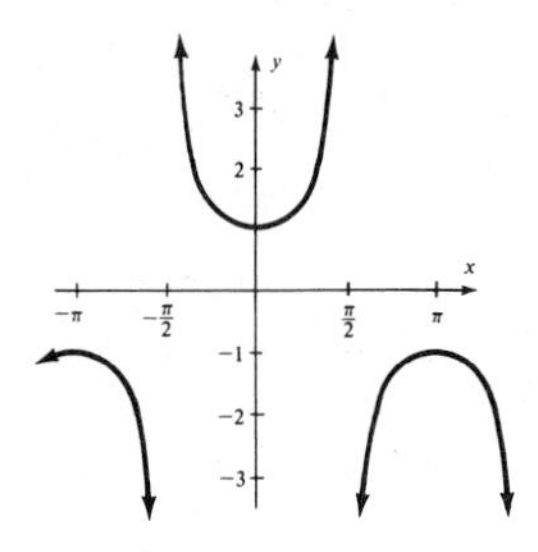

49.
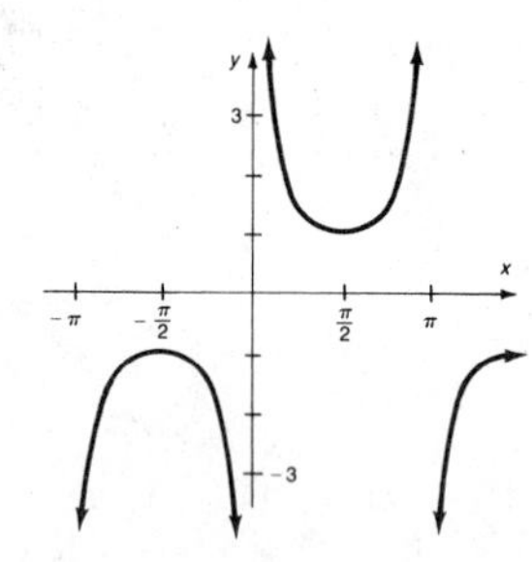

51.
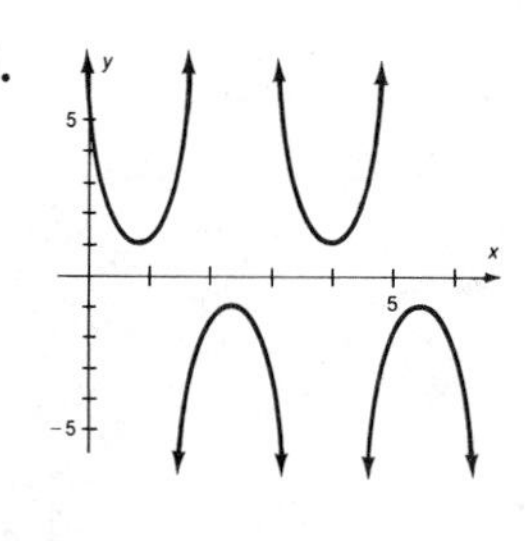

53.
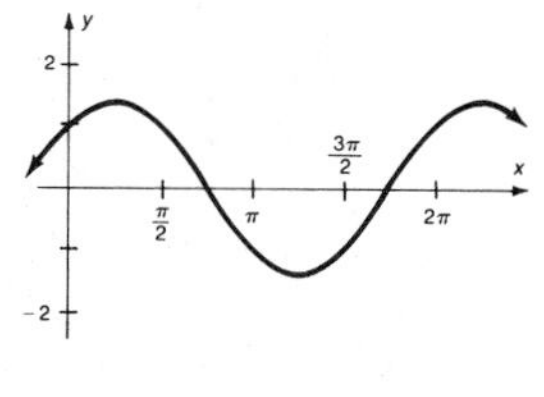

55.
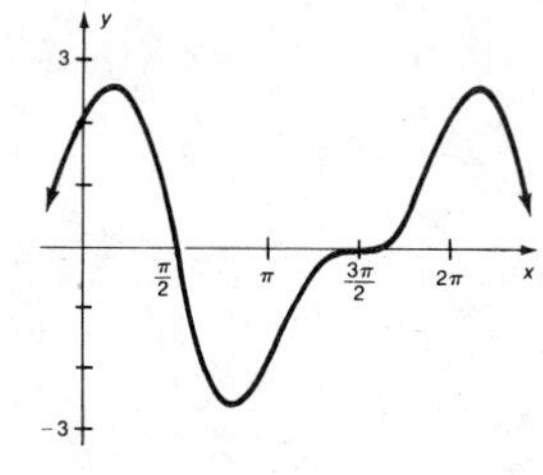

57.
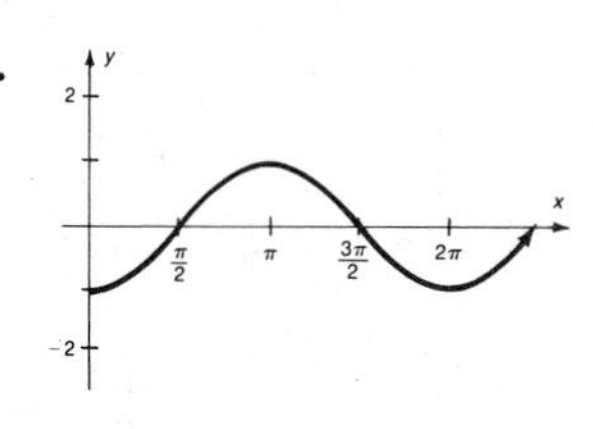

59.
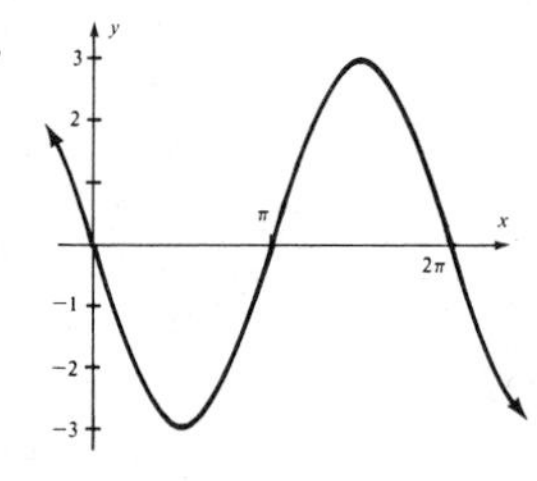

61.
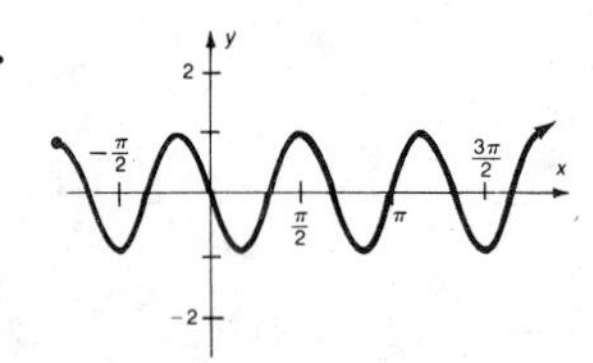

63.
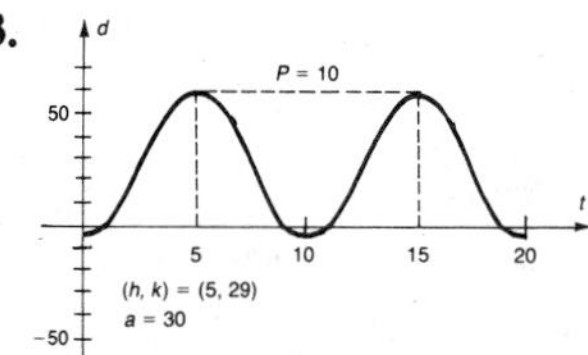

65.
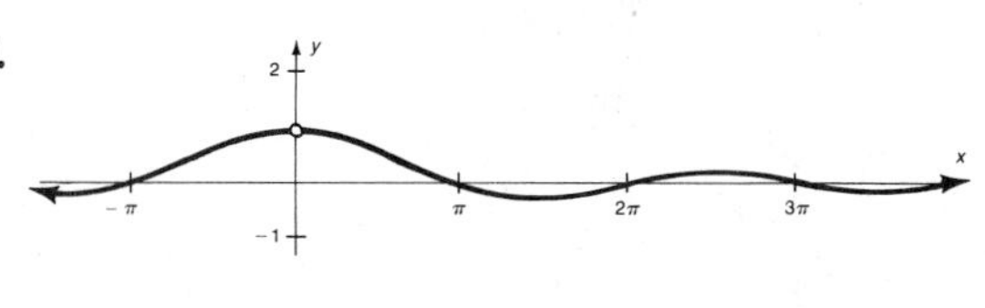

PROBLEM SET 6.5, PAGES 242–243

1. a. $0°$ or 0 **b.** $30°$ or $\frac{\pi}{6}$ **c.** $30°$ or $\frac{\pi}{6}$ **d.** $0°$ or 0 **3. a.** $60°$ or $\frac{\pi}{3}$ **b.** $45°$ or $\frac{\pi}{4}$ **c.** $45°$ or $\frac{\pi}{4}$ **d.** $45°$ or $\frac{\pi}{4}$ **5. a.** $150°$ or $\frac{5\pi}{6}$ **b.** $-45°$ or $-\frac{\pi}{4}$ **c.** $-45°$ or $-\frac{\pi}{4}$ **d.** $120°$ or $\frac{2\pi}{3}$ **7. a.** $0°$ or 0 **b.** $60°$ or $\frac{\pi}{3}$ **c.** $60°$ or $\frac{\pi}{3}$ **d.** $60°$ or $\frac{\pi}{3}$ **9.** 0.58 **11.** 1.65 **13.** 2.81 **15.** 1.32 **17.** 0.98 **19.** 1.34 **21.** 0.34 **23.** 2.90 **25.** 1.77 **27.** $69°$ **29.** $-28°$ **31.** $46°$ **33.** $85°$ **35.** $54°$ **37.** $141°$ **39.** $135°$ **41.** $153.43°$ **43.** $-71.57°$ **45.** $\frac{\pi}{6}$ **47.** $\frac{\pi}{15}$ **49.** $\frac{2\pi}{15}$ **51.** 0.4163 **53.** 1.28 **55.** 0.2836 **57.** -0.64

59.

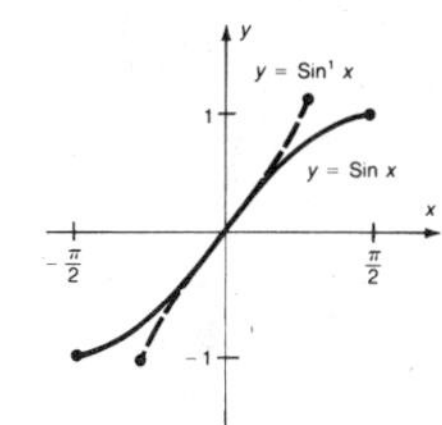

61.

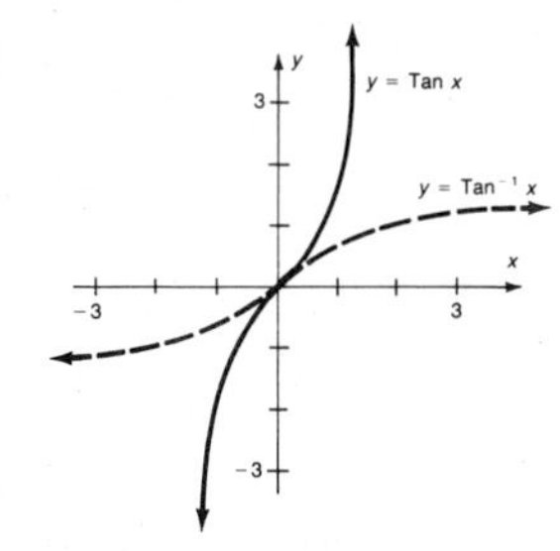

63.

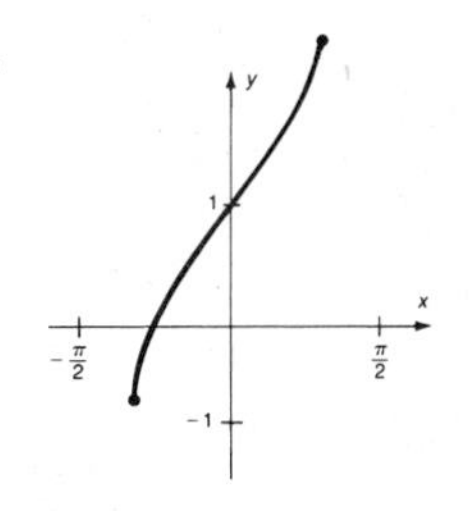

65.

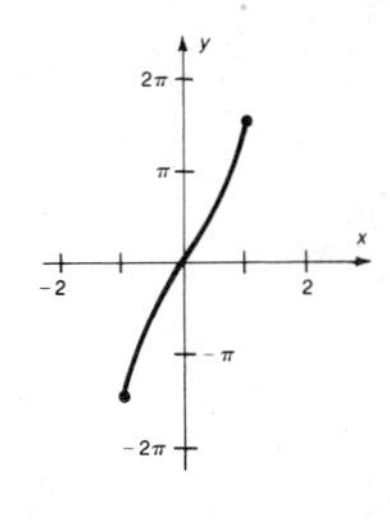

67.

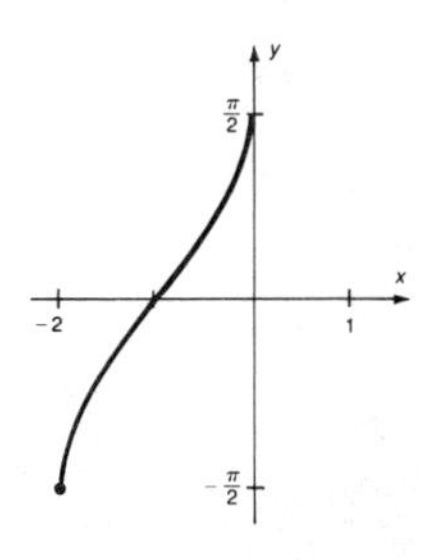

CHAPTER 6 SUMMARY, PAGES 243–245

1. a. lambda **b.** theta **c.** phi **d.** alpha **e.** beta **3.** $x^2 + y^2 = 1$ **5.** 145° **7.** $\frac{7\pi}{6}$

9. 180° **11.**

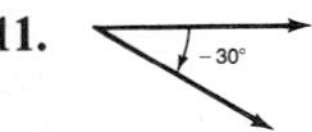

13.

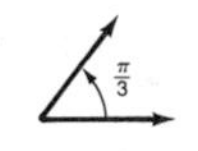

15.

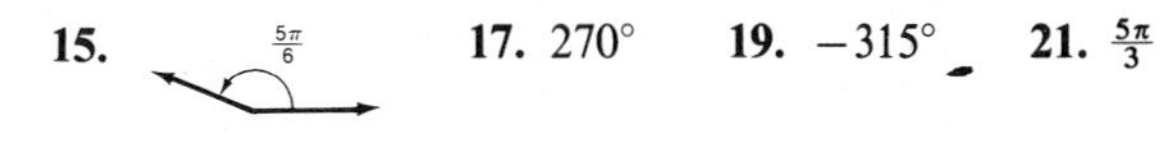

17. 270° **19.** −315° **21.** $\frac{5\pi}{3}$

23. $\frac{3\pi}{10}$ or 0.9425 **25.** $s = r\theta$; radius; angle measured in radians **27.** $5\pi \approx 15.7$ cm **29.** 60° **31.** $4 - \pi$

33. an angle in standard position **35.** $\sin\theta = b$; $\csc\theta = \dfrac{1}{b}$, $b \neq 0$ **37.** all **39.** tangent and cotangent **41.** $\frac{4}{3}$

43. secant **45.** 1.0896 **47.** 1.4587

49. θ is an angle in standard position with a point $P(x, y)$ on the terminal side of θ a nonzero distance of r from the origin

51. $\sin\theta = \frac{y}{r}$; $\csc\theta = \frac{r}{y}$, $y \neq 0$ **53.** $\cos\theta = \frac{5}{13}$; $\sin\theta = -\frac{12}{13}$; $\tan\theta = -\frac{12}{5}$; $\sec\theta = \frac{13}{5}$; $\csc\theta = -\frac{13}{12}$; $\cot\theta = -\frac{5}{12}$

55. $\cos\theta = -\frac{5}{\sqrt{29}}$; $\sin\theta = \frac{2}{\sqrt{29}}$; $\tan\theta = -\frac{2}{5}$; $\sec\theta = -\frac{\sqrt{29}}{5}$; $\csc\theta = \frac{\sqrt{29}}{2}$; $\cot\theta = -\frac{5}{2}$ **57. a.** 1 **b.** $\frac{\sqrt{3}}{2}$ **c.** $\frac{\sqrt{2}}{2}$ **d.** $\frac{1}{2}$ **e.** 0 **f.** −1 **g.** 0 **59. a.** 0 **b.** $\frac{\sqrt{3}}{3}$ **c.** 1 **d.** $\sqrt{3}$ **e.** undefined **f.** 0 **g.** undefined

61. $\frac{1}{2}$ **63.** −1 **65.** 1.4587 **67.** 1.0896 **69.**

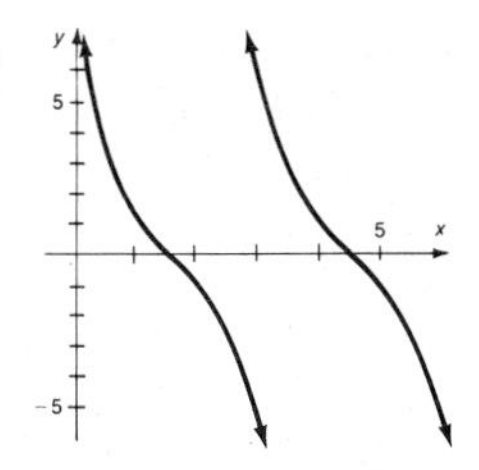

71.

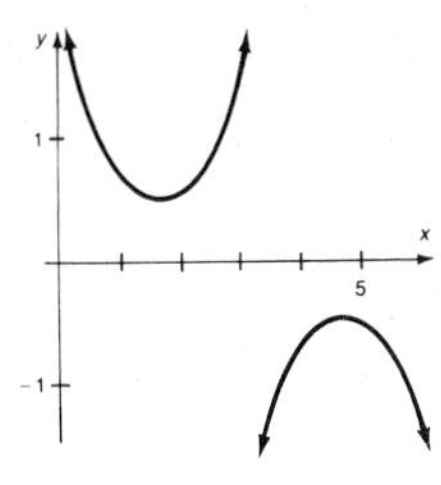

73.

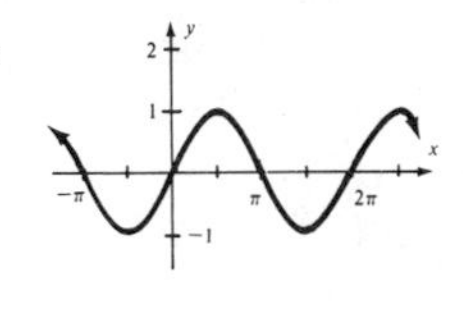

75.

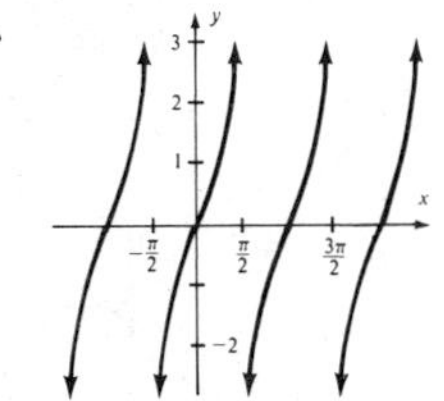

77.

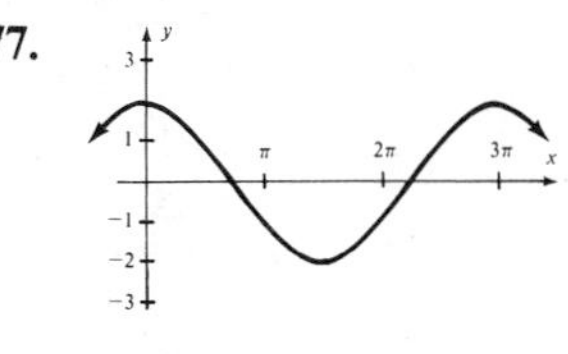

79.

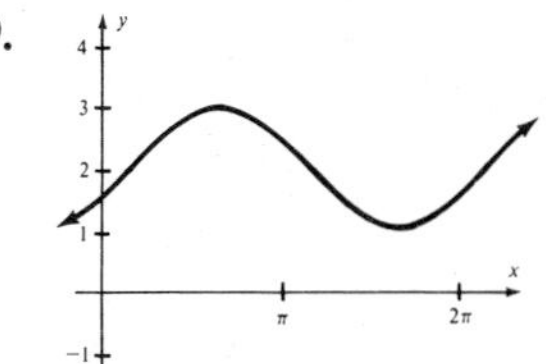

81. $-\frac{\pi}{2} < y < \frac{\pi}{2}$ **83.** $0 < y < \pi$ **85.** $\frac{\pi}{6}$ or 30° **87.** $\frac{\pi}{6}$ or 30° **89.** 0.3194 **91.** 1.2741

93.

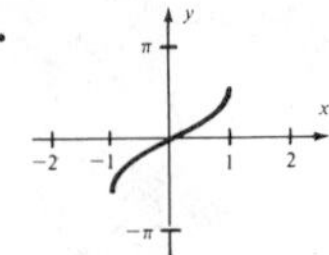

95.

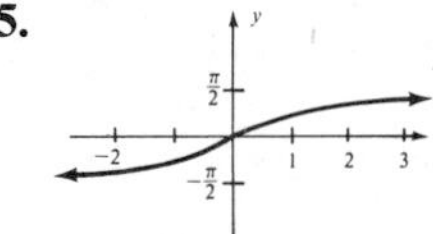

97. 1 **99.** -0.4292

PROBLEM SET 7.1, PAGES 252–253

1. $\frac{\pi}{6}$ **3.** $-\frac{\pi}{6}$ **5.** $\begin{cases}\frac{\pi}{6}+2n\pi\\ \frac{5\pi}{6}+2n\pi\end{cases}$ **7.** $\frac{\pi}{3}$ **9.** $120°$ **11.** $\begin{cases}\frac{\pi}{3}+2n\pi\\ \frac{5\pi}{3}+2n\pi\end{cases}$ **13.** $\frac{\pi}{6}, \frac{5\pi}{6}, \frac{7\pi}{6}, \frac{11\pi}{6}$ **15.** $\frac{\pi}{3}, \frac{2\pi}{3}, \frac{4\pi}{3}, \frac{5\pi}{3}$
17. $\frac{2\pi}{3}, \frac{5\pi}{6}, \frac{5\pi}{3}, \frac{11\pi}{6}$ **19.** $\frac{\pi}{12}, \frac{5\pi}{12}, \frac{3\pi}{4}, \frac{13\pi}{12}, \frac{17\pi}{12}, \frac{7\pi}{4}$ **21.** $\frac{5\pi}{12}, \frac{7\pi}{12}, \frac{17\pi}{12}, \frac{19\pi}{12}$ **23.** $0, \pi$ **25.** $\frac{\pi}{2}, \frac{3\pi}{2}$
27. $\frac{\pi}{6}, \frac{\pi}{3}, \frac{5\pi}{6}, \frac{5\pi}{3}$ **29.** $0, \pi, \frac{\pi}{4}, \frac{5\pi}{4}$ **31.** $\frac{\pi}{4}, \frac{3\pi}{4}, \frac{5\pi}{4}, \frac{7\pi}{4}$ **33.** $0, \pi, \frac{\pi}{3}, \frac{5\pi}{3}$ **35.** 2.2370, 4.0461
37. 0.3649, 1.2059, 3.5065, 4.3475 **39.** 0.6662, 2.4754 **41.** 0.8213, 2.3203 **43.** 0.00, 1.57, 2.09, 3.14, 4.19, 4.71
45. 0.41, 1.16, 3.55, 4.30 **47.** 0.32, 1.89, 3.46, 5.03 **49.** 0.00, 2.09, 4.19 **51.** 0.00, 0.52, 2.62, 3.14
53. 0.00, 1.05, 1.22, 1.92, 2.09, 3.14, 3.32, 4.01, 4.19, 5.24, 5.41, 6.11 **55.** 0.21, 1.05, 1.47, 2.30, 2.72, 3.56, 3.98, 4.82, 5.24, 6.07
57. $\begin{cases}\frac{5\pi}{6}+2n\pi\\ \frac{7\pi}{6}+2n\pi\end{cases}$ **59.** $\begin{cases}0.4014+2n\pi\\ 2.7402+2n\pi\end{cases}$ **61.** $0.9423+n\pi$ **63.** 0.185

PROBLEM SET 7.2, PAGES 256–257

1. See inside back cover. **3.** III, IV **5.** I, IV **7.** II **9.** I **11.** $\cot(A+B)$ **13.** $\tan(\frac{\pi}{15})$
15. $\cos(\frac{\pi}{8})$ **17.** $\cos 127°$ **19.** 1 **21.** 1 **23.** -1

25.
$\frac{1}{\sin\theta} = \frac{1}{y/r}$ By definition of the trig functions
$= \frac{r}{y}$ Dividing fractions
$= \csc\theta$ By definition of the trig functions

27.
$\frac{\cos\theta}{\sin\theta} = \frac{x/r}{y/r}$ By definition of the trigonometric functions
$= \frac{x}{r}\cdot\frac{r}{y}$ Division of fractions
$= \frac{x}{y}$ Multiplication and simplification of fractions
$= \cot\theta$ By definition of cotangent

29.
$x^2+y^2=r^2$ By the Pythagorean Theorem
$1+\frac{y^2}{x^2}=\frac{r^2}{x^2}$ Dividing both sides by x^2, $x\neq 0$
$1+\left(\frac{y}{x}\right)^2=\left(\frac{r}{x}\right)^2$ Properties of exponents
$1+\tan^2\theta=\sec^2\theta$ By definition of the trig functions

31.
$\cot\theta=\cot\theta$
$\tan\theta=\frac{1}{\cot\theta}$
$\csc\theta=\pm\sqrt{1+\cot^2\theta}$
$\sin\theta=\frac{1}{\pm\sqrt{1+\cot^2\theta}}$
$\cos\theta=\frac{\cot\theta}{\pm\sqrt{1+\cot^2\theta}}$
$\sec\theta=\frac{\pm\sqrt{1+\cot^2\theta}}{\cot\theta}$

33.
$\csc\theta=\csc\theta$
$\sin\theta=\frac{1}{\csc\theta}$
$\cot\theta=\pm\sqrt{\csc^2\theta-1}$
$\tan\theta=\frac{1}{\pm\sqrt{\csc^2\theta-1}}$
$\cos\theta=\frac{\pm\sqrt{\csc^2\theta-1}}{\csc\theta}$
$\sec\theta=\frac{\csc\theta}{\pm\sqrt{\csc^2\theta-1}}$

35. $\sin\theta=-\frac{4}{5}$
$\tan\theta=-\frac{4}{3}$ $\cot\theta=-\frac{3}{4}$
$\sec\theta=\frac{5}{3}$ $\csc\theta=-\frac{5}{4}$

	$\cos\theta$	$\sin\theta$	$\tan\theta$	$\sec\theta$	$\csc\theta$	$\cot\theta$
37.	$\frac{5}{13}$	$\frac{12}{13}$	$\frac{12}{5}$	$\frac{13}{5}$	$\frac{13}{12}$	$\frac{5}{12}$
39.	$\frac{-12}{13}$	$\frac{-5}{13}$	$\frac{5}{12}$	$\frac{-13}{12}$	$\frac{-13}{5}$	$\frac{12}{5}$
41.	$\frac{-\sqrt{5}}{3}$	$\frac{2}{3}$	$\frac{-2\sqrt{5}}{5}$	$\frac{-3\sqrt{5}}{5}$	$\frac{3}{2}$	$\frac{-\sqrt{5}}{2}$
43.	$\frac{5\sqrt{34}}{34}$	$\frac{3\sqrt{34}}{34}$	$\frac{3}{5}$	$\frac{\sqrt{34}}{5}$	$\frac{\sqrt{34}}{3}$	$\frac{5}{3}$
45.	$\frac{-\sqrt{10}}{10}$	$\frac{-3\sqrt{10}}{10}$	3	$-\sqrt{10}$	$\frac{-\sqrt{10}}{3}$	$\frac{1}{3}$

47. $\sin\theta$ **49.** $\dfrac{2}{1-\cos^2\theta}$ or $\dfrac{2}{\sin^2\theta}$ **51.** $\dfrac{1}{\sin\theta}$ **53.** $\dfrac{\sin^2\theta-\cos^2\theta}{\sin\theta\cos\theta}$ **55.** $\dfrac{\sin^2\theta+\cos\theta}{\sin\theta}$ **57.** 1 **59.** $\dfrac{1+\sin^2\theta}{\cos^2\theta}$
61. $\cos^2\theta-\sin\theta$

PROBLEM SET 7.3, PAGES 261–262

Proofs of trigonometric identities vary.

PROBLEM SET 7.4, PAGES 269–270

1. $\dfrac{\sqrt{3}\cos\theta-\sin\theta}{2}$ **3.** $\dfrac{1+\tan\theta}{1-\tan\theta}$ **5.** $\dfrac{\sqrt{2}}{2}(\cos\theta+\sin\theta)$ **7.** $\cos^2\theta-\sin^2\theta$ **9.** $\dfrac{2\tan\theta}{1-\tan^2\theta}$ **11.** $\cos 52°$
13. $\cos\frac{\pi}{3}$ **15.** $-\tan\frac{\pi}{6}$ **17.** $-\tan 49°$ **19.** $-\sin 31°$ **21.** $-\tan 24°$ **23.** 0.8746 **25.** 0.6561 **27.** 0.6745

	angle θ	$\cos\theta$	$\sin\theta$	$\tan\theta$
29.	$-15°$	$\frac{\sqrt{6}+\sqrt{2}}{4}$	$\frac{\sqrt{2}-\sqrt{6}}{4}$	$-2+\sqrt{3}$
31.	$75°$	$\frac{\sqrt{6}-\sqrt{2}}{4}$	$\frac{\sqrt{2}+\sqrt{6}}{4}$	$2+\sqrt{3}$
33.	$105°$	$\frac{\sqrt{2}-\sqrt{6}}{4}$	$\frac{\sqrt{2}+\sqrt{6}}{4}$	$-2-\sqrt{3}$

35. a. $\cos(\theta-\frac{2\pi}{3})$ **b.** $-\sin(\theta-\frac{2\pi}{3})$ **c.** $-\tan(\theta-\frac{2\pi}{3})$

37.
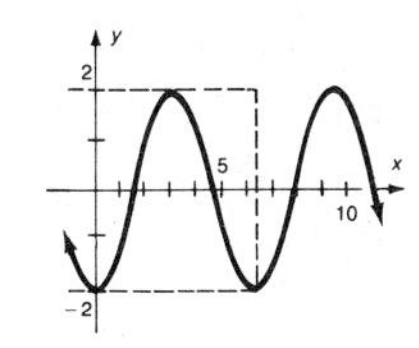

39.
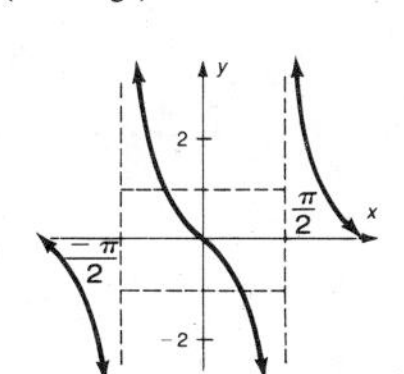

41.
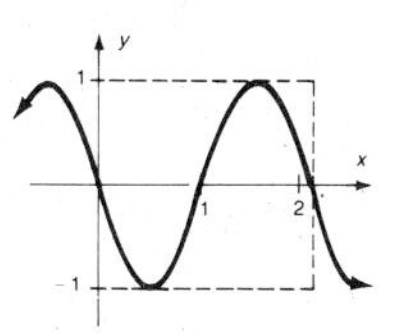

43.
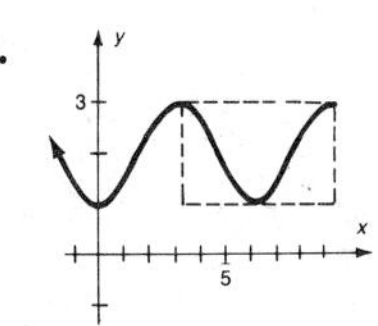

45.
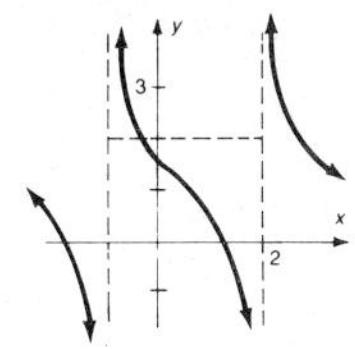

47. a. odd **b.** even **49. a.** even **b.** odd **51–71.** Proofs of identities vary.

PROBLEM SET 7.5, PAGES 277–278

1. $\frac{\sqrt{2}}{2}$ **3.** $\frac{1}{2}$ **5.** -1 **7.** $\frac{1}{2}\sqrt{2-\sqrt{2}}$ **9.** $\sqrt{2}-1$

	$\cos 2\theta$	$\sin 2\theta$	$\tan 2\theta$
11.	$\frac{119}{169}$	$\frac{-120}{169}$	$\frac{-120}{119}$
13.	$\frac{7}{25}$	$\frac{-24}{25}$	$\frac{-24}{7}$
15.	$\frac{-119}{169}$	$\frac{120}{169}$	$-\frac{120}{119}$

	$\cos\frac{1}{2}\theta$	$\sin\frac{1}{2}\theta$	$\tan\frac{1}{2}\theta$
17.	$\frac{\sqrt{26}}{26}$	$\frac{5\sqrt{26}}{26}$	5
19.	$\frac{\sqrt{10}}{10}$	$\frac{3\sqrt{10}}{10}$	3
21.	$\frac{-2\sqrt{13}}{13}$	$\frac{3\sqrt{13}}{13}$	$-\frac{3}{2}$

23. $\cos 28° + \cos 64°$
25. $\frac{1}{2}\sin 315° + \frac{1}{2}\sin 85°$
27. $\frac{1}{2}\cos 3\theta - \frac{1}{2}\cos 7\theta$
29. $\frac{1}{2}\sin 5\theta - \frac{1}{2}\sin\theta$

31. $2\sin 8°\cos 14°$ **33.** $2\cos 51.5°\cos 26.5°$ **35.** $2\sin 62.5°\sin 37.5°$ **37.** $2\cos 4x\cos 2x$ **39.** $0, \frac{\pi}{4}, \frac{\pi}{2}, \frac{3\pi}{4}, \pi, \frac{5\pi}{4}, \frac{3\pi}{2}, \frac{7\pi}{4}$
41. $\frac{\pi}{4}, \frac{\pi}{2}, \frac{3\pi}{4}, \frac{5\pi}{4}, \frac{3\pi}{2}, \frac{7\pi}{4}$ **43.** $0, \frac{\pi}{8}, \frac{3\pi}{8}, \frac{5\pi}{8}, \frac{7\pi}{8}, \pi, \frac{9\pi}{8}, \frac{11\pi}{8}, \frac{13\pi}{8}, \frac{15\pi}{8}$ **45.** $0, \frac{\pi}{4}, \frac{\pi}{2}, \frac{3\pi}{4}, \pi, \frac{5\pi}{4}, \frac{3\pi}{2}, \frac{7\pi}{4}$

	$\cos\theta$	$\sin\theta$	$\tan\theta$
47.	$\frac{\sqrt{2}}{2}$	$\frac{\sqrt{2}}{2}$	1
49.	$\frac{1}{2}$	$\frac{\sqrt{3}}{2}$	$\sqrt{3}$
51.	$\frac{3\sqrt{10}}{10}$	$\frac{\sqrt{10}}{10}$	$\frac{1}{3}$

53–65. Proofs of identities vary.

PROBLEM SET 7.6, PAGES 286–287

The complex numbers in Problems 1–27 should also be plotted.

1. a. $\sqrt{10}$ **b.** $5\sqrt{2}$ **c.** $\sqrt{13}$ **d.** $\sqrt{13}$ **3. a.** $\sqrt{29}$ **b.** $\sqrt{41}$ **c.** $\sqrt{17}$ **d.** $\sqrt{2}$ **5.** $\sqrt{2}$ cis 315°
7. 2 cis 30° **9.** 2 cis 240° **11.** 5 cis 0° **13.** 6 cis 345° **15.** 2 cis 320° **17.** $\frac{3}{2} + \frac{3\sqrt{3}}{2}i$ **19.** $-\frac{5}{2} - \frac{5\sqrt{3}}{2}i$
21. $-5i$ **23.** -2 **25.** $7.3084 + 3.2539i$ **27.** $-8.8633 - 1.5628i$ **29.** 15 cis 140° **31.** $\frac{5}{2}$ cis 267°
33. 3 cis 130° **35.** 81 cis 240° **37.** 64 cis 180° or -64 **39.** 256 cis 120° or $-128 + 128\sqrt{3}i$
41. 2 cis 80°, 2 cis 200°, 2 cis 320° **43.** 2 cis 40°, 2 cis 112°, 2 cis 184°, 2 cis 256°, 2 cis 328°
45. 2 cis 32°, 2 cis 104°, 2 cis 176°, 2 cis 248°, 2 cis 320° **47.** 3 cis 0°, 3 cis 120°, 3 cis 240°
49. cis 22.5°, cis 112.5°, cis 202.5°, cis 292.5° **51.** $\sqrt[8]{2}$ cis 56.25°, $\sqrt[8]{2}$ cis 146.25°, $\sqrt[8]{2}$ cis 236.25°, $\sqrt[8]{2}$ cis 326.25°
53. 2 cis 15°, 2 cis 75°, 2 cis 135°, 2 cis 195°, 2 cis 255°, 2 cis 315°
55. $2^{\frac{1}{18}}$ cis 15°, $2^{\frac{1}{18}}$ cis 55°, $2^{\frac{1}{18}}$ cis 95°, $2^{\frac{1}{18}}$ cis 135°, $2^{\frac{1}{18}}$ cis 175°, $2^{\frac{1}{18}}$ cis 215°, $2^{\frac{1}{18}}$ cis 255°, $2^{\frac{1}{18}}$ cis 295°, $2^{\frac{1}{18}}$ cis 335°
57. cis 0°, cis 36°, cis 72°, cis 108°, cis 144°, cis 180°, cis 216°, cis 252°, cis 288°, cis 324°
Note: Problems 59–63 should also be illustrated graphically.
59. $1, i, -1, -i$ **61.** $-0.6840 + 1.8794i$, $-1.2856 - 1.5321i$, $1.9696 - 0.3473i$
63. $1.9696 + 0.3473i$, $-0.3473 + 1.9696i$, $-1.9696 - 0.3473i$, $0.3473 - 1.9696i$
65. 8 cis 0°, 8 cis 72°, 8 cis 144°, 8 cis 216°, 8 cis 288° **67.** cis 72°, cis 144°, cis 216°, cis 288° **69–71.** Answers vary.

CHAPTER 7 SUMMARY, PAGES 287–289

1. $\frac{\pi}{12}, \frac{7\pi}{12}, \frac{13\pi}{12}, \frac{19\pi}{12}$ **3.** $\frac{\pi}{12}, \frac{\pi}{4}, \frac{3\pi}{4}, \frac{11\pi}{12}, \frac{17\pi}{12}, \frac{19\pi}{12}$ **5.** $\frac{\pi}{3}, \frac{2\pi}{3}, \frac{4\pi}{3}, \frac{5\pi}{3}$ **7.** No solution
9. $\sec\theta = \dfrac{1}{\cos\theta}; \csc\theta = \dfrac{1}{\sin\theta}; \cot\theta = \dfrac{1}{\tan\theta}$ **11.** $\cos^2\theta + \sin^2\theta = 1; \tan^2\theta + 1 = \sec^2\theta; 1 + \cot^2\theta = \csc^2\theta$
13. Quad II: $\csc\delta = \frac{5}{3}; \cos\delta = -\frac{4}{5}; \sec\delta = -\frac{5}{4}; \tan\delta = -\frac{3}{4}; \cot\delta = -\frac{4}{3}$
15. Quad III: $\csc\omega = -\frac{5}{3}; \cos\omega = -\frac{4}{5}; \sec\omega = -\frac{5}{4}; \tan\omega = \frac{3}{4}; \cot\omega = \frac{4}{3}$ **17.** $\dfrac{\sin^2\theta + \cos\theta}{\sin\theta\cos\theta}$
19. $\dfrac{2\sin^2\theta + \cos^2\theta}{\cos^2\theta}$ **21–27.** Answers vary. **29.** cos 52° **31.** sin 1.115 **33.** $\cos(\theta - \frac{\pi}{6})$

35.
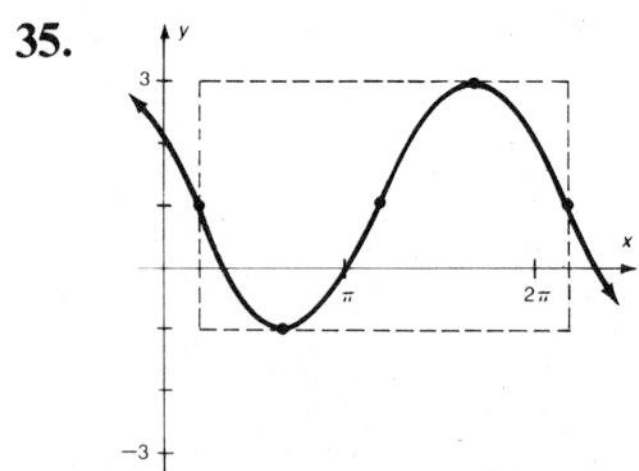

37. $\cos(\theta - 30°) = \frac{1}{2}(\sqrt{3}\cos\theta + \sin\theta)$

39. Write as $\tan(23° - 85°) = \tan(-62°) \approx -1.881$. **41.** Write as $\tan 2(\frac{\pi}{6}) = \tan\frac{\pi}{3} = \sqrt{3}$. **43.** $-\frac{24}{25}$
45. Write as $\cos 120° = -\frac{1}{2}$. **47.** $\frac{\sqrt{10}}{10}$ **49.** $\frac{1}{2}\sin 4\theta + \frac{1}{2}\sin 2\theta$ **51.** $0, \frac{\pi}{4}, \frac{3\pi}{4}, \pi, \frac{5\pi}{4}, \frac{7\pi}{4}$ **53.** $7\sqrt{2}$ cis 315°
55. 7 cis 330° **57.** $2\sqrt{2} - 2\sqrt{2}i$ **59.** $-5i$ **61.** $-128 - 128\sqrt{3}i$ **63.** cis 154°
65. If n is any positive integer, then the nth roots of $r \operatorname{cis}\theta$ are given by $\sqrt[n]{r} \operatorname{cis}\left(\dfrac{\theta + 2\pi k}{n}\right)$ for $k = 0, \ldots, n - 1$.
67. $-2.5556 + 0.6848i$; $2.5556 - 0.6848i$; also plot points using $\sqrt{7}$ cis 165° and $\sqrt{7}$ cis 345°.

PROBLEM SET 8.1, PAGES 296–298

	α	β	γ	a	b	c
1.	30°	60°	90°	80	140	160
3.	45°	45°	90°	9.0	9.0	13
5.	37°	53°	90°	11	15	19
7.	71°	19°	90°	69	24	73
9.	54°	36°	90°	18	13	22
11.	40°	50°	90°	24	29	38
13.	76°	14°	90°	29	7.2	30
15.	45°	45°	90°	49	49	69
17.	77°	13°	90°	390	90	400
19.	50°	40°	90°	98	82	130
21.	6°	84°	90°	3.8	36	36
23.	56.00°	34.00°	90.00°	3484	2350	4202
25.	27.66°	62.34°	90.00°	1625	3100	3500
27.	42°	48°	90°	320	350	470
29.	17.54°	72.46°	90.00°	1296	4100	4300

31. The building is 23 m tall. **33.** The ship is 200 m away. **35.** The car is 340 ft away.
37. The distance is 350 ft. **39.** It is 13 ft above the ground. **41.** The angle of elevation is about 34°.
43. The height is 1251 ft. **45.** 263 m **47.** The distance across the river is 170 m. **49.** The tower is 222.0 ft high.
51. The height of the center is 14.7 ft. **53.** The distance from the earth to Venus is 63.4 million miles (or 6.34×10^7 mi).
55. The radius of the inscribed circle is 633.9 ft. **57.** Answers vary. **59.** The shadow will be about 12 ft.
61. Answers vary.

PROBLEM SET 8.2, PAGES 301–302

1. 54° **3.** 80° **5.** 21° **7.** 13 **9.** 10 **11.** 17 **13.** 14 **15.** 46.6° **17.** 604 **19.** 80° **21.** 22°
23. 78° **25.** $a = 4.2, b = 5.2, c = 3.4, \alpha = 54°, \beta = 85°, \gamma = 41°$
27. $a = 214, b = 320, c = 126, \alpha = 25.8°, \beta = 139.4°, \gamma = 14.8°$ **29.** $a = 140, b = 85.0, c = 105, \alpha = 94.3°, \beta = 37.3°, \gamma = 48.4°$
31. no solution, since no sides are given **33.** $a = 641, b = 520, c = 235, \alpha = 110.5°, \beta = 49.4°, \gamma = 20.1°$
35. $a = 341, b = 340, c = 138, \alpha = 78.7°, \beta = 77.9°, \gamma = 23.4°$ **37.** 537 ft **39.** 362 ft **41.** 293 ft
43–47. Answers vary. **49.** 700 mi **51.** 410 mi
53. $d = 2.25, q = 1.88, p = 2.13, \angle D = 68.1°, \angle P = 61.2°, \angle Q = 50.7°$ **55.** 16 in. **57.** 450 in.^{2}
59–63. Answers vary.

PROBLEM SET 8.3, PAGES 305–307

1. $a = 10, b = 12, c = 13, \alpha = 48°, \beta = 62°, \gamma = 70°$ **3.** $a = 30, b = 46, c = 59, \alpha = 30°, \beta = 50°, \gamma = 100°$
5. $a = 33, b = 40, c = 37, \alpha = 50°, \beta = 70°, \gamma = 60°$ **7.** $a = 310, b = 280, c = 43, \alpha = 120°, \beta = 53°, \gamma = 7°$
9. $a = 107, b = 276, c = 325, \alpha = 18.3°, \beta = 54.0°, \gamma = 107.7°$ **11.** $a = 85, b = 97, c = 110, \alpha = 49°, \beta = 59°, \gamma = 72°$
13. $a = 105, b = 125, c = 131, \alpha = 48.5°, \beta = 62.7°, \gamma = 68.8°$ **15.** no solution (the sum of the angles must be less than 180°)
17. $a = 45.3, b = 145, c = 105, \alpha = 9.5°, \beta = 148.0°, \gamma = 22.5°$
19. $a = 41.0, b = 21.2, c = 53.1, \alpha = 45.2°, \beta = 21.5°, \gamma = 113.3°$
21. no solution (the sum of the angles must be less than 180°) **23.** $a = 26, b = 71, c = 88, \alpha = 14°, \beta = 42°, \gamma = 123°$
25. no solution (at least one side must be given) **27.** $a = 80.6, b = 23.2, c = 83.6, \alpha = 74.7°, \beta = 16.1°, \gamma = 89.2°$
29. $a = 16.90, b = 19.46, c = 28.36, \alpha = 35.60°, \beta = 42.10°, \gamma = 102.30°$
31. $a = 10, b = 48, c = 52, \alpha = 11°, \beta = 61°, \gamma = 108°$ **33.** $a = 48.1, b = 569, c = 574, \alpha = 4.8°, \beta = 82.0°, \gamma = 93.2°$
35. no solution (the sum of the angles must be less than 180°) **37.** $a = 8.1, b = 15, c = 17, \alpha = 29°, \beta = 62°, \gamma = 90°$

39. $a = 6.74, b = 10.9, c = 4.45, \alpha = 16.2°, \beta = 153.2°, \gamma = 10.6°$ **41.** The height of the building is about 260 ft.
43. The tower is about 280 ft tall.
45. Luke's spacecraft is 970.4 m from the first observation point and 1532 m from the second.
47. It is about 60 ft to the top of the building, and the height of the building is about 31 ft. **49.** The boat is about 8 ft long.
51–59. Answers vary.
61. The rate of the car is 38.8428 mph or 119.8128 mph. Since the latter solution is highly unlikely, the published solution of 38.843 is correct to the nearest thousandth.

PROBLEM SET 8.4, PAGES 316–318

Remember, when solving triangles you must work with the given or calculated accuracy and round only when stating your answer.
1. $\alpha > 90°$ and OPP $\leq$ ADJ, so no triangle is formed. **3.** $\alpha < 90°$ and OPP $< h <$ ADJ, so no triangle is formed.
5. $a = 5.0, b = 4.0, c = 1.5, \alpha = 125°, \beta = 41°, \gamma = 14°$ **7.** $a = 9.0, b = 7.0, c = 7.8, \alpha = 75°, \beta = 49°, \gamma = 56°$
9. $\alpha < 90°$ and OPP $< h <$ ADJ, so no triangle is formed. **11.** $a = 8.63, b = 11.8, c = 8.05, \alpha = 47.0°, \beta = 90.0°, \gamma = 43.0°$
13. $a = 14.2, b = 16.3, c = 4.00, \alpha = 52.1°, \beta = 115.0°, \gamma = 12.9°$
15. $a = 49.5, b = 23.5, c = 28.1, \alpha = 147.0°, \beta = 15.0°, \gamma = 18.0°$
17. $h \approx 43.8 <$ OPP $<$ ADJ; ambiguous case: $a = 98.2, a' = 41.6, b = 82.5, c = 52.2, \alpha = 90.8°, \alpha' = 25.0°, \beta = 57.1°, \beta' = 122.9°, \gamma = 32.1°$ **19.** no solution since the sum of the given angles is greater than 180°
21. no solution because the sum of the two smaller sides must be larger than the third side
23. $a = 179, b = 195, c = 196, \alpha = 54.5°, \beta = 62.5°, \gamma = 63.0°$
Answers to Problems 25–47 are in square units and are given to the correct number of significant digits.
25. 37 **27.** 83 **29.** 680 **31.** 560 **33.** 6.4 **35.** 35 **37.** 54.0 **39.** 58.2 and 11.3 **41.** 2150 and 911
43. 180 **45.** no triangle formed **47.** 20,100 **49.** He is 5.39 mi from the target.
51. The distance from the first city is 1.926 mi (10,179 ft) and from the second city is 0.5363 mi (2832 ft). The altitude is 0.3503 mi (1850 ft). **53.** The angle of elevation is 19.0°. **55.** The radius is about 7.82 cm.
57. The area is about 18,000 m^2 (17,872 before rounding). **59.** The volume is about 21,000 ft^3, or 760 yd^3.
61. The height of the tower is 985 ft. **63–65.** Answers vary.

CHAPTER 8 SUMMARY, PAGES 318–320

1. If θ is an acute angle in a right triangle, then $\cos\theta =$ ADJ/HYP, $\sin\theta =$ OPP/HYP, $\tan\theta =$ OPP/ADJ.
3. $a = 475, b = 678, c = 828, \alpha = 35.0°, \beta = 55.0°, \gamma = 90.0°$ **5–7.** Answers vary
9. $\cos\alpha = \dfrac{b^2 + c^2 - a^2}{2bc}; \alpha = \cos^{-1}\left(\dfrac{b^2 + c^2 - a^2}{2bc}\right)$ **11.** $\alpha + \beta + \gamma = 180°; \alpha = 180° - \beta - \gamma$
13. $a = 24, b = 52, c = 61, \alpha = 23°, \beta = 58°, \gamma = 99°$
15. ambiguous case: $a = 51, a' = 14, b = 34, c = 21, \alpha = 137°, \alpha' = 11°, \beta = 27°, \beta' = 153°, \gamma = 16°$ **17.** 460 ft^2
19. 122 $in.^2$ and 17.4 $in.^2$ **21.** 430 ft; 48.9° **23.** 278 ft

CUMULATIVE REVIEW II, PAGE 321

1. Let θ be an angle in standard position with a point $P(x, y)$ on the terminal side a distance of r ($r \neq 0$) from the origin. Then $\cos\theta = x/r$; $\sin\theta = y/r$; $\tan\theta = y/x$ ($x \neq 0$); $\sec\theta = r/x$ ($x \neq 0$); $\csc\theta = r/y$ ($y \neq 0$); $\cot\theta = x/y$ ($y \neq 0$).
2. a. $-\frac{\sqrt{3}}{2}$ **b.** $-\frac{\sqrt{3}}{2}$ **c.** 0 **d.** $\frac{4\pi}{3} + 2n\pi, \frac{5\pi}{3} + 2n\pi$ **e.** $\frac{\pi}{3} + 2n\pi, \frac{5\pi}{3} + 2n\pi$ **f.** no value **g.** 1.162253 **h.** $2\sqrt{2}$
3. a. **b.** 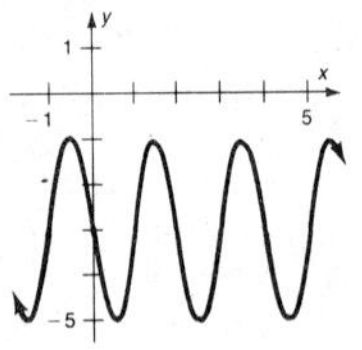 **c.** 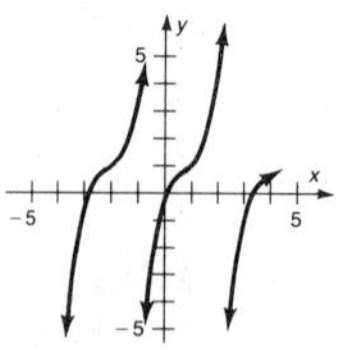 **4.** Answers vary.

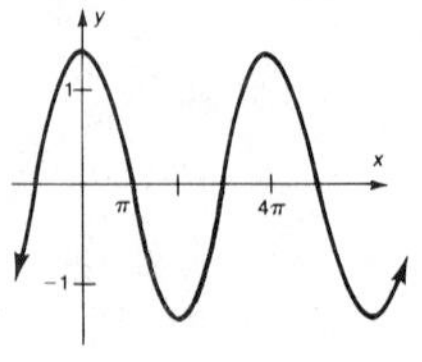

5. a. $\frac{\pi}{9}, \frac{4\pi}{9}, \frac{7\pi}{9}, \frac{10\pi}{9}, \frac{13\pi}{9}, \frac{16\pi}{9}$ **b.** 0.1183, 3.0238 **6.** Answers vary. **7. a.** $\dfrac{\sin\psi}{s} = \dfrac{\sin\theta}{t} = \dfrac{\sin\phi}{u}$

b. $u^2 = s^2 + t^2 - 2st\cos\phi$; $s^2 = u^2 + t^2 - 2ut\cos\psi$; $t^2 = s^2 + u^2 - 2su\cos\theta$
8. a. $a = 14, b = 27, c = 19, \alpha = 29°, \beta = 109°, \gamma = 42°$ **b.** $a = 19, b = 7.2, c = 15, \alpha = 113°, \beta = 20°, \gamma = 47°$
c. $a = 35, b = 45, c = 69, \alpha = 26°, \beta = 35°, \gamma = 119°$ **d.** $a = 92.8, b = 119, c = 85.7, \alpha = 50.8°, \beta = 83.5°, \gamma = 45.7°$
9. a. 320 ft^2 **b.** 51.7 in.^{2} **c.** 1900 cm^2 **10. a.** I will be 22 or 67 mi from Mexico City.
b. The bridge is 4620 ft long.

EXTENDED APPLICATION: SOLAR POWER, PAGES 325–326

1. 16 h 50 m or 4:50 P.M. **3.** 19 h 1 m or 7:01 P.M. **5.** 6 h 18 m or 6:18 A.M.

7.

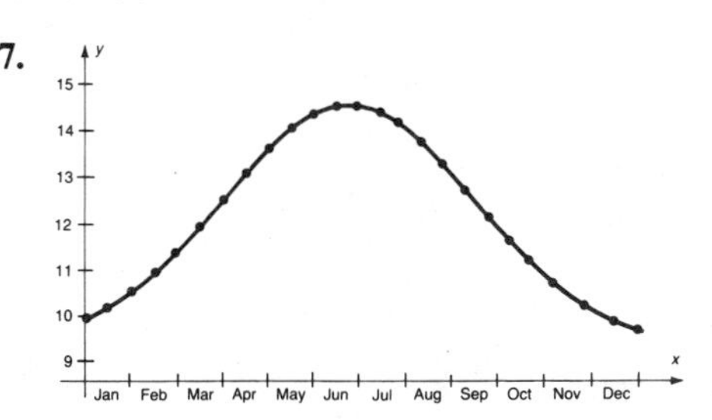

9–11. Answers vary.

PROBLEM SET 9.1, PAGES 336–337

1.

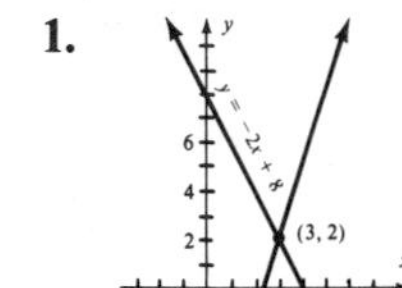

3.

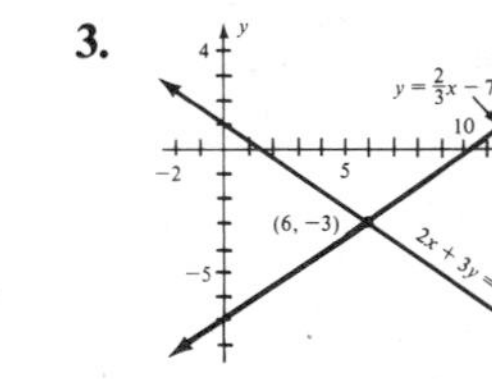

5.

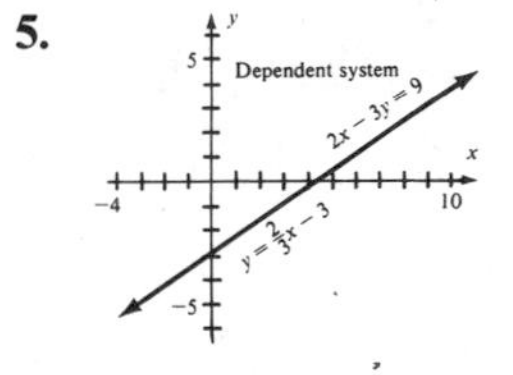

7.

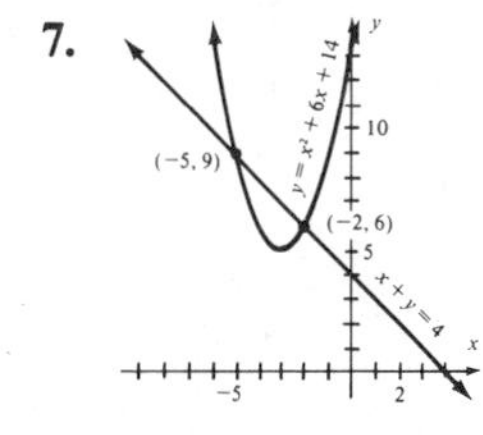

9.

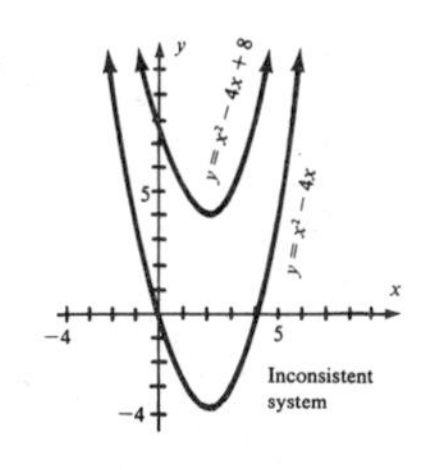

The answers for variables and constants in Problems 11–27 may vary, but the solution to the problems will not vary.

11. The variables are $s = x_1$ and $t = x_2$; the constants are
$a_{11} = 1 \quad a_{12} = 1 \quad b_1 = 1$
$a_{21} = 3 \quad a_{22} = 1 \quad b_2 = -5$
The solution is $(s, t) = (-3, 4)$.

13. The variables are $u = x_1$ and $v = x_2$; the constants are
$a_{11} = 3 \quad a_{22} = -5 \quad b_1 = -10$
$a_{21} = 3 \quad a_{22} = -5 \quad b_2 = 10$
The system is inconsistent.

15. The variables are $t_1 = x_1$ and $t_2 = x_2$; the constants are
$a_{11} = 3 \quad a_{12} = 5 \quad b_1 = 1541$
$a_{21} = 2 \quad a_{22} = -1 \quad b_2 = -160$
The solution is $(t_1, t_2) = (57, 274)$.

17. The variables are $\gamma = x_1$ and $\delta = x_2$; the constants are
$a_{11} = 1 \quad a_{12} = -3 \quad b_1 = -4$
$a_{21} = 5 \quad a_{22} = -4 \quad b_2 = -9$
The solution is $(\gamma, \delta) = (-1, 1)$.

19. The variables are $c = x_1$ and $d = x_2$; the constants are
$a_{11} = 1 \quad a_{12} = 1 \quad b_1 = 2$
$a_{21} = 2 \quad a_{22} = -1 \quad b_2 = 1$
The solution is $(c, d) = (1, 1)$.

21. The variables are $q_1 = x_1$ and $q_2 = x_2$; the constants are
$a_{11} = 3 \quad a_{12} = -4 \quad b_1 = 3$
$a_{21} = 5 \quad a_{22} = 3 \quad b_2 = 5$
The solution is $(q_1, q_2) = (1, 0)$.

23. The variables are $x = x_1$ and $y = x_2$; the constants are
$a_{11} = 7 \quad a_{12} = 1 \quad b_1 = 5$
$a_{21} = 14 \quad a_{22} = -2 \quad b_2 = -2$
The solution is $(\frac{2}{7}, 3)$.

25. The variables are $\alpha = x_1$ and $\beta = x_2$; the constants are
$a_{11} = 1 \quad a_{12} = 1 \quad b_1 = 12$
$a_{21} = 1 \quad a_{22} = -2 \quad b_2 = -4$
The solution is $(\alpha, \beta) = (\frac{20}{3}, \frac{16}{3})$.

27. The variables are $\theta = x_1$ and $\phi = x_2$; the constants are
$a_{11} = 2 \quad a_{12} = 5 \quad b_1 = 7$
$a_{21} = 3 \quad a_{22} = 4 \quad b_2 = 0$
The solution is $(\theta, \phi) = (-4, 3)$.

29. dependent system **31.** $(3, 15)$ **33.** $(6, -1)$ **35.** $(2\gamma - \delta, \gamma - \delta)$ **37.** $(\frac{a+b}{2}, \frac{a-b}{2})$ **39.** $(\frac{c+d}{5}, \frac{2d-3c}{5})$
41. $(-2, 2), (-5, -4)$ **43.** $(1, 2), (1, -2), (-1, 2), (-1, -2)$ **45.** $(\sqrt{13}, 1), (-\sqrt{13}, 1), (5, 2), (-5, 2)$
47. Answers vary. **49.** Answers vary. **51.** The numbers are 1 and -3 or 3 and -1.
53. The sides are 5 and 12 ft. **55.** $(\frac{\pi}{3}, \frac{7\pi}{6}), (\frac{\pi}{3}, \frac{11\pi}{6}), (\frac{5\pi}{3}, \frac{7\pi}{6}), (\frac{5\pi}{3}, \frac{11\pi}{6})$ **57.** inconsistent system
59. $(2\sqrt{2}, 0), (-2\sqrt{2}, 0)$ **61.** $(\frac{17}{21}, -\frac{20}{9})$ **63.** $(\frac{-9+\sqrt{113}}{2}, \frac{-15+\sqrt{113}}{8}), (\frac{-9-\sqrt{113}}{2}, \frac{-15-\sqrt{113}}{8})$
65. $(h, k) = (1, 8)$ **67.** $(\frac{a+\sqrt{a^2-4}}{2}, \frac{a-\sqrt{a^2-4}}{2}), (\frac{a-\sqrt{a^2-4}}{2}, \frac{a+\sqrt{a^2-4}}{2})$

PROBLEM SET 9.2, PAGES 340–341

1. 10 **3.** -10 **5.** 36 **7.** 10 **9.** -28 **11.** -1 **13.** 1 **15.** $\cot^2\theta$ **17.** $(s,t)=(5,-3)$ **19.** $(\frac{2}{13},\frac{3}{13})$
21. $(q_1,q_2)=(\frac{25}{29},-\frac{3}{29})$ **23.** $(\frac{21}{19},-\frac{24}{19})$ **25.** inconsistent system **27.** dependent system **29.** $(3,-3)$
31. $(31,19)$ **33.** $(-54,-31)$ **35.** $(\frac{5a+3b}{13},\frac{a-2b}{13})$ **37.** $(\frac{a}{a^2-b^2},\frac{-b}{a^2-b^2})$ **39.** $(\frac{sf-td}{cf-ed},\frac{ct-es}{cf-ed})$
41. $(\frac{ad-b\beta}{ad-bc},\frac{a\beta-c\alpha}{ad-bc})$ **43.** $(800,200)$ **45.** $(1800,200)$ **47.** $(63,84)$ **49.** $(-\frac{1}{5},\frac{3}{10})$ **51.** $(\frac{11}{15},\frac{1}{5})$
53–55. Answers vary. **57.** $(2,1),(-2,1),(2,-1),(-2,-1)$ **59.** $(3,2)$

PROBLEM SET 9.3, PAGES 347–349

1. -75 **3.** -20 **5.** -44 **7.** 3 **9.** -3 **11.** 38 **13.** -4 **15.** 21 **17.** -260 **19.** -28 **21.** 10
23. -5 **25.** 150 **27.** 0 **29.** 24 **31.** $(5,4,-3)$ **33.** $(3,-1,2)$ **35.** $(4,-3,2)$ **37.** $(-4,0,3)$
39. $(5,\frac{15}{2},\frac{3}{2})$ **41.** $x+4y+11=0$ **43.** $8x-3y+7=0$ **45.** 84 **47.** $(w,x,y,z)=(3,1,-2,1)$
49. $(s,t,u,x)=(-3,1,-2,-1)$ **51–59.** Answers vary.

PROBLEM SET 9.4, PAGES 358–359

1. $\begin{bmatrix}2&4&2\\6&-2&4\\2&2&5\end{bmatrix}$ **3.** $\begin{bmatrix}16&19&10\\29&16&15\\35&9&31\end{bmatrix}$ **5.** $\begin{bmatrix}13&25&20\\25&31&23\\42&24&33\end{bmatrix}$ **7.** $\begin{bmatrix}20&21&34\\29&16&15\\7&48&5\end{bmatrix}$ **9.** $\begin{bmatrix}34&117&44\\45&143&97\\109&100&151\end{bmatrix}$

11. $\begin{bmatrix}2&8\\16&8\\-7&7\\7&7\end{bmatrix}$ **13.** $\begin{bmatrix}14&14\\-7&7\end{bmatrix}$ **15.** $\begin{bmatrix}14&4&8&30\\-7&-1&-9&3\end{bmatrix}$ **17.** not conformable **19.** $\begin{bmatrix}1&0&0&0\\0&1&0&0\\0&0&1&0\\0&0&0&1\end{bmatrix}$

21. $\begin{bmatrix}2&7\\1&4\end{bmatrix}$ **23.** $\begin{bmatrix}0&\frac{1}{2}\\\frac{1}{3}&-\frac{1}{6}\end{bmatrix}$ **25.** $\begin{bmatrix}3&-17&-20\\3&-18&-20\\-1&6&7\end{bmatrix}$ **27.** $\begin{bmatrix}1&0&-1&0\\0&\frac{1}{2}&0&0\\-2&0&2&1\\0&0&1&0\end{bmatrix}$ **29.** $\begin{bmatrix}\frac{3}{5}&0&-\frac{2}{5}&\frac{1}{5}\\\frac{1}{5}&0&\frac{1}{5}&-\frac{1}{10}\\0&1&0&0\\-\frac{6}{5}&0&\frac{4}{5}&\frac{1}{10}\end{bmatrix}$

31. $(-4,7)$ **33.** $(25,14)$ **35.** $(50,29)$ **37.** $(-1,4)$ **39.** $(-5,2)$ **41.** $(-3,-2)$ **43.** $(4,-2)$ **45.** $(12,-5)$
47. $(0,4)$ **49.** $(-2,4,3)$ **51.** $(3,-5,2)$ **53.** $(-5,18,5)$ **55.** $(5,4,-1)$ **57.** $(1,2,3)$ **59.** $(-3,2,7)$
61. $(w,x,y,z)=(3,-2,1,-4)$ **63–67.** Answers vary.

PROBLEM SET 9.5, PAGES 363–365

Answers for Problems 1–27 may vary.

1. Interchange Rows 1 and 3. $\left[\begin{array}{ccc|c}1&3&-4&9\\0&2&4&5\\3&1&2&1\end{array}\right]$

3. Multiply first row by $\frac{1}{2}$. $\left[\begin{array}{ccc|c}1&2&5&-6\\6&3&4&6\\10&-1&0&1\end{array}\right]$

5. Add (-1) times second row to first row. $\left[\begin{array}{ccc|c}1&5&-12&2\\4&1&9&2\\7&6&1&3\end{array}\right]$

7. $\left[\begin{array}{ccc|c}1&2&-3&0\\0&3&1&4\\0&1&7&6\end{array}\right]$ **9.** $\left[\begin{array}{ccc|c}1&2&4&1\\0&9&8&4\\0&13&17&7\end{array}\right]$ **11.** $\left[\begin{array}{cccc|c}1&4&-1&3&3\\0&16&3&13&9\\0&-19&14&-14&-17\\0&-4&3&-3&-3\end{array}\right]$

13. Multiply Row 2 by $\frac{1}{2}$. $\left[\begin{array}{ccc|c}1&3&5&2\\0&1&3&-4\\0&3&4&1\end{array}\right]$

15. Add (-1) times third row to second row. $\left[\begin{array}{ccc|c}1&4&-1&6\\0&1&-5&-2\\0&4&6&5\end{array}\right]$

17. Multiply Row 2 by $\frac{1}{5}$. $\left[\begin{array}{ccc|c}1&3&-2&4\\0&1&\frac{1}{5}&\frac{3}{5}\\0&7&9&2\end{array}\right]$

19. $\left[\begin{array}{ccc|c} 1 & 5 & -3 & 2 \\ 0 & 1 & 4 & 5 \\ 0 & 0 & -8 & -13 \end{array}\right]$ **21.** $\left[\begin{array}{cccc|c} 1 & 7 & 6 & 6 & 2 \\ 0 & 1 & 9 & 2 & 1 \\ 0 & 0 & 42 & 9 & 5 \\ 0 & 0 & -37 & 0 & -2 \end{array}\right]$ **23.** Multiply Row 3 by $\frac{1}{8}$. $\left[\begin{array}{ccc|c} 1 & -3 & 4 & -5 \\ 0 & 1 & 3 & 6 \\ 0 & 0 & 1 & \frac{3}{2} \end{array}\right]$ **25.** $\left[\begin{array}{ccc|c} 1 & 3 & 0 & 9 \\ 0 & 1 & 0 & -2 \\ 0 & 0 & 1 & 4 \end{array}\right]$

27. $\left[\begin{array}{cccc|c} 1 & 6 & 0 & 0 & -2 \\ 0 & 1 & 0 & 0 & 1 \\ 0 & 0 & 1 & 0 & 0 \\ 0 & 0 & 0 & 1 & -6 \end{array}\right]$ **29.** $\left[\begin{array}{ccc|c} 1 & 0 & 0 & 35 \\ 0 & 1 & 0 & 4 \\ 0 & 0 & 1 & -21 \end{array}\right]$ $(x, y, z) = (35, 4, -21)$

31. $(3, 2)$ **33.** $(1, \frac{3}{2})$ **35.** $(3, 1)$ **37.** $(1, 2, 3)$ **39.** $(3, -1, 2)$ **41.** $(5, 6, 1)$ **43.** $(1, 1, 1)$ **45.** $(6, 2, -9)$
47. $(\frac{9}{7}, \frac{1}{7}, -\frac{4}{7})$ **49.** dependent system **51.** $(1, 0, 0)$ **53.** $(w, x, y, z) = (1, 2, -3, -2)$ **55.** $y = x^2 + 4x + 5$
57. $y = x^2 - 10x + 20$ **59.** Mix 7 bars of first alloy and 3 bars of second alloy.
61. 3 units of Type I, 7 units of Type II, and 5 units of Type III

PROBLEM SET 9.6, PAGE 370

1. 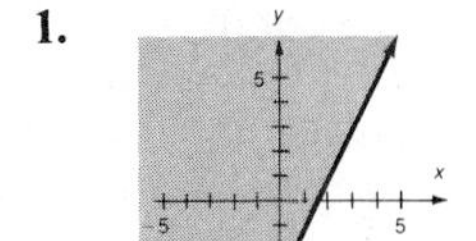

3. 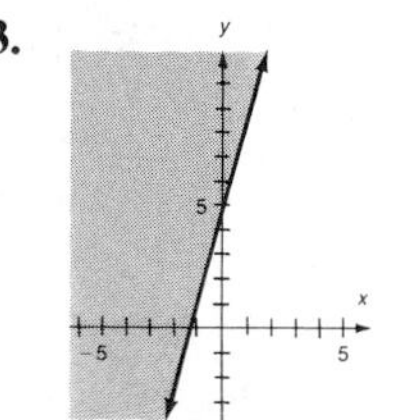

5. 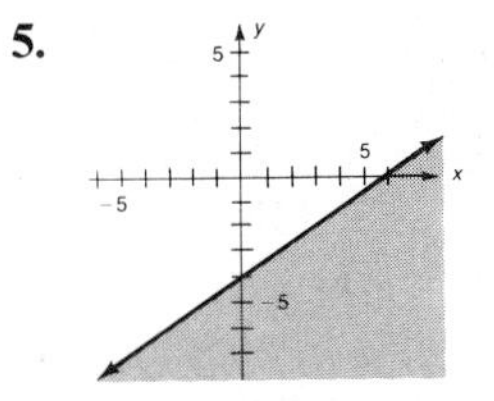

7. 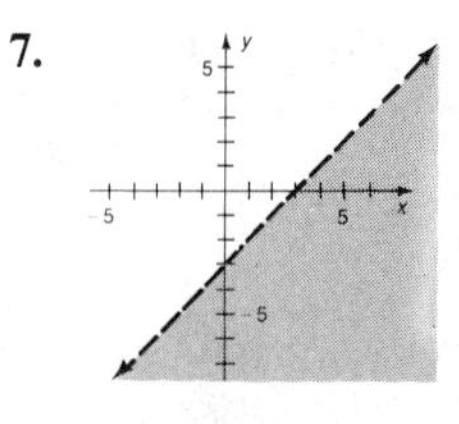

9. 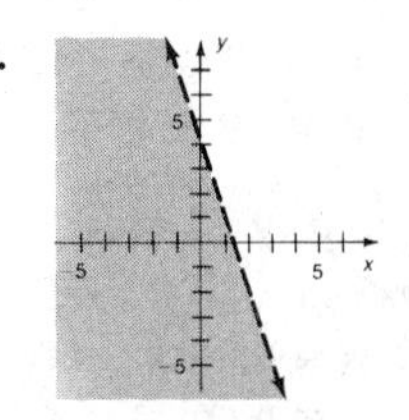

11. 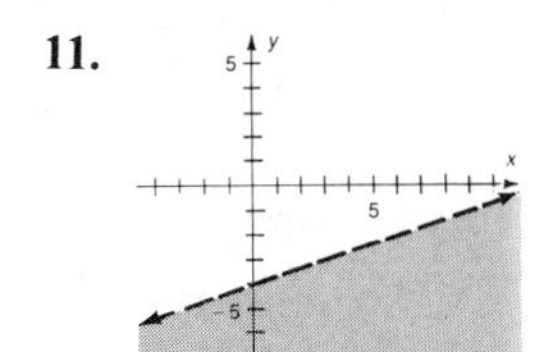

13. 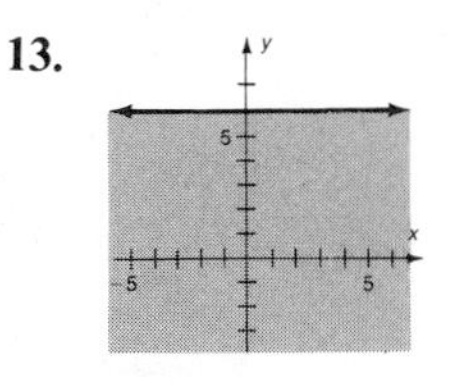

15. 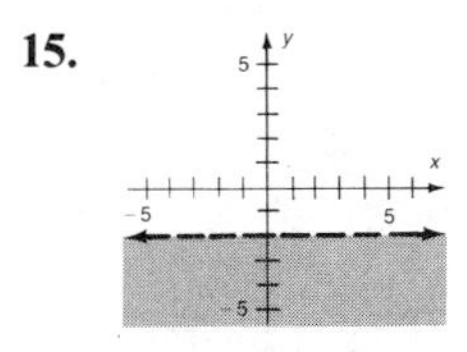

17. 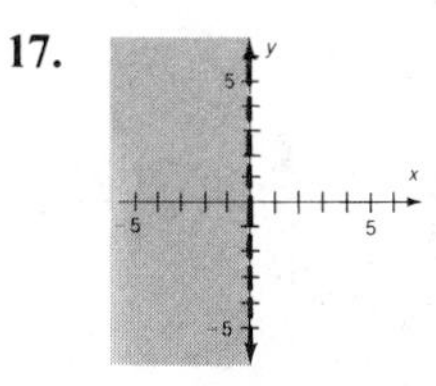

19. 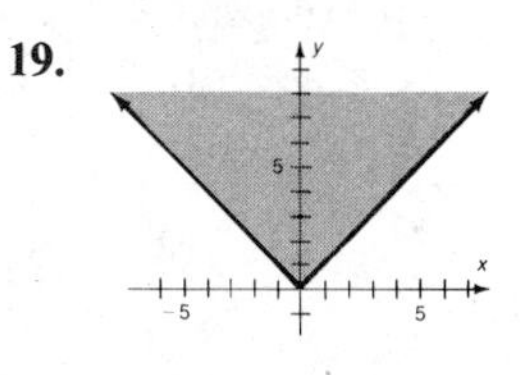

21. 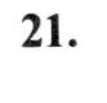 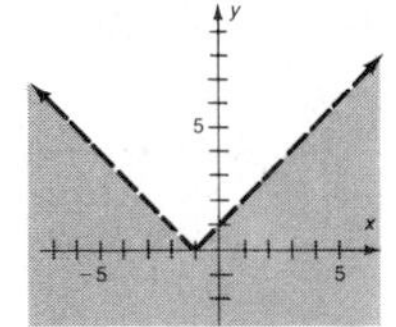

23. 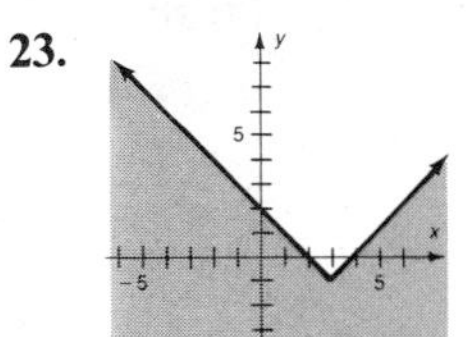

25. 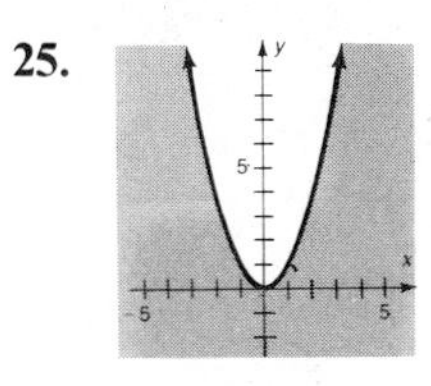

27. 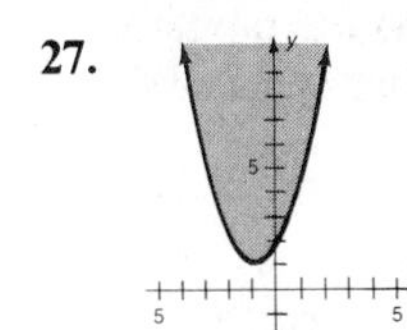

29. 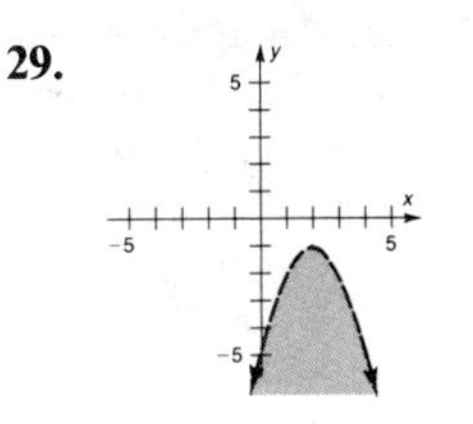

31. 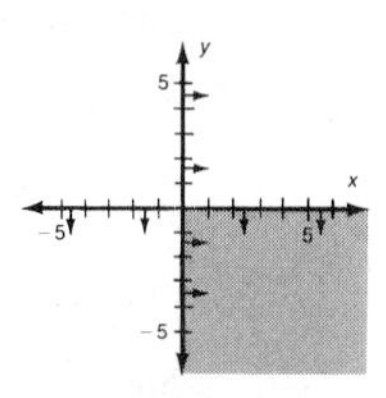 

33. 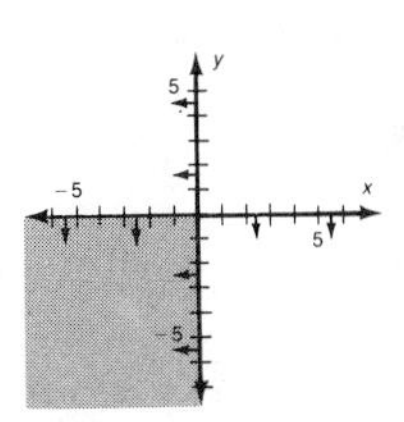

35. 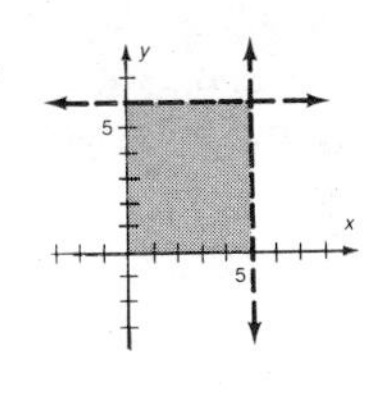 

37. 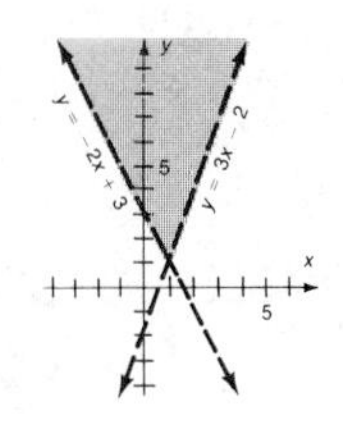

39. 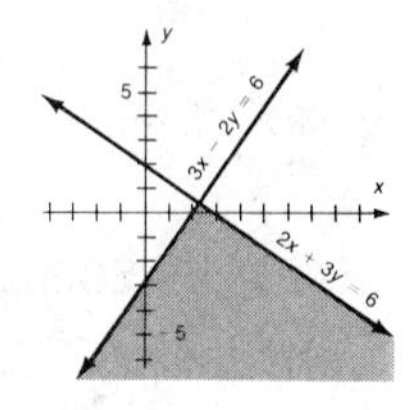

41. 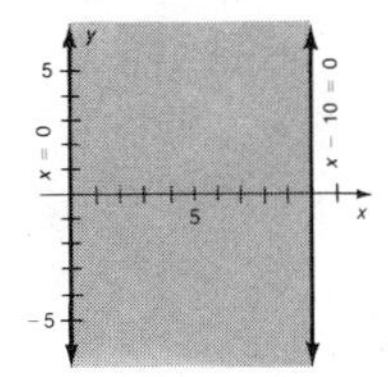

43. 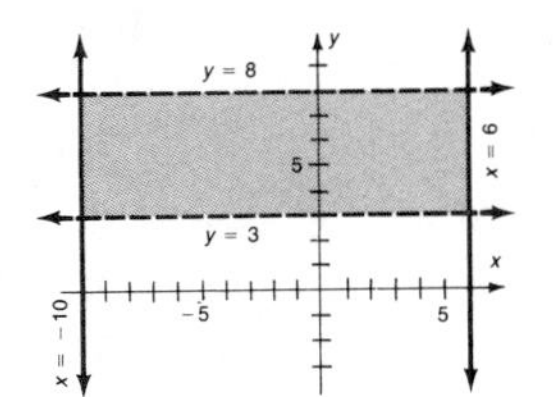

45. 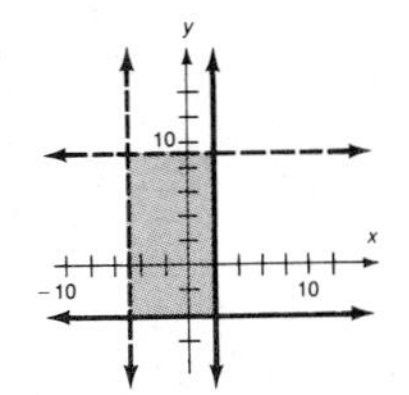

47. 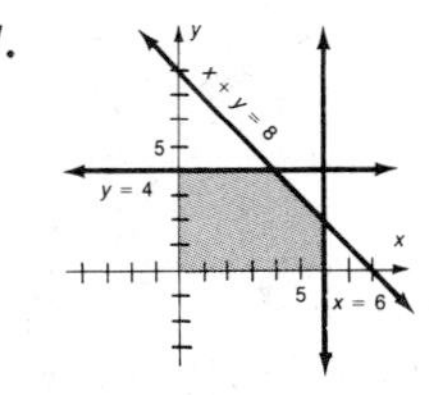

49.

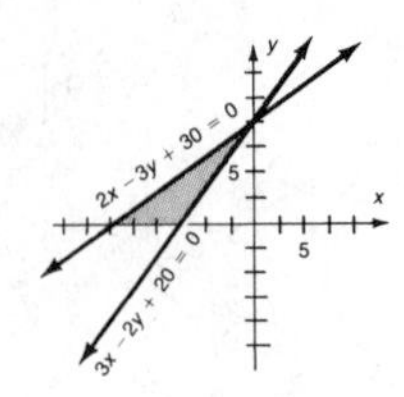

51.

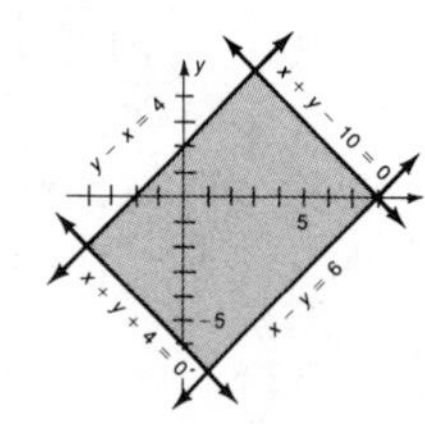

53.

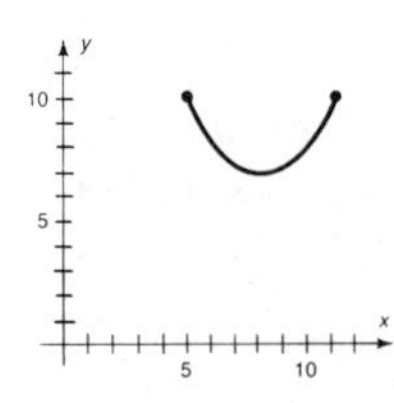

55.

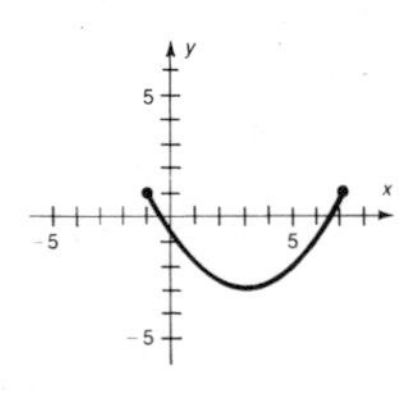

57.

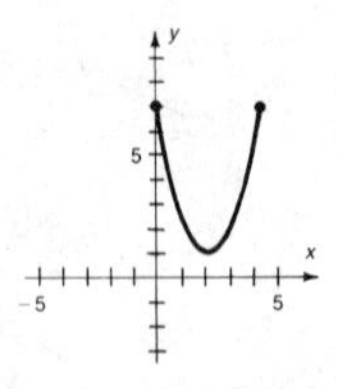

59.

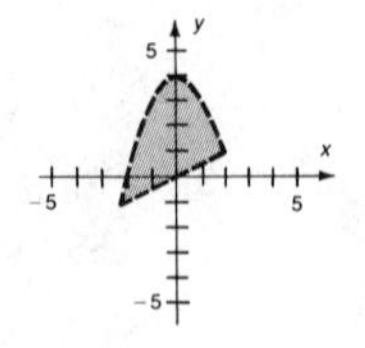

61.

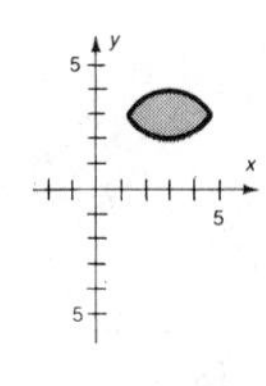

63.

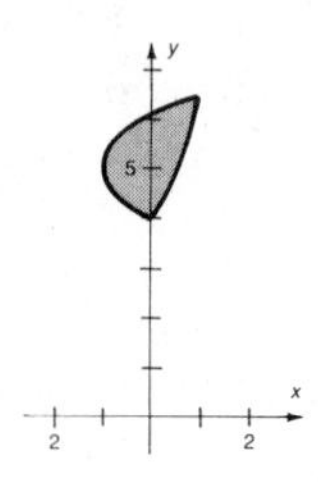

PROBLEM SET 9.7, PAGES 379–382

1. no **3.** no **5.** no **7.** no **9.** yes **11.** yes **13.** $(0,0), (0,\frac{9}{2}), (5,2), (6,0)$ **15.** $(0,0), (0,4), (2,3), (4,0)$
17. $(0,0), (0,4), (4,4), (6,2), (6,0)$ **19.** $(0,0), (0,\frac{8}{5}), (\frac{8}{3},0), (\frac{24}{13},\frac{16}{13})$ **21.** $(50,0), (\frac{200}{7},\frac{60}{7}), (8,24), (0,40)$
23. $(3,2), (5,5), (7,5), (\frac{10}{3},\frac{4}{3})$ **25.** Max $W = 190$ at $(5,2)$. **27.** Max $T = 600$ at $(6,0)$. **29.** Max $P = 500$ at $(2,3)$.
31. Min $C = 72$ at $(0,6)$. **33.** Max $F = 12$ at $(6,0)$. **35.** Min $I = 120$ at $(0,6)$. **37.** Max $P = 701.5$ at $(\frac{25}{2},9)$.
39. Max $P = 45$ at $(6,3)$. **41.** Min $X = \frac{62}{3}$ at $(\frac{10}{3},\frac{4}{3})$. **43.** Min $K = 560$ at $(4,0)$.

45. x = number of regular widgets
y = number of deluxe widgets
Maximize $P = 25x + 30y$
subject to: $\begin{cases} x \geq 0 \\ y \geq 0 \\ 3x + 2y \leq 8 \\ 2x + 4y \leq 8 \end{cases}$

47. x = amount invested in stocks
y = amount invested in bonds
Maximize $R = 0.12x + 0.08y$
subject to: $\begin{cases} x \geq 0 \\ y \geq 0 \\ x \leq 8 \\ y \geq 2 \\ x + y \leq 10 \\ y \leq 3x \end{cases}$

49. x = number of commercial guests
y = number of other guests
Maximize profit $P = 20x + 24y$
subject to: $\begin{cases} x \geq 0 \\ y \geq 0 \\ x + y \leq 200 \\ 0.4x + 0.2y \leq 50 \end{cases}$

51. x = number of Alpha products produced per day
y = number of Beta products produced per day
Maximize profit $P = 5x + 8y$
subject to: $\begin{cases} 0 \leq x \leq 700 \\ y \geq 0 \\ x + 3y \leq 1200 \\ x + 2y \leq 1000 \end{cases}$

53. Max profit $P = 14{,}300$ with all 100 acres planted in corn.

55. Max profit = \$100 with 4 regular widgets and 0 deluxe widgets produced.
57. Minimum cost = \$5.92 obtained with 8 g of food A and 24 g of food B.
59. Maximum return = \$900,000 with \$2,500,000 invested in stocks and \$7,500,000 invested in bonds.

CHAPTER 9 SUMMARY, PAGES 382–385

1.

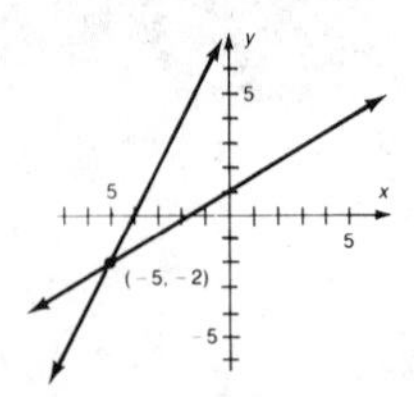

3.

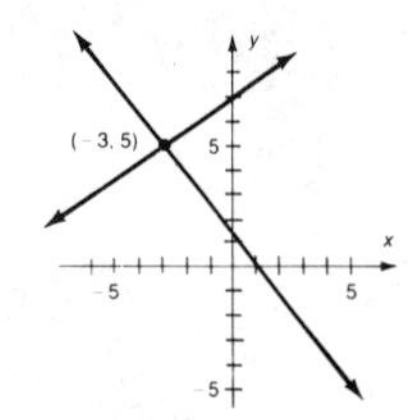

5. $(-6,2)$ **7.** $(-16,21)$ **9.** $(-3,4)$

11. $(5,-9)$ **13.** $(\frac{7}{16}, -\frac{11}{16})$ **15.** dependent system **17.** 29 **19.** -57 **21.** $(4,0,-3)$ **23.** $(-1,-3,5)$

25. $\begin{bmatrix} 5 & -1 & 1 \\ -3 & 8 & 7 \\ 3 & -6 & 6 \end{bmatrix}$ **27.** $\begin{bmatrix} 10 & -14 & -1 \\ 3 & -7 & 23 \\ 7 & -23 & -4 \end{bmatrix}$ **29.** $\begin{bmatrix} 7 & 2 \\ 3 & 1 \end{bmatrix}$ **31.** does not exist **33.** $(1, 7, 0)$ **35.** $(8, -2, 3)$

37. $(7, -5, 1)$ **39.** $(1, 2, -3)$ **41.**

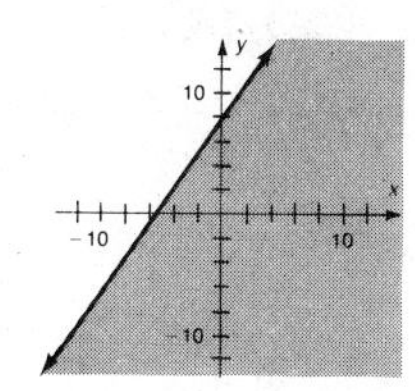

43.

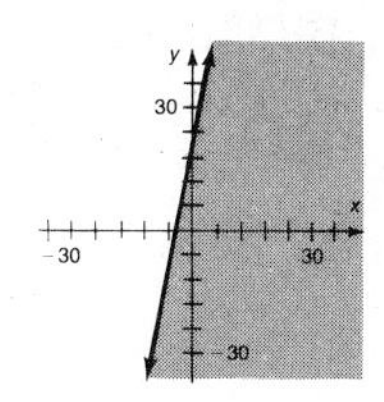

45.

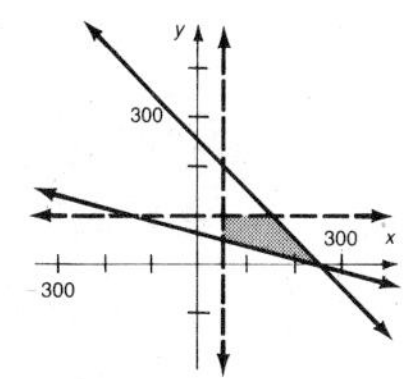

47.

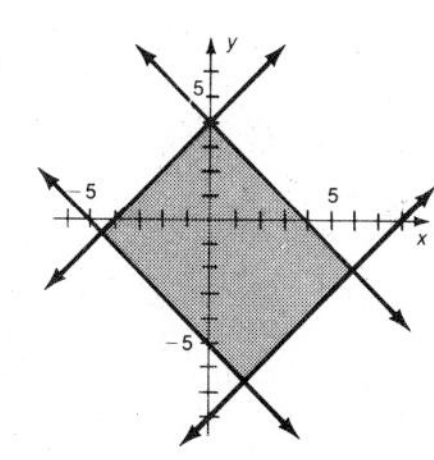

49. Maximum $P = 52.5$ at $(7, \frac{5}{2})$. **51.** Maximum $M = 1170$ at $(170, 80)$.

53. 20 g of I and 40 g of II **55.** Minimum cost $C = \$2.52$ with 12 g of food A and 14 g of food B.

PROBLEM SET 10.1, PAGES 392–393

Proofs for Problems 1–10 vary.

1. true **3.** true **5.** true **7.** true **9.** true **11.** $P(k)$: $5 + 9 + 12 + \cdots + (4k + 1) = k(2k + 3)$
$P(k + 1)$: $5 + 9 + 13 + \cdots + (4k + 5) = (k + 1)(2k + 5)$

13. $P(k)$: $2^2 + 4^2 + 6^2 + \cdots + (2k)^2 = \dfrac{2k(k + 1)(2k + 1)}{3}$
$P(k + 1)$: $2^2 + 4^2 + 6^2 + \cdots + (2k + 2)^2 = \dfrac{2(k + 1)(k + 2)(2k + 3)}{3}$

15. $P(k)$: $\cos(\theta + k\pi) = (-1)^k \cos\theta$
$P(k + 1)$: $\cos(\theta + k\pi + \pi) = (-1)^{k+1} \cos\theta$

17. $P(k)$: $k^2 + k$ is even
$P(k + 1)$: $(k + 1)^2 + (k + 1)$ is even

19. $P(k)$: $(\frac{2}{3})^{k+1} < (\frac{2}{3})^k$
$P(k + 1)$: $(\frac{2}{3})^{k+2} < (\frac{2}{3})^{k+1}$

21–53. Proofs vary.

55. Conjecture: $1^3 + 2^3 + \cdots + n^3 = \left[\dfrac{n(n + 1)}{2}\right]^2$ or $\dfrac{n^2(n + 1)^2}{4}$. Proofs vary.

57. Conjecture: $2 + 6 + 18 + 54 + \cdots + (2 \cdot 3^{n-1}) = 3^n - 1$. Proofs vary. **59–61.** Proofs vary.

PROBLEM SET 10.2, PAGES 399–400

1. 22 **3.** 2 **5.** 72 **7.** 132 **9.** 220 **11.** 1140 **13.** 8 **15.** 28 **17.** 56 **19.** 70 **21.** 1326
23. 1,000 **25.** $(a + b)^6 = a^6 + 6a^5b + 15a^4b^2 + 20a^3b^3 + 15a^2b^4 + 6ab^5 + b^6$ **27.** $(2x + 3)^3 = 8x^3 + 36x^2 + 54x + 27$
29. $(x + y)^5 = x^5 + 5x^4y + 10x^3y^2 + 10x^2y^3 + 5xy^4 + y^5$ **31.** $(3x + 2)^5 = 243x^5 + 810x^4 + 1080x^3 + 720x^2 + 240x + 32$
33. $(x + y)^4 = x^4 + 4x^3y + 6x^2y^2 + 4xy^3 + y^4$ **35.** $(\frac{1}{2}x + y^3)^3 = \frac{1}{8}x^3 + \frac{3}{4}x^2y^3 + \frac{3}{2}xy^6 + y^9$ or $\frac{1}{8}(x^3 + 6x^2y^3 + 12xy^6 + 8y^9)$
37. $(x^{1/2} + y^{1/2})^4 = x^2 + 4x^{3/2}y^{1/2} + 6xy + 4x^{1/2}y^{3/2} + y^2$ **39.** $1 - 8x + 28x^2 - 56x^3 + 70x^4 - 56x^5 + 28x^6 - 8x^7 + x^8$
41. $a^{10} + 10a^9b + 45a^8b^2 + 120a^7b^3$ **43.** $a^{14} + 14a^{13}b + 91a^{12}b^2 + 364a^{11}b^3$
45. $x^{16} + 32x^{15}y + 480x^{14}y^2 + 4480x^{13}y^3$ **47.** $x^{12} - 24x^{11}y + 264x^{10}y^2 - 1760x^9y^3$ **49.** $1 - 0.24 + 0.0264 - 0.00176$
51. -16 **53.** 4032 **55.** 160 **57–63.** Proofs vary.

PROBLEM SET 10.3, PAGES 405–407

1. a. arithmetic **b.** $d = 3$ **c.** 17 **3. a.** geometric **b.** $r = 2$ **c.** 96 **5. a.** neither **b.** Difference increases by one between each term. **c.** 85 **7. a.** geometric **b.** $r = q$ **c.** pq^5 **9. a.** neither **b.** Difference increases by two between each term. **c.** 36 **11. a.** neither **b.** Number of fives between twos increases by one. **c.** 2

13. a. neither **b.** Systematic listing of fractions (excluding those previously listed); that is, $\frac{2}{6}, \frac{3}{6}, \frac{4}{6}$ are previously listed in another form. **c.** $\frac{5}{6}$ **15. a.** neither **b.** perfect cubes **c.** 216 **17.** 1, 5, 9 **19.** a, ar, ar^2 **21.** $-1, 1, -1$ **23.** $2, \frac{3}{2}, \frac{4}{3}$ **25.** 2, 2, 2 **27.** $\cos x, \cos 2x, \cos 3x$ **29.** 20 **31.** 49 **33.** 12 **35.** 11 **37.** 80 **39.** 10 **41.** 57 **43.** 1 **45.** 5^4 or 625 **47.** 2, 6, 18, 54, 162 **49.** 1, 1, 2, 3, 5 **51.** $\sum_{k=1}^{7} \frac{1}{2^k}$

53. $\sum_{k=1}^{5} 6^{k-1}$ **55.** arithmetic **57.** geometric **59.** geometric **61. a.** $a_1b_1 + a_2b_2 + a_3b_3 + \cdots + a_rb_r$ **b.–c.** Answers vary. **63.** 47 (add preceding two terms) **65.** 7 (arranged alphabetically)

PROBLEM SET 10.4, PAGES 411–412

1. 5, 9, 13, 17 **3.** 85, 88, 91, 94 **5.** 100, 95, 90, 85 **7.** $-\frac{5}{2}, -2, -\frac{3}{2}, -1$ **9.** $2\sqrt{3}, 3\sqrt{3}, 4\sqrt{3}, 5\sqrt{3}$ **11.** $x, x + y, x + 2y, x + 3y$ **13.** $a_1 = 5; d = 3$ **15.** $a_1 = 6; d = 5$ **17.** $a_1 = -8; d = 7$ **19.** $a_1 = x; d = x$ **21.** $a_1 = x - 5b; d = 2b$ **23.** $a_n = 5$ **25.** $a_n = 24 + 11n$ **27.** $a_n = -3 + 2n$ **29.** $a_n = x + n\sqrt{3}$ **31.** 101 **33.** 25 **35.** $-10{,}600$ **37.** -2 **39.** 2 **41.** -140 **43.** 4 **45.** 12 **47.** 11,500 **49.** 2030 **51.** n^2 **53.** **a**, **b**, **c**, and **d** are harmonic. **55. a.** 12 **b.** $\frac{19}{2}$ **c.** $\frac{5}{12}$ **d.** -6 **e.** $\frac{1}{15}$ **57.** Option *D*
59. Option *A*: $43,440.
Option *B*: $44,160; $720 more than *A*
Option *C*: $44,520; $1080 more than *A* and $360 more than *B*
Option *D*: $44,760; $1320 more than *A*, $600 more than *B*, and $240 more than *C*.

PROBLEM SET 10.5, PAGES 417–418

1. 5, 15, 45 **3.** 1, -2, 4 **5.** $-15, -3, -\frac{3}{5}$ **7.** $8, 8x, 8x^2$ **9.** x, xy, xy^2 **11.** $g_1 = 7; r = 2$ **13.** $g_1 = 100; r = \frac{1}{2}$ **15.** $g_1 = xyz; r = \frac{1}{z}$ **17.** $g_n = 7 \cdot 2^{n-1}$ **19.** $g_n = 200(\frac{1}{2})^n$ or $200 \cdot 2^{-n}$ or $100(\frac{1}{2})^{n-1}$ **21.** $g_n = xyz^{2-n}$ **23.** 2000 **25.** $-\frac{40}{3}$ **27.** $-\frac{1296}{5}$ **29.** 10^{-7} or 0.0000001 **31.** 2 **33.** $\frac{93}{25}$ **35.** $\frac{1}{3}$ **37.** $\frac{1111}{10{,}000}$ **39.** $\frac{85}{256}$ **41.** $\frac{5}{9}$ **43.** $\frac{3}{11}$ **45.** $\frac{5}{11}$ **47.** $\frac{218}{999}$ **49.** $\frac{27}{11}$ **51.** $\frac{22{,}309}{9900}$ **53.** 9 mailings **55.** 50 ft **57.** She must pay $25 in taxes, and they must pay her $125. **59.** $4 + 2\sqrt{2}$ **61.** $\frac{4 + 3\sqrt{2}}{2}$ **63.** $3 - 3\sqrt{2}, 3, 3 + 3\sqrt{2}$ **65.** It has a limit for $p > 1$. **67.** The area of the shaded portion is $\frac{1}{4}$ the area of the original square.

CHAPTER 10 SUMMARY, PAGES 419–420

1. If a given proposition $P(n)$ is true for $P(1)$ and if the truth of $P(k)$ implies the truth of $P(k + 1)$, then $P(n)$ is true for all positive integers. **3.** $P(k)$: $4 + 8 + 12 + \cdots + 4k = 2k(k + 1)$
$P(k + 1)$: $4 + 8 + 12 + \cdots + 4(k + 1) = 2(k + 1)(k + 2)$
5. $a^5 + 5a^4b + 10a^3b^2 + 10a^2b^3 + 5ab^4 + a^5$ **7.** $32x^5 + 80x^4y + 80x^3y^2 + 40x^2y^3 + 10xy^4 + y^5$ **9.** 2,598,960 **11.** 70 **13.** $(a + b)^n = \sum_{k=0}^{n} \binom{n}{k} a^{n-k}b^k = \binom{n}{0}a^n + \binom{n}{1}a^{n-1}b + \cdots + \binom{n}{n-1}ab^{n-1} + \binom{n}{n}b^n$
15. $\binom{15}{r}x^{15-r}(-y)^r$ **17.** arithmetic; $a_n = -9 + 10n$ **19.** neither; 11111, 111111; number of ones increases by 1. **21.** 40 **23.** $3^{10} - 1$ or 59,048 **25.** $A_{10} = 460$ **27.** $G_{10} = 81 - 3^{-6} \approx 80.99862826$ **29.** $d = 2; A_{10} = 110$ **31.** $g_{10} = 2560; G_5 = 155$ **33.** 2000 **35.** $\frac{24}{11}$

PROBLEM SET 11.1, PAGES 430–432

1. 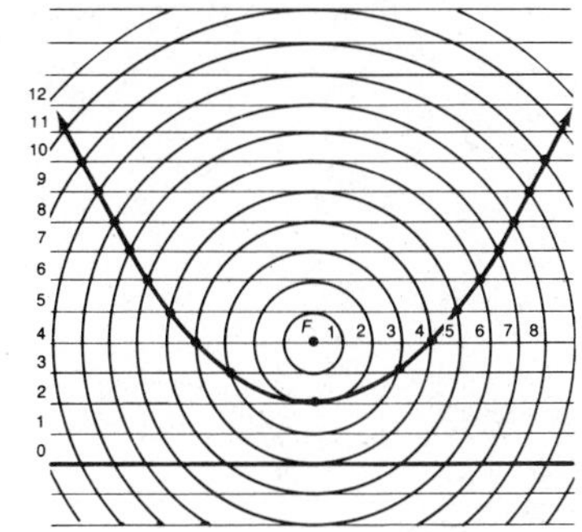

3. 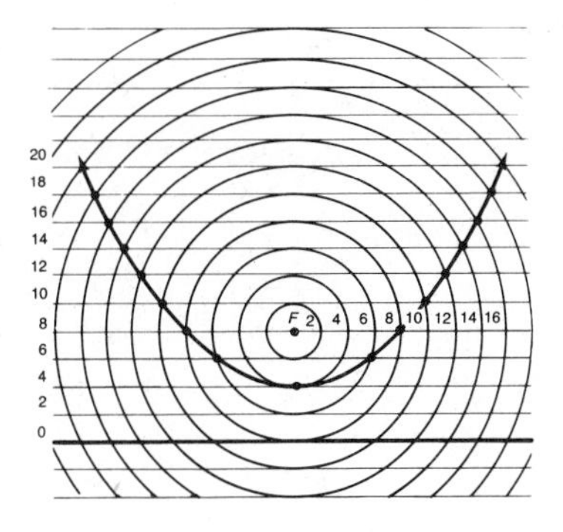

5. 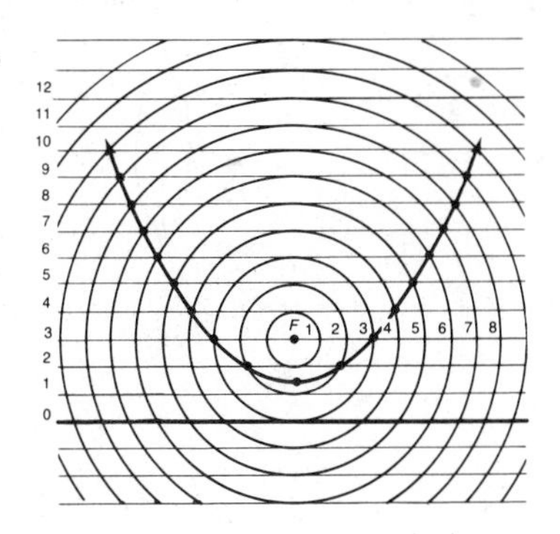

7. 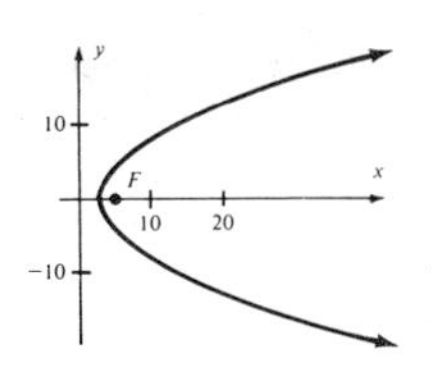

9. 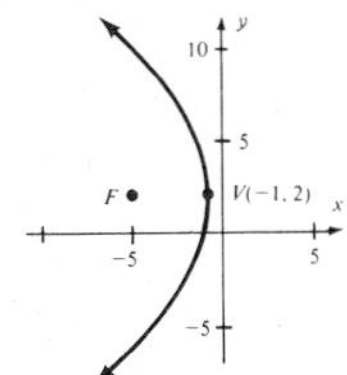

11. 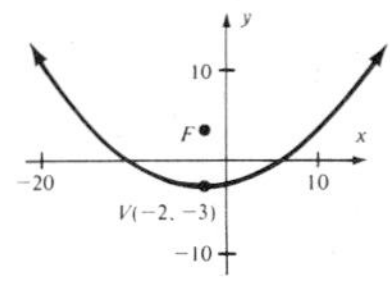

13. 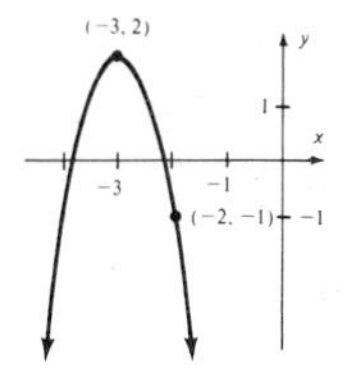

15. 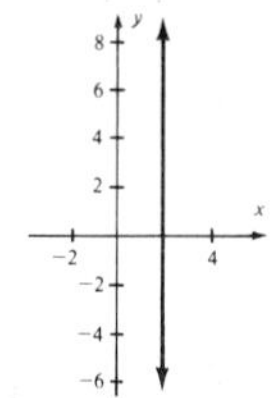

17. 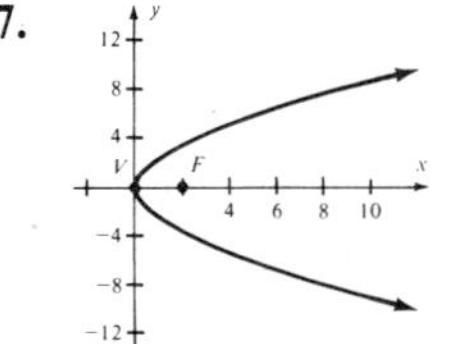

19. 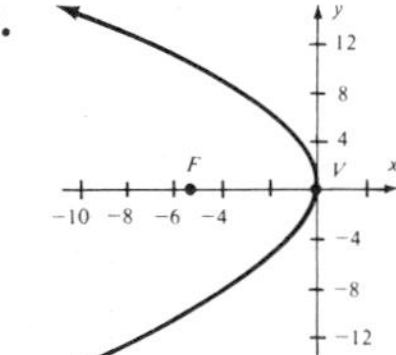

21. 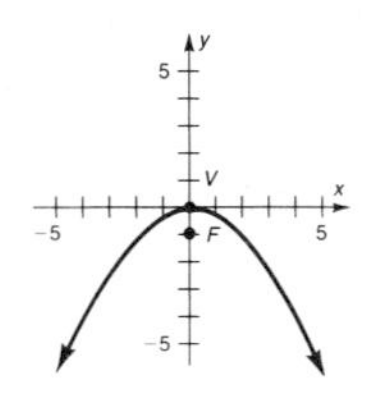

23. 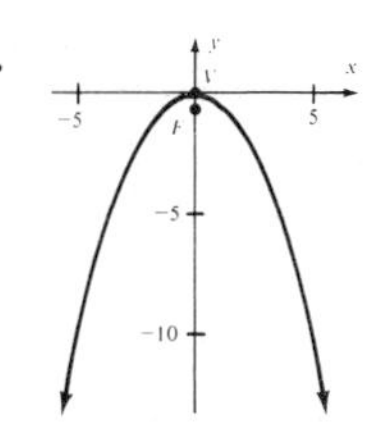

25. 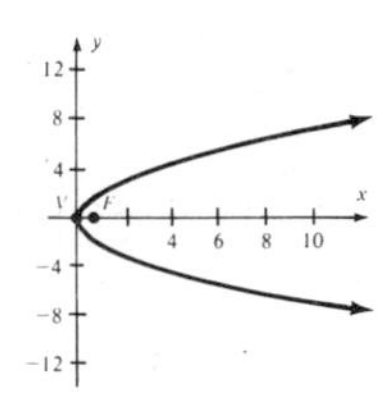

27. 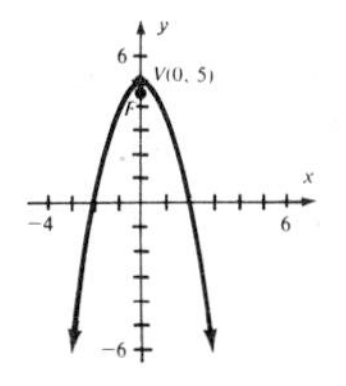

29. 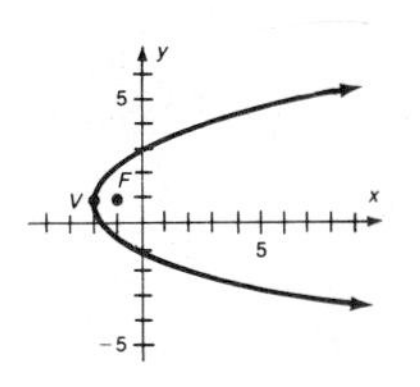

31. 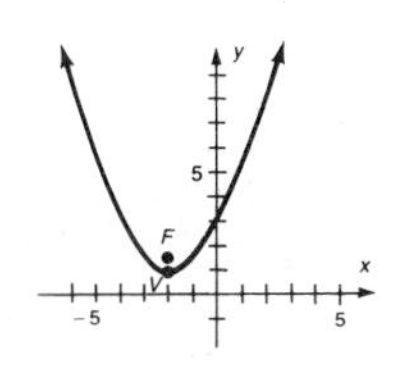

33. 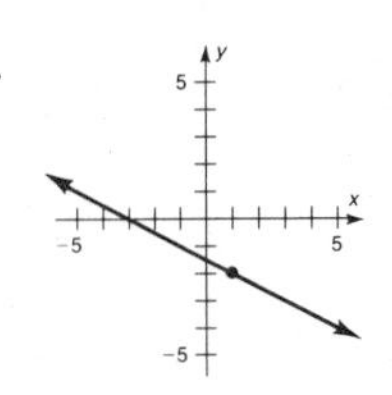

35. 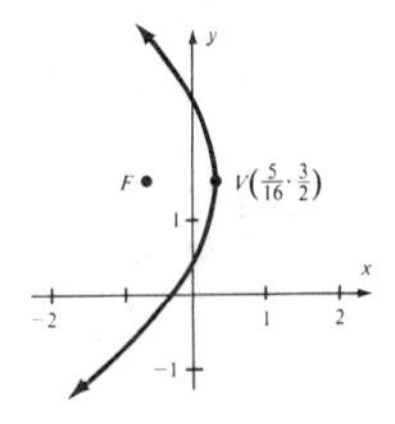

37. 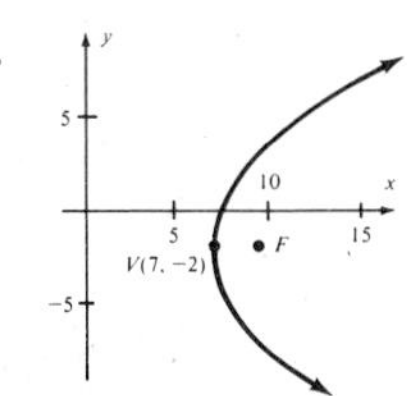

39. 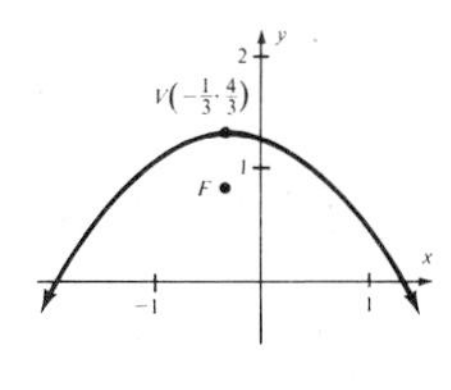

41. $y^2 = 10(x - \frac{5}{2})$ **43.** $(y - 2)^2 = -16(x + 1)$

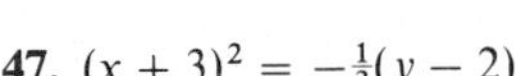

45. $(x + 2)^2 = 24(y + 3)$ **47.** $(x + 3)^2 = -\frac{1}{3}(y - 2)$ **49.** $x = 2$
51. $(x - 100)^2 = -200(y - 50)$. Also may add: $0 \le x \le 200$ **53.** 2.25 m from the vertex on the axis of the parabola
55. $(-3, 4)$ and $(1, 12)$ **57.** $(2, 3)$ and $(3, 1)$ **59.** 21.0 seconds at age 40 **61–63.** Answers vary. **65. a.** $m = -\frac{A}{B}$
b. $m' = \frac{B}{A}$ **c.** $Bx - Ay - Bx_0 + Ay_0 = 0$ **d.** $x = \dfrac{B^2x_0 - ABy_0 - AC}{A^2 + B^2}$, $y = \dfrac{-ABx_0 + A^2y_0 - BC}{A^2 + B^2}$
e. Answers vary. **67.** $25x^2 + 120xy + 144y^2 - 1110x + 1730y + 4209 = 0$

PROBLEM SET 11.2, PAGES 440–441

1. 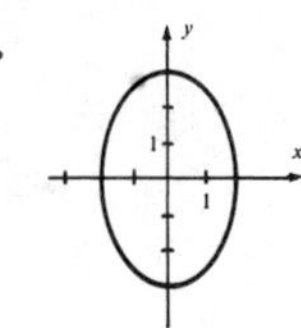

3. 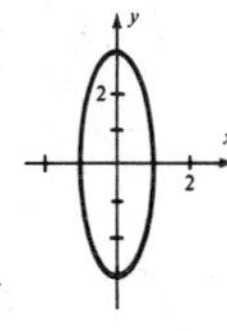

5. 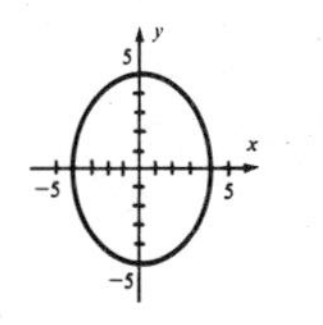

7. 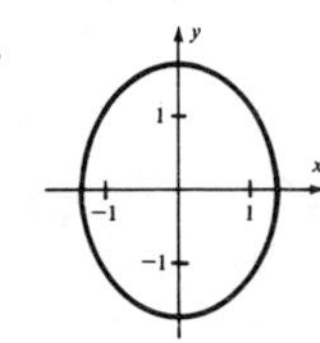

9. 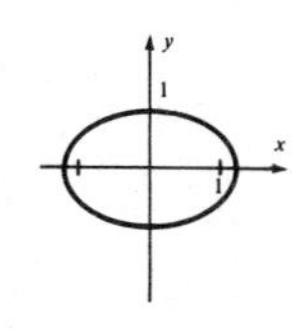

11. 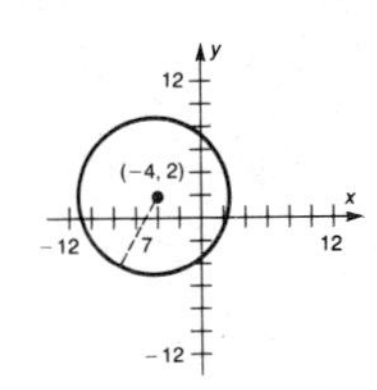

13. 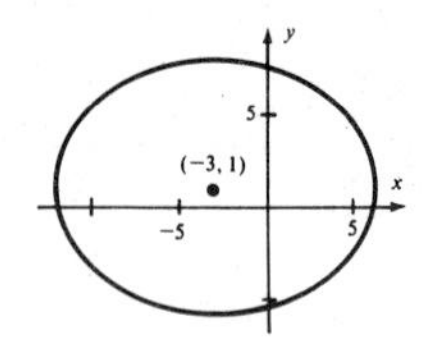

15. 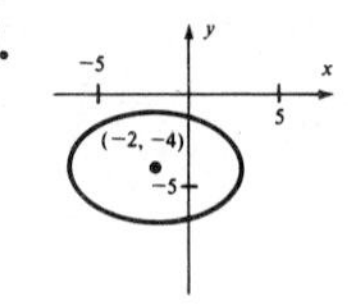

17. 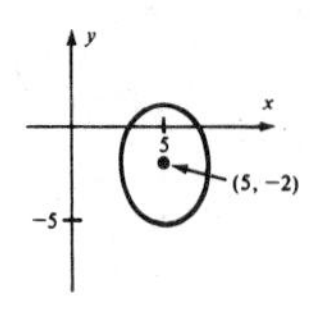

19. 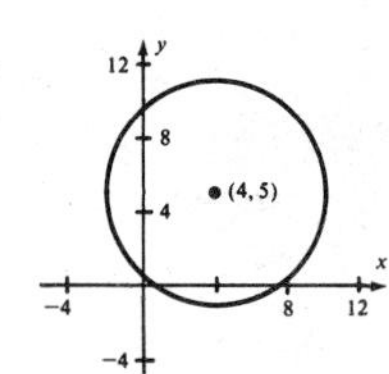

21. 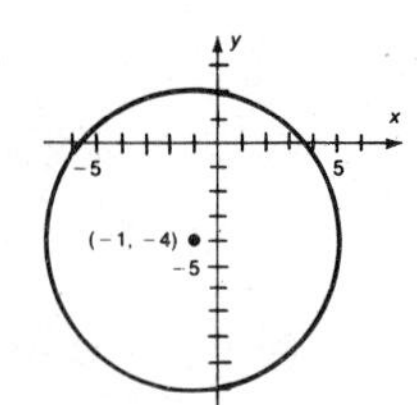

23. 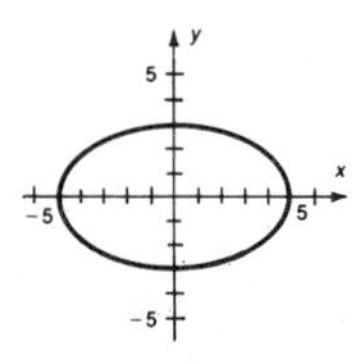

25. 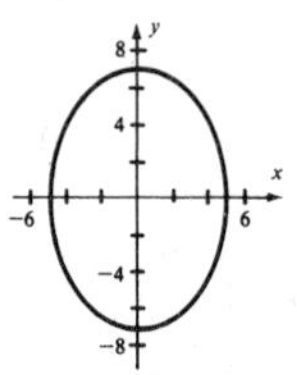

27. 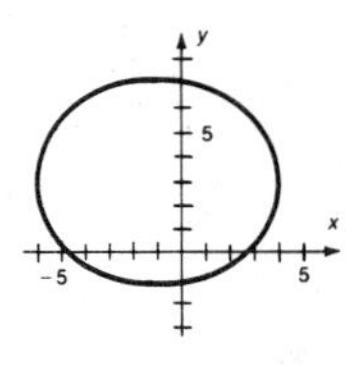

29. 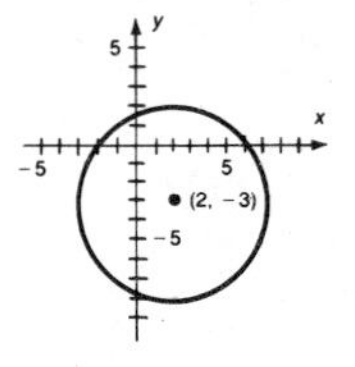

31. $(x-4)^2+(y-5)^2=36$ **33.** $(x+1)^2+(y+4)^2=36$ **35.** $\dfrac{x^2}{25}+\dfrac{y^2}{9}=1$ **37.** $\dfrac{x^2}{24}+\dfrac{y^2}{49}=1$

39. $\dfrac{(x+1)^2}{25}+\dfrac{(y-3)^2}{16}=1$ **41.** $x^2+y^2-4x+6y-12=0$ or $(x-2)^2+(y+3)^2=25$

43. 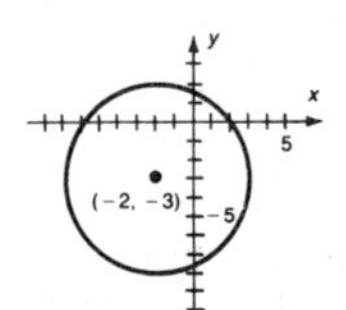

45. 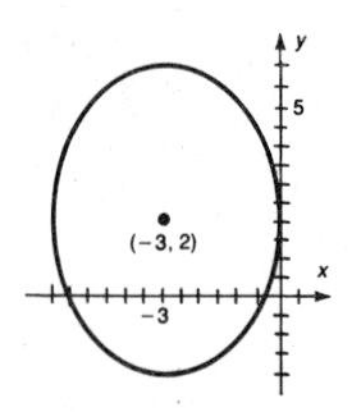

47. 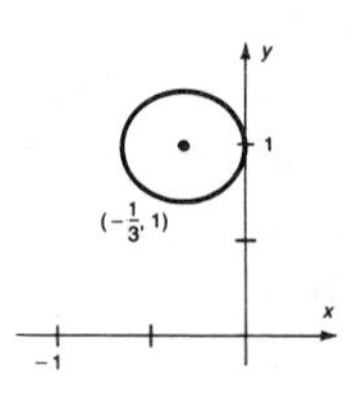

49. 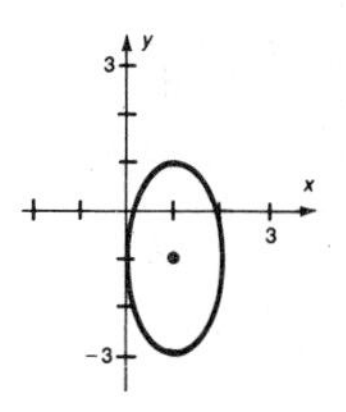

51. 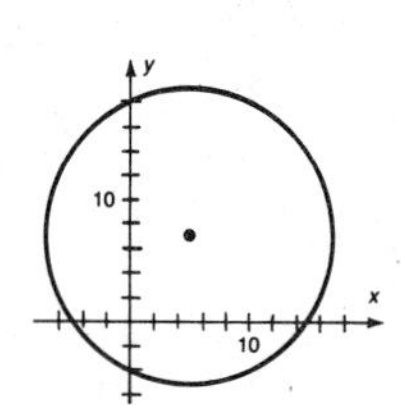

53. 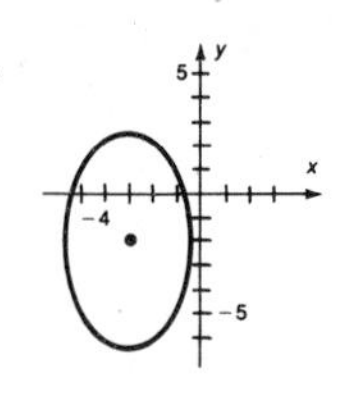

55. Answers vary. **57.** The distance at aphelion is 94,500,000 miles and at perihelion is 91,500,000 miles. **59.** 0.053

61. 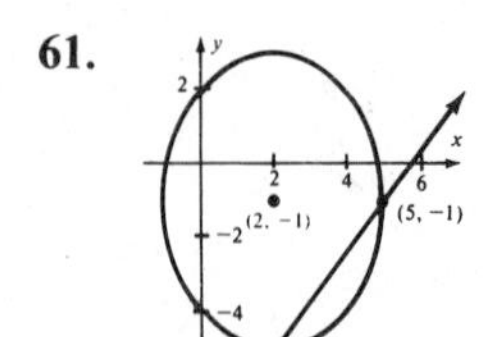

63. Answers vary.

PROBLEM SET 11.3, PAGES 450–451

1. 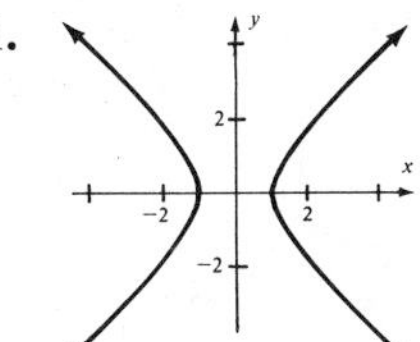

3. 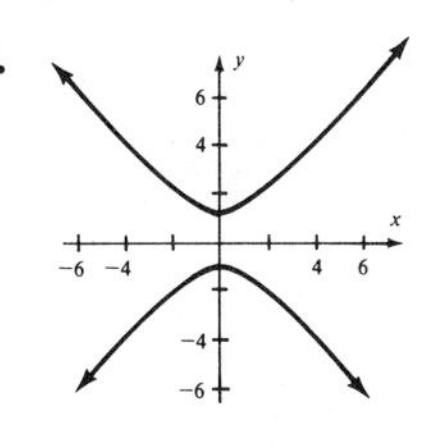

5. 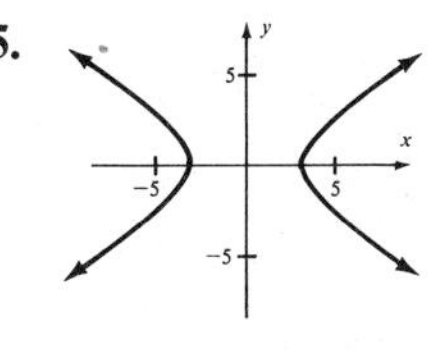

7. 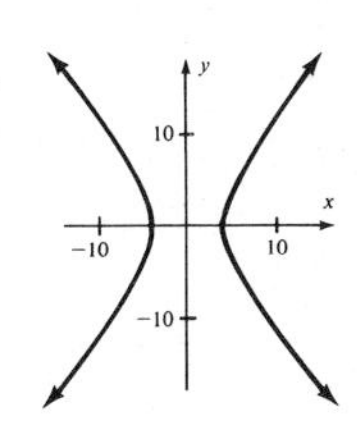

9. 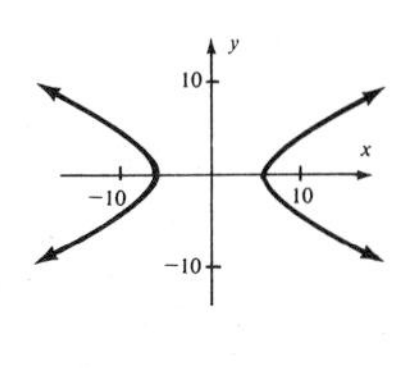

11. 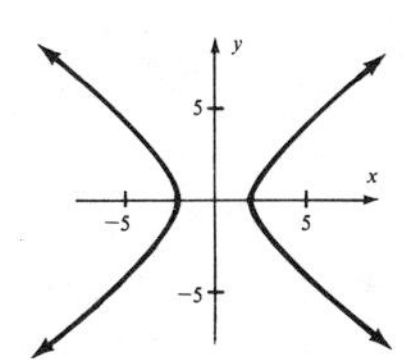

13. 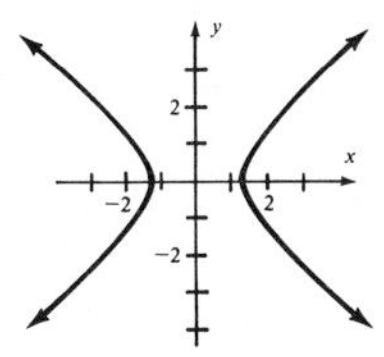

15. 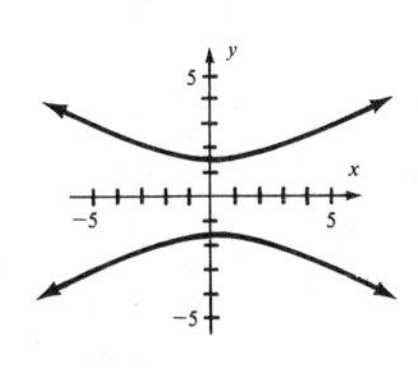

17. 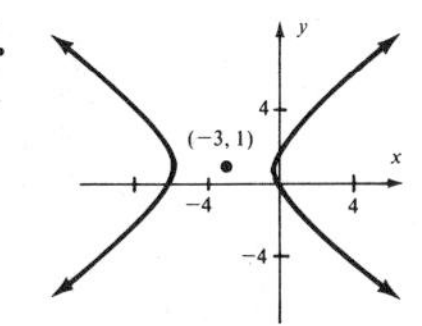

19. 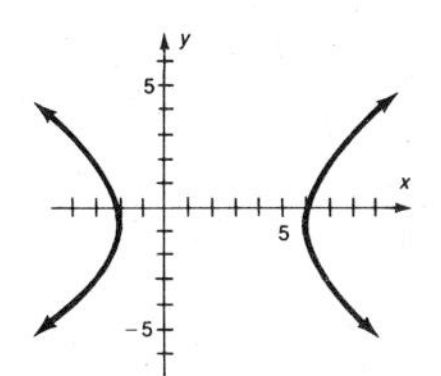

21. 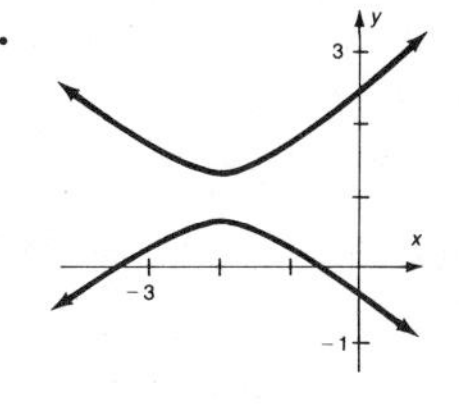

23. line 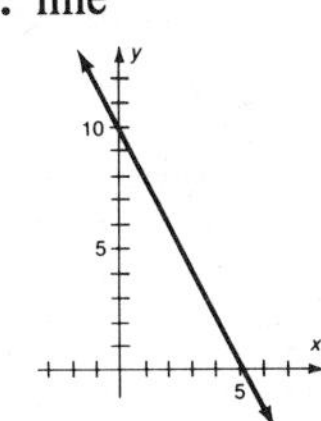

25. parabola 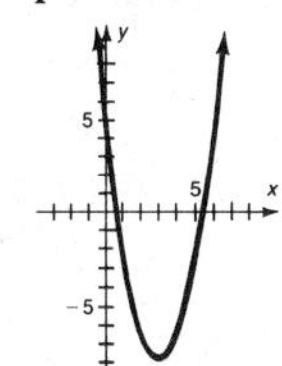

27. line 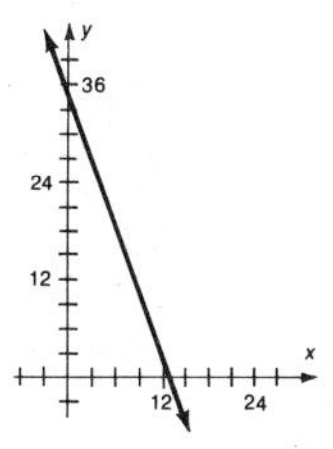

29. parabola 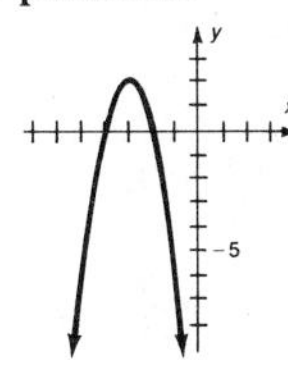

31. ellipse 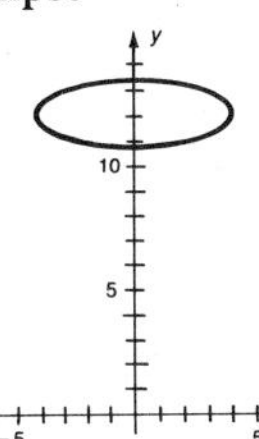

33. circle 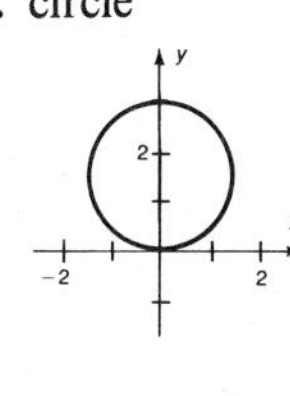

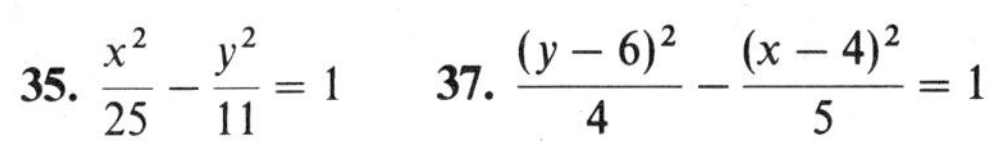

35. $\dfrac{x^2}{25} - \dfrac{y^2}{11} = 1$

37. $\dfrac{(y-6)^2}{4} - \dfrac{(x-4)^2}{5} = 1$

39. $\dfrac{(x-2)^2}{16} - \dfrac{y^2}{3} = 1$ 

41. 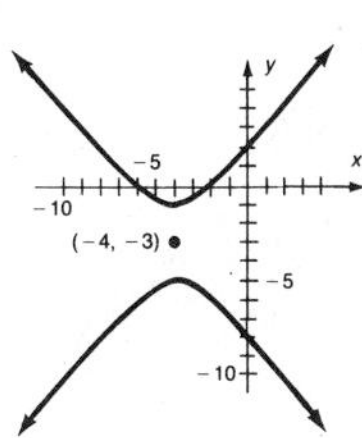

43. 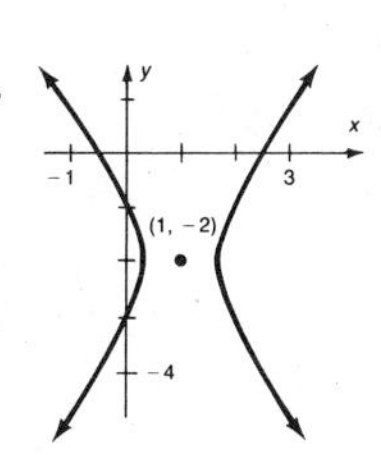

45. 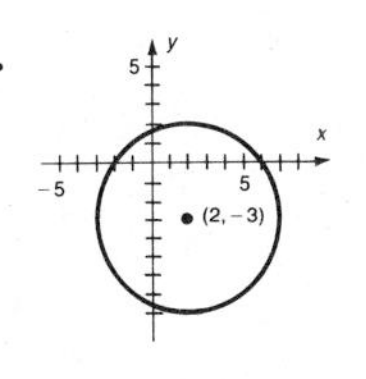

47. 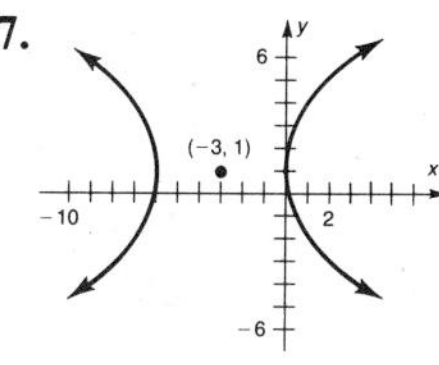

49. parabola 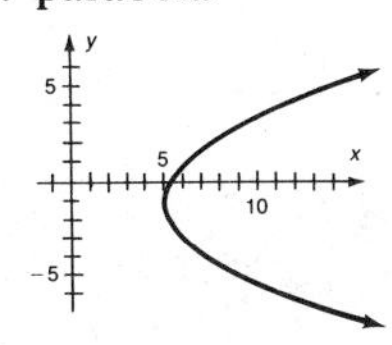

51. parabola 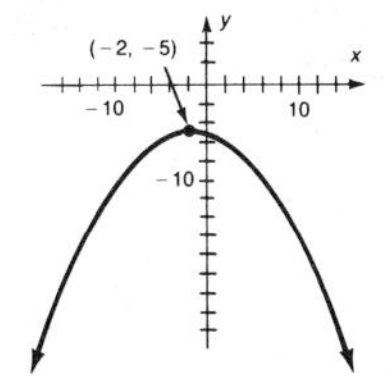

53. hyperbola 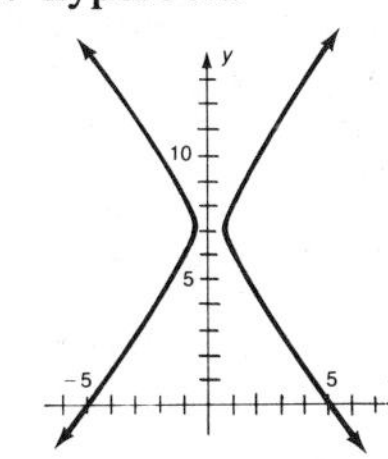

55. circle 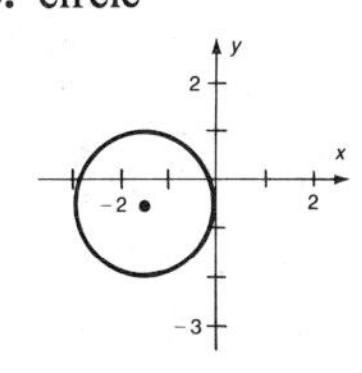

57. ellipse 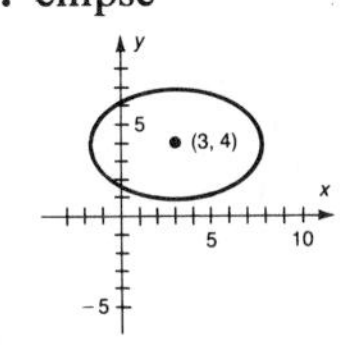

PROBLEM SET 11.4, PAGES 457–458

1.
$$\begin{aligned} xy &= 6 \\ [\tfrac{1}{\sqrt{2}}(x'-y')][\tfrac{1}{\sqrt{2}}(x'+y')] &= 6 \\ \tfrac{1}{2}(x'-y')(x'+y') &= 6 \\ x'^2 - y'^2 &= 12 \end{aligned}$$

3.
$$\begin{aligned} x^2 - 4xy + 4y^2 + 5\sqrt{5}y - 10 &= 0 \\ \tfrac{1}{5}(2x'-y')^2 - 4(\tfrac{1}{5})(2x'-y')(x'+2y') + 4(\tfrac{1}{5})(x'+2y')^2 + 5\sqrt{5}(\tfrac{1}{\sqrt{5}})(x'+2y') - 10 &= 0 \\ 4x'^2 - 4x'y' + y'^2 - 8x'^2 - 12x'y' + 8y'^2 + 4x'^2 + 16x'y' + 16y'^2 + 25x' + 50y' - 50 &= 0 \\ 25y'^2 + 50y' &= -25x' + 50 \\ y'^2 + 2y' &= -x' + 2 \\ y'^2 + 2y' + 1 &= -x' + 3 \\ (y'+1)^2 &= -(x'-3) \end{aligned}$$

5. hyperbola **7.** ellipse **9.** parabola **11.** hyperbola **13.** parabola **15.** ellipse **17.** ellipse
19. ellipse **21.** parabola **23.** $\theta = 45°$; $x = \frac{1}{\sqrt{2}}(x'-y')$; $y = \frac{1}{\sqrt{2}}(x'+y')$ **25.** $\theta = 45°$; $x = \frac{1}{\sqrt{2}}(x'-y')$; $y = \frac{1}{\sqrt{2}}(x'+y')$
27. $\theta \approx 63.4°$; $x = \frac{1}{\sqrt{5}}(x'-2y')$; $y = \frac{1}{\sqrt{5}}(2x + y')$ **29.** $\theta = 30°$; $x = \frac{1}{2}(\sqrt{3}x' - y')$; $y = \frac{1}{2}(x' + \sqrt{3}y')$
31. $\theta = 30°$; $x = \frac{1}{2}(\sqrt{3}x' - y')$; $y = \frac{1}{2}(x' + \sqrt{3}y')$ **33.** $\theta = 60°$; $x = \frac{1}{2}(x' - \sqrt{3}y')$; $y = \frac{1}{2}(\sqrt{3}x' + y')$
35. $\theta \approx 71.6°$; $x = \frac{1}{\sqrt{10}}(x' - 3y')$; $y = \frac{1}{\sqrt{10}}(3x' + y')$

37.

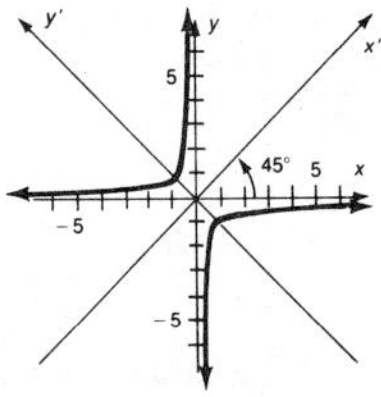

39.

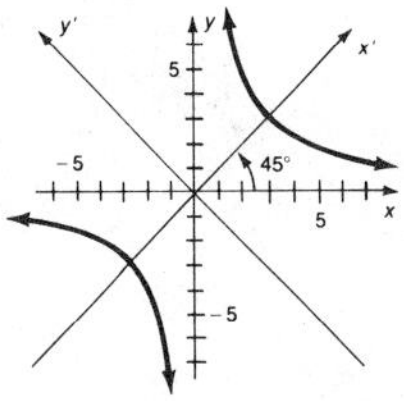

41.

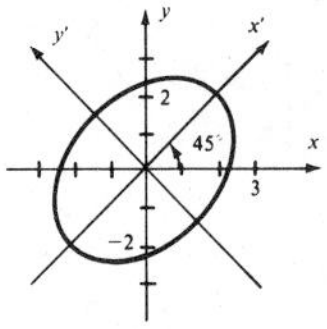

43.

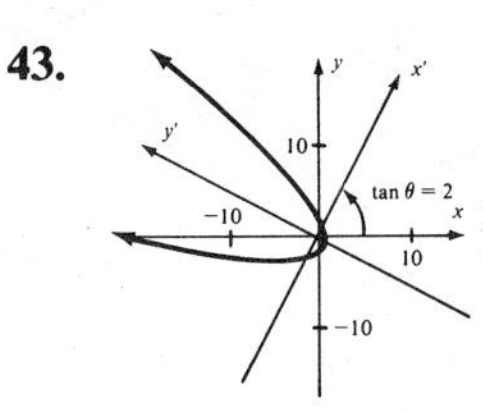

45.

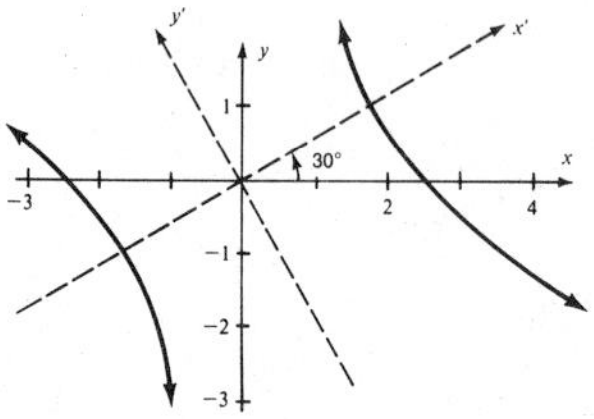

47.

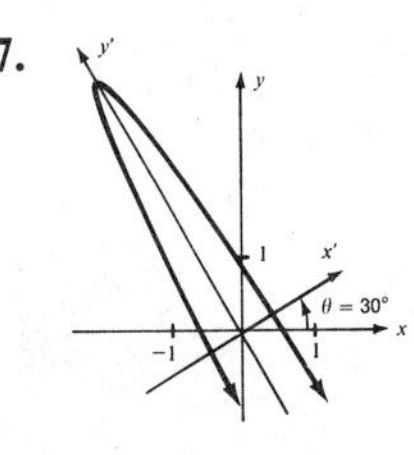

49.

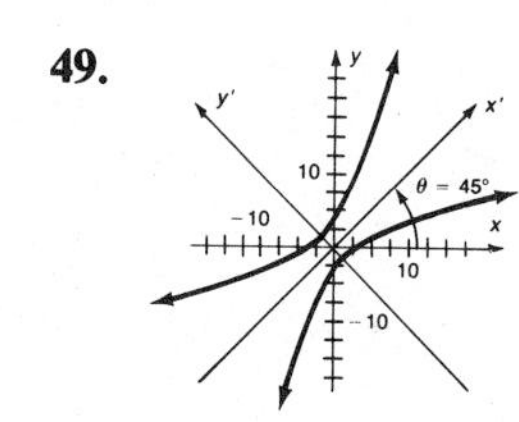

51.

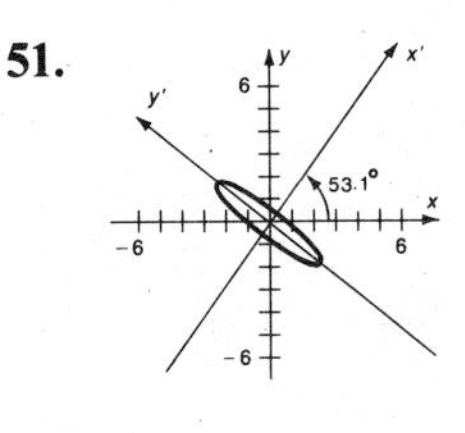

53.

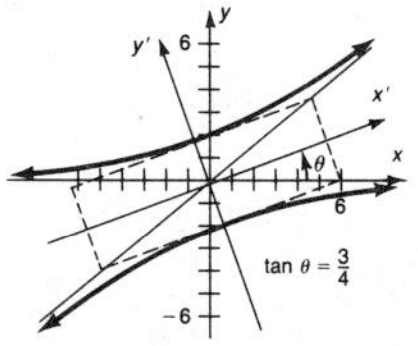

55.

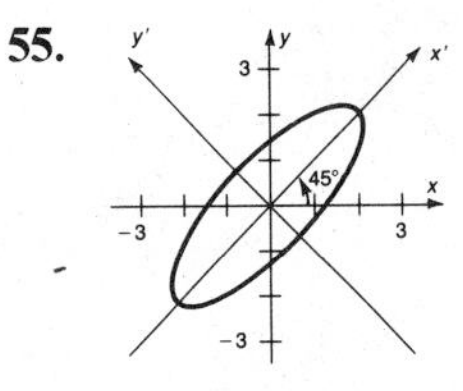

57.

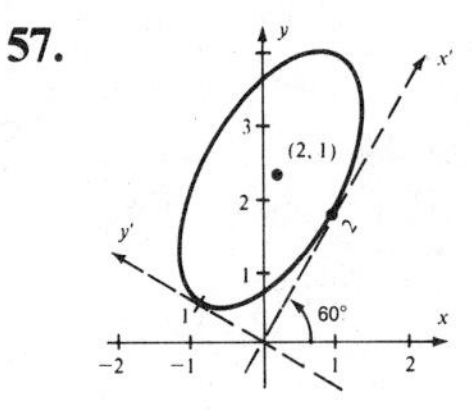

59.

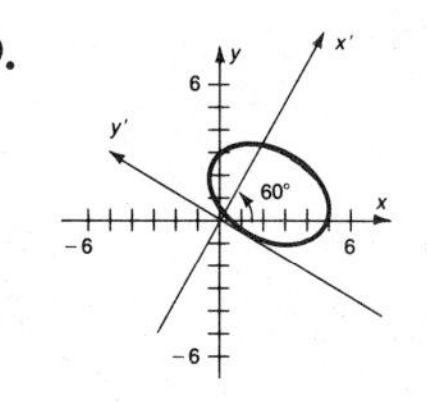

CHAPTER 11 SUMMARY, PAGE 459

1.

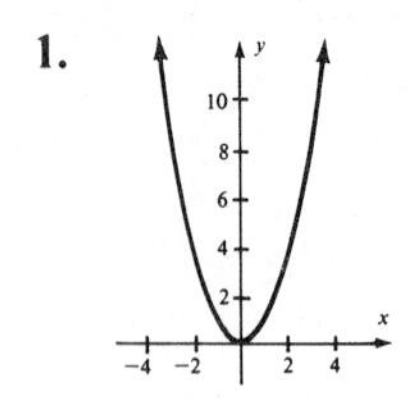

3.

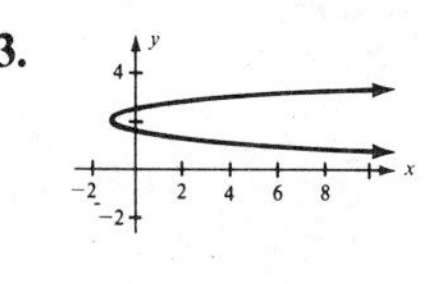

5. $(y-3)^2 = 20(x-6)$ **7.** $(x+3)^2 = -24(y-5)$

9.

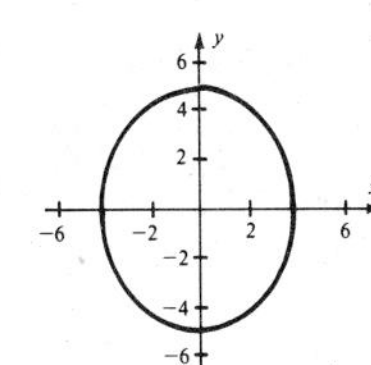

11.

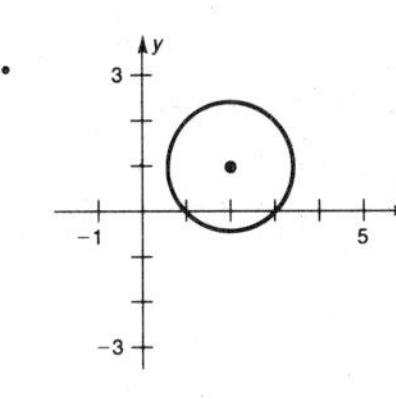

13. $\dfrac{(x-4)^2}{4} + \dfrac{(y-1)^2}{3} = 1$ **15.** $(x+1)^2 + (y+2)^2 = 64$

17.

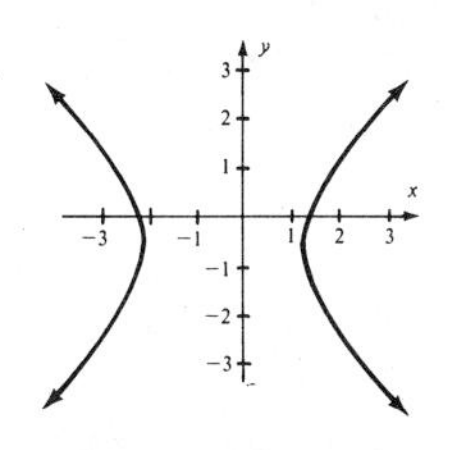

19.

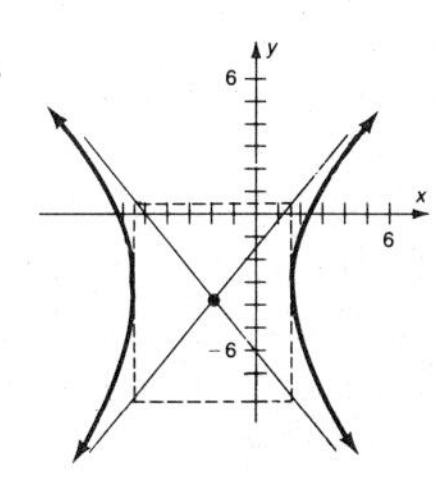

21. $\dfrac{(x+5)^2}{1} - \dfrac{(y-4)^2}{3} = 1$ **23.** $\dfrac{y^2}{9} - \dfrac{x^2}{16} = 1$

25. $\dfrac{(x-h)^2}{a^2} + \dfrac{(y-h)^2}{b^2} = 1$ **27.** $(y-k)^2 = 4c(x-h)^2, c > 0$ **29.** parabola **31.** ellipse

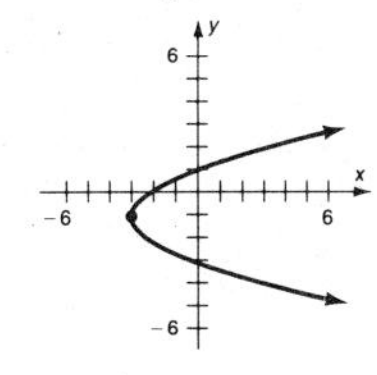

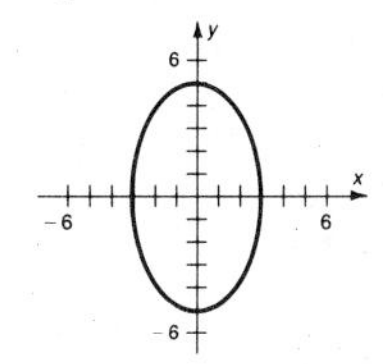

33. 45° **35.** 26.6° **37.** hyperbola **39.** parabola

PROBLEM SET 12.1, PAGES 468–469

Answers to Problems 1–17 should also include a vector diagram.

1. $\mathbf{v} = 6\mathbf{i} + 6\sqrt{3}\mathbf{j}$ **3.** $\mathbf{v} = \mathbf{i} + \mathbf{j}$ **5.** $\mathbf{v} = 7\cos 23°\mathbf{i} + 7\sin 23°\mathbf{j} \approx 6.4435\mathbf{i} + 2.7351\mathbf{j}$
7. $\mathbf{v} = 4\cos 112°\mathbf{i} + 4\sin 112°\mathbf{j} \approx -1.4984\mathbf{i} + 3.7087\mathbf{j}$ **9.** $\mathbf{v} = -2\mathbf{i} + 2\mathbf{j}$ **11.** $\mathbf{v} = -6\mathbf{i} - 5\mathbf{j}$ **13.** $\mathbf{v} = 8\mathbf{i} - 10\mathbf{j}$
15. $\mathbf{v} = -7\mathbf{i} - \mathbf{j}$ **17.** $\mathbf{v} = -9\mathbf{i} - 6\mathbf{j}$ **19.** 5 **21.** $\sqrt{85} \approx 9.2195$ **23.** $2\sqrt{2} \approx 2.8284$ **25.** $\sqrt{10} \approx 3.1623$
27. $\sqrt{41} \approx 6.4031$ **29.** not orthogonal **31.** orthogonal **33.** orthogonal
35. The plane is traveling at 244 mph with a direction of S80°W (a heading of 260°).
37. In 2 hr it has traveled 338 mi east.

	$\mathbf{v} \cdot \mathbf{w}$	$\|\mathbf{v}\|$	$\|\mathbf{w}\|$	$\cos\theta$
39.	−112	10	13	$\frac{-56}{65}$
41.	0	8	16	0
43.	39	$3\sqrt{10}$	$\sqrt{29}$	$\frac{-13\sqrt{290}}{290}$
45.	1	1	1	1
47.	7	$\sqrt{26}$	$\sqrt{13}$	$\frac{7\sqrt{2}}{26}$
49.	1	$\sqrt{2}$	1	$\frac{\sqrt{2}}{2}$

51. 75° **53.** 45° **55.** 90° **57.** $-\frac{8}{5}$
59. The pilot must fly in the direction of N4.9°E; the speed relative to the ground is 240 mph.
61. The weight of the astronaut is resolved into two components, one parallel to the inclined plane with length y and the other perpendicular to it with length x. The simulated weight of the astronaut is $|\mathbf{x}|$,

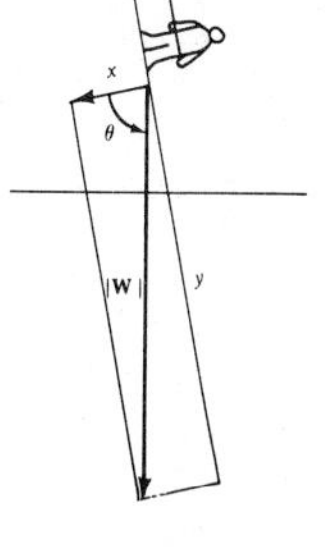

PROBLEM SET 12.2, PAGES 474–475

Answers to Problems 1–24 may vary.

1. $2\mathbf{i} - 3\mathbf{j}$ **3.** $\mathbf{i} - \mathbf{j}$ **5.** $3\mathbf{i} - 2\mathbf{j}$ **7.** $9\mathbf{i} + 7\mathbf{j}$ **9.** $4\mathbf{i} - \mathbf{j}$ **11.** $\mathbf{i} + 2\mathbf{j}$ **13.** $3\mathbf{i} + 2\mathbf{j}$ **15.** $\mathbf{i} + \mathbf{j}$ **17.** $2\mathbf{i} + 3\mathbf{j}$

19. $7\mathbf{i} - 9\mathbf{j}$ **21.** $\mathbf{i} + 4\mathbf{j}$ **23.** $2\mathbf{i} - \mathbf{j}$ **25.** $\frac{63}{13}$ **27.** 0 **29.** $\frac{3}{61}\sqrt{61}$ **31.** $\frac{315}{169}\mathbf{i} + \frac{756}{169}\mathbf{j}$ **33.** **0** **35.** $\frac{18}{61}\mathbf{i} + \frac{15}{61}\mathbf{j}$
37. $d = 0$ **39.** $d = \frac{47}{5}$ **41.** $d = \frac{12}{5}$ **43.** $d = \frac{22}{5}$ **45.** $d = \frac{17}{5}\sqrt{10}$ **47.** $d = \frac{19}{10}\sqrt{10}$ **49.** $d = \frac{17}{29}\sqrt{29}$ **51.** $\frac{21}{2}$
53. $\frac{41}{2}$ **55.** 19 **57–59.** Answers vary.

PROBLEM SET 12.3, PAGES 481–482

1. The area of a rectangle with dimensions 15 by 35 is $K(15, 35) = 525$ square units.
3. The area of a rectangle with dimensions $\sqrt{2}$ by $\sqrt{3}$ is $K(\sqrt{2}, \sqrt{3}) = \sqrt{6} \approx 2.45$ square units.
5. The volume of a box 3 by 5 by 8 is $V(3, 5, 8) = 120$ cubic units.
7. The volume of a box $\sqrt{2}$ by $2\sqrt{5}$ by 4 is $V(\sqrt{2}, 2\sqrt{5}, 4) = 8\sqrt{10} \approx 25.3$ cubic units.
9. The simple interest from a \$500 investment at 5% for 3 years is $I(500, 0.05, 3) = \$575$.
11. The simple interest from a \$1000 investment at 11% for 4 years is $I(1000, 0.11, 4) = \$1440$.
13. The future value of a \$2500 investment at 12% compounded quarterly for 6 years is $A(2500, 0.12, 6, 4) = \$5081.99$.
15. The future value of a \$110,000 investment at 9% compounded monthly for 30 years is $A(110{,}000, 0.09, 30, 12) = \$1{,}620{,}363$.
17. 1 **19.** 49 **21.** $a^2 - 12a + 36$ **23.** 0 **25.** $\frac{11}{17}$ **27.** $2/s$ **29.** 0.2 **31.** $e^6/\sqrt{13} \approx 111.89$

33. $1/\pi \approx 0.3183$ **35.**

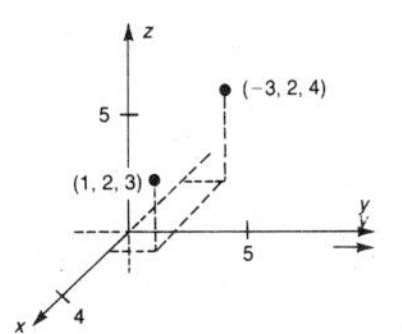

37.

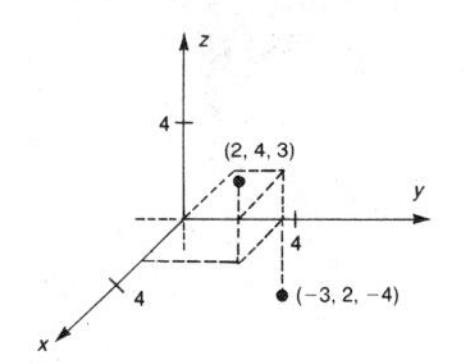

39.

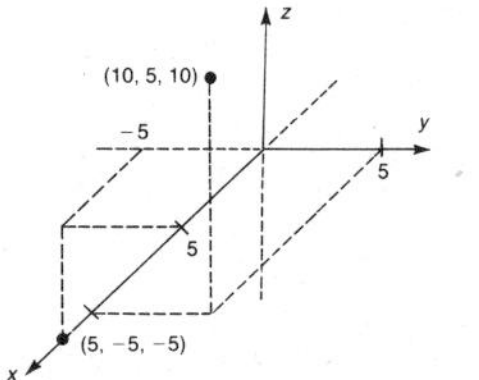

41.

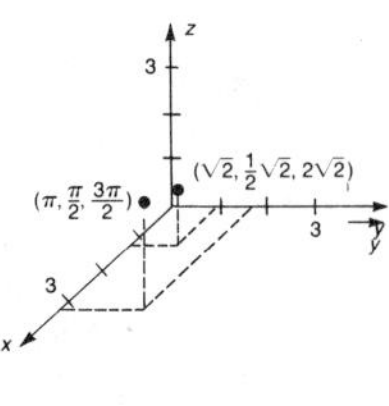

43. plane with interecepts $(10, 0, 0)$, $(0, 5, 0)$, $(0, 0, 2)$

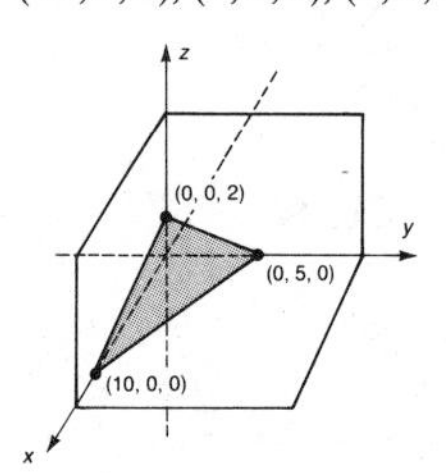

45. plane with intercepts $(4, 0, 0)$, $(0, -6, 0)$, $(0, 0, -12)$

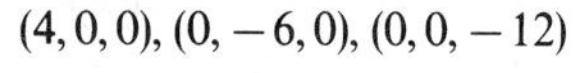

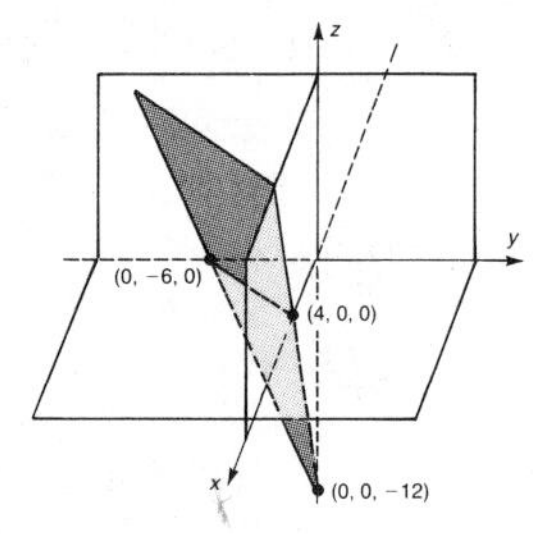

47. elliptic paraboloid

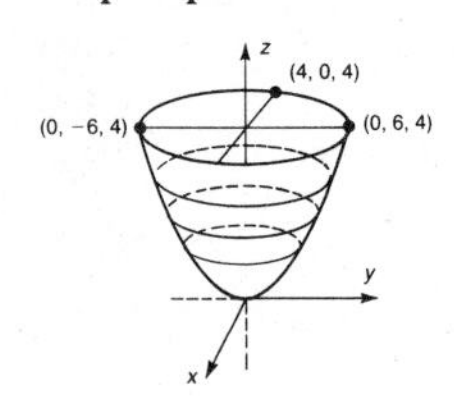

49. ellipsoid

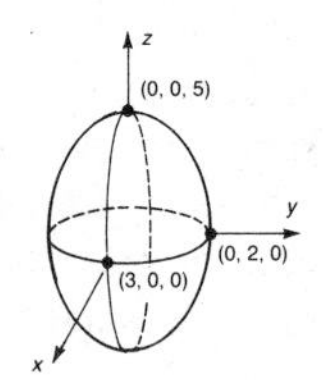

51. elliptic paraboloid

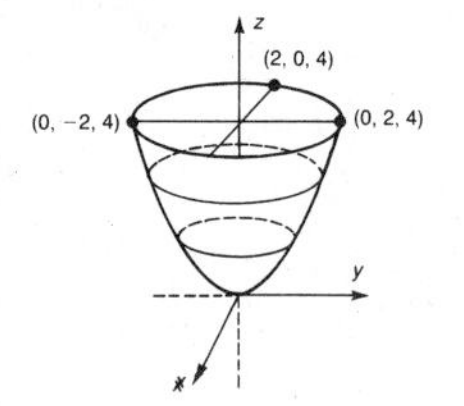

53. circular paraboloid

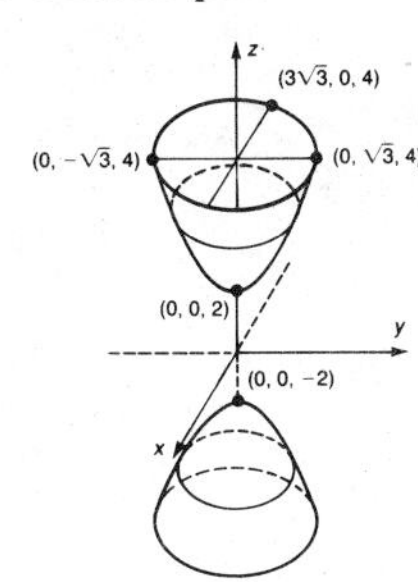

55. circular cylinder

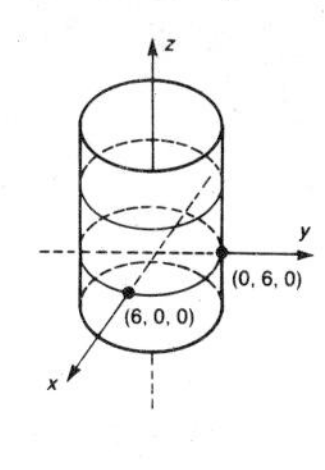

57. circular cylinder

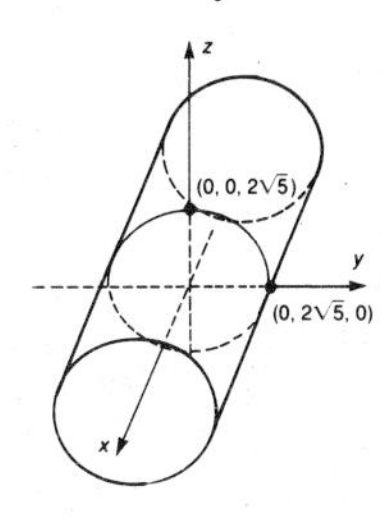

59. $C(x, y) = 3700 + 800x + 550y$; **a.** \$19,950 **b.** \$21,450 **c.** \$21,200 **d.** \$20,200
61. $C(x, y, z) = 3.25xy + 1.5xz + 1.5yz$ **63. a.** 3.875×10^8 ml **b.** 3.06×10^{10} ml **c.** 1.92×10^7 ml

PROBLEM SET 12.4, PAGES 489–490

1. $\mathbf{v} = 2\mathbf{i} + 6\mathbf{j} - 6\mathbf{k}$; $|\mathbf{v}| = 2\sqrt{19}$ **3.** $\mathbf{v} = -5\mathbf{i} + 6\mathbf{j} - 3\mathbf{k}$; $|\mathbf{v}| = \sqrt{70}$ **5.** $\mathbf{v} = -4\mathbf{i} - 3\mathbf{j} + 5\mathbf{k}$; $|\mathbf{v}| = 5\sqrt{2}$ **7.** $\sqrt{14}$
9. $18\mathbf{i} - \mathbf{j} - 7\mathbf{k}$ **11.** $6\mathbf{i} - 5\mathbf{j} + 5\mathbf{k}$ **13.** $5\mathbf{i} + \mathbf{j} + 7\mathbf{k}$ **15.** $25\mathbf{i} - 3\mathbf{j} + 11\mathbf{k}$ **17.** 1 **19.** 1 **21.** 3 **23.** 24
25. $\mathbf{k}$ **27.** $-2\mathbf{i} + 4\mathbf{j} + 3\mathbf{k}$ **29.** $-2\mathbf{i} + 25\mathbf{j} + 14\mathbf{k}$ **31.** $-2\mathbf{i} + 16\mathbf{j} + 11\mathbf{k}$ **33.** $-14\mathbf{i} - 4\mathbf{j} - \mathbf{k}$ **35.** $\frac{5}{13}\mathbf{i} + \frac{12}{13}\mathbf{k}$
37. $\frac{3}{13}\mathbf{i} + \frac{12}{13}\mathbf{j} - \frac{4}{13}\mathbf{k}$ **39.** $\frac{4}{29}\sqrt{29}\mathbf{i} + \frac{2}{29}\sqrt{29}\mathbf{j} - \frac{3}{29}\sqrt{29}\mathbf{k}$ **41.** $\frac{2}{21}\sqrt{7}$ **43.** $s = -\frac{7}{2}t$ **45.** $-\frac{9}{13}$ **47.** $s = \frac{13}{9}t$
49. $\frac{4}{63}\sqrt{42}$ **51.** $s = -\frac{27}{8}t$ **53.** $\frac{26}{35}$ **55.** $s = -\frac{25}{26}t$ **57.** $\frac{2}{3}\mathbf{i} - \frac{2}{3}\mathbf{j} - \frac{2}{3}\mathbf{k}$; $\frac{2}{3}\sqrt{3}$ **59.** $\frac{20}{7}\mathbf{i} + \frac{4}{7}\mathbf{j} - \frac{12}{7}\mathbf{k}$; $\frac{4}{7}\sqrt{35}$
61. $-\frac{24}{29}\mathbf{i} + \frac{32}{29}\mathbf{j} - \frac{16}{29}\mathbf{k}$; $\frac{8}{29}\sqrt{29}$ **63. a.** scalar **b.** vector **c.** scalar

PROBLEM SET 12.5, PAGE 500

Points in Problems 1–19 should also be plotted.

	Polar	Rectangular
1.	$(4, \frac{\pi}{4}) = (-4, \frac{5\pi}{4})$	$(2\sqrt{2}, 2\sqrt{2})$
3.	$(5, \frac{2\pi}{3}) = (-5, \frac{5\pi}{3})$	$(-\frac{5}{2}, \frac{5\sqrt{3}}{2})$
5.	$(\frac{3}{2}, \frac{7\pi}{6}) = (-\frac{3}{2}, \frac{\pi}{6})$	$(\frac{-3\sqrt{3}}{4}, -\frac{3}{4})$
7.	$(-4, 4) = (4, 0.86)$	$(2.61, 3.03)$
9.	$(-4, \pi) = (4, 0)$	$(4, 0)$

11. $(5\sqrt{2}, \frac{\pi}{4}) = (-5\sqrt{2}, \frac{5\pi}{4})$ **13.** $(4, \frac{5\pi}{3}) = (-4, \frac{2\pi}{3})$
15. $(3\sqrt{2}, \frac{7\pi}{4}) = (-3\sqrt{2}, \frac{3\pi}{4})$ **17.** $(2, \frac{5\pi}{6}) = (-2, \frac{11\pi}{6})$
19. $(13, 2.75) = (-13, 5.89)$ **21.** lemniscate **23.** 3-leaved rose
25. cardioid **27.** none **29.** none **31.** none
33. none (line) **35.** cardioid

37.

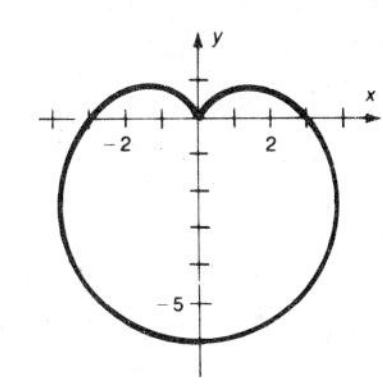

39.

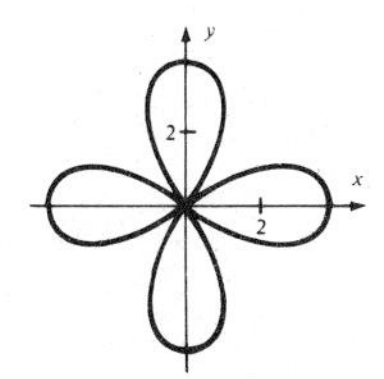

41.

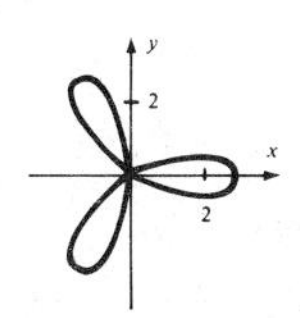

43.

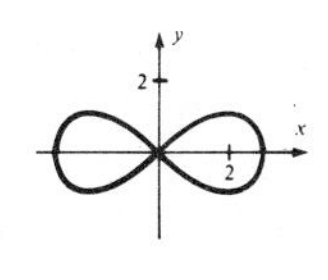

45.

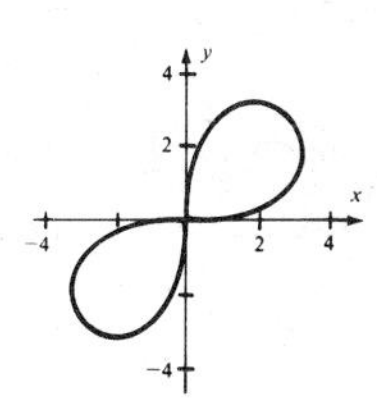

47.

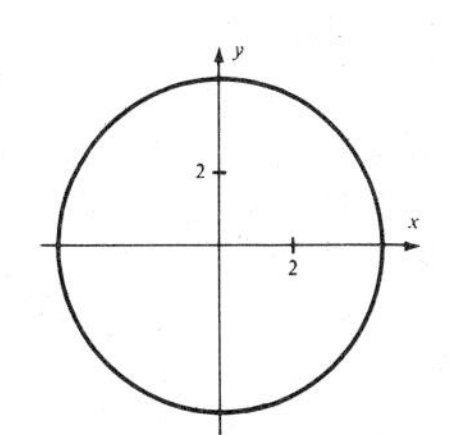

49.

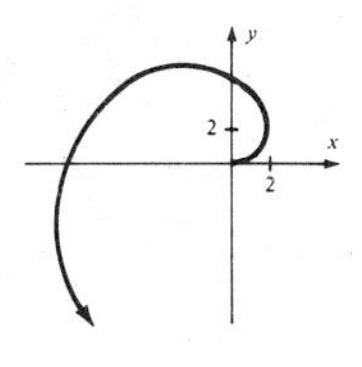

51. Answers vary. **53.** $d \approx 4.1751$

55. a.

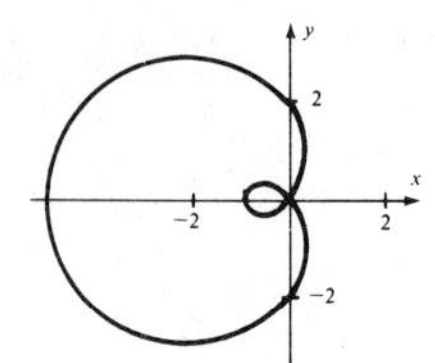

b.

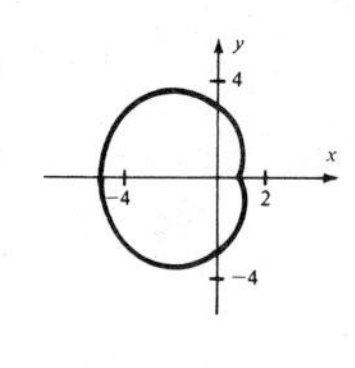

c.

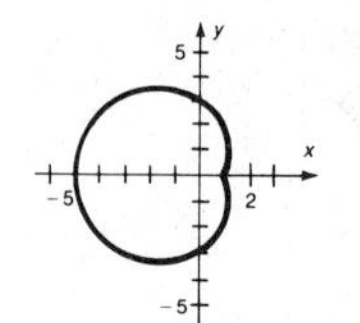

d.

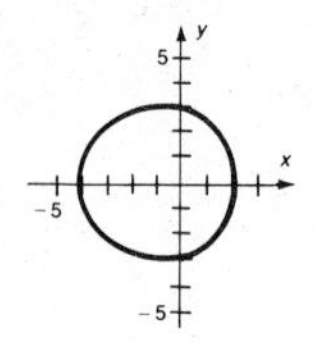

57. a.

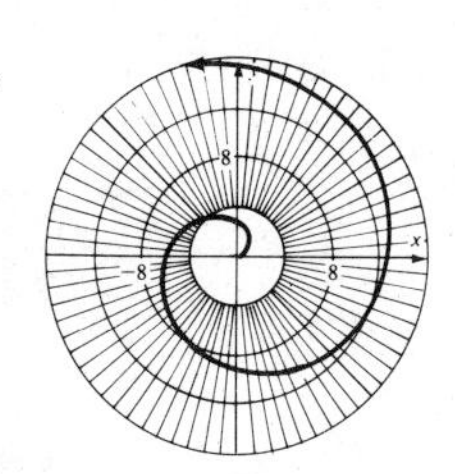

b.

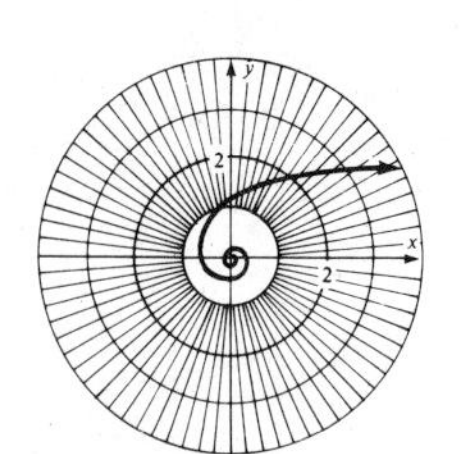

c.

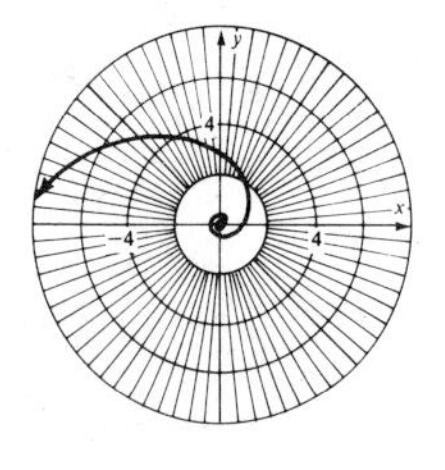

59.

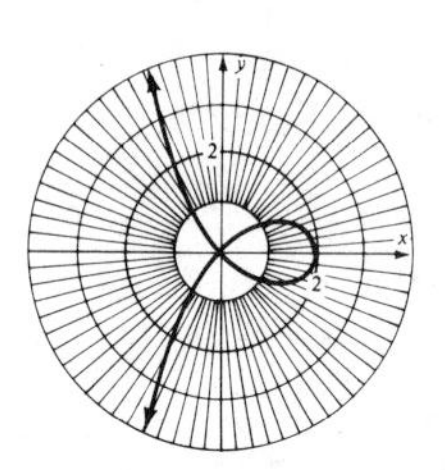

61.

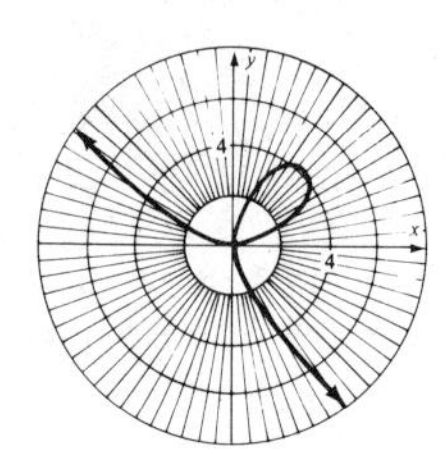

PROBLEM SET 12.6, PAGE 506

1.

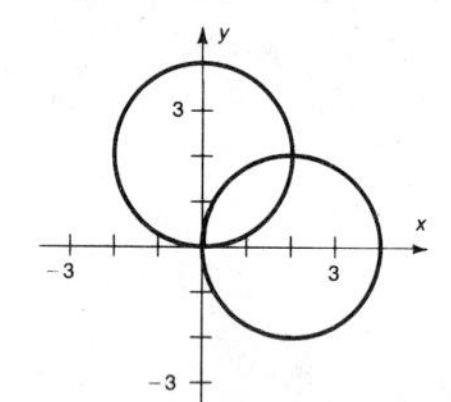

3.

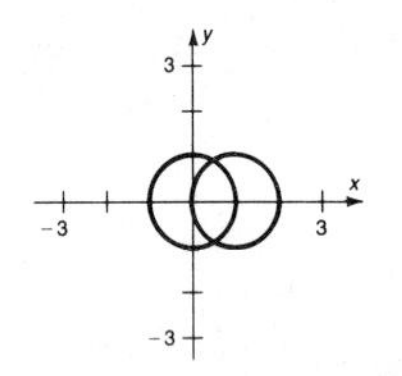

5.

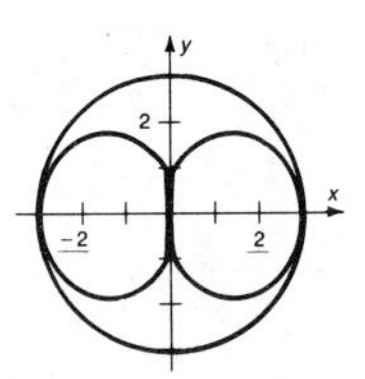

7.

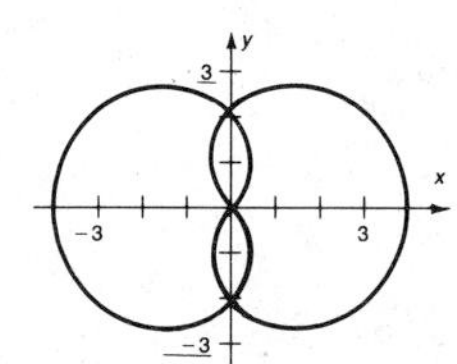

9.

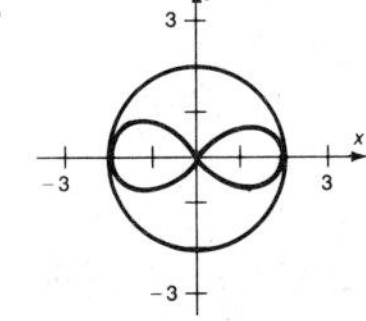

11.

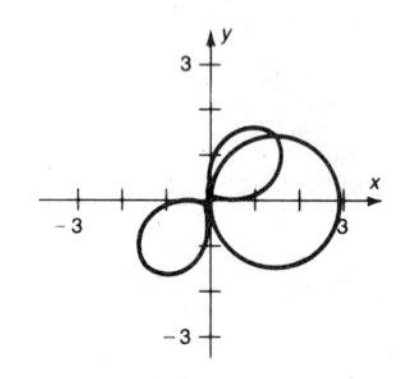

13.

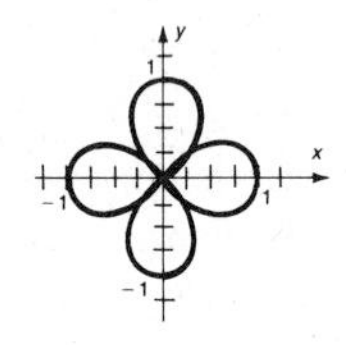

15.

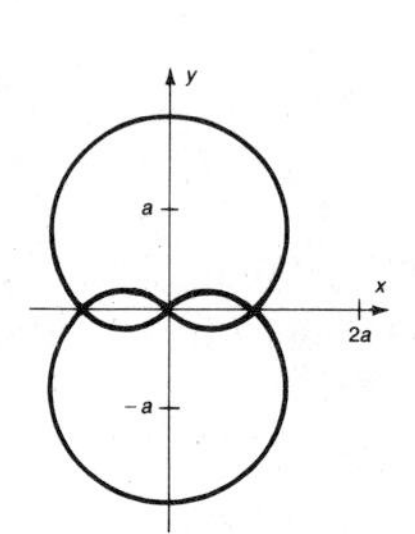

17. $(0, 0), (4\sqrt{2}, \frac{\pi}{4})$ **19.** $(2, \frac{\pi}{3}), (2, \frac{5\pi}{3})$

21. $(2, \frac{\pi}{2}), (2, \frac{3\pi}{2})$ **23.** $(2, 0), (2, \pi), (0, 0)$ **25.** $(3, \frac{\pi}{4}), (3, \frac{5\pi}{4})$ **27.** $(0, 0), (\frac{3}{2}\sqrt{2}, \frac{\pi}{6}), (\frac{3}{2}\sqrt{2}, \frac{5\pi}{6})$ **29.** $(0, 0), (\frac{2+\sqrt{2}}{2}, \frac{\pi}{4}), (\frac{2-\sqrt{2}}{2}, \frac{5\pi}{4})$

31.

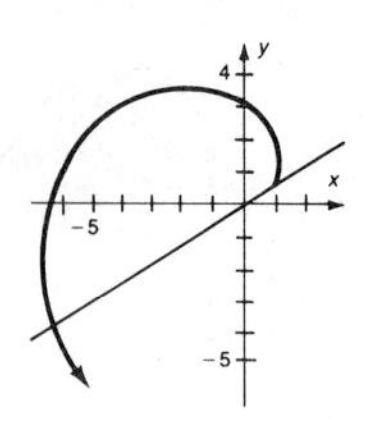

33.

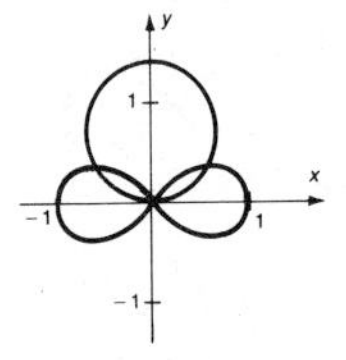

35.

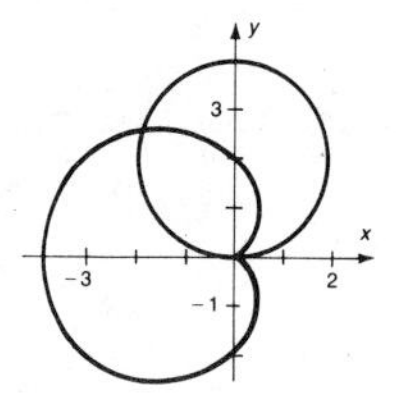

37.

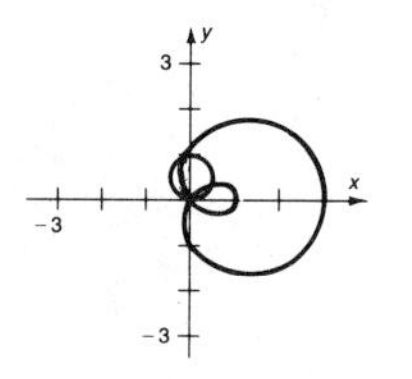

39.

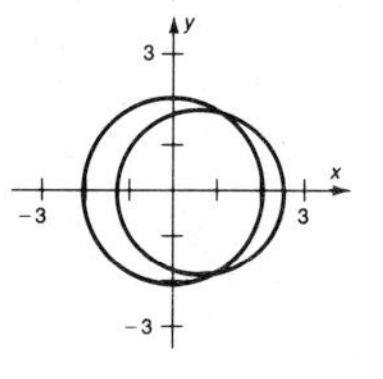

41.

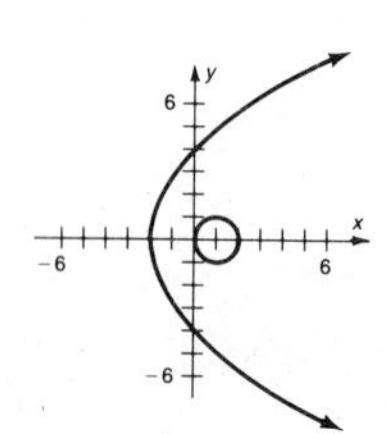

43.

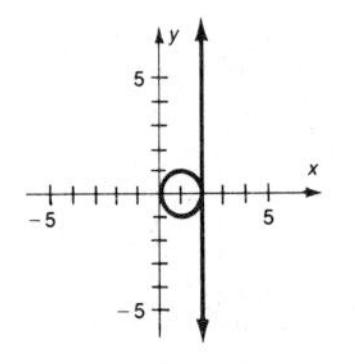

45.

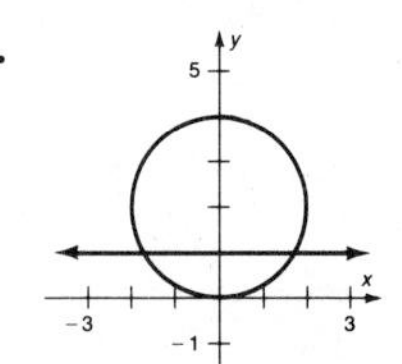

47. $(0, 0), (\pi, \frac{\pi}{3}), (4\pi, \frac{4\pi}{3})$

49. $(0, 0)$, $(\frac{16}{5}, \text{Arccos}(-\frac{3}{5}))$. *Note:* $\text{Arccos}(-\frac{3}{5}) \approx 2.214$ or $126.87°$
51. $(0, 0)$, $(3.2, 2\pi - \text{Arccos}\, 0.8)$. *Note:* $2\pi - \text{Arccos}\, 0.8 \approx 5.640$ or $323.13°$
53. $(0, 0)$, $(1, 0)$, $(0.6, \text{Arcsin}\, 0.8)$. *Note:* $\text{Arcsin}\, 0.8 \approx 0.927$ or $53.1°$ **55.** $(2, \frac{\pi}{2})$, $(2, \frac{3\pi}{2})$
57. $(2 + \sqrt{2}, \frac{5\pi}{4})$, $(2 + \sqrt{2}, \frac{3\pi}{4})$, $(2 - \sqrt{2}, \frac{\pi}{4})$, $(2 - \sqrt{2}, \frac{7\pi}{4})$ **59.** $(a, \frac{\pi}{2})$

PROBLEM SET 12.7, PAGES 510–511

1.
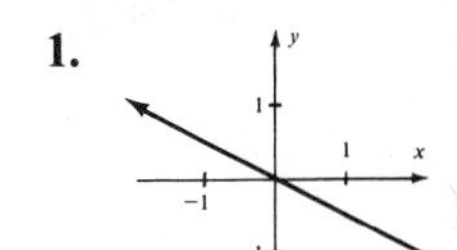

3.
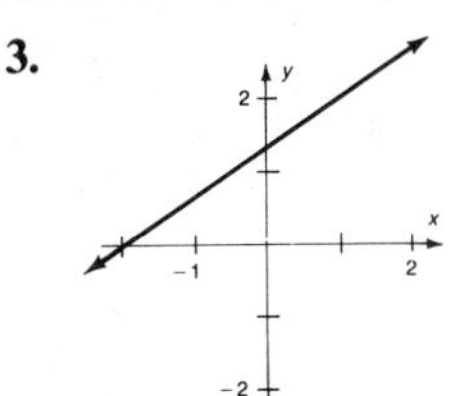

5.
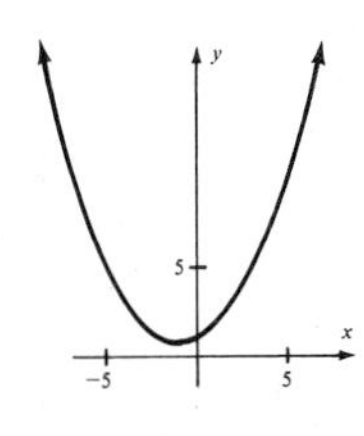

7.
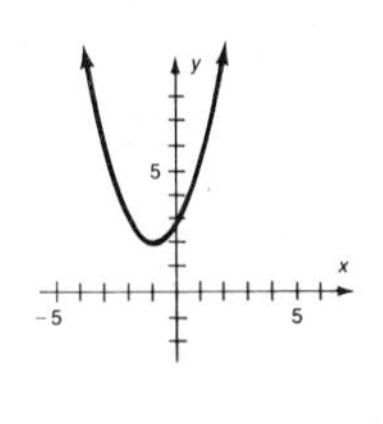

9.
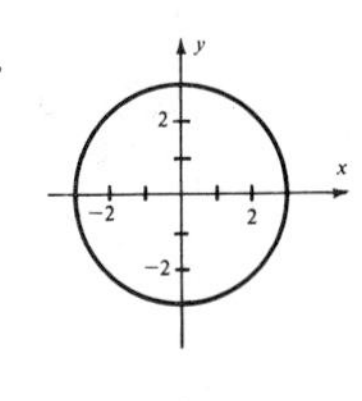

11.
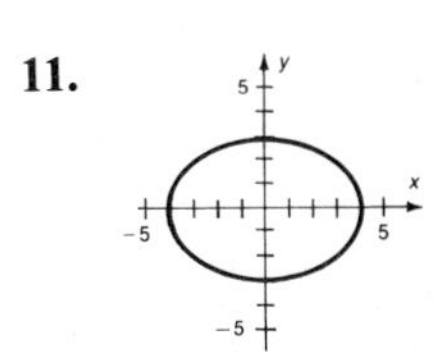

13.
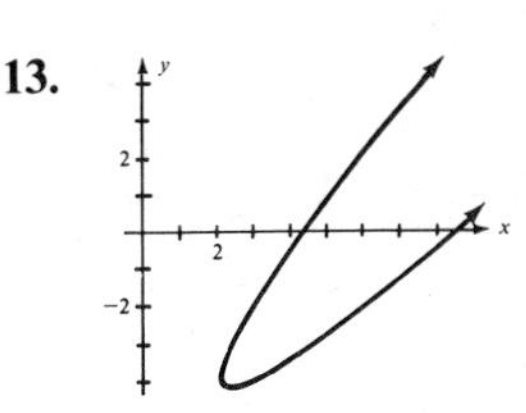

15.
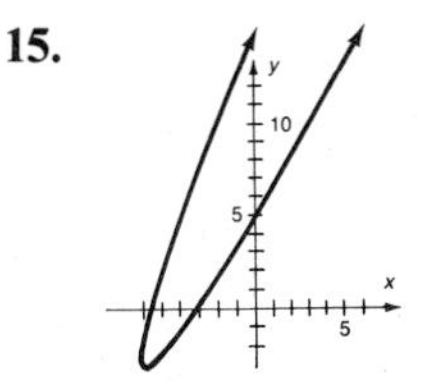

17.
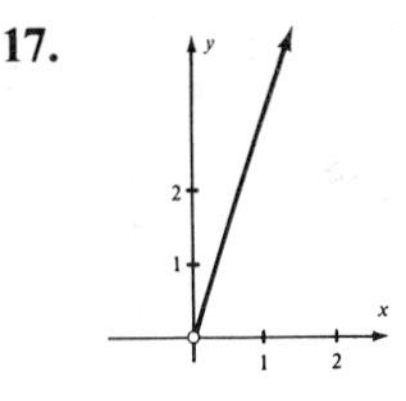

19.
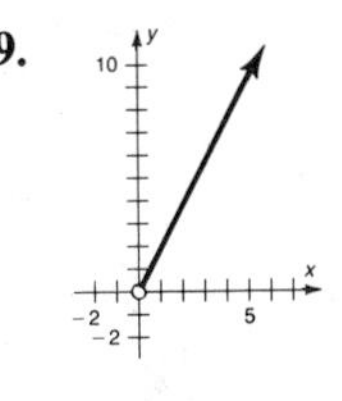

Graphs for Problems 21–39 are the same as for Problems 1–19.

21. $y = -\frac{1}{2}x$ **23.** $y = \frac{2}{3}x + \frac{4}{3}$ **25.** $(x + 1)^2 = 4(y - \frac{3}{4})$ **27.** $y = x^2 + 2x + 3$ **29.** $x^2 + y^2 = 9$

31. $\dfrac{x^2}{16} + \dfrac{y^2}{9} = 1$ **33.** $x^2 - 2xy + y^2 - 13x + 12y + 38 = 0$; $\cot 2\theta = \dfrac{1 - 1}{-2} = 0$; $\theta = 45°$

35. $4x^2 - 4xy + y^2 + 36x - 20y + 75 = 0$; $\cot 2\theta = \dfrac{4 - 1}{-4} = -\dfrac{3}{4}$; $\theta \approx 63.4°$ **37.** $y = 3x, x > 0$

39. $y = ex, x > 0$

41.
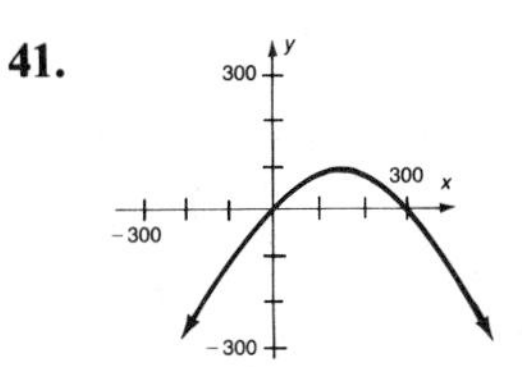

43.
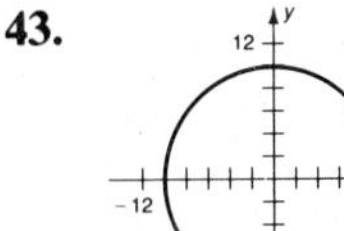

45.
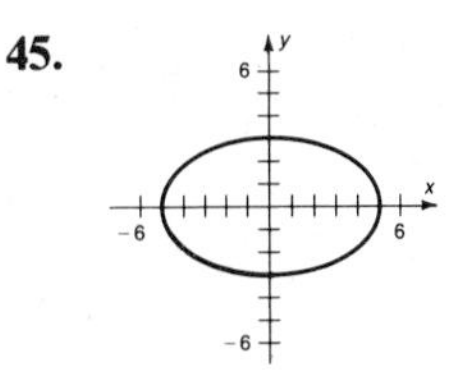

47.
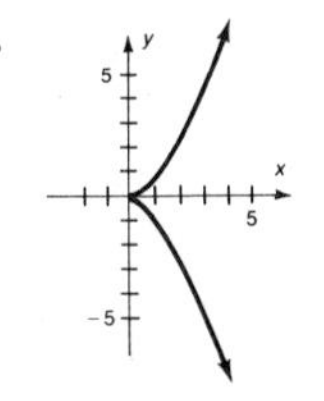

49.

51.
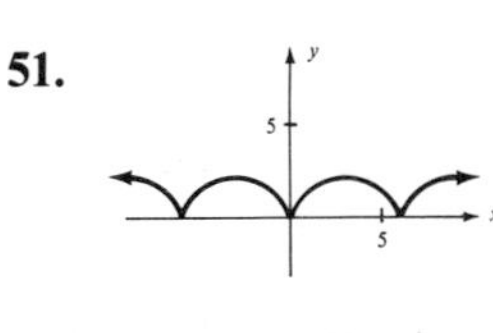

53. **55.**
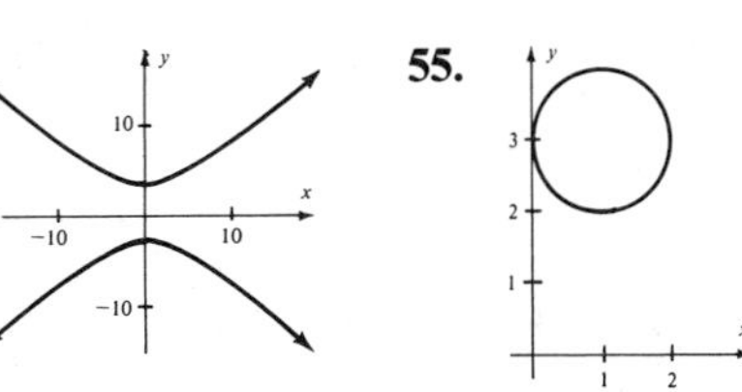

57.
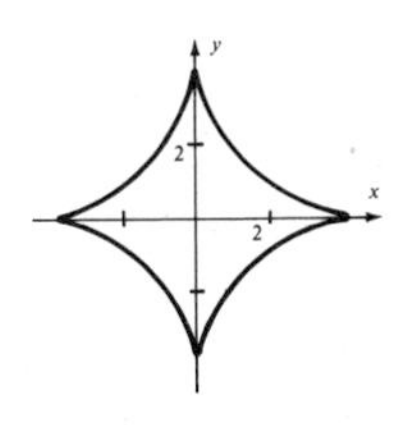

59.

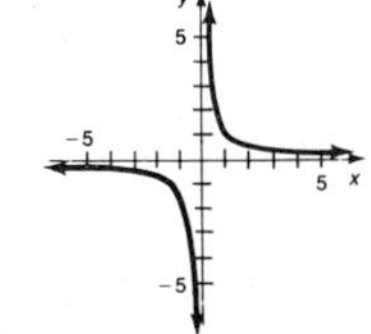

61. $x = a(\theta - \sin\theta)$, $y = a(1 - \cos\theta)$

CHAPTER 12 SUMMARY, PAGES 511–514

1. $-2\mathbf{i} + \mathbf{j}$ **3.** The magnitude is 9.8 and the direction is N80° E. **5.** $\mathbf{v}_x \approx 2.8$; $\mathbf{v}_y \approx 3.5$ **7.** $5\mathbf{i} + 12\mathbf{j}$
9. $\sqrt{29}$ **11.** 6 **13.** Let $\mathbf{v} = a\mathbf{i} + b\mathbf{j}$ and $\mathbf{w} = c\mathbf{i} + d\mathbf{j}$. Then the scalar product is $\mathbf{v} \cdot \mathbf{w} = ac + bd$ **15.** 9

17. $-\frac{9}{169}\sqrt{13}$ **19.** 5° **21.** $\mathbf{N} = 5\mathbf{i} - 12\mathbf{j}$; $\mathbf{v} = -\frac{3}{5}\mathbf{i} - \frac{1}{4}\mathbf{j}$; $\mathbf{N}\cdot\mathbf{v} = -3 + 3 = 0$ **23.** $\mathbf{N} = \mathbf{i} - 5\mathbf{j}$; $\mathbf{v} = -4\mathbf{i} - \frac{4}{5}\mathbf{j}$; $\mathbf{N}\cdot\mathbf{v} = -4 + 4 = 0$ **25.** $\frac{5}{2}\mathbf{i} + \frac{\sqrt{5}}{2}\mathbf{j}; \frac{\sqrt{30}}{2}$

27. 0; 0 **29.** $\frac{73}{13}$ **31.** $\frac{31}{13}\sqrt{13}$ **33.** $4\mathbf{i} + 8\mathbf{j}$ or $4\mathbf{i} + 8\mathbf{j} - 2\mathbf{k}$ **35.** $-8\mathbf{i} - \mathbf{j}$ or $-8\mathbf{i} - \mathbf{j} + 3\mathbf{k}$ **37.** 339 **39.** -60

41.

45.

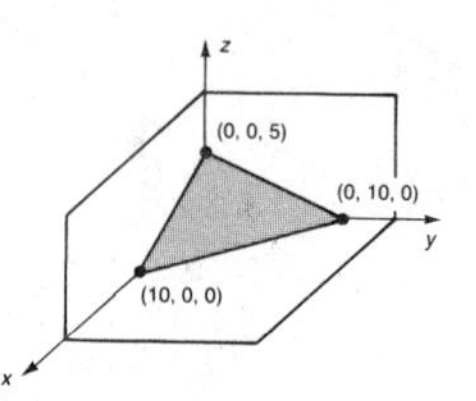

47.

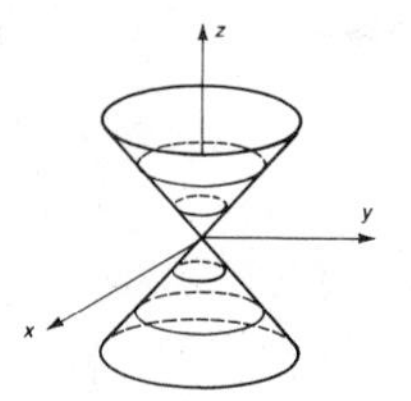

49. $4\mathbf{i} - 7\mathbf{j} + 2\mathbf{k}$ **51.** $3\mathbf{i} + \mathbf{j} - \mathbf{k}$ **53.** 0 **55.** 4 **57.** $2\mathbf{k}$ **59.** $-5\mathbf{j} + 5\mathbf{k}$ **61.** $\frac{1}{3}\mathbf{i} + \frac{1}{3}\mathbf{j} - \frac{\sqrt{7}}{3}\mathbf{k}$ **63.** $\frac{3+\sqrt{7}}{9}$

Points in Problems 65–69 should also be plotted.

65. $(5, 2.3771) = (-5, 5.5187)$ **67.** $(-2, 2) = (2, 5.1416)$ **69.** $(-1.50000, -2.5981)$ **71.** $(4.2426, 5.4978)$

73.

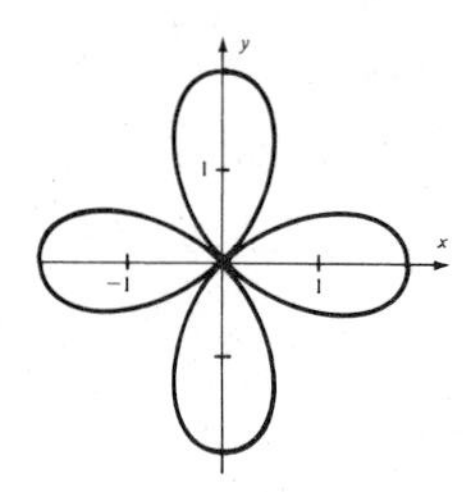

75.

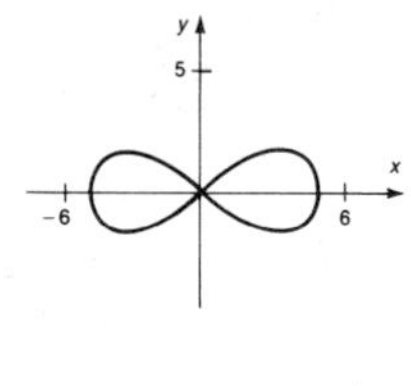

77. a. lemniscate **b.** 4-leaved rose

79. a. none (limaçon) **b.** cardioid **81.** $(0, 0), (4, 0), (4, \pi)$ **83.** $(4, \pi), (0, 0), (4, 0)$

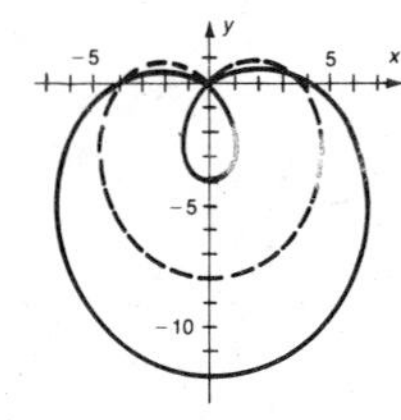

85.

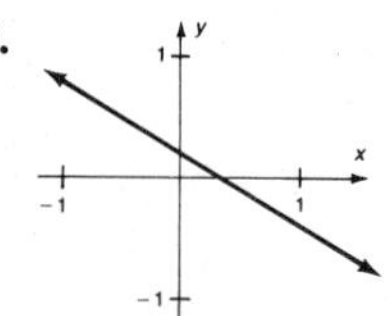

87.

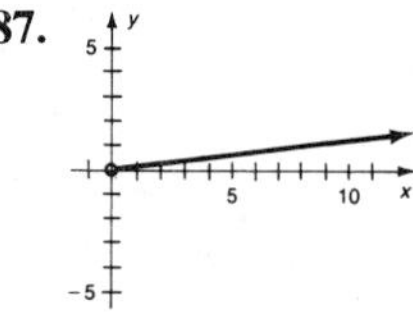

89. $3x + 5y - 1 = 0$ (See Problem 85 for graph.) **91.** $y = e^{-2}x, x > 0$ (See Problem 87 for graph.)

CUMULATIVE REVIEW III, PAGES 515–516

1. a. $\frac{40}{3}$; geometric; $r = \frac{2}{3}$ **b.** 0; arithmetic; $d = -15$ **c.** 30; neither; 45, 30, 25, 30, 5, ... **2. a.** $(-2, 5)$ **b.** $(-1, 2), (-4, 5)$ **c.** $(1, -3)$ **d.** $(2, -1, -1)$ **3. a.** $\begin{bmatrix} -1 & 1 \\ 3 & -2 \end{bmatrix}$ **b.** not conformable **c.** $\begin{bmatrix} 0 & -1 \\ -1 & 1 \end{bmatrix}$

4. a. parabola **b.** hyperbola **c.** hyperbola **d.** circle **e.** line **f.** hyperbola **g.** 4-leaved rose **h.** lemniscate **i.** cardioid **j.** circle **5. a.** $\frac{(x-2)^2}{4} + \frac{(y-1)^2}{3} = 1$ **b.** $\frac{(x-6)^2}{4} - \frac{y^2}{5} = 1$

c. $(y-3)^2 = 12(x-4)$ **6. a.** 

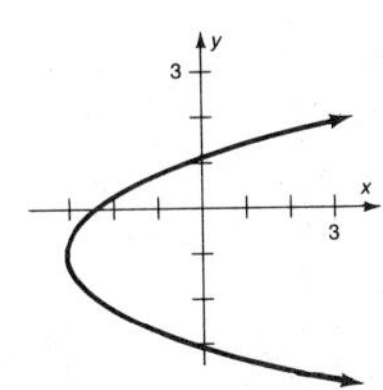

b.

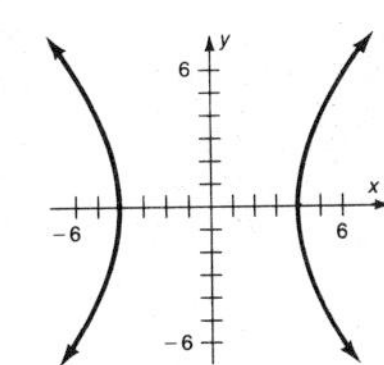

c.

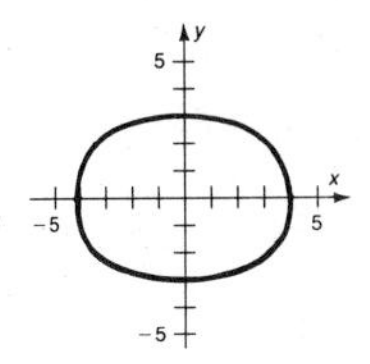

d.

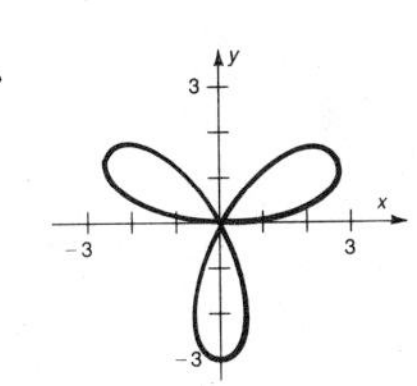

e.

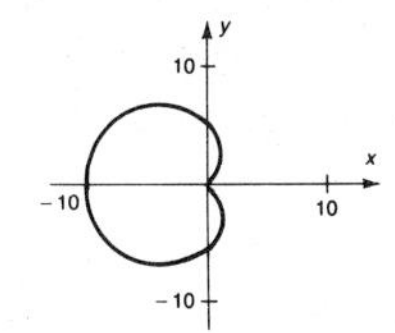

7. a.

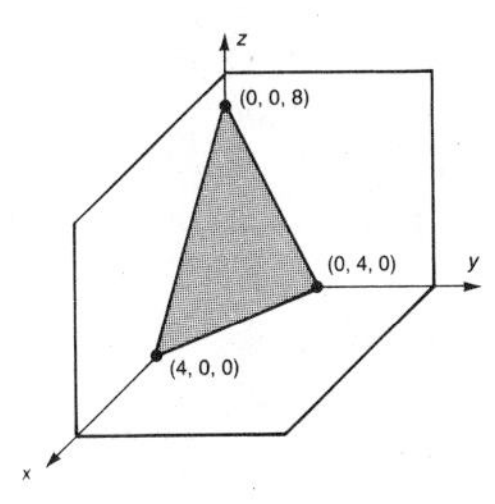

b.

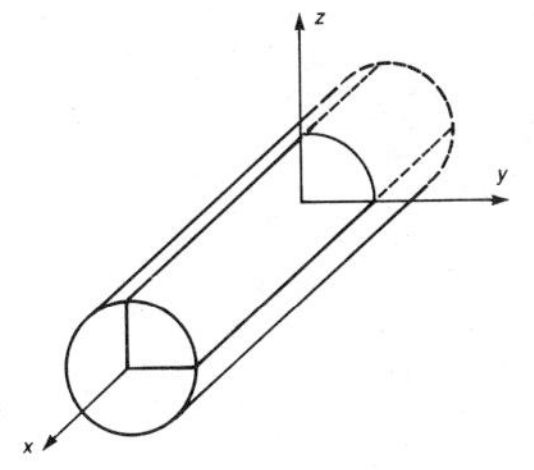

8. a. $(a+b) = \sum_{k=0}^{n} \binom{n}{k} a^{n-k} b^k$ **b.** $(a+b)^n = \binom{n}{0} a^n + \binom{n}{1} a^{n-1} b + \cdots + \binom{n}{n-1} ab^{n-1} + \binom{n}{n} b^n$

c. $x^4 + 4x^3(-3y) + 6x^2(-3y)^2 + 4x(-3y)^3 + (-3y)^4 = x^4 - 12x^3y + 54x^2y^2 - 108xy^3 + 81y^4$ **9.** Answers vary.

10. a. $9\mathbf{i} - 7\sqrt{5}\mathbf{j} + 5\mathbf{k}$ (or, in two dimensions, $9\mathbf{i} - 7\sqrt{5}\mathbf{j}$) **b.** -27 (or -25) **c.** $\sqrt{53}$ (or 7)

d. $\frac{5}{31}\sqrt{31}\,\mathbf{i} - \frac{1}{31}\sqrt{155}\,\mathbf{j} + \frac{1}{31}\sqrt{31}\,\mathbf{k}$ **e.** $-\sqrt{5}\mathbf{i} + 8\mathbf{j} + 13\sqrt{5}\mathbf{k}$

EXTENDED APPLICATION: PLANETARY ORBITS, PAGE 519

1. The greatest distance is $a + c \approx 1.5 \cdot 10^8$, and the least distance is $a - c \approx 1.3 \cdot 10^8$.
3. The greatest distance is $3.7 \cdot 10^9$, and the least distance is $3.6 \cdot 10^9$. **5.** 5500 mi **7.** $1.2x^2 + 1.3y^2 = 1.6 \cdot 10^{15}$
9. $8.802x^2 + 8.805y^2 = 7.751 \cdot 10^{16}$

Applications Index

Calculus

The purpose of this text is preparation for calculus. This list gives examples which are not typical of precalculus textbooks, but which provide practice of concepts that will be particularly useful in the study of calculus.

Chemistry

Consumer

Earth Science

Economics

Engineering

Extended Applications

Each extended application is introduced by a news clipping and examines a topic in more than the usual depth.

General Interest

Index

A

B

C

F

G

H

I

J

K

L

M

T

U

V

Trigonometry

Trigonometric Functions

Let θ be an angle in standard position with a point $P(x, y)$ on the terminal side a distance of r from the origin ($r \neq 0$). Then the trigonometric functions are defined by

$$\cos\theta = \frac{x}{r} \qquad \sec\theta = \frac{r}{x}\ (x \neq 0)$$

$$\sin\theta = \frac{y}{r} \qquad \csc\theta = \frac{r}{y}\ (y \neq 0)$$

$$\tan\theta = \frac{y}{x}\ (x \neq 0) \qquad \cot\theta = \frac{x}{y}\ (y \neq 0)$$

Inverse Trigonometric Functions

Inverse Function	*Domain*	*Range*
$y = \text{Arccos}\, x$ or $y = \text{Cos}^{-1} x$	$-1 \leq x \leq 1$	$0 \leq y \leq \pi$
$y = \text{Arcsin}\, x$ or $y = \text{Sin}^{-1} x$	$-1 \leq x \leq 1$	$-\frac{\pi}{2} \leq y \leq \frac{\pi}{2}$
$y = \text{Arctan}\, x$ or $y = \text{Tan}^{-1} x$	All reals	$-\frac{\pi}{2} < y < \frac{\pi}{2}$
$y = \text{Arccot}\, x$ or $y = \text{Cot}^{-1} x$	All reals	$0 < y < \pi$

Exact Values

Function \ Angle θ	0	$\frac{\pi}{6}$	$\frac{\pi}{4}$	$\frac{\pi}{3}$	$\frac{\pi}{2}$	π	$\frac{3\pi}{2}$
$\cos\theta$	1	$\frac{\sqrt{3}}{2}$	$\frac{\sqrt{2}}{2}$	$\frac{1}{2}$	0	-1	0
$\sin\theta$	0	$\frac{1}{2}$	$\frac{\sqrt{2}}{2}$	$\frac{\sqrt{3}}{2}$	1	0	-1
$\tan\theta$	0	$\frac{\sqrt{3}}{3}$	1	$\sqrt{3}$	undef.	0	undef.
$\sec\theta$	1	$\frac{2}{\sqrt{3}}$	$\frac{2}{\sqrt{2}}$	2	undef.	-1	undef.
$\csc\theta$	undef.	2	$\frac{2}{\sqrt{2}}$	$\frac{2}{\sqrt{3}}$	1	undef.	-1
$\cot\theta$	undef.	$\frac{3}{\sqrt{3}}$	1	$\frac{\sqrt{3}}{3}$	0	undef.	0

Solving Triangles

Given	Conditions on Given Information	Law to Use for Solution
1. SSS	a. The sum of the lengths of the two smaller sides is less than or equal to the length of the larger side.	No solution
	b. The sum of the lengths of the two smaller sides is greater than the length of the larger side.	Law of Cosines
2. SAS	a. The angle is greater than or equal to 180°.	No solution
	b. The angle is less than 180°.	Law of Cosines
3. AAA		No solution
4. ASA or AAS	a. The sum of the angles is greater than or equal to 180°.	No solution
	b. The sum of the angles is less than 180°.	Law of Sines
5. SSA	Let θ be the given angle with adjacent (ADJ) and opposite (OPP) sides given; the height h is found by $h = (\text{ADJ})\sin\theta$.	
	a. $\theta > 90°$	
	i. OPP $\leq$ ADJ	No solution
	ii. OPP $>$ ADJ	Law of Sines
	b. $\theta < 90°$	
	i. OPP $< h <$ ADJ	No solution
	ii. OPP $= h <$ ADJ	Right-triangle solution
	iii. $h <$ OPP $<$ ADJ	*Ambiguous case*: Use the Law of Sines to find *two* solutions.
	iv. OPP $\geq$ ADJ	Law of Sines

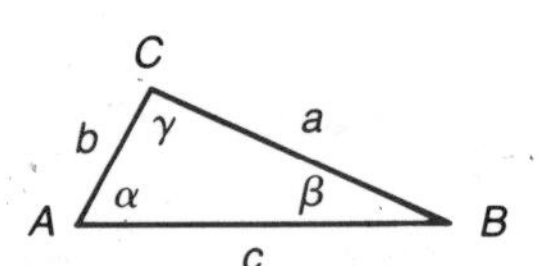

Triangle *ABC*

Pythagorean Theorem

In a right triangle, $c^2 = a^2 + b^2$

Law of Cosines

$$a^2 = b^2 + c^2 - 2bc\cos\alpha$$
$$b^2 = a^2 + c^2 - 2ac\cos\beta$$
$$c^2 = a^2 + b^2 - 2ab\cos\gamma$$

$$\cos\alpha = \frac{b^2 + c^2 - a^2}{2bc}$$

$$\cos\beta = \frac{a^2 + c^2 - b^2}{2ac}$$

$$\cos\gamma = \frac{a^2 + b^2 - c^2}{2ab}$$

Law of Sines

$$\frac{\sin\alpha}{a} = \frac{\sin\beta}{b} = \frac{\sin\gamma}{c}$$

THE COMMISSAR VANISHES
СТРОИМ
ЗА ИНДУСТРИАЛИЗАЦИЮ
На строительных площадках семи союзных республи
и заводах—поставщиках оборудования, на
ИДЕТ БОИ ЗА ПУСК 518
НОВАЯ ОБСТАНОВКА—
НОВЫЕ ЗАДАЧИ
ХОЗЯЙСТВЕННОГО СТРОИТЕЛЬСТВА
ССР
АВГУСТ 18 1933 г. ПЯТНИЦА. № 227 (5753).
Центр. Ком. и Моск. Ком. ВКП(б)
МАКСИМ ГОРЬКИЙ
СССР
ПРИКАЗ
ВПЕРЕДЪ
Россійская Соціальдемократическая Рабочая Партія.
ЖЕНЕВА
ЛЕНИН
В издании газеты приняли участие объединенные литературные организации и группы Москвы:
УДАРНИ

treacherous scheme to enslave the U.S.S.R. and deliver it, bound hand and foot, to the imperialist vultures.
Stalin tore the mask from these despicable capitulators and exposed their Trotskyite-Menshevik souls.
The prime task of the Party, Stalin emphasized at the Fourteenth Congress, was to ensure a durable alliance between the working class and the middle peasantry for the construction of Socialism.
The Congress endorsed Socialist industrialization and the fight for the victory of Socialism in the U.S.S.R. as the fundamental task of the Party.
Shortly after the Congress, at the beginning of 1926, Stalin published his book, On the Problems of Leninism. In this historic work, he demolished the Zinovievite "philosophy" of liquidation and capitulation and proved that the policy adopted by the Fourteenth Party Congress, namely, the Socialist industrialization of the country and the construction of a Socialist society, was the only correct one. He armed the Party and the working class with an indomitable faith in the victory of Socialist construction.
The Bolshevik Party, mustering its forces and re-
ces, and brushing aside all capitulators and
led the country into a new historical phase
se of Socialist industrialization.
ht against the sceptics and capitulators,
Zinovievites, Bukharins and Kame-
88
J. V. Stalin and S. M. Kirov. Leningrad 1926
Photo
В. Сталин, С. М. Киров и Н. М. Шверник в Ленинграде в 1926 г.

DAVID KING

THE COMMISSAR VANISHES

THE FALSIFICATION OF PHOTOGRAPHS AND ART IN STALIN'S RUSSIA

PREFACE BY STEPHEN F. COHEN

PHOTOGRAPHS FROM THE DAVID KING COLLECTION

METROPOLITAN BOOKS

HENRY HOLT AND COMPANY
NEW YORK

Metropolitan Books
Henry Holt and Company, Inc.
Publishers since 1866
115 West 15th Street
New York, New York 10011
Metropolitan Books™ is an imprint of
Henry Holt and Company, Inc.

Published in Canada by Fitzhenry & Whiteside Ltd.
195 Allstate Parkway, Markham,
Ontario L3R 4T8.
All photographs and art are from the David King Collection, London.
Library of Congress Cataloging-in-Publication Data
King, David [*date*]
The commissar vanishes: the falsification of photographs and art in Stalin's Russia/ David King; preface by Stephen F. Cohen; photographs from the David King collection.—1st ed.
p. cm.
Includes bibliographical references and index.
ISBN 0-8050-5294-1 (alk. paper)
1. Photography—Soviet Union—History—20th century. 2. Photographs—Censorship—Soviet Union. 3. Photographs—Political aspects—Soviet Union. 4. Photographs—Forgeries—Soviet Union. 5. Photography—Retouching—Soviet Union. I. Title.
TR85.K55 1997 97-20832
303.3'76—dc21 CIP
Henry Holt books are available for special promotions and premiums.
For details contact: Director, Special Markets.
First Edition—1997
Designed by David King and Judy Groves
Printed in Hong Kong
All first editions are printed on acid-free paper.∞
3 5 7 9 10 8 6 4 2

Opposite: The first portrait of Joseph Stalin, dated April 1, 1922, was drawn by Nikolai Andreyev at the time of Stalin's appointment by Lenin to the post of General Secretary of the Central Committee of the Communist Party of the Soviet Union, a position he was to keep for more than thirty years.

Ominously, the future dictator scrawled over a print of the original: "This ear says that the artist is not well schooled in anatomy. J. Stalin. . . . The ear screams and shouts against anatomy. J.S."

We loved dipping our fingers in the inkwell filled with diluted soot
and were sometimes overzealous. I once inked out Comrade Kaganovich
himself because his name sounded like an exiled one to me.
I was lucky I was only eleven years old.
Sylva Darel, A SPARROW IN THE SNOW

Today a man only talks freely with his wife—
at night, with the blankets pulled over his head.
Isaak Babel

The "Index" grew longer and longer, and the scale of our *auto da fé*
grander and grander. We even had to burn Stalin's *On the Opposition.*
This too had become illegal under the new dispensation.
Evgenia Ginzburg, JOURNEY INTO THE WHIRLWIND

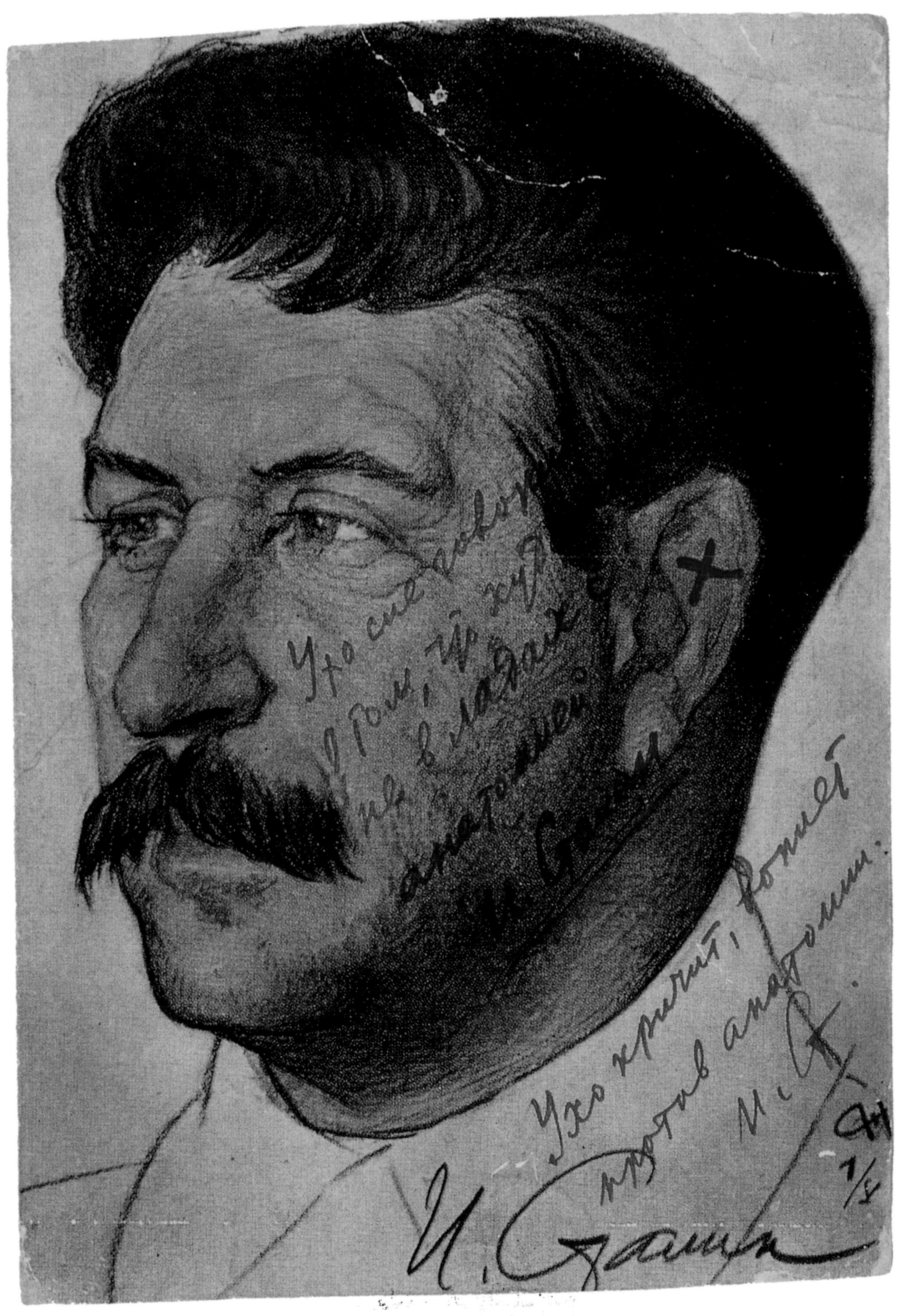
Ухо кричит, вопиет
против анатомии.
И. С.
И. Сталин

ИСКРА

PREFACE

Under Stalin's regime, which ruled the Soviet Union from 1929 to 1953, photographs lied. In David King's unique and revealing book, the same photographs, their original images restored, speak volumes of truth.

Stalinist censorship of photography was, of course, part of a much larger official purpose—the systematic falsification of history itself. From the mid-1930s, almost nothing of significance could be published, exhibited, or publicly uttered in the Soviet Union that failed to glorify every aspect of Stalin's leadership. The cult spanned the three defining chapters of Stalinism, each of which brought the deaths of millions of innocent people: his merciless "collectivization" war against the peasantry from 1929 to 1933; his murderous police terror against Communist officials and ordinary citizens that peaked in the late 1930s but continued until his death in 1953; and his catastrophic political and military maleficence before and after the German invasion in 1941.

Therein lay the calculus of falsification: The greater the Stalinist calamity inflicted on the nation, the more its architect needed to be extolled, and the less any alternatives could be admitted. Everything contrary to Stalin's cult was criminalized or expunged from history, especially all the non-Stalinist and insufficiently Stalinist Communists who had previously led the revolutionary party and new Soviet state. Thus, as David King shows us, were the commissars made to vanish. Why all this happened is still being passionately debated in Russia.

Nor did the consequences end with Stalin. Generations of Soviet children, now Russian adults, were taught his versions of history well into the 1980s. In the 1950s and early 1960s, Nikita Khrushchev, the first post-Stalin leader, did reveal part of the historical truth, but even that was largely re-falsified during the long reign of his successor, Leonid Brezhnev. Unlimited truth-telling began only in the late 1980s, when Mikhail Gorbachev unleashed a tidal wave of historical revelations in his attempt to de-Stalinize the Soviet system.

Considering the five decades of official falsification and zealous concealment, David King's book is heroic—the product of an immense, one-man archaeology. His thirty-year unearthing of pre-falsified Soviet photographs around the world has produced a legendary private collection long known to his many admirers, from history scholars in need of reliable photographs for their books to Hollywood filmmakers in search of visual authenticity. *The Commissar Vanishes* is only a representative sample of his collection, but it is ingeniously selected and presented to re-create a lost political world for both a specialist and general audience.

Readers will understand, I think, that behind all the suppressed and disfigured photographs lay countless family tragedies. They, too, linger on. Having written a biography of Nikolai Bukharin, a Soviet founding father tried and shot in 1938, I searched terror-era archives for years, urged on by his widow, Anna Larina, for even one of the many photographs taken of them in the 1920s and 1930s—without any success. Larina had lost everything when she was arrested shortly after Bukharin in 1937, and she died in 1996 without ever again seeing herself with her husband. Their son, Yuri Larin, who grew up in foster homes and orphanages, has never seen a photograph of his parents together.

A recent event, not unlike ones recounted by David King, saddened me even more. In the tenth year of our close friendship, the elderly daughter of a famous Soviet leader executed by Stalin suddenly handed me her mother's family album. As she studied my reaction intently, I opened the thick, leather-bound book, worn by decades of secrecy, with excitement. I expected to see unknown photographs of her father with other Communist luminaries of the pre-terror years. Some of them were there, at informal family gatherings, but her father had been crudely excised with scissors or a razor. Only his hand, shoulder, elbow, ear, or booted foot remained in a few photographs, left behind so as not to risk touching figures still in Stalin's favor.

I understood immediately that my friend's mother had done the mutilation. Overwhelmed by fear after the father's arrest, she had desperately tried to sanitize herself and her young daughter. My friend knew that I knew, but we spoke not a word about it. She had been devoted to her mother, who survived Stalin's Gulag and died at a very old age in 1989. The daughter kept her secret for sixty years. I interpreted her decision to show me the album as an act of trust. But I wished for her sake that her mother had spared at least one photograph, and that I could forget what I had seen.

Stephen F. Cohen, New York, April 1997

Opposite: "Comrade Stalin and Lado Ketskovali." Irin Shtenberg's quasi-religious drawing was made in 1939 as the Stalin cult reached its apotheosis. It shows the future "leader and teacher" (right) at the turn of the century with the man who is supposed to have taught him the principles of Marxism.

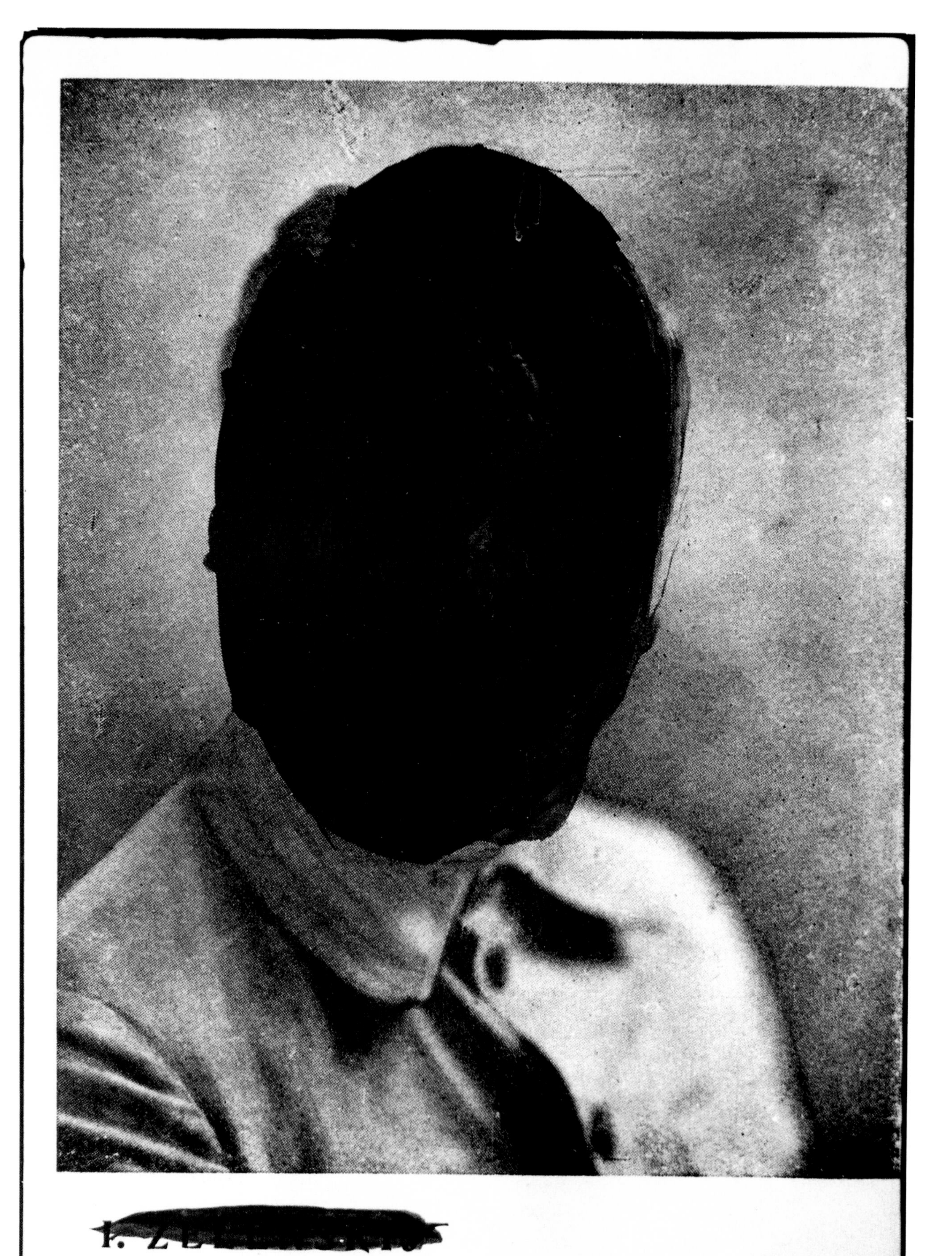

I. Z

VKP (ʙ) MQ Ortaasija ʙjurosiniꞑ 1924-nci jildan 1931-nsi jilgaca ʙolgan katiʙi

INTRODUCTION

HEAVY SOVIET LOSSES

Like their counterparts in Hollywood, photographic retouchers in Soviet Russia spent long hours smoothing out the blemishes of imperfect complexions, helping the camera to falsify reality. Joseph Stalin's pockmarked face, in particular, demanded exceptional skills with the airbrush. But it was during the Great Purges, which raged in the late 1930s, that a new form of falsification emerged. The physical eradication of Stalin's political opponents at the hands of the secret police was swiftly followed by their obliteration from all forms of pictorial existence.

Photographs for publication were retouched and restructured with airbrush and scalpel to make once famous personalities vanish. Paintings, too, were often withdrawn from museums and art galleries so that compromising faces could be blocked out of group portraits. Entire editions of works by denounced politicians and writers were banished to the closed sections of the state libraries and archives or simply destroyed.

At the same time, a parallel industry came into full swing, glorifying Stalin as the "great leader and teacher of the Soviet people" through socialist realist paintings, monumental sculpture, and falsified photographs representing him as the only true friend, comrade, and successor to Lenin, the leader of the Bolshevik Revolution and founder of the USSR. The whole country was subjected to this charade of Stalin-worship.

Soviet citizens, fearful of the consequences of being caught in possession of material considered "anti-Soviet" or "counterrevolutionary," were forced to deface their own copies of books and photographs, often savagely attacking them with scissors or disfiguring them with India ink. There is hardly a publication from the Stalinist period that does not bear the scars of this political vandalism, a haunting example of which can be seen on the opposite page. The grim but not unusual story of the picture's unfortunate subject is worth studying in detail.

Isaak Abramovich Zelensky joined the Bolshevik Party in 1906 at the age of sixteen and took an active part in the October Revolution of 1917. In 1922 he was elected a full member of the Central Committee of the Communist Party. As secretary of the Moscow Party organization, Zelensky served on the commission that arranged the burial of Lenin in 1924. In the fall of that year, however, Stalin attacked Zelensky for "insufficient hostility to Kamenev and Zinoviev," who were at that moment in opposition to the future dictator in the power struggle that followed Lenin's death. Zelensky was packed off to Tashkent for seven years to become secretary of the Central Asian Bureau but was recalled to Moscow in 1931 to run the state consumer distribution network. Zelensky was a conscientious, hardworking revolutionary turned party official, albeit one who, like many others, was deeply concerned about the advance of Stalinism.

In October 1937 Isaak Zelensky was arrested, on Stalin's orders, as an "enemy of the people." The state prosecutor of the time was the universally loathed and feared Andrei Yanuarievich Vyshinsky. Vyshinsky plumbed new depths of cruelty in the late 1930s, when he willingly acted as Stalin's mouthpiece in the three notorious Moscow show trials. At these trials he had many of the "Old Bolsheviks"—those who had created the Revolution that he never took part in—put to death. False confessions to ridiculous charges were extracted from the defendants by sadistic interrogators. Independent defense counsel was unheard of. Confession was sufficient to convict.

Opposite: Isaak Zelensky as he appeared in Alexander Rodchenko's copy of *Ten Years of Uzbekistan.*
Left: A detail from a composite of photographs published in 1928, showing Lenin in the center, flanked by Zelensky (left) and Stalin.

Soon Vyshinsky and Zelensky were to meet face-to-face as prosecutor and victim. The third and last Moscow show trial took place in the House of Trade Unions in March 1938, where Zelensky appeared alongside twenty other defendants, the most prominent of whom was Nikolai Bukharin. Vyshinsky accused Zelensky of having been a tsarist police agent since 1911. He was said to have used his position as head of the state distribution network to sabotage the distribution of food by "spoiling" fifty truckloads of eggs, as well as "throwing nails and broken glass into the masses' butter with a view to undermining Soviet health." With most of his co-defendants, Zelensky was sentenced to death and shot. For his endeavors at the trials, Vyshinsky was rewarded by Stalin with a seat on the Central Committee.

The haunting image of Zelensky's defaced photograph (as seen on page 8) surfaced nearly half a century later. One night in 1984, I made my way up Kirov Street, which was ill-lit even by Moscow standards, to the studio of Alexander Rodchenko. An archway led into a claustrophobic courtyard hemmed in by several tenement blocks. The studio was situated on the tenth story of one of these, and there was no elevator. I began the long climb up.

Rodchenko was one of the heroes of Russian art, design, and photography from the avant-garde period at the time of the Revolution until the late 1930s. He was married to the equally gifted artist and designer Varvara Stepanova. During the 1920s their studio served as the editorial offices of *Lef* magazine, an arts journal under the editorship of the revolutionary poet Vladimir Mayakovsky.

Rodchenko died in 1956. Three generations of his family continue to reside in the apartment, and by 1984 very little had changed. Paintings by the master leaned against the walls as if he had just finished working on them. The dim light of naked bulbs cast dense shadows from tall cupboards and bookcases. The same dust, Rodchenko's dust, occupied the same crevices and the tops of the same books. I was there to see the books—strangely, the first person to have asked to see them.

Rodchenko had divided his working life in the 1930s between photography and his remarkable output of book and magazine design. Huge photographic albums, with titles like *First Cavalry* and *Red Army*, lined the bookshelves of the Kirov Street studio, side by side with pioneering photo-graphics for special issues of the famous magazine *USSR in Construction*. But one book stood out from the others. It was called *Ten Years of Uzbekistan*.

Looking inside Rodchenko's copy of *Ten Years of Uzbekistan* was like opening the door onto the scene of a terrible crime. A major purge of the Uzbek leadership by Stalin in 1937, three years after the book's publication, meant that many of the official portraits of Party functionaries in the album had to be destroyed. The concept of "personal responsibility" had been forced on the whole country by the Stalinists during a vast campaign of vigilance against the regime's enemies. The names of those who had been arrested or had "disappeared" could no longer be mentioned, nor could their pictures be kept without the greatest risk of arrest. Petty informers were everywhere. The walls really did have ears.

Rodchenko's response in brush and ink came close to creating a new art form, a graphic reflection of the real fate of the victims. For example, the notorious secret-police torturer Yakov Peters (page 133) had suffered an ethereal, Rothko-like extinction. The face of party functionary Akmal Ikramov, veiled in ink, had become a terrifying apparition (page 129). And there, suffering a second death, was Isaak Zelensky, his face wiped out in one great blob and his name obliterated in the caption beneath.

This defacing, forced upon Rodchenko, is only one example among thousands of similar actions from the Great Terror and beyond. The libraries of the former Soviet Union still bear these scars of "vigilant" political vandalism. Many volumes—political, cultural, or scientific—published in the first two decades of Soviet rule had whole chapters ripped out by the censors. Reproductions of photographs of future "enemies of the people" were attacked with disturbing violence. In schools across the country, children were actively engaged by their teachers in the "creative" removal of the denounced from their textbooks. A collective paranoia stretched right through the period of Soviet rule.

Huge numbers of publications were banned from the bookshelves altogether. For example, one directive, issued by the Central Committee of the Communist Party on March 7, 1935, ordered the removal of Leon Trotsky's works from libraries throughout the Soviet Union. This ban continued until the late 1980s, but sometime in between it was toughened up to include even some *anti*-Trotsky material. Publications with titles like *Trotskyists: Enemies of the People* and *Trotskyist-Bukharinist Bandits* also became proscribed reading.

The censors published an extraordinary volume entitled *Summary List of Books Not Available in Libraries and the Book Trade Network*. It contained hundreds of pages in small print, "for official use only," listing alphabetically the publications that were banned. A friend of mine, the manager of an antiquarian bookshop in Leningrad in the 1960s, told me that he remembered well the twice-monthly visits of a matronly lady from the censorship bureau, who spent hours rifling through the thousands of books on his shelves, checking them against her latest copy of the *Summary List* (which was always being updated). Those volumes found to be unacceptable were put in a special garbage can at the back of the store.

Opposite: Lenin (center) flanked by Lev Kamenev (left) and Grigorii Zinoviev (right), with delegates to a session of the All-Russian Central Executive Committee in Moscow, October 31, 1922. What makes the photograph remarkable is the presence of Andrei Vyshinsky (center, below Lenin, with three white buttons on the collar of his black shirt), who had been a member of the Bolshevik Party for less than two years. During the next two decades, Vyshinsky would have most of these comrades put to death.

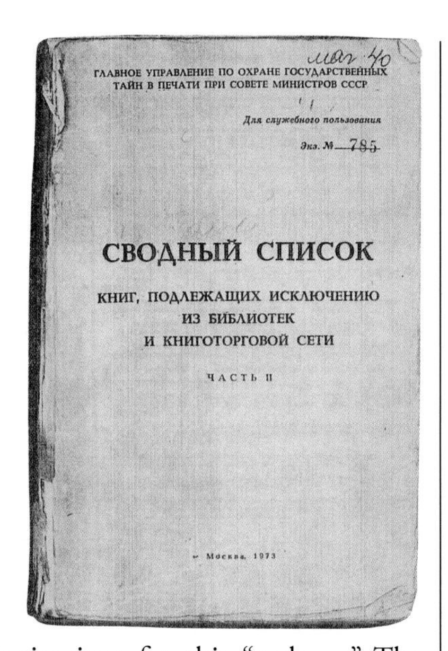

Far left: An edition of the *Summary List*, published in 1973.
Left: A report on the famous Shakhti trial of engineers in 1928. The introduction was written by Vyshinsky. Cover design by Rodchenko.

There were three possible destinations for this "garbage." The happiest one was when rare and interesting volumes found their way quietly into the many fabulous private libraries assembled, sometimes at great personal risk, by bibliophiles or lovers of history. The unhappiest destination was the shredder. So many beautifully produced books and rare manuscripts ended up there during the 1960s because of the boorish campaign of *Makulatura* (book pulping). Ostensibly due to a shortage of newsprint, a system was introduced whereby old books and papers would be weighed and exchanged at a fixed rate for a few rubles or, say, a new copy of an officially approved novel. The heavier the book, the greater the value. This is why many of the giant photographic and graphic albums published in Russia during the 1920s and 1930s are now so scarce.

The third route for material that was considered a threat to Soviet rule led to the official archives. These archives served a dual purpose for the state. The first was to preserve, the second to banish. Sound recordings, the printed word, movie footage, photographs, paintings, drawings and posters, personal effects—any trace of the "disappeared"—vanished into the "closed sections." Whatever the state considered undesirable was withdrawn from view. In Stalin's time, this was a monumental task. Inquiries were met, after hours of waiting, with deflected answers, a shrug of the shoulders, or a harsh "Not possible!"

My first encounter with the open sections of the photographic archives in Moscow took place in the exceptionally bitter Russian winter of 1970, seventeen years after Stalin's death. When I inquired about photographs of Trotsky, the reply would invariably be, "Why do you ask for Trotsky? Trotsky not important in Revolution. Stalin important!" In the dark green metal boxes containing mug shots of subjects starting with "T" were hundreds of photographs of famous Russians: Tolstoy, Turgenev, etc.—but no Trotsky. They had completely wiped him out. It was at this moment that I determined to start my collection.

Nearly three decades later the collection has grown into a working library that documents visually every important aspect of Soviet history, with particular reference to alternatives to Stalinism. Where did all this material come from? Initially from the Soviet Union itself. The worldwide dissemination of photographs and printed matter through the Communist International was a major objective of the Soviet propaganda machine in the 1920s and 1930s. Vast amounts of material reached the West in those times and can still be found even today. With the relaxation by Gorbachev of foreign travel restrictions for Russians in the 1980s, large numbers of books, photographs, and other documents, hidden for many years, arrived in the West. But it is worth pointing out that it is still an offense under Russian law to export almost anything without official permission, except recent publications.

So much falsification took place during the Stalin years that it is possible to tell the story of the Soviet era through retouched photographs. That is the purpose of this book. The photographs are displayed chronologically, at the time they were taken, rather than when they were doctored. The altered versions are usually shown alongside the originals, or on the following pages. A number of key unfalsified photographs and documents are also included to explain important moments in the story. Paintings, graphics, and other examples of Stalinist hero worship appear, as well. Only the most interesting and varied images from a political, cultural, and of course visual point of view are presented here. New examples of falsification are always coming to light. A photograph might appear strange, as a result of heavy retouching. To find the original might take years—and often does. The search continues.

Left: Embossed-metal cover for a 1939 biography of Stalin, based on Sergei Merkurov's 100-foot-tall granite sculpture, which stood at the entrance to the Moscow-Volga canal.

Photographic retouching for publication in books, magazines, and newspapers in Russia started as early as 1917 but did not reach a grand scale until 1935, in the wake of the terror that followed the assassination of Leningrad Party chief Sergei Kirov. No single skyscraper existed where legions of Stakhanovite airbrushers, montagists, and scissormen would labor through the darkest hours of the night, slavishly fulfilling their work norms for some glowering Ministry of Falsification. Rather, photographic manipulation worked very much on an ad hoc basis. Orders were followed, quietly. A word in an editor's ear or a discreet telephone conversation from a "higher authority" was sufficient to eliminate all further reference—visual or literal—to a victim, no matter how famous she or he had been.

Faking photographs was probably considered one of the more enjoyable tasks for the art department of publishing houses during those times. It was certainly much subtler than the "slash-and-burn" approach of the censors. For example, with a sharp scalpel, an incision could be made along the leading edge of the image of the person or object adjacent to the one who had to be removed. With the help of some glue, the first could simply be stuck down on top of the second. A little paint or ink was then carefully brushed around the cut edges and background of the picture to hide the joins. Likewise, two or more photographs could be cannibalized into one using the same method. Alternatively, an airbrush (an ink-jet gun powered by a cylinder of compressed air) could be used to spray clouds of ink or paint onto the unfortunate victim in the picture. The hazy edges achieved by the spray made the elimination of the subject less noticeable than crude knifework.

Many photographic deletions were not the result of retouching at all but of straightforward cropping. Art departments have always cropped photographs on aesthetic grounds, but in the Soviet Union cropping was also used with political objectives in mind. The subtraction of Stalin's enemies, and even some of his friends, was one problem, but for the General Secretary, addition—the addition of himself—was another. From the time of his birth in 1879 until he was appointed General Secretary in 1922, there probably exist fewer than a dozen photographs of him. For a man who claimed to be the standard-bearer of the Communist movement, this caused grave embarrassment, which could only be overcome by painting and sculpture. Impressionism, expressionism, abstraction—for Stalin, none of these artistic movements was capable of showing his image properly. So he made realism—socialist realism—the central foundation of the Stalin cult. A whole art industry painted Stalin into places and events where he had never been, glorifying him, mythologizing him. Sculpture worked well for him, too. The bronze Stalin, the marble Stalin, were invulnerable to the bullets of the "Zinovievite bandits." The flesh and blood Stalin could safely stay out of the public gaze. Sculpture became the real Stalin—heavy, ponderous, immortal.

Skillful photographic retouching for reproduction depended, like any craft before the advent of computer technology, on the skill of the person carrying out the task and the time she or he had to complete it. But why was the standard of retouching in Soviet books and journals often so crude? Did the Stalinists want their readers to see that elimination had taken place, as a fearful and ominous warning? Or could the slightest trace of an almost vanished commissar, deliberately left behind by the retoucher, become a ghostly reminder that the repressed might yet return?

David King, London, 1997

Overleaf: "Stalin and the Masses," 1930, published in a book entitled *Udarniki* (*Shockworkers*). The figure of Stalin is clearly photomontaged onto the picture of the masses. But the masses themselves are also montaged; close inspection reveals that some sections of the crowd appear three times.

A meeting of the St. Petersburg Union of Struggle for the Liberation of the Working Class in February 1897. Shortly after the photograph was taken, the whole group, which had been founded by Lenin and Yuli Martov, was arrested by the Okhrana (tsarist secret police) and sentenced to three years' exile in Siberia. When Martov died from tuberculosis in 1923, exiled in Berlin, he was publicly mourned by the Bolsheviks as their "most sincere and selfless opponent."
Left to right, standing: Alexander Malchenko, Petr Zaporozhets, Anatolii Vanayev. Seated: V. V. Starkov, Gleb Krzhyzhanovsky, Vladimir Ulyanov (Lenin), and Yuli Martov.

Martov survives in this version of the photograph, published in 1939, but Malchenko (standing on the left in the original) has disappeared. At the time the picture was taken, he was an engineering student, and his mother sometimes sheltered Lenin from the St. Petersburg police. Returning from exile in 1900, Malchenko seems to have abandoned revolutionary politics. He moved to Moscow, where he worked as a senior engineer in various state departments after the Revolution. In 1929 he was arrested, wrongfully accused of being a "wrecker," and executed on November 18, 1930. For nearly thirty years, Malchenko was airbrushed out of the photograph whenever it was reproduced. He was rehabilitated in 1958, at which point his presence was allowed to reappear.

A GHOSTLY PILLAR

Right: A game of chess in Capri, Italy, April 1908. Lenin was spending a week's vacation with the writer Maxim Gorky (center, hand on chin), who supported the Bolsheviks and lived in exile on the island from 1906 to 1913.

From left to right, standing: Vladimir Bazarov, Gorky, Zinovii Peshkov, Natalya Bogdanova. Seated: Ivan Ladyzhnikov, Lenin, and Alexander Bogdanov.

Several of those pictured here would later come into conflict with Lenin or Stalin. Gorky opposed the seizure of power in 1917 and formed a non-Bolshevik left-wing group. Bazarov, a former Bolshevik theoretician turned Menshevik, joined the group and wrote articles criticizing the Bolsheviks' plans for insurrection. A mild and yielding personality, Bazarov was arrested and tried with other Mensheviks in an infamous trial in 1931. He was shot in 1937. Zinovii Peshkov (behind Gorky and Bogdanova) was the elder brother of Yakov Sverdlov, the first head of the new Bolshevik state following the October 1917 Revolution. Peshkov became a Jewish convert to Christianity. He looked upon Gorky as a father figure and even took his name (Gorky's real name was Alexei Maximovich Peshkov). Zinovii Peshkov emigrated to France and became a high-ranking officer in World War I, losing an arm at Verdun. Following a spell in the French Foreign Legion, he became a military adviser to Chiang Kai-shek in China in the 1930s. He made friends with General de Gaulle and worked in the Deuxième Bureau (the French secret service) after the war. He died in France in 1966, aged eighty-two.

Lenin's chess opponent in the photograph, the Bolshevik Alexander Bogdanov, was soon to become his political opponent, too. A year later, in 1909, Bogdanov cofounded a school in Capri proposing socialism as a kind of religious experience—a "religion of labor." Lenin's response was "frenzied with indignation." He had Bogdanov expelled from the Bolshevik Party. After 1917 Bogdanov's ideas on proletarian culture (Proletcult) became highly influential. He was arrested for a short time in 1923 as a member of the Workers' Opposition. He founded the first blood-transfusion clinic in the Soviet Union, but died in 1928 as a result of an unsuccessful experiment on himself. In the photograph, his wife Natalya stands behind him. Gorky noted that Lenin, who lost the game, "grew angry and despondent."

Left: Cover of the report on the "Trial of the Counterrevolutionary Organizations of Mensheviks," Moscow, 1931.

Above: Two years after his execution, Vladimir Bazarov was cut off and airbrushed out of the left-hand side of the photograph on the previous page, when it appeared in a giant album entitled *Lenin* (Moscow 1939). Also erased was Zinovii Peshkov, who had been standing behind Maxim Gorky and Natalya Bogdanova.

Opposite: In this version of the picture, which was published in 1960 by the Moscow Institute of Marxism-Leninism, Peshkov has been allowed to reappear. Bazarov is still missing, his place taken by a ghostly pillar. There is no explanation for the appearance, disappearance, and reappearance of a woman's skirted knee in the foreground of the three versions of the photograph.

SUMMER IN SIBERIA

Stalin attending a meeting of Bolshevik exiles in Monastyrskoye village, Turukhansk, Siberia, in the summer of 1915, two years before the Revolution. The Bolsheviks had convened this conference to discuss the trial of their five members of the Fourth State Duma (parliament), who had the previous winter been arrested and exiled for protesting against World War I. Standing in the back row are Suren Spandarian (second from left), Stalin (wearing wide-brimmed black hat), Lev Kamenev (with mustache), Grigorii Petrovsky (hat and beard), Yakov Sverdlov (beard and white shirt), and Filipp Goloshchekin (leather jacket). Seated left to right are Fedor Samoilov, Vera Schweitzer (Spandarian's wife), Alexei Badaev, and Nikolai Shagov.

Stalin's gruff and monosyllabic approach contributed little to the conference, in contrast to the spirited and combative Armenian Marxist Spandarian, who died of a severe illness a year later. In 1937 his widow, writing her memoirs of the time, was made to exaggerate Stalin's role in exile and play down that of her late husband.

Stalin had been in exile in the Turukhansk region since 1913 without once making an attempt to escape. He astonished his fellow deportees there by his total lack of principles, his slyness, and his exceptional cruelty.

Another print (*above*) of the previous photograph, found in the Central State Archives in Moscow, obliterates Kamenev from the center of the picture. Kamenev and his friend Grigorii Zinoviev had been among Lenin's closest lieutenants in spite of condemning the impending October Revolution as an "act of despair." After Lenin's death in 1924, Kamenev was expelled several times from the party, and spent a period of exile in Siberia. He was sentenced to five years in jail in 1935 following the assassination of Stalin's designated successor, Sergei Kirov. He was retried in the first Moscow trial of August 1936, sentenced, and shot. Both his wives—he had first been married to Trotsky's sister, Olga Davidovna—and his two eldest sons were also killed in the purges.

Opposite: Another version of this photograph (*top*) was reproduced in the giant commemorative album *Stalin* published in 1939 to celebrate the tyrant's sixtieth birthday. This time five more members of the group have been replaced by vegetation, fencing, and a better view of the log cabin.

Only Stalin remains (*below left*) in this soft-focus, cropped version of the original photograph when it appeared in another portfolio of 1939, also called *Stalin.*

Never published by the Soviet regime, an alternative and unretouched photograph from the same group (*below right*) lay in the Central State Archives for seventy-five years.

Top: Soldiers demonstrating in the Liteiny Prospekt during the first days of the February Revolution, Petrograd, 1917. Thousands of souvenir postcards like this one were published during the great upheavals of 1917. In the background can be seen a jeweler's shop with its signboard, "Watches—gold and silver." The slogan on the soldier's flag is almost illegible. *Above:* The same photograph, also published as a postcard in the same year, but now the jeweler's signboard has been crudely replaced with the slogan "Struggle for Your Rights!" The soldier's flag has been transformed from black to white and reads "Down with the Monarchy—Long Live the Republic!"

LENIN IN DISGUISE

This famous and unusually glamorous photograph of Lenin (*below*) in the disguise of a farm laborer, with wig and without beard, was taken at the Razliv Station on the Gulf of Finland, July 29, 1917. Lenin was on the run from Kerensky's secret police and needed the photograph for a passport to enter Finland. He lived with Zinoviev in a hut belonging to a worker called Nikolai Emelianov in a field near the station for most of that July. It was here that he wrote his tour de force *State and Revolution*, a strategy for the destruction of all bourgeois institutions. In 1927 the humble fisherman's hut was rebuilt as a granite shrine. Helping Lenin during the Revolution did not guarantee longevity, however. Emelianov's whole family was arrested in 1935, his two

eldest sons were shot and he spent the next twenty years in the Gulag (forced-labor camps).

Opposite: Another photograph taken at the same time as the one above shows Lenin in a far less flattering light and seems to have slipped into print only once, in a book called *Lenin and the Art of Photography*.

A GRAY BLUR

The huge reception organized by the Bolsheviks for Lenin on his return to Russia on the night of April 16, 1917, was graphically described by the distinguished chronicler of the Russian Revolution, Nikolai Sukhanov. "The worldwide socialist revolution has already dawned!" Lenin told the crowd as a thunderous "Marseillaise" was struck up by a band at the Finland Station in Petrograd.

Mikhail Sokolov's painting (*right*), made twenty years later, draws heavily on Sukhanov's eye-witness account of the momentous occasion, but is flawed by one glaring inaccuracy: the figure of Stalin in the doorway of the train, behind Lenin. Stalin did not travel on the train with Lenin, nor was he delegated by the Executive Committee of the Petrograd Soviet to meet him at the station. That task had been given to the Presidium, of which Stalin was not a member.

Sukhanov described Stalin's activity in 1917 as "a gray blur, dimly looming up now and again but not leaving any trace." Years later, at the height of his power, Stalin felt the need to rectify this most embarrassing image problem. He mobilized an army of propagandists, artists, and writers to fabricate material depicting him as the decisive organizer of the Revolution, Lenin's closest collaborator and natural successor, a great orator, and an invincible commander in the Civil War. Meanwhile Sukhanov was arrested in 1931 by the OGPU (secret police), released, and then rearrested in 1939. He died in the Gulag in 1940.

Below: The editors of the famous Russian newsmagazine *Ogonyek* didn't regard the October Revolution as important enough at the time to feature it on their cover, although it was extensively reported inside. Postrevolutionary puritanism ensured that the cover did not see the light of day again. This copy was found in London.

Overleaf, left-hand page: Evgenii Kibrik's odious contribution to Stalinist falsification in art, perpetrated in charcoal in 1947, was entitled "Lenin Arrives at the Smolny during the Night of October 24." The Smolny Institute in Petrograd was the headquarters of the Bolsheviks during the October Revolution. It was from here that Trotsky, as chairman of the Military Revolutionary Committee, directed the insurrection. Stalin was not even there at the time.

Overleaf, right-hand page: In a steel engraving from 1936 by Petr Staronosov, who fittingly became known for his insipid illustrations of fairy tales, a confident Stalin is shown in front of a giant map planning the rout of the counterrevolutionary White armies. Lenin can only look on in dumb bewilderment.

ФЖД.
130052.
III КЛАС.

КАРТА
МОСКВА
СТАРОНОСОВ

PROBLEMS OF PROPAGANDA

Lenin understood well the power of the motion picture for propaganda purposes, and from its earliest days Soviet cinema was highly regarded throughout the world. But there was a lack of documentary footage covering the major revolutionary events. The 1905 and 1917 revolutions went largely unrecorded. Film stock was in short supply. Cameras were unwieldy. Slow film speeds made dramatic movement at night difficult to capture.

The two famous photographs (*left*), which became icons of the revolutionary era not only in the Soviet Union but also in the West, are both falsifications. The photograph (*top*) ostensibly depicts the events of January 9, 1905, when a priest called Father Gapon led thousands of peaceful demonstrators through the streets of St. Petersburg to the Winter Palace to petition Tsar Nicholas II for a constitution. They were met by a hail of bullets. Hundreds were killed and wounded. "Bloody Sunday" sparked off a year of revolt against the autocracy. In reality, no cameras were at the scene. The picture is a still taken from a 1925 feature film directed by Vyacheslav Viskovsky, entitled *The Ninth of January*.

Likewise, the dramatic photograph (*bottom*) passed into history as a record of the storming of the Winter Palace in October 1917. In fact, neither footage nor photographs exist of the "storming," as it took place at night and with a certain amount of stealth. This photograph is actually of one of the many annual street-theater reenactments staged by the Bolsheviks. Artists like Yuri Annenkov and Natan Altman were commissioned to help stage-manage the proceedings.

Feature-film directors often ran into trouble with Stalin, who from the mid-1920s closely followed all their productions. Sergei Eisenstein's masterpiece *October* had been commissioned to celebrate the tenth anniversary of the Revolution. A special preview was screened for Stalin, who immediately ordered all references to Trotsky excluded from the film, which had to be heavily reedited. Stills from the film became a great source of "documentary" photographs. *Opposite:* "October Night," the 1917 Revolution glamorized in paint by the artist Rudolf Frentz. Made before 1932, the work is more expressionistic than the socialist realist style that was to follow.

ОКТЯБРЬСКАЯ НОЧЬ

EXTRAORDINARY COMMISSION

N. Tolkunov's painting *Handing over to F. E. Dzerzhinsky the Resolution of the Soviet of People's Commissars on December 7, 1917, for the organization of the Cheka* is a classic example of 1950s socialist realism. The subject matter glorifies a hierarchical leadership, and the nineteenth-century style of painting leaves no room for creative expression or development. Whether or not the puffed-up figure of Stalin (third from the left) was present at that fateful moment is unknown, as Lenin's new secret police was born in secret and only admitted to in 1958. Initially Dzerzhinsky packed the Cheka (the acronym stands for "Extraordinary Commission to Fight Counterrevolution and Sabotage") with non-Russians, especially Latvians, who were considered disciplined and not at all squeamish. The seated figure is Yakov Sverdlov, first head of the Bolshevik state. Sverdlov died of pneumonia in Moscow in March 1919. Years later, his private safe in the Kremlin was unlocked by the NKVD to reveal, puzzlingly, a horde of gold, silver, diamonds, and faked passports. Sverdlov also left a son, André, who joined the NKVD under Lavrentii Beria and became one of the most vicious interrogators during the purges at the end of the 1930s. *Opposite:* Moisei Nappelbaum's portrait of Felix Edmundovich Dzerzhinsky, 1919.

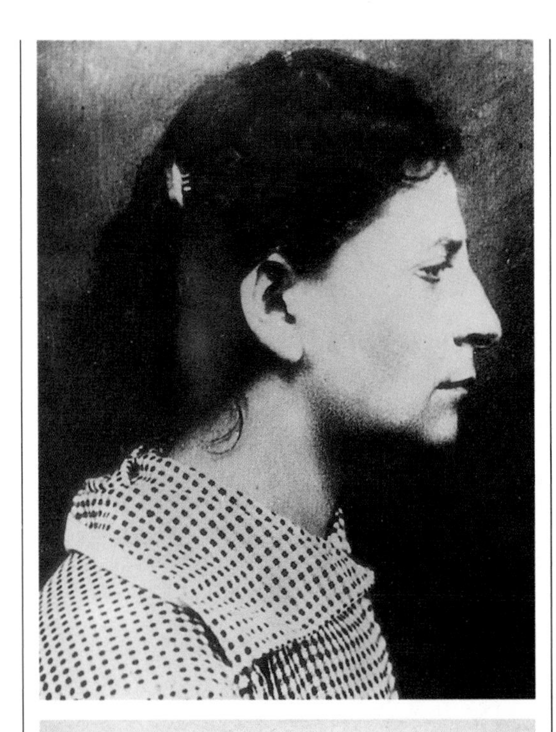

ВСѢМ СОВѢТАМ

РАБОЧИХ, КРЕСТ., КРАСНОАРМ. ДЕПУТ.

ВСѢМ арміям, ВСѢМ, ВСѢМ, ВСѢМ.

Нѣсколько часов тому назад совершено злодѣйское покушеніе на тов. ЛЕНИНА. Роль тов. Ленина, его значеніе для рабочаго движенія Россіи, рабочаго движенія всего міра извѣстны самым широким кругам рабочим всѣх стран.

ПРИЗЫВАЕМ ВСѢХ ТОВАРИЩЕЙ К ПОЛНѢЙШЕМУ СПОКОЙСТВІЮ, К УСИЛЕНІЮ СВОЕЙ РАБОТЫ ПО БОРЬБѢ С КОНТРРЕВОЛЮЦІОННЫМИ ЭЛЕМЕНТАМИ.

НА ПОКУШЕНІЯ, НАПРАВЛЕННЫЯ ПРОТИВ ЕГО ВОЖДЕЙ, РАБОЧІЙ КЛАСС ОТВѢТИТ ЕЩЕ БОЛЬШИМ СПЛОЧЕНІЕМ СВОИХ СИЛ, ОТВѢТИТ БЕЗПОЩАДНЫМ МАССОВЫМ ТЕРРОРОМ ПРОТИВ ВСѢХ ВРАГОВ РЕВОЛЮЦІИ.

Товарищи! Помните, что охрана ваших вождей в ваших собственных руках. Тѣснѣе смыкайте свои ряды и господству буржуазіи вы нанесете рѣшительный, смертельный удар. Побѣда над буржуазіей—лучшая гарантія, лучшее укрѣпленіе всѣх завоеваній октябрьской революціи, лучшая гарантія безопасности вождей рабочаго класса.

Спокойствіе и организація! Всѣ должны стойко оставаться на своих постах! Тѣснѣе ряды!

Предсѣдатель В. Ц. И. К. Я. Свердлов.

30 Августа 1918 года. 10 час. 40 мин. вечера.

TERRIBLE RETRIBUTION

An unretouched photograph of Lenin (*right*) leaving a meeting of the First All-Russian Congress on Education in Moscow, August 28, 1918. Two days later, also in Moscow, he was gunned down while leaving a meeting of factory workers. He was severely wounded in the neck and arm. This was the last photograph taken of Lenin before he was shot.

A Socialist Revolutionary (SR), Fanya Kaplan (*top left*), was arrested for the crime. "I shot Lenin because I believe him to be a traitor . . . he postpones the idea of socialism for decades to come," she is said to have told her interrogators.

The official announcement of the attempted assassination, in the form of a broadsheet poster (*center left*), was written by Yakov Sverdlov. It is dated August 30, 1918, at 10:40 P.M. and blames "Right-wing SRs, hirelings of the English and French," although the suspect's interrogation had yet to begin. It goes on, "The working class will respond to attempts against its leaders with even greater consolidation of its forces, with merciless mass terror against all the enemies of the Revolution."

Some hours before the attempt on Lenin's life, the head of the Petrograd Cheka, Moisei Uritsky, was assassinated. Although these attacks were later found to be unconnected, they brought terrible retribution from the Bolsheviks in the form of the Red Terror. Hundreds of suspects were rounded up and shot. A senior Chekist, Yakov Peters, who led the interrogation of Fanya Kaplan, stated, "Anyone who knows the meaning of Red Terror will grow pale." Peters himself later suffered at the hands of the GPU.

The order was given for Kaplan to be shot without trial. The Kremlin commandant Pavel Malkov (*bottom left*) was directed to carry out the order. Years later Malkov proclaimed: "The execution of a person, especially a woman, was no easy matter. It was a heavy, very heavy responsibility." Malkov was given instructions to shoot Kaplan in a parking lot with an automobile engine running to drown out the screams. The body was to be disposed of without a trace. The order was carried out at 4 A.M. on the morning of September 4, 1918.

Doubt has recently been cast about whether Kaplan really was guilty, since her extremely poor eyesight would have made her a rather unlikely assassin.

СЕЗД

IS IT A PLANE?

Lenin's first public appearance after the attempt on his life was to attend the unveiling of a memorial to Marx and Engels in Revolution Square, Moscow, on November 7, 1918. The unretouched photograph of this event (*right*) was only published in full in 1970.

A retouched version (*above*) was published in the 1930s, confusingly captioned "V. I. Lenin, Y. M. Sverdlov, and V. A. Avanesov watching the flight of an airplane at a demonstration on November 7, 1918." Varlam Avanesov, a high-ranking officer in the Cheka, had two months previously given the order for Fanya Kaplan to be shot without trial.

Another retouched version (*top*) of the original picture was published in 1939 by the Central Lenin Museum in a folder of photographs on the life of Lenin. Avanesov, who died in 1930, has now disappeared from view, leaving only Lenin and Sverdlov.

LENIN'S MAD DOG

Grigorii Zinoviev (*opposite*, holding hat) is shown seated next to Yakov Sverdlov (striped tie) with delegates to the Fifth All-Russian Congress of Soviets in Petrograd, July 1918. Zinoviev has been defaced, as an enemy of the people, in this photograph, which was recently found in Soviet archives.

Although he was regarded by the Mensheviks before the Revolution as "Lenin's mad dog," Zinoviev objected, along with Lev Kamenev, to the policy of armed insurrection in October 1917. "Traitors and strikebreakers of the revolution," Lenin called them. After Uritsky's assassination and Lenin's close call, Zinoviev, as chairman of the Petrograd Soviet, was thrown into blind panic. He feared that the end of the revolution was at hand and responded with terrible cruelty against "bourgeois elements." Launching the Red Terror in Petrograd, he declared: "We must carry along with us 90 million out of the 100 million of Soviet Russia's inhabitants. As for the rest, we have nothing to say to them. They must be annihilated."

Zinoviev was expelled from the Communist Party in 1927 for his opposition to Stalin and was twice exiled. He was one of the chief defendants in the Moscow show trials and was shot in 1936.

The woman in the white hat is Rosa Berzina. She was the wife of the prominent Latvian Communist Yan Antonovich Berzin, who held a number of ambassadorial posts in the 1920s. Berzina's exact fate is unknown, but her husband died in the Gulag in 1938.

GROSS VILIFICATION

As well as being dictator of Petrograd and the whole of northwestern Russia, Grigorii Zinoviev was head of the Comintern (Communist International) from 1919 until 1926. He is seen (*above*) in tense discussion with Lenin on the world-revolutionary situation at a meeting in the Moscow Kremlin in 1920. On the right of the picture can be seen Giacinto Serrati and Angelica Balabanova. Serrati, the Italian socialist leader, was sympathetic to the Communists but objected to their splitting his party for their own ends. He was met with a terrible campaign of vilification and expelled from the Comintern. Balabanova came to Serrati's defense. Zinoviev later justified Serrati's treatment by saying, "We have fought and slandered him because of his great merits. It would not have been possible to alienate the masses from him without resorting to these means." Balabanova emigrated in 1921.

Left: When the same photograph was published in Moscow fifty years later, Zinoviev and Balabanova had gone, leaving only Lenin, Serrati, and the candlestick.

LENIN'S HOLY STATUS

Right: A photograph of the Council of People's Commissars (Sovnarkom) taken to record the return to work of its president, Lenin, following the assassination attempt two months earlier. His survival in the face of death gave him mythic, almost holy, status among his supporters. The picture was taken in the Moscow Kremlin on October 17, 1918, by Petr Otsup, one of the foremost reportage photographers of the revolutionary period. Many of Lenin's comrades in the photograph, including Leon Trotsky (seen at the back of the room with light reflecting from his spectacles), were subsequently killed by Stalin.

Above: In 1970, when the photograph was published in an album on Lenin, only three of the original thirty-three commissars could be shown with him. The rest remained out of favor well into the Gorbachev era.

Two of the "survivors" can be identified. Lev Karakhan (black beard) was a Bolshevik diplomat, killed in the purges of 1937. Petr Stuchka (back left) was Lenin's commissar of justice. He died in 1932 and was buried in the Kremlin Wall. That did not save him in Stalin's eyes; in 1937 he was posthumously purged as an enemy of the people and a deliberate wrecker in the field of jurisprudence. It was Stuchka who, in June 1918, had legalized capital punishment in the new Soviet state.

дных Комиссаров".

Above: Delegates to the Eighth Congress of the Bolshevik Party, Moscow, March 1919. Lenin told the congress that the existence of the Soviet state side by side with the imperialist states was inconceivable: "In the end, one or the other must conquer." For seventy years, this photograph was not shown in its entirety in the Soviet Union. Out of the twenty delegates identified, eleven were killed by Stalin and three more (Mikhail Tomsky, Adolf Yoffe, and Mikhail Lashevich) committed suicide in protest against his policies. The photograph shows (front row, left to right): Ivan Smilga, Vasily Schmidt, and Sergei Zorin. Middle row, from the left: Grigor Yevdokimov, Stalin, Lenin, Mikhail Kalinin, and Petr Smorodin. Top row: Pavel Malkov, Eino Rahja, Said Galiyev, Petr Zalutsky, Yakov Drobnis, Tomsky, Moisei Haritonov, Yoffe, David Ryazanov, Alexei Badaev, Leonid Serebryakov, and Lashevich. Kalinin had just been made chairman of the Central Executive Committee after the death of Sverdlov.

Opposite, top to bottom: The most famous retouched version of the photograph showed only Stalin, Lenin, and Kalinin as a ruling triumvirate. Occasionally Kalinin was made to disappear, falsely suggesting Stalin's extreme closeness to Lenin. Finally, when reproduced in the massive *First Cavalry* album designed by Alexander Rodchenko in 1938, even Lenin has gone.

OLD BOLSHEVIKS

THE COMMISSAR VANISHES

Right: Lenin and Trotsky (center, at top of stairs), the co-leaders of the Russian Revolution, celebrating its second anniversary in Red Square, Moscow, on November 7, 1919. The photograph was taken by L. Y. Leonidov. A heavily retouched version (*top* and *above*, detailed) was published in *V. I. Lenin in the Art of Photography* in Moscow in 1967. Trotsky has been airbrushed out and so has Kamenev (leather cap and glasses, to the left of Lenin in the original).

The man with the black beard (standing center right in front of Trotsky in the unretouched version) has also been eliminated. He was Artashes Khalatov, a Georgian who joined the Bolsheviks in 1917 and became a commissar of Soviet publishing. When Stalin started his assault on the party cadres in 1937, Khalatov perished with thousands of other talented officials.

In front of Khalatov is Maxim Litvinov (hands in pockets). He joined the revolutionary movement in 1898 and became one of the Bolsheviks' chief organizers. He smuggled guns into Russia to advance the 1905 revolution and published the first legal Bolshevik newspaper. When this photograph was taken, he had just returned from London with his English wife, Ivy. As Soviet representative there since the Revolution, he had been arrested and exchanged for the British spy Bruce Lockhart. He was foreign minister from 1930 until 1939. Litvinov died under suspicious circumstances in 1951, possibly a victim of Beria's secret police.

Right: During the 1930s, Trotsky's supporters in Paris, believing that all hope for Marxism in Russia was lost, formed the Fourth International under the leadership of his son Leon Sedov, in opposition to Stalin's policies. They published, somewhat misguidedly in view of Stalin's falsifications, a postcard signed and dated "Leon Trotsky 12/XI/1935," which was itself a retouched version of one of the 1919 photographs (*above*).

Below left: A detail from "World War," panel No. 11 from Diego Rivera's epic series of paintings, *Project of America*, for the New Workers' School in New York, 1933. When the school closed down, eighteen of the twenty-one panels were moved to the headquarters of the International Ladies Garment Workers Union in Forest Park, Pennsylvania. The sight of Lenin and Trotsky proved too much for the union bosses, and four panels were sold to a private collection in Mexico. Thirteen of the other panels were destroyed by fire in 1969.

Opposite: As late as 1987, when Progress Publishers in Moscow issued an album called *Lenin's Portrait: A Few More Touches*, the series of four shots reproduced there were all heavily cropped, leaving only a mysterious elbow—Trotksy's elbow—on the right.

GHOST TRAIN

Opposite: Throughout the Civil War that followed the October Revolution, propaganda trains steamed across the embattled countryside, trying to persuade the local citizens to join the Reds in the struggle against the capitalist-backed White armies. Powerfully painted on the outside by enthusiastic revolutionary designers and caricaturists, the carriages were converted into movie theaters, libraries, telegraph offices, printshops, and meeting rooms. The latest news from the various fronts was turned into type, printed, and published in traveling newspapers. In the cities agitprop trams railed through the streets, and on the rivers there were even agitprop ships.

Right: This photograph shows a young staff member of the famous "October Revolution" train distributing news to an eager crowd. When we look closely at the window on the right of the photograph, a ghostly apparition—the result of the retoucher's inept hand—is all that remains of the person who had been looking out of the carriage.

9

А ЧТОБЫ ПОБЕДИТЬ ГРЯЗЬ,

CHAPAYEV: THE LEGEND CONTINUES

Commanders and commissars of the 25th Infantry Division of Ufa in the Urals, June 9, 1919. In the center of the picture is Vasilii Ivanovich Chapayev (seated, with bandaged head), legendary partisan commander who died in action the same year. Next to him on the left sits Dmitri Andreyevich Furmanov, ex-anarchist, political commissar, and writer, whose novel *Chapayev* was turned into an epic feature film in the 1930s. The photograph in this form was published in the magazine *Projector* in March 1926 as well as in a photographic portfolio entitled *Chapayev* in 1932.

This version (*right*) of the photograph shown on the previous page was recently discovered in Soviet archives. From the original group of forty-one men, only thirty remain, due to airbrushing and cropping. Four faces from the eleven officers erased in the original have been superimposed onto other bodies in this photo (see the three details at bottom for comparison). The soft-focus airbrushing of this version betrays the style favored by Soviet retouchers from the late 1930s onward. The 25 percent reduction in personnel in the picture can be explained by Stalin's purge of the Red Army in 1937, which claimed the lives of 25,000 officers.

Below: A photo-illustration from the *Chapayev* portfolio showing troops from the 25th Division meeting with cavalry from the 217th Pugachevsky Regiment during the Civil War. The slogan on the banner was clearly considered to have been inappropriate by the political censors and was blacked out when the picture was published in 1932.

CIVIL WAR HEROES

Commanders of the Red Army (*left*) in earnest discussion during the Civil War. Left to right: Semyon Budenny, Mikhail Frunze, and Kliment Voroshilov. All three were accorded hero status by Stalin's propaganda machine. The photograph reflects this; in fact, they are lying on a platform in a photographer's studio, and the muddy road behind them is a painted backdrop.
Above: This touching Civil War photograph of a young Red Cavalry soldier was probably taken to be sent to his family back home. It is another studio picture, but what makes the scene disconcerting is that a taxidermist has been at work on the horse.

TROUBLE IN THE CAUCASUS

Above left: Three prominent Party activists, Sergei Kirov, Mikhail Levandovsky, and Konstantin Mekhonoshin, photographed in 1920 in Baku during the Civil War. Kirov organized some of the first systematic purges of the Bolshevik Party there.

Above right: When the picture was published in a biography of Kirov in 1936, Mekhonoshin was cropped out, necessitating some slight retouching to Levandovsky's arm. Mekhonoshin had been appointed commissar on the southern front in 1918 by Trotsky in an attempt to counter Stalin's military shortcomings. Levandovsky was made military commander of the Caucasian Bureau in 1920. Later he commanded the only force that was independent of Moscow—the Red Banner Army of the Caucasus. Stalin was particularly jealous of all this; neither he nor his henchman Voroshilov had been appointed to the Revolutionary Military Soviet during the Civil War. He had a long memory and, in 1937, he took his revenge. He wiped out Mekhonoshin, Levandovsky, and the rest of the commissars who had been unlucky enough to survive.

Top right: Stalin on the Tsaritsyn front, 1918.

Opposite: Back in the photographer's studio, Semyon Budenny (perched on a plastic rock) and Kliment Voroshilov seemingly struggle to find the way to Rostov-on-Don in 1920. The landscape in watercolor deserves mention. This particular confection was published in 1935 in a magazine on the Red Cavalry.

Above: Soldiers of the Eleventh Red Army with political advisors at the train station in Baku, Azerbaijan, in 1920. With the White armies defeated, Baku became the center for the "export of revolution on the point of a bayonet," and the independent Caucasian republics of Georgia, Armenia, and Azerbaijan lived on borrowed time. In the front row can be seen Sergei Kirov (second from left), Sergo Ordjonikidze (third from left), leader of the Caucasian Bureau, and Anastais Mikoyan (right, with dark beard), the sole survivor among the legendary "twenty-six commissars" shot by anti-Communists in 1918.
Opposite: A heavily cropped and airbrushed version (*top*) of the same picture published in *USSR in Construction* magazine in 1939 shows only Kirov and Ordjonikidze.

In another issue of the magazine (*below right*), the two commissars have now been joined by Mikoyan, who has been made to jump at least eight places from his position in the original photograph. The soldiers in the background of the original print have been upstaged by a backdrop drawing of a much more colorful band of stereotypes.

BAND OF STEREOTYPES

Lenin and Mikhail Kalinin, president of the Supreme Soviet, conferring with delegates to the First All-Russian Congress of Working Cossacks in the House of Trade Unions, Moscow, March 1, 1920.

The picture was taken by the distinguished photographer Moisei Nappelbaum. It was reproduced uncropped in the album *V. I. Lenin in the Art of Photography,* published in Moscow in 1967, but a man's head has been blatantly blocked out with a sheet of paper in the top left-hand corner. *Opposite:* When a similar photograph (*top*) of the meeting appeared in *Lenin Album: 100 Photos* in 1927, the missing person was seen to be a tall, bearded comrade listening intently to Lenin's words. When this same photograph was published again in 1970 in another Lenin collection, it was cropped in much closer (*below*), thereby eradicating this unknown "Old Bolshevik" once more.

A large proportion of the photographs in *Lenin Album: 100 Photos* showed Lenin in the company of Stalin's enemies. There was only one picture that included Stalin in the whole book, which was withdrawn from circulation shortly after publication. In 1970 the Institute of Marxism-Leninism in Moscow called it "practically extinct," in spite of the fact that 10,000 copies had been printed. Fortunately, a number of copies reached the West at the time of publication and have been preserved.

PRACTICALLY EXTINCT

THE INFLATABLE LOG

Above: May Day 1920 at the Moscow Kremlin. Lenin gives a hand at a "Subbotnik." To help rebuild the country at the end of the Civil War, hundreds of thousands of workers were urged to perform voluntary work each Saturday. The blurred picture of Lenin (*bottom right*), at the end of the tree trunk, was turned into a laborious painting (*top*) by the artist Mikhail Sokolov in 1927. Widely reproduced throughout the Soviet Union, it soon became known as "The Inflatable Log."

Bottom left: An earlier, more "workerist" painting by Vladimir Meshkov was considered to be too expressionist by the 1930s and disappeared from view.

Opposite: Lenin also found time that day to spread the cement for the foundation stone of a new monument to Karl Marx in Teatralnaya Square. Kamenev has a cold.

ONE STEP FORWARD, FIVE STEPS BACK

Lenin addresses the troops (*left*) from a wooden podium set up outside the Bolshoi Theater in Moscow, May 5, 1920. The soldiers are about to depart for the Polish front to fight Marshal Pilsudski's forces, which had recently invaded the Ukraine. On the steps to the right of Lenin we can see Trotsky, with Kamenev partly obscured behind him. The photograph was taken by G. P. Goldshtein. The subsequent falsification of this photograph (and its variations, which can be seen on the following six pages) is probably the first and certainly the most famous example of Stalinist retouching. The original photograph, which achieved icon status while Lenin was alive and Trotsky still had power, was published throughout the world. It became as much a symbol of revolutionary Russia as the hammer and sickle or the red flag. After Trotsky's downfall, the photograph was never again shown in its entirety in the Soviet Union. Although reproduced in many hundreds of publications, it was always cropped (as in the examples below), even during the Gorbachev period, to eliminate Trotsky and Kamenev.

N 81

... „эта война представляет собой одно из звеньев длинной цепи событий, означающих бешеное сопротивление международной буржуазии по отношению к победоносному пролетариату, бешеную попытку международной буржуазии задушить Советскую Россию, свергнуть первую советскую власть во что бы то ни стало, какими бы то ни было средствами".

В. И. ЛЕНИН. Речь на соединенном заседании ВЦИК, Московского Совета Рабочих, Крестьянских и Красноармейских Депутатов, Профессиональных Союзов и Фабрично-Заводских Комитетов 5 мая 1920 года. (Соч., т. XXV, с. 257).

Фото:

В. И. Ленин произносит речь на Театральной площади в Москве при отправке частей Красной армии на польский фронт.

Парад частей Красной армии в Казани перед отправкой на польский фронт.

Top: Another shot by Goldshtein, taken within seconds of the one on the previous page, was also widely published. Trotsky and Kamenev are now seen in profile.

The same photograph has been montaged into the avant-garde postcard (*above*), published to commemorate the tenth anniversary of the Russian Revolution (1927). This was probably Trotsky's last appearance in published versions of the photograph. He was expelled from the Communist Party on November 14, 1927. Subsequent versions (as shown *right*) have both Trotsky and Kamenev painted out and five wooden steps painted in.

Left: The falsification of Goldshtein's photograph next moved into the world of art. The famous socialist realist painter Isaak Brodsky was commissioned to portray the scene on a giant canvas in 1933. There was no room in this glorification for either Trotsky or Kamenev, their places having been taken by two reporters. The painting became a major attraction at the Central Lenin Museum in Moscow. Large-scale painted copies were made for other Soviet museums, and millions of reproductions were sold.
Top: A photograph taken by A. Garanin for *Pravda* in 1940 shows earnest young soldiers, commanders, and political workers of the Red Army and Navy enthralled by the work.
Above: Isaak Brodsky. He was rewarded for his devotion to socialist realism with the directorship of the Russian Academy of Arts in Leningrad in 1934. He died in 1939.

Left: Another photograph from the May 5 meeting, taken from a different angle. Trotsky is now haranguing the troops while Kamenev and Lenin look on. The picture was found in a special album given to foreign delegates at the Second Congress of the Third International in 1920. The photographer is unknown. In the Stalinist period, being caught in possession of a photograph like this would have meant certain arrest and banishment to a labor camp, or worse. Even to murmur of its existence would have been considered a crime. In the background is a good view of the Bolshoi Theater.
Above: The graphic artist Petr Nikolaievich Staronosov clearly saw the Bolshoi's potential when he made this hideous scraperboard drawing in 1936 for an album entitled *The Life of Lenin.* Trotsky and Kamenev do not feature, of course, and Lenin is back, raging against Pilsudski.

CLOSE TO LENIN

Lenin and a group of delegates on the steps of the Uritsky Palace in Petrograd, July 19, 1920, at the Second Congress of the Communist International. The original photograph (*left*) comes from an album given to foreign delegates at the conference.

Among those who can be identified are, from left to right: Lev Mikhailovich Karakhan (far left, hat and beard), Karl Radek (with cigarette), Nikolai Bukharin (cigarette in hand), Mikhail Lashevich (in uniform), Maxim Peshkov (behind pillar), Lenin, Maxim Gorky, Sergei Zorin (hat), Zinoviev (white tie), M. N. Roy (black tie and jacket), Maria Ulyanova (Lenin's sister), and Abram Belenky (foreground in sunhat).

Gorky moved to Italy a few months after this picture was taken. He returned to the Soviet Union in 1928, four years after Lenin's death. By this time, the photograph was being widely published heavily cropped to an upright (*above*) and retouched to exclude everyone but the writer and the dead leader. Stalin now befriended Gorky, and it was useful to him to show that the friend of the new leader had once been so close to the old one.

Right: As late as the 1980s, only this cropped version of the group on the previous page could be published, albeit slightly modified. Mikhail Lashevich and Sergei Zorin, both prominent supporters of Stalin's rival, Zinoviev, were allowed to return, as was Maxim Peshkov, Gorky's son (peering out from the left of the pillar). Peshkov had been poisoned in 1934 by Genrich Yagoda, then head of the NKVD, who was obsessed with Peshkov's wife. Lashevich committed suicide, and Zorin was shot in 1937. The exact circumstances of Gorky's death are uncertain, but he died in 1936 and was given a state funeral. His body was cremated, and the ashes were placed in the Kremlin Wall.

The fate of some of the others who are identified in the original photograph can be told. Karakhan, an eminent Soviet diplomat, was shot in 1937. Radek, a famous revolutionary and journalist, died in the Gulag in 1939, brutally murdered by a criminal inmate. Bukharin, the immensely popular, liberal, and most humane member of the Bolsheviks, was shot on March 13, 1938, after the third Moscow show trial. Zinoviev, chairman of both the Petrograd Soviet and the Comintern at the time the photograph was taken, was shot in 1936 following the first Moscow trial.

Maria Ulyanova was a member of the editorial board of *Pravda*, as was her friend Bukharin. She died of a brain tumor in 1937.

M. N. Roy, the famous Indian Communist, led the Asian delegates at the Second Congress of the Comintern and clashed sharply with Lenin over colonial issues.

ABSENT

Stalin's relative unimportance during the Revolution and its aftermath is graphically illustrated by his absence from this photomontage of the Bolshevik leaders (*opposite*). It was published in *Oktyabr* (cover, *top*), a spectacular photographic album with a print run of 100,000 copies commemorating the Second Congress of the Communist International in 1920. In its survey of three years of revolution, there is not a single reference to the future dictator. Trotsky, however, as founder and leader of the victorious Red Army during the Civil War, is given star billing (*above*). Few of those pictured in the album were to die later of natural causes.

т. Балабанова
т. Крупская
т. Евдокимов
т. Шляпников
т. Лилина
т. Стасова
т. Зиновьев
т. Троцкий
т. Калинин
т. Луначарский
т. Бухарин
т. Роза Люксембург
т. Свердлов
т. Урицкий

Lenin making a speech in Dvortsovaya Square, Petrograd, on July 19, 1920 (*above*). The picture was taken during the Second Congress of the Communist International by the famous reportage photographer Viktor Bulla. By the time the photograph was used in a memorial issue dedicated to Lenin published by *Krasnaya Niva* (*Red Field*) magazine on February 17, 1924, it must have been judged necessary to boost Lenin's popularity. In spite of the magazine's poor reproduction it can be seen (*opposite, above*) that the audience in front of Lenin in the first picture has been replaced by a much more dramatic crowd scene shot on another occasion (*opposite, below*).

PLANNING THE WORLD REVOLUTION

Isaak Brodsky's massive painting *Lenin's Opening Speech at the Second Congress of the Communist International* was started in 1920 and took four years to complete. Less work of art than historical document, it is a breathtaking panorama portraying more than 300 delegates, Russian and international, who took part in this month-long conference to plan the oncoming world revolution. As the work was not completed until 1924,

Stalin is more prominent in the painting than he would have been were it finished sooner. Trotsky, Bukharin, Radek, and Zinoviev all feature strongly, as do the foreign delegates John Reed (USA), Béla Kun (Hungary), Giacinto Serrati (Italy), Klara Zetkin (Germany), and Alfred Rosmer (France). Since many of the revolutionaries featured in the painting were later murdered by Stalin, and in view of his antipathy to internationalism in general, the work was withdrawn from view in 1927. It was resurrected in 1989 and displayed on the stairs of the Lenin Museum in Moscow, where it had been rolled up, unseen for sixty-two years.

THE CHILDREN OF KASHINO

Above: Lenin and his wife Nadezhda Krupskaya photographed with a group of kulaks (rich peasants) from Kashino village, Moscow region, at the opening of an electric power station. Photograph by F. Feofanov, November 14, 1920.

Left: When the photograph was reproduced nineteen years later at the height of the purges, all the adults from the village and many of the children had been eliminated. The darkened, airbrushed background makes this version doubly sinister.

ENEMY OF THE PEOPLE

Above left: Participants at the Eleventh Congress of the Communist Party, in Moscow, April 1922. From left to right: Emilian Yaroslavsky, Mikhail Kalinin, Stalin, Grigorii Petrovsky, and Sergo Ordjonikidze. When this photo was taken, Stalin had just been made General Secretary, a new post responsible for the organization of the Party apparatus. He was appointed by the Central Committee, on Kamenev's recommendation. Stalin already held several Party and state portfolios, but he worked unobtrusively behind the scenes, and wasn't noticed much. While Lenin was around, there was only one leader.

Yaroslavsky started his political life as a left-wing Bolshevik, and became head of the Union of Atheists, which organized militant anti-religious campaigns from the early 1920s onward.

Above right: In a retouched version of the same photo, published in a portfolio of photographs celebrating Stalin's sixtieth birthday in 1939, there is no longer room for Petrovsky, and Ordjonikidze has been moved closer to his boss by the retoucher. Petrovsky was sacked from his position on the Central Committee in 1938, accused of having connections with "enemies of the people," but he wasn't arrested.

Left: An even closer cropping of the photograph was published at the same time.

THE STALIN SCHOOL OF PETRIFICATION

The photograph (*above*) of Lenin and Stalin in Gorki, near Moscow, in 1922 bears every sign of having been faked. But from the mid-1930s the Stalinist propaganda machine churned out thousands of sculptures, paintings, prints, and drawings to exaggerate the closeness of their relationship in ever more ridiculous degrees. In fact, Lenin had become increasingly alarmed that Stalin was growing too powerful and, in spite of his ill health, tried to break off all relations with him. Lenin had suffered his first stroke in May 1922. The Politburo needed someone to take overall responsibility for him; they chose the recently appointed General Secretary Stalin. More strokes followed, however, leaving Lenin partially paralyzed, and he spent most of the final months of his life resting in Gorki.

The artist Yuri Annenkov went there to draw Lenin in December 1923 for a portfolio of Bolshevik leaders. Annenkov reported later that, "wrapped in a blanket and looking past us with the helpless, twisted, babyish smile of a man in his second infancy, Lenin could serve only as an illustration of his illness, and not as a model for a portrait."

Opposite: A ponderous sculpture, made in 1938, based on the faked photograph above.

Overleaf, left-hand page: Petr Staronosov's inept gravure from 1936, showing Stalin with the disdainful expression of a prizefighter who has just nailed his opponent to the canvas.

Overleaf, right-hand page: Venyamin Pinchuk and R. Taurit's *V. I. Lenin and J. V. Stalin in Gorki* was very well received in 1949 when it was shown in the massive exhibition entitled "Joseph Vissarionovich Stalin in Visual Art" at the State Tretchyakov Gallery in Moscow.

СТАРОНОСОВ

Opposite: Maria Ulyanova, Lenin's sister, who aspired to become a photographer, shot this picture of her brother and sister-in-law, Nadezhda Krupskaya, in Gorki, in late August 1922. Through the viewfinder Ulyanova failed to see the end of a telescope, which looks like a gun pointed at Krupskaya's head.

When the picture was published in a Lenin album in 1960 (*above*) the telescope was slightly retouched, but the adjustment seems to make the threat even greater.

Republished in 1970 (*above*), the threat has become less critical, but a faint and sinister beam now radiates from the barrel toward Krupskaya.

Left: At the close of Soviet rule at the end of the 1980s, the retouchers at last eradicated the threat, and calm was restored.

BASIC RULES OF PHOTOGRAPHY

NOT TO BE REPRODUCED

This extraordinary photograph, never before published, shows Felix Dzerzhinsky, first boss of the Cheka, at his desk in the Lubyanka in 1922. He is surrounded by his closest collaborators, arranged like the twelve disciples at the Last Supper. The photograph is rubber-stamped on the back "Supreme Soviet of the USSR," alongside the ominous command "Not to be reproduced."

Included in the picture are: Yakov Peters (second from the left), with Joseph Unschlicht (hand to chin) and Vyacheslav Menzhinsky flanking Dzerzhinsky. Genrich Yagoda is standing behind Menzhinsky, between Abram Belenky (left) and Martin Latsis and Filipp Goloshchekin (right).

Dzerzhinsky died of heart failure in 1926 and was replaced by Menzhinsky, who died in 1934 after a long illness. Yagoda followed next as chief of security, but he was arrested in 1937 and shot a year later.

Peters was a particularly sadistic Chekist, who was himself liquidated in 1938. Unschlicht, once described by Trotsky as an "ambitious but talentless intriguer" was shot, also in 1938. Belenky was Lenin's personal chief of security. He joined the Leningrad opposition to Stalin in the 1920s and was shot in 1937. Latsis ran a secret section of the Cheka formed after the attempt on Lenin's life. In his book *Two Years' Struggle on the Internal Front* (Moscow, 1920), he wrote: "It [the Cheka] does not try the enemy; it strikes him down." He got struck down himself, shot in 1938. Goloshchekin, who had been in exile with Stalin before the Revolution, masterminded the assassination of the tsar and his family in 1918. Stalin had him shot, along with Trotsky's sister, Olga Kameneva, and many others, in a late purge in 1941.

THE PROPHET PORTRAYED

Yuri Annenkov, the cubo-futurist artist, greatly admired Trotsky and made a large portrait of him in 1923. Trotsky posed for the work in the artist's studio. The painting was so well received that it became the centerpiece (*right*) of the Russian exhibit at the Venice Biennale of the following year. The catalog of the exhibition credited the painting as being the property of the Red Army Museum in Moscow, which may have originally commissioned it. No record of it now exists, and, assuming it was not destroyed, the painting has been in hiding for more than seventy years.
Above: Trotsky with Annenkov at the studio.

21 ЯНВАРЯ 1924 г. В ГОРКАХ, ПОД МОСКВОЙ, УМЕР ВОЖДЬ И УЧИТЕЛЬ УГНЕТЕННЫХ ВСЕГО МИРА, СОЗДАТЕЛЬ БОЛЬШЕВИСТСКОЙ ПАРТИИ И ВЕЛИКОГО СОЦИАЛИСТИЧЕСКОГО ГОСУДАРСТВА — ВЛАДИМИР ИЛЬИЧ ЛЕНИН.
«РАБОЧИЙ КЛАСС ВСЕГО МИРА ВСТРЕТИЛ ВЕСТЬ О СМЕРТИ ЛЕНИНА, КАК САМУЮ ТЯЖЕЛУЮ УТРАТУ. В ДЕНЬ ПОХОРОН ЛЕНИНА МЕЖДУНАРОДНЫЙ ПРОЛЕТАРИАТ ОБЪЯВИЛ ПЯТИМИНУТНУЮ ОСТАНОВКУ ВСЕХ РАБОТ. ОСТАНОВИЛИСЬ ЖЕЛЕЗНЫЕ ДОРОГИ, ОСТАНОВИЛАСЬ РАБОТА НА ЗАВОДАХ И ФАБРИКАХ. ТРУДЯЩИЕСЯ ВСЕГО МИРА С ГЛУБОЧАЙШЕЙ СКОРБЬЮ ПРОВОЖАЛИ В МОГИЛУ СВОЕГО ОТЦА И УЧИТЕЛЯ, ЛУЧШЕГО ДРУГА И ЗАЩИТНИКА — ЛЕНИНА».
ИСТОРИЯ ВКП(б). КРАТКИЙ КУРС.
Коммунистическая партия до сих пор была сильна поддержкой рабочего класса — теперь долг каждого рабочего еще сильнее поддерживать свою партию.
К партии, ко всем трудящимся.
Воззвание Центрального Комитета Коммунистической Партии.
Смерть Ленина не есть смерть его дела.
Ленин—это упорная, все преодолевающая воля.
Ленин—гений рабочего класса.
Ленин всю свою жизнь отдал рабочему классу.
Не уныние и панику должна вселить в нас утеря вождя, а энергию и волю для осуществления великих задач рабочего класса.
Воззвание ЦК РКП(б) о смерти В. И. Ленина.
39

RELIC STATUS

Lenin died in Gorki on January 21, 1924. His body was brought to Moscow, where it lay in state in the Hall of Columns. Soon the body was moved to Red Square, where a large pine boxlike structure was erected to house it (*top right*). On the front of the box a huge sign simply read "LENIN." Throughout February temperatures were subzero and the body lay naturally frozen. But as spring began to melt the ice, and the lines of mourners grew longer, the body started to show signs of decomposition. At this point, the Central Committee hurriedly passed a resolution proposed by Stalin to embalm the body. Krupskaya, Lenin's widow, pleaded to have her late husband simply buried. Stalin, who hated Krupskaya, rejected her request and threatened that "We can always find another widow for Lenin." The "genius of geniuses," as Stalin called Lenin, was about to become a secular saint.

Two embalming experts, Professors V. P. Vorobiev and B. I. Zbarsky (*second from top*), were called in to administer formaldehyde and a glycerin solution. Lenin's internal organs were removed. His brain was taken away to a special laboratory for dissection. The rest of the corpse was then inserted into a glass sarcophagus designed by the constructivist architect Konstantin Melnikov. This in turn was housed in a beautifully designed oak mausoleum (*third from top*).

A heavy granite monster (*bottom right*) was built to replace it in the early 1930s and has remained there ever since, one of Communism's foremost places of pilgrimage. Checks are carried out on the body twice a week, and there is a full overhaul every eighteen months.

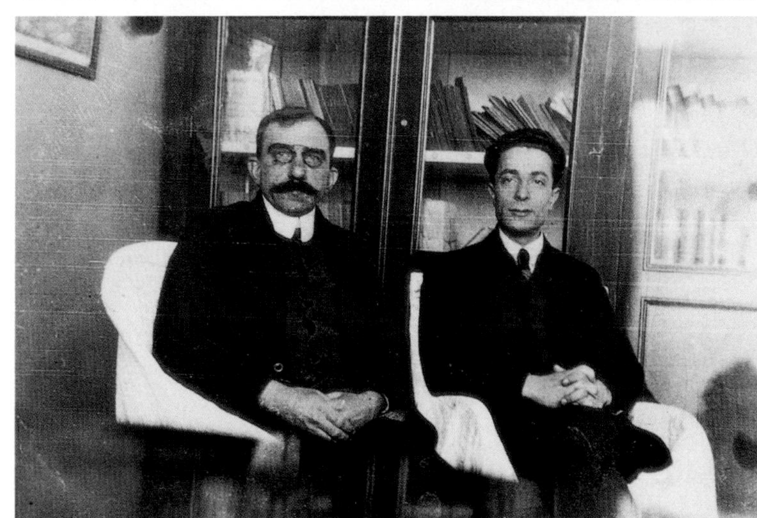

LENIN LEVY

Opposite: A photomontage made in 1939 glorifying Stalin's grief at Lenin's funeral. The two henchmen standing at the coffin are Dzerzhinsky and Voroshilov.

With Lenin sealed safely in the sarcophagus, the internecine savagery could begin in earnest. Stalin's first task was to neutralize Lenin's allies, the "Old Bolsheviks," by creating a network of opportunistic alliances. Stalin also suppressed Lenin's Testament (notes intended for publication after his death concerning the future of the Party, including a demand for Stalin's removal as General Secretary) and sought to make the Party his own. He packed it with hundreds of thousands of new members—the so-called Lenin Levy—political illiterates who shared nothing with the old revolutionaries but owed their sudden rise in status only to Stalin. Trotsky famously referred to this period as the "conspiracy of the epigones." A series of illnesses made it very difficult for Trotsky to mount a properly coordinated opposition.

THE CANONIZATION OF LENIN

The Lenin cult, as orchestrated by Stalin, gained momentum from the day of his death. Plans were set in motion to construct huge museums and monuments all over the Soviet Union. Towns and factories were renamed. Paintings, sculptures, and a huge publishing drive sought to exploit his name and legacy. Among the mass of adulatory books and albums published to commemorate his death, one extraordinary volume more than five hundred pages long contained photographs of every wreath sent to his funeral, with detailed inscriptions and lists of signatories. A four-page newspaper simply titled *Lenin* appeared for one edition and for one day only. The protestations of Lenin's widow—"When he [Lenin] was alive he had no time for such things, he found such things oppressive"—were to no avail. It was absolutely necessary for Stalin's plan: the Lenin cult had to be established in order to lay the foundations for the Stalin cult that was to follow. In the 1930s Stalin became known as "The Lenin of Today." Two early and somewhat bizarre examples of the Lenin cult are shown here.

Opposite: Nikolai Strunnikov's unwittingly sinister portrait of a shrewd and calculating Lenin was painted in 1924 from a photograph taken in 1921.

Above: Design for a Lenin carpet, 1924, from Soviet Central Asia.

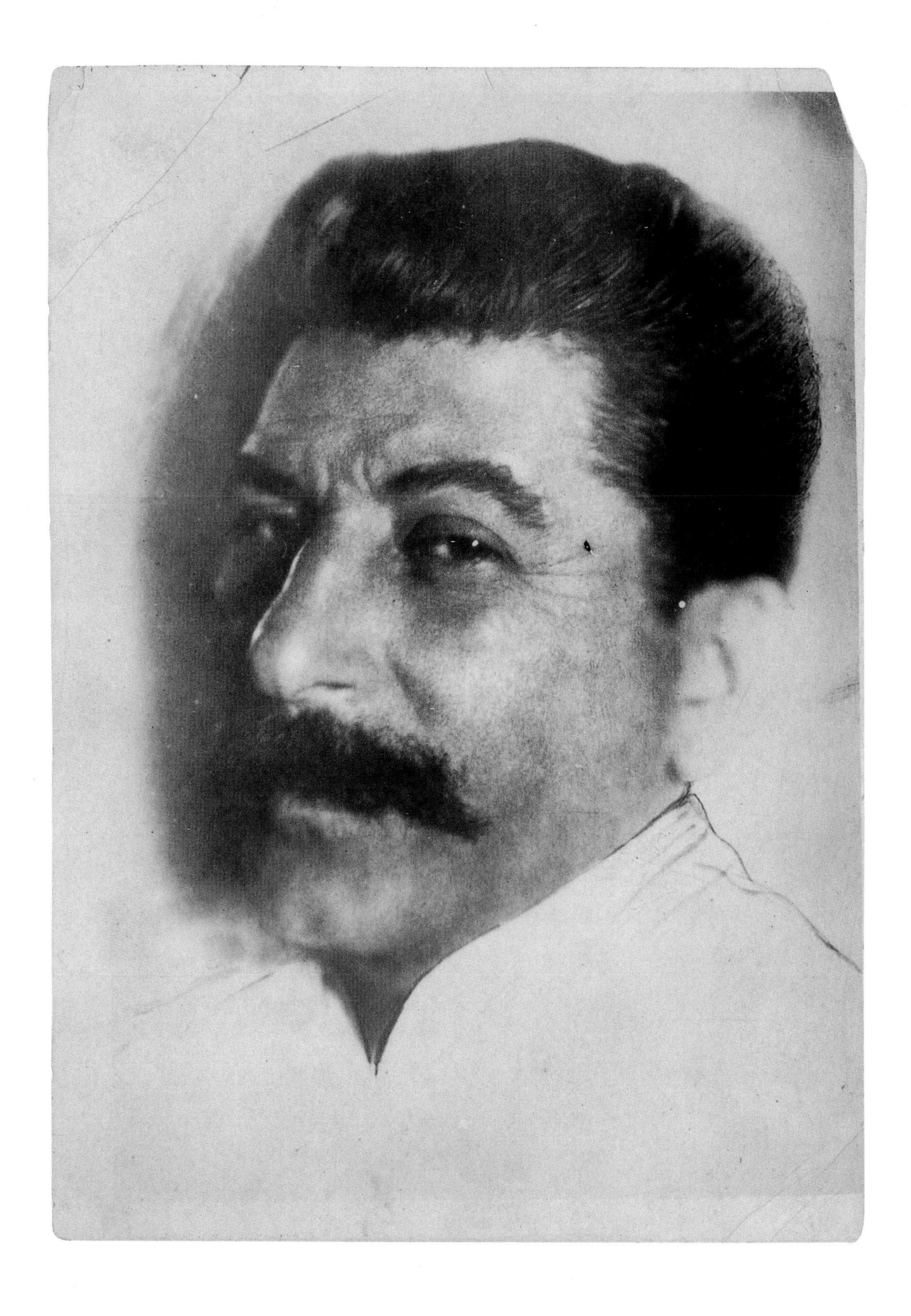

HOLLYWOOD, USSR

Portraits of Party leaders in the Soviet Union were often treated to vignetting, a process used since the beginning of photography. Clouding the edges of the photographic print, achieved with the airbrush, was particularly suited to the Stalinist hierarchy. The subjects took on an ethereal, almost godlike status, quite removed from reality.

Above: This photograph is an original print thought to be by Moisei Nappelbaum and taken in 1924. Although the background has been vignetted, the face has been only lightly retouched. When the same photograph was printed (*opposite*) in 1939 as part of a pictorial biography published to celebrate Stalin's sixtieth birthday, the retoucher's deft hand was there for all to see. Stalin's skin has been positively pancaked, his hair and mustache are now as smooth as a matinee idol's, and the glint in his eye is all that remains of the original.

This sort of vanity was often compounded by the film star's trick of using younger pictures than the captions allege; the birthday album slips the photograph into a section on the early 1930s, seven years after it was actually taken and when Stalin was on the wrong side of fifty.

This group shot, taken at the Fourteenth Party Conference in April 1925, was to become a classic example of Stalinist photographic manipulation. Left to right can be seen: Mikhail Lashevich, Mikhail Frunze, Ivan Nikitich Smirnov, Alexei Rykov, Kliment Voroshilov, Stalin, Nikolai Skrypnik, Andrei Bubnov, Sergo Ordjonikidze, and Joseph Unschlicht.

LONG-TERM FRIENDSHIPS

Only one of Stalin's comrades in this picture would die from natural causes. Lashevich had been an active participant in the October Revolution and a commander in the Red Army during the Civil War. Stalin viciously attacked him in his speech to the conference, calling him a "schemer." Lashevich committed suicide in 1928.

Frunze died on the operating table in October 1925, possibly the victim of a medical murder instigated by the General Secretary.

Stalin had a special hatred for Smirnov, whom Lenin once described as "the conscience of the party." He was accused of Trotskyism and shot after the first Moscow show trial. His wife and daughter were arrested and shot by the GPU in 1937.

Rykov became chairman of the Council of People's Commissars after Lenin's death. He was branded a right-wing opportunist and sacked in 1929. He was shot in 1938 after the third Moscow trial.

Voroshilov, military commissar and friend of Stalin's, survived all the purges. He was criticized by Khrushchev after Stalin's death and forced to retire.

Skrypnik was a veteran Ukrainian Bolshevik who was accused by Stalin of Ukrainian nationalism, and shot himself in 1933. Bubnov's appointment as head of the political administration of the Red Army in the late 1920s did not prevent his arrest and death in the Gulag in 1940.

Ordjonikidze was a political commissar in the Red Army during the Civil War. He was a close supporter of Stalin's and a fellow Georgian. A key figure in the massive Five-Year Plan expansion programs, he committed suicide in 1937, at the height of the Great Purges, in protest against Stalin's murderous policies.

Unschlicht, another active participant in 1917, held various posts both in the Cheka and in Stalin's military command. He was arrested and shot in 1937.

The same photograph, published in two biographies of Stalin that appeared in 1939 and 1949, has been retouched and rearranged to reduce the group to four. Sixty percent of those present at the meeting were erased from history.

SOLEMN OCCASION

The funeral procession (*left*) for the head of the Cheka, Felix Dzerzhinsky, in July 1926. Days earlier, Trotsky had delivered a speech to the Central Committee arguing that Stalin and Nikolai Bukharin had abandoned revolutionary politics. In violent reaction, Dzerzhinsky screamed in his high-pitched voice for almost two hours at Trotsky, Kamenev, and the other oppositionists. As he stepped down from the platform, he suffered a heart attack and collapsed. He died the same day.

In the photograph, Rykov, Kamenev, Molotov, Bukharin, and Stalin bear Dzerzhinsky's coffin through the streets of Moscow. Trotsky is also visible on the opposite side. This was to be Trotsky's last official appearance in the Soviet Union. To the left, a column from the GPU makes its presence felt lining the gutter.

Top: A postcard published in the late 1930s with the caption "J. V. Stalin and F. E. Dzerzhinsky after a meeting of the Central Committee." The photograph shows neither Stalin nor the founder of the secret police but two actors playing their roles in the Soviet feature film *Lenin in 1918*, made in 1939. Stalin's part was played, as ever, by the Georgian Mikhail Gelovani.

A photograph of Nikolai Antipov, Stalin, Sergei Kirov, and Nikolai Shvernik in Leningrad in 1926, celebrating the destruction of Zinoviev's anti-Stalinist opposition. Stalin had just made Kirov first secretary of the Leningrad party.

FOUR, THREE, TWO, ONE . . .

When the photograph *above* was published in the poorly printed *History of the USSR* (Moscow, 1940), Antipov and the chandelier had been deleted. He had joined the Bolsheviks in 1912, and was chairman of the Petrograd Cheka in 1918. He rose to become Prime Minister Molotov's deputy in the 1930s. He was arrested and sent to the infamous Orel prison, where he became the last leading Stalinist cadre to be shot, on August 24, 1941, as the Nazis were advancing toward Moscow.

Overleaf, left-hand page: It is unclear why Shvernik was also eclipsed from the picture, now heavily airbrushed, when it was used in *Joseph Stalin—A Short Biography* (Moscow, 1949). At the time the photograph was taken, he was secretary of the Central Committee of the Communist Party. In 1946, Stalin made Shvernik head of state, a post he held until Stalin's death in 1953.

Overleaf, right-hand page: In this obsequious oil painting by Isaak Brodsky, based on the original photograph and made in 1929, Stalin the executioner alone remains.

Above: Mikhail Tomsky (back row, left) and President Kalinin (seated second from left) in Leningrad with unidentified Party functionaries in 1927. Tomsky, a printer, had been president of the Congress of Trade Unions since 1919. He joined with Bukharin and Rykov in their antipathy toward Trotsky, but in 1929 he boldly criticized Stalin's disastrous program of collectivization of the peasantry. He was accused of "right deviationism" and struck off the Politburo. After the first Moscow show trial, when the press was running a virulent campaign against "hidden Trotskyites," Chief Prosecutor Vyshinsky called for full exposure of these "despicable terrorists." After one particularly unpleasant attack in *Pravda*, Tomsky shot himself. He left a suicide note to Stalin pleading for mercy for his family. Stalin had them arrested. Tomsky was the only genuinely working-class member of Stalin's Politburo.

Kalinin avoided becoming a victim of the purges himself, but his position as head of state did not stop Stalin from arresting his wife, Yekaterina, in 1938. She spent eight years in the Gulag, and was released only in the year of her husband's death.

A SAVAGE ATTACK ON THE WORKING CLASS

Left and (detail) opposite: A most vicious attack in pen and ink on Tomsky and another comrade—by persons unknown—on a copy of the above photograph found in Soviet archives.

CRUEL MOSCOW

Top: Trotsky's persistent ill health during the 1920s necessitated numerous sojourns in climates friendlier than the cruel Moscow winter. On one such visit to Georgia in 1924, he and his wife, Natalya Sedova, were photographed in the backseat of an automobile behind Sergo Ordjonikidze, who was active at this time in installing Soviet power in the Caucasus. A monograph on Ordjonikidze published to celebrate his fiftieth birthday in 1936 reproduces the photograph slightly cropped (*above*), with a certain "Comrade Masnikov" and another man moved in to block out Trotsky and Natalya. At the time the book was printed, Trotsky was in exile in Norway, railing against the Moscow trials.

Right: Trotsky, Kamenev, Zinoviev, and comrades in a rare moment of oppositionist optimism in the mid-1920s.

OGPU

Trotsky's own copy of his expulsion order from the USSR, which has never before been published. Certain of the more outrageous accusations have been underlined by Trotsky, who has also written in the margin, "The scoundrels." The full text follows:

Copy. Extract from the minutes of the special session of the Directorate of the OGPU, January 18, 1929.

We have heard: "The case of citizen Trotsky Lev Davidovich in accordance with article 58 clause 10 of the criminal code accusing him of counterrevolutionary activities, expressed in the organization of an illegal anti-Soviet party whose recent activities have been aimed at provoking anti-Soviet speeches and the preparation of armed struggle against Soviet power."

We have decreed: "To expel citizen Trotsky Lev Davidovich from the territory of the USSR."

Validated: By the chief of the Alma-Ata Section of the OGPU, January 20, 1929.

КОПИЯ

ВЫПИСКА
из протокола Особаго Совещания
при Коллегии О.Г.П.У.
от 18 января 1929 года

СЛУШАЛИ:	ПОСТАНОВИЛИ:
"Дело гражданина Троцкого Льва Давыдовича по ст. 58/10 Уголовнаго Кодекса по обвинению в контр-революционной деятельности, выразившейся в организации нелегальной анти-советской партии, деятельность которой за последнее время направлена к провоцированию анти-советских выступлений и к подготовке вооруженной борьбы против Советской власти".	"Гражданина ТРОЦКОГО Льва Давыдовича – выслать из пределов С.С.С.Р."

Вот прохвосты!

ВЕРНО:

Нач. Алма-Атинского Окротдела Г.П.У.

г. АЛМА-АТА

20 января 1929 года.

GPU

A poster for the GPU dated 1930 by Viktor Deni, a caricaturist who began his political life as a Menshevik. The lightning bolt, which spells out "GPU," strikes the head of a demonic character labeled "Counterrevolutionary wrecker."

Stalin's contempt for the ordinary worker is shown clearly in two versions of the same picture taken at the time of the Sixteenth Party Congress in 1930.

In the original photograph (*above*) an attendant is seen pointing his finger, helpfully directing the "Boss" (*Vhozd*), as he was often known. When the photograph was printed in *Projector* magazine (*above right*) soon afterward, no humble worker was there to tell Stalin where to go.

The Sixteenth Party Congress was characterized by its abject subservience to Stalin and his ruthless policies. There were wholesale expulsions of trade-union Communists whom Stalin attacked for trying to defend the interests of the workers against those of the state. In spite of the devastating assault on the countryside and the massacre of millions of peasants that had been taking place under the slogan "Liquidate the Kulaks as a Class," a long and determined battle was deemed still to be necessary to complete the collectivization of agriculture.

Opposite: Collective farm laborers set off to work in the fields, Uzbekistan, 1930.

Left: A photograph of Stalin and his second wife, Nadezhda Alliluyeva (*center*), with the Voroshilovs and a guard picnicking in Georgia during the summer of 1930. Stalin's coldness and rudeness toward his young wife contributed to her deep depression, and on the night of November 8, 1932, she killed herself. Stalin did not attend the funeral, and lived out the rest of his life alone. The photograph was taken by Lieutenant General Vlasik, Stalin's chief of personal security.

THE BOSS

OLDER AND WISER

Right: A postcard of Friedrich Engels and Karl Marx published by Soyuzfoto in Leningrad in 1933 on the fiftieth anniversary of Marx's death. The picture is a photomontage joining together a portrait of Engels taken in Brighton, England, in 1877 (*above*) with a heavily retouched one of Marx, the original of which had been taken in London on August 24, 1875 (*top*). Both men were fifty-seven years old when the original photographs were taken. The purpose of dramatically aging Marx in the photomontage seems to have been to show him as a wiser and more venerable sage compared to the youthful Engels. In fact, there were only two years between them.

ЛЕНИНГРАДСКОЕ ОТДЕЛЕНИЕ
СОЮЗФОТО
ЛЬС.
ФОТО-КАЛЕНДАРЬ НА 1933 Г.

SOVIET CITIZENS

The lead story in the March 1934 issue of *Projector* magazine concerned the arrival in Moscow of three Bulgarian Communists (welcomed by Stalin and others, *top*). The trio (Georgii Dimitrov, Blagoi Popov, and Vasil Tanev) had been accused of setting fire to the Reichstag in Berlin and prosecuted by Goering at a famous trial in Leipzig. They put on a brilliant defense and were acquitted by the Nazi court. They were each given Soviet citizenship. Left to right can be seen: Sergo Ordjonikidze, Dmitri Manuilsky, Wilhelm Knorin, Vyacheslav Molotov, Lazar Kaganovich, Kliment Voroshilov, Georgii Dimitrov, Valerian Kuibyshev, Joseph Stalin, Popov, and Tanev.

Three years later, Popov and Tanev were condemned by a Soviet court on false charges. This accounts for their departure from this late-1930s retouched version (*center*). Knorin, a Latvian, was also dispensed with. He was later tortured with a hot iron to his neck for "nationalist deviations." In a Bulgarian homage to Stalin published in 1949 (*left*), Ordjonikidze and Manuilsky have also disappeared from view. Dimitrov became premier of Bulgaria in 1946, so perhaps it was on his orders that Manuilsky, a ruthless Stalinist, had to be ejected.

Opposite: A portrait of Stalin in 1935, by Isaak Brodsky.

Всесоюзное Совещание Судебно-Прокур
Работников в Москве 23-IV-34

MURDERERS

Left: Published for the first time, this remarkable photograph shows the prosecutor-general of the Supreme Soviet, Andrei Vyshinsky (seventh from left, front row), with his staff of 228 men and women in the courtyard of their factory of death in Moscow. It was taken on April 23, 1934. These are the faces of the bureaucrats who processed the lies, in the guise of socialist justice, that sent millions of people to be destroyed in the Gulag. Hand in glove with their friends in the NKVD, they invented extremes of interrogation and torture to break their victims' minds and bodies. Less than eight months after this photograph was taken, a law was passed ordering death sentences to be executed immediately. In the Lubyanka, some were shot in the back of the head on the stairs to the basement, others in the basement itself.

Seated second from the left in the front row can be seen Vasilii Ulrikh. This notorious trial judge had been dispatching countless thousands to their deaths since 1920. Sixth from the left, in worker's cap, is Nikolai Krylenko, the people's commissar of justice. His fearful reputation was made at the major political trials of Mensheviks and others at the turn of the decade, where he performed the task of chief public prosecutor. Stalin, who disliked Krylenko personally, had him arrested and shot without trial in 1938.

Vyshinsky and Ulrikh reported regularly to Stalin with lists of the latest executions and requests for advice on further sentencing procedures. During the "Campaign of Vigilance" in 1937, Vyshinsky, with Nikolai Yezhov's help, carried out his own murderous purge on the procuracy itself. Many of those photographed here would not live to see the end of the decade. Both Vyshinsky and Ulrikh survived the purges and died in the 1950s.

ASSASSINS

Sergei Kirov (mourned *above*, lying in state) was assassinated on December 1, 1934 by a student. Mystery still shrouds the real reason for the killing of the Leningrad party boss because the full report of the original investigation has never been released. What is known, however, is that Stalin feared Kirov's immense popularity and considered him a threat to his leadership. Kirov's speech at the Seventeenth Party Congress had called for reconciliation after the terrible upheavals associated with collectivization. The speech was received with great acclaim from the party delegates. No reconciliation followed his death. In a purge labeled "the assassins of Kirov," thousands of Trotskyists, Zinovievists, and even Stalinists were rounded up and sent to the camps. The terror had begun in earnest.

ARCHITECTURAL RETOUCHING

The headquarters of the secret police, which had been, before the 1917 revolution, the offices of the Yakor Insurance Company. During the 1930s the Lubyanka, as it is known, became a nonstop conveyor belt of punishment and death operating within the heart of Moscow.

The transformation of its giant exterior, as prerevolutionary Baroque was swallowed up by Brezhnevite concrete, can be seen here, in photographs taken in 1917 (*top*), 1976, and 1984 (*right*)—retouching on an enormous scale.

Dzerzhinsky's statue stood in the middle of Lubyanka Square. One night in 1991, toward the end of Soviet power, the 40-ton monument was torn down (*right*). The plinth remains. Occasionally a cross is stuck on it.

Control of the secret police was essential for Stalin to tighten his grip over the Party apparatus and the country as a whole. By the mid-1930s the NKVD had become a "state within a state" supported by an acquiescent judiciary who ruled that all opposition to Stalin's policies was "criminal." The prison camp system meanwhile ruthlessly crushed tens of millions of lives in a gigantic network that spanned the whole Soviet Union. *Opposite:* An NKVD officer's badge.

НКВД

WARM WELCOME IN RED SQUARE

Stalin and members of the Politburo *(above)* in Red Square, Moscow, to welcome the return of the captain and crew of the *Chelyuskin*, June 19, 1934. The explorers had camped on a polar reef for two months after their ship had run aground in the Arctic Circle. The entire crew was eventually saved by Soviet fliers in a heroic rescue operation. *Opposite:* When the photograph was printed in a mammoth volume honoring the exercise, some refinements had been made. The onlookers to the left have all disappeared. Lazar Kaganovich, the heavy-industry boss who was obscured by Voroshilov (second from the right in the original), has now been planted behind Kuibyshev (right). His waving hand (see original) has gone, but a lump of his shoulder has been left just behind Voroshilov's. Most thoughtfully, the retouchers have cleared up the litter under the commissar's boots.

СССР

TEN YEARS OF UZBEKISTAN

In 1934 the artist Alexander Rodchenko was commissioned by the state publishing house OGIZ to design the album *Ten Years of Uzbekistan* (photographed *right* and *below*) celebrating a decade of Soviet rule in that state. It was a hefty piece of work, weighed down with mug shots of bureaucrats and endless fake statistics. Nevertheless, the book was interesting due to Rodchenko's skillful design techniques: gatefolds, acetates, embossing, bold photo layouts, full-color and duotone printing. The Russian edition appeared in 1934, and the Uzbek edition, with some politically induced changes, in 1935. But in 1937, at the height of the Great Purges, Stalin ordered a major overhaul of the Uzbek leadership, and heads began to roll.

Many Party bosses photographed in *Ten Years of Uzbekistan* were liquidated. The album suddenly became illegal literature. Using thick black India ink, Rodchenko was compelled to deface his own book. The macabre results are both brutal and terrifying. On the following six pages can be seen the original photographs of five men and one woman alongside the images defaced by Rodchenko's hand.

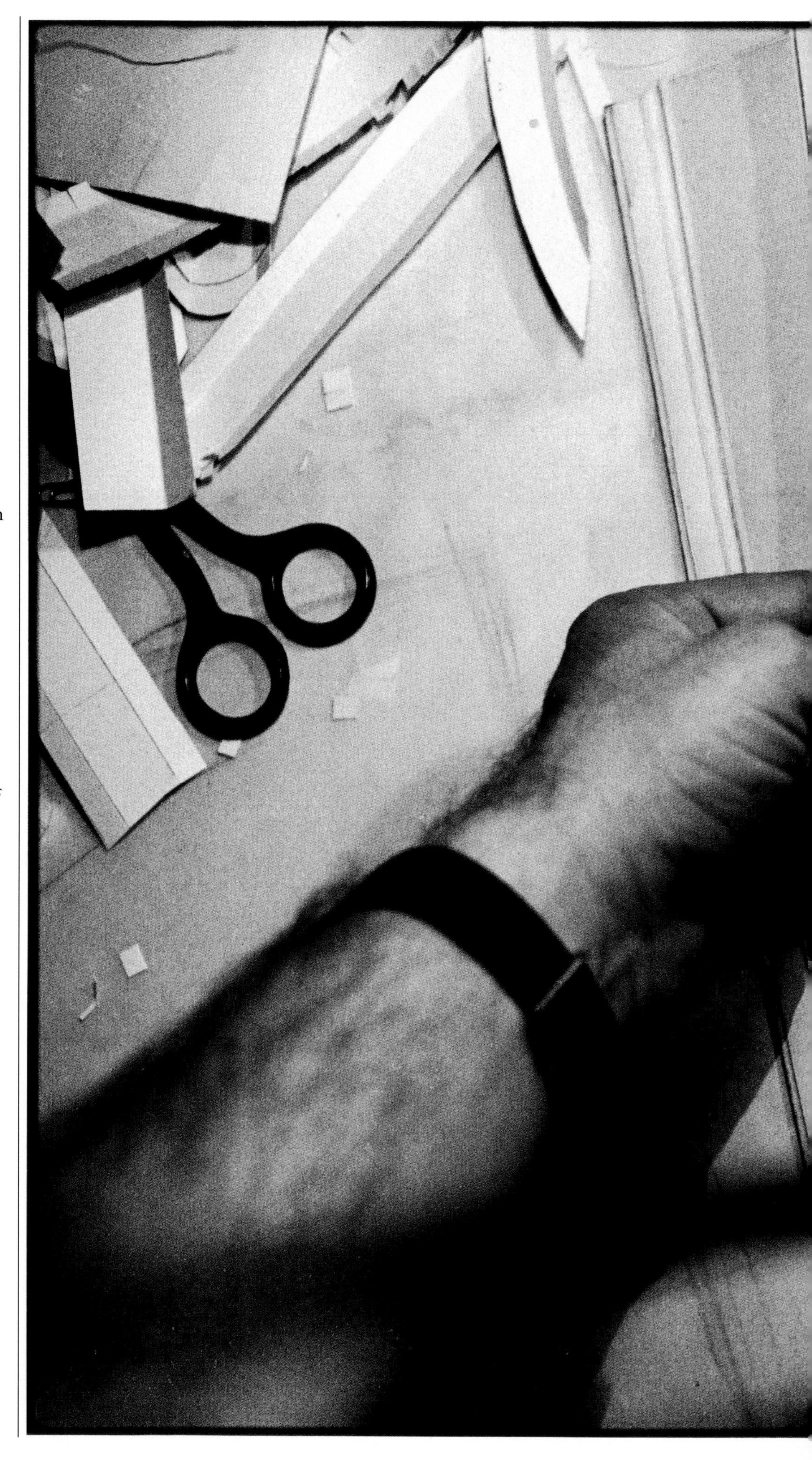

В. КУЙБЫШЕВ
Л. КАГАНОВИЧ

A. A. Tsexer, secretary of the Central Committee of the Communist Party of Uzbekistan. Exact fate unknown.

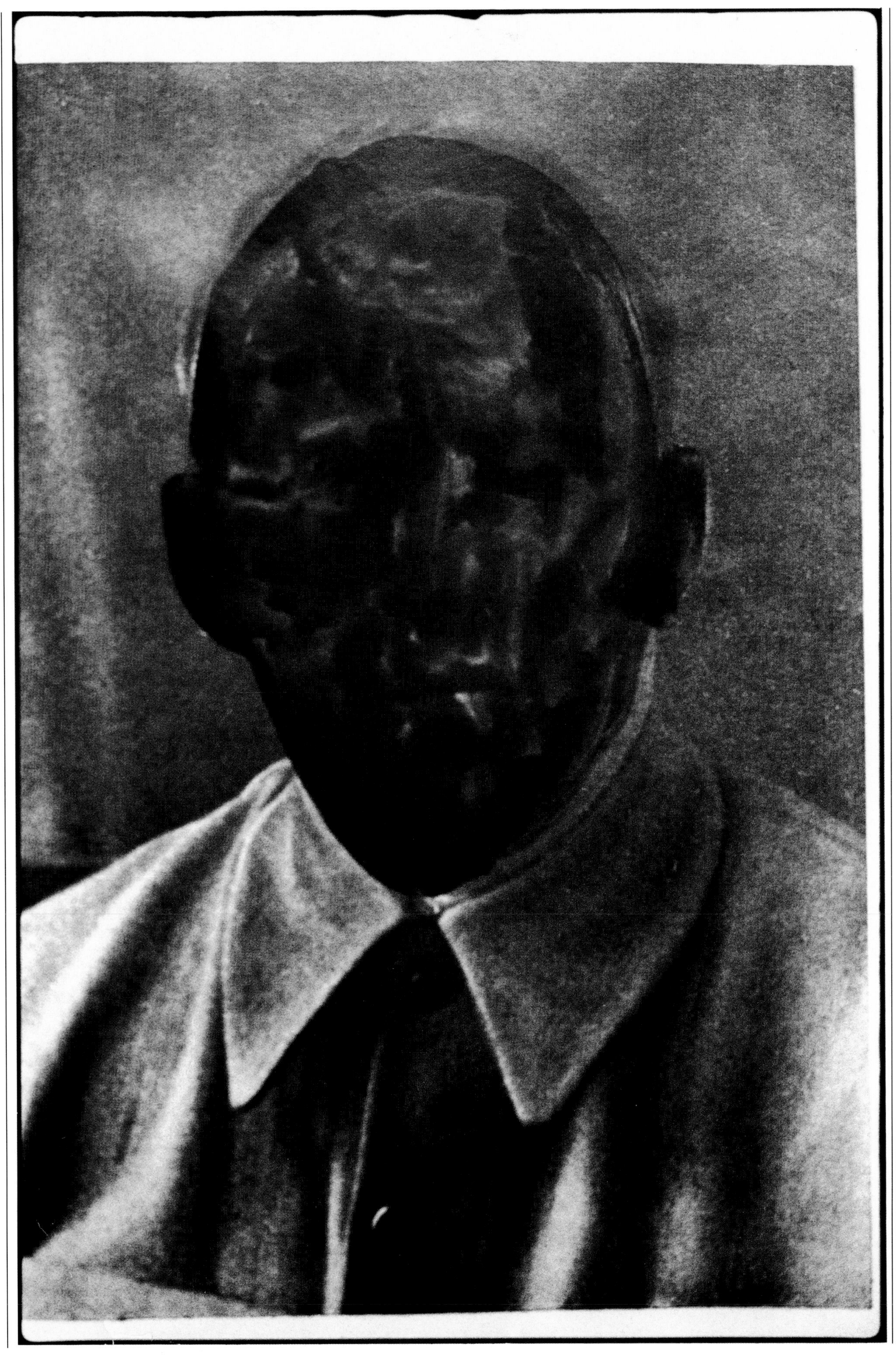

Akmal Ikramov, first secretary of the Communist Party of Uzbekistan. Arrested on Stalin's orders, tried, and shot with Bukharin, Faizulla Khodzhaev, and others after the show trial of 1938. Apparently, Stalin had not forgotten Ikramov's speech at the so-called Congress of Victors in 1934, in which Ikramov had warned of "self conceit." He was expelled from the Party and handed over to the NKVD on September 12, 1937. The Party membership in Tashkent, on hearing of his arrest, greeted the news "with warm applause."

Faizulla Khodzhaev, Soviet chairman of commissars in Uzbekistan. Cofounder of the nationalist Young Bokhara movement. Did not join the Bolsheviks until 1922. Arrested on Stalin's orders, accused of bourgeois nationalism and conspiring to deliver Central Asia into the hands of the British. Tried and shot with Bukharin and others in 1938.

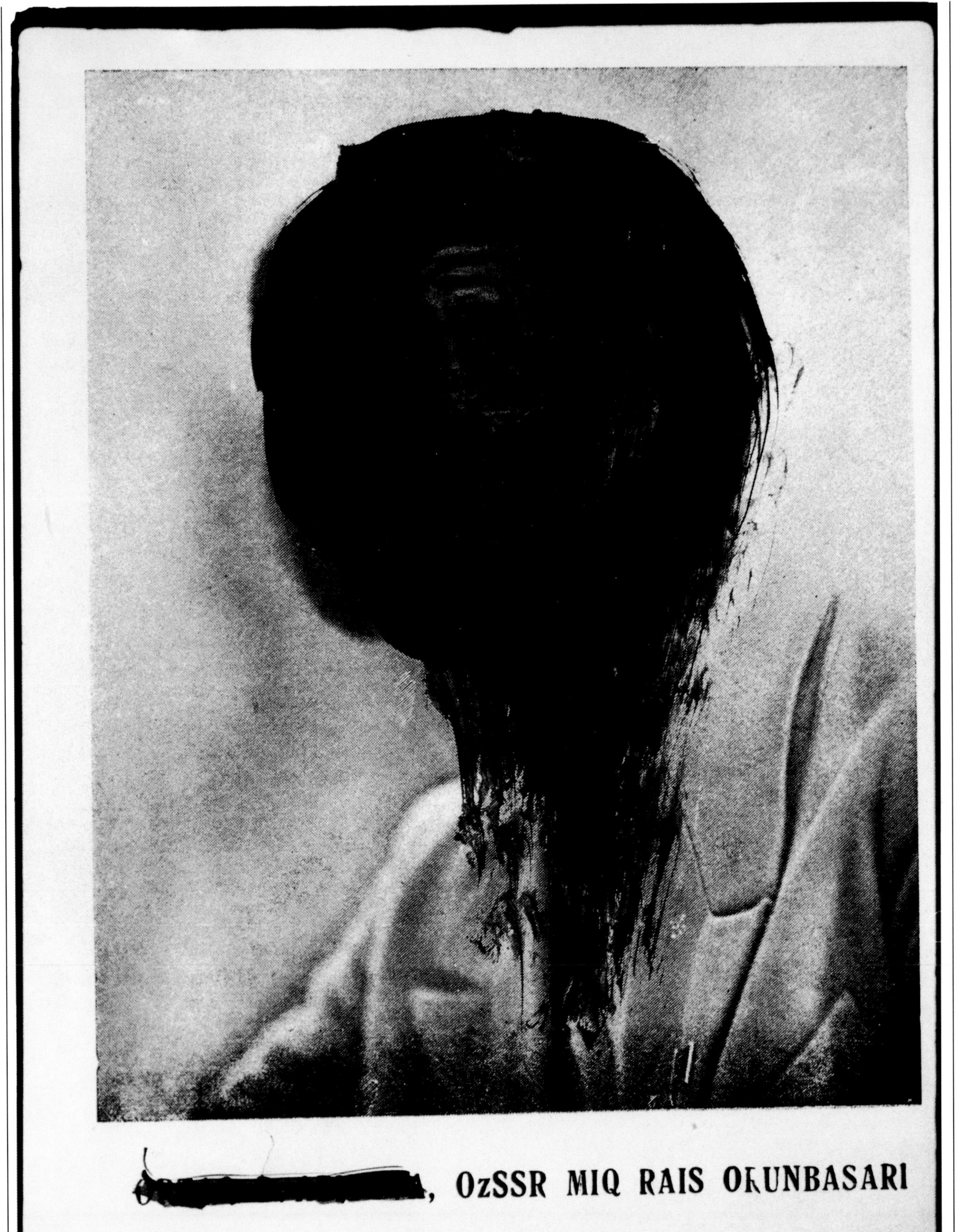

D. Abidova, member of the Council of People's Commissars in Uzbekistan. Her exact fate is unknown.

Yan Rudzutak, a Latvian from the prerevolutionary underground. Party member since 1905. Spent ten years in tsarist prisons. Supported Stalin's rise to power. Became candidate member of the Politburo in 1926. Opposed Stalin's more grandiose proposals during the first Five-Year Plan, but his astute objections were swept aside. He was arrested at a party in May 1937. Among the guests also arrested were four women who were seen, three months later, in the Butyrka prison, still in their bedraggled evening dresses. On July 28, 1937, Yezhov sent Stalin a list of 138 names. Stalin scrawled "Shoot all 138." Rudzutak's name was on the list. The trumped-up charges included "spying for Germany."

Yakov Peters, notoriously sadistic Latvian who was appointed deputy director of the Cheka soon after the Revolution. He worked in the East End of London from 1909 to 1917. Married an Englishwoman, May Freeman. Member of the British Labor Party. Chief of Internal Defense in Petrograd in 1919, where his work included the signing of innumerable death warrants. Key figure in the GPU, elected member of the Party Central Committee. Arrested and shot on Stalin's orders, in 1938. His daughter, who was a famous ballet dancer, was also arrested.

Above: Stalin and members of the Central Committee at the Seventeenth Party Congress, Moscow, late January 1934. Left to right, standing: Abel Yenukidze, Kliment Voroshilov, Lazar Kaganovich, and Valerian Kuibyshev. Seated: Sergo Ordjonikidze, Stalin, Vyacheslav Molotov, and Sergei Kirov.
Right: Abel Yenukidze was expelled from the Party by the Central Committee in June 1935. Not surprisingly, therefore, in the edition of *Pravda* that broke the news of the decision, he was described as a "typical specimen of a corrupted, degenerate, Menshevik-like Communist." Furthermore, he was accused of "rotten liberalism." In reality Yenukidze was an inoffensive character who for many years had organized the difficult task of the day-to-day administration of the Kremlin. His great mistake, as far as his old friend Stalin was concerned, had been his opposition to the terror that had been launched in the aftermath of the Kirov assassination in December 1934. Eighteen months prior to his expulsion, Yenukidze had made a speech to the delegates at the Seventeenth Party Congress saying that "Comrade Stalin was surrounded by the best people in our party."

There had been 1,961 delegates at this congress. No fewer than 1,108 of them were later liquidated.

STALIN'S OLD FRIEND

Above: The seven joint chairmen of the Central Executive Committee of the USSR in 1935, with the secretary to the committee, Abel Yenukidze, standing behind them. Most of the assembled would soon come to grief. Left to right: Faizulla Khodzhaev (Uzbekistan), G. Musabekov (Transcaucasia), Grigorii Petrovsky (Ukraine), Mikhail Kalinin (Russian Federation), A. Chervyakov (Byelorussia), A. Rakhimbayev (Tajikistan), and N. Aitakov (Turkmenistan). Khodzhaev was shot in 1938. Beria personally arrested Musabekov and had him shot, also in 1938. Chervyakov was accused of espionage in 1937 and committed suicide. Rakhimbayev was arrested in September 1937 at a Youth Day parade. He was accused of keeping a harem, and shot. Aitakov was also shot. Petrovksy and Kalinin died peacefully after the war.

Left: Almost exactly the same photograph was published shortly afterward in a portfolio about the congress, but Yenukidze's presence was no longer acceptable, and he has been faded into the background.

THE EXECUTIVES' EXIT

Above: A photograph from the 1934 Russian edition of the album *Ten Years of Uzbekistan* shows, seated left to right: Akhun Babayev, Vyacheslav Molotov, and Abel Yenukidze. Standing: Ortaqlar Blan Birlikda, Zalaridan Avezov, and Tursun Kodzhayev.

Opposite, above: The Uzbek edition of *Ten Years of Uzbekistan* appeared in 1935, after Yenukidze had been sacked from the Central Committee. The Uzbek Party apparatchiks and Molotov remain, but Yenukidze's departure has necessitated major alterations to Tursun Kodzhayev's suit.
Opposite, below: More blocking-out by Alexander Rodchenko from his copy of the first Russian edition.

In 1937 Stalin sanctioned Yenukidze's execution. He was one of the last "Old Bolsheviks" to be murdered by Yezhov, and eight members of his family were to perish with him.

SOVIET TAILORING IN THE 1930s

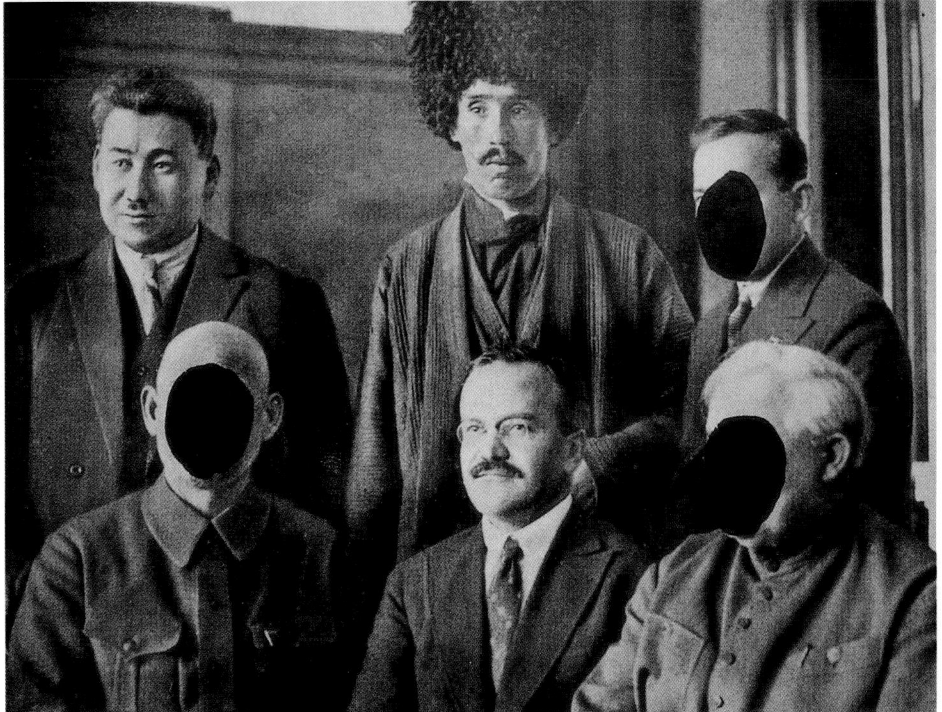

DENOUNCED BY A JEALOUS DESIGNER

Working in Moscow, the Latvian artist Gustav Klutsis (*right*) crossed over from being an avant-garde painter at the time of the Revolution to Stalin's most dedicated political designer. From 1928 onward, his posters for industrialization, collectivization, and in particular the Stalin cult employed all the most powerful weapons in the arsenals of suprematist composition, militant typography, and montage. His posters were published in huge print runs of up to 100,000 copies each, urging the "mass of the people" to "overfill their work norms" and "build the Plan."

In 1938 Klutsis was denounced to the GPU by a jealous designer as an agent of imperialism "working for the British." He was arrested, subjected to intensive interrogation, and shot.

Top: "Under Lenin's Banner," a Five-Year Plan poster that Klutsis designed in 1930. Hindsight reveals the work in a sinister light: with Stalin's shadowy face, partly hidden behind an unsuspecting Lenin, the poster becomes a prophetic icon for the wholesale destruction of the Marxist ideal in the 1930s.

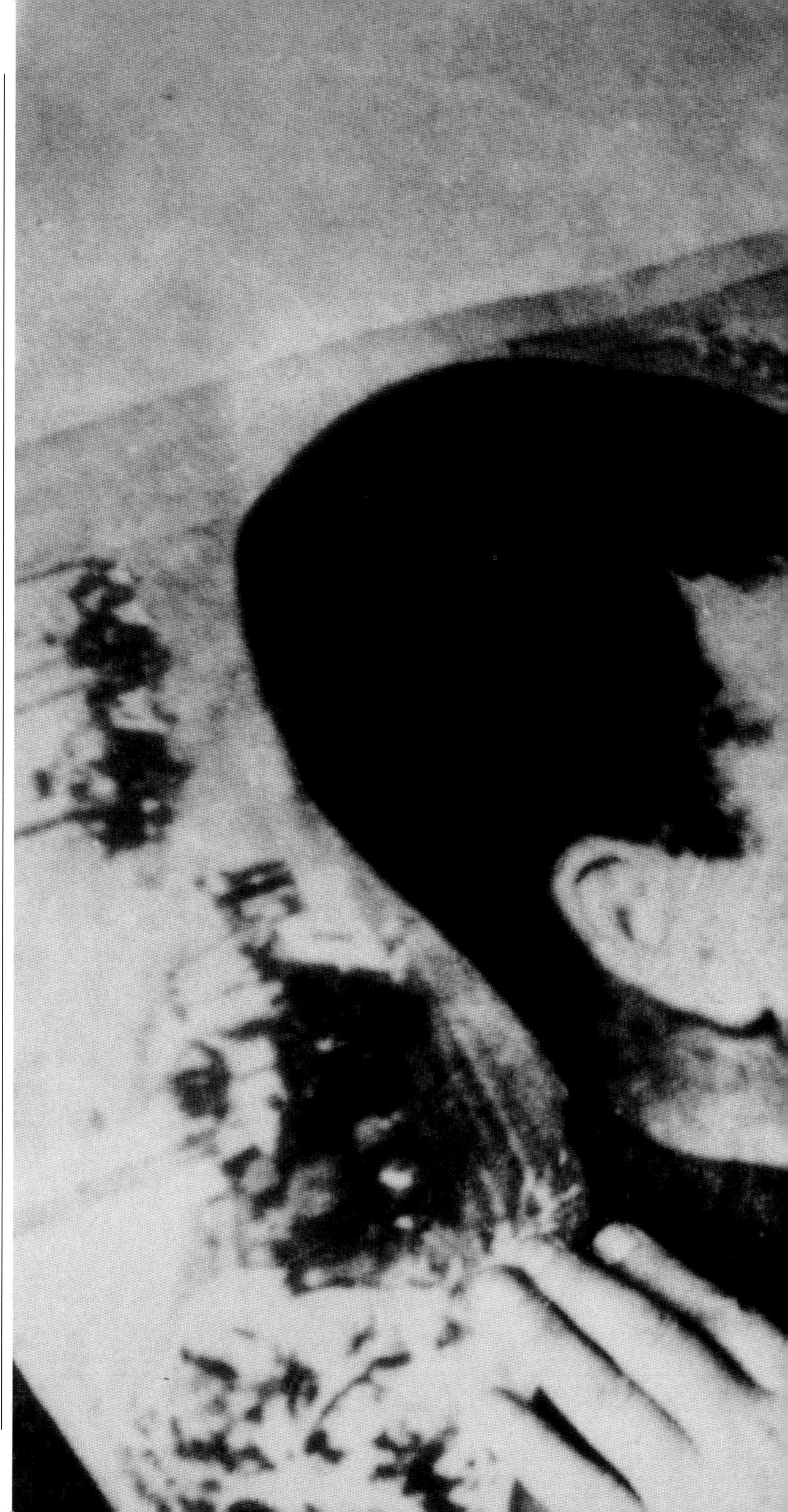

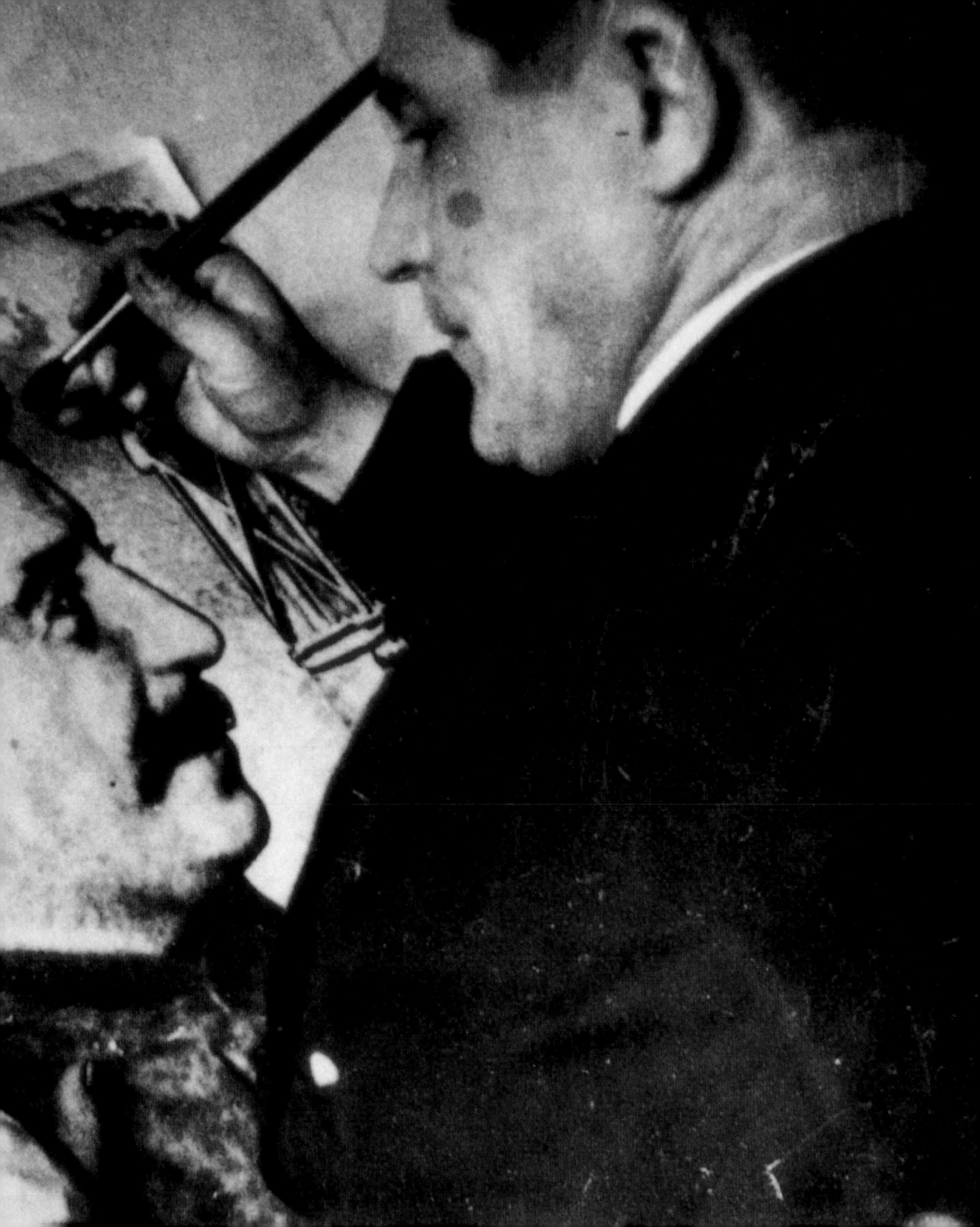

Photomontage was an ideal art form for Stalin. His head could be enlarged to enormous size and juxtaposed with those of the great socialist thinkers. He could also appear alongside the mass of the people, something that in reality he was less than willing to do as the terror deepened. *Above:* A powerful Gustav Klutsis poster from 1933 entitled "Under the Banner of Marx, Engels, Lenin, and Stalin." *Right:* An unused variation by Klutsis for the same poster. *Opposite:* A demonstration by Latvian youth in Riga, 1940, at the time of annexation. A small version of Klutsis's poster can be seen on the placard. The designer of the work had already been shot.

Overleaf, left-hand page: "Long Live Our Happy Socialist Land! Long Live Our Beloved Leader, the Great Stalin!" The apotheosis of his commitment to Stalinism, this breathtaking poster by Klutsis was made in 1935. The mind-bending perspective and wild changes in scale have turned the work, unwittingly perhaps, into a sort of Soviet surrealism. The airplanes over Red Square list the names: "Vladimir Lenin," "Joseph Stalin," "Maxim Gorky," "Mikhail Kalinin," etc., and the formation in the distance on the right spells out "STALIN." *Overleaf, right-hand page:* A maquette for another major Klutsis poster, "Long Live the Red Army of Workers and Peasants!" When Marshals Tukhachevsky and Gamarnik were purged from the Red Army in 1937, Klutsis had to hack them out from their positions in this photomontage, flanking Voroshilov. Marshal Yegorov remained on the right of the picture but was arrested and tortured to death in the depths of the Lubyanka in 1939.

ВЛАДИМИР
ЛЕНИН
ИОСИФ
СТАЛИН
МАКСИМ
ГОРЬКИЙ
МИХАИЛ
КАЛИНИН
ВЯЧЕСЛАВ
КЛУЦИС 35
ЗДРАВСТВУЕТ НАША СЧАСТЛИВАЯ СОЦИАЛИСТИЧЕСКАЯ РОДИНА.
здравствует наш любимый великий Сталин!

Фото=
=ВЦИК

SHOCKWORKER'S SARCOPHAGUS

Left: Tartar delegates at the Second All-Union Conference of Collective Farm Shockworkers, with Stalin and members of the Politburo in Moscow, February 1935. The state of the photograph is such that it looks as if the comrades in the back row have been sprayed with machine-gun fire. But what is really chilling is the afterimage of the departed figure who was seated next to Stalin. Closeness to the leader, physical or ideological, was no guarantee of survival. *Above:* Another photograph from the same conference.

LOOK FOR AN ENEMY!

The photographic composite (*right*) shows graphically how four out of the top seven "specialist worker" graduates from an advanced training course in 1935 later became victims in the campaign against "saboteurs" and "wreckers." Suspicion was endemic. *Pravda* announced that "not one disorder, not one accident, should go unheeded." In the factories, "there can be no breakdowns. . . . Look for an enemy." The slightest mistake, miscalculation, or misprint was judged deliberate. Mass executions of workers, skilled and unskilled, followed.

Top: Fyodor Antonov's painting *Disclosure of the Enemy in the Workshop*, 1938.

Above: Comrade Stalin with Railroad Craftsmen in Tbilisi, 1926. Iraklii Moiseivich Toidze's painting, shown at the Socialist Industry Exhibition of 1939, could just be ignored as bad art were it not for the chronic misrepresentation of its message. In reality, Stalin was completely removed from ordinary workers. He never visited factories or collective farms and lived in constant fear of assassination.

СОВЕРШЕНСТВОВАНИЯ РУКОВОД.
СПЕЦ РАБОТНИКОВ системы
НКЗ СССР
Савицкий.
Сахаров. М.Я.
Балдин. И.В.
Васильев-Новиков
Коряков
Герасименко
Белоусов. А.Р.
Рябов. Г.Г.
Арефьев. Н.К.
Норбей. И.И.
Попов. В.И.
Фалько. Я.Г.
Кондаков. И.Г.
Семенов. В.Г.
Вязовой. А.Н.
Гусаров М.С.
Кулигин. М.Ф.
Иванов. Г.И.
Еюкин. И.С.
1935 г.
обед. фото-снимок Фото №23.
Москва 4. Н.Радищевская д. №2.

Above: Delegates at the Seventh World Congress of the Communist International, July 25, 1935, in Moscow. Left to right: Maurice Thorez (French Communist chief), Stalin, Marcel Cachin, Joseph Pyatnitsky, Georgii Dimitrov, and Wilhelm Pieck. The photograph is heavily retouched, but when it was published fourteen years later (*right*), it was retouched even further: now Pyatnitsky has vanished. Lenin had regarded him as "one of the best Bolsheviks." He had been a key organizer in the Comintern since 1923. On July 7, 1937, he was arrested as a spy and subjected to appalling torture. He had spoken out against a proposal to increase the powers of the NKVD. He was shot later that month with 137 others in a massacre of the Party leadership. His wife was also arrested and never seen again, and his eldest son spent years in the Gulag. "For the Party," he once said, "we must endure everything, just so that it remains alive."

WORKERS OF THE WORLD, UNITE!

Stalin's policy of "Socialism in One Country" effectively destroyed any lingering hope in the 1930s of the world socialist revolution, and in 1943 he finally disbanded the Comintern. *Opposite:* Not many women were in evidence in this photomontage of Stalin among a sea of Party functionaries that appeared on the cover of *Rabotnitsa* (*Woman Worker*), January 1935. The issue was dedicated to the Seventh All-Union Congress of Soviets, which set up a special commission to produce a new Soviet constitution, which was written with painstaking care by Bukharin and Karl Radek. Not surprisingly, it became known as the "Stalin Constitution." Bukharin and Radek were subsequently liquidated.

Overleaf: Vasilii Yakovlev and Petr Shukhmin's "Comrade Stalin amid the Peoples of the USSR," published in a 1937 album entitled *The Stalin Socialist Constitution.*

ДА ЗДРАВСТВУЕТ VII ВСЕСОЮЗНЫЙ
С'ЕЗД СОВЕТОВ!

ПРОЛЕТАРИИ ВСЕХ СТРАН СОЕДИНЯЙТЕСЬ!

РАБОТНИЦА

ЯНВАРЬ 1935 МОСКВА 3

ЖУРНАЛ РАБОТНИЦ и ЖЕН РАБОЧИХ изд. ЦК ВКП(б)

Opposite: This photograph, which became a famous Soviet icon, shows the "Friend of the Little Children" in a joyous mood, having just received a huge bouquet from six-year-old Gelya Markizova. To the right of the picture can be seen M. I. Erbanov, first secretary of the Buryat Mongol ASSR, who also seems very happy. They are attending a reception in the Kremlin in 1936.

A year later Gelya's father, Ardan (*above*), the republic's second secretary, was shot for "spying for Japan" and "making preparations for the assassination of Stalin." Soon after, her mother, Dominica (*top right*), "the wife of an enemy of the people," was murdered under mysterious circumstances—so mysterious that the case was never investigated by the authorities.

Top center: M. I. Erbanov, the first secretary, was also purged and deleted from the photograph when it was reproduced later. *Above:* The "icon" joyously paraded through the streets of Leningrad in 1937 on the twentieth anniversary of the October Revolution. A sculpture glorifying the unforgettable occasion, by Georgii Lavrov, was erected in the Stalinskaya metro station. When rumor began to circulate that one of the subjects in the sculpture had murdered the other's parents, the propaganda value of the sculpture fell, causing the Siberian artist to be denounced himself.

UNFORGETTABLE MOMENT

СТАРОНОСОВ

HIGHER AND HIGHER!

The Lenin cult climbed to its zenith in the mid-1930s with the proposed construction of the Palace of Soviets. Boris Yofan won a lengthy competition that included entries from architects of many countries. Yofan's submission (*above*) was approved in February 1934. The 1,250-foot tower included an enormous 300-foot tall figure of Lenin. The whole building became a pedestal for his figure. The historic Church of Christ the Savior in the center of Moscow was blown up to make way for it, but only the foundations were laid. The palace was never built because war intervened. Subsequently it became the site of the huge Moscow open-air swimming pool. Today the church has been rebuilt on its original site, with new materials. The marble rubble from the old church was recycled into the construction of the Moscow Metro.

Opposite: On a smaller scale, but no less bizarre, Staronosov's engraving from *The Life of Lenin* was published in 1936. It shows Lenin and Krupskaya in a somewhat theatrical setting in the Alps during their three-year exile in Switzerland before the Revolution.

It was at this time that Lenin wrote his famous work, "Imperialism: The Highest Stage of Capitalism."

A SWIFT EXIT

In July 1933, Stalin, Kliment Voroshilov, Sergei Kirov, and Genrich Yagoda took a boat trip up the Stalin-Belomor canal, recently constructed by slave labor at a huge cost to human life. Eight OGPU project chiefs, Yagoda included, received the Order of Lenin for "their work." This painting (*right*) was commissioned from the socialist realist artist Dmitri Nalbandyan, who died in 1994, still an unreconstructed Stalinist. In early 1937, when the work was almost complete, Yagoda and his cadres were arrested as a gang of thieves, embezzlers, and spies and for being "four years late in exposing the Trotskyite-Zinovievite bloc."

Nalbandyan was thrown into confusion, but on Voroshilov's orders retouchers were sent from the Tretyakov Art Gallery to extricate Yagoda from the oil painting, a task requiring more skill than doctoring a photograph. Chillingly, his place in the painting (to the left of Kirov) was taken by an overcoat folded over a handrail—as if he had neither the time nor the need to take it with him.

Below: Construction of the canal by slave labor. Thirty thousand lives are said to have been lost on the project.

Bottom: Genrich Yagoda, who was shot in the Lubyanka, 1938.

Nikolai Yezhov and Joseph Stalin (*above*) conspiring together on the twentieth anniversary of the creation of the Cheka, Moscow, December 20, 1937. With the replacement of Yagoda by Yezhov as secret police chief, Stalin's reign of terror reached its peak. Yezhov was to head the NKVD for less than two years. "Yezhovschina" (the days of Yezhov), as they were known, brought death on a scale unprecedented in Russian history. Stalin would initial the endless lists of names, and Yezhov's special gang of killers—two hundred strong—would do the rest, with the utmost cruelty. In the end, all opposition, real or imagined, was wiped out.

PROCESS OF ELIMINATION

Left: Andrei Vyshinsky, state prosecutor, summing up at the third Moscow show trial of Nikolai Bukharin, Alexei Rykov, Isaak Zelensky, and seventeen others: "I demand that mad dogs be shot!" Vyshinsky screamed, "Every one of them must be shot!"
Opposite: Courtroom scene at one of the Moscow trials.
Leon Trotsky, languishing in exile in Mexico, denounced the trials as Stalin's "conscious and premeditated frame-up."

ФАТЕРЛЯНД

ON GUARD

Above: A photomontage by El Lissitsky from the photographic album *Red Army,* which he designed in 1934. It shows Kliment Voroshilov, people's commissar for defense, with his leading marshals. They are (top row, left to right) Gamarnik, Tukhachevsky, Yegorov, Khalepsky, Orlov, and Yakir. Second row: General S. Kamenev, Ordjonikidze, Budenny, Alksnis, Muklevich, Eideman, and Uborevich.

Stalin's purge of the Red Army took place in mid-1937. Tukhachevsky and other high-ranking officers were condemned to death and executed without trial on June 11. Marshals Voroshilov, Budenny, Blucher, and Yegorov signed the death sentences. Shortly afterward it was the turn of Blucher and Yegorov to face the firing squad. General Kamenev and Yan Gamarnik both committed suicide. Altogether, 25,000 officers were slain. From the photomontage only Voroshilov and Budenny survived, both outdated incompetents, to confront the impending Nazi invasion.

At a Politburo meeting in 1938, Stalin was to praise Tukhachevsky's military talents, his great sense of responsibility, and his striving to keep abreast of the fast-changing theory, technology, and practice of military affairs. This was less than twelve months after he had had him shot. Tukhachevsky's family was also destroyed. His wife, Nina, his mother, and two brothers died in prison, while his daughter and four sisters were all arrested, as well as friends.

Top: The decapitation of the Red Army: a surreal relic from Moscow in the darkest days of 1937. Budenny, Kalinin, and Voroshilov are flanked by a lineup of top-flight Soviet marshals, all later obliterated.

Opposite: Boris Efimov's vicious caricature of Rykov, Sokolnikov, Bukharin, Radek, and Trotsky wallowing in a trough entitled "Vaterland" (German for "Fatherland"), published at the time of the third Moscow show trial of 1938.

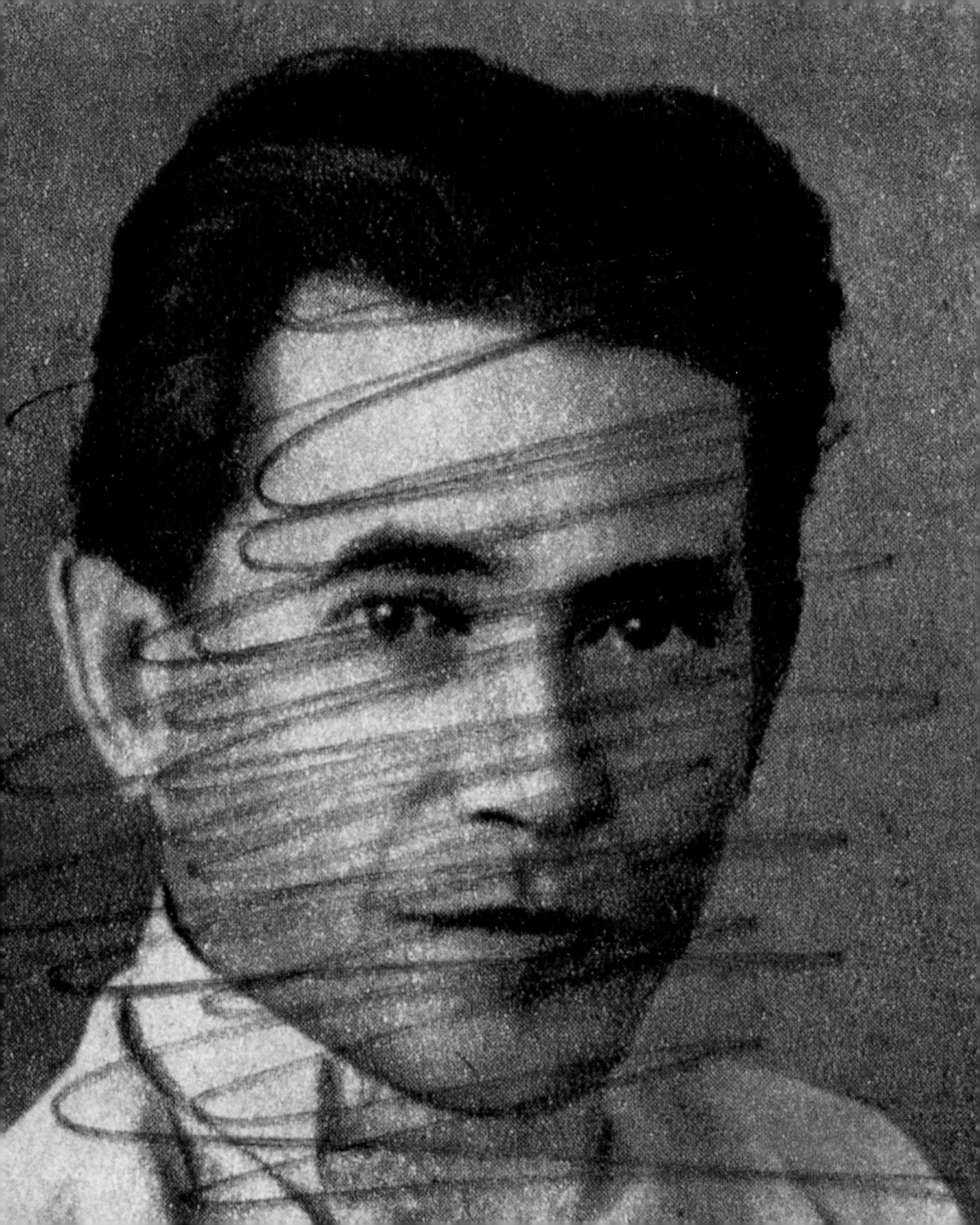

DROWNED IN BLOOD

Opposite: Nikolai Yezhov's portrait, from an album on Kazakhstan published in the late 1930s, has been scribbled out by the owner, following the news of his execution.

Yezhov had been removed from his post as people's commissar for internal affairs on December 8, 1938. His position at the NKVD was taken by Beria, who arrested him on April 10, 1939. Yezhov was shot on February 4, 1940.

Top: Leaders of the Party and government with members of the Moscow Art Theater on its fortieth anniversary, October 27, 1938. Yezhov is standing fifth from the right. Nikita Khrushchev is next to him on the left.

Center: The same photograph published on Stalin's seventieth birthday in 1949. Yezhov and the director of the Art Theater, Y. Boyarsky, have been spirited away by the retouchers, who have carefully extended the dark wooden paneling.

Above, left and right: Yezhov's removal from his other post as commissar of water transport is clearly illustrated in this photograph of a gentle stroll along the banks of the Moscow-Volga canal. In the retouched version, only Voroshilov, Molotov, and Stalin are left to pass the time of day. The canal had been constructed by forced labor at an enormous cost in human lives.

Much later, in 1956, Khrushchev was to tell the Twentieth Party Congress that Yezhov was an alcoholic and drug addict who "got what he deserved."

JEALOUSY AND REVENGE

The last photographs of Isaak Babel, the distinguished writer, were these mug shots (*right*) taken by the NKVD following his arrest on May 16, 1939. Babel was extremely nearsighted, so his glasses must have already been smashed during his interrogation in the Lubyanka. Babel's arrest was Yezhov's final act of jealousy and revenge. In 1927, some years before she married Yezhov, Yevgenia Khayutina had an affair with Babel in Berlin. In 1936 the acquaintance was renewed when Yevgenia, by this time chief editor of the magazine *USSR in Construction*, invited him to work for the journal. Babel rarely met the murderous secret police chief, but Yezhov became suspicious. His wife entered a sanatorium in a depressed state in October 1938 and died under mysterious circumstances the following month. Yezhov, who was arrested the following year, denounced Babel to his interrogators, telling them he had been spying for France. Babel had been working for a long time on a novel about the Cheka, but the book was never completed. The author of *Red Cavalry* (see cover *top*), and other famous stories from the time of the Civil War, was shot on January 27, 1940.

EARTH IN TURMOIL

The famous theatrical director Vsevolod Meyerhold (*top*) was arrested at the same time as Babel and tortured by the same interrogators. He was shot on February 2, 1940. He is seen here in his study standing in front of a portrait of his wife, the actress Zinaida Raikh. At the height of their fame, their apartment in Moscow had been a glamorous meeting place for the international intelligentsia. Shortly after Meyerhold's arrest, his wife was found at home, brutally stabbed to death. It is said that neighbors heard screaming but imagined it was only the sound of Zinaida rehearsing a new play.
Above: Liubov Popova's design for Meyerhold's 1923 production of *Earth in Turmoil* included a portrait of Trotsky, upper left, to whom the production was dedicated. Meyerhold's NKVD inquisitors seized upon this as a key charge against him.

BORN AGAIN

The Stalin cult reached a new, quasi-religious phase with the advent of the dictator's sixtieth birthday on December 21, 1939. His every childhood exploit was rewritten into history like some lost book from the Bible.

In painting and sculpture he was given a prophet-like aura. His image was everywhere—in metal, stone, paint, light, grains of wheat. The replica became the reality; Stalin was made of marble and he stood hundreds of feet high.

Opposite: Z. Snigir's contribution to the "sacred" style was *J. V. Stalin and His Mother* in 1939. Stalin's mother had given birth to three children before Stalin, but they all died before he was born. After 1903 he had little further time for his mother, who stayed in Georgia. She traveled to Moscow on two occasions to visit her son in the Kremlin, and died peacefully in 1937.

Top: Comrade Stalin and the Youth of the Town by the Georgian artist K. K. Gzelishvili. The town is Gori, where Stalin was born.

Above, left and right: In 1939 Stalin's humble birthplace was transformed under the vigilant eye of NKVD chief Lavrentii Beria. A giant marble pavilion was built over it, surrounded by pink flower beds.

STALIN'S SIGNATURE

Above: Stalin shaking hands with von Ribbentrop at the signing of the Stalin-Hitler pact in the Moscow Kremlin, August 1939.

For years the Stalinists had slandered Trotsky and the oppositionists by equating them with the Nazis. Now the Communist parties of the world had to accept that Hitler was not, after all, a threat to world peace. World War II began the following month.

A year later Stalin finally executed his plan to assassinate his last and most hated Communist opponent, Leon Trotsky.

Right: Attacked by a Stalinist agent, Trotsky lies dying in a hospital in Mexico City, August 21, 1940.

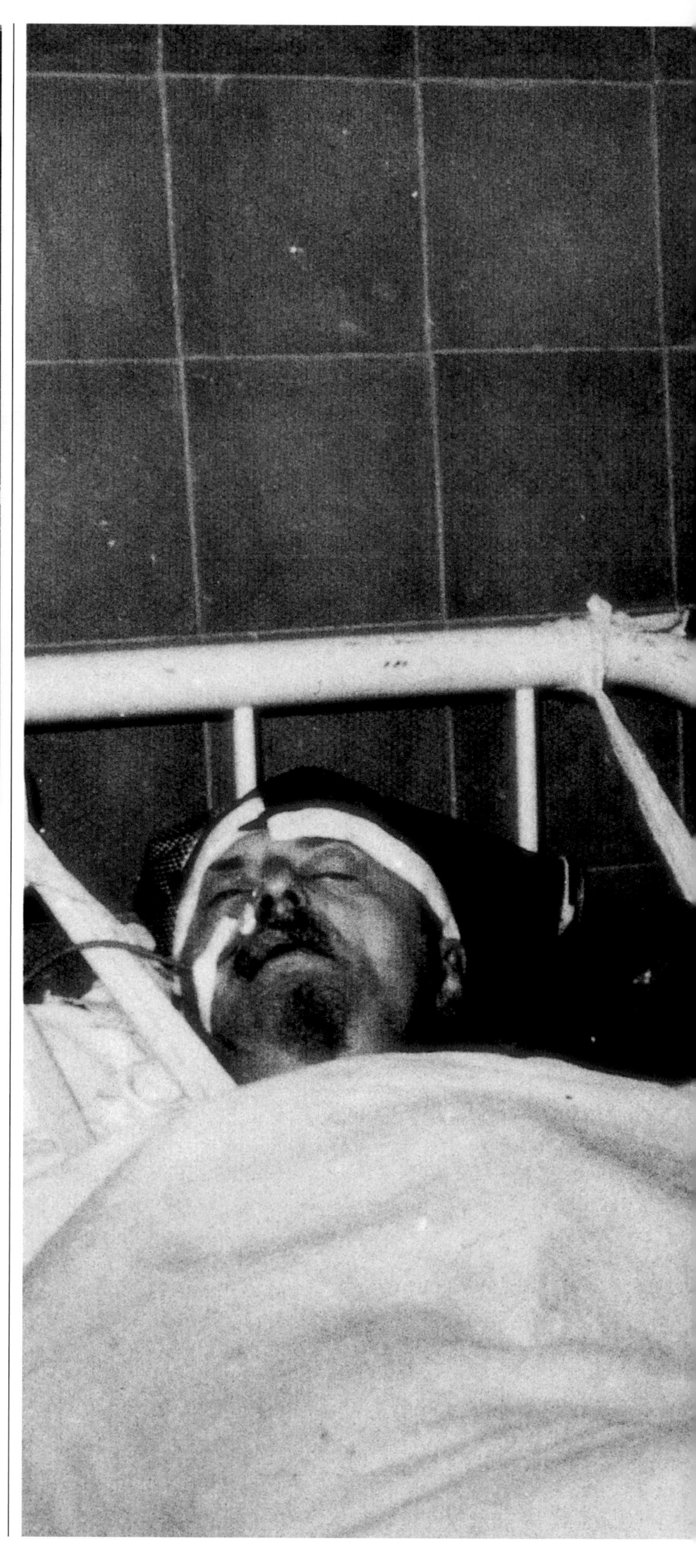

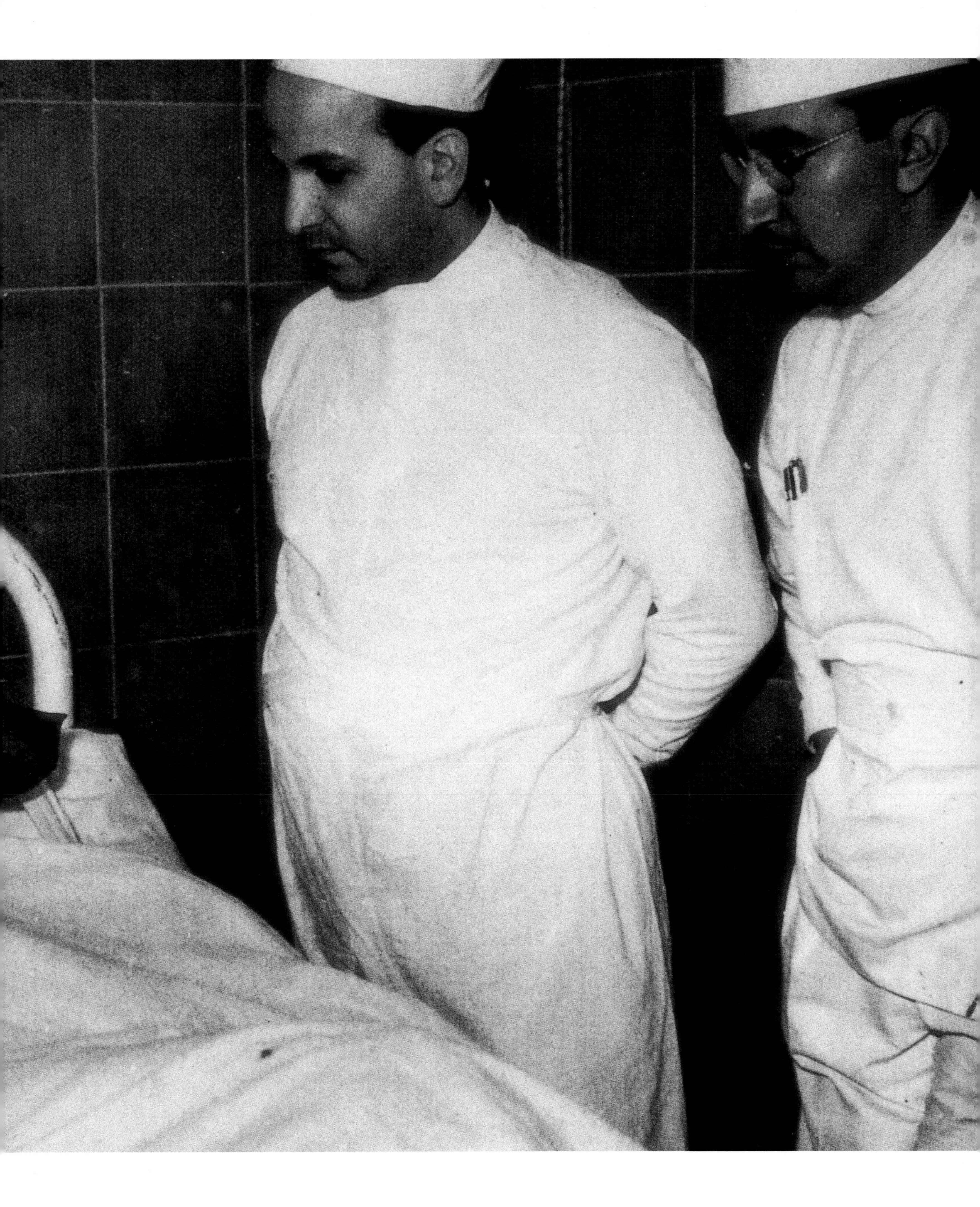

VICTORY!

The heroism and suffering of the Soviet people following the Nazi invasion of June 1941 cannot satisfactorily be conveyed here. By the time the Nazis were finally defeated in May 1945, between 25 million and 30 million Russians had lost their lives.

Stalin's initial response to the invasion was disbelief, followed by indecision. When he recovered his senses, he called upon the people to fight—not for Marx and Lenin—but for Russian nationalism.

Stalin made only two visits to the front, but it is thought that he did not leave Moscow when it looked like the capital was in danger of falling during the winter of 1941.

Above: This drawing, made in 1947 by K. Finogenov entitled *Stalin at the Moscow Front in 1941*, shows that the Great Strategist was less than concerned about placing himself in grave danger.

Right: Another example of Stalinist postwar glorification, a battle painting by P. A. Krivonogov portraying the seizure of the Reichstag in May 1945.

Above: The "Big Three" talks, bringing together the Allied leaders Stalin, Harry Truman, and British prime minister Clement Attlee, got under way in Potsdam in July 1945. Yevgenii Khaldei's photograph of the Soviet team shows Vyshinsky and Molotov gazing at Stalin, with Red Army chief of staff Antonov (reading) and Andrei Gromyko on the right. It was this veteran foreign minister who forty years later, in 1985, nominated Mikhail Gorbachev for the post of General Secretary.

Right: The same photograph, widely published four years later, shows the deletion of two officials between Antonov and Gromyko. Admiral of the Fleet N. G. Kuznetsov had been Supreme Commander of the Soviet Navy throughout the war. Stalin accused him in 1947 of handing over sea maps to the wartime Allies. He was swiftly demoted and sent to the Pacific. After Stalin's death, he was cleared and given his old job back. The figure in the shadows sitting between Kuznetsov and Gromyko in the original photograph is Ivan Maisky. He was the popular—too popular—Soviet ambassador to Britain during the war. In 1947 he was dismissed from the Central Committee, and in 1952 he was arrested, accused of spying for the British. He was released from prison in 1954.

MISSING FROM POTSDAM

Opposite: Besides Stalin, only two other figures in history wished to be referred to as "Generalissimo": the Spanish fascist dictator, Francisco Franco, and Chinese Nationalist leader Chiang Kai-shek. This doesn't seem to have bothered the Ukrainian artist A. Reznichenko, whose work adorned an album entitled "Glory to the Great Stalin," about the victory over fascism.

RED STAR OVER THE KREMLIN

Above left: A poster from the late 1940s titled "Long Live the All-Union Physical Culture Day."
Above right: J. V. Stalin and V. M. Molotov with Children, artist unknown, circa 1951. For a short time after 1945, it seemed that life for the Russian people might improve after three decades of war, revolution, famine, terror, and invasion. The Stalin cult, however, continued to grow, and the specter of Beria's secret police was ever-present.

Opposite: The weekly magazine *Ogonyek* celebrated the despot's seventieth birthday in December 1949 with a cover photomontage showing the "Celestial Stalin" in the night sky over Red Square. The Muscovites below him, however, seem to be looking in the other direction.

Overleaf: High-ranking Party officials at a May Day rally at the end of the war. Left to right: Georgii Malenkov (in silhouette), Andrei Zhdanov (Stalin's hated culture boss), Alexei Kuznetsov, Lavrentii Beria, Vyacheslav Molotov, and A. S. Shcherbakov (a mass slaughterer who drank himself to death nine days after this picture was taken). Kuznetsov, a war hero from the days of the siege of Leningrad, became a victim of Beria's mass purge of Party workers in that city in 1949. A huge anti-semitic purge was rumored to be next on Stalin's agenda, but his own death intervened.

ОГОНЁК
№ 52 ДЕКАБРЬ 1949
ИЗДАТЕЛЬСТВО «ПРАВДА»

DEATH IS NOT THE END

Stalin suffered a stroke and brain hemorrhage at the beginning of March 1953. He was dismissed from his post as prime minister by the other Party leaders just ninety minutes before his death on March 6. The bloodletting that he had inflicted on the Soviet people did not quite end there, however. In what seemed like a last terrible act of malevolence, hundreds of mourners were crushed to death in the struggle to enter the Hall of Columns in Moscow, where the dictator's body was lying in state.

Above: Great apprehension can be seen on the faces of the Politburo in this photomontage published in the Soviet press at the time of the funeral.

Right: High priests of the Russian Orthodox church paying their last respects. *Center right:* Andrei Vyshinsky is devastated by the news of his master's death. He died of a heart attack the following year, amid persistent rumors of an impending commission to investigate the years of terror. *Far right:* Three minutes' silence is observed for Stalin throughout the Soviet Union on March 9, 1953.

Opposite: Secret police chief Lavrentii Beria. His fifteen-year rule came to an end late at night on December 23, 1953. In the power struggle that followed Stalin's death, his bid for the leadership was terminated by Georgii Malenkov and Nikita Khrushchev. A secret trial took place in an underground bunker in Moscow, presided over by four famous World War II military leaders, Ivan Konev, Kirill Moskalenko, Mikhail Mikhailov, and Mikhail Gromov. Beria was found guilty and shot. His last words were "I am innocent."

Beria must have liked the expression on his face in the official portrait because the same head was used again (*above left*), this time montaged onto civilian dress.

Following his execution, his photograph and fawning biography in the *Great Soviet Encyclopedia* were promptly substituted. Subscribers world-wide were sent a four-page insert carrying a series of photographs and "new information" on the Bering Sea (*above right*). Whalers were shown harpooning their catch—a fitting metaphor for the fate of Stalin's blubbery henchman.

Above: Beria and Malenkov waving to the crowds in Red Square, Moscow, 1953.

BLUBBERY HENCHMAN

GHOSTWRITER

Left: This photograph, published for the first time, shows Petr Nikolaievich Pospelov (seated, wearing glasses) with his personnel. A veteran Stalinist, Pospelov was variously an editor of the theoretical magazine of the Central Committee *Bolshevik* from the 1930s, director of the Institute of Marxism-Leninism, and editor in chief of *Pravda.* At the height of the Great Purges, he was Stalin's ghostwriter for the infamous *History of the All-Russian Communist Party (Bolsheviks) Short Course.* Ten years later his adulation knew no bounds as he wrote the *Short Biography of J. V. Stalin.* After Stalin's death, Pospelov proved himself to be more than flexible. He was the man who, in 1956, wrote the first draft of Khrushchev's secret speech acknowledging Stalin's crimes. At the direction of Khrushchev but against the wishes of the likes of Molotov and Voroshilov, Pospelov had headed a commission on the terror. Khrushchev reported the findings in a closed session of the Twentieth Party Congress. Entitled "On the Cult of Personality and its Consequences," the speech was leaked to the West, where it was highly publicized, but it was not published in the Soviet Union until 1989.

Right: Stalin had ordered his body to be embalmed and placed beside Lenin's in Red Square. Thus it became, for eight years, the Lenin-Stalin Mausoleum (*top*). The corpse remained there until the Twenty-second Party Congress voted it out in 1961—five years after Khrushchev's secret speech. The removal of his name and the replacement letters just reading "LENIN" can be seen in the two center photographs. *Bottom:* Stalin's body ended up under a horizontal monolith behind the mausoleum.

SZTALIN
"ESKÜSZÜNK NEKED LENIN
ELVTÁRS, HOGY ÉLETÜNKET SEM KI-
MÉLVE RAJTA LESZÜNK, HOGY ERŐ-
SITSÜK AZ EGÉSZ VILÁG DOLGOZÓI-
NAK SZÖVETSÉGÉT, A KOMMUNISTA
INTERNACIONÁLÉT.

THE LOSING FINALISTS

Stalin's "cult of personality" was denounced in Khrushchev's secret speech of 1956, which also promised much-needed reforms throughout the Soviet bloc. This was not lost on the people of Hungary. Dissatisfaction with the Stalinist government there led thousands of students in Budapest to head for the city park, where they felled and destroyed (*above*) the hated Stalin statue, a massive bronze confection that stood 60 feet high on a pink marble plinth. The student revolt quickly turned into a nationwide struggle for liberation, with workers setting up revolutionary committees to replace the totally discredited Party and state apparatus. The Hungarian revolution was crushed within a month by the swift invasion of the Red Army.

The Stalin statue had been the work of a Hungarian sculptor named Kistaludi-Strobl. It was the winning entry in a competition, sponsored by the Hungarian Workers' Party.

Published here for the first time are photographs of the works that were submitted by the losing finalists in the competition. *Opposite and overleaf:* Eight "Stalins" that made it no further than their maquettes.

SZTÁLIN
Bock Andras

SZTÁL
Bokros B. Dezső

J.V. STALIN

FULL CIRCLE

Vladimir Serov's *Lenin Proclaiming Soviet Power at the Second Congress of Soviets*, painted in 1947, received a Stalin Prize the following year. Behind Lenin can be seen Stalin, Dzerzhinsky, and Sverdlov.

Under pressure from Khrushchev, Serov had to repaint the portrait in full in 1962. De-Stalinization required that revolutionary workers should replace Lenin's henchmen. This tragic and terrible episode in history had now come full circle.

ACKNOWLEDGMENTS

It is almost three decades since I first started collecting photographs, books, posters, and all kinds of graphic and printed matter documenting the political and social history of Russia and the former Soviet Union. During that time I have been fortunate to meet many people, worldwide, who share the same interests and who have generously contributed their knowledge and enthusiasm in the search for new material.

I would particularly like to thank my friends Judy Groves, Doris Rau, Francis Wyndham, Stephen Cohen, and Dr. Yoram Gorlizki. Without their help, this book would not have been possible.

A special thanks is also due to Professor Miklos Kun of Budapest University, a friend who must have taken to heart the title of Isaak Babel's short story "You Must Know Everything."

Sadly missed is another great friend, the late Natan Federovsky, whose infectious humor, generosity, and understanding of Soviet art and photography was, and still is, legendary.

I would like to thank everybody at Metropolitan Books for making work on this project so exciting. I am particularly indebted to Stephen Hubbell (senior editor) for his unfailing wisdom and tireless support, likewise Sara Bershtel (editorial director), Thomas Engelhardt (consulting editor), Lucy Albanese (design director), and Betty Lew.

Many thanks are due to the following people, who have also been so helpful. In alphabetical order, they are:

Dawn Ades, Celia Barnett, Ken Campbell, the late Bruce Chatwin, Susannah Clapp, Ann Creed, Alec Flegon, Dr. Agnes Gereben, the late Walter Goldwater, Ronald Gray, Christabel Gurney, Anthony C. Hall, Adam Hochschild, the late Michael Katanka, Francis King, Joseph King, Josephine King, Robin King, Alexei Ladyzhensky, Alexander Lavrentiev, Sasha Lurye, Gavin MacFadyen, Harold Landry, Tom Maschler, Claudine Meissner, Neil Middleton, Galina Panova, Howard Garfinkel and Larry Zeman of Productive Arts, Sasha Pyshakov, Alexander Rabinovich, Michael Rand, Sylva Rubashova, James Ryan, Wolf Ryzhik, Howard Schickler, Colin Smith, Gabriel Superfin, Sophy Thompson, Dr. Vera Tolz, Richard Weigand, and the late George Weissman.

BIBLIOGRAPHY

Antonov-Ovseyenko, Anton. *The Time of Stalin: Portrait of a Tyranny.* New York, 1980.

Babel, Isaak. *Konarmiya.* Moscow and Leningrad, 1927.

Bown, Matthew Cullerne. *Art Under Stalin.* Oxford, 1991.

Cohen, Stephen F. *Bukharin and the Bolshevik Revolution.* New York, 1973.

Conquest, Robert. *The Great Terror.* London, 1968.

Darel, Sylva. *A Sparrow in the Snow.* New York, 1973.

Deutscher, Isaac. *The Prophet Armed: Trotsky, 1879–1921.* London, 1954.

———. *The Prophet Unarmed: Trotsky, 1921–1929.* London, 1959.

———. *The Prophet Outcast: Trotsky, 1929–1940.* London, 1963.

———, and David King. *The Great Purges.* Oxford, 1984.

Erickson, John. *The Soviet High Command: A Military-Political History, 1918–1941.* London, 1962.

Fischer, Louis. *The Life of Lenin.* London, 1965.

Ginzburg, Evgenia Semyonovna. *Journey into the Whirlwind.* New York, 1967.

Hochschild, Adam. *The Unquiet Ghost.* New York, 1994.

Ivanenko, G. I., et al., eds. *Protsess Kontrrevolutsionnoi Organizatsii Menshevikov.* Moscow, 1931.

Kerr, Alfred, ed. *Russische Filmkunst.* Berlin, 1927.

King, David, et al. *Trotsky: A Photographic Biography.* Oxford, 1986.

Malkin, B. F., ed. *Udarniki.* Leningrad, 1930.

Medvedev, Roy. *Let History Judge: The Origins and Consequences of Stalinism.* New York, 1989.

Nicolaievsky, Boris I. *Power and the Soviet Elite.* New York, 1965.

NKU Soyuza SSR, ed., *Protsess Antisovetskogo Trotskistskogo Tesentra.* Moscow, 1937.

Pankratova, A. M. *Istoriya SSSR.* Moscow, 1940.

Pospelov, P. N., et al. *Joseph Stalin: A Short Biography.* Moscow, 1949.

Rivera, Diego, and Bertram D. Wolfe. *Portrait of America.* New York, 1934.

Schapiro, Leonard. *The Communist Party of the Soviet Union.* London, 1963.

Shentalinsky, Vitaly. *Arrested Voices.* New York, 1996.

Solomon, Peter H., Jr. *Soviet Criminal Justice Under Stalin.* Cambridge, England, 1996.

Sukhanov, N. N. *The Russian Revolution, 1917: A Personal Record.* London, 1955.

Trotsky, Leon. *Stalin: An Appraisal of the Man and His Influence.* London, 1947.

Tucker, Robert C., and Stephen F. Cohen, eds. *The Great Purge Trial.* New York, 1965.

Wyndham, Francis, and David King. *Trotsky: A Documentary.* Harmondsworth, England, 1972.

PHOTOGRAPHIC ALBUMS

Brilliantov, A. V., ed. *15 Let Kazakhskoi ASSR.* Moscow and Leningrad, 1936.

Glebov-Putilovsky, N. N., ed. *Oktyabr—Foto-Ocherk po Istorii Velikoi Oktyabrskoi Revolutsii (1917–1920).* Moscow and Petrograd, 1920.

Goltsev, V. V., ed. *Lenin Albom—Sto Fotograficheskikh Snimkov.* Moscow and Leningrad, 1927.

Gorky, M., et al., eds. *Belomorsko—Baltiiskii Kanal imeni Stalina.* Moscow, 1934.

Gosplan SSSR, ed. *Stalinskaya Konstitutsiya Sotsializma.* Moscow, 1937.

Institut Marksa-Engelsa-Lenina pri TsK VKP (B), ed. *Feliks Edmundovich Dzerzhinskii.* Moscow, 1951.

———, and Tsentralnii Muzei V. I. Lenina, eds. *Iosif Vissarionovich Stalin.* Moscow, 1949.

———, ed. *Karl Marks & Fridrikh Engels—Sobranie Fotografii.* Moscow, 1976.

———, ed. *Lenin—Albom Fotografii 1917–1922.* Moscow, 1957.

———, ed. *Lenin: Sobranie Fotografii.* Moscow, 1970.

———, ed. *V. I. Lenin—Albom Fotografii.* Moscow, 1972.

———, ed. *Vladimir Ilyich Lenin—Albom Fotografii.* Moscow, 1960.

Kerzhentsev, P., ed. *Zhizn Lenina.* Illustrated by P. Staronosov. Moscow, 1936.

Leonidov, O. L., ed. *Pervaya Konnaya.* Designed by Alexander Rodchenko. Moscow, 1938.

Mekhlis, L., et al, eds. *Geroicheskaya Epopeya—Albom Foto-dokumentov.* Moscow, 1935.

Orakhelashvili, M. D., et al., eds. *Sergo Ordzhonikidze—Biograficheskii Ocherk.* Moscow, 1936.

Ot Moskovskogo Komiteta R.K.P., ed. *Tovarishchu Delegatu II Kongressa III Kommunisticheskogo Internatsionala—Albom.* Moscow, July 27, 1920.

Pozern, B. P., ed. *Sergei Mironovich Kirov.* Leningrad, 1936.

Prigorovsky, N. N. et al., eds. *Stalin.* Moscow, 1939.

Ramzin, K., ed. *Revolutsiya v Srednei Azii.* Moscow, 1928.

Reznichenko, A., ed. *Slovo Velikomu Stalinu ot Ukrainskogo Naroda.* Kiev, 1947.

Rodionov, F. E., ed. *Raboche-Krestyanskaya Krasnaya Armiya.* Designed by El Lissitsky. Moscow, 1934.

Teimin, E., et al., ed. *Stalin.* Moscow, 1939.

Tursunkhodzhaev, M., et al., ed. *10 Let Uzbekistana.* Moscow and Leningrad, 1934 (Russian-language edition) and 1935 (Uzbek edition).

Volkov-Lannit, L. F. *V. I. Lenin v Fotoiskusstve.* Moscow, 1967.

Yakovlev, Igor., ed. *Lenin's Portrait.* Moscow, 1987.

Yakovlev, N. M., ed. *Stalin—Zhivopis, Plakat, Grafika, Skulptura.* Moscow, 1934. The first "personality cult" album.

PHOTOGRAPHIC AND GRAPHIC PORTFOLIOS

Soyuzfoto, ed. *Chapaev.* Moscow, 1932.

Muzei V. I. Lenina, ed. *Lenin.* Moscow, 1939.

Institut Marksa-Engelsa-Lenina pri TsK VKP(B), ed. *Lenin.* Moscow, 1934.

Goltsman, B. Ya. et al., eds. *I. V. Stalin v Proizvedeniyakh Zhivopisi Khudozhnikov Gruzii.* Moscow, 1939.

Mitskevich, S. I., ed. *Proletarskaya Revolutsiya v Obrazakh i Kartinakh.* Moscow, n. d. (c. 1924).

———, ed. *Muzei Revolutsii—Revolutsionnoe Dvizhenie v Kartinakh.* Moscow and Leningrad, 1932.

Goskinoizdat, ed. *I. V. Stalin (Fotosnimki).* Moscow, 1939.

———. *I. V. Stalin: Fotoseriya v 72 Listakh.* Moscow, 1939.

INDEX